Quaternion Fusion Packets

CONTEMPORARY MATHEMATICS

765

Quaternion Fusion Packets

Michael Aschbacher

AMERICAN MATHEMATICAL SOCIETY

2020 *Mathematics Subject Classification.* Primary 20D05; Secondary 20C20.

Library of Congress Cataloging-in-Publication Data

Names: Aschbacher, Michael, 1944– author.
Title: Quaternion fusion packets / Michael Aschbacher.
Description: Providence, Rhode Island : American Mathematical Society, 2021. | Series: Contemporary mathematics, 0271-4132 ; volume 765 | Includes bibliographical references and index.
Identifiers: LCCN 2020043155 | ISBN 9781470456658 (paperback) | ISBN 9781470464219 (ebook)
Subjects: LCSH: Finite simple groups. | Sylow subgroups. | AMS: Group theory and generalizations – Abstract finite groups – Finite simple groups and their classification. | Group theory and generalizations – Abstract finite groups – Sylow subgroups, Sylow properties, π-groups, π-structure.
Classification: LCC QA177 .A826 2021 | DDC 512/.23–dc23

LC record available at https://lccn.loc.gov/2020043155

Contemporary Mathematics ISSN: 0271-4132 (print); ISSN: 1098-3627 (online)
DOI: https://doi.org/10.1090/conm/765

∞ The paper used in this book is acid-free and falls within the guidelines
established to ensure permanence and durability.
Visit the AMS home page at `https://www.ams.org/`

10 9 8 7 6 5 4 3 2 1   26 25 24 23 22 21

Contents

Background and overview

CHAPTER 0

Introduction

The notion of fusion in groups first arose in the context of finite group theory. The fusion theorems of Alperin [**Alp67**] and Goldschmidt [**Gol74**] , and Glauberman's Z^*-Theorem [**Gla66**] are highlights of the early work on fusion in finite groups. Saturated fusion systems were first defined and studied by L. Puig, although Puig calls these objects *Frobenius categories* rather than fusion systems; see for example [**Pui09**]. Our introduction to the subject was from [**BLO03**] and we adopt the notation and terminology found there and in [**AKO11**].

Let p be a prime and S a finite p-group. A *fusion system* on S is a category $\mathcal{F}$ whose objects are the subgroups of S and for $P, Q \leq S$, the set $\hom_{\mathcal{F}}(P, Q)$ of morphisms from P to Q consists of injective group homomorphisms from P into Q, and some weak axioms are satisfied. The standard example is a system $\mathcal{F}_S(G)$, where G is a finite group, $S \in Syl_p(G)$, and the morphisms from P to Q are those induced via conjugation in G. The fusion system is *saturated* if it satisfies two additional axioms which are easily checked in $\mathcal{F}_S(G)$ using Sylow's Theorem. There are also *exotic* saturated fusion systems which are not realized in finite groups. See [**BLO03**] or [**AKO11**] for precise definitions, notation, terminology, and basic results on fusion systems. There is also some introductory discussion in section 1.2.

Puig developed his theory as a tool in modular representation theory, while homotopy theorists use fusion systems to study the classifying spaces of finite groups, and more general objects called *p-local finite groups.* However our interest in the subject is motivated by local finite group theory. It is by now clear that techniques from the local theory of finite groups are effective in the investigation of fusion systems, and indeed some arguments are easier to carry out in the category of fusion systems than the category of groups. Thus we are interested in, first, using techniques from the local theory to prove theorems about saturated fusion systems, and perhaps to even classify the simple saturated 2-fusion systems, or the subclass of systems of component type (as defined in Definition II.14.1 of [**AKO11**]). Then, second, we hope to use the theorems about fusion systems to simplify the proof of the theorem classifying the finite simple groups, particularly that part of the proof on groups of component type. See section II.14 in [**AKO11**] and [**Asc15**] for discussion of such a program, begun by the author, in which this manuscript plays a major part.

An obvious early step in a program to classify simple 2-fusion systems of component type is to prove an analogue of the so-called "classical involution theorem" on finite groups (found in [**Asc77**]) for 2-fusion systems. The hypotheses of that theorem are for the most part fusion theoretic, so a translation of the statement on groups into a statement about fusion systems is not difficult. This leads to the following definition.

A *quaternion fusion packet* is a pair $\tau = (\mathcal{F}, \Omega)$ where $\mathcal{F}$ is a saturated fusion system on a finite 2-group S and Ω is an $\mathcal{F}$-invariant set of subgroups of S such that

(QFP1) There exist an integer m such that for all $K \in \Omega$, K has a unique involution $z(K)$ and is nonabelian of order m.

(QFP2) For each pair of distinct $K, J \in \Omega$, $|K \cap J| \leq 2$.

(QFP3) If $K, J \in \Omega$ and $v \in J - Z(J)$, then $v^{\mathcal{F}} \cap C_S(z(K)) \subseteq N_S(K)$.

(QFP4) If $K, J \in \Omega$ with $z = z(K) = z(J)$, $v \in K$, and $\phi \in \hom_{C_{\mathcal{F}}(z)}(v, S)$ then either $v\phi \in J$ or $v\phi$ centralizes J.

We next consider the generic class of examples of quaternion fusion packets. Let p be an odd prime and write $\mathrm{Chev}(p)$ for the collection of finite groups G of Lie type of characteristic p with an irreducible root system and generated by root groups. Let $\mathrm{Chev}^*(p)$ consist of the members of $\mathrm{Chev}(p)$ which are not isomorphic to $L_2(q)$ or ${}^2G_2(q)$ for any power q of p. Thus $G \in \mathrm{Chev}^*(p)$ is quasisimple unless $G \cong SL_2(3)$, and G is an image of the universal group $\tilde{G}$ of type G, defined by the Steinberg relations of type G.

The *fundamental subgroups* of G are the conjugates of groups $\langle Z(U_\alpha), Z(U_{-\alpha})\rangle$, for α a long root in a root system for G, and U_α the root subgroup of G corresponding to α. Each fundamental subgroup is isomorphic to $SL_2(q)$ for a suitable power q of p.

Let $S \in Syl_2(G)$ and let Ω be the set of groups $S \cap L$, as L varies over the fundamental subgroups L of G with $S \cap L \in Syl_2(L)$. Then the pair $(\mathcal{F}_S(G), \Omega)$ is a quaternion fusion packet; for example this follows from 14.5 in [**Asc77**] or 2.4.1 in this paper. We call such packets *Lie fusion packets*.

Write $SL_2[m]$ for the fusion system $\mathcal{F}_S(G)$ of a group G isomorphic to $SL_2(q)$, q odd, with $m = (q^2 - 1)_2$ the 2-share of $|G|$. If $(\mathcal{F}_S(G), \Omega)$ is a Lie fusion packet, $K \in \Omega$, and $z = z(K)$, then the fundamental subgroup L of G containing K is subnormal in $C_G(z)$, so $\mathcal{F}_K(L) \cong SL_2[m]$ is subnormal in $C_{\mathcal{F}}(z)$.

Given a saturated fusion system $\mathcal{F}$ on a finite 2-group S and a collection Σ of subgroups of S, the *normal closure* of Σ in $\mathcal{F}$ is the smallest normal subsystem $[\Sigma]_{\mathcal{F}}$ of $\mathcal{F}$ such that a Sylow group of the subsystem contains Σ.

We can now state our main theorem.

Main Theorem. Assume $\mathcal{F}$ is a saturated fusion system on a finite 2-group S, z is an involution in S, and $\mathcal{E}$ is a subnormal subsystem of $C_{\mathcal{F}}(z)$ containing z and isomorphic to $SL_2[m]$ for some integer m. Let K be the Sylow group of $\mathcal{E}$ and set $\Omega = K^{\mathcal{F}}$. Assume $\mathcal{F}$ is the normal closure of K. Then $(\mathcal{F}, \Omega)$ is a Lie fusion packet.

The Main Theorem is a corollary to Theorems 1 and 2, stated in section 1.1. The Main Theorem provides a characterization of the 2-fusion systems of the groups L in $\mathrm{Chev}^*(p)$ for p an odd prime. By Theorem 5.6.18, if L is simple then its 2-fusion system is also simple.

Section 1.2 contains a brief introduction to the theory of fusion systems, and then section 1.3 supplies an outline of the proof of Theorems 1 and 2, plus the proofs of Theorems 3-8, which are used in the proof of Theorem 1.

We close with a historical remark. In January 1974, John Thompson gave a series of lectures at the winter meeting of the AMS in San Francisco. One of the talks was on a possible approach to proving the B-conjecture. That approach involved a finite group G and a normal subgroup Q of $C_G(z)$ for some involution $z \in K \in Syl_2(Q)$ with K quaternion. It is easy to see that the 2-fusion system of G has an associated quaternion fusion packet. Thompson's idea was to show that $\langle K^G \rangle$ is a group of Lie type. Toward that end he constructed a 2-torus of G and part of the Weyl group of the 2-torus. This idea served as the foundation of the proof of the classical involution theorem, and is also the foundation for the study of quaternion fusion packets.

CHAPTER 1

The major theorems and some background

1.1. Theorems 1 through 8

To state the remainder of our major results we need more notation and terminology.

Let $\tau = (\mathcal{F}, \Omega)$ be a quaternion fusion packet. We will show that τ comes equipped with a collection $\eta(\tau)$ of sets η such that η consists of cyclic subgroups of S each of which is conjugate to a subgroup of index 2 in some member of Ω. For $\eta \in \eta(\tau)$, $W = \langle \eta \rangle$ is abelian and should be viewed as a "2-torus" of $\mathcal{F}$. For example if τ is the Lie packet of some group G of Lie type then W is a 2-torus of G.

Each $K \in \Omega$ acts on W, $K \cap W \in \eta$, and $k \in K - W$ induces an automorphism $d_{K \cap W}$ on W via conjugation that is independent of the choice of k. Indeed for $V \in \eta$, V is conjugate under $Aut_{\mathcal{F}}(W)$ to $K \cap W$ for some $K \in \Omega$, so we get a corresponding map d_V, and hence a set $D_\eta = \{d_V : V \in \eta\}$ of involutions in $Aut_{\mathcal{F}}(W)$. Set $\mu(\tau) = \langle D_\eta \rangle \leq Aut_{\mathcal{F}}(W)$, and call $\mu(\tau)$ the *Thompson group* of τ. It can be shown that usually $\eta(\tau)$ contains a unique member, and in any event $\mu(\tau)$ is independent of the choice of η in $\eta(\tau)$. One should think of the Thompson group as the Weyl group of its 2-torus W, and indeed if τ is the Lie packet of G then the Thompson group is the Weyl group of W, or at least a large subgroup of the Weyl group.

Let $\mathcal{Z} = \mathcal{Z}(\tau)$ be the union of the classes $z(K)^{\mathcal{F}}$ as K ranges over Ω. We consider certain graphs $\mathcal{D} = \mathcal{D}(\tau)$, $\mathcal{D}^* = \mathcal{D}^*(\tau)$, and $\mathcal{A} = \mathcal{A}(\tau)$ on $\mathcal{Z}$. Given such a graph $\mathcal{B}$ and $z \in \mathcal{Z}$, write $\mathcal{B}(z)$ for the vertices adjacent to z in $\mathcal{B}$. Write $\mathcal{B}^c$ for the complementary graph.

Write $\mathcal{F}^\circ$ for the normal closure $[\Omega]_{\mathcal{F}}$ of Ω in $\mathcal{F}$ and set $\tau^\circ = (\mathcal{F}^\circ, \Omega)$.

For $K \in \Omega$, $\Omega(z(K)) = \{J \in \Omega : z(K) = z(J)\}$.

We are now ready to state our eight major theorems. The notation for describing the examples appearing as conclusions in the theorems can be found immediately after the statement of the eight theorems.

Theorem 1. Assume $\tau = (\mathcal{F}, \Omega)$ is a quaternion fusion packet. Let $K \in \Omega$, $z = z(K)$, and assume that $\mathcal{F}$ is the normal closure of K and $\mathcal{F}$ is transitive on Ω. Then one of the following holds:

(1) $z \in O_2(\mathcal{F})$.

(2) τ is a Lie fusion packet.

(3) $\mathcal{F}$ is $L_2[2m]^{(1)}$, $L_3^+[m]^{(1)}$, or M_{12}.

(4) $\mathcal{F}$ is a central factor system of $\hat{S}p_6(2)$ or $\hat{\Omega}_8^+(2)$.

Theorem 2. Assume $\tau = (\mathcal{F}, \Omega)$ is a quaternion fusion packet. Let $K \in \Omega$, $z = z(K)$, and assume that $\mathcal{F}$ is the normal closure of K in $\mathcal{F}$, $\mathcal{F}$ is transitive on Ω, and $z \in O_2(\mathcal{F})$ but $z \notin Z(\mathcal{F})$. Then $\mathcal{F}$ is constrained, $K \leq O_2(C_{\mathcal{F}}(z))$, and a model G for $\mathcal{F}$ satisfies one of the following:

(1) $G \cong L_3(2)/E_8$.
(2) $G/Z(G) \cong L_3(2)/E_{64}$ or $L_3(2)/2^{3+6}$.
(3) $G/Z(G) \cong AE_n$.
(4) For $\eta \in \eta(\tau)$, $\langle \eta \rangle \trianglelefteq G$ and $\mu(\tau)$ is generated by 3-transpositions.

Theorem 3. Assume $\tau = (\mathcal{F}, \Omega)$ is a quaternion fusion packet, $K \in \Omega$, $z = z(K)$, and $\mathcal{A}(z) \neq \emptyset$. Set $\mathcal{F}_1 = [\Omega(z)]_{\mathcal{F}^\circ}$, $\tau_1 = (\mathcal{F}_1, \Omega(z))$, and $\Omega_2 = \Omega - \Omega(z)$. If $\Omega_2 \neq \emptyset$ set $\mathcal{F}_2 = [\Omega_2]_{\mathcal{F}^\circ}$ and $\tau_2 = (\mathcal{F}_2, \Omega_2)$. Then

(1) τ_1 is a quaternion fusion packet with $\mu(\tau_1) \cong D_{12}$.
(2) If $\Omega_2 \neq \emptyset$ then τ_2 is a quaternion fusion packet and $\mathcal{F}^\circ = \mathcal{F}_1 \times \mathcal{F}_2$.
(3) One of the following holds:
(i) $\langle \eta \rangle \trianglelefteq \mathcal{F}_1$ for $\eta \in \eta(\tau_1)$.
(ii) $\mathcal{F}_1 \cong L_3(2)/E_8$.
(iii) $\mathcal{F}_1 \cong L_3^+[m]^{(1)}$.
(iv) $\mathcal{F}_1 \cong G_2[m]$.
(v) $\mathcal{F}_1 \cong M_{12}$.

Theorem 4. Assume $\tau = (\mathcal{F}, \Omega)$ is a quaternion fusion packet such that $\mathcal{F} = \mathcal{F}^\circ$. Then there exists a partition $\{\Omega_i : i \in I\}$ of Ω such that, setting $\mathcal{F}_i = [\Omega_i]_{\mathcal{F}}$ and $\tau_i = (\mathcal{F}_i, \Omega_i)$, each of the following holds:

(1) τ_i is a quaternion fusion packet.
(2) $\mathcal{F}$ is a central product of the $\mathcal{F}_i$, for $i \in I$.
(3) For $i \in I$, either $\mathcal{F}_i$ is transitive on Ω_i or $\mu(\tau_i) \cong D_{12}$.
(4) For $i \in I$, either $\mathcal{D}(\tau_i)^c$ is a connected component of $\mathcal{D}(\tau)^c$ or $\mathcal{F}_i/Z(\mathcal{F}_i) \cong Sp_n[m]$ or $L_n^-[m]$ for some $n \geq 6$.

In the situation described in Theorem 4, we say that τ is the *central product* of the subpackets τ_i and write $\tau = \tau_1 * \cdots * \tau_r$. Moreover we call the subpackets τ_i, $i \in I$, the *coconnected components* of τ. For example from part (4) of Theorem 4, we have the graph $\mathcal{D}(\tau)$ on the set $\mathcal{Z}(\tau)$, and, aside from the exceptional cases appearing in (4), the sets $\mathcal{Z}(\tau_i)$ are the connected components of the complementary graph $\mathcal{D}(\tau)^c$: the coconnected components of $\mathcal{D}(\tau)$. Alternatively from part (3) of Theorem 4, we can view coconnection as corresponding to orbits of $\mathcal{F}$ on Ω.

Theorem 5. Assume $\tau = (\mathcal{F}, \Omega)$ is a quaternion fusion packet such that $\Omega = \{K\}$ is of order 1, $\mathcal{F} = \mathcal{F}^\circ$, and $z(K) \notin Z(\mathcal{F})$. Then one of the following holds:

(1) $\mu(\tau) \cong S_3$ and for $\eta \in \eta(\tau)$, $\langle \eta \rangle \trianglelefteq \mathcal{F}$.
(2) $\mathcal{F} \cong L_3(2)/E_8$.
(3) $\mathcal{F} \cong L_2[2m]^{(1)}$.
(4) $\mathcal{F} \cong L_3^\epsilon[m]$.
(5) $\mathcal{F} \cong L_3^+[m]^{(1)}$.
(6) $\mathcal{F} \cong G_2[m]$.
(7) $\mathcal{F} \cong M_{12}$.

Theorem 6. Assume $\tau = (\mathcal{F}, \Omega)$ is a quaternion fusion packet such that $\mathcal{F} = \mathcal{F}^\circ$ is transitive on Ω, $|\Omega(z(K))| > 1$ for some $K \in \Omega$, and $\mathcal{A}(\tau) = \emptyset$. Then one of the following holds:

(1) $\mathcal{F} \cong L_3(2)/E_{64}$.
(2) $\mathcal{F}/Z(\mathcal{F}) \cong L_3(2)/2^{3+6}$.
(3) $\mathcal{F}/Z(\mathcal{F}) \cong AE_n$ for some $n \geq 5$.
(4) For $\eta \in \eta(\tau)$, $\langle \eta \rangle \trianglelefteq \mathcal{F}$ and $\mu(\tau) \cong \mathrm{Weyl}(D_n)$ for some $n \geq 3$.
(5) $\mathcal{F}/Z(\mathcal{F}) \cong Sp_6(2)$ or $\Omega_8^+(2)$.
(6) $\mathcal{F}/Z(\mathcal{F}) \cong P\Omega_n^\epsilon[m]$ for some $n \geq 5$.

Theorem 7. Assume $\tau = (\mathcal{F}, \Omega)$ is a quaternion fusion packet such that $\mathcal{F} = \mathcal{F}^\circ$ is transitive on Ω, $\mu(\tau)$ is abelian, and for $K \in \Omega$ and $z = z(K)$, $\Omega(z) = \{K\} \neq \Omega$. Then one of the following holds:

(1) $\mathcal{F} \cong \hat{A}E_5$.
(2) $\mathcal{F}/Z(\mathcal{F}) \cong PSp_n[m]$ for some $n \geq 4$.
(3) $\mathcal{F}/Z(\mathcal{F}) \cong L_n^-[m]$ for some $n \geq 4$.

Theorem 8. Assume $\tau = (\mathcal{F}, \Omega)$ is a quaternion fusion packet such that $\mathcal{F} = \mathcal{F}^\circ$ is transitive on Ω, $\mu = \mu(\tau)$ is nonabelian, and for $K \in \Omega$ and $z = z(K)$, $\Omega(z) = \{K\} \neq \Omega$. Then $\mu \cong \mathrm{Weyl}(\Phi)$ where $\Phi = A_n$ for some $n \geq 4$; $\Phi = D_n$ for some $n \geq 3$; or Φ is E_6, E_7, or E_8. Moreover one of the following holds:

(1) For $W \in W(\tau)$, $W \trianglelefteq \mathcal{F}$.
(2) $\Phi = A_n$ for some $n \geq 4$ and $\mathcal{F}$ is a central factor system of $SL_{n+1}^+[m]$.
(3) $\Phi = D_n$ for some $n \geq 3$ and $\mathcal{F} \cong Spin_{2n}^+[m]$, $Spin_{2n+1}[m]$, $Spin_{2n+2}^-[m]$, $\hat{A}E_{2n}$, or $\hat{A}E_{2n+1}$ for some $n \geq 4$; or $n > 4$ is even and $\mathcal{F} \cong HSpin_{2n}^+[m]$ or $H\hat{A}E_{2n}$.
(4) $m = 8$, $\Phi = D_3$, and $\mathcal{F} \cong \hat{L}_3(2)/E_{64}$ or $\hat{S}p_6(2)$.
(5) $\Phi = D_4$ and $\mathcal{F} \cong \hat{\Omega}_8^+(2)$, $\hat{L}_3(2)/2^{3+6}$, $F_4[m]$, or $E_6^-[m]$.
(6) $\Phi = E_6$ and $\mathcal{F} \cong E_6^+[m]$.
(7) $\Phi = E_7$ and $\mathcal{F} \cong E_7[m]$ or $\tilde{E}_7[m]$.
(8) $\Phi = E_8$ and $\mathcal{F} \cong E_8[m]$.

We next define the notation used to describe the fusion systems appearing in our theorems. We begin with a table summarizing our notation for the fusion system $\mathcal{F}_S(G)$ of a Lie packet $\tau = (\mathcal{F}_S(G), \Omega)$, where G is a group of Lie type over $\mathbf{F}_q$. In the table, q is a prime power, $\pi = \pm 1 \equiv q \mod 4$, and $m = (q^2-1)_2$ is the 2-share of $q^2 - 1$. In the last two rows of the table, $\mu = \mu(\tau)$ is the Thompson group of τ and $W(\Phi)$ is an abbreviation for the Weyl group $\mathrm{Weyl}(\Phi)$ of type Φ.

group	$G_2(q), {}^2D_4(q)$	$L_n^\epsilon(q)$	$Sp_n(q)$	$\Omega_{4n}^+(q)$	$\Omega_{2n+1}(q)$
system	$G_2[m]$	$L_n^{\pi\epsilon}[m]$	$Sp_n[m]$	$\Omega_{4n}^+[m]$	$\Omega_{2n+1}[m]$
$\mu : \pi\epsilon = 1$	$W(A_2)$	$W(A_{n-1})$	$E_{2^{n/2}}$	$W(D_{2n})$	$W(D_n)$
$\mu : \pi\epsilon = -1$	$W(A_2)$	$E_{2^{[n/2]}}$	$E_{2^{n/2}}$	$W(D_{2n})$	$W(D_n)$

group	$\Omega_{4n+4}^-(q)$	$\Omega_{4n+2}^\epsilon(q)$	$F_4(q)$	$E_6^\epsilon(q)$	$E_7(q)$	$E_8(q)$
system	$\Omega_{4n+4}^-[m]$	$\Omega_{4n+2}^{\pi\epsilon}[m]$	$F_4[m]$	$E_6^{\pi\epsilon}[m]$	$E_7[m]$	$E_8[m]$
$\mu : \pi\epsilon = 1$	$W(D_{2n+1})$	$W(D_{2n+1})$	$W(D_4)$	$W(E_6)$	$W(E_7)$	$W(E_8)$
$\mu : \pi\epsilon = -1$	$W(D_{2n+1})$	$W(D_{2n})$	$W(D_4)$	$W(D_4)$	$W(E_7)$	$W(E_8)$

Implicit in the table is the fact that the 2-fusion system of $X(q)$ does not depend on q, but on X and $(q^2-1)_2$, and possibly also on π. This fact emerges in the proof of our major theorems, as a consequence of the uniqueness, up to isomorphism, of a quadratic fusion packet τ with suitable invariants, such as $\mu(\tau)$. For example the Thompson groups of the Lie packets are calculated in Chapter 5. The last section in the paper supplies an even stronger uniqueness condition for Lie packets.

Next the notation for the exceptional examples.

Write $L_2[2m]^{(1)}$ for the 2-fusion system $\mathcal{F}_S(G)$ of G, where G is the extension of $L_2(q^2)$ of degree 2 with a semidihedral Sylow 2-subgroup S of order $2m$. Write $L_3^+[m]^{(1)}$ for $\mathcal{F}_S(G)$ where G is the extension of $L \cong L_3^\epsilon(q)$ by a graph automorphism and L has a wreathed Sylow 2-subgroup of order $m^2/2$. The 2-fusion system of L is $L_3^+[m]$, while if L has semidihedral Sylow groups then its 2-fusion system is $L_3^-[m]$. Write M_{12} for the 2-fusion system of M_{12}.

Write $\hat{Sp}_6(2)$, $\hat{\Omega}_8^+(2)$ for the universal covering group of $Sp_6(2)$, $\Omega_8^+(2)$, respectively. We also use that notation for the 2-fusion system of the group.

We write $L_3(2)/E_8$ for both the nonsplit extension of E_8 by its general linear group and for the 2-fusion system of that group. Similarly $L_3(2)/E_{64}$ denotes both the maximal parabolic in $Sp_6(2)$ with this structure and its 2-fusion system, and $L_3(2)/2^{3+6}$ denotes both the parabolic in $\Omega_8^+(2)$ and its 2-fusion system. The groups and systems $\hat{L}_3(2)/2^{3+6}$ and $\check{L}_3(2)/2^{3+6}$ are defined in 5.10.10 and 5.10.12.

The systems AE_n, $\hat{A}E_n$, $H\hat{A}E_n$, and $\check{A}E_n$ are defined and discussed in section 5.1. For example AE_n is the split extension of A_n by its natural module.

A summary of basic concepts, notation, and terminology from the theory of quaternion fusion packets can be found in section 5.11.

1.2. Background

For background involving fusion systems, we refer the reader for the most part to [**BLO03**] or [**AKO11**]. However in this section we recall some notation, terminology, and definitions involving fusion systems used frequently in this paper.

Let $\mathcal{C}$ be a category and A, B objects in $\mathcal{C}$. Write $\hom_{\mathcal{C}}(A, B)$ for the set of morphisms in $\mathcal{C}$ from A to B, and $Aut_{\mathcal{C}}(A)$ for the group of automorphisms of A in $\mathcal{C}$. Given an isomorphism $\alpha : A \to B$ and C a subobject of A, write α^* for the bijection $\alpha^* : \hom_{\mathcal{C}}(C, A) \to \hom_{\mathcal{C}}(C\alpha, B)$ defined by $\alpha^* : \phi \mapsto \alpha^{-1}\phi\alpha$.

Let $\tilde{\mathcal{F}}$ be a fusion system on a p-group $\tilde{S}$. A *morphism* $\alpha : \mathcal{F} \to \tilde{\mathcal{F}}$ of fusion systems is a group homomorphism $\alpha : S \to \tilde{S}$ such that the induced map takes morphisms of $\mathcal{F}$ to morphisms of $\tilde{\mathcal{F}}$, as defined in II.2.2 in [**AKO11**].

For $\mathcal{E}$ a subsystem of $\mathcal{F}$ on $T \leq S$ and $\alpha \in \hom_{\mathcal{F}}(T, S)$, let $\mathcal{E}\alpha^*$ be the fusion system on $T\alpha$ with $\hom_{\mathcal{E}\alpha^*}(P\alpha, Q\alpha) = \hom_{\mathcal{E}}(P, Q)\alpha^*$ for $P, Q \leq T$. Notice that $\alpha : \mathcal{E} \to \mathcal{E}\alpha^*$ is an isomorphism.

Let S be a group. For $x, y \in S$, $x^y = y^{-1}xy$ is the *conjugate* of x by y, and $c_y : S \to S$ defined by $c_y : x \mapsto x^y$ is *conjugation by* y.

A *fusion system* on S is a category $\mathcal{F}$ whose objects are the subgroups of S, and such that for $P, Q \leq S$, $\hom_{\mathcal{F}}(P, Q)$ is a set of injective group homomorphisms from P into Q, and such that:

(1) for each $s \in S$ and $P, Q \leq S$ with $P^s \leq Q$, $c_s : P \to Q$ is in $\hom_{\mathcal{F}}(P, Q)$, and

(2) for each $\phi \in \hom_{\mathcal{F}}(P, Q)$, $\phi : P \to P\phi$ and $\phi^{-1} : P\phi \to P$ are morphisms in $\mathcal{F}$.

From now on let p be a prime, S a finite p-group, and $\mathcal{F}$ a fusion system on S. We call S the *Sylow group* of $\mathcal{F}$.

Given $P \leq S$, let

$$P^{\mathcal{F}} = \{P\phi : \phi \in \hom_{\mathcal{F}}(P, S)\}$$

be the set of *$\mathcal{F}$-conjugates* of P. Define P to be *fully centralized, fully normalized* if for all $Q \in P^{\mathcal{F}}$, $|C_S(P)| \geq |C_S(Q)|$, $N_S(P)| \geq |N_S(Q)|$, respectively. Write $\mathcal{F}^f$ for the set of fully normalized subgroups of S.

Given a subgroup T of S and a family $(\Delta_i : i \in I)$ of morphism in $\mathcal{F}$ between subgroups of T, define the *fusion system generated by the family* to be the intersection of all fusion systems on T containing the morphisms in each Δ_i, and write $\langle \Delta_i : i \in I \rangle_T$ for that fusion system. If the choice of T is clear, we often omit the subscript T.

For $U \leq S$, define $N_{\mathcal{F}}(U)$, $C_{\mathcal{F}}(U)$ to be the subsystems of $\mathcal{F}$ on $T = N_S(U)$, $C_S(U)$ such that morphisms from $P \leq T$ into T consist of those $\phi \in \hom_{\mathcal{F}}(P, T)$ extending to $\varphi \in \hom_{\mathcal{F}}(PU, T)$ such that φ acts on U, centralizes U, respectively. Call $N_{\mathcal{F}}(U)$ the *normalizer in $\mathcal{F}$* of U and $C_{\mathcal{F}}(U)$ the *centralizer in $\mathcal{F}$* of U, respectively.

Our fusion system $\mathcal{F}$ is *saturated* if:

(I) For all $P \in \mathcal{F}^f$, P is fully centralized and $Aut_S(P) \in Syl_p(Aut_{\mathcal{F}}(P))$, and

(II) whenever $P \leq S$ and $\phi \in \hom_{\mathcal{F}}(P, S)$ with $P\phi$ fully centralized, then ϕ extends to a member of $\hom_{\mathcal{F}}(N_\phi, S)$, where

$$N_\phi = \{g \in N_S(P) : c_g^* \in Aut_S(P\phi)\},$$

using the $*$-notation defined above.

In the remainder of the section assume $\mathcal{F}$ is a saturated fusion system on the finite p-group S. We have the following theorem of Puig; see I.5.5 in [**AKO11**] for a proof.

LEMMA 1.2.1. *For $U \in \mathcal{F}^f$, $N_{\mathcal{F}}(U)$ and $C_{\mathcal{F}}(U)$ are saturated fusion systems.*

A subgroup P of S is *centric* if for each $Q \in P^{\mathcal{F}}$, $C_S(Q) \leq Q$. Further P is *radical* if $Inn(P) = O_p(Aut_{\mathcal{F}}(P))$. Write $\mathcal{F}^c$, $\mathcal{F}^r$ for the centric, radical subgroups of S, respectively. For $Y \subseteq X = \{f, r, c\}$, set

$$\mathcal{F}^Y = \bigcap_{y \in Y} \mathcal{F}^y.$$

LEMMA 1.2.2. *(Alperin's Fusion Theorem)* $\mathcal{F} = \langle Aut_{\mathcal{F}}(U) : U \in \mathcal{F}^{frc} \rangle$.

PROOF. See for example I.3.5 in [**AKO11**]. □

DEFINITION 1.2.3. The *Alperin data* for $\mathcal{F}$ consists of

(1) S and

(2) $\mathcal{F}^{frc}$ and

(3) the subgroups $Aut_{\mathcal{F}}(U)$ of $Aut(U)$, as U varies over $\mathcal{F}^{frc}$.

LEMMA 1.2.4. *The Alperin data determines $\mathcal{F}$ up to isomorphism; that is if $\tilde{\mathcal{F}}$ is a saturated fusion systems on $\tilde{S}$ and $\alpha : S \to \tilde{S}$ is an isomorphism of groups* *which induces a bijection* $\alpha : \mathcal{F}^{frc} \to \tilde{\mathcal{F}}^{frc}$ *such that for each* $U \in \mathcal{F}^{frc}$, $\alpha^* : Aut(U) \to Aut(U\alpha)$ *maps* $Aut_{\mathcal{F}}(U)$ *to* $Aut_{\tilde{\mathcal{F}}}(U\alpha)$, *then* α *induces an isomorphism of* $\mathcal{F}$ *with* $\tilde{\mathcal{F}}$.

PROOF. This follows from Remark 9.11 in [**Asc10**]; it is almost immediate from Alperin's Fusion Theorem. □

For $P \leq S$, define $\mathfrak{A}(P) = \mathfrak{A}_{\mathcal{F}}(P)$ to consist of those $\alpha \in \hom_{\mathcal{F}}(N_S(P), S)$ such that $P\alpha \in \mathcal{F}^f$.

LEMMA 1.2.5. *Let* $P \leq S$ *and* $\phi \in \hom_{\mathcal{F}}(P, S)$.

(1) If $P\phi$ *is fully centralized then* ϕ *extends to a member of* $\hom_{\mathcal{F}}(PC_S(P), S)$.

(2) If $P\phi$ *is fully normalized then there exists* $\chi \in Aut_{\mathcal{F}}(P)$ *such that* $\chi\phi$ *extends to a member of* $\mathfrak{A}(P)$.

(3) $\mathfrak{A}(P) \neq \emptyset$.

PROOF. These are special cases of I.5.2.c in [**AKO11**]. □

A subgroup T of S is *strongly closed* in S with respect to $\mathcal{F}$ if for each $t \in T$, $t^{\mathcal{F}} \subseteq T$. The notions of an $\mathcal{F}$*-invariant subsystem* of $\mathcal{F}$, *weakly normal subsystem* and *normal subsystem* of $\mathcal{F}$ are defined in Definition I.6.1 of [**AKO11**]. We write $\mathcal{E} \trianglelefteq \mathcal{F}$ to indicate $\mathcal{E}$ is normal in $\mathcal{F}$. If $\mathcal{E}$ is an $\mathcal{F}$-invariant subsystem on T then T is strongly closed in S with respect to $\mathcal{F}$ and the invariant condition and Frattini condition of I.6.1 in [**AKO11**] are satisfied. An $\mathcal{F}$-invariant system $\mathcal{E}$ is normal if $\mathcal{E}$ is saturated and the extension condition of I.6.1 in [**AKO11**] (called condition (N1) in [**Asc08a**]) is satisfied.

Recall for Σ a collection of subgroups of S, $[\Sigma]_{\mathcal{F}}$ is the smallest normal subsystem $\mathcal{E}$ of $\mathcal{F}$ such that Σ is contained in the Sylow group of $\mathcal{E}$. Such a subsystem exists by Theorem 1 in [**Asc11**]. We call $[\Sigma]_{\mathcal{F}}$ the *normal closure* of Σ in $\mathcal{F}$.

Write $O^{p'}(\mathcal{F})$ for the normal closure of S in $\mathcal{F}$.

LEMMA 1.2.6. *(Craven's Theorem) If* $\mathcal{E}$ *is a saturated* $\mathcal{F}$*-invariant subsystem of* $\mathcal{F}$ *then* $O^{p'}(\mathcal{E}) \trianglelefteq \mathcal{F}$.

PROOF. See [**Cra11**]. □

The normal subsystem $O^p(\mathcal{F})$ of $\mathcal{F}$ is defined in I.7.4 in [**AKO11**]. If $\mathcal{F}_0$ is a normal subsystem of $\mathcal{F}$ on S_0 and $S_0 \leq T \leq S$ with T strongly closed in S with respect to $\mathcal{F}$, then there is a product subsystem $T\mathcal{F}_0$ of $\mathcal{F}$ on T defined in 8.19 of [**Asc11**].

If T is strongly closed in S with respect to $\mathcal{F}$, then from section 8 in [**Asc08a**] we can form the factor system $\mathcal{F}/T$ and then form the surjective morphism $\Theta : \mathcal{F} \to \mathcal{F}/T$ of fusion systems as in Definition 12.13 of [**Asc11**].

LEMMA 1.2.7. *Let* T *be strongly closed in* S *with respect to* $\mathcal{F}$, $\mathcal{F}^+ = \mathcal{F}/T$, *and* $\Theta : \mathcal{F} \to \mathcal{F}^+$ *the natural surjection. Let* $\mathcal{E}$ *be a subsystem of* $\mathcal{F}$ *on* E *and* $\mathcal{E}^+ = \mathcal{E}\Theta$.

(1) If $\mathcal{E}$ *is saturated then so is* $\mathcal{E}^+$.

(2) If $\mathcal{E} \trianglelefteq \mathcal{F}$ *then* $\mathcal{E}^+ \trianglelefteq \mathcal{F}^+$.

(3) If $T \leq Z(\mathcal{F})$ *and* $\mathcal{E}^+ \trianglelefteq \mathcal{F}$ *then the preimage of* $\mathcal{E}^+$ *in* $\mathcal{F}$ *is normal in* $\mathcal{F}$.

PROOF. See 8.5 in [**Asc08a**] or II.5.4 in [**AKO11**] for (1). See 8.9.2 in [**Asc08a**] for (2) and 8.10 in [**Asc08a**] for (3). □

A subgroup Q of S is *normal* in $\mathcal{F}$ if $\mathcal{F} = N_{\mathcal{F}}(Q)$. Write $O_p(\mathcal{F})$ for the largest normal subgroup of $\mathcal{F}$. Further $\mathcal{F}$ is *constrained* if there exists a normal centric subgroup of $\mathcal{F}$.

A *model* for $\mathcal{F}$ is a finite group G such that $S \in Syl_p(G)$, $F^*(G) = O_p(G)$, and $\mathcal{F} = \mathcal{F}_S(G)$. If G is a model for $\mathcal{F}$ then $O_p(G) = O_p(\mathcal{F})$ is centric, so $\mathcal{F}$ is constrained. By a theorem in [**BCG$^+$05**], the converse is true.

LEMMA 1.2.8. *If $\mathcal{F}$ is constrained then*
(1) $\mathcal{F}$ possesses a model G.
(2) G is unique up to an isomorphism which is the identity on S.
(3) If $\alpha : \mathcal{F} \to \tilde{\mathcal{F}}$ is an isomorphism of fusion systems and $\tilde{G}$ is a model for $\tilde{\mathcal{F}}$ then there exists an isomorphism $\breve{\alpha} : G \to \tilde{G}$ extending α, and $\breve{\alpha}$ is determined up an automorphism c_z of G, for $z \in Z(S)$.
(4) The map $H \mapsto \mathcal{F}_{S\cap H}(H)$ is a bijection between the set of normal subgroups of G and the set of normal subsystems of $\mathcal{F}$.

PROOF. See Theorem III.5.10 in [**AKO11**] for (1), and II.4.3 in [**AKO11**] for (3). Part (3) implies (2). See II.7.5 in [**AKO11**] for (4). □

We recall Definition 2.7.2 from [**Asc19**]:

DEFINITION 1.2.9. For $T \leq S$ define $\mathrm{sub}_0(\mathcal{F}, T) = \mathcal{F}$, and, proceeding recursively, given $\mathrm{sub}_i(\mathcal{F}, T)$ define $\mathrm{sub}_{i+1}(\mathcal{F}, T) = [T]_{\mathrm{sub}_i(\mathcal{F},T)}$. Write $n(\mathcal{F}, T)$ for the first n such that $\mathrm{sub}_n(\mathcal{F}, T) = \mathrm{sub}_{n+1}(\mathcal{F}, T)$, and set $\mathrm{sub}(\mathcal{F}, T) = \mathrm{sub}_{n(\mathcal{F},T)}(\mathcal{F}, T)$. We call $\mathrm{sub}(\mathcal{F}, T)$ the *subnormal closure* of T in $\mathcal{F}$, and $\mathrm{sub}_i(\mathcal{F}, T)$, $1 \leq i \leq n(\mathcal{F}, T)$, the *subnormal series* for T in $\mathcal{F}$. From 2.7.3 in [**Asc19**], the members of the subnormal series are indeed subnormal subsystems of $\mathcal{F}$, so in particularly these systems are saturated.

A subgroup $X \in \mathcal{F}^{fc}$ is *essential* in $\mathcal{F}$ if $Out_{\mathcal{F}}(X)$ has a strongly p-embedded subgroup. Write $\mathcal{F}^e$ for the set of essential subgroups of $\mathcal{F}$.

LEMMA 1.2.10. *(Alperin-Goldschmidt Fusion Theorem) If $\mathcal{F}$ is saturated then $\mathcal{F} = \langle Aut_{\mathcal{F}}(R) : R \in \mathcal{F}^e \cup \{S\}\rangle_S$.*

PROOF. See for example I.3.5 in [**AKO11**]. □

LEMMA 1.2.11. *(1) $\mathcal{F}^e \subseteq \mathcal{F}^{frc}$.*
(2) $O_2(\mathcal{F})$ is contained in each member of $\mathcal{F}^{frc}$.

PROOF. See I.3.3.a in [**AKO11**] for (1) and I.4.5 in [**AKO11**] for (2). □

LEMMA 1.2.12. *Let $U \in \mathcal{F}^f$ and $\mathcal{E} \trianglelefteq \mathcal{F}$. Then $N_{\mathcal{E}}(U) \trianglelefteq N_{\mathcal{F}}(U)$.*

PROOF. This is 8.24 in [**Asc11**]. □

1.3. An outline of the proof

In this section $\tau = (\mathcal{F}, \Omega)$ is a quaternion fusion packet.

We begin with a preview of some of the concepts and notation involving τ that will be formally introduced later in the paper. Then, with this background in place, we give an outline of the proof of our main theorems appearing in section 1.1.

Let $\mathcal{Z} = \mathcal{Z}(\tau)$ be the union of the classes $z(K)^{\mathcal{F}}$ as K ranges over Ω, where $z(K)$ is the involution in K. Define $\mathcal{Z}_S = \mathcal{Z}_S(\tau) = \{z(K) : K \in \Omega\}$, and for $z \in \mathcal{Z}_S$ set $\Omega(z) = \{K \in \Omega : z(K) = z\}$. Set $W_S = \langle \mathcal{Z}_S \rangle$. For $K \in \Omega$ we define a fusion system $\mathcal{O}_K$ on K. If $m = 8$ set $\mathcal{O}_K = \mathcal{F}_K(K)$. If $m > 8$ set $\mathcal{O}_K = \langle O^2(Aut_{\mathcal{F}}(Q)) : Q \in \mathcal{Q}(K) \rangle_K$, where $\mathcal{Q}(K)$ is the set of Q_8-subgroups of K.

Let $z \in \mathcal{Z}_S \cap \mathcal{F}^f$, and set $O(z) = \langle \Omega(z) \rangle$. By 2.6.6.1, $O(z)$ is the central product of the members of $\Omega(z)$. Further by 2.6.8, $\mathcal{F}$ contains a central product $\mathcal{O}(z)$ of the subsystems $\mathcal{O}_K$, for $K \in \Omega(z)$, and by 2.6.11, $\mathcal{O}(z) \trianglelefteq C_{\mathcal{F}}(z)$, so each $\mathcal{O}_K$ is subnormal in $C_{\mathcal{F}}(z)$.

Similarly set $O(\tau) = \langle \Omega \rangle$. By 3.1.5, $O(\tau)$ is a central product of the members of Ω, and by 3.3.7, $\mathcal{F}$ contains a central product $\mathcal{O}(\tau) = \prod_{K\in\Omega} \mathcal{O}_K$ that is normal in $N_{\mathcal{F}}(W_S)$.

Next for $t \in \mathcal{Z} = \mathcal{Z}(\tau)$ and $\alpha \in \mathfrak{A}(t)$,

$$\Delta(t) = \{(K \cap C_S(t)\alpha)\alpha^{-1} : K \in \Omega(t\alpha) \text{ and } |K \cap C_S(t)\alpha| > 2\}.$$

By 3.1.3, $\Delta(t)$ is independent of the choice of $\alpha \in \mathfrak{A}(t)$. Set $\mathcal{Z}_\Delta = \mathcal{Z}_\Delta(\tau) = \{t \in \mathcal{Z} : \Delta(t) \neq 1\}$ and

$$\Delta = \Delta(\tau) = \bigcup_{t \in \mathcal{Z}_\Delta} \Delta(t).$$

By 3.1.5, $\Delta(z) = \Omega(z)$ for $z \in \mathcal{Z}_S$, and then by 3.1.11, each member of $\Delta - \Omega$ is cyclic of order $m/2$.

Let $\eta(\tau)$ consist of the sets η of cyclic subgroups of S such that

(i) η is $\langle \eta \rangle$-invariant, and

(ii) for each $V \in \eta$, V is contained in a unique $D(V) \in \Delta$, and either $V = D(V)$ or $D(V) \in \Omega$ and $|D(V) : V| = 2$, and

(iii) the map $V \mapsto D(V)$ is a bijection of η with Δ.

Let $\eta_S(\tau)$ denote the set of S-invariant members of $\eta(\tau)$. By 3.1.10, $\eta_S(\tau)$ is nonempty, and by 3.1.11, for each $\eta \in \eta(\tau)$, $\langle \eta \rangle$ is abelian. We write $W(\tau)$ for $\{\langle \eta \rangle : \eta \in \eta_S(\tau)\rangle$.

By 3.3.14:

LEMMA 1.3.1. *Either $\eta(\tau) = \{\eta\}$ is of order 1, or $m = 8$ and if $\mathcal{F}$ is transitive on Ω then $\Delta = \Omega$.*

Thus, the exceptional case of 1.3.1 aside, $W(\tau) = \{W\}$ is of order 1 and W is weakly closed in S with respect to $\mathcal{F}$.

In any case pick $\eta \in \eta(\tau)$ and set $W = \langle \eta \rangle$. By 3.1.12, for each $K \in \Omega$, K acts on W and W acts on K. Then $k \in K - W$ induces an involutory automorphism $d_{K\cap W}$ on W via conjugation. Then from 3.1.20, for each $V \in \eta$, V is conjugate in $N_{\mathcal{F}}(W)$ to $K \cap W$ for some $K \in \Omega$, so we get a conjugate d_V of $d_{K\cap W}$ in $Aut_{\mathcal{F}}(W)$. Set $D_\eta = \{d_V : V \in \eta\}$ and $\mu_\eta = \langle D_\eta \rangle \leq Aut_{\mathcal{F}}(W)$. If $\eta(\tau) = \{\eta\}$ then $\mu = \mu_\eta$ is uniquely determined. But even if $|\eta(\tau)| > 1$, we find in 3.1.25 that $\mu_\eta \cong \mu_{\eta'}$ for all $\eta, \eta' \in \eta(\tau)$. Thus we write $\mu = \mu(\tau)$ for μ_η where $\eta \in \eta(\tau)$, and we call μ the *Thompson group* of τ.

If $U, V \in \eta$ are distinct then by 3.1.18, $\langle d_U, d_V \rangle$ is isomorphic to E_4, S_3, or D_{12}. In the last case, by definition, $z(U) \in \mathcal{A}(z(V))$, so in particular the graph $\mathcal{A}(\tau)$ is nonempty. Moreover this shows that if $\mathcal{A}(\tau) = \emptyset$ then D_η is a set of 3-transpositions of μ.

Therefore in section 4.1 we consider groups generated by 3-transpositions. In 4.1.25 we determine all groups generated by a conjugacy class of 3-transpositions satisfying Hypothesis 4.1.12. Then in 4.2.14 we find that if D_η is a set of 3-transpositions of μ then Hypothesis 4.1.12 is indeed satisfied.

By 3.1.10 each $W \in W(\tau)$ is normal in S. Therefore $W \in \mathcal{F}^f$ and by 3.3.9, $\mathcal{M} = [\Omega]_{N_\mathcal{F}(C_S(W))}$ is constrained. Hence by 1.2.8, $\mathcal{M}$ has a model M. We write $M(\tau)$ for the set of such groups M as W ranges over $W(\tau)$. By 1.3.1, if $m > 8$ or $\mathcal{F}$ is transitive on Ω and μ is nonabelian, then $W(\tau) = \{W\}$ and $M(\tau) = \{M\}$ are of order 1.

Assume D_η is a conjugacy class of 3-transpositions of μ. By an earlier remark, 4.1.25 supplies a list of possibilities for μ. Then, in 4.3.8, that list is refined to show that:

LEMMA 1.3.2. *If D_η is a conjugacy class of 3-transpositions and μ is nonabelian, then μ is the Weyl group Weyl(Φ), where Φ is a Coxeter diagram of type A_n, D_n, or E_n.*

Moreover in 4.3.1 we define a universal group $\bar{G} = \bar{\omega}(\Phi, m)$ and show in 4.3.8 that there is a surjective homomorphism $\pi : \bar{G} \to M$ with $\ker(\pi) \leq Z(\bar{G})$. In particular a Sylow 2-group T of M is determined up to isomorphism by Φ, m, and $\ker(\pi)$; often in a Lie packet, we have $T = S$ and $\mathcal{F}_T(M) = N_\mathcal{F}(W)$, in which case S and $N_\mathcal{F}(W)$ are determined up to isomorphism.

Further $\xi = (\mathcal{F}_T(M), \Omega)$ is a quaternion fusion packet which decomposes as a central product $\xi = \xi_1 * \cdots * \xi_r$ with ξ_i a coconnected component of ξ such that $\mu(\xi_i) = \mu_i$ and $\mu = \mu_1 \times \cdots \times \mu_r$.

The proof of Theorem 2 is completed in section 6.7, building on work in sections 6.2, 6.3, and 6.4. The proof quickly reduces to the case where $\mathcal{F}$ is constrained, so by 1.2.8, $\mathcal{F}$ has a model G. In particular $F^*(G) = O_2(G)$, and the proof of Theorem 2 can now be carried out in the category of such groups, where things go relatively smoothly and some difficult machinery required to establish later theorems is unnecessary.

Define the *order* of τ be the triple $(|S|, |\Omega|, |Mor(\mathcal{F})|)$ and order such triples lexicographically.

We prove many of our theorems by "induction on the order of τ". Hence, when necessary, we may assume the following *Inductive Hypothesis* : Theorem 1 holds in proper subpackets $\rho = (\mathcal{Y}, \Gamma)$ of τ with $\mathcal{Y} \leq \mathcal{F}$, $\Gamma \subseteq \Omega$, and ρ satisfying the hypothesis of Theorem 1. Sometimes we assume the stronger *Extended Inductive Hypothesis*: For each $Q \leq S$ centralizing $\mathcal{F}^\circ$ with $Q \cap \mathcal{Z}(\tau) = \emptyset$, we have $\tau_Q^+ = (\mathcal{F}^{\circ +}, \Omega^+)$ satisfying the Inductive Hypothesis, where $(Q\mathcal{F}^\circ)^+ = Q\mathcal{F}^\circ/Q$.

Observe that if τ is a counter example to Theorem 1 of minimal order, then τ satisfies the Extended Inductive Hypothesis.

We handle the case $\mathcal{A}(\tau) \neq \emptyset$ later when proving Theorem 3 in section 7.2; so assume $\mathcal{A}(\tau) = \emptyset$. Then by 6.6.6, $\tau^\circ = \tau_1 * \cdots * \tau_r$ is a central product of subpackets $\tau_i = (\mathcal{F}_i, \Omega_i)$, where $\mathcal{F}_i = [\Omega_i]_{\mathcal{F}_i}$ is transitive on Ω_i. In particular the first three parts of Theorem 4 hold, and the fourth part holds by 6.6.6 if either $\tau \neq \tau_i$ or $\tau = \tau_i$ satisfies Theorem 1. Thus Theorems 1 and 3 imply Theorem 4. Similarly, with a bit of work, Theorem 1 (together with Theorem 2) implies Theorems 3-8, although in practice Theorem 1 is established by a series of steps which include Theorems 3-8.

Most of our theorems include the hypotheses that $\mathcal{F} = \mathcal{F}^\circ$ and $\mathcal{F}$ is transitive on Ω. But it is difficult to use these hypotheses early in the game, so in effect we are looking at a system $\mathcal{F}$ with a normal subsystem $\mathcal{F}^\circ = \mathcal{F}_1 * \cdots * \mathcal{F}_r$ where $\tau_i = (\mathcal{F}_i, \Omega_i)$ for $1 \leq i \leq r$ are the coconnected components of τ°. Thus often we produce a subpacket $\rho = \rho_1 * \cdots * \rho_r$ with $\rho = (\mathcal{Y}, \Omega)$, $r > 1$, and $\rho_i = (\mathcal{Y}_i, \Omega_i)$ the coconnected components of ρ. We then show that $\mathcal{Y}_1$ is subnormal in $\mathcal{F}$ with conjugates $\mathcal{Y}_i$, $1 \leq i \leq r$, so as $\mathcal{F} = \mathcal{F}^\circ$ we have $\mathcal{F}^\circ = \mathcal{Y}$, contradicting $\mathcal{F}$ transitive on Ω and $r > 1$. This is often accomplished by showing that $\mathcal{Y}_1$ is tightly embedded in $\mathcal{F}$, and appealing to the theory of tightly embedded subsystems in [**Asc19**]. Such arguments appear in section 8.1.

For example assume μ is nonabelian. Then from earlier remarks, $\xi = \xi_1 * \cdots * \xi_r$ and as $\mathcal{F}$ is transitive on Ω, we have $N_{\mathcal{F}}(W)$ transitive on the ξ_i and the μ_i, $1 \leq i \leq r$, with μ_i generated by a conjugacy class of 3-transpositions. Then by 1.3.2, $\mu_i \cong \mathrm{Weyl}(\Phi)$ for each i and some Coxeter diagram Φ of type A_n, D_n, or E_n. It turns out this forces $\mu_i = \mu(\rho_i)$, and if $r > 1$ and the Inductive Hypothesis holds then ρ_i is a known packet with Thompson group μ_i, facilitating the arguments described in the previous paragraph. So, via this approach and with some effort, we can reduce to the case where either μ is abelian; or $\mathcal{A}(\tau) \neq \emptyset$ and $\mu \cong D_{12}$; or $\mu \cong \mathrm{Weyl}(\Phi)$ with Φ of type A_n, D_n, or E_n.

In order to work effectively with our system $\mathcal{F}$ and eventually identify it as one of our examples, it seems necessary to show that $\mathcal{F}$ is generated by some nice collection of subsystems. In Definition 3.4.2 we are led to define the following subsystem of $\mathcal{F}$:

$$\mathcal{E}(\tau) = \langle S\mathcal{O}(\tau), N_{\mathcal{F}}(O(\theta)), N_{\mathcal{F}}(W) : W \in W(\tau) \text{ and } \emptyset \neq \theta \subseteq \mathcal{Z}_S \text{ with } O(\theta) \in \mathcal{F}^f \rangle.$$

Here $O(\theta) = \prod_{t \in \theta} O(t)$. Sometimes we can prove $\mathcal{F} = \mathcal{E}(\tau)$ but sometimes $\mathcal{F} \neq \mathcal{E}(\tau)$. In general, given a candidate $\mathcal{B}$ for $\mathcal{F}$ containing $Aut_{\mathcal{F}}(S)$, the Alperin-Goldschmidt Fusion Theorem 1.2.10 says that if $\mathcal{B}$ is proper in $\mathcal{F}$ then their exists $R \in \mathcal{F}^e$ such that $Aut_{\mathcal{B}}(R)$ is proper in $Aut_{\mathcal{F}}(R)$. This observation is exploited in section 3.4 and later sections where we prove $\mathcal{F}$ is generated by suitable subsystems: for example often $\mathcal{F} = \mathcal{E}(\tau)$.

As mentioned earlier, most of our theorems are proven under the hypothesis that $\mathcal{F} = \mathcal{F}^\circ$, but it can be difficult to make use of this condition early in our analysis. Therefore in many cases we need to identify a candidate $\mathcal{E}$ for $\mathcal{F}^\circ$ as a subsystem of $\mathcal{F}$, prove $\mathcal{E} = \tilde{\mathcal{F}}$ for some example $\tilde{\mathcal{F}}$, and prove $\mathcal{F}^\circ = \mathcal{E}$. Typically we begin by proving something like $\mathcal{F} = \langle \mathcal{F}_1, \mathcal{F}_2 \rangle$ is generated by a suitable pair of subsystems. Now $\tilde{\mathcal{F}}$ satisfies our hypothesis, so also $\tilde{\mathcal{F}} = \langle \tilde{\mathcal{F}}_1, \tilde{\mathcal{F}}_2 \rangle$. Then we identify subsystems $\mathcal{E}_i$ of $\mathcal{F}_i$, prove $\mathcal{E}_2 \cong \tilde{\mathcal{F}}_2$, so that we may take $\mathcal{E}_2 = \tilde{\mathcal{F}}_2$, and then prove that $\mathcal{E}_1 = \tilde{\mathcal{F}}_1$. Thus $\mathcal{E} = \langle \mathcal{E}_1, \mathcal{E}_2 \rangle = \langle \tilde{\mathcal{F}}_1, \tilde{\mathcal{F}}_2 \rangle = \tilde{\mathcal{F}}$; in particular $\mathcal{E} = \mathcal{E}^\circ$ is saturated. Then we also prove that $\mathcal{E}_i \trianglelefteq \mathcal{F}_i$ for $i = 1, 2$, and perhaps prove one or two other technical conditions. This makes possible an appeal to section 1.4 in [**Asc10**], which implies that $\mathcal{E} \trianglelefteq \mathcal{F}$, so as $\mathcal{E} = \mathcal{E}^\circ$ we have $\tilde{\mathcal{F}} = \mathcal{E} = \mathcal{F}^\circ = \mathcal{F}$, completing our identification of $\mathcal{F}$.

The actual proof of Theorem 1 is begun with the case $n_S = |\mathcal{Z}_S| = 1$. Most of our exceptional examples arise in this case, so it requires special treatment. Thus in the next few paragraphs we consider the case where $n_S = 1$, so that $\mathcal{Z}_S = \{z\}$.

In Theorem 5, $|\Omega| = 1$, so $n_S = 1$ and μ is $\mathbf{Z}_2$ or $S_3 = \mathrm{Weyl}(A_2)$. Theorem 5 is proved in section 7.1; see in particular Remark 7.1.30.

Next in Theorem 3, $\mathcal{A}(\tau) \neq \emptyset$. Then $\mathcal{A}(z) \neq \emptyset$ for some $z \in \mathcal{Z}_S$, so by 7.2.2 we have $|\Omega(z)| = 2$. Then, in the most important subcase where $\mathcal{F} = [O(z)]_{\mathcal{F}}$, we have $\Omega = \Omega(z)$, so indeed $n_S = 1$, and $\mu \cong D_{12}$. Theorem 3 is proved in section 7.2; see Remark 7.2.25.

Given Theorem 3, from now on we may assume μ is generated by a set of 3-transpositions. Then from 1.3.2, in each case to be considered we can eventually reduce to μ abelian or $\mu \cong \mathrm{Weyl}(\Phi)$ with Φ of type A_n, D_n, or E_n.

We next consider the cases where $|\Omega(z)| = 2$. Sections 9.1 and 9.2 contain basic results established under this hypothesis. Section 9.3 considers the case where $\mathcal{Z} \cap O(z) \neq \{z\}$, which turns out to force $n_S = 1$ and $\mu \cong S_4 = \mathrm{Weyl}(A_3) = \mathrm{Weyl}(D_3)$. Section 9.4 considers the case μ abelian; as $|\Omega(z)| = 2$ this eventually forces $\mu \cong E_4$. In addition to identifying $\mathcal{F}$ we prove that $\mathcal{F} = \langle C_{\mathcal{F}}(z), Aut_{\mathcal{F}}(R_i) : 1 \leq i \leq k\rangle$ for suitable $R_i \in \mathcal{F}^e$ and small k. Then in section 9.5 we treat the case where $\mu \cong S_4$ and $\mathcal{Z} \cap O(z) = \{z\}$. Here, in the process of identifying $\mathcal{F}$, we show $\mathcal{F} = \mathcal{E}(\tau) = \langle C_{\mathcal{F}}(z), N_{\mathcal{F}}(W)\rangle$, except in one unusual case where $\mathcal{F} = \langle \mathcal{E}(\tau), C_{\mathcal{F}}(b)\rangle$ for a certain involution b in W. This completes the treatment of the case $n_S = 1$ and $|\Omega(z)| = 2$.

In our examples, $|\Omega(z)|$ is 1, 2, or 4. In section 10.2, we prove this fact in general when $\mathcal{F} = \mathcal{F}^\circ$ is transitive on Ω and τ satisfies the Extended Inductive Hypothesis , reducing to the case $n_S = 1$ and $\mu \cong \mathrm{Weyl}(D_4)$, treated earlier in section 10.1, where we identify $\mathcal{F}$ as one of our examples, each of which is a subsystem of $P\Omega_8^+[m]$. This completes the treatment of the case $n_S = 1$.

We next turn to a proof of Theorem 6, where we assume $\mathcal{F} = \mathcal{F}^\circ$ is transitive on Ω, $|\Omega(z)| > 1$ for $z \in \mathcal{Z}_S$, and $\mathcal{A}(\tau) = \emptyset$. We also assume the Extended Inductive Hypothesis. As discussed in the previous paragraph, from section 10.2 we have $|\Omega(z)| = 2$ or 4, and in the latter case Theorem 6 holds. Therefore we may assume $|\Omega(z)| = 2$.

We interject a generational result proven in sections 11.2 and 11.3 under one of the following two hypotheses:

HYPOTHESIS 1.3.3. *For $z \in \mathcal{Z}_S$, $|\Omega(z)| = 2$ and $\mu \cong$ Weyl(D_n) for some $n \geq 3$.*

HYPOTHESIS 1.3.4. *For $z \in \mathcal{Z}_S$, $|\Omega(z)| = 1$ and either $m(W_S) = n_S$ or $\prod_{t \in \mathcal{Z}_S} t = 1$ and $m(W_S) = n_S - 1$.*

In Theorems 11.3.10 and 11.3.11 it is shown that if 1.3.3 or 1.3.4 hold then, with known exceptions, $\mathcal{F} = \mathcal{B}(\tau)$, where $\mathcal{B}(\tau) = \langle \mathcal{E}(\tau), C_{\mathcal{F}}(x) : x \in \mathcal{W}(\tau) \cap \mathcal{F}^f\rangle$ for a suitable set $\mathcal{W}(\tau)$ of involutions in W or W_S.

Note that Hypothesis 1.3.4 is satisfied if $|\Omega(z)| = 1$ and $\mu \cong \mathrm{Weyl}(\Phi)$ with Φ of type A_n, D_n, or E_6. If Φ is E_7 then $n_S = 7$ is odd, so some member of Ω is normal in S; then 3.4.6 says $\mathcal{F} = \mathcal{E}(\tau)$. If Φ is E_8 then Theorem 14.2.24 supplies the required generational result. When Φ is D_n with $n \neq 4$ we appeal to Theorem 14.1.18.

Next Theorem 11.3.11 is extended in Theorem 11.5.3 to show that if $\mathcal{F}^\circ = O^2(\mathcal{F})$ is the system of an orthogonal group then $\mathcal{F}$ is generated by a small collection of nice subsystems. For example if $m > 8$ and $\mathcal{F}^\circ = P\Omega_{2n}^+[m]$ then $\mathcal{F} = \langle \mathcal{SO}(\tau), \mathcal{F}_S(SM)\rangle$.

Suppose $\mu \cong \mathrm{Weyl}(D_n)$ for some $n \geq 3$. In section 12.1 we treat the case $n = 4$, while the case $n \geq 5$ is handled in section 12.2. Note if $n = 3$ then $\mu \cong S_4$ with $n_S = 1$, a case already dealt with.

Therefore to complete the proof of Theorem 6, modulo the Extended Inductive Hypothesis, it remains to show that, under the hypothesis of Theorem 6 and the Extended Inductive Hypothesis, we have $\mu \cong \mathrm{Weyl}(D_n)$ for some $n \geq 2$. When μ is nonabelian this is accomplished in Theorem 12.3.50. The case μ abelian is handled in section 9.4. The proof of Theorem 6 is wrapped up in sections 12.4 and 16.2.

Assuming the Extended Inductive Hypothesis, we have established Theorem 6 and treated the case $|\Omega(z)| > 1$ for $z \in \mathcal{Z}_S$. Thus in the remainder of the discussion we may assume $|\Omega(z)| = 1$. We turn next to Theorem 7, so we assume $\mathcal{F} = \mathcal{F}^\circ$ is transitive on Ω, μ is abelian, and $n_S > 1$. As usual, we also assume the Extended Inductive Hypothesis.

We begin in section 13.1, where we reduce to the case where $Aut_{\mathcal{F}}(O(\tau))^\Omega = Sym(\Omega)$. Set $\Theta = \mathcal{Z}_S$. As $|\Omega(z)| = 1$ for $z \in \Theta$, also $Aut_{\mathcal{F}}(W_S)^\Theta = Sym(\Theta)$, so it follows that either W_S is the natural module for $Sym(\Theta)$ or its image modulo e_Θ. In particular Hypothesis 1.3.4 is satisfied, so $\mathcal{F} = \mathcal{B}(\tau)$ by an earlier remark. Then this result is extended further in Theorem 13.5.3. Finally the proof of Theorem 7 (modulo the Extended Inductive Hypothesis) is completed in section 13.6; see in particular Remark 13.6.7.

We've proven Theorems 2 through 5, established Theorems 6 and 7 under the Extended Inductive Hypothesis, and reduced the proof of Theorem 1 to the case where $|\Omega(z)| = 1$ for $z \in \mathcal{Z}_S$ and μ is nonabelian. This is the case covered by Theorem 8. After some basic lemmas are established in section 15.1, we prove in Theorem 15.2.2 that D_η is a conjugacy class of 3-transpositions of μ. Therefore by 1.3.2, μ is Weyl(Φ) where Φ is of type A_n, D_n, or E_n. As discussed earlier, Hypothesis 1.3.4 is satisfied unless Φ is D_n with $n \neq 4$, E_7 or E_8, so, except possibly in these three cases, $\mathcal{F} = \mathcal{B}(\tau)$. In the three exceptional cases, a similar result holds by other means. When Φ is of type A_n the result is extended further in Theorem 15.6.3. Finally the various possibilities for Φ are treated in sections 15.3-15.8.

Formal proofs of Theorem 1 and Theorems 6, 7, and 8 appear in sections 16.1 and 16.2, respectively, although these proofs only gather together the various threads that we've been discussing here.

Basics and examples

CHAPTER 2

Some basic results

Chapter 2 contains a variety of basic results, almost all on saturated fusion systems $\mathcal{F}$ on a p-group S, often with $p = 2$.

For example Theorem 2.1.9 establishes an analogue for fusion systems of a theorem of D. Holt for groups, that determines the finite permutation groups containing an involution fixing a unique point.

A *solvable component* of $\mathcal{F}$ is a subnormal subsystem of $\mathcal{F}$ isomorphic to the 2-fusion system of A_4 or $SL_2(3)$. Section 2.2 develops a theory of solvable components similar to the theory in [**Asc11**] on components.

Section 2.3 studies *intrinsic* $SL_2[m]$-components and solvable components $\mathcal{C}$ of $\mathcal{F}_z = C_{\mathcal{F}}(z)$ for involutions z in S: that is $z \in \mathcal{C}$. The results in this section are used in Theorem 2.4.1 to show that intrinsic $SL_2[m]$ components and solvable components give rise to quaternion fusion packets. This makes possible the derivation of the Main Theorem from Theorems 1 and 2.

Sections 2.5 and 2.6 establish basic results on quaternion fusion packets $\tau = (\mathcal{F}, \Omega)$. In particular, given $K \in \Omega$ and $z = z(K)$, we find in Theorem 2.6.11 that K is Sylow in a subnormal subsystem $\mathcal{O}_K$ of $\mathcal{F}_z$, and that the normal closure $\mathcal{O}(z)$ of $\Omega(z)$ in $\mathcal{F}_z$ is the central product of the $\mathcal{O}_J$ for $J \in \Omega(z)$.

In Theorem 2.7.3 we find that if $\mathcal{F}$ is the product $SO^2(\mathcal{F})$ of S with $O^2(\mathcal{F})$ the central product of $SL_2[m]$-components, then $\mathcal{F}$ is uniquely determined by S and the set Ω of Sylow groups of the components.

Finally in section 2.8 we consider a finite group G with a strongly embedded subgroup, and an involution a in G such that $G = \langle a^G \rangle$. Then Theorem 2.8.15 determines all $\mathbf{F}_2G$-modules $V = \langle [V, a]^G \rangle$ with $m([V, a]) = 2$. This result is used in various places to study essential subgroups of S.

2.1. Preliminary lemmas

In this section $\mathcal{F}$ is a fusion system on a finite p-group S.

DEFINITION 2.1.1. For Δ a collection of subgroups of S, define $\Phi(S, \Delta) = \langle O^p(Aut(D)) : D \in \Delta \rangle_S$ to be the fusion system on S generated by these automizers. Set $\mathfrak{D}(\mathcal{F}) = \{D \leq S : O^p(Aut_{\mathcal{F}}(D)) \neq 1\}$.

LEMMA 2.1.2. *Let $\mathcal{F}$ be saturated and $\mathcal{U}$ a set of representatives in $\mathcal{F}^f$ for the orbits of $\mathcal{F}$ on $\mathcal{F}^{rc}$. Set $\Delta = \mathfrak{D}(\mathcal{F}) \cap \mathcal{U}$ and assume for each $D \in \Delta$ that $O^p(Aut_{\mathcal{F}}(D)) = O^p(Aut(D))$. Then $\mathcal{F} = \Phi(S, \Delta)$.*

PROOF. As $\mathcal{F}$ is saturated, $\mathcal{F} = \langle O^p(Aut_{\mathcal{F}}(D)) : D \in \Delta \rangle$ by 1.3.1 in [**Asc19**]. Then the lemma follows. □

EXAMPLE 2.1.3. Let S be semidihedral of order $2m$. Let $\Delta(S) = \{D \leq S : O^2(Aut(D)) \neq 1\}$. Then $\Delta(S) = \Delta_1 \cup \Delta_2$, where Δ_1 is the set of 4-subgroups of S and Δ_2 is the set of Q_8-subgroups of S. Moreover S is transitive on Δ_i for $i = 1, 2$; for each $D \in \Delta_1$, $O^2(Aut(D)) \cong \mathbf{Z}_3$; for each $D \in \Delta_2$, $O^2(Aut(D)) \cong A_4$; and $O^2(Aut(D)$ is the unique subgroup of $Aut(D)$ containing $Inn(D)$ which is not a 2-group.

(1) Let $L_2(q^2)^{(1)}$ be the extension of $L_2(q^2)$ with semidihedral Sylow 2-groups. Let $L_2[2m]^{(1)}$ be the fusion system of $L \cong L_2(q^2)^{(1)}$ where q is an odd prime power with $(q^4-1)_2 = 2m$. We may take $S \in Syl_2(L)$ and observe that $\mathfrak{D}(\mathcal{F}_S(L)) = \Delta_1$. Hence by 2.1.2, $L_2[2m]^{(1)} = \Phi(S, \Delta_1)$.

(2) Next let $SL_2(q)^{(2)}$ be the extension of degree 2 of $SL_2(q)$ with semidihedral Sylow 2-subgroups, where q is an odd prime power. If $(q^2-1)_2 = m$ we write $SL_2[m]^{(2)}$ for $\mathcal{F}_S(L)$ where $L \cong SL_2(q)^{(2)}$ and $S \in Syl_2(L)$. Observe $\mathfrak{D}(\mathcal{F}_S(L)) = \Delta_2$, so by 2.1.2, $SL_2[m]^{(2)} = \Phi(S, \Delta_2)$.

(3) Finally let $L \cong L_3(q)$ for q an odd prime power with $q \equiv -1 \mod 4$. If $(q^2-1)_2 = m$ we may choose $S \in Syl_2(L)$ and we write $L_3^-[m]$ for $\mathcal{F}_S(L)$. Observe that $\mathfrak{D}(\mathcal{F}_S(L)) = \Delta(S)$, so by 2.1.2, $L_3^-[m] = \Phi(S, \Delta(S))$.

LEMMA 2.1.4. *Assume $\mathcal{F}$ is saturated and T is strongly closed in S with respect to $\mathcal{F}$. Then $\mathcal{E} = \langle O^p(Aut_{\mathcal{F}}(D)) : D \leq T\rangle_T$, regarded as a fusion system on T, is $\mathcal{F}$-invariant.*

PROOF. Observe first that, by definition of $\mathcal{E}$, $O^p(Aut_{\mathcal{E}}(D)) = O^p(Aut_{\mathcal{F}}(D))$ for each $D \leq T$.

We next claim that $Aut_{\mathcal{F}}(T) \leq Aut(\mathcal{E})$. Namely for $\alpha \in Aut_{\mathcal{F}}(T)$,

$$\begin{aligned}\mathcal{E}\alpha^* &= \langle O^p(Aut_{\mathcal{F}}(D)) : D \leq T\rangle\alpha^* \\ &= \langle O^p(Aut_{\mathcal{F}}(D))\alpha^* : D \leq T\rangle \\ &= \langle O^p(Aut_{\mathcal{F}}(D\alpha) : D\alpha \leq T\rangle = \mathcal{E},\end{aligned}$$

establishing the claim.

By the claim and 3.3 in [**Asc08a**], to complete the proof it suffices to show that $\mathcal{E}$ is $\mathcal{F}$-Frattini; that is for $P \leq T$ and $\alpha \in \hom_{\mathcal{F}}(P, S)$, $\alpha = \varphi\phi$ for some $\varphi \in Aut_{\mathcal{F}}(T)$ and $\phi \in \hom_{\mathcal{E}}(P\varphi, T)$.

By Alperin's Fusion Theorem and 5.5.3 in [**Asc08a**], $\mathcal{F} = \langle Aut_{\mathcal{F}}(U) : U \in \mathcal{U}\rangle$, for any set $\mathcal{U}$ of representatives in $\mathcal{F}^f$ for the orbits of $\mathcal{F}$ on $\mathcal{F}^{rc}$. We choose $\mathcal{U}$ so that for each $U \in \mathcal{U}$, $U_T = U \cap T$ is in $\mathcal{F}^f$. Now it suffices to take α to be the restriction to P of some $\beta \in Aut_{\mathcal{F}}(U)$ for some $U \in \mathcal{U}$ with $P \leq U$. Further we choose P and α to be a counter example with $n = |T : P|$ minimal. If $P = T$ then $\alpha \in Aut_{\mathcal{F}}(T)$, contrary to the choice of P; therefore $n > 1$. Further by minimality of n, $P = U_T$, so as T is strongly closed, $P \trianglelefteq N_{\mathcal{F}}(U)$ and hence $\alpha \in Aut_{\mathcal{F}}(P)$.

Let $V = PC_S(P)$, G a model for $N_{N_{\mathcal{F}}(V)}(P)$, $c : G \to Aut(P)$ the conjugation map, $Q = N_T(P)$, and $Y = O^p(G)C_G(P)$. Thus $Gc = Aut_{\mathcal{F}}(P)$, so $Yc = O^p(Aut_{\mathcal{F}}(P))$. As $\alpha \in Aut_{\mathcal{F}}(P)$, $\alpha = gc$ for some $g \in G$. As Q is strongly closed in $N_S(P)$ with respect to G, $QY \trianglelefteq G$ and $G = N_G(Q)Y$ by a Frattini argument. Thus $g = xy$ for some $x \in N_G(Q)$ and $y \in Y$. Let $c_x \in Aut(Q)$ be conjugation by x. As $P < Q$, $c_x = \varphi\gamma$ for some $\varphi \in Aut_{\mathcal{F}}(T)$ and $\gamma \in \hom_{\mathcal{E}}(Q\varphi, T)$ by minimality of n. Also $yc \in Yc = O^p(Aut_{\mathcal{F}}(P)) \leq Aut_{\mathcal{E}}(P)$, so $\alpha = gc = \varphi\phi$, where $\phi = \gamma_{|P\varphi} \cdot (yc) \in \hom_{\mathcal{E}}(P\varphi, T)$, contrary to the choice of α. $\square$

LEMMA 2.1.5. *Assume $\mathcal{F}$ is saturated, $T \leq S$, and the following hold:*

(a) There is a normal subgroup Q of $\mathcal{F}$ of index p in T.

(b) Let G be a model for $N_{\mathcal{F}}(QC_S(Q))$ and $H = \langle T^G \rangle$. Assume $T \in Syl_p(H)$ and $O^p(Aut_G(Q)) = O^p(Aut_H(Q)) \neq 1$.

(c) There exists $W \leq T$ such that W is weakly closed in S with respect to $\mathcal{F}$, $W \not\leq Q$, and $Aut_{N_{\mathcal{F}}(W)}(T)$ is a p-group.

Then

(1) Q is the greatest member of $\mathfrak{D}(\mathcal{F})$ contained in T.

(2) $\mathcal{D} = \mathcal{F}_T(H) \trianglelefteq \mathcal{F}$.

(3) $\mathcal{D} = \langle O^p(Aut_{\mathcal{F}}(Q))\rangle_T$ as a fusion system on T.

PROOF. Let $P \leq T$ and $\phi \in \hom_{\mathcal{F}}(P, S)$. As $Q \trianglelefteq \mathcal{F}$ by (a), ϕ extends to $\varphi \in \hom_{\mathcal{F}}(PQ, S)$ acting on Q. If $P \leq Q$ then $P\phi = P\varphi \leq Q$. If $P \not\leq Q$ then as $|T : Q| = p$ by (a), $PQ = T$. Similarly as $W \not\leq Q$, $T = WQ$. Thus $P\phi \leq (PQ)\varphi = T\varphi = W\varphi Q\varphi = WQ = T$ as W is weakly closed by (c). Hence T is strongly closed in S.

Next if $P \in \mathfrak{D}(\mathcal{F})$ we may choose $\phi \in Aut_{\mathcal{F}}(P)$ to be a p'-element. But then if $P \not\leq Q$ then $\varphi \in Aut_{\mathcal{F}}(T)$ is a p-element as $W\varphi = W$ since W is weakly closed, so $\varphi \in Aut_{N_{\mathcal{F}}(W)}(T)$, which is a p-group by (c). Thus each member of $\mathfrak{D}(\mathcal{F})$ contained in T is contained in Q. Further by (b), $Q \in \mathfrak{D}(\mathcal{F})$, so (1) holds.

Let $\mathcal{E} = \langle O^p(Aut_{\mathcal{F}}(D)) : D \leq T\rangle$, regarded as a fusion system on T. By 2.1.4, $\mathcal{E}$ is $\mathcal{F}$-invariant. By (1), $\mathcal{E} = \mathcal{B}$, where $\mathcal{B} = \langle O^p(Aut_{\mathcal{F}}(Q))\rangle_T$. On the other hand $\mathcal{B} = \mathcal{D}$ by (b), so $\mathcal{E} = \mathcal{B} = \mathcal{D}$ and (3) holds. By definition in (2), $\mathcal{D} = \mathcal{F}_T(H)$, while $T \in Syl_p(H)$ by (b), so $\mathcal{D}$ is saturated. Finally by (c), $Aut_{\mathcal{F}}(T)$ is a p-group, so $\mathcal{F}, \mathcal{D}$ satisfies condition (N1) and hence (2) follows. □

LEMMA 2.1.6. *Assume $\mathcal{F}$ is saturated and let $I = \{1, \ldots, n\}$, let $\{\mathcal{C}_i : i \in I\}$ be an $\mathcal{F}$-invariant set of components of $\mathcal{F}$, let $Q_i \in Syl(\mathcal{C}_i)$, and let Q Sylow be in $\mathcal{C} = \mathcal{C}_1 \cdots \mathcal{C}_n$. Define $\mathcal{F}_1 = N_{\mathcal{F}}(\mathcal{C}_1)$ and, proceeding recursively, given $1 \leq i < n$ define $\mathcal{F}_{i+1} = N_{\mathcal{F}_i}(\mathcal{C}_{i+1})$. Set $\mathcal{E} = \mathcal{F}_n$ and $T = \bigcap_{i \in I} N_S(Q_i)$. Then $\mathcal{E}$ is a normal subsystem of $\mathcal{F}$ on T.*

PROOF. Let $\mathcal{A} = N_{\mathcal{F}}(Q)$, $\mathcal{S} = \{Q_i : i \in I\}$, and form $\mathcal{B} = \mathcal{A}(\mathcal{S})_*$ as in Theorem 2.8.11 in [**Asc19**]. By that theorem, $\mathcal{B} \trianglelefteq \mathcal{A}$ with T Sylow in $\mathcal{B}$.

Set $\mathcal{Y} = N_{\mathcal{E}}(Q)$. By 2.8.11.2 in [**Asc19**], for each $V \leq T$, $Aut_{\mathcal{B}}(V) = A(V)$, and by construction of $A(V)$ in section 2.8 of [**Asc19**], each $\alpha \in A(V)$ extends to $\hat{\alpha} \in Aut(VQ)$ acting on each Q_i, so $A(V) \leq Aut_{\mathcal{Y}}(V)$. Therefore $\mathcal{B} \leq \mathcal{Y}$. Conversely $\mathcal{Y} \leq \mathcal{B}$ by 2.8.11.4 in [**Asc19**]. Thus $\mathcal{Y} = \mathcal{B} \trianglelefteq \mathcal{A}$. As $\{\mathcal{C}_i : i \in I\}$ is $\mathcal{F}$-invariant, $\mathcal{C} \trianglelefteq \mathcal{F}$ by 7.4 in [**Asc11**]. Hence $\mathcal{E} \trianglelefteq \mathcal{F}$ by Theorem 1.5.1 in [**Asc19**]. □

LEMMA 2.1.7. *Assume $\mathcal{F}$ is saturated and constrained and let G be a model for $\mathcal{F}$. Then the map $H \mapsto \mathcal{F}_{S\cap H}(H)$ is a bijection between the set of subnormal subgroups of G and the set of subnormal subsystems of $\mathcal{F}$.*

PROOF. Let $\mathcal{S}, \mathcal{T}$ be the set of subnormal subgroups of G, subnormal subsystems of $\mathcal{F}$, respectively, and for $H \in \mathcal{S}$ set $H\varphi = \mathcal{F}_{S\cap H}(H)$.

Let $H \in \mathcal{S}$. As $F^*(G) = O_p(G)$, also $F^*(H) = O_p(H)$. Let M be the normal closure of H in G; then either $H = G$ or M is proper in G. We prove the lemma by induction on the order of G, so pick a counter example G of minimal order.

If $H = G$ then $H\varphi = \mathcal{F} \in \mathcal{T}$. If $H \neq G$ then as $F^*(M) = O_p(M)$, $H\varphi$ is subnormal in $M\varphi$ by minimality of G. By 1.2.8.4, $M\varphi \in \mathcal{T}$, so $H\varphi \in \mathcal{T}$. Thus $\varphi : \mathcal{S} \to \mathcal{T}$.

Similarly if $\mathcal{H} \in \mathcal{T} - \{\mathcal{F}\}$ then $\mathcal{H}$ is subnormal in a proper normal subsystem $\mathcal{N}$ of $\mathcal{F}$ and $\mathcal{N} = N\varphi$ for some proper normal subgroup N of G by 1.2.8.4. Then by minimality of G, $\mathcal{H} = H\varphi$ for some H subnormal in N, so $H \in \mathcal{S}$. That is φ is a surjection, so it remains to show that φ is an injection.

Assume $K \in \mathcal{S}$ with $H\varphi = K\varphi$ but $H \neq K$. Then some $J \in \{H, K\}$ is proper in G, so its normal closure A is proper in G. As $H\varphi = K\varphi$, $T = S \cap K = S \cap H \leq A$, so $B = \langle T^G \rangle \leq A$ and hence also $L = O^{p'}(H)$ and $O^{p'}(K)$ are contained in A. Now by 1.2.8.4, $L\varphi = O^{p'}(H\varphi) = O^{p'}(K\varphi)$, so working in A, by minimality of G, $L = O^{p'}(K)$.

Set $Q = O_p(L)$, $Z = C_G(Q)$, and $P = O_p(Z)$. As $F^*(N_G(Q)) = O_p(N_G(Q))$, if Q is not normal in G, then working in $N_G(Q)$ we conclude $H = K$ by minimality of G. Therefore $Q \trianglelefteq G$, so also $Z \trianglelefteq G$.

As $F^*(H) = Q = F^*(K)$, $C_H(Q) \leq Q \geq C_K(Q)$ so p'-elements of H and K are faithful on Q. As $H\varphi = K\varphi$, $Aut_H(Q) = Aut_K(Q)$, so $HZ = KZ$. Hence by minimality of G, $G = HZ = KZ$.

Let x be a p'-element of H and $W = [P, x]$. Then also $W = [W, x]$ and $W \leq M$. Then by induction on the subnormal length of H, $W \leq H$, so $W \leq Q$. Therefore $[P, x] \leq P \cap Q \leq Z(Z)$, so as $F^*(Z) = P$ we have $[Z, x] \leq C_Z(P/Z(Z)) \leq P$, and then $[Z, x] \leq P \cap Q \leq Q$. Therefore $[Z, O^p(H)] \leq Q$. Then as $G = HZ$, we conclude that $O^p(H)Q \trianglelefteq G$. By symmetry also $O^p(K)Q \trianglelefteq G$. By 7.8.2 in [**Asc11**], $O^p(H)\varphi = O^p(H\varphi) = O^p(K)\varphi$, so if $G \neq O^p(H)O^p(K)Q$ then $O^p(H) = O^p(K)$ by minimality of G. But then $H = O^p(H)L = K$, a contradiction. Therefore $G = O^p(H)O^p(K)Q$, so $L \trianglelefteq G$. Hence $H = LO^p(H)Q \trianglelefteq G$ and by symmetry also $K \trianglelefteq G$. Therefore $H = K$ by 1.2.8.4. This contradicts the choice of H and K and completes the proof. □

LEMMA 2.1.8. *Assume $\mathcal{F}$ is saturated and $\mathcal{E}$ is a subsystem of $\mathcal{F}$ on S. Assume $z \in Z(S)$ such that*

(a) $C_{\mathcal{F}}(z) \leq \mathcal{E}$ and

(b) for each $s \in z^{\mathcal{F}}$ there is $\alpha \in \hom_{\mathcal{E}}(C_S(s), S)$ with $s\alpha = z$.

Then $\mathcal{E} = \mathcal{F}$.

PROOF. By Alperin's Fusion Theorem it suffices to show that if $R \in \mathcal{F}^{frc}$ then $Aut_{\mathcal{F}}(R) = Aut_{\mathcal{E}}(R)$. As $z \in Z(S)$, $z \in Z(R)$. Let $\beta \in Aut_{\mathcal{F}}(R)$ and $s = z\beta$. By (b), there is $\alpha \in \hom_{\mathcal{E}}(C_S(s), S)$ with $s\alpha = z$. As $z \in Z(R)$, $R \leq C_S(s)$ so $\zeta = \beta\alpha \in \hom_{\mathcal{F}}(R, S)$ with $z\zeta = z$, so by (a), $\zeta \in \hom_{\mathcal{E}}(R, S)$. Then $\beta = \zeta\alpha^{-1} \in Aut_{\mathcal{E}}(R)$, completing the proof. □

THEOREM 2.1.9. *(Holt's Theorem for Fusion Systems) Assume $\mathcal{F}$ is saturated and $\mathcal{E}$ is a saturated subsystem of $\mathcal{F}$ on S. Assume $z \in Z(S)$ such that*

(1) $z^{\mathcal{F}} = z^{\mathcal{E}}$ and

(2) $C_{\mathcal{F}}(z) \leq \mathcal{E}$.

Then $\mathcal{E} = \mathcal{F}$.

PROOF. As $z \in Z(S)$, z is fully centralized in $\mathcal{E}$. Let $s \in z^{\mathcal{F}}$; then by (1), $s \in z^{\mathcal{E}}$, so as z is fully centralized in $\mathcal{E}$ and $\mathcal{E}$ is saturated, there is $\alpha \in \mathfrak{A}_{\mathcal{E}}(s)$ with $s\alpha = z$. Now the theorem follows from 2.1.8. □

LEMMA 2.1.10. *Assume $\mathcal{F}$ is saturated, $p = 2$, and S is dihedral or semidihedral. Then either*

(1) $\mathcal{F}$ has one class of involutions, or

(2) $\mathcal{F} \neq O^2(\mathcal{F})$.

PROOF. Assume $\mathcal{F}$ has more than one class of involutions. Then there is a subgroup T of S of index 2 that is strongly closed in S with respect to $\mathcal{F}$. Then by I.7.5 in [**AKO11**], $O^2(\mathcal{F}) \neq \mathcal{F}$. □

LEMMA 2.1.11. *Assume $\mathcal{F}$ is saturated and let $Q = Z(\mathcal{F})$ and $\mathcal{F}^+ = \mathcal{F}/Q$. Assume $\mathcal{C}^+$ is a component of $\mathcal{F}^+$ and let $\mathcal{C}$ be the preimage in $\mathcal{F}$ of $\mathcal{C}^+$. Then $\mathcal{C} = QO^2(\mathcal{C})$ and $O^2(\mathcal{C})$ is a component of $\mathcal{F}$.*

PROOF. By 8.10 in [**Asc08a**], $\mathcal{C}$ is a subnormal subsystem of $\mathcal{F}$. Let T be Sylow in $\mathcal{C}$, $\mathcal{B} = O^2(\mathcal{C})$, and R Sylow in $\mathcal{B}$. By 7.7.3 in [**Asc11**], $\mathcal{B} \trianglelefteq \mathcal{C}$, so $\mathcal{B}$ is subnormal in $\mathcal{F}$. Then, replacing $\mathcal{F}$ by $\mathcal{C}$, we may assume $\mathcal{F} = \mathcal{C}$. Hence $\mathcal{F} = Q\mathcal{B}$ by 7.15 in [**Asc11**]. Therefore $\mathcal{F}^+ = \mathcal{B}^+$. Let $\Theta : \mathcal{F} \to \mathcal{F}^+$ be the natural map and $\varphi = \Theta_{|R} : R \to S^+ = R^+$. Then $\varphi : \mathcal{B} \to \mathcal{B}^+ = \mathcal{B}/(Q \cap R)$ is the natural map with $\mathcal{B}^+$ simple and $Q \cap R \leq Z(\mathcal{B})$. Therefore $\ker(\varphi) = Z(\mathcal{B})$, so $\mathcal{B}$ is quasisimple, completing the proof. □

LEMMA 2.1.12. *Assume $\mathcal{F}$ is a saturated and $U \leq Q \leq S$ with U and Q normal in $\mathcal{F}$. Define $\mathcal{R} = C_{\mathcal{F}}(Q/U)$ to be the fusion system on $R = C_S(Q/U)$ such that for $P \leq R$, $\hom_{\mathcal{R}}(P, R)$ consists of those $\phi \in \hom_{\mathcal{F}}(P, S)$ such that for some $\hat{\phi} \in \hom_{\mathcal{F}}(PQ, S)$ extending ϕ, $[Q, \hat{\phi}] \leq U$. Then $\mathcal{R}$ is weakly normal in $\mathcal{F}$, so $O^{p'}(\mathcal{R}) \trianglelefteq \mathcal{F}$.*

PROOF. First R is strongly closed in S with respect to $\mathcal{F}$: if $x \in R$ and $\alpha \in \hom_{\mathcal{F}}(x, S)$ then $[Q, x] \leq U$ so $[Q, x\alpha] = [Q\alpha, x\alpha] = [Q, x]\alpha \leq U\alpha = U$.

Next by I.5.5 in [**AKO11**], $\mathcal{R}$ is a saturated fusion system on R.

Claim $\mathcal{R}$ is $\mathcal{F}$-invariant. For $x \in R$ and $\phi \in \hom_{\mathcal{R}}(x, S)$ $[x, \phi] \in U$, so $[x, \phi] = u$ for some $u \in U$. Then for $\alpha \in \hom_{\mathcal{F}}(X, S)$ with $x \in X \leq R$ and $x\phi \in X$, we have

$$(x\alpha)(\phi\alpha^*) = (x\alpha)(\alpha^{-1}\phi\alpha) = (x\phi)\alpha = (xu)\alpha = x\alpha u\alpha \in x\alpha U,$$

so $\phi\alpha^* \in \hom_{\mathcal{R}}(x, R)$. This proves the claim.

By the claim and paragraph two, $\mathcal{R}$ is weakly normal in $\mathcal{F}$. Then by Craven's Theorem 1.2.6, $O^{p'}(R) \trianglelefteq \mathcal{F}$. □

DEFINITION 2.1.13. Assume $\mathcal{F}$ is saturated and $p = 2$. Following Definition 2.19 in [**AO16**], define the *reduction* of $\mathcal{F}$ to be

$$\mathfrak{red}(\mathcal{F}) = (C_{\mathcal{F}}(O_2(\mathcal{F}))/Z(O_2(\mathcal{F})))^{\infty},$$

where $\mathcal{E}^{\infty}$ is the last term in the Puig series of $\mathcal{E}$, defined in Definition 2.18 in [**AO16**]: the series obtained by alternating applications of the operators O^2 and $O^{2'}$.

LEMMA 2.1.14. *Assume $\mathcal{F}$ is saturated and $p = 2$. Assume*

(a) $\mathcal{F}_0$ is a quasisimple normal subsystem of $\mathcal{F}$ on S_0 with $Z(\mathcal{F}_0) = U \cong E_4$.

(b) $Aut_{\mathcal{F}}(U) = GL(U)$.

(c) $C_S(S_0) = C_S(\mathcal{F}_0)$.

(d) $O^2(C_{Aut_{\mathcal{F}}(S_0)}(U)) = 1$.

(e) There exists T strongly closed in S with respect to $\mathcal{F}$ containing S_0, such that $C_T(S_0) = U$ and $[U,T] \neq 1$.
Let $\mathcal{E} = [T]_{\mathcal{F}}$ be the normal closure of T in $\mathcal{F}$. Then
(1) T is Sylow in $\mathcal{E}$.
(2) $F^*(\mathcal{E}) = \mathcal{F}_0$.
(3) $C_S(S_0) = U$, so $N_{\mathcal{E}}(S_0)$ has a model G.
(4) $C_G(U) = C_T(U)$ and $G/C_T(U) \cong S_3$.
(5) $\mathfrak{red}(\mathcal{E}) = \mathcal{F}_0/U$.

PROOF. Assume otherwise and choose $\mathcal{F}$ to be a counter example of minimal order. Observe that as $[T,U] \neq 1$, it follows from (b) that $Aut_{\mathcal{E}}(U) = GL(U)$. As $\mathcal{F}_0$ is quasisimple, $\mathcal{F}_0$ is contained in the normal closure of S_0 in $\mathcal{F}$. Thus if $\mathcal{D}$ is a normal subsystem of $\mathcal{F}$ containing T, then as $S_0 \leq T$ we have $\mathcal{F}_0 \leq \mathcal{D}$, and it follows that $\mathcal{D}$ satisfies conditions (a)-(e). Therefore as each of our conclusions is a statement about $\mathcal{E}$, rather than $\mathcal{F}$, we conclude that $\mathcal{F} = \mathcal{D}$ by minimality of $\mathcal{F}$. Then as $\mathcal{E} \trianglelefteq \mathcal{F}$, we have $\mathcal{F} = \mathcal{E}$.

Let $\mathcal{C} = E(C_{\mathcal{F}}(\mathcal{F}_0))$ and S_c Sylow in $\mathcal{C}$. Then $[T, S_c] \leq C_T(S_0) = U$ by (e), so by 9.5.2 in [**Asc11**], T centralizes $\mathcal{C}$. Therefore $\mathcal{E} = [T]_{\mathcal{F}} \leq C_{\mathcal{F}}(\mathcal{C})$, so as $\mathcal{F} = \mathcal{E}$ by paragraph one, we have $\mathcal{F} = C_{\mathcal{F}}(\mathcal{C})$, so $\mathcal{F}_0 = E(\mathcal{F})$.

Set $Q = O_2(\mathcal{F})$ and $\mathcal{R} = C_{\mathcal{F}}(Q/U)$. Now $[T,Q] \leq T \cap Q \leq C_T(S_0) = U$ by (e). Therefore $T \leq C_S(Q/U)$, so $\mathcal{E} \leq O^{2'}(\mathcal{R})$ by 2.1.12. Hence $\mathcal{F} = O^{2'}(\mathcal{R})$ as $\mathcal{F} = \mathcal{E}$ by paragraph one.

Let $Q_0 = S_0Q$. Then $S_1 = C_S(Q_0) \leq C_S(S_0) = C_S(\mathcal{F}_0)$ by (c). As $\mathcal{F}_0 = E(\mathcal{F})$ we have $Q = F^*(C_{\mathcal{F}}(\mathcal{F}_0))$, so $S_1 = C_{S_1}(Q) \leq Q \leq Q_0$. That is $C_S(Q_0) \leq Q_0$, so $N_{\mathcal{F}}(Q_0)$ has a model M. Set $H = \langle T^M \rangle$ and $L = O^2(C_M(U))$.

As $\mathcal{F} = O^{2'}(\mathcal{R})$, L centralizes Q. By (d), L centralizes S_0. Thus L centralizes $Q_0 = QS_0$, so as $F^*(M) = Q_0$, we have $L = 1$. Therefore $C_M(U) = O_2(M)$ and $M/O_2(M) = GL(U) \cong S_3$ by (b). Hence $M = YS$ with $|S : O_2(M)| = 2$ and $Y \cong \mathbf{Z}_3$. Similarly as T is strongly closed in S with respect to M, we have $H = YT$ with $H/C_T(U) \cong S_3$.

As $C_M(U) = O_2(M)$, we conclude that $C_{\mathcal{F}}(\mathcal{F}_0) = Q$ and then that $C_{\mathcal{F}}(Q) = C_S(Q)\mathcal{F}_0$. Now by Lynd transfer [**Lyn14**], $O^2(C_{\mathcal{F}}(Q)) = \mathcal{F}_0$, so that $\mathfrak{red}(\mathcal{F}) = \mathcal{F}_0/U$. Hence (5) holds as $\mathcal{F} = \mathcal{E}$.

As T is strongly closed in S, $TO^2(\mathcal{F}) \trianglelefteq \mathcal{F}$, so $\mathcal{F} = TO^2(\mathcal{F})$ by paragraph one.

Claim $T = S$. Assume otherwise; then $T \leq X < S$ with $|S : X| = 2$. Now $S_0 \leq T$ is Sylow in $\mathcal{F}_0 \trianglelefteq \mathcal{F}$, so $N_{\mathcal{F}}(T) \leq N_{\mathcal{F}}(S_0Q) = N_{\mathcal{F}}(Q_0)$, so $N_{\mathcal{F}}(T) = \mathcal{F}_S(N_M(T)) = \mathcal{F}_S(S)$. Therefore $\mathcal{F}/T \cong N_{\mathcal{F}}(T)/T \cong S/T$. Therefore $\mathfrak{hyp}(\mathcal{F}) \leq T$, so as $\mathcal{F} = TO^2(\mathcal{F})$, we have T Sylow in $\mathcal{F}$, establishing the claim.

We've shown $T = S$; in particular (1) holds. As $C_S(S_0) = C_T(S_0) = U$, we conclude that $Q = U$, so $F^*(\mathcal{F}) = \mathcal{F}_0Q = \mathcal{F}_0$, proving (2) and (3). Also $Q_0 = S_0$, so $H = M = G$ with $C_T(U) = O_2(G)$, and (4) holds. Thus the proof is complete. □

LEMMA 2.1.15. *Assume $\mathcal{F}$ is a saturated, $Z \leq Z(\mathcal{F})$, and set $\mathcal{F}^+ = \mathcal{F}/Z$. Assume $\{\mathcal{F}_i : i \in I\}$ is set of subsystems $\mathcal{F}_i$ of $\mathcal{F}$ on S_i with $Z \leq S_i$ such that $\mathcal{F}^+ = \langle \mathcal{F}_i^+ : i \in I \rangle$. Then $\mathcal{F} = \langle \mathcal{F}_i : i \in I \rangle$.*

PROOF. Let $R \in \mathcal{F}^{frc}$ and $\mathcal{E} = \langle \mathcal{F}_i : i \in I \rangle$. As $R \in \mathcal{F}^{frc}$, we have $Z \leq C_S(R) \leq R$. By Alperin's Fusion Theorem 1.2.2, it suffices to show that for each $\alpha \in Aut_{\mathcal{F}}(R)$, $\alpha \in Aut_{\mathcal{E}}(R)$. As $\mathcal{F}^+ = \langle \mathcal{F}_i^+ : i \in I \rangle$, there exists n and $\mathcal{F}_{i_j}$-maps α_j, $1 \leq j \leq n$, with $\alpha^+ = \alpha_1^+ \cdots \alpha_n^+$. Now $R^+ \leq S_1^+$ and $Z \leq R \cap S_1$, so $R \leq S_1$.

Also $\beta = \alpha_1 \cdots \alpha_n$ is an $\mathcal{E}$-map with $\alpha^+ = \beta^+$. Next $R^+\beta^+ = R^+\alpha^+ = R^+$, so $R\beta = R$. As $\alpha^+ = \beta^+$, $\phi = \alpha\beta^{-1} \in Aut_{\mathcal{F}}(R)$ centralizes R/Z. Then as ϕ also centralizes Z, we have $\phi \in O_p(Aut_{\mathcal{F}}(R)) = Inn(R)$ as $R \in \mathcal{F}^r$. Then as $R \in \mathcal{F}^c$, we have $\phi = c_r$ for some $r \in R$, so $\alpha = c_r\beta$ is an $\mathcal{E}$-map, completing the proof. □

LEMMA 2.1.16. *Let X be a set of order $n = 2^a$ and for $b \leq a$ let $\mathcal{U}_b = \{\Pi_1, \ldots, \Pi_b\}$ be a set of partitions of X such that Π_k as 2^k blocks of size $n/2^k$ and Π_{k+1} refines Π_k. Let G_b be the stabilizer in $Sym(X)$ of $\mathcal{U}_b$ and $G_b^{\mathcal{U}_b}$ the induced action of G_b on $\mathcal{U}_b$. Then $G_b^{\mathcal{U}_b}$ is a 2-group.*

PROOF. Indeed $G_b^{\mathcal{U}_b}$ is an iterated wreath product of $\mathbf{Z}_2$; this follows by induction on b. □

2.2. Solvable components

In this section we assume $\mathcal{F}$ is a saturated fusion system on a finite 2-group S. Define a *solvable component* of $\mathcal{F}$ to be a subnormal subsystem $\mathcal{C}$ of $\mathcal{F}$ on $T \leq S$, such that $\mathcal{C} \cong \mathcal{F}_T(G)$ for some group G isomorphic to $SL_2(3)$ or $L_2(3)$.

Write $\mathrm{Sol}(\mathcal{F})$ for the set of solvable components of $\mathcal{F}$, $\mathrm{Comp}(\mathcal{F})$ for the set of components of $\mathcal{F}$, and set $\mathrm{Comp}_+(\mathcal{F}) = \mathrm{Comp}(\mathcal{F}) \cup \mathrm{Sol}(\mathcal{F})$.

LEMMA 2.2.1. *Assume $\mathcal{C}$ is a solvable component of $\mathcal{F}$ on T. Then*

(1) $T \leq O_2(\mathcal{F})$.

(2) Let $\Gamma = \{\mathcal{C}\phi^* : \phi \in \hom_{\mathcal{F}}(T, S)\}$ be the set of conjugates of $\mathcal{C}$ in $\mathcal{F}$. Then a central product $\mathcal{X}$ of the members of Γ is a normal subsystem of $\mathcal{F}$.

(3) Suppose $T \in \mathcal{F}^f$ and let G be a model for $N_{N_{\mathcal{F}}(V_T)}(T)$, where $V_T = TC_S(T)$. Then there is a normal subgroup H of G that is a model for $\mathcal{C}$. Further $[V_T, H] = T$ and if $T \cong Q_8$ then H centralizes $C_S(T)$.

(4) If $T \in \mathcal{F}^f$ then $\mathcal{C} \trianglelefteq N_{\mathcal{F}}(T)$.

PROOF. As $T \trianglelefteq \mathcal{C} \trianglelefteq \trianglelefteq \mathcal{F}$, T is subnormal in $\mathcal{F}$, so (1) follows from 9.10 in [**Asc11**].

Assume $T \in \mathcal{F}^f$ and let $\mathcal{C} = \mathcal{F}_1 \trianglelefteq \cdots \trianglelefteq \mathcal{F}_n = \mathcal{F}$ be a subnormal series. If $n = 1$ then (2) is trivial, so assume $n > 1$, and let $\mathcal{E} = \mathcal{F}_{n-1}$ be a system on $E \leq S$. Let $\Delta = \{\mathcal{C}\psi^* : \psi \in \hom_{\mathcal{E}}(T, E)\}$. By induction on n, there is a central product $\mathcal{Y}$ of the members of Δ normal in $\mathcal{E}$; let Y be Sylow in $\mathcal{Y}$. For $\alpha \in Aut_{\mathcal{F}}(E)$, $\alpha \in Aut(\mathcal{E})$ by 3.3 in [**Asc08a**], so by 7.3.4 in [**Asc11**], $Y\alpha \trianglelefteq \mathcal{E}$. As $Aut_{\mathcal{E}}(Y)$ is irreducible on $Y/\Phi(Y)$, either $Y = Y\alpha$ or $Y \cap Y\alpha \leq \Phi(Y) \cap \Phi(Y\alpha) \leq Z(\mathcal{Y}) \cap Z(\mathcal{Y}\alpha)$. In the first case $\mathcal{Y} = \mathcal{Y}\alpha$ as $Aut_{\mathcal{Y}}(Y)$ is an $Aut_{\mathcal{F}}(Y)$-invariant normal subgroup of $Aut_{\mathcal{E}}(Y)$, and in the second $\mathcal{Y}\alpha \leq C_{\mathcal{E}}(\mathcal{Y})$ and $\mathcal{Y}\mathcal{Y}\alpha$ is the central product of $\mathcal{Y}$ and $\mathcal{Y}\alpha$ by 9.2 in [**Asc11**]. Set

$$\Omega = \bigcup_{\alpha \in Aut_{\mathcal{F}}(E)} \Delta\alpha.$$

Then by repeated applications of 9.2 in [**Asc11**], there is a central product $\mathcal{Z}$ of the members of Ω that is normal in $\mathcal{E}$ by Theorem 3 in [**Asc11**]; let Z be Sylow in $\mathcal{Z}$. By 7.4 in [**Asc11**], $\mathcal{Z} \trianglelefteq \mathcal{F}$. Let $\phi \in \hom_{\mathcal{F}}(T, S)$. By 3.3 in [**Asc08a**], $\phi = \beta\gamma$, where $\beta \in Aut_{\mathcal{F}}(Z)$ and $\gamma \in \hom_{\mathcal{Z}}(T\beta, Z)$. Then $\beta \in Aut(\mathcal{Z})$, so $\mathcal{C}\beta^* \in \Omega$

and γ fixes $\mathcal{C}\beta^*$. Thus $T\phi = T\beta$ and $\mathcal{C}\phi^* = \mathcal{C}\beta^*$ with $\beta \in Aut_{\mathcal{F}}(Z)$, so $\mathcal{Z} = \mathcal{X}$, completing the proof of (2).

Let X be Sylow in $\mathcal{X}$. Recall $V_X = XC_S(X)$. Let $G(X)$ be a model for $N_{\mathcal{F}}(V_X)$. As $\mathcal{X} \trianglelefteq \mathcal{F}$, there is a model L for $\mathcal{X}$ normal in $G(X)$ by Theorem 2 in [**Asc08a**]. Choose G as in (3). As $V_X \leq V_T$ and X is strongly closed in S, X and V_X are normal in $\mathcal{N} = N_{N_{\mathcal{F}}(V_T)}(T)$, so we may take $G = N_{G(X)}(V_T) \cap N_{G(X)}(T)$. As $\mathcal{C} \trianglelefteq \mathcal{X}$, a similar argument shows there is a model H for $\mathcal{C}$ normal in L, so H is subnormal in $G(X)$. Indeed by (2), L is the central product of copies of H permuted by $G(X)$, so as the Sylow group T of H is normal in G it follows that $G \leq N_{G(X)}(H)$. Hence $[V_T, H] \leq V_T \cap H = T$ and H acts on $C_{V_T}(T) = C_S(T)$. Further if $T \cong Q_8$ then $[H, C_S(T)] \leq C_H(T) = Z(H)$, so as $H = O^2(H)$, we conclude that H centralizes $C_S(T)$, completing the proof of (3).

Finally as $\mathcal{C}$ is the fusion system of a group, $\mathcal{C}$ is saturated, and as $Aut_{\mathcal{C}}(T) = O^2(Aut(T)) \trianglelefteq Aut(T)$, $\mathcal{C}$ is $N_{\mathcal{F}}(T)$-invariant. By (3), $\mathcal{C}$ satisfies (N1) in $N_{\mathcal{F}}(T)$, so (4) holds. □

LEMMA 2.2.2. *(1) $\mathcal{F}$ contains a normal subsystem $\mathcal{E}$ which is a central product of the solvable components of $\mathcal{F}$.*

(2) $\mathcal{E} \leq C_{\mathcal{F}}(E(\mathcal{F}))$.

(3) $C_{\mathcal{F}}(E(\mathcal{F}))$ is constrained, and if G is a model for $C_{\mathcal{F}}(E(\mathcal{F}))$ then the normal model for $\mathcal{E}$ in G is the central product of copies of $SL_2(3)$ and $L_2(3)$.

PROOF. Let $\mathcal{C}$ be a solvable component of $\mathcal{F}$ on $T \in \mathcal{F}^f$. By 2.2.1.1, $T \leq O_2(\mathcal{F})$, so $T \leq \mathcal{D} = C_{\mathcal{F}}(E(\mathcal{F}))$.

Set $\mathcal{N} = N_{\mathcal{F}}(T)$. By 2.2.1.4, $\mathcal{C} \trianglelefteq \mathcal{N}$, so as T centralizes $E(\mathcal{F})$, it follows from 10.3 in [**Asc11**] that $E(\mathcal{F}) = E(\mathcal{N})$. Then as $\mathcal{C}$ is normal in $\mathcal{N}$ and constrained, $\mathcal{C}E(\mathcal{F})$ is the central product of $\mathcal{C}$ and $E(\mathcal{F})$ by 9.14 in [**Asc11**]. Thus $\mathcal{C} \leq \mathcal{D}$, establishing (2).

Define $\mathcal{X}$ as in 2.2.1.2 and let X, E be Sylow in $\mathcal{X}, E(\mathcal{F})$, respectively. As $\mathcal{C} \leq \mathcal{D} \trianglelefteq \mathcal{F}$, $\mathcal{X} \leq \mathcal{D}$. By 3.6.2 in [**Asc08a**], $\mathcal{X}$ is $\mathcal{D}$-invariant, and by 2.2.1.2, $\mathcal{X}$ is saturated. By 2.2.1.3, $\alpha \in Aut_{\mathcal{C}}(T)$ of order 3 lifts to $\beta \in Aut_{\mathcal{F}}(TEC_S(TE))$ with $[\beta, EC_S(TE)] \leq Z(T)$. As E is strongly closed in S, it follows that $[E, \beta] \leq E \cap Z(T) \leq Z(E(\mathcal{F}))$, so $[E, \beta] = 1$ by 9.5.1 in [**Asc11**]. Then $\beta_{|TC_D(T)} \in Aut_{\mathcal{D}}(TC_D(T))$, so $\mathcal{X}$ satisfies (N1) in $\mathcal{D}$. Therefore $\mathcal{X} \trianglelefteq \mathcal{D}$, so that $\mathcal{C} \trianglelefteq \trianglelefteq \mathcal{D}$.

By 9.12.3 in [**Asc11**], $\mathcal{D}$ is constrained, so there exists a model G for $\mathcal{D}$. As $\mathcal{X} \trianglelefteq \mathcal{D}$, there is a model M for $\mathcal{X}$ normal in G by Theorem 1 in [**Asc08a**]. If $\mathcal{C}_1 \in Sol(\mathcal{F})$ is not a conjugate of $\mathcal{C}$ then arguing as in the proof of 2.2.1.2, $\mathcal{C}_1$ centralizes $\mathcal{X}$, so (3) follows. Then (3) implies (1). □

LEMMA 2.2.3. *Let $U \in \mathcal{F}^f$ and $\mathcal{C} \in Comp_+(\mathcal{F})$ be contained in $N_{\mathcal{F}}(U)$. Then $\mathcal{C} \in Comp_+(N_{\mathcal{F}}(U))$.*

PROOF. By 2.2.1.2 there is a normal subsystem $\mathcal{X}$ of $\mathcal{F}$ which is a central product of the conjugates of $\mathcal{C}$. By Theorem 5 in [**Asc11**] we can form the system $U\mathcal{X}$. Let T be Sylow in $\mathcal{C}$.

Suppose $\mathcal{C}$ is a component of $\mathcal{F}$. Claim U acts on T. If not $[T, T^u] = 1$ for some $u \in U$, so $[T, u, T] \not\leq Z(\mathcal{C})$, impossible as $\mathcal{C}$ acts on U and $[T, u] \leq U$. This establishes the claim, and by the claim, $U \leq N_{\mathcal{F}}(\mathcal{C})$. By 8.24 in [**Asc11**], $\mathcal{C} \leq N_{\mathcal{X}}(U) \trianglelefteq N_{\mathcal{F}}(U)$. Then applying 1.1.10 in [**Asc19**] to $\mathcal{C}$, $N_{\mathcal{X}}(U)$, $N_{\mathcal{F}}(\mathcal{C})$ in the role of $\mathcal{E}$, $\mathcal{D}$, $\mathcal{F}$, we conclude $\mathcal{C} \trianglelefteq N_{\mathcal{X}}(U)$. But now the lemma holds in this case, so we may assume that $\mathcal{C}$ is solvable.

As $\mathcal{C}$ is solvable, $U\mathcal{X}$ is constrained, so it has a model G. By Theorem 1 in [**Asc08a**], there is a normal model L for $\mathcal{X}$ in G, and a normal model H for $\mathcal{C}$ in L. Then $N_L(U)$ is a model for $N_{\mathcal{X}}(U)$. As $\mathcal{C}$ normalizes U, so does H, so $H \trianglelefteq N_L(U)$, and hence $\mathcal{C} \trianglelefteq N_{\mathcal{X}}(U)$. Finally $N_{\mathcal{X}}(U) \trianglelefteq N_{\mathcal{F}}(U)$ by 8.24 in [**Asc11**]. □

LEMMA 2.2.4. *Let $\mathcal{C}$ be a solvable component of $\mathcal{F}$ on $T \leq S$, and $U \leq S$. Let $\mathcal{D} \leq \mathcal{F}$ be the central product of the subsystems $\mathcal{C}^u$, $u \in U$ and set $\mathcal{E} = U\mathcal{D}$. Then*

(1) $\mathcal{E}$ is a saturated constrained fusion system on some $E \leq S$, $\mathcal{D} = O^2(\mathcal{E})$, and $\mathcal{C}$ is a solvable component of $\mathcal{E}$. Moreover $\mathcal{E} = \mathcal{F}_E(G)$ for some model G of $\mathcal{E}$, and there exists $H \trianglelefteq K \trianglelefteq G$ with $\mathcal{D} = \mathcal{F}_{S\cap K}(K)$, $\mathcal{C} = \mathcal{F}_T(H)$, and $G = KU$.

(2) Assume U is strongly closed in E with respect to $\mathcal{E}$. Then either $\mathcal{E}$ is the central product of $\mathcal{C}$ and U, or $T \leq U$.

(3) Assume U is strongly closed in S with respect to $\mathcal{F}$ and $W \trianglelefteq U$ such that (*) for each $R \leq W$ and $\phi \in \hom_{\mathcal{F}}(R, U)$ with $R \cap R\phi \neq 1$, we have $R\phi \leq W$. Assume $C_W(H) \neq 1$. Then either $T \leq W$ or $W \trianglelefteq \mathcal{E}$ and $\mathcal{C}$ centralizes W.

(4) $O_2(G) = TC_{O_2(G)}(H)$, so $O_2(\mathcal{F}) = TC_{O_2(\mathcal{F})}(\mathcal{C})$.

PROOF. Let $\mathcal{B}$ be the product of all $\mathcal{F}$-conjugates of $\mathcal{C}$. By 2.2.1.2, $\mathcal{B} \trianglelefteq \mathcal{F}$. Then by Theorem 5 in [**Asc11**], $U\mathcal{B}$ is a saturated fusion system with $\mathcal{B} = O^2(U\mathcal{B})$, so replacing $\mathcal{F}$ by $U\mathcal{B}$, we may assume $\mathcal{F} = U\mathcal{B}$. Then by 7.4 in [**Asc11**], $\mathcal{D} \trianglelefteq \mathcal{F}$, so again by Theorem 5 in [**Asc11**], $U\mathcal{D}$ is saturated with $\mathcal{D} = O^2(U\mathcal{D})$. As $\mathcal{C}$ is a solvable component, $\mathcal{E}$ is constrained, and as $\mathcal{C} \trianglelefteq \mathcal{D} \trianglelefteq \mathcal{E}$, $\mathcal{C}$ is a solvable component of $\mathcal{E}$. Thus (1) holds.

We next prove (4), where we take $U = O_2(\mathcal{F})$, so that also $U = O_2(G)$. If $u \in U - N_U(T)$ then $[H, H^u] = 1$, so $[H, u]$ is not a 2-group, contradicting $[H, u] \leq U$. Therefore $T \trianglelefteq \langle H, U\rangle = G$, so $H \trianglelefteq G$ by 2.2.1.4. Thus $[H, U] = T$ and as H is irreducible on $T/\Phi(T)$, $U = TC_U(T)$. Finally for $A \in Syl_3(H)$, $[C_U(T), A] \leq Z(T)$, so $C_U(T) = Z(T)C_U(H)$, and hence (4) holds.

Thus it remains to prove (2) and (3), where we may assume $\mathcal{F} = \mathcal{E}$ and U is strongly closed in S with respect to $\mathcal{F}$. Set $V = O_2(K) \cap U$; as $\mathcal{F} = \mathcal{E}$ and U is strongly closed in S, we have $V \trianglelefteq G$. Then $[H, V] \leq T \cap V$ and as H is irreducible on $T/\Phi(T)$, either $[H, V] = 1$ or $T \leq V$.

We first prove (2), so assume $\mathcal{F}$ is a counter example to (2). Then $T \not\leq U$, so $[H, V] = 1$. Hence $K = \langle H^G\rangle$ centralizes V. If $[T, U] = 1$ then as $H \trianglelefteq N_G(T)$ by 2.2.1.4 and U is strongly closed in S with respect to G, either $G = HU$ is a central product of H with U, or $T = [H, U] \leq U$, contrary to the choice of $\mathcal{F}$ as a counter example. Therefore $[T, U] \neq 1$. On the other hand, $[T, U] \leq O_2(K) \cap U = V$.

Set $G^* = G/V$, so that G^*, K^*, U^* satisfy the hypothesis of G, K, V with $[T^*, U^*] = 1$ and $T^* \not\leq U^*$. Thus by an earlier case, G^* is a central product of H^* with U^*, so $U \trianglelefteq G$. But then as $K = O^2(K)$ centralizes V and U^*, K centralizes U, completing the proof of (2).

So assume the hypothesis of (3). Thus $C_W(H) \neq 1$, so by (*) applied to $\phi = c_h$ for $h \in H$ and $W \cap Q$ in the role of R:

(**) H acts on $W \cap Q$ for each H-invariant 2-subgroup Q of S containing $C_W(H)$.

Thus if W centralizes T then $[H, W] \leq C_H(T) = Z(T)$, so applying (**) to $Q = WT$ we conclude that H acts on W. Then as $[H, W] \leq Z(T)$, either $T \leq W$ or W centralizes H, and in the latter case as $W \trianglelefteq U$, also $W \trianglelefteq \langle H, U\rangle = G$.

Therefore we may assume $[T, W] \neq 1$. Therefore $[H, U] \neq 1$, so that $T \leq U \leq N_G(W)$ by (2). Thus $[T, W] \leq O_2(K) \cap W = W_V$. Further W_V acts on T and hence also on H, so $[H, W_V] \leq T$. Hence by (**) applied to $Q = TC_W(H)W_V$, H acts on $C_W(H)W_V$, and hence also on W_V. Therefore either $T \leq W_V$ or $T \cap W_V \leq \Phi(T)$ and $[H, W_V] = 1$, and we may assume the latter. As $W \trianglelefteq U$, $W_V \trianglelefteq U$, so $W_V \trianglelefteq \langle H, U\rangle = G$.

Set $\bar{G} = G/W_V$. Then $\bar{H} \in \mathrm{Sol}(\bar{G})$, so as $\bar{W}$ centralizes $\bar{T}$, $\bar{W}$ acts on $\bar{H}$. Hence $[\bar{H}, \bar{W}] \leq C_{\bar{H}}(\bar{T}) \leq \bar{T}$, so H acts on TW. Then by (**) applied to $Q = TW$, H acts on W, so $[H, W] \leq T \cap W \leq \Phi(T)$, and hence $[H, W] = 1$, completing the proof of (3). □

LEMMA 2.2.5. *Assume $z \in \mathcal{F}^f$ is an involution and $\mathcal{C}$ is a solvable component of $\mathcal{F}_z = C_{\mathcal{F}}(z)$ on $T \in \mathcal{F}_z^f$. Let $\alpha \in \mathfrak{A}(T)$ and set $\bar{T} = T\alpha$, $\bar{\mathcal{F}} = N_{\mathcal{F}}(\bar{T})$, etc. Let G be a model for $N_{N_{\mathcal{F}}(V_{\bar{T}})}(\bar{T})$. Then*

(1) There is a normal subgroup H of $C_G(\bar{z})$ that is a model for $\bar{\mathcal{C}}$.

(2) $\bar{T} = [C_{V_{\bar{T}}}(\bar{z}), H]$.

(3) If $T \cong Q_8$ then H centralizes $C_S(\bar{T})$ and $\bar{\mathcal{C}} \trianglelefteq \bar{\mathcal{F}}$.

(4) Suppose $R \leq C_S(\bar{T})$ is $H\langle\bar{z}\rangle$-invariant. Then either H centralizes R or $\bar{T} \leq R$.

PROOF. As $T \in \mathcal{F}_z^f$, it follows from 2.2 in [**Asc10**] that $\bar{z} = z\alpha \in \bar{\mathcal{F}}^f$ and $\alpha : N_{\mathcal{F}_z}(T) \to C_{\bar{\mathcal{F}}}(\bar{z})$ is an isomorphism. Hence as $\mathcal{C}$ is a solvable component of $\mathcal{F}_z$, we have $\bar{\mathcal{C}}$ is a solvable component of $C_{\bar{\mathcal{F}}}(\bar{z})$. Let S_z be Sylow in $\mathcal{F}_z$ and G_0 a model for $N_{N_{\mathcal{F}_z}(TC_{S_z}(T))}(T)$. Then by 2.2 in [**Asc10**], $\check{\alpha} : G_0 \to C_G(z)$ is an isomorphism, so (1) and (2) follow from 2.2.1.3 applied to $C_{\bar{\mathcal{F}}}(\bar{z}), \bar{\mathcal{C}}$ in the role of $\mathcal{F}, \mathcal{C}$. Indeed if $T \cong Q_8$ then H centralizes $C_S(\bar{T}\langle\bar{z}\rangle)$ by 2.2.1.3, so by the Thompson $A\times B$-Lemma, H centralizes $C_S(\bar{T})$. Then the proof of 2.2.1.4 shows $\bar{\mathcal{C}} \trianglelefteq \bar{\mathcal{F}}$, establishing (3).

Assume the setup of (4). Then as H is irreducible on $\bar{T}/\Phi(\bar{T})$, it follows from (2) that either H centralizes $C_R(\bar{z})$ or $\bar{T} \leq R$. In the latter case (4) holds, and in the former H centralizes R by the $A \times B$-Lemma, completing the proof of (4). □

LEMMA 2.2.6. *Let $z \in \mathcal{F}^f$ be an involution, $\mathcal{C} \in Comp_+(C_{\mathcal{F}}(z))$, $T \in Syl(\mathcal{C})$, $\mathcal{L} \in Comp_+(\mathcal{F})$, and $L \in Syl(\mathcal{L})$. Let $\mathcal{E}$ be the product of the $\mathcal{F}$-conjugates of $\mathcal{L}$ and F Sylow in $\mathcal{E}$. Then*

(1) T acts on $\mathcal{L}$.

(2) If $\mathcal{L} \neq \mathcal{L}^z$ then either $\mathcal{C}$ centralizes $\mathcal{L}$ or $\mathcal{C} \leq \mathcal{L}\mathcal{L}^z$ is a morphic image of $\mathcal{L}$.

(3) If $\mathcal{L} = \mathcal{L}^z$ then one of the following hold:

(i) T centralizes $\mathcal{L}$, or

(ii) $T \leq L$, or

(iii) $\mathcal{C} \cong \mathcal{F}_T(L_2(3))$ centralizes $Z(\mathcal{E})$ and, setting $\bar{\mathcal{E}} = \mathcal{E}/Z(\mathcal{E})$, there exist three conjugates $L = L_1, L_2, L_3$ of L under $Aut_{\mathcal{F}}(F)$ such that z acts on each L_i, L_i is dihedral or quaternion, $C_{\bar{L}_i}(z) = \langle\bar{t}_i\rangle \cong \mathbf{Z_2}$ for some $t_i \in L_i$, and each $t \in T^{\#}$ is of the form $t = t_it_j$ for some $1 \leq i < j \leq 3$. In particular each involution in T centralizes some $\mathcal{F}$ conjugate of $\mathcal{L}$, and some involution in T centralizes $\mathcal{L}$.

(4) If $z \in \mathcal{C}$ then either $\mathcal{C}$ and $\mathcal{L}$ commute or $T \leq L$.

(5) If z acts on $\mathcal{L}$ and $\mathcal{C} \cong SL_2[m]$ then either $\mathcal{C}$ and $\mathcal{L}$ commute or $T \leq L$.

PROOF. We begin a series of reductions which will eventually establish (1)-(3). Then at the end of the proof we prove (4) and (5).

(a) The lemma holds if $\mathcal{C}$ is a component, so we may assume $\mathcal{C}$ is a solvable component.

For assume $\mathcal{C}$ is a component. Then by Theorem 7 in [**Asc11**], $\mathcal{C} \leq E(\mathcal{F})$. Thus if $\mathcal{L}$ is a solvable component then $\mathcal{C}$ centralizes $\mathcal{L}$ by parts (1) and (2) of 2.2.2, so the lemma holds trivially. Hence we may assume $\mathcal{L}$ is a component. Therefore as $T \leq E \in Syl(E(\mathcal{F}))$ and E acts on $\mathcal{L}$, (1) holds. Next by 10.11 in [**Asc11**], there exists a component $\mathcal{K}$ of $\mathcal{F}$, such that either $\mathcal{K} \neq \mathcal{K}^z$ and $\mathcal{C} \leq \mathcal{K}\mathcal{K}^z$, or $\mathcal{K} = \mathcal{K}^z$ and $\mathcal{C} \leq \mathcal{K}$. Hence if $\mathcal{L} \notin \{\mathcal{K}, \mathcal{K}^z\}$ then $\mathcal{C}$ centralizes $\mathcal{L}$, so we may take $\mathcal{L} = \mathcal{K}$. But then the lemma holds, so (a) is established.

(b) We may assume z does not centralize $\mathcal{L}$.

For if z centralizes $\mathcal{L}$ then $\mathcal{L} \in \mathrm{Comp}_+(C_{\mathcal{F}}(z))$ by 2.2.3, so $\mathcal{L} = \mathcal{C}$ or $\mathcal{C}$ centralizes $\mathcal{L}$ by 2.2.2. In either case the lemma holds, so (b) is established.

Let $\mathcal{E}$ be the product of the $\mathcal{F}$-conjugates of $\mathcal{L}$, so that $\mathcal{E}$ is a normal subsystem of $\mathcal{F}$, which is the central product of the conjugates of $\mathcal{L}$.

(c) If $\mathcal{L} \neq \mathcal{L}^z$ then $C_{\mathcal{L}\mathcal{L}^z}(z) = \mathcal{W}C_{Z(\mathcal{L}\mathcal{L}^z)}(z)$, where $\mathcal{W} \in Comp_+(C_{\mathcal{F}}(z))$ is a morphic image of $\mathcal{L}$ centralized by $\mathcal{C}$. Further $N_{T\langle z\rangle}(\mathcal{L})$ is of index 2 in $T\langle z\rangle$ and centralizes $\mathcal{L}$.

By 10.11 and 8.24 in [**Asc11**], $C_{\mathcal{L}\mathcal{L}^z}(z) = \mathcal{W}C_{Z(\mathcal{L}\mathcal{L}^z)}(z)$, where $\mathcal{W}$ is a morphic image of $\mathcal{L}$. In particular if $\mathcal{L}$ is quasisimple then so is $\mathcal{W}$, while if $\mathcal{L}$ is solvable then $\mathcal{W}$ is $L_2(3)$ or $SL_2(3)$. In any event as $\mathcal{L}$ is subnormal in $\mathcal{F}$, $\mathcal{W}$ is subnormal in $C_{\mathcal{F}}(z)$ by 8.24 in [**Asc11**]. Therefore $\mathcal{W} \in Comp_+(C_{\mathcal{F}}(z))$. Then by 2.2.2, either $\mathcal{C}$ centralizes $\mathcal{W}$ or $\mathcal{C} = \mathcal{W}$, and the lemma holds in the latter case, so $\mathcal{C}$ centralizes $\mathcal{W}$. Now 2.3.3.2 in [**Asc19**] completes the proof of (c).

(d) T does not centralize $\mathcal{L}$.

Assume otherwise; then (1) and (3i) hold, so it remains to prove (2), and hence we may assume $\mathcal{L} \neq \mathcal{L}^z$, and adopt the notation in (c). We may assume $T \in \mathcal{F}_z^f$ and adopt the notation in 2.2.5. Then $R = \bar{L}\bar{L}^{\bar{z}} \leq C_S(\bar{T})$. Let $\mathcal{Y}$ be the product of the set Γ of $\bar{\mathcal{F}}$-conjugates of $\bar{\mathcal{L}}$ and Y Sylow in $\mathcal{Y}$. Thus $Y \leq C_S(\bar{T}) \leq V_{\bar{T}}$ and Y is strongly closed in $C_S(\bar{T})$ with respect to $\bar{\mathcal{F}}$. Pick h of order 3 in H and set $\gamma = c_h \in Aut_{\bar{\mathcal{F}}}(Y)$. Then γ permutes Γ and centralizes a Sylow group W of $\bar{\mathcal{W}}$, so γ acts on $\{\bar{L}, \bar{L}^{\bar{z}}\}$. Then as γ is of order 3, γ acts on $\bar{L}$. Next as γ centralizes W, $[\bar{L}, \gamma] \leq Z(\bar{\mathcal{L}})$, so γ centralizes $\bar{\mathcal{L}}$ by 9.5.1 in [**Asc11**]. But then $\mathcal{C}$ centralizes $\mathcal{L}$, so (2) holds, completing the proof of (d).

(e) We may take $Z(\mathcal{E}) = 1$.

Let $\gamma \in Aut_{\mathcal{C}}(T)$ be of order 3, and set $E = Z(\mathcal{E})$ and $U = C_E(z)$. Let $\mathcal{X}$ be the product of the $C_{\mathcal{F}}(z)$-conjugates of $\mathcal{C}$, so that $\mathcal{X}$ is a normal subsystem of $C_{\mathcal{F}}(z)$ on some $X \leq S$. Then $UX \trianglelefteq \mathcal{F}_z$, so γ extends to a 3-element $\beta \in Aut_{\mathcal{F}_z}(XU)$ with $[XU, \beta] = T$. Then $[U, \beta] = [U, \beta, \beta] \leq [XU, \beta] \cap U = T \cap U$, so as β is irreducible on $T/\Phi(T)$, either $[U, \beta] = 1$ or $T = [U, \beta] \leq U$. However in the latter case, $\mathcal{E} \leq C_{\mathcal{F}}(T)$, contrary to (d), so we may assume $[U, \beta] = 1$. But β extends to a 3-element $\delta \in Aut_{\mathcal{F}}(XE)$ as $E \trianglelefteq \mathcal{F}$, and then by the Thompson $A \times B$-Lemma, $[E, \delta] = 1$. Therefore $T = [T, \delta]$ also centralizes E, so $\mathcal{C} \leq C_{\mathcal{F}}(E)$.

By 2.2.3, $\mathcal{C} \in \mathrm{Sol}(C_{\mathcal{F}}(U\langle z\rangle))$ and $\mathcal{L} \in \mathrm{Comp}_+(C_{\mathcal{F}}(U))$. Then by induction on the order of $\mathcal{F}$, $U \leq Z(\mathcal{F})$. Now we form the factor system $\mathcal{F}^+ = \mathcal{F}/U$ as in 8.10 of [**Asc08a**]. By 8.10 in [**Asc08a**], the image $\mathcal{C}^+$ of $\mathcal{C}$ is a solvable component of

$C_{\mathcal{F}^+}(z^+)$, and $\mathcal{L}^+ \in \mathrm{Comp}_+(\mathcal{F}^+)$. Thus if $E \neq 1$, then by induction on the order of $\mathcal{F}$, the lemma holds for $\mathcal{C}^+$ and $\mathcal{L}^+$, and hence also for $\mathcal{C}$ and $\mathcal{L}$. Note that, adopting the notation of 2.2.6.3.iii, $T = \langle t_1t_2, t_1t_3\rangle$, so T is abelian, and hence is E_4 rather than Q_8. This completes the proof of (e).

(f) We may assume z acts on $\mathcal{L}$.

Assume otherwise; then we are in the setup of (c), and we write W for the Sylow group of $\mathcal{W}$. If T acts on L then T centralizes $\mathcal{L}$ by (c), contrary to (d). Therefore by (c), $T_0 = N_T(L)$ is of index 2 in T.

Set $Z = \Omega_1(Z(L))$ and let W_0 be the inverse image in W of Z under the projection map, which is an isomorphism by (e). Set $V = TW_0$ and $\bar{V} = V/\Phi(T)$. Then Z acts faithfully on $\bar{V}$ as the group of transvections with axis $\bar{T}_0\bar{W}_0 = \bar{V}_0$ and centers in $\bar{W}_0$. Let $\gamma \in Aut_{\mathcal{C}}(T)$ be of order 3. By 2.2.1.3 applied in $C_{\mathcal{F}}(z)$, γ lifts to a 3-element $\hat{\gamma}$ on V_T with $[C_S(T), \hat{\gamma}] \leq Z(T)$, so $\hat{\gamma}$ induces $\beta \in Aut_{\mathcal{F}}(V)$. As γ is irreducible on $T/\Phi(T)$ and β acts on the strongly closed subgroup $V_1 = V \cap F$, where F is Sylow in the product $\mathcal{E}$, either $T \leq V_1$ or $T \cap V_1 \leq \Phi(T)$, and as $T_0 \neq T$, it must be the latter. Thus $V_1 = W_0$ or $W_0\Phi(T)$, and as V_1 is strongly closed in V, V_1 is β-invariant. Thus $[\beta, W_0] \leq V_1 \cap T \leq \Phi(T)$, so β centralizes W_0. Let $Y = \langle Aut_Z(V), \beta\rangle \leq Aut_{\mathcal{F}}(V)$. Now Z centralizes V_1 and V/V_1, so $Aut_Z(V) \leq O_2(Y)$, and hence β lifts to $\delta \in Aut_{\mathcal{F}}(Q)$ with $ZV \leq Q \leq S$. Next $Z\delta W_0 = (ZW_0)\delta \leq Q \cap F$ with $Z\delta \leq (Q \cap L)\delta \leq \mathcal{K} \in \mathcal{L}^{\mathcal{F}}$ and $(Z\delta)^z \leq (Q \cap L^z)\delta \leq \mathcal{K}^z$. Then $W_0 \leq (ZZ^z)\delta \leq \mathcal{K}\mathcal{K}^z$, so $\{\mathcal{K}, \mathcal{K}^z\} = \{\mathcal{L}, \mathcal{L}^z\}$, and hence $\{Q \cap L, Q \cap L^z\}\gamma = \{Q \cap L, Q \cap L^z\}$. Therefore as δ is of odd order, δ acts on $Q \cap L^z$, so $T = [T, \delta]$ also acts on $Q \cap L^z$, contradicting $T \neq T_0$. This completes the proof of (f), and establishes (2).

(g) We may assume $T \leq F$, so that T acts on $\mathcal{L}$.

Let $\mathcal{P}$ consist of those $P \leq L\langle z\rangle$ with $z \in P \in \mathcal{F}^f$, $C_S(z) \leq N_S(P)$, and $\mathcal{C}$ a solvable component of $\mathcal{F}_P = N_{\mathcal{F}}(P)$. For example $\langle z\rangle \in \mathcal{P}$. As the Sylow group F of $\mathcal{E}$ is strongly closed in S with respect to $\mathcal{F}$, $U = N_F(P)$ is strongly closed in $N_S(P)$ with respect to $\mathcal{F}_P$. Applying 2.2.4.2 to $\mathcal{F}_P$, $\mathcal{C}$, and U, we conclude that either $T \leq U$ or U centralizes $\mathcal{C}$. In the former case (g) holds, so we may assume the latter. When $P = \langle z\rangle$, this shows that $L_z = C_L(z)$ centralizes $\mathcal{C}$.

We can choose a conjugate $\bar{\mathcal{L}}$ of $\mathcal{L}$ such that $\bar{L} \in \mathcal{F}^f$ and $z \in N_{\mathcal{F}}(\bar{\mathcal{L}})^f$. Here if $\mathcal{L}$ is quasisimple then $N_{\mathcal{F}}(\bar{\mathcal{L}})$ is defined in Definition 2.2.1 of [**Asc19**], while if $\mathcal{L}$ is solvable we define $N_{\mathcal{F}}(\bar{\mathcal{L}}) = N_{\mathcal{F}}(\bar{L})$. We may assume $T \not\leq F$, so applying the argument above to $\bar{\mathcal{L}}$ in place of $\mathcal{L}$, we conclude that $\mathcal{C}$ centralizes $\bar{L}_z$. Then, using 2.1.5 in [**Asc19**], $\mathcal{C} \leq N_{\mathcal{F}}(\bar{L}_z) \leq N_{\mathcal{F}}(\bar{L})$. Then by induction on the order of $\mathcal{F}$, either $\bar{\mathcal{L}} \trianglelefteq \mathcal{F}$, or our lemma holds in $N_{\mathcal{F}}(\mathcal{L})$, and hence also in $\mathcal{F}$. Therefore we may assume that $\bar{\mathcal{L}} \trianglelefteq \mathcal{F}$, so $\mathcal{E} = \bar{\mathcal{L}} = \mathcal{L}$. Hence $U = N_L(P)$, and as $C_S(z) \leq N_S(P)$, also $C_S(z) \leq N_S(U)$. Let $V = U\langle z\rangle$ and $\alpha \in \mathfrak{A}(V)$; as $C_S(z) \leq N_S(V)$ and $z \in \mathcal{F}^f$, also $z\alpha \in \mathcal{F}^f$, so replacing z, V by $z\alpha, V\alpha$, we may assume that $V \in \mathcal{F}^f$. As $C_{\mathcal{F}}(V) \leq \mathcal{F}_z$ and $\mathcal{C}$ is a solvable component of $\mathcal{F}_z$, $\mathcal{C}$ is also a solvable component of $C_{\mathcal{F}}(V)$, and hence also of $\mathcal{F}_V$. But now choosing P maximal in $\mathcal{P}$, we conclude that $P = V$, so $L = U$ centralizes $\mathcal{C}$, contrary to (d). This finally completes the proof of (g).

By (g), part (1) of our lemma holds, so it remains to prove (3).

(h) We may assume that $C_L(\mathcal{C}) = 1$.

Suppose not, and set $W = C_L(z)$. Then $C_W(\mathcal{C}) \neq 1$. Further suppose $R \leq L$ and $\gamma \in \hom_{\mathcal{F}}(R, F)$ with $L \cap R\gamma \neq 1$. By 3.3 in [**Asc08a**], $\gamma = \varphi\phi$ with $\varphi \in Aut_{\mathcal{F}}(F)$ and $\phi \in \hom_{\mathcal{E}}(R\varphi, F)$. Thus as $1 \neq L \cap R\gamma$ and $(R\gamma \cap L)\phi^{-1} \leq L$, we have $1 \neq R\varphi \cap L$, so as φ permutes $\mathcal{L}^{\mathcal{F}}$, $R\varphi \leq L$, and then $R\gamma = R\varphi\phi \leq L$. Thus condition (*) of 2.2.4.3 is satisfied by $\mathcal{F}_z$, $C_F(z)$, W in the roles of $\mathcal{F}$, U, W, so applying 2.2.4.3, we conclude either $T \leq W$ or $\mathcal{C}$ centralizes W. In the former case (3ii) holds, so we may assume the latter.

Let $\pi : T \to L$ be the projection map and $T_L = T\pi$. If $T_L = 1$ then T centralizes $\mathcal{L}$, contrary to (d), so $T_L \neq 1$. As π is z-equivariant, $T_L \leq W$, so $\mathcal{C}$ centralizes T_L. But then for $\alpha \in Aut_{\mathcal{C}}(T)$, π is α-equivariant, contradicting $C_T(\alpha) = \Phi(T)$ and $\mathcal{C}$ centralizing T_L. This establishes (h).

Form $C_{\mathcal{E}}(z)$ in $\mathcal{F}$ as in 8.24 in [**Asc11**]; by that lemma, $C_{\mathcal{E}}(z) \trianglelefteq \mathcal{F}_z$. By (f), $\mathcal{L} \trianglelefteq \langle z\rangle\mathcal{E}$, so we can form $C_{\mathcal{L}}(z)$ in $\langle z\rangle\mathcal{E}$ as in 8.24 in [**Asc11**], and $C_{\mathcal{L}}(z) \trianglelefteq \langle z\rangle\mathcal{E}$. Therefore $C_{\mathcal{L}}(z)$ is subnormal in $\mathcal{F}_z$, so either $C_{\mathcal{L}}(z)$ contains a component $\mathcal{D}$, or $C_{\mathcal{L}}(z)$ is constrained. In the former case $\mathcal{D}$ centralizes $\mathcal{C}$ by 2.2.2, contrary to (h). Thus $R = O_2(C_{\mathcal{L}}(z)) = F^*(C_{\mathcal{L}}(z)) \leq O_2(\mathcal{F}_z) = U$ as $C_{\mathcal{L}}(z)$ is subnormal in $\mathcal{F}_z$. Now the hypothesis of 2.2.4 are satisfied with $C_{\mathcal{F}}(z)$ in the role of $\mathcal{F}$, so we can form the group G of 2.2.4.1, and the normal model H for $\mathcal{C}$ in G. By 2.2.4.4, $U = TC_U(H)$, so by (h), $|R| \leq |T : C_T(H)| = 4$. Indeed as $T \leq F \leq N_S(L)$, we can pick an involution $r \in Z(TR)$ in R. Now $[H, R] \leq T$, so if $H \cong SL_2(3)$, then H acts on $Z(\langle r, T\rangle) = \langle r, Z(T)\rangle$, and hence $H = O^2(H)$ centralizes r, contrary to (h). Hence

(i) $\mathcal{C} \cong \mathcal{F}_T(L_2(3))$, so $T \cong E_4$.

Let a be of order 3 in H. Then $R^a = R\varphi\phi$, where $\varphi \in Aut_{\mathcal{F}}(F)$ and $\phi \in \hom_{\mathcal{E}}(R\varphi, F)$. Thus $R^a \leq L\varphi$. By (i), $U = T \times C_U(H)$, so by (h) we have $1 = C_R(H) = C_R(a)$. Also $[R, a] \leq T$. Hence as we may assume (3ii) does not hold, a does not act on R, so $\mathcal{L} \neq \mathcal{L}\varphi^*$. Therefore $RT \geq RR^a = R \times R^a$.

Suppose $|R| = 4$. Then $R \cong E_4$ and $R^{a^2} \leq R \times R^a = RT \leq F$. This is a contradiction as R^{a^i} is contained in a component $\mathcal{J}_i \in \mathcal{L}^{\mathcal{F}}$ and if $\mathcal{J}_i$ is a system on $J_i \leq S$, then as $R^{a^2} \leq LJ_1$, we have $\mathcal{J}_2 \in \{\mathcal{L}, \mathcal{J}_1\}$, so $R^{a^2} \leq L \cap RT = R$ or $J_1 \cap RT = R^a$. Therefore:

(j) $C_L(z) = R = \langle r\rangle \cong \mathbf{Z}_2$.

It follows from (j) that $L\langle z\rangle$ is dihedral or semidihedral, and hence as $Z(\mathcal{L}) = 1$, that L is dihedral. Further $RH \cong \mathbf{Z}_2 \times A_4$ and for each involution $t \in T$, $t = r_1r_2 \in L_1L_2$ for some $r_i \in \{r, r^a, r^{a^2}\}$, with $r_i \in \mathcal{L}_i \in \mathcal{L}^{\mathcal{F}}$ a component on $L_i \leq F$. In particular t centralizes L_3. Therefore (3iii) holds, completing the proof of (3).

Finally we prove (4) and (5), so assume the hypothesis of (4) or (5). In the first case, $z \in \mathcal{C}$, so z acts on $\mathcal{L}$ by (1), while in the second z acts on $\mathcal{L}$ by hypothesis. Thus in either case, z acts on $\mathcal{L}$, so by (3), either T centralizes $\mathcal{L}$ or $T \leq L$. In the latter case we are done, so assume the former. If $z \in T$ then z centralizes $\mathcal{L}$, so $\mathcal{L} \in Comp_+(\mathcal{F}_z)$ by 2.2.3, so as $\mathcal{C} \in Comp_+(\mathcal{F}_z)$, $\mathcal{C}$ and $\mathcal{L}$ commute by 2.2.2. Therefore (4) is established, so we may assume $\mathcal{C} \cong SL_2[m]$.

If $m = 8$ then conjugating in $\mathcal{F}_z$ we may assume $T \in \mathcal{F}_z^f$, and adopt the notation of 2.2.5. By 2.2.5.3, $\bar{\mathcal{C}}$ centralizes $\bar{\mathcal{L}}$, so $\mathcal{L}$ and $\mathcal{C}$ commute. Finally take $m > 8$; then $\mathcal{C}$ is a component of $\mathcal{F}_z$, so $\mathcal{C} \leq E(\mathcal{F})$ by E-balance, and then as T centralizes $\mathcal{L}$, we have $\mathcal{C} = [T]_{E(\mathcal{F})} \leq C_{E(\mathcal{F})}(\mathcal{L})$, completing the proof of (5). □

LEMMA 2.2.7. *Let $z \in \mathcal{F}^f$ be an involution, $\mathcal{C} \in Comp_+(C_{\mathcal{F}}(z))$, and $T \in Syl(\mathcal{C})$. Assume $z \in T$, $\mathcal{C}$ does not centralize $E(\mathcal{F})$, and $\mathcal{C}$ is not a component of $\mathcal{F}$. Then*

(1) there exists a unique component $\mathcal{L}$ of $\mathcal{F}$ such that $\mathcal{C}$ does not centralize $\mathcal{L}$. Moreover $T \leq L \in Syl(\mathcal{L})$.

(2) Let $\mathcal{F}_0 = C_{F^*(\mathcal{F})}(\mathcal{L})$ and $\mathcal{Y} = C_{N_{\mathcal{F}}(\mathcal{L})}(\mathcal{F}_0)$. Then $\mathcal{C} \leq \mathcal{Y}$.

(3) $F^*(\mathcal{Y}) = \mathcal{L}Z(\mathcal{Y})$.

(4) Set $\mathcal{Y}^+ = \mathcal{Y}/Z(\mathcal{Y})$. Then $\mathcal{Y}^+$ is almost simple with $F^*(\mathcal{Y}^+) = \mathcal{L}^+$.

(5) $\mathcal{C}^+ \in Comp_+(C_{\mathcal{Y}^+}(z^+))$ and $z^+ \neq 1$.

PROOF. Let $\mathfrak{L}$ be the set of components of $\mathcal{F}$ not centralized by z. As $\mathcal{C}$ does not centralize $E(\mathcal{F})$ and $z \in T$ it follows from 2.2.3 applied to z in the role of U and 2.2.2 that $\mathfrak{L} \neq \emptyset$. Let $\mathcal{L} \in \mathfrak{L}$ and $L \in Syl(\mathcal{L})$. As $z \in T$ be conclude from 2.2.6.4 that $T \leq L$. Hence for each $\mathcal{L} \neq \mathcal{K} \in Comp(\mathcal{F})$, $z \in T \leq L \leq C_S(\mathcal{K})$, so by 2.2.3 and 2.2.2, $\mathcal{C}$ centralizes $\mathcal{K}$. This completes the proof of (1).

Next $\mathcal{F}_0$ is the product of $O_2(\mathcal{F})$ with the components of $\mathcal{F}$ distinct from $\mathcal{L}$. As $z \in T \leq C_S(O_2(\mathcal{F}))$ and $\mathcal{C}$ and $Comp(\mathcal{F}) - \{\mathcal{L}\}$ are in $Comp_+(\mathcal{F}_z)$ by 2.2.3, it follows that $\mathcal{C}$ centralizes $\mathcal{F}_0$, using 2.2.1.3 when $\mathcal{C}$ is solvable. Therefore $\mathcal{C} \leq \mathcal{Y}$, establishing (2).

By Theorem 6 in [**Asc11**], $C_{\mathcal{F}}(F^*(\mathcal{F})) = Z(F^*(\mathcal{F}))$, so (3) follows.

As $\mathcal{L}$ is quasisimple with $Z(\mathcal{L}) \leq Z(\mathcal{Y})$, $\mathcal{L}^+$ is simple. By (3), $\mathcal{L}^+ = F^*(\mathcal{Y}^+)$, so (4) holds.

If $z \in Z(\mathcal{Y})$ then $\mathcal{L}$ is a component of $\mathcal{F}_z$ by 2.2.3. Then as $T \leq L$ and $\mathcal{C} \in Comp_+(\mathcal{F}_z)$, it follows that $\mathcal{C} = \mathcal{L}$, contradicting the hypothesis that $\mathcal{C} \notin Comp(\mathcal{F})$. Therefore $z^+ \neq 1$.

Let $\mathcal{E}$ be the product of the $\mathcal{F}_z$-conjugates of $\mathcal{C}$. Then $\mathcal{E} \leq C_{\mathcal{Y}}(z)$ with $\mathcal{E} \trianglelefteq \mathcal{F}_z$, so by 1.1.10 in [**Asc19**], $\mathcal{E} \trianglelefteq C_{\mathcal{Y}}(z)$ if $\mathcal{C}$ is quasisimple, and at least $\mathcal{E}$ is weakly normal if $\mathcal{C}$ is solvable. Then in the latter case condition (N1) holds by 2.2.1.3, so in any event $\mathcal{C}$ is subnormal in $C_{\mathcal{Y}}(z)$, and then $\mathcal{C}^+$ is subnormal in $C_{\mathcal{Y}}(z)^+$ by 8.9.2 in [**Asc08a**]. Then as $C_{\mathcal{Y}}(z)^+ \trianglelefteq C_{\mathcal{Y}^+}(z^+)$, we conclude that $\mathcal{C}^+ \in Comp_+(C_{\mathcal{Y}^+}(z^+))$, completing the proof of (5). □

2.3. Intrinsic $SL_2[m]$-components

In this section we assume $\mathcal{F}$ is a saturated fusion system on a finite 2-group S. Often we also assume the following hypothesis:

HYPOTHESIS 2.3.1. *For $i = 1, 2$, $z_i \in \mathcal{F}^f$ is an involution, such that the set Δ_i of components and solvable components of $\mathcal{F}_i = C_{\mathcal{F}}(z_i)$ on quaternion subgroups of S containing z_i of order m is nonempty.*

NOTATION 2.3.2. When assuming Hypothesis 2.3.1, we adopt the following notation: e is an involution in S and $\phi \in \mathfrak{A}(e)$ with $e\phi = z_2$. Pick $\mathcal{C}_i \in \Delta_i$, a system on T_i, set $z = z_1$, $T = T_1$, $\mathcal{C} = \mathcal{C}_1$, $\mathcal{D} = \mathcal{C}_2$, and $D = T_2$. Write $\mathcal{X}_i$ for the central product of the members of Δ_i, which is normal in $\mathcal{F}_i$ by Theorem 6 in [**Asc11**], 2.2.1, and 2.2.2.

LEMMA 2.3.3. *Assume Hypothesis* 2.3.1. *Then* $C_S(T)$ *centralizes* $\mathcal{C}$.

PROOF. Each $R \in \mathcal{C}^{frc}$ (distinct from T if $m > 8$) is isomorphic to Q_8, so the lemma follows from 2.3.5 in [**Asc19**]. □

LEMMA 2.3.4. *Assume* u_i *is an involution in* $\mathcal{F}$ *for* $i = 1, 2$ *with* $u_1 \in \mathcal{F}^f$ *and* $u_2 \in C_{\mathcal{F}}(u_1)^f$. *Let* $\alpha \in \mathfrak{A}(u_2)$. *Then* $u_1\alpha \in C_{\mathcal{F}}(u_2\alpha)^f$ *and* $\alpha : C_{C_{\mathcal{F}}(u_1)}(u_2) \to C_{C_{\mathcal{F}}(u_2\alpha)}(u_1\alpha)$ *is an isomorphism.*

PROOF. This is a special case of 2.2.2 in [**Asc10**]. □

LEMMA 2.3.5. *Assume Hypothesis* 2.3.1 *and* $C_T(e)$ *is quaternion. Then* e *centralizes* $\mathcal{C}$ *and* $z\phi$ *centralizes* $\mathcal{D}$.

PROOF. As $z \in C_T(e)$, $e \in \mathcal{F}_1$, so replacing e by a member of $e^{\mathcal{F}_1} \cap \mathcal{F}_1^f$, we may assume $e \in \mathcal{F}_1^f$. Set $k = |C_T(e)|$. As $Q = C_T(e)$ is quaternion, either e centralizes T or $m = 2k > 8$, $T = \langle v, u\rangle$ with $|v| = k$, $|u| = 4$, $v^u = v^{-1}$, and $Q = \langle w, u\rangle$, where $w = v^2$. Let $\mathcal{R}$ be the set of Q_8-subgroups of Q. For each $R \in \mathcal{R}$, e acts on $\mathcal{C}_R = O^2(N_{\mathcal{C}}(R)) \cong SL_2[8]$. By 2.3.3, e centralizes $\mathcal{C}_R$, so e centralizes $\mathcal{C}_e = \langle \mathcal{C}_R : R \in \mathcal{R}\rangle \cong SL_2[k]$. Observe $\mathcal{C}_e \in \text{Comp}_+(C_{\mathcal{F}_1}(e))$. Then using 2.3.4, $\mathcal{B} = \mathcal{C}_e\phi^* \in \text{Comp}_+(C_{\mathcal{F}_2}(z\phi))$, so by 2.2.6.4, either $\mathcal{B}$ and $\mathcal{D}$ commute or $\mathcal{B} \leq \mathcal{D}$, so that $z = e$. In either case $z\phi$ centralizes $\mathcal{D}$, and in the latter e centralizes $\mathcal{C}$. Finally in the former case, applying the isomorphism $\phi^{-*} : C_{\mathcal{F}_2}(z\phi) \to C_{\mathcal{F}_1}(e)$ of 2.3.4, $\mathcal{C}_e$ commutes with $\mathcal{D}_0 = \mathcal{D}\phi^{-*} \in \text{Comp}_+(C_{\mathcal{F}_1}(e))$, so applying 2.2.6.4 to $\mathcal{F}_1, e, \mathcal{D}_0, \mathcal{C}$ in the roles of $\mathcal{F}, z, \mathcal{C}, \mathcal{L}$, we conclude that $\mathcal{C}$ and $\mathcal{D}_0$ commute, so again e centralizes $\mathcal{C}$. □

THEOREM 2.3.6. *Assume Hypothesis* 2.3.1 *and adopt Notation* 2.3.2. *Then*
(1) If $a \in C_S(e) - \langle e\rangle$ *with* $[a, z] = 1$ *and* $a\phi \in D$, *then* $a \in N_S(T)$.
(2) If $z\phi \in N_S(D)$ *then* $e \in N_S(T)$.

We establish Theorem 2.3.6 via a series of reductions. Thus in the remainder of this section we assume we are working in a counterexample to the theorem. In (1) as $a\phi \in D$ we have $e \in \langle a\rangle \leq C_S(z)$, while in (2) as $z\phi \in N_S(D)$, again $e \in C_S(z)$. Thus in any event $[e, z] = 1$. Hence, replacing e by a member of $e^{\mathcal{F}_1} \cap \mathcal{F}_1^f$, we may assume $e \in \mathcal{F}_1^f$.

For example let $\alpha \in \mathfrak{A}_{\mathcal{F}_1}(e)$, set $\bar{e} = e\alpha$, and in (1) set $\bar{a} = a\alpha$. Let $\varphi \in \mathfrak{A}(\bar{a})$ with $\bar{a}\varphi = z_2$; then $\zeta = \phi^{-1}\alpha\varphi$ is an $\mathcal{F}_2$-map. In (1), $\bar{a}\varphi = (a\phi)\zeta$ and $a\phi \in D$, so as $\mathcal{D} \in Comp_+(\mathcal{F}_2)$, $(a\phi)\zeta \in \bar{D}$ an $\mathcal{F}_2$-conjugate of D. Then if $\bar{a} \in N_S(T_1)$ for each $\mathcal{F}_1$-conjugate T_1 of T, the same holds for $a = \bar{a}\alpha^{-1}$, so (1) holds. Similarly in (2) if $z\phi \in N_S(D)$ then $z\varphi = (z\phi)\zeta \in N_S(\bar{D})$, and hence if $\bar{e} \in N_S(T_1)$ for each T_1, then the same holds for e.

Set $x = z\phi$.

LEMMA 2.3.7. x *acts on* $\mathcal{D}$.

PROOF. If we are working in a counterexample to (1), then x centralizes $a\phi \in D$, and as $a\phi \neq z_2$, $\mathcal{D}$ is the unique member of $\mathrm{Comp}_+(\mathcal{F}_2)$ containing $a\phi$, so x acts on $\mathcal{D}$. On the other hand if we are in a counterexample to (2), then $x \in N_S(D)$ by hypothesis, so x acts on $\mathcal{D}$. $\square$

LEMMA 2.3.8. *e acts on $\mathcal{C}$.*

PROOF. Assume otherwise. Suppose first that $\mathcal{C}$ is a component. Then by 10.11 in [**Asc11**]:

(*) $C_{\mathcal{C}\mathcal{C}^e}(e) = \langle z \rangle \times \mathcal{L}$, where $\mathcal{L} \cong \mathcal{C}/\langle z \rangle$ is a component of $C_{\mathcal{F}_1}(e)$.

On the other hand if $\mathcal{C}$ is solvable, then we apply 2.2.4 to $\mathcal{F}_1$ in the role of $\mathcal{F}$ and $U = \langle e \rangle$, and, working in the group G of 2.2.4.1, $C_{HH^e}(e) = \langle z \rangle J$ where $J \cong A_4$, so again (*) holds, except that this time $\mathcal{L}$ is a solvable component. In any case write L for a Sylow group of $\mathcal{L}$. Now by (*) and 2.3.4, $\mathcal{L}\phi^* \in \mathrm{Comp}_+(C_{\mathcal{F}_2}(x))$. Notice L is dihedral, so as D is quaternion, $L\phi \not\leq D$. By 2.3.7, x acts on $\mathcal{D}$, so we can appeal to 2.2.6.3 to conclude that each involution in $L\phi$ centralizes some member of $\mathcal{D}^{\mathcal{F}}$, and for some involution $l \in L$, $l\phi$ centralizes $\mathcal{D}$.

Suppose we are working in a counterexample to (1). Then $e \in \langle a^2 \rangle$. By the previous paragraph, $a\phi$ centralizes some involution $l\phi \in L\phi$, so a centralizes $l \in \mathcal{L} \leq \mathcal{C}\mathcal{C}^e$. Therefore a acts on $\{\mathcal{C}, \mathcal{C}^e\}$, so $e \in \langle a^2 \rangle$ acts on $\mathcal{C}$, a contradiction.

Therefore no such a exists, and we are working in a counterexample to (2). Suppose x centralizes $d \in D - \langle z_2 \rangle$. Let $\alpha \in \mathfrak{A}(x)$ with $x\alpha = z$ and $\beta \in \mathfrak{A}(z_2\alpha)$ with $z_2\alpha\beta = z_2$. Then $\xi = \alpha\beta$ is an $\mathcal{F}_2$-map, so ξ permutes Δ_2 and hence $d\xi \in T_2'$ Sylow in some member of Δ_2. Then as we've shown that e acts on $\mathcal{C}$ under the hypotheses of (1), and as $z\beta$ centralizes $d\xi \in T_2'$, we conclude that $z_2\alpha$ acts on each member of Δ_1. Hence as $\zeta = \phi\alpha$ is an $\mathcal{F}_1$-map and $e\zeta = z_2\alpha$, we conclude that e acts on $\mathcal{C}$, a contradiction. Thus we've shown that x centralizes no element of $D - \langle z_2 \rangle$. Then as x acts on $\mathcal{D}$ but centralizes no element of $D - \langle z_2 \rangle$, x is conjugate to xz_2 in $D_x = N_D(\langle x, z_2 \rangle)$. Then as $l\phi$ centralizes D, $l\phi xz_2$ is conjugate to $l\phi x$ under D_x, so $(lze)^{\mathcal{F}} = (lz)^{\mathcal{F}}$.

Let $S_z = C_S(z)$ and $\tilde{S}_z = S_z/\langle z \rangle$. Let Q be a Q_8-subgroup of T, and L_0 the inverse image in L of $\tilde{Q}$ under the projection of $\tilde{L}$ on $\tilde{T}$ and $A = \langle z, e \rangle L_0$. Then we may choose $l \in L_0$, $A \cong E_{16}$, $O^2(Aut_{\mathcal{L}}(L)) \cong \mathbf{Z_3}$ lifts to Λ of order 3 in $Aut_{\mathcal{F}}(A)$, and $E_4 \cong Aut_Q(A) = C_{Aut_{\mathcal{C}\mathcal{C}^e}(A)}(\tilde{L}_0)$, so Λ acts on $Aut_Q(A)$, and hence $\Sigma = \Lambda Aut_Q(A) \leq Aut_{\mathcal{F}}(A)$ is isomorphic to A_4. Observe $lze \in e^{\Sigma}$ and $lz \in l^{\Sigma}$. Therefore

$$l^{\mathcal{F}} = (lz)^{\mathcal{F}} = (lze)^{\mathcal{F}} = e^{\mathcal{F}} = z_2^{\mathcal{F}},$$

and indeed by 2.2.6.1, $L\phi$ acts on $\mathcal{D}$ and hence also on D_x, so $[L\phi, D_x] \leq \langle z_2 \rangle$ and hence D_x acts on $A\phi$. Thus $l \in e^{\theta}$, where $\theta = \langle \Sigma, Aut_{D_A}(A) \rangle \leq Aut_{\mathcal{F}}(A)$, and $D_A = Aut_{D_x}(A\phi)\phi^{-*}$.

Hence there exists $\psi \in \mathfrak{A}(l)$ with $l\psi = z_2$. Suppose $\mathcal{C}_0 \in \Delta_1 - \{\mathcal{C}, \mathcal{C}^e\}$ is a system on T_0. As $l \in \mathcal{C}\mathcal{C}^e$, l centralizes $\mathcal{C}_0$. Hence by 2.2.6.4, $z\psi$ centralizes $\mathcal{X}_2$. Now there exists $\varphi \in \mathfrak{A}(z\psi)$ with $z\psi\varphi = z$, and by a second application of 2.2.6.4, $z_2\varphi = l\psi\varphi$ centralizes $\mathcal{X}_1$. However $\psi\varphi \in \hom_{\mathcal{F}_1}(l, C_S(z))$ and $\mathcal{X}_1 \trianglelefteq \mathcal{F}_1$, so as $l\psi\varphi$ centralizes $\mathcal{X}_1$, so does l, a contradiction. Therefore $\Delta_1 = \{\mathcal{C}, \mathcal{C}^e\}$. Then as $\mathcal{X}_1 \trianglelefteq \mathcal{F}_1$, $\Sigma = O^2(Aut_{\mathcal{X}_1}(A)) \trianglelefteq Aut_{\mathcal{F}_1}(A)$.

Let $u = l\phi$, $\mu \in \mathfrak{A}(u)$ with $u\mu = z_2$, and $\eta \in \mathfrak{A}(z_2\mu)$ with $z_2\mu\eta = z_2$. As u centralizes $\mathcal{D}$, the same argument says $u\mu\eta = z_2\eta$ centralizes $\mathcal{X}_2$, and then also u centralizes $\mathcal{X}_2$. By symmetry, $L_0\phi$ centralizes $\mathcal{X}_2$, and hence also D_x, so D_A is generated by a transvection on A with axis $\langle e, L_0\rangle$ and center $\langle e\rangle$. Observe that Σ is indecomposible on A with chief series $1 < \langle z\rangle < \langle z\rangle L_0 < A$, so θ is irreducible on A. As $A_4 \cong \Sigma \trianglelefteq C_\theta(z)$, it follows from the list in [**McL69**] of irreducible subgroups of $GL(A)$ containing a transvection that $\theta \cong O_4^-(2)$ or $Sp_4(2)$ is the isometry group of a quadratic form or symplectic form f on A, in the respective case, and by an earlier observation, $l \in e^\theta$. In either case, z is a singular vector in A, $e \notin z^\perp = \langle z, L_0\rangle$, and $l \in L_0 = \langle z, e\rangle^\perp$, with l nonsingular if f is quadratic. As $l \in e^\theta$, le is singular and hence in z^θ if f is quadratic; then as $l \in \langle z, e\rangle^\perp$, we conclude that $(z, l) \in (le, e)^\theta$. Thus there exists an $\mathcal{F}$-map $\rho : (z, l) \mapsto ((le)\phi, z_2)$ and $\nu \in \mathfrak{A}((le)\phi)$ with $(le)\phi\nu = z$. Further $\langle L_0, e\rangle\phi$ centralizes $\mathcal{X}_2$, so by 2.2.6.4, $z_2\nu = l\rho\nu$ centralizes $\mathcal{X}_1$. However $z\rho\nu = z$, so as $\mathcal{X}_1 \trianglelefteq \mathcal{F}_1$, l centralizes $\mathcal{X}_1$, a contradiction. Thus the proof of 2.3.8 is at last complete. □

REMARK 2.3.9. Note that 2.3.8 establishes part (2) of Theorem 2.3.6, so it remains to prove part (1) of the theorem. Therefore in the remainder of the proof of Theorem 2.3.6, we work in a counter example to part (1) of the theorem. Thus we assume $a \in C_S(\langle e, z\rangle)$ is of order at least 4 with $a\phi \in D$ but $\mathcal{C}^a \neq \mathcal{C}$. As $a\phi \in D$ and $e\phi = z_2$ is the involution in D, e is the involution in $\langle a\rangle$. By 2.3.8, e acts on $\mathcal{C}$, so we may choose a so that a^2 acts on $\mathcal{C}$. Recall we have chosen $e \in \mathcal{F}_1^f$, so by 2.3.4, $x \in \mathcal{F}_2^f$.

LEMMA 2.3.10. *$C_T(e)$ and $C_D(x)$ are cyclic.*

PROOF. Let $(s, Y, \mathcal{Y}) = (e, T, \mathcal{C})$ or $(x, D, \mathcal{D})$. From 2.3.7 and 2.3.8, s acts on Y. Suppose $C_Y(s)$ is noncyclic. Then by 2.3.5, s centralizes $\mathcal{Y}$. Hence x centralizes $\mathcal{D}$ if $s = x$, while if $s = e$ then $\mathcal{C}\phi^*$ is a component of $C_{\mathcal{F}_2}(x)$ by 2.3.4, so applying 2.2.6.4 to $\mathcal{F}_2, x, \mathcal{C}\phi^*, \mathcal{D}$ in the role of $\mathcal{F}, z, \mathcal{C}, \mathcal{L}$, we conclude that x centralizes $\mathcal{D}$. Hence in any event, x centralizes $\mathcal{D}$.

Pick $\varphi \in \mathfrak{A}(x)$ with $x\varphi = z$. By 2.2.3, $\mathcal{D} \in \mathrm{Comp}_+(C_{\mathcal{F}_2}(x))$, so by 2.3.4, $\mathcal{E} = \mathcal{D}\varphi^* \in \mathrm{Comp}_+(C_{\mathcal{F}_1}(z_2\varphi))$. Hence by 2.2.6.4, either $\mathcal{E}$ and $\mathcal{X}_1$ commute or we may take $D\varphi \leq T$. In either case $a\phi\varphi$ acts on T, so as $\phi\varphi \in \hom_{\mathcal{F}_1}(a, C_S(z))$ and $\mathcal{X}_1 \trianglelefteq \mathcal{F}_1$, a acts on $\mathcal{C}$, contrary to the choice of a. □

LEMMA 2.3.11. *Let $E = \langle z, e\rangle$. Then $Aut_{\mathcal{F}}(E) \cong S_3$.*

PROOF. By 2.3.7 and 2.3.8, E acts on T and $E\phi$ acts on D. Then by 2.3.10, $[E, N_T(E)] = \langle z\rangle$ and $[E\phi, N_D(E\phi)] = \langle z_2\rangle$. □

LEMMA 2.3.12. $|C_T(e)| > 2$.

PROOF. Assume $|C_T(e)| = 2$. Then $T\langle e\rangle$ is semidihedral. From 2.3.9, a^2 acts on T and $e \in \langle a^2\rangle$, so as $C_T(e) = \langle z\rangle$, $a^2 = e$. Further $[T, T^a] = 1$ and $T \cap T^a = \langle z\rangle$, with $T^a\langle e\rangle$ semidihedral. Let $v \in T$ be of order 4 inverted by e. Then $u = [a, v] = v^{-a}v \in TT^a$, so u is an involution centralizing e. Then as $e = a^2$ inverts v, $u^a = v^{-a^2}v^a = uz$, so $[u, a] = z$. Set $y = u\phi$ and $b = a\phi$. Then $[y, b] = x \in [y, D] \leq \mathcal{X}_2$, so x induces an inner automorphism on $\mathcal{D}$.

By 2.3.9, $x \in \mathcal{F}_2^f$, so as $x \in \mathcal{X}_2 \trianglelefteq \mathcal{F}_2$, we have $x \in \mathcal{X}_2^f$. Let d be the projection of x on D; as $\mathcal{X}_2$ is a central product of the members of Δ_2, we have $d \in \mathcal{D}^f$. Let $B = C_D(x)$ and observe that $B = C_D(d)$. Claim B is cyclic of index 2 in D. If $m = 8$ this is clear, so take $m > 8$. Then $\mathcal{D}$ is a component of $\mathcal{F}_2$ so $\mathcal{D}$ is transitive on the members of D of order 4. Then as $d \in \mathcal{D}^f$, the claim follows.

Next $x = bb^{-y} \in BB^y$ with $B^y \leq D^y$ and D^y Sylow in $\mathcal{D}^y$. By 2.3.11 there is $\rho \in Aut_{\mathcal{F}}(E\phi)$ interchanging z_2 and x. Indeed as $z_2 \in \mathcal{F}^f$ and $x \in \mathcal{F}_2^f$, $E\phi \in \mathcal{F}_2^f$, and then by 2.3.11, $E\phi \in \mathcal{F}^f$. Thus we can extend ρ to $\chi \in Aut_{\mathcal{F}}(C_S(E\phi))$. Let $\varphi \in \mathfrak{A}(x)$ with $x\varphi = z$. Now $\langle B, y\rangle \leq C_S(E\phi)$, so $\langle B, y\rangle\chi\varphi \leq C_S(E\phi)\varphi \leq C_S(x)\varphi \leq S$, with $z_2\chi\varphi = x\varphi = z$. Therefore as $B \leq \mathcal{D}$, $B\chi\varphi \leq \mathcal{C}' \in \Delta_1$, so $e\phi\varphi \in E\phi\varphi \leq (BB^y)\chi\varphi \leq \mathcal{X}_1$. Then as $\mathcal{X}_1 \trianglelefteq \mathcal{F}_1$, $e \in \mathcal{X}_1$, contradicting $C_T(e) = \langle z\rangle$. □

LEMMA 2.3.13. *e induces an inner automorphism on $\mathcal{C}$ and $|T : C_T(e)| = 2$.*

PROOF. Assume otherwise and let S_1 be Sylow in $\mathcal{X}_1$. By 2.3.9, $e \in \mathcal{X}_1^f$. Then as $\mathcal{X}_1 \trianglelefteq \mathcal{F}_1$, we have $|C_{S_1}(e)| \geq |C_{S_1}(f)|$ for each $f \in e^{\mathcal{X}_1}$. Hence as $\mathcal{X}_1$ is a central product of the members of Δ_1, we have $|C_T(s)| \geq |C_T(g)|$ for each $g \in e^{\mathcal{C}}$.

Let $V = \langle v\rangle$ be an e-invariant cyclic subgroup of T of index 2, $u \in T - V$, and $\pi \in V$ of order 4. Suppose first that $C_V(e) = \langle z\rangle$; then by 2.3.12 we may take $C_T(e) = \langle u\rangle$. Set $c = ue$; then c centralizes $Q = \langle \pi, u\rangle \cong Q_8$, so that lemma holds if $m = 8$, so we may assume $m > 8$. Then $\mathcal{C} \cong SL_2[m]$ is a component of $\mathcal{F}_z$, so (as in the proof of 2.3.3) $\beta \in Aut_{\mathcal{C}}(Q)$ of order 3 with $u\beta = \pi$ extends to γ centralizing $C_S(Q)$. Hence $e\gamma = c \cdot u\beta = c\pi$ centralizes π. Therefore $k = |V : C_V(e\gamma)| \leq 2$. If $k = 1$ then $|C_T(e\gamma)| = m/2 > 4 = |C_T(e)|$, contrary to paragraph one. Thus $k = 2$, where $C_T(e\gamma) = \langle v^2, vu\rangle$, for the same contradiction.

Therefore $|C_V(e)| > 2$, so $j = |V : C_V(e)| \leq 2$. If $j = 1$ then $e\pi$ centralizes T, so the lemma holds; thus we may assume $j = 2$ and hence $m > 8$ and $v^e = vz$. But now $\langle e, z\rangle = \Omega_1(V\langle e\rangle)$ is u-invariant, so $e^u = ez = e^v$, and hence e centralizes uv, contrary to 2.3.10. □

By 2.3.13, $e = \pi\pi^a y$ for some π of order 4 in T and $y \in C_S(TT^a)$. Let $t \in T - C_T(e)$, $u = tt^a$, $w = u\phi$, and $b = a\phi$. Then u is an involution centralizing e. Next

$$z = t^{-1}t^{\pi} = t^{-1}t^{e} = t^{-1}t^{a}t^{-a}t^{e} = u^{-1}u^{a} = [u, a],$$

so $x = [u\phi, a\phi] = [w, b] \in \mathcal{D}\mathcal{D}^w$. As $z \neq e$ it follows that $\mathcal{D} \neq \mathcal{D}^w$.

By 2.3.10 and 2.3.13, $\langle v\rangle = C_T(e)$ is of order $m/2$. Then $t_1 = tv \in T - C_T(e)$, so by symmetry between t and t_1, $\mathcal{D}^{w_1} \neq \mathcal{D}$, where $w_1 = (t_1t^a)\phi$. Then as $x \in DD^w$, $\mathcal{D}^w = \mathcal{D}^{w_1}$, so $v\phi$ acts on $\mathcal{D}$. Then x centralizes $[v, a]\phi \in D$ and $a \notin \langle [v, a]\rangle$, so if $[v, a] \neq e$ then x centralizes a Q_8-subgroup of D, contrary to 2.3.10. Hence $[v, a] = e$ so $|v| = 4$ and hence $m = 8$ and v inverts a.

Now $x = [w, b] = b^{-w}b$ and $y = v\phi$ inverts b. Let $d \in D - \langle b\rangle$; then $[x, d] = z_2$, so $d^y = dbz_2^{\epsilon}$. Then

$$dz_2 = d^x = (dbz_2^{\epsilon})^y = dbz_2^{\epsilon}b^{-1}z_2^{\epsilon} = d,$$

for our final contradiction.

This completes the proof of Theorem 2.3.6.

2.4. A sufficient condition for quaternion fusion packets

In this section we assume $\mathcal{F}$ is a saturated fusion system on a finite 2-group S. We prove:

THEOREM 2.4.1. *Assume z is an involution in S, $z \in \mathcal{F}^f$, and $\mathcal{C}$ is a component or solvable component of $C_{\mathcal{F}}(z)$ on T such that $z \in T$ and T is quaternion. Set $\Omega = T^{\mathcal{F}}$. Then $\tau = (\mathcal{F}, \Omega)$ is a quaternion fusion packet.*

PROOF. By construction, Ω is $\mathcal{F}$-invariant, and condition (QFP1) is satisfied. For $P \in \Omega$, let $z(P)$ be the involution in P, so that:

(a) For $P \in \Omega$, $z(P)$ is the unique involution in P.

Set $\mathcal{F}_z = C_{\mathcal{F}}(z)$, $E = \langle T^{\mathcal{F}_z} \rangle$, and $\Delta = \Delta_z = \{\mathcal{C}\alpha^* : \alpha \in Aut_{\mathcal{F}_z}(E)\}$. Now $\mathcal{C}$ is a component or solvable component of $\mathcal{F}_z$, so by Theorem 6 in [**Asc11**] and 2.2.2:

(b) There exists a central product $\mathcal{E}$ of the members of Δ normal in $\mathcal{F}_z$.

Suppose $D \in \Omega - \{T\}$ with $1 \neq T \cap D$. Then by (a), $z = z(T) = z(D)$. By definition of Ω, there is $\alpha \in \hom_{\mathcal{F}}(T, S)$ with $T\alpha = D$. Then $z\alpha = z(T)\alpha = z(D) = z$, so $\alpha \in \hom_{\mathcal{F}_z}(T, C_S(z))$. Then by (b), $T \cap D = \langle z \rangle$ and $1 = [T, D]$. Hence (QPF2) is satisfied.

We next establish (QFP3). Assume $D \in \Omega$, $d \in D - \langle z(D) \rangle$, and $\mu \in \hom_{\mathcal{F}}(d, S)$ with $d\mu \in C_S(z)$. Replacing D by a suitable conjugate, we may assume $z(D) \in \mathcal{F}^f$. Set $e = z(D)\mu$, $a = d\mu$, and let $\phi \in \mathfrak{A}(e)$ with $e\phi = z(D)$. Thus Hypothesis 2.3.1 is satisfied with $z(D) = z_2$. As $z(D)\mu\phi = e\phi = z(D)$ and $d \in D \leq \mathcal{D} \in \Delta_{z_2}$, $a\phi = d\mu\phi \in D' \leq \mathcal{D}' \in \Delta_{z_2}$. Hence by Theorem 2.3.6, $a \in N_S(T)$, so that (QFP3) is satisfied.

Finally assume $D \in \Omega$ with $z = z(D)$, $v \in T$, and $\alpha \in \hom_{\mathcal{F}_z}(v, S)$. By (b), $\mathcal{E} \trianglelefteq \mathcal{F}_z$, so by 3.3 in [**Asc08a**], $\alpha = \varphi\phi$ with $\varphi \in Aut_{\mathcal{F}_z}(E)$, and $\phi \in \hom_{\mathcal{E}}(v\varphi, E)$. By definition, $\mathcal{C}\varphi^* \in \Delta$, and of course $v\varphi \in T\varphi$. Further as ϕ is a $\mathcal{E}$-map and $v\varphi \in T\varphi$, also $v\alpha = v\varphi\phi \in T\varphi$. Finally by (b), either $D = T\varphi$ or $[D, T\varphi] = 1$, so $v\alpha \in D$ or $[v\alpha, D] = 1$. Thus (QFP4) is satisfied, completing the proof of the theorem. □

2.5. Basic results on fusion packets

In this section we assume that $\tau = (\mathcal{F}, \Omega)$ is a quaternion fusion packet, and we adopt the following notation:

NOTATION 2.5.1. Set $\mathcal{Z}_S = \mathcal{Z}_S(\tau) = \{z(K) : K \in \Omega\}$ and

$$\mathcal{Z} = \mathcal{Z}(\tau) = \bigcup_{z \in \mathcal{Z}_S} z^{\mathcal{F}}.$$

For $z \in \mathcal{Z}_S$, set $\Omega(z) = \{J \in \Omega : z \in J\}$ and $\mathcal{F}_z = C_{\mathcal{F}}(z)$.

LEMMA 2.5.2. *Suppose $\mathcal{E}$ is a saturated subsystem of $\mathcal{F}$ on $T \leq S$, Γ_0 is a nonempty subset of Ω, and for some divisor k of m, $\Gamma = \{K \cap T : K \in \Gamma_0\}$ satisfies:*

(a) For each $J \in \Gamma$, J is nonabelian of order k.

(b) $\Gamma^{\mathcal{E}} = \Gamma$.

Then the pair $(\mathcal{E}, \Gamma)$ is a quaternion fusion packet.

PROOF. The pair satisfies condition (QFP1) by (a). The remaining conditions inherit from τ to the pair. □

LEMMA 2.5.3. *Assume $z \in \mathcal{Z}_S \cap \mathcal{F}^f$. Then*
(1) $(\mathcal{F}_z, \Omega(z))$ is a quaternion fusion packet.
(2) If z centralizes Ω then $(\mathcal{F}_z, \Omega)$ is a quaternion fusion packet.

PROOF. As $z \in \mathcal{F}^f$, $\mathcal{F}_z$ is a saturated fusion system by 1.2.1. Thus the hypotheses of 2.5.2 are satisfied by $\mathcal{F}_z$ in the role of $\mathcal{E}$, m in the role of k, and $\Omega(z)$ in the role of Γ, so (1) follows. A similar argument establishes (2). □

DEFINITION 2.5.4. Define $SL_2[8]$ to be the 2-fusion system of $SL_2(3)$. Suppose $m > 8$ and let $K \in \Omega$. Then K has two classes R_i^K, $i = 1, 2$, of subgroups $R_i \cong Q_8$. Define $SL_2[m]$ to be the fusion system on K generated by $O^2(Aut(R_i))$, $i = 1, 2$. Define $SL_2^1[m/2]$ to be the fusion system on K generated by $O^2(Aut(R_1))$.

LEMMA 2.5.5. *Let $K \in \Omega$ and assume $\mathcal{E}$ is a saturated fusion system on K. Then:*
(1) Either $\mathcal{E}$ is isomorphic to $\mathcal{F}_K(K)$ or $SL_2[m]$, or $m > 8$ and $\mathcal{E} \cong SL_2^1[m/2]$.
(2) Assume K is Sylow in $G \cong SL_2(q)$. Then $\mathcal{F}_K(G) \cong SL_2[m]$, for $m = (q^2 - 1)_2$.
(3) Assume K is Sylow in G such that $F^*(G) \cong SL_2(q)$ and $|G : F^*(G)| = 2$. Then $\mathcal{F}_K(G) \cong SL_2^1[m/2]$, where $m = 2(q^2 - 1)_2$.
(4) $\mathcal{F}_K(K)$, $SL_2[m]$, and $SL_2^1[m/2]$ are saturated.

PROOF. From 1.2.4, $\mathcal{E}$ is determined by its *Alperin data* : A set $\mathcal{Q}$ of representatives for the orbits of $\mathcal{E}$ on $\mathcal{E}^{frc}$, and the systems $Aut_{\mathcal{E}}(Q)$ for $Q \in \mathcal{Q}$. However for $X \leq K$, either $Aut(X)$ is a 2-group or $X \cong Q_8$ and $Aut(X) \cong S_4$. We conclude that $\mathcal{E}$ has one of the following sets of Alperin data:
(i) $Aut_{\mathcal{E}}(X) = Aut_K(X)$ for each $X \leq K$ and $\mathcal{E}^{frc} = \{K\}$.
(ii) $m = 8$, $Aut_{\mathcal{E}}(K) = O^2(Aut(K)) \cong A_4$, and $\mathcal{E}^{frc} = \{K\}$.
(iii) $m > 8$, $Aut_{\mathcal{E}}(R_i) = Aut(R_i) \cong S_4$ for $i = 1, 2$, and $\mathcal{E}^{frc} = \{K, R_1^K, R_2^K\}$.
(iv) For $i = 1$ or 2, $Aut_{\mathcal{E}}(R_i) = Aut(R_i) \cong S_4$, $Aut_{\mathcal{E}}(R_{3-i}) = Aut_K(R_{3-i}) \cong D_8$, and $\mathcal{E}^{frc} = \{K, R_i^K\}$.
It follows that $\mathcal{E}$ is isomorphic to $\mathcal{F}_K(K)$, $SL_2[8]$, $SL_2[m]$, $SL_2^1[m/2]$ in cases (i)-(iv), respectively, since those 2-fusion systems have the respective Alperin data. This establishes (1). Then (1) implies (2) and (3), since the 2-fusion systems for the groups in (2) and (3) are saturated and have the Alperin data of the indicated fusion system. Finally (2) and (3) imply (4). □

DEFINITION 2.5.6. We define a fusion system $\mathcal{O}_K$ on $K \in \Omega$ which is a subsystem of $\mathcal{F}$. Set $z = z(K)$. If $m = 8$, set $\mathcal{O}_K = \mathcal{F}_K(K)$. If $m > 8$, define $\mathcal{Q}_K$ to be the set of Q_8 subgroups of K and define $\mathcal{O}_K = \langle O^2(Aut_{\mathcal{F}}(Q)) : Q \in \mathcal{Q}_K \rangle_K$, regarded as a fusion system on K.

LEMMA 2.5.7. *Let $K \in \Omega$. Then:*
(1) $\mathcal{O}_K$ is isomorphic to $\mathcal{F}_K(K)$, $SL_2[m]$, or $SL_2^1[m/2]$.
(2) $\mathcal{O}_K$ is saturated.
(3) $\mathcal{O}_K = O^{2'}(\mathcal{O}_K)$.
(4) $Aut_{\mathcal{F}}(K) \leq Aut(\mathcal{O}_K)$.

PROOF. From 2.5.6, $\mathcal{O}_K$ is $\mathcal{F}_K(K)$ or one of the three fusion systems defined in 2.5.4, so (1) follows. Then (2) follows from 2.5.5.4. If $m = 8$ then $\mathcal{O}_K =$

$\mathcal{F}_K(K)$, so (3) holds in that case. If $m > 8$ then for $R \leq Q$, $O^2(Aut_{\mathcal{F}}(R)) \leq O^{2'}(Aut_{\mathcal{O}_K}(R))$, so again (3) holds. Part (4) follows as $Aut_{\mathcal{F}}(K)$ permutes $\mathcal{Q}_K$; cf. 1.2.4 for example. □

LEMMA 2.5.8. *Let $m > 8$. Then:*
(1) $L_2[m/2]$ is simple.
(2) $SL_2[m]$ is quasisimple.

PROOF. Suppose $\mathcal{E}$ is a fusion system on T isomorphic to $L_2[m/2]$ or $SL_2[m]$. In the first case, $\mathcal{E}$ has one class of involutions and T is generated by involutions, so there is no proper nontrivial subgroup of T strongly closed in T with respect to $\mathcal{E}$. Then as $N_{\mathcal{F}}(T) = T$, $\mathcal{E}$ is simple by part (4) of Theorem 8 in [**Asc11**].

So assume $\mathcal{E}$ is $SL_2[m]$. Then $Z = Z(\mathcal{E})$ is of order 2, and $\mathcal{E}/Z \cong L_2[m/2]$ is simple, so it remains to show $\mathcal{E} = O^2(\mathcal{E})$. But $O^2(\mathcal{E}) \trianglelefteq \mathcal{E}$ and Z is the unique subgroup of T of order 2, so $Z \leq O^2(\mathcal{E})$, and hence $\mathcal{E} = O^2(\mathcal{E})$ by 7.15 in [**Asc11**]. □

2.6. The case $z \in Z(\mathcal{F})$.

In this section we assume:

HYPOTHESIS 2.6.1. *$\tau = (\mathcal{F}, \Omega)$ is a quaternion fusion packet and $z \in \mathcal{Z}_S$ with $z \in Z(\mathcal{F})$. Set $O = O(z) = \langle \Omega(z) \rangle$ and $\mathcal{F}^+ = \mathcal{F}/\langle z \rangle$. Let $K \in \Omega(z)$.*

LEMMA 2.6.2. *Suppose $V \leq K$ with $|V| > 2$ and let $\phi \in \hom_{\mathcal{F}}(V, S)$. Then*
(1) K is the unique member of Ω containing V.
(2) $V\phi \leq J$ for some $J \in \Omega(z)$.
(3) $J \in K^{\mathcal{F}}$.
(4) If $K \in \mathcal{F}^f$ then $V^{\mathcal{F}} \cap \mathcal{F}^f \cap K \neq \emptyset$.

PROOF. Part (1) follows from (QFP2). Set $X = V\phi$ and let $\alpha \in \mathfrak{A}(X)$.

Assume (2) fails. Suppose first that $U = X \cap J \neq \langle z \rangle$ for some $J \in \Omega(z)$. As the lemma fails, $U < X$, so there exists $v \in V$ with $x = v\phi \notin J$. By (QFP4), x centralizes J. Hence x centralizes U, so U is cyclic and $U \leq \langle x \rangle \leq C_S(J)$, contradicting $U \leq J$.

Therefore for each $J \in \Omega(z)$, $J \cap X = \langle z \rangle$, so $X \leq C_S(O)$ by (QFP4). As $\Omega^{\mathcal{F}} = \Omega$, as $z \in Z(\mathcal{F})$, and as $O \leq C_S(X)$, $\Omega(z)\alpha = \Omega(z)$, so $O\alpha = O$ and hence $X\alpha \leq C_S(O)$. In particular $X \notin \Omega(z)$, so $V \neq K$ and hence $V < N_K(V)$. By 1.2.5.2 there is $\chi \in Aut_{\mathcal{F}}(V)$ such that $\chi\phi\alpha$ extends to $\varphi \in \hom_{\mathcal{F}}(N_K(V), S)$. Then $Y = N_K(V)\varphi$ acts on $V\chi\phi\alpha = X\alpha$, and by induction on the order of V, $Y \leq J \in \Omega(z)$, contradicting $X \leq C_S(O) \leq C_S(J)$. This completes the proof of (2).

Assume (3) fails and choose a counter example with $n = |K : V|$ minimal; then $n > 1$. By (2), $X\alpha \leq L \in \Omega$ and then by (1) and (2), $N_J(X)\alpha \leq L$, so $L \in J^{\mathcal{F}}$ by minimality of n. Thus replacing ϕ by $\phi\alpha$, we may assume $X \in \mathcal{F}^f$. Hence by 1.2.5.2 there is $\chi \in Aut_{\mathcal{F}}(V)$ such that $\chi\phi$ extends to $\beta \in \mathfrak{A}(V)$. Then $N_S(V)\beta \leq J$ by (1) and (2), so (3) follows from the minimality of n.

Assume $K \in \mathcal{F}^f$ and choose $\phi \in \mathfrak{A}(V)$. By (3) there is $\gamma \in \mathfrak{A}(J)$ with $J\gamma = K$. By (1), $N_S(X) \leq N_S(J)$, so $X\gamma \in \mathcal{F}^f$. Thus (4) holds. □

LEMMA 2.6.3. *Assume* $K \in \mathcal{F}^f$, $\langle z \rangle < P \leq K$, *and* $\psi \in \hom_{\mathcal{F}}(P, K)$. *Then* $\psi = \varphi\phi$ *for some* $\varphi \in Aut_{\mathcal{F}}(K)$ *and* $\phi \in \hom_{\mathcal{O}_K}(P\varphi, K)$.

PROOF. Assume otherwise and pick a counter example P with $n = |K : P|$ minimal. Then $\psi \notin Aut_{\mathcal{F}}(K)$, so $n > 1$. By 2.6.2.4 there is a member of $P^{\mathcal{F}} \cap \mathcal{F}^f$ contained in K.

Suppose 2.6.3 holds whenever $P\psi \in \mathcal{F}^f$. Let $\beta \in \mathfrak{A}(P\psi)$ with $P\psi\beta \leq K$. Then $\psi\beta = \varphi_1\phi_1$ and $\beta = \varphi_2\phi_2$, so

$$\psi = (\psi\beta)\beta^{-1} = \varphi_1\phi_1\phi_2^{-1}\varphi_2^{-1} = \varphi\phi,$$

where $\varphi = \varphi_1\varphi_2^{-1} \in Aut_{\mathcal{F}}(K)$ and $\phi = (\phi_1\phi_2^{-1})\varphi_2^{-*} \in \hom_{\mathcal{O}_K}(P\varphi, K)$, using 2.5.7.4. Thus it suffices to assume $P\psi \in \mathcal{F}^f$.

Let $R = N_K(P)$; as $n > 1$, $P < R$. By 1.2.5.2, there exists $\chi \in Aut_{\mathcal{F}}(P)$ such that $\chi\psi$ extends to $\rho \in \hom_{\mathcal{F}}(R, S)$. As $P\psi \leq K$ it follows from 2.6.2 that $R\rho \leq K$. Thus as $P < R$ it follows from minimality of n that $\rho = \varphi_3\phi_3$. Next $\chi\psi = \psi\zeta$ where $\zeta = \chi\psi^* \in Aut_{\mathcal{F}}(P\psi)$, so $\psi = \varphi_3(\phi_3\zeta^{-1})$. If P is not isomorphic to Q_8 then $Aut(P)$ is a 2-group, so $\zeta = c_s$ for some $s \in N_S(P\psi)$ as $P\psi \in \mathcal{F}^f$. Thus as $N_S(P\psi) \leq N_S(K)$ by (QFP2), $\zeta = c_s \in Aut_{\mathcal{F}}(K)$, so $\psi = \varphi_4\phi_4$ with $\varphi_4 = \varphi_3\zeta^{-1} \in Aut_{\mathcal{F}}(K)$ and $\phi_4 = \phi_3\zeta^{-*} \in \hom_{\mathcal{O}_K}(P\varphi_4, K)$, using 2.5.7.4. On the other hand if $P \cong Q_8$ then $\zeta \in Aut_{\mathcal{O}_K}(P\psi)$, so $\psi = \varphi_3\phi_4$ with $\phi_4 = \phi_3\zeta^{-1} \in \hom_{\mathcal{O}_K}(P\varphi_3, K)$. This completes the proof of the lemma. □

LEMMA 2.6.4. *Assume* $K \in \mathcal{F}^f$, $\langle z \rangle < V \leq K$ *with* $V \in \mathcal{O}_K^f$, *and* $\alpha \in \mathfrak{A}(V)$. *Then* $N_{\mathcal{O}_K}(V)\alpha^* \trianglelefteq N_{\mathcal{F}}(V\alpha)$.

PROOF. By 2.6.2, $X = V\alpha$ is contained in a unique member J of Ω, and by 2.6.2.4 we may take $J = K$. Set $\mathcal{N} = N_{\mathcal{F}}(X)$, $T = N_S(X)$, $L = N_K(X)$, and $\mathcal{O} = N_{\mathcal{O}_K}(X)$. By 2.6.3, $\alpha_{|N_K(V)} = \varphi_1\phi_1$ for some $\varphi_1 \in Aut_{\mathcal{F}}(K)$ and $\phi_1 \in \hom_{\mathcal{O}_K}(V\varphi_1, K)$. Therefore as $V \in \mathcal{O}_K^f$, $X \in \mathcal{O}_K^f$ and $N_{\mathcal{O}_K}(V)\alpha^* = \mathcal{O}$ is saturated by 1.2.1.

Let $Y \leq L$ and $\phi \in \hom_{\mathcal{N}}(Y, T)$. Then ϕ lifts to $\varphi \in \hom_{\mathcal{N}}(XY, T)$, and by 2.6.2, $(XY)\varphi = X(Y\varphi) \leq K$ and hence $Y\phi \leq L$. Therefore L is strongly closed in T with respect to $\mathcal{N}$.

By 2.5.2, $\rho = (\mathcal{N}, L)$ is a quaternion fusion packet. Define $\mathcal{O}_L$ as in 2.5.6 with respect to ρ. By definition, $\mathcal{O}_L \leq \mathcal{O}$. Further if $P \in \mathcal{O}^f$ and $\psi \in Aut_{\mathcal{O}}(P)$, then either $\psi \in Aut_L(P)$ or $P \in \mathcal{Q}_L$ and $\psi \in O^2(Aut_{\mathcal{O}}(P)) \leq Aut_{\mathcal{O}_L}(P)$, so indeed $\mathcal{O}_L = \mathcal{O}$. As L is strongly closed in T with respect to $\mathcal{N}$, $\mathcal{O}_L$ is $\mathcal{N}$-invariant by construction of $\mathcal{O}_L$. By 2.5.7.3, $\mathcal{O}_L = O^{2'}(\mathcal{O}_L)$, so by Craven's Theorem 1.2.6, $\mathcal{O} \trianglelefteq \mathcal{N}$. □

LEMMA 2.6.5. *If* $K \in \mathcal{F}^f$ *then* $\mathcal{O}_K^+$ *is tightly embedded in* $\mathcal{F}^+$.

PROOF. As $K \in \mathcal{F}^f$, $K^+ \in \mathcal{F}^{+f}$, and as $\mathcal{O}_K$ is saturated by 2.5.7.2, $\mathcal{O}_K^+$ is saturated. Thus we must verify conditions (T1)-(T3) of Definition 3.1.2 in [**Asc19**] for the embedding of $\mathcal{O}_K^+$ in $\mathcal{F}^+$. Condition (T1) holds by 2.6.4, condition (T2) holds by 2.6.3, and condition (T3) holds by 2.5.7.4. □

LEMMA 2.6.6. *(1)* O *is the central product of the members of* $\Omega(z)$.

(2) $O^+ = \prod_{J \in \Omega(z)} J^+$ *is a direct product.*

(3) Assume $m > 8$ *and let* Q *be a* Q_8*-subgroup of* K *and* $R = C_S(Q)$. *Suppose* $\alpha \in Aut_{\mathcal{F}}(Q)$ *is of order 3. Then* $\alpha \times 1 \in Aut_{\mathcal{F}}(QR)$, *where* $\alpha \times 1$ *is the automorphism of* QR *which acts as* α *on* Q *and 1 on* R.

PROOF. For distinct $K, J \in \Omega(z)$, K and J commute by (QFP4) applied to each $v \in K$ with ϕ the identity map. Note that $K \cap \langle J \neq K \in \Omega \rangle = \langle z \rangle$ as $\langle z \rangle = Z(K)$. Thus (1) holds and then (1) and (QFP2) imply (2). Thus it remains to prove (3), so we may assume the setup of (3).

First suppose $Q \in \mathcal{F}^f$ and let $V = QR$. Then α extends to $\beta \in Aut_{\mathcal{F}}(V)$ by 1.2.5.1. Next as $m > 8$, α is inverted by c_s for some $s \in N_K(Q)$. Then $\gamma = [\beta, c_s] \in C_{Aut(V)}(R)$ as s centralizes R. Further $\gamma_{|Q} = [\alpha, c_s] = \alpha$, so $\gamma = \alpha \times 1$.

In general, let $\beta \in \mathfrak{A}(Q)$ and $Q\beta = P$. Then $\alpha\beta^* = \delta$ is of order 3 in $Aut_{\mathcal{F}}(P)$, so $\delta \times 1 \in Aut_{\mathcal{F}}(PR')$ for $P \leq J \in \Omega(z)$ and $R' = C_S(P)$ by the previous paragraph. Then $\alpha \times 1 = (\delta \times 1)\beta^{-*} \in Aut_{\mathcal{F}}(V)$, completing the proof. □

DEFINITION 2.6.7. We define a fusion system $\mathcal{O} = \mathcal{O}(z)$ on O. By construction, $\mathcal{O}$ is a subsystem of $\mathcal{F}$. If $m = 8$, set $\mathcal{O} = \mathcal{F}_O(O)$. If $m > 8$, define notation as in 2.6.6 and set

$$\mathcal{O} = \langle \alpha \times 1 \in Aut_{\mathcal{F}}(QC_O(J) : Q \in \mathcal{Q}_J,\ J \in \Omega(z),\ \alpha \in Aut_{\mathcal{F}}(Q) \text{ of order } 3\rangle_O,$$

regarded as a fusion system on O. Here $\mathcal{Q}_J$ is the set of Q_8-subgroups of J.

LEMMA 2.6.8. *$\mathcal{O}$ is the central product of the subsystems $\mathcal{O}_K$, for $K \in \Omega(z)$, so $\mathcal{O}$ is saturated.*

PROOF. If $m = 8$ then $\mathcal{O} = \mathcal{F}_O(O)$ is the central product of the systems $\mathcal{O}_K = \mathcal{F}_K(K)$, as O is the central product of the members of $\Omega(z)$ by 2.6.6.1. Thus we may take $m > 8$, where the lemma follows from 2.3 in [**Asc11**]. □

LEMMA 2.6.9. *If $m > 8$ then $\mathcal{O} \trianglelefteq \mathcal{F}$.*

PROOF. By 2.6.5, $\mathcal{O}_K^+$ is tightly embedded in $\mathcal{F}^+$ and as $m > 8$, $\Phi(K^+) \neq 1$. By 2.6.2, the set $\mathcal{T}$ of Notation 3.3.2 in [**Asc19**] is empty. By 2.5.7.3, $\mathcal{O}_K = O^{2'}(\mathcal{O}_K)$. Thus $\mathcal{F}^+, \mathcal{O}_K^+$ satisfies Hypothesis 3.6.1 of [**Asc19**]. Therefore $\mathcal{O}^+ \trianglelefteq \mathcal{F}^+$ by Theorem 3.6.8 in [**Asc19**], so $\mathcal{O} \trianglelefteq \mathcal{F}$ by 8.10 in [**Asc08a**]. □

LEMMA 2.6.10. *If $m = 8$ then $O \trianglelefteq \mathcal{F}$.*

PROOF. Assume otherwise and pick a minimal counter example τ. By minimality of τ and 2.5.2:

(a) $\mathcal{F} = [O]_{\mathcal{F}}$.

Let $\mathcal{Y} = \{P \leq O : \langle z \rangle < P\}$ and $\mathcal{Y}^f = \mathcal{Y} \cap \mathcal{F}^f$. Let Z be the preimage in O of $Z(\mathcal{F}^+) \cap O^+$ and $\mathcal{F}^! = \mathcal{F}/Z$. Set $\mathcal{N} = N_{\mathcal{F}}(O)$.

As O^+ is a weakly closed abelian subgroup of S^+, it follows that

(b) $\mathcal{N}^+$ controls fusion in O^+.

(c) For each $P \in \mathcal{Y}$, $P^{\mathcal{F}} \cap \mathcal{Y}^f \neq \emptyset$.

(d) For $P \in \mathcal{Y}^f$ either $P \trianglelefteq \mathcal{F}$ or $O \trianglelefteq N_{\mathcal{F}}(P)$.

As O^+ is abelian, $\tau_P = (N_{\mathcal{F}}(P), \Omega(z))$ is a quaternion fusion packet by 2.5.2, so (d) follows from the minimality of τ.

(e) For each $P \in \mathcal{Y}$ with $P \trianglelefteq \mathcal{F}$, $P \leq Z$.

By (a) and 8.10 in [**Asc08a**], $\mathcal{F}^+ = [O^+]_{\mathcal{F}^+}$. As O^+ is abelian, $O^+ \leq C_{S^+}(P^+)$, so $\mathcal{F}^+$ centralizes P^+, proving (e).

(f) $O^!$ is tightly embedded in $\mathcal{F}^!$.

We must verify conditions (T1)-(T3) of 3.1.2 in [**Asc19**] for the pair $\mathcal{F}^!$, $O^!$. Condition (T1) follows from (c)-(e), (T2) follows from (b), and (T3) is trivial.

Define the set $\mathcal{P}^*$ as in Definition 3.1.9 of [**Asc19**] with respect to the pair $\mathcal{F}^!$, $O^!$. Let $\mathcal{B}$ be the set of preimages in S of members of $\mathcal{P}^*$. Then for $B \in \mathcal{B}$, $B^{\mathcal{F}} \cap \mathcal{Y} \neq \emptyset$ and B is maximal subject to that constraint. As τ is a counter example to our lemma, it follows from 3.1.7 in [**Asc19**] that

(g) There exists $B \in \mathcal{B} - \{O\}$.

From 3.1.12 in [**Asc19**]:

(h) $B^!$ is a TI-subgroup in $S^!$ and $B \cap O = Z$.

Let A be the preimage in O of $C_{O^!}(B^!)$. As a consequence of (h):

(i) $[O^!, B^!] \leq A^!$, $|A^!| \geq |B^!|$, and for each $1 \neq b^! \in B^!$, $A^! = C_{O^!}(b^!)$.

Pick $B^!$ of maximal order.

(k) We may choose B so that A is conjugate to B in $Aut_{\mathcal{F}}(AB)$.

As $B^{\mathcal{F}} \cap \mathcal{Y} \neq \emptyset$ by (c), it follows that there is $\alpha \in \mathfrak{A}(B)$ with $B\alpha \in \mathcal{Y}$. As $A \leq N_S(B)$, $A\alpha \leq N_S(B\alpha)$ and by 3.1.14 in [**Asc19**], the member B_0 of $\mathcal{B}$ containing $A\alpha$ is not O, so $|A^!| \leq |B_0^!| \leq |B^!|$. Thus $|A^!| = |B_0^!| = |B^!|$ by (i), so $B_0 = A\alpha$ and $(B\alpha)^! = C_{O^!}(A\alpha)$. Further O acts on $(AB)\alpha$ and for $x \in O - C_0(A\alpha)$, $(A\alpha)^x \neq A\alpha$, so $\phi = c_x\alpha^{-*} \in Aut_{\mathcal{F}}(AB)$ with $A\phi \neq A$. Thus replacing B by $A\phi$, (k) holds.

(l) $K \cap Z = \langle z \rangle$, so $K^! \cong E_4$.

For if $V = K \cap Z \neq \langle z \rangle$ then $V \trianglelefteq \mathcal{F}$, so $K \trianglelefteq \mathcal{F}$ by 2.6.5, contrary to the choice of τ.

(m) $K \cap A = \langle z \rangle$.

If $V = K \cap A \neq \langle z \rangle$ then by (h) and (k) there is $\phi \in \hom_{\mathcal{F}}(V, S)$ with $V \not\leq O$, contrary to 2.6.2.

(n) For $b \in B - Z$, $K^b \neq K$.

If (n) fails then $1 \neq C_{K^!}(b^!) \leq A^!$ by (i), contrary to (m).

By minimality of τ:

(o) $\mathcal{F}$ is transitive on $\Omega(z)$.

Let $\Omega(z) = \{K_1, \ldots, K_n\}$. By 2.6.6.2:

(p) $O^+ = K_1^+ \times \cdots \times K_n^+$.

By (p) for each $1 \neq x^+ \in O^+$, $x^+ = x_1^+ \cdots x_n^+$ where x_i^+ is the projection of x^+ on K_i^+. Set $I = \{1, \ldots, n\}$, $I(x^+) = \{i \in I : x_i^+ \neq 1\}$, and $r(x^+) = |I(x^+)|$.

(q) For each $1 \neq x^+ \in Z^+$, $r(x^+) = n$. Thus $|Z^+| \leq 4$.

For if $i \notin I(x^+)$ then x centralizes K_i, and then by (o), x centralizes O, a contradiction. Thus $r(x^+) = n$ and for $y^+ \in Z^+ - \langle x^+ \rangle$, $y_i^+ \neq x_i^+$, so $|Z^+| \leq |K^+| = 4$.

We are now in a position to obtain a contradiction, and hence to establish 2.6.10. By (n), $B^!$ acts semiregularly on $\Omega(z)$, so $k = |B^!|$ divides n. On the other hand it follows from (i) that O/A is semiregular on $b^!A^!$ for $1 \neq b^! \in B^!$, so $j = |O : A| \leq |A^!| = k$. Thus $k^2 \geq kj = |O^!|$. But $|O^+| = 4^n$, so $|O^!| \geq 4^{n-1}$ by (q). Therefore $4^{n-1} \leq k^2 \leq n^2$, while $n \geq 2$ by (n), so we conclude $n = k = j = 2$. This is impossible as $j = |O : A| \geq |K^!| = 4$ by (m). This completes the proof of 2.6.10. □

THEOREM 2.6.11. *(1)* $\mathcal{O} \trianglelefteq \mathcal{F}$.

(2) For each $K \in \Omega(z)$, $\mathcal{O}_K$ *is subnormal in* $\mathcal{F}$.

(3) $Aut_{\mathcal{F}}(O)$ *permutes* $\Omega(z)$ *and* $\{\mathcal{O}_K : K \in \Omega(z)\}$.

PROOF. Part (1) follows from 2.6.9 and 2.6.10. By 2.6.8, $\mathcal{O}_K \trianglelefteq \mathcal{O}$, so (1) implies (2). The first statement in (3) follows as $\Omega(z)^{\mathcal{F}} = \Omega(z)$, and the second statement in (3) follows from the first and the definition of $\mathcal{O}_K$ in 2.6.7. □

LEMMA 2.6.12. *$C_S(O)$ centralizes $\mathcal{O}$.*

PROOF. If $m = 8$ then $\mathcal{O} = O$, so the lemma is trivial. If $m > 8$ then the lemma follows from 2.6.6.3 and the definition of $\mathcal{O}$ in 2.6.7. □

LEMMA 2.6.13. *Either*
(1) $\mathcal{O}_K = \mathcal{F}_K(K)$ and $K \leq O_2(\mathcal{F})$, or
(2) $\mathcal{O}_K \cong SL_2[m]$ or $SL_2^1[m/2]$ and $O^2(\mathcal{O}_K)$ is a component or solvable component of $\mathcal{F}$.

PROOF. This is a consequence of 2.6.11.1, 2.5.7.1, and 2.5.8. □

LEMMA 2.6.14. *(1) $C_{\mathcal{F}}(\mathcal{O}) = C_{\mathcal{F}}(O)$.*
(2) If $|\Omega(z)| \leq 2$ and $m > 8$, or if $Aut_{\mathcal{F}}(O)$ is a 2-group, then $\mathcal{F} = S\mathcal{O}C_{\mathcal{F}}(\mathcal{O})$.
(3) Assume $m = 8$ and $\mathcal{O}_z \trianglelefteq \mathcal{F}$ with O Sylow in $\mathcal{O}_z \cong SL_2[8] * SL_2[8]$. Then $C_{\mathcal{F}}(O) = C_{\mathcal{F}}(\mathcal{O}_z)$ and $\mathcal{F} = S\mathcal{O}_z C_{\mathcal{F}}(\mathcal{O}_z)$.

PROOF. By 2.6.12, $S_0 = C_S(O)$ is Sylow in $\mathcal{Y} = C_{\mathcal{F}}(\mathcal{O})$ and $\mathcal{Y} \leq \mathcal{X} = C_{\mathcal{F}}(O)$. As $\mathcal{Y} \leq \mathcal{X}$ and S_0 is Sylow in both $\mathcal{X}$ and $\mathcal{Y}$, it follows that $\mathcal{X}^{fc} \subseteq \mathcal{Y}^{fc}$. For $U \in \mathcal{Y}^{fc}$, set $A(U) = O^2(Aut_{\mathcal{X}}(U))Aut_{S_0}(U)$; thus $A(U) = Aut_{\mathcal{X}}(U)$. By 6.12 in [**Asc11**], $Aut_{\mathcal{Y}}(U) = A(U)$, so as $\mathcal{X}^{fc} \subseteq \mathcal{Y}^{fc}$, it follows from Alperin's Fusion Theorem that

$$\mathcal{X} = \langle Aut_{\mathcal{X}}(U) : U \in \mathcal{X}^{fc}\rangle \leq \langle Aut_{\mathcal{Y}}(U) : U \in \mathcal{Y}^{fc}\rangle = \mathcal{Y},$$

and then as $\mathcal{Y} \leq \mathcal{X}$, we have $\mathcal{X} = \mathcal{Y}$, establishing (1).

If $m > 8$ then $Aut(K)$ is a 2-group, and then as $Aut_{\mathcal{F}}(O)$ permutes $\Omega(z)$ by 2.6.11.3, if in addition $|\Omega(z)| \leq 2$ it follows that $Aut_{\mathcal{F}}(O)$ is a 2-group. Therefore we may assume $Aut_{\mathcal{F}}(O)$ is a 2-group. By 2.6.11.1, $\mathcal{O} \trianglelefteq \mathcal{F}$. Set $\mathcal{E} = \mathcal{O}C_{\mathcal{F}}(\mathcal{O})$ and let $E = OS_0$ and G a model for $N_{\mathcal{F}}(E)$. By 1.1.11.1 in [**Asc19**], $\mathcal{E} \trianglelefteq \mathcal{F}$. As $Aut_{\mathcal{F}}(O)$ is a 2-group, $O^2(G) \leq C_G(O)$, so $G = SC_G(O)$ and hence $N_{\mathcal{F}}(E) = SN_{C_{\mathcal{F}}(O)}(S_0)$. Finally by 1.3.2 in [**Asc19**], $\mathcal{F} = \langle S\mathcal{E}, N_{\mathcal{F}}(E)\rangle$, so $\mathcal{F} = S\mathcal{E}$, establishing (2).

Assume the hypothesis of (3). Then $\mathcal{O}_z = \mathcal{O}_1 * \mathcal{O}_2$ with $\mathcal{O}_i \cong SL_2[8]$ a solvable component of $\mathcal{F}$. Hence $C_{\mathcal{F}}(O) = C_{\mathcal{F}}(\mathcal{O}_z)$ by 2.2.1.3. Define $\mathcal{E} = \mathcal{O}_z C_{\mathcal{F}}(\mathcal{O}_z)$, $E = OS_0$, and G as in the previous paragraph. Let H be the model for $\mathcal{O}_z$ normal in G. Then $G = SHC_G(O)$, so, as in the previous paragraph, $\mathcal{F} = \langle S\mathcal{E}, N_{\mathcal{F}}(E)\rangle$, completing the proof of (3). □

2.7. $\mathcal{F} = S\mathcal{O}_\tau$

DEFINITION 2.7.1. Write $\mathfrak{T}$ for the set of pairs $\tau = (\mathcal{F}, \Omega)$ such that $\mathcal{F}$ is a saturated fusion system on a finite 2-group S, Ω is a set of subgroups of S, $O^2(\mathcal{F}) = \prod_{K\in\Omega} \mathcal{F}_K$ is a central product, where K is Sylow in $\mathcal{F}_K$, $\mathcal{F}_K \cong SL_2[m]$, and S permutes $\{\mathcal{F}_K : K \in \Omega\}$.

For $\tau = (\mathcal{F}, \Omega) \in \mathfrak{T}$ and S Sylow in $\mathcal{F}$, we say that τ is *uniquely determined* by S and Ω (or $\mathcal{F}$ is *uniquely determined* by S and Ω) if whenever $\tau_0 = (\mathcal{F}_0, \Omega_0) \in \mathfrak{T}$ with S Sylow in $\mathcal{F}_0$ and $\Omega = \Omega_0$, then we have $\mathcal{F}_0 = \mathcal{F}$, so also $\tau_0 = \tau$.

REMARK 2.7.2. Suppose $\tau = (\mathcal{F}, \Omega) \in \mathfrak{T}$. Then τ is a quaternion fusion packet. Namely it is straightforward to see that 2.7.1 implies that the four defining conditions (QFP1)-(QFP4) in the definition of quaternion fusion packet are satisfied.

The object of this section is to prove:

THEOREM 2.7.3. *For each $(\mathcal{F},\Omega) \in \mathfrak{T}$, $\mathcal{F}$ is uniquely determined by S and Ω.*

NOTATION 2.7.4. Let $\mathcal{Y} = O^2(\mathcal{F})$ and, using 2.7.2, set $O = O(\tau)$, and $Z = Z(O)$. Then $Z = \langle \mathcal{Z}_S \rangle = Z(\mathcal{Y})$, so $Z \trianglelefteq \mathcal{F}$. Set $\tilde{\mathcal{F}} = \mathcal{F}/Z$.

LEMMA 2.7.5. $\mathcal{F} = S\mathcal{Y}$.

PROOF. This is Theorem 5 in [**Asc11**]. □

LEMMA 2.7.6. *Theorem 2.7.3 holds if $m = 8$.*

PROOF. Assume $m = 8$. Then $O = O_2(\mathcal{Y})$, so $\mathcal{F}$ is constrained by 2.7.5, and hence $\mathcal{F}$ has a model G. Set $H = O^2(G)$. By 2.7.1, $H = \prod_{K\in\Omega} H_K$ is a central product of subgroups $H_K \cong SL_2(3)$ with $K = O_2(H_K)$. Then $O = O_2(H)$ and $O \leq O_2(G) \leq R$ for each $R \in \mathcal{F}^{frc}$.

Set $S_0 = C_S(O)$. Now $[O_2(G), H_K] \leq O_2(H) = O$, so $O_2(G)$ acts on $H_K = O^2(H_K O)$ and hence also on $K = O_2(H_K)$. Then as H_K is irreducible on $\tilde{K}$, $O_2(G) \leq KC_S(K)$. Therefore $O_2(G) = OS_0$. Set $G^+ = G/S_0$ and $\mathcal{F}^+ = \mathcal{F}/S_0$. Then $G^+ = Aut_G(O) = Aut_{\mathcal{F}}(O)$ and $\mathcal{F}^+ = \mathcal{F}_{S^+}(G^+)$.

Next $[H, S_0] \leq C_O(O) = Z \leq Z(H)$, so $H = O^2(H)$ centralizes S_0. Let A be the subgroup of $Aut(O)$ permuting Ω, A_* the kernel of the action of A on Ω, and $B = O^2(A_*)$. Then $Aut_{\mathcal{F}}(O) = Aut_G(O) \leq A$ and $O^2(Aut_{\mathcal{F}}(O)) = B$ is determined by S and Ω.

As $S_0 \trianglelefteq \mathcal{F}$, we have the bijection $R \mapsto R^+$ of $\mathcal{F}^{frc}$ with $(\mathcal{F}^+)^{frc}$. Therefore as $\mathcal{F}^+$ is determined by S and Ω, so is $\mathcal{F}^{frc}$. By Alperin's Fusion Theorem 1.2.2 , $\mathcal{F} = \langle Aut_{\mathcal{F}}(R) : R \in \mathcal{F}^{frc} \rangle$, so it remains to show $Aut_{\mathcal{F}}(R)$ is determined by S, Ω, and R, for each $R \in \mathcal{F}^{frc}$.

Let $R \in \mathcal{F}^{frc}$, $G_R = N_G(R)$, and $H_R = O^2(G_R)$. Then

$$Aut_{\mathcal{F}}(R) = Aut_{G_R}(R) = Aut_{H_R}(R)Aut_S(R),$$

so it remains to show $Aut_{H_R}(R)$ is determined by R. Observe $H_R^+ = O^2(N_B(R^+))$ is determined by S, Ω, and R.

Let $Y \in Syl_3(H_R)$; thus as $O \leq R$ we have $H_R = YO_R$, where $O_R = O_2(H_R) = [O, Y] \leq O$ is a product $O_R = \prod_{K\in\Gamma} K$, where $\Gamma = \{K \in \Omega : K = [K, Y]\}$. Now $R = O_R C_R(Y)$ and $Aut_{H_R}(R) = Aut_{O_R}(R)Aut_Y(R)$ is determined by S, Ω, and $Aut_Y(R)$, with $Aut_Y(O_R)$ determined by $Y^+ = Aut_Y(O)$. Thus it remains to show that $C_R(Y)$ is determined by S and Ω.

Finally Y centralizes S_0, so $C_R(Y)^+ = C_{R^+}(Y^+)$, and hence $C_R(Y)$ is the preimage in R of $C_{R^+}(Y^+)$. Therefore $C_R(Y)$ is determined by $C_{R^+}(Y^+)$, which is in turn determined by S, Ω, and R. This completes the proof of 2.7.6. □

Because of 2.7.6, we may assume in the remainder of the section:

HYPOTHESIS 2.7.7. $m > 8$.

NOTATION 2.7.8. For $K \in \Omega$ let $\mathcal{Q}_K$ consist of the Q_8-subgroups of K. For $\emptyset \neq \Gamma \subseteq \Omega$ let Ξ_Γ consist of the sets ξ that contain exactly one member of Q_K for each $K \in \Gamma\}$ (so that $|\xi| = |\Gamma|$). For $\xi \in \Xi_\Gamma$ set $Q_\xi = \prod_{X \in \xi} X$, $S_\xi = C_S(Q_\xi)$, and $R_\xi = Q_\xi S_\xi$. Write $K(X)$ for the member of Ω containing X and set $\mathcal{F}_X = O^2(N_{\mathcal{F}_{K(X)}}(X))$. Pick $\emptyset \neq \Gamma \subseteq \Omega$ and $\xi \in \Xi_\Gamma$.

LEMMA 2.7.9. *(1) For $K \in \Omega$ and $Q_K \in \mathcal{Q}_K$, $Q_K^{\mathcal{F}} = Q_K^S$, so $Q_K \in \mathcal{F}^f$.*

(2) $Q_\xi^{\mathcal{F}} = Q_\xi^S$, so $Q_\xi \in \mathcal{F}^f$.

(3) $R_\xi \in \mathcal{F}^{frc}$.

(4) Let G_ξ be a model for $N_{\mathcal{F}}(R_\xi)$ and $H_\xi = O^2(G_\xi)$. Then $H_\xi = \prod_{X \in \xi} H_X$ is a central product where $H_X \cong SL_2(3)$ and $X = O_2(H_X)$.

PROOF. Set $Q = Q_\xi$, $R = R_\xi$, etc. Set $\Gamma' = \Omega - \Gamma$ and $P = \prod_{J \in \Gamma'} J$.

Assume the setup of (1). As $\mathcal{Y}$ is a central product of the $\mathcal{F}_K$, $Q_K^{\mathcal{Y}} = Q_K^{\mathcal{F}_K}$. Further as $\mathcal{F}_K \cong SL_2[m]$, $Q_K^{\mathcal{F}_K} = Q_K^K$. Then (1) follows from 2.7.5.

Next for $\phi \in \hom_{\mathcal{Y}}(Q, S)$,

$$Q\phi = \prod_{X \in \Gamma} X\phi = \prod_{X \in \Gamma} X^{s_X} = Q^s,$$

for some $s_X \in K(X)$ and $s = \prod_{X \in \Gamma} s_X$ by (1). Therefore $Q^{\mathcal{Y}} = Q^O$, so (2) follows.

Let $\alpha \in \mathfrak{A}(R)$. By (2), $Q \in \mathcal{F}^f$, so $R\alpha = Q\alpha C_S(Q\alpha)$. Define $P = \langle J \in \Omega : J \leq R\rangle$ and observe that ξ is the set of Q_8-subgroups of R contained in a member of Γ and centralizing P. Hence $N_S(R\alpha)$ acts on $P\alpha$, and then also on $\xi\alpha$ and hence on $Q\alpha$. Let $\beta \in \mathfrak{A}(Q\alpha)$ with $Q\alpha\beta = Q$. Then $N_S(R\alpha) \leq N_S(Q\alpha)$ and $R\alpha\beta = Q\alpha\beta \cdot C_S(Q\alpha)\beta = R$, so $N_S(R\alpha)\beta \leq N_S(R)$. Therefore $R \in \mathcal{F}^f$.

As $C_S(R) \leq C_S(Q) \leq R$, $R \in \mathcal{F}^c$. Thus to complete the proof of (3), it remains to show $R \in \mathcal{F}^r$.

Let G be a model for $N_{\mathcal{F}}(R)$. Then $O^2(G)$ is a model for $O^2(N_{\mathcal{F}}(R))$ and there is $H \trianglelefteq O^2(G)$ with $H = \prod_{X \in \xi} H_X$, where H_X is a model for $O^2(N_{\mathcal{F}_{K(X)}}(X))$. Set $R_0 = O_2(G)$. Then $QP \leq R \leq R_0$. For $X \in \xi$, since $H_X \leq G$, $K(X) \cap R_0 = X$, so $\Gamma' = \{J \in \Omega : J \leq R_0\}$ and ξ are G-invariant. As $m > 8$, $Aut(J)$ is a 2-group for $J \in \Gamma'$, so $H = O^2(G)$. Then $[H_X, R_0] \leq O_2(H) = Q$, so R_0 acts on $H_X = O^2(H_X Q)$ and then also on $X = O_2(H_X)$. As H_X is irreducible on $\tilde{X}$, $[R_0, \tilde{X}] = 1$. Therefore $[R_0, \tilde{Q}] = 1$, so $R_0 \leq QC_S(Q) = R$. Therefore $R = O_2(G)$, completing the proof of (3). Moreover in the process of proving (3), we also established (4). □

LEMMA 2.7.10. *For each $R \in \mathcal{F}^{frc}$, $O_2(O^2(N_{\mathcal{F}}(R))) = Q_\xi$ for some $\Gamma \subseteq \Omega$ and $\xi \in \Xi_\Gamma$. Hence $N_{\mathcal{F}}(R) \leq N_{\mathcal{F}}(Q_\xi)$.*

PROOF. Let $\mathcal{N} = N_{\mathcal{F}}(R)$, G a model for $\mathcal{N}$, and $H = O^2(G)$. Then $\mathcal{H} = O^2(\mathcal{N}) \leq O^2(\mathcal{F}) = \mathcal{Y}$, so H is model for $\mathcal{H}$ and $Q = O_2(H) \leq O$. For $K \in \Omega$, let $\pi_K : \tilde{O} \to \tilde{K}$ be the projection map. Set $\Gamma = \{K \in \Omega : \tilde{Q}\pi_K \neq 1\}$.

As $\mathcal{H} \leq \mathcal{Y}$ and $\mathcal{Y}$ is the central product of copies of $SL_2[m]$, H is a $\{2,3\}$-group with an elementary abelian Sylow 3-subgroup Y. Further for $y \in Y^\#$, $y = y_1 \times \cdots \times y_r \times 1$ where $y_i \in Aut_{\mathcal{F}_{K_i}}(Q_{K_i})$ is of order 3 for some $K_i \in \Gamma$ and $Q_{K_i} \in \mathcal{Q}_{K_i}$. Thus Q_{K_i} is the preimage in K_i of $\tilde{Q}\pi_{K_i}$. Hence for $K \in \Gamma$, $Q_K \in \mathcal{Q}_K$, where Q_K is the preimage of $\tilde{Q}\pi_K$.

Set $Q_0 = \prod_{K \in \Gamma} Q_K$; then R acts on Q_0, so $[N_{Q_0}(R), R] \leq Q_0 \cap R = R_0$. Now $\tilde{R}_0 = [\tilde{R}_0, Y] \leq \tilde{Q} \leq \tilde{Q}_0$, so $R_0 = Q$. Therefore $N_{Q_0}(R)$ centralizes R/Q,

$\tilde{Q}$, and Z, so $N_{Q_0}(R) \leq R$ and hence $Q_0 \leq R$. Therefore $Q_0 = [Q_0, Y] \leq Q$, so $Q_0 = Q$. Further $\xi = \{Q_K : K \in \Gamma\} \in \Xi_\Gamma$, so $Q = Q_\xi$, completing the proof of the lemma. □

We are now in a position to complete the proof of Theorem 2.7.3. By 2.7.6 we may assume $m > 8$, so Hypothesis 2.7.7 is satisfied. By Alperin's Fusion Theorem, $\mathcal{F} = \langle Aut_{\mathcal{F}}(R) : R \in \mathcal{F}^{frc}\rangle$. Let $R \in \mathcal{F}^{frc}$. By 2.7.10, $N_{\mathcal{F}}(R) \leq N_{\mathcal{F}}(Q_\xi) = \mathcal{F}_\xi$ for some $\emptyset \neq \Gamma \subseteq \Omega$ and $\xi \in \Xi_\Gamma$. Set $\tau_\xi = (\mathcal{F}_\xi, \xi)$. By 2.5.2, τ_ξ is a quaternion fusion packet, and by construction $\tau_\xi \in \mathfrak{T}$ with $m_\xi = |X| = 8$ for $X \in \xi$. As $N_{\mathcal{F}}(R) \leq \mathcal{F}_\xi$, $Aut_{\mathcal{F}}(R) = Aut_{\mathcal{F}_\xi}(R)$ and $R \in \mathcal{F}_\xi^{frc}$. As $m_\xi = 8$, $Aut_{\mathcal{F}_\xi}(R)$ is determined by $N_S(Q_\xi)$ and ξ by 2.7.6. Therefore

$$\mathcal{F} = \langle Aut_{\mathcal{F}_\xi}(R) : R \in \mathcal{F}_\xi^{frc},\ \emptyset \neq \Gamma \subseteq \Omega,\ \xi \in \Xi_\Gamma\rangle$$

is determined by S and Ω, proving Theorem 2.7.3.

2.8. Modules for groups with a strongly embedded subgroup

In this section we assume the following hypothesis:

HYPOTHESIS 2.8.1. *(1) G is a finite group with a strongly embedded subgroup.*
(2) a is an involution in G and $H = \langle a^G\rangle$.
(3) V is a faithful $\mathbf{F}_2G$-module such that $V = \langle [V, a]^G\rangle$.

LEMMA 2.8.2. *Either*
(1) $H = O^{2'}(G)$ and $H/O(H)$ is a Bender group, or
(2) $m_2(G) = 1$ and $H = \langle a\rangle O(H)$ is the subgroup generated by the set a^G of involutions of G.

PROOF. By the Bender-Suzuki classification of groups with a strongly embedded subgroup (cf. Theorem SE in [**GLS99**]) either
(i) $O^{2'}(G)/O(O^{2'}(G))$ is a Bender group, or
(ii) $m_2(G) = 1$.
Suppose (i) holds and set $L = O^{2'}(G)$. Then $L/O(L)$ is simple so as $H \trianglelefteq L$ we conclude that (1) holds. Thus we may assume that (ii) holds. Hence by the Z^*-Theorem , $H = \langle a\rangle O(H)$, so (2) holds. □

In the remainder of the section we assume:

HYPOTHESIS 2.8.3. *Hypothesis 2.8.1 holds with $V_a = [V, a]$ of dimension 2.*

LEMMA 2.8.4. *(1) $V = [V, O^2(H)]$.*
(2) If U is an $O^2(H)$-submodule of V containing V_a then $U = V$.

PROOF. By 2.8.2, $O^2(H)$ is transitive on a^G, so the lemma follows from G.6.2 in [**AS04**]; note that lemma does not require the SQTK-condition. □

LEMMA 2.8.5. *Let U be an $O^2(H)$-submodule of V. Then one of the following holds:*

(1) $[U, H] = 0$.

(2) $U = V$.

(3) a induces a transvection on U and V/U.

PROOF. See G.6.3 in [**AS04**]. □

LEMMA 2.8.6. *Assume a inverts $x \in G$ of odd order $p > 1$. Then*

(1) $p = 3$ or 5.

(2) If $p = 5$ then $m([V,x]) = 4$, $V_a = [V,x,a]$, and a centralizes $C_V(x)$.

(3) If $p = 3$ then either $m([V,x]) = 4$, $V_a = [V,x,a]$ and a centralizes $C_V(x)$, or $m([V,x]) = 2$ and a induces a transvection on $[V,x]$ and $C_V(x)$.

PROOF. See D.2.11 in [**AS04**]. □

LEMMA 2.8.7. *(1) If $H/O(H)$ is Bender then $H/O(H) \cong A_5$ or $U_3(4)$.*

(2) If $H \cong A_5$ or $U_3(4)$ then $H \cong A_5$ and either

(a) $m(V) = 4$ and V is the A_5-module for H, or

(b) $V/C_V(H)$ is the 4-dimensional $L_2(4)$-module for H and $m(C_V(H)) \leq 2$.

PROOF. Let $G^* = G/O(H)$. Then H^* is Bender so $H^* \cong L_2(q)$, $Sz(q)$, or $U_3(q)$ with $q \geq 4$ even. It follows that either H^* is $L_2(4)$ or $U_3(4)$, or a^* inverts some x^* of odd order $m > 5$. But then a inverts a preimage x of x^* of odd order at least $|x^*|$, contrary to 2.8.6.1. This establishes (1).

Assume H is A_5 or $U_3(4)$. In the latter case there is no nontrivial $\mathbf{F}_2H$-module U with $m([U,a]) = 2$. Therefore $H \cong A_5$, so there are two nontrivial irreducible $\mathbf{F}_2H$-modules U_1 and U_2: U_1 is the A_5-module and U_2 the $L_2(4)$-module. In each case, $m([U_i,a]) = 2$, so we conclude from 2.8.4 and 2.8.5 that $V = [V,H]$ with $\bar{V} = V/C_V(H)$ a 4-dimensional irreducible for H. If $\bar{V}$ is the A_5-module then $\bar{V}$ is projective, so $C_V(H) = 0$ and (2a) holds. On the other hand if $\bar{V}$ is the $L_2(4)$-module then $m(H^1(H,\bar{V})) = 2$, so (2b) holds. □

LEMMA 2.8.8. *$O_p(H) \leq Z(H)$ for each prime $p \notin \{3,5\}$.*

PROOF. Let $P = O_p(H)$. If a centralizes P then $P \leq Z(H)$ as $H = \langle a^H \rangle$. On the other hand if a does not centralize P then a inverts some x of order p in P, so $p \in \{3,5\}$ by 2.8.6.1. □

LEMMA 2.8.9. *Assume $O_5(H) \not\leq Z(H)$. Then $m(V) = 4$ and $H \cong D_{10}$.*

PROOF. As in D.2.12 of [**AS04**], write $\mathcal{X}_2$ for the set of subgroups X of $P = O_5(H)$ such that $|X| = 5$ and $m([V,X]) = 4$. As in the proof of 2.8.8, a inverts some X of order 5 in P, and by 2.8.6.2, we have $X \in \mathcal{X}_2$, $V_a = [V,X,a]$ and a centralizes $C_V(X)$. In particular $\mathcal{X}_2 \neq \emptyset$.

Let $Q = \langle \mathcal{X}_2 \rangle$. By D.2.12 in [**AS04**], $Q = X_1 \times \cdots \times X_s$ is the direct product of the members of $\mathcal{X}_2$ and $[V,Q] = V_1 \oplus \cdots \oplus V_s$ where $V_i = [V,X_i]$. As H permutes $\mathcal{X}_2$ it acts on Q and $[V,Q]$. We may take $X = X_1$; then, as we just saw, a acts on V_1 and centralizes $C_V(X) = V_2 + \cdots + V_s$. Thus a acts on each V_i, so also $H = \langle a^H \rangle$ acts on each V_i. Thus by 2.8.4.2, $V = V_1$ is of dimension 4, so $\mathcal{X}_2 = \{X\}$ and hence $X \trianglelefteq H$. Now the lemma follows as $N_{GL(V)}(X)$ is the multiplicative group of $F = \mathbf{F}_{16}$ extended by $Aut(F)$. □

LEMMA 2.8.10. *Assume V is a decomposable H-module. Then $m(V) = 4$, $H = O(H)\langle a\rangle$, and a inverts $O(H) \cong \mathbf{Z}_3$ or E_9.*

PROOF. By assumption, $V = V_1 \oplus V_2$ with V_1 and V_2 nontrivial H-submodules. Set $V^i = V/V_{3-i}$. As $V = \langle V_a^H\rangle$ also $V^i = \langle [V^i, a]^H\rangle$, so as V^i is isomorphic to V_i, we have $V_i = \langle [V_i, a]^H\rangle$. In particular $[V_i, a] \neq 0$, so a induces a transvection on each V_i.

Let H^i be the image of H in $GL(V_i)$. Claim H^i has a strongly embedded subgroup. If $m_2(H) = 1$ then $\langle a^i\rangle$ is such a subgroup. On the other hand if $H/O(H)$ is Bender then $O(H)$ is contained in the strongly embedded subgroup M of H and as $H/O(H)$ is simple, $C_H(V_i) \leq O(H)$, so M^i is strongly embedded in H^i, completing the proof of the claim. Now by 2.4.1 in [**Asc19**], $m(V_i) = 2$ and $H^i \cong S_3$, so the lemma follows using 2.8.2. □

LEMMA 2.8.11. *Assume V is an indecomposable H-module and H preserves a direct sum decomposition $\mathcal{V} = \{V_i : 1 \leq i \leq n\}$ in which $n > 1$ and $\dim(V_i) > 1$ for some i. Then $m(V) = 6$, $H = O(H)\langle a\rangle$ with $O(H) \cong 3^{1+2}$, and a centralizes $Z(O(H))$.*

PROOF. As H preserves $\mathcal{V}$ and is indecomposable on V, it follows that H is transitive on $\mathcal{V}$. Thus for each i, $\dim(V_i) = m > 1$. As $H = \langle a^H\rangle$ is transitive on $\mathcal{V}$, it follows that a moves some member of $\mathcal{V}$, so we may take $V_1^a = V_2$. Then $2 \leq m = m([V_1 + V_2, a]) \leq m(V_a) = 2$, so we conclude that $m = 2$, $V_a = [V_1 + V_2, a]$, and a centralizes V_i for $i > 2$. Then a induces a transposition on $\mathcal{V}$, so as $H = \langle a^H\rangle$ is transitive on $\mathcal{V}$, it follows that $H^{\mathcal{V}} = Sym(\mathcal{V})$. Then by 2.8.2, $n = 2$ or 3, so $m(V) = 4$ or 6. In the first case H is decomposable on V, so the second holds. Then H is contained in the stabilizer $\mathbf{Z}_3$ *wr* S_3 of $\mathcal{V}$ in $GL(V) = GL_6(2)$, so as $O(H) = [O(H), a]$ and $m(V_a) = 2$, the lemma follows. □

DEFINITION 2.8.12. Let p be an odd prime and P a p-group. From Definition A.1.20 in [**AS04**], a *supercritical subgroup* of P is a characteristic subgroup X of P such that $\Phi(X) \leq Z(X) \geq [X, P]$, X is of exponent p, and X contains all elements of order p in $C_P(X)$.

LEMMA 2.8.13. *Let p be an odd prime and P a p-group. Then*
(1) P possesses a supercritical subgroup X.
(2) X is of class at most 2.
(3) Each p'-automorphism of P is faithful on X.

PROOF. See A.1.21 in [**AS04**]. □

LEMMA 2.8.14. *Assume $F^*(H) = O_3(H)Z(H)$. Then either*
(1) V is a decomposable H-module, or
(2) H preserves a direct sum decomposition $\mathcal{V} = \{V_i : 1 \leq i \leq n\}$ with $n > 1$ and $m(V_i) > 1$ for some i.

PROOF. Assume neither (1) nor (2) holds and set $P = O_3(H)$. As $F^*(H) = PZ(H)$, we have $C_H(P) = Z(P)Z(H)$. By 2.8.13.1, $P = O_3(H)$ contains a supercritical subgroup Q. By 2.8.13.3, each $3'$-automorphism of P is nontrivial on Q, so $C_H(Q) = C_P(Q)Z(G)$. In particular a is nontrivial on Q.

Let $X = Z(Q)$; as H is indecomposable on V, we have $V = [V, X]$. By definition of Q, Q is of exponent 3, so $\Phi(X) = 1$. Thus if X is noncyclic then the set $\mathcal{V}$ of weight spaces of Q is a direct sum decomposition of V of size $n > 1$ whose members are noncyclic, so that (2) holds. Hence we may assume $X \cong \mathbf{Z}_3$.

If a inverts X then $m(V) = 2m(V_a) = 4$, contradicting V an indecomposable H-module. Therefore a centralizes X, so $X \neq Q$. Hence Q is of class 2 and exponent 3 with $\Phi(Q) = Z(Q) \cong \mathbf{Z}_3$, so $Q \cong 3^{1+2w}$. Further a inverts some $Y \leq Q$ of order 3, and the set $\mathcal{W}$ of weight spaces for XY on V is $\mathcal{W} = \{W_1, W_2, W_3\}$ of order 3 with $W_1^a = W_2$ and Q transitive on $\mathcal{W}$. Thus $2 = m(V_a) = m(W_1)$, so $m(V) = 6$ and $Q \cong 3^{1+2}$. But now $\mathcal{W}$ satisfies (2), contrary to assumption. □

THEOREM 2.8.15. *Assume Hypothesis* 2.8.3. *Then one of the following holds:*

(1) $m(V) = 4$ *and* $H = O(H)\langle a\rangle$ *and* $O(H) \cong \mathbf{Z}_3$ *or* E_9 *is inverted by* a.

(2) $m(V) = 4$ *and* $H \cong D_{10}$.

(3) $m(V) = 4$ *and* V *is the* A_5*-module for* $H \cong A_5$.

(4) $H \cong A_5$, $V = [V, H]$ *with* $V/C_V(H)$ *the* $L_2(4)$*-module for* H, *and* $m(C_V(H)) \leq 2$.

(5) $m(V) = 6$, $H = O(H)\langle a\rangle$ *with* $O(H) \cong 3^{1+2}$ *and* a *centralizes* $Z(O(H))$.

PROOF. Suppose first that $F^*(H) \neq F(H)$. Then from 2.8.2, $H = E(H)$ with $H/Z(H)$ Bender. Now by 2.8.7.1, $H \cong A_5$ or $U_3(4)$, and then (3) or (4) holds by 2.8.7.2. Therefore we may assume $F^*(H) = F(H)$.

Next by 2.8.8, $F(H) = O_3(H)O_5(H)Z(H)$. Further if $O_5(H) \not\leq Z(H)$ then (2) holds by 2.8.9. Thus we may assume that $F^*(H) = O_3(H)Z(H)$. Therefore by 2.8.14 either H is decomposable on V or H preserves a direct sum decomposition as in 2.8.14.2. In the first case, (1) holds by 2.8.10. Thus we may assume V is an indecomposable H-module and H preserve a direct sum decomposition, so (5) holds by 2.8.11. This completes the proof of the theorem. □

CHAPTER 3

Results on τ

In Chapter 3 we consider quaternion fusion packets $\tau = (\mathcal{F}, \Omega)$ and let S be Sylow in $\mathcal{F}$. We develop and explore some of the most basic notions concerning such objects.

For example in 3.1.5 we show that $O(\tau) = \langle \Omega \rangle$ is a central product of the members of Ω.

The set $\eta(\tau)$ of collections η of cyclic subgroups of S, the subset $\eta_S(\tau)$ of S-invariant members of $\eta(\tau)$, and the set $W(\tau)$ of subgroups $\langle \eta \rangle$ for $\eta \in \eta_S(\tau)$ are defined in Definition 3.1.9. By 3.1.10, $\eta_S(\tau)$ is nonempty, and by 3.1.11, for each $\eta \in \eta(\tau)$, $\langle \eta \rangle$ is abelian and its members are of order $m/2$. Then the *Thompson group* $\mu(\tau)$ is defined in Definition 3.1.26 as a certain subgroup of $Aut_{\mathcal{F}}(W)$ for $W \in W(\tau)$.

From 3.1.25, $\mu(\tau)$ is a direct product of a group generated by 3-transpositions with copies of $D_{12} \cong \mathrm{Weyl}(G_2)$. From 3.2.1, D_{12}-factors appear if and only if $\mathcal{A}(\tau) \neq \emptyset$. The graph $\mathcal{A}(\tau)$ on $\mathcal{Z}(\tau)$ is discussed in section 3.2. The graphs $\mathcal{D}(\tau)$ and $\mathcal{D}^*(\tau)$ are introduced in Definition 3.1.13 and discussed in section 3.1.

Lemma 3.3.2 shows that if T is a normal subgroup of $\mathcal{F}$ such that $\mathcal{Z}(\tau) \cap T = \emptyset$ and $\mathcal{F}^+ = \mathcal{F}/T$ then $\tau^+ = (\mathcal{F}^+, \Omega^+)$ is a quaternion fusion packet; this makes possible many inductive arguments.

In 3.3.7 we find that there is a subsystem $\mathcal{O}(\tau)$ of $\mathcal{F}$ on $O(\tau)$ which is the central product of the subsystems $\mathcal{O}_K$ for $K \in \Omega$, and $\mathcal{O}(\tau) \trianglelefteq N_{\mathcal{F}}(W_S)$, where $W_S = \langle \mathcal{Z}_S(\tau) \rangle$.

Let $W \in W(\tau)$ and $\mathcal{M} = [O(\tau)]_{N_{\mathcal{F}}(W)}$. From 3.3.11, $\mathcal{M}$ is constrained and hence has a model $M(W)$; we write $M(\tau)$ for the collection of groups $M(W)$, $W \in W(\tau)$. From 3.3.14, if $\mathcal{F}$ is transitive on Ω then either $\eta(\tau)$, and hence also $W(\tau)$ and $M(\tau)$, has a unique member, or $m = 8$ and $\Delta(\tau) = \Omega$. Therefore, the exceptional case aside, $W \in W(\tau)$ is weakly closed in S with respect to $\mathcal{F}$ and $M(\tau)$ has a unique member M. Later we will see that $\mu(\tau) \cong M/W$.

Set $Z = Z(\mathcal{F})$ and $\mathcal{F}^+ = \mathcal{F}/Z$; assume $\mathcal{F} = \mathcal{F}^\circ$. In 3.3.16 we find that if $\mathcal{F}^+$ is tamely realized by a simple group L^+ appearing as a conclusion in Theorem 1, then $\mathcal{F} = \mathcal{F}_S(L)$ for some covering group L of L^+. This makes it possible to lift statements about simple factor systems $\mathcal{F}^+$ to $\mathcal{F}$.

Definition 3.4.2 introduces the subsystem $\mathcal{E}(\tau)$ of $\mathcal{F}$. Section 3.4 explores $\mathcal{E}(\tau)$ and its relationship to $\mathcal{F}$. In time we will see that often $\mathcal{F} = \mathcal{E}(\tau)$. Lemma 3.4.5 is a step in that direction; for example it is used to prove 3.4.8, where it is shown that if Ω is of odd order and $|\Omega(z)| = 1$ then generically we have $\mathcal{F} = \mathcal{E}(\tau)$.

3.1. $\Delta(\tau)$, $\eta(\tau)$, and $\mu(\tau)$

In this section we assume $\tau = (\mathcal{F}, \Omega)$ is a quaternion fusion packet and adopt Notation 2.5.1. In addition we adopt the following notation:

NOTATION 3.1.1. For $z \in \mathcal{Z}_S$ define $O(z) = \langle \Omega(z) \rangle$ and for $z \in \mathcal{Z}$ define

$$\Delta(z) = \{(K \cap C_S(z)\alpha)\alpha^{-1} : K \in \Omega(z\alpha) \text{ and } |K \cap C_S(z)\alpha| > 2\}$$

for $\alpha \in \mathfrak{A}(z)$. In 3.1.3 we find that $\Delta(z)$ is independent of the choice of $\alpha \in \mathfrak{A}(z)$. If $z \in \mathcal{Z}_S \cap \mathcal{F}^f$ write $\mathcal{O}(z)$ for the normal subsystem of $\mathcal{F}_z$ on $O(z)$ supplied by Definition 2.6.7 and Theorem 2.6.11. Similarly for $K \in \Omega(z)$, write $\mathcal{O}_K$ for the normal subsystem of $\mathcal{O}(z)$ on K supplied by Definition 2.5.6 and 2.6.11.

Set $O(\tau) = \langle \Omega \rangle$. We find in 3.1.5.1 that $O(\tau)$ is a central product of the members of Ω.

LEMMA 3.1.2. *Let $z \in \mathcal{Z}$ and $\alpha \in \mathfrak{A}(z)$. Then*
(1) If $z \in \mathcal{Z}_S$ then $\Omega(z)\alpha \subseteq \Omega(z\alpha)$, so $\Omega(z) \subseteq \Delta(z)$.
(2) $z\alpha \in \mathcal{Z}_S$, so $\mathcal{Z} \cap \mathcal{F}^f \subseteq \mathcal{Z}_S$.
(3) If $z \in \mathcal{F}^f$ then $\alpha^{-1} \in \mathfrak{A}(z\alpha)$ and $\Omega(z)\alpha = \Omega(z\alpha)$.

PROOF. Suppose $K \in \Omega(z)$. Then $K\alpha \leq C_S(z)\alpha \leq S$, so $K\alpha \in K^{\mathcal{F}} \subseteq \Omega$ as τ is a quaternion fusion packet. Then as $z\alpha \in K\alpha$, $K\alpha \in \Omega(z\alpha)$, establishing (1).

Let $\phi \in \hom_{\mathcal{F}}(z, S)$ with $z\phi \in \mathcal{Z}_S$. Let $\beta \in \mathfrak{A}(z\phi)$ with $z\phi\beta = z\alpha$. By (1), $\Omega(z\phi)\beta \subseteq \Omega(z\phi\beta) = \Omega(z\alpha)$, so (2) holds.

If $z \in \mathcal{F}^f$ then $|C_S(z)| \geq |C_S(z\alpha)|$, so $\alpha^{-1} \in \mathfrak{A}(z\alpha)$, and then (3) follows from (1) and symmetry between (z, α) and $(z\alpha, \alpha^{-1})$. □

LEMMA 3.1.3. *Let $z \in \mathcal{Z}$. Then*
(1) $\Delta(z)$ is independent of the choice of $\alpha \in \mathfrak{A}(z)$.
(2) For $D \in \Delta(z)$, $P \leq D$ with $|P| > 2$, and $\phi \in \hom_{\mathcal{F}}(P, S)$, $P\phi \leq E \in \Delta(z\phi)$.

PROOF. For $i = 1, 2$, let $\alpha_i \in \mathfrak{A}(z)$ and set $z_i = z\alpha_i$, $T_i = C_S(z_i)$, $S_i = C_S(z)\alpha_i$, and $\Delta_i = \{(K \cap S_i)\alpha_i^{-1} : K \in \Omega(z_i)\}$. Then $\psi = \alpha_1^{-1}\alpha_2 : S_1 \to S_2$ is an isomorphism. Let $\varphi \in \mathfrak{A}(z_2)$ with $z_2\varphi = z_1$. Then $\beta = \psi\varphi \in \hom_{\mathcal{F}}(S_1, T_1)$ with $z_1\beta = z_1$, so β is an $\mathcal{F}_{z_1}$-map. By 3.1.2.3, $\Omega(z_2)\varphi = \Omega(z_1)$. Next for $D \in \Delta_1$, $D = (K_1 \cap S_1)\alpha_1^{-1}$ for some $K_1 \in \Omega(z_1)$, and as β is an $\mathcal{F}_{z_1}$-map, $(K_1 \cap S_1)\beta \leq J_1 \in \Omega(z_1)$ by 2.6.2. Thus $(K_1 \cap S_1)\psi = (K_1 \cap S_1)\beta\varphi^{-1} \leq J_1\varphi^{-1} = K_2 \in \Omega(z_2)$, so $D\alpha_2 = D\alpha_1\psi = (K_1 \cap S_1)\psi \leq K_2$. Further $D\alpha_2 = D\alpha_1\psi \leq S_1\psi = S_2$, so $D \leq E \in \Delta_2$. By symmetry, $E \leq D' = (K_1' \cap S_1)\alpha_1^{-1} \in \Delta_1$, so $D = E = D'$ as D and E have order at least 4, so $K_1 = K_1'$ by (QFP2). This shows $\Delta(z)$ is independent of α, establishing (1).

Assume the hypothesis of (2). Let $\alpha \in \mathfrak{A}(z\phi)$ and $\beta \in \mathfrak{A}(z)$ with $z\beta = z\phi\alpha = a$. Then $\gamma = \beta^{-1}\phi\alpha$ is an $\mathcal{F}_a$-map. As $D \in \Delta(z)$, $D\beta \leq K \in \Omega(a)$, so by 2.6.2.2 and as $|P| > 2$, $P\phi\alpha = P\beta\gamma \leq J \in \Omega(a)$. Therefore $P\phi \leq (J \cap C_S(z\phi)\alpha)\alpha^{-1} \in \Delta(z\phi)$, proving (2). □

LEMMA 3.1.4. *Suppose $z \in \mathcal{Z}$ and $D \in \Delta(z)$ is nonabelian and acts on $T \leq S$ with $D \cap T = 1$. Then D centralizes T and $DT = D \times T$.*

PROOF. Let $\alpha \in \mathfrak{A}(z)$, $a = z\alpha$, and $W = \langle D^{C_T(z)} \rangle$. As D acts on T, so does W. Then $D\alpha \leq K \in \Omega(a)$ is invariant under $C_S(z)\alpha \cap N_S(K)$, so $W\alpha$ is the central product of the members of $(D\alpha)^{C_T(z)\alpha}$ by 2.6.6.1. Thus as D is nonabelian, $\langle z \rangle = Z(W)$. Then as $T \cap W \lhd W$ and $z \notin T$, $T \cap W = 1$. Thus it remains to show that D centralizes T.

Let $R = C_T(z)$. Then $[R, W] \leq W \cap T = 1$, so we may assume $R < T$. Thus $R < Y = N_T(WR)$. Now $[W, Y] \leq WR \cap T = (W \cap T)R = R \leq C_S(W)$, so $[Y, W, W] = 1 = [W, Y, W]$. Therefore by the 3-Subgroups Lemma (cf. 8.7 in [**Asc86**]), Y centralizes $[W, W]$. As $z \in [W, W]$, this contradicts $R < Y$. □

LEMMA 3.1.5. *(1) $O(\tau)$ is a central product of the members of Ω.*

(2) $\langle \mathcal{Z}_S \rangle = Z(O(\tau))$.

(3) If $z \in C_{\mathcal{Z}}(O(\tau))$ then $z \in \mathcal{Z}_S$.

(4) If $t \in \mathcal{Z}_S$ and $\alpha \in \mathfrak{A}(t)$, then $\Omega\alpha = \Omega$ and $\Omega(t)\alpha = \Omega(t\alpha)$.

(5) If $t \in \mathcal{Z}$ and some member of $\Delta(t)$ is nonabelian, then $t \in \mathcal{Z}_S$ and $\Delta(t) = \Omega(t)$.

(6) $Aut_{\mathcal{F}}(O(\tau))$ is transitive on each orbit of $\mathcal{F}$ on Ω.

PROOF. We first prove (1). Let Σ be a subset of Ω maximal subject to $T = \langle \Sigma \rangle$ a central product of the members of Σ. Set $E = \langle z(J) : J \in \Sigma \rangle$. We may assume $K \in \Omega$ with $K \notin \Sigma$. If $\{J \in \Omega : J \leq N_S(T)\}$ is contained in T then $N_S(N_S(T)) = N_S(T)$, so $T \lhd S$. Hence we may take $K \leq N_S(T)$. Let $z = z(K)$.

If K centralizes E, then by (QFP3), K acts on each $J \in \Sigma$. Then by 3.1.4, either K centralizes J or $z = z(J)$, in which case $K = J$ or K centralizes J by 2.6.6.1. However this contradicts $K \notin \Sigma$ and the maximality of Σ.

Therefore $[K, E] \neq 1$. Hence by 3.1.4, $z \in E \leq Z(T)$. Then by (QFP3), each $J \in \Sigma$ acts on K. Then by 3.1.4, either K centralizes J or $z = z(J)$, for the same contradiction. This completes the proof of (1).

Observe (1) implies (2) as $\langle z \rangle = Z(K)$ for $K \in \Omega(z)$. Assume the hypothesis of (3) and let $\alpha \in \mathfrak{A}(z)$. Then $z\alpha \in \mathcal{Z}_S$ by 3.1.2.2, and $\Omega\alpha \subseteq C_S(z)\alpha \subseteq S$, so $\Omega\alpha = \Omega$ as $\Omega^{\mathcal{F}} = \Omega$. Thus $\mathcal{Z}_S\alpha = \mathcal{Z}_S$, so as $z\alpha \in \mathcal{Z}_S$, (3) holds. Similarly (4) holds.

Observe next that the proof of (1) (with $\Sigma = \Omega$) shows that if $t \in \mathcal{Z}$ and $D \in \Delta(t)$ is nonabelian, then one of the following holds:

(i) $D \in \Omega$ and hence $t \in \mathcal{Z}_S$.

(ii) D centralizes Ω.

(iii) $t \in E = \langle \mathcal{Z}_S \rangle$.

In case (i), $t \in \mathcal{Z}_S$, while in cases (ii) and (iii), $t \in \mathcal{Z}_S$ by (3). Thus in any event, $t \in \mathcal{Z}_S$. Then for $\alpha \in \mathfrak{A}(t)$, $\Omega\alpha = \Omega$ by (4). Thus for $V \in \Delta(t)$, $V\alpha \leq K \in \Omega(t\alpha)$, so as $\Omega(t)\alpha = \Omega(t\alpha)$ by (4), we conclude that $V = K\alpha^{-1} \in \Omega(t)$ and (5) holds.

Let $K \in \mathcal{F}^f$ and $J \in K^{\mathcal{F}}$. Then there is $\alpha \in \mathfrak{A}(J)$ with $J\alpha = K$. By (4), $O(\tau)\alpha = O(\tau)$, so (6) follows. □

Note by 3.1.5.1 and the weak closure of $O(\tau)$, the orbits of $\mathcal{F}$ on Ω are the same as those of $Aut_{\mathcal{F}}(O(\tau))$. In particular if $\mathcal{F}$ is transitive on Ω then so is $Aut_{\mathcal{F}}(O(\tau))$.

LEMMA 3.1.6. *For $K \in \Omega$, $C_{\mathcal{Z}}(K) = C_{\mathcal{Z}}(O(z(K))$.*

PROOF. Let $u = z(K)$, $J \in \Omega(u)$, $t \in C_{\mathcal{Z}}(K)$, $\beta \in \mathfrak{A}(t)$, and $z = t\beta$. Suppose first that $u \in \mathcal{F}^f$. Then there exists $\alpha \in \mathfrak{A}(u\beta)$ with $u\beta\alpha = u$. Thus $\gamma = \beta\alpha$ is an $\mathcal{F}_u$-map. As $K\beta \leq C_S(t)\beta \leq S$, $u\beta \in \mathcal{Z}_S$. Thus $\Omega(u) = \Omega(u\beta)\alpha$ by 3.1.5.4. By 3.1.2.2, $z \in \mathcal{Z}_S$, so z centralizes $\Omega(u\beta)$ by 3.1.5.2, and hence $z\alpha$ centralizes

$\Omega(u\beta)\alpha = \Omega(u)$. Thus $t\gamma = z\alpha \in C_S(\Omega(u))$, so as $\mathcal{O}(u) \lhd \mathcal{F}_u$ by 2.6.11 and $C_S(O(u)) \leq C_{\mathcal{F}_u}(\mathcal{O}(u))$ by 2.6.12, we have $t \in C_S(\Omega(u))$. Thus the lemma holds when $u \in \mathcal{F}^f$.

In general let $\phi \in \mathfrak{A}(u)$. By 3.1.5.4, $\Omega(u)\phi = \Omega(u\phi)$, so as $t\phi$ centralizes $K\phi$, $t\phi$ centralizes $\Omega(u\phi)$ by the previous paragraph. Then t centralizes $\Omega(u\phi)\phi^{-1} = \Omega(u)$, completing the proof. □

LEMMA 3.1.7. *Suppose $t \in \mathcal{Z} - \mathcal{Z}_S$, $V \leq D \in \Delta(t)$, $K \in \Omega$, and assume V centralizes $z = z(K)$ and $|V| > 2$. Let $\alpha \in \mathfrak{A}(t)$. Then*

(1) V acts on K.

(2) V is cyclic.

(3) $C_K(t) = C_K(V)$ contains each V-invariant cyclic subgroup of K.

(4) V is a TI-subgroup of KV.

(5) Either $C_K(t)$ is a cyclic subgroup of K of index 2, or t centralizes K.

(6) If t centralizes K then t centralizes $O(z)$ and $z\alpha \in \mathcal{Z}_S$.

(7) Assume $[t, K] \neq 1$, let U_0 be the subgroup of $C_K(t)$ of order $|V|$, and $X = U_0V$. Then $z\alpha$ does not centralize $O(t\alpha)$ and $Aut_{\mathcal{F}}(X)$ induces $GL(\langle t, z\rangle)$ on $\langle z, t\rangle$.

PROOF. Part (1) follows from (QFP3). Part (2) follows from 3.1.5.5 and the hypothesis that $t \notin \mathcal{Z}_S$.

Let $U = C_K(t)$. Now V acts on some cyclic subgroup B of K of index 2, while as $|V| > 2$, $t \in \Phi(V)$, and hence $C_B(t)$ is of index at most 2 in B, and is of order at least 4. Now $V\alpha \leq J \in \Omega(t\alpha)$ and $U\alpha \leq N_S(J)$ by (QFP3). Next $[U\alpha, V\alpha] \leq U\alpha \cap J = 1$ as $z \neq t$, so $U = C_K(V)$. Then as t is the unique involution in V, (4) holds. Further $C_B(t) = C_B(V)$, so as $t \in \Phi(V)$, t centralizes B. Thus $U = B$ or K, completing the proof of (3) and establishing (5).

Suppose t centralizes K. Then by 3.1.6, t centralizes $O(z)$. Now $\Omega(z)\alpha \subseteq C_S(t)\alpha \leq S$, so as $\Omega^{\mathcal{F}} = \Omega$, $\Omega(z)\alpha \subseteq \Omega$. Hence $z\alpha \in \mathcal{Z}_S$, establishing (6).

Assume the hypothesis of (7). Let $\gamma \in \mathfrak{A}(z)$ and $\beta \in \mathfrak{A}(z\alpha)$ with $z\alpha\beta = z\gamma$. Set $a = z\gamma$, $b = t\gamma$, and $\phi = \gamma^{-1}\alpha\beta \in \hom_{\mathcal{F}}(C_S(E)\gamma, S)$, where $E = \langle z, t\rangle$. Then ϕ is an $\mathcal{F}_a$-map and $E\gamma = \langle a, b\rangle$.

Suppose $[z\alpha, J] = 1$. Then by (6) applied to the pair $z\alpha, t\alpha$ in the role of t, z, either $z\alpha \in \mathcal{Z}_S$ or $t\alpha\beta \in \mathcal{Z}_S$. In the first case, $O(t\alpha)\beta$ centralizes $O(a)$ so again $t\alpha\beta \in \mathcal{Z}_S$. Thus in any event, $b\phi = t\gamma\phi = t\alpha\beta$ centralizes $\Omega(a)$, so as $\mathcal{O}(a) \lhd \mathcal{F}_a$ by 2.6.11 and $C_S(O(a)) \leq C_S(\mathcal{O}(a))$ by 2.6.12, also b centralizes $\Omega(a)$. By 3.1.5.4, $\Omega(z)\gamma = \Omega(a)$, so $t = b\gamma^{-1}$ centralizes $\Omega(z)$, a contradiction.

Therefore $z\alpha$ does not centralize $\Omega(t\alpha)$. Let $k \in K - U$ and $j \in C_J(z\alpha)$. Then $c_k \in Aut_{\mathcal{F}}(X)$ with $[E, k] = \langle z\rangle$. Similarly $c_j \in Aut_{\mathcal{F}}(X\alpha)$ with $[E\alpha, j] = \langle t\alpha\rangle$, so $c_j\alpha^{-*} \in Aut_{\mathcal{F}}(X)$ with $[E, c_j\alpha^{-*}] = \langle t\rangle$. Therefore $Aut_{\mathcal{F}}(X)$ induces $GL(E)$ on E, completing the proof of (7). □

LEMMA 3.1.8. *Let $T \leq S$ and Σ a T-invariant set of cyclic subgroups of T such that there exists a map D from Σ to the set of subgroups of S such that:*

(a) Each $V \in \Sigma$ is of order at least 4 and contained in $D(V) \in \Delta(z(V))$, where $z(V)$ is the involution in V.

(b) If $U, V \in \Sigma$ are distinct then $D(U) \neq D(V)$.

Then $\langle\Sigma\rangle$ is abelian.

PROOF. Assume otherwise and choose a counter example with T minimal. Then $T = \langle\Sigma\rangle$ and there exists $A, B \in \Sigma$ such that $[A, B] \neq 1$. As T is nonabelian,

$A < T$, so $X = \langle A^T \rangle < T$, and hence X is abelian by minimality of T. Similarly $Y = \langle B^T \rangle$ is abelian. Let a, b be the involutions in A, B, respectively, and let $\beta \in \mathfrak{A}(b)$. As $B \leq D(B)$, $B\beta \leq J \in \Omega(b\beta)$ by 3.1.3.2.

Suppose $b \in N_T(A)$. Then $[b, A] \leq Y \cap A \leq C_A(B)$. If $[b, A] = 1$ then $A\beta \leq N_S(J)$ for each $J \in \Omega(b\beta)$ by (QFP3), so $A\beta$ acts on $Y\beta \cap J$. As $Y\beta \cap J$ is abelian, it is cyclic, so $A\beta$ acts on $B\beta$. As $D(A) \neq D(B)$, $A\beta \not\leq J$. Therefore $[A\beta, B\beta] = 1$ by 3.1.5.1 and 3.1.7.3, contradicting $[A, B] \neq 1$. On the other hand if $[b, A] \neq 1$ then $a \in [b, A] \leq C_A(B)$, so by symmetry we again have a contradiction.

Therefore B acts regularly on A^B. Let x, y be elements of order 4 in A, B, respectively, and set $A_0 = \langle x \rangle$. As Y is abelian, $[x, y] \leq Y \leq C_T(y)$, so $x^{-1}x^y = (x^{-1}x^y)^y = x^{-y}x^b$, and hence $a^y = (x^y)^2 = xx^b$. Now b centralizes xx^b, and hence also a^y. Then b centralizes $(a^y)^{y^{-1}} = a$.

Let $\alpha \in \mathfrak{A}(a)$. Then $A\alpha \leq K \in \Omega(a\alpha)$. As b centralizes a but does not act on A, and as $[A, A^b] = 1$, $[K, K^{b\alpha}] = 1$. Then as $a^y = xx^b$, $(a^y)\alpha \in A_0\alpha(A_0\alpha)^{b\alpha}$ induces the inner automorphism $c_{x\alpha}$ on K, so $(a^y)\alpha$ does not centralize K.

Next $x^b = x^{-1}a^y$, so $A_0^b \leq A_0A_0^y = Q$. Then $Q^y = A_0^yA_0^b \leq Q$, so y acts on Q, and $Q = A_0^yA_0^b$. As $A_0A_0^y = A_0^yA_0^b$ and $(A_0\alpha)^{b\alpha}$ centralize K but $A_0\alpha$ does not, it follows that $(A_0^y)\alpha$ does not centralize K. Therefore by 3.1.7.7, $\Lambda = Aut_{\mathcal{F}}(Q)$ induces $GL(E)$ on $E = \langle a, a^y \rangle$. In particular setting $c = aa^y$, there is $\lambda \in \Lambda$ with $c\lambda = a$, so as $A_0, A_0^b \leq Q$ and as $C = \langle xx^y \rangle$ and $C' = \langle x^{-1}x^y \rangle$ are the two cyclic subgroups of order 4 in Q containing c, $C\lambda = A_0$ or A_0^b. Hence $C \leq D \in \Delta(c)$ by 3.1.3.2.

Let $\gamma \in \mathfrak{A}(c)$. As $D \in \Delta(c)$, $D\gamma \leq J$ for some $J \in \Omega(c\gamma)$. However $(xx^y)^y = x^yx^b = x^yx^{-1}a^y = x^{-y}x^{-1} = (xx^y)^{-1}$, so y inverts C. But $y\gamma \in D' \in \Delta(b\gamma)$, contrary to 3.1.7.3 applied to $b\gamma, D', A_0\gamma, J$ in the role of t, D, V, K. This completes the proof of the lemma. □

DEFINITION 3.1.9. Set $\mathcal{Z}_\Delta = \{z \in \mathcal{Z} : \Delta(z) \neq \emptyset\}$ and

$$\Delta = \Delta(\tau) = \bigcup_{z \in \mathcal{Z}_\Delta} \Delta(z),$$

Define $\eta(\tau)$ to consist of those sets Σ of cyclic subgroups of S such that:

(i) Σ is $\langle \Sigma \rangle$-invariant.

(ii) Each $V \in \Sigma$ is contained in some $D(V) \in \Delta(\tau)$, and either $V = D(V)$, or $D(V) \in \Omega$ and $|D(V) : V| = 2$.

(iii) The map $V \mapsto D(V)$ is a bijection of Σ with $\Delta(\tau)$.

Write $\eta_S(\tau)$ for the set of S-invariant members of $\eta(\tau)$ and $W(\tau)$ for $\{\langle \eta \rangle : \eta \in \eta_S(\tau)\}$. We find in 3.1.11.4 that $|\eta(\tau)| = 1$ when $m > 8$.

LEMMA 3.1.10. $\eta_S(\tau) \neq \emptyset$.

PROOF. Let $\Delta_1 = \Delta - \Omega$. By 3.1.5.5, each member of Δ_1 is cyclic. Pick a set $\{K_i : 1 \leq i \leq r\}$ of representatives for the orbits of S on Ω, and for $1 \leq i \leq r$, pick a $N_S(K_i)$-invariant cyclic subgroup V_i of index 2 in K_i. Define

$$\eta = \Delta_1 \cup \bigcup_{1 \leq i \leq r} V_i^S.$$

Then $\eta \in \eta_S(\tau)$. □

LEMMA 3.1.11. *(1)* $\eta(\tau) \neq \emptyset$.

(2) For each $\eta \in \eta(\tau)$, $\langle \eta \rangle$ *is abelian.*

(3) If $\eta \in \eta(\tau)$ and $\phi \in \hom_{\mathcal{F}}(\eta, S)$ then $\eta\phi \in \eta(\tau)$.
(4) If $m > 8$ then $|\eta(\tau)| = 1$.
(5) For each $\eta \in \eta(\tau)$ and $V \in \eta$, $|V| = m/2$.
(6) Each member of $\Delta(\tau) - \Omega$ is cyclic of order $m/2$.
(7) For $\eta \in \eta(\tau)$, $\mathcal{Z} \cap \langle\eta\rangle = \mathcal{Z}_\Delta$.

PROOF. Part (1) follows from 3.1.10, and (2) follows from 3.1.8.

Let $\eta \in \eta(\tau)$, $V \in \eta$, z the involution in V, and $\phi \in \hom_{\mathcal{F}}(\eta, S)$. Then $V \leq D = D(V) \in \Delta(z)$ and $V\phi \leq E = E(V) \in \Delta(z\phi)$ by 3.1.3.2. If $V' \in \eta - \{V\}$ and $E(V) = E(V')$ then either

(a) $\langle V\phi, V'\phi\rangle$ is abelian, or
(b) $\langle V\phi, V'\phi\rangle$ is nonabelian.

In case (a), either $V\phi \leq V'\phi$ or $V'\phi \leq V\phi$. But then $V \leq V'$ or $V' \leq V$, contrary to the hypothesis in 3.1.9.iii that $D : V \mapsto D(V)$ is injective. Similarly in case (b), $\langle V, V'\rangle$ is nonabelian, so $D(V) = D(V') \in \Omega$, for the same contradiction. Therefore the map $E : V \mapsto E(V)$ is injective, while as $D : \eta \to \Delta$ is a bijection, we have $|\eta| = |\Delta|$. Thus $E : \eta \to \Delta$ is a bijection, so condition (iii) of 3.1.9 is satisfied by $\eta\phi$.

As $\phi : \langle\eta\rangle \to \langle\eta\phi\rangle$ is an isomorphism, and as η satisfies condition (i) of 3.1.9, $\mu = \eta\phi$ also satisfies that condition. Thus it remains to show that μ satisfies condition (ii) of 3.1.9. As $E : \eta \to \Delta$ is a bijection, we must show that either $E(V) \in \Delta - \Omega$ or $E(V) \in \Omega$ and $|E(V) : V\phi| = 2$. For $\nu \in \{\eta, \mu\}$, set $\nu' = \{X \in \nu : |X| = m/2\}$ and $\bar{\nu} = \nu - \nu'$. Then $|\nu'| = |\Omega| + t$, where $t = |\Delta'|$ and $\Delta' = \{X \in \Delta - \Omega : |X| = m/2\}$. As $E : \eta \to \Delta$ is a bijection with $|V| = m/2$ for $D(V) \in \Omega$, it follows that $E(\eta') = \Omega \cup \Delta'$ and $\phi : \eta' \to \mu'$ is a bijection. Thus as $|\eta| = |\mu|$, also $\phi : \bar{\eta} \to \bar{\mu}$ is a bijection, and $E(\bar{\eta}) = \phi(\bar{\eta}) = \bar{\mu} = \Delta - (\Omega \cup \Delta')$, completing the proof of (3).

Suppose $m > 8$. If $D(V) \in \Omega$ then $|D(V) : V| = 2$, so $|V| = m/2$ and V is the unique cyclic subgroup of $D(V)$ of order $m/2$. On the other hand if $D(V) \notin \Omega$ then $D(V)$ is cyclic. In either case V is the unique subgroup of $D(V)$ of its order, so η is the unique member of $\eta(S)$. Thus (4) is established.

Further if $V \in \eta$ with $|V| < m/2$, then $V = D(V) \in \Delta - \Omega$. Let z be the involution in V and $\alpha \in \mathfrak{A}(z)$. Then by (2), $\eta\alpha \subseteq C_S(z)\alpha \leq S$, and hence by (3), $\eta\alpha \in \eta(\tau)$. Therefore by (4), $\eta\alpha = \eta$. In particular $V\alpha \in \eta\alpha = \eta$. However by 3.1.2.2, $z\alpha \in \mathcal{Z}_S$, so $\Delta(z\alpha) \subseteq \Omega$, and hence $|D(V\alpha) : V\alpha| = 2$, contradicting $|V| < m/2$. Therefore (5) holds in this case.

On the other hand if $m = 8$ then as $|V| \geq 4 = m/2$, we have $|V| = m/2$, completing the proof of (5). Observe (5) implies (6).

Let $z \in \mathcal{Z} \cap \langle\eta\rangle$ and $\alpha \in \mathfrak{A}(z)$. By (3), $\eta\alpha \in \eta(\tau)$ and by 3.1.2.2, $z\alpha \in \mathcal{Z}_S$, so $z\alpha \in V \in \eta\alpha$. Then $z \in V\alpha^{-1} \in \eta$, so (7) holds. □

LEMMA 3.1.12. *Let $\eta \in \eta(\tau)$ and $K \in \Omega$. Then*
(1) K acts on η and $\langle\eta\rangle$ acts on K.
(2) If $t \in \mathcal{Z}_\Delta$ centralizes K then so does each member of $\Delta(t)$.

PROOF. As $\eta \in \eta(\tau)$, the member U of η contained in K is of index 2 in K. In particular $U \trianglelefteq K$. Let $z = z(K)$, $V \in \eta - \{U\}$, and t the involution in V. By 3.1.11.2, $W = \langle\eta\rangle$ is abelian, so V centralizes U and V acts on K by (QFP3). If t centralizes K then so does V by 3.1.7.3, so (2) holds and in particular K acts on V. Thus we may assume $[t, K] \neq 1$. Therefore $D(V) \notin \Omega$ by 3.1.5.1, so

$V = D(V) \in \Delta - \Omega$. By 3.1.3.2, $\Delta - \Omega$ is S-invariant, so $V^K \subseteq \Delta - \Omega \subseteq \eta$. Thus in each case, $V^K \subseteq \eta$, completing the proof of (1) and the lemma. □

DEFINITION 3.1.13. Let $\mathcal{D}$, $\mathcal{D}^*$ be the set of pairs $(a, b) \in \mathcal{Z} \times \mathcal{Z}$ such that $a \neq b$ and for some $\phi \in \hom_{\mathcal{F}}(\langle a, b\rangle, S)$, $(a\phi, b\phi) \in \mathcal{Z}_S \times \mathcal{Z}_S$, $\mathcal{Z}_\Delta \times \mathcal{Z}_\Delta$, respectively. Visibly $\mathcal{D}$ and $\mathcal{D}^*$ are symmetric relations on $\mathcal{Z}$, so we can regard $\mathcal{D}$ and $\mathcal{D}^*$ as undirected graphs on $\mathcal{Z}$. Also both $\mathcal{D}$ and $\mathcal{D}^*$ are *$\mathcal{F}$-invariant*: that is if (a, b) is an edge in the graph and $\phi \in \hom_{\mathcal{F}}(\langle a, b\rangle, S)$, then also $(a\phi, b\phi)$ is an edge in the graph.

LEMMA 3.1.14. *Let $(z_1, z_2) \in \mathcal{Z} \times \mathcal{Z}$ and $E = \langle z_1, z_2\rangle$. Then the following are equivalent:*

(1) $(z_1, z_2) \in \mathcal{D}$.

(2) For each $\varphi \in \hom_{\mathcal{F}}(E, S)$ with $E\varphi \in \mathcal{F}^f$, $(z_1\varphi, z_2\varphi) \in \mathcal{Z}_S \times \mathcal{Z}_S$.

(3) For some $i \in \{1, 2\}$ and $\phi \in \hom_{\mathcal{F}}(E, S)$, there exists $K \in \Omega(z_i\phi)$ centralized by $z_{3-i}\phi$.

(4) For each $i \in \{1, 2\}$ and $\phi \in \hom_{\mathcal{F}}(E, S)$ with $z_i\phi \in \mathcal{Z}_S$, $z_{3-i}\phi$ centralizes $O(z_i\phi)$.

PROOF. We first prove (2) implies (4), so assume the setup of (2) and let $i \in \{1, 2\}$ and $\phi \in \hom_{\mathcal{F}}(E, S)$ with $z_i\phi \in \mathcal{Z}_S$. Let $\alpha \in \mathfrak{A}(z_i\phi)$ and $\psi = \phi\alpha$. By 3.1.5.4, $\Omega(z_i\phi)\alpha = \Omega(z_i\psi)$. If $E\psi$ centralizes $\Omega(z_i\psi)$ then $E\phi = E\psi\alpha^{-1}$ centralizes $\Omega(z_i\psi)\alpha^{-1} = \Omega(z_i\phi)$, so that (4) holds. Hence, replacing ϕ by ψ, we may assume $a = z_i\phi \in \mathcal{F}^f$.

Next $\delta = \phi^{-1}\varphi \in \hom_{\mathcal{F}}(E\phi, S)$ and as $a \in \mathcal{F}^f$ there exists $\beta \in \mathfrak{A}(z_i\varphi)$ with $a\delta\beta = a$. Then $\gamma = \delta\beta$ is an $\mathcal{F}_a$-map and as $(z_1, z_2)\varphi \in \mathcal{Z}_S \times \mathcal{Z}_S$, we conclude that $z_{3-i}\phi\gamma = z_{3-i}\varphi\beta$ centralizes $\Omega(a)$ by 3.1.5.4. But $\mathcal{O}(a) \trianglelefteq \mathcal{F}_a$ by 2.6.11 and $C_S(O(a)) \leq C_S(\mathcal{O}(a))$ by 2.6.12. Therefore also $z_{3-i}\phi \in C_S(\Omega(a))$, so (4) holds. That is (2) implies (4).

Trivially (4) implies (3). Assume (3) and let $\alpha \in \mathfrak{A}(z_{3-i}\phi)$. By 3.1.2.2, $z_{3-i}\phi\alpha \in \mathcal{Z}_S$, and by 3.1.7.6, $z_i\phi\alpha \in \mathcal{Z}_S$, so $(z_1, z_2)\phi\alpha \in \mathcal{Z}_S \times \mathcal{Z}_S$, and hence (1) holds. That is (3) implies (1).

Finally assume (1) holds, so that there is $\phi \in \hom_{\mathcal{F}}(E, S)$ with $(z_1, z_2)\phi \in \mathcal{Z}_S \times \mathcal{Z}_S$. Choose φ as in (2), and let $\alpha \in \mathfrak{A}(E\phi)$ with $E\phi\alpha = E\varphi$. Then $\Omega\alpha \subseteq C_S(E\phi)\alpha \leq C_S(E\phi\alpha)$, so $\Omega\alpha = \Omega$ centralizes $E\phi\alpha = E\varphi$. Therefore $(z_1, z_2)\varphi \in \mathcal{Z}_S \times \mathcal{Z}_S$, proving that (1) implies (2). □

LEMMA 3.1.15. *Let $(z_1, z_2) \in \mathcal{Z} \times \mathcal{Z}$ and $E = \langle z_1, z_2\rangle$. Then the following are equivalent:*

(1) $(z_1, z_2) \in \mathcal{D}^*$.

(2) For each $\varphi \in \hom_{\mathcal{F}}(E, S)$ with $E\varphi \in \mathcal{F}^f$, $(z_1\varphi, z_2\varphi) \in \mathcal{Z}_\Delta \times \mathcal{Z}_\Delta$.

(3) For some $i \in \{1, 2\}$ and $\phi \in \hom_{\mathcal{F}}(E, S)$, there exists $K \in \Omega(z_i\phi)$ such that $|C_K(z_{3-i}\phi)| > 2$.

(4) For each $i \in \{1, 2\}$ and $\phi \in \hom_{\mathcal{F}}(E, S)$ such that $z_i\phi \in \mathcal{Z}_S$, $z_{3-i}\phi$ induces an inner automorphism on each $K \in \Omega(z_i\phi)$.

PROOF. The proof is similar to that of 3.1.14. Again, trivially (4) implies (3).

We begin the proof that (2) implies (4) as in 3.1.14; in particular, as in that proof, we may assume $a = z_i\phi \in \mathcal{F}^f$ and define δ, β, and γ as in the proof of 3.1.14. Then as $(z_1, z_2)\varphi \in \mathcal{Z}_\Delta \times \mathcal{Z}_\Delta$, $E\varphi \leq \langle\eta\rangle$ for some $\eta \in \eta(\tau)$, so $E\phi\gamma = E\varphi\beta \leq \langle\eta\beta\rangle$ with $\eta\beta \in \eta(\tau)$ by 3.1.11.3. Let $d = z_{3-i}\phi$, $b = d\gamma$, and $\Omega(a) = \{K_1, \ldots, K_n\}$.

As $\eta\beta \in \eta(\tau)$, $b = ux$ with $u \in C_S(O(a))$ and $x \in O(a)$. By 2.6.12, u centralizes $\mathcal{O}(a)$, so b induces the inner automorphism c_x on $\mathcal{O}(a)$. Hence as $d\gamma = b$ and $\gamma \in \hom_{\mathcal{F}_a}(E\alpha, E\alpha\gamma)$, with $\mathcal{O}(a) \trianglelefteq \mathcal{F}_a$ by 2.6.11, it follows that c_d is also an inner automorphism of $\mathcal{O}(a)$. But then d induces an inner automorphism on each K_i, proving that (2) implies (4).

We also begin the proof that (1) implies (2) as in 3.1.14. Then $E\phi \leq \langle\eta\rangle$ for some $\eta \in \eta(\tau)$, so $E\varphi = E\phi\alpha \leq \langle\eta\alpha\rangle$ and $\eta\alpha \in \eta(\tau)$ by 3.1.11. Therefore (1) implies (2) by 3.1.11.7.

Finally assume (3) and let $\beta \in \mathfrak{A}(z_{3-i}\phi)$. Set $U = C_K(z_{3-i}\phi)\beta$, $a = z_i\phi\beta$, and $b = z_{3-i}\phi\beta$. By 3.1.3.2, $U \leq A \in \Delta(a)$, so $(a, b) \in \mathcal{Z}_\Delta \times \mathcal{Z}_\Delta$ and hence (1) holds. □

DEFINITION 3.1.16. For $i = 1, 2$, let $V_i \leq D_i \in \Delta(\tau)$ with V_i cyclic of order $m/2$. Let z_i be the involution in V_i and assume $\langle V_1, V_2\rangle = V_1 \times V_2$ and $(z_1, z_2) \notin \mathcal{D}$. Let $\alpha_i \in \hom_{\mathcal{F}}(V_1V_2, S)$ with $V_i\alpha_i \leq K_i \in \Omega$. Define

$$A(V_1, V_2, \alpha_1, \alpha_2) = \langle c_{k_i}\alpha_i^{-*} : i = 1, 2\rangle \leq Aut_{\mathcal{F}}(V_1V_2),$$

for $k_i \in K_i - V_i\alpha_i$, where $c_{k_i} \in Aut((V_1V_2)\alpha_i)$ is conjugation by k_i. In the next lemma we see that $c_{k_i}\alpha_i^{-*}$ is independent of the choice of α_i, so also $A(V_1, V_2, \alpha_1, \alpha_2)$ is independent of the choice of α_1 and α_2. Thus we write $A(V_1, V_2)$ for $A(V_1, V_2, \alpha_1, \alpha_2)$ and c_i for $c_{k_i}\alpha_i^{-*}$.

LEMMA 3.1.17. *Assume the hypothesis of* 3.1.16. *Then*
(1) $A(V_1, V_2, \alpha_1, \alpha_2)$ is independent of the choice of α_1 and α_2.
(2) $c_i = c_{k_i}\alpha_i^{-*}$ is independent of α_i.
(3) $Aut_{A(V_1,V_2)}(\langle z_1, z_2\rangle) = GL(\langle z_1, z_2\rangle)$.

PROOF. Let $E = \langle z_1, z_2\rangle$ and $X = \langle V_1, V_2\rangle$. By hypothesis, $X = V_1 \times V_2$, and $V_i\alpha_i \leq K_i \in \Omega$. Then as V_{3-i} centralizes V_i, $V_{3-i}\alpha_i$ acts on K_i by (QFP3). As $(z_1, z_2) \notin \mathcal{D}$, the equivalence of parts (1) and (3) of 3.1.14 says that $z_{3-i}\alpha_i$ does not centralize K_i, so c_{k_i} induces the transvection on $E\alpha_i$ with center $\langle z_i\alpha_i\rangle$. Hence (3) holds.

Let $i \in \{1, 2\}$, $\beta_1 \in \mathfrak{A}(z_i)$, $a = z_i\beta_1$, $\beta_2 \in \mathfrak{A}_{\mathcal{F}_a}(X\beta_1)$, and set $\beta = \beta_1\beta_2$; thus $X\beta \in \mathcal{F}_a^f$. Let $V_i\beta \leq J_i \in \Omega$. Then $\gamma = \alpha_i^{-1}\beta_{|X\alpha_i} \in \hom_{\mathcal{F}}(X\alpha_i, S)$. Observe that $C_{Aut(X)}(z_i)$ is a 2-group as X is abelian of rank 2, so $B = Aut_{\mathcal{F}_a}(X\beta)$ is a 2-group (cf. 24.3 in [**Asc86**]). Further by choice of β, $Aut_S(X\beta)$ is Sylow in B, so $B = Aut_S(X\beta)$. Thus $c_{k_i} \in N_\gamma$ (cf. the definition of saturation in section 2), and hence γ extends to $\hat{\gamma} \in \hom_{\mathcal{F}}(K_iX\alpha_i, S)$. Then as $V_i\alpha_i\gamma = V_i\beta \leq J_i$, $K_i\alpha_i\hat{\gamma} = J_i$, so $c_{k_i}\gamma^* = c_j$ for $j \in J_i - V_i\beta$. Therefore $c_{k_i}\alpha_i^{-*} = c_{k_i}\gamma^*\beta^{-*} = c_j\beta^{-*}$. As this holds for each choice of α_i, (1) and (2) follow. □

LEMMA 3.1.18. *Assume the hypothesis of* 3.1.16, *and let $A = A(V_1, V_2)$, $X = V_1V_2$, and $z = z_2\alpha_2$. Then either*

(1) $A \cong S_3$ and given a generator v_1 for V_1, we can pick a generator v_2 for V_2 such that $[v_{3-i}, c_i] = v_i$ for $i = 1, 2$, or

(2) $A \cong D_{12}$ and there exists $\delta \in A$ of order 3 such that $V_1\delta\alpha_2 \leq J_2 \in \Omega(z) - \{K_2\}$, $\alpha_1^{-1}\delta\alpha_2 \in iso_{\mathcal{F}}(X\alpha_1, X\alpha_2)$ extends to $\rho \in \hom_{\mathcal{F}}(K_1(V_2\alpha_1), S)$ with $K_1\rho = J_2$, $z_1\alpha_2 \in \Phi(V_1\alpha) \leq K_2J_2$, and $\Omega(z) = \{K_2, J_2\}$.

PROOF. Let $E = \langle z_1, z_2\rangle$ and $B = C_A(E)$. As $A = \langle c_1, c_2\rangle$ is generated by a pair of involutions, $A \cong D_{2r}$ is dihedral, where $r = |\gamma|$ and $\gamma = c_1c_2$. By 3.1.17.3,

$A/B \cong Aut_A(E) = GL(E)$. As X is abelian and $E = \Omega_1(X)$, B is a 2-group, (cf. 24.3 in [**Asc86**]) we have $r = 3 \cdot 2^a$.

Pick a generator v_1 for V_1. As A/B acts faithfully as $GL(E)$ on E, $z_1^{c_2} = z_1 z_2$, so $[v_1\alpha_2, k_2] \in V_2\alpha_2$ is inverted by c_2 and of order $m/2$, and hence

(a) $[v_1, c_2] = v_2$ generates V_2, and

(b) $V_i = I(c_i)$, where $I(c_i) = \{x \in X : x^{c_i} = x^{-1}\}$.

Suppose first that $a = 0$, so that $A \cong S_3$. Then $c_2^A = \{c_1, c_2, c_2\gamma\}$ is the set of involutions in A, so it follows from (b) that

(c) $V_1^A = I(c_1)^A = \{I(c) : c \in c_1^A\} = \{V_1, V_2, V_3\}$ is of order 3, where $V_3 = I(c_2\gamma) = I(c_2c_1c_2)$.

But by (a), $v_1^{c_1} = v_1v_2$, so as $V_3 = I(c_1^{c_2}) = I(c_1)^{c_2} = V_1^{c_2}$, we conclude that

(d) $V_3 = \langle v_1v_2\rangle$.

Similarly $[v_2, c_1] = v_1^b$ for some odd integer b. Thus $V_3 = \langle v_2v_1^b\rangle$, and then $b = 1$ by (d). Therefore (1) holds in this case, so we may assume $a > 0$.

As $a > 0$, $A_2 = \langle c_2, c_1^\delta\rangle$ is a Sylow 2-subgroup of A, where $\delta \in A$ is of order 3 with $z_1\delta = z_2$. Then $z_2 \in V_1\delta = I(c_1)\delta = I(c_1^\delta)$. By 3.1.3.2, $V_1\delta$ is of index at most 2 in $D(V_1\delta) \in \Delta(z_2)$. If $V_1\delta = V_2$, then $I(c_1^\delta) = I(c_2)$, so $\zeta = c_1^\delta c_2$ centralizes V_2, and then as ζ centralizes δ, ζ centralizes $X = V_2(V_2\delta)$, contradicting $|A_2| > 2$ so $\zeta \neq 1$. Therefore $V_2 \neq V_1\delta$, so as $D(V_1\delta) \in \Delta(z_2)$, $V_2\alpha_2$ and $V_1\alpha_2(\delta\alpha_2^*)$ are of index 2 in K_2 and $J_2 \in \Omega(z)$, respectively. As V_2 and $V_2\delta$ commute and are distinct, $K_2 \neq J_2$.

As in the proof of 3.1.17, we may choose α_2 so that $X\alpha_2 \in \mathcal{F}_z^f$, and hence $Aut_S(X\alpha_2) = Aut_{\mathcal{F}_z}(X\alpha_2)$. Then as $\xi = \alpha_1^{-1}\delta\alpha_2 \in \mathrm{iso}_{\mathcal{F}}(X\alpha_1, X\alpha_2)$ maps $z_1\alpha_1$ to z, we conclude that $c_{k_1} \in N_\xi$, so ξ extends to $\rho \in \hom_{\mathcal{F}}(K_1(V_2\alpha_1), C_S(z))$. Then as $V_1\alpha_1\xi = V_1\delta\alpha_2 \leq J_2$, $K_1\rho = J_2$. Hence $c_1^\delta\alpha_2^* = c_{k_1}\alpha_1^{-*}\delta^*\alpha_2^* = c_1\rho^* = c_j$ for $j \in J_2 - V_1\delta\alpha_2$. Further as $K_2 \neq J_2$, $[K_2, J_2] = 1$, so $c_1^\delta\alpha_2^* = c_j$ commutes with $c_{k_2} = c_2\alpha_2^*$, and hence $[c_1^\delta, c_2] = 1$. Therefore the Sylow 2-subgroup $\langle c_1^\delta, c_2\rangle$ of A is isomorphic to E_4, so $a = 1$ and $A \cong D_{12}$.

Next $V_1\delta\alpha_2 \cap V_2\alpha_2 \leq K_2 \cap J_2 = \langle z\rangle$, so $V_1\delta V_2$ is of index 2 in X. Therefore $(v_1\alpha_2)^2 \in (V_1\delta V_2)\alpha_2 \leq K_2J_2$, so $z_1\alpha_2 \in K_2J_2$. Therefore $z_1\alpha_2$ centralizes each member of $\Omega(z) - \{K_2, J_2\}$, so $\Omega(z) = \{K_2, J_2\}$ by 3.1.6. Therefore (2) holds in this case. □

DEFINITION 3.1.19. Let $\eta \in \eta(\tau)$ and set $W = \langle\eta\rangle$. For $V \in \eta$ and $\alpha \in \hom_{\mathcal{F}}(W, S)$ with $V\alpha \leq K \in \Omega$, define $d_{V,\eta,\alpha} = c_k\alpha^{-*} \in Aut_{\mathcal{F}}(W)$, for $k \in K - V\alpha$, where $c_k \in Aut(W\alpha)$ is conjugation by k. Observe that by 3.1.11.3, $\eta\alpha \in \eta(\tau)$, so K acts on $\eta\alpha$ by 3.1.12.1. Thus $d_{V,\eta,\alpha}$ acts on η, and is indeed a member of $Aut_{\mathcal{F}}(W)$. In the next lemma we find that $d_{V,\eta,\alpha}$ is independent of the choice of α, so we write $d_{V,\eta}$ for this map. When the choice of η is clear, we also write d_V for $d_{V,\eta}$.

Set $D_\eta = \{d_V : V \in \eta\}$ and $\mu_\eta = \langle D_\eta\rangle$. Thus $\mu_\eta \leq Aut_{\mathcal{F}}(W)$.

LEMMA 3.1.20. *Assume the hypothesis of* 3.1.19, *and for* $U \in \eta$, *write* $z(U)$ *for the involution in* U. *Then*

(1) $d_V = d_{V,\eta,\alpha}$ *is independent of the choice of* α.

(2) d_V *inverts* V *and centralizes* $V' \in \eta - \{V\}$ *whenever* $z(V')$ *centralizes* d_V.

(3) $[W, d_V] \leq V$ *and* $[\langle\mathcal{Z}_\Delta\rangle, d_V] \leq \langle z(V)\rangle$.

(4) For $\varphi \in \hom_{\mathcal{F}}(W, S)$, $d_{V,\eta}\varphi^* = d_{V\varphi,\eta\varphi}$.

(5) $\mu_\eta \trianglelefteq Aut_{\mathcal{F}}(W)$.

(6) The map $V \mapsto d_V$ is an equivalence of the representations of $Aut_{\mathcal{F}}(W)$ on η and D_η via conjugation.

PROOF. Assume the setup and notation of 3.1.19, and let $E = \langle \mathcal{Z}_\Delta \rangle$ and $z = z(V)$. Let $\beta_1 \in \mathfrak{A}(z)$, $a = z\beta_1$, $\beta_2 \in \mathfrak{A}_{\mathcal{F}_a}(W\beta_1)$, and set $\beta = \beta_1\beta_2$; thus $W\beta \in \mathcal{F}_a^f$, so $Aut_{S_a}(W\beta)$ is Sylow in $Aut_{\mathcal{F}_a}(W\beta)$, for S_a Sylow in $\mathcal{F}_a$. Let $\gamma = \alpha^{-1}\beta$, regarded as an $\mathcal{F}$-isomorphism from $W\alpha$ to $W\beta$ mapping $V\alpha$ to $V\beta$. From 3.1.12.1, $[E\alpha, c_k] \le \langle z\alpha \rangle$, so

(a) $[E\beta, c_k\gamma^*] \le \langle a \rangle$.

As W is abelian, $C_{Aut(W)}(E)$ is a 2-group (cf. 24.3 in [**Asc86**]), and from (a), $(c_k\gamma^*)_{|E\beta}$ is in $O_2(Aut_{\mathcal{F}_a}(E\beta))$, so $c_k\gamma^* \in O_2(Aut_{\mathcal{F}_a}(W\beta))$. Therefore γ extends to $\hat{\gamma} \in \hom_{\mathcal{F}}(W\alpha K, S)$. Then $K\hat{\gamma} = J \in \Omega$ and $c_k\gamma^* = c_j$, where $j = k\hat{\gamma} \in J - V\beta$. Thus $c_j\beta^{-*} = c_k\alpha^{-*}$ for each choice of α. This proves (1).

Next $V\alpha \le K \in \Omega(z\alpha)$ with $(d_V\alpha^*)_{|W\alpha}$ conjugation by $k \in K - V\alpha$, so as k inverts $V\alpha$, d_V inverts V. If $z' = z(V')$ centralizes d_V then $z'\alpha$ centralizes $K = \langle V\alpha, k \rangle$, so $V'\alpha$ centralizes K by 3.1.12.2. Thus d_V centralizes V', completing the proof of (2).

Similarly by 3.1.12.1, $[W\alpha, K] \le W\alpha \cap K = V\alpha$, so $[d_V, W] \le V$, proving (3).

Let $\varphi \in \hom_{\mathcal{F}}(W, S)$. Then $\gamma = \varphi^{-1}\alpha \in \hom_{\mathcal{F}}(W\varphi, W\alpha)$ and $d_{V\varphi,\eta\varphi,\gamma} = c_k\gamma^{-*}$. Therefore

$$d_{V,\eta}\varphi^* = c_k\alpha^{-*}\varphi^* = c_k\gamma^{-*} = d_{V\varphi,\eta\varphi},$$

establishing (4). Then (4) implies (6), and by (6), $Aut_{\mathcal{F}}(W)$ acts on D_η, so (5) follows. □

LEMMA 3.1.21. *Let $\eta \in \eta(\tau)$, and for $i = 1, 2$, let $V_i \in \eta$, $z_i = z(V_i)$, and $d_i = d_{V_i}$. Then exactly one of the following holds:*

(1) $z_1 = z_2$ and $[d_1, d_2] = 1$.

(2) $(z_1, z_2) \in \mathcal{D}$ and $\langle d_1, d_2 \rangle \cong E_4$ centralizes $\langle z_1, z_2 \rangle$.

(3) $(z_1, z_2) \in \mathcal{D}^* - \mathcal{D}$ and $\langle d_1, d_2 \rangle \cong A(V_1, V_2) \cong S_3$ or D_{12}.

PROOF. Set $W = \langle \eta \rangle$, $E = \langle z_1, z_2 \rangle$, $X = V_1V_2$, $d_i = d_{V_i}$, and $B = \langle d_1, d_2 \rangle \le Aut_{\mathcal{F}}(W)$.

Suppose first that $z_1 = z_2$ and let $\alpha \in \mathfrak{A}(z_1)$. Then $V_i\alpha \le K_i \in \Omega(z_1)$ and $d_i = c_{k_i}\alpha^{-*}$, where $k_i \in K_i - V_i\alpha$. Thus either $V_1 = V_2$ and $d_1 = d_2$, or $V_1 \ne V_2$, so $K_1 \ne K_2$, and hence $[k_1, k_2] \le [K_1, K_2] = 1$, so that also $[d_1, d_2] = 1$. Thus (1) holds in this case.

Next assume $(z_1, z_2) \in \mathcal{D}$, and let $\varphi \in \hom_{\mathcal{F}}(W, S)$ with $E\varphi \in \mathcal{F}^f$. Then by 3.1.14, $z_i\varphi \in \mathcal{Z}_S$ for $i = 1, 2$, so $V_i\varphi \le K_i \in \Omega(z_i\varphi)$ with $[K_1, K_2] = 1$. Hence, as in the previous paragraph, (2) holds in this case.

Finally suppose $z_1 \ne z_2$ and $(z_1, z_2) \notin \mathcal{D}$. As $V_i \in \eta$, $z_i \in \mathcal{Z}_\Delta$, so $(z_1, z_2) \in \mathcal{D}^*$. Also as $(z_1, z_2) \notin \mathcal{D}$, the hypothesis of 3.1.16 are satisfied for any $\alpha_i \in \hom_{\mathcal{F}}(W, S)$ with $V_i\alpha_i \le K_i \in \Omega$. Then $A = A(V_1, V_2) = \langle \bar{c}_{k_i}\beta_i^{-*} : i = 1, 2 \rangle$, where $k_i \in K_i - V_i\alpha_i$, $\bar{c}_{k_i} \in Aut(X\alpha_i)$ is conjugation by k_i, and $\beta_i = \alpha_{i|X}$. On the other hand $\bar{c}_{k_i} = (c_{k_i})_{|X\alpha_i}$, so $\bar{c}_{k_i}\beta_i^{-*} = d_{i|X}$, and hence $A = Aut_B(X) = B\rho$, where $\rho : B \to A$ is the restriction map $\rho : b \mapsto b_{|X}$.

Now B is dihedral of order $2k$, where k is the number of involutions in B conjugate to d_1 or d_2. By 3.1.20.6, k is also the number of conjugates of V_1 or V_2 under B, and as all these conjugates are in X, this is also the number of conjugates of V_1 or V_2 under A, which, from the proof of 3.1.18, is the number of involutions conjugate to $d_1\rho$ or $d_2\rho$. That is $|B| = 2k = |A|$, so $\rho : B \to A$ is an isomorphism. Hence (3) holds by 3.1.18, completing the proof. $\square$

LEMMA 3.1.22. *Let $\eta \in \eta(\tau)$ and for $i = 1, 2$ let $V_i \in \eta$, $z_i = z(V_i)$, and $d_i = d_{V_i}$. Assume $z_1 \neq z_2$. Then the following are equivalent:*

(1) $[z_1, d_2] = 1$.

(2) $[d_1, d_2] = 1$.

(3) $[d_1, z_2] = 1$.

(4) $[V_1, d_2] = 1$.

(5) $[d_1, V_2] = 1$.

PROOF. By 3.1.20.2, (1) implies (4) and (3) implies (5), while trivially (4) implies (1) and (5) implies (3). By 3.1.21, (2) is equivalent to (1) and to (3), completing the proof. $\square$

LEMMA 3.1.23. *Let $\eta \in \eta(\tau)$, $\mu = \mu_\eta$, $W = \langle \eta \rangle$, and for $i = 1, 2$ let $V_i \in \eta$, $z_i = z(V_i)$, and $d_i = d_{V_i}$. Suppose $\nu = \langle d_1, d_2 \rangle \cong D_{12}$ and set $\eta_\nu = V_1^\nu \cup V_2^\nu$. Then:*

(1) $V_i^\mu = V_i^\nu$ *is of order 3 for $i = 1, 2$, and η_ν is of order 6.*

(2) $|\Omega(z_i\alpha_i)| = 2$ *for $i = 1, 2$ and $\alpha_i \in \mathfrak{A}(z_i)$.*

(3) $\mu = \nu \times \nu'$, *where $\nu' = \langle d_V : V \in \eta - \eta_\nu \rangle$.*

(4) ν' *centralizes V_1V_2, ν centralizes $W' = \langle \eta - \eta_\nu \rangle$, and $W = V_1V_2 \times W'$.*

PROOF. Part (2) follows from 3.1.21 and 3.1.18. Next $V_i^\mu = V_i^\nu$ from (3) and (4), and hence $|\eta_{nu}| = 6$ by 3.1.20.6. Therefore (1) follows from (3) and (4), so it remains to prove those assertions.

Let $\eta' = \eta - \eta_\nu$ and $V \in \eta'$. By 3.1.20.3, d_V centralizes a hyperplane of $F = \langle \mathcal{Z}_\Delta \rangle$, so d_V centralizes some member of $E^\#$, where $E = \langle z_1, z_2 \rangle$, which we may take to be z_2. Then by 3.1.22, d_V centralizes V_2 and the member $V_1\delta$ of V_1^ν with $z_2 \in V_1\delta$. But by 3.1.18, $z_1 \in V_1\delta V_2$, so d_V centralizes E and hence d_V centralizes $X = V_1V_2$ and ν by 3.1.22. Therefore ν' centralizes ν and X. Similarly as ν centralizes ν', ν centralizes W' by 3.1.22, so ν is faithful on X, and then $\mu = \nu \times \nu'$. For δ of order 3 in ν, $C_X(\delta) = 1$, so $W = X \times W'$, completing the proof of (3) and (4) and the lemma. $\square$

DEFINITION 3.1.24. Set $\hat{\Omega} = \{K \in \Omega : \Delta - \{K\} \subseteq C_S(K)\}$, $\hat{k} = |\hat{\Omega}|$, and for $\eta \in \eta(\tau)$, set $\hat{\eta} = \{\langle \eta \rangle \cap K : K \in \hat{\Omega}\}$, $\hat{D}_\eta = \{d_V : V \in \hat{\eta}\}$, and $\hat{\mu}_\eta = \langle \hat{D}_\eta \rangle$.

LEMMA 3.1.25. *Let $\eta \in \eta(\tau)$, $\mu = \mu_\eta$, $W = \langle \eta \rangle$, $D = D_\eta$, $\hat{\mu} = \hat{\mu}_\eta$, and $\hat{D} = \hat{D}_\eta$. Then there exist subsets η_i of η and D_i of D such that:*

(1) We have

$$\eta = \hat{\eta} \amalg \coprod_{0 \le i \le r} \eta_i \text{ and } D = \hat{D} \amalg \coprod_{0 \le i \le r} D_i,$$

such that $\mu = \hat{\mu} \times \mu_0 \times \mu_1 \times \cdots \times \mu_r$, where $\mu_i = \langle D_i \rangle$, $\hat{\mu} \cong E_{2^{\hat{k}}}$, D_0 is a set of 3-transpositions of μ_0, and $\mu_i \cong D_{12}$ for $1 \le i \le r$.

(2) $W = \hat{W}\bar{W}$ *with $\hat{W} \cap \bar{W} \le \Omega_1(C_{W_0}(\mu_0))$, where $\bar{W} = W_0 \times W_1 \times \cdots \times W_r$ and $\hat{W} = \langle \hat{\eta} \rangle$, and for $0 \le i \le r$, $W_i = \langle \eta_i \rangle$, $\hat{\mu}$ centralizes $\bar{W}$, and μ_i centralizes $\hat{W}$ and W_j for $j \neq i$.*

(3) For each $\Sigma \in \eta(\tau)$,

$$\Sigma = \hat{\Sigma} \amalg \coprod_{0 \le i \le r} \eta_i,$$

and $\mu_\Sigma \cong \mu_\eta$.

PROOF. Let $\bar{\eta} = \eta - \hat{\eta}$, $\bar{D} = \{d_V : V \in \bar{\eta}\}$, etc. By definition of $\hat{\Omega}$, each $K \in \hat{\Omega}$ centralizes $W_K = \langle \eta - \{V_K\} \rangle$, where $V_K = W \cap K$. Thus d_{V_K} centralizes W_K, and then by 3.1.22, $d_{V_K} \in Z(\mu)$. Then as d_{V_K} inverts V_K,

$$\hat{\mu} = \prod_{K \in \hat{\Omega}} \langle d_{V_K} \rangle \cong E_{2^{\hat{k}}},$$

and $W = \hat{W}\bar{W}$ with $\hat{W} \cap \bar{W} \leq C_{\hat{W}}(\hat{\mu}) \leq \Omega_1(\hat{W})$. Moreover by 3.1.22, $\bar{\mu}$ centralizes $\hat{W}$, and then $\mu = \hat{\mu} \times \bar{\mu}$.

Next by 3.1.23,

$$\bar{\eta} = \coprod_{0 \leq i \leq r} \eta_i, \ \bar{D} = \coprod_{0 \leq i \leq r} D_i,$$

and $\bar{W} = W_0 \times W_1 \times \cdots \times W_r$, where $D_i = \{d_V : V \in \eta_i\}$, $W_i = \langle \eta_i \rangle$, $\mu_i = \langle D_i \rangle$, and for $i > 0$, $\mu_i \cong D_{12}$ with $W_i = [W, \mu_i]$ and μ_i centralizes W_j for $j \neq i$. Finally by 3.1.18, D_0 is a set of 3-transpositions of μ_0. Thus (1) and (2) hold.

Let $\Sigma \in \eta(\tau)$. Then $\Delta' = \Delta - \Omega \subseteq \Sigma \cap \eta$. Further for $V \in \eta - (\hat{\eta} \cup \Delta')$, $V = C_K(U)$ for some $K \in \Omega$ and $U \in \Delta'$, so by symmetry, $V \in \Sigma$. Thus $\bar{\eta} = \bar{\Sigma}$. As $W = \hat{W}\bar{W}$, as $\langle \Sigma \rangle = \langle \hat{\Sigma} \rangle \langle \bar{\Sigma} \rangle$, and as $\bar{W} = \langle \bar{\Sigma} \rangle$, it follows that

$$\bar{\mu} = \langle d_{V,\eta} : V \in \bar{\eta} \rangle \cong \langle d_{V,\Sigma} : V \in \bar{\Sigma} \rangle = \bar{\mu}_\Sigma,$$

so as $\hat{\mu} \cong E_{2^{\hat{k}}} \cong \hat{\mu}_\Sigma$, also $\mu_\Sigma \cong \mu$, completing the proof of (3) and the lemma. □

DEFINITION 3.1.26. By 3.1.25, μ_η is independent of the choice of $\eta \in \eta(\tau)$. Thus we can define the *Thompson group* of τ to be $\mu(\tau) \cong \mu_\eta$ for any $\eta \in \eta(\tau)$.

LEMMA 3.1.27. *(1)* $\Delta(\tau^\circ) = \Delta(\tau)$.
(2) $\eta(\tau^\circ) = \eta(\tau)$.
(3) $\mu(\tau^\circ) = \mu(\tau)$.

PROOF. Recall $\tau^\circ = (\mathcal{F}^\circ, \Omega)$, where $\mathcal{F}^\circ = [\Omega]_\mathcal{F}$. Let T be Sylow in $\mathcal{F}^\circ$; then T is strongly closed in S with respect to $\mathcal{F}$.

Let $V \in \Delta(\tau)$; then there exists $K \in \Omega$ and $\phi \in \hom_\mathcal{F}(V, K)$. As T is strongly closed, $V \leq T$. Indeed either $V \in \Omega$ or using 3.1.21, $|K : V\phi| = 2$ and we may choose K and ϕ so that $V\phi = C_K(V)$, ϕ acts on $X = V\phi V$, and $S_3 \cong \Sigma = \langle c_k, \phi \rangle \leq Aut_\mathcal{F}(X)$ for $k \in K - V\phi$. As $\mathcal{F}^\circ \trianglelefteq \mathcal{F}$, $Aut_{\mathcal{F}^\circ}(X) \trianglelefteq Aut_\mathcal{F}(X)$, so as $c_k \in Aut_{\mathcal{F}^\circ}(X)$, we have $\Sigma = \langle c_k^\Sigma \rangle \leq Aut_{\mathcal{F}^\circ}(X)$. Thus $V \in \Delta(\mathcal{F}^\circ)$, establishing (1).

Next for $\eta \in \eta(\mathcal{F})$, $\eta = \Delta(\tau) - \Omega \cup \eta_0$, where η_0 is some choice of cyclic subgroups of index 2 from members of Ω. Hence (1) implies (2).

Finally set $W = \langle \eta \rangle$ and $\mu = \mu(\tau)$; by (2), $W \in W(\tau^\circ)$. Now $\mu = \langle D \rangle \leq Aut_\mathcal{F}(W)$, where D is the set of $Aut_\mathcal{F}(W)$-conjugates of the c_k for $k \in K - W$ and $K \in \Omega$. Then as $Aut_{\mathcal{F}^\circ}(W) \trianglelefteq Aut_\mathcal{F}(W)$, $D \subseteq Aut_{\mathcal{F}^\circ}(W)$, so $\mu \in \mu(\tau^\circ)$, proving (3). □

3.2. The graph $\mathcal{A}$

In this section assume $\tau = (\mathcal{F}, \Omega)$ is a quaternion fusion packet, and adopt the notation from earlier sections.

Given a graph $\mathcal{G}$ and a vertex v of $\mathcal{G}$, write $\mathcal{G}(v)$ for the set of vertices adjacent to v in $\mathcal{G}$.

DEFINITION 3.2.1. Given $(z_1, z_2) \in \mathcal{Z} \times \mathcal{Z}$, define $\Phi(z_1, z_2)$ to be the set of $\phi \in \hom_{\mathcal{F}}(\langle z_1, z_2 \rangle, S)$ such that $(z_1, z_2)\phi \in \mathcal{Z}_\Delta \times \mathcal{Z}_\Delta$, and for some $\eta \in \eta(\tau)$, $A(V_1, V_2) \cong D_{12}$, where $z_i\phi \in V_i \in \eta$. Note that by 3.1.21, $\Phi(z_1, z_2) = \emptyset$ when $(z_1, z_2) \in \mathcal{D}$.

Define $\mathcal{A} = \{(z_1, z_2) \in \mathcal{Z} \times \mathcal{Z} : \Phi(z_1, z_2) \neq \emptyset\}$. Visibly $\mathcal{A}$ is a symmetric $\mathcal{F}$-invariant relation on $\mathcal{Z}$, so we can regard $\mathcal{A}$ as a graph on $\mathcal{Z}$. Set $\mathcal{W}(z) = \{t \in \mathcal{Z} : \mathcal{D}(z) = \mathcal{D}(t)\}$.

LEMMA 3.2.2. *Let $z \in \mathcal{Z}_S$ and $t \in \mathcal{A}(z)$. Then*

(1) $\Omega(z) = \{K, \tilde{K}\}$ is of order 2.

(2) $t \in K\tilde{K}$.

(3) $t \in \mathcal{D}^*(z) - \mathcal{D}(z)$.

PROOF. Let $z_1 = z$ and $z_2 = t$. As $t \in \mathcal{A}(z)$ there exists $\phi \in \Phi(z, t)$, and $\eta \in \eta(\tau)$ such that $z_i\phi \in V_i \in \eta$ with $A(V_1, V_2) \cong D_{12}$. By definition of $\eta(\tau)$ in Definition 3.1.9, and by 3.1.11.5, $V_i \leq D_i \in \Delta$ with V_i cyclic of order $m/2$. Thus there is $\alpha_i \in \mathfrak{A}(z_i\phi)$ with $V_i\alpha_i \leq K_i \in \Omega(s_i)$, for $i = 1, 2$, where $s_i = z_i\phi\alpha_i$; that is the hypothesis of 3.1.16 is satisfied. By 3.1.18, $\Omega(s_1) = \{K_1, \tilde{K}_1\}$ and $s = z_2\phi\alpha_1 \in K_1\tilde{K}_1$. Let $\alpha \in \mathfrak{A}(z)$ with $z\alpha = s_1$. By 3.1.2.1, $\Omega(z)\alpha = \Omega(s_1)$, so (1) holds. Also $\zeta = \alpha_1^{-1}\phi^{-1}\alpha \in \hom_{\mathcal{F}_{s_1}}(s, C_S(s_1))$, so as $\mathcal{O}(s_1) \lhd \mathcal{F}_{s_1}$ by 2.6.11, and as $s \in K_1\tilde{K}_1$, it follows that $t\alpha = s\zeta \in K_1\tilde{K}_1$. Hence (2) holds. By (2), t induces a nontrivial inner automorphism on K, so (3) follows from 3.1.14 and 3.1.15. □

LEMMA 3.2.3. *Let $(z_1, z_2) \in \mathcal{Z} \times \mathcal{Z}$ with $E = \langle z_1, z_2 \rangle \cong E_4$. Then the following are equivalent:*

(1) $z_2 \in \mathcal{A}(z_1)$.

(2) For each $i \in \{1, 2\}$, $\alpha \in \mathfrak{A}(z_i)$, and $\gamma \in \mathfrak{A}_{\mathcal{F}_{z_i\alpha}}(E\alpha)$, we have $\alpha\gamma \in \Phi(z_1, z_2)$.

(3) There exists $\phi \in \hom_{\mathcal{F}}(E, S)$ with $(z_1, z_2)\phi \in \mathcal{Z}_\Delta \times \mathcal{Z}_\Delta$, and for each such ϕ we have $\phi \in \Phi(z_1, z_2)$.

PROOF. Assume (1) and let $\alpha \in \mathfrak{A}(z_i)$ and $\gamma \in \mathfrak{A}_{\mathcal{F}_a}(E\alpha)$, where $a = z_i\alpha$. Set $\rho = \alpha\gamma$. Then there exists $\varphi \in \Phi(z_1, z_2)$, we have $\delta = \rho^{-1}\varphi \in \hom_{\mathcal{F}}(E\rho, S)$, and there exists $\beta \in \hom_{\mathcal{F}}(C_S(z_i\varphi), S)$ with $a\delta\beta = a$. As $\varphi \in \Phi(z_1, z_2)$, there is $\eta \in \eta(\tau)$ and for $j = 1, 2$, $z_j\varphi \in V_j \in \eta$ such that $A(V_1, V_2) \cong D_{12}$. Now by 3.1.11.3, $\eta\beta \in \eta(\tau)$, and $z_j\varphi\beta \in V_j\beta \in \eta\beta$. Moreover $D_{12} \cong A(V_1, V_2)\beta^* = A(V_1\beta, V_2\beta)$. Finally by 3.1.17.3, $Aut_{\mathcal{F}}(E\varphi) = GL(E\varphi)$, so by 2.5 and as $E\rho \in \mathcal{F}_a^f$, there is $\chi \in \mathfrak{A}_{\mathcal{F}_a}(E\varphi\beta)$ with $z_2\varphi\beta\chi = z_2\rho$. Then $D_{12} \cong A(V_1\beta, V_2\beta)\chi^* = A(V_1\beta\chi, V_2\beta\chi)$, so $\rho \in \Phi(z_1, z_2)$. That is (1) implies (2).

Trivially (3) implies (1), so it remains to assume (2) and show that (3) holds. Assume the setup of (2), and let $\alpha \in \mathfrak{A}(z_1)$ and $\gamma \in \mathfrak{A}_{\mathcal{F}_a}(E\alpha)$, where $a = z_1\alpha$ and $\rho = \alpha\gamma$. By (2), $\rho \in \Phi(z_1, z_2)$. Next choose ϕ as in (3); then there exists $\beta \in \mathfrak{A}(z_1\phi)$ with $z_1\phi\beta = a$. As $(z_1, z_2)\phi \in \mathcal{Z}_\Delta \times \mathcal{Z}_\Delta$, $z_i\phi \in V_i \in \eta \in \eta(\tau)$. By 3.1.11.3, $\eta\beta \in \eta(\tau)$, and as $Aut_{\mathcal{F}}(E\rho) = GL(E\rho)$ by 3.1.17.3, it follows from 1.2.5 that there is $\chi \in \mathfrak{A}_{\mathcal{F}_a}(E\phi\beta)$ with $z_2\phi\beta\chi = z_2\rho$. Then $A(V_1, V_2)(\beta\chi)^* \cong A(V_1\beta\chi, V_2\beta\chi) \cong D_{12}$, so indeed (2) implies (3). □

LEMMA 3.2.4. *Assume $z \in \mathcal{Z}_S$ with $\mathcal{A}(z) \neq \emptyset$. Then*

(1) $|\Omega(z)| = 2$.

(2) There exists a unique 4-subgroup E of S containing z such that $\mathcal{A}(z) \cap \mathcal{Z}_\Delta = E - \langle z \rangle$.

(3) $E \leq O(z)$.

(4) $\mathcal{A}(z) = \mathcal{D}^*(z) - \mathcal{D}(z)$.
(5) $\mathcal{A}(z) \subseteq O(z)$.
(6) $\mathcal{A}(z) = \mathcal{Z} \cap O(z) - \{z\}$,
(7) $\mathcal{W}(z) \cap \mathcal{D}^*(z) = \mathcal{A}(z)$.
(8) *If* $z \in \mathcal{F}^f$ *then* $\mathcal{F}_z$ *is transitive on* $\mathcal{A}(z)$.

PROOF. Let $t \in \mathcal{A}(z)$, $E = \langle z, t\rangle$, and $\eta \in \eta(\tau)$. Replacing (z,t) by $(z,t)\alpha\gamma$ for $\alpha \in \mathfrak{A}(z)$ and $\gamma \in \mathfrak{A}_{\mathcal{F}_{z\alpha}}(E\alpha)$, and appealing to 3.2.3, we may assume $z \in \mathcal{F}^f$ and $t \in \mathcal{Z}_\Delta$ with $z \in V \in \eta$, $t \in U \in \eta$, and $A(V,U) \cong D_{12}$. By parts (1) and (2) of 3.2.2, (1) and (3) hold with $E = \langle z, t\rangle$ in (3). For $K \in \Omega(z)$, $[E, K] = \langle z\rangle$, so as $t \in \mathcal{A}(z)$, also $tz \in \mathcal{A}(z)$. By 3.1.25.1, $\mu = \mu(\tau) = \mu_1 \times \mu_2$, where $\mu_1 = \langle d_V, d_U\rangle \cong A(V,U)$ and $D = D_1 \cup D_2$, where $D_i = D \cap \mu_i$. In particular for $x \in \mathcal{Z}_\Delta - E$, and $x \in X \in \eta$, d_X centralizes μ_1, so $x \in \mathcal{D}(z)$ by 3.1.21. Therefore (2) holds, and

$$(*) \qquad \mathcal{Z}_\Delta - \mathcal{D}(z) = E^\#$$

By 3.2.2.3, $\mathcal{A}(z) \subseteq \mathcal{D}^*(z) - \mathcal{D}(z)$. Let $y \in \mathcal{D}^*(z) - \mathcal{D}(z)$, $Y = \langle z, y\rangle$, and $\varphi \in \mathfrak{A}_{\mathcal{F}_z}(Y)$. By 3.1.21, $Aut_{\mathcal{F}}(Y) = GL(Y)$, so by 2.5, $Y\varphi \in \mathcal{F}^f$, and then by 3.1.25, $y\varphi \in \mathcal{Z}_\Delta$. Then $y\varphi \in M \in \eta$, so by (*), $y\varphi \in \{t, tz\}$. Therefore as $\{t, tz\} \subseteq \mathcal{A}(z)$, also $y \in \mathcal{A}(z)$, completing the proof of (4).

Let $s \in \mathcal{A}(z)$. From paragraph one there is $\gamma \in \mathfrak{A}_{\mathcal{F}_z}(s)$ with $s\gamma \in V' \in \Delta$, so $s\gamma \in E$ by (2), proving (8). Then as $\mathcal{O}(z) \trianglelefteq \mathcal{F}_z$ by 2.6.11, $s \in O(z)$ by (3), proving (5).

Let $\mathcal{B} = \mathcal{Z} \cap O(z) - \{z\}$; by (5), $\mathcal{A}(z) \subseteq \mathcal{B}$. On the other hand if $b \in \mathcal{B}$ then b induces a nontrivial inner automorphism on K, so by 3.1.15, $b \in \mathcal{D}^*(z) - \mathcal{D}(z)$ and hence $b \in \mathcal{A}(z)$ by (4). This proves (6).

Suppose $w \in \mathcal{D}(z)$. Then w centralizes $\Omega(z)$ by 3.1.14, so for $\beta \in \mathfrak{A}(w)$, $\Omega(z)\beta$ centralizes $\Omega(w\beta)$ by 3.1.5.1. Then by (5), $\mathcal{A}(z)\beta$ centralizes $\Omega(w\beta)$, so $\mathcal{A}(z) \subseteq \mathcal{D}(w)$. That is $\mathcal{D}(z) \subseteq \mathcal{D}(s)$ for $s \in \mathcal{A}(z)$. But as $Aut_{\mathcal{F}}(\langle z, s\rangle)$ is transitive on $\langle z, s\rangle^\#$, $|\mathcal{D}(z)| = |\mathcal{D}(s)|$, so $\mathcal{D}(z) = \mathcal{D}(s)$; that is $\mathcal{A}(z) \subseteq \mathcal{W}(z)$.

On the other hand let $w \in \mathcal{W}(z)$. Then $w \notin \mathcal{D}(z)$ as $w \notin \mathcal{D}(w)$. Hence if $w \in \mathcal{D}^*(z)$ then $w \in \mathcal{A}(z)$ by (4), completing the proof of (7). □

3.3. More basic lemmas

In this section we assume $\tau = (\mathcal{F}, \Omega)$ is a quaternion fusion packet and adopt Notation 2.5.1 and 3.1.1.

LEMMA 3.3.1. *Let* $z \in \mathcal{Z}_S \cap \mathcal{F}^f$. *Then* $\tau_z = (\mathcal{F}_z, \Omega)$ *is a quaternion fusion packet.*

PROOF. By 3.1.5.1, z centralizes Ω, so the lemma follows from 2.5.3.2. □

LEMMA 3.3.2. *Let* T *be a subgroup of* S *strongly closed in* S *with respect to* $\mathcal{F}$ *such that* $T \cap \mathcal{Z} = \emptyset$. *Set* $\mathcal{F}^+ = \mathcal{F}/T$, *let* $\Theta : \mathcal{F} \to \mathcal{F}^+$ *be the natural surjection, and for* $P \leq S$ *and* $\phi \in \hom_{\mathcal{F}}(P, S)$ *set* $P^+ = P\Theta$ *and* $\phi^+ = \phi\Theta$. *Then*
(1) $T \leq C_S(\Omega)$.
(2) $\tau^+ = (\mathcal{F}^+, \Omega^+)$ *is a quaternion fusion packet.*
(3) *The map* $K \mapsto K^+$ *is a bijection of* Ω *with* Ω^+.
(4) *Assume* $T \trianglelefteq \mathcal{F}$ *and* $\mathcal{F} = [\Omega]_{\mathcal{F}}$. *Then* $T \leq Z(\mathcal{F})$ *and* $\mathcal{F}^+ = [\Omega^+]_{\mathcal{F}^+}$.

PROOF. As $T \cap \mathcal{Z} = \emptyset$, (1) follows from 3.1.4.

We next prove (2). By 1.2.7.1, $\mathcal{F}^+$ is saturated, and as Ω is $\mathcal{F}$-invariant, Ω^+ is $\mathcal{F}^+$-invariant. Thus it remains to verify that $\tau^+ = (\mathcal{F}^+, \Omega^+)$ satisfies (QPF1)-(QPF4). Let $z \in \mathcal{Z}_S$ and $K \in \Omega(z)$.

As $\mathcal{Z} \cap T = \emptyset$, $K \cong K^+$, so τ^+ satisfies (QPF1).

Let $t \in \mathcal{Z}$; then $t^+ = z^+$ iff $t \in zT$. Suppose $t^+ = z^+$. Then $t \in zT$ so by (1) and 3.1.5.1, $t \in C_S(\Omega)$, and hence $t \in \mathcal{Z}_S$ by 3.1.5.3. Thus there is $J \in \Omega(t)$, and if $J \neq K$ then by (1), $K \cap JT \leq Z(K) = \langle z \rangle$. Hence (3) holds and $K^+ \cap J^+ \leq \langle z^+ \rangle$, so τ^+ satisfies (QPF2).

Assume $V \leq K$ with $|V| > 2$ and $\phi^+ \in \hom_{\mathcal{F}^+_{z^+}}(V^+, S^+)$. Then $z\phi \in zT$, say $z\phi = t$. By 3.1.3.2, $V\phi \leq D \in \Delta(t)$, and by the previous paragraph, $t \in \mathcal{Z}_S$, so $D \in \Omega(t)$ and hence $D^+ \in \Omega^+(z^+)$. Therefore τ^+ satisfies (QPF4).

Assume $L^+ \in \Omega^+$, $u^+ \in L^+ - Z(L^+)$, and $\psi^+ \in \hom_{\mathcal{F}^+}(u^+, C_{S^+}(z^+))$. We may assume $i = z(L) \in \mathcal{F}^f$, and choose $\alpha \in \mathfrak{A}(i\psi)$ with $i\psi\alpha = i$. Set $w = u\psi$. Then $\psi\alpha$ is an $\mathcal{F}_i$-map, so $w\alpha = u\psi\alpha \in L' \in \Omega(i)$ by 2.6.2. Then $w \in D \in \Delta(i\psi)$, and D acts on K by 3.1.12. Therefore w^+ acts on K^+, so τ^+ satisfies (QPF3), completing the proof of (2).

Assume the hypothesis of (4). As $T \trianglelefteq \mathcal{F}$, $C_{\mathcal{F}}(T) \trianglelefteq \mathcal{F}$ by Theorem 4 in [**Asc11**]. Then by (1) and as $\mathcal{F} = [\Omega]_{\mathcal{F}}$, $T \leq Z(\mathcal{F})$. Let $\mathcal{E}^+ = [\Omega^+]_{\mathcal{F}^+}$. As $T \leq Z(S)$, the preimage $\mathcal{E}$ of $\mathcal{E}^+$ in $\mathcal{F}$ under Θ is normal in $\mathcal{F}$ by 1.2.7. As $\Omega^+ \subseteq \mathcal{E}^+$, $\Omega \subseteq \mathcal{E}$, so $\mathcal{E} = \mathcal{F}$ as $\mathcal{F} = [\Omega]_{\mathcal{F}}$. Therefore $\mathcal{E}^+ = \mathcal{F}^+$, completing the proof of (4). □

LEMMA 3.3.3. *Let T be a subgroup of S strongly closed in S with respect to $\mathcal{F}$, let $K \in \Omega$, and $\Omega_0 = (K \cap T)^{\mathcal{F}}$.*

(1) If $K \cap T$ is nonabelian then $|K : K \cap T| \leq 2$ and $(\mathcal{F}, \Omega_0)$ is a quaternion fusion packet.

(2) Assume $|K \cap T| > 2$ and $\mathcal{O}_K \neq \mathcal{F}_K(K)$. Then $K \cap T$ is nonabelian and if $K \cap T \neq K$ then $\mathcal{O}_K \cong SL_2^1[m/2]$.

PROOF. Set $J = K \cap T$; then J is strongly closed in K with respect to $\mathcal{O}_K$, and in particular $J \trianglelefteq K$.

Suppose J is nonabelian. Then as $J \trianglelefteq K$, $|K : J| \leq 2$, and in any event J is quaternion. Now it is straightforward to verify that $(\mathcal{F}, \Omega_0)$ is a quaternion fusion packet, proving (1).

Assume the hypotheses of (2). As $\mathcal{O}_K$ is not $\mathcal{F}_K(K)$, it follows from 2.5.5.1 that $\mathcal{O}_K \cong SL_2[m]$ or $SL_2^1[m/2]$. Then as J is strongly closed in K with respect to $\mathcal{O}_K$, it follows that either $K = J$ or $\mathcal{O}_K \cong SL_2^1[m/2]$ and $|K : J| = 2$ with J nonabelian. This proves (2). □

Recall the graphs $\mathcal{D}$ and $\mathcal{D}^*$ on $\mathcal{Z}$ defined in Definition 3.1.13. For $z \in \mathcal{Z}$, set $z^\perp = \{z\} \cup \mathcal{D}(z)$.

LEMMA 3.3.4. *Let $z \in \mathcal{Z}_S$. Then $z^\perp = C_{\mathcal{Z}}(O(z))$.*

PROOF. First if $t \in C_{\mathcal{Z}}(O(z))$ then condition (3) of 3.1.14 is satisfied for (z, t) with $\phi = 1$, so $t \in z^\perp$ by 3.1.14. Therefore $C_{\mathcal{Z}}(O(z)) \subseteq z^\perp$.

Next let $t \in \mathcal{D}(z)$. Then $t \in C_{\mathcal{Z}}(O(z))$ by the equivalence of parts (1) and (4) of 3.1.14 with $\phi = 1$. □

LEMMA 3.3.5. *Let $z \in \mathcal{Z}_S$ and $z \neq t \in \mathcal{Z}$. Then the following are equivalent:*
(1) $t \in \mathcal{D}^*(z)$.
(2) $|C_K(t)| > 2$ for some $K \in \Omega(z)$.
(3) t induces an inner automorphism on each member of $\Omega(z)$.

PROOF. Observe (2) holds iff condition (3) of 3.1.15 is satisfied for (z,t) with $\phi = 1$. Thus as conditions (1), (3), and (4) of 3.1.15 are equivalent, the lemma follows. □

LEMMA 3.3.6. *Let $K \in \Omega$, $z = z(K)$, and assume $t \in \mathcal{Z} - \{z\}$. Set $E = \langle z, t \rangle$. Then exactly one of the following holds:*
(1) $E \cong D_8$ and $K \neq K^t$.
(2) $E \cong E_4$ and $K \neq K^t \in \Omega(z)$.
(3) $E \cong E_4$, $[t, K] = 1$ and $t \in \mathcal{D}(z)$.
(4) $E \cong E_4$, t induces a nontrivial inner automorphism on K, and $t \in \mathcal{D}^*(z) - \mathcal{D}(z)$.
(5) $E \cong E_4$ and t induces an outer automorphism on K with $C_K(t) = \langle z \rangle$. In particular $K\langle t \rangle$ is semidihedral.

PROOF. If $K \neq K^t$ then $[K, K^t] = 1$ by 3.1.5.1, so either $z^t \neq z$ and (1) holds, or $z^t = z$ and (2) holds. Thus we may assume t acts on K. If $[K,t] = 1$ then (3) holds by 3.1.14, so assume $[K,t] \neq 1$. If t induces an inner automorphism on K then (4) holds by 3.3.5. So assume t induces an outer automorphism on K. Then as parts (2) and (3) of 3.3.5 are equivalent, (5) holds. □

Recall from the introduction to the chapter that $W_S = \langle \mathcal{Z}_S \rangle$.

LEMMA 3.3.7. *Set $\mathcal{N} = N_{\mathcal{F}}(W_S)$, $\Gamma = \{\mathcal{O}_K : K \in \Omega\}$, and $\mathcal{O}(\tau) = [O(\tau)]_{\mathcal{N}}$. Then*
(1) $\mathcal{O}(\tau)$ is a central product of the members of Γ.
(2) $\mathcal{O}(\tau) \trianglelefteq \mathcal{N}$.
(3) $Aut_{\mathcal{F}}(O(\tau))$ permutes Γ.

PROOF. As $W_S \trianglelefteq S$, $W_S \in \mathcal{F}^f$, so $\mathcal{N}$ is saturated. By 2.5.2, $(\mathcal{N}, \Omega)$ is a quaternion fusion packet, so, replacing $\mathcal{F}$ by $\mathcal{N}$, we may assume $W_S \trianglelefteq \mathcal{F}$. Set $\mathcal{O} = \mathcal{O}(\tau)$ and $O = O(\tau)$.

Part (2) holds by definition of $\mathcal{O}(\tau)$ and (3) follows as Ω is $\mathcal{F}$-invariant and $Aut_{\mathcal{F}}(K) \leq Aut(\mathcal{O}_K)$ by 2.5.7.4.

Let $K \in \Omega$, $z = z(K)$, and $Q \in \mathcal{Q}_K$. If $\alpha \in Aut_{\mathcal{F}}(Q)$ is of order 3 then by 2.6.6.3, $\alpha \times 1 \in Aut_{\mathcal{F}}(QC_S(Q))$, so by 2.3 in [**Asc11**], the central product $\mathcal{E}$ of the members of Γ is a subsystem of $\mathcal{F}$. By 2.5.7.3, $\mathcal{E} = O^{2'}(\mathcal{E})$, so $\mathcal{E} \leq \mathcal{O}$. Also $\mathcal{O} \leq C_{\mathcal{F}}(W_S) = \mathcal{D}$, and $\mathcal{D} \trianglelefteq \mathcal{F}$ by Theorem 4 in [**Asc11**]. By 2.5.2, $\rho = (\mathcal{D}, \Omega)$ is a quaternion fusion packet. By 2.6.11, $\mathcal{O}(z) \trianglelefteq \mathcal{D}$. Then by Theorem 3 in [**Asc11**], $\mathcal{E} \trianglelefteq \mathcal{D}$, so $\mathcal{E} = [O(\tau)]_{\mathcal{D}}$. Now as Ω is $\mathcal{F}$-invariant, $\mathcal{E} \trianglelefteq \mathcal{F}$ by 7.4 in [**Asc11**], so $\mathcal{E} = \mathcal{O}$, completing the proof of (1) and the lemma. □

LEMMA 3.3.8. *Set $S_0 = O_2(\mathcal{F})$ and assume $K \in \Omega$ with $K_0 = K \cap S_0$ nonabelian. Set $\Omega_0 = K_0^{\mathcal{F}}$, $\Omega_1 = K^{\mathcal{F}}$, and let $\mathcal{O}_1$ be the product of $\mathcal{O}_J$, $J \in \Omega_1$ in $\mathcal{O}(\tau)$. Then*
(1) $\tau_i = (\mathcal{F}, \Omega_i)$ is a quaternion fusion packet for $i = 0, 1$.
(2) $z^{\mathcal{F}} = \{z(J) : J \in \Omega_1\}$ and $E = \langle z^{\mathcal{F}} \rangle \trianglelefteq \mathcal{F}$.
(3) $\mathcal{O}_1 \trianglelefteq \mathcal{F}$ is the central product of the $\mathcal{O}_J$ for $J \in \Omega_1$.

(4) $|K : K_0| \leq 2$ *and in case of equality,* $m = 16$ *and* $\mathcal{O}_K \cong SL_2^1[8]$.
(5) $\mathcal{O}_1 = [\Omega_1]_{\mathcal{F}}$ *centralizes* $[\Omega - \Omega_1]_{\mathcal{F}}$.

PROOF. Suppose (3) holds and let $\Omega_2 = \Omega - \Omega_1$. By (3), $\mathcal{O}_1 = [\Omega_1]_{\mathcal{F}}$. By 3.3.7.1, $\Omega_2 \subseteq C_{\mathcal{F}}(\mathcal{O}_1)$, so as $C_{\mathcal{F}}(\mathcal{O}_1) \trianglelefteq \mathcal{F}$ by (3) and Theorem 4 in [**Asc11**], we conclude that $[\Omega_2]_{\mathcal{F}} \leq C_{\mathcal{F}}(\mathcal{O}_1)$, completing the proof of (5). Thus (3) implies (5), so it remains to prove (1)-(4).

First τ_1 is a quaternion fusion packet by 2.5.2. Thus as (1)-(4) are statements about τ_1, replacing Ω by Ω_1, we may assume $\mathcal{F}$ is transitive on Ω. By 3.3.3.1, τ_0 is a quaternion fusion packet, completing the proof of (1); further $|K : K_0| \leq 2$ by 3.3.3.1.

Let $z = z(K)$; as $\Omega = K^{\mathcal{F}}$, $\mathcal{Z} = z^{\mathcal{F}}$. But for $\phi \in \hom_{\mathcal{F}}(z, S)$, ϕ extends to $\varphi \in Aut_{\mathcal{F}}(S_0)$, so $z\phi = z\varphi$ generates $Z(K_0\varphi)$, and hence $\mathcal{Z} = \mathcal{Z}_S$. Therefore $E = \langle \mathcal{Z}_S \rangle = \langle \mathcal{Z} \rangle \trianglelefteq \mathcal{F}$, establishing (2). Then (3) follows from 3.3.7.

Finally suppose $|K : K_0| = 2$. Then by (3), $\mathcal{O}_K \neq \mathcal{F}_K(K)$, so $\mathcal{O}_K \cong SL_2^1[8]$ by 3.3.3.2 and the fact that $K_0 \leq O_2(\mathcal{O}_K)$, completing the proof of (4). □

LEMMA 3.3.9. *Let* $\eta \in \eta(\tau)$ *and assume* $W = \langle \eta \rangle \trianglelefteq \mathcal{F}$. *Then*
(1) For $K \in \Omega$, $\mathcal{O}_K = \mathcal{F}_K(K)$ *and* $Aut_{\mathcal{F}}(K)$ *is a 2-group.*
(2) $\mathcal{M} = [O(\tau)]_{\mathcal{F}}$ *is constrained.*
(3) Let M *be a model for* $\mathcal{M}$. *Then* $W \trianglelefteq M = \langle O(\tau)^M \rangle$ *and* $M/C_M(W) \cong \mu(\tau)$.

PROOF. As $V = K \cap W$ is cyclic of index 2 in K and normal in $\mathcal{O}_K$, (1) follows. Replacing $\mathcal{F}$ by $\mathcal{M}$, we may assume $\mathcal{F} = \mathcal{M}$. Let $z = z(K)$; we may assume $z \in \mathcal{F}^f$. As $z \in W \leq O_2(\mathcal{F})$, $\mathcal{D} = E(\mathcal{F}) = E(\mathcal{F}_z)$ by 10.3 in [**Asc11**]. By (1) and 2.6.11, $K \leq O_2(\mathcal{F}_z) \leq C_{\mathcal{F}_z}(E(\mathcal{F}_z)) = C_{\mathcal{F}}(\mathcal{D})$, so $\mathcal{F} = [O(\tau)]_{\mathcal{F}} \leq C_{\mathcal{F}}(\mathcal{D})$ and hence $\mathcal{D} = 1$. Then (2) follows from 14.2 in [**Asc11**].

As $\mathcal{F} = [O(\tau)]_{\mathcal{F}}$, $M = \langle O(\tau)^M \rangle$. Set $M^* = M/C_M(W)$; then $M^* = \langle O(\tau)^{*M^*} \rangle$, and from Definition 3.1.19, $K^* = \langle d_V \rangle$, so $M^* = \langle d_U : U \in \eta \rangle = \mu(\tau)$, completing the proof of (3). □

LEMMA 3.3.10. *Let* $\eta \in \eta(\tau)$. *Then*
(1) $\eta = (\Delta(\tau) - \Omega) \cup \{V_K : K \in \Omega\}$ *for some choice of cyclic subgroups* V_K *of index 2 in* K.
(2) η *is weakly closed in* $C_S(\eta)$ *with respect to* $\mathcal{F}$.

PROOF. From conditions (ii) and (iii) of 3.1.9, $\eta = \{V_D : D \in \Delta\}$, where $V_D = D$ if $D \notin \Omega$ and $|D : V_D| = 2$ if $D \in \Omega$. Therefore (1) holds.

Suppose $\phi \in \hom_{\mathcal{F}}(\eta, C_S(\eta))$ with $\xi = \eta\phi \neq \eta$. By 3.1.11.3, $\xi \in \eta(\tau)$, so by (1) applied to ξ we have $\xi = (\Delta - \Omega) \cup \{U_K : K \in \Omega\}$ with $|K : U_K| = 2$. Therefore for some $K \in \Omega$, $V_K \neq U_K$. This is a contradiction as $U_K \leq C_S(\eta) \leq C_S(V_K)$. □

NOTATION 3.3.11. Recall from Definition 3.1.9 that $W(\tau) = \{\langle \eta \rangle : \eta \in \eta_S(\tau)\}$. For $W \in W(\tau)$ set $\tau_W = (N_{\mathcal{F}}(W), \Omega)$. As $W = \langle \eta \rangle$ for some $\eta \in \eta_S(\tau)$, we have $W \trianglelefteq S$, so $W \in \mathcal{F}^f$, and hence $N_{\mathcal{F}}(W)$ is saturated by 1.2.1. Hence τ_W is a quaternion fusion packet by 2.5.2. By 3.3.9.2 applied to τ_W in the role of τ, $[O(\tau)]_{N_{\mathcal{F}}(W)}$ is constrained, and hence has a model $M(W)$. By 3.3.9.3, $M(W)/C_{M(W)}(W) \cong \mu(\tau)$. Write $M(\tau)$ for the set of groups $M(W)$, $W \in W(\tau)$.

As W is normal in S, so is $C_S(W)$, so $N_{\mathcal{F}}(C_S(W))$ is saturated and hence has a model G_W. By 3.3.10.2, $W \trianglelefteq N_{\mathcal{F}}(C_S(W))$, so $[O(\tau)]_{N_{\mathcal{F}}(W)} \trianglelefteq N_{\mathcal{F}}(C_S(W))$, so we may take $M(W) \trianglelefteq G_W$ by 1.2.8.4.

LEMMA 3.3.12. *Set $S_0 = O_2(\mathcal{F})$ and assume $K \in \Omega$ such that $K_0 = K \cap S_0$ is cyclic of order at least 4. Set $\Omega_1 = K^{\mathcal{F}}$, $\eta_1 = K_0^{\mathcal{F}}$, and $E = \langle z^{\mathcal{F}} \rangle$. Then*

(1) $\tau_1 = (\mathcal{F}, \Omega_1)$ is a quaternion fusion packet.

(2) $\mathcal{O}_K = \mathcal{F}_K(K)$ and $Aut_{\mathcal{F}}(K)$ is a 2-group.

(3) $z^{\mathcal{F}} \subseteq \mathcal{Z}_\Delta$ but $z^{\mathcal{F}}$ is not contained in $\mathcal{Z}_S$.

(4) $|K : K_0| = 2$, $\{\eta_1\} = \eta(\tau_1)$, and η_1 is contained in each member of $\eta(\tau)$.

(5) If $\mathcal{A}(z) \neq \emptyset$, set $\Omega_2 = \Omega_1 \cup \tilde{K}^{\mathcal{F}}$ where $\Omega(z) = \{K, \tilde{K}\}$; otherwise set $\Omega_2 = \Omega_1$. Then $[\Omega_2]_{\mathcal{F}}$ centralizes $[\Omega - \Omega_2]_{\mathcal{F}}$, and $\tilde{K} \cap S_0$ is cyclic of index 2 in $\tilde{K}$.

(6) $\langle \eta_1 \rangle \trianglelefteq \mathcal{F}$.

PROOF. Part (1) follows from 2.5.2. As K_0 is cyclic of order at least 4 and invariant under $Aut_{\mathcal{F}}(K)$, (2) follows.

By construction, η_1 is an $\mathcal{F}$-invariant subset of S_0, so $W_1 = \langle \eta_1 \rangle \trianglelefteq \mathcal{F}$, proving (6).

Suppose (4) holds; we prove (5) holds, so adopt the notation of (5), let $J \in \Omega - \Omega_2$ with $t = z(J) \in \mathcal{F}^f$, and set $\Omega_3 = J^{\mathcal{F}}$. As (4) holds, η_1 is contained in each $\eta \in \eta(\tau)$, and we can apply 3.1.25 to η. In particular we obtain the direct product decomposition of μ in that lemma, and as J is not fused to K or $\tilde{K}$, it follows that d_{K_0} and d_{J_0} are in different factors of μ, and hence commute. Therefore J centralizes K by 3.1.22. As this holds for each $K \in \Omega_1$, it follows that J centralizes W_1. Then $\mathcal{E} = [\Omega_3]_{\mathcal{F}}$ centralizes W_1, and hence also centralizes $O(z)$ as J centralizes $O(z)$. Therefore $[\Omega_2]$ centralizes $\mathcal{E}$; that is (4) implies (5). Further (3) is a statement about τ_1, so replacing Ω by Ω_1 and appealing to (1), we may assume $\mathcal{F}$ is transitive on Ω, keeping in mind that to prove (4) in this case we must show η_1 is the unique member of $\eta(\tau)$.

Define a map $\xi : \eta_1 \to \Delta$, where for $U \in \eta_1$, $\xi(U)$ is the member of Δ containing U. If U and V are distinct members of η_1 and $\xi(U) = \xi(V)$, then as $|U| = |V| > 2$, $\langle U, V \rangle$ is a nonabelian subgroup of $\xi(U) \cap S_0$. But $\xi(U) \cap S_0$ is conjugate to the cyclic group K_0, a contradiction. Thus ξ is an injection.

As $\Omega = K^{\mathcal{F}}$, $\mathcal{Z} = z^{\mathcal{F}}$. Then for each $t \in \mathcal{Z}$, $t \in K_0\phi$ for some $\phi \in Aut_{\mathcal{F}}(S_0)$, so $\mathcal{Z} = \mathcal{Z}_\Delta$. Hence ξ is a bijection.

Suppose $\mathcal{Z}_\Delta = \mathcal{Z}_S$. Then $W_S = \langle \mathcal{Z}_S \rangle \trianglelefteq \mathcal{F}$, so by 3.3.7, $K \leq S_0$, contrary to hypothesis. This completes the proof of (3).

Set $z = z(K)$. By (3) there is $\psi \in Aut_{\mathcal{F}}(S_0)$ such that $t = z\psi \notin \mathcal{Z}_S$. Let $D_0 = K_0\psi$ and $D = \xi(D_0) \in \Delta(t)$. As $t \notin \mathcal{Z}_S$, $D \in \eta$ for each $\eta \in \eta(\mathcal{F}, \Omega)$. By 3.1.5.3 and the transitivity of $\mathcal{F}$ on Ω, we may choose D_0 so that $[K, t] \neq 1$. Hence by parts (3) and (5) of 3.1.7, $V = C_K(t)$ is cyclic of index 2 in K and the unique such D-invariant subgroup of K. Therefore as $W = \langle \eta \rangle$ is abelian, $V = K \cap W$. Further $K_0 = K \cap S_0$ is KD-invariant, so $C_K(K_0)$ is cyclic of index 2 in K, and hence $V = C_K(K_0)$. It follows that $\eta = \{C_{\xi(X)}(X) : X \in \eta_1\}$, so in particular $\eta(\tau) = \{\eta\}$ and $W_1 \leq W$. Thus $W \leq C_{\mathcal{F}}(W_1) \leq \mathcal{F}_z$, so $W \cap O_2(\mathcal{F}_z) \leq O_2(C_{\mathcal{F}}(W_1)) \leq S_0$. By (2), $V \leq O_2(\mathcal{F}_z)$, so $V \leq K \cap S_0 = K_0$ and hence $V = K_0$. Thus $\eta_1 = \eta$, completing the proof of (4) and the lemma. □

LEMMA 3.3.13. *Assume $K \in \Omega$ such that $z = z(K) \in O_2(\mathcal{F})$ and $\mathcal{O}_K \neq \mathcal{F}_K(K)$. Then $O^2(\mathcal{O}_K)$ is a component or solvable component of $\mathcal{F}$.*

PROOF. As $z \in O_2(\mathcal{F})$, it follows from 10.3 in [**Asc11**] that $E(\mathcal{F}) = E(\mathcal{F}_z)$. As $\mathcal{O}_K \neq \mathcal{F}_K(K)$, $\mathcal{C} = O^2(\mathcal{O}_K)$ is a component or solvable component of $\mathcal{F}_z$ by 2.6.13.2. In the former case as $E(\mathcal{F}) = E(\mathcal{F}_z)$, $\mathcal{C}$ is a component of $\mathcal{F}$, so we may assume $\mathcal{C} \cong SL_2[8]$. Then $\mathcal{O}_K$ centralizes $E(\mathcal{F}_z) = E(\mathcal{F})$, so $\mathcal{O}_K \leq C_{\mathcal{F}}(E(\mathcal{F})) = \mathcal{Y}$. Therefore replacing $\mathcal{F}$ by $\mathcal{Y}$, we may assume $E(\mathcal{F}) = 1$. Therefore $\mathcal{F}$ is constrained by 14.2 in [**Asc11**], so $\mathcal{F}$ has a model G.

Set $Q = O_2(G)$ and $L = N_G(W_S)$. Then $Q \leq S \leq L$. By 3.3.7, a model H for $\mathcal{O}(\tau)$ is normal in L and $H = H_1 \cdots H_n$ is a central product of the models H_i for $\mathcal{O}_{K_i}$, where $\{K_1, \ldots, K_n\} = \Omega$. Take $K = K_1$; then $H_1 \cong SL_2(3)^{(1)}$ by 3.3.3.2. Now if $u \in Q - N_Q(K)$ then H_1^u centralizes H_1, so $[H_1, u]$ is not a 2-group, a contradiction. Therefore $Q \leq N_G(K) \leq C_G(z)$. Then as $L_1 = O^2(H_1)$ is a solvable component of $C_G(z)$, it follows from 2.2.4.4 that $Q_1 = [Q_1 Q, L_1]$, where $Q_1 = O_2(L_1)$. Hence as $C_G(Q) \leq Q$, we have $Q_1 \leq Q$. Therefore by 3.3.8.3, H_1^G is the central product of the conjugates of H_1, so H_1 is a solvable component of G, completing the proof of the lemma. □

LEMMA 3.3.14. *Assume $\mathcal{F}$ is transitive on Ω. Then either*

(1) $\eta(\tau) = \{\eta\}$ is of order 1, or

(2) $m = 8$, $\Delta(\tau) = \Omega$, and for each $\eta \in \eta(\tau)$, $N_{\mathcal{F}}(\eta) \leq N_{\mathcal{F}}(O(\tau))$.

PROOF. By 3.1.11.4, either (1) holds or $m = 8$. Similarly for $\eta, \Sigma \in \eta(\tau)$, it follows from 3.3.10.1 that either $\Sigma = \eta$ or $\hat{\Omega} \neq \emptyset$, where $\hat{\Omega} = \{J \in \Omega : \Delta - \{J\} \subseteq C_S(J)\}$. But if $\hat{\Omega} \neq \emptyset$, then as $\mathcal{F}$ is transitive on Ω we have $\Omega = \hat{\Omega}$, and then $\Delta = \Omega$. Set $W = \langle \eta \rangle$. By 3.3.9 and as $\Delta = \Omega$, $O(\tau) \trianglelefteq N_{\mathcal{F}}(W)$, so (2) holds, completing the proof of the lemma. □

LEMMA 3.3.15. *Assume $\mathcal{F}$ is transitive on Ω, $K \in \Omega$, $z = z(K) \in \mathcal{F}^f$, and $\Omega(z) = \{K\}$.*

(1) Assume $s, t \in \mathcal{Z} - \mathcal{D}^*(z)$ with $E = \langle z, t, s \rangle \cong E_8$. Then $Aut_{\mathcal{F}}(E) = C_{GL(E)}(e)$ for some $e \in E$.

(2) Assume $\mathcal{D}^*(z) = \emptyset$ and $z \neq t \in \mathcal{Z}$. Then $\mathcal{Z} = \{z\} \cup t^K$.

PROOF. Assume the setup of (1). As $\Omega(z) = \{K\}$, E acts on K and then $K\langle t \rangle$ and $K\langle s \rangle$ are semidihedral of order $2m$ by 3.3.6. Let $F = \langle z, t \rangle$; then $N_K(F) = \langle v \rangle \cong \mathbf{Z}_4$ with $[v, t] = z$. As $\mathcal{F}$ is transitive on Ω and $z \in \mathcal{F}^f$ there is $\alpha \in \mathfrak{A}(t)$ with $t\alpha = z$. Then $\langle v \rangle = N_K(E)$ and $[E, v] = \langle [t, v] \rangle$ with $[t, v] = z$, so v induces a transvection $c_v = \gamma$ on E with center $\langle z \rangle$. Similarly $N_K(E\alpha) = \langle v \rangle$ and $c_v \alpha^{-*} = \beta \in Aut_{\mathcal{F}}(E)$ is a transvection with center $\langle t \rangle$ and $[z, \beta] = t$. Thus $\langle \gamma, \beta \rangle$ induces $GL(F)$ on F. By symmetry there is a transvection $\delta \in Aut_{\mathcal{F}}(E)$ with $[z, \delta] = s$. Now $A = \langle \gamma, \beta, \delta \rangle \leq GL(E)$ is generated by transvections, so $A = C_{GL(E)}(e)$ for some $e \in E$, proving (1).

Next assume (2) fails. As $\mathcal{D}^*(z) = \emptyset$ and $\Omega(z) = \{K\}$, also $\Omega = \{K\}$, so $K \trianglelefteq S$. As (2) fails there is $s \in \mathcal{Z} - K\langle t \rangle$ and $D = \langle s, t \rangle$ is dihedral. We may choose s so that s and t induce the same automorphism on K, so $x = st \in C_S(K)$. Set $E = \langle z, t, s \rangle$. If $|D| = 4$ then as $s \notin t^K$, $x \neq z$ so $E \cong E_8$. If $|D| > 4$ but $z \notin D$ replacing D by $C_D(t)$ we may assume $|D| = 4$. Finally if $z \in D$ then we may choose x of order 4. Then vx is an involution commuting with t and $vxt = vs \in s^K \subseteq \mathcal{Z}$, so replacing v by vs, again $E \cong E_8$. But now by (1), $A = Aut_{\mathcal{F}}(E) = C_{GL(V)}(e)$ for some $e \in E$, so v centralizes a member of $z^A - \{z\}$, a contradiction. □

LEMMA 3.3.16. *Assume $\tau = (\mathcal{F}, \Omega)$ is a quaternion fusion packet with $\mathcal{F} = \mathcal{F}^\circ$ transitive on Ω. Let $Z = Z(\mathcal{F})$ and $\mathcal{F}^+ = \mathcal{F}/Z$ and assume that $\mathcal{F}^+ = \mathcal{F}_{S^+}(L^+)$ for some simple group L^+ appearing in Theorem 1. Then*

(1) We may choose L^+ to tamely realize $\mathcal{F}^+$.

(2) $\mathcal{F} = \mathcal{F}_S(L)$ for some quasisimple group L such that $Z(L) = Z$ and $L/Z = L^+$.

PROOF. As L^+ appears in Theorem 1, L^+ is a known simple group, so by Theorems 3.5 and 3.6 in [**AO16**], we may choose L^+ to tamely realize $\mathcal{F}^+$. Hence (1) holds. Then by 2.22 in [**AO16**], $\mathcal{F} = E(\mathcal{F})Z$ and $\mathcal{F} = \mathcal{F}_S(L)$ for some group $L = E(L)Z$.

Let $K \in \Omega$ with $z = z(K) \in \mathcal{F}^f$. As L^+ appears in Theorem 1, $K^+ \leq O^2(C_{L^+}(z^+))$. Then by 3.3.2.3, $K \leq O^2(C_L(z))$. Hence as $\mathcal{F} = \mathcal{F}^\circ$, also $\mathcal{F} = O^2(\mathcal{F})$, so $L = O^2(L)$. Then as $L = E(L)Z$, we conclude that L is quasisimple and (2) holds. □

3.4. Generating $\mathcal{F}$

In this section we assume the following hypothesis:

HYPOTHESIS 3.4.1. *$\tau = (\mathcal{F}, \Omega)$ is a quaternion fusion packet, $K \in \Omega$, $z = z(K) \in \mathcal{F}^f$, and $\mathcal{F}$ is transitive on Ω.*

DEFINITION 3.4.2. For $\emptyset \neq \theta \subseteq \mathcal{Z}_S$ set $O(\theta) = \prod_{t\in\theta} O(t)$. Define

$$\mathcal{E} = \mathcal{E}(\tau) = \langle S\mathcal{O}(\tau), N_\mathcal{F}(O(\theta)), N_\mathcal{F}(\eta) : \eta \in \eta(\tau) \text{ and } \emptyset \neq \theta \subseteq \mathcal{Z}_S \text{ with } O(\theta) \in \mathcal{F}^f\rangle.$$

(Recall the definition of $T\mathcal{F}_0$ from 8.19 in [**Asc11**].) Write $\mathcal{F}^e$ for the set of $\mathcal{F}$-essential subgroups of S. Let $R \in \mathcal{F}^e$, G a model for $N_\mathcal{F}(R)$, and $G^* = G/R$. Then $N_G(N_S(R)) \leq \mathfrak{M} \leq G$ with $\mathfrak{M}^*$ strongly embedded in G^*.

LEMMA 3.4.3. *(1) Either*

(i) $\eta(\tau) = \{\eta\}$ is of order 1, or

(ii) $m = 8$, $\Delta = \Omega$, and for each $\eta \in \eta(\tau)$, $N_\mathcal{F}(\eta) \leq N_\mathcal{F}(O(\tau))$.

(2) Let $1 \neq X \leq \langle \mathcal{Z}_S\rangle = W_S$ and $\alpha \in \mathfrak{A}(X)$. Then $X\alpha \leq W_S$, $\alpha \in N_\mathcal{F}(O(\tau))$, and $\alpha \in \hom_{\mathcal{E}(\tau)}(N_S(X), S)$. Hence if $N_\mathcal{F}(X\alpha) \leq \mathcal{E}(\tau)$ then $N_\mathcal{F}(X) \leq \mathcal{E}(\tau)$.

(3) For each $\emptyset \neq \theta \subseteq \mathcal{Z}_S$, $N_\mathcal{F}(O(\theta)) \leq \mathcal{E}(\tau)$.

(4) For each $\emptyset \neq \theta \subseteq \mathcal{Z}_S$, $N_\mathcal{F}(\langle\theta\rangle) \leq \mathcal{E}(\tau)$.

PROOF. Part (1) is 3.3.14. Set $O = O(\tau)$.

Assume the setup of (2). Then $O \leq C_S(W_S)$, so $O\alpha \leq S$ and hence $O\alpha = O$ as O is weakly closed in S. Further $W_S \trianglelefteq N_\mathcal{F}(O)$, so $X\alpha \leq W_S$. As $N_\mathcal{F}(O) \leq \mathcal{E}$ by 3.4.2, we have $\alpha \in \hom_\mathcal{E}(N_S(X), S)$. Finally suppose $N_\mathcal{F}(X\alpha) \leq \mathcal{E}$; then $N_\mathcal{F}(X)\alpha^* \leq N_\mathcal{F}(X\alpha) \leq \mathcal{E}$, so as α is an $\mathcal{E}$-map, we have $(N_\mathcal{F}(X)\alpha^*)\alpha^{-*} \leq \mathcal{E}$, completing the proof of (2).

Now assume $\emptyset \neq \theta \subseteq \mathcal{Z}_S$ and let $P = O(\theta)$ and $\alpha \in \mathfrak{A}(P)$. Again $P \trianglelefteq O$, so $O\alpha = O$ and α is an $\mathcal{E}$-map. As $O(\theta\alpha) = P\alpha \in \mathcal{F}^f$, $N_\mathcal{F}(P\alpha) \leq \mathcal{E}$ by 3.4.2, so $N_\mathcal{F}(P) \leq \mathcal{E}$ as in the previous paragraph, establishing (3).

Set $X = \langle\theta\rangle$, $\beta \in \mathfrak{A}(X)$, and $\Theta = \mathcal{Z}_S \cap X\beta$. Let Γ be the union of the $\Omega(t)$ for $t \in \Theta$, $\mathcal{N} = N_\mathcal{F}(X\beta)$, and $\rho = (\mathcal{N}, \Gamma)$. Then ρ is a quaternion fusion packet by 2.5.2. Hence $\mathcal{O}(\Theta) \trianglelefteq \mathcal{N}$ by 3.3.7.2 applied to ρ, so by 1.3.2 in [**Asc19**], $\mathcal{N} = \langle N_S(X\beta)\mathcal{O}(\Theta), N_\mathcal{N}(O(\Theta))\rangle$. By (3), $N_\mathcal{F}(O(\Theta)) \leq \mathcal{E}$ and $N_S(X\beta)\mathcal{O}(\Theta) \leq S\mathcal{O}(\tau) \leq \mathcal{E}$, so $\mathcal{N} \leq \mathcal{E}$. Now (4) follows from (2). □

LEMMA 3.4.4. *Let $Q \leq R$ with $Q \trianglelefteq G$ and set $H = O^{2'}(C_G(Q))$.*
(1) If $H \leq R$ then $C_S(Q) \leq R$.
(2) If $H \not\leq R$ then $C_{S\cap G}(O^2(H^*)) = R$ and

$$Aut_{\mathcal{F}}(R) = Aut_H(R)Aut_{N_{\mathfrak{M}}(S\cap H)}(R).$$

PROOF. Suppose $H \leq R$ and let $X = RC_S(Q)$. As $G^* \not\leq \mathfrak{M}^*$, $Z(G^*)$ is of odd order. However $[G, N_X(R)] \leq [G, RH] = [G, R] \leq R$, so $N_X(R)^* \leq Z(G^*)$ and hence $N_X(R) \leq R$, proving (1).

Suppose $H \not\leq R$. Then H^* is normal in G^* of even order, so by a Frattini argument, $G = HN_G(S \cap H)$ and $N_G(S \cap H) = N_{\mathfrak{M}}(S \cap H)$ as $\mathfrak{M}^*$ is strongly embedded in G^*. This establishes the second statement in (2). Also $G^* = H^*\mathfrak{M}^*$, so as $\mathfrak{M}^*$ is proper in G^*, $H^* \not\leq \mathfrak{M}^*$. Thus $O^2(H^*) \not\leq \mathfrak{M}^*$, so as $\mathfrak{M}^*$ is strongly embedded in G^*, $C_{S^*}(O^2(H^*)) = 1$. Hence (2) holds. □

LEMMA 3.4.5. *Assume $P \in \Delta$ with $|P \cap R| > 2$. Then $Aut_{\mathcal{F}}(R) = Aut_{\mathcal{E}(\tau)}(R)$.*

PROOF. Assume otherwise and choose R of maximal order subject to $J = R \cap P$ of order at least 4 for some $P \in \Delta$ and $Aut_{\mathcal{F}}(R) \neq Aut_{\mathcal{E}}(R)$. Observe:

(a) If $R < T \leq S$ then $\hom_{\mathcal{F}}(T, S) = \hom_{\mathcal{E}}(T, S)$.

For let $\alpha \in \hom_{\mathcal{F}}(T, S)$. By the Alperin-Goldschmidt Fusion Theorem, $\alpha = \alpha_1 \cdots \alpha_n$ where $\alpha_i \in Aut_{\mathcal{F}}(R_i)$, $R_i = S$ or $R_i \in \mathcal{F}^e$, $T_0 = T \leq R_1$, and $T_i = T\alpha_1 \cdots \alpha_i \leq R_{i+1}$ for $i < n$. Moreover if $R_i \in \mathcal{F}^e$ then $Aut_{\mathcal{F}}(R_i) = Aut_{\mathcal{E}}(R_i)$ by maximality of $|R|$ and as $R < T$, while $N_{\mathcal{F}}(S) \leq N_{\mathcal{F}}(O(\tau)) \leq \mathcal{E}$. This proves (a).

Let $\Delta_R = \{P \in \Delta : |P \cap R| > 2\}$ and $\mathcal{J} = \{P \cap R : P \in \Delta_R\}$. For $J \in \mathcal{J}$ there is a unique $D_J \in \Delta$ with $J \leq D_J$. Set $Q = \langle \mathcal{J} \rangle$ and $Q_1 = C_R(Q)$.

Suppose for some $J \in \mathcal{J}$ that J is nonabelian and let $j = z(J)$, $\theta = j^G$, and $X = \langle \theta \rangle$. As J is nonabelian, $D_L \in \Omega$ for each $L \in J^G$, so $\theta \subseteq \mathcal{Z}_S$. Hence by 3.4.3.4, $N_{\mathcal{F}}(R) \leq N_{\mathcal{F}}(X) \leq \mathcal{E}$, contrary to the choice of R. Therefore

(b) Each member of $\mathcal{J}$ is cyclic.

By (b) and 3.1.8:

(c) Q is abelian.

(d) $C_S(Q_1) \leq Q_1$.

Let $S_1 = C_S(Q_1)$ and $S_2 = N_{S_1}(R)$. Then $[S_2, R] \leq C_R(Q) \leq Q_1 \leq C_R(S_2)$, so $S_2 \leq R$. Therefore $S_2 = S_1$ so $S_1 \leq Q_1$, establishing (d).

(e) We may assume $Q_1 \in \mathcal{F}^f$.

Let $\alpha \in \mathfrak{A}(Q_1)$; as $Q_1 \trianglelefteq G$, $R\alpha \in \mathcal{F}^e$. By (a), α is an $\mathcal{E}$-map, so if $Aut_{\mathcal{F}}(R\alpha) = Aut_{\mathcal{E}}(R\alpha)$ then also $Aut_{\mathcal{F}}(R) = Aut_{\mathcal{E}}(R)$, contrary to the choice of R. Hence, replacing R by $R\alpha$ if necessary, (e) holds.

By (d) and (e), $Q_1 \in \mathcal{F}^{cr}$, so there is a model G_1 for $N_F(Q_1)$. As $Q_1 \trianglelefteq G$, we may take $G = N_{G_1}(R)$. Set

$$G_2 = \bigcap_{J \in \mathcal{J}} N_{G_1}(J).$$

Set $\mathcal{J}_1 = \{J \in \mathcal{J} : D_J \notin \Omega\}$, $\mathcal{J}_3 = \{J \in \mathcal{J} : D_J \in \Omega$ and $[L, D_J] = 1$ for all $L \in \mathcal{J}_1\}$, and $\mathcal{J}_2 = \mathcal{J} - (\mathcal{J}_1 \cup \mathcal{J}_3)$. As $G = N_{G_1}(R)$ and R is radical:

(f) $O_2(G_1) \leq R$.

By 3.1.11:

(g) $D_J \in \eta$ for all $J \in \mathcal{J}_1$.

By 3.1.5.1:

(h) For distinct $J, L \in \mathcal{J} - \mathcal{J}_1$, we have $[D_J, D_L] = 1$.

(i) $\mathcal{J}_3 = \emptyset$.

Suppose $J \in \mathcal{J}_3$. By (h), $D = D_J$ centralizes each member of $\mathcal{J} - \{J\}$, so $X = N_D(Q_1) \leq G_2$. But $X \trianglelefteq N_{G_1}(J)$, so $X \trianglelefteq G_2$, and hence $X \leq O_2(G_2) \leq O_2(G_1) \leq R$ by (f). Therefore $X = D \cap R = J$, so $D = J \leq R$, contrary to (b). Hence (i) holds.

(j) $J \leq W$ for each $J \in \mathcal{J}$.

If $J \in \mathcal{J}_1$ this is a consequence of (g), so by (i) we may take $J \in \mathcal{J}_2$. Set $D = D_J$; as $J \in \mathcal{J}_2$ there is $L \in \mathcal{J}_1$ with $V = C_D(L) < D$. By 3.1.7.3, $V = C_D(D_L)$ and by 3.1.7.5, V is cyclic of index 2 in D. Thus as $D_L \in \eta$, also $V \in \eta$, so $J \leq V \leq W$, establishing (j).

By (j), $Q \leq W$, so Q centralizes $V_J = D_J \cap W$ for each $J \in \mathcal{J}$. Set $X = N_{V_J}(Q_1)$; then $X \trianglelefteq N_{G_1}(J)$, so $X \trianglelefteq G_2$, and hence $X \leq R$ by (f). Therefore $X = J$, so $X = V_J$, proving:

(k) $\mathcal{J} \subseteq \eta$.

(l) Let $H = O^{2'}(C_G(Q))$; then $H \not\leq R$.

For if $H \leq R$ then $C_S(Q) \leq R$ by 3.4.4.1. But by (k), $W \leq C_S(Q)$, while by (i), $\Delta \neq \Omega$, so W is weakly close in S by 3.4.3.1, so $N_{\mathcal{F}}(R) \leq N_{\mathcal{F}}(W) \leq \mathcal{E}$ by 3.4.2, contrary to the choice of R.

Set $\Omega_1 = \{K \in \Omega \cap N_S(Q_1) : K \cap R \in \mathcal{J}\}$ and $\mathcal{J}_4 = \{K \cap R : K \in \Omega_1\}$.

(m) $\Omega_1 \neq \emptyset$.

Let $J \in \mathcal{J}$ and $\alpha \in \mathfrak{A}(J)$; then $J\alpha$ is of index 2 in some $K \in \Omega$ by (k). Now $[Q_1\alpha, K] \leq J\alpha \leq Q_1\alpha$, so $K \leq N_S(Q_1\alpha)$. By (e) there is $\beta \in \mathfrak{A}(Q_1\alpha)$ with $Q_1\zeta = Q_1$, where $\zeta = \alpha\beta$. Then $K_0 = K\beta \leq G_1$ is in Ω and $K_0 = D_{J\zeta}$, with $J\zeta = K_0 \cap Q_1 = K_0 \cap R \in \mathcal{J}$. That is $K_0 \in \Omega_1$.

Set $O_1 = \langle \Omega_1 \rangle$, $G_4 = \bigcap_{J \in \mathcal{J}_4} N_{G_1}(J)$, and $G_5 = N_{G_1}(\mathcal{J}_4)$. Therefore $G_4 \trianglelefteq G_5$ as G_5 permutes $\mathcal{J}_4$. Also $O_1 \leq G_4$ and for $J \in \mathcal{J}_4$, we have $D_J \trianglelefteq N_{G_1}(J)$, so that $D_J \trianglelefteq G_4$ and hence $D_J \leq O_2(G_4) \leq O_2(G_5)$. Hence

(n) $O_1 \trianglelefteq G_5$.

Next R permutes Ω_1 and hence R acts on O_1. Similarly $H \leq G_4 \leq G_5$, so

(o) $O_1 \trianglelefteq O_1RH \leq G_5$.

Set $\Theta = \{z(J) : J \in \mathcal{J}_4\}$. Then $O_1 = O(\Theta)$, so by 3.4.3.3, $N_{\mathcal{F}}(O_1) \leq \mathcal{E}$. Then by (o):

(p) $Aut_H(R) \leq Aut_{\mathcal{E}}(R)$.

By (a), (l), and the proof of (e), we have $Aut_{N_{\mathfrak{M}}(S \cap H)}(R) \leq Aut_{\mathcal{E}}(R)$, so by (p) and 3.4.4.2, $Aut_{\mathcal{F}}(R) = Aut_{\mathcal{E}}(R)$, contrary to the choice of R. This final contradiction completes the proof of the lemma. □

LEMMA 3.4.6. *Assume $\Omega(z) = \{K\}$ and $z \in Z(R)$. Then either $Aut_{\mathcal{F}}(R) = Aut_{\mathcal{E}}(R)$ or the following hold:*

(1) $\Omega = \{K\}$.

(2) $\mathcal{Z} = \{z\} \cup t^K$ with $K\langle t \rangle$ semidihedral of order $2m$.

(3) $G = H \times C_G(H)$ with $H = \langle N_K(R)^G \rangle \cong S_4$, $O_2(H) = \langle z^G \rangle$, $R = C_S(O_2(H))$, and $C_G(H)$ is 2-closed.

PROOF. Assume R is a counter example. Then by 3.4.5:

(a) $|P \cap R| \leq 2$ for each $P \in \Delta$.

Set $Z = \langle z^G \rangle$. As $z \in Z(R)$, $Z \leq Z(R)$. Let $v \in K$ be of order 4 with $\langle v \rangle \trianglelefteq C_S(z)$. As $z \in Z(R)$ and $\Omega(z) = \{K\}$, R acts on K, so $[R, v] \leq \langle z \rangle \leq R$, and hence $v \in N_S(R)$. By (a), $v \notin R$, so $[R, v] = \langle z \rangle$. Set $H = \langle v^G \rangle$. We conclude from 2.4.3 in [**Asc19**] that the following two results hold:

(b) $H \cong S_4$ and $G = H \times C_G(H)$ with $Z = O_2(H)$.

(c) $R = C_S(Z)$ and $C_G(H)$ is 2-closed.

Let $t \in Z - \langle z \rangle$.

(d) t induces an outer automorphism on K and $K\langle t \rangle \cong SD_{2m}$.

By (a) and (c), $C_K(t) = \langle z \rangle$, so the result follows from 3.3.6.

(e) $\mathcal{D}^*(z) \neq \emptyset$.

Assume $\mathcal{D}^*(z) = \emptyset$. Then $\mathcal{D}(z) = \emptyset$, so (1) holds, and (2) follows from (d) and 3.3.15.2. Finally (3) follows from (b) and (c), contrary to the choice of R as a counter example.

Notice that (e) shows that Lemma 3.4.6 holds when $\mathcal{D}^*(z) = \emptyset$. In the proof of Lemma 7.1.7, we appeal to this special case of Lemma 3.4.6. At the end of the proof of Lemma 3.4.6 we appeal to Lemmas 7.1.12 and 7.1.13, which are proved under the hypothesis that $\mu(\tau) \cong S_3$. These lemmas do not depend on Lemma 3.4.6.

(f) Suppose $s \in t^{C_S(K)} \cap R$ is not in Z, and set $E = \langle z, t, s \rangle$. Then st or zst is in $C_S(H)$.

For from (b) and (c), H acts on E, so $Aut_H(E)$ centralizes e equal to st or zst, so $e \in C_S(H)$.

(g) $C_{\mathcal{Z}_S}(H) = \emptyset$.

Suppose $X \in \Omega$ with $x = z(X) \in C_S(H)$. Then z centralizes $\mathcal{O}(x)$ by 3.1.5.1, so $Z = [Z, H]$ centralizes $\mathcal{O}(x)$ as $\mathcal{O}(x) \trianglelefteq \mathcal{F}_x$. Therefore $X \leq C_S(Z) = R$, contrary to (a).

(h) $\mathcal{Z}_S \cap R = \{z\}$.

Assume $X \in \Omega - \{K\}$ with $x = z(X) \in R$, and set $\theta = x^R$. Then R acts on $O(\theta)$. Let U be a $C_S(x)$-invariant subgroup of X of order 4 and $Y = \langle U^R \rangle$. By (a), t inverts U, so t inverts Y. Thus $Y_0 = \Omega_1(Y) = C_Y(Z) = R \cap Y$. Then from (b) and (c), Y_0 is of index 2 in $Y_1 = N_Y(R)$. Then $\langle z \rangle = [Y_1, t] \leq B = \langle \theta \rangle$. As t inverts Y, $tx \in t^Y \subseteq t^{C_S(K)}$, so by (f), x or xz is in $C_S(H)$; therefore $xz \in C_S(H)$ by (g). Therefore for all $a, b \in \theta$, $ab = (az)(bz) \in C_S(H)$. Also $n = |\theta|$ is even, or else R acts on U and hence $[t, U] \leq Z \cap U = 1$, a contradiction. Therefore $z \in C_B(R) \leq \langle ab : a, b \in \theta \rangle \leq C_S(H)$, a contradiction.

(i) Set $R_0 = C_R(H) \cap C_R(K)$. Then $R_0 \trianglelefteq G$ and we may take $R_0 \in \mathcal{F}^f$.

The first statement in (i) follows from (b). Let $\alpha \in \mathfrak{A}(R_0)$. As $K \leq N_S(R_0)$, 3.4.5 and the proof of statement (a) in the proof of 3.4.5 implies that α is an $\mathcal{E}$-map. Then the proof of statement (e) in the proof of 3.4.5 shows that we may replace R by $R\alpha$ to obtain $R_0 \in \mathcal{F}^f$.

(j) $\Omega = \{K\}$.

Assume otherwise. By (h), t is fixed point free on $\Gamma = \Omega - \{K\}$. Let $K_1, \ldots, K_n$ be a set of representatives for the orbits of t on Γ, $z_i = z(K_i)$, $u_i = z_i z_i^t$, $J_i = C_{K_i K_i^t}(t)$, and $J = \prod_{i=1}^n J_i$. Then $J_i \cong K$ and J is the central product of the J_i, $1 \leq i \leq n$.

Observe that $J \leq C_S(Z) \leq R$, so $|J : J_0| \leq 2$, where $J_0 = C_J(H)$. In particular $u_i \in \Phi(J_i) \leq C_S(H)$. Also J centralizes K, so $J_0 \leq R_0$ and in particular $R_0 \neq 1$.

As $|J : J_0| \leq 2$, K is the unique member of Ω in $C_S(J_0)$, and hence also in $C_S(R_0)$. Let $\mathcal{F}_0 = N_{\mathcal{F}}(R_0)$; by (i), $R_0 \in \mathcal{F}^f$ and then by 2.5.2, $\tau_0 = (\mathcal{F}_0, \{K\})$ is a quaternion fusion packet. Set $\mathcal{F}_1 = [K]_{\mathcal{F}_0}$. As $H \leq C_G(J_0)$, we conclude from (d) and Theorem 2 that $z \notin O_2(\mathcal{F}_1)$, so, applying Theorem 7.1.29 in an inductive setting, we conclude that $\mathcal{F}_1 \cong L_2[2m]^{(1)}$, $L_3^-[m]$, $G_2[m]$, or M_{12}. Let $E = \langle z_i, z_i^t : 1 \le i \le n\rangle$. Observe $E \cap R_0 \le C_E(t) < E$ so there is $e \in N_E(R_0) - R_0$ and $1 \ne [e,t] = f \in \langle u_1, \dots, u_n\rangle \le J_0$. Now $e \in C_S(K)$ acts on R_0, so e acts on the Sylow group S_0 of $\mathcal{F}_1$. Then $f \in C_{S_0}(K)$ impossible as z is the unique involution in $C_{S_0}(K)$ while $f \in J_0$ centralizes $\mathcal{F}_1$. This completes the proof of (j).

We are now in a position to complete the proof of the lemma. By (j), (d), and (e), Hypothesis 7.1.8 is satisfied. Thus we can adopt notation 7.1.10. By 7.1.12, t inverts W, so $E \le C_S(Z) = R$. Let $A = ZE$; by (b), H acts on A, so as $E \trianglelefteq S$, HR acts on A. By 7.1.13.2, $Aut_{\mathcal{F}}(A) = C_{GL(A)}(e)$ for some $e \in E - \langle z\rangle$. Then as $e \in z^{N_{\mathcal{F}}(W)}$ and $N_{\mathcal{F}}(W) \le \mathcal{E} \ge \mathcal{F}_z$, it follows that $Aut_H(R) \le Aut_{\mathcal{E}}(R)$. Similarly $Aut_{C_G(Z)}(R) \le Aut_{\mathcal{F}_z}(R) \le Aut_{\mathcal{E}}(R)$, so $Aut_{\mathcal{F}}(R) = Aut_{\mathcal{E}}(R)$ by (b), completing the proof of the lemma. □

LEMMA 3.4.7. *$z \in Z(S)$ iff $|\Omega|/|\Omega(z)|$ is odd.*

PROOF. As $\mathcal{F}$ is transitive on Ω, $A = Aut_{\mathcal{F}}(O(\tau))$ is transitive on Ω by 3.1.5.6. Hence $|z^A| = n = |\Omega|/|\Omega(z)|$ and $z \in Z(S)$ iff n is odd. □

LEMMA 3.4.8. *Assume $\Omega(z) = \{K\}$ and $|\Omega|$ is odd. Then either $\mathcal{F} = \mathcal{E}(\tau)$ or conclusions (1)-(3) of Lemma 3.4.6 hold.*

PROOF. By 3.4.7, $z \in Z(S)$, so for each $R \in \mathcal{F}^e$, $z \in C_S(R) \le Z(R)$. Then the lemma follows from 3.4.6 and the Alperin-Goldschmidt Fusion Theorem. □

LEMMA 3.4.9. *Assume $H \trianglelefteq G$ with H^* of even order and $Q \trianglelefteq R$.*

(1) If $[C_G(Q), H] \le R$ then $C_S(Q) \le R$.

(2) If Q is H-invariant and $H^* - C_H(Q)^*$ contains an involution then $C_S(Q) \le R$.

PROOF. As $H \trianglelefteq G$ and H^* is of even order, $G^* = H^*\mathfrak{M}^*$, so $H^* \not\le \mathfrak{M}^*$. Suppose $S_1 = C_S(Q) \not\le R$. As $Q \trianglelefteq R$, R acts on S_1 and $S_2 = N_{S_1}(R) \not\le R$, so $S_2^* \ne 1$. Then $N_{G^*}(S_2^*) \le \mathfrak{M}^*$. Now $S_2 \le C_G(Q)$, so if $[C_G(Q), H] \le R$ then $H^* \le C_{G^*}(S_2^*) \le \mathfrak{M}^*$, a contradiction. Thus (1) holds.

It remains to prove (2), so assume Q is H-invariant, $S_1 \not\le R$, and t^* is an involution in $H^* - C_H(Q)^*$. As S_2^* is a nontrivial 2-group, S_2^* contains an involution s^*. Let $L = HS_2$. Then $C_L(Q)^* \trianglelefteq L^*$ as Q is H-invariant. We showed $G^* = H^*\mathfrak{M}^*$ and $H^* \not\le \mathfrak{M}^*$. Thus $(\mathfrak{M} \cap L)^*$ is strongly embedded in L^*, so L^* is transitive on its involutions, and hence $t^* \in s^{*L^*} \subseteq C_L(Q)^*$, a contradiction. □

LEMMA 3.4.10. *Assume $|\Omega(z)| \le 2$, $N_K(R)$ is nonabelian, and $Aut_{\mathcal{F}}(R) \ne Aut_{\mathcal{E}}(R)$. Set $Q = \langle z^G\rangle$, $J = N_K(R)$, $H = \langle J^G\rangle$, and $\Omega_H = J^H \cap S$. Then*

(1) $|J| = 8$, $\Omega_H = \{J, J^h\}$ for some $h \in H - N_H(J)$, $\tau_H = (\mathcal{F}_{S\cap H}(H), \Omega_H)$ is a quaternion fusion packet, $H/Z(H) \cong AE_5$, and $Z(H) = \langle zz^h\rangle$.

(2) $R = QC_S(H)$ with $C_S(Q) \le R$.

(3) If $z \ne z^h$ then $Q \cong Q_8D_8$ and we may take $h \in z^H$.

PROOF. By 2.5.2, τ_H is a quaternion fusion packet. As $F^*(G) = R = O_2(G)$, $z \in R$ by 3.3.2.4. Then $J \cap R = \langle z\rangle$ by 3.4.5. Hence H is described in Theorem 2, so as H^* has a strongly embedded subgroup we conclude that $H/Z(H) \cong AE_5$. In particular (1) holds. Then (3) follows from (1) and 5.1.7. Set $\tilde{G} = G/Z(H)$. Then

$\tilde{Q}$ is a projective H^*-module, so $\tilde{R} = \tilde{Q} \oplus C_{\tilde{R}}(H)$. Hence $R = QC_R(H)$ using (3). Finally $C_S(Q) \leq R$ by 3.4.9.2, so (2) follows. □

LEMMA 3.4.11. *(1) Let $\theta \subseteq \mathcal{Z}_S$ with $Q = O(\theta) \in \mathcal{F}^f$. Then $N_{\mathcal{F}}(Q) \leq \langle N_{\mathcal{F}}(O(\tau)), N_{\mathcal{F}}(O(\xi))\rangle$ for each $\emptyset \neq \xi \subseteq \theta$ such that $N_S(Q)$ is transitive on ξ.*

(2) $\mathcal{E}(\tau) = \langle S\mathcal{O}(\tau), N_{\mathcal{F}}(O(\xi)), N_{\mathcal{F}}(W) : W \in \mathcal{W}(\tau),\ \xi \in \Xi\rangle$, where Ξ consists of those $\emptyset \neq \xi \subseteq \mathcal{Z}_S$ such that $N_S(\xi)$ is transitive on ξ and $O(\xi) \in \mathcal{F}^f$.

PROOF. It follows from the definition of $\mathcal{E}(\tau)$ in 3.4.2 and from 3.4.3.4 that (1) implies (2). So choose θ and ξ as in (1) and set $\mathcal{N} = N_{\mathcal{F}}(Q)$. Observe that $P = O(\xi) \leq Q$. Set $O = O(\tau)$ and $\mathcal{X} = \langle N_{\mathcal{F}}(O), N_{\mathcal{F}}(P)\rangle$; we must show that $\mathcal{N} \leq \mathcal{X}$.

Set $T = N_S(Q)$ and $\mathcal{Y} = C_{\mathcal{F}}(Q)T$. By 1.3.2 in [**Asc19**] and as O is weakly closed in $QC_S(Q)$, we have $\mathcal{N} = \langle \mathcal{Y}, N_{\mathcal{N}}(O)\rangle$, so it remains to show that $\mathcal{Y} \leq \mathcal{X}$. But as $P \leq Q$, $C_{\mathcal{F}}(Q) \leq N_{\mathcal{F}}(P) \leq \mathcal{X}$. As ξ is T-invariant, $P \trianglelefteq T$. Thus indeed $\mathcal{Y} \leq \mathcal{X}$, completing the proof. □

LEMMA 3.4.12. *Assume $\Delta(\tau) = \Omega$. Then*

(1) $\mathcal{E}(\tau) = \langle S\mathcal{O}(\tau), N_{\mathcal{F}}(O(\xi)) : \xi \in \Xi\rangle$, where Ξ is defined in 3.4.11.

(2) If $m = 8$ then $\mathcal{E}(\tau) = \langle N_{\mathcal{F}}(O(\xi)) : \xi \in \Xi\rangle$.

PROOF. If $m = 8$ then $\mathcal{O}(\tau) = O(\tau)$, so (1) implies (2). Thus it remains to prove (1), so by 3.4.11.2, it suffices to show that $N_{\mathcal{F}}(W) \leq N_{\mathcal{F}}(O)$, for $W \in \mathcal{W}(\tau)$ and $O = O(\tau)$. But as $\Delta = \Omega$, this follows from the proof of 3.3.14. □

LEMMA 3.4.13. *Let x be an involution in $\mathcal{F}^f$. Assume $\tau_x^\circ = \tau_1 * \tau_2$ where $\tau_i = (\mathcal{F}_i, \Omega_i)$ with $\mathcal{F}_i \trianglelefteq \mathcal{F}_x$ and $\Omega_i \neq \emptyset$ for $i = 1, 2$. Then $\mathcal{F}_x \leq \mathcal{E}(\tau)$.*

PROOF. Let $O_i = O(\tau_i)$ and T_i Sylow in $\mathcal{F}_i$. As $\mathcal{F}_i \trianglelefteq \mathcal{F}_x$ for $i = 1, 2$, it follows from 1.3.2 in [**Asc19**] that $\mathcal{F}_x = \langle C_S(x)\mathcal{F}_1, N_{\mathcal{F}_x}(T_1)\rangle$. Further $O_2 \trianglelefteq C_S(x)\mathcal{F}_1$ and $O_1 \trianglelefteq N_{\mathcal{F}_x}(T_1)$, and as $\Omega_i \neq \emptyset$, $N_{\mathcal{F}}(O_i) \leq \mathcal{E}$ by 3.4.3.3, so the lemma holds. □

CHAPTER 4

$W(\tau)$ and $M(\tau)$

In Chapter 4, $\tau = (\mathcal{F}, \Omega)$ is a quaternion fusion packet and S is Sylow in $\mathcal{F}$. Recall from 3.1.26 that $\mu = \mu(\tau) = \langle D \rangle$ for a certain set D of involutions in $Aut_{\mathcal{F}}(W)$, for $W \in \mathcal{W}(\tau)$. Let $W \trianglelefteq M \in \mathcal{M}(\tau)$. From 3.3.11, M is a model for $[O(\tau)]_{N_{\mathcal{F}}(W)}$ and $\mu = Aut_M(W)$. Further $\tau_W = (N_{\mathcal{F}}(W), \Omega)$ is a quaternion fusion packet, so τ_W° satisfies Hypothesis 4.2.1, and from 4.2.4, $\mu = \mu(\tau_W)$.

Also from 3.1.25, generically D is a set of 3-transpositions in μ. Therefore in section 4.1 we are led to consider 3-transposition groups. Our reference for 3-transposition groups is [**Asc97**]. In 4.2.14 we find that when D is a set of 3-transpositions of μ that D satisfies Hypothesis 4.1.12 in μ, while 4.1.25 determines the groups μ generated by a conjugacy class of 3-transpositions satisfying 4.1.12. Indeed by that result, $\mu/Z(\mu)$ is isomorphic to Weyl$(\Phi)/Z($Weyl$(\Phi))$ for some Φ Coxeter diagram Φ of type A, D, or E.

In 4.3.1 we define a universal group $\bar{G} = \bar{\omega}(\Phi, m)$ and show in 4.3.8 that there is a surjective homomorphism $\pi : \bar{G} \to W$ with $\ker(\pi) \leq Z(\bar{G})$. Moreover $W = C_M(W)$, so $\mu = M/W$ and $\mu \cong$ Weyl(Φ). Hence M is determined up to isomorphism by Φ, m, and $\ker(\pi)$. Further the structure of M can be retrieved from the structure of $\bar{G}$.

The group $\bar{G} = \bar{\omega}(\Phi, m)$ is isomorphic to $M(\tau)$ for τ the Lie packet of a suitable universal group of Lie type.

4.1. 3-transposition groups

In this section we assume the following hypothesis:

HYPOTHESIS 4.1.1. *G is a finite group and D is a G-invariant set of involutions in G such that $G = \langle D \rangle$.*

In addition we adopt the following notation:

NOTATION 4.1.2. Let $d \in D$ and set $d^\perp = C_D(d)$, $D_d = d^\perp - \{d\}$, and $A_d = D - d^\perp$. Let $\mathcal{D}(D)$ be the commuting graph on D; that is $a, b \in D$ are adjacent in $\mathcal{D}(D)$ iff $b \in D_a$. Define

$$V_d = \{a \in D : a^\perp = d^\perp\} \text{ and } W_d = \{a \in D : D_a = D_d\}.$$

Recall that D is a *set of 3-transpositions* if Hypothesis 4.1.1 is satisfied and $|ab| \leq 3$ for all $a, b \in D$.

In the remainder of the section we assume:

HYPOTHESIS 4.1.3. *D is a set of 3-transpositions of G.*

Our reference for 3-transposition groups is [**Asc97**].

LEMMA 4.1.4. *If $D = D_1 \coprod D_2$ is a G-invariant partition of D then $[D_1, D_2] = 1$.*

PROOF. See 8.2.1 in [**Asc97**]. □

LEMMA 4.1.5. *If $G = \langle a, b, d\rangle$ with $b \in D_d$ and $a \in A_d \cap A_b$ then $G \cong S_4$ and D is the set of ordinary transpositions in G.*

PROOF. This follows as G is an image of the Coxeter group S_4 of type A_3. □

LEMMA 4.1.6. *$\mathcal{D}(D)$ is disconnected iff G is transitive on D and G has a factor group isomorphic to S_3.*

PROOF. Assume G is transitive on D and $H \trianglelefteq G$ with $G^* = G/H \cong S_3$. As $G = \langle D\rangle$, D is not contained in H, so as G is transitive on D, $H \cap D = \emptyset$. We must show that $\mathcal{D}(D)$ is disconnected; it suffices to show $D_d \subseteq dH$, so assume $b \in D_d - dH$. Then $2 = |bd| = |b^*d^*|$, contradicting $G^* \cong S_3$.

Conversely assume $\mathcal{D}(D)$ is disconnected and let Γ be the set of connected components of $\mathcal{D}(D)$, and for $d \in D$ let Σ_d be the member of Γ containing d. By 4.1.4, G is transitive on D. Let K be the kernel of the action of G on Γ and set $G^+ = G/K$. If $\Sigma_d \neq \Sigma_a$ then d does not act on Σ_a, or else $d \in D_a$ by 4.1.4, a contradiction. Thus d is semiregular on $\Gamma - \{\Sigma_d\}$, so in particular $d^+ \neq 1$. Further for $b \in D_d$ and $a \in D - \Sigma_d$, $a \in A_d \cap A_b$, so $X = \langle a, b, d\rangle \cong S_4$ by 4.1.5. In particular $a^{bd} \in D_a$, so $bd \in K$. Therefore $D_{d^+} = \emptyset$, so by 8.6 in [**Asc97**], $G^+ = O_3(G^+)\langle d^+\rangle$. Now let M^+ be a maximal subgroup of $O_3(G^+)$ normal in G^+ and M the preimage of M^+ in G. Then $G/M \cong S_3$, completing the proof. □

LEMMA 4.1.7. *Assume G is transitive on D. Then*
(1) $V_d = d^{O_2(G)}$ and
(2) $W_d = d^{O_3(d)}$.

PROOF. See 9.2 in [**Asc97**]. □

LEMMA 4.1.8. *Assume that $G/O_2(G) \cong S_3$ and $\mathcal{D}(D)$ is disconnected. Then $2^{n-1} = |V_d|$ for some positive integer n, and G is generated by n elements of $d^\perp$ together with an element of A_d.*

PROOF. Let G be a minimal counter example, $Q = O_2(G)$, $a \in A_d$, and $\pi : G \to G/Q$ the natural map. By 4.1.6, G is transitive on D, so for $b \in \pi^{-1}(d)$, $b = d^g$ for some $g \in G$. Then $g\pi \in C_{G\pi}(d\pi) = \langle d\pi\rangle$, so $b \in d^Q$. Hence by 4.1.7.1, $\pi^{-1}(d) = d^Q = V_d$, so indeed $|V_d| = |Q : C_Q(d)| = 2^{n-1}$ for some positive integer n.

If $Q = 1$ then $S_3 \cong G = \langle d, a\rangle$, so the lemma holds with $n = 1$ in this case. Therefore as G is a counter example, $Q \neq 1$. Let M be a minimal normal subgroup of G contained in Q and set $G^* = G/M$. By minimality of G, $G^* = \langle d_1^*, \ldots, d_m^*, a^* \rangle$ where $2^{m-1} = |V_{d^*}|$, $d = d_1$, and $d_i^* \in d^{*\perp}$. Set $H = \langle d_1, \ldots, d_m, a \rangle$, so that $H^* = G^*$ and hence $G = HM$. As D is a set of 3-transpositions of G, $d_i \in d^\perp$. Further $n \geq m$.

If d centralizes M then as $G = \langle D \rangle$ is transitive on D, $M \leq Z(G)$. Then as $G = MH$, $H \trianglelefteq G$, so as $G = \langle D \rangle$ is transitive on D, we have $G = H$. Then as $n \geq m$, we have a contradiction to the choice of G as a counter example.

Therefore $[M, d] \neq 1$. Next G is irreducible on M, so $M \leq Z(Q)$. Then as $[M, d] \neq 1$, we have $Q = C_G(M)$, so $Aut_G(M) \cong S_3$ and hence as G is irreducible on M we have $M \cong E_4$, so that M is transitive on the involutions in dM. Therefore $C_{G^*}(d^*) = C_G(d)^*$, so $n = m + 1$ and $G = \langle H, d^x \rangle$ for $x \in M - C_M(d)$, again contradicting the choice of G. □

REMARK 4.1.9. Denote by Weyl(D_4) the Coxeter group of type D_4; recall Weyl(D_4) is the split extension W of Q_8^2 by S_3 in which $F^*(W) = O_2(W)$ and the two Q_8-subgroups of $O_2(W)$ are conjugate in W.

LEMMA 4.1.10. *Assume $G/O_2(G) \cong S_3$ with $\mathcal{D}(D)$ disconnected and $|V_d| = 4$. Then G is isomorphic to Weyl(D_4) or Weyl(D_4)/Z(Weyl(D_4)).*

PROOF. By 4.1.8, $G = \langle d_1, \ldots, d_3, a \rangle$ with $d = d_1$, $d_i \in d^\perp$, and $a \in A_d$. Then $a \in A_{d_i}$ for each i, so G is an image of the Coxeter group W of type D_4, with D the image of the reflections in W. Then the lemma follows from Remark 4.1.9. □

A *D-subgroup* of G is a subgroup X of G generated by $D \cap X$. The *width* of G is $|D \cap S|$ for $S \in Syl_2(G)$.

At various times we assume one of the following hypotheses:

HYPOTHESIS 4.1.11. *(1) D is a set of 3-transpositions of G.*

(2) If X is a D-subgroup of G of width 1 then $X \cong \mathbf{Z}_2$ or S_3.

HYPOTHESIS 4.1.12. *Hypothesis* 4.1.11 *holds and*

(*) if $X = \langle E \rangle \leq G$ with E an X-invariant subset of D such that $\mathcal{D}(E)$ is disconnected then $|e^{O_2(X)}| \leq 4$ for $e \in E$.

HYPOTHESIS 4.1.13. *Hypothesis* 4.1.11 *holds and*

(**) if $X = \langle E \rangle \leq G$ with E an X-invariant subset of D such that $\mathcal{D}(E)$ is disconnected then $X \cong S_3$ or S_4.

LEMMA 4.1.14. *The following are equivalent:*

(1) Hypothesis 4.1.11 *holds.*

(2) For each D-subgroup X of G isomorphic to S_3 and for each $d \in D$, $d^\perp \cap X \neq \emptyset$.

PROOF. Suppose X and d are as in (2) with $d^\perp \cap X = \emptyset$. Then by 5.5.4 in [**Asc97**], $\langle X, d \rangle$ is of width 1, so (1) fails. Thus (1) implies (2).

Conversely assume (2) holds and X is a D-subgroup of G of width 1 not isomorphic to $\mathbf{Z}_2$ or S_3. Then there exist distinct $a, b \in D \cap X$, and as X is of width 1, $a \in A_b$, so $Y = \langle a, b \rangle \cong S_3$. Then there exists $c \in (D \cap X) - Y$, and by (2), c commutes with a, b, or a^b, contradicting X of width 1. □

LEMMA 4.1.15. *Each of Hypotheses* 4.1.11, 4.1.12, *and* 4.1.13 *inherit to D-subgroups of G.*

PROOF. This is immediate from the definitions. □

LEMMA 4.1.16. *Hypothesis* 4.1.11 *inherits to homomorphic images of* G.

PROOF. Let $H \trianglelefteq G$ and $\bar{G} = G/H$. Suppose Hypothesis 4.1.11 holds and let $\bar{X}$ be a $\bar{D}$-subgroup of $\bar{G}$ isomorphic to S_3. Then $\bar{X} = \langle \bar{a}, \bar{b} \rangle$ with $b \in D$ and $a \in A_b$, so $X = \langle a, b \rangle \cong S_3$. Let $d \in D$; then by 4.1.14, $d^\perp \cap X \neq \emptyset$, so $\bar{d}^\perp \cap \bar{X} \neq \emptyset$. Then another application of 4.1.14 completes the proof. □

LEMMA 4.1.17. *Assume Hypothesis* 4.1.11 *with* G *transitive on* D, *and suppose* $Y \trianglelefteq G$ *such that* $O_3(G/Y) \not\leq Z(G/Y)$. *Then* $Y = O_2(G)$ *and* $G/Y \cong S_3$.

PROOF. Set $\bar{G} = G/Y$ and let L be the preimage in G of $O_3(\bar{G})$ in G, $a \in D$, and $\bar{a}^L = \theta$. Then $\nu = \langle \theta \rangle$ is of width 1, so by 4.1.16 and 4.1.11, $\nu \cong S_3$. But by 4.1.7, $W_{\bar{a}} = \theta$, while by 4.1.14, for each $\bar{b} \in \bar{D} - \theta$, $D_{\bar{b}} \cap \theta \neq \emptyset$, so $\bar{D} - \theta$ centralizes $\bar{\nu}$. Hence as G is transitive on D, it follows that $\bar{G} = \bar{\nu} \cong S_3$.

We next claim that $aY \cap D \subseteq a^\perp$. If not there is $c \in aY \cap D$ with $\langle a, c \rangle \cong S_3$. Let $b \in aL \cap A_a$; as $\bar{c} = \bar{a}$, $|\bar{c}\bar{b}| = 3$, so $|cb| = 3$. As this also holds for c^a, $b^\perp \cap \langle a, c \rangle = \emptyset$, so 4.1.14 supplies a contradiction, completing the proof of the claim.

Set $Q = O_2(G)$ and $G^* = G/Q$; as $\bar{G} \cong S_3$, we have $Q \leq Y$. By the claim and as $\bar{G} \cong S_3$, $aY \cap D = a^\perp = V_a$, so by 4.1.7.1, $aY \cap D = a^Q \subseteq aQ$. Thus $\{a^*\} = a^*Y^* \cap D^*$, so as $\bar{G} \cong S_3$, $|D^*| = |\bar{D}| = 3$ and hence $G^* \cong S_3$, so that $Y^* = 1$ and hence $Y = Q$. This completes the proof. □

LEMMA 4.1.18. *Assume Hypothesis* 4.1.11.
(1) If $\mathcal{D}(D)$ *is disconnected then* $G/O_2(G) \cong S_3$.
(2) If G *is transitive on* D *and* $O_3(G) \not\leq Z(G)$ *then* $G \cong S_3$.

PROOF. Part (1) follows from 4.1.6 and 4.1.17. Assume the hypothesis of (2). Applying 4.1.17 with $Y = 1$, we conclude that $G \cong S_3$, proving (2). □

LEMMA 4.1.19. *Hypotheses* 4.1.12 *and* 4.1.13 *are inherited by homomorphic images of* G.

PROOF. Let $H \trianglelefteq G$ and $\bar{G} = G/H$. Assume Hypothesis 4.1.12 or 4.1.13 holds, and $\bar{X} = \langle \bar{E} \rangle$ for some $\bar{X} \leq \bar{G}$ and some $\bar{X}$-invariant subset $\bar{E}$ of $\bar{D}$ such that $\mathcal{D}(\bar{E})$ is disconnected. By 4.1.15 and 4.1.16, $\bar{X}$ satisfies Hypothesis 4.1.11, so by 4.1.18.1, $\bar{X}/O_2(\bar{X}) \cong S_3$. By 4.1.6, $\bar{X}$ is transitive on $\bar{E}$, so $\bar{X} = \langle \bar{d}^{\bar{X}} \rangle$ for $d \in D$ with $\bar{d} \in \bar{E}$. Let Y be the preimage of $\bar{X}$ in G, $E = d^Y$, and $X = \langle E \rangle$. By 4.1.6, $\mathcal{D}(E)$ is disconnected.

Suppose Hypothesis 4.1.13 holds. Then $X \cong S_3$ or S_4 and hence $\bar{X} \cong S_3$ or S_4, so $\bar{G}$ satisfies Hypothesis 4.1.13.

Finally suppose Hypothesis 4.1.12 holds. By 4.1.18, $X/O_2(X) \cong S_3$, so $O_2(\bar{X}) = \overline{O_2(X)}$, and hence $|\bar{d}^{O_2(\bar{X})}| \leq |d^{O_2(X)}| \leq 4$. This completes the proof. □

LEMMA 4.1.20. *Assume* G *satisfies Hypotheses* 4.1.11 *and* G *is a subgroup of a group* L *transitive on* D *via conjugation. Assume that* G *is of width 2. Then* $G \cong S_3 \times S_3$, E_4, S_4 *or* S_5.

PROOF. Let $d \in D$, let D_i, $1 \leq i \leq n$, be the orbits of G on D and set $G_i = \langle D_i \rangle$. By 4.1.4, $G = \prod_i G_i$ is a central product. As L is transitive on D, L permutes the factors G_i transitively, so $G_i \cong G_j$ for all i, j. Let w be the width of G_i. Then $2 = nw$ so either $n = 1$ or $n = 2$ and $w = 1$. In the latter case by

Hypothesis 4.1.11, $G_i \cong \mathbf{Z}_2$ or S_3, so that the lemma holds in this case. Therefore we may assume that G is transitive on D.

If $\mathcal{D}(D)$ is disconnected then by 4.1.18.1, $G/O_2(G) \cong S_3$. Then $d^\perp = V_d = d^{O_2(G)}$ by 4.1.7, so as G is of width 2, $|V_d| = 2$ and hence $G \cong S_4$. Therefore we may assume $\mathcal{D}(D)$ is connected. Similarly if $V_d \neq \{d\}$ then as G is of width 2, $d^\perp = V_d$, so $\mathcal{D}(D)$ is disconnected, contrary to assumption. Therefore $V_d = \{d\}$, so $O_2(G) \leq Z(G)$ by 4.1.7. Also as $\mathcal{D}(D)$ is connected, $O_3(G) \leq Z(G)$ by 4.1.18.2.

We've shown that $O_2(G) \leq Z(G) \geq O_3(G)$ and $\mathcal{D}(G)$ is connected. It follows from 9.4.4 and 9.5.4 in [**Asc97**] that $H = \langle D_d \rangle$ is transitive on D_d and $|D_d| > 1$. But as G is of width 2, H is of width 1, so $H \cong S_3$ by 4.1.11. Therefore $H = \langle d_3, d_4 \rangle$ with $d_4 \in A_{d_3}$. Set $d_1 = d$. As G is transitive on D and $\mathcal{D}(D)$ is connected, there exists $d_2 \in A_d$ with $d_3 \in D_{d_2}$. If $d_4 \in D_{d_2}$, then $d_2 \in W_d$, contradicting 4.1.7 and $O_3(G) \leq Z(G)$. Therefore setting $\Delta = \{d_1, \dots, d_4\}$ and $G_0 = \langle \Delta \rangle$, (G_0, Δ) is a Coxeter system of type A_4, so $G_0 \cong S_5$. Finally by 9.5.5 in [**Asc97**], $G = G_0$, completing the proof. □

LEMMA 4.1.21. *If Hypotheses* 4.1.13 *holds with G transitive on D, then $G \cong S_n$ for some $n \geq 2$.*

PROOF. Assume otherwise and let G be a minimal counter example. If $|D| = 1$ then $G \cong S_2$, so $|D| > 1$. If $\mathcal{D}(D)$ is disconnected then $G \cong S_3$ or S_4 by 4.1.13, contrary to the choice of G. Therefore $\mathcal{D}(D)$ is connected.

Observe that $O_3(G) \leq Z(G)$ by 4.1.18.2. Claim $O_2(G) \leq Z(G)$. Assume otherwise and set $\bar{G} = G/O_2(G)$. By 4.1.19 and minimality of G, $\bar{G} \cong S_n$ for some n. We will show:

(!) $\bar{G}$ has no $\bar{D}$-subgroups isomorphic to S_4.

It follows from (!) that $n = 3$, so $\mathcal{D}(D)$ is disconnected by 4.1.6, contrary to an earlier reduction. Thus it remains to establish (!). So suppose $\bar{X}$ is a $\bar{D}$-subgroup isomorphic to S_4. Then $\bar{X} = \langle \bar{d}, \bar{a}, \bar{b} \rangle$ with $b \in D_d$ and $a \in A_d \cap A_b$. Let $X = \langle V_d, a, b \rangle$ and $E = d^X$. Then $\mathcal{D}(E)$ is disconnected by 4.1.6, so $X \cong S_4$ by 4.1.13, whereas by 4.1.7, $V_d \cup \{b\} \subseteq d^{O_2(X)}$, so $|d^{O_2(X)}| > 2$, a contradiction. This completes the proof of the claim.

We've shown that $\mathcal{D}(D)$ is connected and $O_2(G) \leq Z(G) \geq O_3(G)$. At this point we could appeal to Theorem Z in section 15 of [**Asc97**] to finish off the proof fairly quickly. Instead we will give most of the details in this special case, as most are fairly easy, and it is only a little more difficult to include these details than to exclude some of the conclusion appearing in Theorem Z.

By 9.4.4 and 9.5.4 in [**Asc97**], $H = \langle D_d \rangle$ is transitive on D_d. By minimality of G, $H \cong S_n$ for some n. If G is of width 2 then $G \cong S_5$ by 4.1.20, so $n \geq 4$.

Let $a \in A_d$ and $X = \langle a, d \rangle$. As $H \cong S_n$ with $n \geq 4$, it follows that for $b \in D_d$, $C_H(b)$ is transitive on $A_b \cap D_d$, so:

(a) G is transitive on triples (u, v, w) with $w \in A_v$ and $u \in D_v \cap D_w$.

Thus $C_G(X)$ is transitive on $E = D_a \cap D_d$. Set $Y = \langle E \rangle$; we will show that:

(b) Y is transitive on E.

Then by (b), $Y \cong S_m$ for some integer m by minimality of G. We will also show:

(c) $m = n - 1$.

By (c), $Y = \langle \Delta \rangle$ where $\Delta = \{d_1, \dots, d_{n-2}\}$ has Coxeter diagram A_{n-2}, and there exists $d_0 \in D_d \cap A_a$ such that $\Delta_0 = \{d_0, d_1, \dots, d_{n-1}\}$ has Coxeter diagram

A_{n-1} and $H = \langle \Delta_0 \rangle$. Then $\Gamma = \{d, a, d_0, \dots, d_{n-2}\}$ has Coxeter diagram A_n, so $G_0 = \langle \Gamma \rangle \cong S_{n+1}$. Further $G_0 = \langle H, d, a \rangle$, so $G = G_0$ by 9.5.5 in [**Asc97**], contrary to the choice of G. Therefore it remains to prove (b) and (c).

Let $b \in D_d$, $c \in D_d \cap A_b$, and $W = \langle b, c \rangle$. Then $\langle C_D(W\langle d \rangle) \rangle = \langle d \rangle \times K$ where $K = \langle C_{D_d}(W) \rangle \cong S_{n-3}$. Suppose Y is transitive on E. Now for $e \in E$, $C_Y(e) = \langle e \rangle \times L$ where $L \cong S_{m-2}$, so by (a), either $m = n - 1$ or $(n, m) = (4, 2)$. However in the latter case, the rank 3 parameters for the graph $\mathcal{D}(D)$ (cf. section 6 in [**Asc97**]) are $k = 6$ and $\lambda = \mu = 1$, so $\delta = (\lambda - \mu)^2 + 4(k - \mu) = 20$ is not a square, contrary to 6.3.3 in [**Asc97**]. Thus if Y is transitive on E then $m = n - 1$; that is (b) implies (c), so it remains to prove (b).

So assume (b) fails. Then by 4.1.4 and as $C_G(X)$ is transitive on E, $E = E_1 \coprod \cdots \coprod E_r$, for some $r > 1$, and $Y = Y_1 \times \cdots \times Y_r$, where $Y_i = \langle E_i \rangle \cong S_t$ for some $t \geq 2$. Then for $e \in E_1$,

$$Y_e = C_Y(e) = \langle e \rangle \times X_1 \times Y_2 \times \cdots \times Y_r,$$

where $X_1 \cong S_{t-2}$. However by (a), $Y_e \cong \mathbf{Z}_2 \times S_{n-3}$, so we conclude that $r = 2$, $t = n - 3$, and $X_1 = 1$. As $X_1 = 1$, $t \leq 3$, so as $2 \leq t = n - 3$, $n = 5$ or 6.

Suppose $n = 5$. Then $k = 10$, $\lambda = 3$, and $\mu = 2$, so $\delta = 33$ is not a square, again a contradiction. Therefore $n = 6$. Here $k = 15$ and $\lambda = \mu = 6$, so $\delta = 6^2$ is a square. Indeed a second group exists in this case: by 15.6 in [**Asc97**], G is either S_8 or $O_6^-(2)$. However in the latter case, the centralizer of a 2-central involution in G contains a D-subgroup isomorphic to Weyl(D_4), contrary to 4.1.13. This completes the proof of the lemma. □

LEMMA 4.1.22. *Assume Hypotheses 4.1.12 with G transitive on D and $|V_d| > 1$ for $d \in D$. Then*

(1) $G \cong$ Weyl(D_n) or Weyl$(D_n)/Z($Weyl$(D_n))$ for some $n \geq 3$.

(2) Either $|V_d| = 2$ or $n = 4$ and $|V_d| = 4$.

PROOF. Let $d \in D$, $Q = O_2(G)$, and $\bar{G} = G/Q$. By 4.1.7, $V_d = d^Q$, so for $v \in D$, $\bar{d} = \bar{v}$ iff $v \in V_d$. Claim:

(a) $\bar{G}$, $\bar{D}$ satisfy Hypothesis 4.1.13.

For suppose $\bar{X}$ is a subgroup of $\bar{G}$ generated by an $\bar{X}$-invariant subset $\bar{E}$ of $\bar{D}$ such that $\mathcal{D}(\bar{E})$ is disconnected. Let X_0 be the preimage of $\bar{X}$ in G, let $d \in D$ with $\bar{d} \in \bar{E}$, and set $E = d^{X_0}$ and $X = \langle E \rangle$. Arguing as in the proof of 4.1.19, $\bar{X}/O_2(\bar{X}) \cong S_3$ and $\mathcal{D}(E)$ is disconnected. Let $q = |\bar{d}^{O_2(\bar{X})}|$. Then

$$|d^{O_2(X)}| \geq q|d^Q| = q|V_d| \geq 2q,$$

so by 4.1.12 either $|V_d| = 4$ and $q = 1$ or $|V_d| = 2$ and $q \leq 2$. It follows that $\bar{X} \cong S_3$ or S_4, establishing (a). It also follows that

(b) Either $|V_d| = 2$, or $|V_d| = 4$ and $\bar{X} \cong S_3$.

By (a) and 4.1.21, $\bar{G} \cong S_n$ for some n, and as $|V_d| > 1$ and G is transitive on D, $n \geq 3$. As $O_2(\bar{G}) = 1$, $n \neq 4$. If $n = 3$ then $\mathcal{D}(D)$ is disconnected by 4.1.6. Then by (b), $|V_d| = 2^m$ for $m = 1$ or 2. In the first case, $G \cong S_4 \cong$ Weyl(D_3), and in the second the lemma follows from 4.1.10.

Therefore we may assume $n \geq 5$, so $|V_d| = 2$ by (b), establishing (2). As $\bar{G} \cong S_n$ there is a Coxeter system $(\bar{G}, \bar{\Delta})$ with $\bar{\Delta} = \{\bar{d}_1, \dots, \bar{d}_{n-1}\}$ of type A_{n-1} and $\bar{d}_i \in \bar{D}$. Let $\Gamma = \{d_1, \dots, d_n\}$ where $d_i \in D$ is a preimage of $\bar{d}_i$ for $1 \leq i < n$ and $\{d_{n-1}, d_n\} = V_{d_{n-1}}$. As D is a set of 3-transpositions of D, (G_0, Γ) is a Coxeter system of type D_n, where $G_0 = \langle \Gamma \rangle$. Further $\bar{G} = \bar{G}_0$ is transitive on $\bar{D}$ and

$S_4 \cong \langle d_{n-2}, d_{n-1}, d_n \rangle = G_1 \leq G_0$, so $d_n \in d_{n-1}^{G_0}$. Hence G_0 is transitive on D, so that $G = G_0$. This completes the proof of the lemma. □

LEMMA 4.1.23. *Assume Hypotheses 4.1.12 with $\mathcal{D}(D)$ disconnected. Then $G \cong S_3$, S_4, Weyl(D_4) or Weyl(D_4)/Z(Weyl(D_4)).*

PROOF. By 4.1.18.1, $G/O_2(G) \cong S_3$, so either $G \cong S_3$ or $O_2(G) \not\leq Z(G)$. In the latter case $|V_d| > 1$ by 4.1.7, so the lemma holds by 4.1.22. □

LEMMA 4.1.24. *Assume Hypotheses 4.1.12 with G transitive on D and $O_2(G) \leq Z(G) \geq O_3(G)$. Then $G/Z(G)$ is one of the following:*

(1) S_n for some $n \geq 5$.

(2) $O_6^-(2)$.

(3) $Sp_6(2)$.

(4) $O_8^+(2)$.

PROOF. Passing to $G/Z(G)$, we may assume $Z(G) = 1$. Thus by 9.4.1 in [**Asc97**], $F^*(G) = L$ is simple. Then by 4.1.18.1, $\mathcal{D}(D)$ is connected, so by 9.4.4 in [**Asc97**], $|G : L| \leq 2$. Now by Fischer's Theorem in the Introduction to [**Asc97**], G is in one of the six classes of groups listed there. We will examine each class.

Let $\mathcal{M}$ be the set of proper D-subgroups M of G such that M is transitive on $M \cap D$ and $O_2(M) \leq Z(M) \geq O_3(M)$. By 4.1.15 and induction on the order of G, each $M \in \mathcal{M}$ appears on the list of this lemma.

Suppose $G \cong O_n^\epsilon(2)$. As G is almost simple, either $n \geq 6$ or $G \cong O_4^-(2) \cong S_5$, and we may assume the former. Now $O_6^+(2) \cong S_8$ and $O_6^-(2)$ and $O_8^+(2)$ are on our list. Further if $n > 8$ then there is $M \in \mathcal{M}$ isomorphic to $O_8^-(2)$. Thus we may assume G is $O_8^-(2)$. Hence there is a maximal parabolic X of G which is the extension of $O_2(X)$ by $O_6^-(2)$ with $O_2(X)$ the natural module for $X/O_2(X)$. Then there is a D-subgroup Y of X such that $Y/O_2(Y) \cong S_3$ and $|d^{O_2(Y)}| = 8$ for $d \in Y \cap D$, contrary to 4.1.12. Therefore G is not $O_n^\epsilon(2)$.

Suppose $G \cong Sp_n(2)$. As $Sp_4(2) \cong S_6$ and $Sp_6(2)$ are on our list, we may assume $n \geq 8$. But now there is a member of $\mathcal{M}$ isomorphic to $O_8^-(2)$, a contradiction.

Suppose $G \cong U_n(2)$. As G is almost simple, $n \geq 4$, and if $n > 4$ there is a member of $\mathcal{M}$ isomorphic to $U_4(2)$, so we may take $n = 4$. Hence there is a D-subgroup $X = O_3(X)\langle d \rangle$ with $O_3(X) \cong 3^{1+2}$, contrary to 4.1.11.

Suppose $G \cong PO_n^{\mu,\pi}(3)$. As G is almost simple either $G \cong PO_4^{-,\pi}(3) \cong S_6$ or $n \geq 5$, and we may assume the latter. If $n = 5$ then G is $PO_5^{\pi,\pi}(3) \cong O_6^-(2)$ or $PO_5^{-\pi,\pi}(3) \cong U_4(2)$, so the lemma holds in this case as we've already eliminated $U_4(2)$. Therefore $n > 5$, so there is a member of $\mathcal{M}$ isomorphic to $PO_5^{-\pi,\pi}(3)$, for our usual contradiction.

This leaves the case in Fischer's Theorem where G is one of the three Fischer groups. But then there is $M \in \mathcal{M}$ with $M/Z(M) \cong U_6(2)$, for our final contradiction. □

LEMMA 4.1.25. *Assume Hypotheses 4.1.12 with G transitive on D. Then one of the following holds:*

(1) $G \cong S_3$.

(2) $|V_d| > 1$ for $d \in D$ and $G \cong$ Weyl(D_n) or Weyl(D_n)/Z(Weyl(D_n) for some $n \geq 3$.

(3) $O_2(G) \leq Z(G) \geq O_3(G)$ *and* $G/Z(G)$ *is isomorphic to* S_n *for some* $n \geq 5$, $O_6^-(2)$, $Sp_6(2)$, *or* $O_8^+(2)$.

PROOF. If $O_3(G) \not\leq Z(G)$ then (1) holds by 4.1.18.2. If $O_2(G) \not\leq Z(G)$ then $d \in D$ does not centralize $O_2(G)$, so $|V_d| > 1$ by 4.1.7.1, and hence (2) holds by 4.1.22. This leaves the case where $O_2(G) \leq Z(G) \geq O_3(G)$, where (3) holds by 4.1.24. □

4.2. The groups in $M(\tau)$

In this section we assume the following hypothesis:

HYPOTHESIS 4.2.1. *(1)* $\tau = (\mathcal{F}, \Omega)$ *is a quaternion fusion packet.*
(2) $\eta \in \eta(\tau)$ *and* $W = \langle\eta\rangle \trianglelefteq \mathcal{F}$.
(3) $\mathcal{F} = \mathcal{F}^\circ$.

LEMMA 4.2.2. *(1) For* $K \in \Omega$, $\mathcal{O}_K = \mathcal{F}_K(K)$ *and* $Aut_{\mathcal{F}}(K)$ *is a 2-group.*
(2) $\mathcal{F}$ *is constrained.*
(3) $\mathcal{Z} = \mathcal{Z}_\Delta$.

PROOF. Parts (1) and (2) follow from 3.3.9. By definition of $\mathcal{Z}$, for each $t \in \mathcal{Z}$ there exists $\phi \in \hom_{\mathcal{F}}(t, S)$ and $K \in \Omega$ with $t\phi = z(K)$. Hence as $W \trianglelefteq \mathcal{F}$ by 4.2.1.2, $t \in W$. Now (3) follows from 3.1.11.7. □

NOTATION 4.2.3. Let $\mu = \mu(\tau)$. By 4.2.2.2 there exists a model G for $\mathcal{F}$, and as $W \trianglelefteq \mathcal{F}$ also $W \trianglelefteq G$. Set $G^* = G/C_G(W)$ and define $D = D_\eta$ as in Definition 3.1.19; thus $D = \{d_V : V \in \eta\} \subseteq G^*$. Set $\Lambda = \{K^g : K \in \Omega,\ g \in G\}$.

LEMMA 4.2.4. *(1)* $G = \langle O(\tau)^G\rangle = \langle\Lambda\rangle$.
(2) For each $V \in \eta$ *there exists a unique* $K(V) \in \Lambda$ *containing* V. *Moreover the map* $L \mapsto L \cap W$ *is a* G*-equivariant bijection of* Λ *with* η *with inverse* $V \mapsto K(V)$.
(3) For $V \in \eta$, $\langle d_V\rangle = K(V)^*$.

PROOF. Part (1) follows from 3.3.9.3. Let $J \in \Omega$. By definition of η, $U = J \cap W \in \eta$, and by (QFP2), J is the unique member of Ω containing U. Set $z = z(J)$. By 4.2.2.1 and 2.6.11, $O(z) \trianglelefteq C_G(z)$, so $J \trianglelefteq N_G(U)$ and hence J is the unique member of Λ containing U. Then (2) follows. By Definition 3.1.19, $d_U = j^*$ for $j \in J - U$, so (3) follows from (2). □

LEMMA 4.2.5. *Let* $K \in \Omega$, $V = K \cap W$, *and* $U \in \eta$ *with* $\langle d_U, d_V\rangle \cong S_3$. *Set* $M = \langle K, K(U)\rangle$. *Then*
(1) Choose a generator v *for* V. *Then we can choose a generator* u *for* U *with* $[v, d_U] = u$ *and* $[u, d_V] = v$.
(2) $C_M(UV) = UV = U \times V = M \cap W$ *and* $M/UV \cong M^* = \langle d_U, d_V\rangle \cong S_3$.
(3) $K^M = \{K, K(U), K(U')\}$, *where* $U' = \langle uv\rangle$.
(4) M *splits over* UV *and is determined up to isomorphism.*
(5) For each $k \in K - V$ *and* $j \in K(U) - U$, $|kj| = 3$.

PROOF. Let t, z be the involutions in U, V, respectively. First, the hypothesis of 3.1.16 is satisfied with $V = V_1$, $U = V_2$, and $A(U,V) \cong S_3$. Therefore (1) follows from 3.1.18. By (1), $v^{d_U} = vu$ and $u^{d_V} = uv$, so $UV \triangleleft M$. As U does not centralize K, $z \neq t$, so $UV = U \times V$. Then $M = \langle k, j, UV \rangle$ with $k \in K - V$, $j \in K(U) - U$, and $k^2, j^2 \in UV$.

Next $v^{d_U d_V} = (vu)^{d_V} = v^{-1}uv = u$, so (3) follows. Set $M^+ = M/UV$. By (3), $K^M = K(U)^M$ is of order 3, so $k^{+M} = j^{+M}$ is of order 3, and hence $M^+ \cong S_3$. Hence (2) follows.

Let $Y \in Syl_3(M)$. Now $X = \Omega_1(UV) = \langle z, t \rangle$ and $C_X(Y) = 1$, so $C_{UV}(Y) = 1$. Therefore $H = N_M(Y)$ is a complement to UV in M. Let a, b be the involutions in H centralizing z, t, respectively. Then $a^* = d_V$ and $b^* = d_U$, so $[u, a] = v$ and $[v, b] = u$ by (1). This gives a presentation for M, so M is determined up to isomorphism and (4) holds.

Indeed $H = \langle a, b \rangle \cong S_3$ with $C_{UV}(H) = 1$, so $(m/2)^2 = |U||V| = |UV| = |(a,b)^{UV}|$. But for $c \in \{a, b\}$, $\langle c \rangle UV \cong \mathbf{Z}_{m/2}$ wr $\mathbf{Z}_2$, so UV is transitive on the set $I(c)$ of involutions in cUV with $|I(c)| = m/2$. Thus $P = I(a) \times I(b)$ is of order $(m/2)^2 = |(a,b)^{UV}|$, so $P = (a,b)^{UV}$. Hence for each $(a_1, b_1) \in P$, $|a_1 b_1| = 3$. Next $I(a) = aV$ as V is the subgroup of UV inverted by a, and similarly $I(b) = bU$. As $k^* = d_V = a^*$, $k = a_1 u_1$ for some $a_1 \in I(a)$ and $u_1 \in U$. Then as $z = k^2$, we have $u_1 = t$. Similarly $j = b_1 z$ for some $(a_1, b_1) \in P$. Hence $kj = a_1 t b_1 z = ys$ where $y = a_1 b_1$ is of order 3 and $s = tz$. Therefore $(kj)^3 = (ys)^3 = s^{y^2} s^y s = tzs = 1$, proving (5). □

LEMMA 4.2.6. *Let $K \in \Omega$, $V = K \cap W$, $z = z(K)$, and assume $t \in \mathcal{A}(z)$. Then*

(1) $\Omega(z) = \{K, \tilde{K}\}$ for some $\tilde{K}$.

(2) $t \in z^G$ is contained in $K\tilde{K}$ and $\tilde{K} \notin K^G$. Let $t \in \tilde{J} \in \tilde{K}^G$ and $U = \tilde{J} \cap W$; then $\langle d_U, d_V \rangle \cong D_{12}$.

(3) Set $M = \langle K, \tilde{J} \rangle$. Then $C_M(UV) = M \cap W = UV = U \times V = U \times (\tilde{K} \cap W)$ and $M/UV \cong M^* = \langle d_U, d_V \rangle \cong D_{12}$.

(4) Let $Y \in Syl_3(M)$. Then $N_M(Y)$ is a complement to UV in M and the involution in $Z(N_M(Y))$ inverts UV.

(5) M is determined up to isomorphism.

PROOF. Part (1) follows from 3.2.2.1 and $t \in K\tilde{K}$ by 3.2.2.2. By definition of $\mathcal{A}(z)$ in Definition 3.2.1, there is $U \in \eta$ with $t \in U$ such that Hypothesis 3.1.16 is satisfied with $V_1 = V$, $V_2 = U$, and $A(V, U) \cong D_{12}$. Then by 3.1.17.3, $t \in z^G$, so there are two members $J, \tilde{J}$ of Λ containing t, and as $A(V,U) \cong D_{12}$ we can choose notation so that $U = \tilde{J} \cap W$ is conjugate to $\tilde{V} = \tilde{K} \cap W$ and $\bar{U} = J \cap W$ is conjugate to V under $A(V,U)$. In particular (2) holds.

Next $\langle d_V, d_{\bar{U}} \rangle \cong S_3$, so $\bar{M} = \langle K, J \rangle$ is described in 4.2.5. In particular $\bar{M} \cap W = \bar{U} \times V$, $N_{\bar{M}}(Y) \cong S_3$ is a complement to $\bar{U}V$ in $\bar{M}$, and $\bar{M}$ is determined up to isomorphism.

By 3.1.12, $[K, W] \leq K \cap W = V$ and then also $[\tilde{J}, W] \leq U$, so M acts on UV. Then as $\bar{U} \in V^{\bar{M}} \subseteq V^M$, $\bar{U} \times V \leq UV$, so $\bar{U} \times V = UV = U \times V$. Set $M^+ = M/UV$ and $B = \langle k, j \rangle$, where $k \in K - V$ and $j \in \tilde{J} - U$. By 3.1.23.1, $K^M \cup \tilde{K}^M$ is of order 6, so $B^+ \cong D_{12}$ and hence $UV = C_M(UV)$, so (3) holds. Then as $C_{UV}(Y) = C_{\bar{U}V}(Y) = 1$, $N_M(Y)$ is a complement to UV in M. Let c be the involution in $Z(N_M(Y))$ and $\tilde{M} = \langle \tilde{K}, \tilde{J} \rangle$. Then $c = ad$ where $a \in N_{\bar{M}}(Y)$ and $d \in N_{\tilde{M}}(Y)$ are involutions centralizing z. Then $a \in KW$ inverts V and $d \in \tilde{K}W$

centralizes V, so c inverts V. Then as $\bar{U} \in V^{N_{\bar{M}}(Y)}$, c inverts $\bar{U}V = UV$, completing the proof of (4). Moreover as $\bar{M}$ is determined up to isomorphism and c centralizes $N_{\bar{M}}(Y)$ and inverts UV, (5) follows. □

DEFINITION 4.2.7. Define $\mathcal{D}(D)$ to be the commuting graph on D; that is the edges in $\mathcal{D}(D)$ are the pairs (d_U, d_V) such that $\langle d_U, d_V\rangle \cong E_4$.

A *bimorphism* $\phi : \mathcal{G} \to \mathcal{H}$ of graphs is a map of vertices such that for each pair of vertices u, v of $\mathcal{G}$ with $u\phi \neq v\phi$, we have u adjacent to v if and only if $u\phi$ is adjacent to $v\phi$.

LEMMA 4.2.8. *(1) The map $\delta : V \mapsto d_V$ is a G-equivariant bijection of η and D.*

(2) For $V \in \eta$ let $z(V)$ be the involution in V. Then the map $\zeta : d_V \mapsto z(V)$ is a G-equivariant surjection of D onto $\mathcal{Z}$.

(3) The fiber $\zeta^{-1}(z)$ for $z \in \mathcal{Z}$ is $\{d_V : z = z(V)\}$ of order $|\Omega(z^g)|$ for $g \in G$ with $z^g \in \mathcal{Z}_S$. In particular $\langle\zeta^{-1}(z)\rangle$ is elementary abelian.

(4) $\zeta : \mathcal{D}(D) \to \mathcal{D}(\tau)$ is a bimorphism of graphs.

(5) $\mathcal{D}(D)$ is connected iff $\mathcal{D}(\tau)$ is connected.

(6) $|\Omega| = |D \cap Aut_S(W)|$ is the width of μ.

(7) If μ is transitive on D then $\mathcal{F}^\circ$ is transitive on Ω.

PROOF. Part (1) follows from 3.1.20.6. Let $\rho : \eta \to \mathcal{Z}$ be the map $\rho : V \mapsto z(V)$. Then ρ is a G-equivariant surjection, so as $\zeta = \delta^{-1}\rho$, (1) implies (2). Part (3) is trivial.

Let $K \in \Omega$, $z = z(K)$, $V = K \cap W$, $V \neq U \in \eta$, and $t = z(U)$. Then $[d_U, d_V] = 1$ iff $[d_V, U] = 1$ by 3.1.22, so as $K^* = \langle d_V\rangle$, it follows that $[d_U, d_V] = 1$ iff K centralizes U iff $t \in z^\perp$ by 3.3.4. Thus (4) holds. Then (5) follows from (3) and (4) and (1) implies (6).

It remains to prove (7), where by 3.1.27 we may assume $\mathcal{F} = \mathcal{F}^\circ$. Assume μ is transitive on D and let $K, J \in \Omega$, $V = K \cap W$, and $U = J \cap W$. As μ is transitive on D, $d_U \in d_V^\mu$, so by (1) there is $g \in G$ with $d_U = d_{V^g}$. Then by 4.2.4.2, $K^g = J$, establishing (7). □

DEFINITION 4.2.9. For $V \in \eta$, we define a subset $\eta(V)$ of η as follows:

(1) If $\mathcal{A}(z(V)) \neq \emptyset$ set $\eta(V) = V^G \cup \tilde{V}^G$, where $\{V, \tilde{V}\} = \{U \in \eta : z(U) = z(V)\}$.

(2) If $\mathcal{A}(z(V)) = \emptyset$, set $\eta(V) = V^G$.

In either case set $G(V) = \langle K(U) : U \in \eta(V)\rangle$ and $D(V) = \{d_U : U \in \eta(V)\}$.

LEMMA 4.2.10. *(1) $\{\eta(V) : V \in \eta\}$ is a partition of η.*

(2) G is the central product of the subgroups $G(V)$, $V \in \eta$.

(3) If $\mathcal{A}(z(V)) = \emptyset$ then $G(V)$ is transitive on $\eta(V)$ and $D(V)$ is a conjugacy class of 3-transpositions of $G(V)^*$.

(4) If $\mathcal{A}(z(V)) \neq \emptyset$ then $G(V)^* \cong D_{12}$ has two orbits $V^{G(V)}$ and $\tilde{V}^{G(V)}$ on $\eta(V)$ and $\{z(U) : U \in \eta(V)\} = \{z(V)\} \cup \mathcal{A}(z(V))$ is of order 3.

PROOF. By 3.1.25 there is a partition $\eta = \eta_0 \coprod \eta_1 \coprod \cdots \coprod \eta_r$ of η such that, setting $D_i = \{d_V : V \in \eta_i\}$ and $G_i^* = \langle D_i\rangle$, we have $G^* = G_0^* \times G_1^* \times \cdots \times G_r^*$; for $i \neq j$, G_i^* centralizes η_j; D_0 is a set of 3-transpositions of G_0^*; and for $i > 0$, $G_i^* \cong D_{12}$. Indeed when $i > 0$, $\{z(V) : V \in \eta_i\} = \{z(V)\} \cup \mathcal{A}(z(V))$ is of order 3 by 3.2.4. In particular η_0 is partitioned by the sets $V^G = V^{G_0}$ for $V \in \eta_0$, while for $i > 0$ and $V \in \eta_i$, we have $\eta_i = \eta(V)$. Therefore (1) follows.

Let $E_1, \ldots, E_s$ be the orbits of G_0^* on D_0 and $H_i^* = \langle E_i \rangle$. By 4.1.4, $G_0^* = H_1^* \cdots H_s^*$ is a central product and H_i^* is transitive on E_i. Set $\xi_i = \{V : d_V \in E_i\}$; then $\xi_1 \coprod \cdots \coprod \xi_s$ is a partition of η_0, and by 4.2.9, for $V \in \xi_i$, $\xi_i = V^{H_i} = \eta(V)$ and $G(V)^* = H_i^*$, so $G(V)$ is transitive on $\eta(V)$. Similarly we saw earlier that for $j > 0$ and $V \in \eta_j$, $\eta_j = \eta(V)$ and $G(V)^* = G_j^*$. This completes the proof of (3) and (4).

Next for $V_i \in \eta$, $i = 1, 2$, such that $\eta(V_1) \neq \eta(V_2)$, $[d_{V_1}, d_{V_2}] = 1$, so d_{V_i} centralizes V_{3-i} by 3.1.22; hence $K(V_i)$ centralizes V_{3-i}. Then as $K = K(V_{3-i})$ is the unique member of Λ containing V_{3-i}, $K(V_i)$ acts on K. Therefore $G(V_i)$ acts on K. We may choose $K(V_j) \in \Omega$ for $j = 1, 2$, so $[K(V_i), K] = 1$ by 3.1.5.1. Then as

$$G(V_j) = \langle K(V)^{G(V_j)} : V \in \eta(V_j) \text{ and } K(V) \in \Omega \rangle,$$

it follows that $G(V_i)$ centralizes K, and then also $G(V_i)$ centralizes $G(V_{3-i})$. Now (2) follows from 4.2.4.1, completing the proof. □

REMARK 4.2.11. Because of 4.2.10, in many situations we may assume either
(1) $\mu \cong D_{12}$ and $\mathcal{A}(z) \neq \emptyset$ for each $z \in \mathcal{Z}$, so that G is described in 4.2.6, or
(2) G is transitive on η and D is a conjugacy class of 3-transpositions of μ.

LEMMA 4.2.12. *(1) Suppose $a_i \in D$ for $i = 1, 2$ such that $\langle a_1, a_2 \rangle = \nu \cong S_3$. Then for each $d \in D$, d commutes with some member of $D \cap \nu$.*
(2) If D is a set of 3-transpositions of μ then μ satisfies Hypothesis 4.1.11.

PROOF. Assume the setup of (1) and let $z_i = \zeta(a_i)$ and $F = \langle z_1, z_2 \rangle$. From 4.2.5, $D \cap \nu = a_1^\nu$ is of order 3 and $F^\# = \{\zeta(d) : d \in D \cap \nu\}$. Let $Z = \langle \mathcal{Z} \rangle$. By 3.1.20.3, $[Z, d] \leq \langle \zeta(d) \rangle$, so d induces a transvection on Z, and hence there exists $f \in C_F(d)$. Then by 4.2.8.1, d centralizes the member of $D \cap \nu$ in $\zeta^{-1}(f)$, proving (1).

Part (2) follows from (1) and 4.1.14. □

LEMMA 4.2.13. *Assume D is a set of 3-transpositions of μ and $\mathcal{D}(D)$ is disconnected. Let $d \in D$. Then G is transitive on η, μ is transitive on D, and one of the following holds:*
(1) $\mu \cong S_3$ and $|V_d| = 1$.
(2) $\mu \cong S_4$ and $|V_d| = 2$.
(3) μ is isomorphic to Weyl(D_4) or Weyl(D_4)/Z(Weyl(D_4)) and $|V_d| = 4$.

PROOF. As $\mathcal{D}(D)$ is disconnected it follows from 4.2.10.2 that $\eta = \eta(U)$ for each $U \in \eta$, and then, from 4.2.10.3 that D is a conjugacy class of 3-transpositions of μ. Hence G is transitive on η by 4.2.8.1.

By 4.2.12.2 and 4.1.18, $\mu/O_2(\mu) \cong S_3$. Let $a = d_V$ and set $2^{n-1} = |V_a|$. By 4.1.8, $\mu = \langle d_i : 0 \leq i \leq n \rangle$, where $a = d_1$, $d_i \in a^\perp = V_a$ for $1 \leq i \leq n$, and $d_0 \in A_a$. Let $V_i \in \eta$ with $d_i = d_{V_i}$. By 3.1.20, $[W, d_i] \leq V_i$, so G acts on $V_0 V_1 \cdots V_n$. Then as G is transitive on η, $W = V_0 V_1 \cdots V_n$.

Let $U_i = \langle u_i \rangle$ be the subgroup of V_i of order 4, $A = V_1 \cdots V_n$, $B = U_1 \cdots U_n$, $z_i = z(V_i)$, $Z = \langle z_i : 1 \leq i \leq n \rangle$ and $A^+ = A/Z$. If $n \leq 2$ then $\mu \cong S_3$ or S_4, using 4.1.5, and (1) or (2) holds. Thus we may assume $n > 2$. Therefore as $|a^\perp| = 2^{n-1}$, there exists $X \in \eta$ with $d = d_X \in a^\perp - \{a_i : 1 \leq i \leq n\}$. Then $K(X)$ centralizes A but not $u \in X$ of order 4. As $W = V_0 \cdots V_n$, $u = by$ with $b \in A$ and $y \in V_0$. Now

$z = u^2 = b^2y^2$, so $y^2 = b^{-2}z \in Az \cap V_0$. But $K(X)$ centralizes $\langle z, A\rangle$ but not z_0, so $y^2 = 1$. Thus $y = z_0^\epsilon$ and $b^2 = z$. Next $[A, K(X)] = 1 \neq [u, K(X)]$, so $y = z_0$. As $K(V_i)$ centralizes V_j for $i \neq j$ and $1 \leq i, j \leq n$, $A^+ = V_1^+ \times \cdots \times V_n^+$, so as $b^2 = z$, $b = u_1 \cdots u_k e$ for some u_i, some $k \leq n$, and some $e \in Z$. Thus $u = z_0 u_1 \cdots u_k e$.

Let $r \in a^\perp - \{d\}$. Then $u = u^r = \zeta(r) z_0 u_1^r \cdots u_k^r e = u$ or $u\zeta(r)$, depending on whether $r = d_i$ for some $1 \leq i \leq k$ or not. It follows that $n = 3$, so that $|V_a| = 2^{n-1} = 4$. Moreover by 4.1.10, μ is $\omega = \mathrm{Weyl}(D_4)$ or $\omega/Z(\omega)$, and hence (3) holds, completing the proof. □

LEMMA 4.2.14. *If D is a set of 3-transpositions of μ then μ satisfies Hypothesis* 4.1.12.

PROOF. By 4.2.12.2, μ satisfies Hypothesis 4.1.11. Then the lemma follows from 4.2.13 and 4.1.7.1. □

4.3. The groups $\bar{\omega}(\Phi, m)$

NOTATION 4.3.1. Let Φ be a connected Coxeter diagram with no multiple bonds. Thus Φ is A_n, D_n, or E_n. Take $n \geq 2$ and if Φ is D_n take $n \geq 3$. Set $I = \{1, \ldots, n\}$ and write $\mathrm{Weyl}(\Phi)$ for the Coxeter group defined by Φ. Let $m \geq 8$ be a power of 2.

Define $\bar{G} = \bar{\omega}(\Phi, m)$ to be the group presented by generators $\bar{k}_i, \bar{v}_i$, $i \in I$, subject to the following relations:

(1) $\bar{v}_i^{m/2} = 1 = [\bar{v}_i, \bar{v}_j]$, $\bar{k}_i^2 = \bar{v}_i^{m/4}$, $\bar{v}_i^{-1} = \bar{v}_i^{\bar{k}_i}$ for all $i, j \in I$.

(2) If (i, j) is an edge in Φ then $(\bar{k}_i \bar{k}_j)^3 = 1$ and $[\bar{v}_i, \bar{k}_j] = \bar{v}_j$.

(3) If $i, j \in I$ are distinct and (i, j) is not an edge in Φ then $[\bar{k}_i, \bar{k}_j] = 1 = [\bar{k}_i, \bar{v}_j]$.

Set $\bar{W} = \langle \bar{v}_i : i \in I\rangle$ and for $i \in I$, set $\bar{V}_i = \langle \bar{v}_i\rangle$, and $\bar{K}_i = \langle \bar{k}_i, \bar{v}_i\rangle$.

From 4.3.8.2, $\bar{K}_i \cong Q_m$; set $\bar{z}_i = z(\bar{K}_i)$. In 5.8.2 we find that if $\Phi = D_n$ then $\bar{u} = \bar{z}_{n-1}\bar{z}_n \in Z(\bar{G})$; then we define $\omega(D_n, m) = \bar{G}/\langle \bar{u}\rangle$.

LEMMA 4.3.2. *Let $(\mathfrak{W}, \mathfrak{S})$ be a Coxeter system of type Φ.*

(1) $\bar{W}$ is an abelian normal subgroup of $\bar{G}$.

(2) Set $\bar{G}^* = \bar{G}/\bar{W}$ and $\mathfrak{S} = (s_i : i \in I)$. Then the map $s_i \mapsto \bar{k}_i^*$ extends to a surjective homomorphism of $\mathfrak{W}$ onto $\bar{G}^*$.

PROOF. From 4.3.1.1, $[\bar{v}_i, \bar{v}_j] = 1$ for all $i, j \in I$, so $\bar{W}$ is abelian. By 4.3.1.1, $\bar{v}_i^{-2} = [\bar{v}_i, \bar{k}_i] \in \bar{W}$ and for $i \neq j$ in I, $[\bar{v}_i, \bar{k}_j] = 1$ or $\bar{v}_j$, so $[\bar{v}_i, \bar{k}_j]$ is in $\bar{W}$, and hence $\bar{W} \trianglelefteq \bar{G}$, completing the proof of (1).

As $\bar{k}_i^2 = \bar{v}_i^{m/4} \in \bar{W}$ we have $\bar{k}_i^{*2} = 1$. If (i, j) is an edge in Φ then $(\bar{k}_i \bar{k}_j)^3 = 1$, so $(\bar{k}_i^* \bar{k}_j^*)^3 = 1$. If (i, j) is not an edge in Φ then $[\bar{k}_i, \bar{k}_j] = 1$, so $[\bar{k}_i^*, \bar{k}_j^*] = 1$. Therefore $(\bar{k}_i^* : i \in I)$ satisfies the defining relations for $\mathfrak{W}$, so (2) follows. □

LEMMA 4.3.3. *Assume Hypothesis* 4.2.1 *and adopt Notation* 4.2.3. *Let $(\mathfrak{W}, \mathfrak{S})$ be a Coxeter system of type Φ with $\mathfrak{S} = (s_i : i \in I)$ and assume there exists a surjective homomorphism $\alpha : \mathfrak{W} \to G^* = G/C_G(W)$ with $\mathfrak{S}\alpha \subseteq D$ and* $\ker(\alpha) \leq Z(\mathfrak{W})$. *For $i \in I$ set $s_i\alpha = d_i = d_{V_i}$ for $V_i \in \eta$, $K_i = K(V_i)$, and $k_i \in K_i - V_i$.*

(1) $G = \langle K_i : i \in I\rangle$ and $W = \langle V_i : i \in I\rangle$.

(2) We can choose generators v_i for V_i such that the maps $\bar{v}_i \mapsto v_i$ and $\bar{k}_i \mapsto k_i$, $i \in I$, extend to a surjective homomorphism $\pi : \bar{G} \to G$ with $\pi : \bar{K}_i \to K_i$ an isomorphism.

(3) $\bar{W}\pi = W$ *and if* $|W| \geq (m/2)^n$ *then* $\pi : \bar{W} \to W$ *is an isomorphism and* $\bar{W} = \bar{V}_1 \times \cdots \times \bar{V}_n$.

(4) If α *is an isomorphism then* π *induces an isomorphism* $\pi^* : \bar{G}^* = \bar{G}/\bar{W} \to G/W$ *via* $\pi^* : \bar{g}\bar{W} \mapsto \bar{g}\pi W$, $C_G(W) = W$, *and* $\bar{G}^* \cong \mathfrak{W}$.

(5) If α *is an isomorphism and* $|W| \geq (m/2)^n$ *then* $\pi : \bar{G} \to G$ *is an isomorphism, so* $G \cong \bar{\omega}(\Phi, m)$ *with* $W = C_G(W) \cong \mathbf{Z}_{m/2}^n$ *and* $G/W \cong$ *Weyl*(Φ).

PROOF. As $\alpha : \mathfrak{W} \to G^*$ is a surjection and $\mathfrak{W} = \langle \mathfrak{S} \rangle$, we have $G^* = \langle d_i : i \in I \rangle$. Therefore as $d_i = k_i^*$, it follows that $G = G_0 C_G(W)$, where $G_0 = \langle K_i : i \in I \rangle$.

As $\mathfrak{W}$ is transitive on its set $\mathfrak{R}$ of reflections, we have $\mathfrak{R} = s_1^{\mathfrak{W}}$, so $\mathfrak{R}\alpha = D$ and G^* is transitive on D. Therefore as $G = G_0 C_G(W)$, we conclude from 4.2.8.1 that G_0 is transitive on η. Then as $V_i \leq K_i \leq G_0$, also $W = \langle \eta \rangle \leq G_0$. Set $H = C_G(W)$. Then $[K_i, H] \leq C_{K_i}(V_i) = V_i \leq W$, so G_0 centralizes H/W. Hence as $G = G_0 H$ with $W \leq G_0$, we have $G_0 \trianglelefteq G$. Thus as $G = \langle K^G \rangle$ by 4.2.4.1, we conclude that $G = G_0$.

Pick generators v_i of V_i. As $K_i \cong Q_m$ and W is abelian, $(k_i, v_i : i \in I)$ satisfies the relations in 4.3.1.1. If (i, j) is not an edge in Φ, then by 3.1.22, $[K_i, K_j] = 1$, so the relations in 4.3.1.3 are satisfied. Finally as the graph of Φ is a tree, we conclude from parts (1) and (5) of 4.2.5 that we can choose the v_i so that the relations in 4.3.1.2 are satisfied. This proves (2).

Next $\bar{W}\pi = \langle \bar{V}_i \pi : i \in I \rangle = \langle V_i : i \in I \rangle \leq W$. As $\bar{W} \trianglelefteq \bar{G}$ by 4.3.2.1 and as $\pi : \bar{G} \to G$ is a surjection, we have $\bar{W}\pi \trianglelefteq G$. Then as $V_1 \leq \bar{W}\pi$ and G is transitive on η, we conclude that $W = \langle \eta \rangle \leq \bar{W}\pi$, so $\bar{W}\pi = W$, completing the proof of (1).

Further as $\bar{W} = \langle \bar{V}_i : i \in I \rangle$ with $|\bar{V}_i| \leq m/2$, and as $\bar{W}$ is abelian by 4.3.2.1, it follows that $|\bar{W}| \leq (m/2)^n$ with $\bar{W} = \bar{V}_1 \times \cdots \times \bar{V}_n$ in case of equality. Then (3) follows.

Next as $\bar{W}\pi = W$, π induces the homomorphism $\pi^* : \bar{G}^* \to G/W$ defined by $(\bar{g}\bar{W})\pi^* = (\bar{g}\pi)W$. Let $\psi : G/W \to G^*$ be the natural map; then $\varphi = \pi^*\psi : \bar{G}^* \to G^*$ is a surjective homomorphism.

Suppose that α is an isomorphism. Then $G^* \cong \mathfrak{W}$, so as φ is a surjection, $|\bar{G}^*| \geq |\mathfrak{W}|$. But by 4.3.2.2, $|\mathfrak{W}| \geq |\bar{G}^*|$, so all inequalities are equalities, and hence in particular $W = C_G(W)$ and $\pi^* : \bar{G}^* \to G^*$ is an isomorphism. Therefore (4) holds. Finally (3) and (4) imply (5). □

REMARK 4.3.4. Let (X, Γ) be a Coxeter system of type D_n with $n \geq 3$; say $\Gamma = \{s_1, \ldots, s_n\}$. Set $x_0 = s_{n-1}s_n$, $X_0 = \langle x_0^X \rangle$, $\Gamma_0 = \{s_1, \ldots, s_{n-1}\}$, and $L = \langle \Gamma_0 \rangle$. Then (L, Γ_0) is a Coxeter system of type A_{n-1}, so $L \cong S_n$ is a complement to X_0 in X, X_0 is the core of the permutation module for L (ie. the vertices of even weight in the permutation module) on an n-set, and x_0 is a vector of weight 2 in the module X_0.

LEMMA 4.3.5. *Assume Hypothesis* 4.2.1 *and adopt Notation* 4.2.3. *Assume* D *is a conjugacy class of 3-transpositions in* μ *and* $O_2(\mu) \not\leq Z(\mu)$. *Then*

(1) $\mu \cong$ *Weyl*(D_n) *for some* $n \geq 3$.

(2) If n *is even then* $Z(\mu) = \langle \iota \rangle \cong \mathbf{Z}_2$ *and* ι *inverts* W.

PROOF. By 4.2.14, μ, D satisfies Hypothesis 4.1.12, so as $O_2(\mu) \not\leq Z(\mu)$ and μ is transitive on D, we conclude from 4.1.22 that there exists an integer $n \geq 3$ such that μ is $\mathfrak{W}$ or $\mathfrak{W}/Z(\mathfrak{W})$, where $\mathfrak{W} = \mathrm{Weyl}(D_n)$. Thus in proving (1) we may assume $Z(\mathfrak{W}) \neq 1$ and $\mu \cong \mathfrak{W}/Z(\mathfrak{W})$, and it remains to derive a contradiction. As $Z(\mathfrak{W}) \neq 1$ it follows that n is even. On the other hand in proving (2), n is even by hypothesis. So, in any event, n is even and there exists a Coxeter system $(\mathfrak{W}, \mathfrak{S})$ and a surjection $\alpha : \mathfrak{W} \to \mu$ with $\mathfrak{S}\alpha \subseteq D$ and $\ker(\alpha) \leq Z(\mathfrak{W})$. As n is even, $Z(\mathfrak{W}) = \langle \iota \rangle$ is of order 2. Observe that the hypotheses of 4.3.3 are satisfied, and adopt the notation of that lemma. In particular $s_i\alpha = d_i$ and $V_i = \langle v_i \rangle \leq W$.

Let $x = s_{n-1}s_n$, $Q = \langle x^{\mathfrak{W}} \rangle$, and $L = \langle s_i : 1 \leq i < n \rangle \leq \mathfrak{W}$. From Remark 4.3.4, Q is the core of the permutation module for $L \cong S_n$. Thus, as in section 5.1, we can identify Q with the set E of vectors of even weight in the power set of I under the operation of symmetric difference, and $x = [Q, s_{n-1}] = e_{n-1,n}$. In particular for $s \in s_1^L$, s is a transposition (i, j) on I and $[Q, s] = e_{i,j}$, so $s^Q = \{s, s'\}$, where $ss' = e_{i,j}$.

Next $C_Q(L) = \langle \iota \rangle$, where $\iota = e_I = e_{1,2}e_{3,4} \cdots e_{n-1,n}$. For $i < n - 1$ odd, $e_{i,i+1} = s_is_i'$ centralizes V_n as $s_i\alpha = d_i$ and $s_i'\alpha = d_i'$ centralize d_n. Similarly $e_{n-1,n} = s_{n-1}s_n$ inverts V_n, as d_n inverts V_n and d_{n-1} centralizes V_n. Thus ι inverts V_n, so as $\iota \in Z(\mathfrak{W})$, ι inverts $W = \langle V_n^{\mu} \rangle$. This shows $\iota \neq 1$, proving (1), and completes the proof of (2). □

LEMMA 4.3.6. *Assume Hypothesis* 4.2.1 *and adopt Notation* 4.2.3. *Assume*

$$\mu/Z(\mu) \cong \mathit{Weyl}(\Phi)/Z(\mathit{Weyl}(\Phi))$$

for Φ of type E_7 or E_8. Then $\mu \cong \mathit{Weyl}(\Phi)$.

PROOF. Let $\mathfrak{W} = \mathrm{Weyl}(\Phi)$ and assume $\mu \cong \mathfrak{W}/Z(\mathfrak{W})$. Then $\mu \cong Sp_6(2)$ or $O_8^+(2)$. Hence $\prod_{d \in D \cap S} d = 1$, so for $d \in D \cap S$, $d \in \langle D_d \cap S \rangle$. But $d = d_V$ for some $V \in \eta$ and by 3.1.22, each $a \in D_d$ centralizes V, a contradiction as d inverts V. □

LEMMA 4.3.7. *(1) $\bar{G}/\bar{W} \cong \mathit{Weyl}(\Phi)$.*

(2) $\bar{W} = \bar{V}_1 \times \cdots \times \bar{V}_n = C_{\bar{G}}(\bar{W})$.

(3) Let $\bar{S} \in Syl_2(\bar{G})$, $\bar{\Omega} = \bar{K}_1^{\bar{G}} \cap \bar{S}$, $\bar{\mathcal{F}} = \mathcal{F}_{\bar{S}}(\bar{G})$, and $\bar{\tau} = (\bar{\mathcal{F}}, \bar{\Omega})$. Then $\bar{\tau}$ is a quaternion fusion packet.

(4) $\bar{\Omega}(z(\bar{K}_i)) = \{\bar{K}_i\}$.

(5) $\bar{G}$ is isomorphic to a member of $M(\tau)$, where τ is the fusion packet of $SL_{n+1}^+[m]$, $Spin_{2n}^+[m]$, $E_6^+[m]$, $\tilde{E}_7[m]$, $E_8[m]$, for Φ of type A_n, D_n, E_6, E_7, E_8, respectively.

PROOF. We first prove (5), and in the process establish (1) and (2). We exhibit a group L of Lie type over $\mathbf{F}_q$, where $(q^2 - 1)_2 = m$, such that, writing $\tau(L)$ for the Lie packet of L, $\mu = \mu(\tau(L)) \cong \mathrm{Weyl}(\Phi) = \mathfrak{W}$ and, for $W \in W(\tau)$, we have $|W| \geq (m/2)^n$. Then by 4.3.3.5, $\pi : \bar{G} \to G$ is an isomorphism for $G \in M(\tau(L))$, $W = C_G(W)$, and $G/W \cong \mathfrak{W}$.

Suppose that $\Phi = D_n$ and let $L = Spin_{2n}^+(q)$. Then by 5.3.4.4, $\mu \cong \mathfrak{W}$, while from the proof of 5.3.5, $|W| \geq (m/2)^n$.

Next suppose $\Phi = A_n$ and let $L = SL_{n+1}^{\pi}(q)$, where $q \equiv \pi \mod 4$. By 5.4.13.4, $\mu \cong S_{n+1} = \mathfrak{W}$, while from the proof of 5.4.13.5, $|W| \geq (m/2)^n$.

Finally suppose $\Phi = E_n$. Then $\mu \cong \mathfrak{W}$ by 5.5.8, while from the proof of 5.5.10, $|W| \geq (m/2)^n$. This completes the proof of (5). During the proof we showed that (1) and (2) hold.

As $\pi : \bar{G} \to G$ is an isomorphism with $\bar{K}_i\pi = K_i$, (3) holds. As L is a universal group, $\Omega(z(K_i)) = \{K_i\}$, so applying the isomorphism π, we conclude that (4) holds. □

LEMMA 4.3.8. *Assume Hypothesis 4.2.1 and adopt Notation 4.2.3. Assume D is a conjugacy class of 3-transpositions in μ. Then*

(1) $\mu \cong Weyl(\Phi)$, where Φ is of type A_n, D_n, or E_n.

(2) Let $\bar{G} = \bar{\omega}(\Phi, m)$. Then there is a surjective homomorphism $\pi : \bar{G} \to G$ with $\bar{W}\pi = W$, $\pi : \bar{K}_i \to K_i = \bar{K}_i\pi \in \Lambda$ is an isomorphism, and $\ker(\pi) \leq \bar{W}$.

(3) $\ker(\pi) \leq Z(\bar{G})$.

(4) $W = C_G(W)$.

PROOF. By 4.2.14, the pair μ, D satisfies Hypothesis 4.1.12. Hence by 4.1.25, $\mu = \mathfrak{W}/Z$ for some $Z \leq Z(\mathfrak{W})$, where $\mathfrak{W} = \text{Weyl}(\Phi)$ and Φ is of type A_n, D_n, or E_n. Claim $Z = 1$, so that (1) holds. If $Z(\mathfrak{W}) = 1$ this is trivial, so assume otherwise. Then $\Phi = D_n$ with n even, E_7, or E_8. Now 4.3.5.1 and 4.3.6 complete the proof of the claim. Thus (1) is established.

By (1) and 4.3.3 there is a surjection $\pi : \bar{G} \to G$ with $\bar{W}\pi = W$ and π induces isomorphisms $\bar{K}_i \to K_i$ and $\pi^* : \bar{G}/\bar{W} \to G/W = \mu$. This last fact says $\ker(\pi) \leq \bar{W}$, so (2) holds.

As $\pi : \bar{K}_i \to K_i$ is an isomorphism, $\bar{K}_i \cap \ker(\pi) = 1$. Therefore $\bar{K}_i$ centralizes $\ker(\pi)$ by 3.1.4, so $\bar{G} = \langle \bar{K}_i : i \in I \rangle$ centralizes $\ker(\pi)$, proving (3).

It remains to prove (4), so assume $W < C_G(W)$ and let Y be the preimage in $\bar{G}$ of $C_G(W)$. Then $Y \not\leq \bar{W}$, so $Y^* \neq 1$ and $[Y, \bar{W}] \leq \ker(\pi) \leq Z(\bar{G})$ by (3). Therefore as $\bar{W} = C_{\bar{G}}(\bar{W})$ by 4.3.7.2, $Y^* \leq O_2(\bar{G}^*)$, so in particular $O_2(\mathfrak{W}) \neq 1$. Hence $\Phi = D_n$, E_7 or E_8.

Suppose Φ is D_n. If n is even then $Z(\mathfrak{W}) = \langle \iota \rangle$ is the unique minimal normal subgroup of $\mathfrak{W}$ by Remark 4.3.4, so $\iota \in Y^*$, contradicting 4.3.5.2. Thus n is odd, so $\bar{G}^*$ is irreducible on $Q = O_2(\bar{G}^*)$ by Remark 4.3.4, and hence $Y^* = Q$. However $x = d_{n-1}d_n \in Q$ and x inverts V_n, a contradiction.

Therefore Φ is E_n for $n \in \{7, 8\}$, so $O_2(\bar{G}^*) = \langle \iota \rangle$ is of order 2, and hence $Y^* = \langle \iota \rangle$. As in the proof of 4.3.6, $\iota = \prod_{d \in D \cap S} d$, so as $D \cap S - \{d_1\} \subseteq D_{d_1} \subseteq C_D(v_1)$, we conclude that $\bar{v}_1^{\iota} = \bar{v}_1^{d_1} = v_1^{-1}$, a contradiction that completes the proof of (4). □

CHAPTER 5

Some examples

Chapter 5 is devoted to our examples τ of quaternion fusion packets appearing as conclusions in Theorems 1 and 2.

Section 5.1 considers the groups and 2-fusion systems AE_n and $\check{A}E_n$. These groups have an associated quaternion fusion packet τ by 5.1.6.3. The groups are essentially the Weyl groups of type D_n, obtained from their embedding in $\Omega_n^{\epsilon}(q)$ for suitable ϵ and q. The quaternion fusion packet of the group $\hat{A}E_n$, defined in 5.1.16 and 5.1.17, is universal among coverings of τ by 5.1.21.

In section 5.2 the 2-share of the order of groups of Lie type over a field $\mathbf{F}_q$ of odd order q is described from a point of view convenient for discussing their quaternion fusion packets.

Sections 5.3; 5.4; and 5.5 discuss the quaternion fusion packets of orthogonal and spin groups; linear, unitary, and symplectic groups; and exceptional groups, respectively.

Section 5.6 contains a proof that the 2-fusion system $\mathcal{F}_S(G)$ of a simple group G of Lie type and odd characteristic (which is not a Goldschmidt group) is simple. This is used in Theorem 5.6.18 to show that if G is a known simple group that is not Goldschmidt, then the 2-fusion system of G is simple. As a consequence, the known simple 2-fusion systems are the 2-fusion systems of the simple groups other than Goldschmidt groups, together with the exotic Solomon systems.

Section 5.7 considers the systems $L_n^{\pi}[m]$, their coverings, and their subsystems $\bar{\omega}(A_{n-1}, m)$. In particular 5.7.4 and 5.7.5 contain detailed information about $\bar{\omega}(A_3, m)$. Similarly section 5.8 contains information on $\bar{\omega}(D_n, m)$, particularly when $n = 4$.

Section 5.9 considers extensions $\bar{\omega}(C_n, m)$ and $2\bar{\omega}(C_n, m)$ of $\bar{\omega}(D_n, m)$, found in the Lie packets of $Spin_{2n+1}(q)$ and $Spin_{2n+2}^{-}(q)$ for suitable q. The factor group of these systems modulo $C_S(W)$ is the Weyl group of type C_n.

Finally section 5.10 studies the constrained systems $L_3(2)/E_8$, $L_3(2)/E_{64}$, and $L_3(2)/2^{3+6}$ and the coverings of such systems.

We close this chapter with a section listing the most basic notions, notation, and results from the theory of quaternion fusion packets that we've generated in the first five chapters of this manuscript. Hopefully this will provide the reader with a convenient place to refer when the need to refresh his or her memory arises.

5.1. AE_n

In this section $B = Sym(I)$ is the symmetric group on $I = \{1, \dots, n\}$ with $n \geq 5$, and V is the power set of I regarded as an $\mathbf{F}_2$-space via $X + Y = (X \cup Y) - (X \cap Y)$ the symmetric difference of subsets X and Y of I. We also write e_X for X. Observe $B \leq GL(V)$ and for $b \in B$, $e_X^b = e_{Xb}$. Indeed V is the n-dimensional permutation module for B over $\mathbf{F}_2$.

For $X \subset V$, write $Mov(X)$ for the set of points in I moved by X.

Next form the semidirect product $H = BV$ and observe that H is the wreath product $\mathbf{Z}_2$ wr S_n; in particular $Z(H) = \langle e_I \rangle$. Define $E = \langle e_{i,j} : i, j \in I \text{ distinct} \rangle$, $A = F^*(B)$, and $G = O^2(H)$. Then $G = AE$ is the split extension of $E \cong E_{2^{n-1}}$ by $A \cong A_n$. If n is odd then $V = E \oplus \langle e_I \rangle$ is a B-invariant decomposition of V, so G has no center. If n is even then $\langle e_I \rangle = Z(G)$.

In any case set $H^* = H/\langle e_I \rangle$ and $H^+ = H/V$. Denote the isomorphism type of G^* by AE_n, the isomorphism type of A^*V^* by AV_n, and the isomorphism type of G by $\breve{A}E_n$.

LEMMA 5.1.1. *Let X be the stabilizer in A of n. Then $XE \cong \mathbf{Z}_2$ wr A_{n-1} and $X^*E^* \cong \mathbf{Z}_2$ wr A_{n-1} or AV_{n-1} for n odd or even, respectively.*

PROOF. Set $v_k = e_{k,n}$ for $1 \leq k < n$. Then $\{v_k : 1 \leq k < n\}$ is a basis for E such that $v_k^x = v_{kx}$ for $x \in X$. Thus $XE \cong \mathbf{Z}_2$ wr A_{n-1}. Further $\sigma = \sum_k v_k = e_I + ne_n$, so $\sigma^* = 0$ iff n is even, completing the proof. □

LEMMA 5.1.2. *(1) $H^1(A, E^*) = 0$, $\mathbf{F}_2$ for n odd, even, respectively.*

(2) If n is even then $H^1(A, V) = 0$.

(3) If n is even and U is an indecomposable $\mathbf{F}_2A$-module of dimension n with $[U, A]$ A-isomorphic to E, then U is isomorphic to V as an A-module.

PROOF. See Exercise 6.3 in [**Asc86**] for (1), and the fact that V^* is the module $U(A, E^*)$ of section 17 of [**Asc86**]. Then by 17.3 in [**Asc86**], V^* is transitive on the complements to V^* in G^*, so V is transitive on the complements to V in AV, and hence (2) holds.

Assume the setup of (3) and let X be the stabilizer in A of n. By 5.1.1 and (1), U splits over $[U, A]$ as an X-module, so U is an image of the permutation module for A on A/X; then (3) follows. □

DEFINITION 5.1.3. Write $\hat{A} = \hat{A}_n$ for the universal 2-covering group of A; that is $\hat{A}$ is the largest perfect group Y such that $Z(Y)$ is a 2-group and $Y/Z(Y) \cong A$. Then (cf. 33.15 in [**Asc86**]), $Z(\hat{A}) = \langle \sigma \rangle$ is of order 2.

If n is even, write $\hat{G}$ for the semidirect product of $\hat{A}$ with E, such that $\sigma \in Z(\hat{G})$ and $\hat{G}/\langle \sigma \rangle \cong G$. As $\hat{A}$ is the universal 2-covering group of A, it follows from 5.1.2.1 that $\hat{G}$ is the largest 2-covering group Λ of AE_n such that $O_2(\Lambda)$ is elementary abelian.

LEMMA 5.1.4. *Assume U is a normal elementary abelian 2-subgroup of a finite group X such that $X/C_U(X) \cong AE_n$ and $U = [U, X]$. Then either $X \cong AE_n$ or n is even and $X \cong \breve{A}E_n$ or $\hat{G}/\langle \sigma e_I \rangle$.*

PROOF. As $U = [U, X]$ it follows from 5.1.2.1 that $Z = C_U(X) = 0$ if n is odd, while if n is even it follows from 5.1.3 that X is an image of $\hat{G}$. In particular if n is odd then the lemma holds, so we may take n even and $X = \hat{G}/\Sigma$ for some $\Sigma \leq Z(\hat{G}) = \langle \sigma, e_I \rangle$. Then $X \cong AE_n$, $\check{A}E_n$ for $\Sigma = Z(\hat{G})$, $\langle \sigma \rangle$, respectively, while Σ is not contained in $\langle e_I \rangle$ as in that event $X/U \cong \hat{A}$ rather than A. Hence the lemma holds. □

NOTATION 5.1.5. Set $F = \langle (1,2)(3,4), (1,3)(2,4) \rangle \leq A$ and $z = e_{1,2,3,4}$. Set $y = (1,2,3) \in A$.

LEMMA 5.1.6. *(1) $F\langle y \rangle$ is a normal subgroup of $N_A(F)$ isomorphic to A_4.*

(2) $O^2(EF\langle y \rangle) = K\tilde{K}\langle y \rangle$, where K and $\tilde{K}$ are the Q_8-subgroups of $K\tilde{K} \cong Q_8^2$. Moreover $K\tilde{K}\langle y \rangle \trianglelefteq C_G(z)$.

(3) Let $F \leq S \in Syl_2(G)$, $\Omega = K^G \cap S$, and $\mathcal{F} = \mathcal{F}_S(G)$. Then $\tau = (\mathcal{F}, \Omega)$ is a quaternion fusion packet with $\Omega(z) = \{K, \tilde{K}\}$.

(4) Set $k = (n-1)/2$, $n/2$ for n odd, even, respectively. Then $\mu(\tau) \cong Weyl(D_k)$.

PROOF. Set $Y = F\langle y \rangle$ and $\alpha = \{1,2,3,4\}$. Observe that $N_A(F) = (Y \times L)\langle \lambda \rangle$, where $L = A_\alpha \cong A_{n-4}$, $\lambda = 1$ if $n = 5$, and $\lambda = (1,2)(5,6)$ if $n > 5$. Hence (1) follows. Further $[E, Y] = \langle e_{i,j} : 1 \leq i < j \leq 4 \rangle \cong E_8$, so $Y_0 = O^2(YE) = Y[E, Y]$, with $O_2(Y_0) = F[E, Y] \cong Q_8^2$. Hence all but the last remark in (2) follows, with

$$K = \langle ae_{1,3}, be_{1,4} \rangle, \ \tilde{K} = \langle ae_{1,4}, be_{3,4} \rangle, \ a = (1,2)(3,4), \quad \text{and } b = (1,3)(2,4).$$

As $C_G(z) = EN_A(F)$, we have $EY \trianglelefteq C_G(z)$ by (1), so $Y_0 \trianglelefteq C_G(z)$, completing the proof of (2).

Observe next that $K^{e_{1,5}} = \tilde{K}$, so $\tilde{K} \in \Omega(z)$. By construction, Ω is an $\mathcal{F}$-invariant set of Q_8-subgroups of S, so τ satisfies (QFP1). By (2), $C_G(z)$ acts on $\theta = \{K, \tilde{K}\}$, so (QFP4) holds. Suppose $J \in \Omega - \{K\}$. Then J^+ is a root 4-subgroup of G^+ (ie. a conjugate of F^+) contained in S^+, so either $J^+ = K^+$ or $J^+ \cap K^+ = 1$. In the former case $z(J) \in z^G \cap \Phi(KE) = \{z\}$, so $J \in \theta$ as $C_G(z)$ acts on θ. In the latter case, $|K \cap J| \leq 2$; therefore (QFP2) is satisfied. Similarly if $x \in C_J(z)$ is of order 4 then as $C_G(z)$ acts on θ, either x acts on K or $K^x = \tilde{K}$, and since it remains only to verify (QFP3), we may assume the latter. Next x^2 acts on K and $x^2 = e_\beta$ for some 4-subset β of I, so as $K^{e_{1,5}} = \tilde{K}$, $|\alpha \cap \beta| = 0$ or 2. But then up to conjugacy under $C_G(z)$, $x = (5,6)(7,8)e_{5,7}$ or $(1,2)(5,6)e_{1,5}$, and in either case x acts on K, completing the proof of (3).

It remains to prove (4). If $n = 5$ we check that $\mu = \mu(\tau) \cong E_4 \cong \text{Weyl}(D_2)$; thus we may take $n > 5$. If $\mathcal{A}(z) \neq \emptyset$ then by 3.2.4.3, $z^G \cap K\tilde{K} \neq \{z\}$, whereas z is the unique member of $K\tilde{K} \cap E$ of weight 4. Thus by 4.2.10.3, D is a set of 3-transpositions of μ.

Set $\Sigma = \{\{2i, 2i-1\} : 1 \leq i \leq k\}$, a partition of $I' = \{1, \ldots, 2k\}$. We may assume that S acts on Σ. For $i \leq k$, set $t_i = (2i, 2i-1)$, and for $i \neq j$ set $a_{i,j} = t_i t_j$. Write $K_{i,j}, \tilde{K}_{i,j}$ for the conjugate of $K, \tilde{K}$ moving $Mov(a_{i,j})$, and $z_{i,j}$ for the involution in $K_{i,j}$. Pick elements $v_{i,j} \in K_{i,j}$ and $\tilde{v}_{i,j} \in \tilde{K}_{i,j}$ of order 4 with $v_{i,j}^+ = a_{i,j} = \tilde{v}_{i,j}^+$. If $K_{i,j} \notin \Omega$ then $\Delta(z_{i,j}) = \{V_{i,j}, \tilde{V}_{i,j}\}$, where $V_{i,j}, \tilde{V}_{i,j}$ are generated by $v_{i,j}, \tilde{v}_{i,j}$, respectively. Further $C_K(v_{i,j}) = V_{1,2}$ for $|\{i,j\} \cap \{1,2\}| = 1$. Thus $\eta = \{V_{i,j}, \tilde{V}_{i,j} : i,j\} \subseteq \eta_0$ for some $\eta_0 \in \eta(\tau)$. Let $W = \langle \eta \rangle$; then $\{z_{i,j} : i,j\} = z^G \cap C_G(W)$, so $\eta = \eta_0$. Also $M = \langle K_{i,j}, \tilde{K}_{i,j} : i,j \rangle$ is transitive on the $\{K_{i,j}, \tilde{K}_{i,j} : i,j\}$, so μ is transitive on D. Thus D is a class of 3-transpositions. If

$d, \tilde{d}$ are the images of $K, \tilde{K}$ in μ then $\tilde{d} \in V_d$, so by 4.1.7 and 4.3.5, $\mu \cong \mathrm{Weyl}(D_l)$ for some l. Then as $N_G(W) = N_G(\Sigma)$, $N_G(W)/WE \cong S_k$, so $l = k$, completing the proof of (4). □

LEMMA 5.1.7. *Assume X is a finite group, $T \in Syl_2(X)$, $J \leq T$, and $\Gamma = J^X \cap T$ such that $\rho = (\mathcal{F}_T(X), \Gamma)$ is a quaternion fusion packet and $X = \langle J^X \rangle$. Suppose $\beta : X \to G^*$ is a surjective homomorphism with $\ker(\beta) \leq Z(X) \cap O_2(X)$ and $J\beta = K^*$. Let $j = z(J)$, $\tilde{J} \in \Gamma$ with $\tilde{J}\beta = \tilde{K}^*$, and $\tilde{j} = z(\tilde{J})$. Set $f = j\tilde{j}$. Then $\langle f \rangle = \Phi(O_2(X))$ and $X/\langle f \rangle \cong AE_n$ or $\check{A}E_n$.*

PROOF. Assume otherwise and let X be a minimal counter example. Set $Q = O_2(X)$ and $Q_0 = \langle j^X \rangle$, so that $Q_0 \leq Q$.

Suppose $x \in X$ with $J^x \in \Gamma - \{J, \tilde{J}\}$. Then J^x centralizes $J\tilde{J}$ by 3.1.5.1, so $J^x\beta \in \Omega$ centralizes $K^*\tilde{K}^*$, and hence $J^x\beta \not\leq K^*E^*$, so $J^x \not\leq JO_2(X)$. Thus $\{J, \tilde{J}\} = \Gamma \cap JO_2(X)$. Hence for $t \in j^X$, $J^t = J$ or $\tilde{J}$, so $[t, j] \in \langle f \rangle$. As $f \in Z(X)$ we conclude that $\langle f \rangle = \Phi(Q_0)$. Then by minimality of X, $f = 1$. Therefore $\tilde{j} = j$ and as $E = \langle z^G \rangle$, we have $Q = Q_0 Z(X)$ with $\Phi(Q_0) = 1$, so $\Phi(Q) = \Phi(Z(X))$.

Suppose $\Phi(Q) \neq 1$. Then $Z(X/\Phi(Q)) \neq 1$, so by minimality of X, $X/\Phi(Q) \cong \check{A}E_n$ with n even. But then $Q/\Phi(Q) = \langle (j\Phi(Q))^X \rangle$, so $Q = \langle Q_0, \Phi(Q) \rangle = Q_0$, so $\Phi(Q) = 1$.

Suppose $X = \hat{G}/Z$, where $Z = \langle \sigma e_I \rangle$. Then $\xi\beta = \zeta$, where $\xi : \hat{G} \to X$ and $\zeta : \hat{G} \to G^*$ are the natural maps $\xi : g \mapsto gZ$ and $\zeta : g \mapsto g\Sigma$ and $\Sigma = \langle \sigma, e_I \rangle$. From the proof of 5.1.6, $K\tilde{K} = FD$ where $D = \langle e_{1,3}, e_{1,4}, z \rangle$. Therefore $U = \zeta^{-1}(K^*\tilde{K}^*) = \hat{F}D\langle e_I \rangle$, where $Q_8 \cong \hat{F} \leq \hat{A}$ and $\Phi(\hat{F}) = \langle \sigma \rangle$. Then $U_0 = J\tilde{J}Z(X) = U\xi = F_0 D$, where $F_0 = \hat{F}\xi$ and $\Phi(F_0) = \langle \sigma\xi \rangle = Z(X)$. This is a contradiction as $\Phi(U_0) = \Phi(J\tilde{J})\Phi(Z(X)) = \Phi(J\tilde{J}) = \langle j \rangle$.

Therefore X is not $\hat{G}/\langle \sigma e_I \rangle$, so by 5.1.4 it suffices to show $Q = [Q, X]$, so assume otherwise.

Claim $Q = Q_0$. As $X = \langle J^X \rangle$ and $J \leq O^2(X)$, we have $X = O^2(X)$. Thus if $Q \neq Q_0$ then $X/Q_0 \cong \hat{A}$. However $\Phi(J) = \langle j \rangle$, so $JQ_0/Q_0 \cong A_4$, where as the preimage $\hat{F}$ of F in $\hat{A}$ is isomorphic to Q_8. This completes the proof of the claim.

As $Q_0 = Q \neq [Q, X]$, $j \in Q - [Q, X]$. But as $j = \tilde{j}$, $[J\tilde{J} \cap Q, J] = \langle z \rangle$, contradicting $j \notin [Q, X]$ and completing the proof. □

LEMMA 5.1.8. *Let $g \in G$, $J = K^g$, $X = \langle K, J \rangle$, $t = z(J)$, $\alpha = \{1, 2, 3, 4\}$, and $\beta = \alpha g$. Then one of the following holds:*

(1) $\alpha = \beta$, $J^+ = K^+$, and $J \in \{K, \tilde{K}\}$.
(2) $|\alpha \cap \beta| = 3$, $X^+ \cong A_5$, and $K^t = \tilde{K}$.
(3) $|\alpha \cap \beta| = 2$, $X^+ \cong S_4$, and $N_J(K) = C_J(z) \cong \mathbf{Z}_4$.
(4) $|\alpha \cap \beta| = 1$, $X^+ \cong A_7$, and $K^t = \tilde{K}$.
(5) $\alpha \cap \beta = \emptyset$, $X^+ \cong E_{16}$, and J centralizes $K\tilde{K}$.

PROOF. The structure of X^+ is a consequence of 3.2.1 in [**Asc08b**]. We saw during the proof of 5.1.6 that $K^t = \tilde{K}$ iff $|\alpha \cap \beta|$ is odd. □

LEMMA 5.1.9. *Let $g \in G$, $\alpha = \{1, 2, 3, 4\}$, $t = z^g$, $J = K^g$ and $\beta = \alpha g$.*

(1) $t \in \mathcal{D}(z)$ iff $\alpha \cap \beta = \emptyset$.
(2) $t \in \mathcal{D}^*(z) - \mathcal{D}(z)$ iff $|\alpha \cap \beta| = 2$.
(3) If $n > 8$ then $\mathcal{D}(\tau)$ is connected.
(4) If $n > 5$ then $\mathcal{D}^*(\tau)$ is connected.

PROOF. Observe that by 3.1.14, $t \in \mathcal{D}(z)$ iff $[K\tilde{K}, t] = 1$ iff $\alpha \cap \beta = \emptyset$ by 5.1.8. Thus (1) holds. Similarly by 3.1.15, $t \in \mathcal{D}^*(z) - \mathcal{D}(z)$ iff $N_J(K) \cong \mathbf{Z}_4$ iff $|\alpha \cap \beta| = 2$ by 5.1.8. Thus (2) holds. Then (1) implies (3) and (2) implies (4). □

LEMMA 5.1.10. *Assume $n = 6$ or 7, $Y = G$ or G^*, and $X = \langle K^X \rangle \leq Y$ with $X \cong AE_5$. Let $\xi \in Aut(Y)$ be of odd order centralizing X. Then $\xi = 1$.*

PROOF. Let $Q = O_2(Y)$ and $Y^! = Y/Q$. Then ξ induces an automorphism $\xi^!$ of odd order on $Y^!$ centralizing $X^!$, so as $|C_{Aut(Y^!)}(X^!)| \leq 2$, we conclude $\xi^! = 1$. Thus $[Y, \xi] \leq Q$, so as $Q = \langle O_2(X)^Y \rangle$, ξ also centralizes Q. Then as ξ is of odd order, $\xi = 1$. □

LEMMA 5.1.11. *Assume $n = 8$ and set $\gamma = \{5, 6, 7, 8\}$. Then*
(1) $z^* = e^*_\gamma$, so $\Omega^* = \Omega(z^*)$ is of order 4 and $O(z^*) \cong Q_8^4$.
(2) $\mathcal{D}^*(z^*) \subseteq O(z^*)$.

PROOF. Let $\alpha = \{1, 2, 3, 4\}$. As $n = 8$, $e_\alpha + e_\gamma = e_I$, so (1) holds. Then (1) and 5.1.9.2 imply (2). □

LEMMA 5.1.12. *Let $X = G$ or G^*, $u = z$ or z^*, $Y_0 = K\tilde{K}\langle y \rangle$, and $Y = Y_0$ or Y_0^*, respectively. Then*
(1) Either $Y \trianglelefteq C_X(u)$ or $X \cong AE_8$.
(2) Either $C_X(u)$ is maximal in X or $X \cong \check{A}E_8$ and $N_X(\Omega)$ is maximal in X.

PROOF. By 5.1.6.2, $Y_0 \trianglelefteq C_G(z)$, while $C_G(z)^* = C_{G^*}(z^*)$ unless $n = 8$. Thus (1) holds. Part (2) is well known; cf. Exercise 5.6.2 in [**Asc86**]. □

LEMMA 5.1.13. *Let M be the $\mathbf{F}_2$-permutation module for A on the set Γ of 2-subsets of I, and set*

$$M_0 = \langle e_{i,j} + e_{i,k} + e_{j,k} : \{i, j, k\} \text{ is a 3-subset of } I \rangle.$$

Then M/M_0 is isomorphic to E as an $\mathbf{F}_2A$-module.

PROOF. Set $M^! = M/M_0$. First observe that $E = \langle e_\gamma : \gamma \in \Gamma \rangle$ and for each 3-subset $\{i, j, k\}$ of I, $e_{i,j} + e_{i,k} + e_{j,k} = 0$ in E. Therefore E is an image of $M^!$, so it suffices to show that $\dim(M^!) \leq n - 1$. But for $i, j > 1$ distinct, $e^!_{i,j} = e^!_{1,i} + e^!_{1,j}$ by definition of $M^!$, so $M^! = \langle e^!_{1,i} : i > 1 \rangle$ is indeed of dimension at most $n - 1$. □

LEMMA 5.1.14. *Let U be an irreducible $\mathbf{F}_2A$-module, set $Y = \langle y \rangle$, and let X be the global stabilizer in A of $\{1, 2\}$. Assume*
(1) there exists $0 \neq u \in C_U(X)$, and
(2) $\dim(\langle u^Y \rangle) = 2$.
Then U is the natural module E^* for A.

PROOF. Adopt the notation of 5.1.13. As A is irreducible on U, $U = \langle u^A \rangle$, so there is an $\mathbf{F}_2A$-surjection $\varphi : M \to U$ with $u\varphi = e_{1,2}$. Further $e^Y_{1,2} = \{e_{1,2}, e_{1,3}, e_{2,3}\}$, so by (2), $e_{1,2} + e_{1,3} + e_{2,3} \in \ker(\varphi)$. Therefore $M_0 \leq \ker(\varphi)$, so the lemma follows from 5.1.13. □

LEMMA 5.1.15. *Let $\tau^* = (\mathcal{F}_{S^*}(G^*), \Omega^*)$ and $O = O(\tau^*)$. Assume T is a 2-group acting on G^* and centralizing O, and set $Q = O_2(G^*T)$. Then $|TQ : Q| \leq 2$.*

PROOF. Write $n = 4r + a$ with $0 \leq a \leq 2$, and for $1 \leq i \leq r$ let $I_i = \{4i - j : 0 \leq j < 4\}$ and $z_i = e_{I_i}$. We may take $\Omega = \{K_i, \tilde{K}_i : 1 \leq i \leq r\}$, where $z(K_i) = z(\tilde{K}_i) = z_i$ and $Mov(K_i\tilde{K}_i) = I_i$.

Next $Aut_{TG^*}(G^{*+}) \leq H^+ \cong S_n$ and $O^+\langle b^+\rangle$ is a Sylow 2-subgroup of $C_{H^+}(O^+)$, where $b^+ = 1$ if $a \in \{0, 1\}$ and $b^+ = (4r+1, 4r+2)$ otherwise. Therefore $T \leq OQ\langle b\rangle$, so $T \cap OQ$ is of index at most 2 in T. Therefore it suffices to show that $C_{OQ}(O) \leq Q$.

Next G^* is irreducible on E^*, so $E^* \leq Z(Q)$. Then by 5.1.2, $|Q : C_Q(G^*)E^*| \leq 2$ with $Aut_Q(G^*) = V^*$. Passing to $G^*T/C_Q(G^*)$, we may assume $C_Q(G^*) = 1$ and then take $V^* = Q$. So it suffices to show $Z(OV^*) \leq V^*$. Let $V_i = \langle I_i\rangle$, so that $V = V_1 \oplus \cdots \oplus V_r \oplus V_0$, where $V_0 = \langle e_{4r+j} : 1 \leq j \leq a\rangle$. Let P be the preimage in AE of O, so that $P = P_1 \times \cdots \times P_r$ with $P_i = K_i\tilde{K}_i$. Then $PV = P_1V_1 \times \cdots \times P_rV_r \times V_0$, and for $x \in PV$ with $x^* \in Z(OV^*)$, we have $x = x_1 \cdots x_r x_0$ with $x_i \in P_iV_i$ for $1 \leq i \leq r$ and $x_0 \in V_0$. Then for $1 \leq i \leq r$, x_i^* centralizes $P_i^* \cong P_i$, so $x_i \in \langle z_i\rangle$, and hence $x^* \in V^* = Q^*$, completing the proof. □

REMARK 5.1.16. Suppose n is odd; then, up to isomorphism, there exists a largest group X satisfying the hypotheses of 5.1.7. We denote this group by $\hat{A}E_n$ and observe that $\rho = (\mathcal{F}_T(X), \Gamma)$ is a quaternion fusion packet, $Q = O_2(X) \cong 2^{1+2w}$ is extraspecial, where $w = (n-1)/2$, $\langle f\rangle = Z(Q) = Z(X)$ with $f = j\tilde{j}$ and $j, \tilde{j} \in \mathcal{Z}_T(\rho)$ with $|\Gamma(j)| = 1$, and $X/Z(X) \cong AE_n$. Also the induced quadratic form on $Q/Z(Q)$ is the form $\mathfrak{q}$ of Exercise 7.7 in [**Asc86**].

Namely we can find a subpacket $\rho = (\mathcal{F}_T(X), \Gamma)$ of the Lie packet $\xi = (\mathcal{F}, \Gamma)$ of $Spin_n(q)$ and a homomorphism $\beta : X \to AE_n$ satisfying the hypothesis of 5.1.7. In ξ, $\Gamma(j) = \{J\}$ is of order 1 for $j \in \mathcal{Z}_S(\xi)$, so from 5.1.7, $f = j\tilde{j}$ is an involution, $\Gamma(j) = \{J\}$ is of order 1, and setting $Q = O_2(X)$ and $n = 2w + 1$, $Q \cong 2^{1+2w}$ is extraspecial with $\langle f\rangle = Z(Q) = Z(X)$ and $X/Z(X) \cong AE_n$. As E is an absolutely irreducible $\mathbf{F}_2A$-module, A preserves a unique quadratic form $\mathfrak{q}$ on $E \cong Q/Z(Q)$ (cf Exercise 9.1 in [**Asc86**]), so from Exercise 7.7 in [**Asc86**], $\mathfrak{q}$ is the form described in that exercise. Thus it remains to show there is a largest group satisfying the hypotheses of 5.1.7, so assume $\beta_2 : X_2 \to G^*$ is such a group with X_2 not isomorphic to G or X. Let M be the universal 2-covering group of G and $X_1 = X$. Recall the definition of $\hat{G}$ in Definition 5.1.3 and let $\hat{Z} = Z(\hat{A}) = Z(\hat{G}) = \langle\sigma\rangle$. By 8.17 in [**Asc94**], $F^*(M) = \hat{Q} = \hat{Z} \times [\hat{Q}, M]$ with $[\hat{Q}, M] = Q$. Thus $X_i = M/Z_i$ with Z_i generated by σ or $f\sigma$ for $i = 1, 2$. Now there exist surjections $\alpha_1 : M \to X$, $\alpha : M \to \hat{G}$, and $\gamma : \hat{G} \to G$ with $\alpha_1\beta = \alpha\gamma$. Therefore by 6.1 in [**Asc94**], $M \leq \hat{G} \times X$ is the fiber product of this diagram. Let $y \in Y = K\tilde{K} \cap A$, $u \in \hat{G}$, and $x \in X$ with $x\beta = y = u\gamma$. Then xu is in the fiber product M with $(xu)^2 = x^2u^2 = f\sigma$ as the preimage in $\hat{G}$ and X of Y is isomorphic to Q_8. But for $i = 1$ or 2, $Z_i = \langle\sigma f\rangle$, so the image in $X_i = M/Z_i$ of xu is an involution, a contradiction as, by 5.1.7, the preimage in X_i of $K\tilde{K}$ is isomorphic to $Q_8 \times Q_8$. This contradiction establishes the uniqueness of X.

DEFINITION 5.1.17. Suppose $n = 2w$ is even and set $n_0 = n + 1$. Form $I_0 = \{1, \ldots, n_0\}$, G_0, E_0, etc. as at the beginning of the section, but with respect to n_0. Form the group X_0 of 5.1.16 with respect to n_0. Regard I as a subset of I_0 invariant under $T_0 \in Syl_2(X_0)$ and let $v \in Q_0$ with $v\beta_0 = e_I \in E_0$. Set $X = C_{X_0}(v)$ and write $\hat{A}E_n$ for X. If w is even, write $H\hat{A}E_n$ for $X/\langle v\rangle$.

LEMMA 5.1.18. *Assume $n = 2w$ is even and adopt the notation in 5.1.17. Set $Q = O_2(X)$, $T = C_{T_0}(v)$, $\Gamma = \Gamma_0$, and $\rho = (\mathcal{F}_T(X), \Gamma)$.*

(1) ρ is a quaternion fusion packet with $|\Gamma(j)| = 1$ for $j \in \mathcal{Z}_T(\rho)$.

(2) $Q = C_{Q_0}(v) = \langle v \rangle P$ where $P \cong 2^{1+2(w-1)}$ is extraspecial.

(3) $\Phi(Q) = \langle f_0 \rangle \cong \mathbf{Z}_2$, $Z(X) = Z(Q) = \langle v, f_0 \rangle$, and $Q = [Q, X]$.

(4) $X/Z(X) \cong AE_n$ and $X/\langle f_0 \rangle \cong \check{A}E_n$.

(5) If w is odd then $Z(X) = \langle v \rangle \cong \mathbf{Z}_4$, while if w is even then $Z(X) \cong E_4$.

PROOF. As $v\beta_0 = e_I$, $C_{X_0}(v)/C_{Q_0}(v) \cong C_{A_0}(e_I) \cong A_n$ and $E_0 \cong V$ as a $C_{A_0}(e_I)$-module by 5.1.1. Thus $Q = C_{Q_0}(v)$ and $X/Q \cong A_n$. As Q_0 is extraspecial of order 2^{1+2w}, part (2) follows. By (2), $\Phi(Q) = \Phi(Q_0) = \langle f_0 \rangle$. As $E_0 \cong V$ as a $C_{A_0}(e_I)$-module, it follow that $Q/\langle f_0 \rangle = [Q_0/\langle f_0 \rangle, X]$ is the unique X-invariant hyperplane E of V. Thus (3) and (4) hold.

From 5.1.16, the induced form $\mathfrak{q}$ on $Q_0/Z(Q_0)$ is the one described in Exercise 7.7 in [**Asc86**], so $vZ(Q_0)$ is nonsingular if w is odd and singular if w is even. Thus $|v| = 4$ if w is odd while $|v| = 2$ if w is even, so (5) holds. Similarly from the description of $\mathfrak{q}$ in the exercise, v commutes with $j = z(J)$ for each $J \in \Gamma_0$, so as $\Gamma_0(j) = \{J\}$, it follows that v acts on J. But as e_I is T_0-invariant, $[v, T_0] \le \langle f_0 \rangle$, so $[v, J] \le \langle j \rangle \cap \langle f_0 \rangle = 1$. Thus v centralizes Γ_0, so (1) follows from 2.5.2. □

LEMMA 5.1.19. *$\hat{A}E_n$ is the largest group X satisfying the hypothesis of* 5.1.7.

PROOF. If n is odd this follows from 5.1.16, so we may assume $n = 2w$ is even, and form the group $X = \hat{A}E_n$ of 5.1.17. By 5.1.17 the natural map $\beta : X \to X/Z(X) \cong G^*$ satisfies the hypothesis of 5.1.7, so it remains to show that if $\beta_2 : X_2 \to G^*$ satisfies that hypothesis then there is a homomorphism $\alpha : X \to X_2$ with $\alpha\beta_2 = \beta$. The proof is similar to that in 5.1.16.

Let M be the universal 2-covering of G^* and recall the group $\hat{G}$ of Definition 5.1.3 and set $\hat{Z} = \langle \sigma \rangle = Z(\hat{A})$. The proof of 8.17 in [**Asc94**] shows that $\hat{Q} = O_2(M) = \hat{Z} \times Q$ and $M \le \hat{G} \times X$ is the fiber product of the diagram $\hat{X} \to G^*$ and $\beta : X \to G^*$. (We need two observations to implement that proof in our situation: First, let $\check{E}$ be the dual of E as an $\mathbf{F}_2A$-module; then by 5.1.2, $H^1(A, \check{E}) = 0$. Second, E is an indecomposable $\mathbf{F}_2A$-module that admits a unique nonzero A-invariant bilinear form; this replaces condition (a) in the hypothesis of 8.17 in [**Asc94**].) Now there is a homomorphism $\alpha : M \to X_2$ with $\ker(\alpha) \le Z(M) = \langle \sigma \rangle \times Z(X) = \langle \sigma, v, f_0 \rangle$, and to complete the proof it suffices to show that $\sigma \in \ker(\alpha)$.

By 5.1.7, $|Z(X_2)| \le 4$, so $\ker(\alpha) \ne 1$. If $f_0 \in \ker(\alpha)$ then $U = O_2(X_2)$ is elementary abelian, so $\sigma \in \ker(\alpha)$ by 5.1.4. Thus we may assume $f_0 \notin \ker(\alpha)$. Similarly if $\sigma f_0 \in \ker(\alpha)$, an argument in the proof of 5.1.16 supplies a contradiction. Hence $f_0\sigma \notin \ker(\alpha)$, so $\ker(\alpha) \cap \langle f_0, \sigma \rangle = 1$ and hence $|\ker(\alpha)| = 2$. Choose y, x, u as in the proof of 5.1.16; as $|\ker(\alpha)| = 2$ it follows from 5.1.7 that $f_0\sigma = x^2u^2 = (xu)^2 \in f_0 \ker(\alpha)$, so $\sigma \in \ker(\alpha)$, completing the proof. □

DEFINITION 5.1.20. Let $\hat{X}$ be the universal group $\hat{A}E_n$ of 5.1.19. Pick $\hat{S} \in Syl_2(\hat{X})$, set $\hat{\mathcal{F}} = \mathcal{F}_{\hat{S}}(\hat{X})$, and denote $\hat{\mathcal{F}}$ by $\hat{A}E_n$ also. By construction in 5.1.16 and 5.1.18, there is a set $\hat{\Omega}$ of Q_8-subgroups of $\hat{S}$ such that $\hat{\tau} = (\hat{\mathcal{F}}, \hat{\Omega})$ is a quaternion fusion packet.

Similarly if $X = G^*$ or G then, from the start of the section, X is AE_n or $\check{A}E_n$, respectively, and by 5.1.6 there is $S \in Syl_2(X)$ and a set Ω of Q_8-subgroups of S

such that $\tau = (\mathcal{F}_S(X), \Omega)$ is a quaternion fusion packet. Denote $\mathcal{F}_S(X)$ by AE_n or $\check{A}E_n$, respectively.

Finally if $n \equiv 0 \mod 4$ then from 5.1.17, $\hat{X}/\langle v\rangle$ is denoted by $H\hat{A}E_n$ and, setting $\hat{\mathcal{F}}^+ = \hat{\mathcal{F}}/\langle v\rangle$, we have $\tau^+ = (\hat{\mathcal{F}}^+, \hat{\Omega}^+)$ a quaternion fusion packet. Denote the fusion system $\hat{\mathcal{F}}^+$ by $H\hat{A}E_n$ also.

LEMMA 5.1.21. *Assume $\tau = (\mathcal{F}, \Omega)$ is a quaternion fusion packet on S with $\mathcal{F} = \mathcal{F}^\circ$ and $\mathcal{F}^+ = \mathcal{F}/Z(\mathcal{F}) \cong AE_n$. Let $K \in \Omega$ and $z = z(K)$.*

(1) $\mathcal{F}$ is constrained, so $\mathcal{F}$ has a model X.

(2) There is a surjective morphism $\pi : \hat{\mathcal{F}} \to \mathcal{F}$ such that $\hat{S}\pi = S$, $\hat{\Omega}\pi = \Omega$, and $\ker(\pi) \leq Z(\hat{\mathcal{F}})$.

(3) Either $|\Omega(z)| \leq 2$ or $\mathcal{F} \cong AE_8$, and $|\Omega(z)| = 4$.

(4) $|\Omega(z)| = 1$ iff either $\mathcal{F} \cong \hat{A}E_n$ or $8 \neq n \equiv 0 \mod 4$ and $\mathcal{F} \cong H\hat{A}E_n$.

(5) $|\Omega(z)| = 2$ iff either

(a) $\mathcal{F} \cong AE_n$ with $n \neq 8$, or

(b) n is even and $\mathcal{F} \cong \check{A}E_n$.

PROOF. As $\mathcal{F}^+ \cong AE_n$, $\mathcal{F}^+$ is constrained, so as $\mathcal{F}^+ = \mathcal{F}/Z(\mathcal{F})$, (1) holds. Then (2) follows from (1) and 5.1.19.

Recall $G = \check{A}E_n$, $G^* = AE_n$, and taking $\alpha = \{1,2,3,4\}$ and $z_0 = e_\alpha$, from 5.1.6.3 there is a quaternion fusion packet $(\mathcal{F}_T(G), \Gamma)$ such that $|\Gamma(z_0)| = 2$. Now for $z_0 \neq e_\beta \in z_0^G$, $z_0^* = e_\beta^*$ iff $n = 8$ and $\beta = I - \alpha$, so $|\Gamma^*(z_0^*)| = 2$ unless $n = 8$ where $|\Gamma^*(z_0^*)| = 8$. Therefore as $\mathcal{F}^+$ is AE_n and $|\Omega^+(z^+)| \geq |\Omega(z)$, (3) follows once we show $|\Omega(z)| \leq 2$ if $n = 8$ and $\mathcal{F} \neq \mathcal{F}^+$, which we do during the proof of (4) and (5).

Let $Q = O_2(\hat{X})$ and $\langle f\rangle = \Phi(Q)$. By 5.1.16 and 5.1.18, $f = \hat{z}\hat{z}_1$ for distinct $\hat{z}$ and $\hat{z}_1$ in $\mathcal{Z}_{\hat{S}}$. Thus if $|\Omega(z)| = 1$ then $f \notin \ker(\pi)$.

But suppose n is not divisible by 4; then from 5.1.16 and 5.1.18, f is the unique involution in $Z(\hat{X})$. It follows that in that event, $|\Omega(z)| = 1$ iff $\ker(\pi) = 1$ iff $\mathcal{F} \cong \hat{A}E_n$. So suppose $f \in \ker(\pi)$. When n is odd, $Z(\hat{\mathcal{F}}) = \langle f\rangle$ so $\mathcal{F} \cong AE_n$. When n is even, $Z(\hat{\mathcal{F}}) = \langle v\rangle \cong \mathbf{Z}_4$ by 5.1.18.5, so $\mathcal{F}$ is $\check{A}E_n$ or AE_n. Therefore (4) and (5) hold in this case, so we may assume $n \equiv 0 \mod 4$.

If $n = 8$ then $\hat{\mathcal{F}}/\langle x\rangle \cong \check{A}E_8$ for each involution $x \in \hat{Z}(\mathcal{F})$, completing the proof of the lemma in this case; thus we may take $n > 8$. By 5.1.18, $Z(\hat{\mathcal{F}}) = \langle f, v\rangle \cong E_4$ with $\hat{\mathcal{F}}/\langle f\rangle \cong \check{A}E_n$, so we may assume $\ker(\pi)$ is generated by v or fv. Then as fv is conjugate to v under an automorphism of $\hat{X}$ acting on $\hat{S}$ and $\hat{\Omega}$, it follows that $\mathcal{F}$ is $H\hat{A}E_n$, completing the proof. □

5.2. The 2-share of the order of some groups

In this section q is an odd prime power. Recall that if n is a positive integer then n_2 denotes the 2-share of n: the highest power of 2 dividing n. Given positive integers n and k, write $n \sim k$ to indicate that $n_2 = k_2$. Set $m = (q^2 - 1)_2$ and define

$$\Pi(n,q) = \prod_{i=1}^{n} q^i - 1.$$

Set $\Pi^+(n,q) = \Pi(n,q)$ and

$$\Pi^-(n,q) = \prod_{i=1}^{n} q^i - (-1)^i.$$

LEMMA 5.2.1. *(1)* $m \geq 8$.
(2) If $\epsilon = \pm 1 \equiv q \mod 4$ *then* $q + \epsilon \sim 2$ *and* $q - \epsilon \sim m/2$.
(3) $q^2 + 1 \sim 2$.

PROOF. For each odd integer n, $n^2 \equiv 1 \mod 8$, so (1) and (3) hold. Assume the setup of (2). Then $q + \epsilon \equiv 2\epsilon \equiv 2 \mod 4$, so $q + \epsilon \sim 2$. Then $q - \epsilon = (q^2 - 1)/(q + \epsilon) \sim m/2$, completing the proof of (2). □

LEMMA 5.2.2. *If n is odd then* $q^n - 1 \sim q - 1$ *and* $q^n + 1 \sim q + 1$.

PROOF. Recall $q^n - 1 = (q-1)(q^{n-1} + \cdots + q + 1)$. As n and q are odd, so is $q^{n-1} + \cdots + q + 1$, so the first statement in the lemma holds. Similarly the second holds. □

LEMMA 5.2.3. *Let* $n = 2^a k$ *with* $a \geq 0$ *and k odd. Then*
(1) If $q \equiv 1 \mod 4$ *then* $q^n - 1 \sim 2^a m/2$.
(2) If $a > 0$ *then* $q^n - 1 \sim 2^a m/2$.
(3) If $q \equiv 1 \mod 4$ *then* $q^i - 1 \sim im/2$ *for each positive integer i.*
(4) $q^{2i} - 1 \sim im$ *for each positive integer i.*

PROOF. We first prove (1). Replacing q by q^k and appealing to 5.2.2, we may assume $k = 1$. We induct on a; the case $a = 0$ follows from 5.2.1.2. Suppose $a > 0$. Then $q^{2^a} - 1 = (q^{2^{a-1}} - 1)(q^{2^{a-1}} + 1) \sim 2(q^{2^{a-1}} - 1)$ by 5.2.1. By induction on a, $q^{2^{a-1}} - 1 \sim 2^{a-1}m/2$, completing the proof of (1). Observe (1) implies (3).

Suppose $a > 0$ and set $q_0 = q^2$ and $m_0 = (q_0^2 - 1)_2$. By 5.2.1, $m_0 = 2m$, so as $q_0 \equiv 1 \mod 4$, it follows from (1) that

$$q^n - 1 = q_0^{2^{a-1}k} - 1 \sim 2^{a-1}m_0/2 = 2^a m/2,$$

proving (2). Then (2) implies (4). □

LEMMA 5.2.4. *(1) If* $q \equiv 1 \mod 4$ *then* $\Pi(n,q) \sim n!(m/2)^n$.
(2) $\Pi(n,q^2) \sim n!m^n$.
(3) If $q \equiv \epsilon \mod 4$ *with* $\epsilon = \pm 1$ *then* $\Pi^\epsilon(n,q) = n!(m/2)^n$.
(4) If $q \equiv -\epsilon \mod 4$ *then* $\Pi^\epsilon(n,q) = r! \cdot m^r \cdot 2^{r+a}$*, where* $n = 2r + a$ *and* $a \in \{0,1\}$.

PROOF. Part (1) follows from 5.2.3.3, while (2) follows from 5.2.3.4. Suppose $n = 2r$. Then

$$\Pi^\epsilon(n,q) = \Pi(r,q^2) \cdot \prod_{i=1}^{r} q^{2i-1} - \epsilon \sim r!m^r \cdot (q-\epsilon)^r$$

by (2) and 5.2.2. Hence if $q \equiv \epsilon \mod 4$ then by 5.2.1.2 we have

$$\Pi^\epsilon(n,q) \sim r!m^r \cdot (m/2)^r \sim r!2^r \cdot (m/2)^n \sim n! \cdot (m/2)^n,$$

establishing (3) when n is even. On the other hand if $q \equiv -\epsilon \mod 4$ then by 5.2.1.2, $\Pi^\epsilon(n,q) \sim r!m^r \cdot 2^r$, establishing (4) when n is even.

Finally assume $n = 2r + 1$ is odd. Then

$$\Pi^\epsilon(n, q) = (q^n - \epsilon^n) \cdot \Pi^\epsilon(n - 1, q) \sim (q - \epsilon) \cdot \Pi^\epsilon(n - 1, q)$$

by 5.2.2. Then if $q \equiv \epsilon \mod 4$, we have

$$\Pi^\epsilon(n, q) \sim (m/2) \cdot \Pi^\epsilon(n - 1, q) \sim n!(m/2)^n$$

by 5.2.1.2 and (3) applied to $2r = n - 1$. Finally if $q \equiv -\epsilon \mod 4$ then $\Pi^\epsilon(n, q) \sim 2 \cdot \Pi^\epsilon(n-1, q) \sim r!m^r \cdot 2^{r+1}$ by 5.2.1.2 and (4) applied at $2r = n-1$. This completes the proof of (3) and (4), and hence completes the proof of the lemma. □

LEMMA 5.2.5. *(1) If $q \equiv \epsilon \mod 4$ then $|SL_n^\epsilon(q)| \sim n!(m/2)^{n-1}$.*

(2) If $q \equiv -\epsilon \mod 4$ and $n = 2r + a$ with $a \in \{0, 1\}$ then $|SL_n^\epsilon(q)| \sim r!m^r \cdot 2^{r+a-1}$.

PROOF. From Table 16.1 in [**Asc86**], $x = |SL_n^\epsilon(q)| \sim \Pi^\epsilon(n, q)/(q - \epsilon)$. Therefore if $q \equiv \epsilon \mod 4$ then $x \sim \Pi^\epsilon(n, q)/(m/2) \sim n!(m/2)^{n-1}$ by 5.2.1.2 and 5.2.4.3, so that (1) holds. Similarly if $q \equiv -\epsilon \mod 4$ then $x \sim \Pi^\epsilon(n, q)/2 \sim r!m^r \cdot 2^{r+a-1}$ by 5.2.1.2 and 5.2.4.4, establishing (2). □

LEMMA 5.2.6. *$|Sp_{2r}(q)| \sim r!m^r$.*

PROOF. From Table 16.1 in [**Asc86**], $|Sp_{2r}(q)| \sim \Pi(r, q^2)$, so the lemma follows from 5.2.4.2. □

LEMMA 5.2.7. *(1) If $r \geq 4$ is even or $r \geq 5$ is odd and $q \equiv 1 \mod 4$, then $|Spin_{2r}^+(q)| \sim r! \cdot (m/2)^r \cdot 2^{r-1}$.*

(2) If $r \geq 5$ is odd and $q \equiv -1 \mod 4$ then $|Spin_{2r}^+(q)| \sim (r-1)! \cdot (m/2)^{r-1} \cdot 2^r$.

(3) If $r \geq 2$ then $|Spin_{2r+1}(q)| \sim r!m^r \sim r! \cdot (m/2)^r \cdot 2^r$.

(4) If $r \geq 4$ is even or $r \geq 5$ is odd and $q \equiv 1 \mod 4$ then $|Spin_{2r}^-(q)| \sim (r-1)! \cdot (m/2)^{r-1} \cdot 2^r$.

(5) If $r \geq 5$ is odd and $q \equiv -1 \mod 4$ then $|Spin_{2r}^-(q)| \sim r! \cdot (m/2)^r \cdot 2^{r-1}$.

PROOF. From Table 16.1 in [**Asc86**], $x = |Spin_{2r}^+(q)| \sim (q^r - 1) \cdot \Pi(r - 1, q^2)$. Suppose first that r is odd. Then $q^r - 1 \sim q - 1$ by 5.2.2, so by 5.2.1.2 and 5.2.4.2, $x = |Spin_{2r}^+(q)| \sim (m/2) \cdot (r - 1)! \cdot m^{r-1}$ or $2 \cdot (r - 1)! \cdot m^{r-1}$ for q congruent to $1, -1$ modulo 4, respectively. Therefore (2) holds, and when r is odd and $q \equiv 1 \mod 4$, (1) also holds.

So assume r is even. Then $q^r - 1 = (q^2)^{r/2} - 1 \sim rm/2$ by 5.2.3.4, so by 5.2.4.2, $x \sim r \cdot (m/2) \cdot (r - 1)! \cdot m^{r-1} = r! \cdot (m/2)^r \cdot 2^{r-1}$, completing the proof of (1).

Next from Table 16.1 in [**Asc86**], $|Spin_{2r+1}(q)| \sim \Pi(r, q^2) \sim r!m^r = r! \cdot (m/2)^r \cdot 2^r$ by 5.2.4.2, establishing (3).

Finally set $z = |Spin_{2r}^-(q)|$. By Table 16.1 in [**Asc86**], $z \sim (q^r + 1)\Pi(r - 1, q^2)$. Suppose r is odd. Then $q^r + 1 \sim q + 1$ by 5.2.2, so by 5.2.4.2, $z \sim 2 \cdot (r - 1)! \cdot m^{r-1}$ or $(m/2) \cdot (r - 1)! \cdot m^{r-1}$ for q congruent to 1 or -1 modulo 4, respectively. Therefore (5) holds, as does (4) in this case when $q \equiv 1 \mod 4$.

Thus we may assume r is even, so that $q^r + 1 = (q^2)^{r/2} + 1 \equiv 2 \mod 4$ by 5.2.1.3. Therefore $x \sim 2 \cdot (r - 1)! \cdot m^{r-1}$, so that (4) holds, completing the proof. □

LEMMA 5.2.8. *$|F_4(q)| \sim (m/2)^4 \cdot 2^7$.*

PROOF. From Table 16.1 in [**Asc86**], $x = |F_4(q)| \sim \prod_{i=1}^4 q^{2d_i} - 1$, where the d_i are $1, 3, 4, 6$. Hence by 5.2.3.4, $x \sim 1 \cdot 3 \cdot 4 \cdot 6 \cdot m^4 \sim (m/2)^4 \cdot 2^7$. □

LEMMA 5.2.9. *$|E_6^\epsilon(q)| \sim e(q,\epsilon)$, where*
(1) $e(q,\epsilon) = (m/2)^6 \cdot 2^7$ if $q \equiv \epsilon \mod 4$, and
(2) $e(q,\epsilon) = (m/2)^4 \cdot 2^9$ if $q \equiv -\epsilon \mod 4$.

PROOF. From Table 16.1 in [**Asc86**], $y = |E_6^\epsilon(q)| \sim xz$ where $z = (q^9-\epsilon)(q^5-\epsilon)$ and $x = |F_4(q)|$. By 5.2.8, $x \sim (m/2)^4 \cdot 2^7$. By 5.2.2, $z \sim (q-\epsilon)^2$. If $q \equiv \epsilon \mod 4$ then by 5.2.1.2, $z \sim (m/2)^2$, so (1) holds. Thus we may assume $q \equiv -\epsilon \mod 4$, so $z \sim 4$ by 5.2.1.2, and hence (2) holds. □

LEMMA 5.2.10. *(1) $|\tilde{E}_7(q)| \sim (m/2)^7 \cdot 2^{10}$.*
(2) $|E_8(q)| \sim (m/2)^8 \cdot 2^{14}$.

PROOF. From Table 16.1 in [**Asc86**], $u = |\tilde{E}_7(q)| \sim \prod_{i=1}^7 q^{2d_i} - 1$, where the d_i are $1,3,4,5,6,7,9$. Therefore by 5.2.3.4, $u \sim 4 \cdot 6 \cdot m^7$, so (1) holds. Similarly $v = |E_8(q)| \sim \prod_{i=1}^8 q^{2d_i} - 1$, where the d_i are $1,4,6,7,9,10,12,15$, so $v \sim 4 \cdot 6 \cdot 10 \cdot 12 \cdot m^8$, so (2) holds. □

5.3. Orthogonal groups and packets

In this section we assume the following hypothesis:

HYPOTHESIS 5.3.1. *Assume q is an odd prime power and let $F = \mathbf{F}_q$. Set $\pi = \pm 1$ where $q \equiv \pi \mod 4$. Let U be an orthogonal space over F of dimension $4r+b$ with $r \geq 1$, $0 \leq b \leq 4$, and quadratic form $\mathfrak{q}$. Let $\epsilon = sgn(U)$ and assume $\epsilon = +1$ if $b = 0$ and $\epsilon = -1$ if $b = 4$. Set $m = (q^2-1)_2$.*

NOTATION 5.3.2. Write $U = \mathfrak{U}_1 \perp \cdots \perp \mathfrak{U}_r \perp \mathfrak{U}_{r+1}$, where $\mathfrak{U}_i$ is nondegenerate of dimension 4 and sign $+1$ for $1 \leq i \leq r$. Thus $\dim(\mathfrak{U}_{r+1}) = b$ and if $b \neq 0$ then $\mathfrak{U}_{r+1}$ is nondegenerate, and of sign ϵ when b is even.

Let $L = \Omega(U,\mathfrak{q})$ be the subgroup of the orthogonal group $O(U,\mathfrak{q})$ generated by root elements. Thus $L \cong \Omega_{4r+b}^\epsilon(q)$. Let $\bar{L} = Spin_{4r+b}^\epsilon(q)$ be the universal Chevalley group covering L.

If $b > 2$ or $b = 2$ and $\epsilon = \pi$, set $n = 2r+1$ and $a = b-2$. Otherwise set $n = 2r$ and $a = b$. Thus $\dim(U) = 2n + a$ with $0 \leq a \leq 2$ and $\epsilon = \pi^n$ if $a = 0$ and $\epsilon = -1$ if $a = 2$.

For $1 \leq i \leq r$, write $\mathfrak{U}_i = U_{i,1} \perp U_{i,2}$ with $U_{i,j}$ of sign π. If n is odd then $\mathfrak{U}_{r+1}$ contains a subspace $U_{r+1,1}$ of dimension 2 and sign π. For $1 \leq i < r$, set $U_{2i-1} = U_{i,1} + U_{i,2} = \mathfrak{U}_i$ and $U_{2i} = U_{i,2} + U_{i+1,1}$. Set $U_{2r-1} = U_{r,1} + U_{r,2} = \mathfrak{U}_r$, and if n is odd set $U_{n-1} = U_{2r} = U_{r,2} + U_{r+1,1}$. Set $\Gamma_r = \{U_{i,j} : 1 \leq i \leq r, j = 1,2\}$. If n is even set $\Gamma = \Gamma_r$, while if n is odd set $\Gamma = \Gamma_r \cup \{U_{r+1,1}\}$. Thus $|\Gamma| = n$ and the members $U_{i,j}$ of Γ are nondegenerate lines of sign π.

Let $\mathfrak{M} = N_L(\Gamma)$ be the global stabilizer in L of Γ. Write $\mathfrak{W}$ for the kernel of the action on $\mathfrak{M}$ on Γ. If $\pi = +1$ then each member of Γ has exactly two singular points and we write Γ_s for the set of singular points in some member of Γ; thus $|\Gamma_s| = 2n$.

LEMMA 5.3.3. *Let $d = \dim(U)$. Then one of the following holds:*
(1) $d = 2n$ is even, $n = 2r$ is even, and $\bar{L} \cong Spin_{2n}^+(q)$.
(2) $d = 2n$ is even, $n = 2r+1$ is odd, and $\bar{L} \cong Spin_{2n}^\pi(q)$.

(3) $d = 2n + 1$ *is odd and* $\bar{L} \cong Spin_{2n+1}(q)$.
(4) $d = 2n + 2$ *is even and* $\bar{L} \cong Spin^{-\pi}_{2n+2}(q)$.

PROOF. First assume that d is even. Then as $d = 4r + b$, also b is even so $b = 0$, 2, or 4. Thus from 5.3.2, one of the following holds:
(i) $b = 0$, $n = 2r$, $d = 2n$, and $\epsilon = +1$.
(ii) $b = 2$, $n = 2r + 1$, $d = 2n$, and $\epsilon = \pi$.
(iii) $b = 2$, $n = 2r$, $d = 2n + 2$, and $\epsilon = -\pi$.
(iv) $b = 4$, $n = 2r + 1$, $d = 2n + 2$, and $\epsilon = -\pi$.
Now (1) holds in case (i), (2) holds in case (ii), and (4) holds in cases (iii) and (iv). Thus we may assume that d is odd, so $d = 4r + b$ with $b = 1$ or 3. If $b = 1$ then $n = 2r$ by 5.3.2, while if $b = 3$ then $n = 2r + 1$, and in either case (3) holds. □

LEMMA 5.3.4. *(1) For each 2-subset* $\{A, B\}$ *of* Γ *there exist unique fundamental subgroups* L_{A+B} *and* L'_{A+B} *of* L *such that* $A + B = [L_{A+B}, U] = [L'_{A+B}, U]$. *Moreover* $z(L_{A+B}) = z(L'_{A+B})$ *and* $A+B = [U, z(L_{A+B})]$. *Set* $L_{2i-1} = L_{U_{2i-1}}$ *and* $L'_{2i-1} = L'_{U_{2i-1}}$.
(2) There exists a Sylow 2-group S_L *of* L *contained in* $\mathfrak{M}$ *such that, setting* $\mathcal{F}_L = \mathcal{F}_{S_L}(L)$, $K_{2i-1} = L_{2i-1} \cap S_L$, $K'_{2i-1} = L'_{2i-1} \cap S_L$, *and* $\Omega_L = \{K_{2i-1}, K'_{2i-1} : 1 \le i \le r\}$, *we have* $\tau_L = (\mathcal{F}_L, \Omega_L)$ *is a Lie packet for* L.
(3) Set $V_{A+B} = L_{A+B} \cap S_L \cap \mathfrak{W}$, $V'_{A+B} = L'_{A+B} \cap S_L \cap \mathfrak{W}$, *and* $\eta_L = \{V_{A+B}, V'_{A+B} : A, B \in \Gamma\}$. *Then* $\eta_L \in \tau_L$, *and if* $n > 2$ *or* $m > 8$ *then* η_L *is the unique member of* $\eta(\tau_L)$.
(4) $\mathcal{F}_L$ *is transitive on* Ω_L, $\Omega_L(z(K_1)) = \{K_1, K'_1\}$ *is of order 2, and* $\mu(\tau_L) \cong Weyl(D_n)$.
(5) Set $W = \langle \eta_L \rangle$. *Then* $N_L(W) = \mathfrak{M}$.

PROOF. Part (1) is well known; see for example 15.8 in [**Asc77**] and the discussion following that lemma.

For $1 \le i \le r$, set $Y_i = L_{2i-1}L'_{2i-1}$. By (1), $U_{2i-1} = [Y_i, U]$ and $C_U(Y_i) = U^{\perp}_{2i-1}$, so $[Y_i, Y_j] = 1$ for distinct i, j. Pick Sylow 2-groups K_{2i-1} and K'_{2i-1} of L_{2i-1} and L'_{2i-1}. Then $O_i = K_{2i-1}K'_{2i-1}$ is the central product of quaternion subgroups, and O_i is irreducible on U_{2i-1}. We conclude that $O = O_1 \cdots O_r$ is a maximal commuting product of such subgroups, so (using 3.1.5.1) $\tau = (\mathcal{F}_S(L), \Omega_L)$ is a Lie packet for L for any Sylow 2-group S of L containing Ω_L. In particular $\mathcal{F}_L$ is transitive on Ω_L. Define $\Delta = \Delta(\tau)$ as in Definition 3.1.9.

Let $O \le S_L \in Syl_2(\mathfrak{M})$ and choose $S_L \le S$. Define V_{A+B} and V'_{A+B} as in (3). Then V_{A+B} is Sylow in the stabilizer in L_{A+B} of A and B, so V_{A+B} is cyclic of order $m/2$. Observe that by Witt's Lemma, $\mathfrak{M}$ induces $Sym(\Gamma)$ on Γ. Hence for each A, B with $A + B \notin \{U_{2i-1} : 1 \le i \le r\}$, V_{A+B} and V'_{A+B} are in Δ. Thus η_L is contained in some $\eta \in \eta(\tau)$. Then as η centralizes η_L, as Γ_s is the set of singular points that are weight spaces for $W = \langle \eta_L \rangle$ when $\pi = +1$, and as Γ is the set of weight spaces for W that are definite lines when $\pi = -1$, it follows that (5) holds and $\eta \subseteq \mathfrak{W}$. But each $V \in \eta$ is fused into O_1, so $[U, V]$ is of dimension 4 and sign $+1$, so as $V \le \mathfrak{W}$ it follows that $V \in \eta_L$. Therefore $\eta = \eta_L$, so as S acts on η, we have $S = S_L$, establishing (2). Indeed by 3.3.14, if $n > 2$ or $m > 8$ then η_L is the unique member of $\eta(\tau)$, so (3) also holds.

From (1), $z_i = z(K_{2i_1}) = z(K'_{2i-1})$ with

$$U_{2i-1} = [U, z_i], \text{ so } \Omega_L(z_i) = \{K_{2i-1}, K'_{2i-1}\}$$

is of order 2. For $\{A, B\}, \{C, D\}$ 2-subsets of Γ, V_{A+B} centralizes L_{C+D} iff $\{A, B\} \cap \{C, D\} = \emptyset$, so $\mu(\tau_L)$ is transitive on it set D_{η_L} of reflections. Then by 4.1.7.1, $O_2(\mu(\tau_L)) \not\leq Z(\mu(\tau_L))$. Now by 4.3.5.1 and as $|\Gamma| = n$, $\mu(\tau_L) \cong Weyl(D_n)$, completing the proof of (4) and the lemma. □

LEMMA 5.3.5. *Let $p = char(F)$ and for $H \leq L$ set $\theta(H) = O^{p'}(H)$.*

(1) $L = \bar{L}/\langle \bar{u} \rangle$, where $\bar{u} \in Z(\bar{L})$ is an involution.

(2) Let X be a fundamental subgroup of L, X' the second member of $\Omega(z(X))$, and $\bar{X}, \bar{X}'$ the preimages of $\bar{X}, \bar{X}'$ in $\bar{L}$ respectively. Then $\bar{X} = \theta(\bar{X}) \times \langle \bar{u} \rangle$ with $\theta(\bar{X}) \cong SL_2(q)$ a fundamental subgroup of $\bar{L}$, and $z(\theta(\bar{X}))z(\theta(\bar{X}')) = \bar{u}$.

(3) Let $\bar{S}_L$ be the preimage of S_L in $\bar{L}$, $\bar{\Omega}_L = \{\theta(\bar{X}) \cap \bar{S}_L : X \cap S_L \in \Omega_L\}$, $\mathcal{F}_{\bar{L}} = \mathcal{F}_{\bar{S}_L}(\bar{L})$, and $\tau_{\bar{L}} = (\mathcal{F}_{\bar{L}}, \bar{\Omega}_L)$. Then $\tau_{\bar{L}}$ is a quaternion fusion packet with $\Omega_{\bar{L}}(\bar{z}(\bar{K})) = \{\bar{K}\}$ for $\bar{K} \in \bar{\Omega}_L$.

(4) Set $M = M(\tau_L)$ and $\bar{M} = M(\tau_{\bar{L}})$. Then $M = \bar{M}/\langle \bar{u} \rangle$, $\bar{M} \cong \bar{\omega}(D_n, m)$, and $M \cong \omega(D_n, m)$.

(5) $|\bar{\omega}(D_n, m)| = n!m^n/2$.

PROOF. Parts (1) and (2) are well known; cf. Table 6.1.2 in [**GLS98**]. Then (3) follows from (1) and (2). Similarly by (1) and (2), $M \cong \bar{M}/\langle \bar{u} \rangle$.

Let $\bar{W}$ be the preimage of $W = \langle \eta_L \rangle$. For $1 \leq i < n$, set $V_i = V_{U_i}$, $V'_i = V'_{U_i}$, $z_i = z(V_i)$, and $W_i = V_1 \cdots V_i$. Observe $[U, W_i] = A_i$, where $A_i = U_1 + \cdots + U_i$. Also $W_i = V_1 \times \cdots \times V_i$ is a direct product. This follows by induction on i and the fact that $[U, z_i] = U_i \not\leq A_{i-1}$. In particular, $|W_{n-1}| = (m/2)^{n-1}$. Further $V_{n-1} \cap V'_{n-1} = \langle z_{n-1} \rangle$ and $V'_{n-1} \cap W_{n-1} = V_{n-1} \cap V'_{n-1}$, so as $W_{n-1}V'_{n-1} \leq W$, we conclude that $|W| \geq (m/2)^n/2$. Therefore $|\bar{W}| \geq (m/2)^n$. But by 5.3.4.4, $\bar{M}/C_{\bar{M}}(\bar{W})\bar{W} \cong Weyl(D_n)$, which is of order $n! \cdot 2^{n-1}$ by 4.3.4. Therefore we have $|\bar{M}| \geq n!m^n/2$, and by 4.3.2 the universal group $\bar{\omega}(D_n, m)$ is at most of that order, so (4) and (5) follow from 4.3.3.2. □

NOTATION 5.3.6. Define $S_L, \bar{S}_L$, $\Omega_L, \bar{\Omega}_L$, $\mathcal{F}_L, \mathcal{F}_{\bar{L}}$, $\tau_L, \tau_{\bar{L}}$, $M = M(\tau_L), \bar{M} = M(\tau_{\bar{L}})$ as in 5.3.4 and 5.3.5. From 5.3.4, $\eta_L \in \eta(\tau_L)$; set $W = \langle \eta_L \rangle \in W(\tau_L)$. Then from 5.3.5 there is $\eta_{\bar{L}} \in \eta(\tau_{\bar{L}})$; set $\bar{W} = \langle \eta_{\bar{L}} \rangle \in W(\tau_{\bar{L}})$. Let G_W and $\bar{G}_W$ be models for $N_{\mathcal{F}_L}(W)$ and $N_{\mathcal{F}_{\bar{L}}}(\bar{W})$, respectively, and notice that $G_W = \bar{G}_W/\langle \bar{u} \rangle$.

LEMMA 5.3.7. *(1) $C_L(W) = O(\mathfrak{M})C_S(W)$.*

(2) $G_W = MS_L$ with $\mathfrak{M}/O(C_L(W)) \cong G_W$.

(3) If $n > 2$ then $Aut_L(S) = Inn(S)$.

PROOF. From the proof of 5.3.4, $C_L(W)$ is the kernel of the action of $\mathfrak{M}$ on Γ, Γ_s for $\pi = -1$, 1, respectively, with $O^2(C_L(W)) = O(C_L(W))$, so (1) holds.

Set $\mathfrak{M}_K = \langle K_1^{\mathfrak{M}} \rangle O(\mathfrak{M})$; thus $\mathfrak{M}_K \trianglelefteq \mathfrak{M}$ with $\mathfrak{M}_K/O(\mathfrak{M}) \cong M$. Next as $M/W \cong Weyl(D_n)$ by 5.3.5.4, we conclude that either $O^2(Aut(M/W)) = O^2(M/W)$ or $n = 4$ and there is an outer automorphism of M/W of order 3. But if $O^2(\mathfrak{M}) \leq \mathfrak{M}_K$ then (2) follows from (1) and the fact that $S_L \leq G_W$. Therefore we may assume that $n = 4$ and some x of order 3 in $\mathfrak{M}$ induces an outer automorphism on $\mathfrak{M}_K/C_L(W)$. However we may choose x to act on K_1 and have a cycle of length 3 on $\Omega - \{K_1\}$. This contradicts $N_{\mathfrak{M}}(K_1) \leq N_{\mathfrak{M}}(K'_1)$. Hence (2) is established.

It remains to prove (3), so assume $n > 2$. Hence W is strongly closed in S with respect to L by 5.3.4.3, so $N_L(S) = N_{\mathfrak{M}}(S)$ by 5.3.4.5. By (2), $\mathfrak{M}^* = \mathfrak{M}/C_L(W) \cong G_W$, so $N_{\mathfrak{M}^*}(S^*) = S^*$. Hence $N_L(S) = N_{\mathfrak{M}}(S) = S(C_L(W) \cap N_L(S))$. Finally $[N_{O(\mathfrak{M})}(S), S] \leq O(\mathfrak{M}) \cap S = 1$, so $Aut_L(S) = Inn(S)$ by (1). □

LEMMA 5.3.8. *Suppose $n > 2$ and $\bar{L}$ is $Spin^+_{2n}(q)$ with n even or $\bar{L}$ is $Spin^{\pi}_{2n}(q)$ with n odd. Then $\bar{M}$ contains the Sylow group $\bar{S}_L$ of $\bar{L}$ and M contains the Sylow group S_L, so $\bar{M} = \bar{G}_W$ and $M = G_W$.*

PROOF. By 5.3.7.2, $G_W = MS_L$, so also $\bar{G}_W = \bar{M}\bar{S}_L$, and it remains to show $\bar{M}$ contains $\bar{S}_L$; thus it suffices to show that $|\bar{M}| \sim |\bar{L}|$. By parts (4) and (5) of 5.3.5, $|\bar{M}| \sim n!m^n/2$.

Suppose n is even or $n \geq 5$ is odd and $\pi = +1$. Then by 5.2.7.1, $|\bar{L}| \sim n!m^n/2$, so the lemma holds in this case. Further if $n = 3$ then $\bar{L} \cong SL^{\pi}_4(q)$, so this time $|\bar{L}| \sim |\bar{M}$ by 5.2.5.1. Thus we can assume n is odd and $\pi = -1$. Then if $n \geq 5$, the lemma follows from 5.2.7.5, while if $n = 3$ it follows from 5.2.5.1. □

LEMMA 5.3.9. *Suppose $\bar{L}$ is $Spin_{2n+1}(q)$ with $n > 2$. Then*
(1) $|\bar{G}_W : \bar{M}| = |G_W : M| = 2$.
(2) $\bar{G}_W \cong \bar{\omega}(C_n, m)$ and $G_W \cong \omega(C_n, m)$.

PROOF. As $G_W = S_LM$, (1) follows from 5.2.7.3 and parts (4) and (5) of 5.3.5.

Let $A = \langle\Gamma\rangle$ and $B = A^{\perp}$. Then B is a nonsingular point. Set $E = B + U_{1,1}$ and $Y = C_L(E^{\perp})$. Then E is nondegenerate of dimension 3 and Y acts faithfully as $\Omega(E)$ on E. Further $S_Y = S_L \cap Y$ is Sylow in Y and isomorphic to $D_{m/2}$. Next $C_{S_Y}(B) = W \cap Y = \langle x\rangle$ is of index 2 in S_Y, acts on $U_{1,1}$, and an involution $r \in S_Y - \langle x\rangle$ acts with eigenvalue -1 on B. Also $\bar{r}^2 = \bar{u}$. By (1), $G_W = M\langle r\rangle$.

Next (cf. 15.8 in [**Asc77**]) $K_1^r = K_1'$ and r centralizes L_{Q+R} for $Q, R \in \Gamma - \{U_{1,1}\}$, so (2) follows from 5.9.2 and 5.3.5.4 and its proof, which say that $\bar{M}$ satisfies the relations in 4.3.1. □

LEMMA 5.3.10. *Suppose $\bar{L}$ is $Spin^{-\pi}_{2n+2}(q)$. Then*
(1) $|G_W : M| = 4$.
(2) $G_W \cong 2\bar{\omega}(C_n, m)$.

PROOF. As in the proof of 5.3.8, to prove (1) it suffices to show that $|\bar{L}| \sim 4|\bar{M}|$, and $|\bar{M}| \sim n!m^n/2$. Now appeal to parts (2) and (4) of 5.2.7 and to 5.2.5.2 when $n = 2$.

Define A, B, E, Y, and S_Y as in the proof of 5.3.9. This time E is of dimension 4 and sign $-\pi$, $Y \cong \Omega^-_4(q) \cong L_2(q^2)$, and S_Y is dihedral of order m. Also $C_{S_Y}(B) = \langle y\rangle$ is of order $m/2$ and centralizes W and L_{Q+R} for $Q, R \in \Gamma - \{U_{1,1}\}$. From (1), $G_W = M\langle y, r\rangle$.

Let v generate V_{U_1}, $k \in K_1 - V_{U_1}$, and $x = [r, v]$. As W acts on E, $x \in S_Y$. As $K_1^r = K_1'$, $x = v^{-r}v$ is of order $m/4$, so as $|y| = m/2$, x generates $\langle y^2\rangle$. Next $k^x = k^{v^{-r}v} = k^v = kv^2$. Then from the computation in the last paragraph of the proof of 5.7.5, we may choose y so that $y^2 = x$ and $k^y = kv$. Now (2) follows from 5.9.3. □

LEMMA 5.3.11. *Assume $\tau = (\mathcal{F}, \Omega)$ is the Lie packet τ_L.*
(1) Suppose $\mathcal{F} \cong \Omega_7[m]$, $\Omega^{\epsilon}_8[m]$, $\Omega_9[m]$, or $\Omega^-_{10}[m]$. Then $\mathcal{F}$ contains no subsystem $\mathcal{Y} = \mathcal{Y}_1 \times \mathcal{Y}_2$ with $\mathcal{Y}_1 \cong \Omega^-_4[m]$ and $\mathcal{Y}_2 = \mathcal{Y}_2^{\circ} \cong \Omega^-_6[m]$.
(2) Suppose $\mathcal{F} \cong \Omega^-_{2n+2}[m]$ or $\Omega_{2n+1}[m]$ with $d \geq 6$, and $\mathcal{F}$ contains a subsystem $\mathcal{Y} = \mathcal{Y}_1 \times \cdots \times \mathcal{Y}_r$ with $r = |\mathcal{Z}_S| \geq 1$ and $\mathcal{Y}_i = \mathcal{Y}_i^{\circ} \cong \Omega^-_6[m]$. Then $r = 1$ so $n = 2$ or 3.

PROOF. We regard $\mathcal{F}$ as $\mathcal{F}_S(L)$ with $L = \Omega^{\epsilon}_d(q)$ and U the defining orthogonal space for L. We first prove (1).

As $\mathcal{Y}_2 = \mathcal{Y}_2^\circ$, we have $\mathcal{Y}_2 = \mathcal{F}_{T_2}(M_2)$ where $V_2 = [U, M_2]$ is nondegenerate of dimension 6 and sign $-\pi$. Then $\mathcal{Y} = \mathcal{F}_T(M)$ where $M = M_1 \times M_2$, $\mathcal{Y}_1 = \mathcal{F}_{T_1}(M_1)$, and $M_1 \cong \Omega_4^-(q)$. As M_2 is irreducible on V_2 we have $M_1 \leq C_L(V_2)$, so $V_1 = [V, M_1] \leq V_2^\perp$. Now $\dim(V_1) \geq 4$ so $V = V_1 + V_2$ is of dimension at least 10. Thus $V = U$, $d = 10$, and $\dim(V_1) = 4$. As M_1 is faithful on V_1, V_1 is of sign -1, so U is of sign $(-1)(-\pi) = \pi$, contrary to the hypothesis of (1). This completes the proof of (1).

The proof of (2) is similar. This time $\mathcal{Y} = \mathcal{F}_T(M)$ where $M = M_1 \times \cdots \times M_r$ with $M_i \cong \Omega_6^{-\pi}(q)$ and $V_i = [V, M_i]$ of dimension 6 and sign $-\pi$. Then $V = \langle V_i : 1 \leq i \leq r\rangle$ is of dimension $6r$. If $d = 2n + 2$ then from 5.3.2 and 5.3.3, $n = 2r$, so $6r \leq d = 2n + 2 = 4r + 2$, and hence $r = 1$. If $d = 2n + 1$ then $n = 2r + 1$, so $6r \leq d = 2n + 1 = 4r + 3$, and again $r = 1$. □

5.4. Linear, unitary, and symplectic groups and packets

In this section we assume the following hypothesis:

HYPOTHESIS 5.4.1. *Assume q is an odd prime power and set $\pi = \pm 1$ where $q \equiv \pi \mod 4$. Set $m = (q^2 - 1)_2$. Let U be an finite dimensional vector space over a field F, and assume one of the following holds:*

(I) $F = \mathbf{F}_q$ and $G = SL(U)$.

(II) $F = \mathbf{F}_{q^2}$ and $G = SU(U)$ is an isometry group of a unitary form $(\cdot, \cdot)$ on U.

(III) $F = \mathbf{F}_q$ and $G = Sp(U)$ is an isometry group of a symplectic form $(\cdot, \cdot)$ on U.

NOTATION 5.4.2. Write $U = \mathfrak{U}_1 \perp \cdots \perp \mathfrak{U}_r \perp \mathfrak{U}_{r+1}$, where $\dim(\mathfrak{U}_i) = 2$ for $1 \leq i \leq r$ and $\dim(\mathfrak{U}_{r+1}) = a \in \{0, 1\}$, with $a = 0$ in (III). Let $\Sigma = \{\mathfrak{U}_i : 1 \leq i \leq r + 1\}$ and $\mathfrak{N} = N_G(\Sigma)$. For $1 \leq i \leq r$, set $\mathfrak{U}_i' = \langle \mathfrak{U}_j : j \neq i\rangle$. $L_i = C_G(\mathfrak{U}_i') \cap N_G(\mathfrak{U}_i)$, and $K_i \in Syl_2(L_i)$. Set $\Omega = \{K_i : 1 \leq i \leq r\}$.

LEMMA 5.4.3. *Let $\mathcal{U}$ be the set of pairs $\mathfrak{u} = (\mathfrak{U}, \mathfrak{U}')$ such that $U = \mathfrak{U} \oplus \mathfrak{U}'$, $\dim(\mathfrak{U}) = 2$, and in (II) and (III) we have $\mathfrak{U}$ nondegenerate and $\mathfrak{U}' = \mathfrak{U}^\perp$. Then*

(1) for each $\mathfrak{u} \in \mathcal{U}$, $L = L(\mathfrak{u}) = C_G(\mathfrak{U}') \cap N_G(\mathfrak{U})$ is a fundamental subgroup of G.

(2) $\mathfrak{U} = [U, L] = [U, z(L)]$ and $\mathfrak{U}' = C_U(L) = C_U(z(L))$.

(3) L acts naturally on $\mathfrak{U}$ as $SL(\mathfrak{U})$, $SU(\mathfrak{U})$, $Sp(\mathfrak{U})$ in (I), (II), (III), respectively.

PROOF. A fundamental subgroup X of G is of the form $X = \langle Z, Z^-\rangle \cong SL_2(q)$, with $Z \cong E_q$ the center of a long root subgroup of G and Z^- an opposite of Z. In particular Z is a group of transvections on U with center $U(Z)$ and axis $A(Z)$, $U(Z^-)$ is a complement to $A(Z)$ in U, and $A(Z^-)$ is a complement to $U(Z)$ in U. Therefore $[U, X] = U(Z) \oplus U(Z^-)$ is a line in U, and $C_U(X) = A(Z) \cap A(Z^-)$ is a complement to $[U, X]$, with $[U, X]$ a natural module for X. Moreover in (II) and (III), $A(Z) = U(Z)^\perp$. Now the lemma follows from Witt's Lemma. □

LEMMA 5.4.4. *Let X_1 be a fundamental subgroup of G, $U_1 = [U, X_1]$, and $z_1 = z(X_1)$. Suppose $z_2 \in z_1^G$ with $U_1 = [U, z_2]$, and in (I) assume $[z_1, z_2] = 1$. Then $z_1 = z_2$.*

PROOF. Claim z_2 acts on $A = C_U(z_1)$. If (I) holds then by hypothesis, $[z_1, z_2] = 1$, so the claim is trivial. In (II) and (III), $A = U_1^{\perp}$ by 5.4.3, so again the claim holds.

By hypothesis, $[U, z_1] = U_1 = [U_1, z_2]$. Then as z_2 acts on A by the claim, it follows that $A = C_U(z_2)$, so $z_1 = z_2$. □

LEMMA 5.4.5. *Let X_i, $i = 1, 2$, be fundamental subgroups of G, $U_i = [U, X_i]$, $S_i \in Syl_2(X_i)$, and $z_i = z(X_i)$. Then the following are equivalent:*

(1) $[X_1, X_2] = 1$.

(2) $[S_1, S_2] = 1$.

(3) z_1 centralizes S_2 and $z_1 \neq z_2$.

(4) $U_i \leq C_U(z_{3-i})$ for $i = 1, 2$.

PROOF. Trivially, (1) implies (2).

Suppose (2) holds. Then $[z_1, S_2] \leq [S_1, S_2] = 1$. Also S_1 is irreducible on U_1, so as $[S_1, S_2] = 1$, it follows that $z_2 \in [S_2, S_2] \leq C_G(U_1)$, and hence $z_1 \neq z_2$. That is (2) implies (3).

Suppose (3) holds. As S_2 is irreducible on U_2, either $U_2 = [U_2, z_1]$ or $U_2 \leq C_U(z_1) = A$, and the latter holds by 5.4.4. Therefore $U_1 \cap U_2 = 0$, so as $[z_1, z_2] = 1$, it follows that (4) holds.

Finally assume (4) holds. Then $[X_i, U] = U_i \leq C_U(X_{3-i}) = A_{3-i}$, so X_{3-i} acts on U_i and on A_i. Then by 5.4.3, X_{3-i} acts on X_i, so $[X_1, X_2] \leq X_1 \cap X_2 = 1$, as X_1 centralizes A_1 and X_2 is faithful on $U_2 \leq A_1$. Thus (4) implies (1). □

LEMMA 5.4.6. *(1) $\langle L_i : 1 \leq i \leq r\rangle = L_1 \times \cdots \times L_r$.*

(2) $O = \langle K_i : 1 \leq i \leq r\rangle = K_1 \times \cdots \times K_r$.

(3) Let $S \in Syl_2(G)$ with $O \leq S$ and set $\mathcal{F} = \mathcal{F}_S(G)$, $\Omega = \{K_i : 1 \leq i \leq r\}$, and $\tau = (\mathcal{F}, \Omega)$. Then τ is a Lie packet of G.

(4) $\mathcal{F}$ is transitive on Ω and $\Omega(z(L_1)) = \{K_1\}$.

PROOF. Set $L_0 = \langle L_i : 1 \leq i \leq r\rangle$. By 5.4.5, L_0 is a central product of the L_i. Set $L_i' = \langle L_j : j \neq i\rangle$. Then $L_i \cap L_i' \leq C_{L_i}(\mathfrak{U}_i) = 1$, so L_0 is a direct product of the L_i, proving (1). Then (1) implies (2).

Set $\Omega_0 = K_1^G \cap S$ and $\tau_0 = (\mathcal{F}, \Omega_0)$; then τ_0 is a Lie packet of G by 5.4.3. By construction, $\Omega \subseteq \Omega_0$, so to complete the proof of (2) we may assume $K \in \Omega_0 - \Omega$, and it remains to produce a contradiction. By 3.1.5, K centralizes O, so by 5.4.5, $W = \mathfrak{U}_1 + \cdots + \mathfrak{U}_r \leq C_U(K)$. But $U = [U, K] \oplus C_U(K)$, so

$$d = \dim(U) \geq \dim(W) + \dim([U, K]) = 2r + 2 > d,$$

for our contradiction. Thus (3) holds.

As G is transitive on it fundamental subgroups, $\mathcal{F}$ is transitive on Ω. By (2), $\Omega(z(K_i)) = \{K_i\}$. □

NOTATION 5.4.7. Define S, $\mathcal{F} = \mathcal{F}_S(G)$, and $\tau = (\mathcal{F}, \Omega)$ as in 5.4.6. Set $\mathcal{Z} = \mathcal{Z}(\tau)$, $\mathcal{Z}_S = \mathcal{Z}_S(\tau)$, and $z_i = z(L_i)$. For $H \leq N_G(O)$, write H^Ω for the image of H in $Sym(\Omega)$ under the conjugation map.

LEMMA 5.4.8. *(1) If $t \in z_1^G \cap N_G(O)$ with $t^\Omega \neq 1$ then $t^\Omega = (K_i, K_j)$ is a transposition and t centralizes each member of $\Omega - \{K_i, K_j\}$.*

(2) For each $i \neq j$ there exists $t \in z_1^G \cap N_G(O)$ with $t^\Omega = (K_i, K_j)$.

(3) $N_G(O)^\Omega = Sym(\Omega)$.

PROOF. Assume the setup of (1). Then for some distinct i,j, t^Ω has a cycle (K_i, K_j). Therefore t interchanges $\mathfrak{U}_i$ and $\mathfrak{U}_j$, so setting $W = \mathfrak{U}_i + \mathfrak{U}_j$, we have $\dim([W,t]) = 2$. Then as $\dim([U,t]) = 2$, we have $[U,t] = [W,t] \leq W$ and hence t centralizes $\mathfrak{U}_k$ for $k \notin \{i,j\}$. Therefore t also centralizes K_k, proving (1).

Assume the setup of (2). There exists an involution $t \in \mathfrak{N}$ interchanging $\mathfrak{U}_i$ and $\mathfrak{U}_j$ and centralizing $\mathfrak{U}_k$ for $k \notin \{i,j\}$. Then $[U,t] = [W,t]$ is a line, so $t \in z_1^G$, and hence (2) holds. Finally (2) implies (3). □

LEMMA 5.4.9. *Assume $G = Sp(U)$.*

(1) $C_{\mathcal{Z}}(z_1) = z_1^\perp = C_{\mathcal{Z}}(K_1)$.

(2) $C_G(K_1) = \langle z_1 \rangle \times C_G(\mathfrak{U}_1)$.

(3) $C_G(O) = \langle \mathcal{Z}_S \rangle$.

(4) $C_S(z_1) = K_1 C_S(K_1)$.

(5) Let X be the kernel of the action of $N_G(O)$ on Ω. Then either $X = O$ or $m = 8$ and $X = X_1 \times \cdots \times X_r$, where $SL_2(3) \cong X_i = N_{L_i}(K_i)$.

(6) If $m > 8$ then $N_G(S) = S$.

PROOF. By 3.3.4, $C_{\mathcal{Z}}(K_1) = z_1^\perp$. Assume $z \in C_{\mathcal{Z}}(z_1) - z_1^\perp$. Then $0 \leq e = \dim([\mathfrak{U}_1, z]) \leq 2$. By 5.4.4, $e \neq 2$. As $[z, K_1] \neq 1$ we conclude from 5.4.5 that $e \neq 0$. Therefore $e = 1$, so $\mathfrak{U}_1 = Fu + Fv$ where z centralizes u and $vz = -v$. Then $0 \neq (u,v) = (uz, vz) = (u, -v) = -(u,v)$, contradicting $z \in G = Sp(U)$. This proves (1).

Let $g \in C_G(K_1)$. As K_1 is absolutely irreducible on $\mathfrak{U}_1$, g acts on $\mathfrak{U}_1$ via scalar multiplication by some $a \in F$. As $g_{|\mathfrak{U}_1} \in Sp(\mathfrak{U}_1)$, we have $a = \pm 1$, so $gz_1^\epsilon \in C_G(\mathfrak{U}_1)$ for some $\epsilon \in \{0,1\}$, establishing (2). Then (2) implies (3).

Next $C_S(z_1)$ induces a subgroup of $Sp(\mathfrak{U}_1) = L_1$ on $\mathfrak{U}_1$, so as K_1 is Sylow in L_1, (4) follows.

Define X as in (5). Suppose $m > 8$. Then $Aut(K_1)$ is a 2-group, so $X = K_1 C_X(K_1)$ by (4), and then $X = O$ by (3). Therefore we may assume that $m = 8$, so $X_i = N_{L_i}(K_i) \cong SL_2(3)$. In particular $Aut_{X_i}(K_i) = O^2(Aut(K_i))$, so (5) follows from (3) and (4).

Let $Y = N_G(S)$ and assume $m > 8$. As $\Omega = K_1^G \cap S$ by 5.4.6, we have $O \trianglelefteq Y$. By (5), $X = O$, so by 5.4.8.3, $N_G(O)/O = N_G(O)^\Omega = Sym(\Omega)$. Then as $N_{Sym(\Omega)}(S^\Omega) = S^\Omega$, we have $Y^\Omega = S^\Omega$, so $Y = S$, establishing (6). □

LEMMA 5.4.10. *Assume $G = SL^{-\pi}(U)$.*

(1) $C_S(K_1) = \langle z_1 \rangle \times C_S(\mathfrak{U}_1)$.

(2) $\Delta(\tau) = \Omega$.

(3) $C_S(O) = \langle \mathcal{Z}_S \rangle$.

(4) For each $i > 1$ there exists $s_{1,i} \in \mathcal{Z}$ inducing an outer automorphism on L_1 in $PGL_2(q)$ with $C_{K_1}(s_{1,i}) = \langle z_1 \rangle$ and $\dim([\mathfrak{U}_1, s_{1,i}] = \dim([\mathfrak{U}_i, s_{1,i}] = 1$. Therefore $s_{1,i}$ centralizes K_j for $j \notin \{1,i\}$, and if $i \leq r$ then $C_{K_i}(s_{1,i}) = \langle z_i \rangle$.

(5) Let $L_0 = L_1 \cdots L_r$ and $L_O = N_{L_0}(O)$. If $m > 8$ then $L_O = O$, while if $m = 8$ then $L_O = X_1 \times \cdots \times X_r$, where $SL_2(3) \cong X_i = N_{L_i}(K_i)$.

(6) Let X be the kernel of the action of $N_G(O)$ on Ω. Then $X = O(X) L_O X_S$, where $X_S = \langle s_{1,i} : 1 < i \leq r+1 \rangle$ and $X/O(X)L_O \cong E_{2^{r-1+a}}$.

(7) $Aut_G(S) = Inn(S)$.

PROOF. Set $Y_1 = C_S(K_1)$. As K_1 is irreducible on $\mathfrak{U}_1$, $\bar{Y}_1 = Aut_{Y_1}(\mathfrak{U}_1)$ is cyclic and as $G = SL^{-\pi}(U)$, we have $\bar{Y}_1 \leq Z(GL^{-\pi}(\mathfrak{U}_1) \cong \mathbf{Z_{q+\pi}}$. But $q \equiv \pi \mod 4$, so $(q+\pi)_2 = 2$, and hence $\bar{Y}_1 = \langle \bar{z}_1 \rangle$, so (1) holds. Then (1) implies (3).

Suppose (2) fails. Then there is $V \in \Delta(\tau)$ such that V is cyclic of order $m/2$, and we may take $[V, K_1] \neq 1$. Then by 4.2.5, $T = VK_1$ is the wreath product of V by $\mathbf{Z_2}$, so $Z(T) = \langle x \rangle$ is of order $m/2$ with $z_1 \in \langle x \rangle$. But then $\bar{x} \in \bar{Y}_1$ is of order $m/2$, contrary to (1). This proves (2).

Let $i > 1$, W_1 a point of $\mathfrak{U}_1$, and W_i a point of $\mathfrak{U}_i$. Choose W_1 and W_i nondegenerate if G is $SU(U)$ and set $W = W_1 + W_i$. Replacing W by a suitable conjugate under L_0, there exists $s \in z_1^G \cap S$ with $W = [U, s]$. In particular as S acts on O, s acts on $\mathfrak{U}_1$ and $\mathfrak{U}_i$, and then s centralizes $\mathfrak{U}_j$ for each $j \notin \{1, i\}$. As $W_1 = [\mathfrak{U}_1, s]$ is a point and L_1 induces $SL^{-\pi}(\mathfrak{U}_1)$ on $\mathfrak{U}_1$, we conclude that s induces an outer automorphism in $PGL_2(q)$ on L_1 with $C_{K_1}(s) = \langle z_1 \rangle$. We now conclude that (4) holds.

The proof of (5) is essentially the same as that of 5.4.9.5. Moreover $Aut_X(\mathfrak{U}_1) = \langle \bar{s} \rangle Aut_{X \cap L_1}(\mathfrak{U}_1)$, so it follows from (3) and (4) that $X = O(X)L_O X_S$, and then that (6) holds.

Set $Y = N_G(S)$. Arguing as in the proof of 5.4.9.6, $Y = SX$. Then as $[S, O(X)] \leq S \cap O(X) = 1$, it follows from (6) that $Aut_G(S) = Inn(S)Aut_X(S) = Inn(S)Aut_{L_O}(S)$. Hence (7) holds if L_0 is a 2-group, so from (5) we may assume $m = 8$. Now for x_i of order 3 in X_i, x_i is inverted by some $s_{1,i}$, so $N_{L_O}(S) = O$. This completes the proof of (7) and the lemma. □

In the remainder of the section we assume the following hypothesis:

HYPOTHESIS 5.4.11. *Hypothesis* 5.4.1 *holds and* $G = SL^{\pi}(U)$ *with* $\dim(U) \geq 3$.

NOTATION 5.4.12. Let $\{V_i : 1 \leq i \leq r\}$ be an S-invariant collection of cyclic subgroups with V_i of index 2 in K_i. As L_i induces $SL^{\pi}(\mathfrak{U}_i)$ on $\mathfrak{U}_i$, V_i has two weight spaces $\{U_{i,1}, U_{i,2}\}$ on $\mathfrak{U}_i$, each of dimension 1, with $U_{i,j}$ nondegenerate if $\pi = -1$. Set $\Gamma_0 = \{U_{i,j} : 1 \leq i \leq r,\ j = 1, 2\}$, and take $\Gamma = \Gamma_0$ if $a = 0$ and $\Gamma = \Gamma_0 \cup \{\mathfrak{U}_{r+1}\}$ if $a = 1$. Let $\lambda \in F^{\#}$ of order $m/2$. Then for distinct $A, B \in \Gamma$, there is $v_{A,B} \in S$ with eigenvalue λ, λ^{-1} on A, B, respectively, and centralizing each member of $\Gamma - \{A, B\}$. Set $V_{A,B} = \langle v_{A,B} \rangle$, $\eta = \{V_{A,B} : A, B \in \Gamma \text{ distinct}\}$, and $W = \langle \eta \rangle$. Set $\mathfrak{M} = N_G(\Gamma)$ and $d = \dim(U)$.

LEMMA 5.4.13. *(1)* η *is the unique member of* $\eta(\tau)$.
(2) $\mathfrak{M} = N_G(W)$.
(3) $C_G(W) = WO(C_G(W))$ *is the kernel of the action of* $\mathfrak{M}$ *on* Γ.
(4) $\mu(\tau) \cong Sym(\Gamma) \cong S_d$.
(5) $\mathfrak{M}/O(\mathfrak{M}) \cong \bar{\omega}(A_{d-1}, m)$.
(6) $Aut_G(S) = Inn(S)$.

PROOF. By construction, S acts on $\mathcal{V} = \{V_i : 1 \leq i \leq r\}$, and hence also on the set Γ of weight spaces for $\langle \mathcal{V} \rangle$. For pairs (A, B), (C, D) of distinct members of Γ, $V_{A,B}$ and $V_{C,D}$ are in the kernel of the action of $\mathfrak{M}$ on Γ, and by construction the induced actions of the subgroups on each member of Γ commute. Therefore W is an abelian 2-group. Then as S acts on W and $S \in Syl_2(G)$, it follows that $W \leq S$.

From 3.3.10.1, $\eta_0 = \eta - \{V_1, \ldots, V_r\} \subseteq \eta_1 \in \eta(\tau)$. As $d \geq 3$ by 5.4.11, Γ is the set of weight spaces of $W_0 = \langle \eta_0 \rangle$. By 3.1.11.2, $W_1 = \langle \eta_1 \rangle$ is abelian, so W_1 acts on each member of the set Γ of weight spaces for W_0. Therefore for $V \in \eta_1$, $V = V_{A,B}$ for some $A, B \in \Gamma$, so $\eta = \eta_1$. Thus $\eta \in \eta(\tau)$, and indeed the proof shows

η is the *unique* member of $\eta(\tau)$, proving (1). As Γ is the set of weight spaces of W, (2) holds. Further the proof shows that the kernel $\mathfrak{M}_0$ of the action of $\mathfrak{M}$ on Γ is abelian with $W \in Syl_2(\mathfrak{M}_0)$, so $\mathfrak{M}_0 = WO(C_G(W))$, establishing (3).

For $k \in K_i - V_i$, $k^\Gamma = (U_{i,1}, U_{i,2})$ is a transposition. Similarly for distinct $A, B \in \Gamma$, a Sylow 2-subgroup $K_{A,B}$ of $N_G(\{A,B\}) \cap C_G(\langle \Gamma - \{A,B\}\rangle)$ is in K_1^G, so $\mathfrak{M}^\Gamma$ contains the transposition (A, B). This proves (4).

Set $\mathfrak{M}^* = \mathfrak{M}/O(\mathfrak{M})$. By (4) and 4.3.3.2, $\mathfrak{M}^*$ is an image of $\bar{\omega}(A_{d-1}, m)$. By (3) and (4), $\mathfrak{M}/O(\mathfrak{M}W) \cong S_d$. Let $W_{2i-1} = V_i$ for $1 \leq i \leq r$, $W_{2i} = V_{U_{i,2},U_{i+1,1}}$ for $1 \leq i < r$, and if $a = 1$ let $W_{2r} = V_{U_{r,2},U_{r+1}}$. Then $W = W_1 \times \cdots W_{2r-1} \times W_0$, where $W_0 = 1$ if $a = 0$ and $W_0 = W_{2r}$ if $a = 1$. Therefore W is the direct sum of $d-1$ copies of $\mathbf{Z_{m/2}}$, so (5) follows from 4.3.3.5.

As W is weakly closed in S with respect to G, we have $N_G(S) \leq N_G(W) = \mathfrak{M}$. As usual $[N_{O(\mathfrak{M})}(S), S] \leq O(\mathfrak{M}) \cap S = 1$, so from (3), $Aut_G(S) = Aut_{\mathfrak{M}^*}(S^*)$. But by (4), $N_{\mathfrak{M}^*/W^*}(S^*/W^*) = S^*/W^*$, so $Aut_G(S) = Inn(S)$, proving (6). □

5.5. Exceptional groups and packets

In this section G is a 2-simply connected exceptional group of Lie type over $\mathbf{F}_q$ with $q \equiv \pi \mod 4$, where $\pi = \pm 1$, and G is of Lie rank at least 4. Thus G is $F_4(q)$, $E_6^\epsilon(q)$, $\tilde{E}_7(q)$, or $E_8(q)$. Let $S \in Syl_2(G)$, $\mathcal{F} = \mathcal{F}_S(G)$, and $\tau = (\mathcal{F}, \Omega)$ a Lie packet for G.

We will see that there is a unique $\eta \in \eta(\tau)$. Set $W = \langle \eta \rangle$ and $\mathfrak{M} = N_G(W)$. Pick $K \in \Omega$ and set $z = z(K)$, $m = (q^2-1)_2$, and $O = O(\tau)$. Set $\mathfrak{M}_K = \langle K^{\mathfrak{M}} \rangle$. Let $K_2 \in \Omega - \{K\}$ and set $z_2 = z(K_2)$, $u = zz_2$, and $X = \langle K_2^{C_G(z)} \rangle$.

Among other things, we wish to update proofs of some of the results in [**Asc80**], so we will prove:

THEOREM 5.5.1. *(1) If G is $F_4(q)$ or $E_6^\epsilon(q)$ then $|\Omega| = 4$ and $N_G(O)^\Omega = Sym(\Omega)$.*

(2) If G is $\tilde{E}_7(q)$ then $|\Omega| = 7$ and $N_G(O)^\Omega = L_3(2)$.

(3) If G is $E_8(q)$ then $|\Omega| = 8$ and $N_G(O)^\Omega = E_8L_3(2)$ is 3-transitive.

THEOREM 5.5.2. *(1) If G is $F_4(q)$ or $E_6^{-\pi}(q)$ then $\mu(\tau) \cong Weyl(D_4)$ and $\mathfrak{M}_K/O(\mathfrak{M}_K) \cong \bar{\omega}(D_4, m)$.*

(2) If G is $E_6^\pi(q)$ then $\mu(\tau) \cong Weyl(E_6) \cong O_6^-(2)$ and $\mathfrak{M}_K/O(\mathfrak{M}_K) \cong \bar{\omega}(E_6, m)$.

(3) If G is $\tilde{E}_7(q)$ then $\mu(\tau) \cong Weyl(E_7) \cong \mathbf{Z}_2 \times Sp_6(2)$ and $\mathfrak{M}_K/O(\mathfrak{M}_K) \cong \bar{\omega}(E_7, m)$.

(4) If G is $E_8(q)$ then $\mu(\tau) \cong Weyl(E_8) \cong O_8^+(2)/\mathbf{Z}_2$ and $\mathfrak{M}_K/O(\mathfrak{M}_K) \cong \bar{\omega}(E_8, m)$.

LEMMA 5.5.3. *(1) $X \cong Sp_6(q)$, $SL_6^\epsilon(q)$, $Spin_{12}^+(q)$, $\tilde{E}_7(q)$ for G equal to $F_4(q)$, $E_6^\epsilon(q)$, $\tilde{E}_7(q)$, $E_8(q)$, respectively.*

(2) $z \in X$.

(3) Assume G is $E_6^\pi(q)$, $\tilde{E}_7(q)$, or $E_8(q)$ and let l be 6, 7, or 8, respectively. Then $|W| \geq (m/2)^l$.

PROOF. As τ is a Lie packet for G, K is Sylow in $SL_2(q) \cong L \lhd C_G(z)$. Inspecting Table 4.5.1 in [**GLS98**] for such a centralizer, we find that (1) and (2) hold.

Assume the setup of (3). Let $K \in \mathcal{F}^f$, $\tau_X = (X, \Omega - \{K\})$, $\eta_X \in \eta(\tau_X)$, and $W_X = \langle \eta_X \rangle$. Then $\eta_X \subseteq \eta$ so $W_X \leq W$. Claim $|W_X| \geq (m/2)^{l-1}$. If $l = 6$ this follows from (1) and the proof of 4.2.13.5. If l is 7 or 8 it follows from (1) and induction on l. Let $V \in \eta$ with $[K, V] \neq 1$. Then $[K, z(V)] \neq 1$, so $V \cap W_X = 1$ and hence $|W| \geq |W_X||V| = (m/2)^L$, completing the proof of (3). □

Recall that $\mu = \mu(\tau) = M/C_M(W)$ is generated by the set $D = \{d_J : J \in K^M\}$, where d_J is the involutory automorphism induced by J on W. Moreover D is a set of 3-transpositions of μ, and we adopt the notation from section 4.1 in discussing this 3-transposition group.

LEMMA 5.5.4. *(1) If G is $F_4(q)$ or $E_6^{-\pi}(q)$ then $|\Omega| = 4$ and $\langle D_{d_K} \rangle \cong E_8$.*
(2) If G is $E_6^{\pi}(q)$ then $|\Omega| = 4$ and $\langle D_{d_K} \rangle \cong S_6$.
(3) If G is $\tilde{E}_7(q)$ then $|\Omega| = 7$ and $\langle D_{d_K} \rangle \cong Weyl(D_6) \cong S_6/E_{32}$.
(4) If G is $E_8(q)$ then $|\Omega| = 8$ and $\langle D_{d_K} \rangle \cong Weyl(E_7) \cong \mathbf{Z}_2 \times Sp_6(2)$.

PROOF. Let $\xi = (\mathcal{E}, \Gamma)$ be the fusion packet for X and write d for d_K, and observe that $\nu = \langle D_d \rangle \cong \mu(\xi)$. However we know $\mu(\xi)$ when X is in the first three cases of 5.5.3, and we know $\mu(\xi)$ by induction on the order of G in the last case when X is $\tilde{E}_7(q)$. In particular we know $|\Gamma| = |\Omega| - 1$, so the lemma follows. □

Observe that we can now prove Theorem 5.5.1. Namely from 5.5.4, $|\Omega|$ is as indicated in Theorem 5.5.1. Further setting $Y = N_G(O)$, Y^Ω is transitive and $N_Y(K)^\Omega$ contains $N_X(O)^\Omega$ as a normal subgroup. In case (1) of 5.5.1, $N_X(O)^\Gamma = Sym(\Gamma)$, so $Y^\Omega = Sym(\Omega)$. In case (2) of 5.5.1, $N_X(O)^\Omega$ acts transitively as S_4 on the six points of Γ, so Y^Ω is $L_3(2)$ on the seven points of the projective plane Ω. Finally in case (3) of 5.5.1, by induction $N_X(O)^\Omega$ acts 2-transitively as $L_3(2)$ on the projective plane Γ, so Y^Ω is the 3-transitive group $E_8L_3(2)$. This completes the proof of Theorem 5.5.1.

LEMMA 5.5.5. *(1) If G is not $\tilde{E}_7(q)$ then $u \notin z^G$.*
(2) If G is $\tilde{E}_7(q)$ then $u = az_3$ with $z_3 \in \mathcal{Z}_S - \{z, z_2\}$ and $Z(G) = \langle a \rangle$.

PROOF. Let $H = N_G(O)$ and observe the actions of H on Ω and $\mathcal{Z}_S$ are equivalent. But if $u \in z^G$ then H_{z,z_2} fixes a third member u of $\mathcal{Z}_S$, so H^Ω is not 3-transitive, and hence G is $\tilde{E}_7(q)$ by 5.5.1. This proves (1), so it remains to prove (2), and hence we may take G to be $\tilde{E}_7(q)$.

Set $Z(G) = \langle a \rangle$ and $G^* = G/Z(G)$. From 5.5.3, X^* is an image of $Spin_{12}^+(q)$ with $\langle z^* \rangle = Z(X^*)$ and there is a pairing $\{v, \bar{v}\}$ on $\mathcal{V} = \mathcal{Z}_S - \{z\}$ such that $v^*\bar{v}^* = z^*$. Then setting $E = \langle \mathcal{Z}_S \rangle$, $E^* \cong E_8$ and from 5.5.1, $Aut_H(E^*) = GL(E^*)$. Therefore $u = zz_2 \in z_3\langle a \rangle$. Finally if $u = z_3$ then $1 = zz_2z_3$ in G. But by 5.5.3, in $B = E_8(q)$ we may take $a = z(L)$ for some fundamental subgroup L centralizing G, and by 5.5.1, $A = N_B(O_B)$ is 3-transitive on Ω_B, whereas A_{z,z_2} fixes z_3, a contradiction. Thus $u = az_3$, proving (2). □

LEMMA 5.5.6. *Assume G is not $\tilde{E}_7(q)$. Then*

(1) G has two classes of involutions with representatives z and u.

(2) $O^{p'}(C_G(u))$ is isomorphic to $Spin_9(q)$, $Spin_{10}^{\epsilon}(q)$, or $HSpin_{16}^{+}(q)$ for G equal to $F_4(q)$, $E_6^{\epsilon}(q)$, or $E_8(q)$, respectively.

(3) $u \in Z(S)$.

PROOF. From Table 4.5.1 in [**GLS98**], G has two classes of involutions, and then by 5.5.5.1, z and u are representatives for the two classes, so (1) holds. Moreover (2) follows from the description of the involution centralizers in Table 4.5.1 of [**GLS98**] together with 5.5.3. By 5.5.1, $|\Omega|$ is even, so $z \notin Z(S)$, and hence (3) holds. □

LEMMA 5.5.7. *Assume G is $F_4(q)$ or $E_6^{-\pi}(q)$. Then*

(1) $\mu(\tau) \cong Weyl(D_4)$.

(2) $\eta(\tau)$ has a unique member η.

(3) Set $E = C_W(\mathfrak{M}_K)$. Then $u \in E \cong E_4$ with $\mathfrak{M}_K/O(\mathfrak{M}_K) \cong \bar{\omega}(D_4, m)$, $O^{p'}(C_G(E)) \cong Spin_8^{+}(q)$, and $Aut_G(E) = GL(E)$.

(4) $C_G(W) = C_S(W)O(\mathfrak{M})$ with $\mathfrak{M}/C_G(W) \cong Weyl(F_4)$.

(5) $Aut_{\mathcal{F}}(S) = Inn(S)$.

PROOF. By 5.5.6.3, $u \in Z(S)$, so $\nu = (\mathcal{F}_S(C_G(u)), \Omega)$ is a quaternion fusion packet with $\nu \leq \mu = \mu(\tau)$. By 5.5.6.2, $G_u = O^{p'}(C_G(u)) \cong Spin_9(q)$ or $Spin_{10}^{-\pi}(q)$, so $\nu \cong$ Weyl(D_4). On the other hand by 5.5.4.1 the subgroup $\langle d_K^{\perp} \rangle$ of $\mu = \langle D \rangle$ is abelian and $|\Omega| = 4$, so $\mathcal{D}(D)$ is disconnected, and hence (1) follows from 4.2.13.1 and 4.3.5.1. Then (1) and 3.3.14 imply (2). Indeed as $\mu \cong \nu$ it follows that $u \in Z(\mathfrak{M}_K)$. Then by 5.3.5.4 and 5.3.7 applied to G_u, $\mathfrak{M}_K/O(\mathfrak{M}_K) \cong \bar{\omega}(D_4, m)$. Hence by 5.8.4, $E = C_W(\mathfrak{M}_K) \cong E_4$. From the structure of G_u, $G_E = O^{p'}(C_G(E)) = O^{p'}(C_{G_u}(E)) \cong Spin_8^{+}(q)$. Then by 5.5.1.1, $Aut_G(E) = GL(E)$, completing the proof of (3).

From Table 4.5.1 in [**GLS98**], $\langle u \rangle$ is Sylow in $C_G(G_u)$ and $|C_G(u) : G_uC_G(G_u)| \leq 2$, so $C_G(W) = C_{C_G(u)}(W) = O(\mathfrak{M})C_S(W)$ by 5.3.7.1. As $\mathfrak{M}_K \trianglelefteq \mathfrak{M}$, we have $\mathfrak{M} \leq N_G(E)$ by (3). As W is weakly closed in S with respect to G by (2) and W is abelian, $Aut_G(E) = Aut_{\mathfrak{M}}(E)$, so $\mathfrak{M}/C_{\mathfrak{M}}(E) = GL(E)$ by (3). Then as $\mathfrak{M}_K/O(\mathfrak{M}_K) \cong \bar{\omega}(D_4, m)$, it follows that $\mathfrak{M}/C_G(W) \cong$ Weyl(F_4), establishing (4).

As W is weakly closed in S, $N_G(S) \leq \mathfrak{M}$, so $N_G(S) = N_{\mathfrak{M}}(S)$. Set $\mathfrak{M}^* = \mathfrak{M}/C_G(W)$. By (4), $N_{\mathfrak{M}^*}(S^*) = S^*$, so $N_G(S) = S(N_G(S) \cap C_G(W))$. Then as $N_{O(\mathfrak{M})}(S)$ centralizes S, (5) follows from the first statement in (4). □

LEMMA 5.5.8. *(1) μ is transitive on D.*

(2) If G is not $F_4(q)$ or $E_6^{-\pi}(q)$ then $O_2(\mu) \leq Z(\mu) \geq O_3(\mu)$.

(3) If G is $E_6^{\pi}(q)$ then $\mu \cong Weyl(E_6) \cong O_6^{-}(2)$.

(4) If G is $\tilde{E}_7(q)$ then $\mu \cong Weyl(E_7) \cong \mathbf{Z}_2 \times Sp_6(2)$.

(5) If G is $E_8(q)$ then $\mu \cong Weyl(E_8) \cong O_8^{+}(2)/\mathbf{Z}_2$.

PROOF. If G is $F_4(q)$ or $E_6^{-\pi}(q)$ then (1) follows from 5.5.7.1, so assume G is neither group. Let $d = d_K \in \mu$. Then by 5.5.4, $\nu = \langle D_d \rangle$ is transitive on D_d, ν is nonabelian, and for $b \in D_d$, $D_d \cap D_b \neq \emptyset$. Now (1) follows from these observations.

To prove (2), it suffices by 4.1.7 to show $V_d = W_d = \{d\}$. But this follows from the description of ν in 5.5.4 together with (1).

Next by (1) and (2) and 4.1.24, $\mu/Z(\mu)$ is isomorphic to S_n for some $n \geq 5$, $O_6^-(2)$, $Sp_6(2)$, or $O_8^+(2)$. Then from the structure of ν in 5.5.4, it follows that $\mu/Z(\mu) \cong O_6^-(2)$ or S_8 if G is $E_6^\pi(q)$, $Sp_6(2)$ if G is $\tilde{E}_7(q)$, and $O_8^+(2)$ if G is $E_8(q)$.

Suppose G is $E_6^\pi(q)$. Then $\mu/Z(\mu)$ is the Weyl group of A_7 or E_6, so μ is also that Weyl group. Thus (3) holds unless μ is S_8. But $\xi = (\mathcal{F}_S(C_G(u)), \Omega)$ is a subpacket of τ by 5.5.6.3 and $\mu(\xi)$ is Weyl$(D_5) \cong S_5/E_{2^4}$ by 5.3.4.4, whereas S_8 has no such D-subgroup. This proves (3).

Similarly if G is $\tilde{E}_7(q)$ or $E_8(q)$ then (4) or (5) holds by 4.3.6. □

LEMMA 5.5.9. $C_G(W) = O(\mathfrak{M})C_S(W)$.

PROOF. If G is $F_4(q)$ or $E_6^{-\pi}(q)$, this follows from 5.5.7.4, so assume otherwise. If G is not $\tilde{E}_7(q)$ then $u \in Z(S)$, so we can form the quaternion fusion packet $\xi = (\mathcal{F}_S(C_G(u)), \Omega)$. In this case set $W_1 = W(\xi)$ and $w = u$. If G is $\tilde{E}_7(q)$ then $|\Omega|$ is odd by 5.5.1.2, so we can pick $z \in Z(S)$, so $\xi = (\mathcal{F}_S(C_G(z)), \Omega)$ is a quaternion fusion packet. Here set $w = z$ and let W_1 be the member of $W(\xi)$ containing $K \cap W$.

Now in any event, $w \in W_1 \leq W$, so $C = C_G(W) \leq C_G(W_1) = C_1 \leq C_G(w)$. Set $G_w = O^{p'}(C_G(w))$. From Table 4.5.1 in [**GLS98**], $C_G(G_w)$ is cyclic, as is $C_G(w)/G_wC_G(G_w)$. From 5.5.6.2, G_w is $Spin_{10}^+(q)$, $HSpin_{16}^+(q)$ for $G = E_6^\pi(q)$, $E_8(q)$, respectively, so by 5.3.7.1, $C_G(W_1) = O(N_{G_w}(W_1))C_{S\cap G_w}(W_1)$. Then as $C \leq C_1 \leq C_G(w)$, the lemma holds in these two cases. On the other hand if G is $\tilde{E}_7(q)$ then G_w is $SL_2(q) * Spin_{12}^+(q)$, so again by 5.3.7.1, $C_{G_w}(W_1) = O(N_{G_w}(W_1))C_{S\cap G_w}(W_1)$, and the lemma holds in this case too. □

LEMMA 5.5.10. *Assume G is not $F_4(q)$ or $E_6^{-\pi}(q)$. Let $l = 6, 7, 8$ for G equal to $E_6^\pi(q)$, $\tilde{E}_7(q)$, $E_8(q)$, respectively. Let $\Phi = E_l$.*

(1) $S \leq \mathfrak{M}_K$, so $\mathfrak{M}_K = O^{2'}(\mathfrak{M})$.

(2) $\mathfrak{M} = O(\mathfrak{M})\mathfrak{M}_K$.

(3) $\mathfrak{M}/O(\mathfrak{M}) \cong \bar{\omega}(\Phi, m)$.

(4) $W_0 = \Omega_1(W)$ is the l-dimensional orthogonal module for $\mu/Z(\mu) \cong O_l^\epsilon(2)$.

PROOF. By 5.5.8, $\mu \cong$ Weyl(Φ) and by 5.5.3.3, $|W| \geq (m/2)^l$. Therefore by 4.3.3.5, $M \in M(\tau)$ is isomorphic to $\bar{\omega}(\Phi, m)$ with $|W| = (m/2)^l$ and $M/W \cong \mu$. Thus $|S| \geq |M|_2 \sim (m/2)^l|\mu|$. However by 5.2.9 and 5.2.10, $|G| \sim (m/2)^l \cdot |$Weyl$(\Phi)|$, so $S \leq \mathfrak{M}_K$ and $S \leq M$. Then as $\mathfrak{M}_K = \langle K^{\mathfrak{M}} \rangle$, (1) holds.

Indeed as $S \leq M$ and $C_M(W) = W$ by 4.3.3.4, we conclude from 5.5.9 that $C_G(W) = O(\mathfrak{M})W$. Set $\mathfrak{M}_K^* = \mathfrak{M}_K/O(\mathfrak{M}_K)$. As $C_G(W) = O(\mathfrak{M})W$, we have $W^* = C_{\mathfrak{M}_K^*}(W^*)$, so $\mathfrak{M}_K^* \cong M \cong \bar{\omega}(\Phi, m)$. Therefore (2) implies (3), so it remains to prove (2) and (4).

Set $\mathfrak{M}^+ = \mathfrak{M}/C_{\mathfrak{M}}(W)$. Then as $C_G(W) = O(\mathfrak{M}W)$, $\mathfrak{M}_K^+ \cong M/W \cong \mu \cong$ Weyl(Φ). By 3.1.20.3, d_K induces a transvection on W_0 with center z, so as $\dim(W_0) = l$ and $\mu/Z(\mu) \cong O_l^\epsilon(2)$, (4) follows.

As $\mathfrak{M}_K \trianglelefteq \mathfrak{M}$, we have $\mathfrak{M}_K^+ \trianglelefteq \mathfrak{M}^+$. But as $\mu \cong$ Weyl(Φ), $Out(\mu) = 1$, so $\mathfrak{M}^+ = \mathfrak{M}_K^+ C_{\mathfrak{M}^+}(\mathfrak{M}_K^+)$. By (1), $Z^+ = Z(\mathfrak{M}_K^+)$ is Sylow in $C_{\mathfrak{M}^+}(\mathfrak{M}_K^+)$, so $C_{\mathfrak{M}^+}(\mathfrak{M}_K^+) = O(\mathfrak{M}^+)Z^+$. By (4), $\mathfrak{M}_K^+$ is absolutely irreducible on $W_0 = \Omega_1(W)$, unless $l = 7$ where $\mathfrak{M}_K^+$ is absolutely irreducible on $W_0/Z(G)$. Therefore $O(\mathfrak{M}_K^+)$ centralizes W_0, and hence also centralizes W by 24.8 in [**Asc86**]. Hence if M_0 is the preimage of $O(\mathfrak{M}_K^+)$ then $O^2(M_0)$ centralizes $\mathfrak{M}_K^*$, so (2) follows. □

We are now in a position to establish Theorem 5.5.2. If G is $F_4(q)$ or $E_6^{-\pi}(q)$ then Theorem 5.5.2 follows from 5.5.7. In the remaining cases the theorem is a consequence of 5.5.10.

THEOREM 5.5.11. $Aut_{\mathcal{F}}(S) = Inn(S)$.

PROOF. If G is $F_4(q)$ or $E_6^{-\pi}(q)$ this is 5.5.7.5, so we may assume otherwise. As W is weakly closed in S with respect to G, we have $N_G(S) = N_{\mathfrak{M}}(S)$. Set $\mathfrak{M}^* = \mathfrak{M}/O(\mathfrak{M})$. By 5.5.10.3, $N_{\mathfrak{M}^*}(S^*) = S^*$, so $N_G(S) = SN_{O(\mathfrak{M})}(S)$. As usual $N_{O(\mathfrak{M})}(S)$ centralizes S, so the theorem holds. □

5.6. $\mathcal{F}_S(G)$ is simple

In this section G is a simple group of Lie type in odd characteristic which is not a Goldschmidt group. (Recall a *Goldschmidt group* is a nonabelian simple group that either has an abelian Sylow 2-subgroup or is a rank one group of Lie type and characteristic 2. Thus G is neither $L_2(q)$ with $q \equiv \pm 3 \mod 8$ nor ${}^2G_2(q)$. The main result of this section is:

THEOREM 5.6.1. *Let G be simple of Lie type in odd characteristic such that G is not Goldschmidt. Let $S \in Syl_2(G)$ and $\mathcal{F} = \mathcal{F}_S(G)$ be the 2-fusion system of G. Then $\mathcal{F}$ is simple.*

Until the proof of Theorem 5.6.1 is complete, assume G is a counter example.

LEMMA 5.6.2. *There is no nontrivial proper subgroup of S that is strongly closed in S with respect to G.*

PROOF. As G is not Goldschmidt, this follows from a theorem of Foote in [**Foo97**]. □

NOTATION 5.6.3. For $P \leq S$ set $B(P) = O^{2'}(Aut_{\mathcal{F}}(P))$. Set

$$\mathcal{B} = \langle B(P) : P \leq S \rangle_S \leq \mathcal{F},$$

and

$$Aut^0_{\mathcal{F}}(S) = \langle \alpha \in Aut_{\mathcal{F}}(S) : \alpha_{|P} \in \hom_{\mathcal{B}}(P, S) \text{ for some } P \in \mathcal{F}^c \rangle.$$

LEMMA 5.6.4. $Aut_{\mathcal{F}}(S) \neq Aut^0_{\mathcal{F}}(S)$.

PROOF. As G is a counter example to Theorem 5.6.1, it follows that $\mathcal{F}$ is not simple. Then the lemma follows from 5.6.2 and part (4) of Theorem 8 in [**Asc11**]. □

LEMMA 5.6.5. *(1) $Inn(S) \leq Aut^0_{\mathcal{F}}(S)$.*
(2) $Aut_G(S) \neq Inn(S)$.

PROOF. For each $\alpha \in Inn(S)$, $\alpha = c_s \in Inn(S)$ for some $s \in S$. Then $\alpha \in B(S)$ and $S \in \mathcal{F}^c$, so $\alpha \in Aut^0_{\mathcal{F}}(S)$, establishing (1). Then 5.6.4 and (1) imply (2). □

LEMMA 5.6.6. *G is not $L_2(q)$.*

PROOF. Assume otherwise. As G is not Goldschmidt, $q \equiv \pm 1 \mod 8$, so S is dihedral of order at least 8. Therefore $Aut(S)$ is a 2-group, contrary to 5.6.5.2. □

NOTATION 5.6.7. By 5.6.6, G is not $L_2(q)$, and as G is not Goldschmidt, G is not ${}^2G_2(q)$. Therefore G has a Lie packet $\tau = (\mathcal{F}, \Omega)$. Let $K \in \Omega$ and $z = z(K)$.

LEMMA 5.6.8. *G is not $G_2(q)$ or ${}^3D_4(q)$.*

PROOF. Assume otherwise; then $\langle z \rangle = Z(S)$, so $N_G(S) \leq C_G(z) = H$. Now $H = (L_1 * L_2)\langle s \rangle$, where $L_i \cong SL_2(q_i)$ and $s \in S$ induces an outer automorphism in $PGL_2(q_i)$ on L_i. Set $S_i = S \cap L_i$ and $X_i = N_{L_i}(S_i)$. By 5.6.5.2, $N_G(S) \neq S$, so $N_G(S_i) \neq S$ for $i = 1, 2$, and hence $X_i \neq S_i$ for some i. It follows that $q \equiv \pm 3 \mod 8$ and $X_i \cong SL_2(3)$ for $i = 1, 2$. Indeed $O^2(N_G(S_1S_2)) = X_1X_2$. But setting $N_G(S_1S_2)^* = N_G(S_1S_2)/S_1S_2$, we have that s^* inverts X_i^* for $i = 1, 2$, so $N_G(S) = S$, a contradiction. □

LEMMA 5.6.9. *(1) G is classical.*
(2) G is not $P\Omega_d(q)$ for $d \geq 7$.

PROOF. Assume (1) fails; then G is exceptional and, by 5.6.8, G is of Lie rank at least 4, so the hypotheses of section 5.5 are satisfied. But now 5.6.5.2 and Theorem 5.5.11 supply a contradiction. This proves (1). Part (2) follows from 5.6.5.2 and 5.3.7.3. □

By 5.6.9, $G = \bar{G}/Z(\bar{G})$ where $\bar{G} = SL^\epsilon(U)$ or $Sp(U)$ satisfies Hypothesis 5.4.1. Let $\bar{S} \in Syl_2(\bar{G})$ map onto S under the projection of $\bar{G}$ onto G. By 5.6.5.2:

LEMMA 5.6.10. *$Aut_{\bar{G}}(\bar{S}) \neq Inn(\bar{S})$.*

LEMMA 5.6.11. *$G \cong PSp_{2r}(q)$ for some $r \geq 2$ and some q with $m = (q^2-1)_2 =$ 8.*

PROOF. By 5.6.10, $Aut_{\bar{G}}(\bar{S}) \neq Inn(\bar{S})$, so by 5.4.10.7 and 5.4.13.6, $\bar{G} = Sp(U)$. Thus G is $PSp_{2r}(q)$. By 5.6.6, $r \geq 2$. By 5.4.9.6, $m = 8$. □

NOTATION 5.6.12. Adopt Notation 5.4.2 and, as in 5.4.6, set $O = \langle K_i : 1 \leq i \leq r \rangle$. By 5.4.6.3, we may choose $\Omega = \{K_i : 1 \leq i \leq r\}$ and $O \leq S$. Take $K = K_1$. Let X be the kernel of the action of $H = N_{\bar{G}}(O)$ on Ω. By 5.4.9.5, $X = X_1 \times \cdots \times X_r$, where $X_i = N_{L_i}(K_i) \cong SL_2(3)$. By 5.4.8.3, $H/X \cong H^\Omega = Sym(\Omega)$.

Let $Y_i \in Syl_3(X_i)$ and $Y = Y_1 \cdots Y_r$; then $Y \in Syl_3(X)$, so by a Frattini argument, $H = XN_H(Y)$. Set $H^* = H/O$, so that $Y^* = X^* \trianglelefteq H^*$. Write Y_S for the preimage in Y of $C_{Y^*}(S^*)$ and let $y_i \in Y_i^\#$.

LEMMA 5.6.13. *$N_G(S) = SY_S$.*

PROOF. As O is weakly closed in S with respect to G, we have $N_G(S) = N_H(S)$. As $H^\Omega = Sym(\Omega)$, we have $N_H(S)^\Omega = N_{H^\Omega}(S^\Omega) = S^\Omega$, so $N_H(S) = SN_X(S)$. Next $N_X(S)^* = N_{X^*}(S^*) = N_{Y^*}(S^*) = C_{Y^*}(S^*) = Y_S^*$, so the lemma holds. □

NOTATION 5.6.14. For $y \in Y_S$, let $c_y \in Aut(S)$ be conjugation by s. From the construction in the proof of 5.4.8, for $1 \leq i < r$ we can choose Y_i and $t_i \in z^G \cap H$ with $t_i^\Omega = (K_i, K_{i+1})$ and $Y_i^{t_i} = Y_{i+1}$. By 5.4.8.1, t_i centralizes K_j for $j \notin \{i, i+1\}$, so t_i acts on Y. Therefore $\Lambda = \langle t_i : 1 \le i < r\rangle \le N_H(Y)$ and $\Lambda_2 = \langle t_i : 1 < i < r\rangle$ centralizes K. Therefore:

LEMMA 5.6.15. *We may choose our generators y_i so that*

(1) Λ acts on $\mathcal{Y} = \{y_i : 1 \le i \le r\}$, so $Y = \langle \mathcal{Y}\rangle$ is the permutation module for $\Lambda^* = Sym(\mathcal{Y})$.

(2) For $y = \prod_i y_i^{a_i} \in Y$ set $\mathfrak{a}(Y) = \sum_i a_i \in \mathbf{F}_3$, and set $Y_0 = \{y \in Y : \mathfrak{a}(Y) = 0\}$. Then $Y_0 = [\Lambda, Y] \le O^{2'}(H)$.

(3) Let $\mathcal{Y}_1, \dots, \mathcal{Y}_k$ be the orbits of $N_S(Y)$ on $\mathcal{Y}$ and $b_i = \prod_{y\in\mathcal{Y}_i} y$. Then $Y_S = \langle b_1, \dots, b_k\rangle$.

(4) Set $Y_{S,0} = Y_S \cap Y_0$. Then $|Y_S : Y_{S,0}| = 3$ with $Y_S = \langle b_j\rangle Y_{S,0}$ for each $1 \le j \le k$.

LEMMA 5.6.16. *$Aut_{Y_{S,0}}(S) \le Aut^0_{\mathcal{F}}(S)$.*

PROOF. Observe that $O \in \mathcal{F}^c$ with $Y_0 \le O^{2'}(H)$ by 5.6.15.2. Thus $Aut_{Y_{S,0}}(O) \le B(O)$, so the lemma follows from 5.6.3. □

LEMMA 5.6.17. *Let $b = b_j$ for some $1 \le j \le k$ such that $l = |\mathcal{Y}_j| > 1$. Then $c_b \in Aut^0_{\mathcal{F}}(S)$.*

PROOF. Choose notation so that $\mathcal{Y}_j = \{y_1, \dots, y_l\}$ and set $\theta = \{K_1, \dots, K_l\}$. As $l > 1$ there exist conjugates $s_1, \dots, s_{l/2}$ of t_1 under Λ such that $\{s_i^\Omega : 1 \le i \le l/2\}$ is the set of transpositions in S^Ω nontrivial on θ. Choose notation so that $s_i^\Omega = (K_{2i-1}, K_{2i})$.

Set $\mathfrak{W}_i = \mathfrak{U}_{2i-1} + \mathfrak{U}_{2i}$ and $J_i = C_{\bar{G}}(\mathfrak{W}_i^\perp)$. Then J_i acts faithfully as $Sp(\mathfrak{W}_i) \cong Sp_4(q)$ on $\mathfrak{W}_i$. Set $P_i = C_{K_{2i-1}K_{2i}}(s_i)\langle s_i, z_i\rangle$ and observe that $P_i \cong Q_8D_8$ with $N_{J_i}(P_i)/P_i$ acting faithfully as $\Omega_4^-(2)$ on $P_i/Z(P_i)$. Moreover $S_i = S \cap J_i = K_{2i-1}K_{2i}\langle s_i\rangle$ is Sylow in J_i and $x_i = y_{2i-1}y_{2i}$ generates a Sylow 3-subgroup of $N_{J_i}(S_i)$ with $x_i \in N_{J_i}(P_i) = O^{2'}(N_{J_i}(P_i))$. Set $P = P_0P_1 \cdots P_{l/2}$, where $P_0 = \langle K_j : j > 2i\rangle$. Then $P \in \mathcal{F}^c$ and $b = x_1 \cdots x_{l/2} \in O^{2'}(N_{\bar{G}}(P))$, so $c_{b|P} \in B(P)$ and hence $c_b \in Aut^0_{\mathcal{F}}(S)$ by 5.6.3. □

Choose j as in 5.6.17. By 5.6.15.4, $Y_S = \langle b\rangle Y_{S,0}$ and by 5.6.16 and 5.6.17, $Aut_{Y_{S,0}}(S)$ and c_{b_j} are in $Aut^0_{\mathcal{F}}(S)$. Therefore $Aut_{Y_S}(S) \le Aut^0_{\mathcal{F}}(S)$. Then by 5.6.13, $Aut_{\mathcal{F}}(S) = Aut_{SY_S}(S) = Aut^0_{\mathcal{F}}(S)$, contrary to 5.6.5.2. This contradiction completes the proof of Theorem 5.6.1.

THEOREM 5.6.18. *Let $G \in \mathcal{K}$ be a known simple group that is not Goldschmidt. Then the 2-fusion system of G is simple.*

PROOF. By 16.3, 16.5, and 16.8 in [**Asc11**], either the theorem holds or G is of Lie type and odd characteristic. Hence the theorem follows from Theorem 5.6.1. □

REMARK 5.6.19. Given a saturated 2-fusion system $\mathcal{F}$ on S, set $\mathcal{B}_*(\mathcal{F}) = \langle O^{2'}(N_{\mathcal{F}}(R)) : R \in \mathcal{F}^{fc}\rangle$. The proof of Theorem 5.6.1 shows that if $\mathcal{F}$ is the 2-fusion system of a quasisimple group of Lie type of odd characteristic that is not Goldschmidt, then

(a_0) Each proper subgroup of S strongly closed in S with respect to $\mathcal{F}$ is contained in $Z(\mathcal{F})$, and

(b) $\mathcal{F} = \mathcal{B}_*(\mathcal{F})$.

More generally if $\tau = (\mathcal{F}, \Omega)$ is a quaternion fusion packet with $\mathcal{F} = \mathcal{F}^\circ$ transitive on Ω and τ appears as a conclusion in Theorem 1, then, inspecting the examples in Theorem 1 that are not Lie packets, and using Theorem 2, we conclude that (a) and (b) hold, where

(a) S is the only subgroup of S strongly closed in S with respect to $\mathcal{F}$ and containing Ω.

5.7. $L_d^\pi[m]$ and $\bar{\omega}(A_{d-1}, m)$

In this section we assume Hypothesis 5.4.1 with $G = SL^\pi(U)$; that is $G = SL(U)$ if $\pi = 1$ and $G = SU(U)$ if $\pi = -1$. Adopt Notation 5.4.2 and 5.4.12.

NOTATION 5.7.1. Let $\hat{G} = GL^\pi(U)$; that is $\hat{G} = GL(U)$ if $\pi = 1$ and $\hat{G} = GU(U)$ if $\pi = -1$. Set $F_\pi = \{b \in F^\# : b^{q-\pi} = 1\}$. For $A \in \Gamma$ and $b \in F_\pi$ let $h_A(b) \in \hat{G}$ have eigenvalue b on A and centralize each member of $\Gamma - \{A\}$. Set $h_A = h_A(\lambda)$, $\mathfrak{H}_A = \{h_A(b) : b \in F_\pi\}$, $V_A = \langle h_A \rangle$, $\mathcal{H} = \{h_A : A \in \Gamma\}$, and $\mathfrak{H} = \langle \mathfrak{H}_A : A \in \Gamma \rangle$. Set $\hat{\eta} = \{V_A : A \in \Gamma\}$ and $\hat{W} = \langle \hat{\eta} \rangle$. For $A \in \Gamma$ pick $u_A \in A^\#$ with $(u_A, u_A) = 1$ if $\pi = -1$. Set $\mathfrak{B} = \{u_A : A \in \Gamma\}$ and observe that $\mathfrak{B}$ is a basis for U. Set $\mathfrak{S} = Sym(\mathfrak{B})$ and embed $\mathfrak{S}$ in $\hat{G}$ via its action on the basis $\mathfrak{B}$. Set $d = \dim(U)$.

LEMMA 5.7.2. *(1) $N_{\hat{G}}(\Gamma) = \mathfrak{H}\mathfrak{S}$ with $\mathfrak{H} = \prod_{A \in \Gamma} \mathfrak{H}_A$ the kernel of the action of $N_{\hat{G}}(\Gamma)$ on Γ.*

(2) $N_{\hat{G}}(\Gamma) = O(\mathfrak{H})(\hat{W}\mathfrak{S})$ with $\hat{W} = \prod_{A \in \Gamma} V_A$. Indeed for $s \in \mathfrak{S}$ and $A \in \Gamma$, $h_A^s = h_{As}$, so $\mathfrak{S}$ acts faithfully as $Sym(\mathcal{H})$ on $\mathcal{H}$.

(3) For distinct $A, B \in \Gamma$, $v_{A,B} = h_A h_B^{-1}$ and $k_{A,B} = (A, B)z_A$ and $v_{A,B}$ generate $K_{A,B} \in K_1^G$, where (A, B) denotes the transposition in $\mathfrak{S}$ and z_A is the involution in V_A.

(4) For $w = \prod_A h_A^{k_A} \in \hat{W}$ set $\mathfrak{a}(w) = \sum_A k_A$. Then $W = \{w \in \hat{W} : \mathfrak{a}(w) \equiv 0$ mod $m/2\}$.

(5) $\bar{M} = \langle K_{A,B} : A, B \in \Gamma$ distinct$\rangle = \hat{W}\mathfrak{S} \cap G \cong \bar{\omega}(A_{d-1}, m)$ with $W \trianglelefteq \bar{M}$ and $\mathfrak{M} = \bar{M}O(\mathfrak{M})$.

(6) Let $f = \prod_{h \in \mathcal{H}} h$ and $k = (m/2)/(m/2, d)$. Then $Z(\bar{M}) = \langle f^k \rangle$ is cyclic of order $(m/2, d)$.

PROOF. The kernel of the action of $\mathfrak{N} = N_{\hat{G}}(\Gamma)$ on Γ is the diagonal subgroup $\mathfrak{H}$ of $\hat{G}$, with Γ the set of weight spaces for $\mathfrak{H}$. By construction, $\mathfrak{S} \leq \hat{G}$ acts as $Sym(\mathfrak{B})$ on $\mathfrak{B}$, and hence also acts faithfully as the symmetric group on Γ and $\mathcal{H}$ via $As = \langle u_A s \rangle$ and $h_A s = h_{As}$. In particular (1) holds. Also $\mathfrak{H} = O(\mathfrak{H})O_2(\mathfrak{H})$ with $O_2(\mathfrak{H}) = \hat{W}$, so (2) holds.

Next for distinct $A, B \in \Gamma$, $K_{A,B}$ is a Sylow 2-group of $L_{A,B}$ the fundamental subgroup $L(A{+}B, (A{+}B)^\perp)$ of 5.4.3. Thus $A{+}B = [U, K_{A,B}]$ and $K_{A,B}$ centralizes $(A + B)^\perp$. Further $K_{A,B} \cap W = \langle v_{A,B} \rangle$, where $v_{A,B}$ is defined in 5.4.12, and in particular from that definition, $v_{A,B} = h_A h_B^{-1}$. Further $k \in K_{A,B} - V_{A,B}$ induces the transposition (A, B) on Γ, so we can choose $K_{A,B}$ to contain $(A, B)z_A$. Thus (3) holds.

As $W = \{w \in \hat{W} : \det(w) = 1\}$, (4) holds.

By construction, $[W, k_{A,B}] = V_{A,B} \leq W$, so $K_{A,B}$ acts on W with $K_{A,B}W = \langle k_{A,B}\rangle W$, so $\bar{M}/W \cong \mathfrak{S}$ with $\bar{M} = \hat{W}\mathfrak{S} \cap G$. Then from 5.4.13, $\bar{M} \cong \bar{\omega}(A_{d-1}, m)$, establishing (5).

By (2), $\mathfrak{S}$ acts faithfully as $Sym(\mathcal{H})$ on $\mathcal{H}$, so $Z(\hat{W}\mathfrak{S}) = C_{\hat{W}}(\mathfrak{S}) = \langle f\rangle$. Then by (4), $C_W(\mathfrak{S})$ consists of those f^i such that $0 \equiv \mathfrak{a}(f^i) = di \mod m/2$. But $m/2$ divides di iff $(m/2)/(d, m/2)$ divides i, so (6) holds. □

LEMMA 5.7.3. *Let $W_0 = \Omega_1(W)$. Then W_0 is the core of the permutation module for $\bar{M}/W \cong \mathfrak{M}/C_{\mathfrak{M}}(W) \cong S_d$ on a d-set, with $\mathcal{Z}_\Delta$ the set of vectors of weight 2.*

PROOF. Let $\hat{W}_0 = \Omega_1(\hat{W})$. By 5.7.2.2, $\hat{W} = \prod_A V_A$, so $\hat{W}_0$ is the permutation module for $\mathfrak{S}$ on $I = \{z_A : A \in \Gamma\}$, where z_A is the involution in V_A. By 5.7.2.4, $W_0 = \{w \in \hat{W}_0 : \mathfrak{a}(w) \equiv 0 \mod 2\}$, so W_0 is the core of $\hat{W}_0$. By 5.7.2.3, the involution $z_{A,B}$ in $V_{A,B}$ is $z_A z_B$, so the $z_{A,B}$ are the vectors of weight 2 in $\hat{W}_0$. Finally $\mathcal{Z}_\Delta$ consists of the $z_{A,B}$. □

LEMMA 5.7.4. *Assume Hypothesis 4.2.1 and adopt Notation 4.2.3, except write M for the model for $\mathcal{F}$. Assume $\mu \cong S_4$ and let $\bar{G} = \bar{\omega}(A_3, m)$ and $\pi : \bar{G} \to M$ the homomorphism supplied by 4.3.8.2. Write V_i for $\bar{V}_i\pi$, v_i for $\bar{v}_i\pi$, etc. Then*

(1) $W = V_1V_2V_3$ and $M/W \cong \mu \cong S_4$.

(2) $Z(M) = \langle x\rangle$ is cyclic of order $\mathfrak{q} \leq 4$, where $\bar{x} = (\bar{v}_1\bar{v}_2^2\bar{v}_3^{-1})^{m/8}$.

(3) $\mathfrak{q} = 4$ if $z_3 \neq z_1$, $\mathfrak{q} = 2$ if $z_3 = z_1$ and $z_2 \notin V_1V_2$, and $\mathfrak{q} = 1$ if $z_3 = z_1$ and $z_2 \in V_1V_1$. If $\mathfrak{q} = 4$ then $W = V_1 \times V_2 \times V_3$.

(4) If $\mathfrak{q} \leq 2$ then M splits over W, while if $\mathfrak{q} = 4$ then M does not split, but $M = LW$ where $L \cong GL_2(3)$ and $L \cap W = Z(L) \leq Z(M)$.

(5) M is determined up to isomorphism by $\mathfrak{q}$.

(6) If $m = 8$ and $\mathfrak{q} = 1, 2$ then M splits over each $A \in \mathcal{A}(O_2(M))$, and $m(A) = 4, 5$, respectively.

PROOF. By 4.3.7.2, $W = V_1V_2V_3$ and by 4.3.8.4, $W = C_M(W)$, so $M/W \cong \mu$. Thus (1) holds.

By 5.7.2.5, $\bar{G}$ is isomorphic to the group $\bar{M}$ of that lemma when $d = 4$. Thus by 5.7.2.6, $Z(\bar{G}) = \langle \bar{x}\rangle$ where $\bar{x} = f^{m/8}$ is of order 4. Now $f = h_1h_2h_3h_4 = (h_1h_2^{-1})(h_2h_3^{-1})^2(h_3h_4^{-1})^3h_4^4$. Therefore $\bar{x} = (\bar{v}_1\bar{v}_2^2\bar{v}_3^3)^{m/8}h_4^{m/2} = (\bar{v}_1\bar{v}_2^2\bar{v}_3^{-1})^{m/8}$ as $|h_i| = m/2$. Therefore $x \in Z(M)$. By 4.3.8.3, $\ker(\pi) \leq \langle \bar{x}\rangle$. Arguing as in the proof of 5.7.2.6, $C_{\bar{W}}(O^2(\bar{G})) = \langle \bar{x}\rangle$, so as $\ker(\pi) \leq Z(\bar{G})$ we have $C_W(O^2(M)) = \langle x\rangle$, completing the proof of (2). Observe that by (2), $\ker(\pi)$ is determined by $\mathfrak{q}$, so (5) also holds.

Next by 4.3.8.3, $\ker(\pi) \leq Z(\bar{G}) = \langle \bar{x}\rangle$, so $\mathfrak{q} = 4$ iff π is an isomorphism. In that event $W = V_1 \times V_2 \times V_3$ by 4.3.7.2, so $z_3 \neq z_1$. Thus we may assume $\mathfrak{q} \leq 2$, so $1 = x^2 = (v_1v_3^{-1})^{m/4}v_2^{m/2} = z_1z_3$. Finally $\mathfrak{q} = 1$ iff $1 = x = w_1w_3^{-1}z_2$, where $w_i = v_i^{m/8}$ is of order 4, iff $z_2 \in V_1V_3$. This completes the proof of (3).

We can represent M on $I = \{1, 2, 3, 4\}$ as S_4 with kernel W, so that $d_1 = (1, 2)$, $d_2 = (2, 3)$, and $d_3 = (3, 4)$. Let $M_{i,j} = \langle K_i, K_j\rangle$, $z = z_1$, $t = z_2$, and $k \in K_1 - V_1$. Now $a = kt$ is an involution, so by the Baer-Suzuki Theorem (cf. 39.6 in [**Asc86**]), a inverts Y of order 3 in $M_{1,2}$. Then aY contains b with $bW = d_2$. Thus $b = il$ for some $i \in K_2 - V_2$ and $l \in V_1V_2$, and $1 = b^2 = i^2l^il = tl^il$, so $l^il = t$. However $z^iz = t$ and $z^iz = l^il$ iff i inverts lz iff $lz \in V_2$ iff $l \in V_2z$, so we may choose $i \in K_2 - V_2$ so that $b = iz$.

Similarly iz_3 inverts Y_1 of order 3 in $M_{2,3}$ and iz_3Y_1 contains c with $cW = d_3$. Arguing as in the previous paragraph, we may choose $j \in K_3 - V_3$ so that $c = jt$. Now

$$c^a = (jt)^{kt} = j^t t^{kt} = jz_3tz = czz_3.$$

In particular if $z = z_3$ then $[c, a] = 1$, so $L = \langle a, b, c\rangle$ is the Coxeter group S_4, and hence a complement to W in M. By (3), this holds precisely when $\mathfrak{q} \leq 2$. Thus (4) holds when $\mathfrak{q} \leq 2$.

So assume $\mathfrak{q} = 4$. Then from (2) and (3), zz_3 is the involution in $Z(M)$ and $L/\langle zz_3\rangle \cong S_4$, so $L \cong GL_2(3)$ as $ac \in O_2(L)$ is of order 4. Thus (4) holds in this case too.

Suppose $m = 8$ and $\mathfrak{q} \leq 2$. Then by (5), M is the normalizer in $H = AE_6$ or $\check{A}E_6$ of $\eta \in \eta(\tau_H)$, since $\mu(\tau_H) \cong S_4$ and $\mathfrak{q}_H = \mathfrak{q}$ from the discussion in paragraph one of section 5.1. Now (6) can be checked in H. □

LEMMA 5.7.5. *Let $\bar{G} = \bar{\omega}(A_3, m)$ be the universal group of 4.3.1 with $\mu \cong Weyl(A_3) \cong S_4$. Set $\bar{z}_i = z(\bar{K}_i)$, $\bar{u} = \bar{z}_1\bar{z}_3$, $\bar{k} = \bar{k}_1\bar{k}_3$, $\bar{k}_4 = \bar{k}_2^{\bar{k}}$, $\bar{K}_4 = \bar{K}_2^{\bar{k}}$, $\bar{v}_4 = (\bar{v}_1\bar{v}_2\bar{v}_3)^{-1}$, and $\bar{h} = \bar{k}_1\bar{k}_2$. Set $M = \bar{G}/\langle\bar{u}\rangle$, let $\pi : \bar{G} \to M$ be the natural map, and set $z_i = \bar{z}_i\pi$, $K_i = \bar{K}_i\pi$, etc. Set $t = z_2$ and $\bar{O} = \bar{K}_1\bar{K}_3$. Then*

(1) $z = z_1 = z_3$, so $\Omega = \Omega(z) = \{K_1, K_3\}$. Also $\bar{v}_2^{\bar{k}} = \bar{v}_4^{-1}$.

(2) $C_{\bar{W}}(\bar{O}) = \langle\bar{x}_0, \bar{z}_1\rangle$, where $\bar{x}_0 = \bar{v}_1\bar{v}_2^2\bar{v}_3$.

(3) $C_W(O) = \langle x_0, z\rangle$.

(4) $\bar{h}$ is of order 3 with $\bar{O}^{\bar{h}^2} = \bar{K}_2\bar{K}_4$ and $\langle\bar{x}_0\rangle = \bar{O}^{\bar{h}} \cap \bar{O}^{\bar{h}^2}$.

(5) Assume $\rho = (\mathcal{E}, \Omega)$ is a quaternion fusion packet with Sylow group P containing S, $\mu(\rho) \cong S_4$, and H is a model for $N_{\mathcal{E}}(C_P(W))$ such that $M = \langle K^H\rangle$. Assume further that r is an involution in P centralizing O and inverting x_0, and that $s = kr \in z^{\mathcal{E}}$. Then

(a) $K_2^r = K_4$, $v_2^r = v_4$, $s \in \mathcal{D}^*(t) \cup \mathcal{D}^o(t)$, and s inverts W.

(b) $\langle x_0, r\rangle$ is dihedral of order $m/2$ and $[v_2, r] = x_0^{-1}$, so $r^{v_2} = rx_0^{-1}$.

(c) Suppose $s \in \mathcal{D}^*(t)$. Then $s = j_3j_4c$ for some $j_i \in K_i - \langle v_i\rangle$, $i = 2, 4$, and some $c \in C_P(O^{h^2})$. Further, replacing k_2 by j_2, $k_2^r = k_4$. Hence the action of r on M is determined up to an automorphism of M centralizing v_i for $1 \leq i \leq 3$ and k_1 and k_3, and acting on K_2 and K_4.

(d) Suppose $y \in P$ centralizes OW with $y^2 = x_0$. Then $k_2^y = k_2v_2t^\epsilon$ for some $\epsilon \in \{0, 1\}$ and, replacing y by yb if necessary, where b is the involution in $\langle x_0\rangle$, $k_2^y = k_2v_2$.

PROOF. By 5.7.4.2, $\bar{u} \in Z(\bar{G})$, so $z = z_1 = z_3$. As $\mu = \mu(\bar{\tau}) \cong S_4$, $\bar{\Omega} = \{\bar{K}_1, \bar{K}_3\}$ is of order 2, so $\Omega = \Omega(z) = \{K_1, K_3\}$.

Next from the defining relations for $\bar{G}$ in 4.3.1,

(*) $\bar{v}_2^{\bar{k}} = \bar{v}_2^{\bar{k}_1\bar{k}_3} = (\bar{v}_2\bar{v}_1)^{\bar{k}_3} = \bar{v}_2\bar{v}_1\bar{v}_3 = \bar{v}_4^{-1}$,

so (1) holds.

Next $\bar{k}_1$ inverts $\bar{v}_1$ and centralizes $\bar{v}_3$, while $\bar{v}_2\bar{v}_2^{\bar{k}_1} = \bar{v}_2^2[\bar{v}_2, \bar{k}_1] = \bar{v}_2^2\bar{v}_1$, so $\bar{k}_1$ centralizes $\bar{v}_2^2\bar{v}_1$. Therefore $C_{\langle\bar{v}_1,\bar{v}_2\rangle}(\bar{k}_1) = \langle\bar{v}_2^2\bar{v}_1\rangle$, so $C_{\bar{W}}(\bar{k}_1) = \langle\bar{v}_2^2\bar{v}_1, \bar{v}_3\rangle$. Similarly $C_{\bar{W}}(\bar{k}_3) = \langle\bar{v}_1, \bar{v}_2^2\bar{v}_3\rangle$. Hence by 4.3.7.2, $\bar{O}$ centralizes $\bar{v}_1^a v_2^b \bar{v}_3^c$ iff $2a = b = 2c$ modulo $m/2$, so (2) holds.

If $w \in C_W(k_1) - \langle x_0\rangle$ then $[\bar{w}, \bar{k}_1] = \bar{u}$, impossible as $[\bar{W}, \bar{k}_1] = \langle\bar{v}_1\rangle$ does not contain $\bar{u}$. Thus (2) implies (3).

From the defining relation for $\bar{G}$ in 4.3.1, $\bar{h}$ is of order 3. Set $\bar{G}^* = \bar{G}/\bar{W}$, so that $\bar{G}^* = \mu$ and $(\bar{G}^*, \{\bar{k}_1^*, \bar{k}_2^*, \bar{k}_3^*\})$ is a Coxeter system of type A_3, so $\bar{k}_2^{*\bar{h}^*} = \bar{k}_1^*$

and hence $\bar{K}_2^{\bar{h}} = \bar{K}_1$. Then $(\bar{K}_2\bar{K}_4)^{\bar{h}} = \bar{K}_1\bar{K}_4^{\bar{k}_1\bar{k}_2} = \bar{K}_1\bar{K}_2^{\bar{k}_1\bar{k}_3\bar{k}_1\bar{k}_2} = \bar{K}_1\bar{K}_2^{\bar{k}_3\bar{k}_2} = \bar{K}_1\bar{K}_3 = \bar{O}$, and therefore $\bar{O}^{\bar{h}^2} = \bar{K}_2\bar{K}_4$. Also by (2), $\bar{k}_1$ centralizes $\bar{x}_0$, so

$$\bar{x}_0^{\bar{h}} = \bar{x}_0^{\bar{k}_1\bar{k}_2} = \bar{x}_0^{\bar{k}_2} = (\bar{v}_1\bar{v}_2^2\bar{v}_3)^{\bar{k}_2} = \bar{v}_1\bar{v}_2 \cdot \bar{v}_2^{-2} \cdot \bar{v}_3\bar{v}_2 = \bar{v}_1\bar{v}_3 \in \bar{O},$$

and then

$$\bar{x}_0^{\bar{h}^2} = (\bar{v}_1\bar{v}_3)^{\bar{h}} = (\bar{v}_1\bar{v}_3)^{\bar{k}_1\bar{k}_2} = (\bar{v}_1^{-1}\bar{v}_3)^{\bar{k}_2} = \bar{v}_1^{-1}\bar{v}_3 \in \bar{O}.$$

That is $\bar{x}_0^{\bar{h}}$ and $\bar{x}_0^{\bar{h}^2}$ are in $\bar{O}$, so $\bar{x}_0 \in \bar{O}^{\bar{h}} \cap \bar{O}^{\bar{h}^2}$. Conversely $\bar{O} \cap \bar{O}^{\bar{h}^2} = \bar{K}_1\bar{K}_3 \cap \bar{K}_2\bar{K}_4 = \langle \bar{v}_1, \bar{v}_3\rangle \cap \langle \bar{v}_2, \bar{v}_4\rangle = \langle \bar{v}_1\bar{v}_3\rangle$ as $\bar{v}_1\bar{v}_3 = (\bar{v}_2\bar{v}_4)^{-1}$. Thus as $|\bar{v}_1\bar{v}_3| = m/2 = |\bar{x}_0|$, we conclude that (4) holds.

Assume the hypothesis of (5). As $O^{h^2} = K_2K_4$ and $\Omega(z) = \{K_1, K_3\}$, we have $\Omega(t) = \{K_2, K_4\}$. Then $x_0^r = x_0^{-1} = (v_1v_2^2v_3)^{-1}$, while as r centralizes O, $x^r = (v_1v_2^2v_3)^r = v_1v_3(v_2^r)^2$, so $(v_2^r)^2 = (v_1v_2v_3)^{-2} = v_4^2$, and hence as $\Omega(t) = \{K_2, K_4\}$ we conclude that $v_2^r = v_4t^\epsilon$ for some $\epsilon \in \{0,1\}$. In particular $K_2^r = K_4$. But also $K_2^k = K_4$, so $s = rk$ acts on K_2. Hence as $\mu(\rho) \cong S_4$ and $s \in z^{\mathcal{E}}$, it follows that $s \in \mathcal{D}^o(t) \cup \mathcal{D}^*(t)$. Also s acts on $K_2 \cap W = V_2$.

Next as r centralizes O and k inverts $\langle v_1, v_3\rangle$, s also inverts $\langle v_1, v_3\rangle$. Now s acts on K_1 and K_2 and hence also on $M_1 = \langle K_1, K_2\rangle$. Further s inverts V_1 and acts on V_2, so s acts on $W_1 = V_1V_2$, and centralizes $E = \langle z, t\rangle$. By 4.2.5, there is X of order 3 in M_1 with $W_1 = [W_1, X]$. Then X acts on $F = W_1\langle s\rangle = C_{M_1\langle s\rangle}(E)$, and $F = W_1 \cdot C_F(X)$. Now $s = fw$ with $f \in C_F(X)$ and $w \in W$, and f inverts V_1, so f inverts $W_1 = \langle V_1^X\rangle$. Thus s also inverts W_1, so s inverts $W = W_1V_3$, completing the proof of (5a).

As r is an involution inverting x_0 of order $m/4$, $\langle x_0, r\rangle$ is dihedral of order $m/2$. By (5a), $[v_2, r] = v_2^{-1}v_2^r = v_2^{-1}v_4 = x_0^{-1}$, so (5b) holds.

Suppose $s \in \mathcal{D}^*(t)$. By (5a), s inverts W so s inverts $\langle v_2, v_4\rangle$. Therefore $s = j_2j_4c$ with $j_i \in K_i - \langle v_i\rangle$ and $c \in C_P(O^{h^2})$. By 4.2.5.5, the presentation for G is still satisfied if we replace k_2 by j_2, so we may assume that $j_2 = k_2$. Thus $k_2 = k_2^s = k_2^{kr} = k_4^r$ by (5a), so $k_2^r = k_4$. Therefore the action of r on the generators k_i, v_i, $1 \leq i \leq 3$ is determined, so the action of r on G is determined. This completes the proof of (5c).

Assume the setup of (5d). Observe that $x_0^{k_2} = (v_1v_2^2v_3)^{k_2} = v_1v_2 \cdot v_3v_2 \cdot v_2^{-2} = v_1v_3$, so $(k_2^{-1})^{x_0} = x_0^{-1}x_0^{k_2}k_2^{-1} = x_0^{-1}v_1v_3k_2^{-1} = v_2^{-2}k_2^{-1}$, and hence $k_2^{x_0} = k_2v_2^2$. As y centralizes W and acts on M, y acts on K_2, so $k_2^y = k_2v_2^a$. Therefore as $y^2 = x_0$ and y centralizes W, we have $k_2^y = k_2v_2t^\epsilon$. Further as $k_2^{x_0} = k_2v_2^2$ and $|x_0| = m/4$, $[k_2, b] = t$, so $k_2^{yb} = (k_2v_2t^\epsilon)^b = k_2v_2t^{\epsilon+1}$, so replacing y by yb if necessary, we may assume $k_2^y = k_2v_2$. This completes the proof of (5d) and the lemma. □

5.8. $\bar{\omega}(D_n, m)$

In this section we assume $\bar{G}$ is the group $\bar{\omega}(D_n, m)$ for some $n \geq 3$ defined in Notation 4.3.1, and adopt the notation found there. For $i \in I$ write d_i for the image of $\bar{k}_i$ in $\mu = \bar{G}/\bar{W}$, and set $Q = \langle (d_{n-1}d_n)^\mu\rangle \leq \mu$ and $L_0 = \langle d_i : i \in I - \{n\}\rangle \leq \mu$. Let $D = d_1^\mu$ be the set of reflections in the Weyl group μ. Set $\bar{z}_i = z(\bar{K}_i)$.

LEMMA 5.8.1. *$L_0 \cong S_n$ is a complement to Q in μ, Q is the core of the permutation module for L_0 on I, and $x = d_{n-1}d_n$ is of weight 2 in Q.*

PROOF. This is remark 4.3.4. □

LEMMA 5.8.2. *Let* $\bar{u} = \bar{z}_{n-1}\bar{z}_n$. *Then* $\bar{u} \in Z(\bar{G})$ *and for each* $d \in D$, $\bar{u} = \zeta(d)\zeta(d')$, *where* $\{d, d'\} = d^Q$.

PROOF. First $\bar{G} = \langle \bar{K}_i : i \in I\rangle$ and $\bar{K}_i$ centralizes $\bar{u}$ for $i \neq n-2$. Let $\bar{M} = \langle \bar{K}_{n-2}, \bar{K}_{n-1}, \bar{K}_n\rangle$. Then $\bar{M}$ is an image of the universal group $\bar{\omega}(D_3, m)$, so as the diagrams A_3 and D_3 are some same, modulo ordering of the vertices, it follows from 5.7.4.2 that $\bar{u} \in Z(\bar{M})$, so that $\bar{u}$ centralizes $\bar{K}_{n-2}$ and hence $\bar{u} \in Z(\bar{G})$. Next dd' is fused to $d_{n-1}d_n$ in μ, so by 4.2.8.2, $\zeta(d)\zeta(d')$ is fused to $\bar{z}_{n-1}\bar{z}_n = \bar{u}$. Then as $\bar{u} \in Z(\bar{G})$, we have $\bar{u} = \zeta(d)\zeta(d')$. □

LEMMA 5.8.3. *Set* $\bar{W}_0 = \Omega_1(\bar{W})$ *and* $\bar{G}^! = \bar{G}/\langle\bar{u}\rangle$.

(1) $Z(\mu) = C_\mu(\bar{W}_0)$.

(2) $Q/Z(\mu)$ *is a group of transvections on* $\bar{W}_0$ *with center* $\bar{u}$ *and* $\bar{W}_0^!$ *is the core of the permutation module for* μ/Q.

(3) $C_{\mu/Q}(\bar{z}_n^!) \cong \mathbf{Z}_2 \times S_{n-2}$ *is the centralizer in* μ/Q *of the transposition* d_nQ/Q.

PROOF. For $d, a \in D$, $[\zeta(a), d] = 1$, $\zeta(d)$ for $a \in D_d$, $a \in A_d$, respectively. Thus d and $x = d_{n-1}d_n$ are faithful on $\bar{W}_0$, so as L_0 is irreducible on $Q/Z(\mu)$ by 5.8.1, we conclude that $C_Q(\bar{W}_0) \leq Z(\mu)$. If n is odd then $Z(\mu) = 1$, so (1) holds in this case. If n is even then by 4.3.5.2, $Z(\mu) = \langle\iota\rangle$ is of order 2 and ι inverts $\bar{W}$, so ι centralizes $\bar{W}_0$, completing the proof of (1).

By 4.3.7.2, $\bar{W} = \bar{V}_1 \times \cdots \times \bar{V}_n$, so $\{\bar{z}_1, \ldots, \bar{z}_n\}$ is a basis for $\bar{W}_0$. From the previous paragraph, for $j \in \{n-1, n\}$, d_j is a transvection on $\bar{W}_0$ with axis $A = \{\bar{z}_i : i \neq n-2\rangle$ and center $\bar{z}_j$, so $x = d_{n-1}d_n$ is a transvection with axis A and center $\bar{u}$. Then as $Q = \langle x^\mu\rangle$, $Q/Z(\mu)$ is a group of transvections on $\bar{W}_0$ with center $\bar{u}$, so Q centralizes $\bar{W}_0^!$. Now by 5.8.1, $\mu/Q \cong L_0$ acts on $\bar{W}_0^!$. As $D_0 = L_0 \cap D$ is the set of transpositions in L_0, for $d \in D_0$ we have $C_{L_0}(d) \cong \mathbf{Z}_2 \times S_{n-1}$. Then by 4.2.8.2, $C_{L_0}(\zeta(d)) = C_{L_0}(d)$ and the action of L_0 on $\Gamma = \{\zeta(d) : d \in D_0\}$ is equivalent to its action on the 2-subsets of I. As $\bar{W}_0 = \langle\bar{z}_i : i \in I\rangle$ and $\bar{z}_{n-1}\bar{z}_n = \bar{u}$, we conclude that $\bar{W}_0^! = \langle\bar{z}_i^! : i \in I - \{n\}\rangle$, so $\bar{W}_0^! = \langle\zeta(d)^! : d \in D_0\rangle$. Also $\zeta(d)\zeta(a)\zeta(d^a) = 1$ for $a \in A_d$ by 4.2.5, and $\prod_{i<n} \bar{z}_i \notin \langle\bar{u}\rangle$. Hence $\bar{W}_0^!$ is the core of the permutation module for L_0 on I by 5.1.13. This proves (2) and (3). □

LEMMA 5.8.4. $Z(\bar{G}) \cong \mathbf{Z}_4$ *or* E_4 *for* n *odd or even, respectively.*

PROOF. By 4.3.8.4, $\bar{W} = C_{\bar{G}}(\bar{W})$, so $Z = Z(\bar{G}) \leq \bar{W}$.

First assume that n is even. Then $Z(\mu) = \langle\iota\rangle$ is of order 2 and by 4.3.5.2, ι inverts $\bar{W}$, so $\bar{W}_0 = \Omega_1(\bar{W}) = C_{\bar{W}}(\iota) \geq Z$. Set $\bar{G}^! = \bar{G}/\langle\bar{u}\rangle$. By 5.8.3.2, $\bar{W}_0^!$ is the core of the permutation module for μ/Q, so $Z \leq Z_0$, where Z_0 is the preimage in $\bar{W}$ of $C_{\bar{W}_0^!}(\mu/Q)$. In particular $Z \cong \mathbf{Z_2}$ or E_4. By 5.8.3.2, $Q/Z(\mu)$ is a group of transvections on $\bar{W}_0$ with center $\bar{u}$, so as $m(Q/Z(\mu)) = n-2$ and $m(\bar{W}_0) = n$, $C_{\bar{W}_0}(Q) \cong E_4$, and hence $Z_0 = C_{\bar{W}_0}(Q)$. Thus $\bar{G} = L_0Q$ centralizes Z_0, so $Z = Z_0 \cong E_4$ and hence the lemma holds when n is even.

Next assume n is odd. By 5.8.3.2, $C_{\bar{W}_0^!}(\mu) = 1$, so $\langle\bar{u}\rangle = Z(\bar{G}) \cap \bar{W}_0$, and hence Z is cyclic as $\bar{W}_0 = \Omega_1(\bar{W})$. Set $G = \langle\bar{K}_{n-2}, \bar{K}_{n-1}, \bar{K}_n\rangle \leq \bar{G}$. Using the isomorphism of $\bar{G}$ with the group $\bar{M} \cong \mathfrak{M}/O(\mathfrak{M})$ in $\bar{L} = Spin_{2n}^+(q)$ from 5.3.5.4, 5.3.7.2, and 5.3.8, we conclude that $G \cong \bar{\omega}(A_3, m)$. Then by 5.7.4.2, $Z(G) = \langle\bar{w}\rangle$, where $\bar{w} = \bar{w}_n\bar{w}_{n-1}^{-1}\bar{z}_{n-2}$ and $\bar{w}_i = \bar{v}_i^{m/8} \in \bar{V}_i$ is of order 4. Therefore the lemma holds when $n = 3$, so we may assume $n \geq 5$. Set $w = \bar{w}\bar{z}_1\bar{z}_3 \cdots \bar{z}_{n-4}$. Now

$\bar{K}_{n-1}\bar{K}_n$ centralizes $\bar{V}_i$ for $i < n-2$ and $\bar{w}$, so $\bar{K}_{n-1}\bar{K}_n$ centralizes w. Similarly for $1 \leq j \leq n-2$, either j is odd and d_j centralizes $\bar{z}_i$ for all $i \in \{1, 3, \ldots, n-2\}$, or j is even and d_j fails to centralize exactly two of the $\bar{z}_i$. In either case as d_j is a transvection, it centralizes $\bar{z}_1\bar{z}_3 \cdots \bar{z}_{n-2}$. Also if $j < n-2$ then d_j centralizes $\bar{w}_{n-1}\bar{w}_n$, so d_j centralizes w. Therefore $\bar{G} = \langle K_1, \ldots, K_{n-2}, K_{n-1}, K_n \rangle$ centralizes w, so $w \in Z$. Thus to complete the proof of the lemma, it remains to show that $|Z| \leq 4$.

Set $G' = \langle K_1', \bar{K}_1, \ldots, \bar{K}_{n-3} \rangle$. Arguing as above, $G' \cong \bar{\omega}(D_{n-2}, m)$, so by 4.3.7.2, $W' = V_1' \times \bar{V}_1 \times \cdots \times \bar{V}_{n-3}$, where $\zeta(d_1') \in V_1' \in \eta$, and by induction on n, $Z(G') \cong \mathbf{Z}_4$. Further $W = W' \times X$, where $X = \bar{V}_{n-2}\bar{V}_n = O_2(H)$ and $H = \langle \bar{K}_{n-2}, \bar{K}_n \rangle$. Therefore as $\bar{G} = \langle G', H \rangle$, we have $Z \leq Z(G') \times Z(H) = C_{W'}(G') \times C_X(H)$. But $Z(H) = 1$ by 4.2.5, so indeed $|Z| \leq |Z(G')| = 4$, completing the proof. □

LEMMA 5.8.5. *Assume Hypothesis* 4.2.1 *and Notation* 4.2.3. *Assume* $\mu \cong$ *Weyl*(D_n) *with* $n \neq 4$. *Let* $\pi : \bar{G} \to G$ *be the homomorphism of* 4.3.8, $K_i = \bar{K}_i\pi$ *and* $z_i = z(\bar{K}_i)\pi$. *Then the following are equivalent:*

(1) $|\Omega(z_n)| > 1$.

(2) $\Omega(z_n) = \{K_{n-1}, K_n\}$.

(3) $\bar{u} \in \ker(\pi)$.

PROOF. If $K \in \Omega(z_n)$ and $V = K \cap W$, then $d_V \in V_{d_n}$ by 3.1.6. Then by 4.1.7, $d_V \in d_n^{O_2(\mu)}$. As $n \neq 4$, $O_2(\mu) = Q$, so $V_{d_n} = d_n^Q = \{d_{n-1}, d_n\}$ and hence $V = V_i$ for $i \in \{n-1, n\}$. Thus if (1) holds then so does (2) and $z_n = z_{n-1}$, so that (3) also holds. Visibly if $\bar{u} \in \ker(\pi)$ then $z_{n-1} = z_n$, so (2) holds. Trivially, (2) implies (1). □

LEMMA 5.8.6. *Assume Hypothesis* 4.2.1 *and Notation* 4.2.3. *Assume* $\mu \cong$ *Weyl*(D_n) *and let* $\pi : \bar{G} \to G$ *be the homomorphism of* 4.3.8. *Set* $K_i = \bar{K}_i\pi$, $u = \bar{u}\pi$, *etc. Assume* $u = 1$ *and let* $w_i \in V_i$ *be of order 4 and set* $w = w_{n-1}w_n$ *and* $W_1 = \langle w^G \rangle$. *Then* Q *centralizes* W_1 *and* W_1 *is the* n*-dimensional permutation module for* $\mu/Q \cong S_n$ *with* $C_\mu(w) = \langle Q, d_1, \ldots, d_{n-2} \rangle$. *Moreover for each* $d \in D$ *with* $[d, w] \neq 1$, $w \in O(\zeta(d))$.

PROOF. Set $\tilde{\mu} = \mu/Q$; we may regard $\tilde{\mu}$ as $L_0 = Sym(I)$ with $\tilde{d}_n = (n-1, n)$. Now $w \in V_{n-1}V_n \leq O(z_n)$, so w centralizes $O(\zeta(d))$ for $d_{n-1} \neq d \in D_{d_n}$ and $[w, d_n] = z_n = [w, d_{n-1}]$. Further by 5.7.4.2 applied to $M = \langle K_{n-2}, K_{n-1}, K_n \rangle$, w centralizes $O(z)$ for some $z \in z_n^M$. Then, replacing w by wz_n if necessary, we may assume w centralizes $O(z_{n-2})$, so that $[w, d_{n-2}] = 1$. Therefore the maximal subgroup $L_w = \langle d_1, \ldots, d_{n-2} \rangle \cong S_{n-1}$ of L_0 centralizes w, so $L_w = C_{L_0}(w)$. Further, setting $D_0 = D \cap L_0$ and observing that $\zeta(D_0) = \zeta(D)$, L_0 is transitive on $D_0 - C_{D_0}(w)$, so as $w \in O(z_n)$, we conclude that $w \in O(\zeta(d))$ for each $d \in D$ with $[d, w] \neq 1$ and $x^G \subseteq C_G(w)$, so $Q = \langle x^G \rangle$ centralizes W_1. As $[w, d_n] = z_n$, $W_0 \leq W_1$, so as $w \notin W_0$ it follows that $\dim(W_1) \geq n$ by 5.8.3.2. But as $|L_0 : L_w| = n$, W_1 is an image of the permutation module for $L_0 = \tilde{\mu}$ on I, so W_1 is that module. This completes the proof of the lemma. □

LEMMA 5.8.7. *Assume* $n = 4$ *and let* $\bar{\Omega} = \{\bar{K}_1, \bar{K}_1', \bar{K}_3, \bar{K}_4\}$.

(1) $\mathfrak{S} = Sym(I)_2$ *is faithfully represented as a group of automorphisms of* $\bar{G}$ *via* $\bar{k}_i^s = \bar{k}_{is}$ *and* $\bar{v}_i^s = \bar{v}_{is}$.

(2) $Z = Z(\bar{G}) = \langle (\bar{z}_3\bar{z}_4, \bar{z}_1\bar{z}_4 \rangle$.

(3) Assume Hypothesis 4.2.1 *and Notation* 4.2.3 *with* $\mu \cong$ *Weyl*(D_4) *and let* $\pi : \bar{G} \to G$ *be the homomorphism of* 4.3.8. *Set* $K_i = \bar{K}_i\pi$ *and* $z_i = z(K_i)$. *Then*

(i) if $|\ker(\pi)| = 2$ *then* $|\Omega(z_4)| = 2$ *and* $G \cong \omega(D_4, m)$, *while*

(ii) if $\ker(\pi) = Z$ *then* $\Omega = \Omega(z_4)$ *is of order 4 and* $E = \langle z_2, z_4\rangle \trianglelefteq G$. *Also for* $u_i \in V_i$ *and* $u_1' \in V_1'$ *of order 4,* $u_1u_1'u_3u_4 \in E$.

(4) Let $w_i \in \bar{V}_i$, $i = 3, 4$, *be of order 4 and set* $M = \langle C_{\bar{G}}(w_3w_4)\rangle$. *Then* $M = \langle \bar{K}_1, \bar{K}_1', \bar{K}_2\rangle \cong \bar{\omega}(A_3, m)$. *Further* $w_3w_4 \in ZM$.

(5) Assume $\ker(\pi) = Z$ *and* $m = 8$, *and set* $G^+ = G/E$. *Then for* $Y \in Syl_3(G)$, $C_{S^+}(Y^+) = Z(G^+) \cong E_8$.

(6) $\bar{z}_2 \in O(\bar{\tau})$.

(7) Assume $\ker(\pi) = Z$ *and let* $Y \in Syl_3(G)$, $R = \bigcap_{g\in G} O(z_4)^g$, $P = R \cap W$, *and* $G^- = G/P$. *Then* $W^- = [W^-, Y] \cong E_4$.

PROOF. Visibly $\mathfrak{S}$ preserves the defining relations for $\bar{G}$ in 4.3.1, so (1) follows.

By 5.8.4, $Z \cong E_4$, and by 5.8.2, $\bar{u} = \bar{z}_3\bar{z}_4 \in Z$. Now $s = (1, 3, 4) \in \mathfrak{S}$ with $\bar{u}^s = \bar{z}_1\bar{z}_4$, so (2) follows from (1).

By definition in 4.3.1, $\omega(D_4, m) = \bar{G}/\langle\bar{u}\rangle$, so (3i) follows from (1) and (2).

Suppose $\ker(\pi) = Z$. Then by (2), $\Omega = \Omega(z_4)$ is of order 4. Also $[K_2, z_4] = \langle z_2\rangle$ and $[K_i, z_2] = \langle z_4\rangle$ for $i \in I - \{2\}$, so $E \trianglelefteq G$. The final statement of (3ii) follows from (6), completing the proof of (3).

Let $g \in \bar{K}_1' - \bar{W}$ and $H = \bar{V}_1'\bar{V}_1\bar{V}_3\bar{V}_4$. Then g centralizes $\bar{V}_1\bar{V}_3\bar{V}_4$ and inverts $\bar{V}_1'$ so $|H| \geq m^4/32$. Thus $|\bar{W} : H| \leq 2$, so as $\bar{z}_2 \in \Phi(\bar{W})$, (6) holds.

By 5.7.4.2, there are elements w_i of order 4 in $\bar{V}_i$, $i = 3, 4$, such that $w_0 = w_3w_4\bar{z}_2$ centralizes $\bar{K}_2$. Thus $w = w_3w_4 = w_0\bar{z}_2$ also centralizes $\bar{K}_2$, and hence w centralizes $M = \langle \bar{K}_1, \bar{K}_1', \bar{K}_2\rangle$. Observe that $M^* \cong S_4$ is a maximal D-subgroup of $\bar{G}^* = \bar{G}/\bar{W}$. Thus $M^* = \langle C_D(w)\rangle$ and $M = \langle C_{\bar{G}}(w)\rangle$. Moreover all products $w = w_3w_4$ with w_i of order 4 in $\bar{V}_i$ are conjugate in $\bar{K}_3\bar{K}_4$, so (4) holds, modulo the claim that $w \in ZM$. But all products $u_1u_1'u_3u_4$ with $u_i \in \bar{V}_i$ and $u_1' \in \bar{V}_1'$ of order 4 are conjugate in $O(\bar{\tau})$, so from (6), $ww_1w_1' = z(J)$ for some $J \in \bar{K}_2^{O_2(\bar{G})}$ and some w_1 and w_1'. Therefore as w_1 and w_1' are in M, $w \in gM$, where $g = z(J)\bar{z}_2$. Finally $z(J)\bar{z}_2 \in Z(\bar{G})$ by (2), completing the proof of (4).

Assume the setup of (5). Observe $Z(\mu) = \langle c\rangle W/W$, where $c \in O(z_4)$. As $Z \leq \ker(\pi)$, $\Omega = \Omega(z_4)$ by (3). Then as $m = 8$, $E = \Phi(W)$, so $W^+ \cong E_{16}$. Similarly $\Phi(O(z_4)) = \langle z_4\rangle$, so $O(z_4)^+ \cong E_{2^8}$ and c^+ is an involution. As c inverts W, $X^+ = \langle c^+, W^+\rangle \cong E_{32}$ and $X^+ \trianglelefteq G^+$. Now $O(z_4)^+$ centralizes the hyperplane $\langle c^+, (O(z_4) \cap W)^+\rangle$ of X^+, so $G^+ = \langle O(z_4)^+, O(z_2)^+\rangle$ centralizes a subgroup X_0^+ of index 4, and $X_0^+ = Z(G^+)$, proving (5).

Assume the setup of (7) and observe that $G^- \cong \bar{G}/\bar{P}$. By 4.2.5.5 we may take $Y = \langle y\rangle$ where $\bar{y} = \bar{k}_1\bar{k}_2$. From 10.1.3.5 (whose proof uses 5.8.7 but not 5.8.7.7), $|W : O(z_4) \cap W| = 2$, so $\Phi(\bar{W}) \leq \bar{P}$. Therefore W^- is an image of $\tilde{W} = \bar{W}/\Phi(\bar{W}) = \tilde{V}_1 \cdots \tilde{V}_4$. Now $\bar{y}$ has cycles $(\tilde{v}_1, \tilde{v}_1\tilde{v}_2, \tilde{v}_2)$ and $(\tilde{v}_i, \tilde{v}_2\tilde{v}_i, \tilde{v}_1\tilde{v}_2\tilde{v}_i)$ for $i = 3, 4$, so $[\tilde{W}, \bar{y}] = \langle \tilde{v}_1, \tilde{v}_2\rangle$ and $C_{\tilde{W}}(\bar{y}) = \langle \tilde{v}_1\tilde{v}_3, \tilde{v}_1\tilde{v}_4\rangle$. But $O(z_4)$ contains $\langle v_1v_3, v_1v_4\rangle$, so $\Phi(W)\langle v_1v_3, v_1v_4\rangle \leq O(z_4)$, and then as Y is transitive on $\langle z_2, z_4\rangle^\#$, we conclude $P = \Phi(W)\langle v_1v_3, v_1v_4\rangle$, so (7) holds. □

5.9. $\bar{\omega}(C_n, m)$ and $2\bar{\omega}(C_n, m)$

DEFINITION 5.9.1. Let $I = \{1, \ldots, n\}$ and write $\bar{\omega}(C_n, m)$ for the group $\bar{H}$ presented by the generators $\{\bar{k}_i, \bar{v}_i, r : i \in I\}$ and the following relations:

(1) The relations on $\{\bar{k}_i, \bar{v}_i : i \in I\}$ from 4.3.1 defining $\bar{\omega}(D_n, m)$.

(2) $r^2 = \bar{u}$, r centralizes $\bar{k}_i, \bar{v}_i$ for $1 \leq i \leq n-2$, $\bar{k}_{n-1}^r = \bar{k}_n$, and $\bar{v}_{n-1}^r = \bar{v}_n$.

As in 4.3.1, set $\bar{K}_i = \langle \bar{k}_i, \bar{v}_i \rangle$. Recall from section 5.8 that $\bar{z}_i = z(\bar{K}_i)$ is the involution in $\bar{V}_i = \langle \bar{v}_i \rangle$ and that $\bar{u} = \bar{z}_{n-1}\bar{z}_n$ is in the center of $\bar{\omega}(D_n, m)$. Observe that as $r^2 = \bar{u}$, we have r centralizing $\bar{u}$, so $\bar{u} \in Z(\bar{H})$. Set $\omega(C_n, m) = \bar{\omega}(C_n, m)/\langle \bar{u} \rangle$.

LEMMA 5.9.2. *Let $\bar{H} = \bar{\omega}(C_n, m)$. Then*

(1) The subgroup $\langle \bar{k}_i, \bar{v}_i : i \in I \rangle$ of $\bar{H}$ is normal in $\bar{H}$ and isomorphic to the universal group $\bar{\omega}(D_n, m)$. Set $\bar{G} = \langle \bar{k}_i, \bar{v}_i : i \in I \rangle$.

(2) $\bar{H} = \bar{G}\langle r \rangle$.

(3) The subgroup $\bar{W} = \langle \bar{v}_i : i \in I \rangle$ is normal in $\bar{H}$. Set $\bar{H}^* = \bar{H}/\bar{W}$.

(4) $(\bar{H}^*, \{\bar{k}_i^*, r^* : 1 \leq i < n\})$ is a Coxeter system of type C_n.

(5) $\bar{\omega}(C_n, m)$ is also presented as $Grp(\bar{k}_i, \bar{v}_i, r : 1 \leq i < n\})$ subject to the defining relations from 4.3.1 on $\{\bar{k}_i, \bar{v}_i : 1 \leq i < n\}$ for $\bar{\omega}(A_{n-1}, m)$, together with the relations $\langle \bar{k}_{n-1}, \bar{v}_{n-1} \rangle^r$ commutes with $\langle \bar{k}_{n-1}, \bar{v}_{n-1} \rangle$, r centralizes $\{\bar{k}_i, \bar{v}_i : 1 \leq i \leq n-2\}$, and $r^2 = [r, \bar{v}_{n-1}]^{m/4}$.

PROOF. Write $\bar{G}$ for the subgroup of $\bar{H}$ defined in (1); we first prove that $\bar{G} \trianglelefteq \bar{H}$. By definition, $\bar{H} = \langle \bar{G}, r \rangle$. Further r centralizes $\langle \bar{K}_i : 1 \leq i \leq n-2 \rangle$ and $\bar{K}_{n-1}^r = \bar{K}_n$, so r acts on $\bar{G}$. Thus $\bar{G} \trianglelefteq \bar{H}$ and (2) follows.

Next the action of r on the generators of $L = \bar{\omega}(D_n, m)$ preserves the defining relations for that group, so r induces an automorphism of L of order 2. Let s of order 4 induce that automorphism and form the semidirect product D of L by $\langle s \rangle$. Set $D^+ = D/\langle \bar{u}s^2 \rangle$. Then D^+ satisfies the defining relations for $\bar{H}$ with $L^+ \cong L$. Then as D^+ is an image of $\bar{H}$, it follows that $\bar{G} \cong L$, completing the proof of (1).

By construction, r acts on $\bar{W} = \bar{V}_1 \times \cdots \times \bar{V}_n$ as it permutes $\{\bar{v}_i : i \in I\}$. Thus as $\bar{W} \trianglelefteq \bar{G}$ by 4.3.2.1, (3) follows from (2).

From 4.3.2, $\{\bar{k}_i^* : 1 \leq i < n\}$ satisfies the Coxeter relations of type A_{n-1}, and as r centralizes $\bar{k}_i$ for $i \leq n-2$, r^* centralizes $\bar{k}_i^*$. Thus to prove (4), it remains to show $(\bar{k}_{n-1}^* r^*)^4 = 1$. But $(\bar{k}_{n-1}r)^2 = \bar{u}\bar{k}_{n-1}\bar{k}_{n-1}^r = \bar{u}\bar{k}_{n-1}\bar{k}_n$, which squares to $\bar{u} \in \bar{W}$. Thus (4) holds.

Observe that as $\bar{K}_{n-1}^r = \bar{K}_n$, $\bar{H}$ is generated by $\{\bar{k}_i, \bar{v}_i, r : 1 \leq i < n\}$. From 5.9.1, $\bar{K}_{n-1}$ commutes with $\bar{K}_n$, r centralizes $\bar{K}_i$ for $1 \leq i \leq n-2$, and $r^2 = \bar{u} = (\bar{v}_n^{-1}\bar{v}_{n-1})^{m/4}$, so the relations in (5) are satisfied in $\bar{H}$.

Conversely let $H = Grp(k_i, v_i, r : 1 \leq i < n)$ subject to the relations in (5). Set $v_n = v_{n-1}^r$, $k_n = k_{n-1}^r$, and $K_n = \langle k_n, v_n \rangle$; we must show that the relations in 5.9.1 are satisfied in H. By definition the relations among the $K_i = \langle k_i, v_i \rangle$ are satisfied for $1 \leq i < n$, K_n commutes with K_{n-1}, and r centralizes K_i for $1 \leq i \leq n-2$. Therefore as K_{n-1} commutes with K_j for $j < n-2$, so does $K_n = K_{n-1}^r$. Similarly $1 = (k_{n-2}k_{n-1})^3$, so $1 = (k_{n-2}^r k_{n-1}^r)^3 = (k_{n-2}k_n)^3$, and $v_{n-2} = v_{n-2}^r$ commutes with $v_{n-1}^r = v_n$, so the relations defining $\bar{\omega}(D_n, m)$ in 5.9.1.1 are satisfied. In particular $u = (v_n^{-1}v_{n-1})^{m/4} = [r, v_{n-1}]^{m/4} = r^2$, establishing the last relation in $\bar{H}$ and completing the proof of (5). □

DEFINITION 5.9.3. Write $2\bar{\omega}(C_n, m)$ for the group $\bar{Y}$ presented by the generators $\{\bar{k}_i, \bar{v}_i, y, r : 1 \le i < n\}$ together with the following relations:

(a) The relations on $\bar{k}_i, \bar{v}_i, r$ for $1 \le i \le n$ for $\bar{\omega}(C_n, m)$ appearing in 5.9.1, where $\bar{k}_n = \bar{k}_{n-1}^r$ and $\bar{v}_n = \bar{v}_{n-1}^r$.

(b) y centralizes $\bar{k}_i, \bar{v}_i$ for $1 \le i \le n-2$ and $\bar{v}_{n-1}$, r inverts y, $y^2 = x = [r, \bar{v}_{n-1}]$, and $[\bar{k}_{n-1}, y] = \bar{v}_{n-1}$.

We will see in a moment that $\bar{u} = r^2 \in Z(\bar{Y})$, and we set $2\omega(C_n, m) = \bar{Y}/\langle \bar{u} \rangle$.

LEMMA 5.9.4. *Let $\bar{Y} = 2\bar{\omega}(C_n, m)$. Then*

(1) The subgroup $\langle \bar{k}_i, \bar{v}_i, r : 1 \le i < n \rangle$ of $\bar{Y}$ is normal in $\bar{Y}$ and isomorphic to the universal group $\bar{\omega}(C_n, m)$. Set $\bar{H} = \langle \bar{k}_i, \bar{v}_i, r : 1 \le i < n \rangle$.

(2) $\bar{Y} = \bar{H}\langle y \rangle$ and $|\bar{Y} : \bar{H}| = 2$.

(3) Set $\bar{W}_Y = \bar{W}\langle y \rangle$. Then $\bar{W}_Y = \bar{V}_1 \times \cdot \times \bar{V}_{n-1} \times \langle y \rangle$ is normal in $\bar{Y}$ with $|Y| = m$. Set $\bar{Y}^* = \bar{Y}/\bar{W}_Y$.

(4) $(\bar{Y}^*, \{\bar{k}_i^*, r^* : 1 \le i < n\})$ is a Coxeter system of type C_n.

PROOF. Write $\bar{H}$ for the subgroup of $\bar{Y}$ defined in (1) and set $\bar{K}_i = \langle \bar{k}_i, \bar{v}_i \rangle$ for $1 \le i < n$ and $\bar{K}_n = \bar{K}_{n-1}^r$. By definition in 5.9.3, y commutes with $\bar{K}_i$ for $i \le n-2$ and with $\bar{v}_{n-1}$, and $[\bar{k}_{n-1}, y] = \bar{v}_{n-1}$. Therefore y acts on $\bar{K}_{n-1}$. Then as r inverts y and $\bar{K}_{n-1}^r = \bar{K}_n$, y also acts on $\bar{K}_n$. Finally as r inverts y, we have $[r, y] = y^2 = [r, \bar{v}_{n-1}] \in \bar{H}$, so y acts on $\bar{H}$. Thus $\bar{H} \trianglelefteq \bar{Y} = \langle \bar{H}, y \rangle$ and (2) holds.

Let $\rho = (\mathcal{L}, \Gamma)$ be the Lie packet of $L \cong Spin_{2n+2}^-(q)$ for $(q-1)_2 = m/2$, let $\eta_L \in \eta(\rho)$, $W_L = \langle \eta_L \rangle$, and G_L a model for $N_{\mathcal{L}}(W_L)$. Then from the proof of 5.3.10, G_L satisfies the relations defining $2\bar{\omega}(C_n, m)$ and $|G_L| = 4|\bar{\omega}(D_n, m)|$. But by (2) and 5.9.2, $|\bar{Y}| \le 4|\bar{\omega}(D_n, m)|$, so we conclude that $\bar{H} \cong \bar{\omega}(C_n, m)$, completing the proof of (1).

By 4.3.7.2, $\bar{W} = \bar{V}_1 \times \cdots \times \bar{V}_n$. We've seen that y centralizes $\bar{V}_i$ for $1 \le i < n$, so as y is inverted by r and $\bar{V}_{n-1}^r = \bar{V}_n$, it follows that y centralizes $\bar{W}$. Thus $\bar{W}_Y = \bar{W}\langle y \rangle$ is abelian. Then by 5.9.2.3, $\bar{W}$ is normal in $\bar{H}\langle y \rangle = \bar{Y}$. Also y centralizes $\bar{k}_i$ for $i \le n-2$ and $[\bar{k}_{n-1}, y] = \bar{v}_{n-1} \in \bar{W}$, so also $[\bar{k}_n, y^{-1}] = [\bar{k}_{n-1}^r, y^r]$ is in $\bar{W}$, and hence $\bar{W}_Y \trianglelefteq \bar{Y}$. Finally $y^2 = x = [r, \bar{v}_{n-1}] = \bar{v}_n^{-1}\bar{v}_{n-1}$, so $y^2 \cap \bar{W}_0 = 1$ and $\bar{W} = \bar{W}_0 \times \langle x \rangle$, where $\bar{W}_0 = \bar{V}_1 \times \cdots \times \bar{V}_{n-1}$. This completes the proof of (3).

Part (4) follows from (3) and 5.9.2.4. □

5.10. Some constrained examples

In this section we assume:

HYPOTHESIS 5.10.1. *$\tau = (\mathcal{F}, \Omega)$ is a quaternion fusion packet such that $\mathcal{F} = \mathcal{F}^\circ$ is constrained with Sylow group S and model G with $G/Z(G) \cong L_3(2)/E_8$, $L_3(2)/E_{64}$, or $L_3(2)/2^{3+6}$.*

In three instances in this section we appeal to results from section 6.2. We observe that section 6.2 does not depend upon this section.

NOTATION 5.10.2. Set $Z = Z(G)$, $Q = O_2(G)$, $G^* = G/Q$, and $G^+ = G/Z$. Let $z \in \mathcal{Z}_S$, $K \in \Omega(z)$, and $E = \langle z^G \rangle$. Notice that $z \notin Z$ as $\mathcal{O}(z)$ is not normal in $\mathcal{F}$. Set $\mathcal{F}^+ = \mathcal{F}/Z$ and $\tau^+ = (\mathcal{F}^+, \Omega^+)$. By 3.3.2.2, τ^+ is a quaternion fusion packet with model G^+.

LEMMA 5.10.3. $G = \langle O(\tau)^G \rangle$.

PROOF. This follows as $\mathcal{F} = \mathcal{F}^\circ$. □

LEMMA 5.10.4. *(1) K is the unique member of Ω in KZ.*
(2) $N_G(K^+) = N_G(K) \leq C_G(z)$.
(3) $\Phi(E) = 1$.
(4) If $\Omega = \Omega(z)$ then $E_8 \cong E \leq Z(Q)$ and E is the natural module for G^*.

PROOF. Part (1) follows from 3.3.2.3. Then (1) implies (2). From 6.2.8, $E \leq N_G(K^+)$, so E centralizes z by (2), and then as $E = \langle z^G \rangle$, part (3) follows.

Suppose $\Omega = \Omega(z)$. Then as S acts on Ω, $z \in Z(S)$, so $E \leq Z(Q)$. Let $g \in G - C_G(z)$; by 6.2.10.2, $z^g \in \mathcal{D}^*(z) - z^\perp$, so by 4.2.5, $zz^g \in z^G$. Therefore $E^\# = z^G$, so as $E^+ \cong E_8$ is the natural module for G^*, part (4) follows. □

LEMMA 5.10.5. $G = O^2(G)$.

PROOF. By 5.10.3, $G = \langle O(\tau)^G \rangle$, so it suffices to show that $K \leq O^2(N_G(K))$. But $K^+ \leq O^2(N_{G^+}(K^+))$ (cf. the proof of 6.2.14), so $K \leq O^2(N_G(K))$ by 5.10.4.2. □

LEMMA 5.10.6. *If $G/Z(G) \cong L_3(2)/E_8$ then $Z(G) = 1$.*

PROOF. Assume otherwise. As $E^+ = Q^+$ and $[K^+, E^+] \leq \langle z^+ \rangle$, we have $Q \leq N_G(K^+)$, so $K \trianglelefteq KQ$ and $z \in Z(Q)$ by 5.10.4.2. Indeed $[K, Q] \leq \langle z \rangle \leq E$, so $Q/E \leq Z(G/E)$ by 5.10.3. But K^* is a 4-subgroup of $G^* \cong L_3(2)$ and $KE/E \cong K/\langle z \rangle \cong E_4$, so G/E is not $SL_2(7)$. Therefore as $G = O^2(G)$ by 5.10.5, it follows that $E = Q$ and $E = [E, G]$. Then as $H^1(G^*, E^+) \cong \mathbf{F}_2$, we have $m(E) = 4$.

Let $\hat{E}$ be the dual of E as a G^*-module; then $\hat{Z}$ is a hyperplane of $\hat{E}$ and S^* is regular on $\hat{E} - \hat{Z}$, so for s^* an involution in S^* we have $C_{\hat{E}}(s^*) = [\hat{E}, s^*] \cong E_4$. Hence also $C_E(s) = [E, s] \cong E_4$. Therefore for t^* an involution in $S^* - K^*$, $[z, t] \neq 1$. But t^* acts on K^+, so $t \in C_G(z)$ by 5.10.4.2, a contradiction. □

LEMMA 5.10.7. *(1) If $\Omega = \Omega(z)$ then $Z(G) = 1$.*
(2) If $G/Z \cong L_3(2)/E_{64}$ and $Z \neq 1$ then $Z = \langle zz^s \rangle \cong \mathbf{Z}_2$ for $s \in S - C_S(z)$.
(3) If $G/Z \cong L_3(2)/2^{3+6}$ then $\Phi(Z) = 1$, $m(Z) \leq 2$, and $Z = \langle zz^s : s \in S \rangle$. Further $|\Omega(z)| = 4/|Z|$.

PROOF. We first prove (1), so assume $\Omega = \Omega(z)$ but $Z \neq 1$. By 5.10.4.4, $E \cap Z = 1$. By 5.10.6 we may assume Q^+ is E_{64} or 2^{3+6}. Assume first that $Q^+ \cong E_{64}$. From the proof of 6.2.16, Q^+ is the core of the permutation module for G on G/SM, where $M \in M(\tau)$, and $E \cap O(z) = \langle z \rangle$. Set $G^! = G/E$ and $F = O(z) \cap Q$. Then as $\Omega = \Omega(z)$ is of order 2, we have $E_8 \cong F \trianglelefteq S$ with $F \cap ZE = \langle z \rangle$, so $F_0^! = Z(S^!) \cap F^! \neq 1$. Therefore $Q_0^! = \langle F_0^{!G} \rangle \leq Z(Q^!)$, and by 5.10.5, $Q_0^! = Q^!$. As in the proof of 5.10.6, $m(Q^!) = 4$ and for s^* an involution in S^*, $C_{Q^!}(s^*)/Z^! = [Q^!, s]/Z^!$. This is a contradiction as K^* centralizes $F^! \cong E_4$.

Therefore (1) holds when $Q^+ \cong E_{64}$, so assume instead that $Q^+ \cong 2^{3+6}$. From the proof of 6.2.19, for $k \in K - \langle z \rangle$, k^+E^+ is contained in a complement H^+/E^+ to Q^+/E^+ in G^+/E^+; H^+ has three irreducibles Q_i^+/E^+, $1 \leq i \leq 3$, on Q^+/E^+; and setting $H_i = \langle K^{Q_i H} \rangle$ and $\Omega_i = K^{Q_i}$, we have $S_i = S \cap H_i \in Syl_2(H_i)$ with $\tau_i = (\mathcal{F}_i, \Omega_i)$ a quaternion fusion packet with model H_i, where $\mathcal{F}_i = \mathcal{F}_{S_i}(H_i)$ and $H_i/Z(H_i) \cong L_3(2)/E_{64}$. Therefore from the previous paragraph, $Z(H_i) = 1$, so

$Z(H_i^!) = 1$. Therefore $[Q^!, H] = O_2(H_1^!)O_2(H_2^!) \cong E_{64}$, so by 5.10.5, we have $Z^! = 1$, contradicting $1 \neq Z$ and $Z \cap E = 1$. This finally completes the proof of (1).

Assume the setup of (2). Then $\Omega(z) = \{K\}$ by (1), so $\Omega = \{K_1, K_2\}$ where $z_1 \neq z_2$ and $z_i = z(K_i)$. Hence as $\Omega(z^+) = \Omega^+$, we have $u = z_1 z_2 \in Z$. Set $G^- = G/\langle u\rangle$. Then $\Omega(z^-) = \Omega^-$, so $Z(G^-) = 1$ by (1). Therefore $Z = \langle u\rangle$, establishing (2).

Finally assume the setup of (3) and let $U = \langle zz^s : s \in S\rangle$. As $\mu(\tau) \cong Weyl(D_4)$, U is a subgroup of E_4 by 5.8.7.2. Then (3) follows as in the previous paragraph. □

LEMMA 5.10.8. *Assume y is an element of order 3 acting on G and centralizing $O(z)$. Assume $|\Omega(z)| > 1$. Then y centralizes G.*

PROOF. Assume otherwise and set $O = O(\tau)$. Now y acts on $KQ \geq O$, so y acts on O. Then as $|\Omega|$ is 1, 2, or 4, either y acts on each $J \in \Omega$ or y acts on a unique member of Ω. As y centralizes $O(z)$ and $|\Omega(z)| > 1$, the latter is impossible, so y acts on each $J \in \Omega$. Then as y centralizes $K^* = J^*$ it follows that y centralizes J so y centralizes O.

Suppose $[G^+, y] = 1$. Then $G = \langle C_G(y)Z\rangle$, so y centralizes $\langle O^{C_G(y)}\rangle = \langle O^G\rangle = G$ by 5.10.3, contrary to assumption. Therefore $[G^+, y] \neq 1$, so replacing G by G^+, we may assume $Z = 1$.

Next $G^* \cong L_3(2)$ and y centralizes $K^* \cong E_4$, so y centralizes G^*. As $F^*(G) = Q$, y does not centralize Q.

As G^* is irreducible on E and y centralizes $z \in E$, we conclude that y centralizes E. Therefore $E \neq Q$. Set $G^! = G/E$.

Suppose $Q \cong E_{64}$. As we saw, $O \cap Q \not\leq E$ and G^* is irreducible on $Q^!$, so y centralizes $Q^!$. But then y centralizes Q, a contradiction.

This leaves $Q \cong 2^{3+6}$. Now, as we saw during the proof of 5.10.7, G^* has three irreducibles $Q_i^!$, $1 \leq i \leq 3$, on $Q^!$, and $[Q_i, O] \leq Q_i \cap O \not\leq E$, so as usual y centralizes Q_i. But then y centralizes $Q = Q_1Q_2$ for our final contradiction. □

LEMMA 5.10.9. *(1) $Z \leq E = \langle \mathcal{Z}\rangle \leq O^2(C_G(z))$.*
(2) $G = \langle O(z)^g : z^g \in O^2(C_G(z))\rangle$.

PROOF. By 5.10.6 and 5.10.7, $Z \leq E = \langle \mathcal{Z}\rangle$. Let y of order 3 in $C_G(z)$. As $O = O(\tau) \lhd C_G(z)$, y acts on O. Then as $|\Omega| = 1$, 2 or 4, either y acts on each $J \in \Omega$ or $|\Omega| = 4$ and y acts on a unique $K \in \Omega$. The latter does not hold as for each $K_i \in \Omega - \{K\}$, $\{K, K_i\} = \Omega_i$ and $y \in H_i$, as in the proof of 5.10.7. Therefore y acts on each $J \in \Omega$, so as $J^* = K^* = [K^*, y]$, we conclude that $J = [J, y]$. Hence $O = [O, y] \leq O^2(C_G(z))$. Then as $Z \leq O$ by 5.10.6 and 5.10.7 and $E = Z[E, K\langle y\rangle]$, we have $E \leq O^2(C_G(z))$, completing the proof of (1). Then (1) and 5.10.3 imply (2). □

DEFINITION 5.10.10. Let $\hat{L}$ be the covering group of $Sp_6(2)$, $\hat{S} \in Syl_2(\hat{L})$, and $\mathcal{G} = \mathcal{F}_{\hat{S}}(\hat{L})$. Then there exists $\hat{\Omega} \subseteq \hat{S}$ with $\rho = (\mathcal{G}, \hat{\Omega})$ a quaternion fusion packet. This can be seen as $\hat{L} \leq \tilde{L} \cong Spin_7(3)$ with $\hat{S} \in Syl_2(\tilde{L})$, so the Lie packet for $\tilde{L}$ induces the packet ρ. Next let $\hat{P}$ be the preimage in $\hat{L}$ of the maximal parabolic of $\hat{L}.Z(\hat{L})$ isomorphic to $L_3(2)/E_{64}$, and set $\hat{\mathcal{F}} = \mathcal{F}_{\hat{S}}(\hat{P})$; by the same argument, $\hat{\tau} = (\hat{\mathcal{F}}, \hat{\Omega})$ is a quaternion fusion packet. Denote both $\hat{P}$ and $\hat{\mathcal{F}}$ by $\hat{L}_3(2)/E_{64}$.

LEMMA 5.10.11. *Let $\hat{\tau} = (\hat{\mathcal{F}}, \hat{\Omega})$ be a quaternion fusion packet with $\hat{\mathcal{F}} \cong \hat{L}_3(2)/E_{64}$. Then*

(1) $O_2(\hat{\mathcal{F}}) \cong D_8^3$.

(2) $\hat{\tau}$ is the largest quaternion fusion packet $(\mathcal{Y}, \Gamma)$ with $\mathcal{Y} = \mathcal{Y}^\circ$ and $\mathcal{Y}/Z(\mathcal{Y}) \cong L_3(2)/E_{64}$.

PROOF. Part (1) follows from the embedding of $\hat{P}$ in $\hat{L}$.

Suppose for $i = 1, 2$, $(\mathcal{Y}_i, \Omega_i)$ are quaternion fusion packets with $\mathcal{Y}_i = \mathcal{Y}_i^\circ$ and $\mathcal{Y}_i/U_i \cong L_3(2)/E_{64}$, where $U_i = Z(\mathcal{Y}_i) \neq 1$. Let L_i be a model for $\mathcal{Y}_i$. Then by 5.10.7.2, $U_i \cong \mathbf{Z}_2$. As $P = P^\infty$ also $L_i = L_i^\infty$, so $L_i = M/V_i$ for some $V_i \leq Z(M) = V$, and M the universal covering group of P. Then $|V : V_i| = 2$ and we may assume $V_1 \neq V_2$, so setting $V_0 = V_1 \cap V_2$, we have $V/V_0 \cong E_4$. Set $L = M/V_0$ and $Q = O_2(L)$; we abuse notation and regard $V_0 = 1$. Let $J/V \in \Omega$ and $z \in Q$ with $zV = z(J/V)$. Set $Y = O^2(C_L(zV)$; then $Y = XJR$ where $X \cong \mathbf{Z_3}$ with $J/V = [J/V, X]$ and $R = [Q, JX]$ of index 2 in Q. As Y centralizes zV and V and $Y = O^2(Y)$, it follows that Y centralizes $U = V\langle z\rangle$. Let $j \in J - U$; then $j^2 \in zV$ so we may take $z = j^2$. Then as X centralizes U and is transitive on $(J/U)^\#$, also $z = i^2$ for each $i \in J - U$. Further for $i \in J - U\langle j\rangle$, $z = (ij)^2 = zj^ij$, so i inverts j and hence $\langle i, j\rangle \cong Q_8$. Therefore $J = K \times V$ where $K = [J, X] \cong Q_8$ with $z \in K \cong Q_8$. Next $KV_i/V_i \trianglelefteq Y/V_i$, so $[K, JR] \leq V_1Z \cap V_2Z = Z$, and hence $K \trianglelefteq XJR = Y$. Now either $Y = C_L(z)$ or $C_L(z) = Y\rangle t\rangle$ where t inverts X, and in the latter case t acts on $[J, X] = K$. Thus in any event, $K \trianglelefteq C_L(z)$. Let $K \leq S \in Syl_2(L)$, $\mathcal{F} = \mathcal{F}_S(L)$, and $\Gamma = K^{\mathcal{F}}$. As $K \trianglelefteq C_L(z)$, $\zeta = (\mathcal{F}, \Gamma)$ is a quaternion fusion packet with $V = Z(\mathcal{F}) \cong E_4$ and $\mathcal{F}/V \cong L_3(4)/E_{64}$, contrary to 5.10.7.2. This contradiction completes the proof of (2). □

DEFINITION 5.10.12. We denote by $\hat{L}_3(2)/2^{3+6}$ a 2-fusion system $\hat{\mathcal{F}}$ such that there exists $\hat{\Omega} \subseteq \hat{S}$ with $(\hat{\mathcal{F}}, \hat{\Omega})$ a quaternion fusion packet such that $\hat{\mathcal{F}} = \hat{\mathcal{F}}^\circ$, $Z(\hat{\mathcal{F}}) \cong E_4$, and $\hat{\mathcal{F}}/Z(\hat{\mathcal{F}}) \cong L_3(2)/2^{3+6}$. Such a system exists. Namely, arguing as in 5.10.10, writing $\hat{L}$ for the covering group of $\Omega_8^+(2)$, we have $\hat{L} \leq Spin_8^+(3)$, so there is a quaternion fusion packet $(\mathcal{G}, \hat{\Omega})$ with $\mathcal{G}$ the 2-fusion system of $\hat{L}$. Then choosing a maximal parabolic P of $\Omega_8^+(2)$ of the form $L_3(2)/2^{3+6}$, the preimage $\hat{P}$ of P in $\hat{L}$ is of type $\hat{L}_3(2)/2^{3+6}$. Similarly we denote by $\check{L}_3(2)/2^{3+6}$ a 2-fusion system $\check{\mathcal{F}}$ such that there exists $\check{\Omega} \subseteq \check{S}$ such that $(\check{\mathcal{F}}, \check{\Omega})$ is a quaternion fusion packet such that $\check{\mathcal{F}} = \check{\mathcal{F}}^\circ$, $Z(\check{\mathcal{F}})$ is of order 2, and $\check{\mathcal{F}}/Z(\check{\mathcal{F}}) \cong L_3(2)/2^{3+6}$. For example $\hat{\mathcal{F}}$ has such factor systems.

Observe that by 5.10.7.3, if $(\mathcal{F}, \Omega)$ is a quaternion fusion packet with $\mathcal{F} = \mathcal{F}^\circ$, $Z = Z(\mathcal{F}) \neq 1$, and $\mathcal{F}/Z \cong L_3(2)/2^{3+6}$, then $\mathcal{F}$ is of type $\hat{\mathcal{F}}$ or $\check{\mathcal{F}}$.

In this last lemma of the section we do not assume Hypothesis 5.10.1.

LEMMA 5.10.13. *Assume $\tau = (\mathcal{F}, \Omega)$ is a quaternion fusion packet with $\mathcal{F}^\circ$ isomorphic to $\hat{A}E_6$, $\hat{L}_3(2)/E_{64}$, M_{12}, or $L_3(2)/E_8$, and $\mathcal{F}^\circ$ transitive on Ω. Let T be Sylow in $\mathcal{F}^\circ$. Then*

(1) In the first two cases $\Omega = \{K_1, K_2\}$ is of order 2.

(2) In the latter two cases $\Omega = \{K_1\}$ is of order 1 and $K_2 = C_T(K_1) \cong Q_8$.

(3) There exists no $\alpha \in Aut_{\mathcal{F}}(K_1K_2)$ of order 3 with $[K_1, \alpha] = 1$ and $K_2 = [K_2, \alpha]$.

PROOF. Parts (1) and (2) follow by construction of $\mathcal{F}^\circ$. Set $Q = K_1K_2$ and suppose α is as in (3). Observe $\mathcal{F}^\circ$, and hence also $P = O_2(\mathcal{F}^\circ)$, is normal in $\mathcal{F}$, so α extends to PQ. But if $\mathcal{F}^\circ$ is not M_{12} then $PQ = PK_i$ for $i = 1,2$, so as $[K_1,\alpha] = 1$ we have α centralizing PQ/P. This is a contradiction as $K_2 = [K_2,\alpha]$, so $PQ/P = [PQ/P,\alpha]$. Thus we may assume $\mathcal{F}^\circ \cong M_{12}$.

Let $z = z(K_1)$; as $\mathcal{F}^\circ \trianglelefteq \mathcal{F}$ we have $\mathcal{Y} = C_{\mathcal{F}^\circ}(z) \trianglelefteq C_{\mathcal{F}}(z)$. Now a model Y for $\mathcal{Y}$ has $Q = O_2(Y)$ and $Y/Q \cong S_3$ faithful on $K_1/\langle z\rangle$ for $i = 1,2$. But then $\alpha \in [\alpha, Y] \leq Aut_Y(Q)$, a contradiction. □

5.11. Summary of basics

We began this manuscript with statements of our major theorems, followed by a survey of basic notions, notation, and results on fusion systems. The remainder of the first five chapters is devoted to the construction of the foundation of the theory of quaternion fusion packets. In this final section of Chapter 5, we pause to collect in one place a list of the most frequently used concepts, results, and notation from that foundation. Hopefully this list will provide a convenient resource when, in later sections, the reader needs to refresh his or her memory of this theory.

Throughout this section, $\mathcal{F}$ is a saturated fusion system on a 2-group S. Recall $\mathcal{F}^f$ denotes the set of fully normalized subgroups (or involutions) in S. Given $X \leq S$ (or an involution X in S), $\mathfrak{A}(X)$ denotes the set of $\alpha \in \hom_{\mathcal{F}}(N_S(X), S)$ such that $X\alpha \in \mathcal{F}^f$.

For x an involution in S, $\mathcal{F}_x = C_{\mathcal{F}}(x)$.

For Σ a set of subgroups of S, $[\Sigma]_{\mathcal{F}}$ denotes the *normal closure* of Σ in $\mathcal{F}$: the smallest normal subsystem of $\mathcal{F}$ containing Σ. The *subnormal closure* of $T \leq S$ in $\mathcal{F}$ is $\mathrm{sub}(\mathcal{F},T)$: the limit of the sequence $\mathrm{sub}_i(\mathcal{F},T)$, where $\mathrm{sub}_0(\mathcal{F},T) = \mathcal{F}$ and $\mathrm{sub}_{i+1}(\mathcal{F},T) = [T]_{\mathrm{sub}_i(\mathcal{F},T)}$.

The notion of a set D of 3-transpositions of a finite group G is defined in section 4.1. Let D be a set of 3-transpositions of G. For $d \in D$, $d^\perp = C_D(d)$, $D_d = d^\perp - \{d\}$, and $A_d = D - d^\perp$. $\mathcal{D}(D)$ denotes the commuting graph on D. Set $V_d = \{a \in D : a^\perp = d^\perp\}$. By 4.1.7.1, if G is transitive on D then $V_d = d^{O_2(G)}$.

Let $\mathcal{G}$ be a graph. For v a vertex of $\mathcal{G}$, $\mathcal{G}(v)$ denotes the set of vertices adjacent to v in $\mathcal{G}$. Also $\mathcal{G}^c$ denotes the complementary graph to $\mathcal{G}$: the graph in which x and y are adjacent if and only if $x \neq y$ and $y \notin \mathcal{G}(x)$.

From the Introduction, a quaternion fusion packet is a pair $\tau = (\mathcal{F}, \Omega)$, where $\mathcal{F}$ is a saturated fusion system on a 2-group S, Ω is an $\mathcal{F}$-invariant set of subgroups of S, and

(QFP1) There exists an integer m such that for all $K \in \Omega$, K has a unique involution $z(K)$ and K is nonabelian of order m.

(QFP2) For each pair of distinct $K, J \in \Omega$, $|K \cap J| \leq 2$.

(QFP3) If $K, J \in \Omega$ and $v \in J - Z(J)$, then $v^{\mathcal{F}} \cap C_S(z(K)) \subseteq N_S(K)$.

(QFP4) If $K, J \in \Omega$ with $z(J) = z(K)$, $v \in K$, and $\phi \in \hom_{C_{\mathcal{F}}(z(K))}(v, S)$, then either $v\phi \in J$ or $v\phi$ centralizes J.

In the remainder of the section we assume $\tau = (\mathcal{F}, \Omega)$ is a quaternion fusion packet.

Set $\mathcal{F}^\circ = [\Omega]_{\mathcal{F}}$ and $\tau^\circ = (\mathcal{F}^\circ, \Omega)$. By 2.5.2, τ° is a quaternion fusion packet.

Set $\mathcal{Z} = \bigcup_{K\in\Omega} z(K)^{\mathcal{F}}$ and $\mathcal{Z}_S = \{z(K) : K \in \Omega\}$. Set $W_S = \langle \mathcal{Z}_S \rangle$. For $z \in \mathcal{Z}_S$, $\Omega(z) = \{K \in \Omega : z(K) = z\}$.

By 2.5.2, if $\mathcal{E}$ is a saturated subsystem of $\mathcal{F}$ on T, and $\Gamma \subseteq \Omega$ is contained in T and is $\mathcal{E}$-invariant, then $(\mathcal{E}, \Gamma)$ is a quaternion fusion packet. As a special case, we find in 3.1.1: For $z \in \mathcal{Z} \cap \mathcal{F}^f$, $\tau_z = (\mathcal{F}_z, \Omega)$ is a quaternion fusion packet.

For $z \in \mathcal{Z}_S$ we set $O(z) = \langle \Omega(z) \rangle$, while $O(\tau) = \langle \Omega \rangle$. By 3.1.5.1, $O(\tau)$ is a central product of the members of Ω.

For $K \in \Omega$ we define a subsystem $\mathcal{O}_K$ on K. If $m = 8$ set $\mathcal{O}_K = \mathcal{F}_K(K)$. If $m > 8$, set $\mathcal{O}_K = \langle O^2(Aut_{\mathcal{F}}(Q)) : Q \in \mathcal{Q}_K \rangle_K$, where $\mathcal{Q}_K$ is the set of Q_8-subgroups of K. For $z \in \mathcal{Z}_S \cap \mathcal{F}^f$, $\mathcal{O}(z)$ is the central product of the $\mathcal{O}_K$, for $K \in \Omega(z)$. By 2.6.11, $\mathcal{O}(z) \trianglelefteq \mathcal{F}_z$. Write $\mathcal{O}(\tau)$ for the normal closure of $O(\tau)$ in $N_{\mathcal{F}}(W_S)$. By 3.3.7, $\mathcal{O}(\tau)$ is a central product of the $\mathcal{O}_K$, for $K \in \Omega$.

For $z \in \mathcal{Z}$, $\Delta(z) = \{(K \cap C_S(z)\alpha)\alpha^{-1} : K \in \Omega(z\alpha) \text{ and } |K \cap C_S(z)\alpha| > 2\}$, for $\alpha \in \mathfrak{A}(z)$. By 3.1.3, $\Delta(z)$ is independent of the choice of α. Set $\Delta(\tau) = \bigcup_{z\in\mathcal{Z}} \Delta(z)$ and $\mathcal{Z}_\Delta = \{z \in \mathcal{Z} : \Delta(z) \neq \emptyset\}$. By 3.1.11.6, each member of $\Delta(\tau) - \Omega$ is cyclic of order $m/2$.

By 3.1.3 if $P \leq D \in \Delta(\tau)$ with $|P| > 2$, then each conjugate of P is contained in a member of $\Delta(\tau)$.

Define $\eta(\tau)$ to consist of the sets Σ of cyclic subgroups of S such that

(i) Σ is $\langle \Sigma \rangle$-invariant.

(ii) Each $V \in \Sigma$ is contained in some $D(V) \in \Delta(\tau)$ such that either $V = D(V)$ or $D(V) \in \Omega$ and $|D(V) : V| = 2$.

(iii) The map $V \mapsto D(V)$ is a bijection of Σ with $\Delta(\tau)$.

Write $\eta_S(\tau)$ for the set of S-invariant members of $\eta(\tau)$; by 3.1.10, $\eta_S(\tau) \neq \emptyset$. Set $W(\tau) = \{\langle \eta \rangle : \eta \in \eta_S(\tau)\}$. By 3.1.11 each member of $W(\tau)$ is abelian, and for each $\eta \in \eta(\tau)$, all members of η are of order $m/2$.

By 3.3.14, if $\mathcal{F}$ is transitive on Ω then either $|\eta(\tau)| = 1$ or $m = 8$ and $\Delta(\tau) = \Omega$. Observe that if $\eta(\tau) = \{\eta\}$ then $W(\tau) = \{W\}$ is of order 1 and W is weakly closed in S.

Let $\eta \in \eta(\tau)$ and $W = \langle \eta \rangle$. For $V \in \eta$, define $d_V \in Aut_{\mathcal{F}}(W)$ by $d_V = c_k \alpha^{-*}$, where $\alpha \in \hom_{\mathcal{F}}(W, S)$ with $V\alpha \leq K \in \Omega$ and $k \in K - V\alpha$. By 3.1.20, d_V is independent of the choice of α and k. Set $D_\eta = \{d_V : V \in \eta\}$ and $\mu_\eta = \langle D_\eta \rangle \leq Aut_{\mathcal{F}}(W)$. By 3.1.10, $\mu_\eta \trianglelefteq Aut_{\mathcal{F}}(W)$.

By 3.1.26, μ_η is independent of the choice of $\eta \in \eta(\tau)$, so we write $\mu = \mu(\tau)$ for μ_η. We call μ the *Thompson group* of τ. By 3.1.25, μ is a direct product of a 3-transposition group and copies of D_8.

Let $D = D_\eta \leq \mu_\eta = \mu$. By 4.2.8.6, $|\Omega| = |D \cap Aut_S(W)|$ is the *width* of μ. If μ is transitive on D then $\mathcal{F}^\circ$ is transitive on Ω by 4.2.8.7.

Let $\mathcal{D}, \mathcal{D}^*$ be the set of pairs (z,t) in $\mathcal{Z} \times \mathcal{Z}$ such that for some $\phi \in \hom_{\mathcal{F}}(\langle z, t \rangle, S)$, $(z,t)\phi$ is in $\mathcal{Z}_S \times \mathcal{Z}_S$, $\mathcal{Z}_\Delta \times \mathcal{Z}_\Delta$, respectively. Regard $\mathcal{D}$ and $\mathcal{D}^*$ as graphs. There are necessary and sufficient conditions for (z,t) to be an edge in the graphs in 3.1.14 and 3.1.15. The subgraph $\mathcal{A}$ of $\mathcal{D}^*$ is defined in 3.2.1.

Let $\eta \in \eta_S(\tau)$, $W = \langle \eta \rangle$, and $\mu = \mu_\eta$. Then $W \trianglelefteq S$, so $W \in \mathcal{F}^f$. By 3.3.11, $\tau_W = (N_{\mathcal{F}}(W), \Omega)$ is a quaternion fusion packet with $\mu(\tau_W) = \mu$. Further

$N_{\mathcal{F}}(W)^{\circ} = [O(\tau)]_{N_{\mathcal{F}}(W)}$ is constrained, and hence has a model $M(W)$. Set $M(\tau) = \{M(W) : W \in W(\tau)\}$.

Let Φ be a Coxeter diagram of type A_n, D_n, or E_n with $n \geq 2$ and $n \geq 4$ if Φ is of type D_n. In 4.3.1 we defined a certain universal group $\bar{G} = \bar{\omega}(\Phi, m)$ via generators and relations. By 4.3.8, if D_η is a class of 3-transposition of μ, then μ is the Weyl group of type Φ, for some Φ, and $M(W)$ is an image $\pi : \bar{G} \to M(W)$ of $\bar{G} = \bar{\omega}(\Phi, m)$ with $\ker(\pi) \leq Z(\bar{G})$.

For $\theta \subseteq \mathcal{Z}_S$ define $O(\theta) = \langle O(z) : z \in \theta \rangle$. Define $\mathcal{E}(\tau) = \langle S\mathcal{O}(\tau), N_{\mathcal{F}}(O(\theta)), N_{\mathcal{F}}(W) : W \in W(\tau), \emptyset \neq \theta \subseteq \mathcal{Z}_S$ with $O(\theta) \in \mathcal{F}^f \rangle_S$. Often $\mathcal{E}(\tau) = \mathcal{F}$.

Finally the reader is directed to the list of examples appearing as conclusions to our major theorems; recall this list appears at the end of section 1.1.

Theorems 2 through 5

CHAPTER 6

Theorems 2 and 4

In Chapter 6 we prove Theorem 2 and reduce Theorem 4 to Theorems 1 and 3.

Let $\tau = (\mathcal{F}, \Omega)$ be a quaternion fusion packet and S Sylow in $\mathcal{F}$. We begin by showing in Theorem 6.1.5 that if $\mathcal{F} = \mathcal{F}^{\circ}$ is transitive on Ω and τ satisfies one of the conclusions of Theorem 1, then, with the exception of the two classes of examples in case (4) of Theorem 4, τ is coconnected: that is $\mathcal{D}(\tau)^c$ is connected.

Then in section 6.2 we begin the proof of Theorem 2. We operate under Hypothesis 6.2.1, which, in addition to the hypotheses of Theorem 2, assumes that for $K \in \Omega$ we have $K \cap O_2(\mathcal{F}) = Z(K)$, and for each nontrivial subgroup Q of S normal in $\mathcal{F}$, we have $Q \cap \mathcal{Z}(\tau) \neq \emptyset$. These hypotheses isolate the interesting minimal cases from Theorem 2.

Set $E = \langle \mathcal{Z}(\tau) \rangle$. Immediately in 6.2.2 we show that E is the unique minimal normal subgroup of $\mathcal{F}$, and then show in 6.2.4 that $\mathcal{F}$ is constrained. Hence $\mathcal{F}$ has a model G, and for the most part we work with G rather than with $\mathcal{F}$.

There are two cases: E acts on K and E does not act on K. We treat the first case in section 6.2, where we show that G is $L_3(2)/E_8$, $L_3(2)/E_{64}$, or $L_3(2)/2^{3+6}$. The second case is treated in section 6.3, where we show G is AE_n.

In section 6.4 we introduce the *subnormal closure* $\tau(K) = (s(K), K^{s(K)})$ of K in τ; here $s(K)$ is the subnormal closure of K in $\mathcal{F}$, defined in 1.2.9. For the most part section 6.4 is devoted to a description of $s(K)$ in the case where $z = z(K) \in O_2(\mathcal{F})$, building on the earlier work in sections 6.2 and 6.3. Then in Theorem 6.5.4 we prove that if $z \notin O_2(\mathcal{F})$ then z is contained in a unique component $\mathcal{C}$ of $\mathcal{F}$, and if $\mathcal{O}_K$ is not solvable then $O^2(\mathcal{O}_K) \leq \mathcal{C}$. Building on these results, we show in 6.6.6 that if Theorem 1 holds in all proper subpackets of the form $\tau(K)$ for $K \in \Omega$, then a weak version of Theorem 4 holds. In particular, as discussed in Remark 6.6.10, Theorems 1 and 3 imply Theorem 4.

Lemma 6.6.11 gives information about the embedding of subpackets of τ in τ.

Finally Theorem 2 is established in the short section 6.7, using the earlier sections in Chapter 6.

6.1. $\mathcal{D}(\tau)^c$

In this section $\tau = (\mathcal{F}, \Omega)$ is a quaternion fusion packet and S is Sylow in $\mathcal{F}$. Let $\mathcal{D} = \mathcal{D}(\tau)$ be the graph of τ, and for $z \in \mathcal{Z}$, write Σ_z for the connected component of $\mathcal{D}^c$ containing z.

LEMMA 6.1.1. *If all members of $\mathcal{Z}_S$ are in the same connected component of $\mathcal{D}^c$ then $\mathcal{D}^c$ is connected.*

PROOF. Suppose $t \in \mathcal{Z} - \mathcal{Z}_S$. Then by 3.1.5.3, there is $K \in \Omega$ such that t does not centralize K. Set $z = z(K)$; then by 3.3.4, $t \notin z^\perp$, so $\Sigma_t = \Sigma_z$. The lemma follows. □

LEMMA 6.1.2. *If $\Omega = \Omega(z)$ for some $z \in \mathcal{Z}_S$ then $\mathcal{D}^c$ is connected.*

PROOF. This is immediate from 6.1.1. □

LEMMA 6.1.3. *Assume $\mu(\tau) \cong Weyl(\Phi)$, where Φ is A_n for some $n \geq 2$, D_n for some $n \geq 3$, or E_n for some $n \in \{6, 7, 8\}$. Then $\mathcal{D}^c$ is connected.*

PROOF. Let $\eta \in \eta(\tau)$, $W = \langle \eta \rangle$, $M \in M(\tau)$, and $D = D_\eta$. Then D is the set of reflections in the Weyl group $\mu = \mu(\tau) = M/W$. Let $\mathcal{G}$ be the full subgraph of $\mathcal{D}$ on $\mathcal{Z}_\Delta$. By 6.1.1 it suffices to show that $\mathcal{G}^c$ is connected. By 4.2.8.4, $\zeta : \mathcal{D}(D) \to \mathcal{G}$ is a bimorphism of graphs, where $\zeta : d_V \mapsto z(V)$ for $V \in \eta$. Thus it suffices to show that $\mathcal{D}(D)^c$ is connected. But by the hypothesis on μ, μ is transitive on D, so $\mathcal{D}(D)^c$ is connected by 4.1.4. □

LEMMA 6.1.4. *Assume $\mathcal{F} = \mathcal{F}^\circ$ and $\mathcal{Z} \cap Z(\mathcal{F}) = \emptyset$. Set $\mathcal{F}^+ = \mathcal{F}/Z(\mathcal{F})$. Then $\tau^+ = (\mathcal{F}^+, \Omega^+)$ is a quaternion fusion packet, $\mathcal{F}^+ = (\mathcal{F}^+)^\circ$, and $\xi : \mathcal{D} \to \mathcal{D}(\tau^+)$ is a morphism of graphs, where $\xi : z \mapsto z^+$. Therefore if $\mathcal{D}(\tau^+)^c$ is connected then so is $\mathcal{D}^c$.*

PROOF. Let $\Theta : \mathcal{F} \to \mathcal{F}^+$ be the natural map $\Theta : s \mapsto s^+$; then $\xi = \Theta_{|\mathcal{Z}}$. By 3.3.2.2, τ^+ is a quaternion fusion packet, and by 3.3.2.4, $(\mathcal{F}^+)^\circ = \mathcal{F}^+$.

Recall from 3.1.13 that (z, t) is an edge in $\mathcal{D}$ iff $z \neq t$ and $(z, t)\phi \in \mathcal{Z}_S \times \mathcal{Z}_S$ for some $\phi \in \hom_{\mathcal{F}}(\langle t, z \rangle, S)$. Therefore if (z, t) is an edge in $\mathcal{D}$ then $(z, t)\phi = (z(K), z(J))$ for some $K, J \in \Omega$. Then $K^+, J^+ \in \Omega^+$, with

$$(z^+, t^+)\phi^+ = (z(K^+), z(J^+)),$$

so either (z^+, t^+) is an edge in $\mathcal{D}(\tau^+)$, or $z^+ = t^+$. That is $\xi : \mathcal{D} \to \mathcal{D}(\tau^+)$ is a morphism of graphs.

Suppose $\mathcal{D}(\tau^+)^c$ is connected and let $(u, v) \in \mathcal{Z} \times \mathcal{Z}$, $a = u^+$, and $b = v^+$. Then there exists a path $a = a_0, \ldots, a_n = b$ in $\mathcal{D}(\tau^+)^c$. As $\xi : \mathcal{Z} \to \mathcal{Z}(\tau^+) = \mathcal{Z}^+$ is a surjection, there is a sequence $u = u_0, \ldots, u_n = v$ in $\mathcal{Z}$ such that $u_i\xi = a_i$ for each i. As ξ is a morphism and (a_i, a_{i+1}) is not an edge in $\mathcal{D}(\tau^+)$, also (u_i, u_{i+1}) is not an edge in $\mathcal{D}$. Thus $\Sigma_u = \Sigma_v$, completing the proof. □

THEOREM 6.1.5. *Assume $\mathcal{F} = \mathcal{F}^\circ$ is transitive on Ω and τ satisfies one of the conclusions of Theorem 1. Then one of the following holds:*

(1) $\mathcal{D}^c$ is connected.

(2) $\mathcal{F}/Z(\mathcal{F}) \cong PSp_n[m]$ for some $n \geq 6$.

(3) $\mathcal{F}/Z(\mathcal{F}) \cong PSL_n^-[m]$ for some $n \geq 6$.

PROOF. Assume τ is a counter example. Then by 6.1.4 and induction on the order of τ, replacing τ by τ^+ we have:

(a) $Z(\mathcal{F}) = 1$.

By 6.1.1:

(b) There exist $z, t \in \mathcal{Z}_S$ such that $\Sigma_z \neq \Sigma_t$.

Similarly by 6.1.2:

(c) $\Omega \neq \Omega(z)$.

Let $K \in \Omega(z)$; we next show:

(d) $z \notin O_2(\mathcal{F})$.

Assume otherwise. The proof of Theorem 2 in section 6.7 does not involve Theorem 6.1.5, so we may appeal to Theorem 2. We conclude that $\mathcal{F}$ is constrained with model G, $K \leq O_2(\mathcal{F}_z)$, and G satisfies one of the conclusions of Theorem 2. By (c), G satisfies neither conclusion (1) nor (2) of Theorem 2 (cf. 6.2.9.5) and if G is AE_n then $n \geq 9$. Therefore if conclusion (3) of Theorem 2 holds then $\mu = \mu(\tau) \cong \mathrm{Weyl}(D_k)$ for some $k \geq 4$, contrary to 6.1.3.

Therefore conclusion (4) of Theorem 2 is satisfied by G. Thus $G \in M(\tau)$ and μ is generated by a set D of 3-transpositions. As $\mathcal{F} = \mathcal{F}^\circ$ is transitive on Ω, μ is transitive on D by 4.1.4, so 4.3.8.1 and 5.4.3 supply a contradiction, completing the proof of (d).

(e) τ is a Lie fusion packet.

By (d) and the hypothesis of our theorem, τ satisfies one of conclusions (2), (3), or (4) of Theorem 1. But by (c), neither conclusion (3) nor (4) holds, establishing (e).

(f) $\mathcal{F}$ is not the 2-fusion system of an exceptional group.

Assume otherwise, By (c), $\mathcal{F}$ is not $G_2[m]$. Therefore by (e), $\mathcal{F}$ is $F_4[m]$, $E_6^\epsilon[m]$, $E_7[m]$, or $E_8[m]$, contrary to 6.1.3.

(g) $\mathcal{F}$ is $P\Omega_n^\epsilon[m]$ for some $n \geq 9$ or $PSL_n^+[m]$ for some $n \geq 5$.

Assume otherwise. By (f), $\mathcal{F}$ is the 2-fusion system of a classical group L. If $\mathcal{F} = PSp_n[m]$ is symplectic, then $n \geq 6$ by (c), so $\mathcal{F}$ appears in conclusion (2) of our theorem, contrary to the choice of τ as a counter example. Similarly if $\mathcal{F}$ is $PSL_n^-[m]$ then $n \geq 5$ by (c), so $n = 5$ as τ is a counter example. But then by 5.4.6, $\Omega = \{K_1, K_2\}$ is of order 2, and by 5.4.10.4 there exists $s \in \mathcal{Z}$ nontrivial on K_1 and K_2, contrary to (b).

We've shown that $\mathcal{F}$ is $P\Omega_n^\epsilon[m]$ or $PSL_n^+[m]$. Then by (c), $n \geq 9$ in the first case, while $n \geq 5$ in the second. Therefore (g) holds.

We are now in a position to obtain a contradiction, establishing our theorem. For if $\mathcal{F}$ is $P\Omega_n^\epsilon[m]$ with $n \geq 9$ then $\mu \cong \mathrm{Weyl}(D_k)$ for some $k \geq 4$, contrary to 6.1.3. Therefore by (g), $\mathcal{F}$ is $PSL_n^+[m]$ for some $n \geq 5$. Hence μ is $\mathrm{Weyl}(A_{n-1})$, again contrary to 6.1.3. □

LEMMA 6.1.6. *Let $\eta \in \eta(\tau)$, $W = \langle \eta \rangle$, $\mu = \mu(\tau)$, and G a model for $N_{\mathcal{F}}(C_S(W))$.*

(1) $Aut_{\mathcal{F}}(O(\tau))^\Omega = Aut_{\mathcal{F}}(O(\tau)W)^\Omega$.

(2) If $\mathcal{F}$ is transitive on Ω and D_η is a set of 3-transpositions of μ, then either

(i) $m = 8$ and $\Delta = \Omega$, or

(ii) $\eta(\tau) = \{\eta\}$, $N_G(O(\tau))$ is transitive on Ω, and G is transitive on the connected components of $\mathcal{D}(D_\eta)$.

PROOF. Let $O = O(\tau)$, $V = OC_S(O)$, Y a model for $N_{\mathcal{F}}(V)$, and X the kernel of the action of Y on Ω. Set $R = S \cap X$. Then $W \leq R$ and by a Frattini argument, $Y = XN_Y(R) = XN_Y(\Delta)$. Therefore $Y^\Omega = N_Y(\Delta)^\Omega$, so (1) holds.

Assume $\mathcal{F}$ is transitive on Ω. Then by 3.3.14, either (2i) holds or $\eta(\tau) = \{\eta\}$, and we may assume the latter. Hence W is weakly closed in S. As $\mathcal{F}$ is transitive on Ω, so is $Aut_{\mathcal{F}}(O)$, and hence also $Aut_{\mathcal{F}}(OW)$ by (1). Then as W is weakly closed, $N_{\mathcal{N}}(O)$ is transitive on Ω, where $\mathcal{N} = N_{\mathcal{F}}(W)$. Therefore $N_G(O)$ is transitive on Ω.

Now G induces a group of automorphisms on μ via $(xC_M(W))^g = x^g C_M(W)$ for $x \in M$ and $g \in G$. Further $D = D_\eta = \{d_U : U \in \eta\}$, so as G permutes η, it also permutes D, and hence the set Σ of connected components of $\mathcal{D}(D)^c$. Then as $N_G(O)$ is transitive on Ω, also G is transitive on Σ, completing the proof of (2). □

6.2. Beginning the case $z \in O_2(\mathcal{F})$

In this section we assume:

HYPOTHESIS 6.2.1. *$\tau = (\mathcal{F}, \Omega)$ is a quaternion fusion packet and, setting $S_0 = O_2(\mathcal{F})$:*
(1) $\mathcal{F}$ is transitive on Ω.
(2) $\mathcal{F} = \mathcal{F}^\circ$.
(3) For some $K \in \Omega$, $K \cap S_0 = Z(K)$.
(4) For each nontrivial normal subgroup Q of $\mathcal{F}$, $Q \cap \mathcal{Z} \neq \emptyset$.

Set $z = z(K)$ and $E = \langle \mathcal{Z} \rangle$.

LEMMA 6.2.2. *(1) $\mathcal{F}$ is transitive on $\mathcal{Z}$.*
(2) E is the unique minimal normal subgroup of $\mathcal{F}$.
(3) $E \leq Z(S_0)$.
(4) $Aut_{\mathcal{F}}(E)$ is irreducible on E.
(5) K does not centralize E.

PROOF. Part (1) follows from 6.2.1.1. By 6.2.1.3, $z \in S_0$, so $E \leq S_0$ by (1). Then as $\mathcal{Z}$ is $\mathcal{F}$ invariant, $E \trianglelefteq \mathcal{F}$. If M is a minimal normal subgroup of $\mathcal{F}$ then $M \cap \mathcal{Z} \neq \emptyset$ by 6.2.1.4, so $E \leq M$ by (1). Hence (2) holds, and (2) implies (3) and (4).

Suppose K centralizes E. Then by (1), $\Omega \subseteq C_{\mathcal{F}}(E) \trianglelefteq \mathcal{F}$, so $E \leq Z(\mathcal{F})$ by 6.2.1.2, contrary to 6.2.1.4. □

LEMMA 6.2.3. *(1) $[K, N_{S_0}(K)] \leq \langle z \rangle$.*
(2) If $t \in N_{\mathcal{Z}}(K)$ then either $[K, t] = 1$ or $C_K(t)$ is cyclic of index 2 in K.

PROOF. First $[K, N_{S_0}(K)] \leq K \cap S_0 = \langle z \rangle$ by 6.2.1.3, establishing (1). Thus if $t \in N_{\mathcal{Z}}(K)$ then $|K : C_K(t)| \leq 2$ by 6.2.2.3, so (2) follows from 3.3.6. □

LEMMA 6.2.4. *(1) $\mathcal{O}_K = \mathcal{F}_K(K)$.*
(2) $C_K(E) = \langle z \rangle$.
(3) $\mathcal{F}$ is constrained.

PROOF. First $E \leq S_0 \leq C_{\mathcal{F}}(E) \leq \mathcal{F}_z$, so

(*) $S_0 \leq O_2(\mathcal{F}_z)$ and $C_{O_2(\mathcal{F}_z)}(E) = O_2(C_{\mathcal{F}}(E)) = S_0$.

Suppose (1) fails. Then $\mathcal{C} = O^2(\mathcal{O}_K)$ is a component or solvable component of $\mathcal{F}_z$ by 2.6.13, so either $\mathcal{C} \leq C_{\mathcal{F}}(O_2(\mathcal{F}_z)) \leq C_{\mathcal{F}}(S_0)$ by (*), or $\mathcal{C} \cong SL_2[8]$. In the latter case as $K \cap S_0 = \langle z \rangle$, again $\mathcal{C}$ centralizes S_0. But the Sylow subgroup Q of $\mathcal{C}$ is quaternion and $E \leq C_S(Q) \leq N_{S_0}(K)$, so E centralizes K by 6.2.3.2, contrary to 6.2.2.5. Thus (1) holds.

By (1), $K \leq O_2(\mathcal{F}_z)$, so $C_K(E) \leq K \cap S_0 = \langle z \rangle$ by (*), proving (2).

Next as $z \in S_0$, $E(\mathcal{F}) = E(\mathcal{F}_z)$ by 10.3 in [**Asc11**], so $E(\mathcal{F}) \leq C_{\mathcal{F}_z}(K)$ as $K \leq O_2(\mathcal{F}_z)$. But now $\mathcal{F} = \mathcal{F}^\circ \leq C_{\mathcal{F}}(E(\mathcal{F})$ by 6.2.1.2, so $E(\mathcal{F}) = 1$. Therefore (3) follows from 14.2 in [**Asc11**]. □

NOTATION 6.2.5. By 6.2.4.3, $\mathcal{F}$ is constrained, so by 1.2.8.1 there exists a model G for $\mathcal{F}$.

LEMMA 6.2.6. *(1) E is the unique minimal normal subgroup of G.*
(2) $G = \langle K^G \rangle$.

PROOF. Part (1) follows from 6.2.2.2. By 1.2.8.4, there is a bijection between the normal subsystems of $\mathcal{F}$ and the normal subgroups of G, so (2) follows from 6.2.1.2. □

LEMMA 6.2.7. *(1) $m = 8$.*
(2) $O(z)$ is extraspecial of width $|\Omega(z)|$.
(3) $|E : N_E(K)| \leq 2$, and in case of equality, $E_8 \cong E \cap KK^e = [K, e]$ for $e \in E - N_E(K)$.
(4) For $t \in \mathcal{Z}$, one of the following holds:
(i) $t \in z^{\perp}$ and t centralizes $O(z)$.
(ii) $t \in \mathcal{D}^*(z)$ and for each $J \in \Omega(z)$, $|C_J(t)| = 4$.
(ii) t is fixed-point-free on $\Omega(z)$.

PROOF. Suppose that $E \leq N_S(K)$. Then by 6.2.3.2, for each $t \in \mathcal{Z}$, $C_K(t)$ is K or cyclic of index 2 in K. Thus if $m > 8$ then $\mathcal{Z}$ centralizes the unique cyclic subgroup V of K of index 2, contrary to 6.2.4.2.

On the other hand suppose $e \in E - N_E(K)$. Then $[K, e] \leq KK^e \cap E$, so $\Phi([K, e]) = 1$. Hence $m = 8$ in this case too, completing the proof of (1). Moreover as $m = 8$, $KK^e \cap E = [K, e] \cong E_8$. Then as E is abelian and $O(z)$ is the central product of the members of $\Omega(z)$, which are permuted by E, it follows that E acts on KK^e, so $|E : N_E(K)| = 2$, completing the proof of (3).

Observe (1) implies (2).

Next if t centralizes K then $t \in z^{\perp}$ and $[O(z), t] = 1$ by 3.3.4, while if $|C_K(t)| = 4$ then $t \in \mathcal{D}^*(z)$ and $|C_J(t)| = 4$ for each $J \in \Omega(z)$ by 3.3.4 and 3.3.5. Thus (4) holds. □

THEOREM 6.2.8. *Assume $E \leq N_G(K)$ and let $\mu = \mu(\tau)$. Then*
(1) $m(E) = 3$, $S_0 = C_G(E)$, and $Aut_G(E) = GL(E)$.
(2) One of the following holds:
(a) $\mu \cong S_3$ and $G \cong L_3(2)/E_8$.
(b) $\mu \cong S_4$ and $G \cong L_3(2)/E_{64}$.
(c) $\mu \cong$ Weyl(D_4) and $G \cong L_3(2)/2^{3+6}$.

Until the proof of Theorem 6.2.8 is complete, assume $E \leq N_G(K)$. Set $O_0(z) = O(z) \cap S_0$ and for $g \in G$, set $O_0(z^g) = O_0(z)^g$.

LEMMA 6.2.9. *(1) $E \cong E_8$.*
(2) Let $E = \langle z, t, s \rangle$ with $t, s \in \mathcal{Z}$. Then $G = \langle O(z), O(t), O(s) \rangle$ and $C_G(E) = S_0 = O_0(z)O_0(t)O_0(s)$.
(3) $Aut_G(E) = GL(E)$.
(4) $Aut_{O(z)}(E) = Aut_K(E) \cong E_4$ is the group of transvections with center z.
(5) $\Omega = \Omega(z)$.

PROOF. Let R be the group of transvections on E with center z. By 6.2.3.1, $Aut_{O(z)}(E) \leq R$, and by 6.2.4.2, $Aut_K(E) \cong E_4$. Then as G is irreducible on E, it follows from [**McL69**] that $Aut_G(E) = GL(E)$, proving (3). In particular G is 2-transitive on $E^{\#}$, so if (5) fails then $\mathcal{Z} \subseteq z^{\perp}$. But then E centralizes K, contrary to 6.2.2.5. Thus (5) holds and more generally $C_E(K) = \langle z \rangle$.

By (5) and 6.2.7.4, E induces inner automorphisms on K, so $|E : C_E(K)| \leq 4$. Therefore as $C_E(K) = \langle z \rangle$, we have $m(E) \leq 3$, while $m(E) > 2$ by 6.2.4.2, so (1) holds. Then as $E_4 \cong Aut_K(E)$ and $Aut_{O(z)}(E) \leq R$, (4) follows.

Let $E = \langle z, t, s \rangle$ with $t = z^g$ and $s = z^h$, and set $X = C_G(E)$ and $G^* = G/X = Aut_G(E)$. By (4), $G^* = H^*$, where $H = \langle O(z), O(z)^g, O(z)^h \rangle$. Further X acts on $O(z)$ so $[X, O(z)] \leq C_{O(z)}(E) \leq O_2(X) = S_0$, so $[X, O(z)] \leq O_0(z)$. Hence H and X, and then also G, acts on $T = O_0(z)O_0(t)O_0(s)$ and H^+ centralizes X^+, where $G^+ = G/T$. Hence $G = \langle K^G \rangle = \langle O(z)^H \rangle$ by 6.2.6.2, so $H = G$. Further $G^+ = H^+$ is a central extension of $X^+ = Z(G^+)$ by $L_3(2)$ with $E_4 \cong K^+$ normal in $C_{G^+}(z)$, so we conclude that $G^+ = \langle K^{+G^+} \rangle \cong L_3(2)$. This establishes (2) and completes the proof of the lemma. □

Notice 6.2.9 says that part (1) of Theorem 6.2.8 holds, so it remains to establish part (2) of the theorem. Choose notation so that $E_0 = \langle z, t \rangle \trianglelefteq S$.

Recall the definition of the *essential subgroups* of S from section 1.2.

LEMMA 6.2.10. *(1)* $z^\perp = \{z\}$.

(2) $E - \langle z \rangle = \mathcal{D}^*(z) - z^\perp$.

(3) $\mathcal{Z}_\Delta = E_0^\#$.

(4) $\mu \cong S_3$, D_{12}, S_4, *or Weyl*(D_4) *with* $|\Omega(z)| = 2^n$ *for* $n = 0, 1, 1, 2$, *respectively.*

(5) $O(z) \cong 2^{1+2^{n+1}}$.

(6) $O_2(C_G(z))$ *and* $O_2(N_G(E_0))$ *are the essential subgroups of* S.

PROOF. Part (1) follows from 6.2.9.5. Then (2) follows from (1) and 6.2.7.4. By 3.1.19, $\mu = \langle D \rangle$, where $D = D_\eta = \{d_V : V \in \eta\}$ for $\eta \in \eta(\tau)$. By (1) and 4.2.10, either $\mathcal{A}(z) \neq \emptyset$ and $\mu \cong D_{12}$, or D is a conjugacy class of 3-transpositions of μ. In the latter case by (1), 4.2.13, and 4.3.5.1, we have $\mu \cong S_3$, S_4, or Weyl(D_4). Next $|\Omega(z)| = |d_V^\perp|$ for $V \in \eta$ by (1) and 4.2.8, so (4) holds. Then (4) and 6.2.7.2 imply (5).

Let $M \in M(\tau)$. By (4), $M/O_2(M) \cong S_3$, so as $Aut_G(E) = GL(E)$ and S acts on M, we conclude that $MS = N_G(E_0)$. Therefore (3) holds.

Let R be an essential subgroup of S. By 1.2.11, $S_0 \leq R$, so as the radical subgroups of G/S_0 are S/S_0 and the minimal parabolics $C_G(z)/S_0$ and $N_G(E_0)/S_0$, (6) follows. □

LEMMA 6.2.11. *If* $\mu \cong S_3$ *or* D_{12} *then* $\mu \cong S_3$ *and* $G \cong L_3(2)/E_8$.

PROOF. Assume first that $\mu \cong D_{12}$. Then $t \in \mathcal{A}(z)$ and $t \in O(z)$ by 3.2.2.2. Next as $X \in Syl_3(C_G(z))$ acts nontrivially on K, $E = \langle z \rangle [E, X] \leq O(z)$, so as X is irreducible on $O(z)/E$, $O_0(z) = E$. Then as G is transitive on $E^\#$, $O_0(t) = E = O_0(s)$, so $E = S_0 = C_G(E)$ by 6.2.9.2, and then by 6.2.9.3, G is an extension of E_8 by $L_3(2)$. Further for $U \in \Delta(t)$, $S = UKE$, so the two members of $\Omega(z)$ are normal in S, and hence not conjugate in G, contradicting 6.2.1.1.

Therefore $\mu \cong S_3$. By 6.2.10.4, $O(z) = K$, so $O_0(z) = \langle z \rangle$. Thus $S_0 = E = C_G(E)$ by 6.2.9.2, so, as in the previous paragraph, G is an extension of E_8 by $L_3(2)$. Let $\eta \in \eta(\tau)$, $W = \langle \eta \rangle$, and $M = \langle K^{N_G(W)} \rangle$. By 4.2.5.4, M is determined up to isomorphism and $S_M = S \cap M$ is wreathed of order 32. Therefore $|S : S_M| = 2$

and $m(S_M) = 2$, so $m(S) \leq 3$. Hence G does not split over E, so by an unpublished result of Alperin, G is determined up to isomorphism. Hence the lemma holds. $\square$

Because of 6.2.10.4 and 6.2.11, for the remainder of the proof of Theorem 6.2.8 we assume:

HYPOTHESIS 6.2.12. *$E \leq N_G(K)$ and $\mu \cong S_4$ or Weyl(D_4). Let $\eta \in \eta(\tau)$ and set $W = \langle \eta \rangle$ and $M = \langle K^{N_G(W)} \rangle$. Thus $W = C_M(W)$ by 4.3.8.4, so $M/W \cong \mu$ and $M \leq N_G(E_0)$ by 6.2.10.3.*

LEMMA 6.2.13. *If $E \leq F \leq N_{S_0}(K)$ with $F \trianglelefteq G$, then $F = E$.*

PROOF. By 6.2.3.1, $[F, K] \leq \langle z \rangle$, so as $[E, K] = \langle z \rangle$, we have $F = EC_F(K)$. Thus by 6.2.6.2, $G = \langle K^G \rangle$ centralizes F/E. Suppose $f \in F - E$; without loss $F = E\langle f \rangle$. Then $F \cong E_{16}$ and $|C_F(K)| = 4$ but $C_E(K) = \langle z \rangle$, so K centralizes two members of fE. By 6.2.6.1, $C_F(G) = 1$, so by 6.2.9, either $Aut_S(F)$ is regular on fE and $Aut_G(F) = Aut_{S_0}(F)C_{Aut_G(F)}(f)$, or S_0 centralizes F and $C_{Aut_G(F)}(f) \cong$ Frob(21). In either case, G is 2-transitive on fE and $C_G(f)$ is transitive on $E^{\#}$. In particular as K centralizes two members of fE, we may choose those members to be f and fz. Therefore $C_{Aut_G(F)}(f)$ is of even order, so $Aut_{S_0}(F)$ is regular on fE.

For $x \in fE$, set $\mathcal{J}(x) = \{J \in K^G : [J, x] = 1\}$ and $r = |\mathcal{J}(f)|$. As $C_G(f)$ is transitive on $E^{\#}$, $r = 7j$ where $j = |\Omega(z) \cap \mathcal{J}(f)|$. Then counting the set Γ of pairs (x, J) with $x \in fE$ and $J \in \mathcal{J}(x)$ in two ways, we conclude that

$$2 \cdot 7 \cdot |\Omega(z)| = 2|K^G| = |\Gamma| = 8r = 56 \cdot j.$$

Then as $|\Omega(z)| = 2$ or 4 by 6.2.12, we conclude $|\Omega(z)| = 4$ and $j = 1$.

Let $S_1 = C_{S_0}(F)$. As $j = 1$, S_1 acts on K, so as $[N_{S_0}(K), K] \leq \langle z \rangle$, we conclude that $S_1 = EC_{S_1}(K)$. Thus $\Phi(S_1) = \Phi(C_{S_1}(K))$ centralizes K, and hence $G = \langle K^G \rangle$ centralizes $\Phi(S_1)$, so $\Phi(S_1) = 1$ by 6.2.6.1. Next as $Aut_{S_0}(F)$ is regular on fE, the map $s_0S_1 \mapsto [s_0, f]$ is a K-isomorphism of S_0/S_1 with E. Therefore $S_2 = [S_0, K]S_1 = \langle s \rangle S_1$ with $f^s = fz$, so s acts on K. Thus $S_2 = EC_{S_2}(K)$, so $KS_2 = PS_3$, where $P = KE \cong Q_8^2$ and $S_3 = C_{S_2}(P)$. Now $O(z) = \langle K^{S_0} \rangle \leq KS_2$ as $[S_0, K] \leq S_2$. But by 6.2.10.5, $O(z) \cong 2^{1+8}$, so S_3 has a Q_8^2-subgroup, impossible as $|S_3 : S_3 \cap S_1| = 2$ and $\Phi(S_1) = 1$. $\square$

LEMMA 6.2.14. *(1) $G = O^2(G)$.*
(2) $G/[G, S_0] \cong L_3(2)$.
(3) $[G, S_0] \not\leq E$.

PROOF. By 6.2.12 and 6.2.10.4, $|\Omega(z)| = 2$ or 4, so some $X \in Syl_3(C_G(z))$ acts on K. Then from 6.2.9, $K = [K, X]$, so $K \leq O^2(G)$. Then (1) follows from 6.2.6.2.

Let $G^* = G/[G, S_0]$. From 6.2.9, $G^*/Z(G^*) \cong L_3(2)$, so by (1), $G^* \cong L_3(2)$ or $SL_2(7)$. However by 6.2.1.3, $K^* \cong E_4$, so (2) follows.

If (3) fails then $E = S_0$ by (2), a contradiction since by 6.2.12, $\mu = M/W \cong S_4$ or Weyl(D_4). $\square$

LEMMA 6.2.15. *Set*

$$T = \bigcap_{J \in \Omega(z)} N_{S_0}(J)$$

and let $z^g = t$, $z^h = s$. Then

(1) $[O(z), T] \leq \langle z \rangle \leq E$ *and*
(2) $E = T \cap T^g \cap T^h$.

PROOF. Set $F = T \cap T^g \cap T^h$. First $[F, O(z)] \leq [T, O(z)] \leq \langle z \rangle \leq E$ by 6.2.3.1, establishing (1). Similarly $O(z)^g$ and $O(z)^h$ centralizes F/E, so, using 6.2.9.2, $G = \langle O(z), O(z)^g, O(z)^h \rangle$ centralizes F/E. Therefore $F = E$ by 6.2.13. □

LEMMA 6.2.16. *If* $\mu \cong S_4$ *then* $G \cong L_3(2)/E_{64}$.

PROOF. Assume $\mu \cong S_4$. Then $n = |\Omega(z)| = 2$ by 6.2.10.4, so $T = N_{S_0}(K)$ is of index at most 2 in S_0. Set $F = T \cap T^g \cap T^h$. As $|S_0 : T| \leq 2$, $|S_0 : F| \leq 8$. But $F = E$ by 6.2.15, so $|S_0 : E| \leq 8$. Then by 6.2.14.3, S_0/E is a natural module for G/S_0.

Suppose $e \in E \cap O(z) - \langle z \rangle$. Then as $X \in Syl_3(C_G(z))$ acts on $O(z)$, $E = [e, X]\langle z \rangle \leq O(z)$, so $E = O_0(z)$ as X is irreducible on $O(z)/E$. But then $O_0(t) = O_0(z)^g = E = O_0(s)$, so $S_0 = E$ by 6.2.9.2, a contradiction. Therefore $E \cap O(z) = \langle z \rangle$. Hence by 5.7.4.3, $Z(M)$ is generated by an involution i, and by 5.7.4.5, M is determined up to isomorphism. By 6.2.6.1, $i \notin Z(G)$, and as $M \trianglelefteq MS_0$, S_0 centralizes i, so $H = MS_0 = C_G(i)$ is of index 7 in G. Therefore $S_0 = \langle i^G \rangle$ is the core of the permutation module for G on G/H, using 6.2.6.1 and the fact that E and S_0/E are natural modules for G/S_0. In particular $S_0 = J(S)$ is determined in S.

Now $H = MS_0$ and M splits over $S_0 \cap M$ by 5.7.4.6, so H splits over S_0. Then as $S \leq H$, G splits over S_0 by Gaschutz' Theorem (cf. 10.4 in [**Asc86**]). So as S_0 is the core of the permutation module for G on H, G is determined up to isomorphism, so $G \cong L_3(2)/E_{64}$. □

Because of 6.2.16, during the remainder of the proof of Theorem 6.2.8 we assume that $\mu \cong$ Weyl(D_4).

LEMMA 6.2.17. *(1) There is a surjective homomorphism* π *of* $\bar{\omega}(D_4, 8)$ *onto* M *with kernel* $Z(\bar{\omega}(D_4, 8))$.
(2) $E \leq O(z)$.
(3) $S \leq M$ *and* $|S| = 2^{12}$.

PROOF. As $\mu \cong$ Weyl(D_4), there is a surjective homomorphism $\pi : \bar{M} = \bar{\omega}(D_4, 8)$ onto M by 4.3.8. As $\Omega = \Omega(z)$, $\ker(\pi) = Z(\bar{M})$ is of order 4 by 5.8.7.3, establishing (1). Then as $m = 8$, $|W| = 2^6$ by 4.3.7.2. Hence as $|\text{Weyl}(D_4)|_2 = 2^6$, $S_M = S \cap M$ is of order 2^{12}. By 5.8.7.6, $t \in O(z)$, so $E = [t, Y]\langle z \rangle \leq O(z)$, where $Y \in Syl_3(C_G(z))$. Thus (2) holds.

Define T, g, and h as in 6.2.15, and set $R = T \cap T^g$. Let T_1 be the kernel of the action of S on Ω and $R_1 = T_1 \cap T_1^g$. Then S/T_1 acts faithfully on Ω of order 4, so $|S : T_1| \leq 8$.

Set $H = MS$ and $H^+ = H/W$. Observe that $R_1 = C_S(O(z)^+)$ as $C_S(M^+)$ acts on the preimage KW of K^+ in S, and hence also acts on $K = KW \cap \Omega$. Then $R_1 = C_S(M^+)$ as $M^+ = \langle O(z)^+, O(z)^{g+} \rangle$. Hence as $H/S_0 \cong S_4$, we have $R_1 \leq S_0$, so $R_1 = R$ as R is the kernel of the action of S_0 on $\Omega(z) \cup \Omega(z)^g$. Similarly $T_1 = C_S(O(z)^+) = O(z)C_S(M^+) = O(z)R$, so $|T_1 : R_1| = |O(z) : C_{O(z)}(M^+)| = 8$. Hence $|T_1 : R| \leq 8$, so $|S : R| \leq 2^6$.

By 6.2.15, $E = R \cap T^h$, while $|T : T \cap T^h| = |T : R| \leq 8$, so $|R : E| = |RT^h : T^h| \leq |TT^h : T^h| = |T : T \cap T^h| \leq 8$. Thus $|S : E| \leq 2^9$, so we have $|S| \leq 2^{12}$. Therefore (3) holds as $|S_M| = 2^{12}$. □

LEMMA 6.2.18. *(1) S_0/E is the direct sum of two natural modules for G/S_0 dual to E.*

(2) S_0 is special with center E.

PROOF. Let $X \in Syl_3(M)$, $M^* = M/E_0$, and $G^+ = G/E$. By 6.2.17.3, $S \leq M$. Then from 6.2.17.1 and 5.8.7.5, $C_{S^*}(X^*) = Z(M^*) \cong E_8$. So as $C_E(X) \cong \mathbf{Z_2}$ it follows that $C_{S^+}(X^+) = C_{S^+}(M^+) \cong E_4$. Thus for u^+ an involution in $C_{S^+}(M^+)$, $U^+ = \langle u^{+G} \rangle$ is an image of the permutation module for G/S_0 on G/M in which $C_{U^+}(X^+) = C_{U^+}(M^+)$. Using 6.2.14 and 6.2.17.3, it follows that U^+ is a natural module for G/S_0, and as M is the stabilizer in G of E_0, U^+ is dual to E. Then (1) follows.

By 6.2.10.5, $O(z) \cong 2^{1+8}$ and by 6.2.9.4, $Aut_{O(z)}(E) \cong E_4$, so $O_0(z)$ is nonabelian. Thus S_0 is nonabelian. But by (1), $\Phi(S_0) \leq E$, so $\Phi(S_0) = [S_0, S_0] = E$ by 6.2.6.1, establishing (2). □

LEMMA 6.2.19. *If $\mu \cong Weyl(D_4)$ then $G \cong L_3(2)/2^{3+6}$.*

PROOF. Let $G^* = G/S_0$, $G^+ = G/E$, and $U \in \Delta(t)$. Then U^+K^+ is a complement to S_0^+ in S^+, so by Gaschutz' Theorem (cf. 10.4 in [**Asc86**]) there exists a complement H_0^+ to S_0^+ in G^+. By 6.2.18.1, $S_0^+ = S_1^+ \oplus S_2^+$ with $S_1^+ \cong S_2^+$ dual to E as H_0^+-modules. Let S_i be the preimage of S_i^+ in G and set $G^! = G/S_1$.

Claim each involution in $G^+ - S_0^+$ is in a complement to S_0^+ in G^+. First for i^+ an involution in H_0^+, $i^!S_0^!$ contains two $S_0^!$-classes of involutions. Moreover $H^1(G^*, S_0^!) \cong \mathbf{Z_2}$ with S^* regular on the involutions in $V - S_0^!$ for V a 4-dimensional indecomposable $\mathbf{F}_2G^*$-module with $[V, G^*] = S_0^!$. Thus V fuses the two classes and each involution in $i^!S_0^!$ is contained in a complement. Then a similar argument in $H_0^+S_1^+$ completes the proof of the claim.

Let $k \in K$ be of order 4; by the claim we may take $k \in H_0$. Then as $K = [k, X]$ for suitable $X \in Syl_3(C_{H_0}(z))$, $K \leq H_0$. Further as H_0 is transitive on the seven Sylow 2-subgroups of G, $S \cap H_0 \in Syl_2(H_0)$. For $i = 1, 2$, set $G_i = H_0S_i$ and $\Omega_i = K^{G_i} \cap S$. Then $T_i = S \cap G_i \in Syl_2(G_i)$ and we set $\mathcal{F}_i = \mathcal{F}_{T_i}(G_i)$. By 2.5.2, $\tau_i = (\mathcal{F}_i, \Omega_i)$ is a quaternion fusion packet.

Claim τ_i satisfies Hypothesis 6.2.1. By construction, τ_i satisfies 6.2.1.1. As $G_i = \langle K^{G_i} \rangle$, 6.2.1.2 is satisfied. Condition 6.2.1.3 is inherited from τ. By 6.2.18.2, either 6.2.1.4 holds or $S_i = E \oplus F$ where F is dual to E as an H_0^*-module, and we may assume the latter. But then by 3.3.2.4, $F \leq Z(G_i)$, a contradiction. This establishes the second claim.

By the second claim and our treatment of earlier cases, $G_i \cong L_2(3)/E_{64}$. In particular there is a complement L to S_0 in G. Also $S_i = \langle \Gamma_i \rangle$, where $\Gamma_i = v_i^L$ is of order 7 and $M_i = M \cap G_i$ centralizes v_i. Then as L is 2-transitive on Γ_i, the isomorphism type of G is determined by $e = [v_1, v_2]$ and $f = [v_1, w]$ for $w \in \Gamma_2 - \{v_2\}$ with $[w^+, S^+] = \langle v_2^+ \rangle$.

Now M centralizes v_i^+ for $i = 1, 2$, so M centralizes e. Then as $C_E(M) = 1$, $e = 1$. Now as $C_L(v_1)$ is irreducible on $S_2^+/\langle v_2^+ \rangle$, it follows that $S_1\langle v_2 \rangle = C_{S_0}(v_1)$, so $f \neq 1$. Next for $g \in S$, $w^g = wv_2a$ and $v_1^g = v_1b$ for some $a, b \in E$. Therefore

$$f^g = [v_1, w]^g = [v_1^g, w^g] = [v_1b, wv_2a] = [v_1, w] = f$$

as v_1 centralizes v_2 and $E \leq Z(S_0)$. Then as $f \neq 1$ and $C_E(S) = \langle z \rangle$, we have $f = z$. Therefore G is determined up to isomorphism by an earlier remark, completing the proof of the lemma. □

We are now in a position to complete the proof of Theorem 6.2.8. By 6.2.10.4 and 6.2.11, $\mu \cong S_3$, S_4, or Weyl(D_4), and in the first case, $G \cong L_3(2)/E_8$. In the remaining two cases, G is $L_3(2)/E_{64}$ or $L_3(2)/2^{3+6}$ by 6.2.16 and 6.2.19. This completes the proof of Theorem 6.2.8.

6.3. The case $E \not\leq N_G(K)$

In this section we assume the following hypothesis:

HYPOTHESIS 6.3.1. *Hypothesis 6.2.1 holds, and setting $E = \langle \mathcal{Z} \rangle$, E does not normalize K.*

In addition we adopt the notation of section 6.2. In particular set $z = z(K)$ and, appealing to 6.2.4.3, let G be a model for $\mathcal{F}$. Set $E_z = N_E(K)$ and let $\eta \in \eta_S(\tau)$. Set $W = \langle \eta \rangle$ and $M = \langle O(z)^{N_G(W)} \rangle$. Then $\mu = \mu_\eta = M/C_M(W)$. Set $G^* = G/S_0$.

LEMMA 6.3.2. *(1) $|E : E_z| = 2$, $E_z = N_E(J)$ for each $J \in \Omega(z)$, and E/E_z is semiregular on $\Omega(z)$.*
(2) $[K, E] \cong E_8$ and $K[K, E] \cong Q_8^2$.
(3) $K^E = \{K, \tilde{K}\}$ is the set of quaternion subgroups of $K[K, E]$.
(4) $\mathcal{Z} \cap E_z = \{z\} \cup \mathcal{D}^*(z)$ and for each $t \in \mathcal{Z} \cap E_z$, $|K : C_K(t)| \leq 2$.
(5) If $t \in \mathcal{Z} \cap E_z$ then $z \in E_t$.

PROOF. By Hypothesis 6.3.1, $E \neq E_z$, so $|E : E_z| = 2$ by 6.2.7.3.

Part (4) follows from 6.2.7.4. As the relation $\mathcal{D}^*$ is symmetric, (4) implies (5).

By 6.2.6.1, G is irreducible on E, so there is $t \in \mathcal{Z} - E_z$. By 6.2.7.3, $[K, t] = E \cap KK^t \cong E_8$. As $[K, E_z] \leq \langle z \rangle$ and $E = E_z\langle t \rangle$ since $|E : E_z| = 2$, it follows that $[K, E] = E \cap KK^t \cong E_8$, and $K^E = \{K, K^t\}$. Then $K[K, E] \cong Q_8^2$, and hence contains exactly two Q_8-subgroups. Thus (2) and (3) hold.

By 6.2.7.4, t is semiregular on $\Omega(z)$. Let $e \in E$. Then $e = t_1 \cdots t_n$ with $t_i \in \mathcal{Z}$, and from (4), $e \in E_z$ iff $I(e, K) = \{i : t_i \notin \{z\} \cup \mathcal{D}^*(z)\}$ is of even order. From 6.2.7.4, $I(e, K) = I(e, J)$ for each $J \in \Omega(z)$, so $E_z = N_E(J)$, completing the proof of (1). □

LEMMA 6.3.3. *(1) $\mathcal{A}(z) = \emptyset$.*
(2) D_η is a set of 3-transposition of μ.

PROOF. Part (1) and 4.2.10.3 imply (2), so it remains to prove (1). Suppose $t \in \mathcal{Z}_\Delta$ with $t \in \mathcal{A}(z)$ and let $U \in \Delta(t)$ and $V = C_K(t)$. By 3.2.2.1, $|\Omega(z)| = 2$, so $\Omega(z) = \{K, \tilde{K}\}$ by 6.3.2.3. By 4.2.6, $t = z^g$ for some $g \in G$ and $X = \langle K, \tilde{K}^g \rangle$ is the split extension of $U \times V \cong \mathbf{Z}_4^2$ by $Y \cong D_{12}$ with the involution $f \in Z(Y)$ inverting UV, and $\tilde{K} \leq X$. But then by 6.3.2.2, $E_8 \cong [K, E] \leq K\tilde{K} \leq X$, and $[K, E]$ is contained in the elementary abelian normal subgroup $E \cap X$ of X. We conclude from the description of X above that $E \cap X = [K, E] = E_0\langle f\rangle$, where $E_0 = \langle z, t\rangle$. Let $s \in \mathcal{Z} - E_t$; then $1 \neq E_s \cap E_0$, so without loss $z \in E_s$. Now s acts on $\langle O(z), O(t)\rangle = X$ by 4.2.10.4; set $M_0 = X\langle s\rangle$. Then $F = [E, K]\langle s\rangle = E \cap M_0 \trianglelefteq M_0$ and $F = E_0 \times F_0$, where $F_0 = C_F(O(Y)) = \langle f, e\rangle$. As s and $[E, K]$ act on K, so does F. Thus as F_0 centralizes $O(Y)$ which is transitive on $E_0^{\#}$, it follows that F_0 acts on K^g. But now $F = E_0F_0$ acts on K^g, so $s \in F \leq N_G(K^g)$, contrary to the choice of s. □

LEMMA 6.3.4. *(1)* $[G, S_0] \leq E$.
(2) $S_0 = C_G(E)$.
(3) Let $\{K_1, \ldots, K_n\}$ *be a set of representatives for the equivalence classes* $\{J, \tilde{J}\}$, $J \in \Omega$. *Then* $O(\tau)^* = K_1^* \times \cdots \times K_n^*$ *with* $K_i^* \cong E_4$.
(4) If $t \in \mathcal{Z}_\Delta - z^\perp$ *and* $U \in \Delta(t)$, *then* $K^*U^* \cong D_8$.
(5) $K^* \cong E_4$ *is a TI-subgroup of* G^* *with* $N_{G^*}(K^*) \leq N_G(K\tilde{K})^* \leq C_G(z)^*$.
(6) K^* *acts faithfully as a group of transvections on* E_z *with center* $\langle z\rangle$.
(7) For $k \in K - \langle z\rangle$, $[E, k] \cong E_4$.
(8) Suppose $G^* \cong L_3(2)$ *or* $U_3(4)$, *or* $G^* \cong A_n$ *for some* $n \geq 5$ *and* K^* *is a root 4-subgroup of* G^*. *Then* $G = O^2(G)$ *and* $E = S_0$.

PROOF. By 6.2.2.3, $E \leq Z(S_0)$, so S_0 acts on $[K, E]$. By parts (2) and (3) of 6.3.2, $[K, E] \leq K\tilde{K}$. Therefore each member of $\Omega - \{K, \tilde{K}\}$ centralizes $[K, E]$, so S_0 acts on $\{K, \tilde{K}\}$. On the other hand by 6.2.1.3, $K \cap S_0 = \langle z\rangle$, so $K\tilde{K} \cap S_0 = [K, E]$. Therefore $[K, S_0] \leq K\tilde{K} \cap S_0 = [K, E] \leq E$. Then (1) follows from 6.2.6.2.

As G is a model for $\mathcal{F}$, $F^*(G) = S_0$. Thus (2) follows from (1).

By 6.2.1.3, $K^* \cong E_4$. Adopt the notation of (3) and let $K = K_1$ and $X = \langle K_i : i > 1\rangle$. Then X centralizes $K\tilde{K}$ and hence also $[E, K]$. Thus $X \cap KS_0 \leq C_{KS_0}([E, K]) = S_0$, so $X^* \cap K^* = 1$, completing the proof of (3).

Assume the hypothesis of (4). Then by 4.2.5.2, $KU/\langle z, t\rangle \cong D_8$, so as $\langle z, t\rangle \leq E$ and $K \cap S_0 = \langle z\rangle$, (4) follows.

Suppose $g \in G - N_G(K^*)$ with $1 \neq K^* \cap K^{g*}$. By (3), $K^* \cong E_4$, so $K^* \cap K^{g*} = V^{g*}$ for some V of index 2 in K. Then, using (3), $V^g \in \Delta(z^g)$, contrary to (4). Thus K^* is a TI-subgroup of G^*. By 6.3.2.2, $K^* = \tilde{K}^*$, so $N_{G^*}(K^*) \leq N_G(K\tilde{K})^*$, so (5) holds.

By 6.2.3.1, $[E_z, K] \leq \langle z\rangle$, and by 6.3.2.2, K^* is faithful on E_z, so (6) and (7) hold.

Assume the hypothesis of (8) and set $G^+ = G/E$ and $L = O^2(G)$. By (1), $S_0^+ \leq Z(G^+)$, so as G^* is simple in (8) we conclude that G^+ is the central product of L^+ with S_0^+ and L^+ is quasisimple with $Z(L^+)$ a 2-group and $L^+/Z(L^+) \cong G^*$. Moreover $K^* \leq X^* \leq G^*$ with $X^* \cong A_4$, so it follows from (5) that $K \leq O^2(G) = L$. Thus $G = L$ by 6.2.6.2. As $K^+ \cong E_4$, it follows from the structure of the covering group of G^* that $Z(L^+) = 1$, so $E = S_0$, completing the proof of (8). □

LEMMA 6.3.5. *If* $\mu \cong E_4$ *then* $G \cong AE_5$.

PROOF. Assume $\mu \cong E_4$; then $\Delta(\tau) = \{K, \tilde{K}\}$. Hence K^* is strongly closed in S^* with respect to G^* and by 6.3.4.5, $K^* \cong E_4$ is a TI-subgroup of G^*. As $G = \langle K^G\rangle$ by 6.2.6.2, it follows from Theorem 1 in [**Asc73**] that $G^* \cong A_5$ or $U_3(4)$. Thus $S_0 = E$ by 6.3.4.8.

Suppose $G^* \cong U_3(4)$. Then $N_G(S) = SX$ where $X \cong \mathbf{Z}_{15}$. Set $Y = O_5(X)$ and $T = [S, Y]$. As $O = K\tilde{K} \cong Q_8^2$ is Y-invariant, Y centralizes O, so T centralizes

O. But $K^* = Z(T^*)$, so for $k \in K$ of order 4, $k = te$ for some $t \in T$ and $e \in E$. Then $e = t^{-1}k \in C_E(\tilde{K}) = C_E(K)$ as $KE = \tilde{K}E$. But then $k = te$ centralizes K, a contradiction.

Therefore $G^* \cong A_5$. Further $N_G(K^*) = C_G(z)$ is of index 5 in G, so $|\mathcal{Z}| = 5$ by 6.2.2.1 and indeed E is an image of the 5-dimensional permutation module for G^*. Then $E \cong E_{16}$ is the core of that module by 6.2.6.1. As O splits over $[K, E]$, G splits over E, so G is determined up to isomorphism, completing the proof. □

LEMMA 6.3.6. *Assume $g \in G$ with $z^g \in \mathcal{Z}_\Delta - z^\perp$, and let $U \in \Delta(z^g)$ and $J \in K^G$ with $U \leq J$. Set $Q = K\tilde{K}$, $X = \langle Q, J\rangle$ and let $V = K \cap W$, $\tilde{V} = \tilde{K} \cap W$, $E_X = E \cap X$, and $W_X = W \cap X$. Then*

(1) $W_X = V\tilde{V}U = C_X(W_X)$ and $X/W_X \cong S_4$.

(2) $O_2(X) = E_X W_X$ and $S \cap X = KE_X W_X$.

(3) $S_0 \cap X = E_X$ and either

(i) $z^g \in Q$, $\Omega(z) = \{K, \tilde{K}\}$, $W_X = V \times U$, and $E_X \cong E_{16}$, or

(ii) $z^g \notin Q$, $W_X = V\tilde{V} \times U$, and $E_X \cong E_{32}$ with $E_X \geq Z(X) = \langle v\tilde{v}z^g\rangle \cong \mathbf{Z}_2$ for suitable generators $v, \tilde{v}$ of V and $\tilde{V}$.

(4) For $Y \in Syl_3(X)$, $C_{W_X}(Y) = Z(X)$.

(5) In each of the two cases in (3), $X/E_X \cong S_4$, X is determined up to isomorphism, and X splits over E_X.

PROOF. Set $M^+ = M/C_M(W)$. By 6.3.3.2 and 4.1.5, $X^+ \leq M^+ = \mu$ is isomorphic to S_4. Let $\tau_X = (X, \{K, \tilde{K}\})$ and $\eta_X = V^X$. Then τ_X is a quaternion fusion packet with $\eta_X \in \eta(\tau_X)$, $W_0 = \langle \eta_X\rangle \trianglelefteq X$, and $\mu(\tau_X) \cong S_4$. By 4.3.8.4, $W_0 = C_X(W_0)$ so as W_X centralizes W_0 we conclude that $W_0 = W_X$. Then by parts (1) and (2) of 4.3.7, $W_X = V\tilde{V}U$ and $X/W_X \cong S_4$, so (1) holds.

By 6.3.2.3, $\tilde{K} \in \Omega(z)$, so by 5.7.4.3, $q = |Z(X)| = 1$ or 2, with $q = 1$ iff $z^g \in Q$. Further if $z^g \in Q$ then $\Omega(z) = \{K, \tilde{K}\}$ by 3.1.6, and as $q = 1$, $|W_X| = 16$, so $W_X = V \times U$. In particular $C_{W_X}(Y) = 1$, so (4) holds in this case. On the other hand if $q = 2$ then $z^g \notin Q$, so $V\tilde{V} \cap U = 1$ and $|W_X| = 32$, so $W_X = V\tilde{V} \times U$. Here $C_{W_X}(Y) = Z(X)$ is of order 2 in this case, so again (4) holds.

Next $Q^+ \cong E_4$ with $Q \cap W = V\tilde{V}$, so $[E, K]^+ \neq 1$ and hence as $E_X^+ \trianglelefteq X^+$, we have $E_X^+ = O_2(X^+)$, so that (2) holds. Moreover $E_X \leq S_0 \cap X \leq O_2(X) = E_X W_X$, so $S_0 \cap X = E_X(S_0 \cap W_X)$, and then $|(S_0 \cap X) : (S_0 \cap W_X)| = |E_X : E \cap W_X| = 4$ and $E_0 = \langle z, z^g\rangle \leq E \cap W_X$. As $S_0 \cap X = E_X(S_0 \cap W_X)$ it follows from (4) that $C_{S_0 \cap X}(Y) \leq E_X Z(X)$. By 6.3.4.1, $[X, S_0 \cap X] \leq E_X$, so we conclude that $S_0 \cap X \leq E_X Z(X)$.

Suppose $q = 1$. Then $Z(X) = 1$ so $S_0 \cap X = E_X$. Also $\Phi(E) = 1$ and $E_0 = \Omega_1(W_X)$, so $E_X \cong E_{16}$ and hence (3i) holds. Observe also that $X/E_X \cong S_4$.

So assume $q = 2$. Then $u = v\tilde{v}z^g \in E$ as $v\tilde{v} \in [E, K] \leq E$ and z^g are in E. Thus $E \cap W_X$ contains $E_0\langle u\rangle \cong E_8$, and hence $S_0 \cap X = E_X \cong E_{32}$ as $\langle u\rangle = Z(X)$ by 5.7.4.3. This completes the proof of (3). Observe that again $X/E_X \cong S_4$.

By 5.7.4.5, in each of the two cases in (3), X is determined up to isomorphism with $m_2(X) = m_2(E_X)$ and X splits over each member of $\mathcal{A}(O_2(X))$ by 5.7.4.6, so (5) holds. □

LEMMA 6.3.7. *If $\mu \cong S_4$ then $G \cong AE_6$ or AE_7.*

PROOF. Assume $\mu \cong S_4$. Then M is the group X of 6.3.6, with $Q = O(\tau) = K\tilde{K}$ and $T = \langle \Delta\rangle = QW$. By 6.3.6.2, $S_M = S \cap M = KWE_M$; by 6.3.6.3,

$S_0 \cap M = E_M$; and then by 6.3.6.5, $M^* = MS_0/S_0 \cong M/(M \cap S_0) = M/E_M \cong S_4$, with $D_8 \cong S_M^* = (KWE_M)^* = K^*U^* = T^*$ for $z^g \in \mathcal{Z}_\Delta - \{z\}$ and $U \in \Delta(z^g)$. Then as $M^* \cong S_4$, each subgroup of T^* of order 2 is of the form V_1^* for some V_1 of index 2 in some member of K^M, so as $T = OW$, T^* is the weak closure of its subgroups of order 2 in S^*.

Claim T^* is strongly closed in S^* with respect to G^*. As T^* is the weak closure of its involutions, if the claim is false then the $\mathbf{Z}_4$-subgroup B^* of T^* is not weakly closed in S^*, so by Alperin's Fusion Theorem there is $R^* \in \mathcal{F}_{S^*}(G^*)^{frc}$ containing B^* with B^* not normal in $N^* = N_{G^*}(R^*)$. As T^* is weakly closed in S^* and B^* char T^*, $T^* \not\leq R^*$, so $B^* = T^* \cap R^*$. Thus as T^* is the weak closure of its involutions, $Z(T^*) = Z^*$ is strongly closed in R^*, so $Z^* \trianglelefteq N^*$. But $[R^*, K^*] \leq Z^*$ so $K^* \leq R^*$, a contradiction. This completes the proof of the claim.

By 6.2.6.2, $G = \langle K^G \rangle$, so by Corollary B3 in [**Gol75**], $S^* = T^* \cong D_8$. As $S_0 = O_2(G)$, we have $O_2(G^*) = 1$. Therefore, appealing to the classification of groups with D_8-Sylow groups, G^* is A_7 or $L_2(q)$ for some odd q. In the latter case as K^* is a TI-subgroup of G^*, $q = 7$ or 9. However if $q = 7$ then $C_G(z)^* = N_{G^*}(K^*)$ is of index 7 in G^*, so G is 2-transitive on $\mathcal{Z}$ of order 7. But then $\mathcal{Z} = \{z\} \cup z^{gC_G(z)} \subseteq E_z$, a contradiction. Therefore $G^* \cong A_6$ or A_7, and up to conjugacy in $Aut(G^*)$, K^* is a root 4-subgroup of G^*. Then by 6.3.4.8, $S_0 = E$.

Next let $\mathcal{X}^*$ be the set of A_5-subgroups of G^* containing K^* and $\mathcal{X}$ consist of the subgroups $X = \langle Q, Q^x \rangle$ for some $x \in X$ with $X^* \in \mathcal{X}$. Then X is 2-transitive on $\mathcal{Z}_X = z^X$ of order 5, so setting $E_X = \langle \mathcal{Z}_X \rangle$, E_X is an image of the permutation module for X^* on $\mathcal{Z}_X$. Thus $m(E_X) = 4$ or 5. As $X = \langle Q^X \rangle$ and $Q \leq O^2(N_X(Q))$, $X = O^2(X)$, so $m(E_X) = 4$. Let $S_X = S \cap X$, $\mathcal{F}_X = \mathcal{F}_{S_X}(X)$, and $\Omega_X = K^X \cap S$. Then $\tau_X = (\mathcal{F}_X, \Omega_X)$ is a quaternion fusion packet by 2.5.2. As $O_2(X^*) = 1$, $O_2(X) = X \cap E$ and as E_X is a projective X^*-module, $E \cap X = E_X \oplus F$ for some X-submodule F of $E \cap X$. By 3.3.2.4, X centralizes F, so as $X = O^2(X)$ and $K^* \cong E_4$, $F = 1$. Thus τ_X satisfies Hypothesis 6.2.1. Observe $[E, K] = E_X \cap E_z$, so τ_X satisfies Hypothesis 6.3.1. Therefore $X \cong AE_5$ by 6.3.5.

Next let $X_1 \in \mathcal{X} - \{X\}$ and $H = \langle X, X_1 \rangle$. Then $H^* \cong A_6$. Set $S_H = S \cap H$, $\mathcal{Z}_H = z^H$, and $E_H = \langle \mathcal{Z}_H \rangle$. Then $|\mathcal{Z}_H| = |H : C_H(z)| = 15$, and by G.5.3 in [**AS04**] either $m(E_H) = 4$ or E_H is the core of the permutation module for H^* on H^*/X^*. As in the previous paragraph, $O_2(H) = E \cap H$. As $[E, K] \leq E_H$, $[E \cap H, H] = E_H$. Then as $H = O^2(H)$ and $K^* \cong E_4$, we conclude that $E_H = O_2(H)$. Next $M \leq H$ and by 6.3.6.5, M is determined up to isomorphism with $|M|_2 = |H|_2$, so $S_H \in Syl_2(M)$. Then by 6.3.6.5, S_H splits over E_H, so H splits over E_H. Therefore H is determined up to isomorphism, so $H \cong AE_6$ or $\check{A}E_6$. In particular if $G^* \cong A_6$ then $G = H$ and the lemma holds, so we may assume $G^* \cong A_7$.

Next as K^* is a root 4-subgroup of G^*, $|\mathcal{Z}| = |G : C_G(z)| = |G^* : N_{G^*}(K^*)| = 35$. Therefore $m(E) \geq 6$. But $G = \langle H, X_2 \rangle$ for $X_2 \in \mathcal{X}$ with $X_2^* \not\leq H^*$, so $E = [E, G] = [E, H][E, X_2]$ with $m([E, X_2]) = 4$ and $[E, K] \leq [E, X_2] \cap E_H$. Thus $m(E) \leq m(E_H) + 1$, so $m(E_H) = 5$ and $m(E) = 6$. Let $\langle e \rangle = Z(H)$. Then $E = \langle e^G \rangle$ is an image of the permutation module for G on G/H, so as $m(E) = 6$, E is the core of that module. As H splits over E_H and H^* contains a Sylow 2-subgroup of G^*, S splits over E, so G splits over E. Therefore G is determined up to isomorphism, so $G \cong AE_7$, completing the proof of the lemma. □

LEMMA 6.3.8. *Assume $\Omega = \Omega(z)$ is of order 4. Then μ is nonabelian.*

PROOF. Assume μ is abelian and set $O = O(z)$. By 6.3.4.3, $O^* = K_1^* \times K_2^* \cong E_{16}$ where $K = K_1$ and $K_2 \in \Omega - \{K, \tilde{K}\}$. As μ is abelian, $\Delta(\tau) = \Omega(z)$, so $\Sigma = K^* \cup K_2^*$ is strongly closed in S^* with respect to G^*. Claim O^* is strongly closed in S^* with respect to G^*. Assume otherwise, so that there is $x \in O$ and $g \in G$ with $y^* = x^{g*} \in S^* - O^*$. Thus $x^* \notin \Sigma$.

Suppose first that y^* acts on K^*. Then y^* centralizes k^* for some $k \in K$ of order 4, so x^* centralizes $k^{g^{-1}*}$. Then as O^* is abelian and weakly closed in S^*, there is $h \in C_G(x^*)$ with $j^* = k^{g^{-1}h*} \in S^*$, so $j^* \in \Sigma$ and hence $k^{g^{-1}h} \in J \in \Omega(z)$ as $\Delta(\tau) = \Omega(z)$. Hence $g^{-1}h \in C_G(z)$, so as $O \trianglelefteq C_G(z)$, $x^{g*} = x^{*h^{-1}g} \in O^*$, contrary to the choice of y^*.

Thus $K^{*y} = K_2^*$, so $B_0^* = C_{O^*}(y^*) \cong E_4$. Set $B^* = B_0^*\langle y^* \rangle$; then $[B^*, O^*] = B_0^*$, so $Aut_O(B^*)$ is the group of transvections on B^* with axis B_0^*. Next, arguing as above, there exists $h \in C_G(x^*)$ with $D^* = B^{g^{-1}h*} \leq S^*$. If $D^* \leq O^*$ then $D^* \cap K^* \neq 1$, contradicting Σ strongly closed. Further by the previous paragraph, $D_0^* = N_{D^*}(K^*) \leq O^*$, so there exists $d \in D$ with $K^{d*} = K_2^*$. Let $\theta = O^{g^{-1}h}$ and $Y = \langle \theta, O \rangle$. Then $Aut_\theta(D^*)$ is the group of transvections with axis $B_0^{g^{-1}h*}$ and $Aut_O(D^*)$ is the group of transvections with axis D_0^*, so as $x^* \in D_0^* - B^{g^{-1}h*}$, it follows that $Aut_Y(D^*)$ is the stabilizer in $GL(D^*)$ of the point $D_0^* \cap B^{g^{-1}h*}$. In particular there is $k \in K$ of order 4 with $k^{g^{-1}h} \in N_Y(O)$, contradicting $\Delta(\tau) = \Omega(z)$. This completes the proof of the claim.

By 6.2.6.2, $G = \langle K^G \rangle$, so by the claim and Goldschmidt's Theorem [**Gol74**], $O^* \in Syl_2(G^*)$. This is a contradiction as K^* and K_2^* are conjugate in G^*. □

THEOREM 6.3.9. *$G \cong AE_n$ for some $5 \leq n$.*

During the remainder of the section assume G is a minimal counter example to Theorem 6.3.9.

LEMMA 6.3.10. *Assume $K_i \in K^G$ for $i = 1, 2$ with $Y = \langle K_1^Y, K_2^Y \rangle < G$. Let $S_Y \in Syl_2(Y)$ and set $\Omega_Y = (K_1^Y \cup K_2^Y) \cap S_Y$, $\mathcal{F}_Y = \mathcal{F}_{S_Y}(Y)$, and $\tau_Y = (\mathcal{F}_Y, \Omega_Y)$. Take $K = K_1 \leq S_Y$, set $z_i = z(K_i)$, and assume $z_2 \notin E_z$. Then*

(1) τ_Y is a quaternion fusion packet.

(2) $K_2 \in K^Y$ and $Y \cong AE_n$ or $\check{A}E_n$ for some $n \geq 5$.

PROOF. Conjugating in G, we may assume $S_Y \leq S$. Then (1) follows from 2.5.2.

Claim $K_i \cap O_2(Y) = \langle z_i \rangle$. For if $|K_2 \cap O_2(Y)| > 2$ then $K_2 \cap O_2(Y) \leq N_G(K)$, contrary to the choice of z_2. This establishes the claim.

Set $Y_1 = \langle K^Y \rangle$ and $E_1 = \langle z^{Y_1} \rangle$. Assume $K^Y = K^{Y_1}$ and let T be maximal subject to $z \notin T \leq S$ and $T \trianglelefteq Y_1$. Set $Y_1^+ = Y_1/T$, $S_1 = S \cap Y_1$ and let $\mathcal{F}_1^+ = \mathcal{F}_{S_1^+}(Y_1^+)$, $\Omega_1 = K^{Y_1} \cap S$, and $\tau_1^+ = (\mathcal{F}_1^+, \Omega_1^+)$. Then by 3.3.2, $T \leq Z(Y_1)$ and τ_1^+ is a quaternion fusion packet. By construction and the claim, τ_1^+ satisfies Hypothesis 6.2.1, so as $Y \neq G$ and G is a minimal counter example to Theorem 6.3.9, it follows that either $E_1 \not\leq N_G(K)$ and $Y_1^+ \cong AE_n$ for some $n \geq 5$, or $E_1 \leq N_G(K)$ and from Theorem 6.2.8, $Y_1^+ \cong L_3(2)/E_8$, $L_3(2)/E_{64}$, or $L_3(2)/2^{3+6}$. Then as $T \leq Z(Y_1)$, it follows from 5.1.7 that $Y_1 \cong AE_n$ or $\check{A}E_n$ for some $n \geq 5$ when $E_1 \not\leq N_G(K)$.

In particular if $K_2 \in K^Y$ then $Y_1 = Y$ and $K^{Y_1} = K^Y$, and as $K_2 \in K^Y$, $E_1 \not\leq N_G(K)$, so that (2) holds. Thus we may assume $K_2 \notin K^Y$.

Suppose $z_2 \notin E_1$. Then by 3.3.2.1, K_2 centralizes E_1, so $Y_0 = \langle K_2^Y \rangle$ centralizes E_1. But then Y_0 acts on $O(z)$ and $\Omega_2 = K_2^Y \cap S$ centralizes $O(z)$, so $Y_0 = \langle \Omega_2^{Y_0} \rangle$ centralizes $O(z)$, contradicting $z_2 \notin E_z$.

Thus $z_2 \in E_1$. Then as z_2 does not act on K, $E_1 \not\leq N_G(K)$, so $Y_1 \cong AE_n$ or $\check{A}E_n$. Next $\Omega_2 = \Omega \cap Y - \Omega_1$ centralizes $\Sigma = K^{M \cap Y}$ by 6.3.3 and 4.2.10.2, so for $J \in \Omega_2$, $j = z(J) \in C_{E_1}(\Sigma)$. But if $n = 5$ then $\langle z \rangle = C_{E_1}(K)$, so as $j \neq z$, $n > 5$. Then by 5.1.9.4 the graph $\mathcal{D}^*$ in that lemma is connected, so $j \in C_{E_1}(Y_1)$, and we get a contradiction as above. This completes the proof in the case $K^Y = K^{Y_1}$.

Finally we treat the case where $K^Y \neq K^{Y_1}$. Consider K_2Y_1. There is $y \in Y_1$ with $J = K_2^y \leq S$, so $J \in \Omega$ acts on K. Now $K_2Y_1 = JY_1$ and $K^{K_2Y_1} = K^{JY_1} = K^{Y_1}$. If $Y = K_2Y_1$ then $K^Y = K^{Y_1}$, contrary to assumption. Therefore $K_2Y_1 < Y$. Set $Y_2 = \langle K^{Y_1} \rangle$. If $Y_2 = Y_1$ then by induction on $|Y|$, $K_2 \in K^{Y_1}$, a contradiction. Proceeding in this manner, we arrive at Y_r subnormal in Y with Y_r K_2-invariant and $Y_r = \langle K^{Y_r} \rangle$. But now by induction, $K_2 \in K^{Y_r}$, a contradiction. $\square$

LEMMA 6.3.11. $m(E) > 6$.

PROOF. Assume $m(E) \leq 6$. Suppose first that $|\Omega| = 2$. By 6.3.2.3, $\Omega = \{K, \tilde{K}\} = \Omega(z)$, and by 6.3.5, μ is not E_4, so $\mathcal{D}(D_\eta)$ is disconnected. Then by 4.2.13 and 6.3.3, $\mu \cong S_4$. But then 6.3.7 contradict the choice of G as a counter example.

Thus there exists $K_3 \in \Omega - \{K, \tilde{K}\}$; set $z_3 = z(K_3)$ and $E_0 = [E, K][E, K_3]$. Then $E_0 \leq E_z$ and K centralizes $[E, K_3]$ as $[E, K_3] \leq K_3\tilde{K}_3$. Then as $\langle z \rangle = C_{[E,K]}(K)$, $E_1 = [E, K] \cap [E, K_3] \leq \langle z \rangle$. Therefore $m(E_0) \geq 5$ with $\langle z \rangle = E_1$ in case of equality. As $m(E) \leq 6$ and $|E : E_z| = 2$ by 6.3.2.1, we conclude that $E_0 = E_z$ is of rank 5, $m(E) = 6$, and $E_1 = \langle z \rangle$. By symmetry, $E_1 = \langle z_3 \rangle$, so $z_3 = z$ and $\Omega = \Omega(z)$.

Set $K_1 = K$, $K_2 = \tilde{K}$, and $K_4 = \tilde{K}_3$. If $K_5 \in \Omega - \{K_1, \ldots, K_4\}$ then K_5 centralizes $E_0 = E_z$, a contradiction. Thus $\Omega = \{K_1, \ldots, K_4\}$. By 6.3.8, μ is nonabelian, so $\mu \cong \mathrm{Weyl}(D_4)$ by 4.2.13 and 4.3.5.1.

Let $s = z^h \in \mathcal{Z} - E_z$ and set $O = O(z)$. Then $O \cong Q_8^4$, $K^s = \tilde{K}$, and $K_3^s = K_4$, so $[O, s] \cong E_{32}$. Hence as $m(E) = 6$, $E = [O, s]\langle s \rangle$ and $|s^O| = |O : C_O(s)| = 16$ with $zs \notin s^O$. Therefore $E - E_z = s^O \cup (sz)^O$. By 6.2.6, G is irreducible on E, so each member of E_z is fused into $E - E_z$, and hence G has at most two orbits on $E^\#$ with representatives s and sz.

Let $e \in [E, K_3] - \langle z \rangle$, $\Omega_e = \{K, \tilde{K}\}$, $Y = \langle K^{C_G(e)} \rangle$, $S_e = S \cap Y$, $E_e = \langle z^Y \rangle$, and $\mathcal{F}_e = \mathcal{F}_{S_e}(Y)$. We may assume S_e is Sylow in Y. By 2.5.2, $\tau_e = (\mathcal{F}_e, \Omega_e)$ is a quaternion fusion packet. By 5.8.7.4, $\mu(\tau_e) \cong S_4$, and setting $M_e = \langle K^{C_M(e)} \rangle$, $e \in M_e$.

Choose $e \in E_s$ and set $J = K^h$. Then $|C_J(e)| > 2$ and by Sylow there is $y \in C_G(e)$ with $C_J(e)^y \leq S$. Then $C_J(e)^y$ is contained in a member of Δ, while as $\Omega = \Omega(z)$, $F = \langle z(V) : V \in \Delta \rangle \cong E_4$ and M_e is transitive on $F^\#$ from 5.8.7.3.ii, so $s^y \in z^{C_G(e)}$ and hence we may take $h \in Y$. Therefore $Y \cong AE_n$ or $\check{A}E_n$ for some $n \geq 5$ by 6.3.10.2. Then as $\mu(\tau_e) \cong S_4$, $n = 6$ or 7, and as $e \in M_e$, also $e \in Z(Y)$, so it follows that $Y \cong \check{A}E_6$.

If Y centralizes $f \in E - E_e$ then f and e are fused in G, since by an earlier observation, G has two orbits on $E^{\#}$ and neither e nor f is in $\mathcal{Z}$ since the centralizer of neither involution has a normal subgroup isomorphic to $O(z)$. This is a contradiction as $e \in Y$ but $f \notin Y$.

Therefore Y is indecomposable on E, so by 5.1.2.3, E is the 6-dimensional permutation module for $Y/E_e \cong A_6$. In particular Y has orbits Γ_i of length $6, 15, 20, 15, 6, 1$ on $E^{\#}$, where Γ_i is the set of vectors in E of weight i, $1 \leq i \leq 6$.

We've seen that G has two orbits on $E^{\#}$ with representatives e and z, and that $E - E_z = E_1 \cup E_2$, where $E_1 = \mathcal{Z} \cap (E - E_z)$ and $E_2 = e^G \cap (E - E_z)$ are of order 16. Next there is $X \in Syl_3(C_Y(z))$ with $K\tilde{K} = [K\tilde{K}, X]$. As X acts on K_3K_4 and centralizes e, $[X, K_3K_4] = 1$. Similarly there is X_1 of order 3 centralizing $K\tilde{K}$ with $[K_3K_4, X_1] = K_3K_4$. Then XX_1O has orbits of length $1, 6, 6, 18$ on $E_z^{\#}$ with the orbits $[E, K] - \langle z \rangle$ and $[E, K_3] - \langle z \rangle$ of length 6 in e^G. We conclude that

$$|\mathcal{Z}| = 1 + 18 + 16 = 35 \text{ and } |e^G| = 6 + 6 + 16 = 28.$$

Then it follows that $\mathcal{Z} = \Gamma_3 \cup \Gamma_4$ and $e^G = \Gamma_1 \cup \Gamma_2 \cup \Gamma_5 \cup \Gamma_6$. In particular, writing E additively, for $f \in e^G - \{e\}$, $f \in E_e$ iff $f + e \in \mathcal{Z}$.

Define a form $\varphi : E \times E \to \mathbf{F_2}$ by $\varphi(x, y) = 0$ if $y \in E_x$ and $\varphi(x, y) = 1$ otherwise. The form is symmetric on $\mathcal{Z} \times \mathcal{Z}$ by 6.3.2.5. Similarly $z \in E_e$ and $e \in E_z$, so the form is symmetric on $(\mathcal{Z} \times e^G) \cup (e^G \times \mathcal{Z})$. If $f \in e^G - \{e\}$, we saw that $f \in E_e$ iff $f + e \in \mathcal{Z}$, so φ is symmetric on $e^G \times e^G$. Therefore φ is symmetric. Then as $|E : E_x| = 2$ for each $x \in E^{\#}$, φ is bilinear. As G is irreducible on E, φ is nondegenerate, so φ is a symplectic form.

Define $q : E^{\#} \to \mathbf{F_2}$ by $q(x) = 0$ if $x \in \mathcal{Z}$ and $q(x) = 1$ otherwise. We check that:

(*) for $x, y \in E$, $q(x + y) = q(x) + q(y) + \varphi(x, y)$,

so that q is a quadratic form on E. For example $x + z \in x^O$ for $x \in E_z - \langle z \rangle$, so $q(x + z) = q(x)$. Also $z + s \in e^G$ so $q(z + s) = 1$. Finally $f \in E_e$ iff $f + e \in \mathcal{Z}$, so $q(f + e) = 0$ iff $f \in E_e$, completing the verification of (*).

We've shown that $G^* \leq O(E, q) \cong O_6^{\epsilon}(2)$. As $|e^G| = 28$, 7 divides the order of G^*, so $\epsilon = +1$ and $|G^*| \geq 28|Y^*| = |O_6^+(2)|/4$. Then as $G = \langle K^G \rangle$, $G^* \cong A_8 \cong L_4(2)$. By 6.3.4.8, $S_0 = E$.

Let H be the preimage in G of a parabolic H^* of G^* containing S^* and with $H^*/O_2(H^*) \cong L_3(2)$. Set $\mathcal{F}_H = \mathcal{F}_S(H)$ and $\tau_H = (\mathcal{F}_H, \Omega)$. By 2.5.2, τ_H is a quaternion fusion packet; further $\mu(\tau_H) \cong S_4$ and $E_H = \langle z^H \rangle \cong E_8$. Indeed τ_H satisfies Hypothesis 6.2.1 with $E_H \leq N_G(K)$, so $H \cong L_3(2)/E_{64}$ by 6.2.16. Therefore S is determined up to isomorphism as is the embedding of $E = [O, s]\langle s \rangle$ in S. Hence S splits over E, as this is the case in the overgroup AE_8 of H. Thus G splits over E, and hence is determined up to isomorphism. But then $G \cong AE_8$, contrary to the choice of G as a counter example to Theorem 6.3.9. □

LEMMA 6.3.12. *Let $J \in K^G$ such that $z(J) \notin E_z$ and set $Y = \langle K, J \rangle$. Then $J \in K^Y$ and $Y \cong AE_5$ or AE_7.*

PROOF. Set $t = z(J)$, $E_0 = [E, K][E, J]$, and $E_Y = \langle z^Y \rangle$. Then $m(E_0) \leq 2m([E, K]) = 6$ and $E_0 = [E, Y] \trianglelefteq Y$ with $z \in [E, K] \leq E_0$. Thus $E_Y \leq E_0$, so $m(E_Y) \leq 6$. Therefore $Y \neq G$ by 6.3.11, so by 6.3.10.2, $J \in K^Y$ and $Y \cong AE_n$ or $\check{A}E_n$ for some $n \geq 5$. Let $Y^+ = Y/E_Y$, so that $Y^+ \cong A_n$ and K^+ is a root 4-subgroup of Y^+. Then $Y^+ = \langle K^+, J^+ \rangle$ and as $t \notin E_z$, it follows from 5.1.8 that $Y^+ \cong A_5$ or A_7, so the lemma holds. □

NOTATION 6.3.13. Set $\theta(K) = \{J \in K^G : \langle K, J\rangle \cong AE_5\}$. Thus for $J \in \theta(K)$, $z(J) \notin E_z$ and $m([E,K][E,J]) = 4$ with $[E,K] \cap [E,J]$ a complement to $\langle z(X)\rangle$ in $[E,X]$ for $X \in \{K,J\}$.

LEMMA 6.3.14. *(1) For $t \in \mathcal{Z}$, $t \in \mathcal{D}(z)$ iff $[O(z), O(t)] = 1$.*
(2) For $t \in \mathcal{Z}$, $t \in \mathcal{D}^*(z)$ iff $t \in E_z - \langle z\rangle$.
(3) $\mathcal{D}^*$ is connected.
(4) $G = \langle C_G(z), N_G(O), M\rangle$.

PROOF. By 3.3.4, $t \in \mathcal{D}(z)$ iff $[t, O(z)] = 1$, and then as $O(t) \trianglelefteq C_G(t)$, (1) follows. Similarly by 3.3.5, $t \in \mathcal{D}^*(z)$ iff $|C_K(t)| > 2$ iff $t \in E_z - \langle z\rangle$, so (2) holds.

Assume (3) fails and let Σ and Σ' be distinct connected components of $\mathcal{D}^*$ with $z \in \Sigma$. Let $t = z^g \in \Sigma'$ and $J = K^g$. By (2), $t \notin E_z$, so by 6.3.12, $Y = \langle K, J\rangle \cong AE_5$ or AE_7, and as $\Sigma \neq \Sigma'$, the former holds by 5.1.9.4. Thus $J \in \theta(K)$ for each $t \in \Sigma'$ and $J \in K^G$ with $t = z(J)$.

Suppose $|\Omega| > 2$. Then there is $J' \in \Omega^g - \{J, \tilde{J}\}$; set $Y' = \langle Y, J'\rangle$. As $J' \in \theta(K)$, $(E \cap Y)[E, J']$ is of rank 5 and normal in Y', so $Y' \neq G$ by 6.3.11. Then $Y' \cong \check{A}E_6$ by 6.3.10, so 5.1.9.4 contradicts $\mathcal{D}^*$ disconnected.

Thus $|\Omega| = 2$. Then by 6.3.3 and 4.2.13, $\mu \cong E_4$ or S_4, contrary to 6.3.5 and 6.3.7. This completes the proof of (3).

As G is transitive on $\mathcal{Z}$ and $\mathcal{D}^*$ is connected, $G = \langle C_G(z), \Gamma\rangle$, where $\{z^g : g \in \Gamma\}$ is a set of representatives for the orbits of $C_G(z)$ on $\mathcal{D}^*(z)$. Next for $t \in \mathcal{D}^*(z)$, t is conjugate under $C_G(z)$ to a member of $\mathcal{Z}_\Delta$. Now we appeal to 6.1.6.2. If $\Delta = \Omega$ then (4) follows as $N_G(O)$ is transitive on $\mathcal{Z}_S$. On the other hand if $\Delta \neq \Omega$ then $M(N_G(W) \cap N_G(O))$ is transitive on $\mathcal{Z}_\Delta$ by 6.1.6.2, so again (4) follows. □

LEMMA 6.3.15. *Suppose $\mathcal{D}$ is disconnected. Then*
(1) $\mu \cong$ Weyl(D_4) and
(2) $\Omega = \Omega(z)$.

PROOF. By 4.2.13 and 6.3.3, μ is S_4 or Weyl(D_4), so (1) follows from 6.3.7.

Suppose (2) fails and let $K_3 \in \Omega - \Omega(z)$ and set $z_3 = z(K_3)$. By 5.8.7.2, $e = zz_3 \in Z(M)$. Thus if $O = O(\tau) \trianglelefteq C_G(z)$ then e is centralized by $\langle C_G(z), N_G(O), M\rangle$, contrary to 6.3.14.4. Therefore O is not normal in $C_G(z)$, so $X = \langle O(z_3)^{C_G(z)}\rangle \neq O(z_3)$. But $X \leq C_G(O(z))$, so $\tau_X = (\mathcal{F}_X, \Omega(z_3))$ is a quaternion fusion packet by 2.5.2, with $\mu(\tau_X) \cong E_4$, where $\mathcal{F}_X = \mathcal{F}_{S\cap X}(X)$. As $\mu(\tau_X) \cong E_4$ it follows from 3.3.12.3 that $K_3 \cap O_2(X) = \langle z_3\rangle$. Let $T \leq S$ be maximal subject to $T \trianglelefteq X$ and $z_3 \notin T$, and set $X^+ = X/T$ and $\tau_X^+ = (\mathcal{F}_X/T, \Omega^+)$. By 3.3.2, τ_X^+ satisfies Hypothesis 6.2.1. Then by 6.2.8, τ_X^+ satisfies Hypothesis 6.3.1, so it follow from 6.3.5 that $X \cong AE_5$.

Let Σ be the connected component of $\mathcal{D}$ containing z, $H = N_G(\Sigma)$, and $\mathcal{F}_\Sigma = \mathcal{F}_S(H)$. By 2.5.2, $\tau_\Sigma = (\mathcal{F}_\Sigma, \Omega)$ is a quaternion fusion packet. Observe that $\mu(\tau_\Sigma) \cong E_{16}$ by 6.3.14.4. Let $\mathcal{E} = [K_3]_{\mathcal{F}_\Sigma}$, $\Gamma = K_3^{\mathcal{E}}$, $\rho = (\mathcal{E}, \Gamma)$, T_0 maximal subject to $T_0 \trianglelefteq \mathcal{E}$ with $z_3 \notin T_0$, and $\mathcal{E}^+ = \mathcal{E}/T_0$. By 3.3.2, $\rho^+ = (\mathcal{E}^+, \Gamma^+)$ is a quaternion fusion packet satisfying Hypothesis 6.3.1, so by minimality of G, $X \trianglelefteq \langle K^H\rangle$. It follows that $\langle K^H\rangle = X \times X^g$ for $g \in H$ with $z_3^g = z$.

Let $f \in [E, K_3] - \langle z_3\rangle$ and $Y = \langle K^{C_G(f)}\rangle$. Set $S_Y = S \cap Y$, $\mathcal{F}_Y = \mathcal{F}_{S_Y}(Y)$, and $\tau_Y = (\mathcal{F}_Y, \Omega(z))$. By 6.3.10, τ_Y is a quaternion fusion packet, and by 5.8.7.4, $\mu(\tau_Y) \cong S_4$, so by 6.3.10, $Y \cong AE_n$ for $n = 6$ or 7, or $\check{A}E_6$. Let $b \in C_X(f)$ be of order 3. Then b acts on Y and centralizes X^g, so we conclude from 5.1.10 that

b centralizes Y. Therefore $F = \langle f \rangle [E \cap X, b]$ centralizes Y, a contradiction as F contains a member of $\mathcal{Z}$. This completes the proof of (2). □

LEMMA 6.3.16. *$\mathcal{D}$ is connected.*

PROOF. Assume otherwise. By 6.3.15, $\mu \cong \mathrm{Weyl}(D_4)$ and $O = O(z)$. Let $J \in \theta(K)$ and $Y = \langle O, J \rangle$. As in the proof of 6.3.12, $[E, Y] = [E, O][E, J]$ is of rank 6, so by 6.3.10 and 6.3.11, $Y \cong AE_8$. From 5.1.11.2, $\mathcal{Z}_\Delta \subseteq O$. Further if $t = z^g \in \mathcal{D}^*(z)$ then $|C_{K^g}(z)| > 2$, so for $h \in C_G(z)$ with $C_{K^g}(z)^h \leq S$, $t^h \in \mathcal{Z}_\Delta$. Hence as $O \trianglelefteq C_G(z)$, $\mathcal{D}^*(z) \subseteq O$. Indeed from the proof of 6.3.11, $C_Y(z)$ is transitive on $\mathcal{Z} \cap O - \{z\}$, so $C_Y(z)$ is transitive on $\mathcal{D}^*(z)$. But then as $\mathcal{D}^*$ is connected by 6.3.14, $z^Y = \mathcal{Z}$. Thus $E = E \cap Y$ is of rank 6, contrary to 6.3.11. □

LEMMA 6.3.17. *Either*
(1) $\Omega(z) = \{K, \tilde{K}\}$, or
(2) $|\Omega(z)| = 4$ and for $J \in \theta(K)$, $\langle O(z), J \rangle \cong AE_8$.

PROOF. Assume $K_3 \in \Omega(z) - \{K, \tilde{K}\}$, and let $J \in \theta(K)$ and $Y = \langle K, K_3, J \rangle$ and $E_Y = [E, K][E, K_3][E, J]$. Then $m(E_Y) = 6$, so by 6.3.10 and 6.3.11, $Y \cong AE_8$. By 5.1.11.2, there is $t \in \mathcal{Z}_\Delta \cap K\tilde{K}K_3\tilde{K}_3$, so by 3.1.6, $\Omega(z) = \{K, \tilde{K}, K_3, \tilde{K}_3\}$, completing the proof. □

LEMMA 6.3.18. *Let $z^g \in \mathcal{Z} - z^\perp$ and set $Y = \langle O(z), O(z^g) \rangle$. Then*
(1) $O(z)$ and $O(z)^g$ are conjugate in Y.
(2) $K \not\leq O_2(Y)$.
(3) $C_Y(z)$ is transitive on $\Omega(z)$.

PROOF. Suppose first that $t = z^g \in \mathcal{D}^*(z)$. Then t is fused under $C_G(z)$ into $\mathcal{Z}_\Delta$, so without loss, $t \in \mathcal{Z}_\Delta$. Then by 6.3.3 and 4.2.5, (2) and (3) hold and $Y \leq M$ is transitive on $\langle z, t \rangle^\#$, so that (1) holds.

Thus we may assume that $t \notin E_z$, so by 6.3.12, $Y_0 = \langle K, K^g \rangle \cong AE_5$ or AE_7. In particular $K \not\leq O_2(Y_0)$ and $t \in z^{Y_0}$. Thus (1) and (2) hold in this case too. Further $K^t = \tilde{K}$, so if $|\Omega(z)| = 2$ then (3) holds. On the other hand if $|\Omega(z)| > 2$ then (3) follows from 6.3.17. □

NOTATION 6.3.19. Set $D = O(z)^G$, and for $P \in D$ let $z(P)$ be the involution generating $Z(P)$ and $D_P = C_D(P)$. For α a set of imprimitivity for G on D define D_α to be the set of conjugates of α centralizing α and $D^*_\alpha = \{z^g : O(z^g) \in D_\alpha\}$. Define $\mathcal{Y}(K)$ to consist of those elements y of order 3 with $[K\tilde{K}, y] = K\tilde{K}$ and $m([E, y]) = 2$. Observe for each $J \in \theta(K)$ that $\langle K, J \rangle \cap \mathcal{Y}(K) \neq \emptyset$. Define $\mathcal{D}(D)$ to be the commuting graph on D.

LEMMA 6.3.20. *(1) The map $t \mapsto O(t)$ is an isomorphism of $\mathcal{D}(\tau)$ with $\mathcal{D}(D)$.*
(2) For $P, Q \in D$, either $[P, Q] = 1$ or $Q \in P^{\langle P, Q \rangle}$.
(3) $\mathcal{D}(D)$ is connected.
(4) G is transitive on D.
(5) There exists a unique maximal set of imprimitivity α for G on D containing $O(z)$.
(6) $D_\alpha \neq \emptyset$ and $\langle D_\alpha \rangle$ is transitive on D^*_α by conjugation.
(7) $G = \langle D_\alpha, D_\beta \rangle$ for each $\beta \in D_\alpha$.

PROOF. Part (1) follows from 6.3.14.1, (2) follows from (1) and 6.3.18.1, and (3) follows from (1) and 6.3.16. By 6.2.1.1, $\mathcal{F}$ is transitive on Ω, so G is transitive on $\mathcal{Z}$, and hence (4) holds. Finally (5)-(7) follow from (2)-(4) and 7.8 in [**Asc97**]. $\square$

NOTATION 6.3.21. Choose α as in 6.3.20.5 and set $X = \langle D_\alpha \rangle$. Replacing S by a suitable conjugate, we may assume $S_X = S \cap X \in Syl_2(X)$. Set $\Omega_X = \Omega \cap X$, $\mathcal{F}_X = \mathcal{F}_{S_X}(X)$, and $\tau_X = (\mathcal{F}_X, \Omega_X)$. Set $E_X = \langle D_\alpha^* \rangle$. Let $z^g \in D_\alpha^*$.

LEMMA 6.3.22. *(1) τ_X is a quaternion fusion packet.*
(2) $X \cong AE_n$ or $\check{A}E_n$ for some $n \geq 5$.

PROOF. Part (1) follows from 2.5.2. If $D_\alpha^* = \{z^g\}$ then $\alpha = \{O(z)\}$ and $\{O(z), O(z^g)\}$ is a set of imprimitivity for G on D, contrary to 6.3.20.5. Hence $|D_\alpha^*| > 1$. Then as X is transitive on D_α^* by 6.3.20.6, it follows that there exists $x \in X$ with $z^x \notin z^{g\perp}$. In particular $K^g \not\leq O_2(X)$ and $C_X(z^g)$ is transitive on $\Omega(z^g)$ by 6.3.18. Therefore $\mathcal{F}_X$ is transitive on Ω_X.

Suppose $K^g \cap O_2(X) \neq \langle z^g \rangle$. Then by 3.3.12, $W_X = \langle \eta_X \rangle \trianglelefteq X$ for some $\eta_X \in \eta(\tau_X)$. Now $\eta_X \subseteq \eta \in \eta(\tau)$, so for $J \in K^{gX}$, $J \leq N_G(W)$. Thus $X = \langle K^{gX} \rangle \leq N_G(W)$. As $W \leq C_G(z^g) \leq N_G(X^g)$, by symmetry, $X^g \leq N_G(W)$, so by 6.3.20.7, $G = \langle X, X^g \rangle \leq N_G(W)$, contrary to 6.2.1.3. Therefore $K^g \cap O_2(X) = \langle z \rangle$.

Let $T \leq S_X$ be maximal subject to $z^g \notin T$ and $T \trianglelefteq X$. Set $X^+ = X/T$, $\mathcal{F}_X = \mathcal{F}_{S_X^+}(X^+)$, and $\tau_X^+ = (\mathcal{F}_X^+, \Omega_X^+)$. By 3.3.2, τ_X^+ is a quaternion fusion packet and $T \leq Z(X)$. We saw that τ_X^+ satisfies conditions (1) and (3) of Hypothesis yy.1. By the choice of T, 6.2.1.4 is satisfied, and as X is transitive on D_α^*, 6.2.1.5 holds. As $X = \langle K^{gX} \rangle$, 6.2.1.2 holds. Thus τ_X^+ satisfies Hypothesis 6.2.1.

Suppose $E_X \leq N_G(K^g)$. Then by Theorem 6.2.8, $X^+ \cong L_3(2)/E_8$, $L_3(2)/E_{64}$, or $L_3(2)/2^{3+6}$. Hence there is y of order 3 in $C_{X^g}(z)$ with $K\tilde{K} = [K\tilde{K}, y]$. Now $y \in C_G(O(z^g))$ acts on X, so by 5.10.8, y centralizes X. Then $P = [O_2(C_{X^g}(z)), y] \leq C_G(X)$. But by 5.10.9.2, $X^g = \langle K^h : z^h \in P, h \in X^g \rangle$, and so as $X \leq C_G(z^h) \leq N_G(O(z^h))$, X centralizes K^h, and hence X centralizes X^g. Thus $X \trianglelefteq \langle X, X^g \rangle = G$ by 6.3.20.7, a contradiction.

Therefore $E_X \not\leq N_G(K^g)$, so (2) follows from 6.3.10. $\square$

LEMMA 6.3.23. *(1) $\alpha = \{O(z)\}$.*
(2) $\Omega(z) = \{K, \tilde{K}\}$.
(3) For $y \in \mathcal{Y}(K) \cap X^g$, $K\tilde{K}\langle y \rangle \trianglelefteq C_{X^g}(z)$ and y centralizes X.

PROOF. From a remark in 6.3.19, there exists $y \in \mathcal{Y}(K) \cap X^g$, and from 5.1.12.1, either $K\tilde{K}\langle y \rangle \trianglelefteq C_{X^g}(z)$ or $X \cong AE_8$. In any event $y \in C_G(z) \leq N_G(X)$. As $[E, y] = [E, K, y]$, y centralizes E_X, so $[X, y] \leq C_X(E_X) = E_X$, and then (3) holds unless $X \cong AE_8$.

From 6.3.22 and 5.1.12.2, $C_{X^g}(z)$ is a maximal subgroup of X^g unless $X \cong \check{A}E_8$, where $N_G(\Omega_{X^g})$ is maximal. In the first case $O(z)$ is a maximal set of imprimitivity for X^g on $D \cap X^g$, so that (1) holds in this case as $\alpha \subseteq D_{\alpha^g}$. In the second case (1) holds unless $\alpha = \{K\tilde{K}, K_3\tilde{K}_3\}$ for $K_3 \in \Omega_{X^g} - \{K, \tilde{K}\}$. Similarly, working in X^g, either $|\Omega(z)| = 2$ or $X \cong AE_8$ and $|\Omega(z)| = 4$, and in the first case (2) holds. Thus the lemma holds unless $n = 8$, which we now assume.

Now $L = O^{2'}(C_{X^g}(y)) \cong AE_5$ contains $K_3\tilde{K}_3$ for $K_3 \in \Omega_{X^g} - \{K, \tilde{K}\}$. Set $B = \langle L, X \rangle$. Then $B \leq C_G(y)$ and L does not centralize X by 7.8.5 in [**Asc97**], as $\langle \alpha \rangle \leq \langle \Omega_{X^g} \rangle$ and α is maximal. Thus by 6.3.10, $B \cong AE_k$ or $\check{A}E_k$ for some $k > 8$.

As $k > 8$, $C_B(z^g)$ is maximal in B with $\{K^g, \tilde{K}^g\} = \Omega(z^g) \cap B$, so it follows that (1) and (2) hold. Also $X \cong \check{A}E_8$, so (3) also holds. □

LEMMA 6.3.24. *For* $y \in \mathcal{Y}(K)$, $O(z)\langle y\rangle = \langle \mathcal{Y}(K)\rangle \trianglelefteq C_G(z)$.

PROOF. Choose y as in 6.3.23.3 and set $B = O(z)\langle y\rangle$. By 6.3.23.2, X is not AE_8. Then by 5.1.12.1, B char $C_{X^g}(z) \trianglelefteq C_G(\langle z, z^g\rangle)$. Further $X \trianglelefteq C_G(z)$ and X is transitive on $\mathcal{D}(z)$, so by a Frattini argument, $C_G(z) = XC_G(\langle z, z^g\rangle)$. Therefore $B \trianglelefteq C_G(z)$.

Now let $x \in \mathcal{Y}(K)$. Conjugating in $O(z)$, we may take $x \in yC_G(B)$; in particular x centralizes y so $F = \langle x, y\rangle$ is a 3-group. As $x, y \in \mathcal{Y}(K)$, $[E, x] = [E, y] \leq [E, K]$, so xy^{-1} centralizes $[E, y]$ and $E/[E, y]$, so as F is a 3-group, $xy^{-1} \in C_G(E) = S_0$ by 6.3.4.2. Hence $x = y \in B$, completing the proof. □

LEMMA 6.3.25. *Let* $e \in [E, K^g] - \langle z^g\rangle$ *and* $Y_e = \langle K^{C_G(e)}\rangle$. *Then*

(1) $Y_e \cong AE_{n+2}$ *or* $\check{A}E_{n+2}$.

(2) Let $x \in C_{K^g}(e)$ *be of order 4. Then* x *acts on* Y_e *and induces the transposition on* Y_e^* *centralizing* X^{g*}.

(3) Let $y \in \mathcal{Y}(K^g)$ *with* $e \in [E, y]$, *and set* $Y_y = \langle K^{C_G(y)}\rangle$. *Then* $X^g \leq Y_y \leq Y_e$ *with* $Y_y \cong AE_{n+1}$ *or* $\check{A}E_{n+1}$.

PROOF. Set $B = C_X(e)$ and $Y = \langle X^{gB}\rangle$. Observe that X^g centralizes $O(z^g)$, and hence also e, so $Y \leq C_G(e)$. By 6.3.22.2, $X^g = \langle K^{X^g}\rangle$. As X centralizes $O(z)$, for each $b \in B$, $O(z) \leq X^g \cap X^{gb}$, so $Y = \langle K^Y\rangle$. Hence by 6.3.10, $Y \cong AE_p$ or $\check{A}E_p$ for some $p \geq 5$.

Suppose $n > 5$ and let Γ consist of those $O(t) \in D_\alpha$ centralizing e and set $L = \langle\Gamma\rangle$. Then L is isomorphic to Q_8^2 when $n = 6$ and to $\check{A}E_{n-2}$ when $n > 6$. In particular L is transitive on Γ. By construction, L acts on Y, so from 6.3.20.2, either $\Gamma \subseteq O(z)^Y$ or L centralizes Y. But in the latter case as $X^g \leq Y$ it follows from maximality of α that $Y = X^g$ and $\Gamma = \{O(z^g)\}$, a contradiction as e does not centralize $O(z^g)$. Therefore $\Gamma \subseteq O(z)^Y$. We conclude that Γ consists of those $O(t) \in O(z)^Y$ centralizing $O(z)$, and then that $p = n + 2$.

So take $n = 5$. Then $\Omega \cap Y = \Omega(z)$, so $5 \leq p \leq 7$. For $b \in B$ of order 3, $z^{gb} \neq z^g$ so $X^{gb} \neq X^g$ and hence b acts nontrivially on Y and centralizes $O(z)$, so $p = 7$. Thus in any event $Y \cong AE_{n+2}$ or $\check{A}E_{n+2}$.

Next if $J \in K^{C_G(e)} - K^Y$ then $z(J) \notin z^\perp$, so J is conjugate to K in Y_e by 6.3.18. Thus $Y_e = \langle K^{Y_e}\rangle$. Then by 6.3.10, $Y_e \cong AE_k$ or $\check{A}E_k$ for some $k \geq 5$. As $Y \leq Y_e$ with $\Gamma \subseteq Y$ we conclude that $Y = Y_e$, completing the proof of (1).

Next $x \in K^g$ so x centralizes X^g, and as x centralizes e, x acts on Y_e. Thus either (2) holds or x centralizes Y_e^*. But in the latter case x acts on $J \in K^{Y_e} - K^{X^g}$, so $z(J) \in \mathcal{D}^*(z^g) - (z^g)^\perp$. Thus $[x, J^*] \neq 1$, contradicting $[x, Y_e^*] = 1$. Hence (2) holds.

Choose y as in (3). By 6.3.23.3, y centralizes X^g. By 6.3.24, $y \in X$ so $y \in Y_f$ for $f \in [E, K] - \langle z\rangle$. Then by (1), $Y_0 = \langle K^{gY_f} \cap C_G(y)\rangle$ is isomorphic to AE_{n-1} or $\check{A}E_{n-1}$ if $n > 5$ and isomorphic to Q_8^2 if $n = 5$. Further Y_0 centralizes $[E, y]$, $Y_0 \neq \langle K^{gY_f} \cap C_G(e)\rangle$, and Y_0 contains $O(z^h)$ with $z^h \notin E_{z^g}$. Then as Y_0 centralizes $[E, y]$ and $e \in [E, y]$, we have $z^h \in z^{Y_e}$ by 6.3.18.1, so $Y_1 = \langle K^{Y_e} \cap C_G(y)\rangle \neq X^g$ or Y_e, and hence $Y_1 \cong AE_{n+1}$ or $\check{A}E_{n+1}$. As Y_y acts on $[E, y]$ and is perfect, Y_y centralizes $[E, y]$, so $Y_y = Y_1$, completing the proof of (3). □

LEMMA 6.3.26. *The group presented by generators y_i, $1 \leq i \leq n+2$, and the following set of relations, is the alternation group A_{n+4}:*
(1) $y_1^3 = y_i^2 = 1$ for $1 < i \leq n+2$.
(2) $(y_iy_{i+1})^3 = 1$ for $1 \leq i < n+2$.
(3) $(y_iy_j)^2 = 1$ for $i \geq 1$ and $j > i+1$.

PROOF. See for example Theorem 265 in [**Dic58**]. □

Define e and Y_e as in 6.3.25, and let $\langle x\rangle = C_{K^g}(e)$. By 6.3.25 we can represent $B = Y_e\langle x\rangle S_0$ on $I = \{1,\ldots,n+4\}$ so that S_0 is the kernel of the representation, $Y_e^* = Alt(I)_{1,2}$ is the stabilizer in the alternation group on I of $1,2$, $X^{g*} = Alt(I)_{1,2,3,4}$, and $x^* = (1,2)(3,4)$.

Next choose y and Y_y as in 6.3.25.3. Then $X^g \leq Y_y \leq Y_e$ with $Y_y^* \cong A_{n+1}$ generated by root 4-subgroups of Y_e^*, so we may choose $Y_y^* = Alt(I)_{1,2,3}$. As $K \leq X^g$, we can choose notation so that K^* is the root 4-subgroup moving $\{n+1,\ldots,n+4\}$. Choose $f \in [K,E] - \langle z\rangle$; thus $Y_f^* \cong A_{n+2}$ by 6.3.25.1. Let $x_0 \in x^{Y_e}$ with $x_0^* = (1,2)(n+3,n+4)$.. Then x_0^* acts on K^*, so x_0 acts on K, and we choose f so that x_0 centralizes f. As $x_0 \notin X^g$, x_0^* is nontrivial on K^{g*}, so, using 6.3.24, x_0^* inverts two subgroups of $K^{g*}\langle y^*\rangle$ of order 3, and we choose y so that x_0^* inverts y^*. For $3 \leq i \leq n+2$, let $x_i \in x_0^{Y_y}$ with $x_i^* = (1,2)(i+1,i+2)$, and set $y_i = x_i^*$ and $y_1 = y^*$. As x_0^* inverts y^*, y_i inverts y_1. Set $y_2 = x^*$. Then by direct computation:

LEMMA 6.3.27. *y_i, $1 < i \leq n+2$, satisfy the relations in 6.3.26.*

Next we verify that the remaining relations in 6.3.26, involving y_1, are also satisfied. As $|y| = 3$ and $y_1 = y^*$, $y_1^3 = 1$. As $y_2 = x^* \in K^{g*}$ and $K^{g*}\langle y^*\rangle \cong A_4$, $(y_1y_2)^2 = 1$. Recall that for $i \geq 3$, y_i inverts y_1, so $(y_1y_i)^2 = 1$. This completes the verification. Thus it follows from 6.3.26 that $G_0^* = \langle y_i : 1 \leq i \leq n+2\rangle \cong A_{n+4}$.

Next each y_i is in G^*, so $G_0^* \leq G$. Further $X^{g*} \leq Y_e^* \leq G_0^*$ and $X^* = \langle K^{g*}, (Y_e \cap X)^*\rangle \leq \langle x^*, y^*, (Y_e \cap X)^*\rangle \leq G_0^*$, so by 6.3.20.7, $G_0^* = G^*$. Thus we have shown:

LEMMA 6.3.28. *$G^* \cong A_{n+4}$ with K^* a root 4-subgroup of G^*.*

Then by 6.3.28 and 6.3.4.8:

LEMMA 6.3.29. *$E = S_0$ and $G = O^2(G)$.*

REMARK 6.3.30. By 6.3.28 and 6.3.29 we can represent G on $I = \{1,\ldots,n+4\}$ with kernel E as the alternating group on I; that is $G^* = Alt(I)$. Moreover we can choose notation so that K^*, K^{g*} are the root 4-subgroups moving $\{n+1,\ldots,n+4\}$ and $\{1,\ldots,4\}$, respectively, and X^*, X^{g*} are $G^*_{n+1,\ldots,n+4}$, $G^*_{1,\ldots,4}$, respectively. Choose e, x, and Y_e as in 6.3.25. From that lemma, we may choose $Y_e^* = G^*_{1,2}$ to be the stabilizer of 1 and 2, and $x^* = (1,2)(3,4)$. Then $Y_e^*\langle x^*\rangle$ is the global stabilizer of $\{1,2\}$ and centralizes $e \in E$. Choose $y \in \mathcal{Y}(K^g)$ as in 6.3.25.3; from that lemma we may choose $y^* = (1,3,2)$ and $\langle e^{\langle y\rangle}\rangle \cong E_4$.

LEMMA 6.3.31. *E is the natural module for G^*.*

PROOF. By 6.2.6.1, G is irreducible on E. From 6.3.30, the global stabilizer in $G^* = Alt(I)$ of $\{1,2\}$ centralizes $e \in E^\#$ and $\langle e^{\langle y\rangle}\rangle \cong E_4$. Hence the lemma follows from 5.1.14. □

LEMMA 6.3.32. *Let* $H_{n+4} = \langle K^{gG_{n+4}}\rangle$. *Then* $H_{n+4} \cong \check{A}E_{n+3}$.

PROOF. By 6.3.10, $H_{n+4} \cong AE_{n+3}$ or $\check{A}E_{n+3}$, and the latter holds by 6.3.31. □

LEMMA 6.3.33. *n is even.*

PROOF. Assume n is odd and set $H = H_{n+4}$. By 6.3.32, HE splits over E. But as n is odd, H^* contains a Sylow 2-subgroup S^* of G^*, so S splits over E. Therefore by Gaschutz' Theorem (cf. 10.4 in [**Asc86**]) G splits over E. Hence $G \cong AE_{n+4}$, contrary to the choice of G as a counter example. □

By 6.3.31, E is the natural module for G^*, so $E = \bar{V} = V/\langle v_I\rangle$, where V is the core of the permutation module for G^* on I, and for $J \subseteq I$ of even order, v_J is the vector in V with support J. Then $e = \bar{v}_{1,2}$ and $f = \bar{v}_{n+3,n+4} \in [E,K] - \langle z\rangle$, so Y_f is described in 6.3.25.1.

Next from 6.3.32, there is a complement H to E in H_{n+4}. Then $C_H(f)$ is a complement to E in EY_f. By 6.3.25.1, and as E is the natural module for G^*, $Y_f \cong \check{A}E_{n+2}$ and E is the permutation module for Y_f^* on $n+2$ points. Therefore by 5.1.2.2, $H^1(C_H(f), E) = 0$, so E is transitive on the complements to E in Y_fE. Let $t^* = (1,2)(n+3,n+4)$. Then $C_H(f)^* = Y_f^*\langle t^*\rangle \cong S_{n+2}$, and as E is transitive on complements to E in Y_f, we may choose t to be an involution acting on $C_H(f)$, so that $H_f = C_H(f)\langle t\rangle$ is a complement to E in $C_G(f)$.

Define y as in 6.3.30 and set $G_y = N_G(\langle y\rangle)$. Then, using 6.3.30 and 6.3.25.3, $G_y = (\langle y\rangle \times Y_y)\langle t\rangle$, with t inverting y and t^* inducing a transposition on $Y_y^* \cong A_{n+1}$, and $Y_y \cong AE_{n+1}$. Let $H_1 = N_H(\langle y\rangle)$, so that H_1 is a complement to E in the stabilizer $G_{y,n+4}$ of $n+4$ in G_y. Next $O_2(Y_y)$ is the permutation module for $C_H(y) \cong A_n$ on n points, so by 5.1.2.2, $O_2(Y_y)$ is transitive on its complements in $C_H(y)O_2(Y_y)$. Thus H_1 is contained in a complement H_2 to $O_2(Y_y)$ in G_y. By 5.1.2.2, $O_2(Y_y)$ is regular on such complements, so H_1 is contained in exactly $|C_{O_2(Y_y)}(H_1)| = 2$ such complements.

Similarly $F = C_{O_2(Y_y)}(C_H(f)) \cong E_4$ is regular on the four complements to $O_2(Y_y)$ in G_y containing $C_H(f)$, and $F\langle t\rangle \cong D_8$, so each such complement contains an F-conjugate of $H_f = C_H(f)\langle t\rangle$. Thus replacing H_f by a suitable F-conjugate, we may assume $H_f \leq H_2$.

Set $y_1 = y$ and for $1 < i < n+2$, let y_i be the element of H with $y_i^* = (1,2)(i+1,i+2)$. Observe that by construction, $\{y_1, \ldots, y_{n+1}\}$ satisfies the relations in 6.3.26. Next set $y_{n+2} = t$. As $y_i \in C_H(f)$ for $1 \leq i \leq n$ and $C_H(f) \leq H_f$, also $y_i \in H_f$. But also $t \in H_f$, so the relations in 6.3.26 involving t and y_i, $1 \leq i \leq n$ are satisfied. Finally t and y_{n+1} are contained in H_2, so $(y_{n+1}t)^3 = 1$, verifying the last of the relations in 6.3.26. Thus we conclude from 6.3.26 that:

LEMMA 6.3.34. $B = \langle y_i : 1 \leq i \leq n+2\rangle \cong A_{n+4}$, *so* B *is a complement to* E *in* G.

It follows from 6.3.31 and 6.3.34 that $G \cong AE_{n+4}$, contrary to the choice of G. This final contradiction completes the proof of Theorem 6.3.9.

6.4. Subnormal closure

In this section we assume $\tau = (\mathcal{F}, \Omega)$ is a quaternion fusion packet and adopt Notation 2.5.1 and 3.1.1. Moreover pick $K \in \Omega$ and set $z = z(K)$.

DEFINITION 6.4.1. Recall Definition 1.2.9 where the *subnormal series* $\mathrm{sub}_i(\mathcal{F}, K)$, $0 \leq i \leq n(\mathcal{F}, K)$, and the *subnormal closure* $sub(\mathcal{F}, K)$ of K in $\mathcal{F}$ are defined. Set $s(K) = \mathrm{sub}(\mathcal{F}, K)$, $\tau_i(K) = (\mathrm{sub}_i(\mathcal{F}, K), K^{\mathrm{sub}_i(\mathcal{F},K)})$, and $\tau(K) = (s(K), K^{s(K)})$. Write $S(K)$ for the Sylow group of $s(K)$.

LEMMA 6.4.2. *(1) $\tau(K)$ and $\tau_i(K)$ are quaternion fusion packets and $sub_i(\mathcal{F}, K)$ and $s(K)$ are subnormal subsystems of $\mathcal{F}$.*
(2) $O_2(s(K)) = S(K) \cap O_2(\mathcal{F})$.
(3) $z \in O_2(\mathcal{F})$ iff $z \in O_2(s(K))$.
(4) $\langle z \rangle < K \cap O_2(\mathcal{F})$ iff $\langle z \rangle < K \cap O_2(s(K))$.
(5) $K \cap O_2(\mathcal{F})$ is nonabelian iff $K \cap O_2(s(K))$ is nonabelian.
(6) $\mathcal{O}_K \leq s(K)$.

PROOF. By 2.7.3.1 in [**Asc19**], $\mathrm{sub}_i(\mathcal{F}, K)$ is subnormal in $\mathcal{F}$; then 2.5.2 completes the proof of (1). Part (2) is 2.7.3.6 in [**Asc19**]. Then (2) implies (3)-(5). Finally set $s_i = \mathrm{sub}_i(\mathcal{F}, K)$. If (6) fails then $\mathcal{O}_K$ is not contained in s_i for some i, and we choose i minimal subject to that constraint. Then $i > 0$ and $\mathcal{O}_K \leq s_{i-1}$, so replacing τ by $\tau_{i-1}(K)$, we may assume $i = 1$, so $s_i = s_1 \trianglelefteq \mathcal{F}$. As $\mathcal{F}_K(K) \leq s_1$, $\mathcal{O}_K \neq \mathcal{F}_K(K)$, so $\mathcal{C} = O^2(\mathcal{O}_K)$ is a component or solvable component of $\mathcal{F}_z$ by 2.6.13. Next $C_{s_1}(z) \trianglelefteq \mathcal{F}_z$ by 8.23.2 in [**Asc11**], so as $K \leq s_1$, we conclude from 7.2.2 in [**Asc11**] that $\mathcal{O}_K \wedge C_{s_1}(z)$ is a normal subsystem of $\mathcal{O}_K$ on K, and hence that $\mathcal{O}_K \leq C_{s_1}(z)$. But then $\mathcal{O}_K \leq s_1$, contrary to the choice of i. □

LEMMA 6.4.3. *Assume $z \in O_2(\mathcal{F})$ and $\mathcal{O}_K \neq \mathcal{F}_K(K)$. Then*
(1) $[K]_{\mathcal{F}} = \mathcal{O}(\mathcal{F}, K^{\mathcal{F}})$ is the central product of the members of $\{\mathcal{O}_J : J \in K^{\mathcal{F}}\}$.
(2) $\tau(K) = (\mathcal{O}_K, \{K\})$.
(3) $[K]_{\mathcal{F}}$ centralizes $\mathcal{O}_J$ for $J \in \Omega - K^{\mathcal{F}}$.

PROOF. By 3.3.13, $\mathcal{C} = O^2(\mathcal{O}_K)$ is a component or solvable component of $\mathcal{F}$. Let $\Phi = \{O^2(\mathcal{O}_J) : J \in K^{\mathcal{F}}\}$ and $T \in Syl(E_+(\mathcal{F}))$. Then $E_+(\mathcal{F}) \trianglelefteq \mathcal{F}$ and $Aut_{\mathcal{F}}(T)$ permutes Φ transitively as $\mathcal{F}$ is transitive on $K^{\mathcal{F}}$, and

$$\mathcal{E} = \prod_{\mathcal{D} \in \Phi} \mathcal{D}$$

is the central product of the members of Φ by 2.2.2. Therefore $\mathcal{E} \trianglelefteq \mathcal{F}$ by 7.4 in [**Asc11**]. Hence $Z = \langle z(J) : J \in \Phi \rangle \trianglelefteq \mathcal{F}$, so by 3.3.7.2, $\mathcal{O}(\mathcal{F}, K^{\mathcal{F}}) \trianglelefteq \mathcal{F}$. Thus, by definition of $[K]_{\mathcal{F}}$, (1) holds. Moreover $\mathcal{O}_K \trianglelefteq \mathcal{O}(\mathcal{F}, K^{\mathcal{F}})$, so $s_2(K) = \mathcal{O}_K$. Hence (2) also holds. Part (3) follows from (1) and 3.3.7.1. □

LEMMA 6.4.4. *Set $S_0 = O_2(\mathcal{F})$ and assume $K_0 = K \cap S_0$ is nonabelian. Then $[K]_{\mathcal{F}} = \mathcal{O}(\mathcal{F}, K^{\mathcal{F}})$ is the central product of the members of $\{\mathcal{O}_J : J \in K^{\mathcal{F}}\}$, and $\tau(K) = (\mathcal{O}_K, \{K\})$.*

PROOF. By 3.3.8.3, $\mathcal{O}(\mathcal{F}, K^{\mathcal{F}})$ is the central product of the members of $\{\mathcal{O}_J : J \in K^{\mathcal{F}}\}$ and is normal in $\mathcal{F}$. Then, as in the proof of the previous lemma, $[K]_{\mathcal{F}} = \mathcal{O}(\mathcal{F}, K^{\mathcal{F}})$ and $\tau(K) = (\mathcal{O}_K, \{K\})$. □

LEMMA 6.4.5. *$\mathcal{E} = N_S(K)s(K)$ is a saturated fusion system with $s(K) \trianglelefteq \mathcal{E}$ and $O^2(\mathcal{E}) = O^2(s(K))$. Further $(\mathcal{E}, K^{s(K)})$ is a quaternion fusion packet.*

PROOF. The first few statements follow from 2.7.3.7 in [**Asc19**], while the last statement follows from the first and 2.5.2. □

LEMMA 6.4.6. *Assume $z \in O_2(\mathcal{F})$ and $\mathcal{O}_K = \mathcal{F}_K(K)$. Then*
(1) $[K]_{\mathcal{F}} \leq C_{\mathcal{F}}(E(\mathcal{F}))$.
(2) $[K]_{\mathcal{F}}$, $s(K)$, $N_S(K)[K]_{\mathcal{F}}$, and $N_S(K)s(K)$ are constrained.
(3) Let G be a model for $N_S(K)[K]_{\mathcal{F}}$. Then there is a unique subnormal subgroup H of G such that H is a model for $s(K)$. Moreover $N_S(K)$ acts on H, $HN_S(K)$ is a model for $N_S(K)s(K)$, and $H = \langle K^H \rangle$.
(4) Suppose $J \in \Omega - K^{s(K)}$ and $J \leq O_2(Js(K))$. Then J centralizes $s(K)$.

PROOF. As $z \in O_2(\mathcal{F})$, $E(\mathcal{F}) = E(\mathcal{F}_z)$ by 10.3 in [**Asc11**]. As $\mathcal{O}_K = \mathcal{F}_K(K)$, $K \leq O_2(\mathcal{F}_z) \leq C_{\mathcal{F}_z}(E(\mathcal{F}_z))$, so (1) holds. As $s_1 = [K]_{\mathcal{F}} \trianglelefteq \mathcal{F}$, $E(s_1) \leq E(\mathcal{F})$, so $K \leq C_{s_1}(E(s_1))$ by (1). Therefore $s_1 \leq C_{\mathcal{F}}(E(s_1))$, so $E(s_1) = 1$. Hence s_1 is constrained by 14.2 in [**Asc11**], establishing the first statement in (2), while the second statement follows by induction on $n(\mathcal{F}, K)$.

By 6.4.5, $\mathcal{E} = N_S(K)s(K)$ is a saturated fusion system with $s(K) \trianglelefteq \mathcal{E}$ and $O^2(\mathcal{E}) = O^2(s(K))$. Let $\mathcal{C} = E(\mathcal{E})$. By 7.17.3 in [**Asc11**], $\mathcal{C} = O^2(\mathcal{C}) \leq O^2(\mathcal{E}) = O^2(s(K))$. Then by 1.11 in [**Asc19**], $\mathcal{C} \trianglelefteq s(K)$, contradicting $E(s(K)) = 1$. A similar argument shows $N_S(K)s_1$ is constrained. This completes the proof of (2).

Part (3) follows from (2) and 2.1.7.

Assume $J \in \Omega - K^{s(K)}$ and let $T \in Syl(Js(K))$. Suppose that $J \leq O_2(Js(K))$. Then as $O_2(Js(K)) = O_2(HJ)$, $Q = \langle J^H \rangle$ is a normal 2-subgroup of HJ. As $K \notin J^H$, K centralizes Q by 3.1.5.1, so $H = \langle K^H \rangle$ centralizes Q. This establishes (4). □

LEMMA 6.4.7. *Assume $K \cap O_2(\mathcal{F}) = \langle z \rangle$ and $\mathcal{O}_K = \mathcal{F}_K(K)$. Let H be a model for $s(K)$. Then*
(1) $H/Z(H) \cong L_3(2)/E_8$, $L_3(2)/E_{64}$, $L_3(2)/2^{3+6}$, or AE_n for some $n \geq 5$.
(2) If $H/Z(H) \cong L_3(2)/E_8$ then $Z(H) = 1$.

PROOF. By parts (1), (4), and (6) of 6.4.2, we may assume that $\mathcal{F} = s(K)$. Thus $\mathcal{F} = [K]_{\mathcal{F}}$, and replacing Ω by $K^{\mathcal{F}}$, we may assume that $\mathcal{F}$ is transitive on Ω. As $K \cap O_2(\mathcal{F}) = \langle z \rangle$ and $\mathcal{O}_K = \mathcal{F}_K(K)$, $z \notin Z(\mathcal{F})$.

Let T be maximal subject to $T \leq S$, $z \notin T$, and $T \trianglelefteq \mathcal{F}$. Set $\mathcal{F}^+ = \mathcal{F}/T$. By 3.3.2, $T \leq Z(\mathcal{F})$, $\tau^+ = (\mathcal{F}^+, \Omega^+)$ is a quaternion fusion packet, $\mathcal{F}^+$ is transitive on Ω^+, and $\mathcal{F}^+ = [K^+]_{\mathcal{F}^+}$. By maximality of T, $z^+ \in Q^+$ for each $1 \neq Q^+ \leq S^+$ with $Q^+ \trianglelefteq \mathcal{F}^+$. Therefore τ^+ satisfies Hypothesis 6.2.1, so a model H^+ for $\mathcal{F}^+$ satisfies (1) by Theorems 6.2.8 and 6.3.9. As $T \leq Z(\mathcal{F})$, $T \leq Z(H)$ and $H^+ = H/T$. Thus (1) holds. Part (2) follows from 5.10.6. □

LEMMA 6.4.8. *Assume $K \cap O_2(\mathcal{F}) = \langle z \rangle$ and $\mathcal{O}_K = \mathcal{F}_K(K)$. Assume $J \in \Omega - K^{s(K)}$ and let H be a model for $s(K)$. Then either*
(1) J centralizes $s(K)$, or

(2) $H \cong L_3(2)/E_8$ *and* $\tau_0 = (s(K), \{K, J\})$ *is a quaternion fusion packet with* $z = z(J)$ *and* $z^H - \{z\} \subseteq \mathcal{A}(z)$.

PROOF. As in 6.4.6.3, let G be a model for $N_S(K)[K]_{\mathcal{F}}$ and choose H subnormal in G. By 6.4.6.3, $J \leq N_G(H)$ and $H_0 = HJ$ is a model for $Js(K)$. Let $T \in Syl(s(K))$, $T_0 = TJ$, $\Omega_0 = K^H \cup J^H$, and $\tau_0 = (Js(K), \Omega_0)$. By 2.5.2, τ_0 is a quaternion fusion packet. Set $t = z(J)$ and assume J does not centralize $s(K)$; equivalently, J does not centralize H.

Set $E = \langle z^H \rangle$. As J acts on H and centralizes K, $E \trianglelefteq H_0$. Set $Z = Z(H)$, $Q = O_2(H)$, and $\tilde{H}_0 = H_0/Z$. Then $\tilde{H}$ is described in 6.4.7.1. In particular $C_T(\tilde{E}) = Q$. As J does not centralize H, it follows from 6.4.6.4 that $J \not\leq O_2(H_0) = Q_0$.

Suppose $\tilde{E} \cong E_8$ and $H/C_H(\tilde{E}) = GL(\tilde{E})$. Then by 6.2.8 there is $Q_8 \cong \tilde{K}_0$ subnormal in $C_{\tilde{H}}(\tilde{z})$ with $\tilde{X} = \tilde{K}\tilde{K}_0 \cong Q_8^2$, and either $K_0 \in K^H$ and $\tilde{K} \in \Omega(\tilde{z})$, or $H \cong L_3(2)/E_8$. Further $\tilde{E} \leq \tilde{X}$, so if J centralizes $\tilde{K}\tilde{K}_0$ then $J \leq C_{H_0}(\tilde{E}) \leq Q_0$, a contradiction. Therefore J does not centralizes $\tilde{K}_0$, so $\tilde{K}_0 \notin \Omega(\tilde{z})$ by 3.3.4, and hence $H \cong L_3(2)/E_8$. If $J = K_0$ then (2) holds, so we may assume otherwise. Now J acts on but does not centralize K_0, so $J \cap K_0 \neq 1$ by 3.1.4. Thus $z \in J$ so $J \in \Omega(z)$. Also $JK_0 = C_{H_0}(K) \trianglelefteq C_{H_0}(z)$, so as $C_H(z)$ is irreducible on $K_0/\langle z \rangle$ we have $[J, K_0] \leq \langle z \rangle$. Hence $JK_0 \cong Q_8^2$ or D_8Q_8 and as $J \cong Q_8$ does not centralize K_0 it is the latter. Then $\Omega_0(z) = \{K, J\}$ by 3.1.5.1. But $K_0 = [JK_0, Y]$ for Y of order 3 in $C_H(z)$ and $C_{JK_0}(Y) \cong D_8$, so Y does not act on J. This contradicts $\Omega_0(z) = \{K, J\}$.

Hence we may assume $\tilde{H} \cong AE_n$ for some $n \geq 5$. Now J centralizes $R = O(\tau(K))$, while by 5.1.15, $|JQ_0 : Q_0| \leq 2$, so $J_0 = J \cap Q_0$ is of order at least 4. Therefore $\tilde{t}$ is the unique involution in $\Phi(\tilde{Q}_0)$, so $[\tilde{H}, \tilde{t}] = 1$ and hence $J \leq Q_0$, a contradiction. □

LEMMA 6.4.9. *Assume* $K \cap O_2(\mathcal{F}) = \langle z \rangle$ *and* $\mathcal{O}_K = \mathcal{F}_K(K)$. *Set* $n = n(\mathcal{F}, K)$, $s_i = sub_i(\mathcal{F}, K)$, *and let* $S_i \in Syl(s_i)$. *Then*

(1) Either $s(K)$ *is transitive on* $\Omega \cap S(K)$ *or* $H \cong L_3(2)/E_8$ *and* $\Omega \cap S(K) = \{K, K_0\}$ *with* $z^H - \{z\} \subseteq \mathcal{A}(z)$, *where* H *is a model for* $s(K)$.

(2) If $n > 1$ *then* s_{n-1} *is the central product of the members of* $\Phi = \{s(K)\alpha : \alpha \in Aut_{\mathcal{F}}(S_{n-1})\}$.

(3) $n \leq 2$ *and if* $\mathcal{F} = \mathcal{F}^\circ$ *then* $n \leq 1$.

PROOF. Choose G and H as in 6.4.6.3. Then $H = \langle K^H \rangle$ by 6.4.6.3. Set $\Omega_1 = \Omega \cap S(K)$.

If $K_1 \in \Omega_1 - K^H$ then $K_1 \in \Omega_1 \subseteq H$. Therefore as K_1 is nonabelian, K_1 does not centralize H, so (1) follows from 6.4.8.

If $n \leq 1$ then (2) and (3) hold, so we may assume $n > 1$. By 2.1.7, for $i \geq 1$, there is a model X_i for s_i that is subnormal in G. Let $X = X_{n-1}$; as $s(K) \trianglelefteq s_{n-1}$, we have $H \trianglelefteq X$ and for each $\alpha \in Aut_{\mathcal{F}}(S_{n-1})$, $s(K)\alpha \trianglelefteq s_{n-1}$ by 7.3.4 in [**Asc11**]. Observe that $s(K) = [K]_{s_{n-1}} = [\Omega_1]_{s_{n-1}}$ as $\Omega_1 \subseteq S(K)$. By 2.8.3, α extends to $\breve{\alpha} \in Aut(X)$. If $\Omega_1\alpha = \Omega_1$ then using (1),

$$s(K) = [\Omega_1]_{s_{n-1}} = [\Omega_1\alpha]_{s_{n-1}} = s(K)\alpha.$$

On the other hand if $\Omega_1\alpha \neq \Omega_1$ then $\Omega_1 \cap \Omega_1\alpha = \emptyset$ by (1), so $\Omega_1\alpha$ centralizes $s(K)$ by 6.4.8, and hence as $H \trianglelefteq X$, $H\breve{\alpha} = \langle K\alpha^{H\breve{\alpha}} \rangle$ centralizes H. Thus the 2-fusion system $\mathcal{Y} = \mathcal{F}_{S\cap Y}(Y)$ of the central product Y of the conjugates of H is normal in s_{n-2}. Therefore $\mathcal{Y} = s_{n-1}$, so (2) holds.

It remains to prove (3). For each $J \in \Omega - S_{n-1}$, J centralizes $s(K)\alpha$ for $\alpha \in Aut_{\mathcal{F}}(S_{n-1})$ by 6.4.8. Thus J centralizes s_{n-1} by (2), so if $n > 2$ then $L = \langle X^{Aut_{s_{n-3}}(S_{n-2})}\rangle$ is the central product of the $Aut_{s_{n-3}}(S_{n-2})$-conjugates of X, so $L = X_{n-2}$. But then $H \trianglelefteq X_{n-2}$, so $s(K) \trianglelefteq s_{n-2}$, and hence $s(K) = s_{n-1}$, contrary to the definition of n. Thus $n \leq 2$. Similarly if $\mathcal{F} = \mathcal{F}^\circ$ then $n \leq 1$, completing the proof. □

LEMMA 6.4.10. *Assume $\mathcal{F} = \mathcal{F}^\circ$ and set*

$$\Omega_0 = \{J \in \Omega : z(J) \notin O_2(\mathcal{F})\}.$$

Then $\tau = \tau_0 * \tau_1 * \cdots * \tau_k$, where $\mathcal{F}_i = [\Omega_i]_{\mathcal{F}}$, $\tau_i = (\mathcal{F}_i, \Omega_i)$ is a quaternion fusion packet, and for each $i > 0$ and $K \in \Omega_i$, either

(1) $\mathcal{F}_i = s(K)$ and $\Omega_i = K^{\mathcal{F}} = K^{\mathcal{F}_i}$, or

(2) $\Omega_i = \{K, \tilde{K}\} = \Omega(z)$, $\mathcal{F}_i$ is constrained with model H, $z^H - \{z\} \subseteq \mathcal{A}(z)$, and either $H \cong L_3(2)/E_8$ or $W = \langle\eta\rangle \trianglelefteq H$ for $\eta \in \eta(\tau_i)$ and $H/W \cong \mu(\tau_i) \cong D_{12}$.

PROOF. Set $Q = O_2(\mathcal{F})$. If $K \cap Q$ is nonabelian then by 3.3.8, $[K]_{\mathcal{F}}$ is the central product of the $\mathcal{O}_J$ for each $J \in K^{\mathcal{F}}$, and centralizes $[\Omega - K^{\mathcal{F}}]_{\mathcal{F}}$, so as $\mathcal{F} = \mathcal{F}^\circ$ we conclude that $\{K\}$ is an orbit of $\mathcal{F}$ on Ω and $\mathcal{O}_K = [K]_{\mathcal{F}} = s(K)$. Thus we may assume $K \cap Q$ is cyclic for each $K \in \Omega$.

Suppose next that $|K \cap Q| > 2$ for some $K \in \Omega$. Then by 3.3.12, there is $\Gamma \subseteq \Omega$ containing K such that $\mathcal{F}_\Gamma = [\Gamma]_{\mathcal{F}}$ centralizes $[\Omega - \Gamma]_{\mathcal{F}}$, $\tau_\Gamma = (\mathcal{F}_\Gamma, \Gamma)$ is a quaternion fusion packet, and $\eta \in \eta(\tau_\Gamma)$ is normal in $\mathcal{F}_\Gamma$. Moreover either $\mathcal{F}_\Gamma$ is transitive on Γ or $\Omega(z) = \{K, \tilde{K}\}$ is of order 2 and $\mathcal{A}(z) \neq \emptyset$. Then by 3.3.9, $\mathcal{F}_\Gamma$ is constrained with model H and for $\eta \in \eta(\tau_\Gamma)$, $W = \langle\eta\rangle \trianglelefteq H$. Finally if $\mathcal{A}(z) \neq \emptyset$ then $H/W \cong \mu(\tau_\Gamma) \cong D_{12}$ by 4.2.10. Thus we may assume $|K \cap Q| \leq 2$ for each $K \in \Omega$, and there exists some K with $z = z(K) \in Q$.

Suppose $\mathcal{O}_K \neq \mathcal{F}_K(K)$. Then by 6.4.3.3, $s_1 = [K]_{\mathcal{F}}$ centralizes $[\Omega - K^{\mathcal{F}}]_{\mathcal{F}}$ and then by 6.4.3.1, $K^{\mathcal{F}} = \{K\}$. Therefore we may assume $\mathcal{O}_K = \mathcal{F}_K(K)$ for each $K \in \Omega$ with $z(K) \in Q$.

As $\mathcal{F} = \mathcal{F}^\circ$, we conclude from 6.4.9.3 that $n(\mathcal{F}, K) \leq 1$, so $s(K) = s_1$. Now $s_1 \trianglelefteq \mathcal{F}$, so $O_2(s_1) \leq Q$. If $J \in \Omega - S(K)$ then by 6.4.8, J centralizes s_1, so that $[J]_{\mathcal{F}} \leq C_{\mathcal{F}}(s_1)$. In particular $[\Omega_0]_{\mathcal{F}}$ centralizes $[L]_{\mathcal{F}}$ for each $L \in \Omega - \Omega_0$, so we may assume $\Omega_0 = \emptyset$. Then as $[J]_{\mathcal{F}}$ centralizes s_1 for each $J \in \Omega - S_1$, and as $\mathcal{F} = \mathcal{F}^\circ$, it follows that $\mathcal{F} = \mathcal{F}_1 * \cdots * \mathcal{F}_k$ is a central product, where $\mathcal{F}_i = s(K_i)$ for some $K_i \in \Omega$. Now the lemma follows from 6.4.9.1. □

6.5. $F^*(\mathcal{F})$

In this section we assume $\tau = (\mathcal{F}, \Omega)$ is a quaternion fusion packet and adopt Notation 2.5.1 and 3.1.1. Moreover pick $K \in \Omega$ with $z = z(K) \in \mathcal{F}^f$.

LEMMA 6.5.1. *(1) $\mathcal{O}(z) \trianglelefteq \mathcal{F}_z$.*

(2) For each $K \in \Omega(z)$, $\mathcal{O}_K$ is subnormal in $\mathcal{F}_z$.

(3) $\mathcal{O}(z)$ is the central product of the systems $\mathcal{O}_K$, $K \in \Omega(z)$.

(4) For each $K \in \Omega(z)$, $\mathcal{O}_K = \mathcal{F}_K(K)$ or $\mathcal{O}_K \cong SL_2[m]$ or $SL_2^1[m]$.

PROOF. Parts (1) and (2) are a restatement of 2.6.11, part (3) is a restatement of 2.6.8, and part (4) was established in 2.5.7. □

LEMMA 6.5.2. *Suppose $\mathcal{C}$ is a component of $\mathcal{F}$ centralized by z. Then*
(1) $\mathcal{C}$ is a component of $\mathcal{F}_z$.
(2) One of the following holds:
(i) $\mathcal{O}_K$ centralizes $\mathcal{C}$.
(ii) $\mathcal{O}_K = \mathcal{C}$.
(iii) $\mathcal{O}_K \cong SL_2^1[m]$, $\mathcal{C} = O^2(\mathcal{O}_K))$, and $\mathcal{O}_K$ centralizes each component of $\mathcal{F}$ distinct from $\mathcal{C}$.

PROOF. Part (1) follows from 2.2.3, so it remains to prove (2). Appealing to (1) and 2.5.3.1, we may assume $\mathcal{F} = \mathcal{F}_z$. Thus $\mathcal{O}_K$ is subnormal in $\mathcal{F}$ by 6.5.1.2. Therefore if $\mathcal{O}_K = \mathcal{F}_K(K)$, then by 9.10 and Theorem 6 in [**Asc11**], $K \leq O_2(\mathcal{F}) \leq C_S(\mathcal{C})$, so that (2i) holds. Thus we may assume that $m > 8$. Similarly if $\mathcal{O}_K \cong SL_2[m]$, then as $m > 8$, $\mathcal{O}_K$ is quasisimple, so $\mathcal{O}_K$ is a component of $\mathcal{F}$, and hence (2i) or (2ii) holds as $E(\mathcal{F})$ is the central product of its components.

Therefore by 6.5.1.4, we may assume $\mathcal{O}_K \cong SL_2^1[m]$. Set $\mathcal{B} = O^2(\mathcal{O}_K)$, so that $\mathcal{B} \cong SL_2[m/2]$, and let B be Sylow in $\mathcal{B}$. Then either $\mathcal{B}$ is a component of $\mathcal{F}$, or $m = 16$ and $\mathcal{B} \cong SL_2[8]$ is a solvable component. In either case by 2.2.2, either $\mathcal{B} = \mathcal{C}$ or $\mathcal{B}$ centralizes $\mathcal{C}$.

In the first case, for $\mathcal{E} \in Com(\mathcal{F}) - \{\mathcal{B}\}$ and E Sylow in $\mathcal{E}$, we have $E \leq C_S(B) \leq N_S(K)$ by 2.6.6. Thus $[K, E] \leq C_K(B) = \langle z \rangle$, so K acts on $\mathcal{E}$ by 9.7.2 in [**Asc11**], so $[K, E] \leq \langle z \rangle \cap E \leq Z(\mathcal{E})$. Hence K centralizes $\mathcal{E}$ by 9.5 in [**Asc11**], so $\mathcal{O}_K = K\mathcal{B}$ centralizes $\mathcal{E}$, and therefore (2iii) holds.

Finally assume $\mathcal{B}$ centralizes $\mathcal{C}$. Then the argument in the previous paragraph applied to $\mathcal{C}$ in the role of $\mathcal{E}$, shows that $\mathcal{O}_K$ centralizes $\mathcal{C}$, so that (2i) holds, completing the proof. □

LEMMA 6.5.3. *If $z \notin O_2(\mathcal{F})$ then*
(1) $\mathcal{O}_K \leq C_{\mathcal{F}}(O_2(\mathcal{F}))$ and
(2) $z \in E(\mathcal{F}) - Z(E(\mathcal{F}))$.

PROOF. As $z \notin Q = O_2(\mathcal{F})$, K centralizes Q by 3.1.4. Then 2.6.6 implies (1).

Let $\mathcal{E} = E(\mathcal{F})$ and $E \in Syl(\mathcal{E})$. Then $E \trianglelefteq S$, so K acts on E. Hence by 3.1.4, either $z \in E$ or $[K, E] = 1$, and we may assume the latter. By 2.3.1 in [**Asc19**], z centralizes $\mathcal{E}$. Then by 6.5.2.2, $\mathcal{O}_K \leq C_{\mathcal{F}}(E(\mathcal{F})) = \mathcal{B}$. By 9.12.1 in [**Asc11**], $Q = F^*(\mathcal{B})$. But then $K \leq C_{\mathcal{B}}(Q) = Z(Q)$ by Theorem 6 in [**Asc11**], contradicting K nonabelian. □

THEOREM 6.5.4. *Assume $z \notin O_2(\mathcal{F})$. Then*
(1) z is contained in a unique component $\mathcal{C} = \mathcal{C}(z)$ of $\mathcal{F}$.
(2) $\mathcal{O}_K$ centralizes each component of $\mathcal{F}$ distinct from $\mathcal{C}$.
(3) If $\mathcal{O}_K$ is not solvable then $O^2(\mathcal{O}_K) \leq \mathcal{C}$.

We prove Theorem 6.5.4 in a series of reductions. Assume $\tau = (\mathcal{F}, \Omega)$ is a minimal counter example. That is $\mathcal{F}$, Ω is a counter example, but Theorem 6.5.4 holds for all quaternion fusion packets $\tau_1 = (\mathcal{F}_1, \Omega_1)$ such that $\mathcal{F}_1$ is of smaller order than $\mathcal{F}$, or Ω_1 is of smaller order than Ω.

LEMMA 6.5.5. *z is contained in no component of $\mathcal{F}$.*

PROOF. Suppose z is contained in a component $\mathcal{C}$ of $\mathcal{F}$. As $z \notin O_2(\mathcal{F})$, $\mathcal{C}$ is the unique component containing z, so K acts on $\mathcal{C}$. Then K also acts on the product $\mathcal{E}$ of the remaining components of $\mathcal{F}$. Let $E \in Syl(\mathcal{E})$. Then K acts on E and $z \notin E$, so $[K, E] = 1$ by 3.1.4. Claim $\mathcal{O}_K$ centralizes $\mathcal{E}$. For by 6.5.3.2 applied to $K\mathcal{E}$, we have $z \in O_2(K\mathcal{E})$, so that z centralizes $\mathcal{E}$ and the claim follows from 6.5.2. Further if $\mathcal{O}_K$ is not solvable, then $\mathcal{L} = O^2(\mathcal{O}_K)$ is a component of $\mathcal{F}_z$ by 6.5.1, so by L-balance (cf. Theorem 7 in [**Asc11**]), $\mathcal{L} \leq E(\mathcal{F})$. Hence $\mathcal{L} \leq C_{E(\mathcal{F})}(\mathcal{E}) = \mathcal{X} = Z(E(\mathcal{F}))\mathcal{C}$, so by 7.17.3 in [**Asc11**], $\mathcal{L} = O^2(\mathcal{L}) \leq O^2(\mathcal{X}) = \mathcal{C}$. But now all conclusions of Theorem 6.5.4 hold, contrary to the choice of τ as a counter example. □

LEMMA 6.5.6. *$\mathcal{O}_K = \mathcal{F}_K(K)$.*

PROOF. Assume otherwise. Then $\mathcal{L} = O^2(\mathcal{O}_K) \cong SL_2[k]$ with $k = m$ or $m/2$. In particular $\mathcal{L}$ is a component or solvable component of $\mathcal{F}_z$ containing z, on some $L \leq K$, so by 2.2.6.4, either $L \leq \mathcal{C}$ for some component $\mathcal{C}$ of $\mathcal{F}$, or L centralizes $E(\mathcal{F})$. The former is impossible by 6.5.5, and the latter contradicts 6.5.3. □

LEMMA 6.5.7. *$\mathcal{F} = E(\mathcal{F})K$ and $\mathcal{F}$ is transitive on Ω.*

PROOF. Let $\mathcal{E} = KE(\mathcal{F})$. By Theorem 5 in [**Asc11**], $\mathcal{E}$ is a saturated fusion system with $O^2(\mathcal{E}) = E(\mathcal{F})$. Thus $E(\mathcal{E}) = E(\mathcal{F})$. Let $\Sigma = K^{\mathcal{E}}$. By 2.5.2, $\rho = (\mathcal{E}, \Sigma)$ is a quaternion fusion packet. Thus if $\mathcal{F} \neq \mathcal{E}$ or $\Omega \neq \Sigma$, then by minimality of τ, the theorem holds for ρ. But then z is contained in some component of $\mathcal{F}$, contrary to 6.5.5. □

LEMMA 6.5.8. *$O_2(\mathcal{F}) = 1$.*

PROOF. Suppose $Q = O_2(\mathcal{F}) \neq 1$. By hypothesis, $z \notin Q$, so K centralizes Q by 3.1.4. By Theorem 6 in [**Asc11**], $E(\mathcal{F})$ centralizes Q, so $Q = Z(\mathcal{F})$ by 6.5.7.

Set $\mathcal{F}^+ = \mathcal{F}/Q$. By 3.3.2.2, $\tau^+ = (\mathcal{F}^+, \Omega^+)$ is a quaternion fusion packet, so by minimality of τ, z^+ is contained in some component $\mathcal{C}^+$ of $\mathcal{F}^+$, and K^+ acts faithfully on $\mathcal{C}^+$. By 2.1.11, the preimage $\mathcal{C}$ of $\mathcal{C}^+$ in $\mathcal{F}$ is $Q\mathcal{B}$, with $\mathcal{B} = O^2(\mathcal{C})$ a component of $\mathcal{F}$. Now K acts on $\mathcal{C}$, and hence also on $\mathcal{B}$. By 6.5.5, $z \notin \mathcal{B}$, contrary to 6.5.3.2 applied to $K\mathcal{B}$. □

LEMMA 6.5.9. *K is transitive on the components of $\mathcal{F}$.*

PROOF. Let $\mathcal{C}$ be a component of $\mathcal{F}$), $\mathcal{E} = \langle \mathcal{C}^K \rangle$, and $\bar{\mathcal{E}}$ the product of the components not in $\mathcal{C}^K$. By 7.4 in [**Asc11**] and 6.5.7, $\mathcal{E}$ and $\bar{\mathcal{E}}$ are normal in $\mathcal{F}$. We may assume $\bar{\mathcal{E}} \neq 1$. By Theorem 5 in [**Asc11**], $\mathcal{B} = K\mathcal{E}$ and $\bar{\mathcal{B}} = K\bar{\mathcal{E}}$ are saturated subsystems with $\mathcal{E} = E(\mathcal{B})$ and $\bar{\mathcal{E}} = E(\bar{\mathcal{B}})$. By 2.5.2, $(\mathcal{B}, K^{\mathcal{B}})$ and $(\bar{\mathcal{B}}, K^{\bar{\mathcal{B}}})$ are quaternion fusion packets, so by minimality of $\mathcal{F}$, z is contained in some component of $\mathcal{B}$ or $\bar{\mathcal{B}}$, contrary to 6.5.5. □

Let $\{\mathcal{C}_i : i \in I)$ be the set of components of $\mathcal{F}$, with $I = \{1, \ldots, s\}$ and $\mathcal{C}_i$ a system on T_i. Then $E(\mathcal{F})$ is a system on $T = T_1 \times \cdots \times T_s$ by 6.5.8. As $z \in T$, $z = z_1 \cdots z_s$, where z_i is the projection of z on T_i, and by 6.5.9, $z_i \neq 1$ for each $i \in I$, and K is transitive on $\mathcal{U} = \{z_i : i \in I\}$. By 6.5.5, $s > 1$.

Set $R = C_T(z)$, $Q = N_{T_1}(K)$, and $X = K \cap T_1T_2$.

LEMMA 6.5.10. *If $X \neq 1$ then $s = 2$ and $z_1 \in Q$.*

PROOF. Assume $X \neq 1$. Then as z is the unique involution in K, $z \in X$, so $z = z_1z_2$, and then as $z_i \neq 1$ for all $i \in I$ and as $s > 1$, it follows that $s = 2$. As K is transitive on $\mathcal{U}$ of order 2, $|C_K(z_1)| = m/2 > 2$, so $z_1 \in Q$. □

LEMMA 6.5.11. *X is cyclic.*

PROOF. Assume X is noncyclic. Then by 2.5.2, $(E(\mathcal{F}), X^{E(\mathcal{F})})$ is a quaternion fusion packet, so by minimality of $\mathcal{F}$, z is contained in a component of $\mathcal{F}$, contrary to 6.5.5. □

LEMMA 6.5.12. *Choose notation so that there is $k \in K$ of order 4 with $\mathcal{C}_1^k = \mathcal{C}_2$. Then the map $\varphi : y \mapsto [y, k]$ is an isomorphism of Q with X.*

PROOF. Let $u, v \in Q$. Then $[u, k] \in X$, so $\varphi : Q \to X$. Further $u\varphi = v\varphi$ iff $uv^{-1} = u^kv^{-k} = (uv^{-1})^k$ iff k centralizes uv^{-1}. However $uv^{-1} \in T_1$ and $T_1^k = T_2$ with $T_1 \cap T_2 = 1$, so k centralizes uv^{-1} iff $u = v$. Thus φ is an injection. Hence if $|X| \leq 2$ then the lemma follows from 6.5.10, so we may assume $X = \langle x \rangle$ is of order greater than 2. Now $x = x_1x_2$ with $x_i \in T_i$. Further $x_1 \in C_{T_1}(x) \leq N_{T_1}(K) = Q$ as $|x| > 2$. Therefore $|Q| \geq |X|$, so φ is a bijection and $Q = \langle x_1 \rangle$. Then $QX = \langle x_1, x_2 \rangle$ is abelian, so from 8.5.4 in [**Asc86**], $\varphi : Q \to X$ is a homomorphism, completing the proof. □

LEMMA 6.5.13. *$R_1 = R \cap T_1$ is noncyclic, and either $\Phi(R_1) = 1$, or R_1 is nonabelian dihedral and $|R_1 : Q| = 2$.*

PROOF. By 9.1.2 in [**Asc11**], T_1 is not cyclic, so $R_1 = C_{T_1}(z) = C_{T_1}(z_1)$ is noncyclic. But Q is cyclic by 6.5.11 and 6.5.12, so there exists $r \in R_1 - Q$. Then $u = [r, k] = k^{-r}k$, and as r does not act on K, $[k, k^{-r}] \in [K, K^r] = 1$, so $[u, k] = 1$. But k inverts u, so $u^2 = 1$, and hence r is an involution.

We have shown that each element in $R_1 - Q$ is an involution, so each such element inverts Q, and hence as Q is cyclic either $|Q| \leq 2$ or $|R_1 : Q| = 2$ and R_1 is nonabelian dihedral. Thus the lemma holds. □

LEMMA 6.5.14. *$|R_1 : Q| = 2$.*

PROOF. Assume otherwise. Then by 6.5.13, $\Phi(R_1) = 1$, so as $R = \prod_i C_{T_i}(z)$ and K is transitive on the components of $\mathcal{F}$), $R = R_1 \times \cdots \times R_s$ with $R_i = C_{T_i}(z) \cong R_1$. Thus $\Phi(R) = 1$, so $|X| \leq 2$ by 6.5.11,

Let $W = O(z)$. If $K = W$ then $R_1 = Q$, whereas R_1 is noncyclic by 6.5.13, while Q is cyclic by 6.5.11 and 6.5.12. Therefore $W \neq K$. But $W = KW_R$, where $W_R = W \cap R$, so $W/K \cong (W \cap R)/(K \cap R)$ is elementary abelian, and hence as $W \neq K$ we conclude that $m = 8$. Then $W \cong 2^{1+2w}$ is extraspecial, as W is the central product of w copies of $K \cong Q_8$. As $\Phi(W_R) = 1$, $m(W_R) \leq w + 1$. But $|W : W_R| = |K : K \cap R| = 4$, so $m(W_R) = 2w - 1$. Then $2w - 1 \leq w + 1$, so $w = 2$. As R permutes the two members of $\Omega(z)$, $|R_1 : Q| \leq 2$. By 6.5.13, $|R_1| > 2$, so $Q \neq 1$ and by 6.5.12, $X \cong Q$ so $X \neq 1$. By the previous paragraph, $|X| \leq 2$ so $|Q| = 2$ and $|R_1 : Q| = 2$. Thus the lemma holds. □

We are now in a position to establish Theorem 6.5.4. By 6.5.11 and 6.5.12, Q is cyclic, so by 6.5.13 and 6.5.14, R_1 is dihedral and $|R_1 : Q| = 2$. In particular $Q \neq 1$, while $X \cong Q$ by 6.5.12, so $X \neq 1$. Therefore $s = 2$ by 6.5.10.

If R_1 is nonabelian, then $\langle z_1\rangle = Z(R_1)$, so as $R_1 = C_{T_1}(z_1)$, we have $T_1 = R_1$ dihedral. On the other hand if $R_1 \cong E_4$, then by a lemma of Suzuki (cf. Exercise 8.6 in [**Asc86**]), T_1 is dihedral or semidihedral. So in any event, T_1 is dihedral or semidihedral. Then as $\mathcal{C}_1$ is simple, it follows from 2.1.10 that $\mathcal{C}_1$ is transitive on its involutions. But as $z \in \mathcal{F}^f$, $z_1 \in \mathcal{C}_1^f$, so $R_1 = T_1$ is dihedral and z is in the center of $S = T_1T_2\langle k\rangle \cong R_1$ wr $\mathbf{Z_2}$. By 9.1.2 in [**Asc11**], R_1 is nonabelian, so $|X| = |Q| > 2$.

Let $\mathcal{E} = \mathcal{C}_1\mathcal{C}_2$. As $|X| > 2$, $K_0 = X\langle k\rangle$ is quaternion, so replacing τ by $\tau_0 = (\langle k\rangle\mathcal{E}, K_0^{\mathcal{E}})$, we may assume $\tau = \tau_0$, so $K = K_0$. As $R_1 = T_1$, $|T_1 : Q| = 2$ by 6.5.14, so as $Q \cong X$ is cyclic, Q is the cyclic subgroup of T_1 of index 2. Let $a_1 \in T_1 - Q$; then $a = a_1a_1^k$ is in $z^{\mathcal{E}}$ and centralizes k, so $a \in \mathcal{D}^*(z)$. Thus there is $z \neq t \in z^{\mathcal{E}}$ such that there is $U = \langle u\rangle \in \Delta(t)$. As z is weakly closed in QX, $t \in T - QX$, so $t \in aQX$. Then as $S/QX \cong D_8$ it follows that $m = 8$ and $u \in a_1kQX$. But then $X^u \neq X$, where is u acts on K by condition (QFP3) in the definition of quaternion fusion packets, and hence also on $X = K \cap T$. This contradiction completes the proof of Theorem 6.5.4.

6.6. z not in $O_2(\mathcal{F})$

In this section we assume $\tau = (\mathcal{F}, \Omega)$ is a quaternion fusion packet and adopt Notation 2.5.1 and 3.1.1. Moreover pick $K \in \Omega$ with $z = z(K) \in \mathcal{F}^f$.

LEMMA 6.6.1. *Assume $\mathcal{F} = \mathcal{F}^\circ$ and $\mathcal{Z} \cap O_2(\mathcal{F}) = \emptyset$. Set $Q = O_2(\mathcal{F})$ and $\mathcal{F}^+ = \mathcal{F}/Q$. Let $I = \{1, \ldots, r\}$, $\{\mathcal{C}_i : i \in I\}$ the set of components of $\mathcal{F}$, $S_i \in Syl(\mathcal{C}_i)$, $\Omega_i = \{J \in \Omega : z(J) \in S_i\}$, $\mathcal{F}_i = [\Omega_i]_{\mathcal{F}}$, and $\tau_i = (\mathcal{F}_i, \Omega_i)$.*

(1) $Q = Z(\mathcal{F})$.

(2) $\tau^+ = (\mathcal{F}^+, \Omega^+)$ is a quaternion fusion packet with $\mathcal{F}^+ = [\Omega^+]_{\mathcal{F}^+}$ and $O_2(\mathcal{F}^+) = 1$.

(3) For each $i \in I$, $\mathcal{C}_i \trianglelefteq \mathcal{F}$ and $\Omega_i \neq \emptyset$.

(4) $\Omega = \coprod_{i\in I} \Omega_i$ is a partition of Ω.

(5) $\mathcal{F} = \mathcal{F}_1 \cdots \mathcal{F}_r$ is a central product of the subsystems $\mathcal{F}_i$, so $\tau = \tau_1 * \cdots * \tau_r$ is a central product of quaternion fusion packets.

(6) $F^*(\mathcal{F}_i) = \mathcal{C}_iZ(\mathcal{F}_i)$.

PROOF. By hypothesis, $\mathcal{Z} \cap Q = \emptyset$, so Ω centralizes Q by 3.1.4. Then (1) follows as $\mathcal{F} = \mathcal{F}^\circ$. Now (2) follows from (1) and 3.3.2.

Next by Theorem 6.5.4, $z \in \mathcal{C}_j$ for a unique $j \in I$ and $\mathcal{O}_K$ centralizes $\mathcal{C}_i$ for $i \in I - \{j\}$. In particular K acts on each component of $\mathcal{F}$, so as $\mathcal{F} = \mathcal{F}^\circ$, each component of $\mathcal{F}$ is normal in $\mathcal{F}$ by 2.1.6. Further if $\Omega_i = \emptyset$ then each member of Ω centralizes $\mathcal{C}_i$, so $\mathcal{F} = [\Omega]_{\mathcal{F}}$ centralizes $\mathcal{C}_i$, a contradiction. Hence (3) and (4) hold.

For $i \in I$, set

$$\mathcal{E}_i = \prod_{j \in I - \{i\}} \mathcal{C}_j \text{ and } \mathcal{D}_i = C_{\mathcal{F}}(\mathcal{E}_i),$$

and let $T_i \in Syl(\mathcal{D}_i)$. From the previous paragraph, $\Omega_i \subseteq T_i$ and $\mathcal{D}_i \trianglelefteq \mathcal{F}$, so $\mathcal{F}_i \leq \mathcal{D}_i$. Observe $F^*(\mathcal{D}_i) = \mathcal{C}_iQ$. Also for $i \neq j$, $C_{KQT_i}(\mathcal{C}_i) = KQ$ and $C_{\mathcal{D}_i}(\mathcal{C}_i) = Q$, so $KQ = O_2(K\mathcal{D}_i)$ and hence $\{K\} = \Omega \cap O_2(K\mathcal{D}_i)$ so $K \trianglelefteq K\mathcal{D}_i$. Hence $\mathcal{F}_i = [\Omega_i]_{\mathcal{F}_i}$ centralizes K. Then $\mathcal{F}_j = [\Omega_j]_{\mathcal{F}}$ centralizes $\mathcal{F}_i$, proving (5). As $F^*(\mathcal{D}_i) = \mathcal{C}_iQ$, (6) holds. □

LEMMA 6.6.2. *Assume $O_2(\mathcal{F}) = 1$ and $\mathcal{F} = [K]_{\mathcal{F}}$. Then*
(1) $\mathcal{F}$ is almost simple.
(2) $z \in F^*(\mathcal{F})$.
(3) If $\mathcal{O}_K \neq \mathcal{F}_K(K)$ then $O^2(\mathcal{O}_K) \leq F^*(\mathcal{F})$.

PROOF. By 6.6.1, there exists a unique component $\mathcal{C}$ of $\mathcal{F}$ such that $z \in T \in Syl(\mathcal{C})$, K centralizes all other components of $\mathcal{F}$, and each component of $\mathcal{F}$ is normal in $\mathcal{F}$. Hence as $\mathcal{F} = [K]_{\mathcal{F}}$, $\mathcal{C}$ is the unique component of $\mathcal{F}$, so $\mathcal{C} = F^*(\mathcal{F})$ since $O_2(\mathcal{F}) = 1$. In particular (1) and (2) hold, while (3) follows from 6.5.4.3. □

Recall the definition of $s(K) = \text{sub}(\mathcal{F}, K)$ and $\tau(K) = (s(K), K^{s(K)})$ from Definitions 1.2.9 and 6.4.1.

LEMMA 6.6.3. *Assume $F^*(\mathcal{F}) = Z(\mathcal{F})\mathcal{C}$ for some component $\mathcal{C}$ of $\mathcal{F}$, and $z \notin O_2(\mathcal{F})$. Then*
(1) $F^*(s(K)) = Z(s(K))\mathcal{C}$ and $z \notin O_2(s(K))$.
(2) $\tau(K) = (s(K), K^{s(K)})$ is a quaternion fusion packet with $s(K) = [K]_{s(K)}$.

PROOF. By 6.4.2.1, $\tau(K)$ is a quaternion fusion packet, and by definition of $s(K)$, we have $s(K) = [K]_{s(K)}$. Thus (2) holds.

As $z \notin O_2(\mathcal{F})$, also $z \notin O_2(s(K))$ by 6.4.2.3. By 6.4.2.1, $s(K)$ is subnormal in $\mathcal{F}$, so as $F^*(\mathcal{F}) = Z(\mathcal{F})\mathcal{C}$, it follows that either $\mathcal{C} \leq s(K)$ or $F^*(s(K)) \leq Z(\mathcal{F})$. But in the latter case as $C_{s(K)}(F^*(s(K)) \leq F^*(s(K))$, we have $s(K)$ abelian, contradicting $K \leq s(K)$. Therefore $\mathcal{C} \leq s(K)$, so $F^*(s(K)) = Z(s(K))\mathcal{C}$. Thus (1) holds. □

LEMMA 6.6.4. *Assume $\mathcal{A}(\tau) = \emptyset$. Then $\tau^\circ = \tau_1 * \cdots * \tau_r$ is a central product of quaternion fusion packets $\tau_i = (\mathcal{F}_i, \Omega_i)$ such that $\mathcal{F}_i = [\Omega_i]_{\mathcal{F}_i}$ and for $K_i \in \Omega_i$ and $z_i = z(K_i)$,*
(1) if $z_i \in O_2(\mathcal{F}_i)$ then $\mathcal{F}_i$ is transitive on Ω_i, $\mathcal{D}(\tau_i)^c$ is connected, and either $\mathcal{F}_i$ is constrained or $\mathcal{F}_i = \mathcal{O}_{K_i}$, and
(2) if $z_i \notin O_2(\mathcal{F}_i)$ then $F^*(\mathcal{F}_i) = Z(\mathcal{F}_i)\mathcal{C}_i$ for some component $\mathcal{C}_i$ of $\mathcal{F}$ such that $\mathcal{C}_i \leq [K_i]_{\mathcal{F}_i}$.

PROOF. We may assume $\tau = \tau^\circ$. Then by 6.4.10, $\tau = \tau_0 * \tau_1 * \cdots * \tau_s$, where $\tau_i = (\mathcal{G}_i, \Gamma_i)$ is a quaternion fusion packet, $\Gamma_0 = \{J \in \Omega : z(J) \notin O_2(\mathcal{F})\}$, and for $j > 0$ and $J_j \in \Gamma_j$, we have $z(J_j) \in O_2(\mathcal{G}_j)$ and $\mathcal{G}_j$ is transitive on Γ_j. By Theorem 6.1.5, $\mathcal{D}(\tau_j)^c$ is connected. Further if $\mathcal{O}_{J_j} \neq \mathcal{F}_{J_j}(J_j)$ then $\mathcal{G}_j = \mathcal{O}_{J_j}$ by 6.4.3. On the other hand if $\mathcal{O}_{J_j} = \mathcal{F}_{J_j}(J_j)$ then $\mathcal{G}_j$ is constrained by 6.4.6.3. Therefore, replacing τ by τ_0, we may assume $\Omega = \Gamma_0$. Now the lemma follows from 6.6.1 and 6.6.3. □

LEMMA 6.6.5. *Assume*
(a) $\mathcal{A}(\tau) = \emptyset$.
(b) $F^*(\mathcal{F}) = Z(\mathcal{F})\mathcal{C}$ for some component $\mathcal{C}$ of $\mathcal{F}$, and $\mathcal{Z} \cap O_2(\mathcal{F}) = \emptyset$.
(c) If $\tau(K) \neq \tau^\circ$ then $\tau(K)$ satisfies one of the conclusions of Theorem 1.
Then
(1) $\mathcal{F}^\circ$ is transitive on Ω.
(2) $\mathcal{F}^\circ = [K]_{\mathcal{F}}$.
(3) If $\tau^\circ = (\mathcal{F}^\circ, \Omega)$ satisfies one of the conclusions of Theorem 1 then either $\mathcal{D}(\tau)^c$ is connected or $\mathcal{F}/Z(\mathcal{F})$ is $PSp_n[m]$ or $L_n^-[m]$ for some $n \geq 6$.

PROOF. Set $\mathcal{E} = s(K)$, $\Gamma = K^{\mathcal{E}}$, and $\rho = \tau(K)$. By (b) and 6.6.3, ρ satisfies the hypotheses of Theorem 1, so in particular ρ satisfies (1) and (2). Suppose (1) or (2) fails; then as ρ satisfies (1) and (2), we conclude $\tau^\circ \neq \rho$. Therefore by (c), ρ satisfies one the conclusions of Theorem 1. By 6.6.3.1, $z \notin O_2(\mathcal{E})$, so ρ satisfies one of conclusions (2)-(4) of Theorem 1. Further by 6.6.3.1, $F^*(\mathcal{E}) = Z(\mathcal{E})\mathcal{C}$, so $\mathcal{C}$ is described in Theorem 1.

If $\mathcal{E}$ is $L_2[2m]^{(1)}$ then there is a unique quaternion subgroup Q of order m normal in a Sylow group T of $\mathcal{E}$. Therefore $K = Q$ and $\Gamma = \{K\}$. Indeed if $J \in \Omega - \Gamma$ then $J \leq C_S(Q)$, so from the structure of $Aut(L_2(q))$, $J \leq C_S(\mathcal{C})$, so $J \leq C_S(F^*(\mathcal{F})) = Z(\mathcal{F})$, contradicting J nonabelian. Therefore $\Omega = \{K\}$, contradicting the assumption that (1) or (2) fails.

Similarly if $\mathcal{E}$ is $L_3^+[m]^{(1)}$, then there is a unique quaternion subgroup Q of KT of order m such that $KT = QT$, so $K = Q$ and $\Gamma = \{K\}$. Then $J \leq C_S(K) = C_S(Q) = Z(\mathcal{F})P$, where $P \cong K$ is Sylow in a fundamental subgroup of $L_3^\pi(q)$. Also $J \leq O^2(Z(\mathcal{F})\mathcal{C}) = \mathcal{C}$, so $J = P$ is determined. But now $z = z(J)$ and $\mathcal{A}(z) \neq \emptyset$, contrary to (a).

Therefore ρ satisfies conclusion (2) or (4) of Theorem 1, or $\mathcal{E} \cong M_{12}$. In particular each member of Γ is in $Z(\mathcal{F})\mathcal{C}$, so $\mathcal{F}^\circ = Z(\mathcal{F})\mathcal{C}$. But also $\Gamma \leq O^2(\mathcal{F}^\circ)$, so $\mathcal{F}^\circ = \mathcal{C}$.

Suppose $\mathcal{C}$ is M_{12} or $G_2[m]$. Then $\Gamma = \{K\}$, and if $J \in \Omega - \Gamma$ then $z(J) = z$ and $\mathcal{A}(z) \neq \emptyset$, contrary to (a). Thus ρ satisfies conclusion (2) or (4) of Theorem 1, but $\mathcal{C}$ is not $G_2[m]$. Therefore Γ is uniquely determined in $\mathcal{E} = \mathcal{C}$, so $\Omega = \Gamma$, contrary to the assumption that (1) or (2) fails. This completes the proof of (1) and (2). Further (3) follows from 6.1.5. □

LEMMA 6.6.6. *Assume $\mathcal{A}(\tau) = \emptyset$ and Theorem 1 holds in all proper subpackets of τ of the form $\tau(K)$ for some $K \in \Omega$ with $z(K) \in \mathcal{F}^f$. Then*

(1) $\tau^\circ = \tau_1 * \cdots * \tau_r$ is a central product of quaternion fusion packets $\tau_i = (\mathcal{F}_i, \Omega_i)$, where $\mathcal{F}_i = [\Omega_i]_{\mathcal{F}_i}$ is transitive on Ω_i.

(2) If either $\tau \neq \tau_i$ or τ_i satisfies one of the conclusions of Theorem 1, then either $\mathcal{D}(\tau_i)^c$ is connected or $\mathcal{F}_i/Z(\mathcal{F}_i)$ is $PSp_n[m]$ or $L_n^-[m]$ for some $n \geq 6$.

PROOF. We may assume $\tau = \tau^\circ$. Then by 6.6.4, $\tau = \tau_1 * \cdots * \tau_r$ is a central product of quaternion fusion packets $\tau_i = (\mathcal{F}_i, \Omega_i)$, where $\mathcal{F}_i = [\Omega_i]_{\mathcal{F}_i}$ satisfies (1) or (2) of 6.6.4. In 6.6.4.1, $\mathcal{F}_i$ is transitive on Ω_i and $\mathcal{D}(\tau_i)^c$ is connected, so we may assume τ_i appears in 6.6.4.2. Let $K \in \Omega_i$. If $\tau_i = \tau(K)$, then we have transitivity by 6.6.3.2, so assume otherwise. Then $\tau(K)$ is a proper subpacket of τ, so by hypothesis, $\tau(K)$ satisfies one of the conclusions of Theorem 1. But now we have transitivity by 6.6.5.1. This completes the proof of (1).

It remains to prove (2). By 6.6.4 we may assume $z_i \notin O_2(\mathcal{F}_i)$. Observe that $\tau_i = \tau(K)$ for $K \in \Omega_i$ with $z(K) \in \mathcal{F}^f$. Thus if $\tau \neq \tau_i$, then τ_i is proper in τ, so τ_i satisfies one of the conclusions of Theorem 1 by hypothesis. On the other hand if $\tau = \tau_i$ this also holds by the hypotheses of (2). Now 6.6.5.3 completes the proof of (2). □

DEFINITION 6.6.7. We often assume the following *Inductive Hypothesis* : Theorem 1 holds in each proper subpacket of τ satisfying the hypothesis of Theorem 1 and of the form $(\mathcal{Y}, \Gamma)$ with $\mathcal{Y} \leq \mathcal{F}$ and $\Gamma \subseteq \Omega$.

Sometimes we instead assume the stronger *Extended Inductive Hypothesis* : For each $Q \leq S$ centralizing $\mathcal{F}^\circ$ with $Q \cap \mathcal{Z}(\tau) = \emptyset$, the Inductive Hypothesis is satisfied by $\tau_Q^+ = (\mathcal{F}^{\circ +}, \Omega^+)$, where $(Q\mathcal{F}^\circ)^+ = Q\mathcal{F}^c/Q$.

Observe that if τ is a counter example to Theorem 1 of minimal order, then τ satisfies the Extended Inductive Hypothesis. Rather than work in such a minimal counter example, we often assume instead the Inductive Hypothesis or the Extended Inductive Hypothesis is satisfied. This approach has the virtue, first, of keeping track of which properties of a minimal counter example are actually needed in a given proof, and, second, once Theorem 1 is established, results established under either of the two hypotheses become real theorems.

LEMMA 6.6.8. *Assume $\mathcal{A}(\tau) = \emptyset$ and the Inductive Hypothesis holds. Then Theorem 4 holds in each proper subpacket of τ of the form $(\mathcal{Y}, \Gamma)$ with $\mathcal{Y} \leq \mathcal{F}$ and $\Gamma \subseteq \Omega$.*

PROOF. For $K \in \Omega$ with $z(K) \in \mathcal{F}^f$, $\tau(K)$ is a subpacket of τ satisfying the hypothesis of Theorem 1. Then if $\tau(K)$ is proper Theorem 1 holds for $\tau(K)$ by the Inductive Hypothesis. Hence the lemma follows from 6.6.6. □

LEMMA 6.6.9. *Assume τ satisfies the Extended Inductive Hypothesis.*

(1) If $\mathcal{F}^\circ \leq \mathcal{Y} \leq \mathcal{F}$ with $\mathcal{Y}$ saturated then $(\mathcal{Y}, \Omega)$ is a quaternion fusion packet satisfying the Extended Inductive Hypothesis.

(2) Suppose $Q \in \mathcal{F}^f$ with $\mathcal{F}^\circ \leq \mathcal{N} = N_{\mathcal{F}}(Q)$ and $Q \cap \mathcal{Z}(\tau) = \emptyset$. Set $\mathcal{N}^+ = \mathcal{N}/Q$ and $\rho^+ = (\mathcal{N}^+, \Omega^+)$. Then Q centralizes $\mathcal{F}^\circ$ and ρ^+ is a quaternion fusion packet satisfying the Extended Inductive Hypothesis.

PROOF. Part (1) is straightforward, so it remains to prove (2). Assume the setup of (2). Replacing $\mathcal{F}$ by $\mathcal{N}$ and appealing to (1), we may assume $\mathcal{F} = \mathcal{N}$. By 3.3.2, Q centralizes $\mathcal{F}^\circ$ and τ^+ is a quaternion fusion packet. Suppose $P^+ \leq S^+$ centralizes $\mathcal{F}^{\circ +}$ with $P^+ \cap \mathcal{Z}(\tau^+) = \emptyset$. We may assume $Q \leq P \leq S$. Then as $P^+ \cap \mathcal{Z}(\tau^+) = \emptyset = Q \cap \mathcal{Z}(\tau)$, also $P \cap \mathcal{Z}(\tau) = \emptyset$. Then P centralizes $\mathcal{F}^\circ$ by 3.3.2. Set $(P\mathcal{F}^\circ)^! = P\mathcal{F}^\circ/P$. As $\mathcal{F}$ satisfies the Extended Inductive Hypothesis, we have $\tau_P^! = (\mathcal{F}^{\circ !}, \Omega^!)$ satisfies the Inductive Hypothesis, completing the proof of (2). □

REMARK 6.6.10. Observe that 6.6.6 says that Theorems 1 and 3 imply Theorem 4. Theorem 3 treats the case where $\mathcal{A}(\tau) \neq \emptyset$, and will be proved later in section 7.2. Similarly, with a bit of work, Theorem 1 (together with Theorem 2) implies Theorems 3 through 8, although in practice Theorem 1 is established by partitioning the class of quaternion fusion packets $\tau = (\mathcal{F}, \Omega)$ with $\mathcal{F} = \mathcal{F}^\circ$ transitive on Ω into a number of subclasses corresponding to the hypotheses of Theorems 3 through 8, and treating each subclass by proving a version of the theorem corresponding to that subclass.

From time to time we appeal to the following observation, often without comment:

LEMMA 6.6.11. *Assume $\rho = (\mathcal{Y}, \Gamma)$ is a subpacket of τ with $\mathcal{Y} = \mathcal{Y}^\circ$ and $\Gamma \subseteq \Omega$. Assume $\mathcal{A}(\rho) = \emptyset$ and $\mathcal{E} \trianglelefteq \mathcal{F}$. Then*

(1) If ρ satisfies one of the conclusions to Theorem 1 and $\mathcal{E}$ contains a member of Γ then $\mathcal{Y} \leq \mathcal{E}$.

(2) If ρ satisfies one of the conclusions of Theorem 1 then $\mathcal{Y} \leq \mathcal{F}^\circ$.

(3) If τ satisfies the Inductive Hypothesis and $\mathcal{E}$ contains a member of each orbit of $\mathcal{Y}$ on Γ then $\mathcal{Y} \leq \mathcal{E}$.

(4) If τ satisfies the Inductive Hypothesis then $\mathcal{Y} \leq \mathcal{F}^\circ$.

(5) If τ satisfies the Inductive Hypothesis and ρ is coconnected then $\mathcal{Y} \leq \mathcal{X}$ for some coconnected component $(\mathcal{X}, \Sigma)$ of τ.

PROOF. We first prove (1). Let T be Sylow in $\mathcal{Y}$ and E Sylow in $\mathcal{E}$. Then some $K \in \Gamma$ is contained in $T \cap E$ and as $\mathcal{E} \trianglelefteq \mathcal{F}$ we have $T \cap E$ strongly closed in T with respect to $\mathcal{Y}$. Hence by condition (a)of Remark 5.6.19, we have $T \leq E$. Now by Remark 5.6.19 it suffices to show $\mathcal{B}_*(\mathcal{Y}) \leq \mathcal{E}$. Let $R \in \mathcal{Y}^{fc}$ and $\mathcal{N} = O^{2'}(N_{\mathcal{Y}}(R))$; it suffices to show $\mathcal{N} \leq \mathcal{E}$. Let G be a model for $N_{\mathcal{F}}(R)$; then $O^{2'}(G) = H$ is a model for $\mathcal{N}$. Now $Q = N_T(R)$ is Sylow in G and $H = \langle Q^H \rangle$. As $Q \leq T \leq E$ and $\Lambda = Aut_{\mathcal{E}}(R) \trianglelefteq Aut_{\mathcal{F}}(R)$, we have $Aut_H(R) \leq \Lambda$. Then as $C_G(R) \leq R$ it follows that $\mathcal{N} = \mathcal{F}_Q(H) \leq \mathcal{E}$, completing the proof of (1).

Part (2) follows from (1) with $\mathcal{E} = \mathcal{F}^\circ$.

It remains to prove (3)-(5), so we may assume the Inductive Hypothesis. Suppose $\rho = \tau$. Then (4) and (5) are trivial. Further in (3), $\mathcal{F} = \mathcal{Y} = \mathcal{Y}^\circ$, so as $\mathcal{E} \trianglelefteq \mathcal{F}$ we have $\mathcal{Y} \leq \mathcal{E}$ and hence (3) holds. Therefore we may assume $\tau \neq \rho$.

Next (3) implies (4). Assume the setup of (5). Then $\mathcal{Y} \leq \mathcal{F}^\circ$ by (3). By 6.6.6, $\tau^\circ = \tau_1 * \cdots * \tau_r$ is the central product of its coconnected components $\tau_i = (\mathcal{X}_i, \Sigma_i)$. Now $K \in \Sigma_i$ for some i and $\mathcal{X}_i \trianglelefteq \mathcal{F}^\circ$, so $\mathcal{Y} \leq \mathcal{X}_i$ by (1), completing the proof of (5) and the lemma. □

LEMMA 6.6.12. *Assume τ satisfies the Inductive Hypothesis and $\rho = (\mathcal{Y}, \Gamma)$ is a proper subpacket of τ with $\Gamma \subseteq \Omega$. Then ρ satisfies the Extended Inductive Hypothesis.*

PROOF. Let T be Sylow in $\mathcal{Y}$ and assume $Q \leq T$ centralizes $\mathcal{Y}^\circ$ with $Q \cap \mathcal{Z}(\rho) = \emptyset$. Replacing $\mathcal{Y}$ by $Q\mathcal{Y}^\circ$ we may assume $\mathcal{Y} = Q\mathcal{Y}^\circ$. Set $\mathcal{Y}^+ = \mathcal{Y}/Q$. By 3.3.2, ρ^+ is a quaternion fusion packet, $\mathcal{Y}^+ = (\mathcal{Y}^+)^\circ$, and the map $K \mapsto K^+$ is a bijection of Γ with Γ^+.

We must show ρ^+ satisfies the Inductive Hypothesis, so assume $\xi^+ = (\mathcal{X}^+, \Sigma^+)$ is a quaternion fusion packet with $\Sigma^+ \subseteq \Gamma^+$ and satisfying the hypothesis of Theorem 1. Let $\mathcal{X}$ be the preimage of $\mathcal{X}^+$ and Σ the inverse image of Σ^+ in Γ. Then $\mathcal{X}^+ = (\mathcal{X}^+)^\circ$ so $\mathcal{X}^+ = (\mathcal{X}^\circ)^+$ by 8.10.2 in [**Asc08a**]. Set $\zeta = (\mathcal{X}^\circ, \Sigma)$. As $K \mapsto K^+$ is a bijection of Σ with Σ^+, and $\mathcal{X}^+ = (\mathcal{X}^\circ)^+$ is transitive on Σ^+, we have $\mathcal{X}^\circ$ transitive on Σ, so ζ satisfies the hypohtesis of Theorem 1. Hence as ρ satisfies the Inductive HYpothesis, ζ satisfies one of the conclusions of Theorem 1. But then ζ^+ also satisfies one of the conclusions, so ρ^+ satisfies the Inductive HYpothesis, completing the proof. □

6.7. The proof of Theorem 2

In this section we assume the hypothesis of Theorem 2 from the introduction. Thus $\tau = (\mathcal{F}, \Omega)$ is a quaternion fusion packet, $K \in \Omega$, and $z = z(K) \in O_2(\mathcal{F})$ but $z \notin Z(\mathcal{F})$. In addition $\mathcal{F} = [K]_{\mathcal{F}}$ and $\mathcal{F}$ is transitive on Ω.

LEMMA 6.7.1. $\mathcal{O}_K = \mathcal{F}_K(K)$.

PROOF. Assume otherwise. By hypothesis, $\mathcal{F} = [K]_{\mathcal{F}}$, so by 6.4.3.1, $\mathcal{F}$ is the central product of the members of $\{\mathcal{O}_J : J \in K^{\mathcal{F}}\}$. As $\mathcal{F} = [K]_{\mathcal{F}}$, it follows that $\mathcal{F} = \mathcal{O}_K$, contradicting $z \notin Z(\mathcal{F})$. □

LEMMA 6.7.2. *$K \leq O_2(\mathcal{F}_z)$.*

PROOF. By 2.6.11, $\mathcal{O}_K$ is subnormal in $\mathcal{F}_z$, and by 6.7.1, $\mathcal{O}_K = K$, so $K \leq O_2(\mathcal{F}_z)$. □

LEMMA 6.7.3. *$\mathcal{F}$ is constrained.*

PROOF. This is a consequence of 6.7.1 and 6.4.6.2, given the hypothesis that $\mathcal{F} = [K]_{\mathcal{F}}$. □

By 6.7.3 and 2.8.1, $\mathcal{F}$ has a model G. Set $S_0 = O_2(G)$. As $\mathcal{F} = [K]_{\mathcal{F}}$ we have:

LEMMA 6.7.4. *$G = \langle K^G \rangle$.*

LEMMA 6.7.5. *$K \cap S_0$ is cyclic.*

PROOF. If not by 6.4.4, G is the central product of the members of K^G, so $G = K$ by 6.7.4, contradicting $z \notin Z(G)$. □

LEMMA 6.7.6. *Assume $K \cap S_0 \neq \langle z \rangle$. Then for $\eta \in \eta(\tau)$, $\langle \eta \rangle \trianglelefteq G$ and $\mu(\tau)$ is generated by 3-transpositions.*

PROOF. By 6.7.5, $V = K \cap S_0$ is cyclic, and by assumption, $|V| > 2$. Then by 3.3.12.4, $|K : V| = 2$ and, as $\mathcal{F}$ is transitive on Ω, for $\eta \in \eta(\tau)$, we have $\langle \eta \rangle \trianglelefteq G$. Therefore Hypothesis 4.2.1 is satisfied. As $\mathcal{F}$ is transitive on Ω, G is transitive on η, so by 4.2.10.2, $G = G(V)$, and by 4.2.10.4, $\mathcal{A}(z) = \emptyset$. Then by 4.2.10.3, $\mu(\tau)$ is generated by 3-transpositions. □

LEMMA 6.7.7. *Assume $K \cap S_0 = \langle z \rangle$. Then one of the following holds:*
(1) $G \cong L_3(2)/E_8$.
(2) $G/Z(G) \cong L_3(2)/E_{64}$ or $L_3(2)/2^{3+6}$.
(3) $G/Z(G) \cong AE_n$ for some $n \geq 5$.

PROOF. As $\mathcal{F} = [K]_{\mathcal{F}}$ is transitive on Ω it follows from 6.4.10 that $\mathcal{F} = s(K)$. By 6.7.1, $\mathcal{O}_K = \mathcal{F}_K(K)$. Now 6.4.7 completes the proof. □

Observe that by 6.7.2 and 6.7.3, $K \leq O_2(\mathcal{F}_z)$ and $\mathcal{F}$ is constrained, so the initial statements in Theorem 2 hold. Then G satisfies one of the four conclusions of Theorem 2 by 6.7.6 and 6.7.7.

CHAPTER 7

Theorems 3 and 5

In Chapter 7 we prove Theorems 3 and 5. Let $\tau = (\mathcal{F}, \Omega)$ be a quaternion fusion packet. Recall that Theorem 3 deals with the case where $\mathcal{A}(\tau) \neq \emptyset$, while Theorem 5 determines those τ with $\Omega = \{K\}$ of order 1. If $z \in \mathcal{Z}_S(\tau)$ with $\mathcal{A}(z) \neq \emptyset$ and $\mathcal{F} = [\Omega(z)]_{\mathcal{F}^\circ}$, then either a member of $W(\tau)$ is normal in $\mathcal{F}$ or $\mathcal{F}$ is one of the systems appearing as a conclusion to Theorem 5. The difference is that in the first case, $\mathcal{F}$ is intransitive on $\Omega = \Omega(z) = \{K_1, K_2\}$, while in the second $\Omega = \{K_i\}$ is of order 1 for $i = 1$ or 2. Thus the two results are related. We prove Theorem 5 first in section 7.1, and then reduce Theorem 3 to Theorem 5 in lemma 7.2.18. Indeed at other points in the proof of Theorem 3 we essentially repeat corresponding arguments from the proof of Theorem 5.

As we observed in Remark 6.6.10, given Theorem 3 it can be shown that Theorem 4 follows from Theorem 1; this is Corollary 7.2.27.

7.1. Packets of width 1

In this section we assume the following hypothesis:

HYPOTHESIS 7.1.1. *(1) $\tau = (\mathcal{F}, \Omega)$ is a quaternion fusion packet with $\Omega = \{K\}$ of order 1.*

(2) $z = z(K) \notin Z(\mathcal{F})$.

NOTATION 7.1.2. Let $\mathcal{Z}_{in}$, $\mathcal{Z}_{out}$ consist of those $s \in \mathcal{Z}$ inducing a nontrivial inner, outer automorphism on K, respectively. Set $\mathcal{F}_z = C_{\mathcal{F}}(z)$. Let $\eta \in \eta(\tau)$ be S-invariant and set $W = \langle \eta \rangle$. Let G_W be a model for $N_{\mathcal{F}}(C_S(W))$ and $M = \langle K^{G_W} \rangle$. Let $\mathcal{O} = \mathcal{O}_K$, $\mu = \mu(\tau)$, $V = K \cap W$, and let v be an element of order 4 in V.

LEMMA 7.1.3. *$\mathcal{O} \trianglelefteq \mathcal{F}_z$, so K is strongly closed in S with respect to $\mathcal{F}_z$.*

PROOF. This follows from 7.1.1.1 and 2.6.11. □

LEMMA 7.1.4. *Either*

(1) $\mu \cong \mathbf{Z}_2$ and $W = V$ is of index 2 in K, or

(2) $\mu \cong S_3$ and W is weakly closed in S with respect to $\mathcal{F}$.

PROOF. If $\Delta = \{K\}$ then (1) holds, so assume otherwise. By 7.1.1.1 the graph $\mathcal{D}$ is disconnected, so $\mu \cong S_3$ by 4.2.13. Thus (2) holds by 3.3.14. □

Recall the subsystem $\mathcal{E}(\tau)$ from Definition 3.4.2.

LEMMA 7.1.5. $\mathcal{E}(\tau) = \langle S\mathcal{O}, N_{\mathcal{F}}(K), \mathcal{F}_S(SM) \rangle$.

PROOF. By Definition 3.4.2 and 7.1.1.1,

$$\mathcal{E} = \mathcal{E}(\tau) = \langle S\mathcal{O}, N_{\mathcal{F}}(K), N_{\mathcal{F}}(W)\rangle.$$

If $\mu \cong \mathbf{Z}_2$ then $W = V$ is of index 2 in K by 7.1.4, so $N_{\mathcal{F}}(W) \leq N_{\mathcal{F}}(K)$. If $\mu \cong S_3$ then by 3.3.9, $\mathcal{F}_{S\cap M}(M) \trianglelefteq N_{\mathcal{F}}(W)$, so as $KW = S \cap M$ and K is weakly closed in KW, we conclude that $N_{\mathcal{F}}(W) = \langle \mathcal{F}_S(SM), N_{N_{\mathcal{F}}(W)}(K)\rangle\rangle$ from 1.3.2 in [**Asc19**]. This completes the proof. □

THEOREM 7.1.6. *Assume $z \in O_2(\mathcal{F})$. Then $\mathcal{F}^\circ$ is constrained, $\mu \cong S_3$, and for G a model of $\mathcal{F}^\circ$, either*

(1) $G \cong L_3(2)/E_8$, or

(2) $G = M$ is the split extension of $W \cong \mathbf{Z}^2_{m/2}$ by S_3.

PROOF. By 7.1.1 and the assumption that $z \in O_2(\mathcal{F})$, the hypotheses of Theorem 2 are satisfied by τ°. Then the lemma follows from that theorem. □

THEOREM 7.1.7. *Assume $z \notin O_2(\mathcal{F})$ and $\mu \cong \mathbf{Z}_2$. Then $\mathcal{F}^\circ \cong L_2[2m]^{(1)}$ or $L_3^-[m]$.*

PROOF. As $\mu \cong \mathbf{Z}_2$, $SM = S$ by 7.1.4, so $\mathcal{E} = \mathcal{E}(\tau) = \langle S\mathcal{O}, N_{\mathcal{F}}(K)\rangle$ by 7.1.5. In particular, $z \in Z(\mathcal{E})$, so $\mathcal{E} \neq \mathcal{F}$ by 7.1.1.2. Therefore by 3.4.8 and 3.4.6, $\mathcal{Z} = \{z\} \cup t^K$ with $T = K\langle t\rangle \cong SD_{2m}$, and for $E_4 \cong B \leq T$ we have $Aut_{\mathcal{F}}(B) = Aut(B)$.

Also as $\mu \cong \mathbf{Z}_2$, $\Delta = \{K\}$, so $x^{\mathcal{F}} \subseteq K$ for each $x \in K - \{z\}$.

Claim T is strongly closed in S with respect to $\mathcal{F}$. As T is semidihedral and $\mathcal{Z} = \{z\} \cup t^K$, $\mathcal{Z}$ is the set of involutions in T. Let $x \in T$ with $2 < |x| < m$. Then $x \in K$ and hence $x^{\mathcal{F}} \subseteq K$ by the previous paragraph. Finally let $Y = \langle y\rangle \leq T$ be of order m and $\phi \in \hom_{\mathcal{F}}(Y, S)$. Then $\Phi(Y) \leq K$, so $\Phi(Y)\phi \leq K$ and hence $z\phi = z$, so ϕ is an $\mathcal{F}_z$-map. But $T = K\langle \mathcal{Z}\rangle$, so T is weakly closed in S. Then as K is strongly closed in S with respect to $\mathcal{F}_z$, T/K is weakly closed in S/K with respect to $\mathcal{F}_z/K$ by 8.9.1 in [**Asc08a**], so T is strongly closed in S with respect to $\mathcal{F}_z$. Therefore $Y\phi \leq T$ as ϕ is an $\mathcal{F}_z$-map, completing the proof of the claim.

Let $Q_8 \cong Q \leq T$ and $\mathcal{B} = \langle Aut(B), Aut_{\mathcal{F}}(Q)\rangle_T$. Observe $Aut_{\mathcal{F}}(Q) = Aut(Q)$ if $\mathcal{O} \neq \mathcal{F}_K(K)$ or $m = 8$ and $Aut_{\mathcal{F}}(Q)$ is not a 2-group, while $Aut_{\mathcal{F}}(Q) = Aut_T(Q)$ otherwise. Therefore $\mathcal{B}$ is $\mathcal{F}$-invariant by 2.1.4. By Example 2.1.3, $\mathcal{B} \cong L_2[2m]^{(1)}$ or $L_3^-[m]$. In particular, $\mathcal{B}$ is saturated. Therefore $\mathcal{B}$ is weakly normal in $\mathcal{F}$, so $\mathcal{B} \trianglelefteq \mathcal{F}$ by 1.2.6. Hence $\mathcal{B} = \mathcal{F}^\circ$. □

Given Theorems 7.1.6 and 7.1.7, in the remainder of the section we assume:

HYPOTHESIS 7.1.8. *Hypothesis* 7.1.1 *holds with $z \notin O_2(\mathcal{F})$ and $\mu \cong S_3$.*

LEMMA 7.1.9. *$\mathcal{F} = \mathcal{E}(\tau)$.*

PROOF. If not then by 3.4.8 and 3.4.6, $\mathcal{Z}_{in} = \emptyset$, contrary to $\mu \cong S_3$ in 7.1.8. □

NOTATION 7.1.10. Set $E = \Omega_1(W)$ and let $t \in E - \langle z\rangle$.

LEMMA 7.1.11. *(1) $KW \in Syl_2(M)$ and KW is the wreath product of $\mathbf{Z_{m/2}}$ by $\mathbf{Z_2}$.*

(2) If $s = kc \in \mathcal{Z}_{in}$ with $k \in K$ and $c \in C_S(K)$ then there exists $\gamma \in \hom_{\mathcal{F}_z}(\langle s, k\rangle, W)$.

(3) $\mathcal{F}_z$ *is transitive on* $\mathcal{Z}_{in}$.

(4) M *is determined up to isomorphism.*

(5) For each $y \in C_S(E)$, *there exists an integer* $n(y)$ *such that* $c_y : W \to W$ *is the map* $w \mapsto w^{n(y)}$.

(6) $W \trianglelefteq N_{\mathcal{F}}(E)$.

PROOF. Parts (1) and (4) follow from 4.2.5.

Choose s as in (2). Then $C_K(s) = U = C_K(k)$ is cyclic of order at least 4. Let $\beta \in \mathfrak{A}(s)$. Then $s\beta = z$ and $U\beta \leq D \in \Delta(z\beta)$, so there is $\delta \in Aut_{\mathcal{F}}(W)$ with $D\delta = V$. Then $\gamma = \beta\delta \in \hom_{\mathcal{F}_z}(U\langle s \rangle, W)$, so that (2) holds. Part (3) follows from (1) and (2).

By 7.1.6 there is a complement X to W in M. Let $y \in C_S(E)$. Then y acts on V, so y acts as $u \mapsto u^n$ for some integer n. Then if y centralizes X, y also has this action on W as $W = V \times V^x$ for suitable $x \in X$. Finally $C_S(E) = W \times C_S(X)$ by a Frattini argument, so $y = ws$ for some $w \in W$ and $s \in C_S(X)$, and y and s have the same action on W, completing the proof of (5).

Finally $V = C_K(E) \trianglelefteq C_{\mathcal{F}_z}(E) = C_{\mathcal{F}}(E)$, so $W = \langle V^M \rangle \trianglelefteq C_{\mathcal{F}}(E)$. Now (6) follows from 7.1.4. □

LEMMA 7.1.12. *If* $s \in \mathcal{Z}_{out}$ *then* s *inverts* W.

PROOF. As $s \in \mathcal{Z}_{out}$, s inverts V by 3.3.6, so if s centralizes E then s inverts W by 7.1.11.5, and hence the lemma holds in this case. So assume $[E, s] \neq 1$. Then for $k \in K - V$, $y = sk$ centralizes E and y^2 generates V. Thus by 7.1.11.5, y centralizes W, so $W\langle y \rangle = C_{M\langle y \rangle}(E) \trianglelefteq M\langle y \rangle$, a contradiction as $\langle z \rangle = (W\langle y \rangle)^m$ while M is transitive on $E^{\#}$. □

LEMMA 7.1.13. *Assume* $s \in \mathcal{Z} - E$ *centralizes* E *and set* $A = E\langle s \rangle$. *Then*

(1) s *inverts* W.

(2) One of the following holds:

(i) $s \in \mathcal{Z}_{in}$, $Aut_{\mathcal{F}}(A) = C_{GL(A)}(z)$, $N_K(A)A \cong Q_8^2$ *is normal in* $N_{\mathcal{F}}(A)$, *and writing* $s = kc$ *with* $k \in K$ *and* $c \in C_S(K)$, $k \notin V$ *but* $k^{\mathcal{F}_z} \cap V \neq \emptyset$.

(ii) $s \in \mathcal{Z}_{in}$, $Aut_{\mathcal{F}}(A) = GL(A)$, *and writing* $s = kc$ *with* $k \in K$ *and* $c \in C_S(K)$, $k \notin V$ *but* $k^{\mathcal{F}_z} \cap V \neq \emptyset$.

(iii) $s \in \mathcal{Z}_{out}$, $Aut_{\mathcal{F}}(A) = C_{GL(E)}(e)$ *for some* $e \in \{t, tz\}$, *and for* $h \in M$ *with* $e^h = z$, *the pair* s^h, A^h *is described in (i).*

(3) $A^{\#} \subseteq \mathcal{Z}$.

PROOF. Let $\alpha \in \mathfrak{A}(s)$ with $s\alpha = z$. By 3.3.6, s inverts or centralizes V, so by 7.1.11.5, s inverts or centralizes W. But if s centralizes W then $W\alpha = W$ by 7.1.4.2, contradicting $s \notin W$. Thus (1) holds.

Observe (2) implies (3), so it remains to prove (2). By (1), $Aut_W(A)$ is the group of transvections on A with axis E. Further $A\alpha$ acts on K with $E\alpha$ faithful on K by 7.1.1.1 and as $z \notin E\alpha$, so $\zeta = c_v\alpha^{-*} \in Aut(A)$ is a transvection with center $\langle s \rangle$. Therefore $B = \langle \zeta, Aut_W(A) \rangle$ is the stabilizer in $GL(A)$ of some $e \in E^{\#}$.

Suppose $s = kc \in \mathcal{Z}_{in}$. Then $k \notin V$ by (1) but $k\alpha \in D \in \Delta - \{K\}$, so there exists $\beta \in \mathfrak{A}(k\alpha)$ with $D\alpha\beta = V$. Then $\gamma = \alpha\beta$ is a $\mathcal{F}_z$-map with $k\gamma \in V$, so $k^{\mathcal{F}_z} \cap V \neq \emptyset$. Further k induces a transvection ξ on A with center $\langle z \rangle$ such that ξ centralizes s, so either $\xi \in B$ and $C_E(\xi) = \langle z \rangle$, so that $e = z$, or $\xi \notin B$ and (2ii) holds. In the first case, (2i) or (2ii) holds.

Thus we may assume $s \in \mathcal{Z}_{out}$. If B centralizes z then $\langle s \rangle = [E, \zeta] \leq KC_S(K)$, contradicting $s \in \mathcal{Z}_{out}$. Therefore we may assume B is the stabilizer of t in $GL(A)$. Now there is $h \in M$ with $t^h = z$; replacing s, A by s^h, A^h we are in the case $B = C_{GL(A)}(z)$. As B is transitive on $A - \langle z \rangle$ and $t \in \mathcal{Z}_{in}$, also $s^h \in \mathcal{Z}_{in}$, so (2i) holds after conjugation by h. Thus A satisfies (2iii), completing the proof. □

LEMMA 7.1.14. *If $E_8 \cong A \leq S$ with $A^\# \subseteq \mathcal{Z}$ then $C_{\mathcal{Z}}(A) = A^\#$, so $z \in A$.*

PROOF. Suppose $r \in C_{\mathcal{Z}}(A) - A$. Let $\gamma \in \mathfrak{A}(r)$ and $B = A\gamma$. As $B^\# \subseteq \mathcal{Z}$ and $z \notin B$, B is faithful on K by 7.1.1.1. Now $B_v = C_B(v)$ is of index at most 2 in B and as $B^\# \subseteq \mathcal{Z}$, B_v induces a group of inner automorphism on K. This is a contradiction as $C_K(v) = V$ is cyclic. □

LEMMA 7.1.15. *$\mathcal{Z}_{in} \neq E^\#$.*

PROOF. Assume otherwise. Then if $s \in \mathcal{Z} - E$, we have $s \in \mathcal{Z}_{out}$. Hence by 7.1.12, s centralizes E, so by 7.1.13.2, $\mathcal{Z}_{in}$ is *not* contained in E, a contradiction. Therefore $\mathcal{Z} = E^\#$, so E is a strongly closed abelian subgroup of S, and hence $E \trianglelefteq \mathcal{F}$ by 14.1 in [**Asc11**], contrary to 7.1.8. □

LEMMA 7.1.16. *Set $W_C = C_W(K)$. Then*

(1) W_C is cyclic of order $m/2$ with $\langle z \rangle = W_C \cap V$.

(2) $W_C \leq C_S(\mathcal{O})$.

(3) $C_S(K)$ is strongly closed in S with respect to $\mathcal{F}_z$.

(4) Let $P \leq W_C$ and $\phi \in \hom_{\mathcal{F}_z}(P, S)$. Then if $|P| > 4$ or $\mathcal{Z}_{in} \subseteq KW$, then $P\phi \leq W_C$.

PROOF. Let $k \in K - V$. By 7.1.11.1, KW is a wreath product so $V = [W, k]$ is inverted by k and $V \cong C_W(k) = W_C$, so (1) holds.

From 2.6.12, $\mathcal{C} = C_{\mathcal{F}}(K) = C_{\mathcal{F}}(\mathcal{O})$, so (2) follows and $\mathcal{C} \trianglelefteq \mathcal{F}_z$. Thus (3) holds.

Assume the setup of (4). By (2), $P\phi \leq C_S(K)$. As W_C is cyclic and normal in S, so is P, so $P \in \mathcal{F}_z^f$, and hence there is $\alpha \in \mathfrak{A}_{\mathcal{F}_z}(P\phi)$ with $P\phi\alpha = P$. As $P\phi \leq C_S(K)$, $K \leq C_S(P\phi)$, so as K is weakly closed in S, $K\alpha = K$. Further $PV = V \times P'$ where $P' \leq D \in \Delta(t)$ is of order $|P|/2$, so if $|P| > 4$ then $P'\alpha^{-1} \leq W$ as $W = \langle \Delta - \{K\} \rangle$. Hence

$$P\phi = P\alpha^{-1} \leq (PV)\alpha^{-1} = (P'V)\alpha^{-1} \leq KW.$$

Therefore if $|P| > 4$ then $P\phi \leq C_{KW}(K) = W_C$. Similarly if $|P| = 4$ and $\mathcal{Z} \subseteq KW$ then $P\phi \leq (K\langle \mathcal{Z} \cap KP \rangle)\alpha^{-1} \leq KW$, so $P\phi \leq W_C$, completing the proof of (4). □

LEMMA 7.1.17. *Assume $m = 8$ and set $S_0 = KC_S(K)$ and $T = KW$. Let G_0 be a model for $\mathcal{N} = N_{\mathcal{F}}(S_0)$, and $H_0 = \langle T^{G_0} \rangle$.*

(1) If $W_C \trianglelefteq G_0$ then $T \in Syl_2(H_0)$ and either

(i) $Aut_{\mathcal{F}}(K)$ is a 2-group and $T \trianglelefteq G_0$, or

(ii) $Aut_{\mathcal{F}}(K) \cong S_4$, $L = O^2(H_0) \cong SL_2(3)$ with $K = O_2(L)$ and $H_0 = L\langle x \rangle$ where $x \in T$ of order 8 induces an involutory outer automorphism on L in $PGL_2(3)$.

(2) Set $K_0 = \langle W_C^{G_0} \rangle$ and assume $K_0 \neq W_C$. Then $K_0 \cong Q_8$, $T_0 = TK_0 \in Syl_2(H_0)$, and $T_0 = T\langle a \rangle$, where a is an involution inverting W in the center of the group $C_S(X_0)$ for X_0 a complement to W in M. Moreover one of the following holds:

(i) $Aut_{\mathcal{F}}(K)$ is a 2-group and $H_0 = T_1L_0$ where $L_0 = O^2(H_0) \cong SL_2(3)$, $K_0 = O_2(L_0)$, and T_1 is semidihedral of order 16 with $K = C_{T_1}(L_0)$ and $H_0/K \cong PGL_2(3)$.

(ii) $Aut_{\mathcal{F}}(K) \cong S_4$, $O^2(H_0) = L * L_0$ *where* $L \cong L_0 \cong SL_2(3)$, $K = O_2(L)$, $K_0 = O_2(L_0)$, *and* $H_0 = O^2(H_0)W$ *with* $y \in W - O^2(H_0)$ *inducing an outer automorphism in* $PGL_2(3)$ *on* L *and* L_0.

(iii) $Aut_{\mathcal{F}}(K) \cong S_4$, $O_2(H_0) = KK_0 \cong Q_8^2$, $H_0/O_2(H_0) \cong S_3$, *and* $A = \langle E^{G_0}\rangle \cong E_8$ *with* $A^\# \subseteq \mathcal{Z}$.

PROOF. As $\mathcal{F}$ is saturated, $Aut_{\mathcal{F}}(K) = Aut_{\mathcal{N}}(K) = Aut_{G_0}(K)$. Observe that $S = S_0W$.

As K is weakly closed in S, K is normal in G_0 and then also $C_S(K) \trianglelefteq G_0$. Set $X = N_{G_0}(V) \cap N_{G_0}(W_C)$. Then $E = \Omega_1(VW_C) \trianglelefteq X$, so $W \trianglelefteq X$ by 7.1.11.6. Thus $T = KW \trianglelefteq X$. In particular in (1), if $Aut_{\mathcal{F}}(K)$ is a 2-group then (1i) holds.

Next let $Y = N_{G_0}(W_C)$, set $H_Y = \langle T^Y\rangle$, and assume $Y \neq X$. Then $Aut_Y(K) \cong S_4$. Set $Y^* = Y/W_C$ and observe that

$$[W, C_S(K)] \leq C_W(K) = W_C,$$

so $[W^*, C_S(K)^*] = 1$. Further $K^*W^* \cong D_8$ and as $S = S_0W$ we have $[S^*, W^*] = [K^*, W^*] = V^*$. As $Aut_Y(K) \cong S_4$ and $O_2(Y^*) = S_0^*$ is of index 2 in S^*, it follows that $H_Y^* = K^*\langle W^{*Y^*}\rangle \cong S_4$. Thus H_Y satisfies the conclusions of (1ii), completing the proof of (1).

So assume the hypotheses and notation of (2). Set $\bar{G}_0 = G_0/\langle z\rangle$, $I_0 = C_{G_0}(\bar{K})$, and $I = C_{G_0}(K)$. Observe $I_0 = KI$ with $C_S(K)$ Sylow and normal in I, so $S_0 = KC_S(K)$ is Sylow in I_0. Further $[T, C_S(K)] \leq C_T(K) = W_C$ with $\bar{W}_C \leq Z(\bar{S}_0)$ of order 2, so $\bar{K}_0 = \langle \bar{W}_C^{G_0}\rangle \leq Z(\bar{S}_0)$. By hypothesis, $K_0 \neq W_C$, so $\bar{K}_0 \neq \bar{W}_C$.

Set $G_0^! = G_0/C_{G_0}(\bar{K}_0)$, and let t_1 be an involution in $T - W$. Then $T^! = \langle t_1^!\rangle$ is of order 2 and $[t_1^!, \bar{K}_0] \leq \bar{W}_C$. If $t_1^!$ centralizes $\bar{K}_0$ then $T = KW \leq I_0$, so $G_0 = I_0N_{G_0}(T)$ as T is weakly closed in S. But now $W_C \trianglelefteq G_0$, contrary to assumption. Thus $t_1^!$ induces a transvection on $\bar{K}_0$ with center $\bar{W}_C$. Further $H_0^! = \langle t_1^{!G_0}\rangle$. As $\bar{K}_0 \leq Z(\bar{S}_0)$, $S_0 \leq C_{G_0}(\bar{K}_0)$, so I/S_0, and hence also $I^!$ is of odd order. Next $G_0/I = Aut_{G_0}(K) = Aut_{H_0}(K)$ is either $Aut(K) \cong S_4$ or a Sylow group T/W_C of $Aut(K)$. Therefore $\hat{G}_0 = G_0/S_0 = \langle \hat{t}_1\rangle O(\hat{G}_0)$ with $|O(\hat{G}_0) : \hat{I}| = 1$ or 3. Hence $G_0^! = \langle t_1^!\rangle O(G_0^!)$, and as $t_1^!$ is a transvection on $\bar{K}_0 \neq \bar{W}_C$, we conclude that $H_0^! = G_0^! \cong S_3$ and $m(\bar{K}_0) = 2$. Then as $W_C \cong \mathbf{Z}_4$, we have $K_0 \cong Q_8$ and $K_1 = KK_0 \cong Q_8^2$.

As S_0 centralizes $\bar{K}_1$, $S_0 = K_1S_1$, where $S_1 = C_S(K_1) \trianglelefteq G_0$. Now $[T, S_1] \leq C_T(K_1) = \langle z\rangle$, so T centralizes $\bar{S}_1$.

Set $G_0^+ = G_0/K_1$. Then T^+ is of order 2 and as T is weakly closed in S with respect to G_0, it follows from Glauberman's Z^*-Theorem that $H_0^+ = T^+O(H_0^+)$. Further as T centralizes $\bar{S}_1$, so does H_0, so $O^2(C_{H_0}(K_1)) = O^2(C_{H_0}(S_0)) \leq O^2(S_0) = 1$. Therefore $C_{H_0}(K_1) = \langle z\rangle$. Hence $H_0^+ \leq Out(K_1) \cong O_4^+(2)$.

In particular one of the three cases listed in (2) hold. For example if $Aut_{\mathcal{F}}(K)$ is a 2-group take $T_1 = K\langle t_1\rangle$; as t_1 induces an outer automorphism on K, T_1 is semidihedral.

Finally for w a generator for W_C, $t = wv$, so for $j \in K_0 - W_C$, $[t, j] = [w, j] = z$, so for $k \in K - V$, $b = kj$ is an involution inverting VW_C. Hence by 7.1.11.5, b inverts W. Now a complement X_0 to W in M acts on $\langle b\rangle W = C_{T_0M}(E)$ and $C_{\langle b\rangle W}(X_0) = \langle a\rangle$ with a an involution inverting W. Further $[a, C_S(X)] \leq C_{T_0}(M) = 1$, completing the proof of (2). □

THEOREM 7.1.18. *Assume* $\mathcal{Z}_{in} \subseteq KW$. *Then* $S = KW$ *is wreathed and* $\mathcal{F}^\circ \cong L_3^+[m]$.

PROOF. Let $T = KW$. By 7.1.16.4, W_C is strongly closed in S with respect to $\mathcal{F}_z$, so by 14.1 in [**Asc11**], $W_C \trianglelefteq \mathcal{F}_z$. Thus $W_C\mathcal{O} \trianglelefteq \mathcal{F}_z$.

By 7.1.15 there is $s \in \mathcal{Z}_{in} - E$, but by hypothesis, $\mathcal{Z}_{in} \subseteq KW_C$. Then as W is transitive on the involutions in $T - W$ since T is a wreath product, it follows that $E^{\#} \cup s^W$ is the set I of involutions in T and $I = \{z\} \cup \mathcal{Z}_{in}$. Further $\mathcal{F}_z$ is transitive on $I - \{z\}$ by 7.1.11.3, so V is not normal in $\mathcal{F}_z$ since E centralizes V but s inverts V. Therefore $\mathcal{O} \cong SL_2[m]$ if $m > 8$, while $Aut_{\mathcal{F}}(K) \cong S_4$ if $m = 8$.

Assume for the moment that $m = 8$ and adopt the notation of 7.1.17. It follows from 7.1.17.1 that $L = O^2(H_0) \cong SL_2(3)$ with $K = O_2(L)$ and $H_0 = L\langle x\rangle$ where $x \in T$ of order 8 induces an involutory outer automorphism on L in $PGL_2(3)$. Set $\mathcal{H} = \mathcal{F}_T(H_0)$.

Observe that the hypotheses of 2.1.5 are satisfied by $\mathcal{F}_z, T, KW_C, W$ in the role of "$\mathcal{F}, T, Q, W$". For example $O^2(Aut(K)) \cong A_4 \cong O^2(Aut_{H_0}(K))$ and $\mathcal{N} = N_{\mathcal{F}_z}(W) = C_{N_{\mathcal{F}}(W)}(z)$, so $O^2(Aut_{\mathcal{N}}(T))$ centralizes E, and hence also W, and centralizes T/W of order 2, so the group is indeed trivial. Therefore $\mathcal{H} \trianglelefteq \mathcal{F}_z$ by 2.1.5.

Next assume $m > 8$. In this case set $\mathcal{H} = T\mathcal{O}$. We claim $\mathcal{H} \trianglelefteq \mathcal{F}_z$ in this case too. For as $m > 8$, $T = N_{\mathcal{H}}(KW_C)$. Next as T is weakly closed in S with respect to $\mathcal{F}_z$ and T/KW_C is of order 2, T is strongly closed in $N_{\mathcal{F}_z}(KW_C)$, and hence $T \trianglelefteq N_{\mathcal{F}_z}(KW_C)$ by 14.1 in [**Asc19**]. Hence the claim follows from Theorem 1.5.2 in [**Asc19**].

Note $\mathcal{H} \cong SL_2[m]^{(1)}$ in both cases, as by 2.1.2:

(*) $\mathcal{H} = \langle O^2(Aut(QW_C))\rangle_T$, where Q is any Q_8-subgroup of K.

This follows since, up to conjugation in $\mathcal{H}$, QW_C is the unique member of $\mathcal{H}^{rc}$ in $\mathfrak{D}(\mathcal{H})$.

Next we claim K is the unique quaternion subgroup of its order in T. Namely suppose K_1 is a second such subgroup. As T/E is wreathed, it is transitive on it involutions in $T/E - W/E$, so as $K \trianglelefteq T$, all elements of order 4 in $T - W$ are in K. As K and K_1 are generated by such elements, the claim follows.

By (*) and the second claim, it follows from 2.1.2 that $Aut(\mathcal{H}) = Aut(T)$.

Let $\mathcal{E} = \langle \mathcal{H}, \mathcal{F}_T(M)\rangle$. We next claim that $\mathcal{E}$ is determined up to isomorphism. We just saw $\mathcal{H}$ is determined, and $\mathcal{F}_T(M)$ is determined by 7.1.11.4. Hence as $Aut(\mathcal{H}) = Aut(T)$, the claim follows from 1.6.8 in [**Asc19**]. In particular as $L_3^+[m]$ has such a fusion system, $\mathcal{E} \cong SL_3^+[m]$, so $\mathcal{E}$ is saturated.

From 7.1.5 and 7.1.9, $\mathcal{F} = \langle \mathcal{F}_z, \mathcal{F}_S(SM)\rangle$. Then by reductions above, Hypotheses 1.4.7 in [**Asc19**] is satisfied by $\mathcal{F}$, $\mathcal{E}$, and the families $(\mathcal{F}_z, \mathcal{F}_S(SM))$ and $(\mathcal{H}, \mathcal{F}_T(M))$. Therefore by 1.4.8 in [**Asc19**], $\mathcal{E} \trianglelefteq \mathcal{F}$. But then $\mathcal{F}^\circ = \mathcal{E}$, completing the proof. □

LEMMA 7.1.19. *Let $\mathcal{Z}_V = (\mathcal{Z} - E) \cap VC_S(K)$ and assume $s \in \mathcal{Z}_V$. Then*

(1) s inverts W_C and W is transitive on sW_C.

(2) $\mathcal{Z}_V = sW_C$.

(3) Let $v \in V$ be of order 4 and set $K_0 = W_C\langle sv\rangle$ and $\Omega_0 = \{K_0\}$. Then $\tau_0 = (\mathcal{F}, \Omega_0)$ is a quaternion fusion packet.

(4) If $m > 8$ then $\mathcal{O}_{K_0} \cong SL_2[m]$.

(5) If $m = 8$ then K_0 is Sylow in $\mathcal{E}_0 \trianglelefteq \mathcal{F}_z$ with $\mathcal{E}_0 \cong SL_2[8]$. Further if $Aut_{\mathcal{F}}(K)$ is not a 2-group then K is Sylow in $\mathcal{E}_1 \trianglelefteq \mathcal{F}_z$ with $\mathcal{E}_1 \cong SL_2[8]$.

(6) For $\eta_0 \in \eta(\tau_0)$, $\langle \eta_0\rangle = W$.

(7) Set $T = KW$ and $T_0 = TK_0$. Then $T_0 = T\langle a\rangle$, where a is an involution inverting W in the center of $C_S(X_0)$, for X_0 a complement to W in M.

(8) K and K_0 are the quaternion subgroups of T_0 of order m.

PROOF. As $s \in \mathcal{Z}_V$, $s = vc$ with $v \in V$ of order 4 and $c \in C_S(K)$ with $c^2 = z$. If s centralizes E then conclusion (i) or (ii) of 7.1.13.2 holds by that lemma, so $s = kc$ with $k \notin V$, a contradiction. Thus $[t, c] = [t, s] = z$. Let $k \in K - V$ and $i = kc$. Then i is an involution centralizing E and inverting V, so i inverts W by 7.1.11.5. Therefore as k centralizes W_C, c and s invert W_C. Therefore $K_0 = W_C\langle c\rangle$ is quaternion of order m and $D = W_C\langle s\rangle \cong D_m$. Moreover as $[t, s] = z$, $W\langle s\rangle$ is wreathed, so as s inverts W_C, $W_C = [W, s]$ and W is transitive on sW_C. That is (1) holds.

Suppose $\mathcal{Z}_V \neq sW_C$. As K, V, and E are normal in S, and S acts on $\mathcal{Z}$, S also acts on $\mathcal{Z}_V$. Thus there exists $r \in (N_S(D) \cap \mathcal{Z}_V) - D$. Hence $[s, r] \in W_C$. If $[s, r]$ generates W_C then setting $f = sr$, we have f^2 generating W_C, so by 7.1.11.5, f centralizes W. But now $\langle z\rangle = (W\langle f\rangle)^m$ is characteristic in $W\langle f\rangle$, a contradiction as M acts on $C_{M\langle f\rangle}(W) = W\langle f\rangle$ and is transitive on $E^\#$. Therefore $s^r = s^w$ for some $w \in W_C$, so $wr \in C_S(s)$. But from (1), $wr \in r^W$, so $wr \in \mathcal{Z}$. Also $r = vc_1$ for some $c_1 \in C_S(K)$, so $swr = vcwvc_1 \in C_S(K)$. By 7.1.11.3 there is $\alpha \in \mathfrak{A}_{\mathcal{F}_z}(s)$ with $s\alpha = t$. Then by 7.1.13.3, $(\langle wr, s, z\rangle\alpha)^\# \subseteq \mathcal{Z}$, so $swr \in \mathcal{Z}$, contradicting $swr \in C_S(K)$. Hence (2) holds.

Claim K_0 is strongly closed in S with respect to $\mathcal{F}_z$. If not, by 7.1.16.4=AA.16.4 there is $X = \langle x\rangle$ of order 4 in K_0 and $\phi \in \hom_{\mathcal{F}_z}(X, S)$ with $X\phi \not\leq K_0$. Observe that we can extend ϕ to KX. For by 1.3.2 in [**Asc19**], $\mathcal{F}_z = \langle S\mathcal{O}, N_{\mathcal{F}}(K)\rangle$, and as $X \leq C_S(\mathcal{O})$, each $\theta \in \hom_{S\mathcal{O}}(X, S)$ is induced by conjugation in S, while of course $K \trianglelefteq N_{\mathcal{F}}(K)$. By 7.1.16.3, $X\phi \leq C_S(K)$. By (1), W is transitive on sW_C, so as W centralizes v, also W is transitive on svW_C. Then as $x \in svW_C \cup W_C$, $r = xv \in E \cup sW_C = E \cup \mathcal{Z}_V$ by (2).

Suppose $V\phi = V$. Then $r\phi \in V(x\phi) \leq VC_S(K)$, so $r\phi \in E \cup \mathcal{Z}_V = E \cup sW_C$ by (2), so $x\phi = r\phi(v^{-1}\phi) \in vE \cup vSW_C$, and hence $x\phi \in W_C \cup cW_C = K_0$, contrary to the choice of ϕ. Thus $V\phi \neq V$, so in particular $m = 8$ and $Aut_{\mathcal{F}}(K) \cong S_4$.

Adopt the notation of 7.1.17 and let $W_C = \langle w\rangle$. We showed that $N_{G_0}(V)$ acts on $K_0 \cong Q_8$. Further by (2), $(wc)v = ws = r_0 \in \mathcal{Z}$, so by 7.1.11.2, there is $\gamma \in \hom_{\mathcal{F}_z}(\langle r_0, v\rangle, W)$. Thus γ acts on V and W_C is generated by $(r_0v^{-1})\gamma = (wc)\gamma$. As $W_C \trianglelefteq S$, we have $W_C \in \mathcal{F}^f$, so there is $\alpha \in \mathfrak{A}_{N_{\mathcal{F}_z}(V)}(\langle wc\rangle)$ with $W_C = \langle (wc)\alpha\rangle$. We can choose $\alpha = c_g$ for some $g \in N_{G_0}(V)$ with $(wc)^g \in W_C$. Therefore by 7.1.17.2, $\langle W_C^{N_{G_0}(V)}\rangle = K_0 \cong Q_8$. Thus conclusion (ii) of 7.1.17.2 holds as $Aut_{\mathcal{F}}(K) \cong S_4$ and $\langle W_C^{N_{G_0}(V)}\rangle \cong Q_8$. Hence there is $l \in L$ with $\varphi = c_l\phi$ fixing v and $x\varphi = x\phi$, so replacing ϕ by φ, we obtain the contradiction $V\varphi = V$. This completes the proof of the claim. Further (3) follows from the claim.

Again by 7.1.11.2 applied to $s = vc$, $c^{\mathcal{F}_z} \cap W_C \neq \emptyset$, so $Aut_{\mathcal{F}}(K_0)$ is transitive on its elements of order 4. Then (4) follows from (3) and 2.5.4 and 2.5.7 applied to K_0. So assume $m = 8$. Then $K_0 \trianglelefteq \mathcal{F}_z$ and by another application of 7.1.11.2, there exists $\gamma \in Aut_{\mathcal{F}}(KK_0)$ with $v\gamma = v$ and $c\gamma = w$ a generator of W_C. Again the hypotheses of 7.1.17.2 are satisfied with γ induced by some g in the group G_0 of that lemma. If follows that conclusion (i) or (ii) of the lemma is satisfied, so in particular there is $SL_2(3) \cong L_0 \trianglelefteq H_0$ with $K_0 = O_2(L_0)$. Further L_0 centralizes $C_S(K_0)$, and therefore also $C_{\mathcal{F}_z}(K_0)$ by 2.3 in [**Asc11**]. Thus $\mathcal{E}_0 = \mathcal{F}_{K_0}(L_0) \trianglelefteq \mathcal{F}_z$. Similarly if $Aut_{\mathcal{F}}(K) \cong S_4$ then conclusion (ii) of 7.1.17.2 holds and $\mathcal{E}_1 = \mathcal{F}_K(L) \triangleleft \mathcal{F}_z$, completing the proof of (5).

Observe W_C is the member of $\eta(\tau_0)$ containing z, so as $W = \langle W_C^M \rangle$, (6) follows from 3.4.3.1 applied to the quaternion fusion packet τ_0.

The proof of (7) is the same as that of the corresponding statement in 7.1.17, appearing in the last paragraph of the proof of 7.1.17.

We next prove (8). Suppose Q is a quaternion subgroup of T_0 of order m. As W is abelian, $Q \not\leq W$ and $Q \cap W$ is cyclic. From (7), all members of aW are involutions, so $Q \cap \langle a \rangle W = Q \cap W$. From (7), $T_0/W \cong E_4$, so $|Q : Q \cap W| = 2$ and $QW = KW$ or K_0W. Thus we may take $QW = KW$, and now (8) follows from the step in the proof of 7.1.18 showing K is the unique quaternion subgroup of KW of order m. □

NOTATION 7.1.20. Suppose $\mathcal{Z}_V \neq \emptyset$ and define K_0 and Ω_0 as in 7.1.19.3. If $m > 8$ define $\mathcal{E}_0 = \mathcal{O}_{K_0}$ and $\mathcal{E}_1 = \mathcal{O}_K$. If $m = 8$ define $\mathcal{E}_0$ as in 7.1.19.5, set $\mathcal{E}_1 = K$ if $Aut_{\mathcal{F}}(K)$ is a 2-group, and if $Aut_{\mathcal{F}}(K)$ is not a 2-group, define $\mathcal{E}_1$ as in 7.1.19.5. Thus in any case K_0 is Sylow in $\mathcal{E}_0 \cong SL_2[m]$, K is Sylow in $\mathcal{E}_1 \cong SL_2[m]$ or K, and $\mathcal{E}_0$ and $\mathcal{E}_1$ are normal in $\mathcal{F}_z$.

Also define $T_0 = KK_0W$ as in 7.1.19.7 and set $M_0 = T_0M$ and $\mathcal{H} = T_0\mathcal{E}_0\mathcal{E}_1$. Sometimes we partition the proof into two cases: In Case I, $\mathcal{E}_1 \cong SL_2[m]$, while in Case II, $\mathcal{E}_1 = K$.

LEMMA 7.1.21. *Assume $\mathcal{Z}_V \neq \emptyset$ and let $L = G_2(q)$ for some odd prime power q with $(q^2-1)_2 = m$. Then*

(1) T_0 is isomorphic to a Sylow 2-subgroup of L, so we may take $T_0 \in Syl_2(L)$. Set $\mathcal{F}(L) = \mathcal{F}_{T_0}(L)$.

(2) $\tau_L = (\mathcal{F}(L), \Omega)$ is a quaternion fusion packet with $\mathcal{Z}_V \neq \emptyset$ appearing in Case I, and $\mathcal{F}(L) \cong G_2[m]$.

(3) We may choose q to be a power of 3. In that event there is a graph-field automorphism γ of L acting on T_0 inducing an automorphism of $\mathcal{F}(L)$ interchanging K and K_0.

(4) $T_0 \leq X \leq L$ with X the extension of $SL_3^\epsilon(q)$ by a graph automorphism, where $q \equiv \epsilon \mod 4$. Further, choosing K_0 Sylow in a fundamental subgroup of L, $\tau_X = (\mathcal{F}(X), \Omega)$ is a quaternion fusion packet with $\mathcal{Z}_V \neq \emptyset$, $K \triangleleft C_{\mathcal{F}(X)}(z)$, and $Aut_{\mathcal{F}(X)}(K)$ a 2-group, so the packet appears in Case II. Moreover $\mathcal{F}(X) \cong L_3^+[m]^{(1)}$.

PROOF. Let $T_L \in Syl_2(L)$ and $\langle z_L \rangle = Z(T_L)$. Then $O^2(C_L(z_L)) = L_0 * L_1$ with $L_i \cong SL_2(q)$ normal in $C_L(z_L)$, so by 2.4.1, $\tau_L = (\mathcal{F}_{T_L}(L), \{K_L\})$ is a quaternion fusion packet, where $K_L \in Syl_2(L_1)$. As L has one class of involutions, $\mathcal{Z}_V \neq \emptyset$ in this example, and as $L_i \cong SL_2(q)$, the packet appears in Case I. By 7.1.11.4 and 7.1.19.7, T_L is determined up to isomorphism by these conditions, so $T_0 \cong T_L$. Thus we have proved (1) and (2).

Observe for $r \equiv 1 \mod 4$, $(r^2-1)_2 = (r+1)_2 \cdot (r-1)_2 = 2(r-1)_2$. Thus $(3^{2^{k+1}} - 1)_2 = (9^{2^k} - 1) = 2^k \cdot 8$, proving the first statement in (3). From 2.5.13 in [**GLS98**], L admits a graph-field automorphism γ which we may choose to be a 2-element, and hence acting on T_0 by Sylow's Theorem. From the discussion in [**GLS98**], γ interchanges long and short root subgroups of L, and hence K and K_0, completing the proof of (3).

Next from Example 2.6.3 in [**GLS98**] (and with more detail in [**Asc87**]), the subgroup X of (4) exists. As K_0 is Sylow in a fundamental subgroup of X, $K \trianglelefteq C_{\mathcal{F}(X)}(z)$ and $Aut_{\mathcal{F}(X)}(K)$ is a 2-group. By 2.5.2, $\tau_X = (\mathcal{F}(X), \{K\})$ is a quaternion fusion packet, and as $E(X)$ has one class of involutions, $\mathcal{Z}_V \neq \emptyset$ in this example. By definition, $\mathcal{F}(X) \cong L_3^+[m]^{(1)}$, completing the proof of the lemma. □

LEMMA 7.1.22. *Assume* $\mathcal{Z}_V \neq \emptyset$.

(1) $\mathcal{F}_{T_0}(M_0) \trianglelefteq N_{\mathcal{F}}(E)$.

(2) $T_0 = C_{M_0}(z) \trianglelefteq N_{\mathcal{F}_z}(E)$.

(3) $Aut_{\mathcal{F}_z}(T_0)$ *is a 2-group.*

(4) $\mathcal{H} \trianglelefteq \mathcal{F}_z$.

(5) In Case I, $\mathcal{H} \cong C_{\mathcal{F}(L)}(z)$ *and* $Aut(T_0) = Aut(\mathcal{H})$.

(6) In Case II, $\mathcal{H} \cong C_{\mathcal{F}(X)}(z)$ *and* $Aut(T_0) = Aut(\mathcal{H})Aut(\mathcal{F}_{T_0}(M_0))$.

(7) Let $\mathcal{E} = \langle \mathcal{H}, \mathcal{F}_{T_0}(M_0)\rangle$. *Then* $\mathcal{E}$ *is isomorphic to* $G_2[m]$ *in Case I, and to* $L_3[m]^{(1)}$ *in Case II.*

PROOF. Recall from 7.1.11 that $T \in Syl_2(M)$ and $\mathcal{F}_T(M) \trianglelefteq N_{\mathcal{F}}(E) = \mathcal{M}$. As $\langle z\rangle = Z(T)$, $N_M(T) \leq C_M(z)$. As $W \trianglelefteq \mathcal{M}$ and $\mathcal{E}_0$ and $\mathcal{E}_1$ are normal in $\mathcal{F}_z$, we have $KK_0 \trianglelefteq C_{\mathcal{M}}(z)$, so $T_0 = KK_0W \trianglelefteq C_{\mathcal{M}}(z)$, proving (2). Then (1) follows from Theorem 1.5.2 in [**Asc19**].

Set $A = O^2(Aut_{\mathcal{F}_z}(T_0))$. By 7.1.4.2, A acts on W. Then as A centralizes z, A also centralizes E, so A centralizes W (cf. 24.3 in [**Asc86**]). Also A acts on K and K_0 and centralizes $K \cap W$ and $K_0 \cap W$, so A centralizes K and K_0. Therefore A centralizes $KK_0W = T_0$, establishing (3).

Assume for the moment that $m = 8$. Claim the hypotheses of 2.1.5 are satisfied on T_0 in $\mathcal{F}_z$ with $Q = KK_0$. For example hypothesis (b) follows from 7.1.17.2 with $O^2(Aut_{H_0}(Q)) = O^2(Aut(Q))$ in Case I and isomorphic to A_4 in Case II. Hypothesis (c) follows from (3). Then (4) follows when $m = 8$ from the claim and 2.1.5, which also says that Q is the greatest member of $\mathfrak{D}(T_0, \mathcal{F}_z)$ and $\mathcal{H} = \langle O^2(Aut_{\mathcal{F}}(Q))\rangle$ as $\langle z\rangle \trianglelefteq N_{\mathcal{F}}(Q)$.

Next suppose that $m > 8$. Then $N_{\mathcal{F}_z}(Q) \leq N_{\mathcal{F}_z}(E)$, so $T_0 \trianglelefteq N_{\mathcal{F}_z}(Q)$ by (2). Hence (4) follows from 1.5.2 in [**Asc19**] when $m > 8$. This completes the proof of (4).

We next prove (5) and (6). Suppose $m = 8$. We saw Q is the greatest member of $\mathfrak{D}(T_0, \mathcal{F}_z)$ and $\mathcal{D} = \langle O^2(Aut_{\mathcal{F}}(Q))$. In Case I, $O^2(Aut_{\mathcal{F}}(Q)) = O^2(Aut(Q))$, so $\mathcal{H}$ is determined up to isomorphism, and hence is isomorphic to $C_{\mathcal{F}(L)}(z)$ by 7.1.21.2. Moreover $Aut(T_0) = Aut(\mathcal{H})$ by 2.1.2. Similarly in Case II, given 7.1.19.8, 2.1.2 says that $Aut(\mathcal{D}) = N_{Aut(T_0)}(K)$ is of index 2 in $Aut(T_0)$, so by 7.1.21.4, $\mathcal{D} \cong C_{\mathcal{F}(X)}(z)$ or $C_{\mathcal{F}(X)}(z)\rho$ where ρ is the automorphism of $\mathcal{F}(L)$ in 7.1.21.3. As $\rho \in Aut(\mathcal{F}_{T_0}(M_0))$, we have $Aut(T_0) = Aut(\mathcal{H})Aut(\mathcal{F}_{T_0}(M_0))$. This completes the proof of (5) and (6) when $m = 8$.

So assume $m > 8$. This time in Case I a set $\mathcal{U}$ of representatives for the $\mathcal{H}$-orbits on $\mathcal{H}^{rc}$ are $\{KQ_0, QK_0, QQ_0\}$, where Q, Q_0 are Q_8-subgroups of K, K_0, respectively. By 2.1.2, $\mathcal{H} = \langle O^2(Aut(U)) : U \in \mathcal{U}\rangle$, so once again $\mathcal{H} \cong C_{\mathcal{F}(L)}(z)$ by 7.1.21.2 and $Aut(T_0) = Aut(\mathcal{H})$ by 2.1.2 and 7.1.19.8. This completes the proof of (5). Finally in Case II, $\mathcal{U} = \{KQ_0\}$, and then arguing as in the case $m = 8$, (6) also holds when $m > 8$.

Finally we prove (7) via an appeal to 1.6.8 in [**Asc19**], applied to $\mathcal{E}$ in the role of $\mathcal{F}$, $\mathcal{H}$ in the role of $\mathcal{F}_1$, and $\mathcal{F}_{T_0}(M_0)$ in the role of $\mathcal{F}_2$. By definition, $\mathcal{E} = \langle \mathcal{H}, \mathcal{F}_{T_0}(M_0)\rangle$, so Hypothesis 1.6.1 of [**Asc19**] is satisfied and $\mathcal{E}$ is a member of the set Λ defined there. By 7.1.11.4 and 7.1.17.2, $\mathcal{F}_{T_0}(M_0)$ is determined up to isomorphism, while by (5) and (6), $\mathcal{H}$ is also determined up to isomorphism. Further by 7.1.21, $\mathcal{F}(L)$, $\mathcal{F}(X)$ is in Λ in Case I, Case II, respectively. Thus it suffices to show all members of Λ are isomorphic, which is the conclusion of 1.6.8 in [**Asc19**] under either of its two hypotheses. Thus it remains to observe that one of those hypotheses does indeed hold by (5) and (6). □

THEOREM 7.1.23. *Assume $\mathcal{Z}_V \neq \emptyset$. Then either*
(1) Case I holds and $\mathcal{F}^\circ \cong G_2[m]$, or
(2) Case II holds and $\mathcal{F}^\circ \cong L_3^+[m]^{(1)}$.

PROOF. We will show that Hypothesis 1.4.7 in [**Asc19**] is satisfied by $\mathcal{F}$, $\mathcal{E}$, and the families $(\mathcal{F}_z, \mathcal{M})$ and $(\mathcal{H}, \mathcal{F}_{T_0}(M_0))$. Then by Lemma 1.4.8 in [**Asc19**] and 1.2.6, $\mathcal{E} \trianglelefteq \mathcal{F}$. But then $\mathcal{F}^\circ = \mathcal{E}$, so that 7.1.22.7 completes the proof. So it remains to verify Hypothesis 1.4.7 in [**Asc19**].

By 7.1.5 and 7.1.9, $\mathcal{F} = \langle \mathcal{F}_z, \mathcal{M}\rangle$, so Hypothesis 1.4.1 in [**Asc19**] is satisfied. By 7.1.22.4, $\mathcal{H} \trianglelefteq \mathcal{F}_z$ and by 7.1.21.1, $\mathcal{F}_{T_0}(M_0) \trianglelefteq \mathcal{M}$. As E and $\langle z\rangle$ are characteristic in T_0, $Aut_{\mathcal{F}}(T_0) = Aut_{\mathcal{M}}(T_0) = Aut_{\mathcal{F}_z}(T_0)$, so $Aut_{\mathcal{F}}(T_0)$ is $Aut(\mathcal{H})$ and $Aut(\mathcal{F}_{T_0}(M_0))$ by parts (1) and (4) of 7.1.22. By construction, $\mathcal{E} = \langle \mathcal{H}, \mathcal{F}_{T_0}(M_0)\rangle$. By 7.1.22.7, $\mathcal{E}$ is saturated. This completes our verification of Hypothesis 1.4.7 in [**Asc19**] and hence also the proof of the lemma. □

Because of 7.1.18, and 7.1.23, in the remainder of the section we assume:

HYPOTHESIS 7.1.24. *Hypothesis* 7.1.8 *holds with $\mathcal{Z}_{in}$ not contained in KW and $\mathcal{Z}_V = \emptyset$.*

In addition we adopt the following notation:

NOTATION 7.1.25. Set $S_0 = KC_S(K)$ and $T = KW$. Let G_0 be a model for $N_{\mathcal{F}}(S_0)$, $H_0 = \langle T^{G_0}\rangle$, $K_0 = \langle W_C^{G_0}\rangle$, $Q = KK_0$, and $T_0 = TK_0$. Let X_0 be a complement to W in M. Set $\mathcal{H}_0 = \mathcal{F}_{T_0}(M_0)$, $\mathcal{H} = \mathcal{F}_{T_0}(H_0)$, and $\mathcal{E} = \langle \mathcal{H}, \mathcal{H}_0\rangle$.

LEMMA 7.1.26. *(1) $W_C \trianglelefteq C_{\mathcal{F}}(K)$.*
(2) V is not normal in $\mathcal{F}_z$.
(3) $m = 8$.
(4) $T_0 \in Syl_2(H_0)$ and $T_0 = T\langle a\rangle$, where a is an involution inverting W and centralizing X_0. Further $K_0 \cong Q_8$, $O_2(H_0) = Q \cong Q_8^2$, $H_0/Q \cong S_3$, and $A = \langle E^{G_0}\rangle \cong E_8$.
(5) $K_0 \trianglelefteq \mathcal{F}_z$.
(6) $\mathcal{H} = \mathcal{F}_{T_0}(H_0) \trianglelefteq \mathcal{F}_z$.

PROOF. Suppose (1) fails and let $P \leq W_C$ and $\beta \in \hom_{C_{\mathcal{F}}(K)}(P, C_S(K))$ with $P\beta \not\leq W_C$. As β is a $C_{\mathcal{F}}(K)$-map, β extends to $\rho \in \hom_{\mathcal{F}_z}(KP, S)$ centralizing K. Now $t = vw$ with $v \in V$ and $w \in P$, so $t\rho = v\rho w\beta = v(w\beta) \in VC_S(K) - VW_C$, contradicting the condition in Hypothesis 7.1.24 that $\mathcal{Z}_V = \emptyset$. So (1) holds.

By Hypothesis 7.1.24, $\mathcal{Z}_{in}$ is not contained in KW, so there is $s = kc \in \mathcal{Z}_{in}$ with $k \in K$ and $c \in C_S(K) - W_C$. By 7.1.11.2 there is $\gamma \in \hom_{\mathcal{F}_z}(\langle s,k\rangle, W)$; in particular $t\gamma^{-1} = s$ or sz.

Suppose $V \trianglelefteq \mathcal{F}_z$. Then $t\gamma^{-1} = v\gamma^{-1}w\gamma^{-1} \in VC_S(K)$, so as $t\gamma^{-1} \in \{s, sz\}$ and $s = kc$, it follows that $k \in V$. Hence $s \in VC_S(K)$, so as $\mathcal{Z}_V$ is empty it follows that $s \in E$, contradicting the choice of s. This proves (2).

Suppose $m > 8$. Then by (2) and as W interchanges the two classes of Q_8-subgroups of K, $\mathcal{O} \cong SL_2[m]$. From 2.6.12, $C_S(K) = C_S(\mathcal{O})$, so as $\mathcal{O} \cong SL_2[m]$, there is $\psi \in Aut_{\mathcal{F}_z}(K\langle c\rangle)$ centralizing c with $k\psi \in V$. But then $s\psi = k\psi c\psi = k\psi c \in \mathcal{Z}_V$, a contradiction. This establishes (3).

By (3), $KC_{\mathcal{F}}(K) \trianglelefteq \mathcal{F}_z$, so $\gamma = \varphi\phi$ with $\varphi \in Aut_{\mathcal{F}_z}(S_0)$ and

$$\phi \in \hom_{KC_{\mathcal{F}}(K)}(\langle s,k\rangle\varphi, W).$$

By (1), $VW_C \trianglelefteq KC_{\mathcal{F}}(K)$, so $\langle s,k\rangle\varphi \leq VW_C$. In particular $c\varphi \in C_W(K) = W_C$, so $W_C\varphi \neq W_C$. But $\varphi = c_{g|\langle s,k\rangle}$ for some $g \in G_0$, so W_C is not normal in G_0; that is $K_0 \neq W_C$, so the hypotheses of 7.1.17.2 are satisfied. Then by (1) and (2), conclusion (iii) of that lemma holds, establishing (4).

By (1), $W_C\mathcal{F}_z = W_C^{N_{\mathcal{F}_z}(S_0)} = W_C^{G_0}$, so as $K_0 \trianglelefteq G_0$ it follows that $K_0 = \langle W_C^{\mathcal{F}_z}\rangle$, so (5) holds.

Next from (1)-(5), the hypotheses of 2.1.5 are satisfied with $\mathcal{F}_z, T_0, KK_0, W$ in the roles of "$\mathcal{F}, T, Q, W$". It follows from 2.1.5 that (6) holds. □

LEMMA 7.1.27. *(1) $\mathcal{H}$ and $\mathcal{H}_0$ are determined up to isomorphism.*

(2) $\mathcal{H}_0 \trianglelefteq N_{\mathcal{F}}(E)$.

(3) $Aut(\mathcal{H}_0) \leq Aut(\mathcal{H})$ and $Aut(\mathcal{H})$ is of index 2 in $Aut(T_0)$.

(4) Either

(i) $a \in A$, $A^\# = z^{\mathcal{E}}$, and $\mathcal{E} = \mathcal{F}_{T_0}(L) \cong L_3(2)/E_8$, where L is a nonsplit extension of E by $L_3(2)$ with Sylow group T_0, or

(ii) $a \notin A$ and $\mathcal{E} \cong M_{12}$.

PROOF. First M is determined up to isomorphism by 7.1.11.4. Then $M\langle a\rangle$ is determined as a inverts W and centralizes X_0. Thus $\mathcal{H}_0$ is determined. Then the proof of 7.1.22.1 establishes (2).

Let $\tilde{H}_0 = H_0/\langle z\rangle$ and $H_0^+ = H_0/Q$. Recall $\mathcal{H} = \mathcal{F}_{T_0}(H_0)$. As there exists an involution in $aW - Q$, H_0 splits over Q. Further $H_0^+ \leq Out(Q) \cong O_4^+(2)$, and H_0^+ is determined up to conjugacy as an S_3-subgroup acting on each of the Q_8-subgroups K and K_0 of Q, and with $C_{\tilde{Q}}(O(H_0^+)) = 1$. Therefore H_0, and hence also $\mathcal{H}$, is determined up to isomorphism, completing the proof of (1).

Observe $\tilde{Q} = J(\tilde{T}_0)$, so Q is characteristic in T_0. Let $D = Aut(T_0)$ and $B = Aut(Q)$. Then $[C_D(Q), T_0] \leq C_{T_0}(Q) = \langle z\rangle$, so $C_D(Q) = \langle\alpha\rangle$ where $y\alpha = yz$ for $y \in T_0 - Q$. Observe α extends to $\hat{\alpha} \in Aut(H_0)$, where $\hat{\alpha}$ centralizes $O^2(H_0)$ and $h\hat{\alpha} = hz$ for $h \in H_0 - O^2(H_0)$.

Next the restriction $\zeta : \beta \mapsto \beta_{|Q}$ is a homomorphism of D into B with kernel $\langle\alpha\rangle$. Let $c : H_0 \to B$ be the conjugation map. Then $Inn(T_0)\zeta = T_0c$ is of order 2^5 with $N_B(T_0c) = B_0$ a Sylow group of B. Further $B_1 = N_{B_0}(H_0c)$ is of index 2 in B_0. We conclude that $|D : Aut(\mathcal{H})| = 2$.

We now apply results in section 1.6 of [**Asc19**] to $\mathcal{E}, \mathcal{H}, \mathcal{H}_0, T_0$ in the roles of "$\mathcal{F}, \mathcal{F}_1, \mathcal{F}_2, S$" in Hypothesis 1.6.1 of [**Asc19**]. By the previous paragraph, $A_1 = Aut(\mathcal{H})$ is of index 2 in $D = Aut(T_0)$, so either $A_2 = Aut(\mathcal{H}_0) \leq A_1$ and $\Delta =$

$A_1 \backslash D/A_2 = \{A_1, A_1\gamma\}$ for $\gamma \in D - A_1$, or $A_2 \not\leq A_1$, so that $D = A_1A_2$ and $|\Delta| = 1$. In the second case, $\mathcal{E}$ is determined up to isomorphism by 1.6.8 in [**Asc19**], while in the first case there are at most two possibilities for $\mathcal{E}$ by 1.6.5.4 and 1.6.6 in [**Asc19**]. Thus to complete the proof of (3), it suffices to exhibit two nonisomorphic systems in the set Λ of 1.6.2 of [**Asc19**]. Hence (4) implies (3).

In $\bar{G} = G_2(3)$ (cf. [**Asc87**]) there is a subgroup Y which is an extension of E_8 by $L_3(2)$ and contains a Sylow 2-group of $\bar{G}$, which we may take to be T_0 by 7.1.21.1. Then by 2.5.2, $\tau_Y = (\mathcal{F}_{T_0}(Y), \{K\})$ is a quaternion fusion packet satisfying Hypothesis 7.1.24, supplying our first example.

Next let $\hat{G}$ be M_{12} and $\hat{S}$ a Sylow 2-subgroup of $\hat{G}$. Then for $\hat{K}$ a Q_8-subgroup of $O_2(C_{\hat{G}}(Z(\hat{S})))$, $\hat{\tau} = (\mathcal{F}_{\hat{S}}(\hat{G}), \{\hat{K}\})$ is a quaternion fusion packet satisfying Hypothesis 7.1.24, supplying us with our second example. This completes the proof of (4) and the lemma. □

THEOREM 7.1.28. *Assume Hypothesis* 7.1.24. *Then* $\mathcal{F}^\circ \cong M_{12}$.

PROOF. We apply 1.4.8 in [**Asc19**] to $\mathcal{F}$ and the families $(\mathcal{F}_z, \mathcal{M})$, $(\mathcal{H}, \mathcal{H}_0)$ of Hypothesis 1.4.1 in [**Asc19**] to show that $\mathcal{E} \trianglelefteq \mathcal{F}$. Then $\mathcal{F}^\circ = \mathcal{E}$. Now by 7.1.27.4, either $z \in O_2(\mathcal{F})$ or the lemma holds, and by 7.1.8 it is the latter.

By 7.1.5 and 7.1.9, $\mathcal{F} = \langle \mathcal{F}_z, \mathcal{M}\rangle$, so Hypothesis 1.4.1 in [**Asc19**] is indeed satisfied.

By 7.1.26.6 and 7.1.27.2, 1.4.7.1 in [**Asc10**] is satisfied. As $\langle z\rangle$ and M are characteristic in T_0, 1.4.7.2 holds. By definition, $\mathcal{E} = \langle \mathcal{D}, \mathcal{H}_0\rangle$, so 1.4.7.3 holds. By 7.1.27.4, 1.4.7.4 holds. Thus we have verified Hypothesis 1.4.7 in [**Asc19**]. Therefore $\mathcal{E} \trianglelefteq \mathcal{F}$ by 1.2.6 and 1.4.8 in [**Asc19**], completing the proof. □

THEOREM 7.1.29. *Assume Hypothesis* 7.1.1 *with* $z \notin O_2(\mathcal{F})$. *Then* $\mathcal{F}^\circ$ *is isomorphic to one of* $L_2[2m]^{(1)}$, $L_3^\epsilon[m]$, $L_3^+[m]^{(1)}$, $G_2[m]$, *or* M_{12}.

PROOF. By Theorem 7.1.7, we may assume that $\mu \cong S_3$, so Hypothesis 7.1.8 is satisfied. Then by Theorems 7.1.18 and 7.1.23, we may assume Hypothesis 7.1.24 holds. But now Theorem 7.1.28 completes the proof. □

REMARK 7.1.30. Observe that Theorems 7.1.6 and 7.1.29 imply Theorem 5. For the hypothesis of Theorem 5 is just Hypothesis 7.1.1 together with the condition that $\mathcal{F} = \mathcal{F}^\circ$, so we can appeal to the results in this section. If $z \in O_2(\mathcal{F})$ then conclusion (1) or (2) of Theorem 5 holds by Theorem 7.1.6. Therefore we may assume $z \notin O_2(\mathcal{F})$. Then Theorem 5 follows from Theorem 7.1.29.

7.2. $\mathcal{A}(z) \neq \emptyset$

In this section we assume the following hypothesis:

HYPOTHESIS 7.2.1. $\tau = (\mathcal{F}, \Omega)$ *is a quaternion fusion packet,* $K \in \Omega$, $z = z(K) \in \mathcal{F}^f$, *and* $\mathcal{A}(z) \neq \emptyset$.

LEMMA 7.2.2. *(1)* $\Omega(z) = \{K, \tilde{K}\}$ *is of order 2.*

(2) There exists a unique 4-subgroup E *of* S *containing* z *such that* $\mathcal{A}(z) \cap \mathcal{Z}_\Delta = E - \langle z\rangle$.

(3) $\mathcal{D}^*(z) - \mathcal{D}(z) = \mathcal{A}(z) = \mathcal{Z} \cap O(z) - \{z\}$.

(4) $\mathcal{F}_z$ *is transitive on* $\mathcal{A}(z)$.

PROOF. Part (1) is 3.2.2.1, while parts (2)-(4) appear in 3.2.4. □

The following theorem is the main result of this section:

THEOREM 7.2.3. *Assume Hypothesis 7.2.1 and let* $\mathcal{F}_0 = [O(z)]_{\mathcal{F}}$, $\Omega_0 = K^{\mathcal{F}} \cup \tilde{K}^{\mathcal{F}}$, *and* $z^{\mathcal{F}} \cap \mathcal{Z}_S = \{z_i : i \in I\}$. *For* $i \in I$, *set* $\mathcal{F}_i = [O(z_i)]_{\mathcal{F}_0}$ *and* $\tau_i = (\mathcal{F}_i, \Omega(z_i))$. *Then*

(1) For each $i \in I$, τ_i *is a quaternion fusion packet with* $\mathcal{A}(z_i) \neq \emptyset$ *and* $\mu(\tau_i) \cong D_{12}$.

(2) $\mathcal{F}_0 = \prod_{i \in I} \mathcal{F}_i$ *is a direct product.*

(3) For each $i \in I$, *one of the following holds:*

(i) $W_i \trianglelefteq \mathcal{F}_i$ *for* $W_i \in W(\tau_i)$.

(ii) $\mathcal{F}_i \cong L_3(2)/E_8$.

(iii) $\mathcal{F}_i \cong L_3[m]^{(1)}$.

(iv) $\mathcal{F}_i \cong G_2[m]$.

(v) $\mathcal{F}_i \cong M_{12}$.

(4) Suppose $\Omega \neq \Omega_0$ *and set* $\Omega_* = \Omega - \Omega_0$ *and* $\mathcal{F}_* = [\Omega_*]_{\mathcal{F}}$. *Then* $\mathcal{F}^\circ = \mathcal{F}_0 \times \mathcal{F}_*$.

We prove Theorem 7.2.3 via a series of reductions. Assume the theorem is false and choose a counter example τ with τ minimal. Adopt the notation from Theorem 7.2.3.

LEMMA 7.2.4. $\Omega = \Omega_0$.

PROOF. Assume otherwise. By 2.5.2, $\tau_0 = (\mathcal{F}, \Omega_0)$ is a quaternion fusion packet, and of course τ_0 satisfies Hypothesis 7.2.1, so by minimality of $|\Omega|$, τ_0 satisfies the conclusions of Theorem 7.2.3. Hence τ also satisfies conclusions (1)-(3) of the theorem.

By 2.5.2, $\tau_* = (\mathcal{F}_*, \Omega_*)$ is a quaternion fusion packet. Let $J \in \Omega_*$ and $t = z(J)$. By 3.1.5, J centralizes $O(z_i)$ so J acts on the factor $\mathcal{F}_i$ of $\mathcal{F}_0$. Let $T_i \in Syl(\mathcal{F}_i)$. If $t \in T_i$ then $t \in C_{T_i}(O(z_i)) = \langle z_i \rangle$ as 7.2.3.3 holds. But this contradicts the choice of t and 7.2.2.1. Therefore $t \notin T_i$, so J centralizes T_i by 3.1.4. Then by 2.3.1 in [**Asc19**], t centralizes $\mathcal{F}_i$. As T_i centralizes J it centralizes $\mathcal{O}(t)$ by 2.6.12. Therefore $\mathcal{F}_i = [O(z_i)]_{\mathcal{F}_i} \leq C_{\mathcal{F}_t}(\mathcal{O}(t))$. Thus as $\mathcal{F}_0 \trianglelefteq \mathcal{F}$, we have $\mathcal{F}_* = [\Omega_*]_{\mathcal{F}} \leq C_{\mathcal{F}}(\mathcal{F}_0)$, so (4) holds, contradicting the choice of τ as a counter example to Theorem 7.2.3. □

LEMMA 7.2.5. *(1) Set* $O = O(\tau)$, $O_i = O(z_i)$, $\mathcal{O} = \mathcal{O}(\tau)$, *and* $\mathcal{O}_i = \mathcal{O}(z_i)$. *Then* $O = \prod_{i \in I} O_i$, $\mathcal{O} = \prod_{I \in I} \mathcal{O}_i$, *and* $Aut_{\mathcal{F}}(O)$ *is transitive on* $\{O_i : i \in I\}$ *and* $\{\mathcal{O}_i : i \in I\}$.

(2) There is a unique $\eta \in \eta(\tau)$. *Let* $W = \langle \eta \rangle$, G_W *a model for* $N_{\mathcal{F}}(C_S(W))$, $M = \langle O(z)^{G_W} \rangle$, *and* $\mu = \mu(\tau)$.

(3) Set $M_i = \langle O_i^M \rangle$ *and* $W_i = W \cap M_i$. *Then* $M = \prod_{i \in I} M_i$, $W = \prod_{i \in I} W_i$, *and* $\mu = \prod_{i \in I} \mu_i$, *where* $\mu_i = Aut_{M_i}(W_i) \cong D_{12}$.

(4) $N_{G_W}(O)$ *is transitive on* $K^{\mathcal{F}}$ *and* $\tilde{K}^{\mathcal{F}}$, *and* G_W *is transitive on* $\{M_i : i \in I\}$.

PROOF. By 7.2.4 and 7.2.2.1, $\mathcal{F}$ is transitive on $\mathcal{Z}_S = \{z_i : i \in I\}$ and hence also on $\{O_i : i \in I\}$, so (1) follows from 3.3.7. Part (2) follows from the proof of 3.3.14. By (1) and as $\mathcal{A}(z) \neq \emptyset$, 6.1.6.1 says that $N_{G_W}(O)$ is transitive on $K^{\mathcal{F}}$ and $\tilde{K}^{\mathcal{F}}$. Now (3) follows from 4.2.10. Then (3) and the transitivity of G_W on $O(z)^{\mathcal{F}}$ says G_W is transitive on $\{M_i : i \in I\}$, completing the proof of the lemma. □

NOTATION 7.2.6. Adopt the notation in 7.2.5. Set $T_i = S \cap M_i$ and $E_i = \Omega_1(W_i)$. Take $z = z_1$ and set $T = T_1$, $Q = O_1$, $V = K \cap W$, and $\tilde{V} = \tilde{K} \cap W$. Set $\mathcal{F}_z = C_{\mathcal{F}}(z)$ and $\mathcal{H} = [T]_{\mathcal{F}_z}$.

LEMMA 7.2.7. $\mathcal{F} = \mathcal{F}_0 = [Q]_{\mathcal{F}}$.

PROOF. Set $\mathcal{Y} = \mathcal{F}^\circ$. If $\mathcal{F} = \mathcal{Y}$ then $\mathcal{F} = \mathcal{F}_0 = [Q]_{\mathcal{F}}$ by 7.2.4, so assume $\mathcal{F} \neq \mathcal{Y}$.

By 2.5.2, $\xi = (\mathcal{Y}, \Omega)$ is a quaternion fusion packet, and ξ satisfies Hypothesis 7.2.1, so by minimality of $|\tau|$, ξ satisfies the conclusions of Theorem 7.2.3. Let $\{z_j : j \in J\} = z^{\mathcal{Y}} \cap \mathcal{Z}_S$, $\mathcal{Y}_0 = [Q]_{\mathcal{Y}}$, and $\mathcal{Y}_j = [O_j]_{\mathcal{Y}_0}$. Then $\mathcal{Y}_0 = \prod_{j \in J} \mathcal{Y}_j$ with $\mathcal{Y}_j$ satisfying 7.2.3.3.

Now $\mathcal{Y}_0 \leq \mathcal{Y}$, and as $\mathcal{F}_0 \trianglelefteq \mathcal{F}$, also $\mathcal{F}_0 \trianglelefteq \mathcal{Y}$ by 1.2.6. If $z^{\mathcal{Y}} = z^{\mathcal{F}}$ it follows that $\mathcal{Y}_0 = \mathcal{F}_0$, while if $z^{\mathcal{Y}} \neq z^{\mathcal{F}}$ then as ξ satisfies 7.2.3.4, we have $\mathcal{F}_0 = \mathcal{Y}_0 \times \mathcal{Y}_*$. Further by minimality of $|\tau|$, $\mathcal{Y}_*$ satisfies conclusions (1)-(3) of 7.2.3, so τ satisfies those conclusions too. By 7.2.4, 7.2.3.4 holds vacuously. Thus we have a contradiction to the choice of τ as a counter example. □

LEMMA 7.2.8. *(1) $T = QW$ and $|T : Q| = 2$.*
(2) $\mathcal{F}_T(M_1) \trianglelefteq N_{\mathcal{F}}(E_1) = N_{\mathcal{F}}(W_1)$, and $T = C_{M_1}(z) \trianglelefteq N_{\mathcal{F}_z}(E_1)$.
(3) $\mathcal{O}_K \cong SL_2[m]$ or $\mathcal{O}_K = K$.
(4) T is Sylow in $\mathcal{H}$ and either
(a) $\mathcal{H} = T\mathcal{O}(z)$, or
(b) $m = 8$, $\mathcal{H}$ is constrained with $Q = O_2(\mathcal{H}) \cong Q_8^2$, and, taking H to be a model for $\mathcal{H}$, $H/Q \leq Out(Q)$ is isomorphic to E_{3^k} extended by $\mathbf{Z}_2$, where $3^k = |Aut_{\mathcal{F}}(Q)|_3$.

PROOF. The group M_1 is described in 4.2.6, from which (1) follows.

Let $E = E_1$ and $\mathcal{M} = N_{\mathcal{F}}(E)$. By 2.5.2, $\xi = (\mathcal{M}, \Omega)$ is a quaternion fusion packet satisfying Hypothesis 7.2.1 by 7.2.5, so by minimality of $|\tau|$, ξ satisfies the conclusion of Theorem 7.2.3. As E centralizes O_i for $i > 1$ but not Q, we conclude $\mathcal{M}_1 = [Q]_{\mathcal{M}^\circ} \trianglelefteq \mathcal{M}$. Next $C_{\mathcal{O}(z)}(E) \geq V\tilde{V}$ is normal in $C_{\mathcal{F}_z}(E)$ by 8.23.2 in [**Asc11**]. Then as $C_{\mathcal{F}}(E) \trianglelefteq \mathcal{M}$, $V\tilde{V} \leq O_2(\mathcal{M})$, so $W_1 = \langle (V\tilde{V})^M \rangle \leq O_2(\mathcal{M})$. Hence as ξ satisfies 7.2.3, we get $\mathcal{M}_1 = \mathcal{F}_T(M_1)$. From the structure of M_1 in 4.2.6, $T = C_{M_1}(z)$, completing the proof of (2).

If $m > 8$ then from 4.2.6, W_1 interchanges the two K-classes of Q_8-subgroups of K, so (3) holds in this case. Further E is characteristic in Q, so by (2), W and T are normal in $N_{\mathcal{F}_z}(Q)$. Therefore $\mathcal{H} = W\mathcal{O}(z) = T\mathcal{O}(z)$ by 1.5.2 in [**Asc19**], establishing (4) when $m > 8$.

Thus we may assume that $m = 8$. Then $\mathcal{O}_K = K$, so once again (3) holds, completing the proof of (3).

We now argue as in the proof of 7.1.17 to show that (4) holds. Let $S_0 = QC_S(Q)$, G a model for $N_{\mathcal{F}}(S_0)$, and $H = \langle T^G \rangle$. Set $G^* = G/Q$, $G^+ = G/\langle z \rangle$, and $G^! = Out_G(Q)$. Now S acts on W by 7.2.5.2, so $C_S(z)$ acts on T and E by (2). Therefore $[C_S(Q), T] \leq C_T(Q) = \langle z \rangle$, so $[H, C_S(Q)] \leq \langle z \rangle$. Thus $O^2(H) \leq C_{G_z}(C_S(Q))$, so a Sylow 3-subgroup X of H is faithful on Q. If $G^!$ is a 2-group then $E \trianglelefteq \mathcal{F}_z$, so $\mathcal{H} = T$ by (2). Thus we may assume $G^!$ is not a 2-group. As X is faithful on Q, $X \cong X^!$. As $t \in T - Q$ is nontrivial on K^+ and $\tilde{K}^+$, $t^!$ inverts $O_3(G_z^!)$, so $X^! \leq H^!$, so $H^!$ is $X^! \cong E_{3^k}$ extended by the involution $t^!$, where $k \in \{1, 2\}$.

Let $\mathcal{Y} = C_{\mathcal{F}_z}(Q)$. By (2), $\mathcal{Y}$ acts on T and centralizes Q and T/Q, so $O^2(\mathcal{Y})$ centralizes T. Therefore $\mathcal{F}_T(H) \leq \mathcal{H} \leq C_{\mathcal{F}_z}(O^2(\mathcal{Y})) = \mathcal{X}$. But $\mathcal{X}$ is constrained, so $\mathcal{H} = \mathcal{F}_T(H)$, completing the proof of (4) and the lemma. □

THEOREM 7.2.9. $\Omega = \{K, \tilde{K}\}$.

We prove Theorem 7.2.9 via a series of reductions. Assume the theorem is false. Then $|I| = n > 1$.

LEMMA 7.2.10. *(1)* $\mathcal{F}_z^\circ = \mathcal{O}(z) \times \mathcal{F}_2 \times \cdots \times \mathcal{F}_n$, *where* $\mathcal{F}_i = [O_i]_{\mathcal{F}_z^\circ}$ *and* $\tau_i = (\mathcal{F}_i, \Omega(z_i))$ *is a quaternion fusion packet.*

(2) T_i *is Sylow in* $\mathcal{F}_i$ *and* $N_{\mathcal{F}_i}(E_i) = \mathcal{F}_{T_i}(M_i)$.

(3) Let $\alpha_i \in \mathfrak{A}(z_i)$ *with* $z_i\alpha_i = z$. *Then* $\mathcal{F}_1 = \mathcal{F}_i\alpha_i^*$ *is independent of* i *and* α_i.

(4) $\mathcal{F}_1 = \langle \mathcal{H}, \mathcal{F}_T(M_1)\rangle$ *and* $\tau_1 = (\mathcal{F}_1, \Omega(z)\}$ *is a quaternion fusion packet appearing in one of the cases in 7.2.3.3.*

(5) $\mathcal{F}_1 \times \cdots \times \mathcal{F}_n \leq \mathcal{F}$.

PROOF. By 7.2.5.3, $\mathcal{F}_{T_i}(M_i) \leq \mathcal{F}_z^\circ$ for $1 < i \leq n$. Therefore the quaternion fusion packet $\tau_z = (\mathcal{F}_z, \Omega)$ satisfies Hypothesis 7.2.1, so by minimality of $|\mathcal{F}|$, τ satisfies the conclusions of Theorem 7.2.3. In particular (1) and (2) hold.

As $\mathcal{F}_z^\circ$ satisfies 7.2.3, we conclude from 7.1.5 and 7.1.9 that $\mathcal{F}_i = \langle \mathcal{H}_i, \mathcal{F}_{T_i}(M_i)\rangle$, where $\mathcal{H}_i = C_{\mathcal{F}_i}(z_i)$. By 7.2.5.2, $W\alpha_i = W$, so by 7.2.5.4, $T_i\alpha_i = T$ and $\mathcal{F}_{T_i}(M_i)\alpha_i^* = \mathcal{F}_T(M_1)$. As $\mathcal{H}_i = [T_i]_{\mathcal{H}_i}$ we have $\mathcal{H}_i\alpha_i^* \leq \mathcal{H}$. Also by 7.2.5.1, $\mathcal{O}_i\alpha^* = \mathcal{O}(z)$, so by 7.2.8.4, either $\mathcal{H}_i = T_i\mathcal{O}_i$ and $\mathcal{H}_i\alpha^* = \mathcal{H}$ or $m = 8$. In the latter case as $Aut_{\mathcal{F}}(O_i)\alpha^* = Aut_{\mathcal{F}}(Q)$, again $\mathcal{H}_i\alpha^* = \mathcal{H}$. Therefore $\mathcal{F}_i\alpha_i^* = \langle \mathcal{H}, \mathcal{F}_T(M_1)\rangle$, proving (3) and (4).

Next T centralizes T_i and is strongly closed in $C_S(z)$ with respect to $\mathcal{F}_z$ as T is Sylow in $\mathcal{H} \trianglelefteq \mathcal{F}_z$ by 7.2.8.4. Therefore T centralizes $\mathcal{F}_i$ by 9.5 in [**Asc11**]. Let $T_{i'} = \langle T_j : j \neq i\rangle$. Then $T_{i'}\alpha_i = T_{1'}$ and $\mathcal{F}_i\alpha_i^* = \mathcal{F}_1$, so as $T_{i'}$ centralizes $\mathcal{F}_i$, also $T_{1'}$ centralizes $\mathcal{F}_1$. Therefore by 2.3 in [**Asc11**], $\mathcal{F}_1 \times (\mathcal{F}_2 \times \cdots \times \mathcal{F}_m) \leq \mathcal{F}$, proving (5). □

LEMMA 7.2.11. *Let* $1 \neq P \in \mathcal{F}_1^f$, $\alpha \in \mathfrak{A}(P)$, *and* $\mathcal{L} = \mathcal{F}_2 \cdots \mathcal{F}_n$. *Then* $\mathcal{L}\alpha^* \trianglelefteq N_{\mathcal{F}}(P\alpha)$.

PROOF. Set $R = P\alpha$, $\mathcal{Y} = N_{\mathcal{F}}(R)$, and $\Gamma = \{J \in \Omega : J \leq N_S(R)\}$. By 2.5.2, $\xi = (\mathcal{Y}, \Gamma)$ is a quaternion fusion packet. As $\mathcal{F}_i\alpha^* \leq \mathcal{Y}$ for $1 < i \leq n$, it follows that ξ satisfies Hypothesis 7.2.1, and hence 7.2.3. Set $\mathcal{B} = [T_i\alpha]_{\mathcal{Y}^\circ}$; then $\mathcal{F}_i\alpha^* \leq \mathcal{B}$. From 7.2.10.4, $\mathcal{B} = \langle \mathcal{H}_B, \mathcal{F}_{T_i}(M_i)\alpha^*\rangle$, where $\mathcal{H}_B = C_{\mathcal{B}}(z_i\alpha)$. Now $z\alpha$ centralizes $\mathcal{F}_{T_i}(M_i)\alpha^*$, and as $z\alpha$ centralizes $T_i\alpha \in Syl(\mathcal{B})$, $z\alpha$ centralizes $\mathcal{H}_B$ from the structure of $\mathcal{H}_B$ in 7.2.8. Thus $z\alpha$ centralizes $\mathcal{B}$. Therefore for $\beta \in \mathfrak{A}(z\alpha)$ with $z\alpha\beta = z$, we have $\mathcal{E} = \mathcal{B}\beta^* \leq \mathcal{F}_z$ and $\mathcal{E} = [O_i\alpha\beta]_{\mathcal{E}}$, so as $\mathcal{F}_i\alpha^*\beta^* \leq \mathcal{E}$ we conclude from 7.2.10 that $\mathcal{E} = \mathcal{F}_j$ for some j, and hence $\mathcal{E} \cong \mathcal{F}_i$ by 7.2.10.3. Therefore $\mathcal{F}_i\alpha^* = \mathcal{B}$, so $\mathcal{L}\alpha^* \trianglelefteq \mathcal{Y}$, completing the proof. □

LEMMA 7.2.12. *Let* $T' = \langle T_i : 1 < i \leq n\rangle$. *Then* $T' \in \mathcal{F}^f$ *and* $\mathcal{F}_1 \trianglelefteq N_{\mathcal{F}}(T')$.

PROOF. As $z \in \mathcal{F}^f$ also $T' \in \mathcal{F}^f$. Set $\mathcal{Y} = N_{\mathcal{F}}(T')$; then $\xi = (\mathcal{Y}, \Omega(z))$ is a quaternion fusion packet by 2.5.2 and T' centralizes $\mathcal{F}_1$ by 7.2.10.5, so ξ satisfies Hypothesis 7.2.1. Now complete the proof as in the previous lemma. □

LEMMA 7.2.13. *For each* $P \in \mathcal{F}_1^f$ *and* $\alpha \in \mathfrak{A}(P)$, $O^{2'}(N_{\mathcal{F}_1}(P))\alpha^* \trianglelefteq N_{\mathcal{F}}(P\alpha)$.

PROOF. Adopt the notation used in 7.2.11 and its proof. By 7.2.11, $\mathcal{L}\alpha^* \trianglelefteq \mathcal{Y}$. Set $\mathcal{P} = O^{2'}(N_{\mathcal{F}_1}(P))$ and $\mathcal{Q} = \mathcal{P}\alpha^*$. By 7.2.10.5, $\mathcal{Q} = [N_T(P)\alpha]_{\mathcal{Q}} \leq C_{\mathcal{Y}}(\mathcal{L}\alpha^*) = \mathcal{X}$. By 7.2.12, $\mathcal{P} \trianglelefteq N_{N_{\mathcal{F}}(T')}(P)$. Then as $\mathcal{X} \leq N_{\mathcal{Y}}(T'\alpha)$, it follows from 1.2.6 that $\mathcal{Q} \trianglelefteq \mathcal{X}$. Let S_0 be Sylow in $\mathcal{X}\mathcal{L}\alpha^*$. Then $N_{\mathcal{Y}}(S_0) \leq N_{\mathcal{Y}}(T'\alpha)$, so $N_{\mathcal{Y}}(S_0)$ acts on $\mathcal{Q}$ and hence $\mathcal{Q} \trianglelefteq \mathcal{Y}$ by 7.4 in [**Asc11**]. The proof is complete. □

LEMMA 7.2.14. *$\mathcal{F}_1$ is tightly embedded in $\mathcal{F}$.*

PROOF. We must verify conditions (T1)-(T3) if Definition 3.1.2 in [**Asc19**]. Condition (T1) is satisfied by 7.2.13. Condition (T3) is satisfied as $\mathcal{H} \trianglelefteq \mathcal{F}_z$, as $\mathcal{F}_T(M_1) \trianglelefteq N_{\mathcal{F}}(E_1)$, and as $\langle z \rangle$ and E_1 are characteristic in T, so that $Aut_{\mathcal{F}}(T) \leq Aut(\mathcal{F}_1)$ using 7.2.10.4.

It remains to verify (T2), so we must show that if $u \in T$ is an involution and $\psi \in \hom_{\mathcal{F}}(u, T)$, then $\psi = \varphi\phi$ for some $\varphi \in Aut_{\mathcal{F}}(T)$ and $\phi \in \hom_{\mathcal{F}_1}(u\varphi, T)$. As condition (T1) is satisfied, it follows from 3.1.5 in [**Asc19**] that $u^{\mathcal{F}} \cap T \cap \mathcal{F}^f \neq \emptyset$, so we may take $u\psi \in \mathcal{F}^f$. Then ψ extends to $\zeta \in \hom_{\mathcal{F}}(\langle u \rangle T', S)$ acting on T'. Then (T2) follows from 7.2.12. □

LEMMA 7.2.15. *$\mathcal{F}_1$ is subnormal in $\mathcal{F}$ and $[T]_{\mathcal{F}} = \mathcal{F}_1 \times \cdots \times \mathcal{F}_n$.*

PROOF. By 7.2.10, $\mathcal{F}_1$ appears in 7.2.3.3. In particular $\mathcal{F}_1 \neq T$, $m(T) > 1$, $\Phi(T) \neq 1$, and T is not dihedral, so the lemma follows from Theorem 3.0.3 in [**Asc19**]. □

We can now complete the proof of Theorem 7.2.9. For parts (1)-(3) of Theorem 7.2.3 hold by 7.2.15 and 7.2.10.4, while part (4) holds by 7.2.4. This contradicts the choice of τ as a counterexample, establishing Theorem 7.2.9.

REMARK 7.2.16. By Theorem 7.2.9, $|I| = 1$, so by 7.2.5.3, $\mu(\tau) \cong D_{12}$. By 7.2.7, $\mathcal{F} = [O(z)]_{\mathcal{F}}$. Therefore parts (1) and (2) of 7.2.3 hold with $\mathcal{F}_1 = \mathcal{F}$. By 7.2.4, part (4) of 7.2.3 holds vacuously. Therefore it remains only to prove part (3) of Theorem 7.2.3.

LEMMA 7.2.17. *$\mathcal{H}$ is transitive on $\mathcal{D}^*(z) = \mathcal{A}(z) = \mathcal{Z} \cap O(z) - \{z\}$.*

PROOF. By 7.2.9, $\mathcal{D}(z) = \emptyset$, so $\mathcal{D}^*(z) = \mathcal{A}(z) = \mathcal{Z} \cap O(z) - \{z\}$ by 7.2.2.3. By 7.2.2.4, $\mathcal{A}(z) = t^{\mathcal{F}_z}$ for $t \in E - \langle z \rangle$, so it remains to show $t^{\mathcal{F}_z} = t^{\mathcal{H}}$. By 7.2.8.4, T is Sylow in $\mathcal{H}$, so as $\mathcal{H} \trianglelefteq \mathcal{F}_z$, $t^{\mathcal{F}_z} = \{t\varphi\phi : \varphi \in Aut_{\mathcal{F}_z}(T) \text{ and } \phi \in \hom_{\mathcal{H}}(t\varphi, T)\}$. Hence as E_1 is characteristic in T, $t^{\mathcal{F}_z} = t^{\mathcal{H}}$, as desired. □

LEMMA 7.2.18. *$\mathcal{F}$ is transitive on Ω.*

PROOF. Assume otherwise and let $K_1 = K$, $K_2 = \tilde{K}$, $\Omega_i = \{K_i\}$, $M_i = \langle K_i^M \rangle$, and $T_i = K_i W$. By 2.5.2, $\tau_i = (\mathcal{F}, \Omega_i)$ is a quaternion fusion packet.

Observe $\mu_i = \mu(\tau_i) \cong S_3$ with $\eta_i = V_i^{M_i}$, where $V_i = K_i \cap W$. By 4.2.5, $W = \langle \eta_i \rangle$. Set $\mathcal{F}_i = [K_i]_{\mathcal{F}}$ and let S_i be Sylow in $\mathcal{F}_i$. Thus $W \leq T_i \leq S_i$; in particular $E = \Omega_1(W)$ is determined.

Suppose first that $z \in O_2(\mathcal{F})$. Then $\mathcal{F}_i$ appears in Theorem 7.1.6. If 7.1.6.2 holds then $W = \langle \eta_i \rangle \trianglelefteq \mathcal{F}_i = \mathcal{F}_{T_i}(M_i)$, so 7.2.3.3.i holds in this case, contrary to Remark 7.2.16 and the choice of τ as a counter example. So assume 7.1.6.1 holds. Then $O_2(\mathcal{F}_i) = \langle E^{C_{\mathcal{F}_i}(z)} \rangle \leq O$ by 7.2.2.3, so $O = KO_2(\mathcal{F}_i) \leq S_i$ and hence $\mathcal{F} = [O]_{\mathcal{F}} = \mathcal{F}_i$. Therefore 7.2.3.3.ii holds in this case, again a contradiction.

Therefore $z \notin O_2(\mathcal{F})$, so $\mathcal{F}_i$ appears in Theorem 7.1.29. Then as $\mu_i \cong S_3$ we conclude that $\mathcal{F}_i$ is $L_3^+[m]$, $L_3[m]^{(1)}$, $G_2[m]$, or M_{12}. In the last three cases $T_i = K_i\langle E^{C_{\mathcal{F}_i}(z)}\rangle W = T$, so $\mathcal{F}_i = [O]_{\mathcal{F}} = \mathcal{F}$ and hence 7.2.3.3 holds, a contradiction. Therefore $\mathcal{F}_i$ is $L_3^+[m]$. Here $T_i = S_i$. But now as $W \leq T_1 \cap T_2$ and $L_3^+[m]$ is quasisimple, $\mathcal{F}_1 = [W]_{\mathcal{F}} = \mathcal{F}_2$, so $T = T_1T_2 \leq S_i = T_i$, a contradiction. □

NOTATION 7.2.19. Set $\mathcal{E} = \langle \mathcal{H}, \mathcal{F}_T(M)\rangle$.

LEMMA 7.2.20. *T is Sylow in $\mathcal{E}$, $\mathcal{H} = C_{\mathcal{E}}(z)$, and $\mathcal{F}_T(M) = N_{\mathcal{E}}(W)$. Moreover one of the following holds:*

(1) $\mathcal{H} = T$, $\mathcal{E} = \mathcal{F}_T(M)$, and E and T are normal in $\mathcal{F}_z$.

(2) $O^2(\mathcal{H}) \cong SL_2[m] * SL_2[m]$ and $\mathcal{E} \cong G_2[m]$.

(3) $m = 8$, $\mathcal{H}$ is constrained with model $H \cong S_3/Q_8^2$, and $\mathcal{E} \cong L_3(2)/E_8$ or M_{12}.

PROOF. Observe that by 7.2.18, $\tilde{K} \in K^{\mathcal{F}_z}$. Set $\mathcal{M} = \mathcal{F}_T(M)$.

First suppose that $m > 8$. Then by 7.2.8, T is Sylow in $\mathcal{H}$ and either E and T are normal in $\mathcal{F}_z$ or (using the fact that $\tilde{K} \in K^{\mathcal{F}_z}$) $O^2(\mathcal{H}) \cong SL_2[m] * SL_2[m]$. Further in the first case, $\mathcal{H} = T$ and $\mathcal{E} = \mathcal{M}$, so $C_{\mathcal{E}}(z) = C_{\mathcal{M}}(z) = T$, and hence (1) holds.

So assume that $m = 8$. Then by 7.2.8, $\mathcal{H}$ is constrained, T is Sylow in $\mathcal{H} = \mathcal{F}_T(H)$, and either $\mathcal{H} = T$, or $O^2(\mathcal{H}) \cong SL_2[8] * SL_2[8]$, or $H \cong S_3/Q_8^2$. Of course in the first case, (1) holds.

Suppose that $O^2(\mathcal{H}) \cong SL_2[m] * SL_2[m]$. Here we argue as in the proof of parts (5) and (7) of 7.1.22 to show that $\mathcal{E} \cong G_2[m]$; hence (2) holds in this case.

Thus we may assume $m = 8$ and $\mathcal{H}$ is constrained with model $H \cong S_3/Q_8^2$. Here we argue as in the proof of 7.1.27 to show that (3) holds. □

LEMMA 7.2.21. *$Aut_{\mathcal{F}}(T) = Aut_{\mathcal{F}_z}(T) = Aut_{N_{\mathcal{F}}(E)}(T) \leq Aut(\mathcal{E})$.*

PROOF. This follows as $\langle z\rangle$ and E are characteristic in T; for example $\langle z\rangle = Z(T)$ and $E = \Omega_1(Z_2(T))$. □

LEMMA 7.2.22. *$\mathcal{Z} \subseteq N_S(K)$.*

PROOF. Suppose $s \in \mathcal{Z}$ with $K^s = \tilde{K}$. Let v be a generator of V. Then s centralizes $vv^s = w \in W$ with $E \leq \langle z, w\rangle$, so s centralizes E. Set $L = M\langle s\rangle \leq G_W$. Then by 4.2.6, $U = \langle s, a, W\rangle = C_L(E) \trianglelefteq L$, where a is an involution in M inverting W. Thus $U = O_2(L) = W \times C_U(O(X))$, where X is a complement to W in M. Let $\{u\} = sW \cap C_U(O(X))$. Then u is an involution centralizing $W_0 = \langle w^{O(X)}\rangle = \Phi(W)$, so $s \in uW$ centralizes W_0. But s inverts $v^{-1}v^s \in \Phi(W)$, so $\Phi(W_0) = 1$ and hence $m = 8$ and $W_0 = E$ is the set of elements of W inverted by s. Therefore $\langle s\rangle E = \Omega_1(\langle s\rangle W)$ and hence $u \in sE$. But W is transitive on sE, so we may take $s = u$. Therefore $O(X)$ centralizes s and $E = [E, O(X)]$. Let $\alpha \in \mathfrak{A}(s)$. Then $s\alpha = z$ and $E\alpha = [O(X)\alpha^*, E\alpha] \leq O(z)C_S(O(z))$, so $E\alpha$ induces inner automorphisms on $O(z)$. Therefore $(E\alpha)^{\#} \subseteq \mathcal{D}^*(z)$ by 3.3.5, so $E\alpha \leq O(z)$ by 7.2.17. Then as the relation $\mathcal{D}^*$ is symmetric, $s \in \mathcal{D}^*(z)$, a contradiction. □

LEMMA 7.2.23. *(1) $\mathcal{Z} = z^{\mathcal{E}}$.*

(2) $\mathcal{E}$ is isomorphic to $G_2[m]$ or M_{12}.

(3) $\mathcal{F} = \langle \mathcal{F}_z, N_{\mathcal{F}}(W)\rangle$.

PROOF. By 7.2.22 and 3.3.6, and adopting the notation of 7.1.2, $\mathcal{Z} = \mathcal{Z}_{out} \cup \mathcal{Z}_{in}$. Arguing as in 7.1.12:

(a) If $s \in \mathcal{Z}_{out}$ then s inverts W.

Then arguing as in 7.1.13:

(b) Suppose $s \in \mathcal{Z} - E$ centralizes E and set $A = E\langle s\rangle$. Then s inverts W, 7.1.13.2 holds, and $A^{\#} \subseteq \mathcal{Z}$.

Suppose $s \in \mathcal{Z}_{out}$. Then by (a), (b), and 7.1.13.2, there is $h \in M$ with $s^h \in O(z)$, so $s \in z^{\mathcal{E}}$. On the other hand if $s \in \mathcal{Z}_{in}$ then $s = z$ or $s \in \mathcal{D}^*(z)$ by 3.3.5 and 7.2.17, and hence again $s \in z^{\mathcal{E}}$ by 7.2.17. This proves (1).

Suppose $\mathcal{E} \cong L_3(2)/E_8$ or $\mathcal{E} = \mathcal{F}_T(M)$. Then $\mathcal{Z} = O_2(\mathcal{E})^{\#}$ by (1), so $O_2(\mathcal{E}) \trianglelefteq \mathcal{F}$ by 14.1 in [**Asc11**]. But then 7.2.3.3 holds by Theorem 2, contrary to Remark 7.2.16. Together with 7.2.20, this proves (2).

It remains to prove (3). Recall the subsystem $\mathcal{E}(\tau)$ from Definition 3.4.2, and set $\mathcal{Y} = \mathcal{E}(\tau)$. In our setup, $\mathcal{Y} = \langle \mathcal{F}_z, N_{\mathcal{F}}(W)\rangle$, so we may assume $\mathcal{Y} \neq \mathcal{F}$, and it remains to produce a contradiction. By the Alperin-Goldschmidt Fusion Theorem 2.10, there exists $R \in \mathcal{F}^e$ with $Aut_{\mathcal{F}}(R) \neq Aut_{\mathcal{Y}}(R)$. As in section 3.4 let G be a model for $N_{\mathcal{F}}(R)$, $S_0 = S \cap G$, and $G^* = G/R$; then G^* has a strongly embedded subgroup M_0^* with $S_0 \leq M_0$.

As $z \in Z(S)$, $z \in R$ and then $U = \langle z^G\rangle \leq Z(R)$. As $\mathcal{F}_z \leq \mathcal{Y}$ and $Aut_G(R) \neq Aut_{\mathcal{Y}}(R)$, we have $U \neq \langle z\rangle = Z$. By 3.4.5:

(c) For each $P \in \Delta$, $|P \cap R| \leq 2$.

By 7.2.22, $U \leq N_S(K)$, so from an earlier remark:

(d) $z^G - \{z\} \subset \mathcal{Z}_{in} \cup \mathcal{Z}_{out}$.

Next $[S, E] = Z \leq R$, so $E \leq G$. Claim:

(e) $E \leq R$.

Assume otherwise and set $H_0 = \langle E^G\rangle$. As $[R, E] \leq Z$, it follows from 2.4.3 in [**Asc19**] that $H_0 \cong S_4$, $U = O_2(H_0)$, and $C_S(U) = R$. Let $u \in U - Z$; then $C_S(u) = C_S(U) = R$, so by (c), $C_K(u) = Z$ and hence $u \in \mathcal{Z}_{out}$ by (d). Similarly u is outer on $\tilde{K}$; but now u centralizes E, contradicting $[U, E] = Z$. This completes the proof of (e).

Let $v \in V$ of order 4. Then $[S, v] = E \leq R$, so $v \in G$ and then by (c):

(f) $v \in G - R$.

Set $H_1 = \langle v^G\rangle$ and $\bar{R} = R/U$. Then $[\bar{R}, v] \leq \bar{E}$ of order at most 2, so if H_1 does not centralize $\bar{R}$ then by 2.4.1 in [**Asc19**], $\bar{U}_1 = \langle \bar{E}^{H_1}\rangle \cong E_4$ and $Aut_{H_1}(\bar{U}_1) = GL(\bar{U}_1)$. But then for $e \in E - Z$ there is $h \in H_1$ with $[\bar{e}^h, v] = \bar{e}$, so e^h does not act on K, contrary to 7.2.22. Thus

(g) $[H_1, R] \leq U$ so $[H_1, U] \neq 1$.

By (g) and (d), $[U, v] = Z$, so by 2.4.1 in [**Asc19**], $m(U) = 2$ and $Aut_{H_1}(U) = GL(U)$. Then as $[H_1, R] \leq U$ it follows that $U = [H_1, R]$ and then $G = H_1 \times C_G(H_1)$ with $H_1 \cong S_4$. Now E centralizes the Sylow group $\langle v, U\rangle$ of H_1, so there is $e \in C_E(H_1) - Z$. Then $[R, e] \leq Z \cap C_R(H_1) = 1$, so $e \in Z(R)$.

As $\mathcal{F}_z \leq \mathcal{Y}$, $Aut_{C_G(H_1)}(R) \leq Aut_{\mathcal{Y}}(R)$, so $Aut_{H_1}(R) \not\leq Aut_{\mathcal{Y}}(R)$. But $z \in e^{N_{\mathcal{F}}(W)}$ and $\mathcal{F}_z$ and $N_{\mathcal{F}}(W) \leq \mathcal{Y}$, so $\mathcal{F}_e \leq \mathcal{Y}$. Therefore as RH_1 centralizes e, we have $Aut_{H_1}(R) \leq Aut_{\mathcal{F}_e}(R) \leq Aut_{\mathcal{Y}}(R)$, a contradiction. □

LEMMA 7.2.24. $\mathcal{F} = \mathcal{E}$.

PROOF. By 7.2.13, $\mathcal{E}$ is $G_2[m]$ or M_{12}. In the first case the proof of 7.1.23.1 shows that $\mathcal{E} = \mathcal{F}$, while this follows from the proof of 7.1.28 when $\mathcal{E} \cong M_{12}$. Of course the proof uses 7.2.23.3 in place of 7.1.5 and 7.1.9, and also makes use of 7.2.21 and the fact that $\mathcal{H} \trianglelefteq \mathcal{F}_z$ and $\mathcal{F}_T(M) \trianglelefteq N_{\mathcal{F}}(W)$. □

Observe that 7.2.16, 7.2.23.2, and 7.2.24 contradict the choice of τ as a counter example and complete the proof of Theorem 7.2.3.

REMARK 7.2.25. Observe that Theorem 3 follows from Theorem 7.2.3.

THEOREM 7.2.26. *Assume $\tau = (\mathcal{F}, \Omega)$ is a quaternion fusion packet and Theorem 1 holds for each subpacket $\tau(K)$ of τ with $K \in \Omega$ and $z(K) \in \mathcal{F}^f$. Then the conclusions of Theorem 4 hold for τ°.*

PROOF. We may assume $\tau = \tau^\circ$. Suppose first that $\mathcal{A}(\tau) \neq \emptyset$. Then by Theorem 7.2.3, $\tau = \tau_0 * \tau_*$ where $\tau_0 = \tau_1 * \cdots * \tau_r$ with $\tau_i = (\mathcal{F}_i, \Omega_i)$ satisfying one of the conclusions of 7.2.3.3 and with $\mu(\tau_i) \cong D_{12}$ for each $1 \leq i \leq r$. By 6.1.2, $\mathcal{D}(\tau_i)^c$ is connected. Therefore it remains to show that τ_* satisfies the conclusions of Theorem 4, so, replacing τ by τ_*, we may assume $\mathcal{A}(\tau) = \emptyset$. Now by 6.6.6.1, $\tau = \rho_1 * \cdots * \rho_n$ where $\rho_i = (\mathcal{Y}_i, \Gamma_i)$ and $\mathcal{Y}_i = [\Gamma_i]_{\mathcal{Y}_i}$ is transitive on Γ_i. Hence the theorem holds unless conclusion (4) of Theorem 4 fails to hold for some $i \in I$. However $\rho_i = \tau(K)$ for $K \in \Omega_i$ with $z(K) \in \mathcal{F}^f$, so by hypothesis, τ_i satisfies one of the conclusions of Theorem 1. Now 6.6.6.2 completes the proof. □

COROLLARY 7.2.27. *Theorem 1 implies Theorem 4.*

Coconnectedness

CHAPTER 8

τ° not coconnected

Let $\tau = (\mathcal{F}, \Omega)$ be a quaternion fusion packet. Given Theorem 3, we may assume that $\mathcal{A}(\tau) = \emptyset$. We will prove Theorem 1, and many of our other theorems, by induction on the order of τ, so with little loss of generality we may assume the following *Inductive Hypothesis* : Theorem 1 holds in all proper subpackets ρ of τ satisfying the hypothesis of Theorem 1 and of the form $(\mathcal{Y}, \Gamma)$ with $\mathcal{Y} \leq \mathcal{F}$ and $\Gamma \subseteq \Omega$. We assume the Inductive Hypothesis repeatedly during the remainder of the paper.

In any event if $\mathcal{A}(\tau) = \emptyset$ and the Inductive Hypothesis holds, then by 6.6.8 if $\rho = \rho^\circ = (\mathcal{Y}, \Gamma)$ is a proper subpacket of τ with $\mathcal{Y} \leq \mathcal{F}$ and $\Gamma \subseteq \Omega$, then ρ is a central product of subpackets, as described in section 1.1 after the statement of Theorem 4; that is to say Theorem 4 holds in ρ.

At many points in the paper we encounter the following setup: $\{\Omega_1, \ldots, \Omega_r\}$ is a nontrivial partition of Ω, $\rho_i = (\mathcal{B}_i, \Omega_i)$ is a quaternion fusion packet, and $\mathcal{F}$ contains a central product $\mathcal{B} = \mathcal{B}_1 * \cdots * \mathcal{B}_r$. Under suitable extra conditions we wish to show that $\mathcal{B} \trianglelefteq \mathcal{F}$; then if $\mathcal{F} = \mathcal{F}^\circ$ is transitive on Ω, we have $\mathcal{B} = \mathcal{F}^\circ$, contradicting $r > 1$ and $\mathcal{F}$ transitive on Ω. In Chapter 8 we consider various hypotheses building on this setup, and show that each such hypothesis implies that $\mathcal{B} = \mathcal{F}^\circ$. This is accomplished by showing that $\mathcal{B}_i$ is tightly embedded in $\mathcal{F}$, and appealing to the theory of tightly embedded subsystems from [**Asc19**].

Among the various such hypotheses, Hypothesis 8.1.2 is the most general. For example from 8.1.1 we find that Hypothesis 8.1.8 implies Hypothesis 8.1.2. Then later hypotheses are shown to reduce to 8.1.8.

8.1. $\mathcal{D}^c$ disconnected

In this section we assume the following hypothesis:

HYPOTHESIS 8.1.1. *$\tau = (\mathcal{F}, \Omega)$ is a quaternion fusion packet satisfying the Inductive Hypothesis.*

Sometimes we also assume the following hypothesis:

HYPOTHESIS 8.1.2. *(1) $\Omega = \Gamma_1 \coprod \Gamma_2$ is a nontrivial partition of Ω, and for $i = 1, 2$, $\rho_i = (\mathcal{B}_i, \Gamma_i)$ is a quaternion fusion packet such that $\mathcal{B}_i = \mathcal{B}_i^\circ$ and $\mathcal{F}$ contains a central product $\mathcal{B} = \mathcal{B}_1 * \mathcal{B}_2$. Let T_i be Sylow in $\mathcal{B}_i$ and assume $T_1 \in \mathcal{F}^f$.*

(2) For each involution $x \in T_1$ with $x \in \mathcal{B}_1^f$, and for $\alpha \in \mathfrak{A}(x)$, $\mathcal{B}_2\alpha^* \trianglelefteq \mathcal{F}_{x\alpha}$. Further if ρ is a coconnected component of $\tau_{x\alpha}^\circ = (\mathcal{F}_{x\alpha}^\circ, C_\Omega(x\alpha))$ not contained in $\rho_2\alpha$ then $\rho = \rho'\alpha$ for some coconnected component ρ' of $(C_{\mathcal{B}_1}(x)^\circ, C_{\Gamma_1}(x))$.

(3) $Z(\mathcal{F}) \cap T_1 = 1$.

(4) For $\alpha \in \mathfrak{A}(T_2)$, $\mathcal{B}_1\alpha^* \trianglelefteq N_{\mathcal{F}}(T_2\alpha)$.

(5) If ρ *is a coconnected component of* $\mathcal{B}_2$ *with Sylow group* T *then* $T_1 \cap T = 1$, *and if* $\phi \in \hom_{\mathcal{F}}(T, T_1)$ *then either* $\rho\phi = \rho_1$ *or* $\rho = (\mathcal{O}_T, T)$ *for some* $T \in \Gamma_2$ *and* $\mathcal{B}_1 = \mathcal{O}(z_1)$ *for some* $z_1 \in \mathcal{Z}_S$.

REMARK 8.1.3. The condition in 8.1.2.3 that $Z(\mathcal{F}) \cap T_1 = 1$ insures that, in the setup of 8.1.2.2, $\tau_{x\alpha}$ is proper in τ, and hence Theorem 4 holds in $\tau_{x\alpha}$ by Hypothesis 8.1.1. This in turn insures that the notion of a coconnected component of $\tau_{x\alpha}$ in 8.1.2.2 makes sense.

LEMMA 8.1.4. *Assume Hypothesis* 8.1.2. *Then*

(1) $T_2 \in \mathcal{F}^f$ *and* $\mathcal{B}_i \trianglelefteq N_{\mathcal{F}}(T_{3-i})$ *for* $i = 1, 2$.

(2) For each $1 \neq P \in \mathcal{C}_1^f$ *and* $\alpha \in \mathfrak{A}(P)$, $O^{2'}(C_{\mathcal{B}_1}(P))\alpha^*$ *and* $\mathcal{B}_2\alpha^*$ *are normal in* $N_{\mathcal{F}}(P\alpha)$.

(3) $Aut_{\mathcal{F}}(T_1) \leq Aut(\mathcal{B}_1)$.

(4) For each involution $u \in T_1$, $u^{\mathcal{F}} \cap T_1 = u^{Aut_{\mathcal{F}}(T_1)\mathcal{B}_1}$.

PROOF. Suppose that $\mathcal{B}_2 \trianglelefteq N_{\mathcal{F}}(T_1)$. Then $N_S(T_1) \leq N_S(T_2)$. Let $\alpha \in \mathfrak{A}(T_2)$. Then as $T_1 \in \mathcal{F}^f$ by 8.1.2.1, also $T_1\alpha \in \mathcal{F}^f$ and there exists $\beta \in \mathfrak{A}(T_1\alpha)$ such that $T_1\xi = T_1$, where $\xi = \alpha\beta$. As $\mathcal{B}_2 \trianglelefteq N_{\mathcal{F}}(T_1)$, $T_2\zeta = T_2$, so $N_S(T_2\alpha)\beta \leq N_S(T_2)$, proving that $T_2 \in \mathcal{F}^f$. Then by 8.1.2.4 applied to the identity map in $\mathfrak{A}(T_2)$, $\mathcal{B}_1 \trianglelefteq N_{\mathcal{F}}(T_2)$. Thus (1) will follow once we show that $\mathcal{B}_2 \trianglelefteq N_{\mathcal{F}}(T_1)$.

Choose P and α as in (2), set $\mathcal{Y} = N_{\mathcal{F}}(P\alpha)$, $\mathcal{X} = C_{\mathcal{F}}(P\alpha)$, and $\mathcal{A} = \mathcal{B}_2\alpha^*$. Let x be an involution in $P \cap Z(N_{T_1}(P))$ and $\theta \in \mathfrak{A}_{\mathcal{B}_1}(x)$. Replacing x, P by $x\theta$, $P\theta$, we may assume $x \in \mathcal{B}_1^f$. Let $\beta \in \mathfrak{A}(x\alpha)$, and $\gamma = \alpha\beta$. Then $\mathcal{A} \leq \mathcal{X} \leq C_{\mathcal{F}}(x\alpha)$, so $\mathcal{X}\beta^* \leq C_{\mathcal{F}}(x\gamma)$. Let $\delta \in \mathfrak{A}(x)$ with $x\delta = x\gamma$; then $\zeta = \delta^{-1}\gamma \in \hom_{\mathcal{F}_{x\gamma}}((PC_S(P))\delta, C_S(x\gamma))$. By 8.1.2.2, $\mathcal{B}_2\delta^* \trianglelefteq \mathcal{F}_{x\gamma}$, so $\mathcal{B}_2\gamma^* = (\mathcal{B}_2\delta^*)\zeta^* = \mathcal{B}_2\delta^* \trianglelefteq \mathcal{F}_{x\gamma}$. Then as $\mathcal{X}\beta^* \leq \mathcal{F}_{x\gamma}$, $\mathcal{A} \trianglelefteq \mathcal{X}$ by 1.2.6. In particular, choosing $P = T_1$ and α the identity map, $\mathcal{B}_2 \trianglelefteq C_{\mathcal{F}}(T_1)$. Then as $\Omega = \Gamma_1 \cup \Gamma_2$ with $\Gamma_1 \subseteq T_1$ but no member of Γ_2 is contained in T_1 as T_1 centralizes T_2, it follows that $\mathcal{B}_2 \trianglelefteq N_{\mathcal{F}}(T_1)$, completing the proof of (1).

As $\mathcal{B}_2\delta^* \trianglelefteq \mathcal{F}_{x\gamma}$, ζ acts on $T_2\delta$. Thus $\zeta\delta^{-*}$ acts on T_2, so as $\mathcal{B}_1 \trianglelefteq N_{\mathcal{F}}(T_2)$, we have $P\gamma\delta^{-1} = P\zeta\delta^{-*} \leq C_{T_1}(x)$. Therefore $P\gamma \leq C_{T_1}(x)\delta$.

Suppose $\mathcal{A}$ is not normal in $\mathcal{Y}$ and let R be Sylow in $\mathcal{Y}^\circ$. By an earlier reduction, $\mathcal{A} \trianglelefteq \mathcal{X}$, so there exists a coconnected component $\rho = (\mathcal{C}, \Gamma)$ of ρ_2 with Sylow group T, and $\phi \in Aut_{\mathcal{Y}}(R)$ such that $\rho\alpha\phi$ is not contained in $\rho_2\alpha$. Therefore $\rho\alpha\phi\beta$ is not contained in $\rho_2\gamma = \rho_2\delta$. Then as ρ is coconnected, $\rho\alpha\phi\beta$ is contained in a coconnected component θ of $\tau_{x\delta}$ not contained in $\rho_2\delta$ by 6.6.11, and by 8.1.2.2, $\theta = \rho'\delta$ for some coconnected component ρ' of $(C_{\mathcal{B}_1}(x), C_{\Gamma_1}(x))$. In particular $T\alpha\phi\beta \leq C_{T_1}(x)\delta$. Now by 8.1.2.5, either $T\alpha\phi\beta\delta^{-1} = T_1$ or $\mathcal{C} = \mathcal{O}_T$ with $T \in \Gamma_2$ and $\mathcal{B}_1 = \mathcal{O}(z_1)$ for some $z_1 \in \mathcal{Z}_S$. Suppose the former. Then as $T\alpha\phi\beta\delta^{-1} \leq C_{T_1}(x)$, we have $x \in Z(T_1)$, so $T\alpha\phi\beta = T_1\delta$, and by the previous paragraph, $P\gamma \leq T_1\delta$, so $P\alpha \leq T_1\delta\beta^{-1} = T\alpha\phi$. Thus $P\alpha = P\alpha\phi^{-1} \leq T\alpha \leq T_2\alpha$, so $1 \neq P \leq T_1 \cap T$, contrary to 8.1.2.5.

So assume the latter and let $z_2 = z(T)$. As $Q = T\alpha\phi\beta\delta^{-1} \leq T_1$ and $\mathcal{B}_1 = \mathcal{O}(z_1)$, we have $Q \in \Omega(z_1)$, so $z_2\alpha\phi\beta\delta^{-1} = z_1$. Next $Q\gamma$ centralizes $\mathcal{B}_2\gamma^* = \mathcal{B}_1\delta^* \trianglelefteq \mathcal{F}_{x\gamma}$, so $Q\gamma \in C_{\Gamma_1}(x)\delta$, and hence $z_1\gamma = z_1\delta$. Therefore $z_2\alpha\phi = z_1\delta\beta^{-1} = z_1\gamma\beta^{-1} = z_1\alpha$. As $\mathcal{B}_1 = \mathcal{O}(z_1)$, $T_1 = O(z_1)$. Also $\Omega(z_2) \subseteq \Gamma_2$, so $O(z_2)\alpha = O(z_2\alpha)$, so as ϕ acts on R, $O(z_2)\alpha\phi = O(z_2\alpha\phi) = O(z_1\alpha)$. Therefore $P\alpha \geq Z(O(z_1\alpha)) = \langle z_1\alpha\rangle$, so

$P = \langle z_1 \rangle$. By 8.1.2.5, $T \cap T_1 = 1$, so $z_2 \neq z_1$, contradicting $z_2\alpha\phi = z_1\alpha \in P\alpha \leq T_1\alpha$ and ϕ acts on $P\alpha$. This finally proves that $\mathcal{A} = \mathcal{B}_2\alpha^* \trianglelefteq \mathcal{Y}$.

We now complete the proof of (2). Let $\mathcal{L} = O^{2'}(N_{\mathcal{B}_1}(P))$. Then $\mathcal{L}\alpha^* \leq C_{\mathcal{Y}}(\mathcal{A}) \leq C_{\mathcal{Y}}(T_2\alpha)$. Let $\varphi \in \mathfrak{A}(T_2\alpha)$ with $T_2\psi = T_2$, where $\psi = \alpha\varphi$. But $\mathcal{B}_1 \trianglelefteq N_{\mathcal{F}}(T_2)$ by (1), so $\mathcal{L}\psi^* = O^2(N_{\mathcal{B}_1}(P\psi)) \trianglelefteq N_{N_{\mathcal{F}}(T_2)}(P\psi)$, and hence $\mathcal{L}\alpha^* \trianglelefteq N_{\mathcal{Y}}(T_2\alpha)$ by 2.2 in [**Asc10**]. Then as $\mathcal{L}\alpha^* \leq C_{\mathcal{Y}}(\mathcal{A}) \leq C_{\mathcal{Y}}(T_2\alpha)$, $\mathcal{L}\alpha^* \trianglelefteq C_{\mathcal{Y}}(\mathcal{A})$ by 2.6. As $\mathcal{L}\alpha^*$ is normal in $\mathcal{A}C_{\mathcal{Y}}(A)$ and $N_{\mathcal{Y}}(T_2\alpha)$, we conclude from 7.4 in [**Asc11**] that $\mathcal{L}\alpha^* \trianglelefteq \mathcal{Y}$. This completes the proof of (2).

By (1), $\mathcal{B}_2 \trianglelefteq N_{\mathcal{F}}(T_1)$, so $T_2 \trianglelefteq N_{N_{\mathcal{F}}(T_1C_S(T_1))}(T_1)$. Hence for each $\varphi \in Aut_{\mathcal{F}}(T_1)$, φ extend to $\psi \in N_{\mathcal{F}}(T_1T_2)$ acting on T_2. Then as $\mathcal{B}_1 \trianglelefteq N_{\mathcal{F}}(T_2)$ by (1), $\varphi \in Aut_{\mathcal{F}}(\mathcal{B}_1)$. Therefore (3) holds.

It remains to prove (4), so let u_i, $i = 1, 2$ be an involutions in T_1 conjugate in $\mathcal{F}$; we must show there is $\varphi \in Aut_{\mathcal{F}}(T_1)$ and $\phi \in \hom_{\mathcal{B}_1}(u_1\varphi, T_1)$ with $u_1\varphi\phi = u_2$. Let $\psi \in \hom_{\mathcal{F}}(u_1, T_1)$ with $u_1\psi = u_2$. Let $\beta \in \mathfrak{A}_{\mathcal{B}_1}(u_1)$. If $\beta^{-1}\psi = \varphi_0\phi_0$ then $\psi = \beta\varphi_0\phi_0 = \varphi_0 \cdot \beta\varphi_0^*\phi_0$, and as $Aut_{\mathcal{F}}(T_1) \leq Aut_{\mathcal{F}}(\mathcal{B}_1)$ by (3), we have $\beta\varphi_0^*\phi_0$ a $\mathcal{B}_1$-map. Thus we may assume $u_i \in \mathcal{B}_1^f$ for $i = 1, 2$.

By (2) and 3.1.5 in [**Asc19**], there is $v \in u_i^{\mathcal{F}} \cap T_1 \cap \mathcal{F}^f$; let $\gamma_i \in \mathfrak{A}(u_i)$ with $u_i\gamma = v$. If $\gamma_i = \varphi_i\phi_i$, then arguing as in the previous paragraph, (4) holds. Thus it suffices to assume $u \in \mathcal{B}_1^f$, and to show that if $\gamma \in \mathfrak{A}(u)$ with $u\gamma = v$ then $\gamma = \varphi\phi$ with $\varphi \in Aut_{\mathcal{F}}(T_1)$ and $\phi \in \hom_{\mathcal{B}_1}(u\varphi, T_1)$.

Choose such a γ. Applying 8.1.2.2 to u, γ and v, id in the role of x, α, we conclude $\mathcal{B}_2\gamma^*$ and $\mathcal{B}_2$ are normal in $\mathcal{F}_v$, and if ρ is a coconnected component of ρ_2 with Sylow group T such that $\rho\gamma$ is not contained in ρ_2, then it follows from 8.1.2.5 that either $\rho\gamma = \rho_1$ or $\mathcal{B}_1 = \mathcal{O}(z_1)$ and $\rho = (\mathcal{O}_T, T)$ for some $z_1 \in \mathcal{Z}_S$ and $T \in \Gamma_2$ with $T\gamma \in \Gamma_1$. In the former case $T\gamma = T_1$, so as $v \in T_1$ we have $t\gamma = v = u\gamma$ for some $t \in T$, and then $u = t \in T \cap T_1$, contrary to 8.1.2.5. In the latter case, setting $z_2 = z(T)$, as $T \in \Gamma_2$ we have $z_2\gamma = z_1$. Then $O(z_2)\gamma = O(z_2\gamma) = O(z_1) = T_1$, so as u centralizes $O(z_2)$, also $v = u\gamma$ centralizes T_1. Thus $v \in Z(T_1) = \langle z_1 \rangle$, so $u\gamma = v = z_1$. But now $u = z_2$ contradicting $\Omega = \Gamma_1 \cup \Gamma_2$ a partition.

Therefore $\mathcal{B}_2\gamma^* = \mathcal{B}_2$ and hence $T_2\gamma = T_2$. By (1), $\mathcal{B}_1 \trianglelefteq N_{\mathcal{F}}(T_2)$, so $\gamma = \varphi\phi$, where $\varphi \in Aut_{\mathcal{F}}(T_1)$ and $\phi \in \hom_{\mathcal{B}_1}(u\varphi, T_1)$, completing the proof of (4). $\square$

LEMMA 8.1.5. *Assume Hypothesis* 8.1.2. *Then $\mathcal{B}_1$ is tightly embedded in $\mathcal{F}$.*

PROOF. Observe that conditions (T1)-(T3) of Definition 3.1.2 of [**Asc19**] follow from parts (2)-(4) of 8.1.4. $\square$

LEMMA 8.1.6. *Assume Hypothesis* 8.1.2, *$\mathcal{B}_1 \neq \mathcal{O}_K$ for any $K \in \Omega$, and either ρ_1 is coconnected or $\mathcal{B}_1 = \mathcal{O}(z_1)$ for some $z_1 \in \mathcal{Z}_S$. Then*

(1) $\mathcal{B}_1$ is subnormal in $\mathcal{F}$.

(2) $\mathcal{F}^\circ = \mathcal{B}_1 * \mathcal{B}_2$.

(3) The coconnected components of τ are the coconnected components of ρ_1 and ρ_2.

PROOF. Let $K \in \Gamma_1$ and $z = z(K)$. By 8.1.5, $\mathcal{B}_1$ is tightly embedded in $\mathcal{F}$. By hypothesis $\mathcal{B} \neq \mathcal{O}_K$ for any $K \in \Omega$, so $m_2(T_1) > 1$. Also either ρ_1 is coconnected and hence $\mathcal{B}_1 \neq \mathcal{F}_{T_1}(T_1)$, or $\mathcal{B}_1 = \mathcal{O}(z)$. As $K \leq T_1$, T_1 is not dihedral. Thus if $\mathcal{B}_1 \neq \mathcal{F}_{T_1}(T_1)$, then (1) follows from Theorem 3.0.3 in [**Asc19**]. So assume that $\mathcal{B}_1 = \mathcal{O}(z) = T_1$; in this case we verify that none of the exceptional cases (2)-(8) of Theorem 3.0.2 of [**Asc19**] are satisfied, and then (1) follows from that theorem.

As $\mathcal{B}_1 \neq \mathcal{O}_K$, $n = |\Omega(z)| > 1$ and T_1 is a central product of n quaternion groups. As $m_2(T_1) > 1$, case (3) of Theorem 3.0.2 in [**Asc19**] does not hold. As T_1 is not dihedral, case (4) does not hold, and as $K \leq T_1$, $\Phi(T_1) \neq 1$ so case (2) does not hold. As T_1 has no abelian subgroup of index 2, cases (5), (7) and (8) are not satisfied. As $T_1 = \Omega_1(T_1)$, case (6) does not hold. This completes the proof of (1).

Let $\mathcal{B}_0 = [T_1]_{\mathcal{F}}$, $\Gamma_0 = \Gamma_1^{\mathcal{F}}$, $\Gamma_3 = \Gamma_0 - \Gamma_1$, and $\Gamma_4 = \Gamma_2 - \Gamma_3$. Observe that $C_{\mathcal{F}}(T_1)^\circ = \mathcal{B}_2$, since $\mathcal{B}_2^\circ = \mathcal{B}_2 \trianglelefteq N_{\mathcal{F}}(T_1)$ by 8.1.4.1, and since $\Gamma_2 = \Omega - \Gamma_1$.

By (1) and Theorem 1 in [**Asc19**], $\mathcal{B}_0$ is the central product of the conjugates of $\mathcal{B}_1$. Let $\mathcal{B}_1 = \mathcal{A}_1, \ldots, \mathcal{A}_n$ and $\Gamma_1 = \Sigma_1, \ldots, \Sigma_n$ be the conjugates of $\mathcal{B}_1$ and Γ_1, respectively, and set $\xi_i = (\mathcal{A}_i, \Sigma_i)$. Then $\xi_2 * \cdots * \xi_n \leq \rho_2$, so as ρ_2 satisfies Theorem 4, the coconnected components of each ξ_i are components of ρ_2. Let ρ be the product of the remaining components of ρ_2; then $\rho = ([\Gamma_4]_{\mathcal{B}_2}, \Gamma_4)$. In particular Γ_4 centralizes $\mathcal{B}_0$, so $[\Gamma_4]_{\mathcal{F}} \leq C_{\mathcal{F}}(\mathcal{B}_0)$, and hence $\mathcal{F}^\circ = \mathcal{B}_0 * C_{\mathcal{F}}(\mathcal{B}_0)^\circ$ with $C_{\mathcal{F}}(\mathcal{B}_0)^\circ = [\Gamma_4]_{\mathcal{F}}$. Then as $C_{\mathcal{F}}(\mathcal{B}_0)^\circ \leq C_{\mathcal{F}}(T_1)^\circ = \mathcal{B}_2$, (2) and (3) follow. □

HYPOTHESIS 8.1.7. *$r > 1$ is an integer, $\Omega = \Omega_1 \coprod \cdots \coprod \Omega_r$ is a partition of Ω, and for $1 \leq i \leq r$, $\tau_i = (\mathcal{C}_i, \Omega_i)$ is a quaternion fusion packet such that τ_i is coconnected and $\mathcal{F}$ contains a central product $\mathcal{C} = \mathcal{C}_1 * \cdots * \mathcal{C}_r$. Let S_i be Sylow in $\mathcal{C}_i$, $S_i' = \prod_{j \neq i} S_j$, and $\mathcal{C}_i' = \prod_{j \neq i} \mathcal{C}_j$. Assume $S_1 \in \mathcal{F}^f$.*

HYPOTHESIS 8.1.8. *Hypothesis 8.1.7 holds, as do the following conditions:*

(1) τ_1 is minimal among the τ_i, $1 \leq i \leq r$; that is τ_i is not conjugate in $\mathcal{F}$ to a proper subpacket of τ_1.

(2) For each involution $x \in S_1$ with $x \in \mathcal{C}_1^f$, for each $1 < i \leq r$, and for $\alpha \in \mathfrak{A}(x)$, $\tau_i\alpha$ is a coconnected component of $\tau_{x\alpha} = (C_{\mathcal{F}}(x\alpha), C_\Omega(x\alpha))$, and any other coconnected component is a subpacket of $(C_{\mathcal{C}_1}(x), C_{\Omega_1}(x))\alpha$.

(3) For $\alpha \in \mathfrak{A}(S_1')$, $\tau_1\alpha$ is a coconnected component of $\tau_1' = (C_{\mathcal{F}}(S_1'\alpha), \Omega_1\alpha)$.

(4) For $1 < i \leq r$, $S_1 \cap S_i = 1$.

LEMMA 8.1.9. *Assume Hypothesis 8.1.8. Then*

(1) For $\alpha \in \mathfrak{A}(S_1')$, $\mathcal{C}_1\alpha^* \trianglelefteq N_{\mathcal{F}}(S_1'\alpha)$.

(2) For each involution $x \in S_1$ with $x \in \mathcal{C}_1^f$, and for $\alpha \in \mathfrak{A}(x)$, $\mathcal{C}_1'\alpha^* \trianglelefteq \mathcal{F}_{x\alpha}$.

PROOF. In (1) take $P = S_1'$ and in (2) take $P = \langle x \rangle$. In each case let $\alpha \in \mathfrak{A}(P)$, $\mathcal{Y} = N_{\mathcal{F}}(P\alpha)$, and $\mathcal{X} = C_{\mathcal{F}}(P\alpha)$.

Suppose first that $P = \langle x \rangle$. Then for $i > 1$, $\mathcal{C}_i$ centralizes P as $\mathcal{C}$ is a central product. Then $\mathcal{C}_1'\alpha^* \leq \mathcal{X}$. By 8.1.8.2, $\tau_i\alpha$ is a coconnected component of $\tau_{P\alpha}$ for each $i > 1$.

Suppose $\mathcal{C}_1'\alpha^*$ is not normal in $\mathcal{Y}$; then there exists i and $\phi \in \hom_{\mathcal{Y}}(S_i, C_S(P\alpha))$ such that $\mathcal{C}_i\alpha^*\phi^* \neq \mathcal{C}_j\alpha^*$ for any j. Then by 8.1.8.2, $\tau_i\alpha\phi = \rho\alpha$ for some coconnected component ρ of $(C_{\mathcal{C}_1}(P), C_{\Omega_1}(P))$. Then by minimality of τ_1 in 8.1.8.1, $\rho = \tau_1$, so $S_i\alpha\phi = S_1\alpha$. In particular, $P\alpha \leq S_i\alpha\phi$, so $P\alpha = P\alpha\phi^{-1} \leq S_i\alpha$, and hence $1 \neq P \leq S_1 \cap S_i$, contrary to 8.1.8.4. Thus (2) holds.

Now assume $P = S_1'$. Then by 8.1.8.3, $\tau_1\alpha$ is a coconnected component of τ_1'. As $\Omega(\tau_1') = \Omega_1\alpha = \Omega(\tau_1\alpha)$, it follows that $\tau_1\alpha$ is the unique such component. Thus $\mathcal{C}_1\alpha^* \trianglelefteq N_{\mathcal{F}}(S_1'\alpha)$, establishing (1). □

LEMMA 8.1.10. *Assume Hypothesis 8.1.8 with $\mathcal{C}_1 \neq \mathcal{O}_K$ for $K \in \Omega_1$ and $S_1 \cap Z(\mathcal{F}) = 1$. Then*

(1) $\mathcal{C}_1$ is tightly embedded in $\mathcal{F}$.

(2) $\mathcal{C}_1$ is subnormal in $\mathcal{F}$.

(3) $\mathcal{F}^\circ = \mathcal{C}$ *and* $\{\tau_i : 1 \leq i \leq r\}$ *is the set of coconnected components of* τ°.

PROOF. We verify that Hypothesis 8.1.2 is satisfied with $\rho_1 = \tau_1$, $\mathcal{B}_2 = \mathcal{C}_1'$, and $\Gamma_2 = \Omega - \Omega_1$; then we verify the extra condition in the hypothesis of 8.1.6, and appeal to 8.1.5 and 8.1.6 to complete the proof.

Condition (1) of Hypothesis 8.1.2 holds by Hypothesis 8.1.7. Indeed by 8.1.7, $\rho_1 = \tau_1$ is coconnected, verifying the condition in 8.1.6. The first constraint in 8.1.2.2 holds by 8.1.9.2, while the second holds by 8.1.8.2. Condition 8.1.2.3 holds by assumption, and 8.1.2.4 follows from 8.1.9.1. The first constraint in 8.1.2.5 holds by 8.1.8.4, and the second follows from 8.1.8.1. □

LEMMA 8.1.11. *Assume Hypothesis* 8.1.7 *and choose notation so that* τ_1 *is minimal among the* τ_i. *In addition assume*

(a) $\{\mathcal{D}(\tau_i)^c : 1 \leq i \leq r)$ *are the connected components of* $\mathcal{D}(\tau)^c$.

(b) $S_1 \cap Z(\mathcal{F}) = 1$.

(c) For $1 < i \leq r$, $S_1 \cap S_i = 1$.

(d) $\mathcal{C}_1 \neq \mathcal{O}_K$ *for* $K \in \Omega_1$.

Then $\mathcal{F}^\circ = \mathcal{C}$ *and* $\{\tau_i : 1 \leq i \leq r\}$ *is the set of coconnected components of* τ.

PROOF. We show that Hypothesis 8.1.8 is satisfied, and then appeal to 8.1.10 and (b) and (d) to complete the proof.

Observe hypotheses 8.1.8.1 holds by the choice of τ_1 and 8.1.8.4 holds by (c). Thus it remains to verify conditions (2) and (3) of Hypothesis 8.1.8.

For $1 \leq j \leq r$, set $\mathcal{Z}_j = \mathcal{Z}(\tau_j)$. Assume either

(i) $P \in \mathcal{C}_1^f$ is of order 2 and $i > 1$, or

(ii) $P = S_1'$ and $i = 1$.

Let $\alpha \in \mathfrak{A}(P)$, and set $\mathcal{Y} = N_{\mathcal{F}}(P\alpha)$ and $\rho = (\mathcal{Y}, N_\Omega(P\alpha))$. By (a), $\mathcal{Z}_i$ is a connected component of $\mathcal{D}(\tau)^c$, so as P centralizes $\mathcal{C}_i$, $\mathcal{Z}_i\alpha$ is contained in a connected component $\mathcal{U}_i$ of $\mathcal{D}(\rho)^c$, and then $\mathcal{U}_i$ is contained in a connected component $\mathcal{Z}(\tau_{j(i)})$ of $\mathcal{D}(\tau)^c$ for some $j(i)$.

Suppose first that (ii) holds. As $\mathcal{Z} \subseteq T = S_1S_1'$, $\mathcal{Z}\alpha = \mathcal{Z}$ and then $\{\mathcal{Z}_j\alpha : 1 \leq j \leq r\}$ is the set $\{\mathcal{Z}_k : 1 \leq k \leq r\}$ of connected components of $\mathcal{D}(\tau)^c$. Then $\mathcal{Z}_1\alpha$ is the unique coconnected component of ρ of order greater than 1, so $\mathcal{C}_1\alpha^* \trianglelefteq \mathcal{Y}$, verifying hypothesis 8.1.8.3.

So assume that (i) holds. Then $\mathcal{Z}_i\alpha \subseteq \mathcal{Z}_{j(i)}$, so $\mathcal{C}_i\alpha \subseteq \mathcal{C}_{j(i)}$. Proceeding recursively, define $j^n(i) = j(j^{n-1}(i))$ whenever $j^l(i) \neq 1$ for $1 \leq l < n$. If $j(i) = 1$ then $\mathcal{C}_i\alpha = \mathcal{C}_1$ by minimality of τ_1, and in particular $\tau_i\alpha$ is a coconnected component of ρ. Similarly if $j(i) > 1$ then either $\mathcal{C}_i\alpha^* = \mathcal{C}_{j(i)}$ and $\tau_i\alpha$ is a coconnected component of ρ, or $\mathcal{C}_i\alpha^* < \mathcal{C}_{j(i)}$. Assume the latter; then by minimality of τ_1, $j^n(i) \neq 1$ for any n. Now there exists m such that $\mathcal{C}_{j^m(i)}\alpha^* < \mathcal{C}_{j^{m+1}(i)}$, and for each $n > m$, $\mathcal{C}_{j^n(i)}\alpha^* = \mathcal{C}_{j^{n+1}(i)}$. Let $\mathcal{I} = \{j^n(i) : n \geq m\}$. Claim j is a permutation of $\mathcal{I}$. For if not there are distinct $k, l \in \mathcal{I}$ with $j(k) = j(l)$, and we may assume $\mathcal{C}_k\alpha^* = \mathcal{C}_{j(k)}$. Thus $\mathcal{C}_l\alpha^* \leq \mathcal{C}_k\alpha^*$, so $\mathcal{C}_l \leq \mathcal{C}_k$, contradicting Ω_l not contained in Ω_k. This establishes the claim. Now as $\mathcal{C}_{j^m(i)}\alpha^* \neq \mathcal{C}_{j^{m+1}(i)}$, $j^m(i)$ is not a fixed point of j, so for some $n > m$, $\mathcal{C}_{j^n(i)}\alpha^* = \mathcal{C}_{j^m(i)}$, a contradiction.

We have shown that for each $i > 1$, $\tau_i\alpha = \tau_{j(i)}$ is a coconnected component of ρ, verifying the first condition of 8.1.8.2. Then all other coconnected components of ρ are contained in τ_k, where $\{k\} = \{1, \ldots, r\} - \{j(i) : 1 < i \leq r\}$, completing the proof. □

HYPOTHESIS 8.1.12. *Hypothesis* 8.1.1 *holds with* $\mathcal{F}$ *transitive on* Ω. *Let* $\eta \in \eta(\tau)$, $W = \langle \eta \rangle$, G *a model for* $N_{\mathcal{F}}(C_S(W))$, *and* $M = \langle K^G \rangle$. *Assume*

(1) For some $r > 1$, $\eta = \eta_1 \coprod \cdots \coprod \eta_r$, $\Omega = \Omega_1 \coprod \cdots \coprod \Omega_r$, $K_i \in \Omega_i$, *and* $M = M_1 * \cdots * M_r$ *is a central product, where* $M_i = \langle K_i^M \rangle$, $\Omega_i = K_i^M \cap S$, *and* $\eta_i = (K_i \cap W)^M$. *Further* D_η *is a set of 3-transpositions of* $\mu(\tau)$.

(2) Let $K = K_1$ *and* $z = z(K) \in \mathcal{F}^f$. *Then* $\mathcal{F}_z^\circ$ *is a central product* $\mathcal{F}_z^\circ = \mathcal{C}_z \prod_{1<i\leq r} \mathcal{C}_i$ *with* $\tau_i = (\mathcal{C}_i, \Omega_i)$ *a coconnected component of* $\mathcal{F}_z$ *and* $(\mathcal{C}_z, \Omega_1)$ *is a subpacket of* τ_z.

(3) For $i > 1$, $S_i = S \cap M_i$ *is Sylow in* $\mathcal{C}_i$ *and* $\mathcal{C}_i = \langle S_i \mathcal{O}_i, \mathcal{F}_{S_i}(M_i) \rangle$, *where* $Z_i = Z(O(\tau_i))$ *and* $\mathcal{O}_i = \mathcal{O}(\tau_i)$ *if* $m > 8$ *while if* $m = 8$ *then* $O(\tau_i) = O^{2'}(\mathcal{O}_i)$ *and* $\mathcal{O}_i = \mathcal{O}(\tau_i)$ *or* $O^2(C_{\mathcal{C}_i}(Z_i))$ *acts on each member of* Ω_i.

LEMMA 8.1.13. *Assume Hypothesis* 8.1.12. *Then*

(1) $N_G(O(\tau))$ *is transitive on* Ω *and* G *is transitive on* $\{M_1, \ldots, M_r\}$. *Set* $T = S \cap M$, $\mathcal{C}_1 = \mathcal{C}_2^g$ *and* $\tau_1 = \tau_2^g$ *for* $g \in N_G(T)$ *with* $M_2^g = M_1$.

(2) $\mathcal{C} = \mathcal{C}_1 * \cdots * \mathcal{C}_r \leq \mathcal{F}$ *with* τ_i *a coconnected component of* $\mathcal{C}$ *and* $Aut_{\mathcal{F}}(T) \leq Aut_{\mathcal{F}}(\mathcal{C})$ *is transitive on* $\Lambda = \{\tau_i : 1 \leq i \leq r\}$.

(3) $S_1 \in \mathcal{F}^f$.

PROOF. As $\mathcal{F}$ is transitive on Ω, G is transitive on the M_i and $N_G(O(\tau))$ is transitive on Ω by 6.1.6, and hence (1) holds.

Observe that by 8.1.12.3, $\mathcal{C}_1 = \langle S_1 \mathcal{O}_1, \mathcal{F}_{S_1}(M_1) \rangle$, where $\mathcal{O}_1 = \mathcal{O}_2^g \trianglelefteq N_{\mathcal{F}}(Z_1)$. Now for $i \neq j$, M_i centralizes S_j by 8.1.12.1. Also if $m > 8$ then $\mathcal{O}_i = \mathcal{O}(\tau_i)$, and S_j centralizes S_i and hence also $\mathcal{O}_i$ by 2.6.12. If $m = 8$ then we work in a model G_i for $N_{\mathcal{F}}(S_i C_S(S_i))$, where by 8.1.12.3 there is a normal subgroup H_i modeling $\mathcal{O}_i$ with Sylow group $O(\tau_i)$ and $[S_j, H_i] \leq C_{H_i}(S_i) = Z_i$, so again S_j centralizes $\mathcal{O}_i$. Therefore as $\mathcal{C}_i = \langle S_i \mathcal{O}_i, \mathcal{F}_{S_i}(M_i) \rangle$ by 8.1.12.3, we conclude that $\mathcal{C}_i$ centralizes S_j. Hence by 2.3 in [**Asc11**], $\mathcal{C} = \mathcal{C}_1 * \cdots * \mathcal{C}_r \leq \mathcal{F}$.

Let $\rho = (\mathcal{C}, \Omega)$, so that ρ is a quaternion fusion packet and T is Sylow in $\mathcal{C}$ since S_i is Sylow in $\mathcal{C}_i$ for each i. Then by construction, Λ is the set of coconnected components of ρ. As $Aut_{\mathcal{F}}(T)$ permutes the M_i and $\mathcal{O}_i$ transitively, it permutes Λ transitively, and then $Aut_{\mathcal{F}}(T) \leq Aut(\mathcal{C})$, completing the proof of (2).

As $z \in \mathcal{F}^f$ and $S \in Syl_2(G)$, the transitivity of G on the M_i implies that $S_1 \in \mathcal{F}^f$, proving (3). □

HYPOTHESIS 8.1.14. *Hypothesis* 8.1.12 *holds. Choose* g *as in* 8.1.13.*1 and assume for each involution* $x \in S_1 - z^{\mathcal{C}_1}$ *with* $x \in \mathcal{C}_1^f$, *for each* $1 < i \leq r$, *and for* $\alpha \in \mathfrak{A}(x)$, $\tau_i\alpha$ *is a coconnected component of* $\tau_{x\alpha} = (C_{\mathcal{F}}(x\alpha), C_\Omega(x\alpha))$, *and any other coconnected component is a subpacket of* $(C_{\mathcal{C}_1}(x), C_{\Omega_1}(x))\alpha$.

LEMMA 8.1.15. *Assume Hypothesis* 8.1.14 *with* $M_1 \neq K$ *and* $S_1 \cap Z(\mathcal{C}_i) = 1$ *for* $i > 1$. *Then* $\mathcal{F}^\circ = \mathcal{C}$ *and* Λ *is the set of coconnected components of* τ.

PROOF. We verify that Hypothesis 8.1.8 is satisfied and appeal to 8.1.10. Hypothesis 8.1.7 is satisfied by parts (2) and (3) of 8.1.13. By the transitivity of $Aut_{\mathcal{F}}(\mathcal{C})$ on Λ, τ_1 is minimal, so that 8.1.8.1 is satisfied. If $Z = S_1 \cap Z(\mathcal{F}) \neq 1$ then as $Aut(T)$ is transitive on Λ, $Z \leq S_1 \cap Z(\mathcal{C}_i)$ for $i > 1$, contrary to assumption. Similarly $S_1 \cap S_i \leq S_1 \cap Z(\mathcal{C}_i)$, so $S_1 \cap S_i = 1$, establishing 8.1.8.4.

Observe 8.1.8.3 follows from the transitivity of $Aut_{\mathcal{F}}(T)$ on Λ in 8.1.13.2, and the fact that τ_2 is a coconnected component of $\mathcal{F}_z$. Thus it remains to verify 8.1.8.2.

Let $x \in \mathcal{C}_1^f$ be an involution. If $x \notin z^{\mathcal{C}_1}$ then the conditions in 8.1.8.2 follow from 8.1.14. If $x \in z^{\mathcal{C}_1}$ then we may take $x = z$, and then the first condition in 8.1.8.2 follow from 8.1.12.2. If $\xi = (\mathcal{X}, \Gamma)$ is a coconnected component of $\mathcal{F}_z$ other than τ_i, then $\Gamma \subseteq \Omega_1$ by 8.1.12.2, and then $[\Gamma]_{\mathcal{F}_z^\circ} \leq [\Omega_1]_{C_\mathcal{F}(S_1')} \leq \mathcal{C}_1$ from our verification of 8.1.8.3. This completes the proof. □

LEMMA 8.1.16. *Assume Hypothesis* 8.1.12 *with* $M_1 \neq K$ *and* $\mathcal{C}_2$ *has one class of involutions. Then* $\mathcal{F}^\circ = \mathcal{C}$ *and* Λ *is the set of coconnected components of* τ.

PROOF. As $\mathcal{C}_2$ has one class of involutions, so does $\mathcal{C}_1$, and that class is $z^{\mathcal{C}_1}$. Therefore the extra condition in Hypothesis 8.1.14 is vacuously satisfied. Similarly as $M_1 \neq K$, $z \notin Z(M_1)$, so as $\mathcal{C}_1$ as one class of involutions, we conclude $Z(\mathcal{C}_1) = 1$, and hence $S_2 \cap Z(\mathcal{C}_1) = 1$. Therefore the lemma follows from 8.1.15. □

HYPOTHESIS 8.1.17. *(1)* $\Omega = \Omega_1 \cup \Omega_2$ *is a partition of* Ω, *and for* $i = 1, 2$, $\tau_i = (\mathcal{C}_i, \Omega_i)$ *is a quaternion fusion packet such that* τ_i *is coconnected,* $\mathcal{A}(\tau_i) = \emptyset$, *and* $\mathcal{F}$ *contains a central product* $\mathcal{C} = \mathcal{C}_1 * \mathcal{C}_2$. *Let* S_i *be Sylow in* $\mathcal{C}_i$.

(2) $K \in \Omega_2$ *with* $z = z(K) \in \mathcal{F}^f$ *and* $\mathcal{F}_z^\circ = \mathcal{C}_1 * C_{\mathcal{C}_2}(z)^\circ$.

(3) $\mathcal{C}_2 \neq \mathcal{O}_K$, $|\mathcal{C}_1| = |\mathcal{C}_2|$, *and* $\mathcal{Z}(\tau_1) \subseteq z^\mathcal{F}$.

LEMMA 8.1.18. *Assume Hypothesis* 8.1.17. *Then:*

(1) $\mathcal{C}_1 \trianglelefteq \mathcal{F}_z$.

(2) Conjugating in $\mathcal{F}$, *we may take* $S_1 \in \mathcal{F}^f$.

(3) $\mathcal{C}_2 = C_\mathcal{F}(S_1)^\circ \trianglelefteq N_\mathcal{F}(S_1)$.

(4) $S_2 \in \mathcal{F}^f$.

(5) $P = N_S(S_1) = N_S(S_2)$ *and there is* $\alpha \in Aut_\mathcal{F}(P)$ *interchanging* S_1 *and* S_2 *and* $\mathcal{C}_1$ *and* $\mathcal{C}_2$.

(6) Suppose $x \in \mathcal{C}_2^f$ *is an involution and for* $\beta \in \mathfrak{A}(x)$, $\mathcal{C}_1\beta^*$ *is a coconnected component of* $\mathcal{F}_{x\beta}^\circ$. *Assume* $S_1 \cap S_2 = 1$. *Then* $\mathcal{F}_{x\beta}^\circ = (C_{\mathcal{C}_2}(x)^\circ * \mathcal{C}_1)\beta^*$.

PROOF. By 8.1.17.2, $\mathcal{F}_z^\circ = C_{\mathcal{C}_2}(z) \circ \mathcal{C}_1$, and by 8.1.17.1, τ_2 is coconnected, while by 8.1.17.3, $\mathcal{C}_2 \neq \mathcal{O}_K$, so we conclude $|C_{\mathcal{C}_2}(z)| < |\mathcal{C}_2|$. On the other hand by 8.1.17.3, $|\mathcal{C}_1| = |\mathcal{C}_2|$, so it follows that $\mathcal{C}_1$ is not fused into $C_{\mathcal{C}_2}(z)$ in $\mathcal{F}_z$, and then (1) follows.

By (1), $S_1 \trianglelefteq C_S(z)$, so (2) follows.

Set $P = N_S(S_1)$ and let $t \in \mathcal{Z}(\tau_1) \cap \mathcal{C}_1^f$. By 8.1.17.3, $t \in z^\mathcal{F}$, so as $z \in \mathcal{F}^f$ there is $\alpha \in \mathfrak{A}(t)$ with $t\alpha = z$. Then $\mathcal{C}_2\alpha^* \leq \mathcal{F}_z^\circ$, so as $|C_{\mathcal{C}_2}(z)| < |\mathcal{C}_2| = |\mathcal{C}_1|$ and as $\mathcal{C}_2$ is coconnected, it follows that $\mathcal{C}_2\alpha^* = \mathcal{C}_1$ and $S_2\alpha = S_1$.

Observe $\mathcal{C}_1$ is transitive on Ω_1 as $\mathcal{C}_1$ is coconnected, $\mathcal{A}(\tau_1) = \emptyset$, and τ_1 satisfies Theorem 4. Hence $N_{\mathcal{C}_1}(O(\tau_1))$ is transitive on Ω_1. Then as P acts on Ω_1, and hence also $O(\tau_1)$, it follows from a Frattini argument that $P = S_1C_P(t)$ and $|P : C_P(t)| = |S_1 : C_{S_1}(t)|$. But we saw that $C_S(z)$ acts on S_1, so $C_S(z) = C_P(z)$ and $|P : C_P(z)| \geq |S_2 : C_{S_2}(z)| = |S_1 : C_{S_1}(t)|$ by the previous paragraph. Then as $|P : C_P(t)| = |S_1 : C_{S_1}(t)|$ and $z \in \mathcal{F}^f$, it follows that $t \in \mathcal{F}^f$. Then we have symmetry between $z, \mathcal{C}_1$ and $t, \mathcal{C}_2$.

By 2.5.2, $\rho = (N_\mathcal{F}(S_1), \Omega)$ is a quaternion fusion packet. Also $\Omega_2 = C_\Omega(S_1) \subseteq C_\Omega(t)$ and $\mathcal{C}_2 \leq C_\mathcal{F}(S_1)^\circ \leq \mathcal{F}_t^\circ$ by 6.6.11, so $\mathcal{C}_2 = C_\mathcal{F}(S_1)^\circ \trianglelefteq N_\mathcal{F}(S_1)$, proving (3).

By (3), $S_2 \trianglelefteq P$, and we saw earlier that $S_2\alpha = S_1$, so $P = N_S(S_2)$ and hence (4) holds. Then $P\alpha \leq N_S(S_1) = P$, so $\alpha \in Aut_\mathcal{F}(P)$. Further by symmetry between S_1 and S_2, $S_1\alpha = S_2$, establishing (5).

Assume the setup of (6). Set $b = x\beta$ and suppose $\mathcal{A} = \mathcal{C}_1\beta^*$ is not normal in $\mathcal{Y} = \mathcal{F}_b$. Then as $\mathcal{A}$ is a coconnected component of $\mathcal{Y}$ there is $\phi \in \mathrm{hom}_{\mathcal{Y}}(S_1\beta, C_S(b))$, such that $\mathcal{A}\phi^*$ is a second component. Then by (3), $\mathcal{A}\phi = C_{\mathcal{F}}(S_1\beta)^\circ = \mathcal{C}_2\beta^*$, so $b\phi = b \in S_2\beta$. Then as $b = (b\phi)\phi^{-1} \in S_2\beta\phi^{-1} = S_1\beta$, we have $x \in S_1 \cap S_2$, contradicting $S_1 \cap S_2 = 1$.

Therefore $\mathcal{A} \trianglelefteq \mathcal{Y}$. Then $\mathcal{Y}^\circ = \mathcal{X} * \mathcal{A}$, with $\mathcal{X} \leq C_{\mathcal{F}}(S_1)^\circ\beta^* = \mathcal{C}_2\beta^*$ by (3), completing the proof of (6). □

HYPOTHESIS 8.1.19. *Hypothesis* 8.1.17 *holds and for each involution* $x \in \mathcal{C}_2^f$ *not in* $z^{\mathcal{C}_2}$ *and for* $\beta \in \mathfrak{A}(x)$, $\mathcal{C}_1\beta^*$ *is a coconnected component of* $C_{\mathcal{F}}(x\beta)^\circ$.

LEMMA 8.1.20. *Assume Hypothesis* 8.1.19 *with* $S_1 \cap S_2 = 1$. *Then* $\mathcal{F}^\circ = \mathcal{C}$.

PROOF. We show that Hypothesis 8.1.8 is satisfied and then appeal to 8.1.10 to complete the proof.

By 8.1.17.1 and 8.1.18.4, Hypothesis 8.1.7 is satisfied with $r = 2$. By 8.1.18.4, $\mathcal{C}_1$ and $\mathcal{C}_2$ are conjugate, so 8.1.8.1 holds. By 8.1.17.2, 8.1.18.6, and 8.1.19, 8.1.8.2 holds. By 8.1.18.3, 8.1.8.3 holds. Finally by 8.1.18.5, S_1 and S_2 are interchanged in $Aut(N_S(S_1))$, so as $S_1 \cap S_2 = 1$, 8.1.8.4 holds and $S_1 \cap Z(\mathcal{F}) = 1$. This completes the proof. □

LEMMA 8.1.21. *Assume Hypothesis* 8.1.17 *and* $\mathcal{C}_1$ *has one class of involutions. Then* $\mathcal{F}^\circ = \mathcal{C}$.

PROOF. By 8.1.18.4, $\mathcal{C}_1$ and $\mathcal{C}_2$ are conjugate, so $\mathcal{C}_2$ has one class of involutions. Thus the extra condition in Hypothesis 8.1.19 is vacuously satisfied. If $1 \neq S_1 \cap S_2$ then there is an involution in $Z(\mathcal{C}_1) \cap Z(\mathcal{C}_2)$ so $z \in Z(\mathcal{C}_2)$ as $\mathcal{C}_2$ has one class of involutions. This is a contradiction as τ_2 is coconnected and $\mathcal{C}_2 \neq \mathcal{O}_K$ by 8.1.17.3. Now the lemma follows from 8.1.20. □

Theorem 6

CHAPTER 9

$\Omega = \Omega(z)$ of order 2

Let $\tau = (\mathcal{F}, \Omega)$ be a quaternion fusion packet with Sylow group S. The next four chapters consider the case where $\mathcal{A}(\tau) = \emptyset$ and $|\Omega(z)| > 1$ for $z \in \mathcal{Z}_S(\tau)$. This thread culminates in a proof of Theorem 6 under the assumption that τ satisfies the Extended Inductive Hypothesis. Chapter 9 focuses on the case where $|\Omega(z)| = 2$.

We begin in section 9.1 with some basic notions and results for $|\Omega(z)| = 2$. Then in section 9.2 we prove results on the normalizers of essential subgroups R of S such that $Aut_{\mathcal{F}}(R) \neq Aut_{\mathcal{E}(\tau)}(R)$; such results will later allow us to pin down the generation of $\mathcal{F}$ when $\mathcal{F} \neq \mathcal{E}(\tau)$.

In sections 9.3 and 9.4 we assume the Inductive Hypothesis and that $\mathcal{F}^{\circ}$ is transitive on Ω; in the final section 9.5 of Chapter 9 these conditions follow from other parts of Hypothesis 9.5.1. Section 9.3 treats the case where $\mathcal{Z}(\tau) \cap O(z) \neq \{z\}$. In section 9.4 we determine the packets with μ abelian. Finally section 9.5 deals with the case $\mu \cong S_4$ and $\mathcal{Z}(\tau) \cap O(z) = \{z\}$.

Using inductive arguments, we remove the Inductive Hypothesis from our treatment in sections 9.3 and 9.4 when $|\Omega| = 2$ in Theorems 9.4.34 and 9.4.35. Then in Theorem 9.5.34 we determine all packets τ with $\mathcal{A}(\tau) = \emptyset$, $\Omega = \Omega(z)$ of order 2, and $\mathcal{F} = \mathcal{F}^{\circ}$ transitive on Ω.

Another important byproduct of Chapter 9 is the description of the generation of $\mathcal{F}$ in these cases, appearing in 9.4.34, 9.4.35 and 9.5.9.

9.1. $|\Omega(z)| = 2$

In this section we assume the following hypothesis:

HYPOTHESIS 9.1.1. *$\tau = (\mathcal{F}, \Omega)$ is a quaternion fusion packet such that $\mathcal{A}(\tau) = \emptyset$ and for all $z \in \mathcal{Z}_S$, $|\Omega(z)| = 2$.*

NOTATION 9.1.2. Pick $K \in \Omega$ such that $z = z(K) \in \mathcal{F}^f$, and let $\Omega(z) = \{K_1, K_2\}$ with $K = K_1$. Define $\mathcal{D}^o(z)$ to consist of those $t \in \mathcal{Z}$ such that t induces an outer automorphism on K, and set $\mathcal{D}^m(z) = \{t \in \mathcal{Z} : K_1^t = K_2\}$. For $s \in \mathcal{Z}$ and $x \in \{o, m\}$, write $\mathcal{D}^x(s)$ for the set of $t \in C_{\mathcal{Z}}(s)$ such that $t\alpha \in \mathcal{D}^x(s\alpha)$ for $\alpha \in \mathfrak{A}(s)$. Observe that the definition is independent of α.

LEMMA 9.1.3. *For each $t \in C_{\mathcal{Z}}(z)$, one of the following holds:*

(1) $t \in z^{\perp}$ and t centralizes $O(z)$.

(2) $t \in \mathcal{D}^*(z) - \mathcal{D}(z)$ and t induces a nontrivial inner automorphism on K_i for $i = 1, 2$.

(3) $t \in \mathcal{D}^o(z)$, and for $i = 1, 2$, t induces an outer automorphism on K_i with $C_{K_i}(t) = \langle z \rangle$, and t inverts the unique t-invariant cyclic subgroup of K_i of index 2.

(4) $t \in \mathcal{D}^m(z)$ and $K_1^t = K_2$.

PROOF. We may assume $t \neq z$. As t centralizes z, t permutes $\Omega(z) = \{K_1, K_2\}$, so either t acts on K or $K_1^t = K_2$. In the latter case, (4) holds, so we may assume the former. Then by 3.3.6, one of cases (3)-(5) of that lemma hold. If 3.3.6.3 holds then $t \in z^\perp$ centralizes K, and then t centralizes $O(z)$ by 3.3.4, so that (1) holds. In case 3.3.6.4, (2) holds by 3.3.5. Finally in case 3.3.6.5, t induces an outer automorphism on K, so $t \notin \mathcal{D}^*(z)$, and hence t induces an outer automorphism on K_i for $i = 1, 2$, and hence (3) holds. □

LEMMA 9.1.4. *For each $t \in \mathcal{Z}$ acting nontrivially on K, $tz \in t^K$, so $tz \in \mathcal{Z}$.*

PROOF. Observe $\Omega_1(C_K(t)\langle t\rangle) = \langle z, t\rangle$, so $t^u = tz$ for $u \in N_K(C_K(t)) - C_K(t)$. □

LEMMA 9.1.5. *Let $t \in \mathcal{Z}$ and $E = \langle t, z\rangle$.*
(1) If $t \in \mathcal{D}(z)$ then $z \in \mathcal{D}(t)$.
(2) If $t \in \mathcal{D}^*(z) - \mathcal{D}(z)$ then $z \in \mathcal{D}^*(t) - \mathcal{D}(t)$ and $Aut_{\mathcal{F}}(E) = GL(E)$.

PROOF. From 3.1.13 the relations $\mathcal{D}$ and $\mathcal{D}^*$ are symmetric, so the first remark in (1) and (2) holds. In (2), t is nontrivial on K, so by 9.1.4 there is $k \in K$ with $t^k = tz$. Then by symmetry there is $\beta \in Aut_{\mathcal{F}}(\langle z, t\rangle$ with $z\beta = tz$, so (2) follows. □

NOTATION 9.1.6. For $i = 1, 2$, let $\langle v_i\rangle$ be a $N_S(K_i)$-invariant subgroup of K_i of order 4, and choose $v_2 \in v_1^{C_S(z)}$ if $K_2 \in K^{\mathcal{F}}$. Set $v = v_1v_2$. Let $\eta \in \eta_S(\tau)$, $W = \langle \eta\rangle$, G a model for $N_{\mathcal{F}}(C_S(W))$, and $M = \langle K^G\rangle$. Set $\mu = \mu(\tau)$ and $D = D_\eta$.

LEMMA 9.1.7. *(1) Each member of $\mathcal{D}^o(z)$ inverts $\langle v_1, v_2\rangle$.*
(2) Each member of $\mathcal{D}^o(z) \cup \mathcal{D}^m(z)$ centralizes $\langle v, z\rangle$.

PROOF. Suppose $t \in \mathcal{D}^o(z)$. From 9.1.3, t inverts v_i for $i = 1, 2$, so (1) holds and t centralizes v. On the other hand if $s \in \mathcal{D}^m(z)$ then $v_1^s = v_2^\epsilon$ for some $\epsilon = \pm 1$, so s centralizes v. □

HYPOTHESIS 9.1.8. *(1) Hypothesis* 9.1.1 *holds.*
(2) $\mathcal{F}$ is transitive on Ω.
(3) $\mathcal{D}^*(z) \neq \mathcal{D}(z)$.

Note that (by 3.3.14) Hypothesis 9.1.8 implies that $|\eta(\tau)| = 1$, and hence that W is weakly closed in S with respect to $\mathcal{F}$.

LEMMA 9.1.9. *Assume Hypothesis* 9.1.8. *Then*

(1) Let $\{\eta_i : 1 \leq i \leq r\}$ be the orbits of M on η, $\Omega_i = \{J \in \Omega : J \cap W \in \eta_i\}$, $M_i = \langle J_i^M\rangle$ for $J_i \in \Omega_i$, $W_i = \langle \eta_i\rangle$, and $D_i = \{d_V : V \in \eta_i\}$. Then D is a set of 3-transpositions of μ, $M/W = \mu$, D_i is a conjugacy class of 3-transpositions of M_i/W_i, and $M = M_1 \cdots M_r$ is a central product. Choose notation so that $K \in \Omega_1$; then $K_2 \in \Omega_1$.

(2) $N_G(O(\tau))$ is transitive on Ω and $\{\Omega_1, \ldots, \Omega_r\}$, and G is transitive on $\{M_1, \ldots, M_r\}$ and on η.

(3) $\tau_i = (\mathcal{F}_{S\cap M_i}(M_i), \Omega_i)$ is a quaternion fusion packet with $\mu_i = \mu(\tau_i) = M_i/W_i$, $W_i = W \cap M_i$, and $\mu = \mu_1 \times \cdots \times \mu_r$.

(4) For each $1 \leq i \leq r$, $\mu_i \cong$ Weyl(D_n) for some $n \geq 3$.

(5) Let $V_i = W \cap K_i$ and $d_i = d_{V_i} \in D$. Then $d_2 \in V_{d_1}$ in the 3-transposition groups μ and μ_1.

PROOF. By 9.1.1, $\mathcal{A}(\tau) = \emptyset$, so D is a set of 3-transpositions of μ by 3.1.25. Next $M = M_1 \cdots M_r$ is a central product by 4.2.10.2, so as M is transitive on η_i, so is M_i. Then D_i is a conjugacy class of 3-transpositions of μ_i by 4.2.8.1. Choose notation so that $K \in \Omega_1$; then (1) is established, modulo the assertions that $K_2 \in \Omega_1$, $W_i = C_{M_i}(W_i)$, and $W = C_M(W)$, which we prove in a moment. For example by 3.1.19, $\mu = Aut_M(W) = M/C_M(W)$, so if $W = C_M(W)$ then $\mu = M/W$.

By 9.1.8.2, $\mathcal{F}$ is transitive on Ω, so $N_G(O(\tau))$ is transitive on Ω by 6.1.6.1. Then (2) follows from (1). By 2.5.2, τ_i is a quaternion fusion packet, and by definition $\mu_i = \mu(\tau_i)$. We prove the remaining two assertions in (3) below.

By 3.1.21, $d_1^\perp = d_2^\perp$ in the 3-transposition group μ on D; that is $d_2 \in V_{d_1}$ in μ. By 9.1.8.3, $\mathcal{D}^*(z) \neq \mathcal{D}(z)$, so M_1 is nonabelian, and hence the groups M_i are nonabelian for each i by (2). Then as K centralizes M_i for $i > 1$ and $d_2 \in V_{d_1}$, K_2 also centralizes M_i, and hence $K_2 \in \Omega_1$ as M_i is nonabelian and transitive on D_i. This also shows $d_2 \in V_{d_1}$ in the 3-transposition group μ_1, completing the proof of (5).

By (1), μ_i is transitive on D_i, and by (5), $d_2 \in V_{d_1}$ in μ_1, so it follows from 4.1.7 that $d_2 \in d_1^{O_2(\mu_1)}$. Then by 4.3.5, $\mu_1 \cong \mathrm{Weyl}(D_n)$ for some $n \geq 3$. By (2), G is transitive on the τ_i, so $\mu_i \cong \mu_1$ for each i, establishing (4).

As M is a central product of the M_i, μ is a central product of the μ_i. By 4.3.8.4, $C_{M_i}(W_i) = W_i$, so $C_M(W) = W$, completing the proof of (1). Also $Z(M_i) \leq W_i$, so $\mu = \mu_1 \times \cdots \times \mu_r$. Let $W_{i'} = \prod_{j \neq i} W_j$. Now $W = W_1 \cdots W_r$ and $W_i \leq W \cap M_i = U_i$, so $U_i = W_i(M_i \cap W_{i'}) \leq W_i Z(M_i)$, so as $Z(M_i) \leq W_i$, $U_i = W_i$, completing the proof of (3) and the lemma. □

LEMMA 9.1.10. *Assume Hypothesis* 9.1.8 *and let* $t \in \mathcal{D}^*(z) - \mathcal{D}(z)$. *Then*

(1) $t = k_1 k_2 c$ *where* $k_i \in K_i$ *is of order 4 and* $c \in C_S(O(z))$ *with* $c^2 = 1$.

(2) $v_i \in W$ *ans there exists* $\phi \in \hom_{\mathcal{F}_z}(\langle k_1, k_2, t\rangle, W)$ *with* $k_i\phi = v_i$, $c\phi \in C_S(O(z))$, *and* $t\phi \in \mathcal{Z}_\Delta$.

(3) $\mathcal{F}_z$ *is transitive on* $\mathcal{D}^*(z) - \mathcal{D}(z)$.

(4) The following are equivalent:

(a) $v \in \mathcal{Z}$.

(b) $\mathcal{Z} \cap O(z) \neq \{z\}$.

(c) $n = 3$ *and* $Z(M_1) = 1$, *where* M_i *is as in* 9.1.9.*1 and* n *as in* 9.1.9.*4*.

PROOF. Part (1) follows from 9.1.3. Let $V_i = K_i \cap W$. As $\mathcal{D}^*(z) \neq \mathcal{D}(z)$, $\Omega_2(V_i)$ is the unique W-invariant subgroup of K_i of order 4, ao $\langle v_i\rangle = V_i \leq W$. As $\mathcal{F}$ is transitive on Ω and $z \in \mathcal{F}^f$, there is $\beta \in \mathfrak{A}(t)$ with $t\beta = z$. Then $k_i\beta \in X_i \in \Delta(z\beta)$, and as $t \notin \mathcal{D}(z)$, X_i is cyclic and nontrivial on K_1 and K_2. Then there exists $\delta \in \mathfrak{A}(z\beta)$ with $z\beta\delta = z$. Set $\phi = \beta\delta$. Now $\langle k_1, k_2, t\rangle\beta \leq W$ and $W\delta = W$, so $\phi \in \hom_{\mathcal{F}_z}(\langle k_1, k_2, t\rangle, W)$. As $\Omega(z) = \{K_1, K_2\}$ and $\langle v_i\rangle$ is the subgroup of order 4 in $K_i \cap W$, $\langle k_i\phi\rangle = \langle v_{j(i)}\rangle$. Then conjugating in M_1 we may take $k_i\phi = v_i$, completing the proof of (2).

By (2), there is $s \in t^{\mathcal{F}_z}$ such that s is in the set $\mathcal{S}$ of $s' \in \mathcal{Z}_\Delta$ such that for $s' \in V \in \eta$, $d_V \in A_{d_1}$. By 3.1.2.1, $\mathcal{S} = \mathcal{Z}_\Delta \cap \mathcal{D}^*(z) - \mathcal{D}(z)$. But by 9.1.9.4 and 7.9 in [**Asc97**], $C_{\mu_1}(d_1)$ is transitive on A_{d_1}, so $C_M(z)$ is transitive on $\mathcal{S}$, completing the proof of (3).

Suppose $s \in O(z) \cap \mathcal{Z} - \{z\}$. Then s induces a nontrivial inner automorphism on K, so by 9.1.3, $s \in \mathcal{D}^*(z) - \mathcal{D}(z)$. Then by (1) and (2), there is $\zeta \in \hom_{\mathcal{F}_z}(s, W)$

with $s\zeta = v_1v_2x$ for some $x \in C_S(O(z))$. As $\mathcal{O}(z) \lhd \mathcal{F}_z$ and $s \in O(z)$, also $s\zeta \in O(z)$, so $s\zeta = v$ or vz. Then by 9.1.4, $v \in \mathcal{Z}$. Therefore (4b) implies (4a).

Assume (4a). Then by 3.1.11.7, $v \in \mathcal{Z}_\Delta$, and then by paragraph two of this proof, $v \in V \in \eta$ such that $d_V \in A_{d_1}$.

By 9.1.9.4, $\mu_1 \cong \text{Weyl}(D_n)$, so by 4.3.8.2, M_1 is an image of the universal group $\bar{M}_1 = \bar{\omega}(D_n, m)$ described in 4.3.1. Adopt the notation of 4.3.1, let $\bar{z}_i$ be the involution in $\bar{K}_i$, and z_i the image of $\bar{z}_i$ in M_1. As $|\Omega(z)| = 2$, by 5.8.5 we may take $z = z_{n-1} = z_n$, so $u = z_{n-1}z_n = 1$. As $v \in \mathcal{Z}_\Delta$ with $d_V \in A_{d_1}$, we may take $v = z_{n-2}$. Then as $v \in O(z)$, 5.7.4.3 says $Z(M_0) = 1$, where $M_0 = \langle K, K_2, J\rangle$ and $J \in K^M$ with $z(J) = v$; equivalently $v_1v_2z_{n-2} = 1$ in M_0 and hence also in M_1. Thus $\bar{v}_1\bar{v}_2\bar{z}_{n-2} \in Z(\bar{M}_1)$, so by 5.8.4 and its proof, $n = 3$. That is $M_1 = M_0$, so $Z(M_1) = Z(M_0) = 1$. Therefore (4a) implies (4c).

Finally if (4c) holds then from the discussion above, $v_1v_2z_{n-2} = 1$, so $v \in \mathcal{Z}_\Delta$ and hence (4b) holds, completing the proof of (4) and the lemma. □

LEMMA 9.1.11. *Assume Hypothesis* 9.1.8.

(1) If $m > 8$ then $\mathcal{O}_K = K$ or $\mathcal{O}_K \cong SL_2[m]$.

(2) Assume $m = 8$ and if $n = 3$ assume $v \in \mathcal{Z}$. Let G_0 be a model for $N_{\mathcal{F}_z}(O(z)C_S(O(z)))$. Recall W acts on $O(z)$ by 3.1.12, and set $H_0 = \langle W^{G_0}\rangle O(z)$. Then either

(a) $C_{G_0}(O(z)) \leq N_{G_0}(W)$, so $O^2(H_0)$ centralizes $C_S(O(z))$ and is either trivial or of index 1 or 3 in $SL_2(3) * SL_2(3)$, or

(b) $n = 4$ and $O^2(H_0)$ centralizes $C_S(O(\tau_1))$ and is of index 3^e in $(SL_2(3) * SL_2(3))^2$ for some $0 \leq e \leq 3$.

PROOF. Part (1) follows from 2.5.4 as $WO(z)$ is transitive on the Q_8-subgroups of K, using 4.2.5.1. Thus we may assume $m = 8$ and if $n = 3$ that $v \in \mathcal{Z}$. Set $S_0 = C_S(O(z))$. Arguing as in the proof of 9.1.10, from 9.1.9.4, M_1 is an image of $\bar{M}_1 = \bar{\omega}(D_n, m)$, described in 4.3.1. Adopt the notation of 4.3.1 and 4.3.3; in that notation, we may take $K = K_n$, take $\Omega(z) = \{K, K_{n-1}\}$, and let $V_1' \in \eta_1 - \{V_1\}$ with $z(V_1) = z(V_1')$. Set $W_- = W_2 \cdots W_r$, $A = V_1 \cdots V_{n-3}W_-$, and $B = V_{n-2}V_{n-1}V_n$. If $n > 3$ set $W_0 = V_1'A$, while if $n = 3$ set $W_0 = A$. Then $W = AB$ with $A \leq S_0$, so $C_W(O(z)) = AC_B(O(z))$. Set $M_0 = \langle K_{n-2}, K_{n-1}, K_n\rangle$; then M_0 is an image of $\bar{\omega}(D_3, m)$, so from 5.7.5.3, $C_B(O(z)) = \langle x_0, z\rangle$, where $x_0 = z_{n-2}v$. Then $C_W(O(z)) = A\langle x_0, z\rangle$.

Suppose $n = 3$. Then $W_0 = A = W_-$ and by hypothesis $v \in \mathcal{Z}$, so $v = z^\epsilon z_1$ by 9.1.10.4 and 5.7.4, and hence $x_0 \in \langle z\rangle$. Hence in this case $C_W(O(z)) = W_0\langle z\rangle = W_-\langle z\rangle$.

So assume $n > 3$; then $W_0 = V_1'A$. Let $L = \Omega_{2n}^+(q)$ and define τ_L as in 5.3.4; then τ_L satisfies Hypothesis 9.1.8 with $r = 1$. Thus we can use the uniqueness of M_1, and can check in $\Omega_{2n}^+(q)$ (cf. 5.3.4) that $x_0z \in W_0$, so again $C_W(O(z)) = W_0\langle z\rangle$.

Observe that $W_0 = \langle \eta_0\rangle$, where $\eta_0 = \eta \cap S_0$. Let $\Omega_0 = \Omega - \{K, K_{n-1}\}$, $g \in G_0$, and $V_0 \in \eta_0$; then $V_0 \leq W_0 \leq S_0$. Now $\Omega_0 \subseteq S_0$ and G_0 acts on Ω and $\Omega(z)$, and hence on S_0 and Ω_0. Therefore if $V_0 \in \Delta$ then as $V_0 \leq S_0$, also $V_0^g \in \Delta$, so $V_0^g \in \eta_0$. If $V_0 \notin \Delta$ then $V_0 \leq K_0 \in \Omega_0$ and, as $W_0 \leq S_0$, either V_0 is the unique S_0-invariant subgroup of K_0 of order 4 or $n = 4$ and $V_0 \in \{V_1, V_1'\}$. In the first case $V_0^g \in \eta_0$, so if $n \neq 4$ then $W_0 \lhd G_0$. If $n = 4$ let $W_+ = O(\tau_1)W_-$. Thus $O(\tau_1)$, W_+, and W_- are normal in G_0.

Now $[W, S_0] \leq W \cap S_0 = C_W(O(z)) = W_0\langle z\rangle$ and W centralizes $W_0\langle z\rangle$, so if $W_0 \lhd G_0$ then $W \leq O_2(WC_{G_0}(O(z)))$ and hence $C_{G_0}(O(z)) \leq N_{G_0}(W)$, so

$O^2(H_0)$ centralizes S_0, and hence $C_{O^2(H_0)}(O(z)) \leq Z(S_0)$. Then as $O(z) \cong Q_8^2$ and as $Aut_{\mathcal{F}}(O(z))$ is transitive on $\Omega(z)$, (2a) holds in this case.

So assume $n = 4$ and set $S_1 = C_S(O(\tau_1))$. Here $[W, S_1] \leq [W, S_0] \leq W_0\langle z\rangle \leq W_+$ and W centralizes $W_* = W_- Z(O(\tau_1))$. Then as $O(\tau_1)$ and W_* are normal in G_0, $O^2(H_0)$ centralizes S_1. Therefore each element of odd order in H_0 is faithful on $O(\tau_1)$. Then as $O(\tau_1) = O(z) \times O(z_1) \cong Q_8^2 \times Q_8^2$, as $Aut_{\mathcal{F}}(O(z))$ is transitive on $\Omega(z)$, and as $Aut_{\mathcal{F}}(O(\tau_1))$ is transitive on Ω_1, either $O^2(H) = 1$, and hence (2a) holds, or (2b) holds. □

DEFINITION 9.1.12. Given $t \in \mathcal{D}^m(z)$ and a Q_8-subgroup Q_1 of K, set $E(t, Q_1) = C_{Q_1Q_1^t}(t)\langle t\rangle \leq S$.

LEMMA 9.1.13. *Assume $z \in E \leq S$ with $E \cong E_{16}$, $t \in \mathcal{D}^m(z) \cap E$, and $N_E(K) \leq O(z)$. Let Q_1 be the preimage of the projection of $N_E(K)/\langle z\rangle$ on $K/\langle z\rangle$. Then $E = E(t, Q_1)$ and $Q_1 = N_K(E)$.*

PROOF. Set $F = N_E(K)$, $O = O(z)$, and $\tilde{O} = O/\langle z\rangle$; by hypothesis $E_8 \cong F \leq O(z)$. As $m(K) = 1$, $F \cap K = \langle z\rangle$, so the projection $\tilde{Q}_1$ of $\tilde{F}$ on $\tilde{K}$ is isomorphic to $\tilde{F} \cong E_4$. Hence $Q_1 \cong Q_8$ and then $F \leq Q = Q_1Q_1^t$. Also $[\tilde{Q}, t] = C_{\tilde{Q}}(t) \cong E_4$, so $\tilde{F} = [\tilde{Q}, t]$, so $Q_1 \leq N_K(E)$. Then as $N_{\tilde{K}}(\tilde{F}) = \tilde{Q}_1$, we have $Q_1 = N_K(E)$. □

LEMMA 9.1.14. *Let $t \in \mathcal{D}^m(z)$, Q_1 a Q_8-subgroup of K, $Q = Q_1Q_1^t$, and $E = E(t, Q_1)$.*

(1) $Q \cong Q_8^2$, $E \cong E_{16}$, and Q acts on E with $QE/E \cong E_4$.

(2) t^Q is a basis for E (viewed as an $\mathbf{F}_2$-space) of order 4 and $\prod_{s\in t^Q} s = z$.

(3) There exists $\gamma \in \mathfrak{A}(E)$ with $z\gamma = z$ and $C_Q(t)\gamma \leq O(z)$.

(4) Suppose $\mathcal{F}$ is transitive on Ω and $A = Aut_{\mathcal{F}}(E)$ is irreducible on E. Then either

(a) A preserves a symplectic form f on E, $E^\# = z^A \subseteq \mathcal{Z}$, and A is of index at most 2 in $O(E, f) = Sp(E)$, or

(b) A preserves a quadratic form q on E of sign -1, A is of index at most 2 in $O(E, q) \cong O_4^-(2)$, and $E \cap \mathcal{Z}$ is the set of singular points of E.

PROOF. Part (1) is straightforward and shows that $|t^Q| = |Q : C_Q(t)| = 4$. Set $F = C_Q(t)$, $X = QE$, $\tilde{X} = X/\langle z\rangle$, and let $t^Q = \{t_1, \ldots, t_4\}$ with $t = t_4$. Set $x = t_1 \cdots t_4$. Then $\tilde{t}^Q = \tilde{t}\tilde{F}$, so $\tilde{x} = 1$ and hence $x = 1$ or z. However if $x = 1$ then $t = t_1t_2t_3$ and $E = T \times \langle z\rangle$, where $T = \langle t_1, t_2, t_3\rangle = \langle t^Q\rangle$ is X-invariant. This is a contradiction as $T \cong \tilde{T}$, so $z \notin T$, whereas $\langle z\rangle = Z(X)$. Similarly $E = \langle t^Q\rangle$, so (2) holds.

Let $\delta \in \mathfrak{A}(E)$ and $\bar{E} = E\delta$. As $C_E(Q) = \langle z\rangle$, $N_S(\bar{E})$ centralizes $z\delta$. Let $\beta \in \mathfrak{A}(z\delta)$ such that $z\zeta = z$, where $\zeta = \delta\beta$. Then $N_S(\bar{E})\beta \leq N_S(E\zeta)$, so as $\bar{E} \in \mathcal{F}^f$, also $E\zeta \in \mathcal{F}^f$. Thus replacing $\bar{E}$ by $E\zeta$, we may assume $z\delta = z$, and hence $\delta \in \hom_{\mathcal{F}_z}(N_S(E), C_S(z))$. As $\mathcal{O}(z) \lhd \mathcal{F}_z$ by 2.6.11, $F\delta = (E \cap O(z))\delta \leq O(z)$, completing the proof of (3).

Assume the setup of (4) and set $O = O(z)$. Replacing E by $E\delta$, we may assume $E \in \mathcal{F}^f$. Let $\Gamma = \{Q_1, Q_1^t\}$ and observe that Γ consists of the subgroups $N_J(E)$ for $J \in \Omega$ with $z(J) \in E$ and $N_J(E)$ nonabelian. For example given such a J, with $j = z(J) \neq z$, J centralizes $O(z)$ but j does not. Therefore by 2.5.2, $\rho = (N_{\mathcal{F}}(E), \Gamma)$ is a quaternion fusion packet. Then as A is irreducible on E it follows from Theorem 2 that $N_{\mathcal{F}}(E)^\circ$ is AE_5 or AE_6. Note that conclusion (4) of

Theorem 2 does not hold as $Q_1^e = Q_1^t$ for $e \in z^{N_{\mathcal{F}}(z)} - Q$. Hence (4) holds, modulo the statement in (4b) that $E \cap \mathcal{Z}$ is the set of singular points of E, which we now prove. Suppose the statement fails; then $E^\# \subseteq \mathcal{Z}$.

Let $e \in Q \cap E - \langle z \rangle$ and $\alpha \in \mathfrak{A}(e)$ with $e\alpha = z$. Then e is nonsingular in E. As $e \in \mathcal{D}^*(z)$, $z \in \mathcal{D}^*(e)$ by 9.1.5, so $z\alpha \in O$, since $\mathcal{D}^*(z) - \mathcal{D}(z) \subseteq O$ by 9.1.10.3. Then $X = \langle z^{C_A(e)} \rangle \alpha \leq O$. But as e is nonsingular, $m(X) = 3$, so $m(X) = m(Q) = 3$ and hence $X = E\alpha \cap O$. Let $y \in E\alpha - X$ and set $y_0 = y\alpha^{-1}$. If $y \in \mathcal{D}(z)$ then $Aut_O(E\alpha)$ contains a 4-group of transvections on $E\alpha$ with center z, so $C_A(y_0)$ contains a 4-group of transvections with center e, contradicting $A \leq O(E, q)$. Thus $y \notin \mathcal{D}(z)$, so as y centralizes X and $\mathcal{D}^*(z) - \mathcal{D}(z) \subseteq O(z)$, $y \in \mathcal{D}^m(z)$. Then by 9.1.13, $E\alpha = E(y, P_1)$ for some Q_8-subgroup P_1 of K, so $Aut_O(E\alpha)$ contains a 4-group free on $E\alpha$. But now $C_A(e)$ contains such a 4-group, contradicting e nonsingular. This completes the proof. □

LEMMA 9.1.15. *Assume $O(z) \cap \mathcal{Z} = \{z\}$ and either*

(a) $|\mathcal{Z}_S| = 1$, or

(b) μ is abelian.

Then

(1) If $t \in \mathcal{D}^m(z)$ then $z \in \mathcal{D}^m(t)$.

(2) If $t \in \mathcal{D}^o(z)$ then $z \in \mathcal{D}^o(t)$ and $Aut_{\mathcal{F}}(\langle z, t \rangle) = GL(\langle z, t \rangle)$.

PROOF. If $t \in \mathcal{D}^o(z)$ then by 9.1.5, $z \notin \mathcal{D}^*(t)$, so by 9.1.3, $z \in \mathcal{D}^o(t) \cup \mathcal{D}^m(t)$. Hence (2) follows from (1), using 9.1.4 to establish the last statement in (2), so it remains to prove (1).

Assume $t \in \mathcal{D}^m(z)$, let Q_1 be a Q_8-subgroup of K, $Q = Q_1 Q_1^t$, and $E = E(t, Q_1)$. Adopt the notation in the proof of 9.1.14 and assume $z \notin \mathcal{D}^m(t)$. Then by 9.1.3 and 9.1.5, $z \in \mathcal{D}^o(t)$. Let $\alpha \in \mathfrak{A}(t)$ with $t\alpha = z$; as $z \in \mathcal{D}^o(t)$, we have $z\alpha \in \mathcal{D}^o(z)$. Set $s_i = t_i\alpha$ for $1 \leq i \leq 4$, and $s = z\alpha$. By 9.1.14.2, $t_1t_2t_3t_4 = z$, so $s_1s_2s_3s_4 = s$. Hence as $s \in \mathcal{D}^o(z)$, an even number r of the s_i are in $\mathcal{D}^m(s_4)$, so $r = 0$ or 2. For $i \leq 3$, $tt_i \in F - \langle z \rangle$, so as $\mathcal{Z} \cap O = \{z\}$, $tt_i \notin \mathcal{Z}$, and hence by 9.1.3 and 9.1.4, $s_i \in \mathcal{D}(s_4) \cup \mathcal{D}^m(s_4)$. Thus we may assume $s_3 \in \mathcal{D}(s_4)$. In particular $|\mathcal{Z}_S| > 1$, so the lemma holds when (a) is satisfied. Thus we may assume (b) is satisfied, so μ is abelian.

Let $\delta \in \mathfrak{A}(s_3)$. By 3.1.14, s_3 centralizes $O(s_4)$, so $s_4\delta \in \mathcal{Z}_S$ while $s_3\delta \in \mathcal{Z}_S$ by 3.1.2.2. Next $f = tt_3 \in F$, so $f = u_1u_2$ with $u_i \in K_i$ of order 4. Also f centralizes each member of $\Omega_0 = \Omega - \Omega(z)$. Let $\beta \in \mathfrak{A}(f)$ and $\gamma \in \mathfrak{A}(f\alpha\delta)$ with $f\alpha\delta\gamma = f\beta = b$. Now $f\alpha\delta = s_3\delta s_4\delta \in Z(O(\tau))$, so $O(\tau)\gamma = O(\tau)$ and hence $b \in Z(O(\tau))$.

Let $\rho \in \mathfrak{A}(z\beta)$ with $z\xi = z$, where $\xi = \beta\rho$. If $z\beta \in \mathcal{Z}_S$ then $O(\tau)\rho = O(\tau)$, so $f\xi = f\beta\rho \in Z(O(\tau))$. This is impossible as ξ is an $\mathcal{F}_z$-map and $\mathcal{O}(z) \trianglelefteq \mathcal{F}_z$ but $f \notin Z(O(z))$. Therefore $z\beta \notin \mathcal{Z}_S$. But $u_i\beta \in Y_i \in \Delta(z_i\beta)$, so $z\beta \in \mathcal{Z}_\Delta - \mathcal{Z}_S$, contrary to (b) and 3.1.21. □

LEMMA 9.1.16. *Let $U = \langle z, v \rangle$ and $B = \langle U, v_1 \rangle$ Then*

(1) $B \trianglelefteq N_{\mathcal{F}_z}(U)$.

(2) Suppose $B \leq P \leq C_S(U)$ and $\alpha \in Aut_{\mathcal{F}}(P)$ is a 3-element with $U = [U, \alpha]$. Then $n = 3$, $v \in \mathcal{Z}$, and $B \leq [B, \alpha] \cong \mathbf{Z}_4^2$.

PROOF. As $U \trianglelefteq C_S(z)$, $U \in \mathcal{F}_z^f$. Then as $z \in \mathcal{F}^f$, also $U \in \mathcal{F}_f$. Let $\mathcal{E} = N_{\mathcal{F}_z}(U)$ and $\tilde{O}(z) = O(z)/\langle z \rangle$. Now $N_{O(z)}(U) = O(z)$, so as $\mathcal{O}(z) \trianglelefteq \mathcal{F}_z$, we have $O(z) \trianglelefteq \mathcal{E}$ by 1.2.12. Therefore $\mathcal{E}$ permutes $\Omega(z)$, and hence also the set $\{\tilde{v}_1, \tilde{v}_2\}$ of projections of $\tilde{v}$ on $\tilde{K}_i$. Therefore $\mathcal{E}$ permutes $\{\langle v_1 \rangle, \langle v_2 \rangle\}$, so (1) holds.

Assume the setup of (2). Then α is transitive on $U^{\#}$, so $v \in z^{\mathcal{F}}$. Therefore $n = 3$ and $Z(M_1) = 1$ by 9.1.10.4. By (1), $A = \langle v_1^{N_{\mathcal{F}}(U)} \rangle = B \cdot B\alpha \cdot B\alpha^2 \leq O_2(C_{\mathcal{F}}(U))$. By 3.1.3.2, $v_1\alpha \in D \in \Delta(z\alpha)$, so $D \in \eta_1$. Thus $A \leq W_1$ and as $n = 3$ and $Z(M_1) = 1$, $A = \langle v_1, v_1\alpha \rangle \cong \mathbf{Z}_4^2$ by 5.7.4, and $A = [A, \alpha] = [B, \alpha]$. □

9.2. Generation when $|\Omega(z)| = 2$

In this section we assume the following hypothesis:

HYPOTHESIS 9.2.1. *$\tau = (\mathcal{F}, \Omega)$ is a quaternion fusion packet such that $\mathcal{A}(\tau) = \emptyset$, $\mathcal{F}$ is transitive on Ω, and $z \in \mathcal{Z}_S \cap \mathcal{F}^f$ with $|\Omega(z)| = 2$. Assume $R \in \mathcal{F}^e$ with $z \in Z(N_S(R))$ and let G be a model for $N_{\mathcal{F}}(R)$, and set $Z = \langle z^G \rangle$ and $G^* = G/R$. After the first lemma we assume in the remainder of the section that $Aut_{\mathcal{F}}(R) \neq Aut_{\mathcal{E}(\tau)}(R)$.*

NOTATION 9.2.2. We adopt the notation of section 9.1, particularly Notation 9.1.2 and 9.1.6. Set $H = \langle v_1^G \rangle$; this definition makes sense as $v_1 \in G$ from 9.2.3.3.

LEMMA 9.2.3. *(1) $Z \leq Z(R)$.*
(2) $v \in R$.
(3) $v_1 \in N_S(R)$.
(4) If $Aut_{\mathcal{E}(\tau)}(R) \neq Aut_{\mathcal{F}}(R)$ then $v_1 \notin R$.

PROOF. As $z \in Z(N_S(R))$, (1) holds. Further $[R, v] \leq [C_S(z), v] = \langle z \rangle$, so $v \in N_S(R) \leq G$.

Suppose v centralizes Z. Then as $[R, v] \leq \langle z \rangle \leq Z$ and $F^*(G) = R$, (2) holds. Then $[R, v_1] \leq [C_S(z), v_1] \leq \langle v, z \rangle \leq R$, so (3) holds. Part (4) follows from 3.4.5.

Thus we may assume $[Z, v] \neq 1$ and it remains to derive a contradiction. Hence there exists $g \in G$ such that $t = z^g$ does not centralize v. Set $H = \langle v^G \rangle$. As $[R, v] = \langle z \rangle$, it follows from 2.4.3 in [**Asc19**] that $H \cong S_4$, $G = H \times C_G(H)$, and $R = C_S(t)$.

As $[t, v] \neq 1$, it follows from 9.1.3 and 9.1.7.2 that $t \in \mathcal{D}^*(z) - \mathcal{D}(z)$. Then by 9.1.10.1, $t = k_1k_2c$ with $k_i \in K_i$ of order 4 and $c \in C_S(O(z))$ with $c^2 = 1$. As $[t, v] \neq 1$ there exists a unique i such that k_i centralizes v_i, say $i = 1$. Then $v_1 \in C_S(t) = R$, so $v_1^g \in R$. But then as $\langle v, z \rangle \trianglelefteq C_S(z)$, $t = (v_1^g)^2$ centralizes v, a contradiction. □

LEMMA 9.2.4. *(1) $[H, Z] \neq 1$.*
(2) $R = C_S(Z)$.
(3) $v_1 \in O_2(C_G(z))$.

PROOF. Let $\mathcal{Y} = C_{N_{\mathcal{F}}(R)}(z)$. As $z \in Z(N_S(R))$, $z \in N_{\mathcal{F}}(R)^f$, so $\mathcal{Y} = N_{\mathcal{F}_z}(R)$. Observe $C_G(z)$ is the model for $\mathcal{Y}$ in G. By 2.6.11.1 and 1.2.12, $N_{\mathcal{O}(z)}(R) \trianglelefteq N_{\mathcal{F}_z}(R)$, so the model X for $N_{\mathcal{O}(z)}(R)$ is normal in $C_G(z)$. Note $N_{O(z)}(R)$ is Sylow in X. As $F^*(X) = O_2(X)$ and $[v_1, O(z)] = \langle z \rangle$, it follows that $v_1 \in O_2(X)$, so as $X \trianglelefteq C_G(z)$, (3) follows.

By 9.2.3.4, $H^* \neq 1$. Suppose (1) fails. Then $H \leq C_G(z)$, so $v_1 \in O_2(H)$ by (3). Thus $v_1 \in R$ as $H \trianglelefteq G$, contrary to 9.2.3.4. This establishes (1). Now (2) follows from (1) and 3.4.9.2 applied to Z in the role of Q. □

LEMMA 9.2.5. *Assume $Z \leq N_S(K)$. Then*

(1) $Z = \langle z, t\rangle$ for some $t \in \mathcal{D}^o(z)$, $O_2(H) = C_H(Z)$, and $H^* = GL(Z)$.

(2) $O_2(H) = U \cong E_{16}$ with $U = Z \times E$, where $E = \langle v^H\rangle$ does not act on K and $H/U \cong S_3$.

(3) $R_0 = C_R(U)$ is of index 2 in R, with $E' = E^r$ the third irreducible for H on U, for $r \in R - R_0$.

(4) $R_0 = U \times C_R(H)$.

(5) There exists $y \in O(z)$ inducing a transvection on U with center v.

(6) Let G_U be a model for the stabilizer in $N_{\mathcal{F}}(R_0)$ of U containing H, and set $H_U = \langle H^{G_U}, y\rangle \leq G_U$. Then $U = O_2(H_U)$, $H_U R_0 = H_U \times C_R(H)$, and there is an H_U-invariant quadratic form q on U of sign $+1$ such that H_U/U is the subgroup of $O(U,q)$ generated by transvections. Further $G_U = H_U R$, $G_U/R_0 = O(U,q)$, and z^{H_U}, v^{G_U} is the set of singular, nonsingular points in U, respectively.

(7) If $J \in \Omega$ then $t \notin \mathcal{D}^*(z(J))$.

PROOF. As Z acts on K, and H does not centralize Z by 9.2.4.1, v_1 induces a transvection on Z with center z. Then by 2.4.1 in [**Asc19**], $Z \cong E_4$ and $H/C_H(Z) = GL(Z)$. By 9.2.4.3, $v_1 \in O_2(C_G(z))$, so $[v_1, C_H(Z)] \leq O_2(C_H(Z)) \leq R$ by 9.2.4.2, so v_1 centralizes $C_H(Z)^*$. Then as $H = \langle v_1^H\rangle$ we have $C_H(Z)^* \leq Z(H^*)$. But by 9.2.4.2, $C_H(Z)^*$ is of odd order, so as $H^*/C_H(Z)^* \cong S_3$, we have $O_2(H) = R \cap H = C_H(Z)$. Therefore $H^* = H/C_H(Z)$.

Let $t \in Z - \langle z\rangle$, and recall $t \in Z \leq N_S(K)$ and $Z^\# \subseteq Z^H$. If $t \in \mathcal{D}^*(z)$ then by 9.1.10.1, $t = k_1k_2c$ with $k_i \in K_i$ of order 4 and $c \in C_S(O(z))$. But then $k_i \in C_S(Z) = R$ by 9.2.4.2, contrary to 3.4.5. Thus $t \in \mathcal{D}^o(z)$ by 9.1.3, completing the proof of (1). As $t \in \mathcal{D}^o(z)$, there is $y \in O(z) - \langle v_1, v_2\rangle$ with $[t, y] = v$.

If R acts on K then $[R, v_1] = \langle z\rangle$, so $H \cong S_4$ and $G = H \times C_G(H)$ by 2.4.3 in [**Asc19**]. In particular H acts on each overgroup of Z in R. In this case set $U = \langle Z, v\rangle$, so that HR acts on U.

Suppose on the other hand that R does not act on K. Then for $r \in R - N_R(K)$, $[v_1, r] = v$ or vz. Set $G^+ = G/Z$ and $E = \langle v^H\rangle$. Then v_1 induces a transvection on R^+ with center v^+, so $E^+ \cong E_4$ and H induces $GL(E^+)$ on E^+. In this case set $U = ZE$. As H is transitive on $E^{+\#}$, as $v \in E$ is an involution, and as $Z \leq Z(U)$, it follows that $U \cong E_{16}$. Let $H_0 = \langle v_1, v_1^h\rangle$ for $h \in H$ of order 3. As $|U : C_U(v_1)| = 4 = |R : C_R(v_1)|$, it follows that $R = UC_R(H_0)$ and then $H = H_0U$ with $H_0 \cong S_4$. Notice that we have established (2) in the cae $R \not\leq N_S(K)$.

Now remove the assumption that R does not act on K. By construction, HR acts on U. By 9.2.4.2, $C_S(U) \leq C_S(Z) = R$, so H acts on $C_S(U)$ and hence we, replacing U by a fully normalized conjugate, we may assume $U$is fully normalized. Now pick G_U to be the stabilizeer of U in a model for $N_{\mathcal{F}}(C_S(U))$. We may choose G_U so that $H \leq G_U$.

Suppose $U = \langle Z, v\rangle$. Then y acts on U as $[t, y] = v$. Set $Y = \langle HU, y\rangle \leq G_U$ and $U_0 = Z(HU)$. Now as $G = H \times C_G(H)$, we have $HU \cong \mathbf{Z}_2 \times S_4$ and $C_U(y) = \langle v, z\rangle \geq U_0$, so Y/U is the stabilizer in $GL(U)$ of U_0. Further R acts on HU and hence on U_0 with $O_2(Y) = [O_2(Y), H]$ and H is irreducible on $O_2(Y)/U$. Thus $[R, O_2(Y)] \leq U \leq R$, so $O_2(Y) \leq N_S(R) \leq G$, and then as $O_2(Y) = [O_2(Y), H]$, $O_2(Y) \leq O_2(H) \leq R$. This is a contradiction as $O_2(Y)$ does not act on Z, and this contradiction completes the proof of (2).

Set $u = v^h$. Then $U = \langle u, v, Z\rangle$ and as E does not act on K, we have $K^u = K_2$. Now $[u, v_1] \in E \cap \langle v, z\rangle$, so $[u, v_1] = v$ and $v_i^u = v_{3-i}^{-1}$. Next as $[t, y] = v$, $y = y_1y_2$

with $y_i \in K_i$ and $[t, y_i] = v_i$. Thus $y_1^u = y_2 v_2^\beta$ for some β, so $y^u = yv^\beta$. If $\beta \equiv 0$ mod 4 then u centralizes y, so y centralizes $E\langle z\rangle$ and (5) holds in this case. If $\beta \equiv 2 \mod 4$ then replacing y by yz and y_1 by $y_1 z$, we reduce to the previous case. Finally suppose $\beta \equiv \pm 1 \mod 4$. If $\beta = -1$, then replacing y by yz and y_1 by $y_1 z$, we reduce to the case $\beta = 1$, where y centralizes $\langle z, v, tu\rangle = E'\langle z\rangle$. Thus in each case, $Y = \langle H, y\rangle \leq G_U$ and (5) holds. Moreover y centralizes $F\langle z\rangle$ for some $F \in \{E, E'\}$.

Set $R_0 = C_R(U)$. Now H has three irreducibles Z, E, and $E' = \langle (vz)^H\rangle$ on U, so either $R = R_0$ and $R = U \times C_R(H)$, or $|R : R_0| = 2$, $R_0 = U \times C_R(H)$, and $E^r = E'$ for $r \in R - R_0$. Also $[E, v_1] \not\leq Z$ so E does not act on K.

As y centralizes F and H acts on F, Y acts on F. If $R = R_0$ then R is Sylow in $C_{G_U}(U)$, so y acts on R and hence $R \trianglelefteq Y$, so $Y \leq G$, a contradiction as y does not act on Z. Therefore $R \neq R_0$ and in particular $E^r = E'$ for $r \in R - R_0$, so B is irreducible on U. Hence (3) and (4) hold.

Set $X = \langle Y, R\rangle$, $B = Aut_X(U)$, and $D = \langle \alpha^B\rangle$, where α is the transvection induced by y on U. As B is irreducible on U, it follows from [**McL69**] that either D is irreducible on U and $B = D = GL(U)$, $Sp(U)$, or $O(U, q)$ for some quadratic form q on U of sign -1, or D is not irreducible and $B = O(U, q)$ with q of sign $+1$. Further setting $\Sigma = Aut_{\mathcal{F}}(U)$, as $Aut_H(Z) = GL(Z)$, $Aut_B(Z) = Aut_\Sigma(Z) = GL(Z)$. Also $O_2(N_\Sigma(Z)) = Aut_R(U) \cong \mathbf{Z_2}$ with $Z = C_U(R)$. We conclude that $B = O(U, q) = \Sigma = G_U/R_0$ with (U, q) of sign $+1$ and Z is a totally singular line in this orthogonal space. Therefore (6) holds.

Finally suppose $J \in \Omega$, $j = z(J)$, and $t \in \mathcal{D}^*(j)$. As $t \in \mathcal{D}^o(z)$, $j \neq z$, so $O(z)$ centralizes J. Now there is $x \in J$ of order 4 centralized by t. Then $x \in C_S(Z) = R$, contrary to 3.4.5. This establishes (7). □

LEMMA 9.2.6. $\mathcal{D}^m(z) \neq \emptyset$.

PROOF. If $Z \not\leq N_S(K)$ then there is $g \in G$ with $z^g \in \mathcal{D}^m(z)$, so we may assume $Z \leq N_S(K)$. Thus the hypotheses of 9.2.5 are satisfied; adopt the notation of that lemma. By 9.2.5.2, U does not act on K, so by 9.2.5.6, there is $g \in G_U$ with $z^g \notin N_U(K)$, so $z^g \in \mathcal{D}^m(z)$. □

LEMMA 9.2.7. *Assume $Z \not\leq N_S(K)$ and set $U = \langle z, v\rangle$. Then*

(1) Either

(a) $[Z, v_1] = U$, or

(b) $Z = Z_0 \times C_Z(H)$; $O^2(H) = XZ_0 \cong A_4$, where $X \cong \mathbf{Z}_3$; $Z_0 = \langle v^G\rangle$; $C_Z(H) = \langle v_1 v_1^t\rangle \cong \mathbf{Z}_2$ for $t \in z^G \cap \mathcal{D}^m(z)$; $H = O^2(H)\langle w\rangle$, where $w = v_1 t$ inverts X and $w^2 = v_1 v_1^t$.

(2) Assume $[Z, v_1] = U$. Then $C_H(Z) = R \cap H$ and $O^2(G^*) = O^2(H^*)$. Moreover one of the following holds:

(i) $m(Z) = 4$ and $H^* \cong D_{10}$, A_5, or $\mathbf{Z}_2/E_9$. If $H^* \cong A_5$ then Z is the A_5-module or $L_2(4)$-module for H^*, and in the latter case $v \in \mathcal{Z}$.

(ii) $m(Z) = 5$ or 6 and $Z/C_Z(H)$ is the $L_2(4)$-module for $H^* \cong L_2(4)$.

(3) Suppose $m(Z) = 4$ and $N_Z(K) \leq O(z)$. Let Q_1 be the preimage of the projection of $N_Z(K)/\langle z\rangle$ on $K/\langle z\rangle$ and $t \in z^G - N_Z(K)$. Then $Z = E(t, Q_1)$ is the A_5-module for $H^* \cong A_5$ and $Q_1 = N_K(Z)$.

(4) If Z is the A_5-module for $H^* \cong A_5$ then z is singular in the othogonal space Z and $N_Z(K) \leq O(z)$.

(5) Assume $[Z, v_1] = U$ *and* Z *is the* A_5*-module if* $H^* \cong A_5$. *Then* $C_H(Z) = Z$ *and there is a complement* L *to* Z *in* H, $G = ZN_G(L)$, *and* $R = Z \times C_R(H)$. *In particular* $\Phi(C_S(Z)) \leq C_R(H)$.

(6) Assume $m(Z) = 4$. *Then* $m = 8$ *and for* $t \in z^G - N_Z(K)$ *and* $a \in N_Z(K) - U$ *there is* $x \in C_{O(z)}(t) - U$ *such that* $a = xc$ *for some* $c \in C_S(O(z))$ *centralizing* Z.

(7) Suppose $Z/C_Z(H)$ *is the* $L_2(4)$*-module for* $H^* \cong L_2(4)$. *Then* $Z = O_2(O^2(H))$ *and* $H = O^2(H)\langle v_1 \rangle$.

PROOF. Sete $G^+ = G/Z$. As $Z \not\leq N_S(K)$ there is $g \in G$ such that $t = z^g \in \mathcal{D}^m(z)$. Thus $[v_1, t] = v \in Z$ and $[v_1, C_H(Z)] \leq U \leq Z$ by 9.1.16.1, so H centralizes $C_H(Z)^+$. In particular $O^2(C_H(Z)) = 1$, so $C_H(Z) = O_2(H) = R \cap H$. Hence $H^* = H/C_H(Z)$ and $C_H(Z)^+ \leq Z(H^+)$.

Suppose that $[v_1, Z] = \langle v \rangle$. Then v_1 induces a transvection on Z with center v, so by 2.4.1 in [**Asc19**], $Z = \langle v, z, t \rangle \cong E_8$, $Aut_H(Z) \cong S_3$, and as an H-module, $Z = \langle v^G \rangle \times C_Z(H)$, with $C_Z(H) = \langle v_1 v_1^t \rangle$ as $v_1 v_1^t = z[v_1, t]$.

Let $H_0 = O^2(H)\langle v_1 \rangle$. As $C_H(Z)^+ = Z(H^+)$ we conclude $O^2(H) = XZ_0$, where $Z_0 = \langle v^G \rangle$. Now $O^2(H)$ is transitive on U^G, and hence as $N_G(U) = C_G(U)$ and $v_1^t = v_2$, $O^2(H)$ is also transitive on $\langle v_1 \rangle^G$, using 9.1.16.1. Now (1b) holds and the proof of (1) is complete.

So assume that $[Z, v_1] = U$. Then by 2.8.15, H^* and its action on Z are described in that lemma. We claim

(*) There exists no 3-element $g \in G$ centralizing v_1^* with $U = [U, g]$.

For if such a g exists then $[v_1, g] \in R \leq C_G(Z)$. We apply 9.1.16.2 to $P = \langle v_1 \rangle R$ and $\alpha = c_g$, to conclude that $B = U\langle v_1 \rangle \leq [B, g]$. This is a contradiction as $B \not\leq R \geq [P, g]$.

By (*), H^*, Z does not appear in case (5) of 2.8.15. Similarly:

(**) In cases (1), (2), and (4) of 2.8.15, there is no element g^* of order 3 in $C_{G^*}(H^*)$ with $Z = [Z, g]$.

Suppose that case (1) of 2.8.15 holds with $H^* \cong S_3$. Then $H^* = Aut(H^*)$ so $G^* = H^* \times C_{G^*}(H^*)$. As G^* has a strongly embedded subgroup, $C_{G^*}(H^*)$ is of odd order, and then $C_{G^*}(H^*) = 1$ by (**). Therefore $G^* = H^*$. But now for each $x \in Z^\#$ with $C_{G^*}(x)$ of even order, $\langle x^G \rangle \cong E_4$, contradicting $Z = \langle z^G \rangle$ of rank 4.

We've shown that either $m(Z) = 4$ and case (1), (2), or (3) of 2.8.15 holds with $H^* \cong D_{10}$, A_5, or $\mathbf{Z}_2/E_9$; or case (4) of 2.8.15 holds. This establishes (2).

Assume the setup of (3). Then $Z = E(t, Q_1)$ and $Q_1 = N_K(Z)$ by 9.1.13, and then (3) follows from comparing the description of $Aut_{\mathcal{F}}(Z)$ in 9.1.14.4 to that in (2). Conversely assume the setup of (4). As $z \in Z(N_S(R))$, z is singular in Z. Then $Z_0 = \langle v^{C_H(z)} \rangle$ is a hyperplane of Z, and as $v \in O(z)$ and $\mathcal{O}(z) \trianglelefteq \mathcal{F}_z$, also $Z_0 \leq O(z)$. Thus $Z_0 = N_Z(K)$, proving (4).

Let $H_0 = O^2(H)\langle v_1 \rangle$. As H^* has a strongly embedded subgroup, $O^2(H^*)$ is transitive on the involutions in H^*. By (2) and (**), $C_{G^*}(v_1^*)$ acts on $[Z, v_1] = U$ as the 2-group $Aut_{N_S(R)}(U)$, so as $z \in Z(N_S(R))$, we have $C_G(v_1^*) \leq C_G(z)$. Then by 9.1.16.1, $C_G(v_1^*)$ permutes $\{\langle v_1 \rangle, \langle v_2 \rangle\}$. Hence as $t \in Z$ with $v_1^t = v_2$, it follows that H_0 is transitive on $\langle v_1 \rangle^G$. Therefore $G = N_G(\langle v_1 \rangle)H_0$, so $H = \langle v_1^G \rangle = \langle v_1 \rangle^{H_0} = H_0$.

Recall that $C_H(Z)^+ \leq Z(H^+)$. Therefore if H^* is D_{10} or $\mathbf{Z}_2/E_9$ then $O^2(H) = XZ$, where X is a Hall $2'$-subgroup of H; thus $H = XZ\langle v_1 \rangle$ and $Z = C_H(Z)$. If Z is the A_5-module for $H^* \cong A_5$ then $O^2(C_H(z))^* \cong A_4$, so as $O^2(C_H(z))$ acts on $K_H = K \cap O_2(C_H(z))$ we have $v_1 \in K_H \leq O^2(C_H(z))$, so $H = O^2(H)$. As

$(R \cap H)^+ \leq Z(H^+)$ we have $H^+ \cong A_5$ or $SL_2(5)$, and the former holds as v_1^+ is an involution in H^+.

We next prove (5), so assume the setup of (5). By the previous paragraph, $C_H(Z) = Z$, so $H^* \cong H^+$. Moreover Z is a projective H^*-module, so there is a complement L to Z in H, and Z is transitive on complements to Z in H. As $[v_1, R] = Z_1 \leq Z$, v_1 centralizes R^+, and hence $H = \langle v_1^G \rangle$ centralizes R^+. Then as H is transitive on complements to Z in H, by a Frattini argument, $G = ZN_G(L)$ and as H centralizes R^+, $R = Z \times C_R(H)$. By 9.2.4.2, $R = C_S(Z)$ so as $R = Z \times C_R(H)$ and $\Phi(Z) = 1$, it follows that $\Phi(C_S(Z)) \leq C_R(H)$, establishing (5).

We next prove (6), so assume $m(Z) = 4$ and let $O = O(z)$ and $a \in N_Z(K) - U$. Then $C_O(t) = \langle z \rangle \times X$ where $X = \{x_1 x_1^t : x_1 \in K\} \cong D_{m/2}$. As $U = [Z, v_1]$, a inverts v_1. Thus if $m > 8$ then for $x_1 \in K$ of order 8 with $x_1^2 = v_1$, $x_1^a = x_1^{-1} z^\epsilon$. Set $x = x_1 x_1^t$; then x is of order 4 with $x^a = x^{-1}$, so x induces a transvection on Z with center $v = v_1 v_1^t$ and axis $U\langle t \rangle$. But now (2) supplies a contradiction, since G^* has a strongly embedded subgroup and $H^* \trianglelefteq G^*$.

Therefore $m = 8$. Suppose next that a induces an outer automorphism on K. Then for $x_1 \in K - \langle v_1 \rangle$ and $x = x_1 x_1^t$, $[x, u] = v$, so again x induces a transvection on Z with center v and axis $U\langle t \rangle$, for the same contradiction. Hence a induces an inner automorphism on K inverting v_1 so $a = x_1 c_1$ for some $x_1 \in K - \langle v_1 \rangle$ and $c_1 \in C_S(K)$. Then as a centralizes t and $K^t = K_2$, $a = xc$ where $x = x_1 x_1^t$ and $c \in C_S(O)$ centralizes t and $c^2 = 1$. Further c centralizes $\langle U, a, t \rangle = Z$. This completes the proof of (6).

Finally assume the setup of (7). Then as v_1^+ is an involution and $C_H(Z)^+ \leq Z(H^+)$, it follows that $H^+ = O^2(H^+)\langle v_1^+ \rangle$ is isomorphic to $L_2(4)$, $L_2(4) \times \mathbf{Z}_2$, or $SL_2(5) * \mathbf{Z}_4$. In the first two cases, (7) holds. In the third case, setting $S_H = S \cap H$, we have $S_H^+ \cong Q_8 * \mathbf{Z}_4$ does not centralize v_1^+, contrary to 9.1.16.1 and the fact that S_H centralizes U. This completes the proof of (7) and the lemma. □

9.3. $|\Omega(z)| = 2$ and $\mathcal{Z} \cap O(z) \neq \{z\}$

In this section we assume the following hypothesis:

HYPOTHESIS 9.3.1. *(1) $\tau = (\mathcal{F}, \Omega)$ is a quaternion fusion packet such that $\mathcal{A}(\tau) = \emptyset$ and for $z \in \mathcal{Z}_S$, $|\Omega(z)| = 2$.*

(2) $\mathcal{F}^\circ$ is transitive on Ω.

(3) For $z \in \mathcal{Z}_S$, $\mathcal{Z} \cap O(z) \neq \{z\}$.

(4) The Inductive Hypothesis holds: Theorem 1 holds in each proper subpacket of τ satisfying the hypotheses of Theorem 1 and of the form $(\mathcal{Y}, \Gamma)$ such that $\mathcal{Y} \leq \mathcal{F}$ and $\Gamma \subseteq \Omega$.

LEMMA 9.3.2. *Hypothesis* 9.1.8 *is satisfied.*

PROOF. Hypothesis 9.1.1 holds by 9.3.1.1. Condition (2) of 9.1.8 holds by 9.3.1.2. Let $z \in \mathcal{Z}_S$ with $z \in \mathcal{F}^f$ and $K \in \Omega(z)$. By 9.3.1.3, there is $u \in \mathcal{Z} \cap O(z) - \{z\}$. Then u induces a nontrivial inner automorphism on K, so by 9.1.3, $u \in \mathcal{D}^*(z) - \mathcal{D}(z)$, verifying condition (3) of 9.1.8. □

NOTATION 9.3.3. Adopt Notation 9.1.2 and 9.1.6, plus the notation in 9.1.9. Set $O = O(z)$.

LEMMA 9.3.4. *(1)* $M = M_1 \times \cdots \times M_r$ *with* $Z(M_i) = 1$.
(2) $\mu = \mu_1 \times \cdots \times \mu_r$ *with* $\mu_i = M_i/W_i \cong S_4$.

PROOF. By 9.1.9, $M = M_1 \cdots M_r$ is a central product, $\mu = \mu_1 \times \cdots \times \mu_r$, and $\mu_i = M_i/W_i \cong \mathrm{Weyl}(D_n)$ for some n. By 9.3.1.3 and 9.1.10.4, $n = 3$ and $Z(M_i) = 1$, so the central product $M = M_1 \cdots \times M_r$ is direct. Finally $\mathrm{Weyl}(D_3) = \mathrm{Weyl}(A_3) \cong S_4$. □

LEMMA 9.3.5. $\mathcal{F}_z$ *is transitive on* $O \cap \mathcal{Z} - \{z\} = \mathcal{D}^*(z) - \mathcal{D}(z)$.

PROOF. By 9.1.10.3, $\mathcal{F}_z$ is transitive on $\mathcal{D}^*(z) - \mathcal{D}(z)$, so as $\mathcal{O}(z) \trianglelefteq \mathcal{F}_z$, it follows from 9.3.1.3 and 9.1.3 that $\mathcal{D}^*(z) - \mathcal{D}(z) = O \cap \mathcal{Z} - \{z\}$. □

Recall from 9.1.9 that $W_i = \langle \eta_i \rangle = W \cap M_i$, and let S_i be Sylow in M_i and $J_i \in \Omega \cap M_i$.

LEMMA 9.3.6. *Suppose* $m = 8$. *Then*
(1) $W_1 \cong \mathbf{Z}_4^2$ *and* $\mu_1 = M_1/W_1 \cong S_4$.
(2) $O \cap W_1$ *is of index 2 in* W_1 *and* $w \in W_1 - O$ *induces an outer automorphism on* K_i *for* $i = 1, 2$.
(3) $N_{S_1}(K) = OW_1$ *and each involution in* $N_{S_1}(K)$ *is in* O.
(4) Each involution s *in* S_1 *with* $K_1^s = K_2$ *is conjugate in* M_1 *to an element of* $O - W_1$.

PROOF. By 9.3.4.1, $Z(M_1) = 1$, so $W_1 \cong \mathbf{Z}_4^2$ from 5.7.4, which says that $W_1 \cong \mathbf{Z}_4^3/\mathbf{Z}_4$. By 9.3.4.2, $\mu_1 = M_1/W_1 \cong S_4$, completing the proof of (1).

Next $O_1 = O \cap W_1 = \langle v_1, v_2 \rangle \cong \mathbf{Z}_2 \times \mathbf{Z}_4$, so O_1 is of index 2 in W_1 by (1). From 4.2.5.1, $w \in W_1 - O$ induces an outer automorphism on K_i for $i = 1, 2$, proving (2).

Next from (1) and (2), $N_{S_1}(K) = OW_1$ and $|OW_1 : O| = 2$. For $k_i \in K_i - W_1$, $a_i = k_i v$ is an involution in $k_i W_1$ and $[W_1, a_i] = [W_1, k_i] = \langle v_i \rangle$ is the subgroup of W_1 inverted by a_i, so the involutions in $k_i W_1$ are the members of $a_i \langle v_i \rangle \subseteq k_i O \subseteq O$. On the other hand $k = k_1 k_2$ is an involution centralizing $\Omega_1(W_1)$ and inverting v_1 with $y_1^k = y_2$, where $\langle y_i \rangle$, $i = 1, 2$, are the cyclic subgroups of order 4 of W_1 containing v. Thus $\langle v_1, v_2 \rangle$ is the subgroup of W_1 inverted by k, so the set of involutions in kW_1 is $k\langle v_1, v_2 \rangle \subseteq O$. This establishes (3).

Choose s as in (4). By (3), $N_{S_1}(K) = OW_1$, so as $\mu_1 \cong S_4$, $s \in O_2(M_1) - W_1$ and then $s^g \in kW_1$ for some $g \in M_1$. But we just saw that all involutions in kW_1 are in O, so (4) holds. □

LEMMA 9.3.7. *(1)* $C_{S_1}(O) = C_{W_1}(O)$ *is cyclic of order* $m/4$, $|W_1 : W_1 \cap OC_{S_1}(O)| = 2$, *and* $w \in W_1 - OC_{S_1}(O)$ *induces an outer automorphism on* K_i *for* $i = 1, 2$.
(2) $|S_1 : OW_1| = 2$, $N_{S_1}(K) = OW_1$ *and each involution in* $N_{S_1}(K)$ *is in* $OC_{W_1}(O)$.
(3) Each involution s *in* S_1 *with* $K_1^s = K_2$ *is conjugate in* M_1 *to an element of* $O - W_1$.
(4) $m(S_1) \leq 4$.

PROOF. Let $\bar{M}_1$ be the universal group $\bar{\omega}(A_3, m)$ of type M_1 described in 4.3.1 and 5.7.4; from 5.7.4 there is a surjective homomorphism $\pi : \bar{M}_1 \to M_1$ and, modifying the notation in 4.3.1 and 5.7.4 slightly so that 3 indexes the middle node of the Coxeter diagram, $\bar{W}_1 = \bar{V}_1 \times \bar{V}_2 \times \bar{U}$, where $\bar{V}_i = \bar{K}_i \cap \bar{W}_1$ for $i = 1, 2$, and $\bar{V}_3 = \bar{U} \in \bar{\eta}$ with involution $\bar{z}_3$. Let $\bar{y}_i$ be the canonical generator of $\bar{V}_i$ and $\bar{v}_i = \bar{y}_i^{m/8}$. Write V_i, U, etc. for $\bar{V}_i\pi$, $\bar{U}\pi$, etc. Set $M_1^* = M_1/W_1$, so that $M_1^* = \mu_1 \cong S_4$.

By 9.3.4.1, $Z(M_1) = 1$, so as $Z(\bar{M}_1) = \langle \bar{v}_1\bar{v}_2^{-1}\bar{z}_3\rangle$ by 5.7.4.2, we conclude that $(y_1y_2^{-1}y_3^2)^{m/8} = 1$.

Now $E_4 \cong O^* = C_{M_1^*}(O^*)$ so $C_{S_1}(O) \leq OW_1$. As $O \cap W_1 = C_O(O \cap W_1)$, we have $C_{S_1}(O) = C_{OW_1}(O) \leq C_{OW_1}(O \cap W_1) = W_1$, so $C_{S_1}(O) = C_{W_1}(O)$. By 5.7.5.3, $C_{W_1}(O) = \langle x_0, z\rangle$, with $\bar{x}_0 = \bar{y}_1\bar{y}_2\bar{y}_3^2$, so as $(y_1y_2^{-1}y_3^2)^{m/8} = 1$ it follows that $\langle x_0\rangle = \langle x_0, z\rangle$ is of order $m/4$. Set $W_0 = OC_{S_1}(O)$. As $Z(O) = \langle z\rangle$, $|W_0| = |x_0||O|/2 = m^3/16 = |OW_1|/2$, so $|OW_1 : W_0| = 2$ and hence $|W_1 : W_1 \cap W_0| = 2$. From 4.2.5.1, $u \in U - \Phi(U)$ induces an outer automorphism on K_i, completing the proof of (1).

Now if $m = 8$ then the remainder of the lemma follows from 9.3.6; for example $m(S_1) \leq m(W_1) + m(M_1^*) = 4$. Thus we may assume $m > 8$.

As $\bar{W}_1 \cong \mathbf{Z}_{m/2}^3$ is of rank 3, so is W_1, and hence $\Omega_1(W_1) = \langle z, v, w\rangle$, where $\bar{w}^2 = \bar{v}_1\bar{v}_2^{-1}\bar{z}_3$, so we may take $w = (y_1y_2^{-1}y_3^2)^{m/16} = x_0^{m/16}y_2^{-m/8} \in W_0$. That is $\Omega_1(W_1) \leq W_0$. Also setting $k = k_1k_2 \in O$, $w^k = x_0^{m/16}y_2^{m/8} \neq w$, so $W_* = W_1\langle k\rangle$ is of 2-rank 3, and hence $m(S_1) \leq m(W_*) + m(F^*/W_*) = 4$ for each 4-subgroup F^* of S_1^*, proving (4).

Let $k_i \in K_i - V_i$; then $a_i = k_iv$ is an involution in k_iW_1 with $I(k_i) = I(a_i) = I_i$, where $I(x)$ is the subgroup of W_1 inverted by x. Hence the set of involutions in k_iW_1 is a_iI_i. Calculating using 5.7.4.2 and the defining relations for $\bar{M}_1$,we find $I_1 = V_1\langle w\rangle$, where w is defined above, so that $a_1I_1 \subseteq W_0$. Similarly $a_2I_2 \subseteq W_0$. On the other hand $k = k_1k_2$ is an involution inverting $V_1V_2 = [W_1, k]$, so the involutions in kW_1 are in $kV_1V_2 \subseteq O$. This completes the proof of (2).

Finally choose s as in (3). Then $s^h \in kW_1$ for some $h \in M_1$, and we saw earlier that all involutions in kW_1 are in O, so (3) holds. □

LEMMA 9.3.8. *One of the following holds:*

(1) $m > 8$, $\mathcal{O}_K = K$, *and* $\mathcal{Z} \cap O = \{z, v, vz\}$.

(2) $m > 8$, $\mathcal{O}_K \cong SL_2[m]$, *and* $\mathcal{Z} \cap O$ *is the set of involutions in* O.

(3) $m = 8$, $O^2(Aut_{\mathcal{F}}(O)) = 1$, *and* $\mathcal{Z} \cap O = \{z, v, vz\}$.

(4) $m = 8$ *and there is* $\mathcal{O} \trianglelefteq \mathcal{F}_z$ *of index 3 in* $SL_2[8] * SL_2[8]$ *with Sylow group* O *and* $O \cap \mathcal{Z} = E_0^\#$ *for some* $E_8 \cong E_0 \trianglelefteq \mathcal{F}_z$.

(5) $m = 8$ *and there is* $\mathcal{O} \trianglelefteq \mathcal{F}_z$ *isomorphic to* $SL_2[8] * SL_2[8]$ *with Sylow group* O *and* $O \cap \mathcal{Z}$ *is the set of involutions in* O.

PROOF. Suppose $m > 8$. Then by 9.1.11.1, $\mathcal{O}_K = K$ or $\mathcal{O}_K \cong SL_2[m]$. In the first case $O \trianglelefteq \mathcal{F}_z$, so $\langle v, z\rangle = \Omega_1(Z_2(O)) \trianglelefteq \mathcal{F}_z$. Hence (1) holds in this case by 9.3.5. Similarly if $\mathcal{O}_K \cong SL_2[m]$ then all involutions in $O - \langle z\rangle$ are fused in $\mathcal{O}(z)$, so (2) holds.

So assume $m = 8$. Then $O = \mathcal{O}(z) \trianglelefteq \mathcal{F}_z$. From 9.3.6.2, $\langle v_i\rangle$ is the unique $N_S(K)$-invariant subgroup of K_i of order 4, so if $A = Aut_{\mathcal{F}}(O)$ is a 2-group then $\langle v_1, v_2\rangle \trianglelefteq \mathcal{F}_z$, and then (3) holds by 9.3.5. Thus we may assume A is not a 2-group.

Adopt the notation of 9.1.11.2. By that lemma, $O^2(H_0)$ is of index 1 or 3 in $SL_2(3) * SL_2(3)$, and centralizes $S_0 = C_S(O)$. Then by 2.3 in [**Asc11**], $\mathcal{E} = \mathcal{O} * C_{\mathcal{F}}(O)$ is a subsystem of $\mathcal{F}_z$, where $\mathcal{O} = \mathcal{F}_O(O^2(H_0))$. Further G_0 is a model for $N_{\mathcal{F}_z}(OS_0) = N_{\mathcal{F}_z}(S_0)$, so as $O^2(H_0) \trianglelefteq G_0$, $N_{\mathcal{E}}(S_0) \trianglelefteq N_{\mathcal{F}_z}(S_0)$, and hence

$\mathcal{E} \trianglelefteq \mathcal{F}_z$ by 1.5.1 in [**Asc19**] applied to $\mathcal{F}_2, C_{\mathcal{F}}(O)$ in the role of $\mathcal{F}, \mathcal{F}_0$. Then as $\mathcal{F}_z = C_S(z)\mathcal{E}$ and $C_S(z)$ acts on $O^2(H_0)$, $\mathcal{O} \trianglelefteq \mathcal{F}_z$. Now (4) or (5) holds for $|O^2(H_0)|_3 = 3$ or 9, respectively. □

NOTATION 9.3.9. If $m > 8$ set $\mathcal{O} = \mathcal{O}(z)$. If $m = 8$ and $Aut_{\mathcal{F}}(O)$ is a 2-group, set $\mathcal{O} = O$. Finally if $m = 8$ and $Aut_{\mathcal{F}}(O)$ is not a 2-group, define $\mathcal{O}$ as in part (4) or (5) of 9.3.8. Then from 9.3.8, $\mathcal{O} \trianglelefteq \mathcal{F}_z$ with Sylow group O, and $\mathcal{O} = O$ or $\mathcal{O} \cong SL_2[m] * SL_2[m]$, or $m = 8$ and $\mathcal{O}$ is of index 3 in $SL_2[8] * SL_2[8]$. Further $\mathcal{O}$ centralizes $C_S(O)$.

In the next lemma, two statements are proved by induction on the order of τ. The order of τ is defined in section 1.3. I suppose technically this process involves proving results first for small packets, then second for larger packets, and so on. But of course reading is much easier when the proof has a different organization.

In each of the two results, $\rho = (\mathcal{Y}, \Gamma)$ is a proper subpacket of τ with $\Gamma \subseteq \Omega$, satisfuing Hypothesis 9.3.1 By induction on the order of τ, ρ satisfies one of the conclusions of Theorem 9.3.24, and has the generation property of 9.3.20. Actually we could avoid the appeal to induction (as we will see in a moment), but I think this approach is easier and more attractive.

To avoid the inductive appeal: in the first case by 9.3.1.4, ρ satisfies one of the conclusions of Theorem 1; then working with that list of conclusions, we could show that ρ appears in 9.3.24. For example $\mu(\rho) \cong S_4$ provides a first sieve. In the second case as ρ appears in 9.3.24, we can check the generation property directly; for example it is trivial in 9.3.24.1 and in 9.3.24.3 it holds by 9.3.20.2.

LEMMA 9.3.10. *Suppose $r > 1$. Then:*

(1) $\mathcal{F}_z^\circ = \mathcal{O}(z) \times \mathcal{C}_2 \times \cdots \times \mathcal{C}_r$ where $\rho_i = (\mathcal{C}_i, \Omega_i)$ is a quaternion fusion packet, S_i is Sylow in $\mathcal{C}_i$, $\eta_i \in \eta(\rho_i)$, and M_i is a model for $\langle J_i^{N_{\mathcal{C}_i}(W_i)} \rangle$ for $J_i \in \Omega_i$. Moreover $(\mathcal{O}_{K_j}, K_j)$, $j = 1, 2$, and ρ_i, $2 \le i \le r$, are the coconnected components of τ_z°.

(2) One of the following holds for each $2 \le i \le r$:

(a) $\mathcal{O} = O$ and $\mathcal{C}_i = \mathcal{F}_{S_i}(M_i)$.

(b) $\mathcal{O} \cong SL_2[m] * SL_2[m]$ and $\mathcal{C}_i \cong L_4^+[m]$.

(c) $m = 8$, $\mathcal{O}$ is of index 3 in $SL_2[8] * SL_2[8]$, and $\mathcal{C}_i \cong AE_6$.

(3) $\mathcal{C}_i = \langle S_i\mathcal{O}_i, \mathcal{F}_{S_i}(M_i)\rangle$, where $\mathcal{O}_i \trianglelefteq N_{\mathcal{C}_i}(z_i)$, $z_i = z(J_i)$, and $\mathcal{O}_i \cong \mathcal{O}$.

PROOF. By 9.3.1.4 and 6.6.8 applied to τ_z in the role of $(\mathcal{Y}, \Gamma)$, τ_z° has coconnected components $\xi_j = (\mathcal{F}_j, \Gamma_j)$, $1 \le j \le a$. We may take $\mathcal{F}_j = \mathcal{O}_{K_j}$ and $\Gamma_j = \{K_j\}$ for $j = 1, 2$. Further $\mathcal{F}_{S_i}(M_i)$ is coconnected for $2 \le i \le r$ and centralizes z, so, using 6.6.11.5, $\mathcal{F}_{S_i}(M_i) \le \mathcal{F}_{j(i)}$ for some $j = j(i) > 2$. Let $z_i = z(J_i)$ for $J_i \in \Omega_i$. Then by induction on the order of τ (cf. section 1.3 for the definition of the order of τ, and apply 9.3.24) in an inductive context to ξ_j), as ξ_j is coconnected and $\mathcal{Z}(\xi_j) \cap O(z_j) \ne \{z_j\}$, it follows that $\mu(\xi_j) \cong S_4$, $\eta_i \in \eta(\xi_j)$, and M_i is a model for $\langle J_i^{N_{\mathcal{F}_j}(W_i)} \rangle$. Moreover by induction, $\mathcal{F}_j = \mathcal{F}_{S_i}(M_i)$ or $\mathcal{F}_j \cong L_4^+[m]$, or $m = 8$ and $\mathcal{F}_j \cong AE_6$. In particular $\Omega_i = \Gamma_j$, so the map $i \mapsto j(i)$ is a bijection of $\{2, \ldots, r\}$ with $\{3, \ldots, a\}$, so we may relabel so that $\xi_1, \xi_2, \rho_2, \ldots, \rho_r$ are the coconnected components of $\mathcal{F}_z$ and (1) holds.

We next prove (2). Suppose $\mathcal{C}_2 \cong L_4^+[m]$. Then if $m > 8$, $\mathcal{O}(z_2) \cong SL_2[m] * SL_2[m]$, while if $m = 8$ then $|O^2(Aut_{\mathcal{C}_2}(O(z_2)))|_3 = 9$. It follows from 9.3.9 that $\mathcal{O} \cong SL_2[m] * SL_2[m]$.

Conversely suppose $\mathcal{O} \cong SL_2[m] * SL_2[m]$. By 9.3.9, $\mathcal{O}$ centralizes $S_2 \cdots S_r$, so as $\mathcal{F}$ is transitive on Ω, $\mathcal{O}(z_2) \cong SL_2[m] * SL_2[m]$ when $m > 8$, and when $m = 8$ we have $|O^2(Aut_{\mathcal{F}_z}(O(z_2))|_3 = 9$. We conclude, using our proof of (1), that $\mathcal{C}_2 \cong L_4^+[m]$. Therefore we may assume $\mathcal{O}$ is a 2-group or $m = 8$ and $\mathcal{O}$ is of index 3 in $SL_2[8] * SL_2[8]$, and we may also assume that $\mathcal{C}_2$ is $\mathcal{F}_{S_2}(M_2)$ or AE_6.

Suppose $\mathcal{C}_2 \cong AE_6$. Then $m = 8$ and $|O^2(Aut_{\mathcal{C}_2}(O(z_2)))|_3 = 3$, so $\mathcal{O}$ is of index 3 in $SL_2[8] * SL_2[8]$. Conversely if $\mathcal{O}$ is of index 2 in $SL_2[8] * SL_2[8]$ then, as in the previous paragraph, $O^2(Aut_{\mathcal{F}_z}(O(z_2))) \neq 1$, so $\mathcal{C}_2 \cong AE_6$. This completes the proof of (2).

Part (3) follows by induction on the order of τ. That is we apply 9.3.20.1 to ρ_i in an inductive setting, observing in each case that $C_{\mathcal{C}_i}(z_i) = S_i\mathcal{O}_i$ amd $N_{\mathcal{C}_I}(E_i) = \mathcal{F}_{S_i}(M_i)$ □

LEMMA 9.3.11. $r = 1$.

PROOF. Assume $r > 1$ and adopt the notation of 9.3.10. We verify Hypothesis 8.1.14, and then appeal to 8.1.15 to conclude that $\mathcal{C}_2$ is subnormal in $\mathcal{F}$ and $\mathcal{F}^\circ$ is the direct product of the conjugates of $\mathcal{C}_2$ under $\mathcal{F}$. (The constraint in 8.1.15 that $Z(\mathcal{C}_i) = 1$ holds as $Z(M_i) = 1$ by 9.3.4.1 and as S_i is Sylow in $\mathcal{C}_i$ by 9.3.10.1.) Then we have a contradiction to the transitivity of $\mathcal{F}^\circ$ on Ω. Hence it remains to verify Hypothesis 8.1.14.

First, Hypothesis 8.1.1 holds by conditions (1) and (4) in Hypothesis 9.3.1. Then condition 8.1.12.1 holds by 9.3.2 and 9.1.9.1. Condition 8.1.12.2 holds by 9.3.10.1, and condition 8.1.12.3 holds by parts (1) and (3) of 9.3.10. Therefore Hypothesis 8.1.12 holds, so it remains to check the condition in 8.1.14: setting $\mathcal{C}_1 = \mathcal{C}_2^g$ for $g \in G$ with $z_2^g = z$, if $x \in S_1$ is an involution in $\mathcal{C}_1^f$, then for each $1 < i \leq r$ and $\alpha \in \mathfrak{A}(x)$, $\rho_i\alpha$ is a coconnected component of $\tau_{x\alpha} = (C_{\mathcal{F}}(x\alpha), C_\Omega(x\alpha))$, and any other coconnected component is a subpacket of $(C_{\mathcal{C}_1}(x)\alpha^*, C_\Omega(x)\alpha)$. The first constraint follows from the argument in paragraph one of the proof of 9.3.10, given that x centralizes $\mathcal{C}_i$ for $i > 1$. Further the same argument shows that for $R = S_2 \cdots S_r$, $C_{\mathcal{F}}(R)^\circ = \mathcal{C}_1$. Then if $\xi = (\mathcal{E}, \Gamma)$ is any other coconnected component, we have $\Gamma \subseteq \Omega(z)\alpha$ and $\mathcal{E}$ centralizes $R\alpha$, so for $\beta \in \mathfrak{A}(R\alpha)$ with $R\alpha\beta = R$, $\mathcal{E}\beta^* \leq C_{\mathcal{F}}(R)^\circ = \mathcal{C}_1$. As $\zeta = \alpha\beta \in \hom_{N_{\mathcal{F}}(R)}(C_S(x), N_S(R))$ and $\mathcal{C}_1 \trianglelefteq N_{\mathcal{F}}(R)$, $\mathcal{E}\alpha^{-*} = \mathcal{E}\beta^*\zeta^{-*} \leq \mathcal{C}_1$, so $\mathcal{E}$ is a subsystem of $C_{\mathcal{C}_1}(x)\alpha^*$, completing the proof. □

NOTATION 9.3.12. Set $T = S \cap M$ and $E = \langle z, v \rangle$. Let G be a model for $N_{\mathcal{F}}(C_S(W))$.

LEMMA 9.3.13. *(1)* $\Omega = \{K_1, K_2\}$ *so* $\mathcal{D}(z) = \emptyset$, $M = M_1$, $W = W_1$, *and* $\mu \cong S_4$.

(2) $z \in Z(S)$ *and* O, T, *and* E *are normal in* S.

(3) W *and* $\mathcal{F}_T(M)$ *are normal in* $N_{\mathcal{F}}(E)$.

PROOF. By 9.3.11, $r = 1$, so as Ω is partitioned by the Ω_i, $1 \leq i \leq r$, by 9.1.9.1, we have $\Omega = \Omega_1$. By 9.3.4.2, $\mu = \mu_1 \cong S_4$, so $\Omega = \Omega_1$ is of order 2 by 4.2.8.6, and hence $\Omega = \{K_1, K_2\}$. By 9.3.4.1, $M = M_1 \times \cdots \times M_r$, so as $r = 1$, $M = M_1$ and $W = W_1$, completing the proof of (1).

By (1) $\Omega = \{K_1, K_2\}$, so $O = K_1K_2 \trianglelefteq S$. Then as $\langle z \rangle = Z(O)$, $z \in Z(S)$. As S acts on W, $S \leq G$, so as $M \trianglelefteq G$ by 9.1.6, S acts on $S \cap M = T$. From the proof of 9.3.8, $E \trianglelefteq C_S(z)$, so $E \trianglelefteq S$, completing the proof of (2).

Let $\mathcal{Y} = N_{\mathcal{F}}(E)$ and $\xi = (\mathcal{Y}, \Omega)$; by 2.5.2, ξ is a quaternion fusion packet. Then $\mu(\xi) \cong S_4$ with $\mathcal{Z}(\xi) \cap O \neq \{z\}$ and $z \in E \trianglelefteq \mathcal{Y}$, so by Theorem 2 and as $E \cong E_4$, $\mathcal{Y}^\circ = \mathcal{F}_T(M)$, proving (3). □

LEMMA 9.3.14. *Assume* $s \in \mathcal{D}^o(z) \cup \mathcal{D}^m(z)$. *Then*
(1) $\mathcal{D}^*(s) \cap E \neq \emptyset$.
(2) $s \in \mathcal{D}^m(z)$.
(3) $s \in T$.
(4) s *is conjugate in* M *to a member of* $O - W$.

PROOF. Suppose that $\mathcal{D}^*(s) \cap E \neq \emptyset$. By hypothesis, $s \notin \mathcal{D}^*(z)$. Thus we may take $v \in \mathcal{D}^*(s)$. Let $g \in M$ with $z^g = v$. By 9.3.5, $s \in O^g$ in M, so $s \in C_{O^g}(z) \leq T$, proving (3). As $s \in T$, 9.3.7.2 implies that $s \notin \mathcal{D}^o(z)$, so (2) holds. Then (4) follows from 9.3.7.3. Observe that we've shown that (1) implies (2)-(4).

It remains to prove (1), so assume $\mathcal{D}^*(s) \cap E = \emptyset$. By 9.1.7.2, s centralizes E. Let $\alpha \in \mathfrak{A}(s)$, so that $s\alpha = z$. Suppose $m > 8$ and $s \in \mathcal{D}^m(z)$. Then $C_O(s) = \langle z \rangle \times X$ where $X \cong D_{m/2}$ and $Z(X) = \langle x \rangle$ for some $x \in \{v, vz\}$. As $X \cong D_{m/2}$ and $m > 8$, $x\alpha \in \Phi(X\alpha) \cap \mathcal{Z} \subseteq \mathcal{D}^*(z)$ and hence $x \in \mathcal{D}^*(s)$, contrary to assumption. Thus:

(a) If $m > 8$ then $s \in \mathcal{D}^o(z)$.

Next as the graph $\mathcal{D}^*$ is $\mathcal{F}$-invariant, $E\alpha \cap \mathcal{D}^*(z) = \emptyset$, so $E\alpha \cap OC_S(O) = 1$ by 9.1.3. Thus, using 9.1.3.3, $E\alpha = \langle y_1, y_2 \rangle$ with $y_1 \in \mathcal{D}^o(z)$ and $y_2 \in \mathcal{D}^m(z)$, so by 9.1.7.2, $E\alpha$ centralizes E. Hence $\langle E, E\alpha \rangle = E \times E\alpha \cong E_{16}$. Also $[y_1, v_1] = z$ and $[y_2, v_1] \in v\langle z \rangle$, so $[E\alpha, v_1] = E$.

Let $\beta \in \mathfrak{A}(E\alpha)$ with $E\zeta = E$, where $\zeta = \alpha\beta$. If $z\zeta \neq z$ then as $z \in Z(S)$ we have $N_S(E\alpha) = C_S(E\alpha)$, and we adjust β by composing with a member of $Aut_M(C_S(E))$ to set $z\zeta = z$. Then $\zeta \in \hom_{\mathcal{F}_z}(C_S(s), C_S(z))$, so as $s \notin \mathcal{D}^*(z)$, $s_0 = s\zeta \notin \mathcal{D}^*(z)$. Set $A = EE\beta$, so that $A = E \times E\beta \cong E_{16}$. Let $\gamma = c_{v_1}\beta^* \in Aut_{\mathcal{F}}(A) = \Sigma$, where c_{v_1} is the automorphism induced on $EE\alpha$ by v_1. Hence $E\beta = [EE\alpha, v_1]\beta = [A, \gamma]$.

Suppose there is $y \in A - E$ with $y \in \mathcal{D}^*(z)$. Then $y \in O$, so $|C_O(s_0)| \geq 8$ and hence $s_0 \in \mathcal{D}^m(z)$ by 9.1.3.3. Then by 9.1.13, $A = E(s_0, Q_1)$ for some Q_8-subgroup Q_1 of K. Set $Q = Q_1Q_1^{s_0}$. By 9.1.14.3 we may assume $A \in \mathcal{F}^f$. Then $\rho = (N_{\mathcal{F}}(A), \{Q_1, Q_2\})$ is a quaternion fusion packet by 2.5.2.

Claim that Σ is irreducible on A; assume otherwise and let B be a proper nonzero Σ-submodule of A. Then $\langle z \rangle = C_A(Q) \leq B$. If $x \in z^\Sigma - Q$ then $A = \langle x^Q \rangle \leq B$, contrary to the choice of B. Thus $z^\Sigma \subseteq Q$. Hence there is $\sigma \in \Sigma$ with $z \neq z\sigma \in Q$, so $z\sigma \in \mathcal{D} * (z)$. Then as $W(\rho) \leq W$, we have $E \leq B$, so as $[E, \gamma] = [A, \gamma] = E\beta$, we have $A \leq B$, contrary to the choice of B. This completes the proof of the claim.

By the claim and 9.1.14.4, and as $v \in A \cap \mathcal{Z}$, Σ is of index at most 2 in $Sp(A)$. This is a contradiction as $C_\Sigma(z\beta)$ acts on $E\beta = (O \cap EE\alpha)\beta$. Therefore $\mathcal{D}^*(z) \cap A \subseteq E$. In particular

(b) $E\beta = \langle a, b \rangle$ with $a \in \mathcal{D}^o(z)$ and $b \in \mathcal{D}^m(z)$.

By 9.3.7.2, $a \notin T$, so as we showed that (1) implies (3), it follows that $\mathcal{D}^*(a) \cap E = \emptyset$. Let $e \in E - \langle z \rangle$ and $g \in M$ with $z^g = e$. As $a \notin \mathcal{D}^*(e)$, an argument above using the action of A on O^g shows that $\mathcal{D}^*(e) \cap A \subseteq E$. Hence

(c) For each $e \in E^\#$, $\mathcal{D}^*(e) \cap A \subseteq E$.

Suppose $m > 8$. Then by (a), $E \cap \mathcal{D}^*(b) \neq \emptyset$, contrary to the previous paragraph. Hence

(d) $m = 8$.

By (c), for $c \in E\beta^{\#}$, $E \cap \mathcal{D}^*(c) = \emptyset$, so c centralizes no element of order 4 in W. Thus

(e) For $1 \neq c \in E\beta$, $C_W(c) = E$.

Set $G^+ = G/E$ and $P = O_2(M)$. By 9.1.3.3, $C_O(a) = E$, so it follows from 9.3.6.3 that $m(C_P(a)) < 4$. In particular $A \not\leq P$.

As M is irreducible on W^+, $C_S(E)$ centralizes W^+, so W acts on A. From (e), for each $x \in W - E$, $C_A(x) = E$. Thus $Aut_W(A) \cong E_4$ and $\Sigma_0 = \langle Aut_W(A), \gamma\rangle \leq \Sigma$ is irreducible on A, as $E = [c, W]$ for each $c \in A - E$ and $[E, \gamma] = E\beta$. Further $E = A \cap O$, so $C_\Sigma(z) \leq N_\Sigma(E)$. It follows from the list of irreducible subgroups of $GL(A)$ that $L_2(4) \cong \Sigma_0 \trianglelefteq \Sigma$ and A is the $L_2(4)$-module for Σ_0.

In particular there is a subgroup of order 3 in $N_{\Sigma_0}(E)$ which is induced by X_1 of order 3 in $N_G(A)$. Then X_1 centralizes a Sylow 3-subgroup X_2 of M. Set $X = X_1X_2$ and $X_0 = C_X(W^+)$. As $[X_2, C_S(E)] = P$ and $A \not\leq P$, we have $X_1 \neq X_2$, so $X \cong E_9$.

Now by 5.7.4.4, $P^+ \cong E_{16}$ and M has three irreducibles W^+ and P_i^+, $i = 1, 2$ on P^+, and M is also irreducible on E. By parts (4) and (5) of 5.7.4, $P_i \cong E_{16}$. For each of these four irreducibles P_0, $End_M(P_0) = \mathbf{F_2}$, so there is $X_0 \cong \mathbf{Z}_3$ centralizes M/P and W, and then also centralizes each of the four irreducibles, so X_0 centralizes M. Thus $Aut_{X_1}(P) = Aut_{X_2}(P)$. Then $A = [A, X_1]$ centralizes P_i^+ for $i = 1, 2$, so A centralizes P^+ and hence P acts on A with $Aut_P(A) = [Aut_P(A), X_1] \leq \Sigma_0$ from the definition of X_1. Therefore $P = WC_P(A)$ and then $C_P(A) = P_i$ for $i = 1$ or 2, contradicting $m(C_P(a)) < 4$. The proof is at last complete. □

LEMMA 9.3.15. *(1)* $\mathcal{D}^o(z) = \emptyset$.
(2) $\mathcal{Z} \subseteq T$.

PROOF. Part (1) follows from 9.3.14.2. By 9.3.13.1, $\mathcal{D}(z) = \emptyset$. By 9.3.5, $\mathcal{D}^*(z) \subseteq O \leq T$, while by 9.3.14.3, $\mathcal{D}^m(z) \subseteq T$. Now (2) follows from 9.1.3. □

LEMMA 9.3.16. *The following are equivalent:*
(1) $\mathcal{F}^\circ = \mathcal{F}_T(M)$.
(2) $E^{\#} = \mathcal{Z}$.
(3) $\mathcal{D}^m(z) = \emptyset$.

PROOF. If (1) holds then $\mathcal{Z} \subseteq W \leq N_S(K)$, so (1) implies (3).

Assume (3). Then by 9.1.3 and 9.3.15.1, $\mathcal{Z} = \{z\} \cup \mathcal{D}^*(z)$, and then by 9.3.5, $\mathcal{Z} \subseteq O$. Suppose $s \in \mathcal{Z} - E$ and let $g \in M - C_M(z)$. Now $s = k_1k_2$ with k_i of order 4 in K_i. By 9.3.8 we may choose $k_i \notin W$ for $i = 1, 2$. Therefore $s \in C_T(E) = O_2(M)$ with $K_1^{gs} = K_2^g$, so $s^g \in O_2(M)$ with $K_1^{s^g} = K_2$. Hence $s^g \in \mathcal{D}^m(z)$, contrary to (3). Thus (3) implies (2).

Assume (2) holds. Then E is a strongly closed abelian subgroup of S, so by 14.1 in [**Asc11**], $E \trianglelefteq \mathcal{F}$. Thus $\mathcal{F}_T(M) \trianglelefteq \mathcal{F}$ by 9.3.13.3, so (1) follows. □

In light of 9.3.16, during the remainder of the section, except in Theorem 9.3.24, we assume the following hypothesis:

HYPOTHESIS 9.3.17. *Hypothesis* 9.3.1 *holds with* $\mathcal{D}^m(z) \neq \emptyset$.

LEMMA 9.3.18. *Either*

(1) $\mathcal{O} \cong SL_2[m] * SL_2[m]$ *and* $O \cap \mathcal{Z}$ *is the set of involutions in* O, *or*

(2) $m = 8$, $\mathcal{O}$ *is of index 3 in* $SL_2[8] * SL_2[8]$, *and* $O \cap \mathcal{Z} = E_0^\#$ *for some* $E_8 \cong E_0 \trianglelefteq \mathcal{F}_z$.

PROOF. By 9.3.17 there exists $s \in \mathcal{D}^m(z)$. Then by 9.3.14.4, s is conjugate to some $x \in O - E$. Hence $\mathcal{Z} \cap O \neq E^\#$, so the lemma follows from 9.3.8. □

LEMMA 9.3.19. $T\mathcal{O} \trianglelefteq \mathcal{F}_z$.

PROOF. As $C_{\mathcal{F}_T(M)}(z) = T$, we conclude from 9.3.13.3 and 1.2.12 that $T \trianglelefteq N_{\mathcal{F}_z}(E)$. Next if $m > 8$ then $E = \Omega_1(Z_2(O))$, so $N_{\mathcal{F}_z}(O) \leq N_{\mathcal{F}_z}(E)$ and $N_{\mathcal{O}}(O) = O$. Thus $TN_{\mathcal{O}}(O) = T \trianglelefteq N_{\mathcal{F}_z}(O)$. Then $T\mathcal{O} \trianglelefteq \mathcal{F}_z$ by Theorem 1.5.2 in [**Asc19**]. Therefore we may assume $m = 8$.

Let $S_0 = C_S(O)$ and $\mathcal{F}_0 = C_{\mathcal{F}}(O)$; by Theorem 4 in [**Asc11**], $\mathcal{F}_0 \trianglelefteq \mathcal{F}_z$. By 1.3.2 in [**Asc19**], $\mathcal{F}_z = \langle \mathcal{F}_1, \mathcal{F}_2 \rangle$, where $\mathcal{F}_1 = S\mathcal{F}_0$ and $\mathcal{F}_2 = N_{\mathcal{F}_z}(S_0)$. Set $\mathcal{E}_1 = T$ and $\mathcal{E}_2 = T\mathcal{O}$. As $T \trianglelefteq N_{\mathcal{F}_z}(E)$ with $E \trianglelefteq \mathcal{F}_1$, we have $T \trianglelefteq \mathcal{F}_1$. Further from the proof of 9.3.8, $W\mathcal{O} = \mathcal{F}_T(O^2(H_0)W) \trianglelefteq \mathcal{F}_S(G) = \mathcal{F}_2$, so $\mathcal{E}_2 \trianglelefteq \mathcal{F}_2$. We have verified Hypothesis 1.4.7 in [**Asc19**], so by 1.4.8 in [**Asc19**], $T\mathcal{O} = \langle \mathcal{E}_1, \mathcal{E}_2 \rangle \trianglelefteq \mathcal{F}_z$, completing the proof. □

LEMMA 9.3.20. *(1)* $\mathcal{F} = \langle \mathcal{F}_z, N_{\mathcal{F}}(E) \rangle$.

(2) Let $\tilde{\mathcal{F}} = L_4^+[m]$, $\tilde{\tau}$ *the Lie packet for* $\tilde{\mathcal{F}}$, *etc. Then* $\tilde{\mathcal{F}} = \langle \tilde{S}\tilde{\mathcal{O}}, \mathcal{F}_{\tilde{S}}(\tilde{M}) \rangle$.

PROOF. We first prove (1). Let $\mathcal{E} = \langle \mathcal{F}_z, N_{\mathcal{F}}(E) \rangle$. We verify that $\mathcal{E}$ satisfies conditions (a) and (b) of 2.1.8, and then appeal to that lemma to complete the proof. By definition of $\mathcal{E}$, 2.1.8.a holds, so it remains to verify 2.1.8.b. Let $s \in \mathcal{Z} - \{z\}$ and $S_s = C_S(s)$.

Suppose first that $s \in \mathcal{D}^*(z)$. By 9.3.5 there is $\alpha \in \mathfrak{A}_{\mathcal{F}_z}(s)$ with $s\alpha = v$. Then $S_s\alpha \leq C_S(E) \leq O_2(SM) = Q$ in G, and there is $g \in M$ with $v^g = z$. Then $\gamma = \alpha c_g \in \hom_{\mathcal{E}}(S_s, S)$ with $s\gamma = z$, so 2.1.8.b holds in this case.

Therefore we may assume $s \notin \mathcal{D}^*(z)$. Then by 9.1.3, 9.3.14.1, and 9.3.15 we may assume $\mathcal{D}^*(s) \cap E = \{v\}$. Therefore $s \in C_S(E) \leq Q$. Then $S_s^g \leq Q$ with $s^g \in \mathcal{D}^*(z)$, so by the previous paragraph there is $\delta \in \hom_{\mathcal{E}}(S_s^g, S)$ with $s^g\delta = z$. Now $\beta = c_g\delta \in \hom_{\mathcal{E}}(S_s, S)$ with $s\beta = z$, completing the proof of (1).

Assume the setup of (2); observe that $\tilde{\tau}$ satisfies Hypothesis 9.3.1 (except possibly the Inductive Hypothesis), so the proof of (1) (based on 2.1.8) works here too, once we make several observations that follow from the description of $\tilde{\mathcal{F}}$ as $\mathcal{F}_{\tilde{S}}(\tilde{L})$ for $\tilde{L} \cong L_4^\pi(q) \cong P\Omega_6^\pi(q)$ in sections 5.7 and 5.3. First, $\tilde{\mathcal{F}}_z = \tilde{S}\tilde{\mathcal{O}}$ and $N_{\tilde{\mathcal{F}}}(\tilde{E}) = \mathcal{F}_{\tilde{S}}(\tilde{M})$. Second, $C_{\tilde{\mathcal{Z}}}(z) = \{z\} \cup \mathcal{D}^*(z) \cup \mathcal{D}^m(z)$. Third, if $s \in \mathcal{D}^m(z)$ then by 9.1.7.2, s centralizes $\tilde{E}$ and a generator for $N_{\tilde{E}}(L_s)$ is in $\mathcal{D}^*(s)$, where $s \in L_s$ and L_s is a fundamental subgroup of $\tilde{L}$. □

LEMMA 9.3.21. *Assume* $m = 8$ *and* $\mathcal{O}$ *is of index 3 in* $SL_2[m] * SL_2[m]$. *Then* $\mathcal{F}^\circ \cong AE_6$.

PROOF. Recall from 9.3.6 that $W \cong \mathbf{Z}_4^2$ and $E = \Omega_1(W)$. By 9.3.18, $O \cap \mathcal{Z} = E_0^\#$ for some $E_8 \cong E_0 \trianglelefteq \mathcal{F}_z$. Set $M^+ = M/E$, $P = O_2(M)$, and let $X \in Syl_3(M)$. Observe that $E_0 \leq C_T(E) = P$. Next by 5.7.4.4, $P^+ = [P^+, X] \cong E_{16}$, so for each $\mathbf{Z}_2 \cong U^+ \leq P^+$, $\langle U^{+X} \rangle \cong E_4$. Let A be the preimage in P of $\langle E_0^{+X} \rangle$. Then A is the union of conjugates of E_0, so as $E_0^\# \subseteq \mathcal{Z}$, $A^\# \subseteq \mathcal{Z}$. Now $\mathcal{D}^*(z) = E_0 - \langle z \rangle$ and by 9.3.14, $\mathcal{Z} = \{z\} \cup \mathcal{D}^*(z) \cup \mathcal{D}^m(z)$. Then by 9.3.14.4, $A^\# = \mathcal{Z}$. Therefore A is a strongly closed abelian subgroup of S, so by 14.1 in [**Asc11**], $A \trianglelefteq \mathcal{F}$. Then the lemma follows from Theorem 2 and the fact that $E_{16} \cong A$ and $\mu \cong S_4$. □

LEMMA 9.3.22. *Assume $\mathcal{O} \cong SL_2[m] * SL_2[m]$ and set $\mathcal{E} = \langle T\mathcal{O}, \mathcal{F}_T(M)\rangle_T$, regarded as a fusion system on T. Then $\mathcal{E} \cong L_4^+[m]$.*

PROOF. By 9.3.4.1, $Z(M) = 1$, so by 5.7.4.5 and 4.3.7.3, the pair M, Ω is determined up to isomorphism. Let $\tilde{\mathcal{F}} = L_4^+[m]$. Then by 5.3.5.4 and 5.3.8, $\tilde{\mathcal{F}}$ satisfies the Hypotheses 9.3.1 and 9.3.17 of $\mathcal{F}$ (except possibly the Inductive Hypothesis) with $\tilde{S} \leq \tilde{M}$ and $\tilde{\mathcal{F}}_{\tilde{z}} = \tilde{S}\tilde{\mathcal{O}}$, so we may take $\tilde{M} = M$, $\tilde{S} = T$, and $\tilde{\Omega} = \Omega$. Further by 9.3.20.2, $\tilde{\mathcal{F}} = \langle T\tilde{\mathcal{O}}, \mathcal{F}_T(M)\rangle$, so to show that $\tilde{\mathcal{F}} = \mathcal{E}$, it suffices to show $\tilde{\mathcal{X}} = \mathcal{X}$, where $\mathcal{X} = T\mathcal{O}$ and $\tilde{\mathcal{X}} = T\tilde{\mathcal{O}}$. But this follows from Theorem 2.7.3 applied to the pairs $(\mathcal{X}, \Omega)$ and $(\tilde{\mathcal{X}}, \Omega)$. □

LEMMA 9.3.23. *Assume $\mathcal{O} \cong SL_2[m] * SL_2[m]$. Then $\mathcal{F}^\circ \cong L_4^+[m]$.*

PROOF. Let $\mathcal{E}_1 = T\mathcal{O}$, $\mathcal{E}_2 = \mathcal{F}_T(M)$, $\mathcal{E} = \langle \mathcal{E}_1, \mathcal{E}_2\rangle$, $\mathcal{F}_1 = \mathcal{F}_z$, and $\mathcal{F}_2 = N_{\mathcal{F}}(E)$. We verify Hypothesis 1.4.7 in [**Asc19**], and then appeal to 1.4.8 in [**Asc19**] to show that $\mathcal{E} \trianglelefteq \mathcal{F}$. Then as $\mathcal{F}^\circ = [O]_{\mathcal{F}} = \mathcal{E}$, the lemma follows from 9.3.22. Thus it remains to establish Hypothesis 1.4.7 in [**Asc19**].

By 9.3.20, Hypothesis 1.4.1 in [**Asc19**] holds. By 9.3.13.3 and 9.3.19, $\mathcal{E}_i \trianglelefteq \mathcal{F}_i$ for $i = 1, 2$. As $\langle z\rangle = Z(T)$ and $E = Z_2(T)$ are characteristic in T, $Aut_{\mathcal{F}}(T) = Aut_{\mathcal{F}_i}(T)$. Finally by 9.3.22, $\mathcal{E}$ is saturated, so Hypothesis 1.4.7 in [**Asc19**] is satisfied, so as $\mathcal{E} = O^{2'}(\mathcal{E})$, 1.4.8 in [**Asc19**] shows $\mathcal{E} \trianglelefteq \mathcal{F}$. □

THEOREM 9.3.24. *Assume Hypothesis 9.3.1, let $\eta \in \eta(\tau)$, $W = \langle \eta\rangle$, G a model for $N_{\mathcal{F}}(W)$, and $M = \langle K^G\rangle$ for $K \in \Omega$. Set $z = z(K)$. Then $\mu(\tau) \cong S_4$, $\mathcal{F} = \langle \mathcal{F}_z, N_{\mathcal{F}}(W)\rangle$, and one of the following holds:*

(1) $\mathcal{F}^\circ = \mathcal{F}_{S\cap M}(M)$.

(2) $m = 8$ and $\mathcal{F}^\circ \cong AE_6$.

(3) $\mathcal{F}^\circ \cong L_4^+[m]$.

PROOF. By 9.3.4.2 and 9.3.11, $\mu \cong S_4$. If $\mathcal{D}^m(z) = \emptyset$ then (1) and the theorem hold by 9.3.16, so we may assume Hypothesis 9.3.17 is satisfied. Therefore $\mathcal{F} = \langle \mathcal{F}_z, N_{\mathcal{F}}(W)\rangle$ by 9.3.20. If 9.3.18.2 holds, then so does (2) by 9.3.21. Thus by 9.3.18, we may assume that 9.3.18.1 holds. Therefore (3) holds by 9.3.23, completing the proof. □

9.4. $|\Omega(z)| = 2$ and $\mathcal{D}^*(z) = \mathcal{D}(z)$

In this section, except in Theorems 9.4.34 and 9.4.35, we assume the following hypothesis:

HYPOTHESIS 9.4.1. *(1) $\tau = (\mathcal{F}, \Omega)$ is a quaternion fusion packet such that $\mathcal{A}(\tau) = \emptyset$ and for $z \in \mathcal{Z}_S$, $|\Omega(z)| = 2$.*

(2) $\mathcal{F}^\circ$ is transitive on Ω.

(3) For $z \in \mathcal{Z}_S$, $\mathcal{D}^*(z) = \mathcal{D}(z)$.

(4) The Inductive Hypothesis is satisfied.

REMARK 9.4.2. Observe that Hypothesis 9.4.1.1 is just a restatement of Hypothesis 9.1.1, so we can appeal to the lemmas in section 9.1 that do not assume Hypothesis 9.1.8, and we adopt Notation 9.1.2 and 9.1.6. Similarly later, after we show $|\Omega| = 2$ in 9.4.8 and essential subgroups enter the picture, such subgroups will satisfy Hypothesis 9.2.1 by parts (1) and (2) of 9.4.1, so we can appeal to results in section 9.2. Set $O = O(z)$.

LEMMA 9.4.3. *Assume $\Omega(z) \subseteq \Gamma \subseteq \Omega$ such that $\rho = (\mathcal{Y}, \Gamma)$ is a proper coconnected subpacket of τ. Then*

(1) If $\mathcal{D}^o_\rho(z) = \emptyset$ then $\mathcal{Y} \cong PSp_4[m]$ or AE_5.

(2) If $\mathcal{D}^o_\rho(z) \neq \emptyset$ then $\mathcal{Y} \cong L_4^-[m]$.

PROOF. Observe that ρ satisfies Hypothesis 9.4.1; most of the conditions in 9.4.1 inherit immediately, while the transitivity of $\mathcal{Y}$ on Γ follows from 9.4.1.4 and 6.6.6. Then the lemma follows by applying Theorem 9.4.33 in an inductive context in the proper subpacket. □

LEMMA 9.4.4. *Let $u \neq z$ be an involution in O. Then*

(1) $u^{\mathcal{F}_z} \cap O \cap \mathcal{F}^f \neq \emptyset$.

(2) If $u \in \mathcal{F}^f$ and $\rho = (\mathcal{Y}, \Gamma)$ is a coconnected subpacket of τ_u° such that $\Omega(t) \subseteq \Gamma$ for some $t \in \mathcal{Z}_S - \{z\}$, then ρ is a coconnected component of τ_z°.

PROOF. As $u \neq z$ is an involution in O, $u = u_1 u_2$ with $u_i \in K_i$ of order 4. Let $\Sigma = \Omega - \Omega(z)$ and $\alpha \in \mathfrak{A}(u)$. Then $u_i\alpha \in D_i \in \Delta(z\alpha)$ centralizes $\Sigma\alpha$, so by 9.4.1.3, $D_i \in \Omega - \Sigma\alpha$ and $Q = D_1 D_2 = O(z\alpha)$. Let $\beta \in \mathfrak{A}(z\alpha)$ with $z\zeta = z$, where $\zeta = \alpha\beta$. As Q is the unique member of $\{O(t) : t \in \mathcal{Z}_S\}$ containing $u\alpha$, $u\zeta = (u\alpha)\beta \in \mathcal{F}^f$. As $\zeta \in \hom_{\mathcal{F}_z}(C_S(u), C_S(z))$, (1) holds.

Assume the hypothesis of (2) and let T be Sylow in $\mathcal{Y}$ and $U = C_O(u)$. Then U is nonabelian with $z \in [U, U]$. As U centralizes $O(t)$, U acts on $\mathcal{Y}$, and hence on T. Indeed T acts on $C_{UT}(O(t)) = U\langle t\rangle$, using 9.4.3, and then acts on $\Phi(U\langle t\rangle) = \Phi(U)$. Hence T centralizes z, so T acts on $U = O \cap U\langle t\rangle$, so $[T, U] \leq T \cap U = 1$. Then by 2.3.1 in [**Asc19**], $z \in [U, U]$ centralizes $\mathcal{Y}$. As $U \trianglelefteq C_{\mathcal{F}_z}(u)$, also U centralizes $\mathcal{Y}$. Indeed if $P = C_O(\mathcal{Y})$ then $P' = N_O(P) \trianglelefteq N_{\mathcal{F}_z}(P)$, and as ρ is a coconnected component of $C_{\mathcal{F}}(P)$ with P' centralizing $O(t)$, P' acts on $\mathcal{Y}$ and then as $P' \trianglelefteq N_{\mathcal{F}_z}(P)$ it follows that P' centralizes $\mathcal{Y}$. That is $O = P'$ centralizes $\mathcal{Y}$. Then $T \leq C_{\mathcal{F}_z}(\mathcal{O}(z))$ by 2.6.12, so $\mathcal{Y} \leq [T]_{\mathcal{F}_z} \leq C_{\mathcal{F}_z}(\mathcal{O}(z))$. Then as $C_{\mathcal{F}_z}(\mathcal{O}(z)) \leq C_{\mathcal{F}_z}(u)$, ρ is a coconnected component of $C_{\mathcal{F}_z}(\mathcal{O}(z))$, and hence of τ_z°. □

LEMMA 9.4.5. $z \notin Z(\mathcal{F}^\circ)$.

PROOF. If $z \in Z(\mathcal{F}^\circ)$, then as $\mathcal{F}^\circ$ is transitive on Ω, $\Omega = \Omega(z)$, so $\mathcal{F}^\circ = \mathcal{F}_z^\circ = \mathcal{O}(z)$. This is a contradiction as $|\Omega(z)| = 2$ and $\mathcal{F}^\circ$ is transitive on Ω. □

LEMMA 9.4.6. *If each coconnected component of $\mathcal{F}_z$ is of the form $\rho_J = (\mathcal{O}_J, J)$ for some $J \in \Omega$, then $\Omega = \Omega(z)$.*

PROOF. Assume otherwise and let $\Gamma_1 = \Omega(z)$, $\Gamma_2 = \Omega - \Gamma_1$, $\rho_1 = (\mathcal{O}(z), \Gamma_1)$, and $\mathcal{B}_2 = \prod_{t \in \mathcal{Z}_S - \{z\}} \mathcal{O}(t)$. We show Hypothesis 8.1.2 is satisfied by ρ_1 and $\rho_2 = (\mathcal{B}_2, \Gamma_2)$. Then by 8.1.6, ρ_1 is a coconnected component of τ, so as $\mathcal{F} = \mathcal{F}^\circ$ and $\mathcal{F}$ is transitive on Ω, $\tau = \mathcal{O}_K$, contrary to 9.4.5.

It remains to verify Hypothesis 8.1.2. First, Hypothesis 8.1.1 holds by 9.4.1.4 and 6.6.6 (cf Remark 6.6.10). By construction, 8.1.2.1 holds. Next $T_1 = O$ and if $x \neq z$ is an involution in $\mathcal{B}_1^f$ then by 9.4.4.1 we may take $x \in \mathcal{F}^f$. Then by 9.4.4.2, $\mathcal{B}_2 \trianglelefteq \mathcal{F}_x$, and as $\Gamma_2 = C_\Omega(x)$, $\mathcal{B}_2$ contains all coconnected components of $\mathcal{F}_x$. Therefore 8.1.2.2 holds.

Next 8.1.2.3 follows from 9.4.5. As in the proof of 8.1.4.1, $T_2 \in \mathcal{F}^f$, and as $\mathcal{F}$ is transitive on Ω and ρ_J is a component of τ_z for each $J \in \Omega$, we have $\mathcal{B}_1 = \mathcal{O}(z) \trianglelefteq N_\mathcal{F}(T_2)$. That is 8.1.2.4 holds.

Assume the setup of 8.1.2.5. Then $T \in \Omega - \Gamma_1$ so $T_1 \cap T = O \cap T = 1$. Hence 8.1.2.5 holds, completing the proof of the lemma. □

LEMMA 9.4.7. *Let $\mathcal{Z}_S = \{z_1, \ldots, z_r\}$ with $z = z_1$, and assume $r > 1$. For $1 \leq i \leq r$, set $\Omega_i = \Omega(z_i)$, $O_i = O(z_i)$,*

$$Q = \bigcap_{i=1}^{r} N_S(O_i), \text{ and } A = Aut_\mathcal{F}(Q).$$

Then $Q = C_S(W_S)$ and:

(1) For each $i > 1$, there exists a coconnected component $\tau_i = (\mathcal{C}_i, \Omega_i)$ of τ_z such that $\mathcal{C}_i \cong L_4^-[m]$, $PSp_4[m]$, or AE_5. Let S_i be Sylow in $\mathcal{C}_i$.

(2) Let $P_i = \prod_{k \neq i} O_k$, $\mathcal{C}_1 = C_\mathcal{F}(P_1)^\circ$, and S_1 Sylow in $\mathcal{C}_1$. Then A is transitive on Ω, on $\mathcal{S} = \{S_1, \ldots, S_r\}$, and on $\{\mathcal{C}_k : 1 \leq k \leq r\}$.

(3) $\mathcal{F}$ contains the central product $\mathcal{C} = \mathcal{C}_1 * \cdots * \mathcal{C}_r$.

PROOF. Set $\mathcal{O}_i = \mathcal{O}(z_i)$. By 9.4.3 applied to the components of τ_z°, for each $i > 1$ there is a quaternion fusion packet $\tau_i = (\mathcal{C}_i, \Omega_i)$ such that either τ_i is a coconnected component of $\mathcal{F}_z$ with $\mathcal{O}_i \leq \mathcal{C}_i$, or $\mathcal{C}_i = \mathcal{O}_i$. Then S_i centralizes $\mathcal{O}_j$ for each $i \neq j$. As $r > 1$, 9.4.6 says there exists a $j > 1$ such that $\mathcal{C}_j \neq \mathcal{O}_j$.

Next as $\mathcal{F}$ is transitive on Ω, A is transitive on Ω, using 3.1.5. Let $1 < i \leq r$ and choose $\alpha \in A$ with $z_j\alpha = z_i$. Then for some $k \neq j$, $z_k\alpha = z$. As z_k centralizes $\mathcal{C}_j$, z centralizes $\mathcal{C}_j\alpha$ and then $\mathcal{C}_j\alpha \leq \mathcal{C}_i$, so $\mathcal{C}_i \neq \mathcal{O}_i$. Then by symmetry between i and j, there is $\beta \in A$ with $\mathcal{C}_i\beta \leq \mathcal{C}_j$, so $\mathcal{C}_j\alpha = \mathcal{C}_i$ and hence $\tau_j\alpha = \tau_i$. Now (1) follows from 9.4.3.

As $z \in \mathcal{F}^f$, also O_1 and P_1 are in $\mathcal{F}^f$. Then by 2.5.2, $\tau_1 = (\mathcal{C}_1, \Omega_1)$ is a quaternion fusion packet. For $i > 1$, $\mathcal{C}_i$ centralizes P_i, and as A is transitive on Ω, also A is transitive on $\{P_k : 1 \leq k \leq r\}$. Thus for $\beta \in A$ with $P_i\beta = P_1$, we have $\mathcal{C}_i\beta^* \leq \mathcal{C}_1$, so as $\mathcal{C}_i$ is transitive on Ω_i, $\mathcal{C}_1$ is coconnected and described in 9.4.3. Further there is $\gamma \in A$ with $O_2\gamma = O_1$, so as $\mathcal{C}_1$ is coconnected, we have $\mathcal{C}_1\gamma^* \leq \mathcal{C}_k$ for some $k > 1$, and hence $\mathcal{C}_1 = \mathcal{C}_i\beta^*$, so that also $S_i\beta = S_1$. Therefore (2) holds.

Next for $i > 1$, Q acts on O_i and hence on S_i, so $S_i \trianglelefteq Q$. Therefore by transitivity of A on $\mathcal{S}$, $S_1 \trianglelefteq Q$. Then $[S_1, S_i] \leq S_1 \cap S_i = 1$ as $\langle z_j \rangle = Z(S_j)$ for each $1 \leq j \leq r$. Thus $S_0 = \prod_k S_k$ is a central product by the transitivity of A on $\mathcal{S}$.

Set $T_1 = \prod_{i>1} S_i$. Claim that $\mathcal{C}_1$ centralizes T_1. Suppose that S_i centralizes $\mathcal{C}_1$ for each $i > 1$. Now $\mathcal{C}_1 \trianglelefteq \mathcal{N} = N_\mathcal{F}(P_1)$ and $T_1 \leq C_\mathcal{N}(\mathcal{C}_1)$, establishing the claim. Therefore it suffices to show that $\mathcal{C}_1$ centralizes S_i for each $i > 1$, or equivalently as A is transitive on $\mathcal{S}$, for all $i \neq j$, $\mathcal{C}_j$ centralizes S_i. Once this claim is established, 2.3 in [**Asc11**] says that $\mathcal{F}$ contains a central product of $\mathcal{C}_1, \ldots, \mathcal{C}_r$, completing the proof of (3) and the lemma.

Suppose first that $r > 2$ and for $1 \leq i \leq r$ set $B_i = \{\beta \in A : z\beta = z_i\}$. Then $B_i^{-1} = \{\alpha_i \in A : z_i\alpha_i = z\}$. Let $j > 1$. Then $\mathcal{C}_j$ centralizes S_i for each $i \notin \{1, j\}$, so $\mathcal{C}_1 = \mathcal{C}_j\alpha_j$ centralizes $S_i\alpha_j$ and hence $\mathcal{C}_1$ centralizes S_k for $k \notin \{1, l\}$, where

$S_1\alpha_j = S_l$. Therefore the claim holds unless for all $j > 1$, $S_1\alpha_j = S_l$ for some fixed l; that is unless for all $j > 1$, $B_j^{-1} = B_l$. But as $r > 2$, there exists $j > 1$ with $j \neq l$, so $B_j^{-1} = B_l = B_l^{-1}$, and hence $B_j = B_l$, so that $z_j = z\beta = z_l$ for $\beta \in B_j$, a contradiction.

Therefore we may assume $r = 2$, and it suffices to show that S_1 centralizes $\mathcal{C}_2$. Recall that O centralizes $\mathcal{C}_2$ and S_1 centralizes S_2. If $m > 8$ set $\mathcal{O}_i = \mathcal{O}(z_i)$. If $m = 8$ then from (1), $C_{\mathcal{C}_i}(z_i)$ has a normal subsystem $\mathcal{O}_i$ of index 1 or 3 in $SL_2[8] * SL_2[8]$ with Sylow group O_i. In either case as S_1 centralizes O_2, it centralizes $\mathcal{O}_2$ by 2.6.12 and its proof. So S_1 centralizes $S_2\mathcal{O}_2$.

Next (cf. 9.4.28 and 9.4.31.3) $\mathcal{C}_i = \langle S_i\mathcal{O}_i, N_{\mathcal{C}_i}(R_i) : R_i \in \mathcal{R}_i\rangle$ for a suitable subset $\mathcal{R}_i$ of $\mathcal{C}_i^e$, and $S_i = O_iV_i$, where $V_i = \langle U_i : R_i \in \mathcal{R}_i\rangle$ and $U_i = O_2(O^2(N_{\mathcal{C}_i}(R_i)))$. Thus $S_1 = O_1V_1$, and as O_1 centralizes $\mathcal{C}_2$, it remains to show that each U_1 centralizes $\mathcal{C}_2$. Then as S_1 centralizes $S_2\mathcal{O}_2$, it suffices to show that U_1 centralizes $N_{\mathcal{C}_2}(R_2)$.

Let G_2 be a model for the stabilizer in $N_{\mathcal{F}}(P)$ of R_2, where $P = R_2C_S(R_2)$. As $r = 2$, O is weakly closed in P with respect to G_2, so $O \trianglelefteq G_2$. Also as $\mathcal{C}_2 \trianglelefteq \mathcal{F}_z$, $N_{\mathcal{C}_2}(R_2) \trianglelefteq N_{\mathcal{F}_z}(R_2)$, so there is a normal subgroup H_2 of G_2 which is a model for $N_{\mathcal{C}_2}(R_2)$. Then $[P, O^2(H_2)] = U_2$, so $O^2(H_2)$ acts on $U = U_1U_2$. Thus we may regard H_2 as a subgroup of a model G_U for the stabilizer of U in $N_{\mathcal{F}}(C_S(U))$. By symmetry there is a corresponding subgroup H_1 of G_U with $[U, H_1] = U_1$. Set $H = \langle H_1, H_2\rangle$. Then (cf. the list of the $\mathcal{R}_i$ in 9.4.22 and 9.4.27, and the structure of the corresponding H_i in 9.4.21. Note that case (2) of 9.4.22 does not occur here as S_i is Sylow in $\mathcal{C}_i \cong L_4^-[m]$ when $\mathcal{D}^o(z) \neq \emptyset$ by 9.4.3.) $O^2(H_i) \cong A_4$ or AE_5, so $m(U) = 4$ or 8, and setting $Y_i = U_i \cap O_i$, Y_i is a hyperplane of U_i such that $Y_i = [U_i, N_{O_i}(R_i)]$. From 9.4.4, for each $y_i \in Y_i^\#$, $C_H(y_i) \leq N_H(U_{3-i}) \leq N_H(H_{3-i})$. Set $H^* = H/C_H(U)$, $L_i = O^2(H_i)$, and $W_i = C_U(L_{3-i})$. As H_i centralizes the hyperplane Y_{3-i} of U_{3-i}, it centralizes U/U_i, and then as U_i is a projective L_i-module, $U = U_i \times W_{3-i}$ with Y_{3-i} a hyperplane of W_{3-i}. As S_i centralizes U_{3-i}, $Y_i = [N_{O_i}(R_i), W_i]$. In particular if $W_0 = W_1 \cap W_2 \neq 1$, as $N_{O_i}(R_i)$ acts on L_1 and L_2 it acts on W_0 and then as $Y_i = [W_i, N_{O_i}(R_i)]$, it follows that $W_0 \leq Y_1 \cap Y_2 = 1$, a contradiction. Therefore $W_1 \cap W_2 = 1$.

Claim H is irreducible on U. For suppose $1 \neq U_0 < U$ is H-invariant. As H_i is irreducible on U_i, each proper nontrivial H_i-submodule is either contained in W_i or contains U_i. Then as $U_1 \cap U_2 = W_1 \cap W_2 = 1$, it follows that $U_0 = U_1$ or U_2, say the former. Set $Q_1 = C_{H^*}(U_1) \cap C_{H^*}(U/U_1)$. Then $Q_1 \leq C_{H^*}(Y_1) \leq N_{H^*}(U_2)$, so Q_1 centralizes U_2 and hence $Q_1 = 1$. Thus $X = C_{H^*}(U/U_1)$ is faithful on U_1 and $H_1^* \leq X$. If $m(U) = 4$ then $H_1^* = GL(U_1)$ so $H_1^* = X \trianglelefteq H^*$ and then $H_1 \trianglelefteq H$. If $m(U) = 8$ then $\rho_1 = (\mathcal{F}_{S \cap L_1}(L_1), \{N_{K_1}(R_1), N_{K_2}(R_1)\})$ is a quaternion fusion packet with $U_1 = O_2(L_1)$ so $L_1 \trianglelefteq H$. Thus in any event, $L_1 \trianglelefteq H$, so $W_1 \trianglelefteq H$, and hence H_2 acts on W_1, so $U_2 = U_1 \trianglelefteq H$. Thus $U_i = W_i$ centralizes H_{3-i}, completing the proof in this case. So the claim is established.

Suppose $m(U) = 8$ and set $\Gamma = \{N_J(U) : J \in \Omega\})$. As in the previous paragraph, $\rho = (\mathcal{F}_{S \cap H}(H), \Gamma)$ is a quaternion fusion packet with $z, z_2 \in U$, $\mu(\Gamma)$ is abelian, and H irreducible on U, so Theorem 2 supplies a contradiction as $|\Gamma| = 4$. Therefore $m(U) = 4$ and H^* is an irreducible subgroup of $GL(U) = L_4(2)$. Then from the list of such subgroups, and as $S_3 \cong H_2^* \trianglelefteq C_{H^*}(Y_1)$, we conclude that $H^* \leq O_4^+(2)$ preserves an orthogonal space structure on U, and Y_1 is a nonsingular point in that orthogonal space. But now L_1^* and L_2^* are the normal subgroups of order 3 centralizing the two definite lines $U_i = [L_i, U]$, so indeed U_{3-i} centralizes L_i, and hence also $H_i = L_i N_{S_i}(R_i)$, completing the proof. $\square$

LEMMA 9.4.8. *$\Omega = \Omega(z)$ is of order 2 so $\mathcal{D}^*(z) = \emptyset$.*

PROOF. Assume otherwise and adopt the notation of 9.4.7; in particular, in that notation, $r > 1$, and Hypothesis 8.1.7 is satisfied by 9.4.7. We will verify Hypothesis 8.1.8. Then by 8.1.10.3, $\mathcal{F} = \mathcal{F}^\circ = \mathcal{C}$, a contradiction as $\mathcal{F}$ is transitive on Ω and $r > 1$. Thus it remains to verify Hypothesis 8.1.8.

As A is transitive on $\mathcal{S}$, condition (1) of 8.1.8 is satisfied. For each involution $x \in \mathcal{C}_1^f$, $x \in O$, so by 9.4.4.1 we may take $x \in \mathcal{F}^f$. Then by 9.4.4.2, τ_i is a coconnected component of τ_x. Further if $x \neq z$ then τ_x has no other components, while if $x = z$ then the remaining components are $\mathcal{O}_{K_i}$ for $i = 1, 2$. Therefore 8.1.8.2 holds.

As in the proof of 8.1.4.1, $S_1' \in \mathcal{F}^f$. Then as $\mathcal{C}_1' \trianglelefteq \mathcal{F}_z$ and A is transitive on $\mathcal{Z}_S$, $\mathcal{C}_1 \trianglelefteq N_{\mathcal{F}}(S_1')$, verifying 8.1.8.3. Finally as $Z(\mathcal{C}_1) = 1$, 8.1.8.4 holds. This completes the proof. $\square$

LEMMA 9.4.9. *$O \trianglelefteq S$ and $z \in Z(S)$.*

PROOF. By 9.4.8, $O = O(\tau)$, so $O \trianglelefteq S$. Then as $\langle z \rangle = Z(O)$, $z \in Z(S)$. $\square$

LEMMA 9.4.10. *Assume $t \in \mathcal{D}^m(z)$, let Q_1 be a Q_8-subgroup of K, and set $Q = Q_1 Q_1^t$, $E = E(t, Q_1)$, and $\mathcal{U} = \mathcal{Z} \cap E$.*

(1) $\mathcal{U} = \{z\} \cup t^Q$ is of order 5.

(2) $C_{\mathcal{Z}}(E) \subseteq E$.

(3) $Aut_{\mathcal{F}}(E)$ acts transitively as $Alt(\mathcal{U})$ or $Sym(\mathcal{U})$ on $\mathcal{U}$.

(4) $\mathcal{Z} \cap tC_S(Q) = \{t\}$.

(5) t centralizes $C_S(Q)$.

(6) $\mathcal{Z} \cap tC_S(Q)Q \subseteq E$.

PROOF. Set $F = E \cap Q$. We first prove (1). By 9.4.1.3, $\mathcal{Z} \cap F = \{z\}$, so $\mathcal{U} - \{z\} \subseteq E - F$. By 9.1.14.2, Q has two orbits t^Q and $(tz)^Q$ on $E - F$ of order 4, so we may assume $tz \in \mathcal{Z}$ and it remains to derive a contradiction. Let $\alpha \in \mathfrak{A}(t)$; thus $t\alpha = z$. Set $t^Q = \{t_1, \ldots, t_4\}$ with $t = t_4$. For $1 \leq i \leq 3$, tt_i and tt_iz are in $F - \langle z \rangle$, so neither is in $\mathcal{Z}$. Hence by 9.1.4, neither $t_i\alpha$ nor $(t_iz)\alpha$ acts on K, so $t_i\alpha$ and $(t_iz)\alpha$ are in $\mathcal{D}^m(z)$. Therefore $z\alpha = (t_i\alpha) \cdot (t_iz)\alpha$ acts on K, contrary to 9.1.15.1. Thus (1) holds.

Suppose $s \in C_{\mathcal{Z}}(E) - E$. If $s \in \mathcal{D}^o(z)$ then $\langle v, z \rangle = C_O(s)$, contradicting $F \leq C_O(s)$. Therefore $s \in \mathcal{D}^m(z)$, so by 9.1.13, $E' = \langle F, s \rangle = E(s, Q_1)$, and then in particular $Q_1^s = Q_1^t$. Now $st \in N_S(Q_1)$ centralizes F, so $[Q, ts] \leq \langle z \rangle$, and hence $ts = fc$ for some $f \in F$ and $c \in C_S(Q)$.

Next let $\beta \in \mathfrak{A}(s)$. As $s \in \mathcal{D}^m(z)$, $z\beta \in \mathcal{D}^m(z)$ by 9.1.15.1. If $\mathcal{U}\beta \subseteq \mathcal{D}^m(z)$, then by 9.1.14.2, $z\beta = \prod_i t_i\beta$ acts on K, contradicting $z\beta \in \mathcal{D}^m(z)$. Therefore we may assume $t\beta \in \mathcal{D}^o(z)$, so by 9.1.4, $ts \in \mathcal{Z}$. Then as $ts \in FC_S(Q)$, ts induces an inner automorphism on O, contradicting 9.4.1.3. This completes the proof of (2).

By (1), $E = \langle \mathcal{U} \rangle$ and by 9.4.8, $\mathcal{Z} \subseteq \mathcal{D}^o(z) \cup \mathcal{D}^m(z)$. Hence by 9.1.7.2, $E\alpha$ centralizes v, so $E\alpha$ centralizes $r = (z\alpha)v$. As $t \in \mathcal{D}^m(z)$, we have $z \in \mathcal{D}^m(t)$ by 9.1.15.1; hence $z\alpha \in \mathcal{D}^m(z)$. Then by (1) applied to $z\alpha$ in the role of t, either $z\alpha v$ or $z\alpha vz$ is in $\mathcal{Z}$. Thus replacing v by vz if necessary, $r \in \mathcal{Z}$. By 9.1.14.3 we may assume $E \in \mathcal{F}^f$, so there is $\gamma \in \mathfrak{A}(E\alpha)$ with $\alpha\gamma \in Aut_{\mathcal{F}}(E)$. Then $r\gamma \in C_{\mathcal{Z}}(E)$, so

$r\gamma \in E$ by (2), and hence $r \in E\alpha$. Thus $v = r(z\alpha) \in E\alpha$, so $[E\alpha, v_1] \le \langle z, v\rangle \le E\alpha$, so v_1 induces $\xi \in Aut_{\mathcal{F}}(E\alpha)$. Then $\zeta = \xi\alpha^{-*} \in Aut_{\mathcal{F}}(E)$ with $1 \ne [z,\zeta]$, so as $Aut_Q(E)$ is regular on $\mathcal{U} - \{z\}$, (3) follows.

Suppose $u \in \mathcal{Z} \cap tC_S(Q) \cap C_S(t)$. Then $u \in C_{\mathcal{Z}}(E)$, so by (2), $u \in tC_S(Q) \cap E = tC_E(Q) = t\langle z\rangle$. Therefore $u = t$ by (1), so $\{t\} = \mathcal{Z} \cap tC_S(Q) \cap C_S(t)$. Now if $t \ne w \in \mathcal{Z} \cap tC_S(Q)$ then $1 \ne tw \in C_S(Q)$, so the involution $d \in \langle tw\rangle$ is in $C_S(Q)$, and td is a conjugate under $\langle t, w\rangle$ of t or w, so $t \ne td \in \mathcal{Z} \cap tC_S(Q) \cap C_S(t)$, a contradiction. This completes the proof of (4). As $C_S(Q)$ acts on $\mathcal{Z} \cap tC_S(Q)$, (4) implies (5).

It remains to prove (6), so suppose $t \ne s \in tC_S(Q)Q \cap \mathcal{Z}$. Set $y = st$ and let y_0 be the involution in $\langle y\rangle$. As above, ty_0 is conjugate to t or s, so $ty_0 \in \mathcal{Z}$. Hence by (4), $y_0 \notin C_S(Q)$, so if $|y| > 2$ then y induces an outer automorphism on Q as $\Phi(Inn(Q)) = 1$, contrary to the choice of s. Therefore $y = y_0$ so $[t,s] = 1$. Next $y = cx$ with $c \in C_S(Q)$ and $x \in Q$, and $\langle s,t\rangle$ centralizes c by (5), so $\langle s,t\rangle$ centralizes $x = c^{-1}y$. Therefore $x \in C_Q(t) = E \cap Q$ so $s = cxt \in C_{\mathcal{Z}}(E) \subseteq E$ by (2), completing the proof. □

NOTATION 9.4.11. Set $\mathcal{R} = \{R \in \mathcal{F}^e : Aut_{\mathcal{F}}(R) \ne Aut_{\mathcal{E}(\tau)}(R)\}$, and for $R \in \mathcal{R}$, let G_R be a model for $N_{\mathcal{F}}(R)$, $Z_R = \langle z^{G_R}\rangle$, and $H_R = \langle v_1^{G_R}\rangle$. Set $T = \langle O, \mathcal{D}^m(z)\rangle$.

LEMMA 9.4.12. *(1)* $\mathcal{E}(\tau) = \mathcal{F}_z$.
(2) $\mathcal{R} \ne \emptyset$ *and* $\mathcal{F} = \langle \mathcal{F}_z, Aut_{\mathcal{F}}(R) : R \in \mathcal{R}\rangle$.
(3) For each $R \in \mathcal{R}$, $z \in Z(R)$, $v \in R$, *and* $v_1 \in N_S(R) - R$.
(4) $\mathcal{D}^m(z) \ne \emptyset$.

PROOF. Part (1) follows from the definition of $\mathcal{E}(\tau)$ in 3.4.2, and from 9.4.8; for example by 1.3.2 in [**Asc19**], $\mathcal{F}_z = \langle SO(z), N_{\mathcal{F}}(O)\rangle$. Then (1), 9.4.5, and the Alperin-Goldschmidt Fusion Theorem imply (2).

Let $R \in \mathcal{R}$. By 9.4.9, $z \in Z(S)$, so $z \in Z(R)$. Then 9.2.3 completes the proof of (3). Part (4) follows from (2), (3), and 9.2.6. □

LEMMA 9.4.13. *(1) If* $m > 8$ *then* $\mathcal{O}_K \cong SL_2[m]$.
(2) If $m = 8$ *then* $O^2(Aut_{\mathcal{F}}(O)) \ne 1$.

PROOF. By 9.4.12.4 there is $t \in \mathcal{D}^m(z)$. Next for each Q_8-subgroup Q_1 of K, setting $E = E(t, Q_1)$, it follows from 9.4.10.3 that $Aut_{N_{\mathcal{F}}(E)}(Q_1)$ contains an automorphism of order 3. Hence (2) follows and (1) follows from 2.5.4. □

LEMMA 9.4.14. *Either*
(1) O *is transitive on* $\mathcal{D}^m(z)$, *so that* $T = O\langle t\rangle$ *for* $t \in \mathcal{D}^m(z)$ *and* $|T : O| = 2$,
or
(2) O *has two orbits on* $\mathcal{D}^m(z)$ *with representatives* t_1 *and* t_2, *so that* $T = O\langle t_1, t_2\rangle$ *with* $T/O \cong E_4$. *Moreover if* $m > 8$ *then* $Q_1^{t_2} \notin Q_1^{t_1 O}$ *for* Q_1 *a* Q_8-*subgroup of* K.

PROOF. If O is transitive on $\mathcal{D}^m(z)$ then (1) holds, so assume s and t are in distinct orbits of O on $\mathcal{D}^m(z)$. Let Q_1 be a Q_8-subgroup of K and set $Q = Q_1Q_1^t$. Suppose $Q_1^t = Q_1^s$. Then st acts on Q_1 and hence also on Q as t inverts st. As $s \notin t^O$ we conclude from 9.4.10.6 that $s \notin tC_S(Q)Q$, so ts induces an outer automorphism on Q_1 and Q_1^t. Now $\Sigma = Aut(Q)$ is the split extension of $I = Inn(Q) \cong E_{16}$ by $O_4^+(2)$, so there are two orbits α_i^I, $i = 1, 2$, of I on the set of involutions in a Sylow 2-subgroup Λ of Σ which do not act on Q_1, and α_1 and α_2 are fused in Λ.

Suppose $m = 8$. Then $O = Q$ and we may take $c_t = \alpha_1$ and $c_s = \alpha_2$, where c_x is the automorphism induced by x on Q via conjugation. In particular if $r \in \mathcal{D}^m(z)$ then $c_r \in \alpha_i^I$ for $i = 1$ or 2, so $r \in s^O \cup t^O$ by 9.4.10.4, and hence O has two orbits on $\mathcal{D}^m(z)$. Further s acts on t^O, so s actx on $O\langle t\rangle$, so $T/O \cong E_4$ and (2) holds in this case.

Therefore we may assume $m > 8$. Then $Aut_O(Q)\langle t\rangle$ is Sylow in Σ, so by an earlier observation, $s \in rC_S(Q)$ for some $r \in t^{N_O(Q)}$, contrary to an earlier reduction. Thus $Q_1^t \neq Q_1^s$. However if $r \in \mathcal{D}^m(z)$ then as K_2 has two orbits on its Q_8-subgroups, $Q_1^r \in Q_1^{xK_2}$ for some $x \in \{t, s\}$, so replacing x by a suitable conjugate under K_2, $Q_1^r = Q_1^x$, and hence $r \in x^O$. Thus O has two orbits on $\mathcal{D}^m(z)$. Again s acts on $O\langle t\rangle$, so $T/O \cong E_4$, completing the proof of the lemma. □

LEMMA 9.4.15. $C_S(O) = C_S(T)$.

PROOF. This follows from 9.4.10.5. □

LEMMA 9.4.16. *Suppose* $t \in \mathcal{D}^o(z)$ *and set* $Z = \langle z, t\rangle$ *and* $R = C_S(Z)$; *then* $Z \in \mathcal{F}^f$ *so there is a model* G_Z *for the stabilizer of* Z *in* $N_{\mathcal{F}}(R)$. *Set* $H_Z = \langle v_1^{G_Z}\rangle$.

(1) *Set* $U = O_2(H_Z)$. *Then* $U \cong E_{16}$ *with* $H_Z/U \cong S_3$. *Set* $R_0 = C_S(U)$ *and embed* G_Z *in a model* G_U *for the stabilizer of* U *in* $N_{\mathcal{F}}(R_0)$.

(2) *There exists* $y \in O - \langle v, z\rangle$ *with* $[t, y] = v$ *such that* y *induces a transvection on* U *with center* v. *Set* $Y = \langle H_Z, y\rangle \leq G_U$.

(3) G_U *preserves a quadratic form* q *on* U *of sign* $+1$ *such that* Y/U *is the subgroup of* $O(U, q)$ *generated by transvections,* $z^Y = \mathcal{Z} \cap U$ *is the set of singular points of* U, *and* $v^Y \cup (vz)^Y$ *is the set of nonsingular points of* U. *Either* $R = R_0$ *and* $G_U = YR$ *or* $|R : R_0| = 2$ *and* $G_U/R_0 = O(U, q)$.

(4) $\mathcal{D}^o(z) \cap U$ *and* $\mathcal{D}^m(z) \cap U$ *are of order 4 and* $U = \langle \mathcal{D}^m(z) \cap U\rangle \leq T$.

(5) $\mathcal{D}^o(z) \subseteq T$.

(6) O *has two orbits on* $\mathcal{D}^m(z)$, *so* $T/O \cong E_4$.

(7) $U = C_T(t)$.

(8) O *is transitive on* $\mathcal{D}^o(z)$.

PROOF. Let $\alpha \in \mathfrak{A}(Z)$; as $v_1 \in N_S(Z)$ we have $v_1\alpha \in D \in \Delta$, so $z\alpha = (v_1\alpha)^2 = z$ and hence α is an $\mathcal{F}_z$-map and $t\alpha \in \mathcal{D}^o(z)$. Therefore, replacing Z by $Z\alpha$, we may assume for the moment that $Z \in \mathcal{F}^f$, and postpone the proof that $\langle z, t'\rangle \in \mathcal{F}^f$ for all $t' \in \mathcal{D}^o(z)$ until later.

Much of the rest of the proof is similar to that of 9.2.5. By 9.1.15.2, $Aut_{\mathcal{F}}(Z) = GL(Z)$. Then as v_1 induces a transvection on Z with center z, there is $h \in H_Z$ of order 3 with $z^h = t$ and $Aut_{H_Z}(Z) = GL(Z)$. Further $[R, v_1] \leq \langle z, v\rangle$ and $[N_R(K), v_1] = \langle z\rangle$, so, as in the proof of 9.2.5, either

(a) R acts on K, $H_Z \cong S_4$, and $G_Z = H_Z \times C_{G_Z}(H_Z)$, or

(b) $U = O_2(H_Z) = Z \times E \cong E_{16}$, where $E = \langle v, v^h\rangle$ does not act on K, $E \triangleleft H_Z$, and $H_Z/U \cong S_3$.

Next as $t \in \mathcal{D}^o(z)$ there is $y \in O - \langle v_1, v_2\rangle$ with $[t, y] = v$. Suppose (a) holds and set $U = Z\langle v\rangle$ and form G_U and $Y = \langle H_Z, y\rangle \leq G_U$ as in (1). Then $C_U(y) = \langle v, z\rangle \geq Z(H_Z U) = U_0$, so $Aut_Y(U)$ is the stabilizer in $GL(U)$ of U_0. But then no Sylow 2-subgroup of Y centralizes z, impossible as $v_1 \in Y$, while by 9.4.8 $\Delta(\tau) = \Omega$, so by 3.1.3.2, z is in the center of any Sylow group of Y containing v_1. Therefore (b) holds, as does (1).

Set $u = v^h$. Then $U = \langle u, v, Z\rangle$, and as E does not act on K, we have $K^u = K_2$. Now $[u, v_1] = v$ so $v_i^u = v_{3-i}^{-1}$. Next $y = y_1y_2$ with $y_i \in K_i$ and $y_1^u = y_2v_2^\beta$ for some β, so $y^u = yv^\beta$. Therefore $(z, t, v, u)^y = (z, tv, v, uv^\beta)$. If β is even then u centralizes y, so y centralizes $E\langle z\rangle$ and (2) holds in this case. If β is odd then y centralizes $\langle z, v, tu\rangle$ and again (2) holds.

Suppose β is even. Then Y acts on E and y centralizes E and U/E, so $c_y \in O_2(Aut_Y(U))$. Then as H_Z is irreducible on U/E it follows that $O_2(Aut_Y(U))$ does not centralize z. Let $v_1 \in S_Y \in Syl_2(Y)$ and $g \in G_U$ with $S_Y^g \leq S$. Then $v_1^g \in D \in \Delta$, so $z = (v_1^g)^2 = z^g$ by 9.4.8. Therefore $z^g \in Z(S_Y^g)$, a contradiction.

Therefore β is odd, so y acts on E and centralizes the third irreducible $E' = \langle zv, tu\rangle$ for H_Z on U. Hence Y acts on E and E', so Y preserves a quadratic form q on U of sign $+1$ with $Y/C_Y(U)$ the subgroup of $O(U, q)$ generated by transvections. As $[y, R_0] \leq C_O(U) \leq C_O(t) = \langle z, v\rangle \leq U$, we have $U = C_Y(U)$. Set $S_U = S \cap G_U$. As $z \in Z(S_U)$ and $U = \langle z^Y\rangle$, we have $R_0 = O_2(G_U)$. Thus by **[McL69]**, either G_U acts on E and E', $R = R_0$, and $G_U = YR$, or G_U is irreducible on U and either $G_U/R_0 = O(U, q)$ with $|R : R_0| = 2$, or $G_U/R_0 = GL(U)$ or $Sp(U)$. In the first two cases (3) holds, while in the last two cases G_U is transitive on $U^\#$ so that $v \in z^{\mathcal{F}}$, a contradiction. This completes the proof of (3).

Further, writing $z^\perp$ for the subspace of U orthogonal to z with respect to q, $z^\perp = N_U(K)$, so $\mathcal{D}^o(z) \cap U = N_U(K) - (O \cap U)$ and $\mathcal{D}^m(z) \cap U = z^G \cap U - N_U(K)$ are of order 4. In particular $U = \langle \mathcal{D}^m(z) \cap U\rangle$, so $U \leq T$ by definition of T. That is (4) holds.

By (4), $t \in T$, so (5) follows from the symmetry among the members of $\mathcal{D}^o(z)$.

Next if O had just one orbit on $\mathcal{D}^m(z)$, then by 9.4.14, $|T : O| = 2$, so $O = N_T(K)$ and hence $N_{\mathcal{Z}}(K) = \{z\}$, a contradiction. Now 9.4.14 completes the proof of (6).

As $C_O(t) = \langle v, z\rangle$ and $|U : U \cap O| = |T : O| = 4$, (7) follows. As $\langle t\rangle K_i$ is semidihedral for $i = 1, 2$, O is transitive on the involutions in tO, so (8) follows from (5). Also (8) implies that for each $s \in \mathcal{D}^o(z)$, we have $\langle z, s\rangle \in \mathcal{F}^f$. This completes the proof of the lemma. □

LEMMA 9.4.17. *Suppose $T/O \cong E_4$. Then*

(1) $\mathcal{D}^o(z) \neq \emptyset$.

(2) There exists $t, s \in \mathcal{D}^m(z)$ such that $s \notin t^O$, $[s, t] = 1$, and ts inverts v_1. Set $A = \langle z, v, t, s\rangle$.

(3) There exists $w \in \mathcal{D}^o(z) \cap A$ and for each such w, $A = C_T(w)$.

PROOF. Let $t \in \mathcal{D}^m(z)$, $\alpha \in \mathfrak{A}(t)$, and $\tilde{T} = T/\langle z\rangle$. Then $O\langle t\rangle$ is of index 2 in T and O is transitive on $tO \cap \mathcal{Z}$ by 9.4.14, so by a Frattini argument there is $a \in N_T(K) \cap C_T(t) = T_t$ such that $a \notin O$.

Next the map $\varphi : y \mapsto yy^t$ is a T_t-equivariant homomorphism of K into $C_O(t)$ with image X such that $C_O(t) = \langle z\rangle \times X$, and φ induces an isomorphism $\tilde{\varphi} : \tilde{K} \to X$ via $\tilde{\varphi} : \tilde{y} \mapsto y\varphi$. Therefore $X \cong \tilde{K} \cong D_{m/2}$. Observe $v = v_1v_1^t$ generates $Z(X)$ if $m > 8$.

We first prove (2). To do so we wish to show:

(a) We may choose a to be an involution.

For if a is an involution, then as O has two orbits on involutions in taO with representatives ta and taz, and as one of these orbits is $\mathcal{Z} \cap taO$, replacing a by az

if necessary, we have $s = ta \in \mathcal{Z}$, so that $s \in \mathcal{D}^m(z) - t^O$. Then if a inverts v_1 we have established (2); that is:

(b) If (a) holds and a inverts v_1 then (2) holds.

As $|N_T(K) : O| = 2$, $a^2 \in \langle z\rangle X$. Thus $a^2 \in C_{X\langle z\rangle}(a)$. Let b be the involution in $X \cap \Phi(C_T(t))$. Now $b\alpha$ acts on K while $z\alpha \in \mathcal{D}^m(z)$ by 9.1.15.1, so $(zb)\alpha$ does not act on K. Therefore as $(a\alpha)^2$ acts on K, we have $(a\alpha)^2 \in X\alpha$, so $a^2 \in X$. Thus if $m = 8$ then $X\langle a\rangle \cong D_8$, so we may choose a to be an involution. As a induces an outer automorphism on K, $K\langle a\rangle \cong SD_{16}$, so a inverts v_1. Thus (2) holds by (b) in this case, so we may assume $m > 8$ during the remainder of the proof of (2). Then by 9.4.13, $\mathcal{O}_K \cong SL_2[m]$, so as $\mathcal{O}_K^t = \mathcal{O}_{K_2}$ it follows from 2.12 in [**Asc11**] that $C_{\mathcal{O}_z}(t) = \langle z\rangle \times \mathcal{X}$, where X is Sylow in $\mathcal{X} \cong \mathcal{O}_K/\langle z\rangle \cong L_2[m]$.

Let $s \in \mathcal{D}^m(z) - t^O$, $\langle v_0\rangle$ the cyclic subgroup of K of index 2, $k_1 \in K - \langle v_0\rangle$, $k = k_1k_1^t$, Q_1 a Q_8-subgroup of K, and $E = E(t, Q_1)$. By 9.4.14.2, $Q_1^s \neq Q_1^{ty}$ for $y \in K$, so st, and hence also a, interchanges the two classes of non-2-central involutions of $\tilde{K}$. Hence applying the isomorphism $\tilde{\varphi}$:

(c) a interchanges the two classes of non-2-central involutions of X.

Write $I(X)$ for the set of involutions of X. Then $v \in I(X)$ with $tzv = t^{v_1} \in \mathcal{Z}$, so as $\mathcal{X}$ is transitive on $I(X)$:

(d) $\{tzi : i \in I(X)\} = \mathcal{Z} \cap tzX$ is of order $m/4$.

Set $v_+ = v_0v_0^t$, $v_- = v_0v_0^{-t}$, $r_+ = tk$, $r_- = tz$, $V_\epsilon = \langle v_\epsilon\rangle$, and $X_\epsilon = \langle v_\epsilon, r_\epsilon\rangle$. Then $r_\epsilon \in I(X_\epsilon)$ inverts v_ϵ, so

(e) $X_\epsilon \cong D_{m/2}$.

Also $v_-^{k_1} = v_+$ and $r_-^{k_1} = r_+$, so

(f) k_1 interchanges X_+ and X_-, $v \in X_+$, and $vz \in X_-$.

Next $k \in X$ inverts v_0, so $j = kv_0$ is an involution which inverts $\langle v_0, v_0^t\rangle$ such that $r_-^j = r_-v_-z$ and

$$r_+^j = (tk)^{kv_0} = (tk)^{v_0} = (tv_-)(kv_0^2) = tkv_-^{-1}v_0^2 = tkv_+ = r_+v_+.$$

Also $(X_+X_-)\langle j\rangle$ is of order $m^2/2 = |C_{T_t}(v)|$, so

(g) $C_{T_t}(v) = (X_+ \times X_-)\langle j\rangle$, where $j \in O$ is an involution inverting v_ϵ with $r_\epsilon^j = r_\epsilon v_\epsilon$.

(h) $\mathcal{Z} \cap T_t = \mathcal{Z}_- \cup \mathcal{Z}_+ \cup \{z\}$ is of order $(m/2) + 1$, where $\mathcal{Z}_\epsilon = v(\epsilon)I(X_\epsilon)$ is of order $m/4$, $v(-) = v$, and $v(+) = vz$.

Namely $I(X_+) = \{v\} \cup r_+^{V_+} \cup (r_+v_+)^{V_+}$ with $r_+v_+ = r_+^j$. Further $v(+) = vz$ centralizes X_+ and $vzr_+ = vztk = t^{v_1k_1} \in \mathcal{Z}$, so also $vzr_+v_+ = vzr_+^j = (vzr_+)^j \in \mathcal{Z}$. Therefore $\mathcal{Z}_+ \subseteq \mathcal{Z}$ is of order $m/4$, so by (f), also $\mathcal{Z}_- \subseteq \mathcal{Z}$ is of order 4. Then as $|\mathcal{Z} \cap T_t| = |t^O| + 1 = (m/2) + 1$, (h) holds.

(i) $Z_\epsilon = \{r \in t^O : rv(-\epsilon) \in \mathcal{Z}\}$.

This is because for $i \in I(\epsilon)$, $v(-\epsilon)i \in I(\epsilon)$.

(j) $X_\epsilon \trianglelefteq C_T(v)$.

By (i) and the fact that $C_T(v)$ centralizes $\langle v, z\rangle = \langle v(+), v(-)\rangle$, $C_T(v)$ acts on $\mathcal{Z}_\epsilon$, and hence also on $I(X_\epsilon) = v(\epsilon)\mathcal{Z}_\epsilon$.

By symmetry between t and s and by (j), we have normal subgroups $\bar{X}_\epsilon$ of $C_T(v)$ defined with respect to s^O rather than t^O. Set $Y_\epsilon = X_\epsilon\bar{X}_\epsilon$. Then $X_\epsilon \cap X_{-\epsilon} = V_\epsilon$, and $\bar{r}_\epsilon$ interchanges the two classes of non-2-central involutions of X_ϵ by (c). Therefore $Y_\epsilon = \langle r_\epsilon, \bar{r}_\epsilon\rangle \cong D_m$. Further $Z(Y_\epsilon) = \langle v(-\epsilon)\rangle$, so $Y_+ \cap Y_- = 1$ and hence $\langle Y_+, Y_-\rangle = Y_+ \times Y_-$ is of order m^2. Therefore as $|C_T(v)| = m^2$ we conclude:

(k) $C_T(v) = Y_+ \times Y_-$ with $Y_\epsilon \cong D_m$.

Next $t, tvz = t^{v_1} \in \mathcal{Z}$, so $t \in \mathcal{Z}_-$ by (i). Thus $t = vi$ for some $i \in i(X_-)$. Now for $i' \in I(\bar{X}_+)$, $vzi' \in s^O$ and $t = vi$ centralizes vzi' by (k). Thus we may take $s = vzi'$, so that $[s,t] = 1$ and (a) holds. As $s \in \bar{\mathcal{Z}}_+$, $sv \in \mathcal{Z}$ by (i), so $svz \notin \mathcal{Z}$ and hence $v_1^s \neq v_1^t$, so $a = ts$ inverts v_1. Therefore (b) finally completes the proof of (2).

Notice (3) implies (1), so it remains to prove (3); thus we choose A as in (2). If $w \in \mathcal{D}^o(z) \cap A$ then $A = C_T(w)$ by 9.4.16.7, so we may assume $\mathcal{D}^o(z) \cap A = \emptyset$. Let $\Sigma = Aut_{\mathcal{F}}(A)$, $\mathcal{U} = \mathcal{Z} \cap A$, $\mathcal{V} = A^{\#} - \mathcal{U}$, $\mathcal{U}_0 = \{t, tzv, s, sv\}$, $\mathcal{V}_0 = \{v, vz\} \cup z\mathcal{U}_0$, and $\mathcal{W} = \{st, stz, stv, stvz\}$. Then $\{z\} \cup \mathcal{U}_0 \subseteq \mathcal{U}$ as we've seen that $t^{v_1} = tvz$, and then as ts inverts v_1, $s^{v_1} = sv$. Also $O \cap \mathcal{Z} = \{z\}$ and for $r \in \mathcal{D}^m(z)$, $rz \notin \mathcal{Z}$, so $\mathcal{V}_0 \subseteq \mathcal{V}$. Next $\mathcal{W} \subseteq N_T(K)$, so if $w \in \mathcal{W} \cap \mathcal{U}$ then $w \in \mathcal{D}^o(z)$, contrary to assumption. Therefore:

(l) $\mathcal{U} = \{z\} \cup \mathcal{U}_0$ is of order 5.

Let $\zeta \in \Sigma$ be the automorphism c_{v_1} induced on A by v_1 and $\mathcal{S}$ the set of $u \in \mathcal{U}$ such that u is the unique fixed point of a Sylow 2-subgroup of $C_\Sigma(u)$. By Gleason's Lemma:

(m) Σ is transitive on $\mathcal{S}$.

As $[t, v_1] = vz$ and $[s, v_1] = v$:

(n) $C_A(v_1) = [A, v_1] = \langle v, z\rangle$, so z is the unique fixed point of ζ on $\mathcal{U}$ and hence $z \in \mathcal{S}$.

Let $\gamma \in \mathfrak{A}(A)$. As $v_1 \in N_S(A)$ it follows from 9.4.1.3 and 9.4.8 that $z\gamma = (v_1\gamma)^2 = z$. Thus, replacing A by $A\gamma$:

(o) We may assume $A \in \mathcal{F}^f$.

(p) $C_{\mathcal{Z}}(A) = \mathcal{U}$.

Namely $C_{\mathcal{Z}}(A) \subseteq C_T(\langle t, z\rangle) = \langle z\rangle \times X\langle a\rangle$ and $C_X(a) = \langle v\rangle$ so $C_T(A) = A$. Indeed there exists $x \in X$ with $[a, x] = v$, so writing ξ for c_x:

(q) $\xi \in \Sigma$ is a transvection with center v and axis $\langle z, t, v\rangle$.

(r) Σ is not transitive on $\mathcal{U}$.

Suppose Σ is transitive on $\mathcal{U}$. By (l), $\mathcal{U} = \{z\} \cup \mathcal{U}_0$ is of order 5. Then as $\langle \zeta, \xi\rangle \leq \Sigma$, Σ acts faithfully on $\mathcal{U}$ as $Sym(\mathcal{U})$. But now $N_A(K) = [N_A(K), C_\Sigma(z)]$, so as $v \in N_A(K)$, we have $N_A(K) \leq O$, contradicting $a \in N_A(K) - O$.

We now obtain a contradiction, establishing (3) and completing the proof of the lemma. By (r) and (m), we may assume $t \in \mathcal{U} - \mathcal{S}$. Let $\alpha \in \mathfrak{A}(t)$. As $t \in \mathcal{D}^m(z)$, $z\alpha \in \mathcal{D}^m(z)$ by 9.1.15.1, so $r = (z\alpha)^{c_{v_1}} = z\alpha v$ or $z\alpha vz$. If $v \notin A\alpha$ then $r \in C_{\mathcal{Z}}(A\alpha) - A\alpha$. By (o), there is $\beta \in \mathfrak{A}(A\alpha)$ with $A\alpha\beta = A$. Then $r\beta \in C_{\mathcal{Z}}(A) - A$, contrary to (p). Thus $v \in A\alpha$, so v_1 acts on $A\alpha$. Let $\theta = c_{v_1}\alpha^{-*} \in \Sigma$. As $t \notin \mathcal{S}$, $\theta = (z, u)$ is a transposition. Then as $\Lambda = \langle \zeta, \theta\rangle$ is not transitive on $\mathcal{U}$, $\{z, t, tzv\}$ is an orbit of Λ. Thus we have shown that if $u \in \mathcal{U} - \mathcal{S}$ then $u \in z^\Sigma$, so Σ is transitive on $\mathcal{U}$ by (m), contradicting (r) and completing the proof of the lemma. $\square$

DEFINITION 9.4.18. Set $\mathfrak{Z}_1 = \{\langle z, t, v\rangle : t \in \mathcal{D}^m(z)\}$, let $\mathcal{Q}(K)$ be the set of Q_8-subgroups of K, define

$$\mathfrak{Z}_2 = \{E(t, Q_1) : t \in \mathcal{D}^m(z) \text{ and } Q_1 \in \mathcal{Q}(K)\},$$

and set $\mathfrak{Z}_3 = \{\langle z, t\rangle : t \in \mathcal{D}^o(z)\}$ and $\mathfrak{Z}_4 = \{C_T(t) : t \in \mathcal{D}^o(z)\}$. Set $\mathcal{R}_i = \{C_S(Z) : Z \in \mathfrak{Z}_i\}$.

LEMMA 9.4.19. *(1) O has $|T : O|/2$ orbits on $\mathfrak{Z}_1$.*

(2) If $m = 8$ and $|T : O| = 2$ then $|\mathfrak{Z}_2| = 1$.

(3) If $|T : O| = 4$ then O is transitive on $\mathfrak{Z}_3$ and $\mathfrak{Z}_4$.

(4) If $i = 2$ assume $m = 8$ and $|T : O| = 2$. Then $\mathfrak{Z}_i$ and $\mathcal{R}_i$ are contained in $\mathcal{F}^f$.

PROOF. By 9.4.14, O has $|T : O|/2$ orbits on $\mathcal{D}^m(z)$. Also for $t \in \mathcal{D}^m(z)$, $\mathcal{Z} \cap \langle z, t, v\rangle = \{z, t, t^{v_1}\}$, so (1) follows. If $m = 8$ then $\mathcal{Q}(K) = \{K\}$, and if $|T : O| = 2$ then O is transitive on $\mathcal{D}^m(z)$ by 9.4.14. Hence (2) holds. Finally if $|T : O| = 4$ then O is transitive on $\mathcal{D}^o(z)$ by 9.4.16.8, so (3) follows.

Choose i as in (4), let $X \in \mathfrak{Z}_i$ and $\alpha \in \mathfrak{A}(X)$. As $v_1 \in N_S(X)$, the argument in the first paragraph of the proof of 9.4.16 shows that $z\alpha = z$. Hence $X\alpha \in \mathfrak{Z}_i$. Now by (1)-(3), $X\alpha \in X^S$, so as $X\alpha \in \mathcal{F}^f$, also $X \in \mathcal{F}^f$. Similarly let $Y = C_S(X)$ and $\beta \in \mathfrak{A}(Y)$, Then $\beta_{|X} \in \mathfrak{A}(X)$ so there is $s \in S$ with $(X\beta)^s = X$. Then $(Y\beta)^s = Y$, so $Y \in \mathcal{F}^f$, completing the proof of (4). □

LEMMA 9.4.20. *Assume $|T : O| = 4$ and let $U \in \mathfrak{Z}_4$. Then*
(1) $C_S(U) \notin \mathcal{R}$.
(2) If $N_S(U) = N_T(U)C_S(U)$ then $\mathcal{R}_3 \cap \mathcal{R} = \emptyset$.
(3) If $N_S(U) \neq N_T(U)C_S(U)$ then $\mathcal{R}_3 \subseteq \mathcal{R}$.

PROOF. As $|T : O| = 4$, there is $t \in \mathcal{D}^o(z)$ by 9.4.17.1. Thus $U = C_T(t) \in \mathfrak{Z}_4$ and we adopt the notation of 9.4.16. By 9.4.16.3, G_U preserves a quadratic form q on U of sign $+1$ and either

(i) $N_S(U) = N_T(U)C_S(U)$, $R = R_0$, where $R_0 = C_S(U)$, $G_U = YR$, and G_U/R is the subgroup of $O(U, q)$ generated by transvections, or

(ii) $N_S(U) \neq N_T(U)C_S(U)$, $|R : R_0| = 2$, and $G_U/R_0 = O(U, q)$.

Therefore $R_0 \in \mathcal{R}_4$ and G_U/R_0 has no strongly embedded subgroup, so (1) holds. Also $R \in \mathcal{R}_3$ and in case (i), $R = R_0 \notin \mathcal{R}$, so (2) holds. On the other hand in case (ii), $G/R \cong S_3$ and $R \in \mathcal{F}^f$ by 9.4.19.4, so $R \in \mathcal{R}$, establishing (3). □

LEMMA 9.4.21. *Let $t \in \mathcal{D}^m(z)$, $Z = \langle t, z, v\rangle$, $R = C_S(Z)$, G_R a model for $N_{\mathcal{F}}(R)$, $G_Z = N_{G_R}(Z)$, and $H_Z = \langle v_1^{G_Z}\rangle$. Then*

(1) $H_Z \cong S_4 \times \mathbf{Z}_2$, $Z = O_2(H_Z)$, $G_Z = H_Z C_{G_Z}(Z)$, and $C_{G_Z}(Z)/R$ is of odd order.

(2) $R \in \mathcal{R}$ and $R \in \mathcal{R}_1$, so $\mathcal{R}_1 \subseteq \mathcal{R}$.

(3) Either

(a) $G_R = G_Z$ and $\mathcal{R}_2 \cap \mathcal{R} = \emptyset$, or

(b) $m = 8$, $|T : O| = 2$, $S = TC_S(O)$, and for $E \in \mathfrak{Z}_2$, G_E a model of $N_{\mathcal{F}}(E)$, and $H_E = \langle v_1^{G_E}\rangle$, we have $H_E \cong AE_5$, $R = EC_S(O) = C_S(E) \in \mathcal{R}$, and $G_Z \leq G_E$.

PROOF. As in the proof of 9.2.7, $H_Z \cong S_4 \times \mathbf{Z_2}$ with $v = v_1 v_1^t$ generating $Z(H_Z)$, and $G_Z = H_Z C_{G_Z}(Z)$. As $Z \trianglelefteq G_Z$, $O_2(H_Z) = \langle z^{H_Z}\rangle = Z$. In particular (1) holds and G_Z/R has a strongly embedded subgroup, so if $Z \trianglelefteq N_{\mathcal{F}}(R)$ then $G_R = G_Z$ and $R \in \mathcal{R}$. Hence

(i) If $Z \trianglelefteq N_{\mathcal{F}}(R)$ then (2) holds.

But, as in the proof of 9.4.17, $C_T(\langle z, t\rangle) = \langle z, t\rangle \times X\langle s\rangle$, where $X = \{x_1 x_1^t : x_1 \in K\} \cong D_{m/2}$, $s = 1$ if $|T : O| = 2$, and $s \in \mathcal{D}^m(z)$ with $ts \in \mathcal{D}^o(z)$ if $|T : O| = 4$. Thus $Z \leq Z(C_T(\langle z, t\rangle)$ so $C_T(Z) = C_T(\langle z, t\rangle)$. Further $X\langle z, t\rangle = \langle E(t, Q_1) : Q_1 \in \mathcal{Q}(K)\rangle$, and if $|T : O| = 4$ then $C_T(Z) = UX$, where $U = C_T(st) \in \mathfrak{Z}_4$. Thus $C_T(Z) = \langle C_{\mathcal{Z}}(Z)\rangle \trianglelefteq N_{\mathcal{F}}(R)$. Further unless $m = 8$ and $|T : O| = 2$, $Z = Z(C_T(Z))$, so indeed $Z \trianglelefteq N_{\mathcal{F}}(R)$. Therefore by (i):

(ii) If $m > 8$ or $|T : O| = 4$ then (2) holds.

Moreover it follows from 9.4.10.3 that if $m > 8$ or $|T : O| = 4$, then for $E \in \mathfrak{Z}_2$, $Aut_{\mathcal{F}}(E) \cong S_5$ has no strongly embedded subgroup, so $\mathcal{R}_2 \cap \mathcal{R} = \emptyset$. Thus (2) and (3a) hold is this case.

Finally suppose $m = 8$ and $|T : O| = 2$. Then $E = C_T(Z) = E(t, K)$, so taking G_E to be a model for $N_{\mathcal{F}}(E)$, $R_0 = C_S(E)$, and $H_E = \langle v_1^{G_E} \rangle$, we have $H_E \cong AE_5$ and $R_0 = EC_S(O) \in \mathcal{R}_2$ by 9.4.15. If $S = TC_S(O)$ then $G_E/R_0 \cong A_5$ has a strongly embedded subgroup, so $R_0 \in \mathcal{R}$. Further $R = R_0$ and $G_Z/R \cong S_3$, so (2) and (3b) hold in this case. Finally if $S \neq TC_S(O)$ then $G_E/R_0 \cong S_5$ has no strongly embedded subgroup, so $R_0 \notin \mathcal{R}$, but $|R : R_0| = 2$ and $G_Z/R \cong S_3$, so $R \in \mathcal{R}$ and (2) and (3a) hold. □

LEMMA 9.4.22. *(1) If $|T : O| = 4$ and $N_S(U) = N_T(U)C_S(U)$ for $U \in \mathfrak{Z}_4$ then $\mathcal{R} = \mathcal{R}_1$ and O has two orbits on $\mathcal{R}$.*

(2) If $|T : O| = 4$ and $N_S(U) \neq N_T(U)C_S(U)$ for $U \in \mathfrak{Z}_4$ then $\mathcal{R} = \mathcal{R}_1 \cup \mathcal{R}_3$ and O has two orbits on $\mathcal{R}_1$ and is transitive on $\mathcal{R}_3$.

(3) If $|T : O| = 2$ and $m > 8$ then $\mathcal{R} = \mathcal{R}_1$ and O is transitive on $\mathcal{R}$.

(4) If $|T : O| = 2$ and $m = 8$, then $\mathcal{R}_2 = \{E\}$ is of order 1, O is transitive on $\mathcal{R}_1$, and for each $R \in \mathcal{R}_1$, $N_{\mathcal{F}}(E) = \langle N_{\mathcal{F}}(R), N_{\mathcal{F}_z}(E) \rangle$. Further either

(a) $S \neq TC_S(O)$ and $\mathcal{R} = \mathcal{R}_1$, or

(b) $S = TC_S(O)$ and $\mathcal{R} = \mathcal{R}_1 \cup \mathcal{R}_2$.

PROOF. Let $R \in \mathcal{R}$ and adopt Notation 9.4.11. Suppose first that $Z_R \leq N_S(K)$. Then by 9.2.5.1, $Z_R = \langle z, t \rangle$ for some $t \in \mathcal{D}^o(z)$. Therefore $|T : O| = 4$ by 9.4.16.6, $Z_R \in \mathfrak{Z}_3$, and $R \in \mathcal{R}_3$ by 9.2.4.2. Further G_R is described in 9.2.5, and in particular $U = O_2(H_R) \in \mathfrak{Z}_4$ by 9.4.16.7. By 9.4.20.2, $N_S(U) \neq N_T(U)C_S(U)$ and by 9.4.19.3, O is transitive on $\mathcal{R}_3$ so $\mathcal{R}_3 \subseteq \mathcal{R}$. Conversely if $|T : O| = 4$ and $N_S(U) \neq N_T(U)C_S(U)$ for $U \in \mathfrak{Z}_3$, then $\mathcal{R}_3 \subseteq \mathcal{R}$ by 9.4.20.3.

Therefore we may assume $Z_R \not\leq N_S(K)$. Then G_R is discussed in 9.2.7, where there are two cases: cases (a) and (b) of 9.2.7.1. In case (b), $Z_R = \langle z, v, t \rangle$ for some $t \in \mathcal{D}^m(z)$ and hence $R = C_S(Z_R) \in \mathcal{R}_1$ by 9.2.4.2. Further by 9.4.21.2, $\mathcal{R}_1 \subseteq \mathcal{R}$ and by 9.4.19.1, O has $|T : O|/2$ orbits on $\mathcal{R}_1$.

So assume that case (a) holds. Then G_R is described in 9.2.7.2. By 9.4.8, $\mathcal{D}^*(z) = \emptyset$, so neither case (i) nor (ii) of 9.2.7.2 holds, and in case (iii), Z_R is not the $L_2(4)$-module. In case (iv), $m(Z_R) > 4$ and $Z_R = \langle \mathcal{Z} \cap Z_r \rangle$, so $Z_R \leq T$, using 9.4.16.5. This contradicts $m(T) = 4$. Therefore $m(Z_R) = 4$, so $m = 8$ by 9.2.7.6. Observe this completes the proof of (3).

Suppose $H_R/Z_R \cong \mathbf{Z}_2/E_9$. Then H_R preserves a quadratic form q on Z_R of sign $+1$ and z^{H_R} is the set of singular points of this form. In particular there is $z \neq r \in z^{H_R} \cap N_{Z_R}(K)$, so $r \in \mathcal{D}^o(z)$. Therefore $|T : O| = 4$ by 9.4.16.6 and $Z_R \in \mathfrak{Z}_4$ by 9.4.16.7, so $R \in \mathcal{R}_4$, contrary to 9.4.20.1. Thus H_R/Z_R is not $\mathbf{Z}_2/E_9$.

Suppose next that $H_R/Z_R \cong D_{10}$. Then $[Z_R, v_1] = \langle z, v \rangle = Z_0$ and there are three hyperplanes of Z_R containing Z_0, all conjugate under $Aut(H_R)$. Therefore $N_{Z_R}(K) \cap z^{H_R} = \{z\}$, as this holds in AE_5 which contains a copy of H_R. Thus $z^{H_R} - \{z\} \subseteq \mathcal{D}^m(z)$. If $|T : O| = 2$ then $N_{Z_R}(K) \leq O$, so $Z_R = E(t, K)$ for $t \in z^{H_R} - \{z\}$, and hence $Aut_{\mathcal{F}}(Z_R) \cong A_5$ or S_5 by 9.4.10, contradicting $H_R/Z_R \cong D_{10}$. Thus $|T : O| = 4$ and $N_{Z_R}(K) \not\leq O$, so Z_R satisfies the hypothesis of the group A in 9.4.17. Hence $Z_R \in \mathfrak{Z}_3$ by 9.4.17.3. But then Z_R has no automorphism of order 5 by parts (3) and (7) of 9.4.16. Thus H_R/Z_R is not D_{10}.

We have eliminated all but one of the cases in 9.2.7.2, so we conclude from that lemma that $H_R/Z_R \cong A_5$ and Z_R is the A_5-module. Then by parts (3) and (4) of 9.2.7+QQ.7, $Z_R = E(t, K)$ for $t \in z^{H_R} - \{z\}$. As G_R/R has a strongly embedded subgroup, $G_R = H_R C_{G_R}(H_R)$. In particular T is Sylow in H_R and $|T : O| = 2$. This completes the proof of (1) and (2), and (4) follows from 9.4.21.3. □

LEMMA 9.4.23. *Assume $m = 8$ and $|Aut_{\mathcal{F}}(O)|_3 = 3$. Then $|T : O| = 2$ and $\mathcal{F}^\circ \cong AE_5$.*

PROOF. Let $t \in \mathcal{D}^m(z)$ and $E = E(t, K)$. By 9.4.10.3, $\Sigma = Aut_{\mathcal{F}}(E)$ is A_5 or S_5.

Claim $|T : O| = 2$. If not then $\Sigma \cong S_5$ and by 9.4.14 there is $s \in \mathcal{D}^m(z) - t^O$. Set $\mathcal{X} = N_{\mathcal{F}_z}(E)$; then $s \notin O_2(C_\Sigma(z))$, so $s \notin O_2(\mathcal{F})$. Set $F = E(s, K)$ and $\mathcal{Y} = N_{\mathcal{F}_z}(F)$. As $s \in O_2(\mathcal{Y})$ but $s \notin O_2(\mathcal{X})$, $O^2(Aut_{\mathcal{X}}(O)) \neq O^2(Aut_{\mathcal{Y}}(O))$, contradicting $|Aut_{\mathcal{F}}(O)|_3 = 3$. This contradiction establishes the claim.

By the claim and 9.4.19.2, $\mathfrak{Z}_2 = \{E\}$ so $E \trianglelefteq \mathcal{F}_z$. By 9.4.22.4, $N_{\mathcal{F}}(R) \leq N_{\mathcal{F}}(E)$ for each $R \in \mathcal{R}$. Then by 9.4.12.2, $\mathcal{F} = \langle \mathcal{F}_z, N_{\mathcal{F}}(R) : R \in \mathcal{R} \rangle \leq N_{\mathcal{F}}(E)$, so $E \trianglelefteq \mathcal{F}$. Then the lemma follows from Theorem 2. □

Because of 9.4.23, during the remainder of the section until Theorem 9.4.33 we assume:

HYPOTHESIS 9.4.24. *Hypothesis* 9.4.1 *holds and if $m = 8$ then $|Aut_{\mathcal{F}}(O)|_3 = 9$.*

LEMMA 9.4.25. *T centralizes $C_{\mathcal{F}}(O)$.*

PROOF. As $T = \langle O, \mathcal{D}^m(z) \rangle$ it suffices to show $t \in \mathcal{D}^m(z)$ centralizes $\mathcal{Y} = C_{\mathcal{F}}(O)$. By 9.4.15, t centralizes $Y = C_S(O)$, so t acts on R and $N_{\mathcal{Y}}(R)$ for each $R \in \mathcal{Y}^e$. Let G be a model for $N_{\mathcal{Y}\langle t \rangle}(R)$; it suffices to show $t \in Z(G)$. But $[G, t] \leq C_Y(R) = Z(R)$, so G acts on $tZ(R)$, and then G centralizes t by 9.4.10.4, since Y centralizes each Q_8-subgroup of K. □

LEMMA 9.4.26. *There exists a normal subsystem $\mathcal{O}$ of $\mathcal{F}_z$ with Sylow group O such that $\mathcal{O} \cong SL_2[m] * SL_2[m]$ and $T\mathcal{O} \trianglelefteq \mathcal{F}_z$.*

PROOF. Suppose first that $m > 8$ and set $\mathcal{O} = \mathcal{O}(z)$; then $\mathcal{O} \cong SL_2[m] * SL_2[m]$ by 9.4.13, so it remains to show $T\mathcal{O}(z) \trianglelefteq \mathcal{F}_z$. By 2.6.11, $\mathcal{O} \trianglelefteq \mathcal{F}_z$. Applying Theorem 1.5.2 in [**Asc19**] to $\mathcal{F}_0 = \mathcal{O}$ and $E = T$, it suffices to show $TN_{\mathcal{O}}(O) \trianglelefteq N_{\mathcal{F}}(O)$. But $O = N_{\mathcal{O}}(O)$ and $Aut(O)$ is a 2-group, so $N_{\mathcal{F}}(O) = SC_{\mathcal{F}}(O)$, and we must show $T \trianglelefteq SC_{\mathcal{F}}(O)$. As $T = \langle O, \mathcal{D}^m(z) \rangle$, $T \trianglelefteq S$, and by 9.4.25, T centralizes $C_{\mathcal{F}}(O)$, completing the proof when $m > 8$.

So assume $m = 8$, let $S_0 = C_S(O)$, $R = OS_0$, G a model for $N_{\mathcal{F}}(R)$, and $H = \langle T^G \rangle$. By 9.4.25, T centralizes $B = C_G(O)$ so H centralizes B. Set $G^* = G/B$. By 9.4.24, $O^2(G^*) = G_1^* \times G_2^*$ where $G_i^* \cong A_4$ and the preimage $\bar{G}_i$ of G_i^* in G centralizes K_{3-i}. Let $t \in \mathcal{D}^m(z)$ and $G_i = O^2(\bar{G}_i)$; then $G_1^t = G_2$ and $H_0 = [G_1 G_2, t] \leq H \leq C_G(B)$, so $O = O_2(H_0)$ and $H_0/O \cong \mathbf{Z}_3$.

Next let $E \in \mathfrak{Z}_2$, $R_E = C_S(E)$, G_E a model for $N_{\mathcal{F}}(R_E)$, and $H_E = \langle {v_1}^{G_E} \rangle$. Then $H_E/E \cong A_5$ by 9.4.10 and H_E has one class of complements to E, so $R_E = E \times C_{R_E}(H_E)$. Then $X = O^2(C_{H_E}(z))$ centralizes S_0 as S_0 centralizes T and then $S_0 = C_{R_E}(H_E)\langle z \rangle$. Hence we may take $X \leq G$ and then $X \leq C_{G_1 G_2}(S_0) = Y$. We conclude that $Y = H_0 X = G_1 * G_2$ with $G_i \cong SL_2(3)$. As T centralizes B, so does

$TY = T[T,Y]$. Then as $Aut_Y(O) = O^2(Aut(O))$, $G = SYB$, and as $T \trianglelefteq S$, we have $TY \trianglelefteq G$.

Set $\mathcal{O} = \mathcal{F}_O(Y)$, $\mathcal{F}_1 = SC_{\mathcal{F}}(O)$, $\mathcal{F}_2 = N_{\mathcal{F}_z}(S_0)$, $\mathcal{E}_1 = T$, and $\mathcal{E}_2 = T\mathcal{O}$. Observe that this 4-tuple satisfies Hypothesis 1.4.7 in [**Asc19**]. For example $\mathcal{F}_z = \langle \mathcal{F}_1, \mathcal{F}_2 \rangle$ by 1.3.2 in [**Asc19**], $\mathcal{E}_1 \trianglelefteq \mathcal{F}_1$ by 9.4.25, and $\mathcal{E}_2 \trianglelefteq \mathcal{F}_2$ as $TY \trianglelefteq G$. Therefore $T\mathcal{O} = \langle \mathcal{E}_1, \mathcal{E}_2 \rangle \trianglelefteq \mathcal{F}_z$ by 1.4.8 in [**Asc19**], completing the proof. □

DEFINITION 9.4.27. We define a subset $\mathcal{R}_0$ of $\mathcal{R}$, depending on which case in 9.4.22 holds. In case (1), let $\mathcal{R}_0 = \{R_1, R_2\}$ be representatives of the orbits of O on $\mathcal{R}$, so that $R_i \in \mathcal{R}_1$ for $i = 1, 2$. In case (2) let $\mathcal{R}_0 = \{R_1, R_2, R_3\}$ be representatives for the orbits on $\mathcal{R}_1$ for $i = 1, 2$, and $R_3 \in \mathcal{R}_3$. In cases (3) and (4) let $\mathcal{R}_0 = \{R_1\}$ with $R_1 \in \mathcal{R}_1$.

Set $k = |\mathcal{R}_0|$, $\mathcal{F}_0 = \mathcal{F}_z$, and for $1 \leq i \leq k$ set $\mathcal{F}_i = N_{\mathcal{F}}(R_i)$ and $S_i = N_S(R_i)$. Set $\mathcal{E}_0 = T\mathcal{O}$ and $T_0 = T$. For $1 \leq i \leq k$ let $T_i = N_T(R_i)$, let G_i be a model for $\mathcal{F}_i$, and set $H_i = \langle T_i^{G_i} \rangle$. We will see that $T_i \in Syl_2(H_i)$, so we can define $\mathcal{E}_i = \mathcal{F}_{T_i}(H_i)$ and we have $\mathcal{E}_i \trianglelefteq \mathcal{F}_i$.

Finally define $\mathcal{E} = \langle \mathcal{E}_0, Aut_{\mathcal{E}_i}(T_i \cap R_i) : 1 \leq i \leq k \rangle_T$, regarded as a fusion system on T. Note $\mathcal{E}$ need not be $\mathcal{E}(\tau)$.

LEMMA 9.4.28. $\mathcal{F} = \langle \mathcal{F}_z, Aut_{\mathcal{F}}(R_i) : 1 \leq i \leq k \rangle$.

PROOF. By 9.4.12.2, $\mathcal{F} = \langle \mathcal{F}_z, Aut_{\mathcal{F}}(R) : R \in \mathcal{R} \rangle$. Therefore also $\mathcal{F} = \langle \mathcal{F}_z, Aut_{\mathcal{F}}(R_i) : 1 \leq i \leq l \rangle$ for any set $\{R_1, \dots, R_l\}$ of representatives for the orbits of O on $\mathcal{R}$. Hence the lemma follows from 9.4.22 and 9.4.27. For example in case (4b) of 9.4.22, $N_{\mathcal{F}}(E) = \langle N_{\mathcal{F}}(R), N_{\mathcal{F}_z}(E) \rangle$, so $Aut_{\mathcal{F}}(E) = \langle Aut_{N_{\mathcal{F}}(R)}(E), Aut_{\mathcal{F}_z}(E) \rangle$ with $Aut_{N_{\mathcal{F}}(R)}(E)$ induced in $Aut_{\mathcal{F}}(R)$ and $Aut_{\mathcal{F}_z}(E)$ induced in $\mathcal{F}_z$. □

LEMMA 9.4.29. *(1) For $1 \leq i \leq k$, $T_i \in Syl_2(H_i)$.*
(2) For $0 \leq i \leq k$, $\mathcal{E}_i \trianglelefteq \mathcal{F}_i$.

PROOF. By 9.4.26, $\mathcal{E}_0 \trianglelefteq \mathcal{F}_0$, so we may take $1 \leq i \leq k$, and from the discussion in 9.4.27, it remains to prove (1).

Suppose $R_i \in \mathcal{R}_1$. Then $R_i = C_S(Z_i)$ where $Z_i = \langle z, v, t_i \rangle$ for some $t_i \in \mathcal{D}^m(z)$. Suppose 9.4.21.2.a holds. Then by 9.4.21, $\bar{T}_i = \langle v_1, v, t_i \rangle \in Syl_2(H_{Z_i})$ and $\bar{T}_i \leq T_i$. Then by 9.4.21, $S_i = R_i\bar{T}_i$ and $G_i = H_{Z_i}N_{G_i}(S_i)$. Finally from the proof of 9.4.21, $T_i = \langle S_i \cap \mathcal{Z} \rangle$, so $T_i \trianglelefteq N_{G_i}(S_i)$, and hence $H_i = H_{Z_i}T_i$, so in particular $T_i \in Syl_2(H_i)$.

So assume instead that 9.4.21.3.b holds. Then by 9.4.21, $H_i \cong AE_5$ with $T_i \in Syl_2(H_i)$.

This leaves the case $R_3 \in \mathcal{R}_3$ in 9.4.22.2. In that event $R_3 = C_S(Z)$ for $Z = \langle z, t \rangle \in \mathfrak{Z}_3$, and from 9.4.16, $U = C_T(t) \in \mathfrak{Z}_4$ and, setting $R_U = C_S(U)$ and G_U a model for $N_{\mathcal{F}}(R_U)$, we have $G_U/R_U = O(U, q)$ for a quadratic form q on U of sign $+1$, and with $G_3 = N_{G_U}(Z)$. Then from 9.4.16, $H_3/U \cong S_3$ with $\bar{T}_3 = U\langle v_1 \rangle \in Syl_2(H_3)$ and $\bar{T}_3 \leq T_3$. Further by 9.4.16.7, $U = C_T(Z)$ so $T_3 = \bar{T}_3$, completing the proof. □

LEMMA 9.4.30. *(1) Assume $|T : O| = 2$; then:*
(1) $\mathcal{E} \cong PSp_4[m]$.
(2) $\mathcal{E} = \langle \mathcal{E}_0, \mathcal{E}_1 \rangle$.

PROOF. As $|T : O| = 2$, case (3) or (4) of 9.4.22 holds, so from 9.4.27, $\mathcal{R}_0 = \{R_1\}$ with $R_1 \in \mathcal{R}_1$. Thus $\mathcal{E} = \langle \mathcal{E}_0, Aut_{\mathcal{E}_1}(T_1 \cap R_1)\rangle$. Let $\tilde{\mathcal{F}} = PSp_4[m]$ with Sylow group $\tilde{S}$ and Lie packet $\tilde{\tau} = (\tilde{\mathcal{F}}, \tilde{\Omega})$.

First, $\mathcal{E}_0 = T\mathcal{O}$ is determined up to isomorphism as $\bar{\mathcal{E}}_0/Z(\bar{\mathcal{E}}_0)$, where $\bar{\mathcal{E}}_0 = SL_2[m]$ wr $\mathbf{Z}_2$. Therefore as $\tilde{\mathcal{F}}$ satisfies our hypothesis with $|\tilde{S}| = |T|$ by 5.2.6 (the Inductive Hypothesis is removed in Theorem 9.4.34), we may take $T = \tilde{S}$, $\tilde{\Omega} = \Omega$, and $\mathcal{E}_0 = \tilde{\mathcal{F}}_z = \tilde{\mathcal{F}}_0$. Using 9.4.28 applied to $\tilde{\mathcal{F}}$, we conclude that $\tilde{\mathcal{F}} = \langle \tilde{\mathcal{F}}_z, Aut_{\tilde{\mathcal{F}}}(\tilde{R}_1)\rangle$. Therefore to prove (1) it suffices to show that, when $m > 8$ we have $Aut_{\mathcal{E}_1}(T_1 \cap R_1) = Aut_{\tilde{\mathcal{F}}}(\tilde{R}_1)$, while when $m = 8$ we have $Aut_{\tilde{\mathcal{F}}}(\tilde{R}_1) = \langle Aut_{\mathcal{E}_0}(T_1 \cap R_1), Aut_{\mathcal{E}_1}(T_1 \cap R_1)\rangle$.

Pick $t \in \mathcal{D}^m(z)$. Then, in our identification of $\mathcal{E}_0$ with $\tilde{\mathcal{F}}_z$, we may choose t to be an image of an involution $\bar{t}$ in a Sylow group $\bar{T}$ of $\bar{\mathcal{E}}_0$ interchanging the components $\mathcal{C}_1$ and $\mathcal{C}_2$ of $\bar{\mathcal{E}}_0$, while tz is the image of the element $\bar{t}\bar{z}_1$ of order 4, where $\bar{z}_1 = z(\mathcal{C}_1)$. We conclude that also $t \in \mathcal{D}^m_{\tilde{\mathcal{F}}}(z)$, and then that $\tilde{\mathcal{Z}} = t^{\mathcal{O}} \cup \{z\} = \mathcal{Z}$.

Suppose $m = 8$ and set $E = E(t, K)$. Then $E = J(T)$ so also $E = E_{\tilde{\mathcal{F}}}(t, K)$ and of course $E \cap \mathcal{Z} = E \cap \tilde{\mathcal{Z}}$. Also $E = R_1 \cap T_1 = \tilde{R}_1$. Observe that $\tilde{\mathcal{F}}$ satisfies 9.4.21.3.b. Thus if $\mathcal{F}$ also satisfies 9.4.21.3.b then

$$Aut_{\mathcal{E}_1}(E) = Alt(\mathcal{Z} \cap E) = Alt(\tilde{\mathcal{Z}} \cap E) = Aut_{\tilde{\mathcal{F}}}(\tilde{R}_1),$$

proving (1) in this case. On the other hand if $\mathcal{F}$ satisfies 9.4.21.3.a, then at least $Aut_{\tilde{\mathcal{F}}}(\tilde{R}_1) = \langle Aut_{\mathcal{E}_0}(T_1 \cap R_1), Aut_{\mathcal{E}_1}(T_1 \cap R_1)\rangle$ by 9.4.22.4, so (1) holds in this case too. Also $\mathcal{E}_1^e = \{E\}$, so by the Alperin-Goldschmidt Fusion Theorem, $\mathcal{E}_1 = \langle Aut_{\mathcal{E}_1}(T_1), Aut_{\mathcal{E}_1}(E)\rangle \leq \mathcal{E}$ since $Aut_{\mathcal{E}_1}(T_1) \leq Aut_{\mathcal{E}_0}(T_1)$. Thus (2) also holds when $m = 8$.

So assume $m > 8$. Then $R_1 = C_S(Z)$ where $Z = \langle z, v, t\rangle$ is a hyperplane of E and $v = v_1 v_1^t$. From the proof of 9.4.21, $\tilde{R}_1 = \langle z, t\rangle \times X$, where $X = \{x_1 x_1^t : x_1 \in K\}$. Let $V = \langle w\rangle$ be of index 2 in K, $k_1 \in K - V$, $Q_1 = \langle v_1, k_1\rangle$, $k = k_1 k_1^t$, and $E = E_{k_1} = E(t, Q_1)$. Set $\tilde{\Sigma} = O^2(Aut_{\tilde{\mathcal{F}}}(\tilde{R}_1))$. Then $Aut_{\tilde{\Sigma}}(E) = O^2(N_{Aut_{\tilde{\mathcal{F}}}(E)}(Z))$ centralizes $\langle v, tzk\rangle$, as tzk and $vtzk$ are the two singular points in the orthogonal space E distinct from the points z, t, tz in Z. Hence $C_{\tilde{R}_1}(\tilde{\Sigma})$ is the subgroup of $\tilde{R}_1$ generated by such 4-subgroups E_{k_1} as k_1 varies over $K - V$. Similarly setting $\Sigma = O^2(Aut_{\mathcal{E}_1}(\tilde{R}_1))$, $C_{\tilde{R}_1}(\Sigma)$ is also generated by such 4-subgroups. Hence $C_{\tilde{R}_1}(\tilde{\Sigma}) = C_{\tilde{R}_1}(\Sigma)$, so $Aut_{\tilde{\mathcal{F}}}(\tilde{R}_1) = Aut_{\mathcal{E}_1}(\tilde{R}_1)$, and again $\mathcal{E} = \tilde{\mathcal{F}}$ and (1) is established. Similarly $\mathcal{E}_1 \leq \mathcal{E}$, so (2) holds. □

LEMMA 9.4.31. *Assume* $|T : O| = 4$. *Then*
(1) $\mathcal{E} \cong L_4^-[m]$.
(2) $\mathcal{E} = \langle \mathcal{E}_i : 1 \leq i \leq k\rangle$.
(3) In case (2) of 9.4.22, $\mathcal{E} = \langle \mathcal{E}_0, Aut_{\mathcal{E}_i}(T_i \cap R_i) : i = 1, 2\rangle$.

PROOF. As $|T : O| = 4$, case (1) or (2) of 9.4.22 holds; in particular $R_i \in \mathcal{R}_1$ for $i = 1, 2$, and $R_3 \in \mathcal{R}_3$ in case (2). This time set $\tilde{\mathcal{F}} = L_4^-[m]$ with Sylow group $\tilde{S}$ and Lie packet $\tilde{\tau} = (\tilde{\mathcal{F}}, \tilde{\Omega})$. Again $\mathcal{E}_0 = T\mathcal{O}$ is determined up to isomorphism: this time $\mathcal{E}_0 = \hat{\mathcal{E}}_0/Z(\hat{\mathcal{E}}_0)$, where $\hat{\mathcal{E}}_0 = SL_2[m]$ *wr* $\mathbf{Z}_2$ is $\bar{\mathcal{E}}_0$ extended by an involution $\bar{a}$ commuting with $\bar{t}$ and inducing an outer automorphism on the components $\mathcal{C}_i$ such that $\bar{K}_i\langle \bar{a}\rangle$ is semidihedral for $i = 1, 2$ and $\bar{K}_i$ is Sylow in $\mathcal{C}_i$. Therefore as $\tilde{\mathcal{F}}$ satisfies our hypothesis with $|\tilde{S}| = |T|$ by 5.2.5.2 (again the Inductive Hypothesis is removed in 9.4.34), we again may take $T = \tilde{S}$, $\tilde{\Omega} = \Omega$, and $\mathcal{E}_0 = \tilde{\mathcal{F}}_z = \tilde{\mathcal{F}}_0$, and once again to prove (1) it suffices to show that $Aut_{\mathcal{E}_i}(T_i \cap R_i) = Aut_{\tilde{\mathcal{F}}}(\tilde{R}_i)$ for $1 \leq i \leq k$.

As in the proof of the previous lemma, we may choose an identification of $\mathcal{E}$ with $\tilde{\mathcal{F}}_z$ so that $\mathcal{D}^m(z) = \mathcal{D}^m_{\tilde{\mathcal{F}}}(z)$. Further, from the proof of 9.4.16.8, O is transitive on the involutions in tsO, so $\mathcal{D}^o(z) = \mathcal{D}^o_{\tilde{\mathcal{F}}}(z)$, and hence $\tilde{\mathcal{Z}} = \mathcal{Z}$.

Note that as $\tilde{S} = T$, for $U \in (\mathfrak{Z}_{\tilde{\mathcal{F}}})_4$ we have $N_{\tilde{S}}(U) = N_T(U)$. Similarly if case (2) of 9.4.22 holds and $R_3 \in \mathcal{R}_3$ then $R_3 = C_S(Z_3)$ for $Z_3 = \langle z, t\rangle$ and $t \in \mathcal{D}^o(z)$; and from 9.4.16, $U = C_T(t) \in \mathfrak{Z}_4$ with $\mathcal{E}_3$ a model for H_3 and $G_3 = N_{G_U}(R_3)$, where G_U is a model for $N_{\mathcal{F}}(U)$. Then $H_3 \leq H_U = \langle T_3^{G_U}\rangle$ with $H_U = X_1 \times X_2$ and $X_i \cong S_4$. But $R_3 \cap T_3 = U$ and $Aut_{H_3}(U) \leq Aut_{H_U}(U) = \langle Aut_{X_1}(U), Aut_{X_2}(U)\rangle$, while $Aut_{X_i}(U) \leq Aut_{N_{\mathcal{F}}(P_i)}(U)$ for $P_i = C_S(Z_i) \in \mathcal{R}_1$ and Z_i a suitable hyperplane of U in $\mathfrak{Z}_1$. Then as $P_i \in \mathcal{R}^O_{j(i)}$ for some $j(i) \in \{1, 2\}$, $Aut_{H_3}(H) \leq Aut_{\mathcal{E}'}(U)$, where $\mathcal{E}' = \langle \mathcal{E}_0, Aut_{\mathcal{E}_i}(T_i \cap R_i) : i = 1, 2\rangle$; that is $\mathcal{E} = \mathcal{E}'$, so (3) holds and it suffices to show $Aut_{\mathcal{E}_i}(T_i \cap R_i) = Aut_{\tilde{\mathcal{F}}}(R_i \cap T_i)$ and $\mathcal{E}_i \leq \mathcal{E}$ for $i = 1, 2$.

If $m = 8$ then $N_{\mathcal{F}}(R_i) \leq N_{\mathcal{F}}(E_i)$ for $i = 1, 2$, where t_1 and t_2 are representatives for the orbits of O on $\mathcal{D}^m(z)$, and $E_i = E(t_i, K)$. Indeed $R_i \cap T_i = E_i\langle r_i\rangle$ for $r_i \in T_i$ inducing a transvection on E_i. Then we can argue as in the proof of the previous lemma to complete the proof.

So assume $m > 8$. Recall $R_i = C_S(Z_i)$ with $Z_i = \langle z, v, t_i\rangle$, and from 9.4.21.1, $H_{Z_i} \cong S_4 \times \mathbf{Z_2}$, while from the proof of 9.4.29, $H_i = H_{Z_i}T_i$, so $R_i \cap T_i = C_{T_i}(Z_i)$. Further from the proof of 9.4.29, $C_{T_i}(Z_i) = UX_i$ where $X_i = \langle E(t_i, Q_1) : Q_1 \in \mathcal{Q}(K)\rangle$ and $U = C_T(t_1t_2)$. Arguing as in the proof of the previous lemma, the action of H_{Z_i} on $E_i = E(t_i, Q_1)$ is determined: $O^2(H_{Z_i})$ fixes the two singular points of E_i not contained in Z_i. Similarly the action of H_{Z_i} on U is determined: H_{Z_i} centralizes the definite line of U not contained in Z_i. Therefore

$$Aut_{\mathcal{E}_i}(R_i \cap T_i) = Aut_{H_i}(R_i \cap T_i) = Aut_{\tilde{\mathcal{F}}}(\tilde{R}_i T_i),$$

establishing (1). Also $R_i \cap T_i$ is the unique member of $\mathcal{E}_i^e$ and $T_i = N_{\mathcal{E}_i}(T_i)$, so $\mathcal{E}_i \leq \mathcal{E}$, completing the proof. □

LEMMA 9.4.32. $\mathcal{F}^\circ = \mathcal{E}$.

PROOF. We verify that the family $\mathcal{F}_i$, $\mathcal{E}_i$, $0 \leq i \leq k$, satisfies Hypothesis 1.4.3 in [**Asc19**]. Then by 1.4.5 in [**Asc19**], $\mathcal{E} \trianglelefteq \mathcal{F}$, so $\mathcal{F}^\circ = \mathcal{E}$.

By 9.4.28, $\mathcal{F} = \langle \mathcal{F}_i : 0 \leq i \leq k\rangle$, and then by construction of our tuple, Hypothesis 1.4.1 in [**Asc19**] is satisfied. By 9.4.29, $\mathcal{E}_i \trianglelefteq \mathcal{F}_i$ for each i. By 9.4.30 and 9.4.31, $\mathcal{E}$ is saturated and $\mathcal{E} = O^{2'}(\mathcal{E})$. As $Aut_{\mathcal{F}}(T)$ permutes the orbits of O on $\mathcal{R}$ contained in $\mathcal{R}_0$, $Aut_{\mathcal{F}}(T) \leq Aut(\mathcal{E})$ by 1.4.6 in [**Asc19**]. Finally we verify the extension condition in 1.4.3.3 in [**Asc19**]: In case (4b) of 9.4.22, $T_1 = T$, so the lifting condition is trivially satisfied. In the remaining cases for $R_i \in \mathcal{R}_0$, $T = T_iO$ and $K_i \cap T_i = \langle v_i\rangle$, so $Aut_{\mathcal{F}}(T_i)$ acts on $\langle v_1, v_2\rangle = X$. But $O \trianglelefteq N_{\mathcal{F}}(X)$, so $Aut_{\mathcal{F}}(T_i)$ lifts to $OT_i = T$, completing the proof. □

THEOREM 9.4.33. *Assume Hypothesis* 9.4.1. *Then either*
(1) $|T : O| = 2$, $\mathcal{D}^o(z) = \emptyset$, *and* $\mathcal{F}^\circ \cong AE_5$ *or* $PSp_4[m]$, *or*
(2) $|T : O| = 4$, $\mathcal{D}^o(z) \neq \emptyset$, *and* $\mathcal{F}^\circ \cong L_4^-[m]$.

PROOF. By 9.4.14, $|T : O| = 2k$, where $k = 1$ or 2 is the number of orbits of O on $\mathcal{D}^m(z)$. Then by 9.4.16.6 and 9.4.17, $k = 1$ iff $\mathcal{D}^o(z) = \emptyset$. By 9.4.23 we may assume that Hypothesis 9.4.24 is satisfied. By 9.4.32, $\mathcal{F}^\circ = \mathcal{E}$, and then $\mathcal{F}^\circ \cong PSp_4[m]$ when $k = 1$ by 9.4.30.1, while $\mathcal{F}^\circ \cong L_4^-[m]$ when $k = 2$ by 9.4.31.1. $\square$

THEOREM 9.4.34. *Assume $\tau = (\mathcal{F}, \Omega)$ is a quaternion fusion packet such that for all $z \in \mathcal{Z}_S$ we have $|\Omega(z)| = 2$. Assume also that $\mathcal{F}^\circ$ is transitive on Ω and $\mu(\tau) \cong E_4$. Then*

(1) $\mathcal{F}^\circ \cong AE_5$, $PSp_4[m]$, or $L_4^-[m]$.

(2) $\mathcal{F} = \langle \mathcal{F}_z, Aut_{\mathcal{F}}(R_i) : 1 \le i \le k \rangle$, where the R_i are defined in 9.4.27

PROOF. We verify Hypothesis 9.4.1; then the theorem follows from 9.4.33 and 9.4.28.

By hypothesis, conditions (1) and (2) of 9.4.1 are satisfied; for example as $\mu \cong E_4$, μ has no D_{12}-subgroup, so $\mathcal{A}(\tau) = \emptyset$ by 3.2.1. As $\mu(\tau) \cong E_4$, 9.4.1.3 holds by 3.1.21. Suppose $\rho = (\mathcal{Y}, \Gamma)$ satisfies the hypothesis of 9.4.1.4. As $\mu(\tau) \cong E_4$, we conclude that $\mu(\rho) \cong \mathbf{Z}_2$ or E_4. In the first case, ρ satisfies one of the conclusions of Theorem 1 by 7.1.29. In the second, ρ satisfies the hypothesis of this theorem, so as ρ is proper in τ, also ρ satisfies one of its conclusions by induction. This completes the proof of the theorem. $\square$

THEOREM 9.4.35. *Assume $\tau = (\mathcal{F}, \Omega)$ is a quaternion fusion packet such that for $z \in \mathcal{Z}_S$, $|\Omega(z)| = 2$ and $\mathcal{Z} \cap O(z) \neq \{z\}$. Assume $\mu(\tau) \cong S_4$. Then $\mathcal{F} = \langle \mathcal{F}_z, N_{\mathcal{F}}(W) \rangle$ for $W \in W(\tau)$, and one of the following holds:*

(1) $\mathcal{F}^\circ = \mathcal{F}_{S \cap M}(M)$ for $M \in M(\tau)$.

(2) $m = 8$ and $\mathcal{F}^\circ \cong AE_6$.

(3) $\mathcal{F}^\circ \cong P\Omega_6^+[m]$.

PROOF. We verify Hypothesis 9.3.1; then the theorem follows from 9.3.24.

By hypothesis, $\mu = \mu(\tau) \cong S_4$, so $\mathcal{A}(\tau) = \emptyset$. By 3.1.27, $\mu = \mu(\tau^\circ)$. Let $\eta \in \eta(\tau^\circ)$; as $\mu \cong S_4$, μ is transitive on D_η, so by 4.2.8.7, $\mathcal{F}^\circ$ is transitive on Ω. Thus conditions (1)-(3) of 9.3.1 are satisfied. Let $\rho = (\mathcal{Y}, \Omega)$ be as in the Inductive Hypothesis; we must show that ρ appears as a conclusion to Theorem 1. Now $\nu = \mu(\rho) \le \mu \cong S_4$, so ν is $\mathbf{Z}_2$, E_4, S_3, or S_4. If ν is $\mathbf{Z}_2$ or S_3 then ρ appears in Theorem 1 by 7.1.29. If $\mu \cong E_4$ this follows from 9.4.34. Finally if μ is S_4 then as $W(\rho) \le W(\tau)$ and $M(\rho) \le M(\tau)$, we conclude that $M(\rho) = M(\tau)$, so $O(z) \cap \mathcal{Z}(\rho) \neq \emptyset$ by 9.1.10.4. Thus ρ satisfies the hypothesis of this theorem, so as ρ is proper in τ, it satisfies one of the conclusions of the theorem by induction, completing the proof. $\square$

The last two lemmas in this section will be applied in section 11.2.

LEMMA 9.4.36. *Assume $\tau = (\mathcal{F}, \Omega)$ is the Lie packet of $\mathcal{F} \cong L_4^-[m]$. Let $z \in \mathcal{Z}_S$, $K \in \Omega$, v_1 of order 4 in K with $\langle v_1 \rangle \lhd N_S(K)$, and set $v = v_1 v_1^s$ for some $s \in S - N_S(K)$. Let $u_1 = v$ and $u_2 = vz$. Let $t \in \mathcal{D}^o(z)$, $U = C_S(t)$, and G_U a model for $N_{\mathcal{F}}(U)$.*

(1) $U \cong E_{16}$ and $G_U^+ = G_U/U$ is the subgroup of $O(U, \mathfrak{q})$ generated by its transvections, for some quadratic form $\mathfrak{q}$ on U of sign $+1$. Further v_1^+ inverts $O(G_U^+)$.

(2) For h of order 3 in G_U with $C_U(h) = 1$ we have $G_U = \langle h, UN_{O(z)}(U) \rangle$.

(3) For $i = 1, 2$ let $U(i) = \langle z^{C_{G_U}(u_i)} \rangle$. Then $\mathcal{F} = \langle C_{\mathcal{F}}(z), N_{\mathcal{F}}(U(i)) : i = 1, 2 \rangle$.

PROOF. First τ satisfies the hypothesis of 9.4.34, and hence from the proof of 9.4.34, τ also satisfies Hypothesis 9.4.1. Then (1) follows from 9.4.16 and (1) implies (2). Part (3) follows from 9.4.22, 9.4.27, and 9.4.28, or see 9.4.34.2. $\square$

LEMMA 9.4.37. *Assume $\tau = (\mathcal{F}, \Omega)$ is the Lie packet of $\mathcal{F} \cong \Omega_7[m]$ or $\Omega_8^-[m]$. Let $z \in \mathcal{Z}_S$, $K \in \Omega$, v_1 of order 4 in K with $\langle v_1 \rangle \lhd N_S(K)$, and $v = v_1 v_1^s$ for some $s \in S - N_S(K)$, Set $u_1 = v$ and $u_2 = vz$. Suppose $E_{16} \cong U \leq S$ such that $U = \langle z, z\alpha, v, v\alpha \rangle$ for some $\alpha \in Aut_{\mathcal{F}}(U)$ of order 3 with $C_U(\alpha) = 1$, $z\alpha \in \mathcal{D}^o(z)$, $K^{v\alpha} \neq K$, and c_{v_1} inverts α. Set $\mathcal{X} = \langle \alpha, N_{O(z)}(U)U \rangle_{UN_{O(z)}U}$, and for $i = 1, 2$ let $U(i) = \langle z^{C_{\mathcal{X}}(u_i)} \rangle$. Set $\mathcal{Y} = \langle U\mathcal{O}_z, N_{O(z)}(U_i) N_{\mathcal{X}}(U_i) : i = 1, 2 \rangle$. Then $\mathcal{Y} \cong L_4^-[m]$.*

PROOF. Regard $\mathcal{F}$ as $\mathcal{F}_S(L)$ where $L = \Omega_d^\epsilon(q)$, and let V be the orthogonal space defining L. Set $z_1 = z$, $z_2 = z\alpha$, $V_i = [V, z_i]$, $Y_i = O^{p'}(C_L(V_i^\perp))$, $Y = \langle Y_1, Y_2 \rangle$, and $V_0 = V_1 + V_2$. By definition, $\mathcal{O}_{z_i} = \mathcal{F}_{O(z_i)}(Y_i)$. As $z_2 \in \mathcal{D}^o(z)$, V_0 is nondegenerate of dimension 6 and sign $-\pi$. Further Y_i centralizes $V_i^\perp$, so $Y \leq C_L(V_0^\perp) \cong \Omega_6^{-\pi}(q)$. Indeed $Y = C_L(V_0^\perp)$ with $T = UO(z) \in Syl_2(Y)$, so we can apply 9.4.36 to $\mathcal{Y}_0 = \mathcal{F}_Y(\mathcal{Y})$ to conclude that $\mathcal{Y} = \mathcal{Y}_0$, completing the proof. $\square$

9.5. $|\Omega(z)| = 2$ and μ isomorphic to S_4

In this section we assume the following hypothesis:

HYPOTHESIS 9.5.1. *$\tau = (\mathcal{F}, \Omega)$ is a quaternion fusion packet such that*
(1) For $z \in \mathcal{Z}_S$, $|\Omega(z)| = 2$ and $\mathcal{Z} \cap O(z) = \{z\}$.
(2) $\mu \cong S_4$.

NOTATION 9.5.2. Observe that Hypotheses 9.1.1 and 9.1.8 are satisfied, and adopt Notation 9.1.2 and 9.1.6. For example as $\mu \cong S_4$, $\mathcal{F}$ is transitive on Ω by 4.2.8.7. Let $t \in z^M - \{z\}$, $E = \langle z, t \rangle$, $a = vzt$, $b = vt$, and $O = O(z)$. For $c \in \{a, b\}$, let T_c be the Sylow group of $\mathcal{F}_c^\circ$. The choice of a presupposes notation is chosen so that $v = v_1 v_2$ and $Z(M) = \langle v_1 v_2 z t \rangle$, as in the proof of 9.5.3.

LEMMA 9.5.3. *(1) $E^\# = \mathcal{Z}_\Delta$.*
(2) $Z(M) = \langle a \rangle$ is of order 2.
(3) b centralizes O and $b \in O^g$ for each $g \in M - C_M(z)$. Moreover M is transitive on $\{b, vz, v\}$.
(4) $a, b, z \in Z(S)$.
(5) $C_S(v) = C_S(t) = C_S(E)$.
(6) Let i be an involution in S.
(i) If $K^i \neq K$ then $i \in C_S(E)$.
(ii) If $i \in \mathcal{D}^o(z)$ then $i \in C_S(E)$.
(iii) If $i \in \mathcal{Z} - C_{\mathcal{Z}}(E)$ then $i \in \mathcal{D}^*(z)$.
(7) The Inductive Hypothesis is satisfied.
(8) $\Omega = \Omega(z)$.

PROOF. By 9.5.1.2, $\mu \cong S_4$, so $|\Omega| = 2$ by 4.2.8.6. By 9.5.1.1, $|\Omega(z)| = 2$, so (8) holds.

As $\mu \cong S_4$, M is described in 5.7.4. In particular $S \cap M = C_M(z)$ is of index 3 in M, so $|z^M| = 3$. Applying 4.2.5 to $\langle K, K^g \rangle$ for $g \in M - C_M(z)$, (1) follows.

We conclude from 5.7.4.2 that $Z(M) = (\bar{v}_1\bar{v}_2^{-1})^{m/8}t$ for suitable choice of generators $\bar{v}_i$ for $K_i \cap W$ and $i = 1, 2$. We choose notation so that $v_i = \bar{v}_i^{m/8}$. By (8), $v_2^{-1} = v_2 z$ so that $Z(M)$ is generated by $v_1v_2zt = a$. As $\mathcal{Z} \cap O(z) = \{z\}$, $t \notin O(z)$, so $a \neq 1$, establishing (2).

By (2), $b = az$ centralizes O. Similarly as M is transitive on $E^\#$, M is also transitive on $aE^\# = \{b, vz, v\}$. Then as $v \in O$, $b \in O^g$ for some $g \in M - O$, and then as $O \leq C_S(b)$ is transitive on $E - \langle z \rangle$, (3) follows.

As $\Omega = \Omega(z)$, $z \in Z(S)$. As S acts on M, $a \in Z(S)$ by (2). Then $b = az \in Z(S)$, proving (4).

By (4), $vt = b$ and z are in $Z(S)$, so (5) follows.

Assume i is an involution in S. If $K^i \neq K$ then $v_1v_1^i \in C_S(i)$ with $v_1v_1^i = vz^\epsilon$, so i centralizes v, and hence (6i) follows from (5). If $i \in \mathcal{D}^o(z)$ then i inverts v_1 and v_2, so $i \in C_S(v) = C_S(E)$, proving (6ii). Finally assume $i \in \mathcal{Z} - C_{\mathcal{Z}}(E)$. Then by (6i), (6ii), and 9.1.3, $i \in \mathcal{D}^*(z)$, completing the proof of (6).

Assume the setup of the Inductive Hypothesis. Then, as in the proof of 9.4.35, ν is $\mathbf{Z}_2$, E_4, S_3, or S_4, and when ν is $\mathbf{Z}_2$ or S_3 then ρ satisfies one of the conclusions of Theorem 1 by 7.1.29. If ν is E_4 the same holds by 9.4.34, so we may assume ν is S_4. Then ρ satisfies Hypothesis 9.5.1, so as ρ is proper in τ, it satisfies one of the conclusions of Theorem 9.5.33 by induction, and hence also appears in Theorem 1. Hence (7) holds. □

For x an involution in S, we write $\mathcal{F}_x^\circ$ for $(\mathcal{F}_x)^\circ$. Recall $PSp_4(q) \cong \Omega_5(q)$ and $L_4^\epsilon(q) \cong P\Omega_6^\epsilon(q)$.

LEMMA 9.5.4. *(1) $\tau_b = (\mathcal{F}_b, \Omega)$ is a quaternion fusion packet with $\mu(\tau_b) \cong E_4$.*
(2) Either $\mathcal{F}_b^\circ = \mathcal{O}(z)$ or $\mathcal{F}_b^\circ \cong AE_5$, $PSp_4[m]$, or $L_4^-[m]$.

PROOF. From 9.5.3.3, b centralizes Ω and $O \trianglelefteq C_M(b)$. Therefore τ_b is a quaternion fusion packet by 2.5.2, and as $O \trianglelefteq C_M(b)$, $\mu(\tau_b) \cong E_4$. That is (1) holds. By 9.5.3.8, S is transitive on Ω, so by 9.5.3.4, $\mathcal{F}_b$ is transitive on Ω. Suppose $\mathcal{F}_b^\circ$ is transitive on Ω. Then (1) and Theorem 9.4.34 imply (2).

So assume $\mathcal{F}_b^\circ$ is not transitive on Ω. If $z \in Z(\mathcal{F}_b^\circ$ then $\mathcal{F}_b^\circ = \mathcal{O}(z)$ and (2) holds, so assume otherwise. Let $\mathcal{D} = [K]_{\mathcal{F}_b^\circ}$; then $\delta = (\mathcal{D}, K)$ is a quaternion fusion packet. Further as $\mu(\tau_b) \cong E_4$, $\mu(\delta) = \mathbf{Z}_2$, so by 7.1.6 and 7.1.7, $\mathcal{D}$ is $L_2[2m]^{(1)}$ or $L_3^-[m]$. Further $\mathcal{C} = E(\mathcal{D})$ is the unique component of $\mathcal{F}_b$ containing z, so $s \in S - N_S(K)$ acts on $\mathcal{C}$. Set $J = K \cap \mathcal{C}$ and let $s \in S - N_S(K)$; then $J^s \leq \mathcal{C}$ centralizes J, a contradiction to the structure of $\mathcal{D}$. □

LEMMA 9.5.5. *(1) $\tau_a = (\mathcal{F}_a, \Omega)$ is a quaternion fusion packet with $\mu(\tau_a) \cong S_4$.*
(2) One of the following holds:
(i) $\mathcal{F}_a^\circ = \mathcal{F}_{T_a}(M)$.
(ii) $m = 8$ and $\mathcal{F}_a^\circ \cong \check{A}E_6$.
(iii) $\mathcal{F}_a^\circ \cong \Omega_6^+[m]$.

PROOF. Part (1) follows from 9.5.3.2 and 2.5.2. Replacing τ by τ_a, we may assume $a \in Z(\mathcal{F})$ and set $\mathcal{F}^+ = \mathcal{F}/\langle a \rangle$ and $\tau^+ = (\mathcal{F}^+, \Omega^+)$. Then by 3.3.2, τ^+ is a quaternion fusion packet with $\mu(\tau^+) \cong S_4$. As $\mu(\tau^+) \cong S_4$, $(\mathcal{F}^+)^\circ$ is transitive on Ω^+ by 3.1.27 and 4.2.8.7. Then using 9.5.3.7 and observing that $t^+ = (Vz)^+ \in O^+$, τ^+ satisfies Hypothesis 9.3.1. Thus $\mathcal{F}^+$ is described in Theorem 9.3.24. Then $\mathcal{F}$ appears in (2) by 5.1.21 and 3.3.16. □

LEMMA 9.5.6. *Let* $\mathcal{E} = \mathcal{E}(\tau)$ *and* $\mathcal{D}^*_E = \bigcup_{e \in E^\#} \mathcal{D}^*(e)$.
(1) $\mathcal{F}_e \leq \mathcal{E}$ *for* $e \in E^\#$.
(2) $\mathcal{F}_a \leq \mathcal{E}$.
(3) For each $s \in \mathcal{D}^*_E$, *there exists* $\alpha \in \hom_{\mathcal{E}}(C_S(s), S)$ *with* $s\alpha = z$.

PROOF. Let $g \in M$ with $t^g = z$; then $\beta = c_g \in \hom_{\mathcal{E}}(C_S(t), S)$ with $t\beta = z$ as $N_{\mathcal{F}}(W) \leq \mathcal{E}$. By 3.4.3.4, $\mathcal{F}_z \leq \mathcal{E}$, so $\mathcal{F}_t\beta^* \leq \mathcal{F}_z \leq \mathcal{E}$. Then as β is an $\mathcal{E}$-map, $\mathcal{F}_t = (\mathcal{F}_t\beta^*)\beta^{-*} \leq \mathcal{E}$, establishing (1).

By 1.3.2 in [**Asc19**], $\mathcal{F}_a = \langle S\mathcal{F}_a^\circ, N_{\mathcal{F}_a}(T_a)\rangle$. Further $W \lhd N_{\mathcal{F}}(T_a)$, so $N_{\mathcal{F}_a}(T_a) \leq N_{\mathcal{F}}(W) \leq \mathcal{E}$. By 9.3.20.1 applied in $\mathcal{F}_a/\langle a\rangle$ and by 2.1.15,

$$S\mathcal{F}_a^\circ = \langle SC_{\mathcal{F}_a^\circ}(z), N_{S\mathcal{F}_a^\circ}(W)\rangle,$$

so as $\langle \mathcal{F}_z, N_{\mathcal{F}}(W)\rangle \leq \mathcal{E}$, (2) follows.

Part (3) follows from an argument in 10.1.15. For $s \in \mathcal{Z}$ let $\mathfrak{B}(s) = \{\alpha \in \hom_{\mathcal{E}}(C_S(s), S) : s\alpha = z\}$. If $s \in \mathcal{D}^*(z)$ then by 9.1.10.3 there is $\gamma \in \mathfrak{A}_{\mathcal{F}_z}(s)$ with $s\gamma = t$. Then as $\mathcal{F}_z \leq \mathcal{E}$, $\gamma\beta \in \mathfrak{B}(s)$. Finally suppose $r \in \mathcal{D}^*(t) - (\{z\} \cup \mathcal{D}^*(z))$. Claim $C_S(r) \leq C_S(E)$. For if not, $\langle C_S(r), C_{O^g}(r)\rangle$ induces $GL(E)$ on E, so $r \in \mathcal{D}^*(z)$, contrary to the choice of r. As $C_S(E) \lhd C_S(E)M$, by the claim, $C_S(r)\beta \leq S$, so $\beta \in \hom_{\mathcal{E}}(C_S(r), S)$. From the previous case there is $\delta \in \mathfrak{B}(r\beta)$. Then $\beta\delta \in \mathfrak{B}(r)$, completing the proof of (3). □

LEMMA 9.5.7. *Let* $s \in \mathcal{Z}$ *such that* $s^{\mathcal{F}_b} \cap E \neq \emptyset$. *Then*
(1) Either $\mathcal{D}^*(s) \cap E \neq \emptyset$ *or* $s \in z^{\mathcal{F}_b} \cap \mathcal{D}^o(z)$ *and* $\mathcal{F}_b^\circ \cong L_4^-[m]$.
(2) Let $\mathcal{D}^*(s) \cap E = \emptyset$ *and set* $U = C_{T_b}(s)$. *Then either* $\mathcal{F}_b \leq \mathcal{E}(\tau)$ *or* $Aut_{T_b}(U) \neq Aut_S(U)$.

PROOF. Assume $\mathcal{D}^*(s) \cap E = \emptyset$ and let $\delta \in \mathfrak{A}(s)$. By 9.5.3.6.iii, s centralizes E. Thus $[s, M] \leq C_M(E) = O_2(M)$. Therefore s induces an automorphism σ on $M/W = \mu \cong S_4$ centralizing $\mu/O_2(\mu)$. Then as $\mu = Aut(\mu)$ it follows that $\sigma \in O_2(M/W)$. Therefore $s = cy$ for some $y \in O_2(M)$ and $c \in C_S(M/W)$. Now c acts on each member of K^M and y acts on some $J \in K^M$, so s acts on $V = J \cap W$. Set $e = z(J)$. As $\mathcal{D}^*(s) \cap E = \emptyset$, it follows that $s \in \mathcal{D}^o(e)$, so s inverts V. Thus $e \in [V, s]$. But if $s \in z^{\mathcal{F}_b}$ then $s \in T_b$ so $[V, s] \leq T_b$. On the other hand if $s \in t^{\mathcal{F}_b}$ then $s = bu$ for some $u \in T_b$, so again $[V, s] = [V, u] \leq T_b$. Hence $e \in T_b \cap E = \langle z\rangle$, so $e = z$. Therefore $s \in \mathcal{D}^o(z)$, and by hypothesis, s is fused in $\mathcal{F}_b$ to z or $t = bv$. Hence there exists $i \in z^{\mathcal{F}_b} \cup v^{\mathcal{F}_b}$ inducing an outer automorphism on K. Then as, among the systems listed in 9.5.4.2, there exists such an involution in T_b inducing an outer automorphism on K iff $\mathcal{F}_b^\circ \cong L_4^-[m]$, and in that case all such involutions are in $z^{\mathcal{F}_b}$, it follows that (1) holds.

Now assume the setup of (2). By (1), $\mathcal{F}_b^\circ \cong L_4^-[m]$, so from 9.4.16.7, $U \cong E_{16}$ and from 9.4.18, U is in the set $\mathfrak{Z}_4$ defined there. We may assume $Aut_{T_b}(U) = Aut_S(U)$, so by 9.4.22.1 and 9.4.28,

$$\mathcal{F}_b = \langle C_{\mathcal{F}_b}(z), Aut_{\mathcal{F}_b}(R_i) : i = 1, 2\rangle,$$

where $R_i \in \mathcal{R}_1$ for $i = 1, 2$. However from 9.4.18 and 9.4.21, $Aut_{\mathcal{F}_b}(R_i)$ centralizes v or vz, and hence also centralizes $t = bv$ or $tz = bvz$. Therefore $Aut_{\mathcal{F}_b}(R_i) \leq Aut_{\mathcal{E}(\tau)}(R_i)$ by 9.5.6.1. Then as $\mathcal{F}_z \leq \mathcal{E}(\tau)$, $\mathcal{F}_b \leq \mathcal{E}(\tau)$, completing the proof of (2). □

LEMMA 9.5.8. *Let $s \in \mathcal{Z}$. Then either $s \in \mathcal{D}_E^*$ or the following hold:*

(1) For some $g \in M$, $s^g \in z^{\mathcal{F}_b} \cap \mathcal{D}^o(z)$. Further either $C_S(s) \leq C_S(E)$ or we may take $g = 1$, and in either case $\gamma = c_g \in \hom_{\mathcal{E}(\tau)}(C_S(s), S)$.

(2) $\mathcal{F}_b^\circ \cong L_4^-[m]$.

(3) There exists $\phi \in \hom_{\mathcal{F}_b}(C_S(s^g), S)$ with $s^g\phi = z$, and if $Aut_{T_b}(C_{T_b}(s^g)) = Aut_S(C_{T_b}(s^g))$ then $\phi \in \hom_{\mathcal{E}(\tau)}(C_S(s^g), S)$ and $\gamma\phi \in \hom_{\mathcal{E}(\tau)}(C_S(s), S)$.

PROOF. Assume $s \notin \mathcal{D}_E^*$ and let $\alpha \in \mathfrak{A}(s)$ and $F = E\langle a \rangle$. By 9.5.3.6.iii, s centralizes F. Then for some $f \in F$, $f\alpha$ centralizes v_1. As $s \notin \mathcal{D}_E^*$, $f \in F - E$. If $f = a$ then for $\delta \in \mathfrak{A}(a\alpha)$, $v_1\delta \in D_1 \in \Delta$, so either $v_1\delta \in W$ or $z\delta = z$, and in either case $z\delta \in E$. As $\xi = \alpha\delta \in \hom_{\mathcal{F}_a}(s, S)$ and $s\xi = z\delta \in E$, we have $s \in z^{\mathcal{F}_a}$. This is a contradiction as $z^{\mathcal{F}_a} \subseteq \mathcal{D}_E^*$ (cf. 9.3.14.1).

Therefore $f \in b^M$, so there is $g \in M$ with $f^g = b$. Now $\langle f\alpha \rangle = C_{F\alpha}(v_1)$, so f is $C_S(s)$-invariant. But if $[C_S(s), E] \neq 1$ then b is the unique member of b^M centralized by $C_S(s)$, so we may take $g = 1$. On the other hand if $C_S(s) \leq C_S(E)$ then $C_S(s)^g \leq S$, as in the proof of 9.5.6.3. Therefore in either case, $\gamma = c_g \in \hom_{\mathcal{E}}(C_S(s), S)$. Let $\beta \in \mathfrak{A}(f\alpha)$, $\zeta = \gamma^{-1}\alpha\beta$, and $r = s\gamma$. Then $\zeta \in \hom_{\mathcal{F}_b}(C_S(s)\gamma, S)$ with $v_1\beta \in W$ or $z\beta = z$, so $r\zeta \in E$. Now as $s \notin \mathcal{D}_E^*$ and $E\gamma = E$, $r \notin \mathcal{D}_E^*$, so (1) and (2) hold by 9.5.7.1 applied to $r = s^g$ in the role of s. Let $\phi \in \mathfrak{A}_{\mathcal{F}_b}(r)$; then ϕ satisfies the conditions of (3), using 9.5.7.2 to verify the second condition. □

LEMMA 9.5.9. *(1) $\mathcal{E}(\tau) = \langle \mathcal{F}_z, N_{\mathcal{F}}(W) \rangle$.*

(2) Set $\mathcal{B}(\tau) = \mathcal{E}(\tau)$ unless $\mathcal{F}_b^\circ \cong L_4^-[m]$ and $Aut_S(C_{T_b}(s)) \neq Aut_{T_b}(C_{T_b}(s))$ for $s \in z^{\mathcal{F}_b} \cap \mathcal{D}^o(z)$, where we set $\mathcal{B}(\tau) = \langle \mathcal{E}(\tau), \mathcal{F}_b \rangle$. Then $\mathcal{F} = \mathcal{B}(\tau)$.

PROOF. Part (1) follows from the definition of $\mathcal{E}(\tau)$ in Definition 3.4.2, and the fact that $\Omega = \Omega(z)$. To establish (2), we argue as in the proof of 10.1.15. Set $\mathcal{B} = \mathcal{B}(\tau)$. We first observe that:

(a) For each $s \in \mathcal{Z}$ there exists $\alpha \in \hom_{\mathcal{B}}(C_S(s), S)$ with $s\alpha = z$.

If $s \in \mathcal{D}_E^*$ this is 9.5.6.3, while if $s \notin \mathcal{D}_E^*$ it is 9.5.8.3.

Now (2) follows from (a) and 2.1.8. □

NOTATION 9.5.10. Let $S_0 = C_S(O)$, $R_0 = OS_0$, and G_z a model for $N_{\mathcal{F}}(R_0)$. Set $H_z = \langle W^{G_z} \rangle$, $L_z = O^2(H_z)$, and $B_z = \langle b^{L_z} \rangle$. If $m > 8$ set $\mathcal{O} = \mathcal{O}(z)$ and for $c \in \{a, b\}$ set $\mathcal{O}_c = C_{\mathcal{F}_c^\circ}(z)^\circ$. If $m = 8$ set $\mathcal{O} = \mathcal{F}_O(OO^2(C_{L_z}(S_0))$ and $\mathcal{O}_c = O^2(C_{\mathcal{F}_c^\circ}(z))O$, except when $\mathcal{F}_b^\circ = O$ set $\mathcal{O}_b = \mathcal{O}$.

LEMMA 9.5.11. *(1) $[W, S_0] \leq C_{O(t)}(O)$, $C_{O(t)}(O)$ is cyclic, and if $m = 8$ then $C_{O(t)}(O) = \langle b \rangle$.*

(2) If $m > 8$ then either $\mathcal{O}(z) = O$, or $\mathcal{O}_K \cong SL_2[m]$ and $\mathcal{O}(z) \cong SL_2[m] * SL_2[m]$.

(3) If $m = 8$ then one of the following holds:

(i) $H_z = W$.

(ii) L_z centralizes S_0, and is of index 1 or 3 in $SL_2(3) * SL_2(3)$.

(iii) $O_2(L_z) = O \times B_z$ where $B_z = \langle b^{L_z} \rangle \cong E_4$, and $L_z = O_2(L_z)X_z$ for $X_z \in Syl_3(L_z)$. Moreover $L_z/C_{L_z}(B_z) \cong \mathbf{Z}_3$, $X_z \cong \mathbf{E}_{3^e}$ for some $1 \leq e \leq 3$, and $Z(L_z) = \langle z \rangle$.

(4) If $m = 8$ then $O^2(Aut_{\mathcal{F}}(O)) = O^2(Aut_{H_z}(O))$ and $O^2(C_{G_z}(S_0)) \leq L_z$.

PROOF. For $g \in M$ set $V_{z^g} = (O \cap W)^g$. From 5.7.4, $W = V_zV_t$, so $[W, S_0] = [V_t, S_0]$. Further $S_0 \leq C_S(v) = C_S(t)$ by 9.5.3.5. Thus $[V_t, S_0] \leq V_t \cap S_0 = C_{V_t}(O)$, while $C_{O(t)}(O)$ is cyclic by 5.7.5.3, with $C_{O(t)}(O) = \langle b \rangle$ when $m = 8$. Thus (1) holds. Part (2) follows from 9.1.11.1. Thus it remains to prove (3) and (4), so we may assume $m = 8$ and set $G_z^+ = G_z/O$.

We first prove (3). Observe that by 5.7.4.1, $W^+ = \langle w^+ \rangle$, where $\mathbf{Z}_4 \cong \langle w \rangle \in \Delta$ with $w^2 = t$. As $b = vt$, $t^+ = b^+$. Set $B_z = \langle b^{L_z} \rangle$. As $b \in Z(S)$, $B_z \leq Z(R_0)$ so $O \leq R_0 \leq C_{G_z}(B_z)$. Set $H_z^! = H_z/C_{H_z}(B_z)$. Then $W^! = \langle w^! \rangle \cong \mathbf{Z}_2$, and as W is weakly closed in S, $W^!$ is weakly closed in $S^!$, so by the Z^*-Theorem, $L_z^! = O(L_z^!)$.

By (1), $[S_0, W] \leq \langle b \rangle$. Suppose L_z centralizes b. Then as $[S_0, W] \leq \langle b \rangle$, it follows that L_z centralizes S_0. Thus $H_z \leq C_G(a)$, so from 9.5.5.2, (3i) or (3ii) holds. Therefore we may assume $B_z^+ = [B_z^+, L_z^+] \neq \langle b^+ \rangle$.

As $w^!$ induces a transvection on B_z^+ with center generated by b^+, and as $L_z^! = O(L_z^!)$, we conclude that $B_z^+ \cong E_4$ and $H_z^! = GL(B_z^+)$. Indeed as $[S_0, W] = \langle b \rangle$, we have $H_z^+ \cong S_4$ with $B_z^+ = O_2(H_z^+)$ and $S_0 = B_z C_{S_0}(H_z)$. Then $w^2 = t \in O_2(H_z) \leq OB_z$.

Suppose L_z centralizes O. Then $L = L_z\langle w \rangle \cong \mathbf{Z}_2 \times S_4$ has center generated by $v = bt$. Now by 9.5.3.3 there is $\alpha \in \mathfrak{A}(v)$ with $v\alpha = b$, and we may assume $v \in C_M(K^g)$, where $g \in M$ with $t = z(K^g)$. Therefore $b = v\alpha \in O_2(L)\alpha = \langle z^L \rangle \leq T_b$, contrary to 9.5.4.2. Thus L_z does not centralize O. Then as $K_2 \in K^S$, $O = [O, L_z] \leq L_z$, so $O_2(L_z) = OB_z$. Now as $O \cong Q_8^2$, $Out(O) \cong O_4^+(2)$ and then (3iii) holds, completing the proof of (3).

Let $\Sigma = Out_{\mathcal{F}}(O)$ and $\tilde{G}_z = G_z/\langle z \rangle$. For $w \in W - OC_S(O)$, w is nontrivial on $\tilde{K}_i$ for $i = 1, 2$, so w inverts $O(\Sigma)$. Thus the first statement in (4) holds. Let $X = O^2(C_{G_z}(S_0))$; as $Out(O)$ is 2-nilpotent, X^+ is of odd order, so $X = X_0 Y$ where X_0 is a Hall $2'$-subgroup of X and $Y = O \cap X$. As $F^*(G_z) = R_0$, X_0 is faithful on Y and then as w inverts $O(\Sigma)$, $X_0^+ = [X_0^+, w] \leq L_z^+$, so $X \leq L_z$, completing the proof of (4). □

LEMMA 9.5.12. *(1) $\mathcal{O} = \mathcal{O}_c$ centralizes S_0 for $c \in \{a, b\}$.*
(2) If $m > 8$ then $\mathcal{O}_K \cong SL_2[m]$ iff $\mathcal{F}_b^\circ \cong PSp_4[m]$ or $L_4^-[m]$ iff $\mathcal{F}_a^\circ \cong \Omega_6^+[m]$.
(3) $\mathcal{O} \trianglelefteq \mathcal{F}_z$.

PROOF. By 2.6.12, S_0 centralizes $\mathcal{O}(z)$. Thus if $m > 8$ then as $\langle a, b \rangle \leq S_0$, $\mathcal{O} \neq O$ iff $\mathcal{O}_a \neq O$ iff $\mathcal{O}_b \neq O$, and then (1) holds in this case by 9.5.11.2. Then (1), 9.5.4, and 9.5.5 imply (2), while (3) holds as $\mathcal{O}(z) \trianglelefteq \mathcal{F}_z$. Therefore we may assume that $m = 8$.

Let $c \in \{a, b\}$. Then from 9.5.4 and 9.5.5, $\mathcal{O}_c$ centralizes $S_0 \cap T_c$, so $\mathcal{O}_c$ centralizes S_0. Therefore $\mathcal{O}_c \leq \mathcal{O}$ by 9.5.11.4. By construction, $\mathcal{O}$ centralizes S_0, so as $c \in S_0$ we have $\mathcal{O} \leq \mathcal{F}_c$.

If $\mathcal{F}_a^\circ \cong \Omega_6^+[m]$ or $\mathcal{F}_b^\circ \cong PSp_4[m]$ or $L_4^+[m]$ then $O^2(Aut_{\mathcal{O}_c}(O)) = O^2(Aut(O))$, so as $\mathcal{O}_c \leq \mathcal{O}$, we have $\mathcal{O}_c = \mathcal{O}$.

Suppose 9.5.5.2.i holds. Then $\mathcal{F}_{T_a}(M) \trianglelefteq \mathcal{F}_a$, so $N_{\mathcal{F}_a}(O) \leq N_{\mathcal{F}_a}(T_a)$, and hence as $O^2(Aut_{\mathcal{F}}(T_a)) = 1$, also $O^2(Aut_{\mathcal{F}_a}(O)) = 1$, so $\mathcal{O} = \mathcal{O}_a = O$.

Suppose $\mathcal{F}_a^\circ = \check{A}E_6$ or $\mathcal{F}_b^\circ \cong AE_5$. Then $O^2(Aut_{\mathcal{F}_c}(O))$ acts on $O_2(\mathcal{F}_c^\circ \cap O)$ and hence $O^2(Aut_{\mathcal{F}_c}(O))) = O^2(Aut_{\mathcal{O}_c}(O))$. Thus again $\mathcal{O} = \mathcal{O}_c$.

This leaves the case $\mathcal{F}_b^\circ = O$, where by definition $\mathcal{O} = \mathcal{O}_b$. This completes the proof of (1), so it remains to prove (3).

By construction, $\mathcal{O}$ centralizes S_0 and O is Sylow in $\mathcal{O}$, so the central product $\mathcal{O} * C_{\mathcal{F}}(O)$ is a subsystem of $\mathcal{F}_z$ by 2.3 in [**Asc11**]. Then as $Aut_{\mathcal{O}}(O) \trianglelefteq Aut_{\mathcal{F}}(O)$ by construction, we have $\mathcal{O} \trianglelefteq \mathcal{F}_z$. This completes the proof of (3) and the lemma. □

LEMMA 9.5.13. *If $a \in Z(\mathcal{F}_z)$ then $a \in Z(\mathcal{F})$.*

PROOF. Assume $a \in Z(\mathcal{F}_z)$. By 9.5.9.1, $\mathcal{E} = \mathcal{E}(\tau) = \langle \mathcal{F}_z, N_{\mathcal{F}}(W)\rangle$, so as $a \in Z(N_{\mathcal{F}}(W))$, also $a \in Z(\mathcal{E})$. Thus we may assume $\mathcal{E} \neq \mathcal{F}$, so by 9.5.9.2, $\mathcal{F}_b^\circ \cong L_4^-[m]$.

Let $s \in \mathcal{D}^m(z) \cap z^{\mathcal{F}_b}$, $Z = \langle s, z, v\rangle$, $R = C_S(Z)$, G_R a model for $N_{\mathcal{F}_b}(R)$, $G_Z = N_{G_R}(Z)$, and $H_Z = \langle v_1^{G_Z}\rangle$. By 9.4.21 there is $h \in H_Z$ with $z^h = s$, and from the proof of 9.4.21, $v_0 = v_1 v_1^s \in Z(H_Z)$. Now $v_0 = v$ or vz, and h centralizes $bv_0 = e = t$ or tz. By 9.5.6.1, $\mathcal{F}_e \leq \mathcal{E}$, so as $a \in Z(\mathcal{E})$ we have $a^h = a$. However $a = bz$, so $a^h = bz^h = bs \neq bz = a$, a contradiction. □

LEMMA 9.5.14. *Assume $\mathcal{F}_b^\circ = \mathcal{O}(z)$. Then*

(1) $t \in Z(\mathcal{F}_{vz})$.

(2) $C_{\mathcal{F}}(O) \leq C_{\mathcal{F}_z}(vz) \leq C_{\mathcal{F}_z}(a)$.

(3) Assume $a \notin Z(\mathcal{F})$. Then $m = 8$, $E_4 \cong B_z \trianglelefteq \mathcal{F}_b$, and $O^2(Aut_{\mathcal{F}}(O)) \neq O^2(Aut_{\mathcal{O}}(O))$.

PROOF. By hypothesis, $\mathcal{O}(z) \trianglelefteq \mathcal{F}_b$, so $z \in Z(\mathcal{F}_b)$. Let $g \in M$ interchange t and z. Then g interchanges $at = vz$ and $az = b$, so (1) holds. By (1), $C_{\mathcal{F}_z}(vz) = C_{\mathcal{F}_z}(\langle vz, t\rangle) \leq C_{\mathcal{F}_z}(a)$, and of course $C_{\mathcal{F}}(O) \leq C_{\mathcal{F}_z}(vz)$, so (2) holds.

Assume $a \notin Z(\mathcal{F})$. Then $a \notin Z(\mathcal{F}_z)$ by 9.5.13. Let $\mathcal{F}_0 = \mathcal{O}(z)C_{\mathcal{F}}(O)$. By 2.6.14, $C_{\mathcal{F}}(O) = C_{\mathcal{F}}(\mathcal{O}(z))$ and $\mathcal{F}_0 \trianglelefteq \mathcal{F}_z$. By (2), a is in the center of $S\mathcal{F}_0$, so $\mathcal{F}_z \neq S\mathcal{F}_0$. But by 1.3.2 in [**Asc19**], $\mathcal{F}_z = \langle S\mathcal{F}_0, N_{\mathcal{F}}(OS_0)\rangle$. Recall that G_z is a model for $N_{\mathcal{F}}(OS_0)$. If $Aut_{\mathcal{F}}(O)$ is a 2-group then $G_z = SC_{G_z}(O)$ and $\mathcal{F}_{S_0}(C_{G_z}(O)) \leq S\mathcal{F}_0$, so $N_{\mathcal{F}}(OS_0) \leq S\mathcal{F}_0$, contradicting $\mathcal{F}_z \neq S\mathcal{F}_0$. It follows that $m = 8$.

By 9.5.11.4, $G_z = SH_zC_{G_z}(O)$, so if $B_z = \langle b\rangle$ then G_z centralizes $\langle a, z\rangle = \langle b, z\rangle$, contradicting $a \notin Z(\mathcal{F}_z)$. Then by 9.5.11.3, $B_z \cong E_4$.

Finally by 9.5.12.1, $\mathcal{O} = \mathcal{O}_a$ centralizes a. By 9.5.11.4,

$$Aut_{L_z}(O) = O^2(Aut_{\mathcal{F}}(O)),$$

so if $O^2(Aut_{\mathcal{F}}(O)) = O^2(Aut_{\mathcal{O}}(O))$ then as $\mathcal{O}$ centralizes S_0 and $b \in S_0$ by 9.5.3.3, we have $L_z = O^2(C_{H_b}(a)C_{L_z}(O)$, so as $C_{L_z}(O) \leq C_{H_z}(a)$ by (2), and as $C_{H_z}(B_z)$ centralizes b, H_z centralizes $\langle a, z\rangle = \langle b, z\rangle$, contradicting $B_z \cong E_4$. □

LEMMA 9.5.15. *Assume $\mathcal{F}_b^\circ = \mathcal{O}(z)$ and $a \notin Z(\mathcal{F})$. Let $k = |Aut_{\mathcal{O}}(O)|_3$ and $T = T_aB_z$. Then $m = 8$ and:*

(1) Either

(a) $k = 1$, $\mathcal{O} = O$, and $\mathcal{F}_a = \mathcal{F}_{T_a}(M)$, or

(b) $k = 3$, $\mathcal{O}$ is of index 3 in $SL_2[8] * SL_2[8]$, and $\mathcal{F}_a \cong \check{A}E_6$.

(2) There exists X of order 3 in H_z such that $C_{OB_z}(X) = \langle z\rangle$ and $XS = SX$.

(3) Let $A = \langle t^{XO}\rangle$ and $\tau_A = (N_{\mathcal{F}}(A), \Omega)$. Then $E_8 \cong A \trianglelefteq S$, τ_A is a quaternion fusion packet, and $N_{\mathcal{F}}(A)^\circ = \langle \mathcal{F}_T(TX), \mathcal{F}_T(TM)\rangle \cong L_3(2)/E_{64}$ has Sylow group T.

(4) $C_{\mathcal{F}}(O) \leq C_{\mathcal{F}_z}(B_z)$.

(5) $\mathcal{F}_z = \langle SC_{\mathcal{F}}(O), \mathcal{F}_S(G_z)\rangle$.

(6) $B_z \trianglelefteq \mathcal{F}_z$.

(7) $\mathcal{F}_T(TM) \trianglelefteq N_{\mathcal{F}}(W)$.

(8) $T \trianglelefteq SC_{\mathcal{F}}(O)$.

(9) $\mathcal{F}_T(TL_z) \trianglelefteq \mathcal{F}_z$.

(10) If $k = 1$ then $\mathcal{F}^\circ = N_{\mathcal{F}}(A)^\circ \cong L_3(2)/E_{64}$.

PROOF. By 9.5.14.3, $m = 8$, $B_z \cong E_4$ and $O^2(Aut_{\mathcal{F}}(O)) \neq O^2(Aut_{\mathcal{O}}(O))$. In particular $O^2(Aut_{\mathcal{O}}(O)) \neq O^2(Aut(O))$ so $k = 1$ or 3, while by 9.5.12.1, $\mathcal{O} = \mathcal{O}_a$, and then (1) follows from 9.5.5.2.

As $b \in B_z \cong E_4$, it follows that $Aut_{H_z}(B_z) \cong S_3$ and then as $b \in S_0$ we conclude that case (iii) of 9.5.11.3 holds, so that $P = O_2(L_z) = OB_z$ and hence $C_P(X_z) = \langle z \rangle$. Set $X_0 = C_{X_z}(B_z)$ and observe that X_0 is faithful on O. As $b \in B_z$ and $\langle a, z \rangle = \langle b, z \rangle$, we have $X_0 = C_{X_z}(a) = C_{X_z}(b)$. As $X_0 \leq \langle W^{C_{G_z}(a)} \rangle$, $Aut_{X_0}(O)$ is induced in $\mathcal{O}_a$, so $|X_0| = k$. Then as $Aut_{L_z}(B_z) \cong S_3$ we have $|X_z| = 3k$. Hence if $k = 1$ then $X_z \cong \mathbf{Z}_3$ and we take $X = X_z$. As $PX = L_z \trianglelefteq G_z$, $SX = XS$, so (2) holds when $k = 1$. If $k = 3$ then X_0 is of order 3 and $PX_0 \trianglelefteq G_z$, so as S interchanges K_1 and K_2, $C_O(X_0) = \langle z \rangle$. As $O^2(Aut_{\mathcal{F}}(O) \neq O^2(Aut_{\mathcal{O}}(O)$, X_z is faithful on O. It follows that X_z has a unique subgroup X of order 3 with $C_P(X) = \langle z \rangle$, so again (2) holds.

Set $\tilde{G}_z = G_z/\langle z \rangle$. As $C_{\tilde{P}}(X) = 1$, $\langle \tilde{t}^X \rangle \cong E_4$, so $A \cong E_8$, and as $PX \trianglelefteq H_zS$ and $\tilde{t} \in Z(\tilde{S})$, $A \trianglelefteq S$. Indeed $Aut_{SX}(A)$ is the stabilizer in $GL(A)$ of z, so either $Aut_{\mathcal{F}}(A) = Aut_{SX}(A)$ or $Aut_{\mathcal{F}}(A) = GL(A)$. Set $\mathcal{Y} = N_{\mathcal{F}}(A)$; in any event $\tau_A = (\mathcal{Y}, \Omega)$ is a quaternion fusion packet by 2.5.2.

Suppose $Aut_{\mathcal{F}}(A) = GL(A)$. Then $A^\# = z^{\mathcal{Y}}$ and $\mathcal{F}_{T_a}(M) \leq \mathcal{Y}$, so $\mu(\tau_A) \cong S_4$. We conclude from Theorem 2 and 5.10.7.1 that one of the first three cases in Theorem 2 holds and $Z(\mathcal{Y}^\circ) = 1$. Then as $\mu(\tau_A) \cong S_4$ it follows from 6.2.8 that $\mathcal{Y}^\circ \cong L_3(2)/E_{64}$. Let T be Sylow in $\mathcal{Y}^\circ$; then $|T| = 2^9$, so as $|T_a| = 2^8$ and $B_z \leq T$ with $B_z \cap T_a = \langle b \rangle$, $|T_aB_z| = 2^9$, so that $T = T_aB_z$. As $\mathcal{Y}^\circ \cong L_3(2)/E_{64}$, $\mathcal{Y}^\circ = \langle \mathcal{F}_T(TX), \mathcal{F}_T(TM) \rangle$, so that (3) holds in this case. Thus to complete the proof of (3), we may assume $Aut_{\mathcal{F}}(A) = Aut_{XS}(A)$, and it remains to derive a contradiction.

Let $x \in X^\#$, $r = t^x$, and $c = b^x$. Then $A = E\langle r \rangle$ and as $t = vb$, $r = v^xc = kc$, where $k = k_1k_2$ with $k_i \in K - W$. Let $g \in M$ have cycle (z, t) and $w \in O^g \cap W$ of order 4. If $r \in \mathcal{D}^*(t)$ then $[A, O^g] \leq A$ and $\langle [z, O^g] \rangle = \langle t \rangle$, contradicting $Aut_{\mathcal{F}}(A) = Aut_{\mathcal{F}_z}(A)$. Therefore $r \notin \mathcal{D}^*(t)$, so $s = r^g \notin \mathcal{D}^*(z)$. As w induces an outer automorphism on K_i for $i = 1, 2$, w inverts XP/P, so $[c, w] = b$ and hence $K^{gc} \neq K^g$. Also $K^{gk} \neq K^g$, so r acts on K^g and hence s acts on K, so $s \in \mathcal{D}^o(z)$. Thus s induces outer automorphisms on K_i, so s inverts XP/P and hence $[c, s] = b$. Therefore $|s^{OB_z}| = |OB_z : C_{OB_z}(s)| = 16$. However $A \trianglelefteq S$, so $A^g \trianglelefteq C_S(E)$. Set $\theta = A^g - E$; it follows that $\theta^S = \theta \cup \theta^k$ is of order at most 8, contradicting $s^{OB_z} \subseteq \theta^S$. This contradiction completes the proof of (3).

By 9.5.14.2, $C_{\mathcal{F}}(O) \leq C_{\mathcal{F}_z}(a) = C_{\mathcal{F}_z}(b)$, so $C_{\mathcal{F}}(O)$ centralizes $B_z = \langle b^{G_z} \rangle$, proving (4).

By 1.3.2 in [**Asc19**], $\mathcal{F}_z = \langle SC_{\mathcal{F}}(O), N_{\mathcal{F}}(R_0) \rangle$, so as $N_{\mathcal{F}}(R_0) = \mathcal{F}_S(G_z)$, (5) holds. As $B_z \trianglelefteq H_z$, (4) and (5) imply (6).

By (1), $T_a \in Syl_2(M)$. Thus by 1.3.2 in [**Asc19**], $N_{\mathcal{F}}(W) = \langle \mathcal{F}_S(SM), N_{\mathcal{F}}(T_a) \rangle$, so as $N_{\mathcal{F}}(T_a) \leq \mathcal{F}_z$, (7) follows from (6) and Theorem 1.5.2 in [**Asc19**]. Similarly by (4), $C_{\mathcal{F}}(O)$ centralizes E, so as $W \trianglelefteq N_{\mathcal{F}}(E)$, $C_{\mathcal{F}}(O) \leq N_{\mathcal{F}}(W)$. Then as $T = C_{\mathcal{F}_T(TM)}(z)$, (8) follows from (7).

Set $\mathcal{F}_1 = \mathcal{F}_S(G_z)$, $\mathcal{E}_1 = \mathcal{F}_T(TL_z)$, $\mathcal{F}_2 = SC_{\mathcal{F}}(O)$, and $\mathcal{E}_2 = T$. Observe that Hypothesis 1.4.7 in [**Asc19**] is satisfied by this tuple, so (9) follows from 1.4.8 in [**Asc19**].

Assume $k = 1$. Then $A \trianglelefteq SH_zC_{G_z}(P) = G_z$, while $A \trianglelefteq SC_{\mathcal{F}}(O)$ by (4). Therefore $A \trianglelefteq \mathcal{F}_z$ by (5). Hence $A - \langle z \rangle = t^{\mathcal{F}_z} = \mathcal{D}^*(z)$ by 9.1.10.3. Then as

$Aut_{\mathcal{F}}(A) = GL(A)$, $A^{\#} = \mathcal{D}_E^*$, so $A \trianglelefteq N_{\mathcal{F}}(W)$. Therefore by 9.5.9, $A \trianglelefteq \mathcal{F}$, so (3) implies (10). □

LEMMA 9.5.16. *Assume* $\mathcal{F}_b^\circ = \mathcal{O}(z)$, $a \notin Z(\mathcal{F})$, *and* $|Aut_{\mathcal{O}}(O)|_3 = 3$. *Then* $\mathcal{F}^c \cong Sp_6(2)$.

PROOF. Adopt the notation of 9.5.15 and set $\mathcal{F}_1 = \mathcal{F}_z$, $\mathcal{E}_1 = \mathcal{F}_T(TL_z)$, $\mathcal{F}_2 = N_{\mathcal{F}}(W)$, $\mathcal{E}_2 = \mathcal{F}_T(TM)$, and $\mathcal{X} = \langle \mathcal{E}_1, \mathcal{E}_2 \rangle$. We will show that $\mathcal{X} \cong Sp_6(2)$; in particular $\mathcal{X}$ is saturated. Then we verify that the tuple satisfies Hypothesis 1.4.7 in [**Asc19**]. Finally we appeal to 1.4.8 in [**Asc19**] to conclude that $\mathcal{X} \trianglelefteq \mathcal{F}$, so it follows that $\mathcal{F}^\circ = \mathcal{X} \cong Sp_6(2)$.

The verification of Hypothesis 1.4.7 in [**Asc19**] has for the most part already been done. By 9.5.9, $\mathcal{F} = \langle \mathcal{F}_1, \mathcal{F}_2 \rangle$. By 9.5.15.9, $\mathcal{E}_1 \trianglelefteq \mathcal{F}_1$ and by 9.5.15.7, $\mathcal{E}_2 \trianglelefteq \mathcal{F}_2$. As $\langle z \rangle$ and W are invariant under $Aut_{\mathcal{F}}(T)$, $Aut_{\mathcal{F}}(T) = Aut_{\mathcal{F}_i}(T)$ for $i = 1, 2$. Thus it remains to show that $\mathcal{X} \cong Sp_6(2)$.

Let $\tilde{G}$ be the group $Sp_6(2)$, $\tilde{S} \in Syl_2(\tilde{G})$, and $\tilde{\mathcal{F}} = \mathcal{F}_{\tilde{S}}(\tilde{G})$. Thus $\tilde{\mathcal{F}} \cong Sp_6(2)$ satisfies the hypothesis of the lemma with $|\tilde{S}| = 2^9 = |T|$. For example $\tilde{G} \leq \hat{G} \cong \Omega_7(3)$ contains a Sylow 2-subgroup of $\hat{G}$, so by 2.5.2, $\tilde{G}$ inherits a quaternion fusion packet $\tilde{\tau}$ from $\hat{G}$. From section 5.3, the Lie packet of $\hat{G}$ satisfies Hypothesis 9.5.1; then $\tilde{\tau}$ inherits that hypothesis. Next $\tilde{G}$ has a maximal parabolic subgroup $\tilde{Y}$ over $\tilde{S}$ such that $\tilde{Y} \cong L_3(2)/E_{64}$. Let $\mathcal{Y} = N_{\mathcal{F}}(A)^\circ$; from 9.5.15.3, $\mathcal{Y} \cong L_3(2)/E_{64} \cong \mathcal{F}_{\tilde{S}}(\tilde{Y})$, so we may take $\tilde{S} = T$, $\tilde{\Omega} = \Omega$, and $\mathcal{Y} = \mathcal{F}_T(\tilde{Y})$. Moreover from 9.5.15.3, $\mathcal{Y} = \langle \mathcal{F}_T(TX), \mathcal{F}_T(TM) \rangle$. Further $\mathcal{E}_1 = \langle \mathcal{F}_T(TX), \mathcal{F}_T(TX_0) \rangle$, where $X_0 = C_{X_z}(B_z)$ appears in the proof of 9.5.15. Thus $\mathcal{X} = \langle \mathcal{Y}, Aut_{\mathcal{E}_0}(P_0) \rangle$, where $\mathcal{E}_0 = \mathcal{F}_T(TX_0)$ and $P_0 = O_2(\mathcal{E}_0)$. Hence it suffices to show that $Aut_{\mathcal{E}_0}(P_0) = Aut_{\tilde{\mathcal{E}}_0}(\tilde{P}_0)$.

Next $TL_z = L_z N_{TL_z}(X_z)$ and $N_{TL_z}(X_z) = X_z I$, where $I = \langle z, u_1, u_2 \rangle \cong E_8$ with $\langle z, u_1 \rangle = C_{O_2(Y)}(X)$ with $u_1 \in z^{\mathcal{Y}} \cap \mathcal{D}^m(z)$ and u_2 is an involution in Y inverting X with $K_1^{u_2} = K_2$. For example Y splits over $O_2(Y)$, which is the core of the permutation module for $Y/O_2(Y)$, and hence $m(C_{O_2(Y)}(u_2)) = 4$. Therefore TL_z is determined up to isomorphism. Recall $P = O_2(L_z)$ and $B_z = [Z(P), X]$ is determined by $\mathcal{F}_T(TX)$ with $X_z = O(N_{TL_z}(X))$ and $X_0 = C_{X_z}(B_z)$. Finally $P_0 = O_2(\mathcal{E}_0) = P\langle u_2 \rangle$ and $O^2(Aut_{\mathcal{E}_0}(P_0)) = O^2(C_{Aut(P_0)}(B_z))$, so indeed $\tilde{P}_0 = P_0$ and $Aut_{\mathcal{E}_0}(P_0) = Aut_{\tilde{\mathcal{E}}_0}(\tilde{P}_0)$. This completes the proof. □

THEOREM 9.5.17. *Assume* $\mathcal{F}_b^\circ = \mathcal{O}(z)$. *Then one of the following holds:*
(1) $a \in Z(\mathcal{F})$ *and* $\mathcal{F}^\circ = \mathcal{F}_{T_a}(M)$ *or* $\mathcal{F}^\circ \cong \check{A}E_6$ *or* $\Omega_6^+[m]$.
(2) $\mathcal{F}^\circ \cong L_3(2)/E_{64}$.
(3) $\mathcal{F}^\circ \cong Sp_6(2)$.

PROOF. If $a \in Z(\mathcal{F})$ then (1) holds by 9.5.5, so assume otherwise. Then by 9.5.15.1, $k = |Aut_{\mathcal{O}}(O)|_3 = 1$ or 3. Therefore (2) or (3) holds by 9.5.15.10 and 9.5.16. □

Because of Theorem 9.5.17, during the remainder of the section we assume the following hypothesis:

HYPOTHESIS 9.5.18. *Hypothesis* 9.5.1 *holds and* $\mathcal{F}_b^\circ \neq \mathcal{O}(z)$.

LEMMA 9.5.19. *Let $g \in M$ interchange z and t.*

(1) $\langle v, z\rangle \trianglelefteq S$.

(2) g fixes v and interchanges b and vz, and $\beta = c_g^* \in \mathfrak{A}(\langle b, t\rangle)$ with $C_S(\langle b,t\rangle) = C_S(E) = C_S(\langle v, z\rangle)$.

(3) There exists a unique (solvable) component $\mathcal{L}_{b,t}$ of $C_{\mathcal{F}_b^\circ}(t)$. Let $S_{b,t}$ be Sylow in $\mathcal{L}_{b,t}$ and set $\mathcal{L}_{vz,z} = \mathcal{L}_{b,t}\beta^*$ and $S_{vz,z} = S_{b,t}\beta$.

(4) $\mathcal{L}_{b,t} \cong L_2[m]$ if $\mathcal{F}_b^\circ \cong AE_5$ or $PSp_4[m]$, and $\mathcal{L}_{b,t} \cong L_2[2m]$ if $\mathcal{F}_b^\circ \cong L_4^-[m]$.

(5) $S_{b,t} = \langle (vz)^{\mathcal{L}_{b,t}}\rangle$.

(6) $\mathcal{L}_{vz,z} \trianglelefteq C_{\mathcal{F}}(O)$ is S-invariant and $S_{vz,z} = \langle b^{\mathcal{L}_{vz,z}}\rangle$.

(7) If $\mathcal{F}_b^\circ \cong PSp_4[m]$ or $L_4^-[m]$ then $\mathcal{F}_z = SOC_{\mathcal{F}}(O)$ and $\mathcal{L}_{vz,z} \trianglelefteq \mathcal{F}_z$.

PROOF. Part (1) is an observation.

As g fixes $a = vzt$ and interchanges z and t, it fixes v, and then interchanges vz and $vt = b$. Also

$$C_S(\langle b,t\rangle) = C_S(t) = C_S(E) = C_S(E\langle b\rangle) = C_S(\langle v,z,b\rangle) = C_S(\langle v,z\rangle),$$

establishing (2).

Parts (3)-(5) follow from the local structure of $\mathcal{X} = \mathcal{F}_b^\circ$, as we see in a moment. By 9.5.18 and 9.5.4, $\mathcal{X}$ is AE_5, $PSp_4[m]$, or $L_4^-[m]$. First, if $\mathcal{X} \cong AE_5$ then $\mathcal{X} = \mathcal{F}_{T_b}(H_b)$ where H_b is the split extension of $U_b \cong E_{16}$ by $Y_b \cong A_5$ with U_b the core of the permutation module for $Y_b = Alt(I)$ and $I = \{1,\ldots,5\}$. We may take $z = e_{1,2,3,4}$ and $v = e_{1,2}$. Then $C_{H_b}(t) = C_{H_b}(v) = U_bC_{Y_b}(v) \cong \mathbf{Z}_2/(E_4 \times A_4)$, so $X_b = O^2(C_{H_b}(v)) \cong A_4$ with $vz = e_{3,4} \in O_2(X_b) = \langle vz, e_{3,5}\rangle$. Here take $\mathcal{L}_{b,t} = \mathcal{F}_{S_{b,t}}(X_b) \cong L_2[8]$. Thus (3)-(5) indeed hold in this case.

Suppose next that $\mathcal{X} \cong \Omega_k^\epsilon[m]$, where $k = 5$ or $(k,\epsilon) = (6,-1)$. Then $\mathcal{X} = \mathcal{F}_{T_b}(H_b)$ where $H_b \cong \Omega(U)$ for some k-dimensional orthogonal space U over $\mathbf{F_q}$ for some suitable prime power q. We can take z to be an involution in H_b with $[U,z]$ 4-dimensional of sign $+1$, while $[U,v]$ is a 2-dimensional subspace of $[U,z]$. Then $C_{H_b}(v)$ has a (solvable) component $X_b \cong \Omega_{k-2}^\epsilon(q) \cong L_2(q)$ or $L_2(q^2)$ for $k = 5$ or 6, respectively. In this case $\mathcal{L}_{b,t} = \mathcal{F}_{S_{b,t}}(X_b)$, and again one can check (3)-(5).

Next as $vz \in O$, $C_{\mathcal{F}}(O) \le C_{\mathcal{F}_z}(vz)$. Also $\mathcal{L}_{b,t}$ is a (solvable) component of $C_{\mathcal{F}_b}(t)$, and hence by (2) and 2.2 in [**Asc10**], $\mathcal{L} = \mathcal{L}_{vz,z} = \mathcal{L}_{b,t}\beta^*$ is a (solvable) component of $C_{\mathcal{F}_z}(vz)$. If $m > 8$ then $\mathcal{L}$ is a component, so by E-balance (cf. Theorem 7 in [**Asc11**]), and as $C_{\mathcal{O}(z)}(vz)$ is a 2-group, $\mathcal{L}$ centralizes $\mathcal{O}(z)$, and in particular $\mathcal{L} \le C_{\mathcal{F}}(O) \le C_{\mathcal{F}_z}(vz)$, so $\mathcal{L} \trianglelefteq C_{\mathcal{F}}(O)$ as $\mathcal{L}_{b,t} \trianglelefteq C_{\mathcal{F}}(\langle b,t\rangle)$ by (3). As $S = OC_S(vz)$, S acts on $\mathcal{L}$, completing the proof of (6) when $m > 8$.

Suppose $m = 8$. As $\langle v,z\rangle \trianglelefteq O$ and $\mathcal{L} \trianglelefteq C_{\mathcal{F}}(\langle v,z\rangle)$ with $z \notin S_{vz,z}$, $\mathcal{L}$ centralizes O, completing the proof of (6).

Suppose $\mathcal{F}_b^\circ \cong PSp_4[m]$ or $L_4^-[m]$. By 9.5.12, $\mathcal{O} \cong SL_2[m] * SL_2[m]$ and $\mathcal{O} \trianglelefteq \mathcal{F}_z$. Then by 2.6.14, $\mathcal{F}_z = SOC_{\mathcal{F}}(O)$, so $\mathcal{L} \trianglelefteq \mathcal{F}_z$ by (6), proving (7). □

LEMMA 9.5.20. *Assume $\mathcal{F}_b^\circ \cong AE_5$ and let L be the subnormal subgroup of G_z which is the model for $\mathcal{L} = \mathcal{L}_{vz,z}$. Then*

(1) $\mathcal{F}_a^\circ \cong \check{A}E_6$.

(2) $L_z = L \times L_O$ where $L \cong A_4$, $L_O = C_{L_z}(L)$ is of index 3 in $SL_2(3)^2$ with $O = O_2(L_O)$ and $\mathcal{F}_O(L_O) = \mathcal{O}$, and $B_z = O_2(L) = S_{vz,z}$.

(3) $\mathcal{F}_z = SOC_{\mathcal{F}}(O)$ and $\mathcal{L} \trianglelefteq \mathcal{F}_z$.

(4) $\mathcal{F}_z = \langle S\mathcal{L}, C_{\mathcal{F}_z}(b)\rangle$.

PROOF. As $\mathcal{F}_b^\circ \cong AE_5$, $\mathcal{O}_b$ is of index 3 in $SL_2(3)^2$. By 9.5.12.1, $\mathcal{O}_b = \mathcal{O} = \mathcal{O}_a$, so $\mathcal{O}_a$ is of index 3 in $SL_2(3)^2$, and hence (1) follows from 9.5.5.2.

By 9.5.19.4, $\mathcal{L} \cong L_2[m]$, so $L \cong A_4$ as $m = 8$. By 9.5.19.6, $O_2(L) = S_{vz,z} = \langle b^L \rangle$. Then as $B_z = \langle b^{L_z} \rangle \trianglelefteq G_z$ is of order 2 or 4 by 9.5.11.3, it follows that $E_4 \cong O_2(L) = B_z$. Let $X_z \in Syl_3(L_z)$. If $x \in X_z^\#$ with $L^x \neq L$ then by 2.2.2.1, $\langle L^{\langle x \rangle} \rangle = L \times L^x \times L^{x^2}$, a contradiction as $E_4 \cong B_z \trianglelefteq G_z$. Therefore L_z acts on L. As $[B_z, W] = \langle b \rangle$ by 9.5.11.1, $L = [L, W] \leq L_z$ and then by 9.5.11.3 and 9.5.19.6, $L = C_{L_z}(O)$ and $L_z = L \times L_0$ where $L_0 = C_{L_z}(L)$. Thus, using 9.5.11.4 and 9.5.12.1, $\mathcal{O} = \mathcal{F}_O(L_0)$, so as $\mathcal{O}$ is of index 3 in $SL_2(3)^2$, the same holds for L_0. This completes the proof of (2).

By 9.5.11.1, $[W, S_0] \leq B_z$, so as L_0 centralizes B_z, it also centralizes S_0. Therefore by 2.3 in [**Asc11**], $\mathcal{F}_z$ contains a central product $\mathcal{B} = \mathcal{O} * C_{\mathcal{F}}(O)$. As $L_0 \trianglelefteq G_z$, we conclude from 1.5.1 in [**Asc19**] that $\mathcal{B} \trianglelefteq \mathcal{F}_z$. By 9.5.19.4, $O^2(Aut_{\mathcal{F}}(O)) = Aut_{L_z}(O)$, and by (2), $Aut_{L_z}(O) = Aut_{L_0}(O) = Aut_{\mathcal{O}}(O)$, so $G_z = SL_zC_{G_z}(O)$ and hence $N_{\mathcal{F}}(OS_0) = SN_{\mathcal{B}}(OS_0)$. Then $\mathcal{F}_z = SOC_{\mathcal{F}}(O)$ by 1.3.2 in [**Asc19**]. By 9.5.19.6, $\mathcal{L} = \mathcal{L}_{vz,z} \trianglelefteq C_{\mathcal{F}}(O)$ is S-invariant, while $\mathcal{O}$ centralizes $\mathcal{L}$ by (2). Therefore $\mathcal{L} \trianglelefteq SOC_{\mathcal{F}}(O) = \mathcal{F}_z$, proving (3).

As $A_4 \cong \mathcal{L} \trianglelefteq C_{\mathcal{F}}(O)$, $O^2(C_{\mathcal{F}}(O)) = \mathcal{L} \times O^2(C_{\mathcal{F}}(P))$, where $P = OB_z$. Then by (3), $\mathcal{F}_z = \langle S\mathcal{L}, SOO^2(C_{\mathcal{F}}(P)) \rangle = \langle S\mathcal{L}, C_{\mathcal{F}_z}(b) \rangle$, proving (4). □

LEMMA 9.5.21. *If $\mathcal{F}_b^\circ \cong AE_5$ then $\mathcal{F}^\circ \cong AE_7$.*

PROOF. For $c \in \{a, b\}$ set $U_c = O_2(\mathcal{F}_c^\circ)$, let G_c be a model for $N_{\mathcal{F}_c}(C_S(U_c))$, and let $H_c \trianglelefteq G_c$ be a model for $\mathcal{F}_c^\circ$. Then H_b is the split extension of $U_b \cong E_{16}$ by A_5 with U_b the A_5-module for H_b/U_b. By 9.5.20.1, $\mathcal{F}_a^\circ \cong \check{A}E_6$, so $U_a \cong E_{32}$. By 9.5.12.1, $\mathcal{O} = \mathcal{O}_c$ for $c \in \{a, b\}$. Now $W\mathcal{O}$ acts on a unique E_8-subgroup F of O, so $F = O \cap U_b = O \cap U_a$. Further U_a centralizes the hyperplane $O \cap U_b = O \cap U_a$ of U_b, and that hyperplane is not the axis of a transvection in $Aut(H_b)$ on U_b, so U_a centralizes U_b.

Set $U = U_aU_b$ and let $\rho = (N_{\mathcal{F}}(U), \Omega)$; by 2.5.2, ρ is a quaternion fusion packet. From the previous paragraph U is elementary abelian, so $U = C_{UH_b}(U_b) \trianglelefteq UH_b$. Thus if $U_b \leq U_a$ then H_b acts on $U_a = U$ and $\mu(\rho) = \mu(\tau_a) \cong S_4$, so by Theorem 2 and 5.1.21, $N_{\mathcal{F}}(U)^\circ \cong \check{A}E_6$. Therefore $\mathcal{F}_b^\circ \leq N_{\mathcal{F}}(U)^\circ = \mathcal{F}_a^\circ$, so H_b centralizes a, a contradiction as $a = bz$. Thus $U_b \not\leq U_a$, so $U_a \cap U_b = O \cap U_b$ and hence $m(U) = 6$.

As $U \trianglelefteq UH_b$, $\mathcal{F}_b^\circ \leq N_{\mathcal{F}}(U)$. As above, $U = C_{UH_a}(U_a) \trianglelefteq UH_a$, so $\mathcal{F}_a^\circ \leq N_{\mathcal{F}}(U)$. Thus $\mu(\rho) \cong S_4$, $U \cong E_{64}$, and $\mathcal{F}_c^\circ \leq N_{\mathcal{F}}(U)^\circ$ for $c \in \{a, b\}$, so we conclude from Theorem 2 and 5.1.21 that $N_{\mathcal{F}}(U)^\circ \cong AE_7$. Thus to complete the proof it remains to show that $U \trianglelefteq \mathcal{F}$.

By 9.5.20.4, $\mathcal{F}_z = \langle S\mathcal{L}, C_{\mathcal{F}_z}(b) \rangle$. As $N_{\mathcal{F}}(U)^\circ \cong AE_7$, $\mathcal{L} \leq \mathcal{F}_{B_zO}(L_z) \leq N_{\mathcal{F}}(U)^\circ$. As $C_{\mathcal{F}_z}(b)$ centralizes $\langle a, b \rangle$, it is contained in $N_{\mathcal{F}}(U)$. Therefore $\mathcal{F}_z \leq N_{\mathcal{F}}(U)$. By 9.5.9, $\mathcal{F} = \langle \mathcal{F}_z, N_{\mathcal{F}}(W) \rangle$, so it remains to show that $N_{\mathcal{F}}(W) \leq N_{\mathcal{F}}(U)$. But $N_{\mathcal{F}}(W) = \langle S\mathcal{F}_S(M), N_{\mathcal{F}_z}(W) \rangle$ and $\mathcal{F}_{T_a}(M) \leq N_{\mathcal{F}}(U)^\circ$, so the proof is complete. □

Because of 9.5.21 and 9.5.4.2, during the remainder of the section we assume the following hypothesis:

HYPOTHESIS 9.5.22. *Hypothesis 9.5.1 holds and $\mathcal{F}_b^\circ \cong PSp_4[m]$ or $L_4^-[m]$. Set $\mathcal{L} = \mathcal{L}_{vz,z}$, $T_l = S_{vz,z}$, and $T = T_aT_l$.*

LEMMA 9.5.23. *(1)* $\mathcal{F}_a^\circ \cong \Omega_6^+[m]$.
(2) $\mathcal{F}_z = \langle S\mathcal{L}, C_{\mathcal{F}_z}(b)\rangle$.
(3) $T \trianglelefteq C_{N_\mathcal{F}(W)}(z) = C_{N_\mathcal{F}(W)}(b)$.
(4) $T\mathcal{O} = C_{T\mathcal{F}_a^\circ}(z) \trianglelefteq C_{\mathcal{F}_a}(z)$.
(5) $T\mathcal{O}\mathcal{L} \trianglelefteq \mathcal{F}_z$.
(6) $\mathcal{F}_T(TM) \trianglelefteq N_\mathcal{F}(W)$.

PROOF. As $\mathcal{F}_b^\circ$ is $PSp_4[m]$ or $L_4^-[m]$, $\mathcal{O}_b \cong SL_2[m] * SL_2[m]$, so by 9.5.12.1, $\mathcal{O}_a \cong SL_2[m] * SL_2[m]$, and hence (1) follows from 9.5.5.2.

By 9.5.19.7, $\mathcal{L} \trianglelefteq \mathcal{F}_z$. Then $O^2(\mathcal{C}_\mathcal{F}(O)) = \mathcal{L} \times O^2(C_\mathcal{F}(P))$, where $P = OT_l$, while by 9.5.19.7, $\mathcal{F}_z = SOC_\mathcal{F}(O)$, so $\mathcal{F}_z = \langle S\mathcal{L}, SOO^2(C_\mathcal{F}(P))\rangle = \langle S\mathcal{L}, C_{\mathcal{F}_z}(b)\rangle$, proving (2).

Next $\mathcal{F}_{T_a}(M) \trianglelefteq N_\mathcal{F}(W)$, so $a \in Z(N_\mathcal{F}(W))$ by 9.5.3.2. Then as $a = bz$, $C_{N_\mathcal{F}(W)}(z) = C_{N_\mathcal{F}(W)}(b)$. Also $T_a = C_M(z)$, and $T_l = C_\mathcal{L}(a)$, so (3) holds, and (3) and Theorem 1.5.2 in [**Asc19**] imply (6). Similarly $N_{\mathcal{F}_a}(T_a) \le N_{\mathcal{F}_z}(W) \le N_\mathcal{F}(T)$ by (3), so (4) follows from Theorem 1.5.2 in [**Asc19**].

Next $\mathcal{O} \trianglelefteq \mathcal{F}_z$ by 9.5.12.3 and $\mathcal{L} \trianglelefteq \mathcal{F}_z$ by 9.5.19.7, so $\mathcal{X} = \mathcal{O}\mathcal{L} \trianglelefteq \mathcal{F}_z$ by Theorem 3 in [**Asc11**]. Suppose $m > 8$. Then $\langle b\rangle = Z(T_l)$ so $N_{\mathcal{F}_z}(P) \le C_{\mathcal{F}_a}(z)$, and then as $N_\mathcal{O}(O) = O$, $T \trianglelefteq N_{\mathcal{F}_z}(P)$ by (4). Then (5) follows from Theorem 1.5.2 in[**Asc19**] in this case.

So assume $m = 8$; we verify Hypothesis 1.4.7 in [**Asc19**] with $\mathcal{F}_1 = S\mathcal{X}$, $\mathcal{E}_1 = T\mathcal{X}$, $\mathcal{F}_2 = C_{\mathcal{F}_z}(b)$, and $\mathcal{E}_2 = T\mathcal{O}$. By (2), $\mathcal{F}_z = \langle \mathcal{F}_1, \mathcal{F}_2\rangle$ and by (4), $\mathcal{E}_2 \trianglelefteq \mathcal{F}_2$. As $T \trianglelefteq S$, $\mathcal{E}_1 \trianglelefteq \mathcal{F}_1$ by 1.3.4 in [**Asc19**]. Therefore Hypothesis 1.4.7 in [**Asc19**] is satisfied, so $T\mathcal{X} = \langle \mathcal{E}_1, \mathcal{E}_2\rangle \trianglelefteq \mathcal{F}_z$ by 1.4.8 in [**Asc19**], proving (5). □

LEMMA 9.5.24. *(1) $\mathcal{O}\mathcal{L}$ is transitive on the set $\mathcal{I}$ of involutions ij such that $i \in O - \langle z\rangle$ and $j \in T_l$. Moreover $\mathcal{I} \subseteq \mathcal{Z}$.*
(2) $C_{O^g}(O) \trianglelefteq T_l$ is cyclic of order $m/4$.
(3) Let r be an involution in $T_l - \langle b\rangle$. Then r inverts $C_{O^g}(O)$.
(4) Choose generators k_i, x_i, $1 \le i \le 3$, for M as in 4.3.1 *and* 5.7.4, *but with x_i in the role of v_i and $[K_1, K_2] = 1$. Set $x = x_1x_2x_3^2$, $k_4 = k_3^{k_1k_2}$, and $s = k_1k_2r$. Then $C_{O^g}(O) = \langle x\rangle$, $K^{gr} \ne K^g$, $x_3^r = x_4$, $s \in \mathcal{D}^*(t) \cup \mathcal{D}^o(t)$, s inverts x_3, and $[x_3, r] = x^{-1}$, so $r^{x_3} = rx$.*
(5) We may choose k_1 and k_2 so that $s \in \mathcal{D}^*(t)$, and k_3 so that $k_3^r = k_4$. In particular $M\langle r\rangle$ is determined up to isomorphism.
(6) $\langle T_b, T_l\rangle = T_b \times T_l$.
(7) If $\mathcal{F}_b^\circ \cong L_4^-[m]$ then $T_l = \langle y, r\rangle$ is dihedral of order m and we may choose y of order $m/2$ with $y^2 = x$, y centralizes OW, and $k_3^y = k_3x_3$. In particular MT is determined up to isomorphism. Moreover $T = T_b \times T_l$.

PROOF. By 9.5.22, $\mathcal{O} \cong SL_2[m] * SL_2[m]$, so $\mathcal{O}$ is transitive on involutions in $O - \langle z\rangle$; by 9.5.19.4, $\mathcal{L}$ is transitive on involutions in T_l; and by parts (6) and (7) of 9.5.19, $\mathcal{L}$ centralizes $\mathcal{O}$. This gives the transitivity in (1). As $t = vb \in \mathcal{Z}$, the second statement in (1) also holds, completing the proof of (1).

By 9.5.11.1, $C_{O^g}(O)$ is cyclic of order $m/4$. Let $K^g \cap W = \langle w\rangle$. By 9.5.19.7, $\mathcal{L} \trianglelefteq \mathcal{F}_z$, so $T_l \trianglelefteq S$. By 9.5.11.1, $[w, r] \in C_{O^g}(O)$, so as b is the involution in $C_{O^g}(O)$, either $K^{gr} \ne K^g$ or r centralizes w. In the latter case $r_0 = r$ or br centralizes K^g, so r_0 centralizes $M = \langle O, K^g\rangle$, so as b does not centralize M, neither does br_0. Then as br_0 satisfies the hypothesis of r, we conclude that $K^{gr_0b} \ne K^g$, and then as b acts on K^g, it follows that $K^{gr} \ne K^g$. Therefore $|[w, r]| = m/4$, so as $|C_{O^g}(O)| = m/4$

and $T_l \trianglelefteq S$, we have $x_0 = [w,r] \in T_l$ a generator of $C_{O^g}(O)$. If $m = 8$ then $x_0 = b \in Z(T_l)$ and (2) and (3) hold. If $m > 8$ then $|x_0| > 2$, so as T_l is dihedral, $\langle x_0 \rangle \trianglelefteq T_l$ and r inverts x_0. This completes the proof of (2) and (3).

It follows from (1)-(3) that the hypotheses of 5.7.5.5 are satisfied with τ_M, τ in the roles of τ, ρ, so (4) follows from parts (a) and (b) of 5.7.5.5.

Let $k = k_1 k_2$ and $k_1' = x_1^{-1} k_1$; then $k_1' \in K_1 - W$ and if $s \in \mathcal{D}^o(t)$ then as x_1 induces an outer automorphism on K^g, $s' = k_1' k_2 r$ induces an inner automorphism on K^g, so $s' \in \mathcal{D}^*(t)$. Thus replacing k_1 by k_1' and appealing to (1), we may assume $s \in \mathcal{D}^*(t)$. Now 5.7.5.5.c completes the proof of (5).

Suppose $\mathcal{F}_b^\circ \cong L_4^-[m]$. By 9.5.19.4, $\mathcal{L} \cong L_2[2m]$ so $T_l \cong D_m$. Therefore $T_l = \langle y, r\rangle$ where y is of order $m/2$, and by (2) and (4) we may choose y so that $y^2 = x$. As T_l centralizes O, so does y. Applying (4) to r and yr, $x_3^r = x_3 x^{-1} = x_3^{yr}$, so y centralizes x_3, and hence also $W = (W \cap O)\langle x_3\rangle$. We have established the hypothesis of 5.7.5.5.d, so by that lemma we may choose y so that $k_3^y = k_3 x_3$. This establishes the first sentence in (7).

From 9.5.22, $T = T_a T_l$ so $TM = T_l M$. As $T_l = \langle y, r\rangle$ with $T_l/\langle x\rangle \cong E_4$, TM is determined by the action of r and y on M, together with the power maps $y^2 = x$ and $r^2 = 1$ and the fact that r inverts y. The action of r is determined by (5), and as y centralizes OW with $y^2 = x$ and $k_3^y = k_3 x_3$, the action of y on M is determined. Thus TM is determined up to isomorphism, establishing the second sentence in (7).

Assume (6). Then $|T_b T_l| = |T_b||T_l| = 2m^2 \cdot m = 2m^3$. Similarly $|T| = |T_a||T_l|/|T_a \cap T_l| = (m^3/2) \cdot m/(m/4) = 2m^3$. Finally $T_b = O\langle s_1, s_2\rangle$ where $s_i \in \mathcal{D}^m(z) \cap z^{\mathcal{F}_b}$, so by 9.5.7.1, we may take $s_i \in \mathcal{D}^*(t)$. Then, choosing g as in 9.5.19, $s_i^g \in \mathcal{D}^*(z) \subseteq OT_l \le T$, so $s_i \in T$ and hence $T_b \le T$. Therefore $T_b T_l \le T$ with $|T_b T_l| = |T|$, so $T = T_b \times T_l$ by (6), completing the proof of (7).

It remains to prove (6). By 9.5.19.7, $T_l \trianglelefteq S$ and by 9.5.3.4, $T_b \trianglelefteq S$. Also $T_l \cap T_b \le C_{T_b}(O) = \langle z\rangle$ and $z \notin T_l$, so $T_b \cap T_l = 1$. Hence (6) holds. □

LEMMA 9.5.25. *If $\mathcal{F}_b^\circ \cong PSp_4[m]$ then $T_l \cong D_{m/2}$, $|T : T_a| = 2$, and TM is determined up to isomorphism.*

PROOF. By 9.5.19.4, $\mathcal{L} \cong L_2[m]$ so $T_l \cong D_{m/2}$. By 9.5.24, $x \in T_l$ is of order $m/4$, so $T_l = \langle x, r\rangle$ and $\langle x\rangle = T_l \cap T_a$ is of index 2 in T_l. Therefore $T = T_a T_l$ is of order $2|T_a|$, so $TM = \langle r\rangle M$ is determined by the action of r on M, and hence is determined up to isomorphism by 9.5.24.5. □

NOTATION 9.5.26. By 9.5.22, $\mathcal{F}_b^\circ$ is $PSp_4[m]$ or $L_4^-[m]$. If $\mathcal{F}_b^\circ \cong PSp_4[m]$ set $\tilde{\mathcal{F}} = \Omega_7[m]$, while if $\mathcal{F}_b^\circ \cong L_4^-[m]$ set $\tilde{\mathcal{F}} = \Omega_8^-[m]$. Let $\tilde{S}$ be Sylow in $\tilde{\mathcal{F}}$ and let $\tilde{\tau} = (\tilde{\mathcal{F}}, \tilde{\Omega})$ be the Lie packet of $\tilde{\mathcal{F}}$. Let $\tilde{\eta} \in \eta(\tilde{\tau})$, $\tilde{W} = \langle \tilde{\eta}\rangle$, and $\tilde{G}$ a model of $N_{\tilde{\mathcal{F}}}(\tilde{W})$.

LEMMA 9.5.27. *(1) $\tilde{G} \cong TM$, so we may take $\tilde{G} = TM$, $\tilde{S} = T$, $\tilde{\Omega} = \Omega$, and $N_{\tilde{\mathcal{F}}}(\tilde{W}) = \mathcal{F}_T(TM)$.*

(2) $\langle a\rangle = Z(\tilde{G})$.

PROOF. By 9.5.24.7 and 9.5.25, TM is determined up to isomorphism by our hypothesis, so as $\tilde{\mathcal{F}}$ satisfies that hypothesis with $|\tilde{S}| = |T|$ by 5.2.7 and 5.3.5.4, we conclude that $\tilde{G} \cong TM$, and then (1) follows. Then as $Z(TM) = \langle a\rangle$, (2) holds. □

LEMMA 9.5.28. *(1) $T\mathcal{F}_a^\circ = \tilde{\mathcal{F}}_a$.*

(2) $TO = T\tilde{O}$.

PROOF. By 9.5.23.1, $\mathcal{X} = \mathcal{F}_a^\circ \cong \Omega_6^+[m]$. Set $\mathcal{Y} = T\mathcal{X}$ and $\tilde{\mathcal{Y}} = \tilde{\mathcal{F}}_a$. By 9.3.20 and 2.1.15, $\mathcal{Y} = \langle \mathcal{Y}_z, \mathcal{F}_T(TM)\rangle$ and $\tilde{\mathcal{Y}} = \langle \tilde{\mathcal{Y}}_z, N_{\tilde{\mathcal{F}}}(\tilde{W})\rangle$. By 9.5.27.1, $N_{\tilde{\mathcal{F}}}(\tilde{W}) = \mathcal{F}_T(TM)$, so to prove (1) it suffices to show $\mathcal{Y}_z = \tilde{\mathcal{Y}}_z$. But $\mathcal{Y}_z = TO$, so (2) implies (1). Finally (2) follows from Theorem 2.7.3. □

LEMMA 9.5.29. *If* $\mathcal{F}_b^\circ \cong PSp_4[m]$ *then* $\mathcal{F}^\circ \cong \Omega_7[m]$.

PROOF. Let $\mathcal{E}_1 = TO\mathcal{L}$, $\mathcal{F}_1 = \mathcal{F}_z$, $\mathcal{E}_2 = \mathcal{F}_T(TM)$, and $\mathcal{F}_2 = N_{\mathcal{F}}(W)$. By 9.5.9, $\mathcal{F} = \langle \mathcal{F}_1, \mathcal{F}_2\rangle$ and $\tilde{\mathcal{F}} = \langle \tilde{\mathcal{F}}_z, \mathcal{F}_T(TM)\rangle$. We claim that $\mathcal{E}_1 = \tilde{\mathcal{F}}_z$, so $\tilde{\mathcal{F}} = \langle \mathcal{E}_1, \mathcal{E}_2\rangle$ is saturated with $\tilde{\mathcal{F}} = O^{2'}(\tilde{\mathcal{F}})$. By 9.5.23.5, $\mathcal{E}_1 \trianglelefteq \mathcal{F}_1$ and by 9.5.23.6, $\mathcal{E}_2 \trianglelefteq \mathcal{F}_2$. Therefore Hypothesis 1.4.7 in [**Asc19**] is satisfied, so by 1.4.8 in [**Asc19**], $\tilde{\mathcal{F}} \trianglelefteq \mathcal{F}$. Then $\mathcal{F}^\circ = \tilde{\mathcal{F}} \cong \Omega_7[m]$, establishing the lemma. Thus it remains to prove the claim.

By 9.5.28.2, $TO = T\tilde{O}$. Further $\mathcal{E}_1 = TO\mathcal{L}$, so $\mathcal{E}_1 = \langle TO, T\mathcal{L}\rangle$. Thus it remains to show $T\mathcal{L} = T\tilde{\mathcal{L}}$. But $\mathcal{L} \cong L_2[m] \cong \tilde{\mathcal{L}}$ and by 9.5.24.4, $r^{x_3} = rx$, so T is transitive on the 4-subgroups of T_l, and hence on $(T\mathcal{L})^e$ and $(T\tilde{\mathcal{L}})^e$, with representative $R = C_T(U)$ for $U = \langle b, r\rangle$. From 9.5.24.4, $R = OU\langle j\rangle$, where $j = k_3k_4$. Also $O^2(Aut_{T\mathcal{L}}(R))$ centralizes O, and we see in a moment that it also centralizes $i = jr$, so $Aut_{T\mathcal{L}}(R) = Aut_{T\tilde{\mathcal{L}}}(R)$, completing the proof of the claim.

To see that i is centralized, we show that i is the unique member of iU in $\mathcal{Z}$ and the unique member in $\tilde{\mathcal{Z}}$. First $k_3^r = k_4$, so $r^{k_3} = k_3^{-1}k_4r = tjr = ti$ and $(br)^{k_3} = b^{k_3}r^{k_3} = bt \cdot ti = bi$ are in $b^{\mathcal{F}} \cap b^{\tilde{\mathcal{F}}}$. Also $b, j \in O^g$, so $bj \in O^g$ and hence $bj \in b^{\mathcal{F}} \cap b^{\tilde{\mathcal{F}}}$. Finally $K_3^i = K_4$ and we showed that $ti \in b^{\mathcal{F}} \cap b^{\tilde{\mathcal{F}}}$, so $i = t \cdot ti \in \mathcal{Z} \cap \tilde{\mathcal{Z}}$. □

It remains to treat the case $\mathcal{F}_b^\circ \cong L_4^-[m]$, so we assume:

HYPOTHESIS 9.5.30. *Hypothesis 9.5.1 holds and* $\mathcal{F}_b^\circ \cong L_4^-[m]$. *Set* $\mathcal{F}_1 = \mathcal{F}_z$, $\mathcal{F}_2 = N_{\mathcal{F}}(W)$, $\mathcal{F}_3 = \mathcal{F}_b$, $\mathcal{E}_1 = TO\mathcal{L}$, $\mathcal{E}_2 = \mathcal{F}_T(TM)$, *and* $\mathcal{E}_3 = T\mathcal{F}_b^\circ$. *Set* $j = k_3k_4$, $i = jr$, $l = x_3y^{-1}$, $i_2 = il^{-1}$, *and* $r_2 = ry$.

LEMMA 9.5.31. *(1)* $\mathcal{F} = \langle \mathcal{F}_i : 1 \le i \le 3\rangle$.
(2) $\tilde{\mathcal{F}} = \langle \tilde{\mathcal{F}}_z, \mathcal{E}_2\rangle$.
(3) $\mathcal{E}_i \trianglelefteq \mathcal{F}_i$ *for* $1 \le i \le 3$.
(4) $T_0 = \langle i, l\rangle \cong D_m$ *centralizes* T_l *with* $i \in \mathcal{Z} \cap \tilde{\mathcal{Z}}$ *inverting* l *of order* $m/2$.
(5) bi_2 *and* i_2r_2 *are in* $b^{\mathcal{F}} \cap b^{\tilde{\mathcal{F}}}$ *and* $i_2 \in \mathcal{Z} \cap \tilde{\mathcal{Z}}$.

PROOF. Part (1) follows from 9.5.9. Similarly by 9.5.24.7, $T = T_b \times T_l$, so as T is Sylow in $\tilde{\mathcal{F}}$ by 9.5.27, applying 9.5.9 to $\tilde{\mathcal{F}}$ we get $\tilde{\mathcal{F}} = \langle \tilde{\mathcal{F}}_z, N_{\tilde{\mathcal{F}}}(W)\rangle$ and then (2) follows from 9.5.27.1.

Next $\mathcal{E}_1 \trianglelefteq \mathcal{F}_1$ by 9.5.23.5 and $\mathcal{E}_2 \trianglelefteq \mathcal{F}_2$ by 9.5.23.6. By 9.5.23.3, $T \trianglelefteq C_{N_{\mathcal{F}}(W)}(b)$, so as $W \trianglelefteq N_{\mathcal{F}_b}(T_b)$, $\mathcal{E}_3 \trianglelefteq \mathcal{F}_3$ by Theorem 1.5.2 in [**Asc19**] applied to $\mathcal{F}_b^\circ \trianglelefteq \mathcal{F}_b$. This proves (3).

As $k_3^r = k_4$, r centralizes $j = k_3k_4$, so $i = jr$ is an involution centralizing r. Next

$$l^r = (x_3y^{-1})^r = x_4y = x_3x_3^{-1}x_4y = x_3x^{-1}y = x_3y^{-2}y = x_3y^{-1} = l,$$

so r centralizes l. Therefore l centralizes $\langle r, y\rangle = T_l$.

Also $k_3^y = k_3x_3$, so $y^{k_3} = yx_3^{-1} = l^{-1}$. Then

$$(y^{-1})^{k_4} = y^{rk_3^r} = (y^{k_3})^r = (l^{-1})^r = l^{-1},$$

so $y^{k_4} = l$. Then

$$y^j = y^{k_3k_4} = (l^{-1})^{k_4} = (yx_3^{-1})^{k_4} = lx_3^{-1} = y^{-1},$$

so j inverts y. Then as r inverts y, i centralizes y, so i centralizes T_l. Thus $T_0 = \langle i, l\rangle$ centralizes T_l. Finally $|l| = |y| = m/2$ and we showed that j inverts y, so as j also inverts x_3, j inverts l. Hence as r centralizes l, i inverts l, so $T_0 \cong D_m$. From the proof of 9.5.29, $i \in \mathcal{Z} \cap \tilde{\mathcal{Z}}$, completing the proof of (4).

As $k_3^r = k_4$, $r^{k_3} = tjr = ti$. Thus $r_2^{k_3} = r^{k_3}y^{k_3} = til^{-1} = ti_2 \in b^{\mathcal{F}} \cap b^{\tilde{\mathcal{F}}}$, so as r_2 interchanges K_3 and K_4, $i_2 = t{\cdot}ti_2 \in \mathcal{Z} \cap \tilde{\mathcal{Z}}$. Also $(br_2)^{k_3} = b^{k_3}r_2^{k_3} = (bt){\cdot}(ti_2) = bi_2$, so $bi_2 \in b^{\mathcal{F}} \cap b^{\tilde{\mathcal{F}}}$. Finally $i_2r_2 = (il^{-1}) \cdot ry = iyx_3^{-1}ry = iyx_3^{-1}y^{-1}r = ix_3^{-1}r = jrx_3^{-1}r = jx_4 \in b^{\mathcal{F}} \cap b^{\tilde{\mathcal{F}}}$ as $jx_4 \in O^g$ is an involution. This completes the proof of (5) and the lemma. □

LEMMA 9.5.32. *If $\mathcal{F}_b^\circ \cong L_4^-[m]$ then $\mathcal{F}^\circ \cong \Omega_8^-[m]$.*

PROOF. We first observe that the tuple $\mathcal{F}_i, \mathcal{E}_i$, $1 \le i \le 3$, satisfies Hypothesis 1.4.3 in [**Asc19**] by parts (1) and (3) of 9.5.31. Therefore by 1.4.4 in [**Asc19**], $\mathcal{Y} = \langle \mathcal{E}_i : 1 \le i \le 3\rangle$ is an $\mathcal{F}$-invariant subsystem of $\mathcal{F}$ on T, so in particular T is strongly closed in S with respect to $\mathcal{F}$. We will show that $\mathcal{Y} = \tilde{\mathcal{F}}$. Thus $\mathcal{Y} = O^{2'}(\mathcal{Y})$ is saturated, and hence $\mathcal{Y} \trianglelefteq \mathcal{F}$ by 1.4.5 in [**Asc19**]. Then $\mathcal{F}^\circ = \mathcal{Y} \cong \Omega_8^-[m]$. It remains to show $\mathcal{Y} = \tilde{\mathcal{F}}$.

We first show that $\mathcal{E}_1 = \tilde{\mathcal{F}}_z$. By 9.5.28.2, $T\mathcal{O} = T\tilde{\mathcal{O}}$. Further $\mathcal{E}_1 = T\mathcal{O}\mathcal{L}$, so $\mathcal{E}_1 = \langle T\mathcal{O}, T\mathcal{L}\rangle$, and similarly $\tilde{\mathcal{F}}_z = \langle T\tilde{\mathcal{O}}, T\tilde{\mathcal{L}}\rangle$, so to show $\mathcal{E}_1 = \tilde{\mathcal{F}}_z$ it suffices to show that $T\mathcal{L} = T\tilde{\mathcal{L}}$. From 9.5.31.4, $T = OT_0 \times T_l$. Let $r_1 = r$, $i_1 = i$, and $U_i = \langle r_i, b\rangle$ for $i = 1, 2$. As $\mathcal{L} \cong L_2[2m]$ and U_1 and U_2 are representatives for the orbits of T_l on its 4-subgroups, $R_i = OT_0U_i$, $i = 1, 2$, are representatives for the orbits of T on $(T\mathcal{L})^e$ and $(T\tilde{\mathcal{L}})^e$, so it suffices to show $Aut_{T\mathcal{L}}(R_i) = Aut_{T\tilde{\mathcal{L}}}(R_i)$ for $i = 1, 2$. Let $G_i, \tilde{G}_i$ be model for $N_{T\mathcal{L}}(U_i)$, $N_{T\tilde{\mathcal{L}}}(U_i)$, respectively. Then $G_i = H_i \times X_i$, where $H_i \cong S_4$ has Sylow 2-group $N_{T_l}(U_i)$ and $X_i = C_{G_i}(H_i) = OC_{T_0}(H_i)$. Moreover $\Phi(T_0) = \langle l^2\rangle = \langle x_3^2x^{-1}\rangle = \langle (x_1x_2)^{-1}\rangle \le O \le C_T(H_i)$, so $X_i = O\langle i_eb^{\epsilon_{i,e}} : e = 1, 2\rangle$. Define similar notation in $\tilde{G}_i$. Thus it remains to show that $\epsilon_{i,e} = \tilde{\epsilon}_{i,e}$ for $i, e \in \{1, 2\}$. From the proof of 9.5.29, i centralizes H_1 and $\tilde{H}_1$, so $\epsilon_{1,1} = 0 = \tilde{\epsilon}_{1,1}$. The same proof together with 9.5.31.5 shows that $\epsilon_{2,2} = 0 = \tilde{\epsilon}_{2,2}$. We claim that i does not centralize $\mathcal{L}$. Given the claim, $\epsilon_{1,2} = 1 = \tilde{\epsilon}_{1,2}$ and similarly $\epsilon_{2,1} = 1 = \tilde{\epsilon}_{2,1}$, showing that $\mathcal{E}_1 = \tilde{\mathcal{F}}_z$. So assume i centralizes $\mathcal{L}$; we produce a contradiction.

By 9.5.31.4, $i \in \mathcal{Z}$ so there is $\alpha \in \mathfrak{A}(i)$ with $i\alpha = z$. Let $\mathcal{X} = O^2(C_{\mathcal{O}}(i))$. As $K^i = K_2$, $\mathcal{X} \cong L_2[m]$. Then $\mathcal{B} = (\mathcal{X}\mathcal{L})\alpha^* \le O^2(\mathcal{F}_z)$ and as T is strongly closed in S with respect to $\mathcal{F}$ and $\mathcal{E}_1 \trianglelefteq \mathcal{F}_z$, $\mathcal{B} \le O^2(\mathcal{E}_1) = \mathcal{O}\mathcal{L}$. Indeed from 9.5.24, $b^{\mathcal{F}} \cap \mathcal{O}\mathcal{L}$ is the set of involutions in $O - \langle z\rangle$ and T_l, so $\mathcal{L}\alpha^*$ is contained in $C_{\mathcal{O}}(z\alpha)$ or $\mathcal{L}$. As $O^2(C_{\mathcal{O}}(z\alpha)) \cong L_2[m]$ it follows that $\mathcal{L}\alpha^* = \mathcal{L}$, so in particular $b\alpha = b$. Therefore $(ib)\alpha = zb = a$, a contradiction as we saw during the proof of 9.5.29 that $ib \in b^{\mathcal{F}}$. This finally completes the proof that $\mathcal{E}_1 = \tilde{\mathcal{F}}_z$.

As $\mathcal{E}_1 = \tilde{\mathcal{F}}_z$, $\langle \mathcal{E}_1, \mathcal{E}_2\rangle = \langle \tilde{\mathcal{F}}_z, \mathcal{E}_2\rangle = \tilde{\mathcal{F}}$ by 9.5.31.2. Thus $\tilde{\mathcal{F}} \le \mathcal{Y}$, so to complete the proof it suffices to show that $\mathcal{E}_3 \le \tilde{\mathcal{F}}$. By 9.4.22 and 9.4.28, $\mathcal{E}_3 = \langle T\mathcal{O}, Aut_{\mathcal{E}_3}(R_i) : i = 1, 2\rangle$, where $R_i = C_T(Z_i)$ and $Z_i = \langle z, v, t_i\rangle$ with $t_i \in \mathcal{D}^m(z) \cap z^{\mathcal{E}_3}$. By 9.4.21, $u_i = v$ or vz is in the center of $N_{\mathcal{E}_3}(R_i)$, so $N_{\mathcal{E}_3}(R_i)$

centralizes $e_i = bu_i = t$ or tz. However there is $\alpha \in \mathfrak{A}_{\mathcal{E}_2}(e_i)$ with $e_i\alpha = z$, so $N_{\mathcal{E}_3}(R_i)\alpha^* \le C_{\mathcal{F}_3}(e_i)\alpha^* \le \mathcal{F}_z$. However $Aut_{\mathcal{F}_3}(R_i) = \langle Aut_T(R_i)^{Aut_{\mathcal{E}_3}(R_i)}\rangle$, so as T is strongly closed in S with respect to $\mathcal{F}$ and $\mathcal{E}_1 \trianglelefteq \mathcal{F}_z$, $Aut_{\mathcal{E}_3}(R_i)\alpha^* \le Aut_{\mathcal{E}_1}(R_i\alpha) \le Aut_{\tilde{\mathcal{F}}}(R_i\alpha)$. Then as α is an $\tilde{\mathcal{F}}$-map, $Aut_{\mathcal{F}_3}(R_i) \le Aut_{\tilde{\mathcal{F}}}(R_i)$, so indeed $\mathcal{E}_3 \le \tilde{\mathcal{F}}$. This finally completes the proof of the lemma. □

Collecting the results in this section, we have proved:

THEOREM 9.5.33. *Assume Hypothesis* 9.5.1. *Then either*
(1) $a \in Z(\mathcal{F})$ *and* $\mathcal{F}^\circ = \mathcal{F}_{S\cap M}(M)$ *or* $\mathcal{F}^\circ \cong \check{A}E_6$ *or* $\Omega_6^+[m]$, *or*
(2) $\mathcal{F}^\circ$ *is isomorphic to* $L_3(2)/E_{64}$, $Sp_6(2)$, AE_7, $\Omega_7[m]$, *or* $\Omega_8^-[m]$.

PROOF. By Theorem 9.5.17 we may assume $\mathcal{F}_b^\circ \ne \mathcal{O}(z)$, so Hypothesis 9.5.18 holds. Then by 9.5.21 and 9.5.4.2 we may assume $\mathcal{F}_b^\circ$ is $PSp_4[m]$ or $L_4^-[m]$; hence Hypothesis 9.5.22 is satisfied. Then 9.5.29 and 9.5.32 complete the proof. □

THEOREM 9.5.34. *Assume* $\tau = (\mathcal{F}, \Omega)$ *is a quaternion fusion packet such that* $\mathcal{A}(z) = \emptyset$, $\Omega = \Omega(z)$ *is of order 2, and* $\mathcal{F}^\circ$ *is transitive on* Ω. *Then either*
(1) $\mu \cong E_4$ *and* $\mathcal{F}^\circ$ *is* AE_5, $PSp_4[m]$, *or* $L_4^-[m]$, *or*
(2) $\mu \cong S_4$ *and* $\mathcal{F}^\circ$ *is* $\mathcal{F}_{S\cap M}(M)$ *for* $M \in \mathcal{M}(\tau)$, AE_6, $P\Omega_6^+[m]$, $\check{A}E_6$, $\Omega_6^+[m]$, $L_3(2)/E_{64}$, $Sp_6(2)$, AE_7, $\Omega_7[m]$, *or* $\Omega_8^-[m]$.

PROOF. Let $W \in \mathcal{W}(\tau)$ and $W \trianglelefteq M \in \mathcal{M}(\tau)$. Then $\tau_W = (\mathcal{F}_{S\cap M}(M), \Omega)$ is a quaternion fusion packet with $\mu(\tau_W) = \mu(\tau)$, satisfying Hypothesis 4.2.1. Adopt Notation 4.2.3. As $\mathcal{A}(\tau) = \emptyset$, D is a set of 3-transpositions of $\mu = M/W$ by 4.2.10. As $\Omega = \Omega(z)$ is of order 2, $V_d = d^\perp$ is of order 2 for $d \in D$, so either $\mathcal{D}(D)$ is disconnected or $D = V_d$. In the first case $\mu \cong S_4$ by 4.2.13 and in the second, $\mu \cong E_4$. In the latter case the theorem follows from 9.4.34, and in the former it follows from 9.4.35 and 9.5.33. □

CHAPTER 10

$|\Omega(z)| > 2$

Let $\tau = (\mathcal{F}, \Omega)$ be a quaternion fusion packet and $z \in \mathcal{Z}_S(\tau)$. Chapter 10 treats the case where $\mathcal{F}^\circ$ is transitive on Ω, $|\Omega(z)| > 2$, and the Inductive Hypothesis holds. We begin in section 10.1 with the case $\mu \cong \mathrm{Weyl}(D_4)$; this is the case where all examples appear; moreover in this case we do not need to assume the Inductive Hypothesis. Then the general case is dealt with in section 10.2. As a corollary, in Theorem 10.2.14 we are then able to determine those τ where $\mathcal{F}^\circ$ is transitive on Ω and $\Omega = \Omega(z)$ is of order at least 3, using an inductive argument to establish the Inductive Hypothesis in a minimal counter example. An important byproduct of 10.2.14 is that in each example τ, $\mathcal{F} = \langle \mathcal{F}_z, \mathcal{F}_S(SM) \rangle$ for $M \in M(\tau)$.

10.1. $|\Omega(z)| = 4$ and μ isomorphic to Weyl(D_4)

In this section we assume the following hypothesis:

HYPOTHESIS 10.1.1. *$\tau = (\mathcal{F}, \Omega)$ is a quaternion fusion packet such that*
(1) For $z \in \mathcal{Z}_S$, $|\Omega(z)| = 4$.
(2) $\mu = \mu(\tau) \cong$ Weyl(D_4).

NOTATION 10.1.2. As $\mu \cong \mathrm{Weyl}(D_4)$, $\mathcal{F}$ is transitive on Ω by 4.2.8.7 and $|\Omega| = 4$ by 4.2.8.6. Hence as $|\Omega(z)| = 4$ for $z \in \mathcal{Z}_S$, we have $\Omega = \Omega(z)$.

Let $\{z\} = \mathcal{Z}_S$ and $\{K_i : 1 \le i \le 4\} = \Omega(z)$. Set $K = K_1$, let $\eta \in \eta(\tau)$, $W = \langle \eta \rangle$, G a model for $N_{\mathcal{F}}(C_S(W))$, and $M = \langle K^G \rangle$. Set $O = O(z)$ and let $\langle v_1 \rangle$ be a $N_S(K)$-invariant subgroup of K of order 4, $v_i \in v_1^S \cap K_i$ for $1 < i \le 4$, and $v = v_1 \cdots v_4$. Set $E = \langle v, z \rangle$ and $T = S \cap M$. Let $\mathcal{D}^m(z)$ consist of those $t \in \mathcal{Z}$ such that t is fixed point free on Ω, and let $\mathcal{D}^o(z)$ consist of those $t \in \mathcal{Z}$ inducing an outer automorphism on some member of Ω.

LEMMA 10.1.3. *(1) $M = \bar{M}/Z(\bar{M}$, where $\bar{M} = \bar{\omega}(D_4, m)$.*
(2) $v \in \mathcal{D}^*(z)$, $E \trianglelefteq MS$, $E^\# = Z_\Delta$, and $Aut_M(E) = GL(E)$.
(3) $|W| = m^4/64$ and $M/W \cong \mu$.
(4) OW is the kernel of the action of T on Ω and $T/OW \cong E_4$ is regular on Ω.
(5) $|W : O \cap W| = 2$ and for each $K \in \Omega$ and $w \in W - O$, w interchanges the two classes of Q_8-subgroups of K.
(6) $\Omega_1(W) = \langle v_1 v_2, v_1 v_3, v, z \rangle \cong E_{16}$,
(7) $\mathcal{D}^m(z) \cup \mathcal{D}^o(z) \subseteq C_S(E)$.
(8) For $s \in \mathcal{Z}$, we have $s \in \mathcal{D}^*(z)$ iff s induces a nontrivial inner automorphism on some K_i iff s induces a nontrivial inner automorphism on each K_i.

PROOF. By parts (1) and (2) of Hypothesis 10.1.1 and 5.8.7.3, $v \in z^M$ and (1) holds; in particular $v \in \mathcal{D}^*(z)$. By 4.3.8.4, $\mu = M/W$, and by 10.1.1.2, $\mu \cong \text{Weyl}(D_4)$, so $M/O_2(M) \cong S_3$, and hence $|z^M| = 3$. Then as $v \in z^M$ and $[v_1, k_1] = z$ for $k_1 \in K - W$, it follows that $E^\# = z^M = \mathcal{Z}_\Delta$, so (2) holds. By 5.8.7 and 4.3.7.2, in the universal group $\bar{M} = \bar{\omega}(D_4, m)$, $|Z(\bar{M})| = 4$, $|\bar{W}| = (m/2)^4$ and $W = \bar{W}/Z(\bar{M})$ by (1), so (3) follows. Indeed $Z(\bar{M}) = \langle \bar{z}_1\bar{z}_2, \bar{z}_1\bar{z}_3\rangle \cong E_4$, so the preimage of $\Omega_1(W)$ is contained in $\Omega_2(\bar{W}) = \langle \bar{v}_1, \ldots, \bar{v}_4, \bar{v}_0\rangle$, where $\bar{v}_0^2 = \bar{v}$. Further from 5.8.7.6, $\bar{v} = \bar{v}_1 \cdots \bar{v}_4$, so $\Omega_1(W) = \langle v_1v_2, v_1v_3, v, z\rangle$, so (6) follows.

As OW acts on $K_i \cap W$ for each $1 \le i \le 4$, OW acts on K_i. Also $M^* = M/W \cong \text{Weyl}(D_4)$ with $E_{16} = O^* \trianglelefteq T^*$, so $T/OW \cong T^*/O^* \cong E_4$. Further T^*/O^* is regular on $d^{\perp}_{K\cap W}$, so T/OW is regular on Ω, establishing (4).

Next $O \cap W = \prod_i (K_i \cap W)$ is of order $(m/2)^4/8 = m^4/128$ as $K_i \cap K_j = \langle z\rangle$ for all $i \neq j$. Therefore $|W : O \cap W| = 2$ by (3). Further for $v \in \langle x\rangle \in \eta$, x induces an automorphism on K interchanging the two classes of Q_8-subgroups by 4.2.5, completing the proof of (5).

Part (8) follows from 3.1.15.

Suppose $t \in \mathcal{D}^m(z)$. Then relabeling if necessary, $v_1^t = v_2 z^\epsilon$ and $v_3^t = v_4 z^\delta$, so t centralizes $\langle v_1v_2, v_3v_4\rangle$ and hence also v. Similarly if $s \in \mathcal{Z}$ induces an outer automorphism on K_1 then by (8) either s is outer on each K_i or we may assume s is outer on K_2 and $K_3^s = K_4$. In either case s centralizes v, so each member of $\mathcal{D}^o(z)$ centralizes v. Thus (7) holds. □

LEMMA 10.1.4. *Let $t \in \mathcal{D}^*(z)$ and $I = \{1, \ldots, 4\}$. Then*

(1) $t = k_1k_2k_3k_4c$ where $k_i \in K_i$ is of order 4 and $c \in C_S(O(z))$ with $c^2 = 1$.

(2) There exists $\phi \in \hom_{\mathcal{F}_z}(\langle k_1, \ldots, k_4, t\rangle, W)$ and $\sigma \in Sym(I)$ with $k_i\phi = v_{\sigma(i)}$, $c\phi \in C_S(O(z))$, and $t\phi = v$.

(3) $\mathcal{F}_z$ is transitive on $\mathcal{D}^*(z)$.

(4) $\mathcal{D}^*(z) \subseteq O$ and we may take $c = 1$.

(5) $\mathcal{D}^*(v) \cup \mathcal{D}^*(vz) \subseteq T$.

PROOF. Part (1) follows as t induces a nontrivial inner automorphism on each K_i. Then the proof of 9.1.10.2 shows we may choose $\phi \in \hom_{\mathcal{F}_z}(\langle k_1, \ldots, k_4, t\rangle, W)$ with $t\phi \in \mathcal{Z}_\Delta$ and $k_i\phi = v_{\sigma(i)}$ for some $\sigma \in Sym(I)$. By 10.1.3.2, $\mathcal{Z}_\Delta = E^\#$, so conjugating in O, we may take $t\phi = v$. That is (2) holds. Then (2) implies (3). As $v \in O$ and $\mathcal{O}(z) \trianglelefteq \mathcal{F}_z$, $\mathcal{D}^*(z) \subseteq O$ by (3). Thus $t \in O$, so $c = k_1 \cdots k_4 t \in C_O(O) = \langle z\rangle$, so replacing k_1 by k_1^{-1} if necessary, we may take $c = 1$, completing the proof of (4). Similarly $\mathcal{D}^*(v) \subseteq O^g$ for $g \in M$ with $z^g = v$, so $\mathcal{D}^*(v) \subseteq C_{O^g}(E) = O^g \cap O_2(M) \le T$, establishing (5). □

LEMMA 10.1.5. *Let $w = v_3v_4$ and $Q = \bigcap_{g\in M} O^g$. Then*

(1) $\tau_w = (\mathcal{F}_w, \Omega_w)$ is a quaternion fusion packet, where $\Omega_w = \{K_1, K_2\}$.

(2) $\eta_w = (K \cap W)^{C_M(w)} \in \eta(\tau_w)$, and setting $W_w = \langle \eta_w\rangle$ and $M_w = \langle K^{C_M(w)}\rangle$, $\mu_w = M_w/W_w = \mu(\tau_w) \cong S_4$ with $\langle w\rangle = Z(M_w)$.

(3) $\mathcal{F}_w^\circ = \mathcal{F}_{S\cap M_w}(M_w)$ or $\mathcal{F}_w^\circ \cong \check{A}E_6$ or $\Omega_6^+[m]$.

(4) There is $g \in C_M(w)$ with $z^g = v$ and $w \in K_3^g K_4^g$.

(5) $\Omega_1(W) \le Q$.

PROOF. Part (1) follows from 2.5.2. Then (2) follows from 5.8.7, while (3) follows from (1), (2), and Theorem 9.5.33.

By (2), $Aut_{M_w}(E) = GL(E)$, so there exists $g \in M_w$ with $z^g = v$. As $w \in K_3K_4$, (4) follows. As $v \in O$ and $Aut_M(E) = GL(E)$, we have $E \leq Q$. By (4), $w \in Q$ and then by symmetry among the v_i and 10.1.3.6, (5) follows. □

LEMMA 10.1.6. *(1)* $\mathcal{D}^m(z) \subseteq \mathcal{D}^*(v) \cup \mathcal{D}^*(vz)$.
(2) $\mathcal{D}^m(z) \subseteq T$.

PROOF. Let $t \in \mathcal{D}^m(z)$ and $w = v_3v_4$. Without loss t acts as $(K_1, K_3)(K_2, K_4)$ on Ω, so $[w, t] = v$ or vz, say the former. Choose g as in 10.1.5.4 and let $J_i = K_i^g$. Then t acts on O^g by 10.1.3.7, so as $w \in J_3J_4$ and $[w, t] = v = [w^t, t]$, t acts on each J_i, and then by 10.1.3.8, either $t \in \mathcal{D}^*(v)$ or t induces an outer automorphism on each J_i. But in the latter case, $C_{O^g}(t) = \langle v_1^g v_i^g, v : 1 < i \leq 4\rangle$, so $C_{O^g}(t) = \Omega_1(W)$ by 10.1.3.6 and symmetry between z and v. Then t centralizes v_1v_2 by 10.1.3.6, a contradiction. Therefore $t \in \mathcal{D}^*(v)$. This proves (1), and (1) and 10.1.4.5 imply (2). □

LEMMA 10.1.7. *Write $\mathcal{D}^a(z)$ for the set of $t \in \mathcal{D}^o(z)$ such that $t \in T$ acts on each K_i and inverts W and set*

$$\mathcal{D}^*(E) = \bigcap_{e \in E^\#} \mathcal{D}^*(e).$$

Then
(1) If $t \in \mathcal{D}^o(z)$ is in T then $t \in \mathcal{D}^a(z)$.
(2) We have

$$T \cap \mathcal{Z} = \bigcup_{e \in E^\#} \mathcal{D}^*(e).$$

(3) If $t \in \mathcal{D}^*(E)$ then $t = x_1x_2x_3x_4$ for some $x_i \in K_i - W$.

PROOF. Suppose $t \in \mathcal{D}^o(z)$ is in T. By 10.1.3.4, $t \in OW$ and t acts on each member of Ω. Then as t induces an outer automorphism on some member of Ω, $t \notin \mathcal{D}^*(z)$ by 10.1.3.8, so by another application of 10.1.4.8, t induces an outer automorphism on each K_i, and hence t inverts $O \cap W$. Therefore $t \in kW$ where $k = k_1 \cdots k_4$ with $k_i \in K_i - W$. Now k inverts W by 4.3.5.2, establishing (1) and showing $\mathcal{D}^a(z) \subseteq \mathcal{Z} \cap kW$.

Set

$$\mathcal{U} = \bigcup_{e \in E^\#} \mathcal{D}^*(e),$$

and assume $t \in T \cap \mathcal{Z}$ but $t \notin \mathcal{U}$. Then by 10.1.6.1, $t \notin \mathcal{D}^m(z) \cup O$, so $t \in \mathcal{D}^o(z)$ and hence $t \in \mathcal{D}^a(z)$ by (1). Define Q as in 10.1.5 and set $P = W \cap Q$, $U = \langle k\rangle W$, and $M^+ = M/P$. Let $X \in Syl_3(M)$. Then X acts on U^+ and by 5.8.7.7, $W^+ = [U^+, X] \cong E_4$ so $U^+ = C_{U^+}(X) \times W^+$ with $|C_{U^+}(X)| = 2$. We may take k_1^+ to invert X^+, so $C_{U^+}(X) \leq C_{U^+}(k_1) = \langle k^+, (K \cap W)^+\rangle \leq O^+$, so each orbit of X on U^+ contains a member of $U \cap O$, and in particular the preimage U_0 in U of $C_{U^+}(X)$ is contained in Q. Thus $\mathcal{Z} \cap U_0 \subseteq \mathcal{D}^*(E) \cup E^\#$ and $\mathcal{D}^a(z) \subseteq \mathcal{Z} \cap kW \subseteq \mathcal{U}$, establishing (2).

Suppose $t \in \mathcal{D}^*(E)$. By 10.1.4, $t = x_1x_2x_3x_4$ with $x_i \in K_i$ of order 4. As $W \cap \mathcal{Z} = E^\#$ by 10.1.3.2, we have $t \notin W$. As $t \in \mathcal{D}^*(E)$, we have $t \in Q \leq QW = \langle k\rangle W$, so $t = kw$ for some $w \in W$. Then $w = kt \in W \cap O$, so as $k_i \in K_i - W$, also $x_i \in K_i - W$, proving (3). □

LEMMA 10.1.8. *(1)* $O \trianglelefteq N_{\mathcal{F}_z}(E)$.
(2) $\mathcal{F}_T(M) \trianglelefteq N_{\mathcal{F}}(E) = N_{\mathcal{F}}(W)$.
(3) $T \trianglelefteq N_{\mathcal{F}_z}(E)$.
(4) $T \trianglelefteq SC_{\mathcal{F}}(O)$.
(5) Either $O \trianglelefteq \mathcal{F}_z$ *or* $m > 8$ *and* $\mathcal{O}_{K_i} \cong SL_2[m]$ *for each* $1 \leq i \leq 4$.

PROOF. As $v = v_1 \cdots v_4$ with $\langle v_i \rangle \trianglelefteq K_i$ of order 4, $N_{\mathcal{O}(z)}(E) = O$. So as $\mathcal{O}(z) \trianglelefteq \mathcal{F}_z$, (1) follows from 8.23.2 in [**Asc11**].

Let $\mathcal{Y} = N_{\mathcal{F}}(E)$. By (1) and 2.2 in [**Asc10**], $O \trianglelefteq C_{\mathcal{Y}}(z)$. Thus $C_O(E) \trianglelefteq C_{\mathcal{Y}}(E) = \mathcal{X}$ by 8.23.2 in [**Asc11**], so $C_O(E) \leq O_2(\mathcal{X})$. But, using 10.1.3.5 for example, $W = \langle (O \cap W)^M \rangle$, so $W \leq O_2(\mathcal{X}) \leq O_2(\mathcal{Y})$. Then as W is weakly closed in S with respect to $\mathcal{F}$, $W \trianglelefteq \mathcal{Y}$. On the other hand by 3.1.11.7 and 10.1.3.2, $E^{\#} = \mathcal{Z} \cap W$, so $E \trianglelefteq N_{\mathcal{F}}(W)$, and hence $N_{\mathcal{F}}(W) = \mathcal{Y}$. Moreover by 3.3.10.2, $\mathcal{M} = [\Omega]_{\mathcal{Y}}$ is constrained, and by 3.3.10.3, M is a model for $\mathcal{M}$, so $\mathcal{F}_T(M) = \mathcal{M} \trianglelefteq \mathcal{Y}$, completing the proof of (2). Further $T = C_{\mathcal{M}}(z) \trianglelefteq C_{\mathcal{Y}}(z) = N_{\mathcal{F}_z}(E)$, so (3) holds.

Set $\mathcal{B} = SC_{\mathcal{F}}(O)$. By 10.1.3.2, $\mathcal{B} \leq N_{\mathcal{F}_z}(E) = C_{\mathcal{Y}}(z)$, while by (3), $T \trianglelefteq C_{\mathcal{Y}}(z)$, so $T \trianglelefteq \mathcal{B}$, establishing (4).

If $m = 8$ then $O = \mathcal{O}(z) \trianglelefteq \mathcal{F}_z$, so (5) holds in this case. Thus we may assume $m > 8$. By 10.1.3.5, $w \in W - O$ induces an outer automorphism on K interchanging the two classes of Q_8-subgroups of K, so if $\mathcal{O}_K \neq K$ then $\mathcal{O}_K \cong SL_2[m]$ by 2.5.7. Then $\mathcal{O}_{K_i} \cong SL_2[m[$ for each $1 \leq i \leq 4$ as T is transitive on Ω by 10.1.3.4. Similarly if $\mathcal{O}_K = K$ then $O = \mathcal{O}(z) \trianglelefteq \mathcal{F}_z$, completing the proof of (5). □

LEMMA 10.1.9. *There exists a normal subsystem* $\mathcal{O}$ *of* $\mathcal{F}_z$ *with Sylow group* O *such that* $T\mathcal{O} \trianglelefteq \mathcal{F}_z$,

$$\Sigma = O^2(\bigcap_{i=1}^{4} N_{Aut_{\mathcal{F}}(O)}(K_i)) = O^2(Aut_{\mathcal{O}}(O)),$$

and one of the following holds:
(1) $\mathcal{O} \cong SL_2[m]^4$.
(2) $\mathcal{O} = O$.
(3) $m = 8$ *and* $\mathcal{O}$ *is of index* 3^a *in* $SL_2[m]^4$ *for some* $1 \leq a \leq 3$.

PROOF. Suppose first that $m > 8$ and set $\mathcal{O} = \mathcal{O}(z)$. From 10.1.8.5, either (1) or (2) holds. Further $Aut(O)$ is a 2-group, so $O^2(Aut_{\mathcal{F}}(O)) = O^2(Aut(O)) = 1$. Thus it remains to show $T\mathcal{O} \trianglelefteq \mathcal{F}_z$. Applying Theorem 1.5.2 in [**Asc19**] to $\mathcal{O}$ in the role of "$\mathcal{F}_0$" and T in the role of "E", it suffices to show $TN_{\mathcal{O}}(O) \trianglelefteq N_{\mathcal{F}}(O)$. As $m > 8$, $N_{\mathcal{O}}(O) = O$ and $E \trianglelefteq N_{\mathcal{F}}(O)$, completing the proof in this case.

So we may assume $m = 8$. Let $S_0 = C_S(O)$, $R = OS_0$, G_0 a model for $N_{\mathcal{F}}(R)$, $H = \langle T^{G_0} \rangle$, $B = C_{G_0}(O)$, and $G_0^* = G_0/B$. Thus $G^* = Aut_{\mathcal{F}}(O)$. By 10.1.8.4, $T \trianglelefteq SB$, so $[T, B] \leq C_T(O) = \langle z \rangle$, and hence $L = O^2(H)$ centralizes B. Let G_1 be the kernel of the action of G_0 on Ω and G_2 the preimage in G of $O_2(Sym(\Omega))$. Then $G_1T = G_2 \trianglelefteq G_0$ by 10.1.3.4. By a Frattini argument, $G_0 = G_2N_{G_0}(TS_0)$. Further as $E = Z_2(TS_0) \cap O$, $N_{G_0}(TS_0) \leq N_{G_0}(E)$, so $T \trianglelefteq N_{G_0}(TS_0)$ by 10.1.8.3. Therefore $T^{G_0} = T^{G_2} = T^{G_1}$, so $H = \langle T^{G_0} \rangle = \langle T^{G_1} \rangle \leq G_2$. Then as L centralizes B, L is of index 3^a in $SL_2(3)^4$ for some $0 \leq a \leq 4$. Further $G_1 \leq SBL$ so $T^{G_1} = T^{SBL} = T^L$, so $H = \langle T^{G_1} \rangle = \langle T^L \rangle = TL$. Let $w \in W - O$; by 10.1.3.5, w induces an outer automorphism on K_i for $1 \leq i \leq 4$. Therefore $Aut_L(O) = \Sigma$.

Set $\mathcal{O} = \mathcal{F}_O(L)$, $\mathcal{F}_1 = SC_{\mathcal{F}}(O)$, $\mathcal{F}_2 = N_{\mathcal{F}_z}(S_0)$, $\mathcal{E}_1 = T$, and $\mathcal{E}_2 = T\mathcal{O}$. Observe that this 4-tuple satisfies Hypothesis 1.4.7 in [**Asc19**]. For example $\mathcal{F}_z = \langle \mathcal{F}_1, \mathcal{F}_2 \rangle$

by 1.5.2 in [**Asc19**], $\mathcal{E}_1 \trianglelefteq \mathcal{F}_1$ by 10.1.8.4, and $\mathcal{E}_2 \trianglelefteq \mathcal{F}_2$ as $TL \trianglelefteq G$. Therefore $T\mathcal{O} = \langle \mathcal{E}_1, \mathcal{E}_2 \rangle \trianglelefteq \mathcal{F}_z$ by 1.4.8 in [**Asc19**], completing the proof. □

NOTATION 10.1.10. Write $\mathcal{O}$ for the normal subsystem of $\mathcal{F}_z$ appearing in 10.1.9. If $m = 8$ then $T\mathcal{O}$ is constrained, so it possess a model H_z, with $L_z = O^2(H_z)$ a model for $\mathcal{O}$. Moreover $L_z = OX_z$ where $O \cong Q_8^4$ and X_z is an elementary abelian 3-group; write $e(\mathcal{O})$ for $m_3(X_z)$.

LEMMA 10.1.11. *(1)* $\mathcal{F}_z = \langle S\mathcal{O}, N_{\mathcal{F}_z}(E) \rangle$.
(2) $\mathcal{D}^*(z) = v^{\mathcal{O}}$.

PROOF. By 10.1.9, $T\mathcal{O} \trianglelefteq \mathcal{F}_z$, so by 1.3.2 in [**Asc19**], $\mathcal{F}_z = \langle S\mathcal{O}, N_{\mathcal{F}_z}(T) \rangle$. As W is weakly closed in S and $W \leq T$, $N_{\mathcal{F}_z}(T) \leq N_{\mathcal{F}_z}(W)$, and then 10.1.8.2 completes the proof of (1).

By 10.1.4.3, $\mathcal{D}^*(z) = v^{\mathcal{F}_z}$. Further if $\alpha \in \mathrm{hom}_{\mathcal{F}_z}(v, O)$ then as $T\mathcal{O} \trianglelefteq \mathcal{F}_z$, $\alpha = \varphi\phi$ for some $\varphi \in Aut_{\mathcal{F}_z}(T)$ and $\phi \in \mathrm{hom}_{T\mathcal{O}}(v\varphi, T)$. As $Aut_{\mathcal{F}}(T)$ acts on E, $v\alpha = v\phi$ or $v\phi z$. Similarly as $\mathcal{O} \trianglelefteq T\mathcal{O}$, $v\phi = v\psi$ for some $\psi \in \mathrm{hom}_{\mathcal{O}}(v, O)$, so (2) holds. □

LEMMA 10.1.12. *Let $t \in \mathcal{Z}$ and $\alpha \in \mathfrak{A}(t)$. Assume that $\mathcal{D}^*(z\alpha) \cap \mathcal{D}^*(z) \neq \emptyset$. Then $t \in T$.*

PROOF. Assume $x \in \mathcal{D}^*(z) \cap \mathcal{D}^*(z\alpha)$, and let $\gamma \in \mathfrak{A}(z\alpha)$ with $z\xi = z$, where $\xi = \alpha\gamma$. Then ξ is a $\mathcal{F}_z$-map and $x\gamma \in \mathcal{D}^*(z\gamma) \cap \mathcal{D}^*(z\xi) = \mathcal{D}^*(t\xi) \cap \mathcal{D}^*(z)$, so by 10.1.4.3 there is $\delta \in \mathfrak{A}_{\mathcal{F}_z}(x\gamma)$ with $x\gamma\delta = v$. Then $t\xi\delta \in \mathcal{D}^*(v) \subseteq T$ by 10.1.4.5, so as $\xi\delta$ is a $\mathcal{F}_z$-map, and as T is strongly closed in S with respect to $\mathcal{F}_z$ by 10.1.9. we conclude that $t \in T$. □

LEMMA 10.1.13. *If $t \in \mathcal{D}^o(z)$ then $t \in \mathcal{D}^*(u)$ for some $u \in \{v, vz\}$.*

PROOF. Assume otherwise; conjugating in $\mathcal{F}_z$, we may assume $t \in \mathcal{F}_z^f$. Let $\alpha \in \mathfrak{A}(t)$ and $S_t = C_S(t)$.

If $t \in T$ then the lemma follows from 10.1.7, so $t \notin T$. Hence by 10.1.12, $\mathcal{D}^*(z) \cap \mathcal{D}^*(z\alpha) = \emptyset$, so $z\alpha \notin T$ by 10.1.7.2, and neither $v\alpha$ nor $(vz)\alpha$ is in $\mathcal{D}^*(z)$. Therefore $z\alpha \in \mathcal{D}^o(z)$ by 10.1.6.2.

As $s = z\alpha \in \mathcal{D}^o(z)$, we may assume s acts on K_1. If $u = v\alpha$ acts on K_1 then as neither u nor us is in $\mathcal{D}^*(z)$, both invert v_1, so $s = u(us)$ centralizes v_1, a contradiction. Thus we may take (K_1, K_2) to be a cycle of u. Then (K_3, K_4) is a cycle in u or su, and we may assume the former. Hence $u \in \mathcal{D}^m(z)$, so there is $y \in \{v, vz\}$ with $y \in \mathcal{D}^*(u)$ by 10.1.6.1.

As $t \in \mathcal{F}_z^f$ it follows from 2.2 in [**Asc10**] that $s \in \mathcal{F}_z^f$ and $S_t\alpha = C_S(s)$. By 10.1.3.7, $y \in S_t\alpha$. Let $\beta \in \mathfrak{A}(s)$ and $\xi = \alpha\beta$. Then $\xi \in \mathrm{hom}_{\mathcal{F}_z}(S_t, S)$ with $t\xi \in \mathcal{F}_z^f$. As $v\xi \in \mathcal{D}^*(z)$, by 10.1.4.3 there is $\gamma \in \mathfrak{A}_{\mathcal{F}_z}(v\xi)$ with $v\xi\gamma = v$. Set $\zeta = \xi\gamma$, $\bar{t} = t\zeta$, and $\bar{y} = y\beta\gamma$.

Let $g \in M$ with $v^g = z$. As $u \in \mathcal{D}^m(z)$ and $y \in E$, u centralizes y by 10.1.3.7, so $\bar{y} = y\beta\gamma$ centralizes $u\beta\gamma = v$, and hence $\bar{y} \in C_S(E) \leq O_2(MS)$. Similarly u centralizes z so $\bar{t} = z\beta\gamma$ centralizes $u\beta\gamma = v$, so $\bar{t} \in O_2(MS)$. As $y \in \mathcal{D}^*(z)$ we have $\bar{y} \in \mathcal{D}^*(\bar{t})$, and hence $\bar{y}^g \in \mathcal{D}^*(\bar{t}^g)$. As $y \in \mathcal{D}^*(u)$, we have $\bar{y} \in \mathcal{D}^*(v)$, so $\bar{y}^g \in \mathcal{D}^*(z)$, and then by 10.1.4.3 there is $\delta \in \mathfrak{A}_{\mathcal{F}_z}(\bar{y}^g)$ with $\bar{y}^g\delta = v$. Then as $\bar{y}^g \in \mathcal{D}^*(\bar{t}^g)$, we have $\bar{t}^g\delta \in T$ by 10.1.4.5, while by 10.1.9, T is strongly closed in S with respect to $\mathcal{F}_z$, so $\bar{t}^g \in T$. Then as T is Sylow in M, $\bar{t} \in T$. As $\bar{t} = t\zeta$ and ζ is an $\mathcal{F}_z$-map, also $t \in T$, contrary to an earlier reduction. This completes the proof. □

LEMMA 10.1.14. *(1)* $\mathcal{Z} \subseteq T$.
(2) We have

$$\mathcal{Z} = \bigcup_{e \in E^{\#}} \mathcal{D}^*(e).$$

(3) $\mathcal{D}^o(z) = \mathcal{D}^a(z)$.

PROOF. Let $t \in \mathcal{Z} - O$. Then $t \in \mathcal{D}^m(z) \cup \mathcal{D}^o(z)$, so $t \in T$ by 10.1.6.2, 10.1.13, and 10.1.4.5. Thus (1) holds. Then (1) and 10.1.7.2 imply (2), while 10.1.13 implies (3). □

LEMMA 10.1.15. *Let* $\mathcal{E} = \mathcal{E}(\tau)$.
(1) For each $s \in \mathcal{Z}$, *there exists* $\alpha \in \hom_{\mathcal{E}}(C_S(s), S)$ *with* $s\alpha = z$.
(2) $\mathcal{F} = \mathcal{E}(\tau) = \langle \mathcal{F}_z, \mathcal{F}_S(SM) \rangle$.

PROOF. We first prove (1). Let $s \in \mathcal{Z}$ and $\mathfrak{B}(s) = \{\alpha \in \hom_{\mathcal{E}}(C_S(s), S) : s\alpha = z\}$. Let $g \in M$ with $v^g = z$. Then $C_S(v) = C_S(E) \trianglelefteq MS$, so $\beta = c_g \in \mathfrak{B}(v)$. Next let $s \in \mathcal{D}^*(z)$. By 10.1.4.3 there is $\gamma \in \mathfrak{A}_{\mathcal{F}_z}(s)$ with $s\gamma = v$. Then $\gamma\beta \in \mathfrak{B}(s)$. Finally let $s \in \mathcal{D}^*(v) - O$. Claim $C_S(s) \leq C_S(E)$. For if not, $\langle C_S(s), C_{O^{g^{-1}}}(s) \rangle$ induces $GL(E)$ on E, so $s \in O$, contrary to the choice of s. By the claim, $C_S(s)\beta \leq S$ with $s\beta \in \mathcal{D}^*(z)$. By the previous case there is $\delta \in \mathfrak{B}(s\beta)$. Then $\beta\delta \in \mathfrak{B}(s)$. Now 10.1.14.2 completes the proof of (1).

Next by definition of $\mathcal{E}(\tau)$ in 3.4.2, $\mathcal{E} = \langle S\mathcal{O}(z), N_{\mathcal{F}}(O), N_{\mathcal{F}}(W) \rangle$. Observe that $\mathcal{F}_z \geq \langle S\mathcal{O}(z), N_{\mathcal{F}}(O) \rangle$ and by 10.1.8.2, $\mathcal{F}_T(M) \trianglelefteq N_{\mathcal{F}}(W)$, so by 1.3.2 in [**Asc19**], $N_{\mathcal{F}}(W) = \langle \mathcal{F}_S(SM), N_{\mathcal{F}}(T) \rangle$. Finally as $\langle z \rangle = Z(T)$, $N_{\mathcal{F}}(T) \leq \mathcal{F}_z$, so $\mathcal{E} = \langle \mathcal{F}_z, \mathcal{F}_S(SM) \rangle$, establishing the second equality in (2).

Also $\mathcal{F} = \mathcal{E}$ by (1) and 2.1.8, completing the proof of (2). □

LEMMA 10.1.16. *Assume* $t_0 \in \mathcal{D}^*(E)$ *and set* $E_0 = E\langle t_0 \rangle$, $\Omega_0 = \{N_{K_i}(E_0) : 1 \leq i \leq 4\}$, *and* $\tau_0 = (N_{\mathcal{F}}(E_0), \Omega_0)$. *Then* τ_0 *is a quaternion fusion packet and* $N_{\mathcal{F}}(E_0)^\circ \cong L_3(2)/2^{3+6}$.

PROOF. Observe that $E_0 \cong E_8$. From 10.1.7.3, $t_0 = k_1 \cdots k_4$ where $k_i \in K_i - W$. It follows that $N_{K_i}(E_0) \cong Q_8$, so $N_O(E_0) \cong Q_8^4$. Then τ_0 is a quaternion fusion packet by 2.5.2. Next $Aut_O(E_0)$ is the group of transvections with center z, so $t_0 z \in \mathcal{Z}$. Further $E_0 \leq Q = \bigcap_{g \in M} O^g$, so for each $e \in E^{\#}$, $E_0 - \langle e \rangle \subseteq \mathcal{D}^*(e)$. Then for each distinct $a, b \in E_0 - E$, there is $e = ab \in \langle a, b \rangle \cap E^{\#}$, so as $e \in \mathcal{D}^*(a)$, also $b = ae \in \mathcal{D}^*(a)$. Hence for each $a \in E_0^{\#}$, $E_0 - \langle a \rangle \subseteq \mathcal{D}^*(a)$. Thus for $\alpha \in \mathfrak{A}(a)$, $Q_a = N_O(E_0\alpha) \cong Q_8^4$ and $Aut_{Q_a}(E_0\alpha)$ is the group of transvections on $E_0\alpha$ with center z. Hence $A_a = Aut_{Q_a}(E_0\alpha)\alpha^{-*}$ is the subgroup of $A = Aut_{\mathcal{F}}(E_0)$ of all transvections with center a. Therefore $GL(E_0) = \langle A_a : a \in E_0^{\#} \rangle$. As $\Omega_0 = \Omega_0(z)$ is of order 4, it follows from Theorem 2 and 5.10.7.1 that $N_{\mathcal{F}}(E_0)^\circ \cong L_3(2)/2^{3+6}$. □

LEMMA 10.1.17. *If* $\mathcal{O} = O$ *then* $\mathcal{F}^\circ = \mathcal{F}_T(M)$.

PROOF. Assume $\mathcal{O} = O$. Then $\mathcal{O}T = T$ is normal in $\mathcal{F}_z$ by 10.1.9, so $W \trianglelefteq \mathcal{F}_z$. Then by 10.1.15.2, $W \trianglelefteq \langle \mathcal{F}_z, \mathcal{F}_S(SM) \rangle = \mathcal{F}$, so $\mathcal{F}_T(M) \trianglelefteq \mathcal{F}$ by 10.1.8.2, the lemma follows. □

LEMMA 10.1.18. *If* $m = 8$ *and* $e(\mathcal{O}) = 1$ *then* $\mathcal{F}^\circ \cong L_3(2)/2^{3+6}$.

PROOF. Recall Notation 10.1.10, and assume $m = 8$ and $e(\mathcal{O}) = 1$, and set $\tilde{H}_z = H_z/\langle z \rangle$. Then $X = X_z$ is of order 3, and as T is transitive on Ω, $C_{\tilde{O}}(X) = 1$.

Therefore $\tilde{E}_0 = \langle \tilde{v}^X \rangle \cong E_4$; let E_0 be the preimage of $\tilde{E}_0$ in O and $t \in E_0 - E$. By 10.1.11.2, $E_0 = \langle v^{\mathcal{O}} \rangle = \langle v^{\mathcal{F}_z} \rangle$, so $E_0 \trianglelefteq \mathcal{F}_z$.

Suppose $t \in \mathcal{D}^*(E)$ and let $\mathcal{F}_0 = N_{\mathcal{F}}(E_0)^\circ$; by 10.1.16, $\mathcal{F}_0 \cong L_3(2)/2^{3+6}$, so $\mathcal{F}_0$ has a model G_0, and $M_0 = N_{G_0}(E) \cong M$, so $E_0 \trianglelefteq M$. Thus $E_0 \trianglelefteq \langle \mathcal{F}_z, \mathcal{F}_S(SM) \rangle = \mathcal{F}$ by 10.1.15.2, so $\mathcal{F}_0 = \mathcal{F}^\circ$. Thus the lemma holds in this case, so we may assume $t \notin \mathcal{D}^*(E)$ and it remains to derive a contradiction.

As $[\tilde{K}_i, X] = \tilde{K}_i$ for each i, $t = k_1 \cdots k_4$ with $k_i \in K_i - W$ for each i. Then from the proof of 10.1.7, t inverts W, so t acts on each member of K^G. Therefore as $t \notin \mathcal{D}^*(E)$, $t \notin \mathcal{D}^*(e)$ for some $e \in E^\#$, so by 10.1.7.1, $s = t^g \in \mathcal{D}^a(z)$ for some $g \in M - O$. Thus $U = \langle v_1, \ldots, v_4 \rangle = [O, s]$ is inverted by s and $\Omega_1(W) = C_O(s)$ is of order 16, so as $|U| = |O|/16$ it follows that $sU = s^O \subseteq \mathcal{Z}$. Then $s\Omega_1(W) \subseteq O^g \cap \mathcal{Z}$ by 10.1.5.5, a contradiction as $O \cap \mathcal{Z} = E_0^\#$ is of order 7 and $|\Omega_1(W)| = 16$. This contradiction completes the proof. □

LEMMA 10.1.19. *If $m = 8$ and $e(\mathcal{O}) = 2$ then $\mathcal{F}^\circ \cong AE_8$.*

PROOF. Assume $m = 8$ and $e(\mathcal{O}) = 2$, and set $H = H_z$ and $\tilde{H} = H/\langle z \rangle$. Then $X = X_z \cong E_9$, and by 10.1.3.4, T is transitive on Ω with WO the kernel of the action and $T/OW \cong E_4$. Therefore $N_T(X)$ is transitive on Ω, so $C_{\tilde{O}}(X) = 1$ and $\tilde{O}$ is self centralizing in $\tilde{T}$. We see below that $T/O \cong E_8$ and T splits over O. Then $N_{\tilde{H}}(\tilde{X}) = \langle \tilde{s} \rangle \times \tilde{X}\langle \tilde{t}, \tilde{r} \rangle$, where $\tilde{A} = \langle \tilde{s}, \tilde{t}, \tilde{r} \rangle \cong E_8$ and $\tilde{s}$ is fixed point free on $\tilde{\Omega}$. Let $\mathcal{Y}$ be the set of hyperplanes Y of X such that $C_{\tilde{O}}(Y) \neq 1$. Then

$$\tilde{O} = \bigoplus_{Y \in \mathcal{Y}} C_{\tilde{O}}(Y),$$

so as T is transitive on Ω it follows that we may chose notation so that $X = X_1 \times X_2$ with $[O, X_1] = K_1K_2 = C_O(X_2)$ and $[O, X_2] = K_3K_4 = C_O(X_1)$, $\tilde{r}$ inverts $\tilde{X}$, $X_1^t = X_2$, $K_1^s = K_2$, and r acts on K_1. As $N_{\tilde{T}}(\tilde{X})$ is abelian and transitive on Ω, r acts on each member of Ω. Then for each $\tilde{a} \in \tilde{A}^\#$, $[\tilde{O}, \tilde{a}] = C_{\tilde{O}}(\tilde{a})$, so $\tilde{O}$ is transitive in the involutions in $\tilde{a}[\tilde{O}, \tilde{a}]$, so in particular if $a[O, a]$ contains an involution then a is an involution. On the other hand recall that T splits over O, so indeed $a[O, a]$ contains an involution. In particular we may choose s to be an involution, so $U_0 = [O, s] = C_O(s) \cong E_{32}$ and then $U = U_0\langle s \rangle \cong E_{64}$. As $\tilde{s} \in Z(N_{\tilde{H}}(\tilde{X}))$, $H = ON_H(X)$ acts on U. Indeed $U_0 = \langle v^H \rangle = \langle D^*(z) \rangle$ by 10.1.11.2 and $U = C_T(U_0)$, so $U \trianglelefteq \mathcal{F}_z$.

We see in a moment that T has four E_{64}-subgroups, and all are normal in M. Therefore $U \trianglelefteq SM$. Then by 10.1.15.2, $U \trianglelefteq \langle \mathcal{F}_z, \mathcal{F}_S(SM) \rangle = \mathcal{F}$. Now as $U \cong E_{64}$ and $\mu \cong$ Weyl(D_4), we conclude from Theorem 2 that $\mathcal{F}^\circ \cong AE_8$.

It remains to check our two claims; to do so we observe that $\bar{G} = AE_8$ satisfies our hypothesis, so as M is determined up to isomorphism by 4.3.8 and 10.1.3.1, we may take $M = N_{\bar{G}}(E)$. Now $\bar{U} = O_2(\bar{G})$ is the natural module for the complement $\bar{Y} = Alt(I)$, where $I = \{1, \ldots, 8\}$ and $M = N_{\bar{Y}}(E)\bar{U}$, where $N_{\bar{Y}}(E)$ is the stabilizer of the partition $\Lambda = \{\{i, i+1\} : i \in \{1, 3, 5, 7\}\}$ of I. We may take $z = e_{1,2,3,4}$ and then

$$O \cap \bar{U} = \langle e_{i,j} : i, j \in \{1, 2, 3, 4\} \text{ or } i, j \in \{5, 6, 7, 8\} \rangle$$

is a hyperplane of U and $O = (O \cap U)\langle (1,2)(3,4), (1,3)(2,4), ((5,6)(7,8), (5,7)(6,8) \rangle$. Then $\langle \bar{s}, \bar{t}, \bar{r} \rangle$ is a complement to O in T isomorphic to E_8, where $\bar{s} = e_{4,8}$, $\bar{t} = (1,5)(2,6)(3,7)(4,8)$, and $\bar{r} = (1,2)(5,6)$.

Next the E_{64} subgroups of T are U, together with the subgroups $U(A) = AC_U(A)$ for $A \leq T_0 = T \cap \bar{Y}$ such that A is an FF-offender on U, in the sense of section B of [**AS04**]. By B.3.2.6 in [**AS04**] there are exactly three choices for A: $A(1) = \langle (1,2)(3,4), (1,2)(5,6), (1,2)(7,8) \rangle$ and the two E_8-subgroups $A(2)$ and $A(3)$ of T_0 which are regular on I. Regard $\bar{Y}$ as $L_4(2)$ with Borel subgroup T_0. Now $N_{\bar{Y}}(A(1)) = N_{\bar{Y}}(E)$ is the middle node parabolic, while $N_{\bar{Y}}(A(i))$, $i = 2, 3$ are the maximal parabolics with Levi factor $L_3(2)$. As all three parabolics contains the middle node parabolic, $U(A(i))$ is normal in M for each $1 \leq i \leq 3$, and of course $\bar{U}$ is normal in M. This completes the proof of the lemma. □

LEMMA 10.1.20. *(1) If* $m = 8$ *and* $e(\mathcal{O}) = 3$ *then* $\mathcal{F}^\circ \cong \Omega_8^+(2)$.
(2) If $\mathcal{O} \cong SL_2[m]^4$ *then* $\mathcal{F}^\circ \cong P\Omega_8^+[m]$.

PROOF. Under the hypothesis of (1), let $\tilde{G}$ be the group $\Omega_8^+(2)$, $\tilde{S} \in Syl_2(\tilde{G})$, and $\tilde{\tau} = (\tilde{\mathcal{F}}, \tilde{\Omega})$ the quaternion fusion packet of $\tilde{G}$ inherited from the embedding of $\tilde{G}$ in $P\Omega_8^+(3)$. On the other hand under the hypothesis of (2) let $\tilde{\mathcal{F}} = P\Omega_8^+[m]$ and $\tilde{\tau} = (\tilde{\mathcal{F}}, \tilde{\Omega})$ its Lie packet. In either case as M is determined up to isomorphism from the proof 4.3.8 and 10.1.3.1, and as $\tilde{\tau}$ satisfies the corresponding hypothesis on τ with $|\tilde{S}| = |T|$ and $\mathcal{F}_{\tilde{S}}(\tilde{M}) = N_{\tilde{\mathcal{F}}}(\tilde{E})$, we may take $\tilde{S} = T$, $\tilde{M} = M$, and $\tilde{\Omega} = \Omega$.

Set $\mathcal{Y} = \langle T\mathcal{O}, \mathcal{F}_T(M) \rangle \leq \mathcal{F}$. We first claim that $\mathcal{Y} = \tilde{\mathcal{F}}$. By 10.1.15.3 applied to $\tilde{\mathcal{F}}$, $\tilde{\mathcal{F}} = \langle T\tilde{\mathcal{O}}, \mathcal{F}_T(M) \rangle$, so it suffices to show that $T\tilde{\mathcal{O}} = T\mathcal{O}$.

Suppose the claim holds. The second step in the proof is to set $\mathcal{F}_1 = \mathcal{F}_z$, $\mathcal{F}_2 = \mathcal{F}_S(SM)$, $\mathcal{E}_1 = T\mathcal{O}$, and $\mathcal{E}_2 = \mathcal{F}_T(M)$. We verify that this tuple satisfies Hypothesis 1.4.7 in [**Asc19**] and then appeal to 1.4.8 in [**Asc19**] to conclude that $\mathcal{Y} \lhd \mathcal{F}$. Then as $\mathcal{F}^\circ = \mathcal{Y}$, proving the lemma.

The verification of Hypothesis 1.4.7 in [**Asc19**] is not difficult. By 10.1.15.2, $\mathcal{F} = \langle \mathcal{F}_1, \mathcal{F}_2 \rangle$. By 10.1.9, $\mathcal{E}_1 \lhd \mathcal{F}_1$ and by 10.1.8.2, $\mathcal{E}_2 \lhd \mathcal{F}_2$. As $\mathcal{Y} \cong \tilde{\mathcal{F}}$, $\mathcal{Y}$ is saturated and $\mathcal{Y} = O^{2'}(\mathcal{Y})$. Thus Hypothesis 1.4.7 in [**Asc19**] is satisfied.

It remains to show $T\tilde{\mathcal{O}} = T\mathcal{O}$, which we do separately in each of the two cases. Assume first that $m = 8$ and $e(\mathcal{O}) = 3$, and set $H = H_z$ and $\tilde{H} = H/\langle z \rangle$. Then $X = X_z \cong E_{27}$. Now $\tilde{H}$ is contained in the stabilizer A in $Out(O)$ of $\tilde{\Omega}$ and $A \cong S_4$ wr S_4; that is $A = (A_1 \times \cdots \times A_4)B$ where $B = Sym(I)$ permutes $\{A_i : i \in I\}$ via $(a_1 \cdots a_4)^b = a_{1b} \cdots a_{4b}$ for $b \in B$, $I = \{1, \ldots, 4\}$, and $A_i \cong S_4$. Set $A^+ = A/O_2(A)$, so that $A^+ \leq Out(O)$ and $A^+ \cong S_3$ wr S_4.

We saw during the proof of 10.1.19 that $T/O \cong E_8$ and T splits over O. Then $E_8 \cong T^+ \leq A^+$ and $E_{27} \cong X^+ \leq O(A^+) \cong E_{81}$. Further the kernel T_0^+ of the action of T^+ on Ω, and hence also on the A_i^+, is of order 2, with T^+/T_0^+ regular on the A_i^+. It follows that T^+ has four 1-dimensional weight spaces on $O(A^+)$. Moreover $N_{A^+}(T^+)$ is transitive on these weight spaces, so the four T^+-invariant hyperplanes of $O(A^+)$ are conjugate under $N_{A^+}(T^+)$. Hence, conjugating in $Aut(H)$, we may indeed choose the model for $T\tilde{\mathcal{O}}$ to be H, so that $T\tilde{\mathcal{O}} = T\mathcal{O}$. (cf. 1.6.7 in [**Asc19**]) This completes the proof of (1).

Next assume $\mathcal{O} \cong SL_2[m]^4$; that is $\mathcal{O} = \mathcal{O}_1 * \cdots * \mathcal{O}_4$ is a central product where $\mathcal{O}_i \cong SL_2[m]$ with K_i Sylow in $\mathcal{O}_i$. Hence $T\mathcal{O} = T\tilde{\mathcal{O}}$ by 2.7.3, completing the proof of (2) and the lemma. □

THEOREM 10.1.21. *Assume Hypothesis* 10.1.1. *Then one of the following holds:*
(1) $\mathcal{O} = O$ *and* $\mathcal{F}^\circ = \mathcal{F}_T(M)$.

(2) $m = 8$, $e(\mathcal{O}) = 1$, *and* $\mathcal{F}^\circ \cong L_3(2)/2^{3+6}$.
(3) $m = 8$, $e(\mathcal{O}) = 2$, *and* $\mathcal{F}^\circ \cong AE_8$.
(4) $m = 8$, $e(\mathcal{O}) = 3$, *and* $\mathcal{F}^\circ \cong \Omega_8^+(2)$.
(5) $\mathcal{O} \cong SL_2[m]^4$ *and* $\mathcal{F}^\circ \cong P\Omega_8^+[m]$.

PROOF. Recall the normal subsystem $\mathcal{O}$ of $\mathcal{F}_z$ is defined in 10.1.9, and by 10.1.9 one of the three cases in that lemma holds. If $\mathcal{O} = O$ then (1) holds by 10.1.17, while if $\mathcal{O} \cong SL_2[m]^4$ then (5) holds by 10.1.20.2. Therefore by 10.1.9 we may assume that $m = 8$ and, as in Notation 10.1.10, $e(\mathcal{O}) = m_3(X_z)$ is 1, 2, or 3. Then 10.1.18, 10.1.19, and 10.1.20.1 say that (2), (3), or (4) holds, completing the proof of the theorem. □

10.2. $|\Omega(z)|$ large

In this section we assume the following hypothesis:

HYPOTHESIS 10.2.1. $\tau = (\mathcal{F}, \Omega)$ *is a quaternion fusion packet such that*
(1) For $z \in \mathcal{Z}_S$, $|\Omega(z)| > 2$.
(2) $\mathcal{F}^\circ$ *is transitive on* Ω.
(3) The Inductive Hypothesis holds.

NOTATION 10.2.2. Let $z \in \mathcal{Z}_S$ with $z \in \mathcal{F}^f$ and set $\{K_i : 1 \leq i \leq k\} = \Omega(z)$. Set $K = K_1$, let $\eta \in \eta(\tau)$, $W = \langle \eta \rangle$, G a model for $N_{\mathcal{F}}(C_S(W))$, $M = \langle K^G \rangle$, and $\mu = \mu(\tau)$. Set $O = O(z)$ and $\mathcal{Z}_S = \{z_1, \ldots, z_r\}$ with $z = z_1$.

LEMMA 10.2.3. *Assume* $\rho = (\mathcal{Y}, \Gamma)$ *is a proper subpacket of* τ *such that* $\Gamma(z) = \Omega(z) \cap \Gamma = \{K_1, \ldots, K_{k_\rho}\}$ *with* $k_\rho > 2$. *Then*
(1) Theorem 4 holds in ρ.
(2) Suppose $\xi = (\mathcal{C}, \Sigma)$ *is a coconnected component of* ρ° *such that* $K \in \Sigma$. *Let* $\theta \in \eta(\xi)$ *and* $U = \langle \theta \rangle$. *Then either*
(a) $\xi = (\mathcal{O}_K, K)$, *and* $(\mathcal{O}_{K_i}, K_i)$ *is a coconnected component of* ρ° *for each* $1 \leq i \leq k_\rho$, *or*
(b) $k_\rho = 4$, $\Sigma = \Gamma(z)$, $\mu(\xi) \cong$ *Weyl*(D_4), *and either* $U \trianglelefteq \mathcal{C}$ *and* $Z(\mathcal{C}) = 1$ *or* $\mathcal{C} \cong L_3(2)/2^{3+6}$, AE_8, $\Omega_8^+(2)$, *or* $P\Omega_8^+[m]$.

PROOF. Part (1) follows from 10.2.1.3 and 6.6.8.

Assume the setup of (2), and suppose first that $K_i \notin \Sigma$ for some i. Then there is a coconnected component $\xi' = (\mathcal{C}', \Sigma')$ of ρ such that $K_i \in \Sigma'$. Then $\mathcal{C}'$ centralizes $\mathcal{C}$, so z centralizes $\mathcal{C}$, and hence $\xi = (\mathcal{O}_K, K)$. Hence (2a) holds in this case, so we may assume $\Gamma(z) \subseteq \Sigma$. Then by induction on the order of τ, $k_\rho = 4$, $\Sigma = \Gamma(z)$, and $\mu(\xi) \cong$ Weyl(D_4). Then (2b) holds by Theorem 10.1.21. □

LEMMA 10.2.4. *Set* $\tau_W = (\mathcal{F}_{S \cap M}(M), \Omega)$. *Then either* μ *is abelian and* τ_W *has* $rk = |\Omega|$ *coconnected components* $(J, \{J\})$ *for* $J \in \Omega$, *or the following hold:*
(1) $k = 4$.
(2) $M = M_1 \times \cdots \times M_r$ *with* $\tau_i = (\mathcal{F}_{S_i}(M_i), \Omega(z_i))$ *a coconnected component of* τ_W, *for* $S_i = S \cap M_i$ *and* $r = |\Omega|/4$.
(3) $\mu = \mu_1 \times \cdots \times \mu_r$ *with* $\mu_i = \mu(\tau_i) \cong$ *Weyl*(D_4).

PROOF. By 6.2.6, G is transitive on Ω. Replacing $\mathcal{F}$ by $N_{\mathcal{F}}(C_S(W))$, we may assume $\mathcal{F} = \mathcal{F}_S(G)$. By 3.2.2.1, $\mathcal{A}(\tau) = \emptyset$. Then by 3.1.25, $\mu = \hat{\mu} \times \mu_1 \times \cdots \times \mu_r$, where $\hat{\mu}$ is described in 3.1.24 and 3.1.25, and each μ_i is generated by 3-transpositions. As G is transitive on Ω and $k > 2$, we conclude from 6.1.6.2 that either $\mu = \hat{\mu}$, so that μ is abelian, or $\mu = \mu_1 \times \cdots \mu_r$ with G transitive on the μ_i, and we may assume the latter. From 4.1.4, each μ_i is generated by a conjugacy class D_i of 3-transposition, and μ_i is nonabelian. Then by 4.2.10.2, $M = M_1 \cdots M_r$ is a central product, where $\tau_i = (\mathcal{F}_{S\cap M_i}(M_i), \Omega_i)$ is a coconnected component of τ. By 10.2.3.2, $\Omega_i = \Omega(z_i)$ for some $z_i \in \mathcal{Z}_S$, $\mu_i \cong \mathrm{Weyl}(D_4)$ and $Z(M_i) = 1$. Thus $k = 4$ and as $Z(M_i) = 1$, the central product is direct. The proof is complete. □

LEMMA 10.2.5. *One of the following holds:*

(1) Either μ is abelian or $r = 1$. Further $\{\tau_J = (\mathcal{O}_J, J) : J \in \Omega\}$ is the set of coconnected components of τ_z°, or

(2) $r > 1$, $k = 4$, μ is nonabelian, and $\{\tau_{K_i} : 1 \le i \le 4\} \cup \{\tau_j = (\mathcal{C}_j, \Omega(z_j) : 2 \le j \le r\}$ is the set of coconnected components of τ_z°, with $\mu(\tau_j) \cong$ Weyl(D_4) and $\mathcal{F}_{S_j}(M_j) \le \mathcal{C}_j$ for $1 < j \le r$.

PROOF. See the proof of 9.3.10. We apply 10.2.3 to τ_z, z_j in the role of ρ, z. By 10.2.3.1, τ_z° has coconnected components $\xi_l = (\mathcal{F}_l, \Gamma_l)$, $1 \le l \le a$. We may take $\mathcal{F}_i = \mathcal{O}_{K_i}$ and $\Gamma_i = \{K_i\}$ for $1 \le i \le k$. By 10.2.3.2, for $l > k$, either $\xi_l = (\mathcal{O}_J, J)$ for some $J \in \Omega - \Omega(z)$, or $k_{\xi_l} = 4 = k$, $\Gamma_l = \Omega(z_{i(l)})$ for some $i(j) > 1$, and $\mu(\xi_l) \cong \mathrm{Weyl}(D_4)$.

If $r = 1$ then $a = k$ and (1) holds, so we may assume $r > 1$.

Suppose μ is nonabelian; then by 10.2.4, $k = 4$ and M and μ are described in 10.2.4. Then, arguing as in the proof of 9.3.10, for each $i > 1$, $\mathcal{F}_{S_i}(M_i) \le \mathcal{C}_{j(i)}$ for some $j = j(i) > k$ with $\mathrm{Weyl}(D_4) \cong \mu_i \le \mu(\xi_j)$, so as $\mu(\xi_j) \cong \mathrm{Weyl}(D_4)$, we conclude that $\mu_i = \mu(\xi_j)$. Therefore the map $i \mapsto i(j)$ is a bijection of $\{2, \ldots, r\}$ with $\{5, \ldots, a\}$, so (2) holds in this case.

So assume that μ is abelian. Then as $\mu(\tau_z) \le \mu$, $\mu(\tau_z)$ is abelian, so that (1) holds in this case. □

LEMMA 10.2.6. *Let $u \in O - \langle z \rangle$ be an involution and assume either*

(a) $r > 1$ and $u \notin \mathcal{Z}$, or

(b) μ is abelian.

Then $u^{\mathcal{F}_z} \cap O \cap \mathcal{F}^f \neq \emptyset$.

PROOF. If (b) holds then the argument establishing 9.4.4.1 works, using the hypothesis that μ is abelian in place of the appeal to 9.4.1.3. So assume (a) holds and μ is nonabelian, and let $\alpha \in \mathfrak{A}(u)$ and $\Sigma = \Omega - \Omega(z)$; we can still argue as in the proof of 9.4.4.1. First $u = u_1 \cdots u_l$ with $l > 0$ even and $u_i \in K_i$ of order 4. Then $u_i\alpha \in D_i \in \Delta(z\alpha)$ centralizes $\Omega - \{K_1, \ldots, K_l\}\alpha$, $S_j\alpha$ for $j > 1$, and $v = u\alpha$, so $D_i \in \Omega(z\alpha) \cup \eta(\tau_1)\alpha$. Therefore either $\mathcal{Z}_S\alpha = \mathcal{Z}_S$ or $l = 4$. In the first case $v \notin \mathcal{Z}$, so $z\alpha$ is the unique member of $\mathcal{Z}_S$ such that v does not centralizes $\Omega(z\alpha)$, and hence $C_S(v) \le C_S(z\alpha)$. Similarly in the second case $z\alpha \in \mathcal{Z}_{\Delta(\tau_1)\alpha}$, and $v \in Q = D_1 \cdots D_4 \le W_1\alpha$, where $W_1 = \langle \eta_1 \rangle$ for $\eta_1 \in \eta(\tau_1)$. However as $v = u_1\alpha \cdots u_4\alpha$ with u_i of order 4, it follows from 10.1.3.2 that $v \in \mathcal{Z}$, contrary to hypothesis.

Therefore $C_S(v) \le C_S(z\alpha)$. Let $\beta \in \mathfrak{A}(z\alpha)$ with $z\zeta = z$, where $\zeta = \alpha\beta$. As $\mathcal{O}(z) \trianglelefteq \mathcal{F}_z$, $v\beta = v\zeta \in O$. As $C_S(v) \le C_S(z\alpha)$, $v\zeta \in \mathcal{F}^f$. □

LEMMA 10.2.7. *Let $u \in O - \langle z \rangle$ be an involution such that $u \in \mathcal{F}^f$.*

(1) For each coconnected subcomponent $\rho = (\mathcal{Y}, \Gamma)$ of τ_u° such that $\Gamma \cap \Omega(z_j) \neq \emptyset$ for some $j > 1$, ρ is a coconnected component of τ_z°.

(2) If $r > 1$ then each coconnected subcomponent of τ_u° is a component of τ_z°, and if u centralizes K_i then τ_{K_i} is a component of τ_u°.

(3) Assume $r = 1$, μ is abelian, and $C_{\Omega(z)}(u) = \{K_1, \ldots, K_l\}$ with $l > 0$. Then for each $1 \leq i \leq l$, $(\mathcal{O}_{K_i}, K_i)$ is a coconnected component of τ_u°.

PROOF. Part (1) can be proved using the argument which establishes 9.4.4.2. Here we use 10.2.3.2 and the fact that if $\mathcal{Y}$ is one of the systems listed there then S_j is Sylow in $\mathcal{Y}$ and $C_{S_j}(O(z_j)) = \langle z_j \rangle$.

Assume $r > 1$ and let $\mathcal{B}_2 = \prod_{j>1} \mathcal{C}_j$ as in 10.2.5. By (1), $\mathcal{B}_2 \lhd \mathcal{F}_u$. Let T_2 be Sylow in $\mathcal{B}_2$. As in 8.1.4.1, $T_2 \in \mathcal{F}^f$. Further for each K_i centralizing u, τ_{K_i} is a coconnected component of $C_{\mathcal{F}_u}(T_2)$, so τ_{K_i} is also a component of $\mathcal{F}_u$. Hence (2) holds.

Assume the setup of (3) and let $\rho_i = (\mathcal{Y}_i, \Gamma_i)$ be the coconnected component of τ_u containing K_i. If $l > 2$ then applying 10.2.3 to τ_u in the role of ρ, and using the assumption that μ is abelian, (3) holds. If $l = 2$ and (3) fails then as μ is abelian, $\mathcal{Y}_i \cong AE_5$, $PSp_4[m]$, or $L_4^-[m]$ by Theorem 9.4.33. In particular for T_i Sylow in $\mathcal{Y}_i$, $C_{T_i}(O_{\mathcal{Y}_i}(z)) = \langle z \rangle$. Similarly if $l = 1$ then by Theorem 7.1.7, $\mathcal{Y}_i \cong L_2[2m]^{(1)}$ or $L_3^-[m]$, so again $C_{T_i}(O_{\mathcal{Y}_i}(z)) = \langle z \rangle$. Using these facts, the proof of 9.4.4.2 shows that z centralizes $\mathcal{Y}_i$, a contradiction. □

LEMMA 10.2.8. *If $r > 1$ then μ is abelian.*

PROOF. Assume $r > 1$ and μ is nonabelian. We check that Hypotheses 8.1.12 and 8.1.14 are satisfied. Then we appeal to 8.1.15 to see that $\mathcal{F}^\circ = \mathcal{C}_1 * \cdots * \mathcal{C}_r$, contradicting $\mathcal{F} = \mathcal{F}^\circ$ transitive on Ω.

Condition 8.1.12.1 follows from 10.2.4. Condition 8.1.12.2 follows from 10.2.5. From Theorem 10.1.21 and its proof, S_j is Sylow in $\mathcal{C}_j$ and by 10.1.15.2, $\mathcal{C}_j = \langle S_j \mathcal{O}_j, \mathcal{F}_{S_j}(M_j) \rangle$. Thus it remains to show that Hypothesis 8.1.14 holds. Here $\mathcal{C}_1 = \langle S_1 \mathcal{O}_1, \mathcal{F}_{S_1}(M_1) \rangle$ is described in 10.2.3.2. Let $\eta_1 \in \eta(\tau_1)$ and $W_1 = \langle \eta_1 \rangle$. Let u be an involution in S_1. We claim that u is fused into O under M_1. First all involutions in S_1/W_1 are fused into OW_1 under M_1. Further from 10.1.3.5, $|OW_1 : O| = 2$ while (cf. 10.1.7.1) each involution in $OW_1 - W_1$ is conjugate under O to a member of O^g for $g \in M_1 - O$. Finally by 10.1.5.5, each involution in W_1 is in O, completing the proof of the claim.

By the claim and 10.2.6, we may choose $u \in O \cap \mathcal{F}^f$, and then 8.1.14 holds by 10.2.7.2. This completes the proof of the lemma. □

LEMMA 10.2.9. *$z \notin Z(\mathcal{F})$.*

PROOF. If $z \in Z(\mathcal{F})$ then by 10.2.1.2, $\mathcal{F}^\circ = \mathcal{O}(z)$, contradicting $\mathcal{F}^\circ$ transitive on Ω. □

LEMMA 10.2.10. *$r = 1$.*

PROOF. Assume $r > 1$; then μ is abelian by 10.2.8. Set $\mathcal{B}_1 = \mathcal{O}(z)$, $\Gamma_1 = \Omega(z)$, $\mathcal{B}_2 = \prod_{j>1} \mathcal{O}(z_i)$, $\Gamma_2 = \Omega - \Gamma_1$, and $T_2 = \prod_{j>1} O(z_j)$. We check that Hypothesis 8.1.2 is satisfied. Then we appeal to 8.1.6.2 to see that $\mathcal{F}^\circ = \mathcal{B}_1 * \mathcal{B}_2$, contradicting $\mathcal{F} = \mathcal{F}^\circ$ transitive on Ω.

Hypothesis 8.1.1 holds by 10.2.1.3 and 6.6.8.

Condition 8.1.2.1 follows from 10.2.5.

Suppose x is an involution in O with $x \in \mathcal{O}^f$. By 10.2.6 we may assume $x \in \mathcal{F}^f$, and then it follows from 10.2.7.2 that all coconnected components of $\mathcal{F}_x$ are of the form τ_J for some $J \in \Omega$, and then in particular that $\mathcal{B}_2 \trianglelefteq \mathcal{F}_x$. That is 8.1.2.2 holds.

Condition 8.1.2.3 follows from 10.2.9, while 8.1.2.4 follows from the transitivity of $\mathcal{F}$ on Ω and 8.1.2.5 follows from the definition of $\mathcal{B}_i$. This completes the proof of the lemma. □

LEMMA 10.2.11. *μ is nonabelian.*

PROOF. Assume μ is abelian. We begin a series of reductions.

Set $\tilde{\mathcal{F}}_z = \mathcal{F}_z/\langle z\rangle$. Write $\mathcal{U}$ for the set of involutions in $O - \langle z\rangle$. Observe that for $u \in \mathcal{U}$, $u = u_1 \cdots u_{l(u)}$ for some even integer $l(u)$, for some $u_i \in K_{\sigma(i)}$ of order 4, and for some injection $\sigma : \{1, \ldots, l(u)\} \to I = \{1, \ldots, k\}$. Moreover as $\tilde{O} = \prod_{i \in I} \tilde{K}_i$ is a direct product, σ is uniquely determined and the elements u_i are determined up to multiplication by z. Further as $\mathcal{F}_z$ permutes the subsystems $\mathcal{O}_{K_i}$, we have:

(a) For $u \in \mathcal{U}$ and $\phi \in \hom_{\mathcal{F}_z}(u, S)$, $u\phi \in O$ and $l(u\phi) = l(u)$.

(b) For each $u \in \mathcal{U}$ and $\delta \in \mathfrak{A}(u)$, $u\delta \in O$ and $\delta \in \hom_{\mathcal{F}_z}(C_S(u), S)$.

This follows from the proof of 9.4.4.1, using the fact that $r = 1$ by 10.2.10 and μ is abelian by hypothesis. As a consequence of (b):

(c) For each $u \in \mathcal{U}$, $u^{\mathcal{F}} \cap O = u^{\mathcal{F}_z}$.

(d) If $u \in \mathcal{U}$ with $l(u) < k$ then $\mathcal{F}_u \leq \mathcal{F}_z$.

For let $\delta \in \mathfrak{A}(u)$; by (b), $v = u\delta \in O$ and $\delta \in \hom_{\mathcal{F}_z}(u, S)$, and by (a), $l = l(u) = l(v)$, so we may assume $C_\Omega(v) = \{K_1, \ldots, K_a\}$, where $a = k - l > 0$. By 10.2.7.3, $\{\tau_{K_i} : 1 \leq i \leq a\}$ is the set of coconnected components of $\mathcal{F}_v$, so $\mathcal{F}_v \leq \mathcal{F}_z$. Then as δ is an $\mathcal{F}_z$-map and $\mathcal{F}_u\delta^* \leq \mathcal{F}_v \leq \mathcal{F}_z$, (d) holds.

Write $\mathcal{D}^m(z)$ for the set of $t \in \mathcal{Z}$ which are fixed point free on Ω, and $\mathcal{D}^o(z)$ for the set of $t \in \mathcal{Z} - \{z\}$ which act on some member of Ω. As $r = 1$ and μ is abelian, $\mathcal{D}^*(z) = \emptyset$, so

(e) If $t \in \mathcal{D}^o(z)$ acts on K then t induces an outer automorphism on K with $K\langle t\rangle \cong SD_{2m}$.

By 10.2.9 there is $t \in \mathcal{Z} - \{z\}$. Let $\alpha \in \mathfrak{A}(t)$.

(f) There exists $u \in C_{\mathcal{U}}(t)$ with $l(u) = 2$.

If t acts on K_1 and K_2 then by (e), t inverts $v_i \in K_i$ of order 4, so $v_1v_2 \in C_{\mathcal{U}}(t)$. If $K_1^t = K_2$ then $v_1v_1^t \in C_{\mathcal{U}}(t)$.

By (f) the set $\mathcal{U}(t)$ of u in $C_{\mathcal{U}}(t)$ with $l(u) < k$ is nonempty. Pick $u \in \mathcal{U}(t)$ and set $B = \langle t, u\rangle$, $s = z\alpha$, and $v = u\alpha$. Let $\gamma \in \mathfrak{A}(u)$ and $\beta \in \mathfrak{A}(v)$ with $v\beta = u\gamma$. Then $\xi = \alpha\beta\gamma^{-1} \in \hom_{\mathcal{F}_u}(C_S(B), S)$, so ξ is an $\mathcal{F}_z$-map by (d). Similarly γ is an $\mathcal{F}_z$-map by (b), so $\alpha\beta = \xi\gamma \in \hom_{\mathcal{F}_z}(C_S(B), S)$. In particular $z\alpha\beta = z \in Z(C_S(u\alpha)\beta)$, so $s = z\alpha \in Z(C_S(v))$. Therefore

(g) $C_S(v) \leq C_S(s)$.

(h) If v acts on K then $C_K(v) = \langle z\rangle$ and $K\langle v\rangle \cong SD_{2m}$.

For if not, $\langle z\rangle < C_K(v)$, so $\langle z\rangle < C_K(s)$ by (g), contrary to (e).

(i) v acts on each member of Ω.

For assume $K^v = K_2$. Then $X = \{x_1x_1^v : x_1 \in K\} \cong \tilde{K} \cong D_{m/2}$ centralizes v, so $X \leq C_{KK_2}(s)$ by (g). Hence s and v act on $\{K_1, K_2\}$, so some $r \in \{s, sv\}$ acts on K_1 and K_2. Further $r \in \langle s, v\rangle \leq C_S(X)$, so r centralizes $\tilde{K}\tilde{K}_2$, contrary to (e) and (h) and the fact that $r\alpha^{-1} = z$ or $zu \in \mathcal{U}(t)$.

(j) $s \in \mathcal{D}^m(z)$.

Suppose s acts on K and let V be a $N_S(K)$-invariant cyclic subgroup of K of index 2. By (e) and (h), s and v invert V, so sv centralizes V. As $zu \in \mathcal{U}(t)$, this contradicts (h).

By (i) and (j) we may take $K_1^s = K_2$ and v acts on K_1 and K_2. Hence $K_1^{sv} = K_2$, contrary to (i) and the fact that $zu \in \mathcal{U}(t)$. This contradiction completes the proof of the lemma. □

LEMMA 10.2.12. *$k = 4$ and $\mu \cong$ Weyl(D_4).*

PROOF. By 10.2.10, $r = 1$. Therefore by 10.2.4 either μ is abelian or the lemma holds, and in the former case 10.2.11 supplies a contradiction. □

THEOREM 10.2.13. *Assume Hypothesis* 10.2.1. *Then $\Omega = \Omega(z)$ is of order 4 for $z \in \mathcal{Z}_S$, $\mu(\tau) \cong$ Weyl(D_4), and taking $W \in W(\tau)$, either $W \lhd \mathcal{F}$ or $\mathcal{F}^\circ \cong L_3(2)/2^{3+6}$, AE_8, $\Omega_8^+(2)$, or $P\Omega_8^+[m]$.*

PROOF. By 10.2.10, $\Omega = \Omega(z)$, and by 10.2.12, $|\Omega(z)| = 4$ and $\mu \cong \mathrm{Weyl}(D_4)$. Therefore Hypothesis 10.1.1 is satisfied, so Theorem 10.1.21 completes the proof. □

THEOREM 10.2.14. *Assume $\tau = (\mathcal{F}, \Omega)$ is a quaternion fusion packet such that $\mathcal{F}^\circ$ is transitive on Ω and for $z \in \mathcal{Z}_S(\tau)$ we have $\Omega = \Omega(z)$ is of order at least 3. Then*

(1) $\mu(\tau) \cong$ Weyl(D_4).

(2) Either $\mathcal{F}^\circ = \mathcal{F}_{S\cap M}(M)$ for $M \in M(\tau)$ or $\mathcal{F}^\circ$ is isomorphic to $L_3(2)/2^{3+6}$, AE_8, $\Omega_8^+(2)$, or $P\Omega_8^+[m]$.

(3) $\mathcal{F} = \langle \mathcal{F}_z, \mathcal{F}_S(SM)\rangle$ for $M \in M(\tau)$.

PROOF. Assume otherwise and pick τ to be a counter example of minimal order. We show that Hypothesis 10.2.1 is satisfied and appeal to Theorem 10.2.13 and lemma 10.1.15 to obtain a contradiction. By hypothesis, conditions (1) and (2) of 10.2.1 are satisfied, so it remains to consider a proper subpacket $\rho = (\mathcal{Y}, \Gamma)$ of τ satisfying the hypothesis of Theorem 1 with $\mathcal{Y} \le \mathcal{F}$ and $\Gamma \subseteq \Omega$, and to show ρ satisfies one of the conclusions of Theorem 1.

As $\Omega = \Omega(z)$ also $\Gamma = \Gamma(z)$. Set $k = |\Gamma|$. If $k > 2$ then by minimality of τ, ρ satisfies one of the conclusions of our theorem, and hence one of the conclusions of Theorem 1. If $k = 1$ then ρ is a conclusion to Theorem 1 by Theorem 5. Finally if $k = 2$ then ρ is a conclusion to Theorem 1 by 9.5.24. This completes the proof. □

CHAPTER 11

Some results on generation

Let $\tau = (\mathcal{F}, \Omega)$ be a quaternion fusion packet with Sylow group S. In Chapter 11 we consider the case where τ satisfies Hypothesis 11.2.1. In particular $\mathcal{F}$ is transitive on Ω and for $z \in \mathcal{Z}_S(\tau)$, we have $|\Omega(z)| \leq 2$; indeed in Chapter 11, usually $|\Omega(z)| = 2$. Our objective is to prove results on the generation of $\mathcal{F}$; that is to say we wish to show that $\mathcal{F} = \mathcal{B}$, where $\mathcal{B}$ is the subsystem of $\mathcal{F}$ generated by some nice collection of subsystems of $\mathcal{F}$.

For example in section 11.3, $\mathcal{B} = \mathcal{B}(\tau)$, where $\mathcal{B}(\tau)$ is generated by $\mathcal{E}(\tau)$ and the centralizers $\mathcal{F}_x$ of fully centralized involutions x in a certain set $\mathcal{W}(\tau)$ of involutions x with $\mathcal{W}(\tau) \subseteq W \in W(\tau)$ when $|\Omega(z)| = 2$ and $\mathcal{W}(\tau) \subseteq W_S$ when $|\Omega(z)| = 1$. The major results in the chapter of this sort are Theorems 11.3.10 and 11.3.11; the heart of the proof of these theorems is the difficult analysis in section 11.2. Theorem 11.3.11 is refined in Theorem 11.5.3 in the case where $\mathcal{F}$ is an extension of the 2-fusion system of an orthogonal group. Similar refinements for all the classical groups appear later in the paper.

Section 11.6 considers generation in the case where $\mathcal{F}^\circ$ is AE_n or $\check{A}E_n$. Section 11.4 contains some results on essential subgroups, used in the proof of 11.5.3 and later results of a similar flavor. Section 11.1 begins the general study of the case $|\Omega(z)| = 2$ and $\mu \cong \mathrm{Weyl}(D_n)$ with $n \geq 4$.

11.1. $|\Omega(z)| = 2$, μ isomorphic to Weyl(D_n), $n \geq 4$

In this section we assume the following hypothesis:

HYPOTHESIS 11.1.1. *$\tau = (\mathcal{F}, \Omega)$ is a quaternion fusion packet such that*
(1) For $z \in \mathcal{Z}_S$, $|\Omega(z)| = 2$.
(2) $\mathcal{F}^\circ$ is transitive on Ω.
(3) $\mu \cong$ Weyl(D_n) for some $n \geq 4$.
(4) The Inductive Hypothesis holds in τ.

NOTATION 11.1.2. Pick $z \in \mathcal{Z}_S \cap \mathcal{F}^f$ and write $\{K, K'\}$ for $\Omega(z)$. Indeed for $J \in \Omega$, write J' for the second member of $\Omega(z(J))$.

By 11.1.1.3 and 3.3.14 there is a unique $\eta \in \eta(\tau)$. Set $W = \langle \eta \rangle$, let G_W be a model for $N_{\mathcal{F}}(C_S(W))$, $\Lambda = K^{G_W}$, $M = \langle \Lambda \rangle$, $T_M = S \cap M$, $D = D_\eta$, and $\mu = \mu(\tau)$.

Let $I = \{1, \ldots, n\}$, $W_\Delta = \langle \mathcal{Z}_\Delta \rangle$, and $n_S = |\mathcal{Z}_S|$; thus $n = 2n_S$ or $2n_S + 1$ by 4.2.8.6. By 5.8.5 there is a surjective homomorphism $\pi : \bar{M} = \bar{\omega}(D_n, m) \to M$ with $\ker(\pi) \leq C_{\bar{W}}(\bar{M})$. As $|\Omega(z)| = 2$, $\bar{u} \in \ker(\pi)$ by 5.8.5.

From 4.3.1, $\bar{M}$ is generated by elements $\bar{k}_i, \bar{v}_i$, $i \in I$, with $\bar{K}_i = \langle \bar{k}_i, \bar{v}_i \rangle$ quaternion of order m with involution $\bar{z}_i$. As in 5.8.5 we set $K_i = \bar{K}_i \pi$, $z_i = \bar{z}_i \pi$, etc, except we write x_i for $\bar{v}_i \pi$. Then $\{K_i : i \in I\} \subseteq \Lambda$ and we set $V_i = K_i \cap W$ and

$k_i \in K_i - V_i$. Indeed $\bar{u} = \bar{z}\bar{z}'$, where $\bar{z} = z(\bar{K})$, and $\bar{z}' = z(\bar{K}')$. Observe that $K'_{n-1} = K_n$. Pick v_i, v'_i of order 4 in V_i, V'_i, respectively, and set $v(i) = v_i v'_i$ and $W_I = \langle v(1)^M \rangle$.

Let $d_i \in D$ be the image k_i^* of k_i in $M^* = M/W$, and write d'_i for $k_i'^*$, where $k'_i \in K'_i - V'_i$. Set $\mu_0 = \langle (d_1 d'_1)^\mu \rangle$; thus (cf. 5.8.3.2) $\mu/\mu_0 = \tilde{\mu} = Sym(I)$, and we may choose notation so that $\tilde{d}_i = (i, i+1)$ for $1 \leq i < n$, and $\tilde{T}_M$ contains $\{\tilde{d}_1, \tilde{d}_3, \ldots, \tilde{d}_{\tilde{n}}\}$, where $\tilde{n}$ is $n-1$, $n-2$ for n even, odd, respectively.

Notice that we write W_I for the module W_1 of 5.8.6 and W_Δ for the module W_0 of 5.8.3. As $\bar{u} \in \ker(\pi)$, it follows from 5.8.3 and 5.8.6 that μ_0 centralizes W_I, and by 5.8.6 and its proof, W_I is the n-dimensional permutation module for $\tilde{\mu} = Sym(I)$, or its image modulo its center $\langle e_I \rangle$, with $v(i)$ of weight 1 in that module. Thus we can view W_I as the power set of I with the group operation symmetric difference. For $\theta \subseteq I$, write e_θ for θ regarded as a member of W_I. By 5.8.3, W_Δ is the core of W_I.

For $w \in W_I$ set $\Lambda_w = \{J \in \Lambda : J \leq C_M(w)\}$, $M_w = \langle \Lambda_w \rangle$, $\eta_w = \{J \cap W : J \in \Lambda_w\}$, and $W_w = \langle \eta_w \rangle$.

LEMMA 11.1.3. *(1)* $z_i = e_{i,i+1}$ *for* $1 \leq i < n$.
(2) $v(n-1) = e_n$.
(3) $\Omega = \{K_1, K'_1, K_3, K'_3, \ldots, K_{\tilde{n}}, K'_{\tilde{n}}\}$.
(4) $N_{\mathcal{F}}(W)$ *controls fusion in* $C_S(W)$ *with* $Aut_{\mathcal{F}}(C_S(W)) = Aut_{G_W}(C_S(W))$. *Further* W_I *and* W_Δ *are* $Aut_{\mathcal{F}}(C_S(W))$*-invariant.*
(5) $Aut_{\mathcal{F}}(W_I) = Aut_{G_W}(W_I) = Aut_M(W_I)$, *so* $Aut_S(W_I) = Aut_{T_M}(W_I)$.
(6) $\mathcal{Z}_\Delta$, $v(n-1)^{\mathcal{F}} \cap W$ *is the set of vectors in* W_I *of weight 2, 1, respectively.*
(7) For $x \in W_I$, $x^{\mathcal{F}} \cap \mathcal{F}^f \subseteq W_I$ *and there exists* $g \in M$ *such that* $c_g^* \in \mathfrak{A}(x)$.
(8) If $x \in W_\Delta \cap \mathcal{F}^f$ *then* $x = \prod_{s \in \sigma} s$ *for some* $\sigma \subseteq \mathcal{Z}_S$.
(9) For each $x = e_\theta \in W_I$, $\tilde{M}_x^* = \langle C_{\tilde{D}}(x) \rangle$ *and* $\tilde{d} \in C_{\tilde{D}}(x)$ *iff* $\tilde{d}$ *acts on* θ.

PROOF. Recall from 3.1.20.3 that $[W_\Delta, d_i] = \langle z_i \rangle$, so (1) follows as $\tilde{d}_i = (i, i+1)$ and W_I is the permutation module W_I or its image $W_I/\langle e_I \rangle$.

Let $w = v(n-1)$. By 5.8.6, $C_{\tilde{\mu}}(w) = \langle \tilde{d}_i : 1 \leq i \leq n-2 \rangle$, so $C_{\tilde{\mu}}(w) = \tilde{\mu}_n$ is the stabilizer in $\tilde{\mu} = Sym(I)$ of e_n. Thus as W_I is the permutation module for $\tilde{\mu}$, $w = e_n$ or $e_n e_I$, and if $e_I = 1$ then $e_n = e_n e_I$. If $e_I \neq 1$ and n is odd then $e_n e_I$ is even, so $w = e_n$ as $w \in W_I - W_\Delta$ and W_Δ is the core of W_I. Finally if n is even and $e_I \neq 1$ then we choose notation so that $w = e_n$. Thus (2) holds.

Let $J \in \Lambda$. Then $J = (J \cap W)\langle j \rangle$ with $j^* = d_{J \cap W} \in D$. Hence $J \leq T_M$ iff $\tilde{j} \in \tilde{T}_M$ iff $J' \leq T_M$. Finally from the notation established in 11.1.2, $\tilde{D} \cap \tilde{T}_M = \langle \tilde{d}_1, \tilde{d}_3, \ldots, \rangle$, so (3) follows. Further for $x \in W_I$, $J \in \Lambda_x$ iff $\tilde{d}_{J \cap W} = \tilde{j}$ centralizes x, so (9) holds.

The first statement in (4) follows as W is weakly closed in S with respect to $\mathcal{F}$, and then the second follows from the first. By construction $M \trianglelefteq G_W$ and Λ is G_W-invariant. By 4.2.8, the map $\zeta : J \mapsto z(J)$ is a surjection of Λ onto $\mathcal{Z}_\Delta$, so $\mathcal{Z}_\Delta$ is the set of elements of W_I of weight 2 by (1) and $W_\Delta = \langle \mathcal{Z}_\Delta \rangle \trianglelefteq G$. Similarly $\mathcal{V} = v(n-1)^M$ is the set of vectors of weight 1 by (2) and consists of the involutions in $JJ' \cap W - \langle z(J) \rangle$ for $J \in \Lambda$, so $\mathcal{V}$ is G_W-invariant, and hence $W_I = \langle \mathcal{V} \rangle \trianglelefteq G_W$. This proves (4) and (6).

Next $W_\Delta/\langle e_I \rangle$ is the natural module for $\tilde{\mu} = Sym(I)$, so $\mathrm{End}_{\mathbf{F}_2\mu}(W_I/\langle e_I \rangle) = \mathbf{F}_2$, and hence $Aut_{\mathcal{F}}(W_I) = Aut_M(W_I)$. This proves (5).

Let $x \in W_I$, and $\alpha \in \mathfrak{A}(x)$. As $W \leq C_S(x)$ is weakly closed, $W\alpha = W$, so $x\alpha \in W_I$ by (4). Further by Sylow's Theorem there is $g \in M$ with $C_S(x^g) \in Syl_2(C_{SM}(x))$, so $c_g^* \in \mathfrak{A}(x)$, proving (7).

Choose x as in (8). Then x is of even weight $2k$ so there is $h \in M$ with $x^h = y = z_1 z_3 \cdots z_{2k-1}$. Thus $y \in \langle \mathcal{Z}_S \rangle$ by (3), while by (7) there is $g \in G$ such that $c_g^* \in \mathfrak{A}(y)$ with $y^g = x$. Then $O(\tau) \leq C_S(y)$, so $O(\tau)^g = O(\tau)$ and (8) follows. □

LEMMA 11.1.4. *Let $1 \neq x = e_\theta \in W_I$ with $x \in \mathcal{F}^f$, and set $k = |\theta|$, $\Omega_x = \Lambda_x \cap \Omega$ and $\tau_x = (\mathcal{F}_x, \Omega_x)$. Set $\theta_1 = \theta$, $\theta_2 = I - \theta$, $\Lambda_i = \{J \in \Lambda : Mov(d_{J\cap W}) \subseteq \theta_i\}$, $\Omega_i = \Omega \cap \Lambda_i$, $M_i = \langle \Lambda_i \rangle$, $\eta_i = \{J \cap W : J \in \Lambda_i\}$, and $W_i = \langle \eta_i \rangle$.*

(1) τ_x is a quaternion fusion packet.

(2) $\eta_x \in \eta(\tau_x)$, M_x is a model for $[\Omega_x]_{N_{\mathcal{F}_x}(W_x)}$, and $\mu(\tau_x) \cong M_x/W_x$.

(3) $\Lambda_x = \Lambda_1 \coprod \Lambda_2$, $M_x = M_1 * M_2$ is a central product, $\eta_x = \eta_1 \coprod \eta_2$, and $\tau_x^\circ = \tau_1 * \tau_2$ with $\tau_i = (\mathcal{F}_i, \Omega_i)$ a quaternion fusion packet, $\eta_i \in \eta(\tau_i)$, and $\mu(\tau_x) = \mu(\tau_1) \times \mu(\tau_2)$, where $\mu(\tau_i) \cong M_i/W_i$.

(4) $x \in W_\Delta$ iff k is even, in which case $\Omega_x = \Omega$.

(5) $M_1 \cong \omega(D_k, m)$, $\mu_1 \cong$ Weyl(D_k), $x \in Z(M_1)$, $M_2 \cong \omega(D_{n-k}, m)$, $\mu_2 \cong$ Weyl(D_{n-k}), and if $M_2 \neq 1$ then $e_I x \in Z(M_2)$.

PROOF. As $x \in \mathcal{F}^f$, $\mathcal{F}_x$ is a saturated fusion system. By definition, Λ_x consists of those $J \in \Lambda$ centralizing x, so Ω_x consists of those members of Ω centralizing x. Then (1) follows from 2.5.2.

Next for $J \in \Lambda_x$, J is conjugate in $C_M(x)$ to a member of Ω_x by Sylow's Theorem, so as $J \cap W \in \eta_x$, $\eta_x \subseteq \nu \in \eta(\tau_x)$. Conversely if $V \in \nu$ then $V \leq X \in \Delta$, and either $X \in \Omega_x$ or $V = X \in \eta$. In the former case as $X \cap W \in \nu$, $V = X \cap W \in \eta$. Thus $\nu = \eta_x$, proving (2).

By 11.1.3.9, Λ_x is the disjoint union of Λ_1 and Λ_2, and hence η_x is the disjoint union of η_1 and η_2. If $J_i \in \Lambda_i$ then $[\tilde{d}_{J_1 \cap W}, \tilde{d}_{J_2 \cap W}] = 1$, so $[J_1, J_2] = 1$ by 3.1.22. Hence $[M_1, M_2] = 1$, so $M_x = M_1 * M_2$ is a central product.

Let $k_1 = k$ and $k_2 = n - k$. As W_Δ is the core of the permutation module W_I, $x \in W_\Delta$ iff k is even. Moreover in that event as $x \in \mathcal{F}^f$, $x \in \langle \mathcal{Z}_S \rangle$ by 11.1.3.8, so $\Omega_x = \Omega$. Hence (4) holds. Further we may choose notation so that $\theta = \{1, \ldots, k\}$, so $M_1 = \langle K_1, \ldots, K_{k-1}, K'_{k-1} \rangle$ and hence (using the fact that $x \in Z(M_1)$ and $e_I x \in Z(M_2)$ if $M_2 \neq 1$) we have $M_i \cong \omega(D_{k_i}, m)$ and $\mu_i = M_i/W_i \cong$ Weyl(D_{k_i}) from 5.8.6. Thus (5) holds.

It remains to prove (3). By 11.1.1.4 and 6.6.6, $\tau_x^\circ = \rho_1 * \cdots * \rho_r$ where $\rho_i = (\mathcal{Y}_i, \Gamma_i)$, $1 \leq i \leq r$ are the coconnected components of τ_x°. Note that if $K \in \Gamma_i$ then by 3.1.6, either $K' \in \Gamma_i$ or $\rho_i = (\mathcal{O}_K, \{K\})$. Hence in the former case for each $z \in \mathcal{Z}_S(\rho_i)$ we have $\Gamma_i(z) = \Omega(z)$ of order 2. Then as ρ_i satisfies Theorem 1, we conclude that $\mu(\rho_i) \cong$ Weyl(D_l) for some $l \geq 2$.

Suppose $k_j > 2$ for some $j \in \{1, 2\}$. Then M_j is transitive on Λ_j so $\Omega_j \subseteq \Gamma_i$ for some i, and $\eta_j \subseteq \eta(\rho_i) \subseteq \eta(\tau_x) = \eta_1 \cup \eta_2$, so $\mu(\rho_i) = \mu_j \times \mu'_{3-j}$ for some $\mu'_{3-j} \leq \mu_{3-j}$. But we saw that $\mu(\rho_i) \cong$ Weyl(D_l), so we conclude that $\mu(\rho_i) = \mu_j$ and hence $\Gamma_i = \Omega_j$. Therefore (3) holds in this case with $\tau_j = \rho_i$ and τ_{3-j} the produce of the remaining ρ_l.

Therefore we may assume that $k_j \leq 2$ for $j = 1, 2$, so as $k_1 + k_2 = n \geq 4$, we have $n = 4$ and $k_1 = k_2 = 2$. Therefore $\mu_x = \mu_1 \times \mu_2$ with $\mu_i \cong E_4$. Hence for each i, $\mu(\rho_i)$ is abelian, and we saw that either $\Gamma_i = \Omega(z)$ is of order 2 for some $z \in \mathcal{Z}_S(\rho_i)$ or $\rho_i = (\mathcal{O}_{K_i}, \{K_i\})$. We conclude that (3) holds. □

LEMMA 11.1.5. *(1) $\mathcal{F}_z^\circ = \mathcal{O}(z) * \mathcal{C}$ where $\rho = (\mathcal{C}, \Omega - \Omega(z))$ is a quaternion fusion packet with $\mathcal{C} = \mathcal{C}^\circ$ and $\mu(\rho) \cong$ Weyl(D_{n-2}).*

(2) Define M_2 as in 11.1.4. Then $Z(M_2) = \langle e_I z\rangle$ is of order 2.

(3) One of the following holds:

(i) $\mathcal{C} = \omega(D_{n-2}, m)$ with $n \geq 5$.

(ii) $\mathcal{C} \cong \check{A}E_{2n-4}$ with $n \geq 5$ or AE_{2n-3}.

(iii) $\mathcal{C} \cong \Omega_{2n-4}^+[m]$ with $n \geq 5$, $\Omega_{2n-3}[m]$, or $\Omega_{2n-2}^-[m]$.

(iv) $n = 4$ and $\mathcal{C} = \mathcal{O}(ze_I)$.

(v) $n = 5$ and $\mathcal{C} \cong L_3(2)/E_{64}$ or $Sp_6(2)$.

(vi) $n = 6$ and $\mathcal{C} \cong \check{L}_3(2)/2^{3+6}$ or $\check{\Omega}_8^+(2)$.

PROOF. As $z \in \mathcal{Z}_S$, $z = e_\theta$ with $|\theta| = 2$ by 11.1.3.1. Therefore by 11.1.4, $\tau_z = (\mathcal{F}_z, \Omega)$ is a quaternion fusion packet and $\tau_z^\circ = \tau_1 * \tau_2$ with $\tau_i = (\mathcal{F}_i, \Omega_i)$ and $\mu_i = \mu(\tau_i)$ satisfying $\mu_1 \cong$ Weyl(D_2) and $\mu_2 \cong$ Weyl(D_{n-2}). From 2.6.11, $\mathcal{O}(z) \trianglelefteq \mathcal{F}_z$, so $\mathcal{F}_1 = \mathcal{O}(z)$. Write $\mathcal{C}$ for $\mathcal{F}_2$; then (1) holds. Similarly (2) follows from 11.1.4.5.

If $n = 4$ then $\mu(\rho) \cong E_4$ by (1), so, using 11.1.1.1, (3) follows from 9.4.33. If $n = 5$ then $\mu(\rho) \cong S_4$ by (1), so as $|Z(M_2)| = 2$ by (2), (3) holds by Theorem 9.5.33. Similarly if $n = 6$ then $\mu(\rho) \cong$ Weyl(D_4), so (3) holds by Theorem 12.1.25, applied in an inductive context. Finally if $n \geq 7$ then $n - 2 \geq 5$, so (3) holds by induction on n (cf. Theorem 12.2.32). □

LEMMA 11.1.6. *Suppose $c = e_I \neq 1$. Then:*

(1) $\tau_c = (\mathcal{F}_c, \Omega)$ is a quaternion fusion packet with $\mu(\tau_c) \cong$ Weyl(D_n).

(2) If $n > 4$ and $c \in Z(\mathcal{F})$ assume the Extended Inductive Hypothesis. Then one of the following holds:

(i) $\mathcal{F}_c^\circ = \mathcal{F}_{T_M}(M)$.

(ii) $\mathcal{F}_c^\circ \cong \check{A}E_{2n}$.

(iii) $\mathcal{F}_c^\circ \cong \Omega_{2n}^+[m]$.

(iv) $n = 4$ and $\mathcal{F}_c \cong \check{L}_3(2)/E_{2^{3+6}}$ or $\check{\Omega}_8^+(2)$.

PROOF. Part (1) follows from 11.1.4. Set $\mathcal{F}_c^+ = \mathcal{F}_c/\langle c\rangle$; by 3.3.2, τ_c^+ is a quaternion fusion packet with $\mu(\tau_c^+) \cong \mu(\tau_c)$. If $n = 4$ then $\mathcal{F}_c^+$ is described in 10.1.21, so that (2) holds, using 5.1.21.5, 5.10.7.1, 5.10.12, and 3.3.16. Thus we may take $n \geq 5$.

If $c \notin Z(\mathcal{F})$ then $\mathcal{F}_c^\circ$ satisfies one of the conclusions of Theorem 1 by 11.1.1.4, so, inspecting the lists of conclusions in Theorems 1 and 2, (2) holds in this case, again using 5.1.21.5 and 3.3.16. Thus we may assume $c \in Z(\mathcal{F})$ and the Extended Inductive Hypothesis holds. Hence by 6.6.9, τ^+ satisfies Hypothesis 11.1.1, so by induction on the order of τ, $\mathcal{F}^+$ is described in Theorem 12.2.32 in the case $c = 1$, and then, arguing as above and as $c \in Z(\mathcal{F})$, $\mathcal{F}$ appears in (2). Formally this approach amounts (for fixed n) to treating the case $c = 1$ in Theorem 12.2.32 before proving 11.1.6 when $c \neq 1$. □

In the next lemma we appeal to Theorems 12.1.25 and 12.2.32 at $n-1$. Formally this amounts to treating the $n - 1$ case to its end, before going on to treat the n case.

LEMMA 11.1.7. *Let $b \in W_I \cap \mathcal{F}^f$ be of weight 1 and $a \in \{b, bc\}$. Then*
(1) $\tau_a = (\mathcal{F}_a, \Omega_a)$ is a quaternion fusion packet with $\mu(\tau_a) \cong Weyl(D_{n-1})$.
(2) If n is odd then $a \in Z(S)$ so $\Omega_a = \Omega$.
(3) One of the following holds:
(i) $\mathcal{F}_a^\circ \cong \omega(D_{n-1}, m)$.
(ii) $\mathcal{F}_a^\circ \cong \check{A}E_{2n-2}$ with $bc \in Z(\mathcal{F}_a^\circ)$.
(iii) $a = b$ and $\mathcal{F}_b^\circ \cong AE_{2n-1}$.
(iv) $\mathcal{F}_a^\circ \cong \Omega_{2n-2}^+[m]$ and $bc \in Z(\mathcal{F}_a^\circ)$.
(v) $a = b$ and $\mathcal{F}_b^\circ \cong \Omega_{2n-1}[m]$ or $\Omega_{2n}^-[m]$.
(vi) $n = 4$, $a = b$, and $\mathcal{F}_a^\circ \cong L_3(2)/E_{64}$ or $Sp_6(2)$.
(vii) $n = 5$, $\mathcal{F}_a^\circ \cong \check{L}_3(2)/2^{3+6}$ or $\check{\Omega}_8^+(2)$, and $bc \in Z(\mathcal{F}_a^\circ)$.

PROOF. Part (1) follows from 11.1.4 and (2) follows from 11.1.3.5.

From 11.1.4, $M_b = M_{bc}$, and from 11.1.4.5 we have $bc \in Z(M_a)$, so if $Z(\mathcal{F}_a^\circ) = Z(M_a)$ then $bc \in Z(\mathcal{F}_a^\circ)$, while if $Z(\mathcal{F}_a) = 1$ then $a \neq bc$ so $a = b$.

Suppose $n = 4$. Then (3) follows from Theorem 9.5.33. Suppose $n = 5$. Then (3) follows from Theorem 12.1.25. Finally assume $n > 5$; then $n - 1 \geq 5$. Now $\mathcal{F}_a$ is proper in $\mathcal{F}$, so by 6.6.12, τ_a satisfies the Extended Inductive Hypothesis. Then (3) holds by Theorem 12.2.32. □

LEMMA 11.1.8. *Let x be an involution in $W_I \cap \mathcal{F}^f$. Then one of the following holds:*
(1) $\mathcal{F}_x \leq \mathcal{E}(\tau)$.
(2) x is of weight 1 or $n - 1$.
(3) $x = e_I$.
(4) n is even, $e_I = 1$ and x is of weight $n/2$.

PROOF. Assume $x \neq c = e_I$ and x is not of weight 1 or $n - 1$. By 11.1.4, $\Omega_x = \Omega_1 \coprod \Omega_2$ and $\mathcal{F}_x^\circ = \mathcal{F}_1 * \mathcal{F}_2$ with $\tau_i = (\mathcal{F}_1, \Omega_i)$ a quaternion fusion packet and $k_i = |\Omega_i| \geq 2$. Observe that from 11.1.4.5, either $\mathcal{F}_1$ is not isomorphic to $\mathcal{F}_2$ or $n - k_1 = k_2 = k_1$. In the former case $\mathcal{F}_i \trianglelefteq \mathcal{F}_x$. In the latter case, n is even and $k_1 = k_2 = n/2$. Moreover if $c \neq 1$ then from 11.1.4.5, $Z(\mathcal{F}_i) = \langle x_i \rangle$ where $x_1 = x$ and $x_2 = cx \neq x$, so again $\mathcal{F}_i \trianglelefteq \mathcal{F}_x$. Thus we may assume $\mathcal{F}_i \trianglelefteq \mathcal{F}_x$ for $i = 1, 2$. Now (1) holds by 3.4.13. □

11.2. Generation

In this section we assume the following hypothesis:

HYPOTHESIS 11.2.1. *$\tau = (\mathcal{F}, \Omega)$ is a quaternion fusion packet such that*
(1) $\mathcal{F}$ is transitive on Ω.
(2) $\mathcal{Z}_S$ is of order n_S and, setting $W_S = \langle \mathcal{Z}_S \rangle$ and $c = \prod_{z \in \mathcal{Z}_S} z$, either $m(W_S) = n_S$ or $m(W_S) = n_S - 1$ and $c = 1$.
(3) For $z \in \mathcal{Z}_S$, $|\Omega(z)| \leq 2$ and in case of equality $\mu = \mu(\tau) \cong Weyl(D_n)$ for some n.

NOTATION 11.2.2. Suppose $|\Omega(z)| = 2$. Then by 11.2.1.3, $\mu \cong \mathrm{Weyl}(D_n)$, so by 3.3.14 there is a unique $M \in M(\tau)$. Moreover $n = 2n_S$ or $2n_S + 1$ by 4.2.8.6, and by 5.8.5, there is a surjective homomorphism $\pi : \bar{G} = \bar{\omega}(D_n, m) \to M$ with

$\ker(\pi) \leq C_{\bar{W}}(\bar{G})$. As $|\Omega(z)| = 2$, $\bar{u} \in \ker(\pi)$ by 5.8.5, so by 5.8.3.2, $W_\Delta = \langle \mathcal{Z}_\Delta \rangle$ is the core of the permutation module for $\tilde{M} = M/C_M(W_S) \cong \mu/\mu_0 = Sym(I) \cong S_n$ on $I = \{1, \ldots, n\}$ (or that module modulo its center) with $\mathcal{Z}_\Delta$ a set of vectors of weight 2, where μ_0 is the subgroup Q of section 5.8 and centralizes W_Δ. Let W_I be the permutation module for $\tilde{M}$, or that module modulo its center $\langle e_I \rangle$; from 5.8.6 we may take $W_\Delta \leq W_I \trianglelefteq M$, such that $C_M(W_\Delta)$ centralizes W_I and for each $z \in \mathcal{Z}_S$, $z = e_{2i-1,2i}$ for suitable $1 \leq i \leq n_S$, and $\Omega(z) = \{K_i, K_i'\}$. Let v_i, v_i' be of order 4 in $K_i \cap W$, $K_i' \cap W$, respectively, and $v(z) = v_i v_i'$. From 5.8.6, $v(z)$ is e_{2i-1} or e_{2i} in W_I.

NOTATION 11.2.3. Let $R \in \mathcal{F}^e$, let G be a model for $N_{\mathcal{F}}(R)$ and $G^* = G/R$. Let Θ be the set of orbits of R on $\mathcal{Z}_S$. We say that R is *exceptional* if $c = 1$ and R is transitive on $\mathcal{Z}_S$, and we say that R is *ordinary* if R is not exceptional. Note that if R is exceptional then $1 < n_S$ is a power of 2. For as R is transitive on $\mathcal{Z}_S$, n_S is a power of 2. Further if $n_S = 1$ then $\mathcal{Z}_S = \{c\}$, so $c \neq 1$. Also note that if R is exceptional and $|\Omega(z)| = 2$ then $n = 2n_S$ as $c = 1$. If R is exceptional write $\hat{\Theta}$ for the set of R-invariant partitions of $\mathcal{Z}_S$ of size 2, and when $|\Omega(z)| = 2$ let $\tilde{\Theta}$ be the set of R-invariant partitions of I of size 2. For $\theta \subseteq \mathcal{Z}_S$, set $f_\theta = \prod_{t \in \theta} t$ and for $\alpha \subseteq I$ set $e_\alpha = \prod_{i \in \alpha} e_i$. If $z \in \mathcal{Z}_S$ and $|\Omega(z)| = 1$, write $v(z)$ for an element of order 4 in $O(z)$ with $\langle v(z) \rangle \trianglelefteq C_S(z)$.

LEMMA 11.2.4. *Let $\theta \in \Theta$. Then*
(1) $f_\theta \in Z(R)$, and
(2) if R is ordinary then $\langle \theta \rangle \cap Z(R) = \langle f_\theta \rangle$.

PROOF. As R permutes θ, R centralizes f_θ. Further as $R \in \mathcal{F}^e$, $C_S(R) = Z(R)$, so (1) holds.

Suppose R is ordinary and let $f_\sigma \in Z(R)$ for some $\emptyset \neq \sigma \subset \theta$. As R is ordinary it follows from 11.2.1.2 that $\langle \theta \rangle$ is the permutation module for R on θ. Therefore as R centralizes f_σ, R acts on σ, contradicting R transitive on θ. Hence (2) holds. □

LEMMA 11.2.5. *Assume R is exceptional.*
(1) $C_{W_S}(R)^\# = \{f_\theta : \theta \in \pi \in \hat{\Theta}\}$ and if $|\Omega(z)| = 2$ then $C_{W_I}(R)^\# = \{e_\alpha : \alpha \in \pi \in \tilde{\Theta}\}$.
(2) Let $\theta_i \in \pi_i \in \hat{\Theta}$ such that $\{f_i = f_{\theta_i} : 1 \leq i \leq \beta\}$ is a basis for $C_{W_S}(R)$. For $1 \leq \gamma \leq \beta$, let

$$\theta(\gamma) = \bigcap_{i=1}^{\gamma} \theta_i \text{ and } \rho_\gamma = \theta(\gamma)^R.$$

Then ρ_γ is an R-invariant partition of $\mathcal{Z}_S$ into 2^γ blocks of size $n_S/2^\gamma$, and $Z_\gamma = \langle f_1, \ldots, f_\gamma \rangle$ consists of the f_ξ such that ξ is the union of blocks in ρ_γ and $\{\xi, \mathcal{Z}_S - \xi\}$ is R-invariant.
(3) Choose β, γ, and f_i, $1 \leq i \leq \beta$ as in (2) and set $E_\gamma = \langle f_\sigma : \sigma \in \rho_\gamma \rangle$. For $1 \leq i \leq \beta$ set $R_i = N_R(\theta_i)$ and $R(\gamma) = R_1 \cap \cdots \cap R_\gamma$. Then $R(\gamma)$ is the kernel of the action of R on ρ_γ, and $R/R(\gamma) \cong E_{2^\gamma}$ is regular on ρ_γ. If $\gamma > 1$ then $R(\gamma) = C_R(E_\gamma)$.
(4) Let $\hat{\Theta}_\gamma$ consist of those $\pi \in \hat{\Theta}$ such that $f_\theta \in Z_\gamma$ for $\theta \in \pi$. Then $\pi \in \hat{\Theta}_\gamma$ iff π is the set of orbits of the preimage in R of some hyperplane of $R/R(\gamma)$ on $\mathcal{Z}_S$. Hence $Z_\gamma^\# = \hat{\Theta}_\gamma$ and $\dim(Z_\gamma) = \gamma$.

PROOF. Let $\theta \subseteq \mathcal{Z}_S$ such that $1 \neq f_\theta$ centralizes R. Then for $r \in R$, $f_\theta = f_\theta^r = f_{\theta^r}$, so $\theta^r = \theta$ or $\bar{\theta} = \mathcal{Z}_S - \theta$. It follows that R acts on $\pi = \{\theta, \bar{\theta}\}$, so $\pi \in \hat{\Theta}$. Similarly $1 \neq e_\alpha$ centralizes R iff $\pi = \{\alpha, \bar{\alpha}\} \in \tilde{\Theta}$. Thus (1) holds.

Choose θ_i, $1 \leq i \leq \beta$ as in (2); we prove (2) by induction on γ; when $\gamma = 1$, the result is trivial. Suppose the result holds at γ, so that $\rho = \theta^R$ is a partition of $\mathcal{Z}_S$, where $\theta = \theta_1 \cap \cdots \cap \theta_\gamma$. Suppose $\beta \neq \gamma$. If $\theta_{\gamma+1}$ is the union of some set of blocks of $\rho = \rho_\gamma$ then $f_{\gamma+1} \in Z_\gamma$ as (2) holds at γ, contrary to the choice of the f_i. Hence $\theta_{\gamma+1}$ is not a union of blocks of ρ, so as $\lambda = \{\theta_{\gamma+1}, \bar{\theta}_{\gamma+1}\}$ is the set of orbits of $R_{\gamma+1}$ on I, we have $R_{\gamma+1}$ transitive on ρ and $|\theta_{\gamma+1} \cap b| = k$ is independent of the choice of $b \in \rho$. Then as R is transitive on λ, also $k = |\bar{\theta}_{\gamma+1} \cap b|$, so $k = |\theta|/2$ and σ^R is a partition of $\mathcal{Z}_S$, where $\sigma = \theta_{\gamma+1} \cap \theta$. This proves the first part of (2), and the second part follows from (4), which we prove in a moment.

Assume the setup of (3). Then R_i is the stabilizer in R of θ_i and of index 2 in R, so the kernel $X(\gamma)$ of the action of R on $\rho = \rho_\gamma$ is contained in $R(\gamma)$ and $R(\gamma) \trianglelefteq R$. Proceeding by induction on γ as in the previous paragraph, $R(\gamma)$ fixes $\theta(\gamma)$, so as $R(\gamma) \trianglelefteq R$, $R(\gamma) = X(\gamma)$. Further for $\gamma > 1$, $r \in R$ centralizes $f_{\theta(\gamma)}$ iff r acts on $\theta(\gamma)$, so $R(\gamma) = C_R(E_\gamma)$.

As $|R : R_i| = 2$, $R/R(\gamma)$ is elementary abelian. Then as $R/R(\gamma)$ is transitive and faithful on ρ, $R/R(\gamma)$ is regular on ρ. In particular $|R/R(\gamma)| = |\rho| = 2^\gamma$, completing the proof of (3).

If P is the preimage of a hyperplane of $R/R(\gamma)$ then as $R/R(\gamma)$ is regular on ρ, the set π of orbits of P on $\mathcal{Z}_S$ is of order 2 and each orbit is the union of blocks of ρ, so $f_\sigma \in Z_\gamma$ for $\sigma \in \pi$. Conversely if $f_\xi \in \mathcal{Z}_\gamma$ then, changing the basis of Z_γ, we may assume $f_\xi = f_1$, so ξ is an orbit of R_1, and $R_1/R(\gamma)$ is a hyperplane of $R/R(\gamma)$. Thus $Z_\gamma^\# = \hat{\Theta}_\gamma$ is the number $2^\gamma - 1$ of hyperplanes of $R/R(\gamma) \cong E_{2^\gamma}$, so (4) holds. □

NOTATION 11.2.6. We consider three cases:

(I) For $z \in \mathcal{Z}_S$, $|\Omega(z)| = 2$; in this case set $W_0 = W_I$.

(II) For $z \in \mathcal{Z}_S$, $|\Omega(z)| = 1$, $n_S = 4$, R is exceptional, $|W_S : W_S \cap R| = 2$, and $W_S \cap R \leq Z(R)$; in this case set $W_0 = \langle \Omega_2(K \cap W) : K \in \Omega \rangle$.

(III) For $z \in \mathcal{Z}_S$, $|\Omega(z)| = 1$ and case (II) does not hold; in this case set $W_0 = W_S$.

In each case let $\mathcal{W}(R)$ consist of those involutions $w \in C_{W_0}(R)$ such that:

(i) In case (I), $w = e_\alpha$ for some $\alpha \subseteq I$ of order a positive power of 2.

(ii) In case (II), $w \in W_0$.

(iii) In case (III), $w = f_\theta$ for some $\theta \subseteq \mathcal{Z}_S$ with $|\theta|$ a power of 2.

Let $\mathcal{U}$ be the set of $U^* \leq G^*$ such that $S^* \cap U^*$ is Sylow in U^*, U^* is not contained in a strongly embedded subgroup of G^* over $N_S(R)^*$, $U^* = \langle u^{*U} \rangle$ for some $u \in S \cap G$ such that u^* is an involution, and $U^* \leq C_{G^*}(w)$ for some $w \in \mathcal{W}(R)$.

Until further notice we assume the following hypothesis:

HYPOTHESIS 11.2.7. *(1)* $O^{2'}(C_G(\langle x^G \rangle)) = R$ *for each* $x \in \mathcal{W}(R)$.

(2) If $\mathcal{U} \neq \emptyset$ *then* $m_2(G^*) = 1$ *and* $\langle \mathcal{U}_0 \rangle$ *is not normal in* G^* *for any nonempty* $\mathcal{U}_0 \subseteq \mathcal{U}$.

LEMMA 11.2.8. *If* R *is exceptional then* $E_\beta \leq R$.

PROOF. Assume R is exceptional but $E_\beta \not\leq R$ and let γ be minimal subject to $E_\gamma \not\leq R$. Observe that from 11.2.5.2, $E_1 = \langle f_1 \rangle \leq C_{W_S}(R) \leq R$, so $\gamma \geq 2$. By minimality of γ, $E_{\gamma-1} \leq R$.

Next E_2 is generated by the four conjugates of $f_{\theta(2)}$, which sum to 0, so $|E_2| = 8$. Thus $|E_2 : Z_2| = 2$, so if $\gamma = 2$ then $f_{\theta(2)} \notin R$ and $[R, f_{\theta(2)}] \leq Z_2 \leq R$, so $f_{\theta(2)} \in G$. In this case set $y = f_{\theta(2)}$. On the other hand suppose $\gamma > 2$. As $Z(R)$ centralizes f_ξ for each $\xi \in \rho_{\gamma-1}$, $Z(R)$ acts on ξ. Further ξ contains two blocks of ρ_γ; for each $\xi \in \rho_{\gamma-1}$, pick one of those blocks ζ_ξ, and define

$$\zeta = \bigcup_{\xi \in \rho_{\gamma-1}} \zeta_\xi \text{ and } y = f_\zeta.$$

If $y \in R$ then $Z(R)$ centralizes y, so as $Z(R)$ acts on ξ it also acts on ζ_ξ, and then $E_\gamma \leq C_S(Z(R)) = R$ by 11.2.7.1, contrary to the choice of γ. Therefore $y \notin R$, but as R acts on $\rho_{\gamma-1}$ and ρ_γ by 11.2.5.3, we have $[R, y] \leq E_{\gamma-1} \leq R$, so $y \in G$.

Thus in any event, $y \in G - R$. Set $Z = \langle f_1^G \rangle$ and $H = \langle y^G \rangle$. By 11.2.7.1, $O^{2'}(C_G(Z)) = R$, so $[Z, y] \neq 1$. Thus $Z \not\leq C_R(E_\gamma) = R(\gamma)$ by 11.2.5.2, so $Z \not\leq R_i$ for some $1 \leq i \leq \gamma$, and then as R is transitive on $\mathcal{Z}_S$, $Z \not\leq R_i$ for any i. Hence Z is nontrivial on ρ_2, so $E_2 \not\leq R$ and therefore $\gamma = 2$ by minimality of γ. Thus $y = f_{\theta(2)}$ by construction. In paragraph two we saw that $|E_2 : Z_2| = 2$, so $E_2 = Z_2\langle y \rangle$ and hence $E_2^* = \langle y^* \rangle$.

Let $h \in H$ with $\langle y^*, y^{*h} \rangle$ not a 2-group, and set $L = \langle E_2, E_2^h \rangle$ and $B = O_2(L)$. Then $L \leq H$ and $[R, E_2] = Z_2$, so $B = Z_2 Z_2^h \leq Z(R)$ as $Z_2 \leq Z(R)$, and $L \trianglelefteq LR$. If $Z_0 = Z_2 \cap Z_2^h \neq 1$ then $m(B) \leq 3$ and y induces a transvection on Z, so $H^* \cong S_3$ by 2.4.1 in [**Asc19**]. But then $Z_0 \leq C_{Z_2}(H)$, contrary to 11.2.7.2. Therefore $B = Z_2 \times Z_2^h$ and $|y^B| = |B : Z_2| = 4$, so B is transitive on ρ_2.

Observe that in case (I), we have $n_S > 2$ (cf. 11.2.15), so $n = 2n_S \geq 8$ is a power of 2. Hence by 11.2.5.1, in case (I) each involution in $C_{W_0}(R)$ is of the form e_α for some $\alpha \subseteq I$ of order a positive power of 2, and in case (III) each involution in $C_{W_0}(R)$ is of the form f_θ for some $\theta \subseteq \mathcal{Z}_S$ of order a power of 2. Therefore in any case, each involution in $C_{W_0}(R)$ is in $\mathcal{W}(R)$.

Next as $[R, y] = Z_2$, $R = B \times C_R(L)$. Moreover if $C_{R \cap W_0}(L) \neq 1$ then $1 \neq C_{Z(R) \cap W_0}(L)$, so that from the previous paragraph, L^* is in the set $\mathcal{U}$ of 11.2.6. Then by 11.2.7.2, $m_2(H^*) = 1$ and $H^* \neq \langle \mathcal{U} \rangle$. Therefore:

(a) We can choose h so that $C_{W_0 \cap R}(L) = 1$.

Replace Z by $\langle Z_2^G \rangle$. As $Z_2 = [R, y]$ and $R = B \times C_R(L)$, we have $|Z : C_Z(y)| = 4$. Set $Z_0 = C_{W_0 \cap R}(y)$. Suppose we can show that:

(b) $Z_2 < Z_0$.

But for $z_0 \in Z_0$, $z_0 = ab$ with $a \in C_Z(y)$ and $b \in C_R(H)$. For $r \in R$, $z_0^r = ab^r$, so either $b = b^r$ or $z_0 z_0^r = bb^r$ is a nontrivial element of $C_{W_0 \cap R}(H)$. The latter contradicts (a), so $b^r = b$ for each $r \in R$, and hence $z_0 \in C_{W_0 \cap Z(R)}(y) = Z_1$. It follows that $Z_2 < Z_1$. But $R = B \times C_R(L)$, so $C_R(y) = Z_2 \times C_R(L)$, and hence as $Z_2 < Z_0$ we have $1 \neq C_{Z_1}(L) \leq C_{W_0 \cap R}(L)$, contrary to (a). Therefore to complete the proof of the lemma, it remains to establish (b).

Set $Z_y = C_Z(y)$, let A be a complement to Z_2 in Z_y, and let D a complement to Z_y in Z. Then D is regular on ρ_2 with Z_y the kernel of the action of Z on ρ_2.

Suppose that Z is transitive on $\mathcal{Z}_S$ and let $s \in \mathcal{Z}_S$, Z_s a complement to $C_Z(s)$ in Z, and $a \in C_Z(s)$. As Z is transitive on $\mathcal{Z}_S$, for each $t \in \mathcal{Z}_S$, $[v(t), a] = 1$ or t, independent of the choice of t. Thus as $\prod_{t \in \mathcal{Z}_S} t = c = 1$ and Z_s is regular on $\mathcal{Z}_S$, in either case a centralizes $w_0 = \prod_{u \in \mathcal{Z}_S} v(s)^u$. Therefore $Z = C_Z(s)Z_s$ centralizes w_0 and w_0 is of weight $|\mathcal{Z}_S| \geq |Z : C_Z(y)| = 4$, so if $w_0 \in W_0$ then by 11.2.7.1, $w_0 \in C_{W_0}(Z) \leq R$ and hence $w_0 \in Z_0 - Z_2$, establishing (b) in this case. Therefore $w_0 \notin W_0$, so case (II) or (III) holds. If $n_S = 4$ then case (II) holds and $w_0 \in W_0$, so (b) holds here too. Thus $n_S > 4$ and case (III) holds.

So we may assume either Z is not transitive on $\mathcal{Z}_S$, or case (III) holds with $n_S > 4$. In case (I), as D is transitive on ρ_2, we have $n_S > 4$.

Let $s \in \theta(2)$ and $w = \prod_{x \in Z} s^x$; then $w \in C_{W_0}(Z) = W_0 \cap R$ by 11.2.7.1. If Z is not transitive on $\mathcal{Z}_S$ then $1 \neq w$ and indeed $w \notin Z_2$, so (b) holds. Therefore we may assume that Z is transitive on $\mathcal{Z}_S$, so $n_S > 4$ and case (III) holds. Then $n_S = |Z : C_Z(s)| = 4|s^A|$.

Let $C_A(s) \leq A_1$ of index 3 in A and set $w_1 = \prod_{x \in A_1} s^x$ and $w_2 = \prod_{d \in D}(w_1^d$. Then w_2 centralizes DA_1 and for $a \in A - DA_1$, $w_2^a = w_2 f_{\mathcal{Z}_S} = w_2$ as R is exceptional, so $w \in Z_0 - Z_2$ and (b) holds. This completes the proof of (b) and the lemma. □

LEMMA 11.2.9. *If R exceptional and $\beta > 1$ then $Z(R)$ acts on each member of ρ_β.*

PROOF. By 11.2.8, $E_\beta \leq R$, so $Z(R) \leq C_R(E_\beta) = R(\beta)$ by 11.2.5.3. But $R(\beta)$ is the kernel of the action of R on ρ_β by 11.2.5.3. □

LEMMA 11.2.10. *If R is ordinary then $\mathcal{Z}_S \subseteq R$.*

PROOF. Assume otherwise; then there exists $\theta \in \Theta$ such that $\theta \not\subseteq R$, and we fix this θ and set $f = f_\theta$. By 11.1.4, $f \in Z(R)$ and $|\theta| = 2^a$ for some $a > 0$. Let $R_a = R$, $\theta_a = \theta$, and, proceeding recursively, let R_i be a subgroup of R_{i+1} of index 2 with an orbit θ_i of length 2^i on θ_{i+1}. Pick b maximal subject to $f_{\theta_b} \notin R$ and set $\sigma = \theta_b$. Then $\pi = \theta_{b+1}^R$ is a partition of θ into 2^{a-b-1} blocks of size 2^{b+1}, and $E = \langle f_p : p \in \pi \rangle \leq R$. Let Γ be a set of coset representatives for R_{b+1} in R,

$$\zeta = \bigcup_{\gamma \in \Gamma} \sigma^\gamma, \text{ and } y = f_\zeta.$$

If $y \in R$ then as $Z(R)$ acts on θ_{b+1}, $Z(R)$ acts on σ, so $f_\sigma \in O^{2'}(C_G(Z(R)) = R$ by 11.2.7.1, contrary to the choice of σ. Therefore $y \notin R$.

Observe that $[R, y] \leq E \leq R$, so $y \in G$. Set $Z = \langle f^G \rangle$ and $H = \langle y^G \rangle$; as $y \notin R$, $H^* \neq 1$. By 11.2.7.1, $R = O^{2'}(C_G(Z))$, so y does not centralize Z. By 11.2.4.2, $E \cap Z \leq E \cap Z(R) = \langle f \rangle$. Therefore $[Z, y] = \langle f \rangle$, so y induces a transvection on Z with center f. Thus by 2.4.1 in [**Asc19**], $Z \cong E_4$ and $H^+ = H/C_H(Z) = GL(Z)$.

Let $h \in H - C_H(f)$, $A = \langle y \rangle E$, and $L = \langle A, A^h \rangle$. Then A is abelian with $[R, y] \leq E$ and $H^+ = \langle y^+, y^{+h} \rangle = L^+$. It follows that $L \trianglelefteq LR$ with $O_2(L) = EE^h$ and $E_0 = E \cap E^h = Z(L)$. Hence if $E_0 \neq 1$ then $1 \neq E_0 \cap Z(R) = E_1$. Then as $E \cap Z(R) = \langle f \rangle$, we conclude that $f \in Z(L)$, contradicting $[f, H^+] \neq 1$. Therefore $E_0 = 1$, so $B = O_2(L) = E \times E^h$. We may take $h \in L$ of order 3 inverted by y; then $C_B(h) = 1$, so for each $e \in E^\#$, $\langle e, e^h \rangle \cong E_4$ is $\langle h, y \rangle$-invariant.

Set $x = f^h$; from the previous paragraph, $[x, y] = f$. Also $\Pi = \{f_p : p \in \pi\}$ is a basis for E, and x acts on each $p \in \pi$, so $p = \sigma^\gamma \cup \sigma^{\gamma x}$, for $\gamma \in \Gamma$ with $\sigma^\gamma \in p$. Therefore $C_{O(p)}(x) = \{ll^x : l \in O(\sigma^\gamma)\} \cong O(\sigma)$. Then $P = C_{O(\theta)}(x)$ is the direct product of 2^{a-b-1} copies of $O(\sigma)$ and $Z(P) = E$. Therefore as $E \cap E^h = 1$, also $P \cap P^h = 1$. Further $P \leq C_G(Z) = R$, and as R acts on $O(\theta)$, $P = C_{O(\theta)}(x) \trianglelefteq R$, so $\langle P, P^h \rangle = P \times P^h$. Thus $[P, h]$ is nonabelian, whereas $[R, L] = EE^h$ is abelian, a contradiction. □

LEMMA 11.2.11. $\mathcal{Z}_S \subseteq R$.

PROOF. Assume otherwise; by 11.2.10, R is exceptional, so $\sigma = \theta(\beta) \not\subseteq R$. The initial treatment of this case is much like that of the case R ordinary.

By construction, $|\sigma| = 2^a$ for some a. Let $P_a = N_R(\sigma)$, $\sigma_a = \sigma$, and, proceeding recursively, let P_i be a subgroup of P_{i+1} of index 2 with an orbit σ_i of length 2^i on σ_{i+1}. Pick b maximal subject to $f_{\sigma_b} \notin R$ and set $\xi = \sigma_b$. Then $\pi = \sigma_{b+1}^R$ is a partition of $\mathcal{Z}_S$ into $n_S/2^{b+1}$ blocks of size 2^{b+1}, and $E = \langle f_p : p \in \pi \rangle \leq R$. Let Γ be a set of coset representatives for P_{b+1} in R,

$$\zeta = \bigcup_{\gamma \in \Gamma} \xi^\gamma, \text{ and } y = f_\zeta.$$

If $y \in R$ then as $Z(R)$ acts on σ_{b+1}, $Z(R)$ acts on ξ, so $f_\xi \in O^{2'}(C_G(Z(R)) = R$ by 11.2.7.1, contrary to the choice of ξ. Therefore $y \notin R$.

Observe that $[R, y] \leq E \leq R$, so $y \in G$. Set $Z = \langle f_1^G \rangle$. By 11.2.7.1, $R = O^{2'}(C_G(Z))$, so as $y \notin R$, y does not centralize Z. Let $x \in Z - C_Z(y)$. Then x acts on each $p \in \pi$, so for $\gamma \in \Gamma$ with $\xi^\gamma \in p$, $\xi^{\gamma x} = \xi^\gamma$ or $p - \xi^\gamma$. As $[x, y] \neq 1$, x moves some ξ^γ, and then as R is transitive on π and $x \in Z(R)$, x moves every ξ^γ. But now $yy^x = c$, so as $c = 1$, x centralizes y, a contradiction. □

LEMMA 11.2.12. *$Z(R)$ acts on $O(s)$ for each $s \in \mathcal{Z}_S$.*

PROOF. By 11.2.11, $Z(R)$ centralizes $\mathcal{Z}_S$, so the remark follows. □

LEMMA 11.2.13. *Assume $|\Omega(z)| = 1$ for $z \in \mathcal{Z}_S$. Then either $|O(z) \cap R| > 2$ for some $z \in \mathcal{Z}_S$ or the following hold:*

(1) R is ordinary.

(2) R is transitive on $\mathcal{Z}_S$.

(3) Assume that τ satisfies the Inductive Hypothesis. Then $\tau^\circ = \prod_{z \in \mathcal{Z}_S} \rho_z$, where $\rho_z = (\mathcal{Y}_z, \Omega(z))$ with $\mathcal{Y}_z \cong L_3^-[m]$, $L_2[2m]^{(1)}$, $G_2[m]$, or M_{12}.

(4) If $\mathcal{F}^\circ$ is $(P)Sp_{2n_S}[m]$, $(P)SL_{2n_S}^\epsilon[m]$, or $L_{2n_S+1}^\epsilon[m]$ then $n_S = 1$ and $\mathcal{F}^\circ$ is $L_3^-[m]$.

(5) 11.2.6.II does not hold.

PROOF. Assume for each $z \in \mathcal{Z}_S$ that $O(z) \cap R = \langle z \rangle$; hence in particular $v(z) \notin R$. If R is ordinary then $z \in \theta \in \Theta$, and we set $f = f_\theta$ and $Z = \langle f^G \rangle$. If R is exceptional let $z \in \theta_1$ and set $\theta = \mathcal{Z}_S$ and $Z = \langle f_1^G \rangle$. In any event set $E = \langle f_j : j \in \theta \rangle$ and $y = \prod_{j \in \theta} v(j)$.

Assume $y \in R$. Then as R is transitive on θ, either Z centralizes the projection $v(s)$ of y on $O(s)$ for each $s \in \mathcal{Z}_S$, so $v(s) \in R$ by 11.2.7.1, or R is exceptional and for some $x \in Z$, $[v(s), x] = s$ for each $s \in \mathcal{Z}_S$. In the former case the lemma holds, so assume the latter.

Let $\sigma = \theta_1$, $u = \prod_{z \in \sigma} v(z)$, $W_1 = \langle W_S, y \rangle$, and $A = \langle W_1, u \rangle$. Then $[R, u] \leq W_1 \leq R$, so $A \leq G$. Also $[x, u] = f_1$, so $u \notin R$ and indeed $[Z(R), u] \leq \langle f_1 \rangle$. Set $H = \langle u^G \rangle$ and $G^+ = G/C_G(Z(R))$. As G^* has a strongly embedded subgroup we conclude from 2.4.1 in [**Asc19**] that $Z \cong E_4$ and $H^+ = GL(Z)$. Let $h \in H$ of order 3 and $L = \langle A, A^h \rangle$. As $A \trianglelefteq S$, $[R, u] \leq R \cap A = W_1$, so $O_2(L) = W_1 W_1^h$

and $L \lhd LR$. Set $W_2 = W_1 \cap W_1^h$; as usual, $W_2 \leq Z(L)$, so if $W_S \cap W_2 \neq 1$ then $1 \neq W_S \cap Z(R)$, and then as $L^+ = H^+$ we obtain a contradiction from 11.2.5.1 and 11.2.7.2. Therefore $W_S \cap W_S^h = 1$ and as $W_S \lhd R$, we have $W_S W_S^h = W_S \times W_S^h$ abelian and either $W_2 = 1$ or we may take $W_2 = \langle y \rangle$ and $O_2(L) = \langle y \rangle \times W_S \times W_S^h$ is abelian. Hence $O_2(L)$ centralizes W_S and $C_{W_S W_S^h}(h) = 1$, so L acts on $W_z = \langle z, z^h \rangle \cong E_4$. Thus $[z^h, uy] = [z^h, u] = z$, a contradiction as z^h acts on $O(s)$ for each $s \in \mathcal{Z}_S$, so $[z^h, uy] \leq \langle f_{\mathcal{Z}_S - \sigma} \rangle$.

Therefore $y \notin R$. By 11.2.7.1, $R = O^{2'}(C_G(Z))$, so $[y, Z] \neq 1$. Let $x \in Z - C_Z(y)$ and $j \in \theta$. By 11.2.12, $[v(j), x] \in \langle j \rangle$, and as $[y, x] \neq 1$ there exists $j \in \theta$ with $[v(j), x] = j$. Then as R is transitive on θ and $x \in Z(R)$, $[v(k), x] = k$ for each $k \in \theta$. But then if R is exceptional, $[y, x] = c = 1$, contrary to $[y, x] \neq 1$. Therefore R is ordinary, establishing (1).

As R is ordinary, $[Z(R), y] \leq E \cap Z = \langle f \rangle$ by 11.2.4.2, so y induces a transvection on $Z(R)$ with center f, and hence $Z \cong E_4$ and $H^+ = H/C_H(Z(R)) = GL(Z)$ by 2.4.1 in [**Asc19**], where $H = \langle y^G \rangle$. Moreover $H^+ = O^{2'}(G^+)$.

Suppose $\theta \neq \mathcal{Z}_S$ and let $\theta \neq \beta \in \Theta$ and $y_0 = \prod_{t \in \beta} t$. By symmetry between θ and β, $\langle f_\beta \rangle = [Z(R), y_0]$ and $1 \neq y_0^+ \in O^{2'}(G^+) = H^+$, so $y_0^+ = y_+$. But then $\langle f_\beta \rangle = [Z(R), y_0^+] = [Z(R), y^+] = \langle f^+ \rangle$ a contradiction. Therefore R is transitive on $\theta = \mathcal{Z}_S$, proving (2).

Set $A = \langle y, W_S \rangle$ and $L = \langle A, A^h \rangle$ for $h \in H$ of order 3 inverted by y. As usual, appealing to 11.2.4.2 and 11.2.7.2, $Z(L) = 1$, $O_2(L) = W_S \times W_S^h$, and $L \lhd LR$. Then h acts on $W_z = \langle z, z^h \rangle$ with $[z^h, y] = z$. As $O_2(L)$ is abelian, z^h acts on $O(s)$ for each $s \in \theta$, and then also on $\langle v(s) \rangle$. Then as $[z^h, y] = z$, z^h centralizes $v(s)$ for $s \neq z$, while z^h inverts $v(z)$. Now h acts on each W_t and $c_{v(z)}$ centralizes $F = \prod_{s \neq z} W_s$ with $\Gamma = \langle c_{v(z)}, c_h \rangle$ inducing $GL(W_z)$ on W_z. Then $\Sigma_z = \langle c_{v(z)}^\Gamma \rangle$ centralizes F and induces $GL(W_z)$ on W_z. As R is transitive on $\mathcal{Z}_S$, Σ_s centralizes W_z and acts as $GL(W_s)$ on W_s. Thus as $z \in \mathcal{D}(s)$, also $z \in \mathcal{D}(s^h)$, so s^h centralizes $O(z)$ by 3.1.14. Therefore $C_{O(z)}(z^h) \leq C_{O(z)}(f^h) \leq C_{O(z)}(Z) = O(z) \cap R = \langle z \rangle$ by 11.2.7.1. Hence $z^h \in \mathcal{D}^o(z)$ by 3.3.6.

Let $\theta_z = \mathcal{Z}_S - \{z\}$ and $\mathcal{E}_z = C_{\mathcal{F}}(\theta_z)$. As $z \in \mathcal{F}^f$ and R is transitive on $\mathcal{Z}_S$, for each $t \in \mathcal{Z}_S$ we have t and $\theta_t \in \mathcal{F}_f$. Then $\mathcal{O}(\theta_z) \lhd \mathcal{E}_z$ with $O(z) \leq C_{\mathcal{E}_z}(\mathcal{O}(\theta_z)) = \mathcal{C}$, so $\mathcal{Y}_z = [O(z)]_{\mathcal{E}_z} \leq \mathcal{C}$. Then as $\Sigma_z \leq Aut_{\mathcal{E}_z}(W_z)$, we have $z^h \in z^{\mathcal{C}}$, so as $z^h \in \mathcal{D}^o(z)$, it follows from Theorem 5 that

(*) $\mathcal{Y}_z \cong L_3^-[m]$, $L_2[2m]^{(1)}$, $G_2[m]$, or M_{12}.

As R is transitive on $\mathcal{Z}_S$ and each $s \in \mathcal{Z}_S$ is in $\mathcal{F}^f$, $\mathcal{Y}_s = [O(s)]_{\mathcal{E}_s} \cong \mathcal{Y}_z$.

Next W_s centralizes $W_z O(z)$, so from the structure of $Aut(K(\mathcal{Y}_z))$ we have W_s inducing automorphisms on $\mathcal{Y}_z$ in $\langle z \rangle$. Then as Σ_z centralizes W_s, W_s centralizes $\mathcal{Y}_z$. But if $\mathcal{Y}_z$ is $L_3^-[m]$ or $L_2[2m]^{(1)}$ then $W_s O(s)$ is Sylow in $\mathcal{Y}_s$, so it follows from 2.3 in [**Asc11**] that $\mathcal{F}$ contains the direct product $\mathcal{Y}_0 = \prod_{t \in \mathcal{Z}_S} \mathcal{Y}_t$ of R-conjugates of $\mathcal{Y}_z$.

Observe that (1) implies (5). If $n_S = 1$ then (3) and (4), and hence also the lemma, holds by (*), so we may assume that $n_S > 1$.

Assume the hypothesis of (4). Then $\mathcal{F}^\circ = \mathcal{F}_S(L)$ for $L = (P)Sp_{2n_S}(q)$, $(P)SL_{2n_S}^{\epsilon\pi}(q)$, or $L_{2n_S+1}^{\epsilon\pi}(q)$. Let $\bar{L}$ be the covering group of L and U its d-dimensional defining module, as in section 5.4. Let $T = \langle \theta_z \rangle$ and observe that $\mathcal{Y}_z = \mathcal{F}_{S_Y}(Y)$ for some $Y \leq \bar{L}$. Now $[U, T]$ is of dimension $2(n_S - 1)$ and $C_U(T)$ is of dimension $d - 2(n_S - 1) = 2$ or 3, so as Y is faithful on $C_U(T)$ we conclude that $d = 2N_S + 1$ and $\mathcal{Y}_z \cong L_3^-[m]$. Next $\mathcal{Y}_0 = \mathcal{F}_{S_0}(Y_0)$ with $Y_0 = \prod s \in \mathcal{Z}_S \mathcal{Y}_s$ and $Y_s \cong L_3^{-\pi}(q)$, so $2N_S + 1 = m(U) \geq 3n_S$, contradicting $n_S > 1$. This proves (4).

It remains to prove (3), so we may assume the hypotheses of (3). Then by 6.6.8, $\tau_z^\circ = \tau_1 * \cdots * \tau_k$ is the central product of packets satisfying Theorem 1. Let $s \in \theta_z$. Then $z \in \theta_s$ and $\rho_s = (\mathcal{Y}_s, O(s))$ is coconnected, so $\rho_s \leq \tau_i = (\mathcal{C}_i, \Omega_i)$ for some i, and then $\mathcal{Y}_s = C_{\mathcal{C}_i}(O(\theta_s))$. As τ_i satisfies Theorem 1, we conclude that either $\tau_i = \rho_s$ or $\mathcal{Y}_s \cong L_3^-[m]$ and $\mathcal{C}_i \cong L_{2k+1}^-[m]$. Let $t \in \theta_s - \{z\}$; then as $O(t) \leq \mathcal{C}_i$, $\langle \mathcal{Y}_t, \mathcal{Y}_s \rangle \leq C_{\mathcal{C}_i}(O(\mathcal{Z}_S - \{t, s\})) \cong L_5^-[m]$. This is a contradiction as we've shown that $\mathcal{Y}_0 \leq \mathcal{F}$. Therefore the direct product $\xi_z = \prod_{s \in \theta_z} \rho_s$ is a subpacket of τ_z°, and $\tau_z^\circ = \mathcal{O}(z) \times \xi_z$.

If $\mathcal{Y}_z$ is $L_3^-[m]$ or $L_2[2m]^{(1)}$ we've shown that $\mathcal{Y}_0 \leq \mathcal{F}$. We will show the same holds when $\mathcal{Y}_z$ is $G_2[m]$ or M_{12}. Then Hypothesis 8.1.8 is satisfied by $(\rho_s : s \in \mathcal{Z}_S)$, so (3) follows from 8.1.10. Hence it remains to assume $\mathcal{Y}_z$ is $G_2[m]$ or M_{12} and to produce $\mathcal{Y}_0$.

Let T_z be Sylow in $\mathcal{Y}_z$. We showed that $X_s = W_s O(s)$ centralizes $\mathcal{Y}_z$, so T_z centralizes X_s. Then, arguing as above, T_z centralizes $\mathcal{Y}_s$, and then by 2.3 in [**Asc11**], $\mathcal{Y}_0$ exists, completing the proof. □

Because of 11.2.13, during the remainder of the section we assume:

HYPOTHESIS 11.2.14. *(1) $|\Omega(z)| = 2$ for each $z \in \mathcal{Z}_S$.*
(2) For each $K \in \Omega$, $|K \cap R| \leq 2$.

For $s \in \mathcal{Z}_S$, let $\Omega(s) = \{K_s, K_s'\}$, $V_s = K_s \cap W$, $V_s' = K_s' \cap W$, v_s, v_s' of order 4 in V_s, V_s', respectively, and $v(s) = v_s v_s'$. For $\theta \subseteq \mathcal{Z}_S$, set $v(\theta) = \prod_{t \in \theta} v(t)$, $W_\theta = \langle v(t), \theta : t \in \theta \rangle$, and $v_\theta = \prod_{t \in \theta} v_t$.

LEMMA 11.2.15. *If $n_S \leq 2$ then R is ordinary and $c \neq 1$.*

PROOF. If $\mathcal{Z}_S = \{z\}$ then $c = z \neq 1$, so R is ordinary. If $\mathcal{Z}_S = \{z_1, z_2\}$ is of order 2 then $z_1 \neq z_2$, so $c = z_1 z_2 \neq 1$, and again R is ordinary. □

LEMMA 11.2.16. *(1) For each $s \in \mathcal{Z}_S$, $v(s) \in R$.*
(2) $W_I \leq R$.

PROOF. By 11.2.2, W_I is the permutation module for $\tilde{\mu} = Sym(I)$, or that module modulo $\langle e_I \rangle$, and the elements $v(s)$, $s \in \mathcal{Z}_S$, are vectors of weight 1 in W_I. Therefore $W_I = \langle v(s), W_S, e_n : s \in \mathcal{Z}_S \rangle$. Observe that $e_n \in Z(S) \leq R$ if n is odd. Therefore (1) and 11.2.11 imply (2).

Assume $s \in \mathcal{Z}_S$ with $v(s) \notin R$. If R is ordinary then $s \in \theta \in \Theta$, and we set $f = f_\theta$ and $Z = \langle f^G \rangle$. If R is exceptional set $\theta = \mathcal{Z}_S$ and $Z = \langle f_1^G \rangle$. In any event set $E = \langle \theta \rangle$ and $y = \prod_{t \in \theta} v(t)$.

Assume $y \notin R$. By 11.2.7.1, $R = O^{2'}(C_G(Z))$, so $[y, Z] \neq 1$. Let $x \in Z - C_Z(y)$ and $t \in \theta$. By 11.2.12, $[v(t), x] \in \langle t \rangle$, and as $[y, x] \neq 1$ there exists $t \in \theta$ with $[v(t), x] = t$. Then as R is transitive on θ and $x \in Z(R)$, $[v(t), x] = t$ for each $t \in \theta$. But then if R is exceptional, $[y, x] = c = 1$, contrary to $[y, x] \neq 1$. Therefore R is ordinary.

Observe $r \in R$ permutes θ and hence also $\{v(t), v(t)t : t \in \theta\}$, so $[y, r] \in E \leq R$ by 11.2.11. Hence $y \in G$. As R is ordinary, $[Z(R), y] \leq E \cap Z(R) = \langle f \rangle$ by 11.2.4.2, so y induces a transvection on Z with center f, and hence $Z \cong E_4$ by 2.4.1 in [**Asc19**]. Next $v(t) = v_t v_t'$ with $v_t \in V_t$, $v_t' \in V_t'$ of order 4. As $[v(t), x] = t$ we may assume x centralizes v_t and inverts v_t'. Thus v_t centralizes $Z = \langle f, x\rangle$, so $v_t \in R$, contrary to 11.2.14.2.

We've shown that $y \in R$. Then either $Z(R)$ centralizes the projection $v(s)$ of y on $O(s)$ for $s \in \theta$, so $v(s) \in R$ by 11.2.7.1, or R is exceptional and for some $x \in Z$, $[v(s), x] = s$ for all $s \in \mathcal{Z}_S$. In the former case the lemma holds, so assume the latter. Then $v(s)$ induces a transvection on Z with center s, so by 2.4.1 in [**Asc19**], we have $Z = \langle f, x\rangle \cong E_4$. As $[v(s), x] = s$ we may assume x inverts v_s' and centralizes v_s. But then $v_s \in C_S(Z) = R$ by 11.2.7.1, contrary to 11.2.14.2. □

LEMMA 11.2.17. *If R is ordinary let $\theta \in \Theta$ and if R is exceptional let $\theta = \mathcal{Z}_S$. Then $v_\theta \in G$ but $v_\theta \notin R$.*

PROOF. As R permutes θ, $[v_\theta, R] \in W_I$, so $v_\theta \in G$ by 11.2.16.2. Suppose $v_\theta \in R$ and let $Z = Z(R)$. Then Z centralizes v_θ and acts on $O(t)$ for each $t \in \theta$ by 11.2.12. Thus if R is ordinary, Z centralizes the projection v_t of v_θ on $O(t)$, so $v_t \in R$ by 11.2.7.1, contrary to 11.2.14.2. Therefore R is exceptional.

Let $P = O(\tau)$ and $P^+ = P/W_S$. Then Z centralizes v_θ^+ and acts on $O(t)$, so Z centralizes the projection v_t^+ of v_θ^+ on $O(t)^+$. Thus for $x \in Z$, $[v_t, x] = t^{\epsilon(x,t)}$ for some $\epsilon(x,t) \in \{0,1\}$. Then $[v_\theta, x] = f_{\sigma(x)}$, where $\sigma(x) = \{t \in \theta : \epsilon(x,t) = 1\}$, so as x centralizes v_θ, $\sigma(x) = \emptyset$ or θ. By 11.2.14.2, $v_t \notin R$, so there exists $x \in Z$ with $\sigma(x) = \theta$. Then $Z_0 = \{x \in Z : \sigma(x) = \emptyset\}$ is a hyperplane of Z centralizing v_t for each $t \in \theta$.

Choose θ_1 as in 11.2.5.2, let $y = v_{\theta_1}$, $E = \langle v_\theta, W_S\rangle$, and $A = \langle y, E\rangle$. Then $[R, y] \leq E \leq R$, so $y \in G$. Suppose $y \in R$. Then, as in the previous paragraph, Z centralizes v_t for $t \in \theta_1$, so $v_t \in R$ by 11.2.7.1, contrary to 11.2.14.2. Therefore $y \notin R$.

Set $H = \langle y^G\rangle$, $f = f_1$, and $F = \langle f^G\rangle$. As $y \notin R$ and $F \leq Z(R)$ by 11.2.5.1, y is nontrivial on F by 11.2.7.1, so there is $x \in F - C_F(y)$. By 11.2.12 either x acts on each V_s or on no V_s. In the first case $[x, y] = f$, while in the second $[x, y] = v(\theta_1)$, contrary to 11.2.5.1. Thus y induces a transvection on F and $Z(R)$ with center f, so by 2.4.1 in [**Asc19**], $F \cong E_4$ and $Aut_H(Z(R)) = GL(F)$. As in the proof of 11.2.13, $H = \langle A, A^h\rangle$ for $h \in H$ of order 3 inverted by y, and setting $E_0 = E \cap E^h$, either $O_2(H) = E \times E^h$ with $C_{O_2(H)}(h) = 1$, or we may take $E_0 = \langle v_\theta\rangle$ and $O_2(H) = W_S \times W_S^h \times E_0$. Hence for each $z \in \mathcal{Z}_S$, $B_z = \langle z, z^h\rangle \cong E_4$ is an H-invariant subspace of $O_2(H)$ and $[z^h, y] = z$. Therefore H acts on $B = \prod_{t \in \mathcal{Z}_S} B_t$. As $[z^h, y] = z$ and z^h centralizes $v_\theta \in O_2(H)$ since $O_2(H)$ is abelian, it follows that z^h inverts v_z and centralizes v_s for $s \in \mathcal{Z}_S - \{z\}$. By symmetry, v_z centralizes s^h, so v_z induces a transvection on B with center z and axis $\langle z, B_s : s \in \mathcal{Z}_S - \{z\}\rangle$. By symmetry, for each $z \neq s \in \mathcal{Z}_S$, v_{z^h} centralizes B_s, so $\Sigma = \langle v_z, v_{z^h}\rangle$ centralizes B_s and induces $GL(B_z)$ on B_z. By symmetry, $C_{\mathcal{F}}(B_z)$ induces $GL(B_s)$ on B_s. Thus as $z \in \mathcal{D}(s)$, also $z \in \mathcal{D}(s^h)$, so s^h centralizes $O(z)$ by 3.1.14. Therefore $C_{K_z}(z^h) \leq C_{K_z}(f^h) \leq C_{K_z}(F) = K_z \cap R = \langle z\rangle$ by 11.2.7.1 and 11.2.14.2. Hence $z^h \in \mathcal{D}^o(z)$ by 3.3.6.

Let $\theta_z = \mathcal{Z}_S - \{z\}$, $X = \langle \theta_z\rangle$, and $\rho_z = (C_{\mathcal{F}}(X), \Omega(z))$. By aa.2, ρ_z is a quaternion fusion packet, and as Σ centralizes X, $z^h \in \mathcal{Z}(\rho_z)$. Then as $z^h \in \mathcal{D}^o(z)$, it follows from 9.4.34, 9.4.35, and 9.5.33 that $\mathcal{Y}_z = C_{\mathcal{F}}(O(\theta_z))^\circ$ is $\Omega_6^-[m]$, $\Omega_7[m]$, or $\Omega_8^-[m]$. This is a contradiction since as R is exceptional, $c = 1$, so we have $z = f_{\theta_z} \in X$, and hence z is in the center of $\mathcal{Y}_z$. □

LEMMA 11.2.18. *R is ordinary.*

PROOF. Assume R is exceptional and let $\theta = \mathcal{Z}_S$ and $y = v_\theta$. By 11.2.17, $y \in G - R$, so $H = \langle y^G \rangle$ is not contained in R. Set $E = Z(R) \cap W_I$. By 11.2.4.1, $C_{W_I}(R)^\# = \{e_\sigma : \sigma \in \pi \in \tilde{\Theta}\}$. Let $\pi = \{\sigma, \bar{\sigma}\} \in \tilde{\Theta}$, and recall from iI.2 that $\mathcal{Z}_S = \{g_i : 1 \leq i \leq n_S\}$, where $g_i = e_{2i,2i-1}$. If $g_i \in \sigma$ for some i, then as R is transitive on $\mathcal{Z}_S$, each g_j is in σ or $\bar{\sigma}$, so π corresponds to a member of $\hat{\Theta}$ and $e_\sigma \in Z_\beta$. Otherwise for each i, σ contains a unique member of $\{2i, 2i-1\}$. We conclude that either $E = Z_\beta$ or $E = Z_\beta\langle e_\xi \rangle$ for some $\xi \subseteq I$ such that ξ contains exactly one member of $\{2i, 2i-1\}$ for each $1 \leq i \leq n/2$. Note in the latter case, $e_\xi = v(\theta)$ for suitable choice of v_t, v_t', $t \in \theta$. so we normalize and take $v(\theta) = e_\xi \in E$.

Set $Z = \langle E^G \rangle$. As $[R, y] \leq W_I$, $[Z, y] \leq Z \cap W_I \leq E$. By 11.2.7.1, y does not centralize Z, so there is $x \in Z - C_Z(y)$. As R is transitive on θ and $x \in Z(R)$, $K_t^x = K_t$ for all $t \in \theta$ or $K_t^x = K_t'$ for all $t \in \theta$. Suppose the first case holds. As $[y, x] \neq 1$, $[v_t, x] \neq 1$ for some $t \in \theta$, and then as R is transitive on θ and $x \in Z(R)$, x inverts v_t for each $t \in \theta$. But then $[y, x] = \prod_{t \in \theta} t = c = 1$, a contradiction. Therefore $K_t^x = K_t'$ for each $t \in \theta$, so it follows that $C_Z(y)$ is a hyperplane of Z. Thus y induces a transvection on Z, so by 2.4.1 in [**Asc19**], $Z = F \times C_Z(H)$ with $F = [Z, H] \cong E_4$ and $Aut_H(Z) = GL(F)$. As R is exceptional, n_S is a power of 2, and then by 11.2.15, $n_S \geq 4$. Then by 11.2.7.2, $W_I \cap C_Z(H)$ contains no vectors of weight n_S. However as $K_t^x = K_t'$, $[v_t, x] = v(t)t^{\epsilon(x,t)}$ for some $\epsilon(x, t) \in \{0, 1\}$, so $[y, x] = v(\theta) f_{\sigma(x)}$, where $\sigma(x) = \{t \in \theta : \epsilon(x, t) = 1\}$. Thus by our choice of notation in paragraph one, $v(\theta) \in E$, so $m(E) > 1$. Thus $C_E(H) \neq 1$, so as each member of $E^\#$ is of weight n_S, we have a contradiction. □

LEMMA 11.2.19. *(1) For each $\theta \in \Theta$, $Z(R) \cap W_\theta = \langle f_\theta \rangle$ or $\langle f_\theta, v(\theta) \rangle$.*

(2) f_θ is of weight $2|\theta|$ in W_I and $v(\theta)$ and $f_\theta v(\theta)$ are of weight $|\theta|$. In particular if $|\theta| > 1$ then each member of $Z(R) \cap W_\theta$ is in $\mathcal{W}(R)$.

PROOF. Set $W_\theta^+ = W_\theta / \langle \theta \rangle$. Then $X = \{v(t)^+ : t \in \theta\}$ is a basis for W_θ^+. Further for $r \in R$ and $t \in \theta$, $v(t)^{+r} = v(t^r)^+$, so as X is a basis and R is transitive on θ, $C_{W_\theta^+}(R) = \langle v(\theta)^+ \rangle$. Hence by 11.2.4.2, $Y = Z(R) \cap W_\theta = \langle f_\theta \rangle$ or $\langle f_\theta, y \rangle$, where $y = v(\theta) f_\sigma$ for some $\sigma \subseteq \theta$. However, as in the proof of the previous lemma, in the latter case for suitable choice of v_t, v_t' for $t \in \theta$, we can choose $\sigma = \emptyset$, and (1) holds. Part (2) is a straightforward observation, given 11.2.2. □

For $\theta \in \Theta$, set $H_\theta = \langle v_\theta^G \rangle \leq G$, $A_\theta = W_\theta \langle v_\theta \rangle$, and $L_\theta = \langle A_\theta, A_\theta^h \rangle$ for some $h \in H_\theta$ of odd order with $\langle v_\theta^*, v_\theta^{*h} \rangle$ not a 2-group. Set $E_\theta = Z(R) \cap W_\theta$ and $F_\theta = \langle E_\theta^G \rangle$.

LEMMA 11.2.20. *Let $H = H_\theta$, $F = F_\theta$, $E = E_\theta$, and $H^+ = H / C_H(Z(R))$.*
(1) $R = C_S(F)$ and v_θ is nontrivial on F.
(2) For $x \in Z(R) - C_S(v_\theta)$, either
(a) for all $t \in \theta$, $[v_t, x] = t$ and $[v_\theta, x] = f_\theta$, or
(b) for all $t \in \theta$, $K_t^x = K_t'$ and $[v_\theta, x] = v(\theta)$ or $v(\theta) f_\theta$.
(3) $C_H(Z(R)) = C_H(F)$.
(4) If $E \cong \mathbf{Z}_2$ then $F \cong E_4$ and $H^+ = GL(F)$.

(5) If $|\theta| > 1$ and $E \cong E_4$ then $E = [F, v_\theta]$ and either $F \cong E_{16}$ and $H^+ \cong S_3$ or D_{10}, or $C_F(H) \neq 1$ and $F/C_F(H)$ is the natural module for $H^+ \cong L_2(4)$.

(6) If $|\theta| = 1$ and $E \cong E_4$ then either (5) holds or $F \cong E_8$ and $H^+ \cong S_3$ or F is the natural module for $H^+ \cong A_5$.

(7) If $|\theta| > 1$ then $L_\theta \trianglelefteq RL_\theta$, $O_2(L_\theta) = W_\theta \times W_\theta^h$, $|h| = p \in \{3, 5\}$, and $C_{O_2(L_\theta)}(h) = 1$.

PROOF. By 11.2.19, F contains $f_\theta \in \mathcal{W}(R)$, so part (1) follows from 11.2.7.1.

Choose x as in (2) and let $f = f_\theta$, $v = v(\theta)$, and $y = v_\theta$. As $x \in Z(R)$ and R is transitive on θ, either x acts on each K_t or x moves each K_t. In the first case as $[y, x] \neq 1$ and $R \leq C_S(x)$ is transitive on θ, $[v_t, x] = t$ for all $t \in \theta$, and hence (2a) holds. In the second case $[v, x] = vf_\sigma$ for some $\sigma \subseteq \theta$, so (2b) holds by 11.2.19. Hence (2) is established.

As $[Z(R), y] \leq F$ and $H = \langle y^G \rangle$, we have $[Z(R), H] \leq F$, so $O^2(C_H(F)) = O^2(C_H(Z(R))$. Then (1) implies (3).

Suppose $E \cong \mathbf{Z}_2$. Then by 11.2.19, $E = \langle f \rangle$ and y induces a transvection on F with center E by (2), so (4) follows from 2.4.1 in [**Asc19**].

So assume $E \cong E_4$. Again by (2) either y induces a transvection on F or $E = [F, y]$. However in the first case $F = [F, H] \times C_E(H)$ with $C_E(H) \neq 1$, so 11.2.19 and 11.2.7.2 supply a contradiction when $|\theta| > 1$. Further if $|\theta| = 1$ then $F \cong E_8$ and (6) holds in this case. Therefore we may assume $E = [F, y]$ is of rank 2. Hence by 2.8.15, H^+ and its action on F are described in that lemma. If $m(F) = 6$ or F is the $L_2(4)$-module for $H^+ \cong L_2(4)$, then $N_H(E)$ is transitive on $E^\#$, a contradiction as $N_{\mathcal{F}}(W)$ controls fusion in W_I while by 11.2.19, f is of weight $2|\theta|$ while v is of weight $|\theta|$. Hence if $F/C_F(H)$ is the natural module for $H^+ \cong L_2(4)$ then $C_F(H) \neq 1$ so (5) holds in this case. Thus we may assume $m(F) = 4$ and $H^+ \cong \mathbf{Z}_2/E_9$, A_5, S_3, or D_{10}. In the first two cases $H^+ = \langle C_{H^+}(v), C_{H^+}(vf) \rangle$, contrary to 11.2.7.2 and 11.2.19 when $|\theta| > 1$. Thus (5) and (6) hold.

Next by (5) and (6), $\langle v^+, v^{+h} \rangle \cong D_{2p}$ for $|h^+| = p \in \{3, 5\}$. The usual argument based on 11.2.19 and 11.2.7.2 show that $L = L_\theta = \langle v, v^h \rangle B \trianglelefteq RL$, where $B = O_2(L) = W_\theta \times W_\theta^h$ and $C_B(h) = 1$. As $|h^+| = p$, $k = h^p$ centralizes F, and $B = [B, k] \times C_B(k)$ with $F \leq C_B(k)$. As $L \trianglelefteq RL$, $R = C_R(k)[B, k]$, so if $k \neq 1$ then there is $1 \neq u \in C_{[B,k]}(R)$, so $u \in C_B(R) = F$, a contradiction. Therefore $k = 1$, so $|h| = p$, completing the proof of (7). □

LEMMA 11.2.21. *R is transitive on $\mathcal{Z}_S$, so $\Theta = \{\mathcal{Z}_S\}$.*

PROOF. Suppose θ_i, $i = 1, 2$, are distinct members of Θ. Let $v_i = v_{\theta_i}$, etc, and $G^+ = G/C_G(Z(R))$. As G^+ has a strongly embedded subgroup and $v_1^+ \in H_1^+ \trianglelefteq G^+$ with v_1^+ an involution, all involutions in G^+ are in H_1^+, so $v_2^+ \in H_1^+$. Then $H_2^+ = \langle v_2^{+G} \rangle \leq H_1^+$, so by symmetry, $H_1^+ = H_2^+ = H^+$.

Suppose first that $v_1^+ = v_2^+$. Then $[Z(R), v_1] = [Z(R), v_1^+] = [Z(R), v_2^+] = [Z(R), v_2]$. But by 11.2.19, $[Z(R), v_i] \leq E_i$, so $E_1 \cap E_2 \neq 1$. By 11.2.19, $E_i \leq \langle f_i, v(i) \rangle$, so as $\theta_1 \neq \theta_2$ we conclude that $f_1 = f_2 = f$ generates $E_1 \cap E_2$ and $\Theta = \{\theta_1, \theta_2\}$. Then we conclude $[Z(R), v_i] = \langle f \rangle$, so by 11.2.20, $H^+ \cong S_3$ and $F = F_i = \langle f^G \rangle \cong E_4$ for $i = 1, 2$.

Suppose $|\theta_i| = 1$ for $i = 1, 2$. Then as $\Theta = \{\theta_1, \theta_2\}$, $n_S = 2$, so $c \neq 1$ by 11.2.15, contradicting $f_1 = f_2$. Thus we may assume $|\theta_1| > 1$. Therefore by 11.2.20.7, $L_1 = B_1\langle v_1, v_1^h \rangle \trianglelefteq RL_1$ where h is of order 3, $B_1 = W_1 \times W_1^h$, and $C_{B_1}(h) = 1$. As $v_1^+ = v_2^+$, $u = v_1 v_2^{-1} \in R$, so as $L_1 \trianglelefteq RL_1$, $[u, L_1] \leq B_1$. Then

$Y = L_1\langle u\rangle = \langle u_0\rangle \times L_1$ where $\langle u_0\rangle = Z(Y) \cong \mathbf{Z}_2$ and $u = u_0 w_1$ for some $w_1 \in W_1$. Now $v_2 = uv_1 \in Y$, so $\langle v_1, v_2^h\rangle$ contains a 3-element k such that $L_2 = \langle A_2, A_2^k\rangle$ with $L_2^+ = H^+$. As usual $L_2 \lhd RL_2$, so as $k \in h^{B_1}$, $B_1 = [B_1, k] \leq L_2$ and then as v_2 centralizes W_1, $W_1 \leq C_{B_2}(v_2) = W_2$. This is a contradiction as $\theta_1 \neq \theta_2$.

Therefore $v_1^+ \neq v_2^+$. But if $H^+ \cong D_{2p}$ for $p \in \{3, 5\}$ then $\langle v_1^+\rangle$ and $\langle v_2^+\rangle$ are Sylow groups of H^+, so $v_1^+ = v_2^+$, a contradiction. Therefore by 11.2.20, $H^+ \cong A_5 \cong L_2(4)$ and $T^+ = \langle v_1^+, v_2^+\rangle \cong E_4$ is Sylow in H^+.

Suppose $E_1 \cap E_2 \neq 1$. Then, as above $\Theta = \{\theta_1, \theta_2\}$ and $f_1 = f_2$ with $c = 1$. Then, again as above, $n_S \neq 2$, so we may choose $|\theta_1| > 1$, so by 11.2.20, $F/C_F(H)$ is the $L_2(4)$-module. Then as $E_1 \cap E_2 = \langle f_1\rangle$, $|C_F(H)| = 2$ and $C_F(v_1) = C_F(H) \times F_0$ where $F_0 \cong E_4$ is $N_{H^+}(T^+)$-invariant. But then as $v(1)$ and $v(1)f_1$ are of weight $|\theta_1|$ and f_1 is of weight $2|\theta_1|$, we conclude $f_1 \in F_0$, so $E_1 \cap E_2 \neq \langle f_1\rangle$.

Therefore $E_1 \cap E_2 = 1$, so it follows that $C_F(H) \cong E_4$ and $F/C_F(H)$ is the $L_2(4)$-module for H^+ with $E = C_F(T) = E_1 \times E_2$. Thus as $E_i \leq W_I$ also $E \leq W_I$. As $E_2 \in E_1^{N_H(E)}$, $k = |\theta_1| = |\theta_2|$. Now E_i has two members $v(i)$ and $v(i)f_i$ of weight k and one member f_i of weight $2k$ in W_I, so the diagonal members of E consist of 4 members of weight $2k$, four of weight $3k$ and one member e of weight $4k$. As $k = 2^a$, e is of weight $4k = 2^{a+2}$, so by 11.2.7.2, $e \notin C_F(H)$. However all members of $E - C_F(H)$ are in orbits of length 3 under $N_L(E)$, so e is also of weight $2k$ or $3k$, and hence $n = 6k$ or $7k$ and $e_I = 1$ in W_I. Thus there are five diagonal members of weight $4k$ and four of weight $3k$ or $2k$, so some member of weight $4k$ is in $C_F(H)$, a contradiction. This completes the proof of the lemma. □

During the remainder of the section we assume:

HYPOTHESIS 11.2.22. $n_S > 1$.

By 11.2.21, $\theta = \mathcal{Z}_S$ is the unique member of Θ. We write H, L, etc. for H_θ, L_θ, etc. and set $f = f_\theta$, $v = v(\theta)$, and $y = v_\theta$. Set $B = W_S W_S^h$.

LEMMA 11.2.23. *(1) B is the direct sum of the irreducible L-modules $B_t = \langle t^L\rangle$ for $t \in \theta$, of dimension $2, 4$ for $p = |h| = 3, 5$, respectively.*

(2) If $p = 3$ then $[f^h, y] = f$ and $[f^h, v_t] = [t^h, y] = t$.

(3) If $p = 5$ then $[f^h, y] = vf^\epsilon$ and $[f^h, v_t] = [t^h, v_t] = v(t)t^\epsilon$ and $K_t^{f^h} = K_t' = K_t^{t^h}$.

PROOF. By 11.2.21, $|\theta| = n_S$, so by 11.2.22, $|\theta| > 1$. Therefore by 11.2.20.7, $B = W_S \times W_S^h$ with $W_S = C_B(y)$ and $C_B(h) = 1$. It follows that each $w \in W_S^\#$ is in a unique irreducible L-submodule B_w of B, so B_w is of dimension $2, 4$ for $p = 3, 5$, respectively. Thus (1) holds.

In particular if $p = 3$ then $B_f \cong E_4$ so $[f^h, y] = f$, and hence $[f^h, v_t] = [t^h, y] = t$, so (2) holds. So take $p = 5$. Then we may view B_f as the restriction of the orthogonal module for $\Omega_4^-(2)$ to L^+ with f^L the set of singular points of B_f. Hence the hyperplane $f^\perp$ contains $C_{B_t}(y)$ with $f^L - \{f\} \subseteq B_f - f^\perp$ and $[f^\perp, y] = \langle f\rangle$. Therefore $[f^h, y] \neq f$, so (3) follows from 11.2.20.2. □

LEMMA 11.2.24. *(1) $\Lambda = Aut_{\mathcal{F}}(B)$ contains a subgroup $\prod_{t\in\theta}\Lambda_t$ such that Λ_t centralizes B_s for $s \in \theta - \{t\}$ and Λ_t acts as $L_{|B_t}$ on B_t.*

(2) B_t centralizes $O(s)$ for $s \in \theta - \{t\}$.

(3) If $|h| = 3$ then $U_t = \langle B_t, v(t), v(t)^h\rangle$ centralizes $O(s)$ for $s \in \theta - \{t\}$, $K_t^{v(t)^h} = K_t'$, and $t^h \in \mathcal{D}^o(t)$.

(4) If $|h| = 5$ then $t^h \in \mathcal{D}^m(t)$.

(5) If $|h| = 5$ then for some $b(t) \in \{v(t), v(t)t\}$, $b(t)^h$ acts on K_t and $C_{K_t}(b(t)^h) = \langle t\rangle$.

PROOF. Set $a = t^h$ and observe that (4) follows from 11.2.23.3.

As B is abelian, a acts on $O(s)$ for each $s \in \theta$. By parts (2) and (3) of 11.2.23, $[a, y] = t,\ v(t)t^{\epsilon}$ for $p = |h| = 3, 5$, respectively. Therefore as $y = \prod_s v_s$ and a acts on $O(s)$, we conclude that a centralizes v_s for $s \in \theta - \{t\}$ and a inverts v_t when $p = 3$. Hence $\langle v_t^L\rangle$ induces $L_{|B_t}$ on B_t and centralizes B_s for $s \in \theta - \{t\}$, so $\Lambda_t \leq \Lambda$ and then (1) holds.

Now $B_t = \langle t^{\Lambda_t}\rangle$, so as Λ_t centralizes B_s and $t \in \mathcal{D}(s)$, also $a \in \mathcal{D}(s)$, and hence (2) follows from 3.1.14.

Suppose $p = 3$. Let $\alpha \in \mathfrak{A}(a)$; then $z = a\alpha \in \mathcal{Z}_S$ and as $O(s)$ centralizes a we have $O(s)\alpha$ centralizing $O(z)$. Then as $v(t)^h \in O(z) \leq C_S(O(s)\alpha)$, we conclude that $v(t)^h$ centralizes $O(s) = O(s\alpha)\alpha^{-1}$, establishing the first statement in (3). By 11.2.20.2, $K_t^{v(t)^h} = K_t'$.

Next if $a \notin \mathcal{D}^o(t)$ then $P = C_{K_t}(a)$ is of order at least 4. But as $O(t)$ centralizes s^h for $s \in \theta - \{t\}$, P centralizes $\prod_{r\in\theta} r^h = f^h$, so $P \leq C_S(Z_0)$, where $Z_0 = \langle f, f^h\rangle$. Note that $Z_0 = \langle f^L\rangle$ while $O^{2'}(C_G(Z_1)) = R$ by 11.2.7.1, where $Z_1 = \langle f^H\rangle$. Thus $O^{2'}(C_G(Z_0)) = R$ if $H = L$. On the other hand if $H \neq L$ then $Aut_H(Z(R)) \cong L_2(4)$ by 11.2.20 and $O^{2'}(C_G(Z_0)) = O^{2'}(C_G(Z_1)) = R$. Thus in any event, $O^{2'}(C_G(Z_0)) = R$, so if $S_1 = C_S(Z_0) \not\leq R$ then $S_2 = N_{S_1}(R) \not\leq R$, a contradiction. Therefore as $P \leq R$ and $|P| > 2$, 11.2.14.2 supplies a contradiction. This completes the proof of (3).

Assume $p = 5$. By (4), for some $b \in \{v, vf\}$, b^h acts on K_t for each $t \in \theta$. Then $[b^h, y] = f$, so $[b(t)^h, y] = t = [b(t)^h, v_t]$, where $b(t)$ is the projection of b on B_t. Set $P = C_{K_t}(b(t)^h)$. By (2), B_s centralizes $O(t)$ for $s \in \theta - \{t\}$, so $P \leq C_S(b^h)$. But P centralizes f and $O^{2'}(C_G(\langle f, b^h\rangle)) = R$, so $P \leq R$ by 11.2.7.1. Therefore $P = \langle t\rangle$ by 11.2.14.2, establishing (5) and the lemma. □

LEMMA 11.2.25. *Let $t \in \theta$, $\Sigma = \theta - \{t\}$, and $\mathcal{Y} = C_{\mathcal{F}}(O(\Sigma))^{\circ}$.*

(1) t and $\langle\Sigma\rangle$ are fully normalized and $\xi = (\mathcal{Y}, \Omega(t))$ is a quaternion fusion packet with $\mu(\xi) \cong E_4, S_4$ for $n = 2n_S,\ 2n_S + 1$, respectively.

(2) $v(t) \notin \mathcal{Z}$.

(3) $|h| = 3$.

(4) $t^h \in \mathcal{D}_{\mathcal{Y}}^o(t)$, $\mathcal{Y} \cong L_4^-[m]$ if n is even, and $\mathcal{Y} \cong \Omega_7[m]$ or $\Omega_8^-[m]$ if n is odd.

(5) Let $\mathcal{X}_t = \langle \Lambda_t, N_{O(t)}(U_t)U_t\rangle_{U_tO(t)}$, $u_1 = v(t)$, $u_2 = tv(t)$, $U(i) = \langle t^{C_{\mathcal{X}_t}(u_i)}\rangle$ for $i = 1, 2$, and let $\mathcal{Y}_t$ be the subsystem of $\mathcal{Y}$ on $U_tO(t)$ generated by $U_t\mathcal{O}_t$ and $N_{O(t)}(U(i))N_{\mathcal{X}_t}(U(i))$ for $i = 1, 2$. Then $\mathcal{Y}_t \cong L_4^-[m]$.

PROOF. As R is transitive on $\mathcal{Z}_S$ by 11.2.21, t and $\langle\Sigma\rangle$ are fully normalized. By construction, $\Omega(t)$ is the subset of Ω centralizing $Q = O(\Sigma)$, so by 2.5.2, ξ is a quaternion fusion packet. From 11.2.2, $\tilde{M} = \mu/\mu_0 \cong Sym(I)$ with W_I the permutation module or the permutation module modulo e_I. Then $\mu(\xi)/\mu(\xi)_0$ is the symmetric group on $n - 2n_S + 2 = 2, 3$ letters, so $\mu(\xi)$ is E_4, S_4, respectively, establishing (1).

As $v(t)$ is of weight 1 in W_I and t is of weight 2, $v(t) \notin t^{N_{\mathcal{F}}(W)}$ and hence (2) follows as $N_{\mathcal{F}}(W)$ controls fusion in W.

Set $a = t^h$ and let Y be Sylow in $\mathcal{Y}$. By parts (1) and (2) of 11.2.24, $B_t \leq C_S(Q)$ and Λ_t acts as $L_{|B_t}$ on B_t and centralizes Q. Therefore $a \in t^{\mathcal{Y}} = \mathcal{Z}(\xi)$.

Assume $|h| = 5$. By 11.2.24.5 for some $b \in \{v(t), v(t)t\}$, b^h acts on K_t and $C_{K_t}(b) = \langle t \rangle$. Suppose $\mu(\xi) \cong E_4$. Then by 9.4.34, $\mathcal{Y}$ is AE_5, $PSp_4[m]$, or $L_4^-[m]$. In the first two cases $N_Y(K_t) = O(t)$ so $b \in O(t)$, contradicting $C_{K_t}(b) = \langle t \rangle$. Similarly in the third case each involution in $N_Y(K_t) - O(t)$ is in $\mathcal{Z}(\xi)$, contrary to (2).

Therefore $\mu(\xi) \cong S_4$, so by 9.4.35 and 9.5.33, $\mathcal{Y}$ is AE_6, $\check{A}E_6$, AE_7, $P\Omega_6^+[m]$, $\Omega_6^+[m]$, $\Omega_7[m]$, $\Omega_8^-[m]$, $Sp_6(2)$, or $L_3(2)/E_{64}$. In the last case, $Aut_{\mathcal{Y}}(B_t)$ is a $5'$-group, a contradiction. If $\mathcal{Y}$ is AE_6, $\check{A}E_6$, or AE_7 then as $D_{10} \cong \Lambda_t \leq Aut_{\mathcal{Y}}(B_t)$, we conclude that $B_t \leq O_2(\mathcal{Y})$. But then $C_{K_t}(b^h) \neq \langle t \rangle$, a contradiction. If $\mathcal{Y}$ is $P\Omega_6^+[m]$ then $\mathcal{Y}$ has one class of involutions, contrary to (2). If $\mathcal{Y}$ is $Sp_6(2)$ then b is of type b_1, so again $C_{K_t}(b^h) \neq \langle t \rangle$. In the remaining cases, $\mathcal{Y}$ is $\Omega_k^\epsilon[m]$ so $\mathcal{Y}$ is the 2-fusion system of $L_0 = \Omega(U, \mathfrak{q})$ as in Notation 5.3.2. Then from 5.3.4, b is of type $i(\pi, 2)$ on U. As b^h acts on K_t with $C_{K_t}(b^h) = \langle t \rangle$, $[U, b^h] \leq [U, t] = \mathfrak{U}_1$ with $[U, b^h]$ of dimension 2 and sign $-\pi$. This is a contradiction as $[U, b^h] = [U, b]^h$, and hence is of sign π. This proves (3).

By (3), $|h| = 3$, so by 11.2.24.3, $a \in \mathcal{D}^o_{\mathcal{Y}}(t)$. Therefore if $\mu(\xi) \cong E_4$ it follows from 9.4.34 that $\mathcal{Y} \cong L_4^-[m]$. Similarly if $\mu(\xi) \cong S_4$ then $\mathcal{Y}$ is $\Omega_7[m]$ or $\Omega_8^-[m]$ by 9.4.35 and 9.5.33, and hence (4) follows from (1).

By 11.2.24.3, $U_t = [U_t, \Lambda_t] \leq C_S(Q)$, so $U_t O(t)$ is contained in the Sylow group Y of $\mathcal{Y}$. If n is even then $\mathcal{Y} \cong L_4^-[m]$, so as $|U_t O(t)| = 2m^2 = |Y|$, we have $U_t O(t) = Y$ and then even $\mathcal{Y} = \mathcal{Y}_t$ by 9.4.36. Similarly if n is odd then $\mathcal{Y}$ is $\Omega_7[m]$ or $\Omega_8^+[m]$, so as $a \in \mathcal{D}^o(t)$ and $K_t^{v(t)^h} = K_t'$, 9.4.37 says that $\mathcal{Y}_t \cong L_4^-[m]$. This proves (5). □

By 11.2.25.3, $|h| = 3$. For $t \in \theta$, define $U_t = \langle t, t^h, v(t), v(t)^h \rangle$ and $\mathcal{Y}_t \leq \mathcal{Y}$ as in 11.2.25.5. Thus $Y_t = O(t)U_t$ is Sylow in $\mathcal{Y}_t$ and $\mathcal{Y}_t \cong L_4^-[m]$. Let X_t be a model for $N_{\mathcal{Y}_t}(U_t)$.

In the remainder of the section we assume the following hypothesis:

HYPOTHESIS 11.2.26. *Either*
(1) $\mathcal{F}^\circ$ *is* $(P)\Omega_{2n}^+[m]$, $\Omega_{2n+1}[m]$, *or* $\Omega_{2n+2}^-[m]$, *or*
(2) The Inductive Hypothesis is satisfied in τ

LEMMA 11.2.27. *Let* $z \in \mathcal{Z} \cap \mathcal{F}^f$. *Then*
(1) $\tau_z^\circ = \tau_1 * \tau_2$, *where* $\tau_1 = (\mathcal{O}(z), O(z))$ *and* $\tau_2 = (\mathcal{C}, \Omega_2)$ *with* $\Omega_2 = \Omega - \Omega(z)$.
(2) If n *is even then* $\mathcal{C} \cong \Omega_{2n-2}^-[m]$.
(3) If n *is odd then either* $\mathcal{Y} \cong \Omega_7[m]$ *and* $\mathcal{C} \cong \Omega_{2n-3}[m]$ *or* $\mathcal{Y} \cong \Omega_8^-[m]$ *and* $\mathcal{C} \cong \Omega_{2n-2}^-[m]$.

PROOF. By 11.2.22, $n_S > 1$, so $n \geq 4$. If 11.2.26.2 holds then by 11.1.5.1, part (1) holds with $\mathcal{C}$ described in 11.1.5.3. If 11.2.26.1 holds this follows from the discussion in section 5.3.

Choose t and $\mathcal{Y}$ as in 11.2.25, with $t \neq z$, and set $\Xi = \Sigma - \{z\}$. Then $\mathcal{Y} = C_{\mathcal{C}}(O(\Xi))^{\circ}$. If n is even then $\mathcal{Y} \cong L_4^-[m]$ by 11.2.25.4, so (2) follows from 11.1.5.3. If n is odd then $\mathcal{Y}$ is $\Omega_7[m]$ or $\Omega_8^-[m]$ by 11.2.25.4, so (3) follows from 11.1.5.3. □

LEMMA 11.2.28. *(1) $U_t = C_{Y_t}(B_t)$ and X_t/U_t is the subgroup of $O(U_t, \mathfrak{q})$ generated by transvections for some quadratic form $\mathfrak{q}$ on U_t of sign $+1$, such that t^{X_t}, $v(t)^{X_t} \cup (v(t)t)^{X_t}$ is the set of singular, nonsingular points of U_t, respectively.*

(2) There is a subsystem $\mathcal{Y}_0 = \prod_{t\in\theta} \mathcal{Y}_t$ of $\mathcal{F}$.

(3) $n = 4$ or 5.

PROOF. Let $\mathfrak{U}_t = C_{Y_t}(B_t)$. By 11.2.25.4, $t^h \in \mathcal{D}^{\circ}_{\mathcal{Y}_t}(t)$ and by 11.2.25.5 $\mathcal{Y}_t \cong L_4^-[m]$, so by 9.4.16, $\mathfrak{U}_t \cong E_{16}$. But by construction, $E_{16} \cong U_t \leq \mathfrak{U}_t$, so we conclude that $\mathfrak{U}_t = U_t$. Then X_t is described in 9.4.16, which establishes (1).

Set $U = O_2(L)$; from 11.2.20.7, $U = W_\theta \times W_\theta^h = \prod_{s\in\theta} U_s$. Let G_U be a model for $N_{\mathcal{F}}(C_S(U)) \cap N_{\mathcal{F}}(U)$. From 9.4.16, $X_t = \langle L_t, N_{O(t)}(U_t)\rangle$, where $U_t = O_2(L_t)$ and $\Lambda_t = Aut_{L_t}(U_t)$, so that $L_t/U_t \cong S_3$. From 11.2.24, $O(t)$ and L_t centralize U_s for $s \in \theta - \{t\}$, so X_t centralizes U_s. Thus we may take X_t to be contained in G_U, and we obtain a subgroup $X = \prod_{s\in\theta} X_s$ of G_U.

Let $T_t = S \cap X_t \leq Y_t$. Then $Y_t = U_tO(t) = T_tO(t)$. As X_s centralizes U_t, $\mathcal{X}_s = \mathcal{F}_{T_s}(X_s) \leq C_{\mathcal{F}_t}(\mathcal{O}_t)$ and as $Y_t = T_tO(t)$, $\mathcal{X}_s$ centralizes Y_t. Let $u_1 = v(t)$, $u_2 = tv(t)$, and $U(i) = \langle t^{C_{X_t}(u_i)}\rangle$. By 9.4.34, 9.4.22, 9.4.27, and 9.4.28, $\mathcal{Y}_t = \langle Y_t\mathcal{O}, N_{\mathcal{Y}}(U(i)) : i = 1, 2\rangle$, where $\mathcal{O} \cong SL_2[m] * SL_2[m]$ with $O(t)$ Sylow in $\mathcal{O}$. As $\mathcal{X}_s$ centralizes Y_t it centralizes $Y_t\mathcal{O}$. From 9.4.21, $N_{O(t)}(U(i))N_{X_t}(U(i))$ is a model for $N_{\mathcal{Y}_t}(U(i))$, so $\mathcal{X}_s$ centralizes $N_{\mathcal{Y}_t}(U(i))$ and hence also $\mathcal{Y}_t$. Let $\Sigma_t = \theta - \{t\}$; by construction $\mathcal{Y}_t \leq C_{\mathcal{F}}(O(\Sigma_t))$ and we just showed $\mathcal{Y}_t$ centralizes T_s, so $\mathcal{Y}_t$ centralizes $S_t = O(\Sigma_t)\prod_{s\neq t} T_s = \prod_{s\neq t} Y_s$. Therefore by 2.2 in [**Asc11**], $\mathcal{F}$ contains a direct product $\prod_{s\in\theta} \mathcal{Y}_s$, proving (2).

Let $\mathcal{X} = \prod_{s\neq z} \mathcal{Y}_s$; then $\mathcal{X} = \mathcal{X}^{\circ}$ is contained in the component $\mathcal{C}$ of $\mathcal{F}_z^{\circ}$ appearing in 11.2.27. By 11.2.21, $n_S = 2^a$ for some a, and $a \geq 1$ by 11.2.22. Therefore $n = 2n_S = 2^{a+1}$ or $2n_S + 1 = 2^{a+1} + 1$. By 11.2.22, $n_S \geq 2$ so $n \geq 4$. Now by 11.2.27, $\mathcal{C}$ is isomorphic to $\Omega_{2n-2}^-[m]$ or $\Omega_{2n-3}[m]$. As $\mathcal{C}$ contains a copy of $\mathcal{X}$, and $\mathcal{X}$ is the product of $n_S - 1 = 2^a - 1$ copies of $L_4^-[m]$, 5.3.11.2 says that $n = 4$ or 5. completing the proof of (3) and the lemma. □

By 11.2.28, $n = 4$ or 5, so $n_S = 2$. Therefore $\mathcal{Z}_S = \{z, t\}$ is of order 2 with $zt = c$. From Notation 11.2.2, we may take $z = e_{1,2}$, $t = e_{3,4}$, and $v(z) = e_1$. Let I_1 be the elements of W_I of weight 1; from 11.2.2, $I_1 = \{e_i : 1 \leq i \leq n\}$. Let $e \in W_I$ be of weight 1 with $e \in \mathcal{F}^f$. Let $\tau_e = (\mathcal{F}_e, \Omega_e)$ where $\Omega_e = \{K \in \Omega : K \leq C_S(e)\}$.

LEMMA 11.2.29. *Assume $n = 4$. Then*

(1) We may take $e = e_1 = v(z)$.

(2) $\mu(\tau_e) \cong S_4$.

(3) $e_i : 2 \leq i \leq 4$ are in $\mathcal{F}_e^{\circ}$.

(4) $\Omega_6^-[m] \cong \mathcal{C} \leq \mathcal{F}_e^{\circ} \cong \Omega_7[m]$ or $\Omega_8^-[m]$.

PROOF. As $n = 4$, $\mu \cong \mathrm{Weyl}(D_4)$, and then from 11.2.2, W_I is the permutation module for $\tilde{M} \cong S_4$. Now $|S : C_S(e_i)| = 4$ for each $i \in I$, so all members of I_1 are fully centralized, so (1) holds. Further $\Omega_e = \{K_t, K_t'\}$ and $C_{\tilde{M}}(e) \cong S_3$, so (2) holds. Next $e_4 \in O(t) \leq \mathcal{F}_e^{\circ}$ and $C_M(e)$ is transitive on $I_1 - \{e\}$, so (3) holds.

As $v(z) \in O(z)$, it follows from 11.2.27 that $e = v(z)$ centralizes $\mathcal{C} \cong \Omega_6^-[m]$, so $\mathcal{C} \leq \mathcal{F}_e^\circ$. Then as $\mathcal{C} \leq \mathcal{F}_e^\circ$, (4) follows from 11.1.7.3 in case 11.2.26.2, and from the discussion in section 5.3 in case 11.2.26.1. □

LEMMA 11.2.30. *Assume $n = 5$. Them*
(1) $e = e_5$.
(2) $\mu(\tau_e) \cong Weyl(D_4)$.
(3) $e_i \in \mathcal{F}_e^\circ$ for $1 \leq i \leq 4$.
(4) $\mathcal{F}_e^\circ \cong \Omega_8^+[m]$, $\Omega_9[m]$, or $\Omega_{10}^-[m]$.

PROOF. The proof is much like that of 11.2.29. As $n = 5$, $\mu \cong$ Weyl(D_5), and then by 11.2.2, W_I is the permutation module for $\tilde{M} \cong S_5$ or that module modulo e_I. Let T be Sylow in M; from the notational conventions of 11.2.2, T fixes e_5 and is transitive on $I_1 - \{e_5\}$, so (1) holds.

Next $\Omega_e = \Omega$ and $C_{\tilde{M}}(e) \cong S_4$, so (2) holds. As $e_i \in O(z)$ or $O(t)$ for $i \leq 4$, and as $\Omega \subseteq \mathcal{F}_e^\circ$, (3) holds. As in the proof of 11.2.29, $e_1 = v(z)$ centralizes $\mathcal{C}$ and by 11.2.27.3, $\mathcal{C}$ is $\Omega_7[m]$ or $\Omega_8^-[m]$. As $e_1 \in e^{\mathcal{F}}$ and $e \in \mathcal{F}^f$, $\mathcal{F}_e^\circ$ contains a copy of $\mathcal{C}$. Then (4) follows from 11.1.7.3 in case 11.2.26.2, and from section 5.3 in case 11.2.26.1. For example as $\mathcal{F}_e^\circ$ contains a copy of $\Omega_7[m]$ or $\Omega_8^-[m]$, only cases (iv) and (v) of 11.1.7.3 can occur. □

LEMMA 11.2.31. *$\mathcal{F}_e^\circ$ contains a copy of $\Omega_4^-[m] \times \Omega_6^-[m]$.*

PROOF. By 11.2.28.2, $\mathcal{F}$ contains the direct product $\mathcal{Y}_0 = \mathcal{Y}_z \times \mathcal{Y}_t$ with $\mathcal{Y}_z \cong \mathcal{Y}_t \cong \Omega_6^-[m]$ by 11.2.25.5. Now $e_1 = v(z)$ is a non-2-central involution in $\mathcal{Y}_z$, so $\mathcal{X} = O^2(C_{\mathcal{Y}_z}(e_1)) \cong \Omega_4^-[m]$. Then $\mathcal{X} \times \mathcal{Y}_t \leq \mathcal{F}_{e_1}$, so for $\alpha \in \mathfrak{A}(e_1)$ with $e_1\alpha = e$, $\mathcal{F}_e$ contains the copy $\bar{\mathcal{X}} \times \bar{\mathcal{Y}}$ of $\mathcal{X} \times \mathcal{Y}_t$, where $\bar{\mathcal{X}} = \mathcal{X}\alpha^*$ and $\bar{\mathcal{Y}} = \mathcal{Y}_t\alpha^*$. As $\mathcal{Y}_t = \mathcal{Y}_t^\circ$, $\bar{\mathcal{Y}} \leq \mathcal{F}_e^\circ$. As $e_1 \in W$, $W\alpha = W$, so $I_1 = I_1\alpha$, and hence by 11.2.29.3 and 11.2.30.3, $I_1 - \{e\} \subseteq \mathcal{F}_e^\circ$. In particular as $e_2 = e_1 z \in \mathcal{X}$ and $e_2\alpha \neq e$, we conclude that $\bar{\mathcal{X}} = [e_2\alpha]_{\mathcal{X}} \leq \mathcal{F}_e^\circ$, completing the proof. □

LEMMA 11.2.32. *$n \neq 4$ or 5.*

PROOF. By 11.2.29.4 or 11.2.30.4, $\mathcal{F}_e^\circ \cong \Omega_7[m]$, $\Omega_8^\epsilon[m]$, $\Omega_9[m]$, or $\Omega_{10}^-[m]$. On the other hand by 11.2.31, $\mathcal{F}_e^\circ$ contains a copy of $\Omega_4^-[m] \times \Omega_6^-[m]$. This contradicts 5.3.11.1. □

We close this section with two theorems that summarize the various results from the section.

THEOREM 11.2.33. *Assume Hypotheses 11.2.1 and $R \in \mathcal{F}^e$ satisfies 11.2.7. In addition assume for each $z \in \mathcal{Z}_S$ that*
(a) $|\Omega(z)| = 1$.
(b) $O(z) \cap R \leq \langle z \rangle$.
(c) One of the following holds:
(1) $\mathcal{F}^\circ$ is $(P)Sp_{2n_S}[m]$, $(P)SL_{2n_S}^\epsilon[m]$, or $L_{2n_S+1}^\epsilon[m]$.
(2) τ satisfies the Inductive Hypothesis.
(3) $n_S = 4$ and $m(W_S) = 3$.
Then R is transitive on $\mathcal{Z}_S$ and $\tau^\circ = \prod_{z \in \mathcal{Z}_S} \rho_z$, where $\rho_z = (\mathcal{Y}_z, \Omega(z))$ and $\mathcal{Y}_z \cong L_3^-[m]$, $L_2[2m]^{(1)}$, $G_2[m]$, or M_{12}. Further (c3) does not hold and if (c1) holds then $n_S = 1$ and $\mathcal{F}^\circ \cong L_3^-[m]$.

PROOF. This follows from 11.2.13. □

THEOREM 11.2.34. *Assume Hypothesis* 11.2.1 *and* $R \in \mathcal{F}^e$ *satisfies* 11.2.7. *In addition assume for each* $z \in \mathcal{Z}_S$ *that*

(a) $|\Omega(z)| = 2$.

(b) For each $K \in \Omega$, $|K \cap R| \leq 2$.

(c) Hypothesis 11.2.26 *holds.*

Then $|\mathcal{Z}_S| = 1$.

PROOF. By (a) and (b), Hypothesis 11.2.14 holds. We may assume that $|\mathcal{Z}_S| > 1$, so Hypothesis 11.2.22 holds. By (c), Hypothesis 11.2.26 holds. Therefore by 11.2.28.3, $n = 4$ or 5. On the other hand $n \neq 4$ or 5 by 11.2.32, completing the proof. □

11.3. More generation

In this section we assume:

HYPOTHESIS 11.3.1. *(1) Hypothesis* 11.2.1 *holds; adopt Notation* 11.2.2 *and* 11.2.3.

(2) If $|\Omega(z)| = 1$ *for* $z \in \mathcal{Z}_S$, $n_S = 4$, *and* $m(W_S) = 3$ *then* μ *is abelian or* $\mu =$ *Weyl*(Φ) *where* Φ *is of type* A_n, D_n, *or* E_n *for some* n.

DEFINITION 11.3.2. Let $\mathcal{W}(\tau)$ consist of those involutions w such that one of the following holds:

(1) $|\Omega(z)| = 2$ for $z \in \mathcal{Z}_S$ and $w = e_\theta$ for some $\theta \subseteq I$ such that $|\theta|$ is a positive power of 2.

(2) $|\Omega(z)| = 1$ for $z \in \mathcal{Z}_S$ and either

(a) $w = f_\theta$ for some $\theta \subseteq \mathcal{Z}_S$ such that $|\theta|$ is a power of 2, or

(b) $n_S = 4$, $m(W_S) = 3$, and $w \in \mathcal{W}_*(\tau)$, where $\mathcal{W}_1(\tau)$ consists of the involutions $w = v_1v_2v_3v_4$ such that v_i generates $\Omega_1(K_i \cap W)$ and $\Omega = \{K_1, \ldots, K_4\}$, and $\mathcal{W}_*(\tau) = \mathcal{W}_1(\tau)$ unless μ is Weyl(A_7), where $\mathcal{W}_*(\tau)$ consists of the M-conjugates of members of $\mathcal{W}_1(\tau)$.

Recall the definition of $\mathcal{E}(\tau)$ from 3.4.2 and define

$$\mathcal{B} = \mathcal{B}(\tau) = \langle \mathcal{E}(\tau), \mathcal{F}_x : x \in \mathcal{W}(\tau) \cap \mathcal{F}^f \rangle.$$

LEMMA 11.3.3. *Assume the setup of* 11.3.2.*2.b and let* $w \in \mathcal{W}_*(\tau)$. *Let* $W \in W(\tau)$ *and* $M \in M(\tau)$.

(1) If μ *is abelian then there is* $\alpha \in \mathfrak{A}(w)$ *such that* α *is a* $N_{\mathcal{F}}(O(\tau))$*-map and* $w\alpha \in \mathcal{W}(\tau)$.

(2) If μ *is Weyl*(Φ) *with* $\Phi = A_n$ *for some* n, *then* $n = 7$ *and* $w \in x^M$, *where* $x = c_0x_0 \in Z(S)$ *for some* $c_0 \in Z(M)$ *with lift* $\bar{c}_0$ *in* $\bar{M}$ *of order 4, and* $x_0 = f_\theta$ *for some* $\theta \subseteq \mathcal{Z}_S$ *of order 2.*

(3) If μ *is Weyl*(Φ) *with* $\Phi = D_n$ *for some* n, *then* $n = 4$ *or* 5 *and* $w^M \cap \mathcal{Z}_S \cap \mathcal{F}^f \neq \emptyset$.

(4) If μ *is Weyl*(E_n) *for some* n, *then* $n = 6$, $w \in W_\Delta$, *and* $w^M \cap W_S \cap \mathcal{F}^f \neq \emptyset$.

PROOF. As we are in the setup of 11.3.2.2.b, we have $\Omega(z_i) = \{K_i\}$ of order 1 and $\mathcal{Z}_S = \{z_1, \ldots, z_4\}$.

Assume first that μ is abelian. Then $\mu \cong E_{16}$ and $\eta = \{V_i : 1 \leq i \leq 4\}$ where $V_i = K_i \cap W$ and v_i is of order 4 in V_i. Set $v = v_1 \cdots v_4$ and $\alpha \in \mathfrak{A}(v)$. If $m > 8$ then $\eta\alpha = \eta$, so $\mathcal{Z}_S\alpha = \mathcal{Z}_S$ and $v\alpha = \prod_i v_i\alpha \in v^{O(\tau)}$. Therefore $v \in \mathcal{F}^f$, so (1) holds in this case.

So take $m = 8$. Still $V_i\alpha \leq K_{j(i)}$, so $\eta\alpha \in \eta(\tau)$ and $\mathcal{Z}_S\alpha = \mathcal{Z}_S$, so α is a $N_{\mathcal{F}}(W_S)$-map and then α is induced in $N_{\mathcal{F}}(O(\tau))$. This completes the proof of (1).

Thus we may assume μ is Weyl(Φ) for Φ of type A_n, D_n, or E_n. Then by 4.3.8 we have a surjective homomorphism $\pi : \bar{G} \to M$ where $\bar{G}$ is the universal group of 4.3.1 and $\ker(\pi) \leq Z(\bar{G})$.

Suppose Φ is D_n; then as $n_S = 4$ we have $n = 4$ or 5. As $|\Omega(z)| = 1$, we have $\ker(\pi) = 1$. Therefore π is an isomorphism. Now $v = v_1 \cdots v_4 \in z^M$ by 5.8.7.6, so (3) holds.

Suppose Φ is A_n. Again as $n_S = 4$, $n = 7$ or 8. As $m(W_S) = 3$, we have $\bar{z}_1 \cdots \bar{z}_4 \in \ker(\pi)$, so $n = 7$ and v is a projective involution. Working in $SL_8(q)$, we find $\bar{v}$ has eigenvalues λ and λ^{-1} of order 4, each of multiplicity 4, so for some $c_0 \in Z(M)$ with lift $\bar{c}_0$ in $\bar{M}$ of order 4, $\bar{x} = \bar{v}\bar{c}_0$ has eigenvalues 1 and -1, each of multiplicity 4. Thus there is $g \in M$ with $x^g = f_\theta$ with $\theta \subseteq \mathcal{Z}_S$ of order 2, and hence $f_\theta \in Z(S)$, so (2) holds.

Finally suppose Φ is E_n. Again as $n_S = 4$ we have $n = 6$. Hence π is an isomorphism and $v \in W_\Delta$. Then as all involutions in W_Δ are M-conjugate to some member of W_S in $\mathcal{F}^f$, (4) holds. □

LEMMA 11.3.4. *For each $x \in \mathcal{W}(\tau)$, $\mathcal{F}_x \leq \mathcal{B}$.*

PROOF. Suppose first that if 11.3.2.2.b holds with $x \in \mathcal{W}_*(\tau)$, then μ is nonabelian. Let $Q = W$ if $|\Omega(z)| = 2$ or 11.3.1.2 holds with μ nonabelian, and let $Q = O(\tau)$ otherwise. By construction of $\mathcal{E}(\tau)$ in 3.4.2, $N_{\mathcal{F}}(Q) \leq \mathcal{E}(\tau) \leq \mathcal{B}$. Observe also that Q is weakly closed in S with respect to S and $x \in \mathcal{W}(\tau) \subseteq Z(Q)$. Let $\alpha \in \mathfrak{A}(x)$. Then $Q\alpha = Q$ by the weak closure of Q, so $x\alpha \in Z(Q)$ and indeed $x\alpha \in \mathcal{W}(x)$. Therefore $\mathcal{F}_{x\alpha} \leq \mathcal{B}$ by 11.3.2. Hence for $P \leq C_S(x)$,

$$\begin{aligned} \hom_{\mathcal{F}_x}(P, S) &= (\hom_{\mathcal{F}_{x\alpha}}(P\alpha, S))\alpha^{-*} \leq \hom_{\mathcal{B}}(P\alpha, S)\alpha^{-*} \\ &= \hom_{\mathcal{B}\alpha^{-*}}(P, S) = \hom_{\mathcal{B}}(P, S), \end{aligned}$$

so indeed $\mathcal{F}_x \leq \mathcal{B}$.

So assume instead that 11.3.2.2.b holds with $x \in \mathcal{W}_*(\tau)$ and μ abelian. Then by 11.3.3.1 there is $\alpha \in \mathfrak{A}(x)$ with α a $N_{\mathcal{F}}(O(\tau))$-map and $x\alpha \in \mathcal{W}(\tau)$. Now complete the argument as in the previous paragraph. □

NOTATION 11.3.5. Let $\mathcal{R}$ consist of those $R \in \mathcal{F}^e$ such that $Aut_{\mathcal{F}}(R) \neq Aut_{\mathcal{B}}(R)$, and write $\mathcal{R}^*$ for the members of $\mathcal{R}$ of maximal order. For $R \in \mathcal{R}$, set $\mathcal{W}(R) = \mathcal{W}(\tau) \cap Z(R)$ and let G be a model for $N_{\mathcal{F}}(R)$ and $G^* = G/R$.

LEMMA 11.3.6. *For each $R \in \mathcal{R}$ and $K \in \Omega$, $|K \cap R| \leq 2$.*

PROOF. As $\mathcal{E}(\tau) \leq \mathcal{B}$ the lemma follows from 3.4.5. □

LEMMA 11.3.7. *Assume $R \in \mathcal{R}^*$.*

(1) If $x \in \mathcal{W}(R)$ and $Z = \langle x^G \rangle$ then $O^{2'}(C_G(Z)) = R$.

(2) Let $\mathcal{U}$ be the set of $U^* \leq G^*$ such that $S^* \cap U^*$ is Sylow in U^*, U^* is not contained in the strongly embedded subgroup of G^* over $N_S(R)^*$, $U^* = \langle u^{*U} \rangle$ for

some $u \in S \cap G$ such that u^* is an involution, and $U \leq C_G(w)$ for some $w \in \mathcal{W}(R)$. Then if $\mathcal{U} \neq \emptyset$ then $m_2(G^*) = 1$ and for $\emptyset \neq \mathcal{U}_0 \subseteq \mathcal{U}$, $\langle \mathcal{U}_0 \rangle$ is not normal in G^*.

PROOF. As $x \in \mathcal{W}(R)$, $x \in Z(R)$, so $Z = \langle x^G \rangle \leq Z(R)$, and hence $R \leq O^{2'}(C_G(Z))$.

Suppose $H = O^{2'}(C_G(Z)) \neq R$ and let $T = S \cap H$. By 3.4.4.2, $Aut_{\mathcal{F}}(R) = Aut_H(R)Aut_{N_G(T)}(R)$. Further $Aut_H(R) \leq Aut_{\mathcal{F}_x}(R) \leq Aut_{\mathcal{B}}(R)$ by 11.3.4, so as $Aut_{\mathcal{F}}(R) \neq Aut_{\mathcal{B}}(R)$, it follows that $Aut_{N_G(T)}(R) \not\leq Aut_{\mathcal{B}}(R)$. But by the Alperin-Goldschmidt Fusion Theorem, there exists $R' \in \mathcal{F}^e$ containing a conjugate of T such that $Aut_{\mathcal{F}}(R') \neq Aut_{\mathcal{B}}(R')$. As $R < T$, this contradicts the maximality of $|R|$, so (1) holds.

Let $L = O^{2'}(G)$ and $P \in Syl_2(L)$; then L^* has a strongly embedded subgroup Y^* and $U^* \leq L^*$ for each $U^* \in \mathcal{U}$. By 11.3.4, $Aut_U(R) \leq Aut_{\mathcal{B}}(R)$ for each $U^* \in \mathcal{U}$. By maximality of $|R|$ and an argument in the proof of (1), $Aut_{N_G(Q)}(R) \leq Aut_{\mathcal{B}}(R)$ for each $R < Q \leq N_S(R)$.

Suppose that $m_2(G^*) > 1$ and $U^* \in \mathcal{U}$. Then $L^*/O(L^*)$ is a Bender group and as $U^* \not\leq Y^*$, we have $L^* = \langle P^*, U^* \rangle$, so $Aut_L(R) = \langle Aut_U(R), Aut_P(R) \rangle \leq Aut_{\mathcal{B}}(R) \rangle$. Then as $G = LN_G(P)$, we obtain a contradiction by arguing as in the proof of (1). Similarly for $\emptyset \neq \mathcal{U}_0 \subseteq \mathcal{U}$, $\langle \mathcal{U}_0 \rangle$ is not normal in G^*, so (2) holds. □

HYPOTHESIS 11.3.8. *Hypothesis* 11.3.1 *holds and for each $z \in \mathcal{Z}_S$, $|\Omega(z)| = 1$. Moreover one of the following holds:*

(1) τ satisfies the Inductive Hypothesis.

(2) $\mathcal{F}^\circ$ is a central image of $Sp_{2n_S}[m]$, $SL^\epsilon_{2n_S}[m]$, or $L^\epsilon_{2n_S+1}[m]$.

(3) $(n_S, m(W_S)) = (4,3)$.

HYPOTHESIS 11.3.9. *Hypothesis* 11.2.1 *holds and for each $z \in \mathcal{Z}_S$, $|\Omega(z)| = 2$. Moreover either*

(1) τ satisfies the Inductive Hypothesis, or

(2) $\mathcal{F}^\circ$ is $(P)\Omega^+_{2n}[m]$, $\Omega_{2n+1}[m]$, or $\Omega^-_{2n+2}[m]$.

We prove the next two theorems together.

THEOREM 11.3.10. *Assume Hypothesis* 11.3.8. *Then either*

(1) $\mathcal{F} = \mathcal{B}(\tau)$, or

(2) S is transitive on $\mathcal{Z}_S$ and $\tau^\circ = \prod_{z \in \mathcal{Z}_S} \rho_z$, where $\rho_z = (\mathcal{Y}_z, \Omega(z))$ and $\mathcal{Y}_z \cong L^-_3[m]$, $L_2[2m]^{(1)}$, $G_2[m]$, or M_{12}. Moreover $(n_S, m(W)) \neq (4,3)$ and if 11.3.8.*2 holds then $\mathcal{F}^\circ \cong L^-_3[m]$.*

THEOREM 11.3.11. *Assume Hypothesis* 11.3.9. *Then either*

(1) $\mathcal{F} = \mathcal{B}(\tau)$, or

(2) $|\mathcal{Z}_S| = 1$.

Assume $\mathcal{F}$ is a counter example to Theorem 11.3.10 or Theorem 11.3.11. Then $\mathcal{F} \neq \mathcal{B}$, so $\mathcal{R} \neq \emptyset$ by the Alperin-Goldschmidt Fusion Theorem. Pick $R \in \mathcal{R}^*$. By 11.3.7, the pair $\mathcal{F}, R$ satisfies Hypothesis 11.2.7. We appeal to Theorems 11.2.33 and 11.2.34, so to complete the proof it remains to verify the hypotheses of those theorems.

By 11.3.6, condition (b) of each of the theorems is satisfied. Hypothesis 11.3.8 says conditions (a) and (c) of Theorem 11.2.33 are satisfied, while Hypothesis 11.3.9

says conditions (a) and (c) of Theorem 11.2.34 are satisfied. This completes the proof of Theorems 11.3.10 and 11.3.11.

11.4. Essentials and normal subsystems

In this section we assume the following hypothesis:

HYPOTHESIS 11.4.1. *$\mathcal{F}$ is a saturated fusion system on a finite 2-group S and $\mathcal{F}_0 \trianglelefteq \mathcal{F}$ with nontrivial Sylow group S_0.*

NOTATION 11.4.2. For $X \leq S$, set $X_0 = X \cap S_0$. Let $R \in \mathcal{F}^{frc}$ and G a model for $\mathcal{X} = N_{\mathcal{F}}(R)$. Set $\mathcal{X}_0 = N_{\mathcal{F}_0}(R_0)$ and $G^* = G/R$. Conjugating in $\mathcal{F}$, we may assume $R_0 \in \mathcal{F}^f$.

LEMMA 11.4.3. *$R_0 \neq 1$.*

PROOF. As $S_0 \neq 1$, also $1 \neq Z(S) \cap S_0$. Moreover as $R \in \mathcal{F}^c$, $C_S(R) \leq R$. □

LEMMA 11.4.4. *(1) $R_0 \in \mathcal{F}_0^{frc}$.*

(2) Set $V = R_0 C_S(R_0)$. Then $V_0 = R_0$ and $V \in \mathcal{F}^{fc}$, so $N_{\mathcal{F}}(V)$ has a model G_V.

(3) There is a model G_0 for $\mathcal{X}_0$ normal in G_V.

(4) $[G_0, C_S(R_0)] \leq Z(R_0)$.

(5) Set $R_c = R_0 C_R(R_0)$. Then, conjugating in $\mathcal{F}$, we may take $R_c \in \mathcal{F}^f$, so $R_c \in \mathcal{F}^{fc}$ and hence $N_{\mathcal{F}}(R_c)$ has a model G_c.

(6) We may take $G_0 \trianglelefteq G_c$ and $G = N_{G_c}(R) \leq G_c$.

(7) $O_2(G_c) \leq R$.

(8) $N_{G_0}(R) \trianglelefteq G$.

(9) $R_0 = O_2(N_{G_0}(R))$, so $N_{G_0}(R)^* \cong N_{G_0}(R)/R_0$ and $O_2(N_{G_0}(R)^*) = 1$.

PROOF. As $R \in \mathcal{F}^{frc}$ we have $O_2(G) = O_2(\mathcal{X}) = R$, and hence $R_0 \trianglelefteq \mathcal{X}$, so $\mathcal{X}$ acts on $O_2(N_{\mathcal{F}}(R_0)) = Q$. Then $N_Q(R) \leq O_2(G) = R$, and hence $Q \leq R$. Therefore $O_2(N_{\mathcal{F}_0}(R_0)) \leq Q_0 \leq R_0$, so $R_0 \in \mathcal{F}_0^r$.

Set $X_0 = C_{S_0}(R_0)$ and $Y_0 = N_{X_0}(R)$. Then $[Y_0, R] \leq R_0 \trianglelefteq G$, so $Y_0 \leq O_2(G) = R$ and hence $Y_0 = Z(R_0)$. Then $X_0 = Z(R_0)$, so as $R_0 \in \mathcal{F}^f$, also $R_0 \in \mathcal{F}_0^c$, completing the proof of (1).

Next $V_0 = S_0 \cap R_0 C_S(R_0) = R_0(S_0 \cap C_S(R_0)) = R_0 C_{S_0}(R_0) = R_0$ by (1). As $R_0 \in \mathcal{F}^f$ and $R_0 = V_0$, also $V \in \mathcal{F}^f$, while by construction $C_S(V) \leq C_S(R_0) \leq V$, so $V \in \mathcal{F}^{fc}$. Thus (2) holds.

As $V_0 = R_0$, from Notation 4.1 in [**Asc08a**], $N_{\mathcal{F}}(V)$ is the system $\mathcal{D}(R_0)$ defined there, and $\mathcal{E}(R_0) = N_{\mathcal{F}_0}(R_0)$. Then by 6.10 in [**Asc08a**], $N_{\mathcal{F}_0}(R_0) \trianglelefteq N_{\mathcal{F}}(V)$, and hence by Theorem 2 in [**Asc08a**], there is a model G_0 for $N_{\mathcal{F}_0}(R_0)$ normal in G_V, establishing (3).

By (3), $G_0 \trianglelefteq G_V$, so $[G_0, C_S(R_0)] \leq C_{G_0}(R_0) = Z(R_0)$ by (1), proving (4).

As $R_0 = V_0$ also $R_c \cap S_0 = R_0$. Then as R_0 and R are in $\mathcal{F}^f$, we may take $R_c \in \mathcal{F}^f$. Set $S_1 = C_S(R_c)$ and $S_2 = N_{S_1}(R)$. Observe that $S_1 \leq C_S(R_0)$, so $[S_2, R] \leq C_R(R_0) \leq R_c$. Then as $R \in \mathcal{F}^{frc}$, we have $S_2 \leq R$, so $S_2 \leq C_R(R_c) \leq R_c \leq R$, and hence $S_2 = S_1$. Therefore $S_1 = S_2 \leq R_c$, so $R_c \in \mathcal{F}^{fc}$, proving (5).

By (4), G_0 acts on R_c, so we may take $G_0 \leq G_c$ and then $G_0 \trianglelefteq G_c$ by Theorem 1 in [**Asc08a**]. As $R_0 \trianglelefteq G$, also $R_c = R_0 C_R(R_0) \trianglelefteq G$, so we may take $G = N_{G_c}(R)$, completing the proof of (6).

As $R \in \mathcal{F}^{frc}$, $R = O_2(G)$, so as $G = N_{G_c}(R)$ by (6), we conclude that (7) holds. Similarly (8) follows from (6).

As G_0 is a model for $N_{\mathcal{F}_0}(R_0)$, we have $N_{S_0}(R_0)$ Sylow in G_0. Therefore $R \cap N_{G_0}(R_0) \leq R \cap S_0 = R_0$. Hence $R_0 = O_2(N_{G_0}(R))$ by (8), so (9) holds. □

LEMMA 11.4.5. *Assume $R \in \mathcal{F}^e$ and $R_0 \neq S_0$. Let $H = N_{G_0}(R)$, K^* the subgroup of G^* generated by the involutions in G^*, and M^* a strongly embedded subgroup of G^* containing $(S \cap G)^*$.*

(1) K^* is the subgroup of G^* generated by the involutions in H^* and K^* is transitive on such involutions.

(2) $M^* \cap H^*$ and $M^* \cap K^*$ are strongly embedded in H^* and K^*, respectively.

(3) Either $O^{2'}(G^*) = O^{2'}(H^*)$ or $m_2(G^*) = 1$.

PROOF. By 11.4.4.8, $H^* \trianglelefteq G^*$. Set $S_1 = N_{S_0}(R_0)$ and $S_2 = N_{S_1}(R)$. As $R_0 \neq S_0$, also $R_0 \neq S_1$, so $R_0 \neq S_2$. Hence as S_2 is Sylow in H, H^* contains an involution. But as $R \in \mathcal{F}^e$, G^* has a strongly embedded subgroup M^*. Therefore G^* is transitive on its involutions, so as $H^* \trianglelefteq G^*$, we have $K^* \trianglelefteq H^*$. As $K^* \trianglelefteq G^*$ is of even order, $M^* \cap K^*$ is strongly embedded in K^*, so K^* is transitive on its involutions, and hence (1) and (2) hold.

As M^* is strongly embedded in G^* either $O^{2'}(G^*) = K^*$ or $m_2(G^*) = 1$, so (3) follows from (1). □

HYPOTHESIS 11.4.6. *$\mathcal{F}_0 = \mathcal{F}_1 * \cdots * \mathcal{F}_k$ is a central product of subsystems permuted by $Aut_{\mathcal{F}}(S_0)$. Let S_i be Sylow in $\mathcal{F}_i$ and for $X \leq S$, set $X_i = X \cap S_i$. Set $I = \{1, \ldots, k\}$.*

LEMMA 11.4.7. *Assume Hypothesis 11.4.6 and define G_0 as in 11.4.4.3. Then*

(1) $R_0 = R_1 \cdots R_k$ is a central product.

(2) $\mathcal{X}_0 = N_{\mathcal{F}_1}(R_1) * \cdots * N_{\mathcal{F}_k}(R_k)$.

(3) For $i \in I$, $R_i \in \mathcal{F}_i^{frc}$.

(4) For $i \in I$ there is a model G_i for $N_{\mathcal{F}_i}(R_i)$ such that $G_0 = G_1 * \cdots * G_k$ is a central product of the G_i for $i \in I$ and G_c permutes the G_i via conjugation.

PROOF. For $i \in I$, set $Z_i = Z(\mathcal{F}_i)$ and set $Z = Z_1 \cdots Z_k$ and $\mathcal{F}_0^+ = \mathcal{F}_0/Z$. As $\mathcal{F}_0$ is a central product, $\mathcal{F}_i \cap \mathcal{F}_i' \leq Z_i$, where $\mathcal{F}_i' = \prod_{j \neq i} \mathcal{F}_j$. Therefore $\mathcal{F}_0^+ = \mathcal{F}_1^+ \times \cdots \times \mathcal{F}_k^+$ is a direct product. Further $Z \leq O_2(\mathcal{F}_0) \leq R_0$ by 11.4.4.1.

Let P_i^+ be the projection of R_0^+ on S_i^+ and P_i the preimage of P_i^+ in S_i. Set $P = \prod_i P_i$. From the definition of the direct product in Chapter 2 of [**Asc11**], we have $N_{\mathcal{F}_0^+}(R_0^+) \leq N_{\mathcal{F}_0^+}(P^+)$, so by 11.4.4.1, $P^+ \leq R_0^+$ and hence $P \leq R_0$. Hence $P_i = R_i$ and then (1) holds.

Next $[N_{\mathcal{F}_i}(R_i), R_0] = [N_{\mathcal{F}_i}(R_i), R_i R_i'] = [N_{\mathcal{F}_i}(R_i), R_i] \leq R_i$, so $N_{\mathcal{F}_i}(R_i) \leq \mathcal{X}_0$. Hence $\mathcal{Y}_0 = \prod_i N_{\mathcal{F}_i}(R_i) \leq \mathcal{X}_0$. As G_0 acts on R_i, also $\mathcal{X}_0 \leq \mathcal{Y}_0$, so (2) holds.

Now $N_{S_i}(R_0) = N_{S_i}(R_i)$, so as $R_0 \in \mathcal{F}_0^f$, also $R_i \in \mathcal{F}_i^f$. Further $C_{S_i}(R_i) = C_{S_i}(R_0)$, so as $R_0 \in \mathcal{F}_0^c$ by 11.4.4.1, we have $R_i \in \mathcal{F}_i^c$. Finally $O_2(N_{\mathcal{F}_i}(R_i)) \leq O_2(N_{\mathcal{F}_i}(R_0)) \leq O_2(N_{\mathcal{F}_0}(R_0)) = R_0$ by 11.4.4.1, so $R_i \in \mathcal{F}_i^{frc}$, completing the proof of (3).

By (1), $R_0 = R_iC_{R_0}(R_i)$. Applying 11.4.4.3 to the pair R_0, R_i in the role of R, R_0, we conclude there is a model G_i for $N_{\mathcal{F}_i}(R_i)$ with $G_i \trianglelefteq G_0$. As R_i centralizes $\mathcal{F}_i'$, R_i centralizes $G_i' = \prod_{j\neq i} G_j$. Now $[G_i, G_i'] \leq G_i \cap G_i' \leq C_{G_0}(R_iR_i') = C_{G_0}(R_0) = Z(R_0)$, so $[G_i, G_i'] = 1$. Therefore the central product X of the subgroups G_i is a normal subgroup of G_0.

Set $T_0 = N_{S_0}(R_0)$ and let T_i be Sylow in G_i. Thus $T_0 = T_1 * \cdots * T_k$ is Sylow in G_0. As $T_0 \leq X$, it follows from a Frattini argument that $G_0 = XY$ where Y is a Hall $2'$-subgroup of $N_{G_0}(T_0)$. For $y \in Y$, $c_y \in Aut_{\mathcal{X}_0}(R_0)$, so by (2), $c_y = \alpha_1 \cdots \alpha_k$ where $\alpha_i = \beta_i \times 1$ centralizes R_i' and $\beta_i \in Aut_{\mathcal{F}_i}(R_0)$. Now $\alpha_i = c_{y_i}$ for some $y_i \in C_{G_0}(R_i')$ of odd order. Then $y_i \in O^2(C_{G_0}(R_i')) \leq G_i$ as $O^2(C_{\mathcal{X}_0}(R_i')) \leq N_{\mathcal{F}_i}(R_i)$ by (2). Therefore $y \in X$, so $G_0 = X$. This also shows that $O^2(C_{G_0}(R_i')) = O^2(G_i)$.

By 11.4.4.6, $G_0 \trianglelefteq G_c$, so by a Frattini argument, $G_c = G_0N_{G_c}(T_0)$. Thus to show that G_c permutes the G_i and complete the proof of (4), it suffices to show that each $g \in N_{G_c}(T_0)$ permutes the G_i.

As $\mathcal{F}_0 \trianglelefteq \mathcal{F}$, $\alpha = c_g = \varphi\phi$ with $\varphi \in Aut_{\mathcal{F}}(S_0)$ and $\phi \in \hom_{\mathcal{F}_0}(T_0\varphi, S_0)$. By 11.4.6, φ permutes the $\mathcal{F}_i$, so $S_i\varphi = S_{i\sigma}$ for some $\sigma \in Sym(I)$. Thus $T_i\alpha = T_i\varphi\phi \leq (S_{i\sigma} \cap T_0\varphi)\phi \leq S_{i\sigma}$, so $T_i\alpha \leq S_{i\sigma} \cap T_0 = T_{i\sigma}$ and hence $T_i\alpha = T_{i\sigma}$. Then also $R_i^g = R_i\alpha = R_{i\sigma}$. Hence as $O^2(C_{G_0}(R_i')) = O^2(G_i)$, we conclude that $O^2(G_i)^g = O^2(G_{i\sigma})$. Also $O^{2'}(G_i) = \langle R_i^{G_0}\rangle$, so $O^{2'}(G_i)^g = O^{2'}(G_{i\sigma})$, and hence g permutes the G_i, completing the proof of (4) and the lemma. □

LEMMA 11.4.8. *Assume Hypothesis* 11.4.6 *with* $R \in \mathcal{F}^e$ *and* $R_0 \neq S_0$. *Let* $H = N_{G_0}(R)$, $I_+ = \{i \in I : R_i = S_i\}$, *and* $I_- = I - I_+$. *Then*

(1) R *is represented on* I *via* $R_i^r = R_{ir}$ *for* $i \in I$ *and* $r \in R$, *and* R *is transitive on* I_-.

(2) Let $1 \in I_-$ *and* $Q = N_R(1)$. *Set* $G_c^! = G_c/R_0$ *and* $\mathcal{E} = Q\mathcal{F}_1$. *Then* $C_{G_1^!}(Q^!) = N_{G_1}(Q)^!$.

(3) Let Σ *be a set of coset representatives for* Q *in* R *and define* $\alpha : N_{G_1}(Q)^! \to C_{G_0^!}(R^!)$ *by* $g^!\alpha = \prod_{\sigma\in\Sigma} g^{!\sigma}$. *Then* α *is an isomorphism and* $H^* = C_{G_0^!}(R^!)$.

(4) $Q \in \mathcal{E}^e$. *Let* G_Q *be a model for* $N_{\mathcal{E}}(Q)$.

(5) $G_Q/Q \cong H^*$.

PROOF. The first statement in (1) is trivial. By construction, R acts on I_-. Suppose $I_- = I_1 \cup I_2$ is a nontrivial R-partition and let $X_j = \prod_{i\in I_j} G_i$ and $H_j = N_{X_j}(R)$. Then $H_j \trianglelefteq H$ and, as in the proof of 11.4.5, H_j^* contains an involution a_j^*. As H_j^*, $j = 1, 2$, are distinct normal subgroups of H^*, a_1^* and a_2^* are not conjugate in H^*, contrary to 11.4.5.1. This completes the proof of (1).

As $T_1 = N_{S_1}(R_0)$ is Sylow in G_1 and $R_0 = R \cap S_0$, we have $Q \cap T_1R_0 = R_0$, so $Q^! \cap T_1^! = 1$. Therefore $C_{G_1^!}(Q^!) = N_{G_1^!}(Q^!) = N_{G_1R_0}(Q)^! = N_{G_1}(Q)^!$, since $R_0 \leq Q$. This proves (2).

By (1) and construction, the map $\beta : g^! \mapsto (\prod_{\sigma\in\Sigma} g^{!\sigma}$ is an isomorphism of $C_{G_1^!}(Q^!)$ with $C_{G_0^!}(R^!)$. Then by an argument in the proof of (2), $C_{G_0^!}(R^!) = N_{G_0}(R)^! = H^*$. Thus (3) holds.

Observe that by 11.4.5.2, H^* has a strongly embedded subgroup, so by (3), $N_{G_1}(Q)^!$ has a strongly embedded subgroup. In particular $N_{S_1}(Q)$ is Sylow in $N_{G_1}(Q)$, so Q is fully normalized in $QN_{\mathcal{F}_1}(R_0)$.

Next $R_0 = Q_0$, so $N_{S_0}(Q) \leq N_{S_0}(R_0)$. Then as $R_0 \in \mathcal{F}^f$ and Q is fully normalized in $QN_{\mathcal{F}_1}(R_0)$, it follows that $Q \in \mathcal{E}^f$. Next $C_{QS_1}(Q) \leq C_{QS_1}(R_1) \leq$

$QC_{S_1}(R_1) \leq QR_1 \leq Q$, so $Q \in \mathcal{E}^{fc}$. Thus $N_{\mathcal{E}}(Q)$ has a model G_Q, and $G_Q = N_{QG_1}(Q) = QN_{G_1}(Q)$. Then $G_Q/Q \cong N_{G_1}(Q)^! \cong H^*$ has a strongly embedded subgroup, so $Q \in \mathcal{E}^e$ and (4) holds. Moreover we have also proved (5). □

11.5. Generating $\Omega_d^\epsilon[m]$

In this section we assume the following hypothesis:

HYPOTHESIS 11.5.1. *$\tau = (\mathcal{F}, \Omega)$ is a quaternion fusion packet such that $\mathcal{F}^\circ = O^2(\mathcal{F})$ is isomorphic to $(P)\Omega_{2n}^+[m]$, $\Omega_{2n+1}[m]$, or $\Omega_{2n+2}^-[m]$ with $n \geq 2$ and $\mathcal{F}^\circ$ is not $(P)\Omega_4^+[m]$. Let S be Sylow in $\mathcal{F}$.*

NOTATION 11.5.2. As in section 5.3, we write $\mathcal{F}^\circ = \mathcal{F}_{S_L}(L)$ where $L = \Omega(U, \mathfrak{q})$ is a group of isometries of an orthogonal space $(U, \mathfrak{q})$ over a finite field F of odd order. (We are abusing notation in the case $\mathcal{F}^\circ = P\Omega_{2n}^+[m]$, where L should be the projective group $P\Omega_{2n}^+(F)$.) We write $\mathcal{F}^\circ = (P)\Omega_{2n+a}^\epsilon[m]$, where $0 \leq a \leq 2$ and $\epsilon = sgn(U, \mathfrak{q})$ if a is even.

Given a nondegenerate subspace A of U, set $L_A = C_L(A^\perp)$. Thus L_A acts faithfully as $\Omega(A, \mathfrak{q})$ on A.

From 5.3.4, $\tau^\circ = (\mathcal{F}_{S_L}(L), \Omega)$ is a Lie packet of L. From that lemma, for $K \in \Omega$, $[U, z(K)]$ is nondegenerate of dimension 4 and sign $+1$, and there is a fundamental subgroup $L(K)$ of L subnormal in $C_L(z(K))$ isomorphic to $SL_2(F)$; set $\mathcal{O}(K) = \mathcal{F}_K(L(K))$, $\mathcal{O}_{z(K)} = \prod_{K \in \Omega(z(K))} \mathcal{O}(K)$, and $\mathcal{O}_\tau = \prod_{K \in \Omega} \mathcal{O}(K)$.

If $a = 0$ set $\Lambda = \emptyset$. So assume for the moment that $a = 1$ or 2.

Suppose first that n is even, and let $\Lambda = \{1\}$ or $\{1, 2\}$ for $a = 1$ or 2, respectively. Now $U = \mathfrak{U}_1 \perp \cdots \perp \mathfrak{U}_{n/2} \perp \mathfrak{U}_0$ is an orthogonal direct sum decomposition, where $\mathfrak{U}_i = [U, s_i]$ is 4-dimensional of sign $+1$, $\mathcal{Z}_S = \{s_1, \ldots, s_{n/2}\}$, and $\dim(\mathfrak{U}_0) = a$ with $sgn(\mathfrak{U}_0) = -1$ if $a = 2$. In this case set $A = \mathfrak{U}_{n/2} + \mathfrak{U}_0$, so A is of dimension $4 + a$ and sign -1 if $a = 2$, so $L_A \cong \Omega_5(F)$ or $\Omega_6^-(F)$. Set $z = s_{n/2}$ and let $S_A = S \cap L_A$ be Sylow in L_A. For $\lambda \in \Lambda$, let $t_\lambda \in \mathcal{D}_{L_A}^m(z)$, $B_\lambda = \langle z, t_\lambda, v \rangle$, and $R_\lambda = C_S(B_\lambda)$. Thus $t_\lambda \in z^{L_A}$ interchanges the two members K_1 and K_2 of $\Omega(z)$, and t_1 and t_2 are not conjugate in $C_{L_A}(z)$ if $a = 2$. As in 9.1.6, $v = v_1 v_2$ where v_i is of order 4 in K_i. Choose notation so that $z \in \mathcal{F}^f$; this insures that B_λ and R_λ are in $\mathcal{F}^f$. If $a = 2$ let $s \in \mathcal{D}_{L_A}^o(z)$ and assume S induces automorphisms in $O(U, \mathfrak{q})$ on L. This insures that, setting $U_s = C_{S_A}(\langle z, s \rangle)$, we have $N_S(U_s) = N_{S_A}(U_s) C_S(U_s)$. Then by 9.4.22, 9.4.27, and 9.4.28, we may choose the t_λ so that $\mathcal{F}_{N_S(L_A)}(N_S(L_A)L_A) = \langle N_S(L_A)\mathcal{O}_z, Aut_{L_A}(R_\lambda) : \lambda \in \Lambda \rangle$. Note that S_A plays the role of T in these lemmas and $|S_A : O(z)| = 2^a$, with $S_A \mathcal{O}_z$ the 2-fusion system of $C_{L_A}(z)$.

Suppose next that n is odd. Here $U = \mathfrak{U}_1 \perp \cdots \perp \mathfrak{U}_{(n-1)/2} \perp \mathfrak{U}_0$ with $\dim(\mathfrak{U}_0) = a + 2$ and $sgn(\mathfrak{U}_0) = -1$ if $a = 2$. This time pick $A = \mathfrak{U}_0$. Then $L_A = \Omega_3(F)$ or $\Omega_4^-(F)$. Take $\Lambda = \{1\}$ and B_λ a 4-subgroup of L_A. Set $R_\lambda = C_S(B_\lambda)$ and observe B_λ and R_λ are in $\mathcal{F}^f$. Again if $a = 2$ assume that S induces automorphisms in $O(U, \mathfrak{q})$ on L, which insures that if we pick A_0 of dimension 5 or 6, s, and U_s as earlier, then $N_S(U_s) = N_{S_{A_0}}(U_s) C_S(U_s)$. Then, setting $B = \mathfrak{U}_{(n-1)/2} + \mathfrak{U}_0$, it follows from 9.5.9 that

$$\mathcal{F}_{N_S(L_B)}(N_S(L_B)L_B) = \langle N_S(L_B)\mathcal{O}, \mathcal{F}_{N_S(L_B)}(N_S(L_B)M(L_B)), Aut_{L_A}(R_\lambda) : \lambda \in \Lambda \rangle,$$

where we apply 9.5.9 to $\tau_B = (\mathcal{F}_B, \Omega(z))$, where $S_B = N_S(L_B)$ and $\mathcal{F}_B = F_{S_B}(S_B L_B)$, observing that $N_{\mathcal{F}_B}(W_B) = \mathcal{F}_{S_B}(S_B M(L_B))$, and when $a = 2$ that the term $\mathcal{F}_b$ is 9.5.9 is contained in $\langle \mathcal{E}(\tau_B), Aut_{L_A}(R_\lambda)\rangle$.

In this section we prove:

THEOREM 11.5.3. *Assume Hypothesis* 11.5.1 *and adopt Notation* 11.5.2. *If* $\mathcal{F}^\circ \cong \Omega^-_{2n+2}[m]$ *assume* S *induces automorphisms in* $O(U, \mathfrak{q})$ *on* L. *Then* $\mathcal{F} = \langle SO_s, \mathcal{F}_S(SM), Aut_{L_A}(R_\lambda) : \lambda \in \Lambda\rangle_S$.

Assume the theorem is false and pick a counter example with $d = \dim(U)$ minimal. As in Notation 11.5.2, $\mathcal{F}^\circ \cong (P)\Omega^\epsilon_{2n+a}[m]$. Set

$$\mathcal{Y} = \langle SO_\tau, \mathcal{F}_S(SM), Aut_{\mathcal{F}}(R_\lambda) : \lambda \in \Lambda\rangle_S.$$

LEMMA 11.5.4. *(1)* $n > 3$.
(2) $\mathcal{F}^\circ$ *is not* $(P)\Omega^+_8[m]$.

PROOF. By Hypothesis 11.5.1, $n \geq 2$ and $n > 2$ if $a = 0$.

Assume first that $n = 2$. Then $\mathcal{F}^\circ = \Omega_5[m]$ or $\Omega^-_6[m]$ for $a = 1, 2$, respectively. Therefore the lemma holds from the discussion in 11.5.2, contrary to the choice of τ as a counter example.

Assume next that $n = 3$. Then $\mathcal{F}^\circ$ is $P\Omega^+_6[m]$, $\Omega_7[m]$, or $\Omega^-_8[m]$. In the latter two cases the lemma holds from the discussion in 11.5.2; in the first case it holds by 9.5.9.

Finally $\mathcal{F}^\circ$ is not $(P)\Omega^+_8[m]$ by 10.2.14.3 and 2.1.15; the first reference handles the $P\Omega^+_8[m]$ case, and then the second lifts the result to $\Omega^+_8[m]$. This completes the proof of the lemma. □

LEMMA 11.5.5. *Let* $\theta \subseteq \mathcal{Z}_S$ *and* $f = f_\theta = \prod_{t\in\theta} t$ *with* $1 \neq f \in \mathcal{F}^f$. *Then either*
(1) $\mathcal{F}_f \leq \mathcal{Y}$, *or*
(2) $|\theta| = n/4$ *and* $L = P\Omega^+_{2n}(F)$.

PROOF. Assume (2) does not hold and let $|\theta| = k$ and $U_1 = [U, f]$ and $U_2 = C_U(f)$. Thus $d_1 = \dim(U_1) = 4k$ and $U_2 = U_1^\perp$ is of dimension $d_2 = 2n + a - 4k = 2(n-2k)+a$. Let $L_i = L_{U_i}$ have Sylow group T_i and $\mathcal{F}_i = \mathcal{F}_{T_i}(L_i)$. Let T be Sylow in $\mathcal{F}_f$. Claim $C_L(f)$ acts on U_i for $i = 1, 2$. If not $d_1 = d_2$, so $2n + a = 8k \equiv 0 \mod 8$. Thus if n is even then $a = 0$ and $k = n/4$. Further $f = (-I)\cdot f$, so $L = P\Omega^+_{2n}(F)$, contrary to the assumption that (2) does not hold. Similarly if n is odd then $a = 2$. But now U_1 is of sign $+1$ while I_2 is of sign -1, contradicting U_1 conjugate to U_2. This completes the proof of the claim.

Now $\mathcal{F}_f = T\mathcal{F}_1\mathcal{F}_2$, and by the claim, $\mathcal{F}_i \lhd \mathcal{F}_f$ for $i = 1, 2$. By minimality of d, $T\mathcal{F}_1 = \langle TO_{\tau_1}, \mathcal{F}_T(TM_1)\rangle_T$, and if $d_2 > 4$ then

$$T\mathcal{F}_2 = \langle TO_{\tau_2}, \mathcal{F}_T(TM_2), Aut_{L_{\bar{A}}}(\bar{R}_\lambda) : \lambda \in \Lambda\rangle,$$

where $\bar{A} \leq U_2$, $g \in M$ such that $A^g = \bar{A}$, $B^g_\lambda = \bar{B}_\lambda$, $\bar{R}_\lambda = C_T(\bar{B}_\lambda)$, and $\bar{R}^{g^{-1}}_\lambda \leq R_\lambda$; this is possible as $B_\lambda \in \mathcal{F}^f$. But $TO_{\tau_i} \leq SO_{\tau_i} \leq \mathcal{Y}$ and $\mathcal{F}_T(TM_2) \leq \mathcal{F}_S(SM) \leq \mathcal{Y}$, and $Aut_{L_{\bar{A}}}(\bar{R}_\lambda) \leq Aut_{\mathcal{Y}}(R_\lambda)$, as $g \in M$, $\mathcal{F}_S(SM) \leq \mathcal{Y}$, and $\bar{R}^{g^{-1}}_\lambda \leq R_\lambda$. Therefore when $d_2 > 4$ we have $\mathcal{F}_f \leq \mathcal{Y}$ by 1.3.2 in [**Asc19**].

So assume $d_2 \leq 4$. If $d_2 = 4$ and $sgn(U_2) = 1$ then $T\mathcal{F}_2 \leq SO_\tau$. If $d_2 \leq 2$ then $T\mathcal{F}_2 = T \leq \mathcal{Y}$. Finally if $d_2 = 3$ or $d_2 = 4$ and $sgn(U_2) = -1$, then n is odd and $a = 1$ or 2. Then $T\mathcal{F}_2 = \langle Aut_{L_A}(R_\lambda)\rangle_T \leq \mathcal{Y}$, completing the proof. □

LEMMA 11.5.6. *(1) $|\mathcal{Z}_S| = n_S$ is a power of 2.*
(2) $L = P\Omega_{2n}^+(F)$.
(3) Let $\theta \subseteq \mathcal{Z}_S$ be of order $n_S/2$ and $f = f_\theta \in \mathcal{F}^f$. Then $\mathcal{F}_f \not\leq \mathcal{Y}$.

PROOF. From 3.4.11.2, $\mathcal{E} = \mathcal{E}(\tau) = \langle N_\mathcal{F}(O(\xi)), SO_\tau, N_\mathcal{F}(W) : \xi \in \Xi\rangle$, where $W \in W(\tau)$ and Ξ consists of those $\xi \subseteq \mathcal{Z}_S$ with $O(\xi) \in \mathcal{F}^f$ and $N_S(O(\xi))$ transitive on ξ. Recall that $\mathcal{B} = \mathcal{B}(\tau) = \langle \mathcal{E}, \mathcal{F}_x : x \in \mathcal{W}(\tau) \cap \mathcal{F}^f\rangle$, where $\mathcal{W}(\tau)$ consists of those f_ζ with $\zeta \subseteq \mathcal{Z}_S$ of order a power of 2. By 11.5.4 and Theorem 11.3.11, we have $\mathcal{B} = \mathcal{F}$. Thus by the choice of τ as a counter example, $\mathcal{B}$ is not contained in $\mathcal{Y}$, so $N_\mathcal{F}(O(\xi))$ or $\mathcal{F}_x$ is not contained in $\mathcal{Y}$ for some $\xi \in \Xi$ or some $x = f_\zeta \in \mathcal{F}^f$ with $|\zeta|$ a power of 2. In either case $\mathcal{F}_y \not\leq \mathcal{Y}$ for some $y = f_\zeta \in \mathcal{F}^f$ with $|\zeta|$ a power of 2. Now the lemma follows from 11.5.5. □

NOTATION 11.5.7. By 11.5.6, $L = P\Omega_{2n}^+(F)$ and n is a power of 2. Hence $a = 0$ so from 11.5.2, $\Lambda = \emptyset$ and hence $\mathcal{Y} = \langle SO_\tau, \mathcal{F}_S(SM)\rangle$.

Choose $\theta = \theta_1$ as in 11.5.6.3; thus $|\theta| = n_S/2$ with $n = 2n_S$ a power of 2. It is convenient to work in $L = \Omega(U, \mathfrak{q})$ rather than in the projective group PL. Thus we regard f_θ as a member of L. Now $\Pi_1 = \{\theta_1, \mathcal{Z}_S - \theta_1\}$ is an S-invariant partition of $\mathcal{Z}_S$, and $U = U_1 \perp U_1^\perp$, where $U_1 = [U, f_\theta]$ and $U_1^\perp = C_U(f_\theta)$ are of dimension $n = d/2$.

Proceeding recursively, assume Π_k is an S-invariant partition of $\mathcal{Z}_S$ into 2^k blocks of size $n_S/2^k$. Pick $\theta_k \in \Pi_k$ and let $\theta_{k+1} \subseteq \theta_k$ be of order $n_S/2^{k+1}$ such that $|N_S(\theta_k) : N_S(\theta_{k+1})| = 2$, and set $\Pi_{k+1} = \theta_{k+1}^S$, so that Π_{k+1} is an S-invariant partition of $\mathcal{Z}_S$ into 2^{k+1} blocks of size $n_S/2^{k+1}$. Continuing this process, we arrive at $\theta_r = \{z\}$ for some $z \in \mathcal{Z}_S$; that is $n_S = 2^r$.

Let $\Pi_k = \{\theta_{k,1}, \ldots, \theta_{k,2^k}\}$ and $U_{k,i} = [U, f_{\theta_{k,i}}]$. Thus $U = U_{k,1} \perp \cdots \perp U_{k,2^k}$ and we set $\mathcal{U}_k = \{U_{k,1}, \ldots, U_{k,2^k}\}$, $L_{k,i} = L_{U_{k,i}}$ and $L_k = \prod_i L_{k,i}$ the direct product of 2^k copies of $\Omega_{4\cdot 2^{r-k}}^+(F)$. Let $\mathcal{U}^k = \{\mathcal{U}_1, \ldots, \mathcal{U}_k\}$ and G_k the stabilizer in SL of $\mathcal{U}^k$. Set $\mathcal{D}_k = \mathcal{F}_S(G_k)$.

LEMMA 11.5.8. *$G_k = SL_k$.*

PROOF. By construction, $SL_k \leq G_k$ and L_k is contained in the kernel X of the action of G_k on $\mathcal{U}_k$. By 2.1.16, $G_k^{\mathcal{U}_k}$ is contained in a Sylow 2-group of $Sym(\mathcal{U}_k)$. For $x \in X$ of odd order, $x_{|U_{k,i}} \in L_{k,i}$ as $L_{k,i}$ acts faithfully as $\Omega(U_{k,i}, \mathfrak{q})$ on $U_{k,i}$. Therefore X/L_k is a 2-group, completing the proof. □

LEMMA 11.5.9. *(1) $L_{r,i} \cong \Omega_4^+(F)$ has Sylow 2-group $O(s_i)$ for that $s_i \in \mathcal{Z}_S$ such that $s_i \in Z(L_{r,i})$ and $U_{t,i} = [U, s_i]$.*
(2) $\mathcal{O}_\tau = \mathcal{F}_{O(\tau)}(L_r)$.
(3) $\mathcal{D}_r = \mathcal{F}_S(G_r) = SO_\tau \leq \mathcal{Y}$.

PROOF. By construction, $U_{r,i} = [U, s_i]$ for some $s_i \in \mathcal{Z}_S$, and then (1) follows. In particular by (1), $O(\tau)$ is Sylow in L_r and (2) holds. By 11.5.8, $G_r = SL_r$, so $\mathcal{F}_S(G_r) = \mathcal{F}_S(SL_r) = SO_\tau$, establishing (3). □

LEMMA 11.5.10. *$\mathcal{E}(\tau) = \mathcal{Y}$.*

PROOF. From 3.4.11.2, $\mathcal{E} = \mathcal{E}(\tau) = \langle N_{\mathcal{F}}(O(\xi)), SO_\tau, N_{\mathcal{F}}(W) : \xi \in \Xi\rangle$, where $W \in W(\tau)$ and Ξ consists of those $\xi \subseteq \mathcal{Z}_S$ with $O(\xi) \in \mathcal{F}^f$ and $N_S(O(\xi))$ transitive on ξ. Now $N_{\mathcal{F}}(W) = \mathcal{F}_S(SM)$, so $\mathcal{Y} = \langle SO_\tau, \mathcal{F}_S(SM)\rangle \leq \mathcal{E}$ with $\mathcal{E} = \langle \mathcal{Y}, N_{\mathcal{F}}(O(\xi)) : \xi \in \Xi\rangle$. Thus it remains to show that $\mathcal{N} = N_{\mathcal{F}}(O(\xi)) \leq \mathcal{Y}$ for $\xi \in \Xi$. But $\mathcal{N} \leq \mathcal{F}_x$, where $x = f_\xi$, and by 11.5.5 either $\mathcal{F}_x \leq \mathcal{Y}$ or $(x,\xi) = (1, \mathcal{Z}_S)$ or (f, θ), and we may assume one of the latter two cases holds.

Suppose $(x,\xi) = (1,\mathcal{Z}_S)$. Then $C_S(O(\tau)) \leq O(\tau)$ so $\mathcal{N}$ has a model G. Let H be the kernel of the action of G on Ω. Then $W \leq T \in Syl_2(H)$ and $O^2(H) = 1$ if $m > 8$, while if $m = 8$ then $\mathcal{F}_{O(\xi)}(O^2(H)) = \mathcal{O}_\tau$. Thus if $m > 8$ then $W \trianglelefteq G$, so $\mathcal{N} \leq \mathcal{F}_S(SM) \leq \mathcal{Y}$. So take $m = 8$. Then $\mathcal{N} = \langle \mathcal{F}_S(SH), \mathcal{F}_S(N_G(T))\rangle$. But $\mathcal{F}_S(SH) = SO_\tau \leq \mathcal{Y}$ and $W \trianglelefteq N_G(T)$, so $\mathcal{F}_S(N_G(T)) \leq \mathcal{F}_S(SM) \leq \mathcal{Y}$. Thus the lemma holds in this case, so we may assume instead that $(x,\xi) = (f,\theta)$.

Now $\mathcal{N} = \mathcal{F}_S(X)$, where $X = N_{L_{1,1}}(O(\xi))L_{1,2}S$. Therefore

$$\mathcal{N} = \langle \mathcal{F}_S(SL_{1,2}), N_{\mathcal{N}}(S_2)\rangle,$$

where S_2 is Sylow in $L_{1,2}$. By minimality of d, $\mathcal{F}_S(SL_{1,2}) \leq \mathcal{Y}$. Further $N_{\mathcal{N}}(S_2) \leq N_{\mathcal{N}}(O(\tau))$, which is contained in $\mathcal{Y}$ by the previous case. □

NOTATION 11.5.11. Observe that $\mathcal{D}_1 = \mathcal{F}_f$. Therefore by 11.5.6.3, $\mathcal{D}_1 \not\leq \mathcal{Y}$. On the other hand by 11.5.9.3, $\mathcal{D}_r \leq \mathcal{Y}$. Thus there exists a least k such that $\mathcal{D}_k \not\leq \mathcal{Y}$. Set $\mathcal{D} = \mathcal{D}_k$, $\rho = (\mathcal{D}, \Omega)$, let $S_0 = S \cap L_k$, and write $\mathcal{D}_0$ for $\mathcal{F}_{S_0}(L_k)$. Let $I = \{1, \ldots, 2^k\}$ and for $i \in I$ set $S_i = S \cap L_{k,i}$ and $\mathcal{L}_i = \mathcal{F}_{S_i}(L_{k,i})$. From 11.5.7, $L_k = \prod_{i\in I} L_{k,i}$, so $\mathcal{D}_0 = \prod_{i \in I} \mathcal{L}_i$ is a direct product. Set $\Omega_i = \Omega \cap S_i$ and $\rho_i = (\mathcal{L}_i, \Omega_i)$.

Observe:

LEMMA 11.5.12. *The pair $\mathcal{D}, \mathcal{D}_0$ satisfies Hypotheses* 11.4.1 *and* 11.4.6.

NOTATION 11.5.13. As in 11.4.2 and 11.4.6, for $X \leq S$ and $0 \leq i \leq 2^k$, write X_i for $X \cap S_i$.

As $\mathcal{D} \not\leq \mathcal{Y}$, it follows from the Alperin-Goldschmidt Fusion Theorem that the set $\mathcal{R}$ of subgroups $R \in \mathcal{D}^e$ such that $Aut_{\mathcal{D}}(R) \not\leq Aut_{\mathcal{Y}}(R)$ is nonempty. Pick $R \in \mathcal{R}$ and let G be a model for $N_{\mathcal{D}}(R)$ and set $G^* = G/R$. By 11.5.12 and 11.4.4.1, $R_0 \in \mathcal{F}^{frc}$, so there is a model G_0 for $N_{\mathcal{D}_0}(R_0)$. By 11.4.4, we can embed G and G_0 in a certain group G_c with $G_0 \trianglelefteq G_c$ and $G = N_{G_c}(R)$. Thus we can form $H = N_{G_0}(R)$ and by 11.4.4 $H \trianglelefteq G$ with $R_0 = O_2(H)$ and $H^* \cong H/R_0$.

Set $Q = N_R(L_{k,1})$ and $\mathcal{X} = \mathcal{F}_{QS_1}(QL_{k,1})$. Let Σ be a set of coset representatives for Q in R with $1 \in \Sigma$. Write $\mathcal{R}^*$ for the set of those $R \in \mathcal{R}$ of maximal order.

LEMMA 11.5.14. *For each $K \in \Omega$, $|K \cap R| \leq 2$.*

PROOF. Assume otherwise; then from 3.4.5, $Aut_{\mathcal{D}}(R) = Aut_{\mathcal{E}(\rho)}(R)$. But $\mathcal{E}(\rho) \leq \mathcal{E}(\tau)$ and by 11.5.10, $\mathcal{E}(\tau) \leq \mathcal{Y}$, contrary to the choice of k. □

LEMMA 11.5.15. *(1) For each $i \in I$, $R_i \neq S_i$.*

(2) R is transitive on $\{L_{k,i} : i \in I\}$.

(3) $Q \in \mathcal{X}^e$. Let G_Q be a model for $N_{\mathcal{X}}(Q)$.

(4) $G_Q/Q \cong H^*$.

(5) Let $W_1 \in W(\rho_1)$ and $W_{1,I}$ the subgroups of W_1 defined in 11.1.2 that is the permutation module for $Sym(I) \cong \mu(\rho_1)/\mu_0(\rho_1)$. Define $\alpha : W_{1,I} \cap Z(Q) \to W \cap R$

by $\alpha : x \to \prod_{\sigma\in\Sigma} x^\sigma$. *Then* $x\alpha \in Z(R)$ *and if* x *is of weight* p *in* $W_{1,I}$ *then* $x\alpha$ *is of weight* $2^k p$ *in* W.

(6) If $p < 2|\mathcal{Z}_S(\rho_1)|$ *then* $C_{\mathcal{D}}(x\alpha) \le \mathcal{Y}$.

PROOF. If $R_i = S_i$ then $\Omega_i \subseteq R$, contrary to 11.5.14. Thus (1) holds.

By 11.5.12, we can appeal to results in section 11.4. In particular (2) follows from (1) and 11.4.8.1, (3) follows from 11.4.8.4, and (4) follows from 11.4.8.5.

From 11.5.13, $1 \in \Sigma$ is a set of coset representatives for Q in R. If $\sigma_i \in \Sigma$ with $R_1\sigma_i = R_i$ then $R_0 = \prod_{i\in I} R_i = \prod_{i\in I} R_1^{\sigma_i}$, so for $x \in W_{1,I} \cap Z(Q)$ of weight p, $x\alpha$ centralizes $\langle Q, \Sigma\rangle = R$ and $x\alpha$ is of weight $|\Sigma|p = |I|p = 2^k p$ in W. Thus (5) holds.

Assume the hypothesis of (6) and set $y = x\alpha$. Observe that $|\mathcal{Z}_S(\rho_1) = n_S/2^k = 2^{r-k}$. As $p < 2|\mathcal{Z}_S(\rho_1)|$, $y \neq 1$ so y is an involution. As $2^k p$ is even, y is fused in M to f_ξ for some $\xi \subseteq \mathcal{Z}_S$ with $|\xi| = 2^{k-1}p$. Then by 11.5.5, either $\mathcal{F}_y \le \mathcal{Y}$ or $2^{k-1}p = n_S/2 = 2^{r-1}$, and we may assume the latter, so that $p = 2^{r-k}$. Then conjugating in M, we may take $x = f_{\theta_{k+1}}$, so $\Pi_{k+1} = \theta_{k+1}^R \cup (\theta_k - \theta_{k+1})^R$. But then $C_{SL_k}(y) = C_S(y)L_{k+1}$, so as $\mathcal{D}_{k+1} \le \mathcal{Y}$, again $C_{\mathcal{D}}(y) \le \mathcal{Y}$. □

LEMMA 11.5.16. $\zeta = (\mathcal{X}, \Omega_1)$ *satisfies Hypothesis* 11.2.1 *and if* $R \in \mathcal{R}^*$ *then* ζ, Q *satisfies Hypothesis* 11.2.7.

PROOF. Visibly ζ satisfies Hypothesis 11.2.1 with $|\Omega(z)| = 2$ and $\mu(\zeta) \cong Weyl(D_{n/2^k})$.

So assume $R \in \mathcal{R}^*$. By 11.5.15.4, $Q \in \mathcal{X}^e$. From 11.2.6, $\mathcal{W}(Q)$ consists of those involutions $x \in Z(Q) \cap W_{1,I}$ such that $x = e_\alpha$ for some $\alpha \subseteq I(\zeta)$ of weight a positive power of 2. By 11.5.15.6, for each $x \in \mathcal{W}(Q)$, $C_{\mathcal{D}}(x\alpha) \le \mathcal{Y}$.

Let $x \in \mathcal{W}(Q)$, $y = x\alpha$ and $Z = \langle y^G\rangle$. As $R \in \mathcal{R}^*$ and $C_{\mathcal{D}}(y) \le \mathcal{Y}$, the proof of 11.3.5.1 shows that $O^{2'}(C_G(Z)) = R$. Set $Z_x = \langle x^{G_Q}\rangle$. Then Z is a full diagonal subgroup of $\prod_{\sigma\in\Sigma} Z_x^\sigma$ and the isomorphism of 11.5.15.4 maps $O^{2'}(C_{G_Q}(Z_x))/Q$ to $O^{2'}(C_G(Z))/R = 1$, so $O^{2'}(C_{G_Q}(Z_x)) = Q$. Hence ζ, Q satisfies 11.2.7.1.

Similarly if ρ, R satisfies the setup of 11.2.7.2 and $\mathcal{U} \neq \emptyset$ then the proof of 11.3.5.2 shows that $m_2(G^*) = 1$ and $\langle\mathcal{U}\rangle$ is not normal in G^*. Let B^* be the subgroup of G^* generated by its involutions. By 11.4.5.2, B^* is the subgroup of H^* generated by its involutions. Thus the isomorphism of 11.5.15.4 maps the subgroup B_Q^* of $G_Q^* = G_Q/Q$ generated by involutions to B^*, so $m_2(G_Q^*) = 1$ and $\langle\mathcal{U}_Q\rangle \neq B_Q^*$. Therefore ζ, Q satisfies 11.2.7.2. This completes the proof of the lemma. □

We are now in a position to obtain a contradiction, completing the proof of Theorem 11.5.3. Namely choose $R \in \mathcal{R}^*$. By 11.5.16, the pair ζ, Q satisfies Hypotheses 11.2.1 and 11.2.7. Further condition (a) of Theorem 11.2.34 is satisfied, as is condition (b) by 11.5.14. Condition (c) of 11.2.34 holds as $\mathcal{X} = \mathcal{F}_{QS_1}(SL_{k,1})$. But then by Theorem 11.2.34, $|\mathcal{Z}_S(\zeta)| = 1$, a contradiction. This completes the proof of Theorem 11.5.3.

In this last lemma of the section we do not assume Hypothesis 11.5.1.

LEMMA 11.5.17. *Assume* $\tau = (\mathcal{F}, \Omega)$ *is a quaternion fusion packet with* $\mathcal{F}$ *isomorphic to* $\hat{A}E_9$, $Spin_9[m]$, *or* $Spin_{10}^-[m]$. *Let* S *be Sylow in* $\mathcal{F}$, $M = M(\tau)$, *and* $U = Z(M)$.

(1) If $\mathcal{F}$ is $\hat{A}E_9$ then $Y = Aut(\mathcal{F})$ is the split extension of $E = O_2(Y)$ by $Sym(I) \cong S_9$ with E the core of the 9-dimensional permutation module for $Sym(I)$.

(2) $U \cong E_4$ and $Z(S) = Z(\mathcal{F}) \cong \mathbf{Z_2}$, so S is transitive on $U - Z(\mathcal{F})$.

(3) There exists no automorphism α of $\mathcal{F}$ of order 4 centralizing M.

PROOF. Observe that $\mu(\tau) \cong$ Weyl(D_4) (cf. 11.6.2 and 5.3.4) and $M \cong \bar{\omega}(D_4, m)$ (cf. 5.1.16, 5.3.9, and 5.3.10), so $U \cong E_4$ by 5.8.4. Then as $Z = Z(\mathcal{F})$ is of order 2, if $Z(S) = Z$ then S is transitive on $U - Z(S)$, so that (2) holds.

Suppose $\mathcal{F}$ is $\hat{A}E_9$ and let $\hat{G}$ be a model for $\mathcal{F}$. By 5.1.16, $Q = O_2(\mathcal{F}) \cong Q_8^3$, so as $F^*(\hat{G}) = Q$ we have $Z(S) = Z(Q) = Z$, and hence (2) holds in this case. As $Q \cong Q_8^3$, $Out(Q) \cong O_6^-(2)$. Now $Y \leq Aut(Q)$ and indeed $Y/Inn(Q) = N_{Out(Q)}(Out_{\hat{G}}(Q))$ with $Out_{\hat{G}}(Q) \cong A_9$ absolutely irreducible on Q/Z, so we have $Y/Inn(Q) \cong S_9$, establishing (1).

Next α induces an automorphism α of $G = \hat{G}/Z \cong AE_9$ of order 4 centralizing M/Z, and by (1), $\alpha \in Y = ESym(I)$. Then ME/E is the stabilizer of the partition $\Lambda = \{\{1,2\},\{3,4\},\{5,6\},\{7,8\},\{9\}\}$ of I, so $C_{Sym(I)}(ME/E) = \langle t \rangle$ where $t = (1,2)(3,4)(5,6)(7,8)$. Hence $\alpha \in tE$, a contradiction as t does not centralize $K \cap E$ for $K \in \Omega$.

So we may take $\mathcal{F}$ to be $Spin_9[m]$ or $Spin_{10}^-[m]$. Then $\mathcal{F}$ is tamely realized by $\hat{L} \cong Spin_9(q)$ or $Spin_{10}^{-\pi}(q)$ for suitable q by 3.5 in [**AO16**], and then by 2.22.3 in [**AO16**], α is induced by an automorphism α of $L = \Omega_9(q)$ or $\Omega_{10}^{-\pi}(q)$. Let V be the defining unitary space for L; then $Aut(L) = \Gamma = \Gamma O(V)$. Now $M = M(\tau) \leq X \leq \hat{L}$ with $X \cong Spin_8^+(q)$. Let $T \in Syl_2(X)$ and $\mathcal{X} = \mathcal{F}_T(X)$. As α centralizes $O(\tau) \leq M$, α centralizes $C_{\mathcal{X}}(z)$ for $z \in \mathcal{Z}_S$ by 2.6.12, so by 11.5.3, α centralizes $\mathcal{X} = \langle C_{\mathcal{X}}(z), M \rangle$. Thus α centralizes X, a contradiction as a Sylow 2-subgroup of $C_\Gamma(X)$ is elementary abelian.

It remains to show that $Z = Z(S)$; as $M \cong \bar{\omega}(D_4, m)$ we have $U = Z(T)$. If $\mathcal{F} \cong Spin_9[m]$ then $S = T\langle x \rangle$ where x induces a reflection on X, and hence $[U, x] = Z$ as $Out(\hat{L})$ induces $GL(U)$ on U and $x\hat{L} \notin O_2(Out(\hat{L}))$. Thus indeed $Z = Z(S)$ in this case. When $\mathcal{F}$ is $Spin_{10}^-[m]$ we have $Z = Z(S)$ by 5.9.4 and its proof. □

11.6. Generating AE_k

In this section we assume the hypothesis and notation of section 5.1, except we write k for the integer "n" in that section, and instead write $k = 2n + a$ for $a \in \{0, 1\}$. Assume $k \geq 7$; thus $n \geq 3$. In particular $G = AE \cong \check{A}E_k$ and $G^+ = G/E = Alt(I) \cong A_k$. For $X \subseteq G$, write $Fix(X)$ for the set of fixed points of X^* on I and write $Mov(X)$ for the set of points moved by X^*.

As in Notation 5.1.5, let $z = e_{1,2,3,4}$. Choose $K, \tilde{K}$, and $S \in Syl_2(G)$ as in 5.1.6, set $\Omega = K^G \cap S$, and $\mathcal{F} = \mathcal{F}_S(G)$. From 6.3.6.3, $\tau = (\mathcal{F}, \Omega)$ is a quaternion fusion packet with $\Omega(z) = \{K, \tilde{K}\}$ of order 2. Choose notation so that $z \in \mathcal{F}^f$, and if k is odd, so that $Fix(S) = \{k\}$.

From 5.1.9 we may choose notation so that $\Omega = \{K_i, \tilde{K}_i : 1 \leq i \leq s\}$, where $n = 2s$, $n = 2s + 1$ for n even, odd, respectively, and $z(K_i) = z_i = z(\tilde{K}_i) = e_{\lambda_i}$ with $\lambda_i \in \bar{P}_2$, where $\bar{P}_2$ is a partition of the set Δ_2 of the first $4s$ points of I into s blocks of size 4, such that $Mov(K_i\tilde{K}_i) = \lambda_i$. Choose notation so that $z = z_1$.

LEMMA 11.6.1. *(1) For $1 \leq i \leq s$, there is $L_{z_i} \trianglelefteq C_G(z_i)$ with $O(z_i) = O_2(L_{z_i})$, $L_{z_i} = O^2(L_{z_i})$, $L_{z_i}/O(z_i) \cong \mathbf{Z}_3$, and $Mov(L_{z_i}) = \lambda_i$. Set $\mathcal{O}_{z_i} = \mathcal{F}_{O(z_i)}(L_{z_i})$.*

(2) Set $L_\tau = \langle L_{z_i} : i\rangle$ and $\mathcal{O}_\tau = \langle \mathcal{O}_{z_i} : i\rangle$. Then $L_\tau = \prod_i L_{z_i}$ and $\mathcal{O}_\tau = \prod_i \mathcal{O}_{z_i}$.

PROOF. Part (1) is 5.1.6.2 and (1) implies (2). □

Again by 5.1.9, there is a partition $\bar{P}_1$ of the set Δ_1 of the first $2n$ points of I into n blocks of size 2, such that $P_1 = \bar{P}_1 \cup (I - \Delta_1)$ is a partition of I that is a refinement of $\bar{P}_2 \cup (I - \Delta_2)$, and $\mathcal{Z}_\Delta = \{e_{\alpha\cup\beta} : \alpha, \beta \in \bar{P}_2 \text{ and } \alpha \neq \beta\}$.

If k and n are even let $P_2 = \bar{P}_2$. If k is odd and n is even, let $P_2 = \bar{P}_2 \cup \{k\}$. If k is even and n is odd, let $P_2 = \bar{P}_2 \cup \{k_1, k_2\}$. If k and n are odd let $P_2 = \bar{P}_2 \cup \{k-1, k-2\} \cup \{k\}$. Thus in each case, P_2 is a partition of I and P_1 is a refinement of P_2. Moreover P_1 and P_2 are S-invariant as Ω and $\mathcal{Z}_\Delta$ are S-invariant.

From the proof of 5.1.6, $K^+ = \tilde{K}^+ = \langle (1,2)(3,4), (1,3)(2,4)\rangle$ is the root 4-group of G^+ moving λ_1. Similarly K_i^+ is the root 4-group with $Mov(K_i^+) = \lambda_i$.

LEMMA 11.6.2. *Let $W = W(\tau)$, and $M_1 = N_G(P_1)$.*

(1) $M_1 = N_G(W)$.

(2) $M_1^+ = M^+$.

(3) $\mu(\tau) \cong Weyl(D_n)$.

PROOF. Observe M_1^+ permutes the set $\mathcal{T}$ of transpositions t in $Sym(I)$ such that $Mov(t) \in \bar{P}_1$, and W^+ is the subgroup of index 2 in $\langle \mathcal{T}\rangle$ that is the core of the permutation module for $M^+/W^+ = Sym(\mathcal{T}) = Sym(\bar{P}_1) \cong S_n$. As W is weakly closed in S, it follows that (1) holds. Next $M = \langle K^{M_1}\rangle \trianglelefteq M_1$ and $M_1^+ = \langle K^{+M_1^+}\rangle = M^+$, establishing (2).

Next define W_I as in Notation 11.1.2, and observe that $W \cap E = W_I = \langle e_\alpha : \alpha \in \bar{P}_1\rangle$ is the permutation module for $M_1/WE \cong S_n$ on $\{e_\alpha : \alpha \in \bar{P}_1\}$, so $|W| = 2^{2n-1}$. If $k = 2n$ is even then E/W_I is the core of the permutation module for M_1/WE, so $E = [E, M] \leq M$ as $M \trianglelefteq M_1$. Then $\mu = \mu(\tau) = M/W$ is an extension of E/W_I by S_n, so (3) holds in this case by 4.3.8.1. On the other hand if $k = 2n+1$ is odd then E/W_I is the permutation module so $M \cap E = [E, M]$ is of index 2 in E and again (3) holds. Moreover we've shown that if k is even then $M = M_1$, which we record as: □

LEMMA 11.6.3. *If k is even then $M = N_G(P_1)$ contains S.*

LEMMA 11.6.4. *Assume k is not congruent to 3 modulo 4. Then*

(1) $N_G(O(\tau)) = L_\tau N_G(O(\tau)W) = N_G(P_2)$.

(2) $M_{\mathcal{F}}(O(\tau)) = \langle S\mathcal{O}_\tau, N_{\mathcal{F}}(O(\tau)W)\rangle$.

PROOF. Let $O = O(\tau)$, $M_2 = N_G(P_2)$, and $\bar{M}_2 = N_G(\bar{P}_2)$. By construction, $\bar{M}_2 = N_G(O)$. But as k is not congruent to 3 modulo 4, $M_2 = \bar{M}_2$, so $M_2 = N_G(O)$. Similarly by construction, L_τ and $L_\tau W$ are normal in M_2, with OW Sylow in $L_\tau W$, so by a Frattini argument, $M_2 = L_\tau N_G(OW)$, completing the proof of (1). Then (1) implies (2) as G is a model for $\mathcal{F}$. □

NOTATION 11.6.5. Set $S_M = S \cap M$, so that S_M is Sylow in M. Set $\mathcal{B} = \langle S_M\mathcal{O}_\tau, \mathcal{F}_{S_M}(M)\rangle$.

LEMMA 11.6.6. *If k is even then S is Sylow in M and $\mathcal{F} = \mathcal{B}$.*

PROOF. Assume k is even. By 11.6.3, S is Sylow in M and $M = N_G(P_1)$. By the Alperin-Goldschmidt Fusion Theorem, $\mathcal{F} = \langle N_{\mathcal{F}}(S), Aut_{\mathcal{F}}(B) : B \in \mathcal{F}^e \rangle$. By 5.2.5 in [**Asc10**], $N_G(S) = S$, so $\mathcal{B} = \langle Aut_{\mathcal{F}}(B) : B \in \mathcal{F}^e \rangle$.

Let $B \in \mathcal{F}^e$, and define $\mathcal{B}_*(G^+, S^+)$ as in section 5 of [**Asc10**]. Then $B^+ \in \mathcal{B}_*(G^+, S^+)$, so by 5.11 in [**Asc10**], $N_G(B) \leq N_G(P_i) = M_i$ for $i = 1, 2$. Therefore $N_{\mathcal{F}}(B) \leq \mathcal{F}_S(M_i)$. If $i = 1$ then $N_{\mathcal{F}}(B) \leq \mathcal{F}_S(M) \leq \mathcal{B}$ by 11.6.3, so it remains to show $\mathcal{F}_S(M_2) \leq \mathcal{B}$. But by 11.6.4, $\mathcal{F}_S(M_2) = \langle S\mathcal{O}_\tau, N_{\mathcal{F}}(O(\tau)W)\rangle \leq \mathcal{B}$ as $N_{\mathcal{F}}(O(\tau)W) \leq N_{\mathcal{F}}(W) = \mathcal{F}_S(M)$. □

LEMMA 11.6.7. *Let $D = \langle (\Omega - \{K, \tilde{K}\})^{C_G(z)} \rangle$ and assume $k \geq 9$. Then $D \cong \breve{A}E_{k-4}$.*

PROOF. Observe that $C_G(z)$ is the stabilizer in G of λ_1 with $\Gamma = \Omega - \{K, \tilde{K}\}$ trivial on λ_1. Therefore $D \leq G_{\lambda_1}$ and $G^+_{\lambda_1} = Alt(I - \lambda_1)$. Further for $J \in \Gamma$, $J \leq O^2(D)$, so $D = O^2(D)$, and hence $D \cap E \leq \langle e_j : j \in I - \lambda_1 \rangle$. The result follows. □

LEMMA 11.6.8. *Assume k is odd and set $L = \langle K^{G_k} \rangle$. Then*

(1) $S_M \in Syl_2(L)$ and $S = S_M E$. Indeed if $k \geq 9$ then $S = S_M(E \cap D)$, where D is as in 11.6.7.

(2) $L \cong \breve{A}E_{2n}$.

(3) $\mathcal{B} = \mathcal{F}_{S_M}(L)$.

PROOF. As $G^+_k = \langle K^{+G^+_k} \rangle$, we have $G_k = LE$. Then as E is the $2n$-dimensional permutation module for $L^+ \cong A_{2n}$, it follows that $L \cap E = [E, L]$ is of index 2 in E, and then that (2) holds. Then as $M \leq L$, we conclude from 11.6.6 that S_M is Sylow in L and (3) holds. Also $|S : S_M| = 2$ with $S = S_M E$. Finally if $k \geq 9$ then $D \cap E \not\leq [L, E]$, so $S = S_M(D \cap E)$, completing the proof of (1) and the lemma. □

LEMMA 11.6.9. *Assume k is odd and let $\lambda \in \bar{P}_1$ such that $R = S_\lambda$ is of maximal order. Then*

(1) $\mathcal{F} = \langle S\mathcal{B}, O^2(Aut_{\mathcal{F}}(R))\rangle$.

(2) $R \in \mathcal{F}^e$.

(3) λ is in the shortest orbit of S on $\bar{P}_1$.

(4) If n is odd then λ is the unique S-invariant member of $\bar{P}_1$.

PROOF. As in the proof of 11.6.6, $\mathcal{F} = \langle Aut_{\mathcal{F}}(B) : B \in \mathcal{F}^e \rangle$, and for each $B \in \mathcal{F}^e$, $B^+ \in \mathcal{B}_*(G^+, S^+)$. By 11.6.8.3, we may choose B with $N_G(B) \not\leq G_k$. Partially order such B by the transitive extension $\lesssim$ of the relation $B_1 \prec B_2$ if $B_1 \leq B_2$ and $\Sigma_1 = Aut_{\mathcal{F}}(B_1) = \langle Aut_{\mathcal{B}}(B_1), N_{\Sigma_1}(Aut_{B_2}(B_1))\rangle$. Pick B maximal with respect to $\lesssim$.

As k is odd, $Fix(B) \neq \emptyset$, so we may appeal to 5.10 in [**Asc10**]. As $N_G(B) \not\leq G_k$, $Fix(B) \neq \{k\}$, so B^+ appears in one of the three cases in 5.10.3; in particular $J = Fix(B)$ is of order 3. If $k = 7$ we check directly that $|\mathcal{F}^e| = 2$ and the lemma holds. Thus we may assume $k > 7$, so that case (ii) or (iii) of 5.10.3 holds.

As $|J| = 3$ and G is 3-transitive on I, we may assume $J = \lambda \cup \{k\}$ for some $\lambda \in \bar{P}_1$. Hence $B \leq S_\lambda$.

Suppose case (ii) holds. Then n is odd and λ satisfies (3) and (4). Let X be of order 3 in G with $Mov(X) = J$. Then $[S_\lambda, X] = Q$ is a 4-subgroup of

E, so $Q \leq E \leq B$ and $N_G(B) = N_{SL}(B)B_2X$, where $B_2 = N_S(B)$. Therefore $Aut_{\mathcal{F}}(B) = Aut_{\mathcal{B}}(B)Aut_{B_2X}(B)$, so by maximality of B, $B = B_2$, so that $B = S_\lambda$. By (4), $|S : B| = 2$ and as $k > 7$, $N_G(B) = SX$, so (2) holds.

Suppose on the other hand that (iii) holds. Then $B \in Syl_2(G_J)$, so $B = S_\lambda$. As $B \in \mathcal{F}^f$ and $N_G(\bar{P}_1)$ is transitive on $\bar{P}_1$, it follows that (3) holds, so in particular if n is odd then we are back in case (ii), so that B is unique and (1), and hence also the lemma holds. Hence we may assume n is even, so all choice for B are in case (iii). Then as $N_G(\bar{P}_1) \leq LE$, B is unique under conjugacy in $S\mathcal{B}$, so (1) holds, and $B = S_\lambda$ with $|S_\lambda|$ maximal. Further (3) implies (2) and (4) holds vacuously, completing the proof. □

LEMMA 11.6.10. *Assume k is odd and $G = AE_k$. Let $S \in Syl_2(G)$. Then $O_2(G) = E = J(S)$.*

PROOF. This follows from B.2.7 and B.3.2.4 in [**AS04**]. □

CHAPTER 12

$|\Omega(z)| = 2$ and the proof of Theorem 6

In Chapter 12 we prove Theorem 6 under the additional hypothesis that the Extended Inductive Hypothesis is satisfied. Then, at the end of the paper when Theorem 1 is established, the extra assumption becomes unnecessary. In any event throughout Chapter 12 we assume $\tau = (\mathcal{F}, \Omega)$ is a quaternion fusion packet such that $\mathcal{F}^\circ$ is transitive on Ω, $|\Omega(z)| \geq 2$ for $z \in \mathcal{Z}_S(\tau)$, and τ satisfies the Inductive Hypothesis. The case where $|\Omega(z)| > 2$ has already been treated in Theorem 10.2.13, so we may assume $|\Omega(z)| = 2$.

We begin in section 12.1 with the case $\mu = \mu(\tau) \cong \mathrm{Weyl}(D_4)$. Then in section 12.2 we determine those packets with $\mu \cong \mathrm{Weyl}(D_n)$ for some $n \geq 5$. By Theorems 9.3.24 and 9.4.33, we may assume $\mathcal{D}^*(z) \neq \mathcal{D}(z)$ and $\mathcal{Z}(\tau) \cap O(z) = \{z\}$. Therefore, strengthening the Inductive Hypothesis to the Extended Inductive Hypothesis, Hypothesis 12.3.1 holds. Hence (cf. 12.3.4) μ is the direct product of r copies of $\mathrm{Weyl}(D_n)$ for some $n \geq 3$. Finally we complete the proof of Theorem 6 (under the additional assumption that the Extended Inductive Hypothesis holds) in Theorem 12.3.8 by proving $r = 1$. Note that the case $\mu \cong \mathrm{Weyl}(D_3) \cong S_4$ was treated in Theorem 9.5.33. These threads are drawn together in Theorem 12.3.50 and the proof of the (weak version) of Theorem 6 formally appears in section 12.4.

12.1. $|\Omega(z)| = 2$, μ isomorphic to Weyl(D_4)

In this section we assume the following hypothesis:

HYPOTHESIS 12.1.1. *$\tau = (\mathcal{F}, \Omega)$ is a quaternion fusion packet such that*
(1) For $z \in \mathcal{Z}_S$, $|\Omega(z)| = 2$.
(2) $\mu \cong$ Weyl(D_4).
(3) τ satisfies the Inductive Hypothesis

NOTATION 12.1.2. By 12.1.1.2 and 4.2.8.7, $\mathcal{F}^\circ$ is transitive on Ω. By 4.2.8.6, $|\Omega| = 4$, so $|\mathcal{Z}_S| = 2$ by 12.1.1.1.

Observe that Hypotheses 9.1.1 and 9.1.8 are satisfied, and adopt Notation 9.1.2 and 9.1.6. Let $t \in \mathcal{Z}_S - \{z\}$, $\Omega(t) = \{K_3, K_4\}$, and set $c = zt$ and $W_I = \langle v^M \rangle$. Let $\mu_0 = \langle (d_{K \cap W} d_{K_2 \cap W})^\mu \rangle$, $O = O(z)$, and $b = v_3 v_4$. For $a \in \{c, b, bc\}$, let T_a be the Sylow group of $\mathcal{F}_a^\circ$. Observe also that Hypothesis 11.1.1 is satisfied.

LEMMA 12.1.3. *(1) $Z(M) = \langle c \rangle$.*
(2) $\mu_0 \cong E_8$ is the core of the 4-dimensional permutation module for $\mu/\mu_0 \cong S_4$, μ_0 centralizes W_I, and W_I is the 4-dimensional permutation module for μ/μ_0.
(3) $c \in Z(S)$.
(4) $C_S(z) = C_S(t)$ is of index 2 in S and $z \in \mathcal{F}^f$.

(5) For each $a \in b^M \cup (bc)^M$, $|S : C_S(a)| = 4$ and $a \in \mathcal{F}^f$.

(6) $\langle z^M \rangle \cong E_8$ with $\mathcal{Z}_\Delta = \langle z^M \rangle - \langle c \rangle$ of order 6.

(7) $z^M \cap \mathcal{D}^*(z) - \mathcal{Z}_S = \{v_1^{\epsilon_1} \cdots v_4^{\epsilon_4} : \epsilon_i \in \{1, -1\}\}$ is of order 4 and $\mathcal{Z}_\Delta = \mathcal{Z}_S \cup (z^M \cap \mathcal{D}^*(z))$.

PROOF. Part (1) follows from 5.8.7.2. Next $\mu_0 \cong E_8$ is the core of the permutation module for $\mu/\mu_0 \cong S_4$ from the structure of the reflection group $M^* = M/W = \mu \cong \text{Weyl}(D_4)$ (cf. Remark 4.3.4). The remaining statements in (2) follow from 11.1.2.

As S acts on M, (3) follows from (1). Part (4) holds as $|\mathcal{Z}_S| = 2$ and $\mathcal{F}$ is transitive on Ω. Similarly (5) and (6) follow from (2). By (2), for some choice of ϵ_i, $x = v_1^{\epsilon_1} \cdots v_4^{\epsilon_4} \in \Sigma = \mathcal{D}^*(z) \cap z^M - \mathcal{Z}_S$, and then as $O(\tau)$ is transitive on such elements, all are in Σ, so (7) holds. □

LEMMA 12.1.4. *(1) $\tau_c = (\mathcal{F}_c, \Omega)$ is a quaternion fusion packet with $\mu(\tau_c) \cong$ Weyl(D_4).*

(2) Either $\mathcal{F}_c^\circ = \mathcal{F}_{T_c}(M)$ or $\mathcal{F}_c^\circ \cong \check{L}_3(2)/2^{3+6}$, $\check{A}E_8$, $\check{\Omega}_8^+(2)$, or $\Omega_8^+[m]$.

PROOF. This is a special case of 11.1.6. □

LEMMA 12.1.5. *For $a \in \{b, bc\}$ let $\eta_a = (K \cap W)^{C_M(a)}$, $W_a = \langle \eta_a \rangle$, and $M_a = \langle K^{C_M(a)} \rangle$. Then*

(1) $\tau_a = (\mathcal{F}_a, \Omega(z))$ is a quaternion fusion packet with $\eta_a \in \eta(\tau_a)$, M_a is a model for $[K]_{N_{\mathcal{F}_a}(W_a)}$, $\mu(\tau_a) \cong S_4$, $Z(M_{bc}) = \langle bc \rangle$, $b \notin M_b$, and $Z(M_b) = \langle bc \rangle$.

(2) Either $\mathcal{F}_{bc}^\circ = \mathcal{F}_{T_{bc}}(M_{bc})$ or $\mathcal{F}_{bc}^\circ \cong \check{A}E_6$ or $\Omega_6^+[m]$.

(3) One of the following holds:

(i) $\mathcal{F}_b^\circ = \mathcal{F}_{T_b}(M_b)$.

(ii) $\mathcal{F}_b^\circ \leq \mathcal{F}_c^\circ$ and $\mathcal{F}_b^\circ \cong \check{A}E_6$ or $\Omega_6^+[m]$.

(iii) $\mathcal{F}_b^\circ \cong L_3(2)/E_{64}$, $Sp_6(2)$, AE_7, $\Omega_7[m]$, or $\Omega_8^-[m]$.

PROOF. From 11.1.7.1, τ_a is a quaternion fusion packet with $\mu(\tau_a) \cong \text{Weyl}(D_3) \cong S_4$. As a centralizes O by construction, $\Omega(z) = C_\Omega(a)$. Therefore $\eta_a \in \eta(\tau_a)$ and M_a is a model for $[K]_{N_{\mathcal{F}_a}(W_a)}$.

By 5.8.7.4, $b \in \langle c \rangle M_b$, so either $b \in M_b$ or $b \in cM_b$. But as $c \in Z(M)$, $M_b = M_{bc}$, so $b \in cM_b$ iff $bc \in M_b = M_{bc}$. By 5.8.7.3, M is determined up to isomorphism by Hypothesis 12.1.1, so to prove $bc \in M_b$ and complete the proof of (1), we may work in the 2-fusion system $\mathcal{F}$ of $L = \Omega(U, \mathfrak{q})$, where $(U, \mathfrak{q})$ is an 8-dimensional orthogonal space of sign +1 over $F = \mathbf{F}_q$, with $q \equiv \pi \mod 4$ and $\pi = \pm 1$. Then $B = [U, b]$ and $A = B^\perp$ are of sign π and dimension 2 and 6, so $L_A \cong \Omega_6^\pi(F)$ has center generated by an involution i of type $i(6, \pi)$. In particular $i \neq b$, so $i = bc$ and hence indeed $bc \in M_b$.

Parts (2) and (3) follow from 11.1.7.3 together with the fact that $bc \in Z(M_b) = Z(M_{bc})$ from (1). □

LEMMA 12.1.6. *(1) $\mathcal{F}_z^\circ = \mathcal{O}(z) * \mathcal{C}$ where $\rho = (\mathcal{C}, \Omega(t))$ is a quaternion fusion packet with $\mathcal{C} = \mathcal{C}^\circ$ and $\mu(\rho) \cong E_4$.*

(2) Either $\mathcal{C} = \mathcal{O}(t)$ or $\mathcal{C} \cong AE_5$, $PSp_4[m]$, or $L_4^-[m]$.

PROOF. This is a special case of 11.1.5. □

LEMMA 12.1.7. *(1) $\mathcal{F}_c \leq \mathcal{E}(\tau)$.*

(2) $\langle \mathcal{F}_b, \mathcal{F}_{bc} \rangle \leq \mathcal{E}(\tau)$.

(3) $\mathcal{E}(\tau) = \langle SO(\tau), N_{\mathcal{F}}(O), N_{\mathcal{F}}(O(\tau)), N_{\mathcal{F}}(W)\rangle = \langle \mathcal{F}_z, \mathcal{F}_c\rangle$.
(4) $\mathcal{F} = \mathcal{E}(\tau) = \langle \mathcal{F}_z, \mathcal{F}_c\rangle$.
(5) Let $\tilde{\mathcal{F}} = \Omega_9[m]$ *or* $\Omega_{10}^-[m]$. *Then* $\tilde{\mathcal{F}} = \langle \tilde{\mathcal{F}}_z, \tilde{\mathcal{F}}_c\rangle$.

PROOF. The first equality in (3) follows from the definition of $\mathcal{E} = \mathcal{E}(\tau)$ in 3.4.2, while $\mathcal{E} \leq \langle \mathcal{F}_z, \mathcal{F}_c\rangle$ as $\langle SO(\tau), N_{\mathcal{F}}(O(\tau)), N_{\mathcal{F}}(W)\rangle \leq \mathcal{F}_c$ while $N_{\mathcal{F}}(O) \leq \mathcal{F}_z$. Thus to complete the proof of (3), it suffices to prove (1) and $\mathcal{F}_z \leq \mathcal{E}$. But $\mathcal{O}(z) \trianglelefteq \mathcal{F}_z$, so by 1.3.2 in [**Asc19**], $\mathcal{F}_z = \langle SO(z), N_{\mathcal{F}}(O)\rangle$. Then as $SO(z) \leq SO(\tau)$, we have $\mathcal{F}_z \leq \mathcal{E}$, so (1) implies (3).

Let $a \in \{b, c, bc\}$. By 1.3.2 in [**Asc19**], $\mathcal{F}_a = \langle C_S(a)\mathcal{F}_a^\circ, N_{\mathcal{F}_a}(T_a)\rangle$. Further $N_{\mathcal{F}_c}(T_c) \leq N_{\mathcal{F}}(O(\tau))$ and $N_{\mathcal{F}_a}(T_a) \leq N_{\mathcal{F}_a}(O)$ for $a \in \{b, bc\}$, so to prove (1) and (2) it remains to show that $\mathcal{X} = C_S(a)\mathcal{F}_a^\circ \leq \mathcal{E}$. But if $a = c$ then by 12.1.4 and 10.1.15.2, $\mathcal{X} = \langle N_{\mathcal{X}}(\langle z, c\rangle), \mathcal{F}_S(SM)\rangle$ and $\mathcal{O}(\tau) \trianglelefteq N_{\mathcal{X}}(\langle z, c\rangle)$, so by 1.3.2 in [**Asc19**], $N_{\mathcal{X}}(\langle z, c\rangle) = \langle SO(\tau), N_{\mathcal{X}}(O(\tau))\rangle \leq \mathcal{E}$, proving (1).

So take $a = b$ or bc. Here by 12.1.5, 9.3.20, and 9.5.9,

$$\mathcal{X} = \langle N_{\mathcal{X}}(\langle z, a\rangle), \mathcal{F}_{C_S(a)}(C_S(a)M_a)\rangle,$$

and $\mathcal{O}(z) \trianglelefteq N_{\mathcal{X}}(\langle z, a\rangle)$, so by 1.3.2 in [**Asc19**], $N_{\mathcal{X}}(\langle z, a\rangle) = \langle C_S(a)\mathcal{O}(z), N_{\mathcal{X}}(O)\rangle \leq \mathcal{E}$, proving (2).

By Theorem 11.3.11, $\mathcal{F} = \mathcal{B}$, where $\mathcal{B} = \langle \mathcal{E}, \mathcal{F}_z, \mathcal{F}_c\rangle$, so (4) follows from (3). A subset of these arguments also proves (5). □

LEMMA 12.1.8. *Assume* $m = 8$ *and set* $S_0 = C_S(O)$, $R_0 = OS_0$, *let* G_z *be a model for* $N_{\mathcal{F}_z}(R_0)$, $H_z = \langle W^{G_z}\rangle$, *and* $L_z = O^2(H_z)$. *Then either* $L_z = 1$ *or* $O(\tau) = O_2(L_z)$, L_z *is of index* 3^e *in* $(SL_2(3) * SL_2(3))^2$ *for some* $0 \leq e \leq 3$, *and* L_z *centralizes* $C_S(O(\tau))$.

PROOF. By 9.1.11.2, either the lemma holds or $O = O_2(L_z)$ and L_z is of index 1 or 3 in $SL_2(3) * SL_2(3)$. As W is nontrivial on $J/\langle z(J)\rangle$ for $J \in \Omega$ and S is transitive on Ω, the latter case does not hold. □

LEMMA 12.1.9. *(1) If* $\mathcal{C} = \mathcal{O}(t)$ *then* $c \in Z(\mathcal{F})$.
(2) If $\mathcal{F}_c^\circ = \mathcal{F}_{T_c}(M)$ *or* $\mathcal{F}_a^\circ = \mathcal{F}_{T_a}(M_a)$ *for some* $a \in \{b, bc\}$ *then* $c \in Z(\mathcal{F})$.
(3) If $\mathcal{F}_c^\circ \cong \hat{L}_3(2)/2^{3+6}$ *or* $\check{\Omega}_8^+(2)$ *then* $c \in Z(\mathcal{F})$.
(4) If $\mathcal{F}_b^\circ \cong L_3(2)/E_{64}$ *or* $Sp_6(2)$ *then* $c \in Z(\mathcal{F})$.
(5) If $m = 8$ *and* $c \notin Z(\mathcal{F})$ *then* $L_z = L_1 \times L_2$ *where* $L_1 \cong L_2$, $O = O_2(L_1)$, $O(t) = O_2(L_2)$, *and* L_1 *is of index* $k = 1$ *or 3 in* $SL_2(3) * SL_2(3)$, *with* $k = 3$ *iff* $\mathcal{C} \cong AE_5$.

PROOF. If $\mathcal{C} = \mathcal{O}(t)$ then $c = zt \in Z(\mathcal{F}_z)$, so by 12.1.7.4, $\mathcal{F} = \langle \mathcal{F}_z, \mathcal{F}_c\rangle = \mathcal{F}_c$, proving (1).

Assume $c \notin Z(\mathcal{F})$. Then by (1) and 12.1.6, $\mathcal{C} \cong AE_5$, $PSp_4[m]$ or $L_4^-[m]$. In particular if $m > 8$ then $\mathcal{O}(z) \cong \mathcal{O}(t) \cong SL_2[m] * SL_2[m]$, while if $m = 8$ then a model L_2 for $O^2(C_{\mathcal{C}}(t))$ is of index $k \in \{1, 3\}$ in $SL_2(8) * SL_2(8)$, with $k = 3$ iff $\mathcal{C} \cong AE_5$, and L_2 centralizes $C_S(O(t))$. Then $L_2 \trianglelefteq L_z$ and if $k = 1$ then $O^2(Aut(O(t)) = O^2(Aut_{\mathcal{F}}(O(t))$. As S is transitive on Ω we conclude, using 12.1.8, that $L_z = L_1 \times L_2$, where $L_1 = L_2^s$ for $s \in S - C_S(z)$. Therefore (5) holds. Notice L_1 centralizes b and bc, and L_z centralizes c.

Suppose $\mathcal{F}_c^\circ = \mathcal{F}_{T_c}(M)$ or $\mathcal{F}_a^\circ = \mathcal{F}_{T_a}(M_a)$ for $a \in \{b, bc\}$. Then as $\mathcal{O}(z)$ centralizes $\langle b, c, z\rangle$, it follows that $\mathcal{O}(z) = O$, so as $\mathcal{O}(z) \cong SL_2[m] * SL_2[m]$ when

$m > 8$, we conclude that $m = 8$. But then as L_1 centralizes $\langle b, c, z\rangle$, L_1 acts on $W \cap O$ and hence centralizes O, again a contradiction. This proves (2).

Assume $\mathcal{F}_c^\circ \cong \hat{L}_3(2)/2^{3+6}$ or $\check{\Omega}_8^+(2)$. Then $m = 8$ and in 12.1.8, $e = 3$ or 1 in the respective case. This is a contradiction as L_z is of index 1 or 3^2 in $(SL_2(3)*SL_2(3))^2$ by (5). Thus (3) holds.

Finally assume $\mathcal{F}_b^\circ \cong L_3(2)/E_{64}$ or $Sp_6(2)$ and let L_b be a model for $O^2(C_{\mathcal{F}_b^\circ}(z))$. Then $m = 8$ and by 9.5.14.3, 9.5.15.10, and 9.5.16, case (iii) of 9.5.11.3 holds, so that $O_2(L_b) = O \times B$ with $B \cong E_4$. This is a contradiction as $L_b \leq L_z$ and $L_z = L_1 \times L_2$ with L_i of index 1 or 3 in $SL_2(3) * SL_2(3)$ by (5). So (4) holds. □

In the remainder of the section we assume:

HYPOTHESIS 12.1.10. *Hypothesis* 12.1.1 *holds and* $c \notin Z(\mathcal{F})$.

LEMMA 12.1.11. *(1)* $\mathcal{C} \cong AE_5$, $PSp_4[m]$, *or* $L_4^-[m]$.
(2) $\mathcal{F}_c^\circ \cong \check{A}E_8$ *or* $\Omega_8^+[m]$.
(3) $\mathcal{F}_{bc}^\circ \cong \check{A}E_6$ *or* $\Omega_6^+[m]$.
(4) $\mathcal{F}_b^\circ \cong AE_7$, $\Omega_7[m]$, *or* $\Omega_8^-[m]$.

PROOF. This follows from the descriptions of the various possible systems in 12.1.4, 12.1.5, and 12.1.6, together with the exclusions in 12.1.9, unless 12.1.5.3.ii holds, and we may assume the latter. Let $s \in S$ interchange z and t, and set $d = b^s$, so that $d \in O$ and $(\mathcal{F}_b^\circ)c_s^* = \mathcal{F}_d^\circ$. Therefore as $c \in Z(S)$ and $\mathcal{F}_b^\circ \leq \mathcal{F}_c^\circ$ in 12.1.5.3.ii, it follows that $\mathcal{F}_d^\circ \leq \mathcal{F}_c^\circ$. Then as $d \in O$, we have $\mathcal{C} \leq \mathcal{F}_d^\circ \leq \mathcal{F}_c^\circ$, a contradiction. □

NOTATION 12.1.12. Set $S_0 = C_S(O)$, $R_0 = OS_0$, and if $m = 8$ define G_z, H_z, and L_z as in 12.1.8 and L_1 as in 12.1.9.5. If $m > 8$ set $\mathcal{O} = \mathcal{O}(z)$ and if $m = 8$ set $\mathcal{O} = \mathcal{F}_O(L_1)$. Set $\mathcal{O}_c = O^2(C_{\mathcal{F}_c^\circ}(O(t)\langle z\rangle))$, and for $a \in \{b, bc\}$ set $\mathcal{O}_a = O^2(C_{\mathcal{F}_a^\circ}(z))$.

LEMMA 12.1.13. *(1)* $\mathcal{O} = \mathcal{O}_a$ *for* $a \in \{b, bc, c\}$.
(2) $\mathcal{O}$ *centralizes* S_0.
(3) Either $\mathcal{O} \cong SL_2[m] * SL_2[m]$ *or* $m = 8$, $\mathcal{C} \cong AE_5$, *and* L_1 *is of index 3 in* $SL_2(3) * SL_2(3)$.
(4) If $m = 8$ *then* $\mathcal{C} \cong AE_5$ *iff* $\mathcal{F}_c^\circ \cong \check{A}E_8$ *iff* $\mathcal{F}_{bc}^\circ \cong \check{A}E_6$ *iff* $\mathcal{F}_b^\circ \cong AE_7$.
(5) $\mathcal{O} \trianglelefteq \mathcal{F}_z$.

PROOF. Suppose $m > 8$. Then by 12.1.11.1, $\mathcal{C} \cong PSp_4[m]$ or $L_4^-[m]$, so $\mathcal{O}(t) \cong SL_2[m] * SL_2[m]$, and hence as $\mathcal{O}(z) \cong \mathcal{O}(t)$, (3) holds. By definition in 12.1.12, $\mathcal{O} = \mathcal{O}_z$. Therefore (5) holds by 2.6.10 and (2) holds by 2.6.11. Part (1) follows from (3) and inspection of the local structure of the systems in 12.1.11 when $m > 8$. Finally (4) is vacuously true.

So we may assume $m = 8$. Then from 12.1.12, $\mathcal{O} = \mathcal{F}_O(L_1)$. But from 12.1.8 and 12.1.9.5, L_1 centralizes S_0, so (2) holds.

Let $a \in \{b, bc, c\}$ and set $W_c = W$. By definition in 12.1.12, $\mathcal{O}_a = O^2(C_{\mathcal{F}_a^\circ}(z))$ if $a \in \{b, bc\}$, so from 12.1.11, either $\mathcal{O}_a \cong SL_2[8]*SL_2[8]$ or $\mathcal{F}_a^\circ \cong \check{A}E_6$ or AE_7 and $\mathcal{O}_a$ is of index 3 in $SL_2[8] * SL_2[8]$. Similarly $\mathcal{O}_c = O^2(C_{\mathcal{F}_c^\circ}(\langle z\rangle O(t))$, so by 12.1.11 either $\mathcal{O}_c \cong SL_2[8] * SL_2[8]$ or $\mathcal{F}_c^\circ \cong \check{A}E_8$ and $\mathcal{O}_c$ is of index 3 in $SL_2[8] * SL_2[8]$. Also $\mathcal{F}_a^\circ \trianglelefteq \mathcal{F}_a$, so $[W_a, C_{S_0}(a)] \leq C_{W_a}(O) \leq C_{\mathcal{F}_a^\circ}(\mathcal{O}_a)$ from 12.1.11, and hence $\mathcal{O}_a$

centralizes $C_{S_0}(a)$. Then as $|S_0 : C_{S_0}(a)| \leq 2$, it follows that $\mathcal{O}_a$ centralizes S_0, so a model L_a for $\mathcal{O}_a$ is contained in $C_{L_z}(S_0) = L_1$. Conversely L_1 centralizes a, so $L_1 = [L_1, W_a] \leq L_a$. Therefore $L_1 = L_a$, so (1) holds, and (3) and (4) follow from (1) and the structure of $\mathcal{O}_a$ just described.

It remains to prove (5). By (2), $\mathcal{O}$ centralizes S_0. Further as W is nontrivial on $J/\langle z(J)\rangle$ for $J \in \Omega$, $O^2(Aut_{\mathcal{F}}(O(\tau))) = Aut_{L_z}(O(\tau))$, so $O^2(Aut_{\mathcal{F}}(O)) = Aut_{L_1}(O) = Aut_{\mathcal{O}}(O)$. Then (5) follows from 2.6.14. □

LEMMA 12.1.14. *If $\mathcal{C} \cong AE_5$ then $\mathcal{F}^\circ \cong AE_9$.*

PROOF. Assume $\mathcal{C} \cong AE_5$. Then $m = 8$ and by 12.1.13.4, $\mathcal{F}_c^\circ \cong \breve{A}E_8$. Set $U_z = O_2(\mathcal{C})$, $U_c = O_2(\mathcal{F}_c^\circ)$, and $U = U_z U_c$. Now $B = O(t) \cap U_z$ is a hyperplane of U_z and the unique $C_S(z)L_2$-invariant E_8-subgroup of $O(t)$. Therefore $B = O(t) \cap U_z \leq U_c$, so $U_c \leq C_S(B)$. However as $\mathcal{C} \cong AE_5$, there is no transvection in $Aut_{\mathcal{F}_z}(U_z)$ centralizing B, so U_c centralizes U_z and hence U is abelian.

Let $\mathcal{Y} = \mathcal{F}_c^\circ$. Now $U = C_{U\mathcal{C}}(U_z) \trianglelefteq U\mathcal{C}$, and similarly $U \trianglelefteq U\mathcal{Y}$. By 2.5.2, $\tau_U = (N_{\mathcal{F}}(U), \Omega)$ is a quaternion fusion packet. As $\mathcal{Y} \leq N_{\mathcal{F}}(U)$, $\mu(\tau_U) \cong Weyl(D_4)$. As $B \leq U_z \cap U_c$ is a hyperplane of U_z, either $U_z \leq U_c$ and $U = U_c$ is of rank 7, or $U_z \not\leq U$ and $m(U) = m(U_c) + 1 = 8$. In either case as a model for $\mathcal{Y}$ is irreducible on $U_c/\langle c\rangle$ and indecomposable on U_c, and as $\mathcal{C}$ does not centralize c, a model for $\mathcal{X} = N_{\mathcal{F}}(U)^\circ$ is irreducible on U and $U = \langle z^{\mathcal{X}}\rangle$. Also $z \in U \leq O_2(\mathcal{X})$ and $K \cap O_2(\mathcal{X}) \leq K \cap O_2(\mathcal{Y}) = \langle z\rangle$, so we conclude from Theorem 2 that $\mathcal{X} \cong AE_9$. Thus to complete the proof of the lemma, it suffices to show that $\mathcal{X} \trianglelefteq \mathcal{F}$, since then $\mathcal{F}^\circ = \mathcal{X}$.

First $\mathcal{L} = \mathcal{OC} \trianglelefteq \mathcal{F}_z$ using 12.1.13.5, so by 1.3.2 in [**Asc19**],

$$\mathcal{F}_z = \langle C_S(z)\mathcal{L}, N_{\mathcal{F}_z}(L)\rangle$$

for L Sylow in $\mathcal{L}$. As $\mathcal{O}, \mathcal{C} \leq \mathcal{X}$, $U \trianglelefteq C_S(z)\mathcal{L}$. As $N_{\mathcal{F}_z}(L) \leq N_{\mathcal{F}_z}(O(\tau)) \leq C_{\mathcal{F}_z}(c)$ and $U_c \trianglelefteq \mathcal{F}_c$, and as $U_z \trianglelefteq \mathcal{F}_z$, it follows that $U = U_z U_c \trianglelefteq N_{\mathcal{F}_z}(L)$. Thus $U \trianglelefteq \mathcal{F}_c$.

Next by 1.3.2 in [**Asc19**], $\mathcal{F}_c = \langle S\mathcal{Y}, N_{\mathcal{F}}(T_c)\rangle$. Also $\mathcal{Y} \leq \mathcal{X}$ and $U = U_c U_z U_z^s$ for $s \in S - C_S(z)$, so $U \trianglelefteq S\mathcal{Y}$. Similarly $N_{\mathcal{F}}(T_c) \leq N_{\mathcal{F}}(O(\tau)) \leq N_{\mathcal{F}}(\langle z, c\rangle)$ and $N_{\mathcal{F}}(\langle z, c\rangle)$ acts on $U = U_c U_z U_z^s$. Therefore $U \trianglelefteq \mathcal{F}_c$. Hence by 12.1.7, $U \trianglelefteq \langle \mathcal{F}_z, \mathcal{F}_c\rangle = \mathcal{F}$, completing the proof. □

Because of 12.1.14, during the remainder of the proof we assume:

HYPOTHESIS 12.1.15. *Hypothesis* 12.1.1 *holds, $c \notin Z(\mathcal{F})$, and $\mathcal{O} \cong SL_2[m] * SL_2[m]$.*

NOTATION 12.1.16. Let T_0 be Sylow in $\mathcal{C}$ and set $T = T_0 T_c$. Set $\mathcal{E}_1 = C_T(z)\mathcal{OC}$, $\mathcal{F}_1 = \mathcal{F}_z$, $\mathcal{E}_2 = T\mathcal{F}_c^\circ$, $\mathcal{F}_2 = \mathcal{F}_c$, and $\mathcal{Y} = \langle \mathcal{E}_1, \mathcal{E}_2\rangle$.

LEMMA 12.1.17. *(1) $\mathcal{E}_i \trianglelefteq \mathcal{F}_i$ for $i = 1, 2$.*

(2) $\mathcal{Y}$ is $\mathcal{F}$-invariant, so in particular T is strongly closed in S with respect to $\mathcal{F}$.

PROOF. We first prove $\mathcal{E}_1 \trianglelefteq \mathcal{F}_1$. Set $T_1 = C_T(z)$, let $\mathcal{L} = \mathcal{OC}$, and let L be Sylow in $\mathcal{L}$. Thus $L = OT_0$ and, using 12.1.13.5, $\mathcal{L} \trianglelefteq \mathcal{F}_1$. Next $N_{\mathcal{F}_1}(L) \leq N_{\mathcal{F}_1}(O(\tau)) \leq C_{\mathcal{F}_1}(c)$, and if $m > 8$ then $N_{\mathcal{F}_c^\circ}(O(\tau)) = T_c$, so $C_{T_c}(z) \trianglelefteq N_{\mathcal{F}_1}(L)$. Therefore $\mathcal{E}_1 \trianglelefteq \mathcal{F}_1$ by Theorem 1.5.2 in [**Asc19**] when $m > 8$.

So take $m = 8$ and let $\mathcal{X} = \mathcal{F}_{O(\tau)}(L_z)$. Then $N_{\mathcal{F}_c^\circ}(O(\tau)) = T_c\mathcal{X}$, so $C_{T_c}(z)\mathcal{X} \trianglelefteq N_{\mathcal{F}_1}(L)$. Then as $N_{\mathcal{L}}(L) = T_0\mathcal{X}$, again $\mathcal{E}_1 \trianglelefteq \mathcal{F}_1$ by Theorem 1.5.2 in [**Asc19**]. So in any event, $\mathcal{E}_1 \trianglelefteq \mathcal{F}_1$.

We next show that $\mathcal{E}_2 \trianglelefteq \mathcal{F}_2$. First $N_{\mathcal{F}_c}(T_c) \leq N_{\mathcal{F}_c}(O(\tau)) \leq N_{\mathcal{F}}(\langle z, c\rangle)$, and from our proof that $\mathcal{E}_1 \trianglelefteq \mathcal{F}_1$, either $C_T(z) \trianglelefteq N_{\mathcal{F}_z}(O(\tau))$ or $m = 8$ and $C_T(z)\mathcal{X} \trianglelefteq N_{\mathcal{F}_z}(O(\tau))$. Then $T \trianglelefteq N_{\mathcal{F}_c}(T_c)$, so $\mathcal{E}_2 \trianglelefteq \mathcal{F}_2$ by Theorem 1.5.2 in [**Asc19**]. This completes the proof of (1).

By 12.1.7, $\mathcal{F} = \langle \mathcal{F}_1, \mathcal{F}_2\rangle$. Hence the tuple $\mathcal{E}_i, \mathcal{F}_i$, $i = 1, 2$, satisfies Hypothesis 1.4.1 in [**Asc19**]. Therefore to establish (2), it suffices by 1.4.4 in [**Asc19**] to verify Hypothesis 1.4.3 in [**Asc19**]. Condition (1) of 1.4.3 is satisfied by (1).

As $N_{\mathcal{F}}(T) \leq N_{\mathcal{F}}(O(\tau)) \leq \mathcal{F}_c$ and T is Sylow in $\mathcal{E}_2 \trianglelefteq \mathcal{F}_2$, $\Sigma = Aut_{\mathcal{F}}(T) \leq Aut(\mathcal{E}_2)$. Similarly Σ permutes $\{z, t\}$ and hence as $\mathcal{E}_1 \trianglelefteq \mathcal{F}_1$, Σ permutes $\mathcal{E}_1, \mathcal{E}_1'$, where $\mathcal{E}_1' = \mathcal{E}_1 c_x^*$ for $x \in S - C_S(z)$. Thus condition (2) of 1.4.3 of [**Asc19**] is satisfied by 1.4.6 in [**Asc19**]. Therefore it remains to verify the extension condition 1.4.3.4 of [**Asc19**]. As T is Sylow in $\mathcal{E}_2$, the extension condition is trivially satisfied by $\mathcal{E}_2$, so it remains to choose $\alpha \in Aut_{\mathcal{F}_1}(T_1)$ and show that α extends to T. But α acts on $O(\tau)$ and hence centralizes c, so as $\mathcal{E}_2 \trianglelefteq \mathcal{F}_c$ and T is Sylow in $\mathcal{E}_2$, $\alpha = \varphi\phi$ for some $\varphi \in Aut_{\mathcal{F}}(T)$ and $\phi \in \hom_{\mathcal{E}_2}(T_1\varphi, T)$. But $T_1 = C_T(\langle z, c\rangle)$ is $Aut_{\mathcal{F}}(T)$-invariant, so $T_1\varphi = T_1$. Then as $O(\tau)$ is weakly closed in S, $\langle z, c\rangle\phi = \langle z, c\rangle$, so $T_1\phi = C_T(\langle z, c\rangle)\phi = T_1$. Finally $T = N_{\mathcal{E}_2}(T_1)$, so ϕ extends to $\psi \in Aut_{\mathcal{E}_2}(T)$, and then $\varphi\psi$ extends α to T, completing the proof. □

LEMMA 12.1.18. *Let $\mathcal{X} = O^2(C_{\mathcal{C}}(b))$ and X Sylow in $\mathcal{X}$.*

(1) $\mathcal{X} \cong L_2[m]$, $L_2[2m]$ for $\mathcal{C} \cong PSp_4[m]$, $L_4^-[m]$, respectively. Further $bt \in Z(X)$.

(2) Set $\bar{b} = bt$. Then $bz = \bar{b}c \in bc^{O(t)}$, $\mathcal{F}_{\bar{b}c}^\circ \cong \Omega_6^+[m]$, and $\mathcal{X} \leq C_{\mathcal{F}_{bz}}(\mathcal{F}_{bz}^\circ)$.

(3) $\mathcal{C} = C_{\mathcal{F}_v}(z)^\circ$ and $\mathcal{F}_v^\circ \cong \Omega_7[m]$, $\Omega_8^-[m]$ for $\mathcal{C} \cong PSp_4[m]$, $L_4^-[m]$, respectively.

(4) $TM \cong \omega(C_4, m)$, $2\omega(C_4, m)$ for $\mathcal{C} \cong PSp_4[m]$, $L_4^-[m]$, respectively. In particular $M_0 = TM$ and T are determined up to isomorphism.

PROOF. By 12.1.11 and 12.1.15, $\mathcal{C}$ is $PSp_4[m]$ or $L_4^-[m]$, so (1) follows from the structure of $\mathcal{C}$.

We next prove (2). Visibly $bz = \bar{b}c \in bc^{O(t)}$. By (1), $\mathcal{X}$ is a (solvable) component of $C_{\mathcal{F}_z}(\langle b\rangle O)$, while $C_{\mathcal{F}_z}(\langle b\rangle O) = C_{\mathcal{F}_{bz}}(O)$ by 2.2 in [**Asc10**]. Now as $bz \in (bz)^{\mathcal{F}}$, we have $\mathcal{F}_{bz}^\circ \cong \Omega_6^+[m]$ by 12.1.11.3, 12.1.13.4, and 12.1.15. Thus from the structure of $\Omega_6^+[m]$, $O^2(C_{\mathcal{F}_{bz}^\circ}(O)) = 1$, so $\mathcal{X} \not\leq \mathcal{F}_{bz}^\circ$. Then by 2.2.6.3, X centralizes $\mathcal{F}_{bz}^\circ$. Indeed unless $\mathcal{X} \cong L_2[8]$, $\mathcal{X}$ is a component of $C_{\mathcal{F}_{bz}}(z)$, so (2) follows from 10.11.3 in [**Asc11**].

So assume $\mathcal{C} \cong PSp_4[8]$. By 2.2.1.4, $\mathcal{X} \trianglelefteq N_{C_{\mathcal{F}_{bz}}(z)}(X)$, so as X centralizes T_{bz} and $T_{bz}\mathcal{O} = N_{\mathcal{F}_{bz}^\circ}(O) \trianglelefteq N_{\mathcal{F}_{bz}}(O)$, $\mathcal{X}$ centralizes T_{bz}. Therefore as X centralizes $\mathcal{F}_{bz}^\circ$, (2) follows from 2.2 in [**Asc11**].

Similarly $\mathcal{C} = C_{\mathcal{F}_z}(v)^\circ$ and $C_{\mathcal{F}_z}(v) = C_{\mathcal{F}_v}(z)$ by 2.2 in [**Asc10**], so $\mathcal{C} = C_{\mathcal{F}_v}(z)^\circ$. As $v \in b^M$, it follows from 12.1.11.4, 12.1.13.4, and 12.1.15 that $\mathcal{F}_v^\circ \cong \Omega_7[m]$ or $\Omega_8^-[m]$. By 12.1.5.1, $vc \in M_v$ so as $O(t) \leq M_v$ and $c = zt$, also $vz \in T_v$. Then $\mathcal{C} = C_{\mathcal{F}_v}(z)^\circ \cong PSp_4[m]$ or $L_4^-[m]$. We conclude that (3) holds.

Let $r \neq bt$ be an involution in X, $J_1 = K$, $J_2 \in K^M$ with $z(J_2) \notin \{z, t\}$, and $J_3 = K_3$. From the structure of $\mathcal{C}$, $K_3^r = K_4$. Therefore from 5.9.2.5, the subgroup

$M_0 = \langle J_1, J_2, J_3, r\rangle$ of G is $M\langle r\rangle$ and isomorphic to $\omega(C_4, m)$. If $\mathcal{C} \cong PSp_4[m]$ then $T = T_c\langle r\rangle$, so $M_0 = TM$ and (4) holds in this case.

So assume $\mathcal{C} \cong L_4^-[m]$, and choose $y \in X$ with $\langle y\rangle$ the cyclic subgroup of X of index 2; thus $X = \langle y, r\rangle$ and $T = T_cT_0 = T_c\rangle y, r\rangle$. Working in $\mathcal{F}_v^\circ \cong \Omega_8^-[m]$ and using 9.5.24.7 and 9.5.29, we may choose y so that $y^2 = [r, \bar{v}_3]$, $k_3^y = k_3\bar{v}_3$ for $k_3 \in K_3 - W$ and $K_3 \cap W = \langle \bar{v}_3\rangle$. Now by 5.9.3, $M_0 = \langle J_1, J_2, J_3, y, r\rangle = M\langle y, r\rangle \cong 2\omega(C_4, m)$ with T Sylow in M_0, so that $TM = M_0$. This completes the proof of (4) and the lemma. □

NOTATION 12.1.19. If $\mathcal{C} \cong PSp_4[m]$ set $\tilde{\mathcal{F}} = \Omega_9[m]$, while if $\mathcal{C} \cong L_4^-[m]$ set $\tilde{\mathcal{F}} = \Omega_{10}^-[m]$. Let $\tilde{S}$ be Sylow in $\tilde{\mathcal{F}}$ and let $\tilde{\tau} = (\tilde{\mathcal{F}}, \tilde{\Omega})$ be the Lie packet of $\tilde{\mathcal{F}}$. Let $\tilde{\eta} \in \eta(\tilde{\tau})$, $\tilde{W} = \langle \tilde{\eta}\rangle$, and $\tilde{G}$ a model of $N_{\tilde{\mathcal{F}}}(\tilde{W})$.

LEMMA 12.1.20. *$\tilde{G} \cong TM$, so we may take $\tilde{G} = TM$, $\tilde{S} = T$, $\tilde{\Omega} = \Omega$, and $N_{\tilde{\mathcal{F}}}(\tilde{W}) = \mathcal{F}_T(TM)$.*

PROOF. By 12.1.18.4, there is an isomorphism $\alpha : TM \to \omega$, where ω is $\omega(C_4, m)$ or $2\omega(C_4, m)$, such that $\Omega\alpha = \Gamma$, where $\rho = (\mathcal{W}, \Gamma)$ is the quaternion fusion packet described in 4.3.7.5 and section 5.9 with ω the model for $\mathcal{W}$. Similarly by 5.3.9.2, and 5.3.10.2, there is an isomorphism $\beta : \tilde{G} \to TM$ with $\tilde{\Omega} = \Gamma$. Thus $\gamma = \beta\alpha^{-1} : \tilde{G} \to TM$ is an isomorphism with $\tilde{\Omega}\gamma = \Omega$, so identifying $\tilde{G}$ with TM via γ, the lemma holds. □

REMARK 12.1.21. We will show that $\tilde{\mathcal{F}} = \mathcal{Y}$. Hence $\mathcal{Y} = O^{2'}(\mathcal{Y})$ is saturated, so as the proof of 12.1.17 showed that the tuple $\mathcal{F}_i, \mathcal{E}_i$, $i = 1, 2$ satisfies Hypothesis 1.4.3 in [**Asc19**], the tuple also satisfies the hypothesis of 1.4.5 in [**Asc19**]. Then by that lemma, $\mathcal{Y} \trianglelefteq \mathcal{F}$, so $\mathcal{F}^\circ = \mathcal{Y} = \tilde{\mathcal{F}}$.

LEMMA 12.1.22. *Let $\mathcal{P} = \mathcal{O} \cdot \mathcal{O}c_s^*$ for $s \in S - C_S(z)$.*
(1) $T\tilde{\mathcal{P}} = T\mathcal{P}$.
(2) $\mathcal{F}_c^\circ \cong \Omega_8^+[m]$ and $\tilde{\mathcal{F}}_c = \mathcal{E}_2$.

PROOF. By 12.1.11.2, 12.1.13.4, and 12.1.15, $\mathcal{F}_c^\circ \cong \Omega_8^+[m]$. Then $N_{\mathcal{E}_1}(\langle z, c\rangle) = C_T(z)\mathcal{P}$ as $\mathcal{O}c_g^* = O^2(C_{\mathcal{C}}(t))$, while by 10.1.15.2, $\mathcal{E}_2 = \langle N_{\mathcal{E}_1}(\langle z, c\rangle), \mathcal{F}_T(TM)\rangle$ and $\tilde{\mathcal{F}}_c = \langle N_{\tilde{\mathcal{F}}_c}(\langle z, c\rangle), N_{\tilde{\mathcal{F}}}(\tilde{W})\rangle$. Therefore given 12.1.20, (1) implies (2). Thus it remains to prove (1), which follows from Theorem 2.7.3. □

LEMMA 12.1.23. *(1) $\mathcal{F}_{bz}^\circ \le \mathcal{F}_c^\circ$, so $\tilde{\mathcal{F}}_{bz}^\circ = \mathcal{F}_{bz}^\circ$.*
(2) $\tilde{\mathcal{F}}_{bz} = \mathcal{F}_{bz}^\circ\mathcal{X}$.
(3) $\tilde{\mathcal{F}}_z = \mathcal{E}_1$.

PROOF. By 12.1.18.2, $\mathcal{B} = \mathcal{F}_{bz}^\circ \cong \Omega_6^+[m]$ and $\mathcal{X}$ centralizes $\mathcal{B}$. Then by 9.3.20, $\mathcal{F}_{bz}^\circ = \langle T_b\mathcal{O}, \mathcal{F}_{T_{bz}}(M_{bz})\rangle$, so as $T_b\mathcal{O}$ and $\mathcal{F}_{T_{bz}}(M_b)$ are contained in $\mathcal{F}_c^\circ$, (1) follows using 12.1.22.2.

Next $T_0 = (T_c \cap T_0)X$, so $C_T(bz) = C_{T_c}(bz)X = T_{bz}X$. Then $C_T(bz)$ is Sylow in $\mathcal{B}\mathcal{X}$. As $\mathcal{X}$ is the fusion system on X generated by $O^2(Aut(U))$ as U varies over the 4-subgroups of X, $\tilde{\mathcal{X}} = \mathcal{X}$. Then by (1),

$$\tilde{\mathcal{F}}_{bz} = \langle C_T(bz)\tilde{\mathcal{F}}_{bz}^\circ, C_T(bz)\tilde{\mathcal{X}}\rangle = \langle C_T(bz)\mathcal{B}, C_T(bz)\mathcal{X}\rangle = \mathcal{B}\mathcal{X},$$

proving (2).

It remains to prove (3). First $\mathcal{E}_1 = C_T(z)\mathcal{O}\mathcal{C} = \langle C_T(z)\mathcal{O}, C_T(z)\mathcal{C}\rangle$ and $C_T(z)\mathcal{O} \le T\mathcal{P}$, so from 12.1.22.1, $C_T(z)\tilde{\mathcal{O}} = C_T(z)\mathcal{O}$. Therefore it suffices to show that $\mathcal{W} =$

$C_T(z)\mathcal{C} = C_T(z)\tilde{\mathcal{C}} = \tilde{\mathcal{W}}$. Now $\mathcal{C} \cong PSp_4[m]$ or $L_4^-[m]$. Therefore by 9.4.22, 9.4.27, and 9.4.28, $\mathcal{W} = \langle C_T(z)\mathcal{O}', Aut_{\mathcal{W}}(R_i) : i \in I\rangle$ where $\mathcal{O}' = \mathcal{O}c_x^*$ for $x \in S - C_S(z)$ and the R_i are suitable members of $\mathcal{W}^e$. By 12.1.24.1, $C_T(z)\tilde{\mathcal{O}}' = C_T(z)\mathcal{O}'$. If $\mathcal{C} \cong PSp_4[m]$ then we are in case (3) or (4a) of 9.4.22, so the R_i are in $\mathcal{R}_1$. If $\mathcal{C} \cong L_4^-[m]$ we are in case (1) of 9.4.22, so again the R_i are in $\mathcal{R}_1$. Therefore $R_i = C_T(\langle z, t, b, s\rangle)$ for suitable $s \in \mathcal{D}^m(t) \cap t^{\mathcal{C}}$. But then by 9.4.21, $Aut_{\mathcal{W}}(R_i)$ centralizes b or bt, and hence also bz or bzt. Therefore by (2), $Aut_{\mathcal{W}}(R_i) = Aut_{\tilde{\mathcal{W}}}(R_i)$, so (3) holds. □

LEMMA 12.1.24. *(1) If $\mathcal{C} \cong PSp_4[m]$ then $\mathcal{F}^\circ \cong \Omega_9[m]$.*
(2) If $\mathcal{C} \cong L_4^-[m]$ then $\mathcal{F}^\circ \cong \Omega_{10}^-[m]$.

PROOF. By Remark 12.1.21 it suffices to show that $\mathcal{Y} = \tilde{\mathcal{F}}$. But this is a consequence of 12.1.7.5, 12.1.22.2, and 12.1.23.3. □

THEOREM 12.1.25. *Assume Hypothesis* 12.1.1. *Then one of the following holds:*
(1) $c \in Z(\mathcal{F})$ and either $\mathcal{F}^\circ = \mathcal{F}_T(M)$ or $\mathcal{F}^\circ \cong \check{L}_3(2)/2^{3+6}$, $\check{A}E_8$, $\check{\Omega}_8^+(2)$, or $\Omega_8^+[m]$.
(2) $\mathcal{F}^\circ \cong AE_9$.
(3) $\mathcal{F}^\circ \cong \Omega_9[m]$.
(4) $\mathcal{F}^\circ \cong \Omega_{10}^-[m]$.

PROOF. If $c \in Z(\mathcal{F})$ then (1) holds by 12.1.4.2. Thus we may assume Hypothesis 12.1.10 holds. Then by 12.1.13.3, either $\mathcal{O} \cong SL_2[m] * SL_2[m]$ or $\mathcal{C} \cong AE_5$, and in the latter case (2) holds by 12.1.14, so we may assume the former. Therefore Hypothesis 12.1.15 is satisfied. Then $\mathcal{C}$ is isomorphic to $PSp_4[m]$ or $L_4^-[m]$ by 12.1.11.1, so (3) or (4) holds by 12.1.24. □

12.2. More $|\Omega(z)| = 2$

In this section we assume Hypothesis 11.1.1 with $n \geq 5$. Adopt Notation 11.1.2. In addition define $\mathcal{C}$ as in 11.1.5.1.

NOTATION 12.2.1. Set $O = O(z)$, $S_0 = C_S(O)$, and if $m = 8$ let G_z be a model for $N_{\mathcal{F}_z}(OS_0)$ and $L_z = O^2(\langle W^{G_z}\rangle)$. If $m > 8$ set $\mathcal{O} = \mathcal{O}(z)$ and if $m = 8$ set $\mathcal{O} = \mathcal{F}_O(OL_z)$. Similarly if $m = 8$ let G_τ be a model for $N_{\mathcal{F}}(O(\tau)C_S(O(\tau))$, identify $N_{G_z}(O(\tau))$ with a subgroup of $N_{G_\tau}(OS_0)$ using 2.2 in [**Asc10**], and set $L_\tau = \prod_{t\in\mathcal{Z}_S}(L_t)$, where $L_t = L_z^g$ for $g \in G_\tau$ with $z^g = t$. If $m > 8$ set $\mathcal{O}_\tau = \mathcal{O}(\tau)$, while if $m = 8$ set $\mathcal{O}_\tau = \mathcal{F}_{O(\tau)}(O(\tau)L_\tau)$. Set $c = e_I$.

LEMMA 12.2.2. *(1) If $m > 8$ then either $\mathcal{O}(z) = O$ or $\mathcal{O}(z) \cong SL_2[m]*SL_2[m]$.*
(2) If $m = 8$ then L_z centralizes S_0 and either $L_z = 1$ or O is Sylow in L_z and L_z is of index 1 or 3 in $SL_2(3) * SL_2(3)$.
(3) If $m = 8$ then $O^2(Aut_{\mathcal{F}}(O)) = Aut_{L_z}(O)$ and $\mathcal{O} \trianglelefteq \mathcal{F}_z$.
(4) If $m = 8$ then for each $t \in \mathcal{Z}_S$, $L_t \cong L_z$, L_τ centralizes $C_S(O(\tau))$, and $Aut_{L_\tau}(O(\tau)) = O^2(\Gamma)$, where

$$\Gamma = \bigcap_{t\in\mathcal{Z}_S} N_{Aut_{\mathcal{F}}(O(\tau))}(O(t)).$$

PROOF. Parts (1) and (2) follow from 9.1.11. Suppose $m = 8$. For $V_j \in \Delta$ with $z_j \in \mathcal{D}^*(z) - \mathcal{D}(z)$ and w a generator of V_j, w induces an outer automorphism on K and K', so the automorphism c_w induced on O by w inverts a Sylow 3-subgroup of $Aut_{\mathcal{F}}(O)$. Therefore the first statement in (3) holds, and the second follows from the first and 2.6.14.2. Then as $\mathcal{O} \trianglelefteq \mathcal{F}_z$, the groups L_t, $t \in \mathcal{Z}_S$ are well defined and isomorphic to L_z, and the remaining statements in (4) follow from (2) and (3). □

LEMMA 12.2.3. *One of the following holds:*

(1) $\mathcal{F} = \mathcal{E}(\tau)$.

(2) $n = 2^k + 1$ *and* $\mathcal{F} = \langle \mathcal{E}(\tau), \mathcal{F}_x \rangle$, *where* x *is the element* $e_{I-\{n\}}$ *of* $W_I \cap \mathcal{F}^f$ *of weight* 2^k.

(3) $n = 2^k$, $c \neq 1$, *and* $\mathcal{F} = \langle \mathcal{E}(\tau), \mathcal{F}_c \rangle$.

(4) $n = 2^k$, $c = 1$, *and* $\mathcal{F} = \langle \mathcal{E}(\tau), \mathcal{F}_x \rangle$, *where* $x \in W_I \cap \mathcal{F}^f$ *is of weight* 2^{k-1}.

PROOF. Assume otherwise. By Theorem 11.3.11, $\mathcal{F} = \mathcal{B}(\tau)$. Therefore by 11.3.2, $\mathcal{F}$ is generated by $\mathcal{E} = \mathcal{E}(\tau)$ and the set of subsystems $\mathcal{F}_x$, as x varies over the set Γ of involutions $x \in W_I \cap F^f$ of weight a positive power of 2. Thus there is $x \in \Gamma$ with $\mathcal{F}_x \not\leq \mathcal{E}$. In particular $\text{wt}(x) = 2^k$ for some $k \geq 1$, while by 11.1.8, x satisfies one of conditions (2), (3), or (4) of that lemma. But now we conclude that the corresponding conclusion of our lemma holds. □

LEMMA 12.2.4. *Let* $\theta \subseteq \mathcal{Z}_S$ *and set* $f_\theta = \prod_{t \in \theta} t$.

(1) $N_{\mathcal{F}}(O(\theta)) \leq C_{\mathcal{F}}(f_\theta)$.

(2) $O(\theta) \in \mathcal{F}^f$ *iff* $f_\theta \in \mathcal{F}^f$.

(3) If $O(\theta) \in \mathcal{F}^f$ *then* $N_{\mathcal{F}}(O(\theta)) \leq \langle N_{\mathcal{F}}(O(\tau)), N_S(O(\theta)) C_{\mathcal{F}}(O(\theta))^\circ \rangle$.

(4) If $f_\theta \in \mathcal{F}^f$ *then* $C_{\mathcal{F}}(f_\theta) \leq \langle C_S(f_\theta) C_{\mathcal{F}}(f_\theta)^\circ, N_{\mathcal{F}}(O(\tau)) \rangle$.

(5) If $m > 8$ *then* $N_{\mathcal{F}}(O(\tau)) = \langle S\mathcal{X}, N_{\mathcal{F}}(XW) \rangle$, *where* $\mathcal{X} = C_{\mathcal{F}}(O(\tau))$ *and* $X = O(\tau) C_S(O(\tau))$.

(6) Suppose $m = 8$. *Then* $N_{\mathcal{F}}(O(\tau)) = \langle S\mathcal{X}, N_{\mathcal{F}}(XW), S\mathcal{O}_\tau \rangle$.

PROOF. Set $P = O(\theta)$ and $f = f_\theta$. As $Aut_{\mathcal{F}}(P)$ permutes θ, (1) holds. Further by 11.1.2, W_I is the core of the permutation module for $\tilde{\mu} = Sym(I)$, or the image of that module modulo $\langle e_I \rangle$, so either $C_S(f) = N_S(P)$, or $n \equiv 0 \mod 4$, $c = 1$ and $|\theta| = n/4$ with $|S : N_S(P)| = 2$. In either case (2) holds by 11.1.3.7.

Assume $P \in \mathcal{F}^f$; then $f \in \mathcal{F}^f$ by (2). Adopt the notation of 11.1.4 with f in the role of x. Then by 11.1.4.3, $\mathcal{F}_f^\circ = \mathcal{F}_1 * \mathcal{F}_2$ with $\mathcal{F}_2 = C_{\mathcal{F}}(P)^\circ$. By 1.3.2 in [**Asc19**], $N_{\mathcal{F}}(P) = \langle N_S(P)\mathcal{F}_2, N_{\mathcal{F}}(T) \rangle$ where T is Sylow in $P\mathcal{F}_2$. Then $N_{\mathcal{F}}(T) \leq N_{\mathcal{F}}(O(\tau))$ as $O(\tau) \leq T$. This proves (3).

Similarly by 1.3.2 in [**Asc19**], $\mathcal{F}_f = \langle C_S(f)\mathcal{F}_f^\circ, N_{\mathcal{F}_f}(R) \rangle$ where R is Sylow in $\mathcal{F}_f^\circ$. Then as $O(\tau) \leq R$, $N_{\mathcal{F}_f}(R) \leq N_{\mathcal{F}}(O(\tau))$, so (4) holds.

Finally let $Q = O(\tau)$, $\mathcal{X} = C_{\mathcal{F}}(Q)$, and $X = QC_S(Q)$. Recall from 12.2.1 that $G = G_\tau$ is a model for $N_{\mathcal{F}}(X)$. Observe that $\mathcal{F}_S(SC_G(Q)) \leq S\mathcal{X}$. By 1.3.2 in [**Asc19**], $N_{\mathcal{F}}(Q) = \langle S\mathcal{X}, N_{\mathcal{F}}(X) \rangle$, so to prove (5) and (6) it suffice to show $W \trianglelefteq G$ in (5) and $N_{\mathcal{F}}(X) = \langle S\mathcal{O}_\tau, N_{\mathcal{F}}(XW) \rangle$ in (6).

Let Y be the kernel of the action of G on $\Sigma = \{O(t) : t \in \mathcal{Z}_S\}$. Then $W \leq Y$, so $G = YN_G(W)$ by a Frattini argument.

Suppose $m > 8$. Then $Aut(O)$ is a 2-group, so $O^2(Y) \leq H = C_G(Q)$. Set $S_Q = C_S(Q)$. Then $[W, S_Q] \leq W \cap S_Q \leq C_{S_Q}(W)$, so W is quadratic on S_Q. Therefore $\Phi(W)$ centralizes $S_Q/\Phi(S_Q)$, so $[\Phi(W), H] \leq S_Q$ and hence $\Phi(W)S_Q \trianglelefteq WH$. Let $\eta_0 = \{\Phi(V) : V \in \Delta - \Omega\}$. Then as $m > 8$, H permutes η_0, so H acts on $W_1 = \langle \eta_0 \rangle$.

Then as W centralizes W_1, we have $[H, W] \leq C_H(W_1)$. But $C_H(W_1)$ acts on each member of Δ, so $O^2(C_H(W_1)) \leq C_H(W)$. Therefore $O^2(Y) \leq H \leq N_G(W)$, so $G = YN_G(W) = N_G(W)$, completing the proof of (5).

So assume $m = 8$. Then by 9.1.11.2, $C_G(Q) \leq N_G(W)$. Therefore $G = L_\tau N_G(W)$, so $N_{\mathcal{F}}(X) = \langle SO_\tau, N_{\mathcal{F}}(XW)\rangle$, establishing (6). □

LEMMA 12.2.5. *The following are equivalent:*
(1) $c \in Z(\mathcal{F}_z)$.
(2) $cz \in Z(\mathcal{C})$.
(3) $\mathcal{C}$ *is isomorphic to* $\omega(D_{n-2}, m)$, $\check{A}E_{2n-4}$, *or* $\Omega^+_{2n-4}[m]$, *or* $n = 6$ *and* $\mathcal{C} \cong \hat{L}_3(2)/2^{3+6}$ *or* $\hat{\Omega}^+_8(2)$.

PROOF. First $c \in Z(\mathcal{F}_z)$ iff $cz \in Z(\mathcal{F}_z)$. Further from 11.1.5.2, $cz \in \mathcal{C}$, so if $c \in Z(\mathcal{F}_z)$ then $cz \in Z(\mathcal{C})$. But from 11.1.5.3, $|Z(\mathcal{C})| \leq 2$, so $cz \in Z(\mathcal{C})$ iff $Z(\mathcal{C}) = \langle cz\rangle$ iff $cz \in Z(\mathcal{F}_z)$. Thus (1) and (2) are equivalent. Also from 11.1.5.3, $Z(\mathcal{C}) \neq 1$ iff $\mathcal{C}$ is listed in (3), in which case $Z(\mathcal{C}) = Z(M_2) = \langle cz\rangle$ by 11.1.5.2 and the structure of the systems in (3). Therefore (2) and (3) are equivalent. □

LEMMA 12.2.6. *If* $c \in Z(\mathcal{F}_z)$ *then* $c \in Z(\mathcal{F})$.

PROOF. We may assume $1 \neq c \in Z(\mathcal{F}_z)$. By definition of c, $\langle c\rangle = Z(M)$, so as $\mathcal{F}_{S\cap M}(M) \lhd N_{\mathcal{F}}(W)$, $c \in Z(N_{\mathcal{F}}(W))$. As $c \in Z(S)$, c centralizes $SO(\tau)$ by 2.6.12.

Let $x = f_\xi \in \mathcal{F}^f$ for some $\emptyset \neq \xi \subseteq \mathcal{Z}_S$. Then $x = e_\theta$ for some $\theta \subseteq I$ with $k = |\theta| = 2|\xi|$. Claim that $\mathcal{F}_x \leq \mathcal{F}_c$. If $k = n$ then $x = c$ and the claim is trivial.

Assume $k \neq n-1$. Then by 11.1.4.3, $\mathcal{F}_x^\circ = \mathcal{F}_1 * \mathcal{F}_2$ with $\langle x\rangle = Z(\mathcal{F}_1)$. Observe that as $c \neq 1$, $cx \neq x$, so even if $k = n/2$, we have $\mathcal{F}_i \lhd \mathcal{F}_x$ for $i = 1, 2$. Now $|\theta_2| = n - k > 1$, so $M_2 \neq 1$ and $\langle cx\rangle = Z(M_2)$ by 11.1.4.5. Let $t \in \xi \cap \mathcal{F}_1^f$. Then $C_{\mathcal{F}_1}(t)^\circ = \mathcal{O}(t) * \mathcal{C}_1$. From 11.1.3.7, there is $g \in M$ with $\alpha = c_g^* \in \mathfrak{A}(t)$ and $t\alpha = z$. Then $(\mathcal{C}_1\mathcal{F}_2)\alpha^* \leq \mathcal{C}$ is centralized by $t\alpha = z$, and hence also by c as $c \in Z(\mathcal{F}_z)$. But $O(\tau)\alpha = O(\tau)$, so α centralizes c. Hence ct centralizes $\mathcal{C}_1\mathcal{F}_2$, so c centralizes $\mathcal{F}_2$ and then as $cx \in M_2$ we have $\langle cx\rangle = Z(\mathcal{F}_2)$. Thus $cx \in Z(\mathcal{F}_x)$, so $c \in Z(\mathcal{F}_x)$, establishing the claim in this case.

We interject a short digression before completing the proof of the claim. As $c \in Z(\mathcal{F}_x)$, also $c \in Z(N_{\mathcal{F}}(O(\xi))$ by 12.2.4.1; that is $c \in Z(N_{\mathcal{F}}(O(\xi)))$ for each $\xi \subseteq \mathcal{Z}_S$ with $|\xi| \neq (n-1)/2$. Therefore by 3.4.11.2, $c \in Z(\mathcal{E}(\tau))$ unless possibly $n = 2^a + 1$ and $c \notin Z(\mathcal{F}_x)$ for $x \in W_S$ of weight $n-1$ in W_I. Hence if $n \neq 2^a + 1$ then $c \in Z(\mathcal{F})$ by 12.2.3, so we may assume $n = 2^a + 1$.

Finally assume $n = k + 1$, and set $\mathcal{Y} = \mathcal{F}_x^\circ$. By 12.2.4.4, we have

$$\mathcal{F}_x \leq \langle C_S(x)\mathcal{Y}, N_{\mathcal{F}}(O(\tau))\rangle.$$

By parts (5) and (6) of 12.2.4, $N_{\mathcal{F}}(O(\tau)) = \langle SC_{\mathcal{F}}(O(\tau)), \mathcal{N}\rangle$, where $\mathcal{N} = N_{\mathcal{F}}(XW)$ and $X = O(\tau)C_S(O(\tau))$, or $m = 8$ and $\mathcal{N} = \langle N_{\mathcal{F}}(XW), SO_\tau\rangle$. Now $C_{\mathcal{F}}(O(\tau)) \leq \mathcal{F}_z \leq \mathcal{F}_c$, $c \in Z(N_{\mathcal{F}}(W))$, and $\mathcal{O}(\tau)$ centralizes $C_S(O(\tau))$ by 12.2.2.4, so $N_{\mathcal{F}}(O(\tau))$ centralizes c. Therefore $\mathcal{E}(\tau) \leq \mathcal{F}_c$ by 3.4.11.2, and it remains to show that $\mathcal{Y}$ centralizes c. Recall that $n - 1 = 2^a$

Set $\mathcal{Y}^+ = \mathcal{Y}/\langle x\rangle$ and $\tau_x^+ = (\mathcal{Y}^+, \Omega^+)$. By 12.2.3, $\mathcal{Y}^+ = \langle \mathcal{E}(\tau_x^+), N_{\mathcal{Y}^+}(y^+)\rangle$ where $y \in W_S$ is of weight $(n-1)/2$ in W_I. Therefore $\mathcal{Y} = \langle \mathcal{E}(\tau_x), N_{\mathcal{Y}}(Y)\rangle$, where $Y = \langle x, y\rangle$. As we saw earlier, $\mathcal{E}(\tau)$ centralizes c, so $\mathcal{E}(\tau_x) \leq \mathcal{E}(\tau) \leq \mathcal{F}_c$. Also $N_{\mathcal{Y}}(Y) = C_{\mathcal{Y}}(Y)^\circ T_x$ for T_x Sylow in $\mathcal{Y}$, and $C_{\mathcal{Y}}(Y)^\circ \leq \mathcal{F}_y^\circ \leq \mathcal{F}_c$, so indeed $\mathcal{Y}$, and hence also $\mathcal{F}_x$, centralizes c.

We've shown that $\mathcal{E}(\tau)$ centralizes c. Hence $c \in Z(\mathcal{F})$ by 12.2.3 unless $n = 2^a + 1$, where $\mathcal{F} = \langle \mathcal{E}(\tau), \mathcal{F}_x \rangle$ for x of weight $n-1$ in W_I. But we also saw that $c \in Z(\mathcal{F}_x)$, completing the proof. □

Because of 12.2.6 and 11.1.6, during the remainder of the section we assume:

HYPOTHESIS 12.2.7. *If $c \in Z(\mathcal{F})$ then $c = 1$.*

LEMMA 12.2.8. *Let $m = 8$, $x = e_\theta \in W_I \cap \mathcal{F}^f$, and define the factorization $\mathcal{F}_x^c = \mathcal{F}_1 * \mathcal{F}_2$ as in 11.1.4.3. Take $z = z_i$ and assume $\{i, i+1\} \subseteq \theta_j$ for some j, so that $O \leq \mathcal{F}_j$. Assume $|\theta_j| \geq 3$.*

(1) $O^2(Aut_{\mathcal{F}}(O)) = O^2(Aut_{\mathcal{F}_j}(O))$.

(2) One of the following holds:

(i) $\mathcal{F}_j \cong \omega(D_k, m)$, $Aut_{\mathcal{F}}(O)$ is a 2-group, and $\mathcal{O} = O$.

(ii) $\mathcal{F}_j \cong \check{A}E_k$ or AE_k and $\mathcal{O}$ is of index 3 in $SL_2[8] * SL_2[8]$.

(iii) $\mathcal{F}_j \cong \Omega_k^\epsilon[8]$ and $\mathcal{O} \cong SL_2[8] * SL_2[8]$.

PROOF. By hypothesis, $x \in S_0$, so L_z centralizes x by 12.2.2.2. Hence $\Sigma = O^2(Aut_{\mathcal{F}}(O)) = Aut_{L_z}(O) = O^2(Aut_{\mathcal{F}_x}(O)) = \Sigma_x$ by 12.2.2.3.

Suppose $\mathcal{F}_j \cong \omega(D_k, m)$. Then $W \cap O \lhd C_{\mathcal{F}_x}(z)$, so Σ_x is a 2-group, and hence $\mathcal{O} = O$, so that (1) holds in this case, as does (2i).

Suppose next that $\mathcal{F}_j \cong \Omega_k^\epsilon[8]$. Then $\Sigma_x \cong SL_2(3) * SL_2(3) = O^2(Aut(O)) = \Gamma$, so (1) holds and $\mathcal{O} \cong SL_2[8] * SL_2[8]$, so (2iii) holds.

Suppose $\mathcal{F}_j$ is $\check{A}E_k$ or AE_k. Then $\Sigma_j = O^2(Aut_{\mathcal{F}_j}(O))$ is of index 3 in $SL_2(3) * SL_2(3)$. Further $E = O \cap O_2(\mathcal{F}_j) \cong E_8$ is Σ_x-invariant, so $\Sigma = \Sigma_x \neq \Gamma$, so $\Sigma = \Sigma_j$ and $\mathcal{O}$ is of index 3 in $SL_2[8] * SL_2[8]$; in particular (1) holds, as does (2ii).

If $a = |\theta_j| \geq 5$ then by induction on n (cf. Theorem 12.2.32), $\mathcal{F}_j$ is in one of the three generic classes of examples considered already, so (1) holds in this case. Therefore we may assume $a = 3$ or 4 and $\mathcal{F}_j$ is not generic.

Suppose $a = 4$. Then $\mathcal{F}_j$ is described in Theorem 12.1.25, and hence as $\mathcal{F}_j$ is not generic, $\mathcal{F}_j \cong \check{L}_3(2)/2^{3+6}$ or $\check{O}_8^+(2)$. Then (cf. 12.1.12 and the proof of 12.1.13) L_z is of index 3 or 3^3 in $(SL_2(3) * SL_2(3))^2$, contrary to 12.2.2.2.

Finally suppose $a = 3$. Then $\mathcal{F}_j$ is described in Theorem 9.5.33, so $\mathcal{F}_j$ is $L_3(2)/E_{64}$ or $Sp_6(2)$. But then (cf. 9.5.11.3, 9.5.15, and 9.5.16) $|\Sigma|_3 = 3$ or 3^3 and O is not Sylow in L_z, contrary to 12.2.2.2.

This completes the proof of (1), and in the process we have also proved (2). □

LEMMA 12.2.9. *Let v_z, v_z' be elements of $K \cap W, K' \cap W$ of order 4, respectively, and set $v = v_z v_z'$.*

(1) If n is odd then $v(n) \in \mathcal{F}^f$ is of weight 1.

(2) If n is even then $v \in \mathcal{F}^f$ is of weight 1.

(3) Let $x \in W_I \cap \mathcal{F}^f$ be of weight 1 and $g \in M$ such that $\alpha = c_g^* \in \mathfrak{A}(v)$ with $v\alpha = x$. Then $\mathcal{C}\alpha^* = C_{\mathcal{F}}(\langle x, z\alpha \rangle)^\circ = C_{\mathcal{F}_x^\circ}(z\alpha)^\circ$ and one of the following holds:

(i) $\mathcal{F}_x^\circ \cong \omega(D_{n-1}, m)$, $\mathcal{C} \cong \omega(D_{n-2}, m)$, $c = 1$, and $z \in \mathcal{C}$.

(ii) $\mathcal{F}_x^\circ \cong \check{A}E_{2n-2}$, $\mathcal{C} \cong \check{A}E_{2n-4}$, $c = 1$, and $z \in \mathcal{C}$.

(iii) $\mathcal{F}_x^\circ \cong AE_{2n-1}$, $\mathcal{C} \cong AE_{2n-3}$, $c \neq 1$, and $z \notin \mathcal{C}$.

(iv) $\mathcal{F}_x^\circ \cong \Omega_{2n-2}^+[m]$, $\mathcal{C} \cong \Omega_{2n-4}^+[m]$, $c = 1$, and $z \in \mathcal{C}$.

(v) $\mathcal{F}_x^\circ \cong \Omega_{2n-1}[m]$, $\mathcal{C} \cong \Omega_{2n-3}[m]$, $c \neq 1$, and $z \notin \mathcal{C}$.

(vi) $\mathcal{F}_x^\circ \cong \Omega_{2n}^-[m]$, $\mathcal{C} \cong \Omega_{2n-2}^-[m]$, $c \neq 1$, and $z \notin \mathcal{C}$.

PROOF. Part (1) follows from 11.1.7.2. Suppose n is even. In Notation 11.1.2 we chose $\tilde{d}_j = (j, j+1) \in \tilde{\mu} = Sym(I)$. Then $\zeta(d_j) = e_{j,j+1}$ with $z = e_{i,i+1}$ for some i by 11.1.3.1, and $v = e_i$ or e_{i+1} by 11.1.3.2. Thus, using 11.1.3.5, $C_S(v)$ is of index 2 in $C_S(z)$, and as $z \in \mathcal{F}^f$, also $v \in \mathcal{F}^f$, proving (2).

Pick x as in (3); by 11.1.3.7 it is possible to pick g and α as in (3). By 2.2 in [**Asc10**],

$$C_{\mathcal{F}_z}(v)\alpha^* = C_{\mathcal{F}}(\langle z, v\rangle)\alpha^* = C_{\mathcal{F}}(\langle z\alpha, x\rangle) = C_{\mathcal{F}_x}(z\alpha),$$

so as $\mathcal{C} = C_{\mathcal{F}}(\langle z, v\rangle)^\circ$, $\mathcal{C}\alpha^* = C_{\mathcal{F}}(\langle x, z\alpha\rangle)^\circ = C_{\mathcal{F}_x^\circ}(z\alpha)^\circ$. Define $\rho = (\mathcal{C}, \Omega - \Omega(z))$ as in 11.1.5; by 11.1.5.1, $\mu(\rho) \cong \mathrm{Weyl}(D_{n-2})$. Now $\mathcal{F}_x^\circ$ is described in 11.1.7.3, and from the structure of the systems listed there, either $C_{\mathcal{F}_x^\circ}(z\alpha)^\circ$ is as described in (3), or one of the following holds:

(vii) $n = 5$, $\mathcal{F}_x^\circ \cong \check{L}_3(2)/2^{3+6}$, and $\mathcal{C} \cong \omega(D_3, m)$.

(viii) $n = 5$, $\mathcal{F}_x^\circ \cong \check{\Omega}_8^+(2)$, and $\mathcal{C} \cong \check{A}E_6$.

By Hypothesis 12.2.7, $c \in Z(\mathcal{F})$ iff $c = 1$, while by 12.2.6, $c \in Z(\mathcal{F})$ iff $c \in Z(\mathcal{F}_z)$. Then by 12.2.5, $c = 1$ iff $cz \in Z(\mathcal{C})$ iff $z \in \mathcal{C}$. Moreover, unless one of the exceptional cases (vii) or (viii) holds, the cases where $cz \in Z(\mathcal{C})$ are listed in 12.2.5.3. Thus to complete the proof of (3) it remains to show that neither (vii) nor (viii) holds. But this follows from 12.2.8.2. □

LEMMA 12.2.10. *If $\mathcal{O} = O$ then $\mathcal{F}^\circ = \mathcal{F}_S(M)$.*

PROOF. Assume $\mathcal{O} = O$ and set $Q = O(\tau)$. By 12.2.8.2 and 12.2.9.3, case (i) of 12.2.9.3 holds. Choose x as in 12.2.9.3; then $\mathcal{F}_x^\circ \cong \omega(D_{n-1}, m)$, $\mathcal{C} \cong \omega(D_{n-2}, m)$, $c = 1$, and $z \in \mathcal{C}$.

As $\mathcal{O} = O$, $\mathcal{O}(\tau) = Q$, so $S\mathcal{O}(\tau) = S \leq N_{\mathcal{F}}(W)$. Also by 12.2.2.3, $Aut_{\mathcal{F}}(O)$ is a 2-group, so by parts (5) and (6) of 12.2.4, $N_{\mathcal{F}}(Q) = \langle SC_{\mathcal{F}}(Q), N_{\mathcal{F}}(QW)\rangle$. Now $C_{\mathcal{F}}(Q) \leq \mathcal{F}_{z_1} \cap \mathcal{F}_{z_{n-1}}$. We may choose notation so that $z = z_1$. As $\mathcal{C} \cong \omega(D_{n-2}, m)$, $\mathcal{F}_z^\circ = O * \mathcal{C} = M_z$ with $W_z = V_1V_1' \cdot V_3V_4 \cdots V_{n-1}V_n$ and $C_{\mathcal{F}}(Q) \leq N_{\mathcal{F}}(W_z)$. Similarly $W_{z_{n-1}} = V_1V_2 \cdots V_{n-3}V_{n-1}V_n$, and $C_{\mathcal{F}}(Q) \leq N_{\mathcal{F}}(W_{z_{n-1}})$. Then as $W = W_zW_{z_{n-1}}$, $C_{\mathcal{F}}(Q) \leq N_{\mathcal{F}}(W)$. Of course $N_{\mathcal{F}}(WQ) \leq N_{\mathcal{F}}(W)$, so $N_{\mathcal{F}}(Q) \leq N_{\mathcal{F}}(W)$.

Let $\theta \subseteq \mathcal{Z}_S$ with $O(\theta) \in \mathcal{F}^f$ and define $f = f_\theta$ as in 12.2.4. By 12.2.4.2, $f \in \mathcal{F}^f$, so by 12.2.4.4, $\mathcal{F}_f \leq \langle C_S(f)\mathcal{F}_f^\circ, N_{\mathcal{F}}(Q)\rangle$. By 12.2.8,

$$C_S(f)\mathcal{F}_f^\circ = \mathcal{F}_{C_S(f)}(C_S(f)M_f) \leq N_{\mathcal{F}}(W),$$

so $\mathcal{F}_f \leq N_{\mathcal{F}}(W)$. Further by 12.2.4.1, $N_{\mathcal{F}}(O(\theta)) \leq \mathcal{F}_f \leq N_{\mathcal{F}}(W)$. Therefore $\mathcal{E}(\tau) \leq N_{\mathcal{F}}(W)$.

If $n = 2^k$ and $x \in W_S \cap \mathcal{F}^f$ is of weight 2^{k-1} in W_I then $x = f_\theta$ where $|\theta| = n/4$, so $\mathcal{F}_x \leq N_{\mathcal{F}}(W)$. Further $c = 1$, so by 12.2.3, $\mathcal{F} = \langle \mathcal{E}(\tau), \mathcal{F}_x\rangle \leq N_{\mathcal{F}}(W)$. Similarly if $n = 2^k + 1$ and $x = e_{I-\{n\}}$ then $x = f_\theta$ where $|\theta| = (n-1)/2$, so $\mathcal{F}_x \leq N_{\mathcal{F}}(W)$ and then $\mathcal{F} \leq N_{\mathcal{F}}(W)$ by 12.2.3. Finally in the remaining cases, $\mathcal{F} = \mathcal{E}(\tau) \leq N_{\mathcal{F}}(W)$ by 12.2.3. Hence in any case $W \trianglelefteq \mathcal{F}$, so the lemma follows from 3.3.13.3. □

LEMMA 12.2.11. *(1) If $\mathcal{C} \cong \check{A}E_{2n-4}$ then $\mathcal{F}^\circ \cong AE_{2n}$.*
(2) If $\mathcal{C} \cong AE_{2n-3}$ then $\mathcal{F}^\circ \cong AE_{2n+1}$.

PROOF. Assume $\mathcal{C}$ is $\check{A}E_{2n-4}$ or AE_{2n-3}. Then $m = 8$ and applying 12.2.8 to $z \neq x \in \mathcal{Z}_S$, $\mathcal{O}$ is of index 3 in $SL_2[8] * SL_2[8]$. Hence there is a unique $W\mathcal{O}$-invariant E_8-subgroup $E(z)$ of O. As $Aut_{W\mathcal{O}}(O) \trianglelefteq Aut_{\mathcal{F}}(O)$, $E(z) \trianglelefteq \mathcal{F}_z$. Thus for $t \in \mathcal{Z}_\Delta$, we can define $E(t) = E(z)^g$ for $g \in M$ with $c_{g^{-1}}^* \in \mathfrak{A}(t)$ and $z^g = t$. Set $U = \langle E(t) : t \in \mathcal{Z}_\Delta\rangle$. Thus $U \trianglelefteq M$, and using 11.1.3.5, also $U \trianglelefteq N_{\mathcal{F}}(W)$.

Next for $x \in W_I \cap \mathcal{F}^f$ we obtain the factorization $\mathcal{F}_x^\circ = \mathcal{F}_1 * \mathcal{F}_2$ of 11.1.4.3, and if $j \in \{1,2\}$ with $|\theta_j| \geq 3$, then by 12.2.8, $\mathcal{F}_j \cong \check{A}E_k$ or AE_k. Hence from the structure of such systems, $O_2(\mathcal{F}_j) = \langle E(t) : t \in \mathcal{Z}(\tau_j)\rangle$. Therefore $O_2(\mathcal{F}_j) \leq U$ and indeed $U = C_{U\mathcal{F}_j}(O_2(\mathcal{F}_j)) \trianglelefteq U\mathcal{F}_j$. Thus for each $\theta \subseteq \mathcal{Z}_S$ and $f_\theta \in \mathcal{F}^f$, $C_\mathcal{F}(f_\theta)^\circ \leq N_\mathcal{F}(U)$. Similarly $U = C_{UL_z}(E(z)) \trianglelefteq UL_z$, so also $U \trianglelefteq UL_\tau$ and hence $U \trianglelefteq U\mathcal{O}_\tau$.

We may choose notation so that $z = z_1$. Observe $\mathcal{Z}_\Delta = \mathcal{Z}(\tau_z) \cup \mathcal{Z}(\tau_z)^g$ for $g \in M$ with $z^g = z_{n-1}$. Thus $U = O_2(\mathcal{C})O_2(\mathcal{C})^g$, so in particular $C_\mathcal{F}(O(\tau)) \leq N_\mathcal{F}(U)$. Then by 12.2.4.6, $N_\mathcal{F}(O(\tau)) \leq N_\mathcal{F}(U)$.

Choose $f = f_\theta$ as above. We've seen that $C_S(f)\mathcal{F}_f^\circ$ and $N_\mathcal{F}(O(\tau))$ act on U, so $\mathcal{F}_f \leq N_\mathcal{F}(U)$ by 12.2.4.4. Then $N_\mathcal{F}(O(\theta)) \leq N_\mathcal{F}(U)$ by 12.2.4.1. Hence $\mathcal{E}(\tau) \leq N_\mathcal{F}(W)$, and then $U \trianglelefteq \mathcal{F}$ by 12.2.3. Now the lemma follows from Theorem 2. □

In light of 12.2.8.2, 12.2.10, and 12.2.11, it remains only to consider cases (iv), (v), and (vi) of 12.2.9.3, where from 12.2.8 we have $\mathcal{O} \cong SL_2[m] * SL_2[m]$. Thus for the remainder of the section we assume:

HYPOTHESIS 12.2.12. *Hypothesis 12.2.7 holds and* $\mathcal{O} \cong SL_2[m] * SL_2[m]$.

Equivalently, as we have already observed, by 12.2.9.3:

LEMMA 12.2.13. *Let $x \in W_I \cap \mathcal{F}^f$ be of weight 1. Then one of the following holds:*

(1) $\mathcal{F}_x^\circ \cong \Omega_{2n-2}^+[m]$, $\mathcal{C} \cong \Omega_{2n-4}^+[m]$, $c = 1$, and $z \in \mathcal{C}$.

(2) $\mathcal{F}_x^\circ \cong \Omega_{2n-1}[m]$, $\mathcal{C} \cong \Omega_{2n-3}[m]$, $c \neq 1$, and $z \notin \mathcal{C}$.

(3) $\mathcal{F}_x^\circ \cong \Omega_{2n}^-[m]$, $\mathcal{C} \cong \Omega_{2n-2}^-[m]$, $c \neq 1$, and $z \notin \mathcal{C}$.

NOTATION 12.2.14. Let Ξ be the set of nonempty subsets ξ of $\mathcal{Z}_S$ such that either

(a) ξ is a union of orbits of S on $\mathcal{Z}_S$, $f_\xi = \prod_{t\in\xi} t \neq 1$, and if $\xi = \mathcal{Z}_S$ then S is transitive on $\mathcal{Z}_S$, or

(b) n is power of 2, $c = 1$, and $|\xi| = n/4$ with $O(\xi) \in \mathcal{F}^f$.

For $\xi \in \Xi$, define $\mathcal{B}_\xi(\tau) = \langle C_\mathcal{F}(f_\xi), N_\mathcal{F}(W)\rangle$.

LEMMA 12.2.15. *Let $\xi \in \Xi$, $f = f_\xi$, and $\mathcal{B} = \mathcal{B}_\xi(\tau)$.*

(1) $f \in Z(S)$.

(2) $N_\mathcal{F}(O(\tau)) \leq \mathcal{B}$.

(3) $SO(\tau) \leq \mathcal{B}$.

(4) Let $\theta \subseteq \mathcal{Z}_S$ with $Q = O(\theta) \in \mathcal{F}^f$, and let T be Sylow in $N_\mathcal{F}(Q)$. Let Σ be a nontrivial orbit of T on θ and $g \in M$ such that $\alpha = c_g^* \in \mathfrak{A}(O(\Sigma))$. Then if $T\alpha C_\mathcal{F}(O(\Sigma)\alpha)^\circ \leq \mathcal{B}$, we have $N_\mathcal{F}(Q) \leq \mathcal{B}$.

PROOF. In case (a) of 12.2.14, ξ is S-invariant, so $f \in Z(S)$. In case (b), $O(\xi) \in \mathcal{F}^f$, so S acts on $\{O(\xi), O(\mathcal{Z}_S - \xi)\}$, and as $c = 1$, we have $f = f_{\mathcal{Z}_S - \xi}$. Hence $f \in Z(S)$, so that (1) holds.

Set $P = O(\tau)$. Then $SC_\mathcal{F}(P) \leq \mathcal{F}_f \leq \mathcal{B}$, and $SO(\tau) \leq \mathcal{F}_f \leq \mathcal{B}$, so (2) follows from parts (5) and (6) of 12.2.4. Similarly $SO(\tau) \leq \mathcal{F}_f$, so (3) holds.

Assume the hypothesis of (4). By 12.2.4.3, $N_{\mathcal{F}}(Q) \leq \langle N_{\mathcal{F}}(P), TC_{\mathcal{F}}(Q)^{\circ}\rangle$ and $N_{\mathcal{F}}(P) \leq \mathcal{B}$ by (2), so to prove (4) it remains to show that $TC_{\mathcal{F}}(Q)^{\circ} \leq \mathcal{B}$. But as α is a $\mathcal{B}$-map and $(TC_{\mathcal{F}}(Q)^{\circ})\alpha^* \leq T\alpha C_{\mathcal{F}}(O(\Sigma)\alpha)^{\circ}$, $TC_{\mathcal{F}}(Q)^{\circ} \leq \mathcal{B}$ if $T\alpha C_{\mathcal{F}}(O(\Sigma)\alpha)^{\circ} \leq \mathcal{B}$. □

LEMMA 12.2.16. *Let $\xi \in \Xi$, $f = f_{\xi}$, and $\mathcal{B} = \mathcal{B}_{\xi}(\tau)$. Then either*

(1) $\mathcal{F} = \mathcal{B}$, or

(2) S is transitive on $\mathcal{Z}_S$, $\xi = \mathcal{Z}_S$, and $\mathcal{F} = \langle \mathcal{B}, N_{\mathcal{F}}(O(\theta))^{\circ}\rangle$, where $\theta \subseteq \mathcal{Z}_S$ is of order $|\mathcal{Z}_S|/2$ and $O(\theta) \in \mathcal{F}^f$.

PROOF. Assume $\mathcal{F} \neq \mathcal{B}$ and set $\Gamma = \mathcal{Z}_S$. Claim $N_{\mathcal{F}}(O(\theta)) \not\leq \mathcal{B}$ for some $\theta \subseteq \Gamma$ with $O(\theta) \in \mathcal{F}^f$. For if not, then $\mathcal{E}(\tau) \leq \mathcal{B}$ by 12.2.15.3. Then by 12.2.3, one of the exceptional cases (2)-(4) of that lemma holds. However in each case S is transitive on Γ, so either $\xi = \Gamma$ or case (4) holds and ξ is in 12.2.14.b. In each case f is the element appearing in 12.2.3 such that $\mathcal{F} = \langle \mathcal{E}(\tau), \mathcal{F}_f\rangle$, so that $\mathcal{F} = \mathcal{B}$, contrary to assumption. This establishes the claim.

By the claim and 12.2.15.4, there is $\theta \subseteq \Gamma$ such that $Q = O(\theta) \in \mathcal{F}^f$, θ is an orbit of a Sylow group T of $N_{\mathcal{F}}(Q)$, and $\mathcal{Y} = C_{\mathcal{F}}(Q)^{\circ} \not\leq \mathcal{B}$. Let $\xi' = \Gamma - \xi$. In case (a) of 12.2.14, $\theta \subseteq \xi$ or ξ', say $\theta \subseteq \xi$. Further if $\theta = \xi$ then $\mathcal{Y} \leq \mathcal{F}_f \leq \mathcal{B}$, a contradiction. So $\theta \subset \xi$ in case (a). Similarly in case (b), conjugating in M we may take $\theta \subset \xi$.

Set $\bar{\theta} = \Gamma - \theta$, $\bar{\Omega} = \{J \in \Omega : z(J) \in \bar{\theta}\}$, and $\bar{\tau} = (T\mathcal{Y}, \bar{\Omega})$. From 11.1.4, $\bar{\tau}$ is a quaternion fusion packet with $\mu(\bar{\tau}) \cong Weyl(D_a)$, where $a = n - 2|\theta|$. Observe that $|\bar{\theta}| = a/2$ or $(a-1)/2$ for n even, odd, respectively.

Set $M^{\Gamma} = N_M(\Gamma)/C_M(\Gamma)$; as W_I is the core of the permutation module for $\tilde{\mu} = Sym(I)$, or the image of that module modulo its center, we conclude that $M^{\Gamma} = Sym(\Gamma)$. Therefore $\Gamma = \Gamma_1 \cup \cdots \cup \Gamma_k$, where $|\Gamma_i| = 2^{a_i}$ with $0 \leq a_1 < \cdots < a_k$ and $S^{\Gamma} = S_1^{\Gamma} \times \cdots \times S_k^{\Gamma}$, where S_i^{Γ} is transitive on Γ_i and fixes the remaining orbits pointwise.

Next $\theta \subseteq \Sigma$ an orbit of S on Γ; thus $\Sigma = \Gamma_i$ for some i. As $Q \in \mathcal{F}^f$, i is minimal subject to $|\theta| \leq 2^{a_i}$, and θ^S is a partition of Σ. Thus in case (a), $S_j \leq T$ for $j \neq i$ and $|\Sigma| = |S : T||\theta|$. In case (b), $|\xi| = |N_S(\xi) : T||\theta|$.

Suppose T is transitive on $\bar{\theta}$. Then as T acts on ξ' and $\xi - \theta$ with $\theta \neq \xi$, we conclude $\xi' = \emptyset$, so $\xi = \Gamma$. Thus by 12.2.14, case (a) of 12.2.14 holds with S transitive on Γ; in particular $\Sigma = \Gamma$. Further T is transitive on $\xi - \theta$, so as $\Sigma = \Gamma$ and $|\Sigma| = |S : T||\theta|$, it follows that $|S : T| = 2 = |\Gamma|/|\theta|$, and hence 12.2.16.2 holds. Therefore we may assume T is not transitive on $\bar{\theta}$.

Suppose $a < 5$. Then as $|\bar{\theta}| = a/2$ or $(a-1)/2$, we conclude that $(a, |\bar{\theta}|) = (4, 2)$ $(3, 1)$, or $(2, 1)$, and in each case the stabilizer T of θ in S is transitive on $\bar{\theta}$, contrary to the previous paragraph. Therefore $a \geq 5$.

Let $\bar{\xi} = \xi \cap \bar{\theta}$. As $a \geq 5$ and as T is not transitive on $\bar{\theta}$, by induction on n, $\mathcal{Y} = \mathcal{B}_{\bar{\xi}}(\bar{\tau})$. However $\mathcal{B}_{\bar{\xi}}(\bar{\tau}) \leq \langle \mathcal{F}_f^{\circ}, N_{\mathcal{F}}(W)\rangle \leq \mathcal{B}$, contradicting $\mathcal{Y} \not\leq \mathcal{B}$. This completes the proof. □

NOTATION 12.2.17. Let $\xi \in \Xi$ and assume either

(I) $\mathcal{F} = \mathcal{B}_{\xi}(\tau)$ and set $f = f_{\xi}$, or

(II) $\mathcal{F} \neq \mathcal{B}_{\xi}(\tau)$, so that (by 12.2.16) S is transitive on $\xi = \mathcal{Z}_S$. Here set $f = f_{\theta}$, where $\theta \subseteq \mathcal{Z}_S$ is of order $|\mathcal{Z}_S|/2$ and $O(\theta) \in \mathcal{F}^f$.

In each case $f = e_{\theta_1}$ for some $\theta_1 \subseteq I$ of even order k; indeed $k = 2|\xi|$ in (I) while $k = 2|\theta|$ in (II). Set $\theta_2 = I - \theta_1$ and define the factorization $\mathcal{F}_f^\circ = \mathcal{F}_1 * \mathcal{F}_2$ as in 11.1.4. Let T_i be Sylow in $\mathcal{F}_i$ for $i = 1, 2$.

LEMMA 12.2.18. *(1) $\mu_1 \cong$ Weyl(D_k) and $\mathcal{F}_1 \cong \Omega_{2k}^+[m]$ with $\langle f \rangle = Z(\mathcal{F}_1)$ and $T_1 \leq M_1$.*
(2) $\mu_2 \cong$ Weyl(D_{n-k}) and one of the following holds:
(i) 12.2.13.1 holds, $\mathcal{F}_2 \cong \Omega_{2(n-k)}^+[m]$, and $T_2 \leq M_2$.
(ii) 12.2.13.2 holds, $\mathcal{F}_2 \cong \Omega_{2(n-k)+1}[m]$, and $|T_2 : T_2 \cap M_2| = 2$.
(iii) 12.2.13.3 holds, $\mathcal{F}_2 \cong \Omega_{2(n-k)+2}^-[m]$, and $|T_2 : T_2 \cap M_2| = 4$.
(3) If $k = 2|\mathcal{Z}_S|$ then k is a power of 2 and $n = k + 1$.
(4) If $k \neq 2|\mathcal{Z}_S|$ then either $n - k \geq 3$ or $n = k + 2$ is even and $k \geq 4$.
(5) Either $f \in Z(S)$ or 12.2.17.II holds and $|S : C_S(f)| = 2$.

PROOF. By 11.1.4.5, $\mu_1 \cong \text{Weyl}(D_k)$, $\mu_2 \cong \text{Weyl}(D_{n-k})$, $\langle f \rangle = Z(M_1)$, $\langle fc \rangle = Z(M_2)$. In particular $f \in Z(\mathcal{F}_1)$, so as $\mathcal{O} \cong SL_2[m] * SL_2[m]$, $\mathcal{F}_1 \cong \Omega_{2k}^+[m]$ by induction on n. By 5.3.8, $T_1 \leq M_1$. Thus (1) holds.

Similarly as $\mu_2 \cong \text{Weyl}(D_{n-k})$, we conclude by induction on n that $\mathcal{F}_2$ is $\Omega_{2(n-k)+a}^\epsilon[m]$ for some $0 \leq a \leq 2$ and suitable ϵ. Let $y \in W_I \cap \mathcal{F}_1^f$ be of weight 1; then $\mathcal{C}_1 = C_{\mathcal{F}_1}(y)^\circ \cong \Omega_{2k-2}^+[m]$ and $C_{\mathcal{F}}(\langle y, f \rangle)^\circ = \mathcal{C}_1 * \mathcal{F}_2$. By 11.1.3.7 there is $g \in M$ such that $\alpha = c_g^* \in \mathfrak{A}(y)$. Set $x = y\alpha$. By 2.2 in [**Asc10**], $C_{\mathcal{F}}(\langle y, f \rangle)\alpha^* = C_{\mathcal{F}_x}(f\alpha)$, so $\mathcal{C}_1\alpha^* * \mathcal{F}_2\alpha^* = C_{\mathcal{F}_x}(f\alpha^*)^\circ = C_{\mathcal{F}_x}(w)^\circ$, where $w = f\alpha x \in \mathcal{F}_x^\circ$ is of weight $k-1$. Therefore as $\mu(\tau_x) \cong \text{Weyl}(D_{n-1})$, $C_{\mathcal{F}_x}(w)^\circ = \mathcal{G}_1 * \mathcal{G}_2$ with $\rho_i = (\mathcal{G}_i, \Gamma_i)$ a quaternion fusion packet, $\mu(\rho_1) \cong \text{Weyl}(D_{k-1})$, and $\mu(\rho_2) \cong \text{Weyl}(D_{n-k})$. Therefore $\mathcal{G}_1 = \mathcal{C}_1\alpha^*$ and $\mathcal{G}_2 = \mathcal{F}_2\alpha^*$. Then $\mathcal{G}_1 \cong \Omega_{2k-2}^+[m]$, so as $\mathcal{F}_x^\circ \cong \Omega_{2(n-2)+a}^\epsilon[m]$ by 12.2.13, it follows that $\mathcal{G}_2 \cong \Omega_{2(n-k)+a}^\epsilon[m]$. By 5.3.8, 5.3.9, and 5.3.10, $|T_2 : T_2 \cap M_2| = 2^a$, establishing (2).

Suppose $k = 2|\mathcal{Z}_S|$. Then 12.2.17.I holds with $\xi = \mathcal{Z}_S$, so by 12.2.14, S is transitive on $\mathcal{Z}_S$. Thus k is a power of 2 and $n = k$ or $k+1$. In the latter case (3) holds. In the former, $c = f \neq 1$, so $c \notin Z(\mathcal{F})$ by 12.2.7. But as 12.2.17.I holds, $\mathcal{F} = \mathcal{B}_\xi(\tau) = \langle \mathcal{F}_f, N_{\mathcal{F}}(W) \rangle$, whereas $f = c$ generates $Z(M) \vartriangleleft N_{\mathcal{F}}(M)$, so $\mathcal{F} = \mathcal{F}_f$, contradicting $c \notin Z(\mathcal{F})$. Hence (3) holds.

Suppose $k \neq 2|\mathcal{Z}_S|$. In 12.2.19.II, $k = |\mathcal{Z}_S|$ and $n = 2k$ or $2k+1$, so $n - k = k$ or $k+1$. If $n = 2k$ then n is a power of 2 so $n \geq 8$ and $n - k = k \geq 4$. If $n = 2k+1$ then as $n \geq 5$, $n - k = k + 1 \geq 3$. Thus (4) holds in 12.2.17.II, so assume 12.2.19.I holds. Then $f = f_\xi$ and by assumption, $\xi \neq \mathcal{Z}_S$, so $|\theta_2| = 2|\mathcal{Z}_S - \xi| \geq 2$. Finally $n - k = |\theta_2|$ or $|\theta_2| + 1$ for n even or odd, respectively, so (4) holds.

In 12.2.17.I, $f = f_\xi$ and ξ is S-invariant, so $f \in Z(S)$. In 12.2.17.II, $f = f_\theta$ with $|S : N_S(\theta)| = 2$. Thus (5) holds. □

NOTATION 12.2.19. Let $T_M = S \cap M$ be Sylow in M, and let T_f be Sylow in $\mathcal{F}_f^\circ$. Set $T = T_M T_f$. Recall from Notation 11.1.2 that $\{K_i : i \in I\} \subseteq K^M$ and $V_i = K_i \cap W \in \eta$ with $W = V_1 \cdots V_n$. Recall from 11.1.4 that for $i = 1, 2$, $W_i = \langle \eta_i \rangle$.

LEMMA 12.2.20. *(1) $|W : W_1W_2| = 2$.*
(2) $C_T(f)\mathcal{F}_f^\circ \vartriangleleft \mathcal{F}_f$.
(3) $\mathcal{F}_f = \langle C_S(f)\mathcal{F}_f^\circ, N_{\mathcal{F}_f}(W) \rangle$.

PROOF. Let $b = 2, 1$ for $c = 1$, $c \neq 1$, respectively and set $U = W_1W_2$. From 4.3.7.2, $|W_1| = (m/2)^k/2$, $|W_2| = (m/2)^{n-k}/2$, and $|W| = (m/2)^n/2b$. Further

$W_1 \cap W_2 \leq Z(\mathcal{F}_1) = \langle f \rangle$, so $W_1 \cap W_2 = \langle f \rangle$, 1, for $b = 2, 1$, respectively. Therefore $|U| = (m/2)^n/4b = |W|/2$, proving (1).

We next prove (2). By 1.5.2 in [**Asc19**], it suffices to show that $C_T(f) \trianglelefteq \mathcal{N} = N_{\mathcal{F}_f}(T_f)$. As U is weakly closed in T_f with respect to $\mathcal{F}_f$, $\mathcal{N}$ acts on U. Observe that there is a unique $i \in I$ with $V_i \not\leq U$, and by (1) $|V_i : U \cap V_i| = 2$, so $s_i = z(K_i) \in U$. Let $\mathcal{X} = C_{\mathcal{F}_f}(\mathcal{F}_f^\circ)$, X Sylow in $\mathcal{X}$, and $\alpha \in \mathfrak{A}(s_i)$ with $s_i\alpha = z$. Then $Y = V_iV_i'$ acts on $\mathcal{X}$ with $|Y : Y \cap U| = 2$ and $Y \cap U$ centralizes $\mathcal{X}$, while $\mathcal{O}_z \trianglelefteq \mathcal{F}_z$, so $[X, Y] \leq Y \cap U \cap X = 1$ and then Y centralizes $\mathcal{X}$ by 9.5.2 in [**Asc11**]. Therefore $W = YU$ centralizes $\mathcal{X}$. Now $F^*(\mathcal{N}) \leq O_2(\mathcal{N})\mathcal{X}$ so as W centralizes $\mathcal{X}$ and $[W, O_2(\mathcal{N})] \leq U \trianglelefteq \mathcal{N}$, it follows that $W \leq O_2(\mathcal{N})$. Then as W is weakly closed in S with respect to $\mathcal{F}$, $W \trianglelefteq \mathcal{N}$. Then as $\mathcal{F}_{T_M}(M) \trianglelefteq N_{\mathcal{F}}(W)$, $O_2(M) \trianglelefteq \mathcal{N}$. Finally $\tilde{\mu} = M/O_2(M) \cong S_n$ and $N_{\tilde{\mu}}(\tilde{T}_f) \leq N_{\tilde{\mu}}(\tilde{T}_M)$, so $C_T(f) \trianglelefteq \mathcal{N}$, completing the proof of (2). Further by 1.3.2 in [**Asc19**], $\mathcal{F}_f = \langle C_S(f)\mathcal{F}_f^\circ, \mathcal{N} \rangle$, and we showed $W \trianglelefteq \mathcal{N}$, so (3) holds too. □

LEMMA 12.2.21. *Assume* 12.2.13.*1 holds and let* $\mathcal{G}_1 = \mathcal{F}_f$, $\mathcal{G}_2 = N_{\mathcal{F}}(W)$, $\mathcal{E}_1 = C_T(f)\mathcal{F}_f^\circ$, $\mathcal{E}_2 = \mathcal{F}_T(M)$, *and* $\mathcal{Y} = \langle T\mathcal{O}_\tau, \mathcal{F}_T(M) \rangle$. *If case II of* 12.2.17 *holds set* $x = f_\xi$, $\mathcal{G}_3 = \mathcal{F}_x$, *and* $\mathcal{E}_3 = T\mathcal{F}_x^\circ$. *Set* $\Sigma = \{1, 2\}$ *in case I and* $\Sigma = \{1, 2, 3\}$ *in case II of* 12.2.17.

(1) $\mathcal{F} = \langle \mathcal{G}_\sigma : \sigma \in \Sigma \rangle$.

(2) $\mathcal{E}_\sigma \trianglelefteq \mathcal{G}_\sigma$ *for each* $\sigma \in \Sigma$.

(3) $\mathcal{Y} \cong P\Omega_{2n}^+[m]$.

(4) $\mathcal{Y} = \langle \mathcal{E}_\sigma : \sigma \in \Sigma \rangle$.

(5) $T = T_M$, *and* T *is Sylow in* $\mathcal{E}_i$ *for* $i = 1, 2$, *unless case II of* 12.2.17 *holds, where* $C_T(f)$ *is of index 2 in* T.

PROOF. As 12.2.13.1 holds, $c = 1$.

By 3.3.13, $\mathcal{E}_2 \trianglelefteq \mathcal{G}_2$, and by 12.2.20.2, $\mathcal{E}_1 \trianglelefteq \mathcal{G}_1$. Thus (2) holds in case I. Further in case I, $f = f_\xi$ and $\mathcal{F} = \mathcal{B}_\xi(\tau) = \langle \mathcal{G}_1, \mathcal{G}_2 \rangle$ by 12.2.16, so (1) also holds in this case.

Suppose case II holds. Then S is transitive on $\xi = \mathcal{Z}_S$ and by 12.2.16, $\mathcal{F} = \langle \mathcal{F}_x, \mathcal{F}_f, N_{\mathcal{F}}(W) \rangle$, so (1) holds. To complete the proof of (2), we must show that $\mathcal{E}_3 \trianglelefteq \mathcal{G}_3$. From 12.2.18.3, $n = 2|\mathcal{Z}_S| + 1$ is odd. As $c = 1$, x is of weight 1, so from 12.2.9.1 we may take $x = v(1) = e_1$. Now $W_x = V_2 \cdots V_n$ and $W = W_xV_1$. Set $W_4 = V_4 \cdots V_n$, $\Gamma = \{K_2, K_2'\}$, and $\rho = (C_{\mathcal{F}}(W_4), \Gamma)$. Then $\Gamma = C_\Omega(W_4)$, so ρ is a quaternion fusion packet by 2.5.2. Also $\mu(\rho) \cong C_\mu(W_4) = \langle d_1, d_2, d_2' \rangle \cong \mathrm{Weyl}(D_3)$, and we may take $K = K_n$, so ρ is a subpacket of $\mathcal{C}$. Therefore as 12.2.13.1 holds, $\mathcal{X} = C_{\mathcal{F}}(W_4)^\circ \cong \Omega_6^+[m]$. Then $C_{\mathcal{X}}(W_x) = C_{\mathcal{X}}(V_2V_2') = V_1V_2V_2'$, so $W = V_1W_x \trianglelefteq C_{\mathcal{F}_x}(W_x)$.

By 1.3.2 in [**Asc19**], $\mathcal{F}_x = \langle S\mathcal{F}_x^\circ, N_{\mathcal{F}_x}(T_x) \rangle$. Also $N_{\mathcal{F}_x}(T_x) \leq N_{\mathcal{F}_x}(W_x) \leq N_{\mathcal{F}_x}(W)$. Then as $T = T_xO_2(M)$, $T \trianglelefteq N_{\mathcal{F}_x}(T_x)$, so $\mathcal{E}_3 = T\mathcal{F}_x^\circ \trianglelefteq \mathcal{F}_x = \mathcal{G}_3$ by 1.5.2 in [**Asc19**]. This completes the proof of (2).

Next $\mathcal{F}_f^\circ = \mathcal{F}_1\mathcal{F}_2$. Then $T_f = T_1T_2$, so by parts (1) and (2) of 12.2.18, $T_f \leq M_1M_2 \leq M$. Hence $T = T_fT_M = T_M$.

Let $\tilde{\mathcal{F}} = P\Omega_{2n}^+[m]$, $\tilde{\tau} = (\tilde{\mathcal{F}}, \tilde{\Omega})$ the Lie packet for $\tilde{\mathcal{F}}$, $\tilde{\eta} \in \eta(\tilde{\tau})$, etc. Let $\omega = \omega(D_n, m)$. By 5.3.5.4, $N_{\tilde{\mathcal{F}}}(\tilde{W}) \cong \omega/Z(\omega)$, and as $c = 1$, $M \cong \omega/Z(\omega)$ from 11.1.2. Thus we may take $M = N_{\tilde{\mathcal{F}}}(\tilde{W})$, and as $T = T_M$ is Sylow in M, $T = \tilde{S}$ and $\Omega = \tilde{\Omega}$, etc. By 11.5.3, $\tilde{\mathcal{F}} = \langle T\tilde{\mathcal{O}}_{\tilde{\tau}}, \mathcal{F}_T(M) \rangle$, and by Theorem 2.7.3, $T\mathcal{O}_{\tilde{\tau}} = T\mathcal{O}_\tau$. Therefore $\mathcal{Y} = \tilde{\mathcal{F}} \cong P\Omega_{2n}^+[m]$, proving (3).

By definition $T = T_M$ is Sylow in M, so T is Sylow in $\mathcal{E}_2$. If case I holds then $f \in Z(S)$ by 12.2.18.5, and as $T_f \leq T$, T is Sylow in $T\mathcal{F}_f^\circ = \mathcal{E}_1$. In case II, $|T_M : C_{T_M}(f)| = 2$, so $|T : C_T(f)| = 2$, completing the proof of (5).

From 12.2.18, $\mathcal{F}_1 \cong \Omega_{2k}^+[m]$ and $\mathcal{F}_2 \cong \Omega_{2(n-k)}^+[m]$. Therefore by 11.5.3, $\mathcal{E}_1 = C_T(f)\mathcal{F}_f^\circ = \langle C_T(f)\mathcal{O}_\tau, \mathcal{F}_{C_T(f)}(C_T(f)M_f)\rangle \leq \mathcal{Y}$. Hence in case I where $T = C_T(f)$, we have $\mathcal{Y} = \langle \mathcal{E}_1, \mathcal{E}_2\rangle$. Therefore (4) holds in case I, so assume case II holds. There $\mathcal{E}_3 = T\mathcal{F}_x^\circ$ with $\mathcal{F}_x^\circ \cong \Omega_{2n-2}^+[m]$, so again by 11.5.3, $\mathcal{E}_3 = \langle T\mathcal{O}_\tau, \mathcal{F}_T(TM_x)\rangle \leq \mathcal{Y}$, and hence $\mathcal{Y} = \langle \mathcal{E}_1, \mathcal{E}_2, \mathcal{E}_3\rangle$, completing the proof. □

LEMMA 12.2.22. *If 12.2.13.1 holds then* $\mathcal{F}^\circ \cong P\Omega_{2n}^+[m]$.

PROOF. We show that $\mathcal{Y} \trianglelefteq \mathcal{F}$. Then $\mathcal{F}^\circ = \mathcal{Y}$, and 12.2.21.3 completes the proof. To prove $\mathcal{Y} \trianglelefteq \mathcal{F}$, we verify that the tuple $\mathcal{E}_\sigma, \mathcal{G}_\sigma$, $\sigma \in \Sigma$, satisfies Hypothesis 1.4.3 in [**Asc19**], and then appeal to 1.4.5 in [**Asc19**].

By 12.2.21.1, $\mathcal{F} = \langle \mathcal{G}_\sigma : \sigma \in \Sigma\rangle$, so Hypothesis 1.4.1 in [**Asc19**] is satisfied. By 12.2.21.2, $\mathcal{E}_\sigma \trianglelefteq \mathcal{G}_\sigma$ for each $\sigma \in \Sigma$, so 1.4.3.1 in [**Asc19**] is satisfied. As $T = T_M$ and W and $\mathcal{O}(\tau)$ are normal in $N_{\mathcal{F}}(T_M)$, $Aut_{\mathcal{F}}(T)$ is contained in $Aut(T\mathcal{O}_\tau)$ and $Aut(\mathcal{F}_T(M))$, so 1.4.3.2 in [**Asc19**] follows from 1.4.6 in [**Asc19**]. Except in case II with $\sigma = 1$, T is Sylow in $\mathcal{E}_\sigma$, so the extension condition 1.4.3.3 in [**Asc19**] is satisfied. Thus Hypothesis 1.4.3 in [**Asc19**] is satisfied except possibly in case II. Finally $\mathcal{Y}$ is saturated with $\mathcal{Y} = O^{2'}(\mathcal{Y})$ by 12.2.21.3. Therefore the lemma follows from 1.4.5 in [**Asc19**], except in case II where we must verify the extension condition when $\sigma = 1$. But if $\alpha \in Aut_{\mathcal{F}_f}(C_T(f))$ then as $W \leq C_T(f)$, $\alpha \in Aut_{\mathcal{F}_2}(C_T(f))$. Further $Aut_{\mathcal{F}_2}(C_T(f)) \leq N_{\mathcal{F}_2}(T)$, so the proof is complete. □

In light of 12.2.22, during the remainder of the proof we assume:

HYPOTHESIS 12.2.23. *Hypothesis* 12.2.12 *holds, as does case (2) or (3) of* 12.2.13.

NOTATION 12.2.24. It is now convenient to reverse our convention in 11.1.2 that Ω is $\{K_1, K_1', K_3, K_3', \ldots\}$; instead we take

$$\Omega = \{K_n, K_{n-1}, K_{n-3}, K_{n-3}', \ldots\}.$$

Subject to this change $x = e_1$ is an element of $\mathcal{F}_f$ of weight 1 in W_I contained in $\mathcal{F}^f$. By 12.2.23 and 12.2.13, $\mathcal{F}_x^\circ \cong \Omega_{2(n-1)+a}^\epsilon[m]$ for $a = 1$ or 2, and with $\epsilon = -1$ if $a = 2$. Set $s_i = z(K_i)$ for $1 \leq i \leq n$. Then $s_1 = e_{1,2}$ so $s_1 x = e_2 \in M_x$ and $e_2 \in V_2V_2'$.

Much as in Notation 11.5.2, we can form the $(2(n-1)+a)$-dimensional orthogonal space $(U, \mathfrak{q})$ (of sign -1 if $a = 2$) over a suitable field of order congruent to 1 modulo 4. Set $L = \Omega(U, \mathfrak{q})$ and regard $\mathcal{F}_x^\circ$ as $\mathcal{F}_{T_x}(L)$. This time we write $U = U_2 \perp U_3 \perp \cdots \perp U_n \perp U_0$, where for $2 \leq i \leq n$, U_i is a 2-dimensional subspace of U of sign $+1$, and U_0 is of dimension a and sign -1 if $a = 2$. Further we choose notation so that $U_i + U_{i+1} = [U, s_i]$ for $2 \leq i < n$. Set $A = U_n + U_0$, $A_0 = U_3 + \cdots + U_0$, and $A_1 = U_{n-1} + U_n + U_0$. Then $s_1 x$ is the involution in M_x with $[U, s_1 x] = U_2$. In particular $\mathcal{C}_1 = C_{\mathcal{F}_x}(s_1)^\circ = C_{\mathcal{F}_x}(s_1 x)^\circ = \mathcal{F}_{T_{A_0}}(L_{A_0}) \cong \Omega_{2(n-2)+a}^\epsilon[m]$.

Set $W_X = V_2V_2' \cdots V_{n-2}V_{n-2}'$. Then $[U, W_X] = A^\perp$, so $\mathcal{X} = O^2(C_{\mathcal{F}_x^\circ}(W_X)) = \mathcal{F}_X(L_A) \cong L_2[m]$, $L_2[2m]$ for $a = 1, 2$, respectively, and for X Sylow in $\mathcal{X}$. Further as $s_1 x = e_2 \in V_2V_2'$, $\mathcal{X} \leq \mathcal{C}_1$, so X centralizes $K_1 \leq M$. By construction, X centralizes $\langle K_2, \ldots, K_{n-2}\rangle \leq M$.

We summarize our observations:

LEMMA 12.2.25. *(1)* $\mathcal{X} = O^2(C_{\mathcal{F}_x^\circ}(W_X)) \cong L_2[m]$, $L_2[2m]$ *for* $a = 1, 2$ *respectively.*

(2) The Sylow group X *of* $\mathcal{X}$ *centralizes* $\langle K_1, \ldots, K_{n-2} \rangle \leq M$.

NOTATION 12.2.26. Suppose $f = f_\xi$ with $\xi = \mathcal{Z}_S$. Then by 12.2.14, S is transitive on ξ and then by 12.2.18.3, $n = 2|\xi| + 1$ is odd. In this case replace $f = f_\xi$ by $f = f_\theta$ with $|\theta| = (n-1)/4$, and regard $\mathcal{F}$ as $\langle f, f_{\mathcal{Z}_S}, N_{\mathcal{F}}(W) \rangle$, a case that must be handled in case II of 12.2.17.

Subject to this convention, we may choose notation so that $f = e_{\theta_1}$ with $\theta_1 \subseteq \{1, \ldots, n-2\}$. Thus f centralizes L_{A_1}, and hence also $\mathcal{X}_1 = \mathcal{F}_{X_1}(L_{A_1}) \cong \Omega_{4+a}^\epsilon[m]$, where X_1 is Sylow in L_{A_1}. Then $\mathcal{X}_1 = \mathcal{X}_1^\circ \leq \mathcal{F}_2$.

As in 11.5.2, let Λ_1 be an index set of order a. From 9.4.22, 9.4.27, and 9.4.28, there are E_8-subgroups $B_{1,\lambda}$ of X_1 such that $\mathcal{X}_1 = \langle X_1 \mathcal{O}_{s_{n-1}}, Aut_{\mathcal{X}}(C_{X_1}(B_{1,\lambda}) : \lambda \in \Lambda_1 \rangle$. If n is even, set $\Lambda = \Lambda_1$ and $B_\lambda = B_{1,\lambda}$. If n is odd, set $\Lambda = \{1\}$ and let B_λ be a 4-subgroup of X. In either case set $R_\lambda = C_T(B_\lambda)$ and $\mathcal{Y} = \langle T\mathcal{O}_\tau, \mathcal{F}_T(TM), Aut_{\mathcal{X}}(R_\lambda) : \lambda \in \Lambda \rangle$.

LEMMA 12.2.27. *(1)* $T_f = T_{M_f}X$ *with* $|T_f : T_{M_f}| = 2^a = |T : T_M|$.

(2) If $a = 1$ *then* $TM \cong \omega(C_n, m)$.

(3) If $a = 2$ *then* $TM \cong 2\omega(C_n, m)$.

PROOF. By 12.2.18.2, $|T_f : T_{M_f}| = 2^a$, and then as $T = T_M T_f$ with $T_{M_f} = T_M \cap T_f$, (1) holds.

Suppose $a = 1$. Then $\mathcal{X} \cong L_2[m]$ so X is dihedral of order $m/2$ and by induction on n, $X \cap T_M$ is cyclic. Let $r \in X - T_M$. From the structure of $\Omega_{2(n-1)+a}[m]$, $K_{n-1}^r = K_n$, while r centralizes $\langle K_1, \ldots, K_{n-2} \rangle$ by 12.2.25.2, so $TM = \langle r \rangle M \cong \omega(C_n m)$ by 5.9.2.5. Therefore (2) holds.

So assume $a = 2$. Then $\mathcal{X} \cong L_2[2m]$, so $X \cong D_m$. By induction on n, the cyclic subgroup $\langle y \rangle$ of X of index 2 centralizes $V_{n-1}V_n$, and we still have $K_{n-1}^r = K_n$. Thus $TM = \langle r, y \rangle M \cong 2\omega(C_n, m)$ by 5.9.3 and an argument in the proof of 5.3.10, establishing (3). □

LEMMA 12.2.28. *(1) For* $\lambda \in \Lambda$, $R_\lambda \leq C_T(f)$.

(2) $Aut_{\mathcal{X}}(R_\lambda)$ *is uniquely determined by* TM *and* Ω.

(3) $\mathcal{Y} \cong \Omega_{2n+a}^\epsilon[m]$.

PROOF. By 12.2.18.5, either $f \in Z(S)$ or 12.2.17.II holds and $|S : C_S(f)| = 2$. In the first case (1) is trivial. In the second, $\mathcal{X}_1 \leq \mathcal{F}_2$ so $C_S(B_\lambda)$ acts on θ_1 and θ_2, and hence $C_S(B_\lambda) \leq C_S(f)$ so (1) holds in this case too.

By induction on n, $\mathcal{F}_x^\circ$ is uniquely determined by $T_x M_x$ and Ω_x. From 12.2.24, we may regard $\mathcal{F}_x^\circ$ as $\mathcal{F}_{T_x}(L)$, and use the representation of L on $(U, \mathfrak{q})$ described in 12.2.24 to calculate in $\mathcal{F}_x^\circ$. From the proof of 12.2.20, $W = W_f V_i V_i'$, where $i = k, k+1$ for n even, odd, respectively. Further by 12.2.18, $n - k \geq 2, 3$ for n even, odd, respectively, so $n \geq i + 2$. Therefore, using the representation of L on $(U, \mathfrak{q})$, we conclude that $Y_i = K_i K_i' \cap S$ centralizes L_A, and hence also $\mathcal{X}$. As $\mathcal{F}_x^\circ$ is determined by $T_x M_x$ and Ω_x, the fact that Y_i centralizes $\mathcal{X}$ is determined by TM and Ω.

Next μ_0 is the core of the permutation module for $\tilde{\mu} \cong S_n$ by 11.1.2, and similarly $\mu_{i,0} = \mu_i \cap \mu_0$ is the core of the permutation module for $\tilde{\mu}_i$, $i = 1, 2$. Therefore $|\mu_0| = 2^{n-1}$, $|\mu_{1,0}| = 2^{k-1}$, and $|\mu_{2,0}| = 2^{n-k-1}$. Then as $\mu_{1,0} \cap \mu_{2,0} = 1$, $\mu(\tau_f)_0 = \mu(\tau_f) \cap \mu_0$ is of order $2^{k-1} \cdot 2^{n-k-1} = |\mu_0|/2$, and indeed $\mu_0 = \mu(\tau_f)_0 Y_i W/W$. Finally $\widetilde{C_T(f)} = \tilde{T}_f$, so $C_T(f) = Y_i T_f$. Therefore $R_\lambda = Y_i T_1 R_2$ where $R_2 = C_{T_2}(B_\lambda)$. As $\mathcal{X}_1 \leq \mathcal{F}_2$, $Aut_{\mathcal{X}}(R_\lambda)$ centralizes T_1, and thus centralizes $Y_i T_1$ by earlier remarks. Further if $n-k \geq 5$, then working in $\mathcal{F}_2$ by induction on n, $Aut_{\mathcal{X}}(R_2)$ is determined by $T_2 M_2$ and Ω_2, so (2) holds in this case. So suppose $n-k \leq 4$. Recall from above that $n-k \geq 2$.

If $n-k = 2$ then $\mathcal{F}_2 = \mathcal{X}_1$ and $Aut_{\mathcal{X}}(R_2)$ is determined by the proofs of 9.4.30 and 9.4.31. If $n-k = 3$ then $\mathcal{F}_2 \cong \Omega_7[m]$ or $\Omega_8^-[m]$ and $Aut_{\mathcal{X}}(R_2)$ is determined by the proofs of 9.5.29 and 9.5.32. Finally if $n-k = 4$ then $\mathcal{F}_2$ is $\Omega_9[m]$ or $\Omega_{10}^-[m]$ and $Aut_{\mathcal{X}}(R_2)$ is determined by the proof of 12.1.25. Therefore (2) holds.

Let $\tilde{\mathcal{F}} = \Omega_{2n+a}^\epsilon[m]$, $\tilde{\tau} = (\tilde{\mathcal{F}}, \tilde{\Omega})$ the Lie packet of $\tilde{\mathcal{F}}$, $\tilde{\eta} \in \eta(\tilde{\tau})$, etc. By 5.3.9 and 5.3.10, a model for $N_{\tilde{\mathcal{F}}}(\tilde{W})$ is isomorphic to $a\omega(C_n, m)$, so by 12.2.27 we may take $T = \tilde{S}$, $\tilde{M} = M$, $\tilde{\Omega} = \Omega$, etc. By 11.5.3, $\tilde{\mathcal{F}} = \langle TO_{\tilde{\tau}}, \mathcal{F}_T(TM), Aut_{\tilde{\mathcal{X}}}(R_\lambda) : \lambda \in \Lambda \rangle$. By Theorem 2.7.3, $TO_{\tilde{\tau}} = TO_\tau$, and by (2), $Aut_{\tilde{\mathcal{X}}}(R_\lambda) = Aut_{\mathcal{X}}(R_\lambda)$, so $\tilde{\mathcal{F}} = \mathcal{Y}$. Thus (3) holds. □

NOTATION 12.2.29. Let $\mathcal{G}_1 = \mathcal{F}_f$ and $\mathcal{E}_1 = C_T(f)\mathcal{F}_f^\circ$. If case I of 12.2.17 holds, set $\mathcal{G}_2 = N_{\mathcal{F}}(W)$ and $\mathcal{E}_2 = \mathcal{F}_T(TM)$. If case II of 12.2.17 holds set $\mathcal{G}_2 = \mathcal{F}_c$ and $\mathcal{E}_2 = T\mathcal{F}_c^\circ$.

LEMMA 12.2.30. *(1)* $\mathcal{F} = \langle \mathcal{G}_1, \mathcal{G}_2 \rangle$.
(2) $\mathcal{E}_i \trianglelefteq \mathcal{G}_i$ *for* $i = 1, 2$.
(3) $\mathcal{Y} = \langle \mathcal{E}_1, \mathcal{E}_2 \rangle$.

PROOF. Suppose case I of 12.2.17 holds. Then (1) follows from 12.2.16. Also $\mathcal{E}_1 \trianglelefteq \mathcal{G}_1$ by 12.2.20.2. By 3.3.13, $\mathcal{F}_{T_M}(M) \trianglelefteq N_{\mathcal{F}}(M) = \mathcal{N}$, so to complete the proof of (2) it suffices by 1.5.2 in [**Asc19**] to show that $T \trianglelefteq N_{\mathcal{N}}(T_M)$. By 11.1.3.5, $Aut_{\mathcal{N}}(W_0) = Aut_G(W_0) = Aut_M(W_0)$, so $Aut_{N_{\mathcal{N}}(T_M)}(W_0) = Aut_{T_M}(W_0)$, and hence $N_{\mathcal{N}}(T_M) \leq T_M C_{\mathcal{N}}(W_0)$. Finally $C_{\mathcal{N}}(W_0) \leq C_{\mathcal{N}}(f)$, and we just saw that $C_T(f) \trianglelefteq C_{N_{\mathcal{N}}(T_M)}(f)$, so $\mathcal{F}_T(TM) \trianglelefteq N_{\mathcal{F}}(W)$, completing the proof of (2) in this case.

Next by 1.3.2 in [**Asc19**], $\mathcal{E}_1 = C_T(f)\mathcal{F}_f^\circ = \langle C_T(f)\mathcal{F}_1, C_T(f)\mathcal{F}_2 \rangle$, and by 11.5.3, $C_T(f)\mathcal{F}_i \leq \mathcal{Y}$, so $\mathcal{E}_1 \leq \mathcal{Y}$. By definition, $\mathcal{E}_2 = \mathcal{F}_T(TM) \leq \mathcal{Y}$, so $\langle \mathcal{E}_1, \mathcal{E}_2 \rangle \leq \mathcal{Y}$. Conversely $\langle TO_\tau, Aut_{\mathcal{X}}(R_\lambda) : \lambda \in \Lambda \rangle \leq \mathcal{E}_1$, so $\mathcal{Y} = \langle \mathcal{E}_1, \mathcal{E}_2 \rangle$, establishing (3) in this case.

So assume case II of 12.2.17 holds. Then $\xi = \mathcal{Z}_S$ and by 12.2.16, $\mathcal{F} = \langle \mathcal{F}_f, \mathcal{F}_y, N_{\mathcal{F}}(W) \rangle$, where $y = f_\xi$. If n is even, $y = c$ generates $Z(\mathcal{F}_S(M)) \trianglelefteq N_{\mathcal{F}}(W)$, so $\mathcal{F} = \langle \mathcal{G}_1, \mathcal{G}_2 \rangle$ and (1) holds. If n is odd then by 1.3.2 in [**Asc19**], $\mathcal{F}_y = \langle S\mathcal{F}_y^\circ, N_{\mathcal{F}}(T_y) \rangle$. By 11.5.3, $S\mathcal{F}_y^\circ = \langle SO_\tau, \mathcal{F}_S(SM_y) \rangle$, so as $SO_\tau \leq \mathcal{F}_c \geq N_{\mathcal{F}}(W) \geq \mathcal{F}_S(SM)$, we have $S\mathcal{F}_y^\circ \leq \mathcal{F}_c$. Thus to prove (1), it suffices to show that $N_{\mathcal{F}_y}(W_y) \leq N_{\mathcal{F}}(W)$. Let $\mathcal{L} = C_{\mathcal{F}_y}(W_y)$ and L Sylow in $\mathcal{L}$. Now $W_y = V_1 \cdots V_n$, so $W = V_1 W_y$. By 1.3.2 in [**Asc19**], $N_{\mathcal{F}_y}(W_y) = \langle S\mathcal{L}, N_{\mathcal{F}_y}(L) \rangle$, so as W is weakly closed in S with respect to $\mathcal{F}$, if $W \trianglelefteq W\mathcal{L}$ then $N_{\mathcal{F}_y}(W_y) \leq N_{\mathcal{F}}(W)$, proving (1). We may take $K = K_n$, so $C_{\mathcal{F}_y}(W_y) \leq \mathcal{F}_z \leq N_{\mathcal{F}}(\mathcal{C})$. Further $W_{\mathcal{C}} = V_1 \cdots V_{n-3} V_{n-3}'$ and $W_{\mathcal{C}} \trianglelefteq N_{N_{\mathcal{F}}(\mathcal{C})}(W_y)$, so indeed $W \trianglelefteq W\mathcal{L}$, completing the proof of (1).

Recall $\mathcal{E}_1 \trianglelefteq \mathcal{G}_1$ by 12.2.20.2, so to prove (2) we must show that $T \trianglelefteq N_{\mathcal{F}_c}(T_c)$. But $W \leq T_M = T_c$, so $N_{\mathcal{F}_c}(T_c) \leq N_{\mathcal{F}}(W)$. Then as we showed that $T \trianglelefteq N_{N_{\mathcal{F}}(W)}(T_M)$, the proof of (2) is complete.

Let $\mathcal{Y}_0 = \langle \mathcal{E}_1, \mathcal{E}_2 \rangle$. First $\langle T\mathcal{O}_\tau, \mathcal{F}_T(TM) \rangle \leq T\mathcal{F}_c^\circ = \mathcal{E}_2$ and $\langle Aut_{\mathcal{X}}(R_\lambda) : \lambda \in \Lambda \rangle \leq \mathcal{E}_1$, so $\mathcal{Y} \leq \mathcal{Y}_0$. By definition of $\mathcal{Y}$, $\mathcal{E}_1 \leq \mathcal{Y}$. By 11.5.3 and 11.1.6, $\mathcal{E}_2 = T\mathcal{F}_c^\circ = \langle T\mathcal{O}_\tau, \mathcal{F}_T(TM) \rangle \leq \mathcal{Y}$, so $\mathcal{Y}_0 = \mathcal{Y}$, establishing (3). □

LEMMA 12.2.31. $\mathcal{F}^\circ \cong \Omega^\epsilon_{2n+a}[m]$.

PROOF. The proof is much like that of 12.2.22. We show that $\mathcal{Y} \trianglelefteq \mathcal{F}$. Then $\mathcal{F}^\circ = \mathcal{Y}$, and 12.2.28.3 completes the proof. To prove $\mathcal{Y} \trianglelefteq \mathcal{F}$, we verify that the tuple $\mathcal{E}_i, \mathcal{G}_i$, $i = 1, 2$, satisfies Hypothesis 1.4.3 in [**Asc19**], and then appeal to 1.4.5 in [**Asc19**].

By 12.2.30.1, $\mathcal{F} = \langle \mathcal{G}_1, \mathcal{G}_2$, so Hypothesis 1.4.1 in [**Asc19**] is satisfied. By 12.2.30.2, $\mathcal{E}_i \trianglelefteq \mathcal{G}_i$ for $i = 1, 2$, so 1.4.3.1 in [**Asc19**] holds. As $T = T_M$ and W and $\mathcal{O}(\tau)$ are normal in $N_{\mathcal{F}}(T_M)$, $Aut_{\mathcal{F}}(T)$ is contained in $Aut(T\mathcal{O}_\tau)$ and $Aut(\mathcal{F}_T(M))$, so 1.4.3.2 follows from 1.4.6 in [**Asc19**]. Except in case II with $i = 1$, T is Sylow in $\mathcal{E}_i$, so the extension condition 1.4.3.3 is satisfied. Thus Hypothesis 1.4.3 in [**Asc19**] is satisfied except possibly in case II. Finally $\mathcal{Y}$ is saturated with $\mathcal{Y} = O^{2'}(\mathcal{Y})$ by 12.2.28.3. Therefore Hypothesis 1.4.3 in [**Asc19**] is satisfied and the lemma holds, except in case II where we must verify the extension condition when $i = 1$. But if $\alpha \in Aut_{\mathcal{F}_f}(C_T(f))$ then as $W \leq C_T(f)$, $\alpha \in Aut_{\mathcal{F}_2}(C_T(f))$. Further $Aut_{\mathcal{F}_2}(C_T(f)) \leq N_{\mathcal{F}_2}(T)$, so the proof is complete. □

We close this section with a theorem summarizing the major lemmas in the section.

THEOREM 12.2.32. *Assume Hypothesis* 11.1.1 *with* $n \geq 5$. *Define* $c = e_I$ *as in* 11.1.2. *If* $1 \neq c \in Z(\mathcal{F})$ *assume the Extended Inductive Hypothesis holds. Then one of the following holds:*

(1) $1 \neq c \in Z(\mathcal{F})$ *and either* $\mathcal{F}^\circ = \mathcal{F}_{S \cap M}(M) \cong \omega(D_n, m)$ *or* $\mathcal{F}^\circ \cong \check{A}E_{2n}$ *or* $\Omega^+_{2n}[m]$.

(2) $\mathcal{F}^\circ = \mathcal{F}_{S \cap M}(M) \cong \omega(D_n, m)/Z(\omega(D_n, m))$.

(3) $\mathcal{F}^\circ \cong AE_{2n}$ *or* AE_{2n+1}.

(4) $\mathcal{F}^\circ \cong P\Omega^+_{2n}[m]$.

(5) $\mathcal{F}^\circ \cong \Omega_{2n+1}[m]$.

(6) $\mathcal{F}^\circ \cong \Omega^-_{2n+2}[m]$.

PROOF. If $1 \neq c \in Z(\mathcal{F})$ then (1) holds by 11.1.6.2, so we may assume Hypothesis 12.2.7. If $\mathcal{O} = O$ then (2) holds by 12.2.8.2 and 12.2.10, while if $\mathcal{C}$ is $\check{A}E_{2n-4}$ or AE_{2n-3} then (3) holds by 12.2.11. Therefore by 12.2.9.3 we may assume Hypothesis 12.2.12. If 12.2.13.1 holds then (4) holds by 12.2.22, so we may assume Hypothesis 12.2.23. Now 12.2.31 completes the proof of the theorem. □

12.3. Completing $|\Omega(z)| = 2$

In this section we assume the following hypothesis:

HYPOTHESIS 12.3.1. *(1) $\tau = (\mathcal{F}, \Omega)$ is a quaternion fusion packet such that $\mathcal{A}(\tau) = \emptyset$ and for $z \in \mathcal{Z}_S$, $|\Omega(z)| = 2$.*

(2) $\mathcal{F}^\circ$ is transitive on Ω.

(3) For $z \in \mathcal{Z}_S \cap \mathcal{F}^f$, $\mathcal{D}^*(z) \neq \mathcal{D}(z)$ and $\mathcal{Z} \cap O(z) = \{z\}$.

(4) τ satisfies the Extended Inductive Hypothesis.

LEMMA 12.3.2. *Hypothesis 9.1.8 is satisfied.*

PROOF. Immediate from the definitions. □

NOTATION 12.3.3. Adopt Notation 9.1.2 and 9.1.6, plus the notation in 9.1.9. Set $O = O(z)$. By 3.3.14 there is a unique member η of $\eta(\tau)$. From 9.1.6, $W = \langle \eta \rangle$, G is a model for $N_{\mathcal{F}}(C_S(W))$, $M = \langle K^G \rangle \in M(\tau)$, and $\mu = \mu(\tau)$.

LEMMA 12.3.4. *Let η_i, $1 \leq i \leq r$, be the orbits of M on η, and for $1 \leq i \leq r$ let $\Omega_i = \{J \in \Omega : J \cap W \in \eta_i\}$, $M_i = \langle J_i^M \rangle$ for $J_i \in \Omega_i$, $S_i = S \cap M_i$, $\zeta_i = (\mathcal{F}_{S_i}(M_i), \Omega_i)$, and $\mu_i = \mu(\zeta_i)$. Then*

(1) $M = M_1 * \cdots * M_r$ is a central product.

(2) $\mu = \mu_1 \times \cdots \times \mu_r$ with $\mu_i \cong Weyl(D_n)$ for some $n \geq 3$.

(3) G is transitive on $\{M_1, \ldots, M_r\}$ and $N_G(S \cap M)$ is transitive on Ω.

PROOF. This follows from 9.1.9. □

NOTATION 12.3.5. Adopt the notation of 12.3.4, and choose notation so that some $z \in \mathcal{Z} \cap \mathcal{F}^f$ is in $\mathcal{Z}(\zeta_1)$ and let $K \in \Omega(z)$. If $m > 8$ set $\mathcal{O}_z = \mathcal{O}(z)$. If $m = 8$ let G_z be a model for $N_{\mathcal{F}_z}(OC_S(O))$, $H_z = \langle W^{G_z} \rangle$, and $L_z = O^2(H_z)$, except when $n = 4$ write L_1 for $O^2(H_z)$. Let $x = z$ when $n \neq 4$, and set $x = 1$ when $n = 4$. When $L_x = 1$ set $\mathcal{O}_x = O$ and when $L_x \neq 1$ set $\mathcal{O}_x = \mathcal{F}_{O_2(L_x)}(L_x)$.

LEMMA 12.3.6. *$\mathcal{F}_z$ is transitive on $\mathcal{D}^*(z) - \mathcal{D}(z)$.*

PROOF. See 9.1.10.3. □

LEMMA 12.3.7. *Suppose $m = 8$, let $x = z$ when $n \neq 4$, and $x = 1$ when $n = 4$. Assume $L_x \neq 1$. Then:*

(1) $L_x = O^2(\langle W_1^{G_z} \rangle)$.

(2) If $n > 4$ then L_z centralizes $C_S(O)$, $O = O_2(L_z)$, and L_z is of index 1 or 3 in $SL_2(3) * SL_2(3)$.

(3) If $n = 4$ then L_1 centralizes $C_S(O(\zeta_1))$, $O(\zeta_1) = O_2(L_1)$, and L_1 is of index 3^e in $(SL_2(3) * SL_2(3))^2$ for some $0 \leq e \leq 3$.

(4) If $n = 3$ then either

(i) L_z centralizes $C_S(O)$, $O = O_2(L_z)$, and L_z is of index 1 or 3 in $SL_2(3) * SL_2(3)$, or

(ii) $O_2(L_z) = O \times B_z$ where $B_z = \langle b^{L_z} \rangle \cong E_4$, $b = za$ for $\langle a \rangle = Z(M_1)$, and $L_z = O_2(L_z)X_z$, where $X_z \in Syl_3(L_z)$, $X_z \cong E_{3^e}$ for some $1 \leq e \leq 3$, and $Z(L_z) = \langle z \rangle$. Moreover $C_S(O) = B_z \times C_S(L_z)$ and $C_S(L_z)$ contains $W_2 \cdots W_r$ and any L_z-invariant subgroup of $C_S(O)$ centralized by W_1.

PROOF. Set $S_0 = C_S(O)$ and observe $W = W_1 * U$ where $U = W_2 \cdots W_r \leq S_0$ and $U = \langle \eta - \eta_1 \rangle = \langle V \in \eta : V \leq S_0 \rangle$. Thus $U \trianglelefteq G_z$, so as W centralizes U, so does H_z. Thus, setting $H_1 = \langle W_1^{G_z} \rangle$, we have $H_z = H_1 U$ with $U \leq Z(H_z)$, so $L_x = O^2(H_z) = O^2(H_1)$, establishing (1).

Parts (2) and (3) follow from 9.1.11.2, so we may take $n = 3$.

Let $\mathcal{A}$ be the set of L_z-invariant subgroups of G_z with $A \leq C_{S_0}(W_1)$. Then H_1 centralizes each $A \in \mathcal{A}$, so L_z centralizes A by (1). In particular the last statement in (4ii) holds. For the remainder of (4), we appeal to (1) and argue as in the proof of 9.5.11. □

THEOREM 12.3.8. *Assume Hypothesis* 12.3.1 *Then* $r = 1$, *so* $\mu \cong Weyl(D_n)$ *for some* $n \geq 3$.

Until the proof of Theorem 12.3.8 is complete, (near the end of the section) we assume that $r > 1$.

LEMMA 12.3.9. $\mathcal{F}_z^\circ = \mathcal{O}_z * \mathcal{C}_z * \mathcal{C}_2 * \cdots * \mathcal{C}_r$ *where*

(1) For $2 \leq i \leq r$, $\rho_i = (\mathcal{C}_i, \Omega_i)$ *is a quaternion fusion packet,* $\eta_i \in \eta(\rho_i)$, *and* $M_i \in M(\rho_i)$, *so* $\mu_i = \mu(\rho_i)$.

(2) If $n = 3$ *then* $\mathcal{C}_z = 1$, *while if* $n > 3$ *then* $\rho_z = (\mathcal{C}_z, \Omega_z)$ *is a quaternion fusion packet, where* $\Omega_z = \Omega_1 - \Omega(z)$, $\eta_z \in \eta(\rho_z)$, *and* $M_z \in M(\rho_z)$, *where* $\eta_z = \{V \in \eta_1 : V \leq C_S(O)\}$ *and* $M_z = \langle \Omega_z^{C_{M_1}(z)} \rangle$. *Further* $\mu_z = \mu(\rho_z) \cong Weyl(D_{n-2})$.

(3) $(\mathcal{O}_{K_j}, K_j)$, $j = 1, 2$, ρ_z, *and* ρ_i, $2 \leq i \leq r$, *are the coconnected components of* τ_z, *unless* $n = 4$ *and* $\mathcal{C}_z = \mathcal{O}(t)$, *where* $\mathcal{Z}_S(\zeta_1) = \{z, t\}$.

PROOF. See the proof of 9.3.10.1. □

NOTATION 12.3.10. Let $I = \{1, \ldots, r\}$ and $I_1 = I - \{1\}$. For $i \in I_1$, let $z_i \in \mathcal{Z}(\rho_i)$ with $z_i \in \mathcal{C}_i^f$, and, when $n > 3$, set $\mathcal{D}_i = C_{\mathcal{C}_i}(O(z_i))^\circ$, $\Gamma_i = \Omega_i - \Omega(z_i)$, and $\delta_i = (\mathcal{D}_i, \Gamma_i)$.

LEMMA 12.3.11. *Assume* $m = 8$, $n \neq 4$, *and if* $n = 3$ *and* $L_z \neq 1$ *assume* 12.3.7.*4.i holds.*

(1) $\mathcal{O}_z \trianglelefteq \mathcal{F}_z$.

(2) For $s \in \mathcal{Z}_S$ *let* $\phi_s \in Aut_{\mathcal{F}}(O(\tau))$ *with* $z\phi_s = z_s$, *and set* $\mathcal{O}_s = \mathcal{O}_z \phi_s^*$. *Then* $\mathcal{O}_s$ *is independent of the choice of* ϕ_s *and* $N_{\mathcal{F}}(O(\tau))$ *contains a normal central product* $\mathcal{O}_\tau = \prod_{s \in \mathcal{Z}_s} \mathcal{O}_s$.

PROOF. Set $S_0 = C_S(O)$. By 12.3.7, L_z centralizes S_0 and O is Sylow in L_z, so by 2.2 in [**Asc11**], $\mathcal{F}_z$ contains a central product $\mathcal{E} = \mathcal{O}_z * C_{\mathcal{F}}(O)$. Then $N_{\mathcal{E}}(S_0)$ has model $L_z C_{G_z}(O) \trianglelefteq G_z$ as $L_z \trianglelefteq G_z$, so $N_{\mathcal{E}}(S_0) \trianglelefteq N_{\mathcal{F}_z}(S_0)$. Therefore $\mathcal{E} \trianglelefteq \mathcal{F}_z$ by 1.5.1 in [**Asc19**]. Next $\mathcal{E}_0 = C_{\mathcal{E}}(C_{\mathcal{F}}(O)) = Z(C_{\mathcal{F}}(O))\mathcal{O}_z$, so $\mathcal{O}_z = O^2(\mathcal{E}_0) \trianglelefteq \mathcal{F}_z$, proving (1).

By (1), $\mathcal{O}_s$ is independent of the choice of ϕ_s in (2). As L_z centralizes S_0, $\mathcal{N} = N_{\mathcal{F}}(O(\tau))$ contains the central product $\mathcal{O}_\tau$ by 2.2 in [**Asc11**]. Then as $Aut_{\mathcal{F}}(O(\tau))$ permutes $\{O(s) : s \in \mathcal{Z}_S\}$, we have $\mathcal{O}_\tau \trianglelefteq \mathcal{N}$. □

LEMMA 12.3.12. *Assume* $m = 8$ *and* $n = 4$. *For* $i \in I$ *set* $O_i = O(\zeta_i)$ *and* $Z_i = Z(O_i)$.

(1) $\mathcal{O}_1 \trianglelefteq C_{\mathcal{F}}(Z_1)$.

(2) $Aut_{\mathcal{F}}(O(\tau))$ *is transitive on* $\{O_i : i \in I\}$.

(3) For $i \in I$ let $\phi_i \in Aut_{\mathcal{F}}(O(\tau))$ with $O_1\phi_i = O_i$, and set $\mathcal{O}_i = \mathcal{O}_1\phi_i^*$. Then $\mathcal{O}_i$ is independent of the choice of ϕ_i and $N_{\mathcal{F}}(O(\tau))$ contains a normal central product $\mathcal{O}_\tau = \prod_{i\in I}\mathcal{O}_i$.

PROOF. The proof of 12.3.11 establishes (1) and (3). Next $A = Aut_{\mathcal{F}}(O(\tau))$ is transitive on $\{O(s) : s \in \mathcal{Z}_S\}$ as $\mathcal{F}$ is transitive on Ω. By 12.3.9, $\{\{O(z_i), O(t_i)\} : i \in I\}$ is a system of imprimitivity for A, where $\mathcal{Z}_S(\zeta_i) = \{z_i, t_i\}$. Therefore (2) holds. □

LEMMA 12.3.13. *If $(m,n) = (8,3)$ and $L_z \neq 1$ assume 12.3.7.4.i holds. Then for $i \in I_1$, one of the following holds:*
(1) $\mathcal{O}_z = O$ and $\mathcal{C}_i = \mathcal{F}_{S_i}(M_i)$.
(2) $m = 8$, $n > 4$, $O = O_2(L_z)$, L_z is of index 3 in $SL_2(3) * SL_2(3)$, and $\mathcal{C}_i$ is AE_{2n}, $\breve{A}E_{2n}$, or AE_{2n+1}.
(3) $n > 4$ if $m = 8$, $\mathcal{O}_z \cong SL_2[m] * SL_2[m]$, O is Sylow in $\mathcal{O}_z$, and $\mathcal{C}_i$ is $(P)\Omega_{2n}^+[m]$, $\Omega_{2n+1}[m]$, or $\Omega_{2n+2}^-[m]$.
(4) $n = 4$, $m = 8$, $O(\zeta_1) = O_2(L_1)$, L_1 is of index 3^e in $(SL_2(3) * SL_2(3))^2$, and one of the following holds:
(i) $\mathcal{C}_i$ is $\breve{A}E_8$ or AE_9 and $e = 2$.
(ii) $\mathcal{C}_i$ is $\Omega_8^+[8]$, $\Omega_9[8]$, or $\Omega_{10}^-[8]$ and $e = 0$.
(iii) $\mathcal{C}_i$ is $\breve{L}_3(2)/2^{3+6}$ and $e = 3$.
(iv) $\mathcal{C}_i$ is $\breve{\Omega}_8^+(2)$ and $e = 1$.
(5) $n = 3$, $m = 8$, 12.3.7.4.i holds, L_z is of index 3^e in $SL_2(3) * SL_2(3)$, and either
(a) $\mathcal{C}_i \cong \breve{A}E_6$ and $e = 1$, or
(b) $\mathcal{C}_i \cong \Omega_6^+[8]$ or $\Omega_8^-[8]$ and $e = 0$.

PROOF. First observe that $\Gamma = \Omega - \Omega_1$ centralizes $\mathcal{O}_z$. If $m > 8$ this follows from 2.6.12, while if $m = 8$ it follows from 12.3.7.

Now for $\phi \in Aut(O(\tau)W)$, $\Gamma\phi$ centralizes $\mathcal{O}_{z\phi}$, and by 12.3.11 we may choose ϕ so that $z\phi = z_i$ and $\mathcal{O}_z\phi^* = \mathcal{O}_{z_i}$, and hence z centralizes $\mathcal{O}_{z_i}$ (or $\mathcal{O}_i$ if $n = 4$ and $m = 8$ by 12.3.12). Also $W_1\phi = W_i$, so $\mathcal{O}_{z_i}$ is contained in the normal closure $\mathcal{P}_i$ of W_i in $C_{\mathcal{C}_i}(z_i)$ (or in $C_{\mathcal{C}_i}(Z_i)$ if $n = 4$ and $m = 8$). Further for $\alpha_i \in \mathfrak{A}(z_i)$ with $z_i\alpha = z$, $\mathcal{P}_i\alpha_i^* \leq \mathcal{O}_{z_i}$. We conclude that $\mathcal{P}_i = \mathcal{O}_{z_i}$. Then the lemma follows from 12.3.1.4 (together with Theorems 9.5.33, 12.1.25, and 12.2.32 for $n = 3$, $n = 4$, and $n > 4$, respectively), since by 12.3.9.1, ρ_i is a quaternion fusion packet with $\mu_i = \mu(\rho_i)$. □

LEMMA 12.3.14. *Assume $n > 3$. Then*
(1) for all $i, j \in I_1$, $\mathcal{C}_i \cong \mathcal{C}_j$.
(2) For all $i \in I_1$ and $\alpha_i \in \mathfrak{A}(z_i)$ with $z_i\alpha_i = z$, we have $\mathcal{D}_i\alpha_i^* = \mathcal{C}_z$, so $\mathcal{D}_i \cong \mathcal{C}_z$.

PROOF. For $\beta, \gamma \in \{\delta_i : i \in I_1\} \cup \{\rho_z\}$, write $\beta \lesssim \gamma$ if β is isomorphic to a subpacket of γ.

Let $i \in I_1$. Then there is $\alpha \in \mathfrak{A}(z_i)$ with $z_i\alpha = z$. For $j \in J = I_1 - \{i\}$, $\rho_j\alpha^* \leq \rho_{j\sigma}$ for some injection $\sigma : J \to I_1$ by 12.3.9 and as ρ_j is coconnected. Similarly $\delta_i\alpha^* \leq \rho_z$ and $\rho_z\alpha^* \leq \delta_k$, where $I_1 = J\sigma \cup \{k\}$. Thus for each $i \in I_1$, $\delta_i \lesssim \rho_z \lesssim \delta_k$.

Let $I_0 = \{i_0 \in I_1 : \delta_{i_0} \cong \rho_z\}$. Suppose that $i \in I - I_0$. Then $I_0\sigma \subseteq I_0$, so as σ is an injection, we have $I_0\sigma = I_0$. But this is a contradiction as also $\rho_z \lesssim \delta_k \lesssim \rho_z$, so that $k \in I_0$. This contradiction proves (2). Then (2) and 12.3.13 imply (1), except perhaps in the case where 12.3.13.4 holds with $\mathcal{C}_i \cong \check{A}E_8$, $\Omega_8^+[8]$, $\check{L}_3(2)/2^{3+6}$, or $\check{\Omega}_8^+(2)$. But in those cases L_1 is of index 3^e in $(SL_2(3) * SL_2(3))^2$ for $e = 2, 0, 3, 1$, respectively, with $\mathcal{D}_i = O(t_i)$ for $\mathcal{Z}_S(\rho_i) = \{z_i, t_i\}$. Then as $\mathcal{O}_1 \cong \mathcal{O}_i$ by 12.3.12, again (1) and (2) hold. □

LEMMA 12.3.15. *Assume $(m, n) = (8, 3)$ and 12.3.7.4.ii holds. Let $X_z \in Syl_3(L_z)$, so that $X_z \cong E_{3^e}$ for some $1 \le e \le 3$.*

(1) L_z centralizes $S_2 \cdots S_r$.

(2) For $i \in I_1$, $\mathcal{O}_z$ centralizes $\mathcal{F}_{S_i}(M_i) \le \mathcal{C}_i$.

(3) For $i \in I_1$, one of the following holds:

(i) $\mathcal{C}_i \cong AE_7$ or $Sp_6(2)$ and $e = 2$.

(ii) $\mathcal{C}_i \cong \Omega_7[8]$ and $e = 3$.

(iii) $\mathcal{C}_i \cong L_3(2)/E_{64}$ and $e = 1$.

(4) For all $i \in I_1$ and $\alpha_i \in \mathfrak{A}(z_i)$ with $z_i\alpha_i = z$, $\mathcal{O}_i\alpha_i^* = \mathcal{O}_z = \mathcal{O}_1$, where $\mathcal{O}_i$ is $O^2(H_i)$ and H_i is the normal closure of W_i in $C_{\mathcal{C}_i}(z_i)$.

(5) For $j \in I$, set $T_j = O_2(\mathcal{O}_j)S_j$. Set $\mathcal{C}_1 = \langle T_1\mathcal{O}_z, \mathcal{F}_{T_1}(T_1M_1)\rangle$. Then for $i \in I_1$, T_i is Sylow in $\mathcal{C}_i$ and $\mathcal{C}_i\alpha_i^* = \mathcal{C}_1$, so $\mathcal{C}_1 \cong \mathcal{C}_i$.

(6) $\mathcal{F}$ contains a central product $\mathcal{C}_0 = \prod_{j \in I} \mathcal{C}_i$ and α_i permutes $\{T_j : j \in I\}$ and $\{\mathcal{C}_j : j \in I\}$.

(7) Set $T' = \prod_{i \in I_1} T_i$. Then $T' \in \mathcal{F}^f$, $\rho = (C_{\mathcal{F}}(T'), \Omega(z))$ is a quaternion fusion packet, and $C_{\mathcal{F}}(T')^\circ = \mathcal{C}_1$.

PROOF. By 12.3.7.4, L_z centralizes $W' = W_2 \cdots W_r$. Also $S' = S_2 \cdots S_r \le S_0 = C_S(O)$. Let $i \in I_1$. As $M_i \trianglelefteq N_G(W_i)$, we have $S_i \trianglelefteq N_{G_z}(W_i)$, so by 12.3.7.4, L_z centralizes S_i and hence also S', establishing (1).

Next $N_{\mathcal{C}_i}(W_i) \trianglelefteq N_{\mathcal{F}_z}(W_i) = \mathcal{N}$ and $\mathcal{M} = \mathcal{F}_{S_i}(M_i)$ is the normal closure of $O(z_i)$ in $N_{\mathcal{C}_i}(W_i)$, so it follows that $\mathcal{M} \trianglelefteq \mathcal{N}$. Observe that $\mathcal{O}_z \le \mathcal{N}$. Let $P = O_2(L_z)$. Then $[P, M_i] \le C_{M_i}(W_i) = W_i$, so $PW_i \trianglelefteq PM_i$ and hence $\mathcal{M} \le N_{\mathcal{N}}(PW_i)$. Let x of order 3 in L_z; thus x acts on P and centralizes S_i, so as $\mathcal{M} \trianglelefteq \mathcal{N} \ge \mathcal{O}_z$, $\beta = c_{x|PS_i}$ induces an automorphism of PM_i acting on M_i. Now $[\beta, M_i] \le C_{M_i}(W_i) = W_i \le C_{M_i}(\beta)$, so β centralizes M_i. Then M_i acts on $[PW_i, \beta] = B_z$, so $M_i = \langle S_i^{M_i}\rangle$ centralizes B_z and hence $\mathcal{M}$ centralizes $\mathcal{O}_z$, completing the proof of (2).

Set $\mathcal{O}_1 = \mathcal{O}_z$ and for $i \in I_1$ define $\mathcal{O}_i$ as in (4). For $\beta, \gamma \in \{\mathcal{O}_i : i \in I\}$, write $\beta \lesssim \gamma$ if β is conjugate to a subsystem of γ.

There is $\alpha \in \mathfrak{A}(z_i)$ with $z_i\alpha = z$. For $j \in J = I_1 - \{i\}$, $\rho_j\alpha^* \le \rho_{j\sigma}$ for some injection $\sigma : J \to I_1$ by 12.3.9, 6.6.11, and as ρ_j is coconnected. Similarly $\mathcal{O}_i\alpha^* \le \mathcal{O}_1$ and $\mathcal{O}_1\alpha^* \le \mathcal{O}_k$, where $I_1 = J\sigma \cup \{k\}$. Thus for each $i \in I_1$, $\mathcal{O}_i \lesssim \mathcal{O}_1 \lesssim \mathcal{O}_k$.

Let $I_0 = \{i_0 \in I_1 : \mathcal{O}_{i_0} \cong \mathcal{O}_1\}$. Suppose that $i \in I - I_0$. Then $I_0\sigma \subseteq I_0$, so as σ is an injection, we have $I_0\sigma = I_0$. But this is a contradiction as also $\mathcal{O}_1 \lesssim \mathcal{O}_k \lesssim \mathcal{O}_1$, so that $k \in I_0$. This contradiction proves (4).

By 12.3.9.1, ρ_i is a quaternion fusion packet with $\mu(\rho_i) \cong S_4$, so by 12.3.1.4, together with Theorem 9.5.33, $\mathcal{C}_i$ is $\mathcal{F}_{S_i}(M_i)$, $\check{A}E_6$, AE_7, $\Omega_6^+[8]$, $\Omega_7[8]$, $\Omega_8^-[8]$, $L_3(2)/E_{64}$, or $Sp_6(2)$. By hypothesis, $\mathcal{L}_z$ satisfies 12.3.7.4.ii, so by (4), $\mathcal{C}_i \ne \mathcal{F}_{S_i}(M_i)$, $\check{A}E_6$, $\Omega_6^+[8]$, or $\Omega_8^-[8]$. Therefore (3) holds.

From (3) a Sylow group U_i of $\mathcal{C}_i$ is of order 2^9 and $|S_i| = 2^8$, so as $O_2(\mathcal{O}_i) \not\le S_i$ we have T_i Sylow in $\mathcal{C}_i$ and $\mathcal{C}_i = \langle T_i\mathcal{O}_i, \mathcal{F}_{T_i}(T_iM_i)\rangle$. By (4), $\mathcal{O}_i\alpha_i^* = \mathcal{O}_z$ and of course $S_i\alpha_i = S_1$ and $\mathcal{F}_{S_i}(M_i)\alpha_i^* = \mathcal{F}_{S_1}(M_1)$. Therefore (5) holds.

We saw that $N_{\mathcal{C}_i}(W_i) \trianglelefteq \mathcal{N}$, so as L_z centralizes W_i, L_z acts on the Sylow group T_i of $\mathcal{C}_i$. Then as $|T_i : S_i| = 2$ and L_z centralizes S_i, also L_z centralizes T_i.

Next $C_S(z_i)\alpha_i \leq C_S(z)$ and $T = \prod_{j\in I} T_j \leq C_S(z_i)$. Therefore α_i permutes $\{S_j : j \in I\}$ and $\{M_j : j \in J\}$. We essentially showed above that α_i permutes $\{\mathcal{O}_j : j \in I\}$, so as $T_j = O_2(\mathcal{O}_j)S_j$, α_i permutes $\{T_j : j \in I\}$, $\{T_j\mathcal{O}_j : j\}$ and $\{T_jM_j : j\}$.

Next S_1 and L_z centralize S_i, so $T_1\mathcal{O}_1 = S_1\mathcal{O}_1$ centralizes S_i. Hence by the previous paragraph, for all distinct j, k in I, $T_j\mathcal{O}_j$ centralizes S_k. Therefore also T_j centralizes S_k. Then T_j centralizes $S_k\mathcal{O}_k = T_k\mathcal{O}_k$, so also T_j centralizes T_k.

As $\mathcal{T}_j$ centralizes T_k it centralizes $\mathcal{F}_{T_k}(T_kM_k)$. Therefore

$$\mathcal{C}_j = \langle T_j\mathcal{O}_j, \mathcal{F}_{T_j}(T_jM_j)\rangle$$

centralizes T_k. Then (6) follows from 2.2 in [**Asc11**].

As $z \in \mathcal{F}^f$, $T' \in \mathcal{F}^f$. By 2.5.2, ρ is a quaternion fusion packet. By (6), $C_{\mathcal{F}}(T_1 \cdots T_{r-1})^\circ = C_{\mathcal{F}_z}(T_1 \cdots T_{r-1}) = \mathcal{C}_r$, so as α_i permutes $\{T_j : j \in I\}$ and $\{\mathcal{C}_j : j \in I\}$, it follows that $\mathcal{C}_1 = C_{\mathcal{F}}(T')^\circ$. Thus (7) holds. □

LEMMA 12.3.16. *If $(m, n) = (8, 3)$ and $L_z \neq 1$ then 12.3.7.4.i holds.*

PROOF. Assume otherwise; then by 12.3.7.4, 12.3.7.4.ii holds, so we can appeal to 12.3.15. Using 12.3.15, we verify that Hypothesis 8.1.8 is satisfied. Then by 8.1.10, $\mathcal{F}^\circ = \mathcal{C}_0$. This contradicts the assumption that $r > 1$ and the transitivity of $\mathcal{F}^\circ$ on Ω in 12.3.1.2. □

LEMMA 12.3.17. *We can choose z, W_1, and $O(\zeta_1)$ in $\mathcal{F}^f$.*

PROOF. Let $\alpha \in \mathfrak{A}(W_1)$. As $W_i \trianglelefteq W$, $W\alpha = W$, so α is a $N_{\mathcal{F}}(W)$-map. Then as $N_{\mathcal{F}}(W)$ permutes $\{W_i : i \in I\}$, it follows that $W_1\alpha = W_j$ for some $j \in I$. Further as S permutes $\{W_i : i \in I\}$, we have $C_S(z) \leq N_S(W_1)$, so $z\alpha \in \mathcal{F}^f$. Hence, replacing z by $z\alpha$ if necessary, we may assume $W_1 \in \mathcal{F}^f$.

Similarly let $Q_i = O(\zeta_i)$, $Q = O(\tau)$, and $\beta \in \mathfrak{A}(Q_1)$. As in the previous paragraph, $Q_1\beta = Q_i$ for some $i \in I$ and $C_S(z) \leq N_S(Q_1) \geq N_S(W_1)$, so $z\beta$ and $W_1\beta \in \mathcal{F}^f$. Hence we may assume z, W_1, Q_1 are in $\mathcal{F}^f$. □

NOTATION 12.3.18. Suppose first that $n = 4$. Then, using 12.3.12 when $m = 8$, for each $j \in I$, $\mathcal{O}_j$ is defined, as is a central product $\mathcal{O}_\tau = \prod_{j\in I} \mathcal{O}_j$. So take $n \neq 4$. Then, using 12.3.16 and 12.3.11.2 when $m = 8$, for $s \in \mathcal{Z}_S$, $\mathcal{O}_s$ is defined, as is a central product $\mathcal{O}_\tau = \prod_{s\in\mathcal{Z}_S} \mathcal{O}_s$. For $j \in I$, let $\mathcal{O}_j = \prod_{s\in\mathcal{Z}_S(\zeta_j)} \mathcal{O}_s$, and observe we also have a central product $\mathcal{O}_\tau = \prod_{j\in I} \mathcal{O}_j$.

For $j \in I$, set $\mathcal{B}_j = \langle S_j\mathcal{O}_j, \mathcal{F}_{S_j}(M_j)\rangle \leq \mathcal{F}$. For $i \in I_1$, by 12.3.17 there exists $\alpha_i \in \mathfrak{A}(W_i)$ with $W_i\alpha_i = W_1$. Let T_i be Sylow in $\mathcal{C}_i$.

LEMMA 12.3.19. *(1) For $i > 1$ and $(n, m) \neq (4, 8)$, $\mathcal{O}_i = \mathcal{P}_i$, where $\mathcal{P}_i = \prod_{s\in\mathcal{Z}_S(\rho_i)} \mathcal{P}_s$, $\mathcal{P}_s$ is a $N_{\mathcal{C}_i}(O(\rho_i))$-conjugate of $\mathcal{P}_{z_i}$, and $\mathcal{P}_{z_i} = O^2(\mathcal{H}_{z_i})$ where $\mathcal{H}_{z_i}$ is the normal closure of W_i in $C_{\mathcal{C}_i}(z_i)$.*

(2) If $(n, m) = (4, 8)$ and $i > 1$ then $\mathcal{O}_i = O^2(\mathcal{H}_i)$ where $\mathcal{H}_i$ is the normal closure of W_i and $C_{\mathcal{C}_i}(Z_i)$.

(3) For $i > 1$, $\mathcal{B}_i = \mathcal{C}_i$ if $\mathcal{C}_i$ is $\mathcal{F}_{S_i}(M_i)$, AE_{2n}, $\check{A}E_{2n}$, $(P)\Omega^+_{2n}[m]$, $\check{L}_3(2)/2^{3+6}$, or $\check{O}^+_8(2)$.

(4) For $i > 1$, $\mathcal{B}_i \leq \mathcal{C}_i$ is isomorphic to $\check{A}E_{2n}$; $\Omega^+_{2n}[m]$ if $\mathcal{C}_i$ is isomorphic to AE_{2n+1}; $\Omega_{2n+1}[m]$ or $\Omega^-_{2n+2}[m]$, respectively.

(5) For $\alpha_i \in \mathfrak{A}(W_i)$ with $W_i\alpha_i = W_1$, we have $\mathcal{B}_i\alpha_i^* = \mathcal{B}_1$.
(6) $\mathcal{F}$ contains a central product $\mathcal{B}_0 = \prod_{i\in I} \mathcal{B}_i$.
(7) α_i permutes $\{S_j : j \in I\}$ and $\{\mathcal{B}_j : j \in I\}$.
(8) Set $\mathcal{D}_1 = \mathcal{C}_z$. Then α_i permutes $\{\mathcal{D}_j : j \in I\}$.

PROOF. Parts (1) and (2) were established during the proof of 12.3.13. Then (3) and (4) follow from 12.3.13, (1), (2), and the following lemmas: 11.6.6 and 11.6.8 when $\mathcal{C}_i$ is AE_k or $\check{A}E_k$; 11.5.3 when $\mathcal{C}_i$ is a Lie system; and 10.1.11.1 and 2.1.15 in the remaining cases.

Choose α_i as in (5). Then $S_i\alpha_i = S_1$, so $\mathcal{F}_{S_i}(M_i)\alpha_i^* = \mathcal{F}_{S_1}(M_1)$. Also $S_i\mathcal{O}_i\alpha_i^* = S_1\mathcal{O}_1$ from the definition of $\mathcal{O}_i$ in 12.3.18. Thus (5) follows.

Let $\mathcal{M}_i = \mathcal{F}_{S_i}(M_i)$. By 12.3.4.1, $\mathcal{F}$ contains a central product $\prod_{i\in I} \mathcal{M}_i$. Observe also that for distinct i, j in I, S_i centralizes $S_j\mathcal{O}_j$, since $O_i = O(\zeta_i)$ centralizes $\mathcal{O}_j$. When $m > 8$ this is 2.6.12, while if $m = 8$ it follows from 12.3.7 and 12.3.16. Therefore $S' = \prod_{i\in I_1}$ centralizes $\mathcal{B}_1$ and S_1 centralizes $\mathcal{B}_i$ for $i \in I_1$, so (6) follows from 2.2 in [**Asc11**].

As in the proof of 12.3.15.6, α_i permutes $\{S_j : j \in I\}$, $\{M_j : j \in I\}$, and $\{S_j\mathcal{O}_j : j \in I\}$, so α_i permutes $\{\mathcal{B}_j : j \in I\}$, establishing (7). Part (8) is essentially established during the proof of 12.3.14. □

LEMMA 12.3.20. *$\mathcal{C}_2$ is not AE_{2n+1}.*

PROOF. Assume otherwise. By 12.3.16 and 12.3.13.5, $n > 3$. Let $i \in I_1$. By 11.6.7, $\mathcal{D}_i \cong AE_{2n-3}$, while by 11.6.8, $|T_i : S_i| = 2$, and $T_i = S_iO_2(\mathcal{D}_i)$. Set $T_1 = S_1O_2(\mathcal{D}_1)$. By 12.3.14.2, $\mathcal{D}_i\alpha_i^* = \mathcal{D}_1$, so $T_i\alpha_i = T_1$. Then by parts (7) and (8) of 12.3.19, α_i permutes $\{T_j : j \in I\}$.

Set $U_i = O_2(\mathcal{C}_i)$ and $U_1 = J(T_1)$. By 11.6.10, $U_i = J(T_i)$, so $U_i\alpha_i = U_1$ and α_i permutes $\{U_j : j \in I\}$.

Next T_i centralizes $\mathcal{D}_1$ and S_i centralizes S_1, so S_i centralizes T_1. As α_i permutes $\{S_j : j \in I\}$, $\{\mathcal{D}_j : j \in I\}$, and $\{T_j : j \in I\}$ for each $i \in I_1$, it follows that for all distinct j, k in I, S_j and $\mathcal{D}_j$ centralize T_k, so $T_j = S_jO_2(\mathcal{D}_j)$ centralizes T_k. Similarly $\mathcal{B}_i$ centralizes S_1 and $O_2(\mathcal{D}_1)$, so $\mathcal{B}_i$ centralizes T_1. Then $T_i\mathcal{B}_i$ centralizes T_1, so $T_j\mathcal{B}_j$ centralizes T_k.

Next T_1 centralizes $\langle Aut_{\mathcal{B}_i}(U_i), Aut_{\mathcal{D}_i}(U_i)\rangle = Aut_{\mathcal{C}_i}(U_i)$, so $[T_1, \mathcal{C}_i] \leq U_i$. Therefore $\langle \mathcal{C}_j : j \in I_1\rangle$ acts on $U = \prod_{k\in I} U_k$.

Claim $U \trianglelefteq S$, so that $U \in \mathcal{F}^f$. For S acts on $\{S_j : j \in I\}$ and for $s \in S$, $T_1^s = S_1^sO_2(\mathcal{D}_1)^s = S_{1^s}O_2(\mathcal{D}_{1^s}) = T_{1^s}$ as $z^s \in \mathcal{F}^f$. Therefore S acts on $\prod_{k\in 1^S} T_k$ and hence on $U' = \prod_k J(T_k) = \prod_k U_k$.

Let $J = I - 1^S$ and $u \in z^S$. Then $u = z$ or $u \in \mathcal{C}_z$ or $u \in \mathcal{C}_i$ for some $i \in 1^S$. In each case u centralize $\mathcal{C}_j$ for each $j \in J$. Then as $u \in \mathcal{F}^f$, ρ_j is the coconnected component of τ_u containing Ω_j. Therefore S permutes $\{\mathcal{C}_j : j \in J\}$, and hence also $U'' = \prod_{j\in J} U_j$. Thus S acts on $U = U'U''$, completing the proof of the claim.

By the claim, $\tau_U = (\mathcal{Y}, \Omega)$ is a quaternion fusion packet, where $\mathcal{Y} = N_{\mathcal{F}}(U)$. We've shown that $\mathcal{C}_i$ and α_i are contained in $\mathcal{Y}$ for each $i \in I_1$, so $\mathcal{Y}$ is transitive on Ω. As $z \in U \leq O_2(\mathcal{Y})$, it follows from Theorem 2 that $\mathcal{C}_1 = \mathcal{C}_2\alpha_2^* = \mathcal{C}_i\alpha_i^*$ for each $i \in I_1$, and $\mathcal{C}_0 = \mathcal{C}_1 \times \cdots \times \mathcal{C}_r = \mathcal{Y}^\circ$. But now Hypothesis 8.1.8 is satisfied, so by 8.1.10, $\mathcal{F}^\circ = \mathcal{C}_0$, contrary to the assumption that $r > 1$ and to the transitivity of $\mathcal{F}^\circ$ on Ω in 12.3.1.2. □

LEMMA 12.3.21. *Assume $n = 3$ and $\mathcal{C}_2$ is $\Omega_7[m]$ or $\Omega_8^-[m]$. For $i \in I_1$, set $\mathcal{D}_i = E(C_{\mathcal{C}_i}(O(z_i)))$. Then*

(1) $\mathcal{D}_1 = \mathcal{D}_i\alpha_i^*$ is a component of $\mathcal{F}_z$ independent of i and α_i.

(2) α_i permutes $\{\mathcal{D}_j : j \in I\}$.

PROOF. First if $m = 8$ then $\mathcal{C}_2$ is $\Omega_8^-[8]$ by 12.3.16 and 12.3.13.5, so $\mathcal{D}_i \cong L_2[2m]$ is indeed quasisimple.

Let D_i be Sylow in $\mathcal{D}_i$ and $U_i = \langle z, z_i\rangle$. Then $\mathcal{D}_i$ is a component of $C_{\mathcal{F}}(U_i) = C_{\mathcal{F}_z}(z_i)$, so $\mathcal{D}_1 = \mathcal{D}_i\alpha_i^*$ is a component of $C_{\mathcal{F}}(U_i\alpha)$ with $U_i\alpha = \langle u_i, z\rangle$ and $u_i = z\alpha_i$. Therefore $\mathcal{D}_1$ is a component of $C_{\mathcal{F}_z}(u_i)$. By 12.3.19.4, $u_i \in \mathcal{B}_k \leq \mathcal{C}_k$ for some $k \in I_1$, so $u_i = z_k$ and either $\mathcal{D}_1$ is a component of $\mathcal{F}_z$ or $\mathcal{D}_1 = \mathcal{D}_k$. Assume the latter. Now $1 \neq V_i = S_i \cap D_i$ is cyclic of index 2 or 4 in D_i, and $V_i\alpha_i \leq S_1 \cap D_k$ by 12.3.19.7. This is a contradiction as $S_1 \cap S_k \leq Z(M_k)$.

Therefore $\mathcal{D}_1$ is a component of $\mathcal{F}_z$. Let $j \in I_1$ and $\alpha \in \mathfrak{A}(z_j)$ with $z_j\alpha = z$, and suppose $\mathcal{D}_1 \neq \mathcal{D}_j\alpha_j^*$. Then $\langle z\rangle \times D_1 \times D_j\alpha \leq C_S(O(z_1))$. Hence $\langle z\rangle \times V_i\alpha_i \times V_j\alpha \leq C_{S_1}(O(z))$, a contradiction as $m(C_{S_1}(O(z)) = 2$. Therefore $\mathcal{D}_1 = \mathcal{D}_j\alpha_j^*$, completing the proof of (1). Then (1) implies (2). □

LEMMA 12.3.22. *$\mathcal{C}_2$ is not $\Omega_{2n+1}[m]$ or $\Omega_{2n+2}^-[m]$.*

PROOF. Assume otherwise. If $n = 3$ define $\mathcal{D}_j$, $j \in I$, as in 12.3.21, and observe that if also $m = 8$ then $\mathcal{C}_2$ is $\Omega_8^-[8]$ by 12.3.16 and 12.3.13.5. If $n > 3$ then $\mathcal{D}_j$ is defined in 12.3.10 when $j \in I_1$ and $\mathcal{D}_1 = \mathcal{C}_z$, as in 12.3.19.8. In any event let D_j be Sylow in $\mathcal{D}_j$.

Let $i \in I_1$. By 5.3.9 and 5.3.10, $T_i = D_iS_i$ with $|T_i : S_i| = 2{,}4$ for $\mathcal{C}_2$ equal to $\Omega_{2n+1}[m]$, $\Omega_{2n+2}^-[m]$, respectively. Set $T_1 = D_1S_1$, Observe that $T_i\alpha_i = T_1$ and α_i permutes $\{T_j : j \in I\}$. For $D_i\alpha_i = D_1$ and α_i permutes $\{\mathcal{D}_j : j \in I\}$ by 12.3.14.2 and 12.3.21.1, while $S_i\alpha_i = S_1$ and α_i permutes $\{S_j : j \in I\}$ by 12.3.19.7.

Next T_i centralizes $\mathcal{D}_1$ and S_i centralizes S_1, so S_i centralizes $T_1 = D_1S_1$. Then, from the previous paragraph, for all distinct $j, k \in I$, S_j centralizes T_k. Now as T_i centralizes D_1 and S_1, T_i centralizes $T_1 = D_1S_1$, so T_j centralizes T_k.

Next we show that $\mathcal{C}_i = \langle \mathcal{B}_i, \Sigma\rangle$, where $\Sigma = \langle O^2(Aut_{\mathcal{C}_i}(R_\lambda) : \lambda \in \Lambda\rangle$ for suitable $R_\lambda \leq T_i$ and $\Sigma \subseteq C_{\mathcal{F}}(T_1)$. Then $\mathcal{C}_i$ centralizes T_1. Set $T' = \prod_{i\in I_1} T_i$ and $Q = \prod_{j\neq i} T_j \in N_{\mathcal{F}}(T_1)^f$. Then $\mathcal{C}_i = C_{\mathcal{F}}(Q)^\circ$ as $z \in Q$, and $T' \in \mathcal{F}^f$ as $W_1 \in \mathcal{F}^f$, so there is $\beta \in \mathfrak{A}(Q)$ with $Q\beta = T'$. Then $\mathcal{C}_1 = \mathcal{C}_i\beta^* = C_{\mathcal{F}}(T')^\circ$. Similarly $\mathcal{C}_1 = C_{\mathcal{F}}(O')^\circ$, where $O' = \prod_{i\in I_1} O(\rho_i)$, so $\mathcal{C}_j\alpha_j^* = \mathcal{C}_1$ for each $j \in I_1$. Therefore by 2.2 in [**Asc11**], there is a direct product $\mathcal{C}_0 = \mathcal{C}_1 \times \cdots \times \mathcal{C}_r$ contained in $\mathcal{F}$. But now we obtain a contradiction from 8.1.10, as in the proof of 12.3.20.

It remains to establish the claim. If n is odd then from 11.5.2, $R_\lambda = C_{T_i}(B_\lambda)$, where B_λ is a 4-subgroup of D_i and $\Sigma \leq Aut_{\mathcal{D}_i}(R_\lambda)$, so indeed $\Sigma \subseteq C_{\mathcal{F}}(T_1)$. Thus we may take n even, so from 11.5.2, $R_\lambda = C_{T_i}(B_\lambda)$ for a suitable 8-subgroup of L_A and $A = \mathfrak{U}n/2 + \mathfrak{U}_0$ with $\mathfrak{U}_{n/2} = [U, s_{n/2}]$ for $s_{n/2} \in \mathcal{Z}_{T_i} \cap \mathcal{C}_i^f$. If $n \equiv 0 \mod 4$ then there is $s_{n/2} \neq s \in \mathcal{Z}_{T_i} \cap \mathcal{C}_i^f$, so we may choose $s = z_i$ and hence z_i centralizes L_A, so that $O^2(Aut_{\mathcal{C}_i}(R_\lambda)) \leq Aut_{\mathcal{D}_i}(R_\lambda) \subseteq C_{\mathcal{F}}(T_1)$. This leaves the case $n \equiv 2 \mod 4$, where $O(z_i) \lhd T_i$, and we define $O' = \langle \Omega_i - \Omega(z_i)\rangle$ and $\mathcal{E} = C_{\mathcal{C}_i}(O')^\circ$. Then $\mathcal{E} \cong \Omega_5[m]$ or $\Omega_6^-[m]$ and Σ is induced in $\mathcal{E}$. Now $Q = O'\alpha_i \in \mathcal{F}^f$ and $\mathcal{E}\alpha_i^*$ is a coconnected component of $N_{\mathcal{F}}(Q)$. Also $Q \leq \mathcal{C}_z$, so $\mathcal{C}_k$ is a coconnected component of $N_{\mathcal{F}}(Q)$. Therefore $\mathcal{E}\alpha_i^*$ centralizes T_k for each $k \in I_1$, so $\mathcal{E}$ centralizes T_1, completing the proof of the claim and the lemma. □

LEMMA 12.3.23. *(1) For $i \in I_1$, $\mathcal{C}_i = \mathcal{B}_i$.*

(2) $\mathcal{C}_i$ is $\mathcal{F}_{S_i}(M_i)$, $\check{A}E_{2n}$, or $\Omega^+_{2n}[m]$, or $(n,m) = (4,8)$ and $\mathcal{C}_i$ is $\check{L}_3(2)/2^{3+6}$ or $\check{O}^+_8(2)$.

(3) Define $\mathcal{C}_1 = \mathcal{B}_1$; then $\mathcal{F}$ contains a central product $\mathcal{C}_0 = \prod_{j\in I} \mathcal{C}_j$.

(4) $Z(\mathcal{C}_i) = \langle c_i \rangle \cong \mathbf{Z}_2$, where $\langle c_i \rangle = Z(M_i)$.

PROOF. First, by 12.3.16, if $(n,m) = (3,8)$ then either $L_z = 1$ or 12.3.7.4.i holds. Therefore the hypothesis of 12.3.13 is satisfied. Further by 12.3.22, $\mathcal{C}_i$ is neither $\Omega_{2n+1}[m]$ nor $\Omega^-_{2n+2}[m]$, while by 12.3.20, $\mathcal{C}_i$ is not AE_{2n+1}. Therefore by 12.3.13, either $\mathcal{C}_i$ appears in (2), or $Z(\mathcal{C}_i) = 1$ and $\mathcal{C}_i$ is AE_{2n} or $P\Omega^+_{2n}[m]$. Thus (4) implies (2). In any event by 12.3.19.3, $\mathcal{C}_i = \mathcal{B}_i$, so (1) holds and (3) follows from 12.3.19.6. By (3) and 8.1.10, $Z(\mathcal{C}_i) \neq 1$, establishing (4) and completing the proof of (2). □

NOTATION 12.3.24. Set $U = \langle c_i : i \in I \rangle$. By 12.3.23.4, $U \neq 1$. Set $\mathcal{U} = N_{\mathcal{F}}(U)$ and $S_M = S_1 \cdots S_r$.

LEMMA 12.3.25. *$Z(\mathcal{F}) = 1$.*

PROOF. Set $\mathcal{F}^+ = \mathcal{F}/Z(\mathcal{F})$. Then, using 6.6.9, $\mathcal{F}^+$ satisfies Hypothesis 12.3.1, so Theorem 12.3.8 holds in $\mathcal{F}^+$ by induction on the order of $\mathcal{F}$, contradicting $r > 1$. □

LEMMA 12.3.26. *$N_{\mathcal{F}}(W) \leq \mathcal{U}$.*

PROOF. As $U = \langle c_i : i \in I \rangle$ and $\langle c_i \rangle = Z(M_i)$, we have $U = Z(\mathcal{F}_{S_M}(M))$. Then as $\mathcal{F}_{S_M}(M) \trianglelefteq N_{\mathcal{F}}(W)$, the lemma follows. □

LEMMA 12.3.27. *(1) $\mathcal{U}$ controls fusion in $C_S(W)$.*

(2) If $X \leq C_S(W)$ and $\alpha \in \mathfrak{A}(X)$ then $X\alpha \leq C_S(W)$ and α is a $\mathcal{U}$-map.

PROOF. As W is weakly closed in S, $N_{\mathcal{F}}(W)$ controls fusion in $C_S(W)$, and then (1) follows from 12.3.26. Similarly in the setup of (2), $W\alpha = W$ as W is weakly closed in S, so (2) follows from 12.3.26. □

LEMMA 12.3.28. *Let $\emptyset \neq J \subseteq I$ and $W_J = \prod_{j\in J} W_j$. Then*

(1) if $W_J \in \mathcal{F}^f$ then $\mathcal{F}_{S_j}(M_j)$, $j \in J$, and $\mathcal{C}_i$, $i \in I - J$ are the coconnected components of $\rho_J = (N_{\mathcal{F}}(W_J), \Omega)$, and are subnormal in $N_{\mathcal{F}}(W_J)$.

(2) $N_{\mathcal{F}}(W_J) \leq \mathcal{U}$.

PROOF. Suppose $W_J \in \mathcal{F}^f$. By 3.3.9, $\mathcal{F}_{S_j}(M_j)$ is a coconnected component of $\mathcal{J} = N_{\mathcal{F}}(W_J)$ for each $j \in J$. Further for $i \in I - J$, $\mathcal{C}_i$ is contained in a coconnected component $\mathcal{E}$ of ρ_J. Then as $\mathcal{E}$ centralizes z_j for $j \in J$, $\mathcal{E}\alpha^*_j$ is contained in a coconnected component $\mathcal{X}$ of τ_z, so as $\mathcal{C}_i\alpha^*_j \leq \mathcal{E}\alpha^*_j \leq \mathcal{X}$ we conclude from 12.3.9 that $\mathcal{C}_i = \mathcal{E}$, completing the proof of (1).

In (2), conjugating in $\mathcal{U}$ and appealing to 12.3.27.2, we may assume $W_J \in \mathcal{F}^f$. By (1), $\mathcal{J}^\circ \leq \mathcal{U}$. Now by 1.3.2 in [**Asc19**], $\mathcal{J} = \langle N_S(W_J)\mathcal{J}^\circ, N_{\mathcal{J}}(W) \rangle \leq \mathcal{U}$, using 12.3.26. That is (2) holds. □

LEMMA 12.3.29. *$\mathcal{F}_z \leq \mathcal{U}$.*

PROOF. By 12.3.9, $\mathcal{F}^\circ_z = \mathcal{O}_z * \mathcal{C}_z * \mathcal{C}_2 * \cdots * \mathcal{C}_r$. Then as $\mathcal{O}_z * \mathcal{C}_z \leq \mathcal{C}_1$, we have $\mathcal{F}^\circ_z \leq \mathcal{U}$. But by 1.3.2 in [**Asc19**] and 12.3.28.2, $\mathcal{F}_z = \langle C_S(z)\mathcal{F}^\circ_z, N_{\mathcal{F}_z}(W_{I_1}) \rangle \leq \mathcal{U}$. □

LEMMA 12.3.30. *Assume $\xi = (\mathcal{Y}, \Gamma)$ is a proper subpacket of τ with $\Gamma \subseteq \Omega$ and $\mathcal{C}_i \leq \mathcal{Y}$ and $\Omega_i \subseteq \Gamma$ for some $i \in I$. Then*
(1) $\mathcal{C}_i$ is a coconnected component of ξ.
(2) $\mathcal{Y} \leq \mathcal{U}$.

PROOF. We first prove (1). As $\mathcal{C}_i$ is coconnected, $\mathcal{C}_i$ is contained in a coconnected component $\zeta = (\mathcal{E}, \Sigma)$ of $\mathcal{Y}$. Arguing as in the proof of 9.3.10, $W_i \in W(\zeta)$ and $\mu_i \cong \mu(\zeta)$. Then by 12.3.1.4 and Theorem 12.3.50, and by 12.3.13, $\mathcal{C}_i = C_{\mathcal{E}}(c_i)$, and then either $\mathcal{C}_i = \mathcal{E}$ or $\mathcal{C}_i \cong \check{A}E_{2n}$ and $\mathcal{E} \cong AE_{2n+1}$ or $\mathcal{C}_i \cong \Omega^+_{2n}[m]$ and $\mathcal{E} \cong \Omega_{2n+1}[m]$ or $\Omega^-_{2n+2}[m]$. Of course we may assume $\mathcal{C}_i \neq \mathcal{E}$, and, conjugating in $\mathcal{U}$ and appealing to 12.3.27, we may assume $i = 1$. Now if $n > 3$ then $C_{\mathcal{E}}(z)^\circ = \mathcal{O}_z * \mathcal{X}$ where $\mathcal{X}$ is AE_{2n-3} or $\Omega_{2n-3}[m]$ or $\Omega^-_{2n-2}[m]$. Then $\mathcal{X}$ is contained in a coconnected component $\mathcal{Y}$ of τ_z, so $\mathcal{Y} = \mathcal{C}_j$ or $\mathcal{C}_z$ by 12.3.9. Then as $\mathcal{C}_z \leq \mathcal{X}$, it follows that $\mathcal{X} = \mathcal{C}_z$, a contradiction as $\mathcal{C}_z \neq \mathcal{X}$.

Therefore $n = 3$, so $\mathcal{C}_1$ is $\check{A}E_6$ or $\Omega^+_6[m]$ and $\mathcal{E}$ is AE_7, $\Omega_7[m]$, or $\Omega^-_8[m]$. However in the first case, and in the second when $m = 8$, we have $\mathcal{E} = \langle \mathcal{C}_1, C_{\mathcal{E}}(z)\rangle$, so $\mathcal{E} \leq \mathcal{U}$ by 12.3.29, so $c_1 \in O_2(\mathcal{E})$ and hence $\mathcal{E}$ is AE_7. Then $O_2(\mathcal{E}) = \langle c_1^{\mathcal{E}} \rangle \leq U \leq C_S(W)$, whereas W_1 does not centralize $O_2(\mathcal{E})$.

Therefore $\mathcal{E}$ is $\Omega_7[m]$ with $m > 8$ or $\Omega^-_8[m]$. Let $\mathcal{X} = E(C_{\mathcal{E}}(O(z)))$. Then $\mathcal{X} \leq \mathcal{U}$ by 12.3.29. But $c_1 \notin O_2(\langle c_1 \rangle \mathcal{X})$ contradicting $c_1 \in O_2(\mathcal{U})$. This finally completes the proof of (1).

If $\mathcal{C}$ is a $\mathcal{Y}$-conjugate of $\mathcal{C}_1$ distinct from $\mathcal{C}_1$, then z centralizes $\mathcal{C}$, so $\mathcal{C} = \mathcal{C}_k$ for some $k \in I$ by 12.3.9. Thus the product $\mathcal{D}$ of the $\mathcal{Y}$-conjugates of $\mathcal{C}_1$ is contained in $\mathcal{U}$. Now, arguing as usual using 1.3.2 in [**Asc19**] and 12.3.28.2, we obtain (2). □

LEMMA 12.3.31. *(1) $\mathcal{U} \neq \mathcal{F}$.*
(2) $\{\mathcal{C}_i : i \in I\}$ is the set of coconnected components of $\tau_{\mathcal{U}} = (\mathcal{U}, \Omega)$.
(3) $\mathcal{C}_0 \trianglelefteq \mathcal{U}$.

PROOF. If $\mathcal{U} = \mathcal{F}$ then $U \trianglelefteq \mathcal{F}$, so $\mathcal{X} = C_{\mathcal{F}}(U) \trianglelefteq \mathcal{F}$. By 12.3.25, $\mathcal{X} \neq \mathcal{F}$, so by 6.6.9 and induction on the order of $\mathcal{F}$, $\mathcal{C}_0 = \mathcal{X}^\circ \trianglelefteq \mathcal{F}$. As usual this contradicts 12.3.1 and $r > 1$. Thus (1) holds.

By (1) and 12.3.30.1, $\{\mathcal{C}_i : i \in I\}$ is contained in the set $\mathfrak{C}$ of coconnected components of $\tau_{\mathcal{U}}$. Further if $\mathcal{E} \in \mathfrak{C}$ is not a $\mathcal{C}_i$ then $\mathcal{E}$ is contained in a coconnected component of τ_z, contrary to 12.3.9. Thus (2) holds, and of course (2) implies (3). □

LEMMA 12.3.32. *U is tightly embedded in $\mathcal{F}$.*

PROOF. Let $1 \neq X \leq U$; we will show $N_{\mathcal{F}}(X) \leq \mathcal{U}$. For $\alpha \in \mathfrak{A}(U)$, α is a $\mathcal{U}$-map by 12.3.27.2, so $X\alpha \leq U$. Thus we may assume $X \in \mathcal{F}^f$. Then $N_{\mathcal{F}}(X) \leq \mathcal{U}$ by 12.3.30.

By 12.3.27, $\mathcal{U}$ controls fusion in U, so the lemma follows from the previous paragraph. □

LEMMA 12.3.33. *Let $P \leq S$ such that for some $i \in I$, $W_i \leq P$.*
(1) $N_{\mathcal{F}}(P) \leq \mathcal{U}$.
(2) For each $\phi \in \hom_{\mathcal{F}}(P, S)$, ϕ is a $\mathcal{U}$-map.

PROOF. Set $J = \{j \in I : W_j \leq P\}$. Then W_J is weakly closed in P, so (1) follows from 12.3.28.2.

Assume the setup of (2). Then $W_J\phi = W_A$ for some $A \subseteq I$. Let $\alpha \in \mathfrak{A}(W_J)$; by 12.3.27.2, α is a $\mathcal{U}$-map and $W_J\alpha = W_B$ for some $B \subseteq I$. Let $\beta \in \mathfrak{A}(W_A)$ with $W_A\alpha = W_B$. By 12.3.27.2, β is a $\mathcal{U}$-map. Also $\zeta = \alpha^{-1}\phi\beta$ is a $N_{\mathcal{F}}(W_B)$-map, and hence also a $\mathcal{U}$-map by 12.3.28.2. Therefore $\phi = \alpha\zeta\beta^{-1}$ is a $\mathcal{U}$-map, proving (2). □

LEMMA 12.3.34. *Let $i \in I$ and $\mathcal{U}_i = N_S(S_i)\mathcal{C}_0$.*

(1) $\mathcal{C}_i \trianglelefteq \mathcal{U}_i$, so $C_S(\mathcal{C}_i)$ exists in $\mathcal{U}_i$ and is the largest subgroup of $C_S(S_i)$ centralizing $\mathcal{C}_i$ in $\mathcal{U}_i$

(2) Let $1 \neq X \leq C_S(\mathcal{C}_i)$ and $\alpha \in \mathfrak{A}(X)$. Then α is a $\mathcal{U}$-map, $\mathcal{C}_i\alpha^* = \mathcal{C}_j$ for some $j \in I$, and $X\alpha \leq C_S(\mathcal{C}_j)$.

(3) If $X \in \mathcal{F}^f$ then $\mathcal{C}_i$ is a coconnected component of $\tau_X = (N_{\mathcal{F}}(X), N_\Omega(X))$.

(4) $N_{\mathcal{F}}(X) \leq \mathcal{U}$.

PROOF. As $\mathcal{C}_0 \trianglelefteq \mathcal{U}$ by 12.3.31, we can form $\mathcal{U}_i$, and by construction, $\mathcal{C}_i \trianglelefteq \mathcal{U}_i$. Then (1) follows from Theorem 4 in [**Asc11**] and a remark preceding that theorem.

Assume the setup of (2). Applying 12.3.33 to $\alpha, N_S(X)$ in the role of ϕ, P, we conclude that α is a $\mathcal{U}$-map, and then (2) follows from 12.3.31. Part (3) follows from 12.3.30, which also says that $N_{\mathcal{F}}(X) \leq \mathcal{U}$ when $X \in \mathcal{F}^f$.

It remains to prove (4), where as α is a $\mathcal{U}$-map, we may assume $X \in \mathcal{F}^f$. Then (4) follows from a remark in the previous paragraph. □

LEMMA 12.3.35. *Either $\mathcal{C}_1$ is isomorphic to $\mathcal{F}_{S_1}(M_1)$ or $\check{A}E_{2n}$, or $(n, m) = (4, 8)$ and $\mathcal{C}_1$ is $\check{L}_3(2)/2^{3+6}$.*

PROOF. Assume otherwise; then by 12.3.23.2, $\mathcal{C}_1$ is $\Omega_{2n}^+[m]$ or $(n, m) = (4, 8)$ and $\mathcal{C}_1$ is $\check{O}_8^+(2)$. In either case, $\mathcal{C}_1$ is quasisimple. Adopt the notation from [**Asc19**], particularly Notation 6.1.1, 6.1.10, and 6.1.12 in [**Asc19**]; thus $\mathcal{C}_1 \in \mathfrak{C}(\mathcal{F})$ the set of quasisimple subsystems $\mathcal{C}$ of $\mathcal{F}$ such that $\mathcal{I}(\mathcal{C}) \neq \emptyset$, where $i \in \mathcal{I}(\mathcal{C})$ if i is an involution centralizing $\mathcal{C}$ such that $\mathcal{C}\alpha^*$ is a component of $\mathcal{F}_{i\alpha}$ for $\alpha \in \mathfrak{A}(i)$.

By 12.3.34, for each involution $i \in C_S(\mathcal{C}_1)$, we have $i \in \mathcal{I}(\mathcal{C})$. Therefore $\mathcal{C}_1$ is maximal in $\mathfrak{C}(\mathcal{F})$ as defined in 6.1.12 of [**Asc19**]. As $W_1 \in \mathcal{F}^f$, also $S_1 \in \mathcal{F}^f$. By 12.3.34, for distinct $j, k \in I$, each involution in S_j is in $\mathcal{I}(\mathcal{C}_k)$, so $\{\mathcal{C}_i : i \in I\} \subseteq \mathcal{C}_1^\perp$. Therefore by Theorem 2 in [**Asc19**], $\mathcal{C}_1$ is a component of $\mathcal{F}$ and $\mathcal{C}_0 \trianglelefteq \mathcal{F}$. But now we obtain our usual contradiction to 12.3.1 and $r > 1$. □

NOTATION 12.3.36. We now follow section 7 in [**Asc19**], beginning with Notation 7.3.14 in [**Asc19**]. By 12.3.22, U is tightly embedded in $\mathcal{F}$, so we can adopt Notation 3.1.9 in [**Asc19**]. In particular

$$\mathcal{P} = \{1 \neq P \leq S : \hom_{\mathcal{F}}(P, U) \neq \emptyset\},$$

$\mathcal{P}^*$ is the set of maximal members of $\mathcal{P}$ under inclusion, and $\hat{\mathcal{P}} = \mathcal{P}^* - \{U\}$. For $P \in \mathcal{P}$ and $\beta \in \mathfrak{A}(P)$ with $P\beta \leq U$, set $U(\beta) = (U \cap N_S(P)\beta)\beta^{-1}$.

From 7.3.15 in [**Asc19**] we have:

LEMMA 12.3.37. *(1) Each member of $\mathcal{P}$ is contained in a unique member $U(P)$ of $\mathcal{P}^*$.*

(2) $U(P) = U(\alpha)$ for each $\alpha \in \mathfrak{A}(P)$ with $P\alpha \leq U$.

(3) $\mathcal{P}^*$ is $Aut_{\mathcal{F}}(S)$-invariant.

(4) For distinct $P_1, P_2 \in \mathcal{P}^*$, $P_1 \cap P_2 = 1$.

(5) Each member of $\mathcal{P}^*$ is a TI-subgroup of S.

LEMMA 12.3.38. $O_2(\mathcal{F}) = 1$.

PROOF. By 6.6.9.1 and induction on the order of τ, we may take $\mathcal{F} = \mathcal{F}^\circ$. By Theorem 2, $z \notin O_2(\mathcal{F})$, so by 3.3.2, $\mathcal{F} = \mathcal{F}^\circ$ centralizes $O_2(\mathcal{F})$, contrary to 12.3.25. □

LEMMA 12.3.39. $\hat{\mathcal{P}} \neq \emptyset$.

PROOF. If not, U is strongly closed in S, so $U \trianglelefteq \mathcal{F}$, contrary to 12.3.38. □

NOTATION 12.3.40. By 12.3.39 we may choose $P \in \hat{\mathcal{P}}$. Define Σ_P to consist of those $1 \neq X \leq N_S(P)$ such that $\mathfrak{T}(P) \neq \emptyset$, where $\mathfrak{T}(P)$ is the set of triples (P_0, ψ, i) such that $1 \neq P_0 \leq N_P(X)$, $\psi \in \hom_{\mathcal{F}}(XP_0, S)$ with $P_0\psi \leq U$, and $X\psi$ centralizes $\mathcal{C}_i$. Write Ξ_P for the set of 4-subgroups A of $N_S(P)$ with $A^\# \subseteq \Sigma_P$ and $A \cap P \neq 1$.

LEMMA 12.3.41. *Assume $X \in \Sigma_P$ and $(P_0, \psi, i) \in \mathfrak{T}(X)$. Then*

(1) There is $\beta \in \mathfrak{A}(X)$ and $\gamma \in \mathfrak{A}(X\psi)$ such that $X\psi\gamma = X\beta$, $\mathcal{C}_i\gamma^* = \mathcal{C}_j$ for some $j \in I$, and $N_{\mathcal{F}}(X\beta) \leq \mathcal{U}$.

(2) $P_0 \leq U(\beta)$.

(3) For each $1 \neq Y \leq XP_0$, $Y \in \Sigma_P$ with $(C_P(Y), \beta_{|YC_P(Y)}, j) \in \mathfrak{T}(Y)$.

PROOF. Part (1) follows from 12.3.34. Then the proof of 7.3.18 in [**Asc19**] establishes (2) and (3). □

LEMMA 12.3.42. *Assume $X \in \Sigma_P$. Then*

(1) $N_S(X) \leq N_S(P)$.

(2) $N_{\mathcal{F}}(X) \leq N_{\mathcal{F}}(U(\beta)) \leq N_{\mathcal{F}}(P)$ for $\beta \in \mathfrak{A}(X)$ as in 12.3.41.1.

(3) X centralizes no $\mathcal{C}_i$.

(4) $U(\beta) \leq P$ and $U(\beta) \trianglelefteq N_S(X)$.

PROOF. See the proof of 7.3.19 in [**Asc19**]. □

LEMMA 12.3.43. *Assume $1 \neq X \leq S$ such that $W_i \leq N_S(X)$ for some $i \in I$, and let $\beta \in \mathfrak{A}(X)$. Then*

(1) β is a $\mathcal{U}$-map.

(2) If $X\beta \leq C_S(\mathcal{C}_i)$ for some $i \in I$ then $X \leq C_S(\mathcal{C}_k)$ for some $k \in I$.

(3) $X \notin \Sigma_P$.

PROOF. Part (1) follows from 12.3.33.2 applied to $N_S(X), \beta$ in the role of P, ϕ. Then (1) implies (2), and (2) and 12.3.42.3 imply (3). □

LEMMA 12.3.44. *Let $X \in \Sigma_P$. Then*

(1) $X \not\leq W$.

(2) $P \cap W = 1$.

(3) $N_W(P) = C_W(P)$.

PROOF. Part (1) follows from 12.3.43.3. Each nontrivial subgroup of P is in Σ_P, so (1) implies (2). Then $[N_W(P), P] \leq P \cap W = 1$ by (2), so (3) holds. □

LEMMA 12.3.45. *Assume* $A \in \Xi_P$. *Then*
(1) A *acts on* W_i *for each* $i \in I$.
(2) $U \leq C_S(PA)$.
(3) $\emptyset \neq U^{\mathcal{F}} - \{U\} = \mathcal{P}^!$.

PROOF. Assume A does not act on W_i for some $i \in I$, say $a \in A$ with $W_i^a = W_j$ with $i \neq j$. If $\langle a \rangle = P \cap A = P_A$ and for each $b \in A - P$, b acts on W_i, then $A = \langle A - P \rangle$ also acts on W_i, a contradiction. Thus we may assume $A = P_A\langle a \rangle$. Let $Y = W_iW_j$, $Y^+ = Y/\langle c_i, c_j \rangle$, and $Y_a = \{ww^a : w \in W_i\}$. Then $Y_a \leq C_Y(a)$ and Y_a^+ is a full diagonal subgroup of $Y^+ = W_i^+ \times W_j^+$. By 12.3.44.3, P centralizes $C_Y(a)$, so $A = P_A\langle a \rangle$ centralizes Y_a, and hence A acts on $\{W_i, W_j\}$, so $B = N_A(W_i)$ is of index 2 in A. As B centralizes the full diagonal subgroup Y_a^+, B centralizes Y^+. Therefore $[W_i, B] \leq \langle c_i \rangle$. But for $V \in \eta_i$, $z(V)$ is the involution in $\Phi(V\langle c_i \rangle)$, so B centralizes $z(V)$ and hence acts on V and then centralizes V. Hence B centralizes W_i, so by 12.3.42.1, $W_i \leq N_S(P)$, contrary to 12.3.43.3. Thus (1) holds.

By (1), A centralizes c_i for each $i \in I$, so $U \leq C_S(A) \leq N_S(P)$, and then (2) follows from 12.3.44.3. Now for $\alpha \in \mathfrak{A}(P)$, $U\alpha \in \hat{\mathcal{P}}$, so $U\alpha \in \mathcal{P}^! = U^{\mathcal{F}} - \{U\}$, establishing (3). □

LEMMA 12.3.46. *Let* $\alpha \in \mathfrak{A}(P)$. *Then*
(1) $P\alpha \leq U$.
(2) $C_U(P)\alpha \leq Q \in \hat{\mathcal{P}}$.
(3) *If* $P \in \mathcal{P}^!$ *then* $C_U(P)\alpha = Q \in \mathcal{P}^!$ *and* $|P| = |C_U(P)$.
(4) *For each* $1 \neq X \leq P$, $X \in \Sigma_P$ *and* $(P, \alpha, i) \in \mathfrak{T}(X)$ *for each* $i \in I$.
(5) *For each* $i \in I$ *and* $1 \neq X \leq C_S(\mathcal{C}_i) \cap N_S(P)$, $X\alpha \in \Sigma_Q$ *and* $(N_P(X)\alpha, \alpha^{-1}, i) \in \mathfrak{T}(X)$.

PROOF. By 12.3.37.2, there is $\beta \in \mathfrak{A}(P)$ with $P\beta \leq U$. Let $\gamma \in \mathfrak{A}(P\beta)$ with $P\beta\gamma = P\alpha$. As $W \leq C_S(P\beta)$, γ is a $\mathcal{U}$-map by 12.3.33.2, so as $P\beta \leq U$, also $P\alpha = P\beta\gamma \leq U$, proving (1).

Let $R = C_U(P)\alpha$. Then $R\alpha^{-1} \leq U$, so $R \in \mathcal{P}$ and then (2) holds. By 12.3.37, P and U are TI-subgroups of S with $P \cap U = 1$, so $|C_U(P)| \geq |P|$; then (3) follows. Parts (4) and (5) are straightforward. □

LEMMA 12.3.47. *For* $i \in I$, *set* $Z_i = \langle \mathcal{Z}_S(\rho_i), c_i \rangle$ *and* $Z = \prod_{i \in I} Z_i$. *Assume* $A \in \Xi_P$ *and let* $\alpha \in \mathfrak{A}(P)$. *Then*
(1) AP *centralizes* Z.
(2) $U\alpha \in \mathcal{P}^!$ *with* $Z\alpha \leq N_S(U\alpha)$.
(3) *For each* $i \in I$, $X_i\alpha \in \Sigma_{U\alpha}$, *where* $X_i = \prod_{i \neq j \in I} Z_i$.
(4) *Let* $E_4 \cong E \leq X_i$ *with* $1 \neq E \cap U$. *Then* $E\alpha \in \Xi_{U\alpha}$.
(5) $X_i\alpha$ *centralizes* Z.
(6) *For* $K \in \Omega$, $|N_K(U\alpha)| = 2$.
(7) $m(UX_i) \leq 2$.
(8) $n = 3$ *or* 4 *and* $r = 2$.
(9) $c_1 = c_2 = c$ *and* $U = \langle c \rangle \cong \mathbf{Z_2}$.

PROOF. Set $P' = U\alpha$; by 12.3.45.3, $P' \in \mathcal{P}^!$.

Suppose (6) fails. Then there is x of order 4 in $N_K(P')$. By 12.3.46.5, $x\alpha \in \Sigma_{P'}$. But $x\alpha \in D \in \Delta$, so $x\alpha \in S_j$ for some $j \in I$, contrary to 12.3.42.3. This proves (6).

By 12.3.46, A acts on each W_i and AP centralizes U.

Assume A does not centralize Z; then for some $t \in \mathcal{Z}_S$ and $a \in A$, $s = t^a \neq t$. As in the proof of 12.3.45, we may assume $A = P_A\langle a\rangle$. Let $Y = Q(t)Q(s)$ and observe that $Y = Q(t) \times Q(s)$ and $Y_a = C_Y(a)$ is a full diagonal subgroup of Y. By 12.3.44.3, $W_a = W \cap Y_a$ centralizes P, so as $W_a \not\leq Z(Y)$ it follows that P acts on $\{Q(t), Q(s)\}$, so $B = N_A(Q(t))$ is of index 2 in A. Also P_A acts on Y_a, and if $[P_A, Y_a] \neq 1$ then $Z(Y_a) \leq [P_A, Y_A] \leq P$, contradicting $Z(Y_a) \leq W$ and 12.3.44.3. Thus $A = P_A\langle a\rangle$ centralizes Y_a, so B centralizes a full diagonal subgroup of Y and therefore centralizes Y. But this contradicts (6), showing that A centralizes Z. Then $Z \leq N_S(P)$, so $[P, Z] \leq P \cap Z \leq P \cap W = 1$, proving (1).

Recall $P' \in \mathcal{P}^!$ and as $Z \leq N_S(P)$, we have $Z\alpha \leq N_S(P')$, proving (2). In (3), X_i centralizes $\mathcal{C}_i$, so (3) follows from 12.3.46.5. Then (3) implies (4). By (4) and symmetry between P and P', $E\alpha$ centralizes Z, so (5) follows.

Set $B = (UX_i)\alpha$. By (5), B centralizes Z, so B acts on $\{K, K_2\} = \Omega(z)$, and then $|B : N_B(K)| \leq 2$. Let V be of order 4 in $K \cap W$. Then $|N_B(K) : C_B(V)| \leq 2$, so $|B : C_B(V)| \leq 4$. Hence to prove (7), it suffices to show that $C_B(V) = 1$. But if $b \in C_B(V)^\#$ then $V \leq C_S(b) \leq N_S(P')$ by (4), contrary to (6). Hence (7) holds.

Next $m(Z_i) \geq n/2$, so $n = 3$ or 4 by (7). Similarly if $r > 2$ then $m(X_i) > 2$, contrary to (7). This proves (8). Indeed if $c_1 \neq c_2$ then $m(UX_i) = 3$, again contrary to (7), so (9) holds. □

Before leaving the setup of this lemma, we derive a contradiction to the assumption that $\Xi_P \neq \emptyset$. Namely by (8) and (9), $Z_1 = \langle z, c\rangle \cong E_4$. Set $b = c\alpha$, $B = Z_1\alpha$, $Y = KK_2$, and $Y^+ = Y/\langle z\rangle$. From the proof of (7), for some $a \in B - \langle b\rangle$, $K^a = K_2$, so setting $Y_a = C_Y(a)$, Y_a^+ is a full diagonal subgroup of $Y^+ = K^+ \times K_2^+$. Now $Y_a \leq N_S(P') = C_S(b)$, so $B = \langle a, b\rangle$ centralizes Y_a. Set $\langle d\rangle = N_B(K)$. As d centralizes the full diagonal subgroup Y_a^+ and acts on K, it follows that d centralizes K^+. But then $|C_K(d)| > 2$, contrary to (6). We've shown:

LEMMA 12.3.48. $\Xi_P = \emptyset$.

LEMMA 12.3.49. *(1)* $P = \langle a\rangle \cong \mathbf{Z}_2$.
(2) $r = 2$.
(3) $W_1^a = W_2$.

PROOF. By 12.3.46.4, $P \in \Sigma_P$, so (1) follows from 12.3.48.

By 12.3.46.3, $P' = C_U(P)\alpha \in \mathcal{P}^!$. Suppose $\emptyset \neq J \subset I$ such that $X = \prod_{j\in J} W_j$ is a-invariant. Then $X \leq C_S(\mathcal{C}_i)$ for $i \in I - J$, so by 12.3.46.5, $A = C_X(a)\alpha \in \Sigma_{P'}$. Then by 12.3.47, $m(A) = 1$, so as X is abelian, $m(X) = 2$. But if $r > 1$ we may choose $J = \{1, 2\}$, so $m(X) \geq 3$, a contradiction. This establishes (2) and shows $J = \{1\}$ and $X = W_1$. As $m(X) = 2$, $n = 3$ or 4. As $m(A) = 1$ and a centralizes c_1, $z^a \neq z$. Thus $Y = C_{O(z)O(z^a)}(z)$ is a full diagonal subgroup of $O(z) \times O(z^a)$, so $Y \cong O(z)$, contradicting $m(A) = 1$. This completes the proof of (3). □

We are now in a position to obtain a contradiction, and hence establish Theorem 12.3.8. Adopt the notation of 12.3.49, and let v be of order 4 in $K \cap W$ and $X = \langle v, v^a\rangle$. By 12.3.49, $W_1^a = W_2$, so $X \cong \mathbf{Z}_4^2$ and $Y = \langle a\rangle X$ is wreathed. Therefore $H = C_X(a) \cong \mathbf{Z_4}$ and the involution in H is $b = zz^a$ with $ab \in a^X$. Let $\alpha \in \mathfrak{A}(P)$, $d = (ab)\alpha$, and $Q = \langle d\rangle\alpha$. As $P \in \mathcal{P}!$ is of order 2 and $ab \in a^X$, we have $Q \in \mathcal{P}^!$, so $W_1^d = W_2$ by 12.3.49. But $H \leq C_S(P)$, so $H\alpha \leq N_S(Q)$ and then $d \in \Phi(H\alpha) \leq \Phi(S)$, contradicting $W_1^d = W_2$. This contradiction completes the proof of Theorem 12.3.8.

THEOREM 12.3.50. *Assume Hypothesis* 12.3.1 *and let* $M \in M(\tau)$ *and* $S_M = S \cap M$. *Then* $\mu(\tau) \cong$ *Weyl*(D_n) *for some* $n \geq 3$, *and one of the following holds:*
(1) $\mathcal{F}^\circ = \mathcal{F}_{S_M}(M)$.
(2) $m = 8$ *and* $\mathcal{F}^\circ \cong AE_{2n}$, $\check{A}E_{2n}$, *or* AE_{2n+1}.
(3) $\mathcal{F}^\circ \cong (P)\Omega_{2n}^+[m]$, $\Omega_{2n+1}[m]$, *or* $\Omega_{2n+2}^-[m]$.
(4) $(n, m) = (4, 8)$ *and* $\mathcal{F}^\circ \cong \check{L}_3(2)/2^{3+6}$ *or* $\check{\Omega}_8^+(2)$.
(5) $(n, m) = (3, 8)$ *and* $\mathcal{F}^\circ \cong L_3(2)/E_{64}$ *or* $Sp_6(2)$.

PROOF. By Theorem 12.3.8 and 12.3.4, $\mu(\tau) \cong \text{Weyl}(D_n)$ for some $n \geq 3$.
If $n = 3$ then the theorem follows from Theorem 9.5.33. If $n = 4$ the theorem follows from Theorem 12.1.25. Thus we may assume $n \geq 5$. Then the hypothesis of Theorem 12.2.32 is satisfied, and that theorem completes the proof. □

12.4. The proof of Theorem 6

In this section we prove Theorem 6 under the additional hypothesis that τ satisfies the Extended Inductive Hypothesis. At the end of the paper, when we will finally be in a position to prove Theorem 1, we will remove the assumption that the Extended Inductive Hypothesis holds, and complete the proof of Theorem 6.

So in this section we assume the hypothesis of Theorem 6, together with the Extended Inductive Hypothesis. Let $z \in \mathcal{Z}_S \cap \mathcal{F}^f$. By the hypothesis of Theorem 6, $|\Omega(z)| > 1$.

Suppose first that $|\Omega(z)| > 2$. Then Hypothesis 10.2.1 is satisfied. But then one of the conclusions of Theorem 6 holds by Theorem 10.2.13. Hence we may assume:

LEMMA 12.4.1. $|\Omega(z)| = 2$.

LEMMA 12.4.2. $\mathcal{Z} \cap O(z) = \{z\}$.

PROOF. If not then one of the conclusions of Theorem 6 holds by Theorem 9.3.24. □

LEMMA 12.4.3. $\mathcal{D}^*(z) \neq \mathcal{D}(z)$.

PROOF. If not then one of the conclusions of Theorem 6 holds by Theorem 9.4.33. □

Observe that by 12.4.1-12.4.3, Hypothesis 12.3.1 is satisfied. But now one of the conclusions of Theorem 6 holds by Theorem 12.3.50. This completes the proof that Theorem 6 holds under the additional assumption that the Extended Inductive Hypothesis is satisfied.

Theorems 7 and 8

CHAPTER 13

$|\Omega(z)| = 1$ and μ abelian

In Chapter 13 we prove Theorem 7, modulo the Extended Inductive Hypothesis. Thus our assumption throughout the chapter is that $\tau = (\mathcal{F}, \Omega)$ is a quaternion fusion packet such that for each $z \in \mathcal{Z}_S(\tau)$ we have $|\Omega(z)| = 1$, and such that $\mu = \mu(\tau)$ is abelian. In section 13.1 we work under Hypothesis 13.1.1, which also assumes that either (i) $\mathcal{F} = S\mathcal{F}^\circ$ and $\mathcal{F}^\circ$ is isomorphic to $(P)Sp_{2n_s}[m]$, $(P)Sp^-_{2n_S}[m]$, or $L^-_{2n_S+1}[m]$, or (ii) τ satisfies the Extended Inductive Hypothesis . Case (i) is included in order to establish various properties of such systems used in the generational results for the systems established in sections 13.3, 13.4, and 13.5.

Section 13.1 proves some basic results and then in Theorem 13.1.15 reduces our treatment of systems satisfying Hypothesis 13.1.1 to the generic case of Hypothesis 13.2.1 where $Aut_{\mathcal{F}}(O(\tau))$ induces $Sym(\Omega)$ on Ω and some member of $\mathcal{Z}(\tau)$ moves some member of Ω. In section 13.2 we establish various results under this hypothesis, which set the stage for the generational results in the middle of the chapter. In the final section we use the theory developed in the earlier sections to determine in Theorem 13.6.6 the systems satisfying Hypothesis 13.1.1 such that $\mathcal{F}^\circ$ is transitive on Ω. Then in Remark 13.6.7 we observe that Theorem 13.6.7 and Theorem 1 imply Theorem 7.

13.1. Systems with μ abelian

In this section we assume the following hypothesis:

HYPOTHESIS 13.1.1. *$\tau = (\mathcal{F}, \Omega)$ is a quaternion fusion packet such that:*
(1) For each $z \in \mathcal{Z}_S$, $|\Omega(z)| = 1$.
(2) $\mu(\tau)$ is abelian.
(3) Either
(i) $\mathcal{F} = S\mathcal{F}^\circ$ and $\mathcal{F}^\circ = O^2(\mathcal{F}) \cong (P)Sp_{2n_S}[m]$, $(P)SL^-_{2n_S}[m]$, or $L^-_{2n_S+1}[m]$,
or
(ii) τ satisfies the Extended Inductive Hypothesis

NOTATION 13.1.2. Pick $z \in \mathcal{F}^f \cap \mathcal{Z}$ and let $K \in \Omega(z)$. Set $\mu = \mu(\tau)$. For $s \in \mathcal{Z}$, define $\mathcal{D}^o(s) = C_{\mathcal{Z}}(s) - s^\perp$. Set $\mathcal{Z}' = \mathcal{Z}'(\tau) = \mathcal{Z} - \mathcal{Z}_S$, $\mathcal{Z}^m = \mathcal{Z}^m(\tau) = \{t \in \mathcal{Z} : \mathcal{Z}_S \not\subseteq C_S(t)\}$, and $O = O(\tau)$. For $\lambda \subseteq \Omega$ set $\lambda' = \Omega - \lambda$ and $O(\lambda) = \prod_{J\in\lambda} J$. For $s \in S$, write s^Ω for the image in $Sym(\Omega)$ of s.

LEMMA 13.1.3. *Suppose $\theta \subseteq \mathcal{Z}$ such that for each $t \in \theta$, $\theta \subseteq t^\perp$. Set $Q = \langle\theta\rangle$. Then for $\alpha \in \mathfrak{A}(Q)$, $\theta\alpha \subseteq \mathcal{Z}_S$.*

PROOF. Replacing θ by $\theta\alpha$, we may assume that $Q \in \mathcal{F}^f$. Let $t \in \theta$, $\Gamma = \theta - \{t\}$, and $\beta \in \mathfrak{A}(\Gamma)$. By induction on $|\theta|$, $\Gamma\beta \subseteq \mathcal{Z}_S$. By hypothesis, $\Gamma\beta \subseteq (t\beta)^\perp$, so for each $s \in \Gamma$, $t\beta$ centralizes $O(s\beta)$ by 3.3.4. Let $\gamma \in \mathfrak{A}(t\beta)$; then $O(s\beta)\gamma \in \Omega$, so $\theta\beta\gamma \subseteq \mathcal{Z}_S$. Finally let $\delta \in \mathfrak{A}(Q\beta\gamma)$ with $Q\beta\gamma\delta = Q$. Then $O = O\delta \leq C_S(Q)$, so $\theta \subseteq \mathcal{Z}_S$ by 3.1.5.3. □

LEMMA 13.1.4. *Suppose $t \in \mathcal{D}^o(z)$ and set $E = \langle z, t\rangle$. Then*
(1) $K\langle t\rangle \cong SD_{2n}$.
(2) $\mathcal{D}^o$ is a symmetric $\mathcal{F}$-invariant relation.
(3) Let $v \in K$ be of order 4 with $\langle v\rangle \lhd C_S(z)$, $\alpha \in \mathfrak{A}(t)$, and $u \in O(t\alpha)$ of order 4 with $\langle u\rangle \lhd C_S(t\alpha)$. Then $GL(E) = \langle c_v, c_u\alpha^{-*}\rangle = Aut_{\mathcal{F}}(E)$.

PROOF. Part (1) follows from 3.3.6. As $\mathcal{D}^o(s) = C_{\mathcal{Z}}(s) - s^\perp$ for $s \in \mathcal{Z}$ and $\mathcal{D}$ is a symmetric $\mathcal{F}$-invariant relation by 3.1.13, (2) follows.

Assume the setup of (3). By (1), t inverts v, so $c_v \in Aut_{\mathcal{F}}(E)$ is the transvection on E with center z. By (2), $z\alpha \in \mathcal{D}^o(t\alpha)$, so by symmetry c_u is the transvection on $E\alpha$ with center $t\alpha$, and hence $c_u\alpha^{-*}$ is the transvection on E with center t. Thus (3) holds. □

LEMMA 13.1.5. *Assume* 13.1.1.*3.ii holds and $\rho = (\mathcal{Y}, \Gamma)$ is a subpacket of τ with $\Gamma \subseteq \Omega$.*
(1) ρ satisfies conditions (1) and (2) of Hypothesis 13.1.1 *and the Inductive Hypothesis.*
(2) $\rho^\circ = \rho_1 * \cdots * \rho_r$ is a central product of quaternion fusion packets $\rho_i = (\mathcal{Y}_i, \Gamma_i)$ such that ρ_i satisfies conditions (1) and (2) of Hypothesis 13.1.1 *and the Inductive Hypothesis, and $\mathcal{Y}_i = \mathcal{Y}_i^\circ$ is transitive on Γ_i.*
(3) Assume ρ is proper in τ and let $1 \leq i \leq r$, $\mathcal{Z}_i = \mathcal{Z}(\rho_i)$, $K_i \in \Gamma_i$, $z_i = z(K_i)$, and $\mu_i = \mu(\rho)_i$. Then one of the following holds:
(i) $\Gamma_i = \{K_i\}$ and $\mathcal{Y}_i = \mathcal{O}_{K_i}$.
(ii) $\mu_i \cong \mathbf{Z}_2$, $\mathcal{Z}_i - \{z_i\} = \mathcal{D}^o(z_i)$, and $\mathcal{Y}_i \cong L_2[2m]^{(1)}$.
(iii) $m = 8$, $\mu_i \cong E_4$, $C_{\mathcal{Z}_i}(z_i) = z_i^\perp$, and $\mathcal{Y}_i \cong \hat{A}E_5$.
(iv) $\mu_i \cong E_{2^{n_i}}$, $C_{\mathcal{Z}_i}(z_i) = z_i^\perp$, and $\mathcal{Y}_i \cong (P)Sp_{2n_i}[m]$ for some integer $n_i \geq 2$.
(v) $\mu_i \cong E_{2^{n_i}}$, $\mathcal{D}^o(z(J)) \neq \emptyset$ for $J \in \Gamma_i$, and $\mathcal{Y}_i \cong (P)SL^-_{2n_i+a}[m]$ for some n_i and $a \in \{0,1\}$ with $2n_i + a \geq 3$.

PROOF. Part (1) is trivial. Part (2) follows from (1), 13.1.1.3.ii, and 6.6.6.

Assume ρ is proper. Then for $1 \leq i \leq r$, ρ_i satisfies the hypothesis of Theorem 1 and the Inductive Hypothesis by (2), so ρ_i satisfies one of the conclusions of Theorem 1. We conclude that one of the following holds:

(a) $\mathcal{Z}_i \subseteq \mathcal{Z}_S$, and (cf. 13.1.11) $\mathcal{Y}_i = \mathcal{O}(\rho_i)$.

(b) $\mathcal{Z}_i' \neq \emptyset = \mathcal{Z}_i^m$ and (cf. 13.1.12) $\mathcal{Y}_i \cong L_2[2m]^{(1)}$ or $L_3^-[m]$.

(c) $\mathcal{Z}_1^m \neq \emptyset$ and (cf. 13.1.15 and 13.6.6) $\mathcal{Y}_i$ is $\hat{A}E_5$, $(P)Sp_{2n}[m]$, $(P)SL^-_{2n}[m]$, or $L^-_{2n+1}[m]$.

In case (a), as $\mathcal{Y}_i = \mathcal{Y}_i^\circ$ is transitive on Γ_i, (3i) holds. In case (b), (3ii) or (3v) holds. Finally in case (c), (3iii), (3iv), or (3v) holds. This completes the proof. □

LEMMA 13.1.6. *$\tau_z^\circ = \tau_1 * \cdots * \tau_r$ where $\tau_i = (\mathcal{F}_i, \Omega_i)$ is a quaternion fusion packet satisfying conditions (1) and (2) of Hypothesis* 13.1.1 *and the Inductive Hypothesis, and τ_i is described in* 13.1.5.*3.*

PROOF. In 13.1.1.3.i the lemma follows from the discussion in section 5.4, so assume that 13.1.1.3.ii holds. Then the lemma is essentially the special case of

13.1.5 with $\rho = \tau_z$. If $z \notin Z(\mathcal{F})$ the application is immediate. If $z \in Z(\mathcal{F})$ then $\mathcal{O}_K \trianglelefteq \mathcal{F}$ and we apply 13.1.5 to $\rho = (\mathcal{F}, \Omega - \{K\})$. □

LEMMA 13.1.7. *Let $t \in \mathcal{D}(z)$.*

(1) For $\alpha \in \mathfrak{A}_{\mathcal{F}_z}(t)$, $t\alpha \in \mathcal{Z}_S$.

(2) Let T be Sylow in $\mathcal{F}_z^\circ$. Then $t \in T$ and $t\alpha = t\varphi\phi$ for some $\varphi \in Aut_{\mathcal{F}_z}(T)$ and $\phi \in \hom_{\mathcal{F}_z^\circ}(t\varphi, T)$. Indeed $t\varphi$ and $t\alpha$ are in T_i Sylow in $\mathcal{F}_i$ for some i, and $t\varphi$ is fused to $t\alpha$ in $\mathcal{F}_i$.

(3) Let $\theta \subseteq \Omega$ with $O(\theta) \in \mathcal{F}^f$ and $t \in \mathcal{D}(z(J))$ for each $J \in \theta$. Let $\rho = (C_{\mathcal{F}}(O(\theta)), \theta')$. Then $t \in \mathcal{Z}(\rho)$.

PROOF. Let $\theta = \{z, t\}$ and $\beta \in \mathfrak{A}(\langle\theta\rangle)$. By 13.1.3, $\theta\beta \subseteq \mathcal{Z}_S$. Let $\gamma \in \mathfrak{A}(z\beta)$ with $z\zeta = z$ where $\zeta = \beta\gamma$. Then $O\gamma = O$, so $t\zeta \in \mathcal{Z}_S$. Let $\delta \in \mathfrak{A}_{\mathcal{F}_z}(t\zeta)$ with $t\zeta\delta = t\alpha$. Then $O\delta = O$, so $t\alpha \in \mathcal{Z}_S$, establishing (1).

By 13.1.6, $\tau_z^\circ = \tau_1 * \cdots * \tau_r$, and by (1), $t\alpha \in \mathcal{Z}_S(\tau_i)$ for some i. As $\mathcal{F}_z^\circ \trianglelefteq \mathcal{F}_z$ and $t\alpha \in T$, also $t \in T$ and φ and ϕ exist as in (2). Then as $\mathcal{F}_z^\circ$ is the central product of the $\mathcal{F}_i$, $t\varphi$ is fused to $t\alpha$ in $\mathcal{F}_i$. Thus (2) holds.

Finally (3) follows from (2) by induction on $|\Omega|$. □

LEMMA 13.1.8. *Assume $K_i \in \Omega$ and $z_i = z(K_i)$ for $i = 1, 2$ with $1 \neq u = z_1z_2 \in Z(\mathcal{F})$. Assume $t \in \mathcal{Z}$ such that either*

(a) $K_1^t = K_2$, or

(b) $t \in \mathcal{D}^o(z_i)$ for $i = 1, 2$.

Set $\mathcal{Y} = [K_1]_{\mathcal{F}}$ and $\Gamma = K_1^{\mathcal{F}}$. Then

(1) $\rho = (\mathcal{Y}, \Gamma)$ is a quaternion fusion packet and $\rho = \rho_1 * \cdots * \rho_r$ is a central product where $\rho_j = (\mathcal{Y}_j, \Gamma_j)$, $|\Gamma_j| = 2$, $\Gamma_1 = \{K_1, K_2\}$, t is fused to z_1 and z_2 in $\mathcal{Y}_1$, and $\mathcal{Y}_j \cong \hat{A}E_5$, $Sp_4[m]$, or $SL_4^-[m]$.

(2) $\tau^\circ = \rho * \xi_1 * \cdots * \xi_k$, where the $\xi_j = (\mathcal{X}_j, \Sigma_j)$ are coconnected components of τ° and $\mathcal{X}_j$ is described in 13.1.5.3.

PROOF. In 13.1.1.3.i, as $u \in Z(\mathcal{F})$ we conclude that $\mathcal{F}^\circ$ is $Sp_4[m]$ or $SL_4^-[m]$, and then the lemma follows. Therefore we may assume that 13.1.1.3.ii holds.

By 13.1.5.2, $\tau^\circ = \tau_1 * \cdots * \tau_n$ is a central product, where $\tau_i = \tau_i^\circ = (\mathcal{T}_i, \Omega_i)$ and $\mathcal{T}_i$ is transitive on Ω_i. By 2.5.2, $\rho = (\mathcal{Y}, \Gamma)$ is a subpacket of τ, and by construction, $\mathcal{Y} \trianglelefteq \mathcal{F}$. Thus we may choose notation so that $\tau_j = \rho_j = (\mathcal{Y}_j, \Gamma_j)$ for $1 \leq j \leq r$ are the conjugates of $\mathcal{Y}_1$, and for $1 \leq j \leq k$, $\tau_{r+j} = \xi_j = (\mathcal{X}_j, \Sigma_j)$. Further we may choose $K_1 \in \Gamma_1$. Now in cases (a) and (b), z_1, t, z_2 is a path in $\mathcal{D}^c$, so it follows that as $K_1 \in \Gamma_1$, also $z_1, t, z_2 \in \mathcal{Z}_1 = \mathcal{Z}(\rho_1)$.

Set $\mathcal{F}^+ = \mathcal{F}/\langle u\rangle$ and $\tau^+ = (\mathcal{F}^+, \Omega^+)$. Then τ^+ is a quaternion fusion packet by 3.3.2.2. Moreover $(\tau^\circ)^+ = \tau_1^+ * \cdots * \tau_n^+$ is a central product with $\tau_j^+ = (\mathcal{T}_j^+, \Omega_j^+)$ and $\mathcal{T}_j^+$ is transitive on Ω_j^+. By 6.6.9.2, τ^+ satisfies the Extended Inductive Hypothesis. Further $\Omega(z_1^+) = \{K_1^+, K_2^+\}$ is of order 2, and from 13.1.1.2, $\mu(\tau^+)$ is abelian. It follows from Theorem 9.4.33 that $\Gamma_1^+ = \{K_1^+, K_2^+\}$, t^+ is fused to z_1^+ in $\mathcal{Y}_1^+$, and $\mathcal{Y}_j^+ \cong AE_5$, $\Omega_5[m]$, or $\Omega_6^-[m]$ for each $1 \leq j \leq r$. Then (1) follows from 5.1.19 and 3.3.16. Part (2) follows from the discussion above together with 13.1.5.3. □

LEMMA 13.1.9. *Assume $K_i \in \Omega$ and $z_i = z(K_i)$ for $i = 1, 2$ with $1 \neq u = z_1z_2 \in \mathcal{F}^f$. Assume $t \in \mathcal{Z}$ such that either*

(a) $K_1^t = K_2$, or

(b) $t \in \mathcal{D}^o(z_i)$ for $i = 1, 2$. Then

(1) τ_u° has a coconnected component $\tau_1 = (\mathcal{Y}, \Gamma)$ where $\Gamma = \{K_1, K_2\}$, t is fused to z_1 and z_2 in $\mathcal{Y}$, and $\mathcal{Y} \cong \hat{A}E_5$, $Sp_4[m]$, or $SL_4^-[m]$.

(2) Suppose $s \in \mathcal{Z}$ centralizes K_1 and K_2. Then $s \in \mathcal{Z}(\tau_u)$ and either s centralizes $\mathcal{Y}$ or $s \in \{z_1, z_2\}$.

PROOF. From 2.5.2, $\tau_u = (\mathcal{F}_u, \Omega)$ is a quaternion fusion packet.

Suppose first that (a) holds. Then $tu \in t^{K_1}$, so $\theta = \{t, tu\} \subseteq \mathcal{Z}$. Claim $tu \in \mathcal{D}(t)$. If not then $tu \in \mathcal{D}^o(t)$, so by 13.1.4.3, $u \in t^{Aut_{\mathcal{F}}(\langle t,u\rangle)} \subseteq \mathcal{Z}$ and $u \notin \mathcal{D}(t)$. Let $\alpha \in \mathfrak{A}(t)$ with $t\alpha = z$. Observe $X = C_{K_1K_2}(t) \cong K_1$ with $u = z(X)$, so $u\alpha$ induces an inner automorphism on K, and hence $u\alpha \in \mathcal{D}(z)$, so $u \in \mathcal{D}(t)$, contradicting and earlier remark. This proves the claim.

Set $Q = \langle\theta\rangle$ and let $\alpha \in \mathfrak{A}(Q)$; by 13.1.3, $\theta\alpha \subseteq \mathcal{Z}_S$. Let $\beta \in \mathfrak{A}(u\alpha)$ with $u\zeta = u$, where $\zeta = \alpha\beta$. Then $O(\theta\alpha)\beta = O(\theta\zeta)$ so $\theta\zeta \subseteq \mathcal{Z}_S$. Therefore as $\zeta \in \hom_{\mathcal{F}_u}(Q, Q\zeta)$, it follows that $t \in \mathcal{Z}(\tau_u)$. Hence (1) follows from 13.1.8 if $u \in Z(\mathcal{F})$, while if $u \notin Z(\mathcal{F})$ then τ_u is proper in τ, and (1) follows from 13.1.5 and the structure of the coconnected components of τ_u in 13.1.5.3; cf. section 5.4.

Next assume that (b) holds. Let $\alpha \in \mathfrak{A}(t)$ with $t\alpha = z$. By 13.1.4.2, $z_i\alpha \in \mathcal{D}^o(z)$ for $i = 1, 2$, so $z_i\alpha$ inverts the $C_S(z)$-invariant cyclic subgroup V of K of index 2. Therefore $u\alpha = z_1\alpha \cdot z_2\alpha$ centralizes V. Let $\beta \in \mathfrak{A}(u\alpha)$ with $u\zeta = u$, where $\zeta = \alpha\beta$. Then $\zeta \in \hom_{\mathcal{F}_u}(C_S(\langle z_1, u, t\rangle), S)$. Now $V\beta \leq J \in \Delta(t\zeta)$, and by 13.1.1.2, $J \in \Omega$, so $t\zeta \in \mathcal{Z}_S$. Hence again $t \in \mathcal{Z}(\tau_u)$, so again (1) follows from 13.1.8 if $u \in Z(\mathcal{F})$ and from 13.1.5 and the structure of the coconnected components of τ_u in 13.1.5 when $u \notin Z(\mathcal{F})$.

Assume the hypothesis of (2) with $s \notin \{z_1, z_2\}$, and let $\alpha \in \mathfrak{A}(s)$ with $s\alpha = z$ and $\beta \in \mathfrak{A}(u\alpha)$ with $u\zeta = u$, where $\zeta = \alpha\beta$. Then $K_i\alpha$ centralizes K, so $K\beta$ centralizes $(K_1K_2)\zeta$ and thus as $u \in (K_1K_2)\zeta$, we have $s\zeta \in \mathcal{Z}_S$ centralizes u. Therefore as $\zeta \in \hom_{\mathcal{F}_u}(K_1K_2\langle s\rangle, S)$, $s \in \mathcal{Z}(\tau_u)$ so $s\zeta \in \mathcal{Z}(\rho')$ for some coconnected component ρ' of τ_u°. As $s \notin \{z_1, z_2\}$ centralizes K_1K_2, $\rho' \neq \rho$, so s centralizes $\mathcal{Y}$, proving (2). □

LEMMA 13.1.10. *Let $t \in \mathcal{Z}$. Then*

(1) $t \in \mathcal{Z}_S$ iff $t \in s^\perp$ for each $s \in \mathcal{Z}_S$.

(2) Suppose $K_i \in \Omega$ and $z_i = z(K_i)$ for $i = 1, 2$ with $1 \neq u = z_1z_2$ and $t \in \mathcal{Z}$ such that either

(a) $K_1^t = K_2$, or

(b) $t \in \mathcal{D}^o(z_i)$ for $i = 1, 2$.

Let $\alpha \in \mathfrak{A}(u)$. Then there exists a coconnected component $\rho = (\mathcal{Y}, \Gamma)$ of $\tau_{u\alpha}^\circ$ such that $\Gamma = \{K_1\alpha, K_2\alpha\}$, $t\alpha$ is fused to $z_1\alpha$ and $z_2\alpha$ in $\mathcal{Y}$, and $\mathcal{Y} \cong \hat{A}E_5$, $Sp_4[m]$, or $SL_4^-[m]$.

(3) If $t \in \mathcal{Z}^m$ then t induces a transposition (K_1, K_2) on Ω and centralizes each member of $\Omega - \{K_1, K_2\}$.

(4) If $\mathcal{D}^o(t) \cap \mathcal{Z}_S \neq \emptyset$ then $|\mathcal{D}^o(t) \cap \mathcal{Z}_S| \leq 2$ and $\mathcal{Z}_S - \mathcal{D}^o(t) \subseteq \mathcal{D}(t)$.

(5) Assume the hypothesis of (2) with $t \in \mathcal{Z}^m$. Then $\mathcal{Z}$ centralizes u and acts on $\{K_1, K_2\}$.

(6) Let $t_i \in \mathcal{Z}'$ and $\Gamma_i = \{J \in \Omega : [t_i, J] \neq 1\}$, for $i = 1, 2$. Assume $\Gamma_1 \cap \Gamma_2 = \emptyset$. Then $[t_1, t_2] = 1$.

(7) If $t, a \in \mathcal{Z}^m$ with $t^\Omega \neq a^\Omega$ then $[t, a] = 1$,

PROOF. Part (1) follows from 13.1.3.

Assume the hypothesis of (2); replacing u by $u\alpha$, we may assume $u \in \mathcal{F}^f$ and $\alpha = 1$. Then (2) follows from 13.1.9.

Suppose t moves some $K_1 \in \Omega$ or is nontrivial on two members K_1 and K_2 of Ω; then we may assume the setup of (2) holds with $u \in \mathcal{F}^f$ and $\alpha = 1$. Therefore for each $J \in \Omega - \Gamma$, $J \in \Gamma'$ where $\rho' = (\mathcal{Y}', \Gamma')$ is a coconnected component of τ_u° distinct from ρ, and hence as $t \in \mathcal{Z}(\rho)$, t centralizes J. It follows that (3) and (4) hold.

Suppose $t \in \mathcal{Z}^m$. As distinct transpositions in S^Ω commute, it follows that all members of $\mathcal{Z}$ act on $\{K_1, K_2\}$ and hence centralize u, establishing (5).

Assume the hypothesis of (6). If $\Gamma_1 = \{K_1, K_2\}$ is of order 2, then, replacing $u = z_1 z_2$ by $u\alpha$ for $\alpha \in \mathfrak{A}(u)$, we may assume $u \in \mathcal{F}^f$. Then (6) follows from 13.1.9.2. Thus we may assume $\Gamma_i = \{K_i\}$ is of order 1. Let $\theta = \Omega - \{K_1, K_2\}$. If $n = |\Omega| > 2$ then appealing to 13.1.7.3 and induction on n, (6) holds, so we may take $n = 2$. Then by 13.1.6 and 13.1.7.2, $\mathcal{F}_{z_i}^\circ = \mathcal{O}(z_i) * \mathcal{Y}_i$ with $\mathcal{Y}_i \cong L_2[2m]^{(1)}$ or $L_3^-[m]$ and $t_{3-i} \in T_i$ Sylow in $\mathcal{Y}_i$ with $T_i \cong SD_{2m}$. Then $[T_1, T_2] \leq C_{T_1}(O(z_1)) \cap C_{T_2}(O(z_2)) = 1$, completing the proof of (6).

Assume the hypothesis of (7). Then $t = (K_1, K_2)$ and $a = (J_1, J_2)$ are distinct cycles on Ω, so (7) follows from (3) and (6). □

LEMMA 13.1.11. *If $\mathcal{Z} = \mathcal{Z}_S$ and $\mathcal{F}$ is transitive on Ω then $\mathcal{F}^\circ = \mathcal{O}(\tau)$.*

PROOF. By 13.1.1.1, $\mathcal{O}_K$ is tightly embedded in $\mathcal{F}$. Thus the lemma follows from Theorem 3.6.8 in [**Asc19**]. □

LEMMA 13.1.12. *Assume $\mathcal{Z}' \neq \emptyset = \mathcal{Z}^m$ then $\mathcal{F}^\circ$ is the central product of conjugates of $L_2[2m]^{(1)}$ or $L_3^-[m]$.*

PROOF. If 13.1.1.3.i holds then as $\mathcal{Z}' \neq \emptyset = \mathcal{Z}^m$, we have $\mathcal{F}^\circ \cong L_3^-[m]$, so the lemma holds. Thus we may assume that 13.1.1.3.ii holds.

Let $n = |\Omega|$. If $n = 1$ the lemma is a consequence of Theorem 7.1.6, so assume $n > 1$.

(a) For each $t \in \mathcal{Z}'$, there exists a unique $J \in \Omega$ with $[J, t] \neq 1$.

As $\mathcal{Z}^m = \emptyset$, t acts on each member of Ω, while as $t \in \mathcal{Z}'$, $[J, t] \neq 1$ for some $J \in \Omega$ by 13.1.10.1. If t is nontrivial on two members of Ω, then by 13.1.10.2, there is a subpacket $\rho = (\mathcal{Y}, \Gamma)$ of τ with $\mathcal{Y}$ isomorphic to $\hat{A}E_5$, $Sp_4[m]$, or $SL_4^-[m]$. But then there is $s \in \mathcal{Z}^m(\rho)$, contradicting $\mathcal{Z}^m = \emptyset$.

From 3.1.5.4 and the transitivity of $\mathcal{F}$ on Ω:

(b) $Aut_{\mathcal{F}}(O)$ is transitive on Ω.

Let $I = \{1, \ldots, n\}$, $\Omega = \{K_i : i \in I\}$, and set $z_i = z(K_i)$. Set $\mathcal{Z}_i' = \{t \in \mathcal{Z}' : [K_i, t] \neq 1\}$ and $T_i = \langle \mathcal{Z}_i', K_i \rangle$. As $\mathcal{Z}^m = \emptyset$, $T = T_1 \cdots T_m$ is in the kernel of the action of S on Ω, so by (b) and a Frattini argument:

(c) $Aut_{\mathcal{F}}(T)$ is transitive on Ω and $\{T_i : i \in I\}$.

By (a), $\mathcal{Z}_i' \neq \emptyset$ for some $i \in I$, and hence $\mathcal{Z}_i' \neq \emptyset$ for each $i \in I$ by (c). Take $z = z_1$.

(d) $\tau_z^\circ = \mathcal{O}(z) * \tau_2 * \cdots * \tau_n$, where for $1 < i \in I$, $\tau_i = (\mathcal{Y}_i, K_i)$ is a quaternion fusion packet, $\mathcal{Y}_i \cong L_2[2m]^{(1)}$ or $L_3^-[m]$, T_i is Sylow in $\mathcal{Y}_i$, and $T_i \cong SD_{2m}$.

By 13.1.6, $\tau_z^\circ = \tau_1 * \cdots * \tau_r$ with $\tau_i = (\mathcal{Y}_i, \Gamma_i)$ and $\mathcal{Y}_i$ is described in 13.1.5.3. Let $t \in \mathcal{Z}_i'$ for $i > 1$. By (a), t centralizes K, so by 13.1.7.2, $t \in \mathcal{Z}(\tau_j)$ for some $j \in I$. Then t centralizes $\mathcal{Y}_k$ for $k \neq j$, so $K_i \in \Gamma_j$. As $\mathcal{Z}^m = \emptyset$, $\mathcal{Z}^m(\tau_l) = \emptyset$ for each l, so from 13.1.5.3, $\mathcal{Y}_l \cong L_2[2m]^{(1)}$ or $L_3^-[m]$. Thus $\Gamma_j = \{K_i\}$ and a Sylow group Q_j of $\mathcal{Y}_j$ is isomorphic to SD_{2m}. Therefore $Q_j = T_i$, establishing (d).

(e) T_1 centralizes $\mathcal{Y}_i$ for $1 < i \in I$.

From (d), K centralizes $\mathcal{Y}_i$, so it remains to show $t \in \mathcal{Z}_1'$ centralizes $\mathcal{Y}_i$. By 13.1.10.6, T_1 centralizes T_i. Hence by 2.6.12, T_1 centralizes $\mathcal{O}_i = C_{\mathcal{Y}_i}(z_i)$. Next $\mathcal{Y}_i = \langle \mathcal{O}_i, Aut_{\mathcal{Y}_i}(R)\rangle$ where R is a 4-subgroup of T_i. Let G be a model for $\langle t\rangle N_{\mathcal{Y}_i}(R)$; then $G = \langle a\rangle \times H$ where $H \cong S_4$ is a model for $N_{\mathcal{Y}_i}(R)$ and $a = t$ or tz_i. If $a = t$ then (e) holds, so assume $a = tz_i$.

Let $\alpha \in Aut_{\mathcal{F}}(T)$ with $z_i\alpha = z$. Then $T_1\alpha = T_l$ for some $1 < l \in I$, so $(t\alpha)z$ is fused to $(z\alpha)z$ in $\mathcal{F}_z$. Therefore $a = tz_i \in u^{\mathcal{F}}$, where $u = z_jz_k \in \mathcal{F}^f$, so there is $\beta \in \mathfrak{A}(a)$ with $a\beta = u$. Now $\mathcal{F}_u^\circ = \mathcal{O}(z_j) * \mathcal{O}(z_k) * \prod_{r\neq j,k} \mathcal{Y}_r$ by 13.1.5, and as $\mathcal{Z}^m = \emptyset$. As $u = t\beta \cdot z_i\beta$ with $K_i\beta \in \Omega$, it follows that $z_i\beta = z_j$ and $t\beta = z_k$. This is a contradiction as $\mathcal{Y}_i\beta^* \leq \mathcal{F}_u^\circ$ and $z_i \notin Z(\mathcal{Y}_i)$. Hence the proof of (e) is complete.

By (c), (e), and 2.2 in [**Asc11**]:

(f) There is $\mathcal{Y}_1 \leq \mathcal{F}$ with Sylow group T_1 such that $\mathcal{Y} = \mathcal{Y}_1 * \cdots * \mathcal{Y}_n \leq \mathcal{F}$ is a central product and $Aut_{\mathcal{F}}(T)$ is transitive on $\{\mathcal{Y}_i : i \in I\}$.

(g) Let $\rho_i = (\mathcal{Y}_i, K_i)$, $i \in I$, and set $\mathcal{Z}_i = \mathcal{Z}_i' \cup \{z_i\}$. Then $\mathcal{Z}_i = \mathcal{Z}(\rho_i)$ and $\{\mathcal{Z}_i : i \in I\}$ is the set of connected components of $\mathcal{D}(\tau)^c$.

From (a), $\mathcal{Z}' = \bigcup_{i\in I} \mathcal{Z}_i'$, and as $\mathcal{Y}_i$ is $L_2[2m]^{(1)}$ or $L_3^-[m]$ with T_i Sylow in $\mathcal{Y}_i$, and as $T_i \cong SD_{2m}$, it follows that $\mathcal{Z}_i$ is the set of involutions in T_i, so $\mathcal{Z}_i = \mathcal{Z}(\rho_i)$. Then from (f), $\{\mathcal{Z}_i : i \in I\}$ is the set of connected components of $\mathcal{D}(\tau)^c$.

By (f) and (g) and 8.1.11, $\{\rho_i : i \in I\}$ is the set of coconnected components of τ°, completing the proof of the lemma. □

In light of 13.1.11 and 13.1.12, during the remainder of the section we assume the following hypothesis:

HYPOTHESIS 13.1.13. *Hypothesis* 13.1.1 *holds and*

(1) $\mathcal{F}$ *is transitive on* Ω, *and*

(2) $\mathcal{Z}^m \neq \emptyset$.

LEMMA 13.1.14. *Let* $A = Aut_{\mathcal{F}}(O)$ *and write* A^Ω *for the image of* A *in* $Sym(\Omega)$. *For* $t \in \mathcal{Z}^m$ *let* c_t *be conjugation by* t. *Set* $B = \langle c_t : t \in \mathcal{Z}^m\rangle \leq A$.

(1) A *is transitive on* Ω.

(2) There is an A*-invariant partition* Λ *of* Ω *such that for* $\lambda \in \Lambda$, $|\lambda| \geq 2$, *and* $B = \prod_{\lambda\in\Lambda} B_\lambda$ *is a direct product, where* $B_\lambda = \langle c_t : t \in \mathcal{Z}_\lambda\rangle$, $\mathcal{Z}_\lambda = \{t \in \mathcal{Z}^m : \lambda' \subseteq Fix_\Omega(c_t)\}$, *and* B_λ^Ω *acts faithfully as* $Sym(\lambda)$ *on* λ.

(3) For each $\lambda \in \Lambda$, B_λ *centralizes* $O(\lambda')$ *and* $\mathcal{Z}_\lambda$ *centralizes* $\mathcal{Z}_{\lambda'} = \bigcup_{\sigma\subseteq\lambda'} \mathcal{Z}_\sigma$.

PROOF. By 13.1.13.1, $\mathcal{F}$ is transitive on Ω, so (1) follows from 3.1.5.4.

Let $t \in \mathcal{Z}^m$. By 13.1.10.3, c_t^Ω is a transposition. Then (2) follows from this fact and (1). In particular $\{\mathcal{Z}_\lambda : \lambda \in \Lambda\}$ is a partition of $\mathcal{Z}^m$ with $\lambda' \subseteq Fix_\Omega(c_t)$ for $t \in \mathcal{Z}_\lambda$. Then by 13.1.10.7, $\mathcal{Z}_\lambda$ centralizes $O(\lambda')$ and $\mathcal{Z}_{\lambda'}$. Finally as B acts on $O(\lambda')$ and $B_\lambda = \langle Z_\lambda^B\rangle$, B_λ centralizes $O(\lambda')$, completing the proof of (3). □

THEOREM 13.1.15. *Assume* $Aut_{\mathcal{F}}(O)$ *does not induce* $Sym(\Omega)$ *on* Ω. *Then* $\mathcal{F}^\circ$ *is the central product of the conjugates of* $\hat{A}E_5$, *(P)*$Sp_n[m]$, *or (P)*$SL_n^-[m]$ *for some* $n \geq 4$.

During the remainder of the section assume that $\mathcal{F}$ is a counter example to Theorem 13.1.15.

As $Aut_{\mathcal{F}}(O)$ does not induce $Sym(\Omega)$ on Ω, 13.1.1.3.i does not hold, so 13.1.1.3.ii holds.

Adopt the notation in 13.1.14. As $A^{\Omega} \neq Sym(\Omega)$, the partition Λ of Ω in 13.1.14 is nontrivial. Set $n = |\lambda|$ for $\lambda \in \Lambda$ and $k = |\Lambda|$. Thus $n, k \geq 2$ and $|\Omega| = nk$.

Pick $\lambda \in \Lambda$ with $O(\lambda) \in \mathcal{F}^f$. As $N_S(O(\lambda)) = N_S(O(\lambda'))$, also $O(\lambda') \in \mathcal{F}^f$. Set $\mathcal{Y} = \mathcal{Y}_\lambda = C_{\mathcal{F}}(O(\lambda'))^\circ$ and $\rho = \rho_\lambda = (\mathcal{Y}, \lambda)$, and let T_λ be Sylow in $\mathcal{Y}$. By 2.5.2, ρ is a quaternion fusion packet. Indeed:

LEMMA 13.1.16. *Either*
(1) $n = 2$ and $\mathcal{Y} \cong \hat{A}E_5$, $Sp_4[m]$, or $SL_4^-[m]$, or
(2) $n \geq 3$ and $\mathcal{Y} \cong (P)Sp_{2n}[m]$ or $(P)SL_{2n+a}[m]$ with $a = 0$ or 1.

PROOF. By 13.1.5, $\rho = \rho_1 * \cdots * \rho_r$ with $\rho_i = (\mathcal{Y}_i, \Gamma_i)$ described in 13.1.5.3. As B acts on $O(\lambda')$, B acts on $O(\lambda) = \prod_i O(\Gamma_i)$, and permutes $\{\Gamma_1, \ldots, \Gamma_r\}$. Hence as $B^\lambda = Sym(\lambda)$ is primitive on λ, we conclude that $r = 1$. If $\mathcal{Y} \cong \hat{A}E_5$ then $n = 2$, while if $\mathcal{Y} \cong (P)Sp_l[m]$ or $(P)SL_l[m]$ then either $l = 2n$ or $\mathcal{Y} \cong (P)SL_{2n+1}[m]$; thus the lemma follows from 13.1.5.3. □

LEMMA 13.1.17. *For $\sigma \in \Lambda$, set $\mathcal{Z}'_\sigma = \{t \in \mathcal{Z}' : \sigma' \subseteq C_S(t)\}$. Then*
(1) $\{\mathcal{Z}'_\sigma : \sigma \in \Lambda\}$ is a partition of $\mathcal{Z}'$.
(2) $\mathcal{Z}'(\rho) = \mathcal{Z}'_\lambda$.

PROOF. We first prove (1). Let $t \in \mathcal{Z}'$. If $t \in \mathcal{Z}^m$ then $t \in \mathcal{Z}'_\sigma$ for a unique $\sigma \in \Lambda$ by 13.1.10.3, so we may assume $\mathcal{Z}_S \subseteq C_S(t)$. By 13.1.10.1, t is nontrivial on some $J \in \Omega$. Let $J \in \sigma \in \Lambda$. If t centralizes $\Omega - \{J\}$ then σ is the unique $\zeta \in \Lambda$ with $t \in \mathcal{Z}'_\zeta$, so assume $L \in \Omega - \{J\}$ with $[L, t] \neq 1$. Set $u = z(J)z(L)$ and let $\alpha \in \mathfrak{A}(u)$. By 13.1.10.2, there exists $s \in \mathcal{Z}^m$ with cycle $(J\alpha, L\alpha)$ on Ω, so $J\alpha, L\alpha \in \zeta \in \Lambda$. As O centralizes u, $\alpha \in Aut_{\mathcal{F}}(O)$, so $\zeta = \sigma\alpha$, and hence $L \in \sigma$, completing the proof of (1).

Choose notation so that $K \in \lambda$. By definition of ρ, $\mathcal{Z}(\rho) = z^{\mathcal{Y}}$ and $\mathcal{Z}'(\rho) = z^{\mathcal{Y}} - \mathcal{Z}_{T_\lambda}$. As z centralizes $O(\lambda')$, $\mathcal{Z}'(\rho) \subseteq \mathcal{Z}'_\lambda$. Conversely if $t \in \mathcal{Z}'_\lambda$ then for each $J \in \Omega$ with $[J, t] \neq 1$, $J \in \lambda$, so by 13.1.7.3, $t \in \mathcal{Z}'(\rho)$. □

Set $S_0 = C_S(O)$ and let G be a model for $N_{\mathcal{F}}(OS_0)$. For $\sigma \in \Lambda$ let $H_\sigma = \langle (\mathcal{Z}'_\sigma)^{N_G(\sigma)} \rangle O(\sigma)$.

LEMMA 13.1.18. *(1) $H_\lambda = \langle (\mathcal{Z}'_\lambda)^{G_\lambda} \rangle O(\lambda) = O^{2'}(G_\lambda)$, where G_λ is model for $N_{\mathcal{Y}}(O(\lambda))$.*
(2) $T_\lambda \in Syl_2(H_\lambda)$.
(3) $H = \langle H_\sigma : \sigma \in \Lambda \rangle$ is a central product of the H_σ, and G acts transitively on $\{H_\sigma : \sigma \in \Lambda\}$ via $H_\sigma^g = H_{\sigma^g}$.

PROOF. Let $\mathcal{X} = N_{\mathcal{F}}(O(\lambda'))$. By construction $\mathcal{Y} \trianglelefteq \mathcal{X}$ and $N_G(\lambda)$ is a model for $N_{\mathcal{X}}(OS_0)$. By 2.6.12, S_0 centralizes $\mathcal{O}(\tau)$, so we conclude from the structure of $\mathcal{Y}$ in 13.1.16 that $OS_0 \trianglelefteq OS_0N_{\mathcal{Y}}(O)$. Then $N_{\mathcal{Y}}(O) \trianglelefteq N_{\mathcal{X}}(OS_0)$, so we may take $G_\lambda \trianglelefteq N_G(\lambda)$. Then as $\mathcal{Z}'_\lambda \subseteq G_\lambda$ by 13.1.17, $H_\lambda \leq G_\lambda$. Indeed from the description of $\mathcal{Y}$ in 13.1.16 and 5.4.8-5.4.10, $O^{2'}(G_\lambda) = \langle (\mathcal{Z}'_\lambda)^{G_\lambda} \rangle O(\lambda)$, so (1) holds. Then (2) follows from (1).

As $H_\lambda \trianglelefteq N_G(\lambda)$, $H_\lambda^g = H_{\lambda^g}$ for $g \in G$. In particular as G is transitive on Λ, G is transitive on $\{H_\sigma : \sigma \in \Lambda\}$. Let G_Λ be the kernel of the action of G on Λ. As $H \leq G_\Lambda \leq N_G(\sigma)$, $H_\sigma \trianglelefteq H$. Set $H_{\lambda'} = \prod_{\sigma \neq \lambda} H_\sigma$. Then $H_\lambda \cap H_{\lambda'} \leq C_{H_\lambda}(O(\lambda)) \cap C_{H_{\lambda'}}(O(\lambda') \leq Z(O(\lambda)) \cap Z(O(\lambda')) \leq Z(H)$. Therefore $[H_\lambda, H_{\lambda'}] \leq Z(H)$, so $[H_\lambda, O^2(H_{\lambda'})] = 1$. Then as $H_{\lambda'} = O^2(H_{\lambda'})\langle \mathcal{Z}'_{\lambda'} \rangle$, where $\mathcal{Z}'_{\lambda'} = \bigcup_{\sigma \in \Lambda - \{\lambda\}} Z'_\sigma$, and as $\mathcal{Z}_\lambda$ centralizes $\mathcal{Z}_\sigma$ for each $\sigma \in \Lambda - \{\lambda\}$ by 13.1.10.6, we have $[H_\lambda, H_{\lambda'}] = 1$. This completes the proof of (3). □

For $\sigma \in \Lambda$, let $T_\sigma = S \cap H_\sigma$ and set $T = \prod_{\sigma \in \Lambda} T_\sigma$. For $\theta \subseteq \Lambda$ set $T_\theta = \prod_{\sigma \in \theta} T_\sigma$. Define $\mathcal{Y}_\sigma = \mathcal{Y} c_g^*$ for $g \in N_G(T)$ with $\lambda^g = \sigma$. By 13.1.18, $N_G(T) \cap N_G(\lambda) \leq N_G(T_\lambda)$, so $\mathcal{Y}_\sigma$ is independent of g.

LEMMA 13.1.19. *(1) $\mathcal{F}$ contains a central product $\mathcal{Y}_0 = \prod_{\sigma \in \Lambda} \mathcal{Y}_\sigma$.*
(2) $Aut_{\mathcal{F}}(T)$ acts on $\mathcal{Y}_0$ and on $Z = Z(\mathcal{Y}_0)$.

PROOF. By 13.1.18.3, $Q = T_{\lambda'}$ centralizes H_λ; therefore Q centralizes $N_{\mathcal{Y}}(O(\lambda))$ using 13.1.18.1, and observing from 13.1.16 that $C_{G_\lambda}(T_\lambda) = Z(G_\lambda)$. Take $K \in \lambda$ and $t \in \mathcal{Z}_\lambda$ with $K^t \neq K$. Set $u = zz^t$. By 13.1.9.2, $\mathcal{Z}'_{\lambda'}$ centralizes $C_{T_\lambda}(u)\mathcal{X}_u$, where $\mathcal{X}_u$ is the coconnected component of $(C_{\mathcal{Y}}(u), \lambda)$ described in 13.1.9.1. Therefore by 1.3.2 in [**Asc19**], $U = \langle \mathcal{Z}'_{\lambda'} \rangle O(\lambda')$ centralizes $\mathcal{Y} = \langle C_{T_\lambda}(u)\mathcal{X}_u, N_{\mathcal{Y}}(O(\lambda)) \rangle$. Let $\mathcal{X} = N_{\mathcal{F}}(O(\lambda'))$, X a model for $N_{\mathcal{X}}(OS_0)$, and $\mathcal{C} = C_{\mathcal{X}}(\mathcal{Y})$. Then $[U]_{\mathcal{X}} \leq \mathcal{C}$. But $H_{\lambda'} \leq X$ with $T_{\lambda'} \leq H_{\lambda'} = \langle U^{H_{\lambda'}} \rangle$ by 13.1.18, so $T_{\lambda'} \leq \mathcal{C}$. Therefore $T_{\lambda'}$ centralizes $\mathcal{Y}$, so $T_{\lambda'^g}$ centralizes $\mathcal{Y}_{\lambda^g}$ for $g \in N_G(T)$. Then (1) follows from 2.2 in [**Asc11**].

As $Aut_{\mathcal{F}}(T)$ permutes Λ, it permutes the T_σ and $\mathcal{Y}_\sigma$ by construction. Thus $Aut_{\mathcal{F}}(T)$ acts on $\mathcal{Y}_0$, and hence also on $Z(\mathcal{Y}_0)$, proving (2). □

LEMMA 13.1.20. $Z(\mathcal{F}) = 1$.

PROOF. Assume otherwise and let $y \in Z(\mathcal{F})$ be an involution and set $\mathcal{F}^+ = \mathcal{F}/\langle y \rangle$. Suppose first that $y = zz^t$ for some $z \in \mathcal{Z}_S$ and $t \in \mathcal{Z}$. Then by 13.1.8, $\mathcal{F}^\circ$ is the central product of the conjugates of $\mathcal{Y} \cong \hat{A}E_5$, $Sp_4[m]$, or $SL_4^-[m]$, contrary to the choice of $\mathcal{F}$.

Hence y is not of that type, so, using 6.6.9, $\tau^+ = (\mathcal{F}^+, \Omega^+)$ satisfies Hypothesis 13.1.1. Then by induction on the order of τ, $(\mathcal{F}^+)^\circ$ is the central product of copies of $\mathcal{Y}^+$, again contrary to 5.1.19, 3.3.16, and the choice of $\mathcal{F}$. □

LEMMA 13.1.21. *(1) Suppose $\theta = (\mathcal{X}, \Gamma)$ is a proper subpacket of τ with $\Gamma \subseteq \Omega$. Then each coconnected component of θ° is contained in $\tau_\sigma = (\mathcal{Y}_\sigma, \sigma)$ for some $\sigma \in \Lambda$, so if τ_σ is a subpacket of θ then it is a component of θ°.*
(2) For some $\sigma \in \Lambda - \{\lambda\}$, $T_\lambda \cap T_\sigma \neq 1$.
(3) $Z = Z(\mathcal{Y}_0) \neq 1$.
(4) Z is not normal in $\mathcal{F}$.
(5) $\mathcal{Y}_\lambda \cong \hat{A}E_5$, $Sp_{2n}[m]$, or $SL_{2n}^-[m]$.

PROOF. Let $\xi = (\mathcal{U}, \Sigma)$ be a coconnected component of θ°. Then ξ is described in 13.1.5.3. In particular taking G_ξ to be a model for $N_{\mathcal{U}}(O(\xi))$ and setting $H =$

$\langle \mathcal{Z}(\xi)^{G_\xi} \rangle$, H is transitive on Σ, so $\Sigma \subseteq \sigma$ for some $\sigma \in \Lambda$. Therefore (1) follows from 13.1.18.

Next by 13.1.20, $Z(\mathcal{F}) = 1$. Hence if $T_\lambda \cap T_\sigma = 1$ for each $\sigma \in \Lambda - \{\lambda\}$, then we can verify the hypotheses of 8.1.8 using (1) and arguments in the proof of 8.1.11. Hence $\mathcal{F}^\circ = \mathcal{Y}_0$ by 8.1.10, contrary to the choice of $\mathcal{F}$. This proves (2).

As $T_\lambda \cap T_\sigma \le Z(\mathcal{Y})$, (2) implies (3). Suppose $Z \trianglelefteq \mathcal{F}$. By 13.1.20, $C_{\mathcal{F}}(Z) \neq \mathcal{F}$ so by induction on the order of τ, $\mathcal{Y}_0 = \mathcal{F}^\circ$, contrary to the choice of $\mathcal{F}$. Thus (4) holds.

Part (5) follows from (3) and 13.1.16. □

Set $\mathcal{B} = N_{\mathcal{F}}(Z)$, where $Z = Z(\mathcal{Y}_0)$, and set $\tau_0 = (\mathcal{B}, \Omega)$. From 13.1.21.3, $Z \neq 1$ and $\mathcal{B} \neq \mathcal{F}$ by 13.1.21.4, so τ_0 is a proper subpacket of τ. Then from 13.1.5 and 13.1.21.1, $\tau_0^\circ = \prod_{\sigma \in \Lambda} \tau_\sigma$ is a central product where $\tau_\sigma = (\mathcal{Y}_\sigma, \sigma)$. In particular $\mathcal{Y}_0 \trianglelefteq \mathcal{B}$. For $\theta \subseteq \Lambda$, set $\mathcal{Y}_\theta = \prod_{\sigma \in \theta} \mathcal{Y}_\sigma$.

LEMMA 13.1.22. *(1) $N_{\mathcal{F}}(T) \le \mathcal{B}$.*

(2) For each $\emptyset \neq \theta \subseteq \Lambda$ with $T_\theta \in \mathcal{F}^f$, $N_{\mathcal{F}}(T_\theta) \le \mathcal{B}$.

(3) Let $\emptyset \neq \theta \subset \Lambda$ with $T_\theta \in \mathcal{F}^f$, $1 \neq P \in \mathcal{Y}_\theta^f$ with $P \not\le Z$, and $\alpha \in \mathfrak{A}(P)$. Then $\mathcal{Y}_{\theta'}\alpha^* \trianglelefteq N_{\mathcal{F}}(P\alpha)$ and $P^{\mathcal{B}} \cap \mathcal{F}^f \neq \emptyset$. If $P \in \mathcal{F}^f$ then $N_{\mathcal{F}}(P) \le \mathcal{B}$.

PROOF. By 13.1.19.2, $Aut_{\mathcal{F}}(T)$ acts on Z, so $Aut_{\mathcal{F}}(T) = Aut_{\mathcal{B}}(T)$. Next $C_{\mathcal{F}}(T) \le C_{\mathcal{F}}(Z) \le \mathcal{B}$. Let $R = TC_S(T)$ and $\mathcal{N} = N_{\mathcal{F}}(T)$. By 1.3.2 in [**Asc19**], $\mathcal{N} = \langle SC_{\mathcal{F}}(T), N_{\mathcal{N}}(R) \rangle$, and as $Aut_{\mathcal{F}}(T) = Aut_{\mathcal{B}}(T)$, $N_{\mathcal{N}}(R) \le \mathcal{B}$, establishing (1).

Pick θ as in (2) and let $Q = T_\theta$ and $\xi = (C_{\mathcal{F}}(Q), \Gamma)$, where Γ consists of those $J \in \Omega$ contained in a member of $\Lambda - \theta$. Then ξ is a quaternion fusion packet, and from 13.1.21.1, $\mathcal{Y}_{\theta'} = C_{\mathcal{F}}(Q)^\circ \trianglelefteq \mathcal{M} = N_{\mathcal{F}}(Q)$. By 1.3.2 in [**Asc19**], $\mathcal{M} = \langle N_S(Q)\mathcal{Y}_{\theta'}, N_{\mathcal{M}}(T) \rangle$, so (2) follows from (1).

Pick P and α as in (3) and let $\mathcal{P} = N_{\mathcal{F}}(P\alpha)$. As usual by 13.1.21.1, $\mathcal{X} = \mathcal{Y}_{\theta'}\alpha^*$ is a product of coconnected components of $\mathcal{P}^\circ$. As $P \not\le Z$, all other coconnected components are of smaller order, using 13.1.21.1, so $\mathcal{X} \trianglelefteq \mathcal{P}$. Let $\beta \in \mathfrak{A}(T_{\theta'}\alpha)$ with $T_{\theta'}\zeta = T_{\theta'}$, where $\zeta = \alpha\beta$. By (2), $\zeta \in \hom_{\mathcal{B}}(N_S(P), S)$. Also $N_S(P\alpha)\beta \le N_S(P\zeta)$, so $P\zeta \in \mathcal{F}^f$. Finally if $P \in \mathcal{F}^f$ we may take $\alpha = 1$ to obtain $\mathcal{Y}_{\theta'} \trianglelefteq \mathcal{P} = N_{\mathcal{F}}(P)$. Then by 1.3.2 in [**Asc19**], $\mathcal{P} = \langle N_S(P)\mathcal{Y}_{\theta'}, N_{\mathcal{P}}(T_{\theta'}) \rangle$, so $\mathcal{P} \le \mathcal{B}$ by (2). □

LEMMA 13.1.23. *(1) $\mathcal{B}$ is transitive on $\mathcal{Z}$ and $\mathcal{F}_t \le \mathcal{B}$ for each $t \in \mathcal{Z}$.*

(2) Suppose $P \le S$ with $\mathcal{Z} \cap Z(P) \neq \emptyset$. Then $\hom_{\mathcal{F}}(P,S) = \hom_{\mathcal{B}}(P,S)$.

(3) Suppose $Q \in \mathcal{F}^f$ and $\phi \in \hom_{\mathcal{F}}(Q,S)$ such that $C_{\mathcal{Z}}(Q\phi) \neq \emptyset$. Then $Q\phi \in Q^{\mathcal{B}}$.

PROOF. As $Aut_{\mathcal{F}}(T)$ is transitive on Λ and $\mathcal{Y}$ is transitive on $\mathcal{Z}(\rho)$, $\mathcal{B}$ is transitive on $\mathcal{Z}$. By 13.1.22.3, $\mathcal{F}_z \le \mathcal{B}$, completing the proof of (1).

Assume $u \in Z(P) \cap \mathcal{Z}$ and let $\psi \in \hom_{\mathcal{F}}(P,S)$. Let $\beta \in \mathfrak{A}_{\mathcal{B}}(u)$. By (1), $u^{\mathcal{B}} = \mathcal{Z}$, so there is $\alpha \in \mathfrak{A}_{\mathcal{B}}(u\psi)$ with $u\psi\alpha = u\beta = v$. Now $\zeta = \beta^{-1}\psi\alpha \in \hom_{\mathcal{F}}(P\beta, S)$ with $v\zeta = v$, so as $\mathcal{F}_v \le \mathcal{B}$ by (1), $\zeta \in \hom_{\mathcal{B}}(P\beta, S)$. Then as α, β, and ζ are $\mathcal{B}$-maps, so is ψ, proving (2).

Assume the hypothesis of (3) and let $u \in C_{\mathcal{Z}}(Q\phi)$ and set $P = \langle u, Q\phi \rangle$ and let $\alpha \in \mathfrak{A}(Q\phi)$ with $Q\phi\alpha = Q$. Then $u \in Z(P)$, so by (2), $\alpha_{|P} \in \hom_{\mathcal{B}}(P,S)$. Hence $\alpha_{|Q\phi} \in \hom_{\mathcal{B}}(Q\phi, S)$, so $Q\phi \in Q^{\mathcal{B}}$, proving (3). □

LEMMA 13.1.24. *(1) Let* $1 \neq P \leq Z$ *and* $\alpha \in \mathfrak{A}(P)$. *Then* $P\alpha \in P^{\mathcal{B}}$ *and* $P\alpha \leq Z$. *If* $P \in \mathcal{F}^f$ *is of order 2 then* $N_{\mathcal{F}}(P) \leq \mathcal{B}$.

(2) Z *is not strongly closed in* S *with respect to* $\mathcal{F}$.

PROOF. As $\mathcal{Z} \subset T \leq C_S(P)$, it follows from 13.1.23.3 that $P\alpha \in P^{\mathcal{B}}$. Then as $Z \trianglelefteq \mathcal{B}$, $P\alpha \leq Z$. Suppose $P \in \mathcal{F}^f$ is of order 2. Then as P centralizes $\mathcal{Y}_0$, 13.1.20 and our usual argument using 13.1.21.1 shows that $\mathcal{Y}_0 \trianglelefteq \mathcal{N} = N_{\mathcal{F}}(P)$. By 1.3.2 in [**Asc19**], $\mathcal{N} = \langle S\mathcal{Y}_0, N_{\mathcal{N}}(T)\rangle$ and $N_{\mathcal{F}}(T) \leq \mathcal{B}$ by 13.1.22.1, so $\mathcal{N} \leq \mathcal{B}$, completing the proof of (1).

Suppose (2) fails. Then Z is a strongly closed abelian subgroup of $\mathcal{F}$, so by 14.1 in [**Asc11**], $Z \trianglelefteq \mathcal{F}$, contrary to 13.1.21.4. □

By 13.1.24.2 there is an involution $a \in Z$ with $a^{\mathcal{F}}$ not contained in Z, and by 13.1.24.1 we may take $a \in \mathcal{F}^f$. Let $b \in a^{\mathcal{F}} - Z$ and $\alpha \in \mathfrak{A}(b)$ with $b\alpha = a$. By 13.1.23.3, $C_{\mathcal{Z}}(b) = \emptyset$, so:

LEMMA 13.1.25. *b is has no fixed points on* Ω.

Suppose for the moment that b acts on some nonempty proper subset θ of Λ and set $P = C_{T_\theta}(b)$.

LEMMA 13.1.26. *For each involution* $x \in P$, $C_{\mathcal{Z}}(x\alpha) = \emptyset$.

PROOF. Assume $C_{\mathcal{Z}}(x\alpha) \neq \emptyset$. By 13.1.22.3, there is $\beta \in \mathfrak{A}_{\mathcal{B}}(x)$ with $y = x\beta \in \mathcal{F}^f$, and $\mathcal{F}_y \leq \mathcal{B}$ by 13.1.22.3. Now $\phi = \beta^{-1}\alpha \in \hom_{\mathcal{F}}(y, S)$, so by 13.1.23.3, $x\alpha = y\phi \in y^{\mathcal{B}}$. Thus there is $\gamma \in \mathfrak{A}_{\mathcal{B}}(x\alpha)$ with $x\alpha\gamma = y$. Therefore $\zeta = \phi\gamma \in \hom_{\mathcal{F}_y}(\langle b\beta, y\rangle, S)$. As $\mathcal{F}_y \leq \mathcal{B}$, ζ, β, and γ are $\mathcal{B}$-maps, so α is also a $\mathcal{B}$-map, contradicting $b \notin a^{\mathcal{B}}$. □

By 13.1.26, $P\alpha$ is semiregular on Ω, so $nk = |\Omega| \geq |P|$. Enlarging θ if necessary, we may assume $|\theta| = k - \epsilon$, where $\epsilon = 1$ or 2, and $\epsilon = 1$ if k is odd. Thus the set Γ of $J \in \Omega$ contained in a member of θ is of order $n|\theta| = n(k-\epsilon)$.

For $J \in \Gamma$, $J^b \neq J$ by 13.1.25, so $|C_{JJ^b}(b)| = m$. Thus $|P \cap O(\theta)| = m^{n(k-\epsilon)/2}/e$, where $e = |Z(\mathcal{Y}_\theta) : C_{Z(\mathcal{Y}_\theta)}(b)| \leq 2^{k-\epsilon-1}$. Therefore as $nk \geq |P| \geq |P \cap O(\theta)|$,

$$2^{k-\epsilon-1}nk \geq enk \geq m^{n(k-\epsilon)/2}.$$

In particular if $\epsilon = 1$ then $2^{k-2}nk \geq m^{n(k-1)/2}$, which is impossible as $k \geq 2$, $n \geq 2$, and $m \geq 8$. Thus $\epsilon = 2$, so k is even and hence $k \geq 4$ as $\theta \neq \emptyset$. But then $2^{k-3}nk \geq m^{n(k-2)/2}$, impossible as $k \geq 4$, $n \geq 2$, and $m \geq 8$.

We've shown that b acts on no nonempty proper subset of Λ, so:

LEMMA 13.1.27. $k = 2$ *and* $\lambda^b \neq \lambda$.

We are now in a position to derive a contradiction, and hence establish Theorem 13.1.15. As $|Z(\mathcal{Y})| = 2$ by 13.1.16 and 13.1.21.3, and as $k = 2$, it follows from 13.1.21.2 that $Z = \langle a \rangle = Z(\mathcal{Y})$. Set $\mathcal{X} = O^2(C_{\mathcal{Y}_0}(b))$ and let X be Sylow in $\mathcal{X}$. As $\lambda^b \neq \lambda$, it follows from 10.11 in [**Asc11**] that $\mathcal{X} \cong \mathcal{Y}/\langle a \rangle \cong AE_5$, $PSp_{2n}[m]$, or $L^-_{2n}[m]$. Then $\mathcal{X}\alpha^* \leq \mathcal{B}$ and as $\mathcal{X} = O^2(\mathcal{X})$, $\mathcal{X}\alpha^*$ acts on $\mathcal{Y}$.

Next $X = \{tt^b : t \in T_\lambda\}$ and for t an involution in T_λ distinct from a, $b \neq b^t = btt^b \in bX$. Therefore $a \neq (b^t)\alpha \in a(X\alpha) \subseteq N_S(\lambda)$. As $c = (b^t)\alpha$ acts on λ, we have $c \in Z$ by 13.1.27. Thus as $Z = \langle a \rangle$, we conclude that $a = c$, contradicting $a \neq (b^t)\alpha$. This contradiction completes the proof of Theorem 13.1.15.

13.2. Generic systems with μ abelian

In this section we assume the following hypothesis:

HYPOTHESIS 13.2.1. *Hypothesis* 13.1.1 *holds and*
(1) $Aut_{\mathcal{F}}(O(\tau))$ *induces* $Sym(\Omega)$ *on* Ω, *and*
(2) $\mathcal{Z}^m = \{t \in \mathcal{Z} : \mathcal{Z}_S \not\subseteq C_S(t)\} \neq \emptyset$.

NOTATION 13.2.2. Adopt Notation 13.1.2, set $S_0 = C_S(O)$, let G be a model for $N_{\mathcal{F}}(OS_0)$, and set $H_0 = \langle (\mathcal{Z}^m)^G \rangle \leq G$. Write G^Ω for the image of G in $Sym(\Omega)$. Set $n = |\Omega|$ and $W = \langle \mathcal{Z}_S \rangle$. Let $c = \prod_{s \in \mathcal{Z}_S} s$.

LEMMA 13.2.3. *(1)* $G^\Omega = H_0^\Omega = Sym(\Omega)$.
(2) $\{t^\Omega : t \in \mathcal{Z}^m\}$ *is the set of transpositions in* S^Ω.
(3) The map $J \mapsto z(J)$ *is a* G*-equivariant bijection between* Ω *and* $\mathcal{Z}_S$.
(4) Either $c \neq 1$ *and* W *is the permutation module for* G^Ω, *or* $c = 1$ *and* W *is the image of the permutation module modulo its center.*

PROOF. Recall that $Aut_{\mathcal{F}}(O) = Aut_G(O)$, so $G^\Omega = Sym(\Omega)$ by 13.2.1.1. By 13.2.1.2 there is $t \in \mathcal{Z}^m$, and by 13.1.10.3, t^Ω is a transposition. Further as $G^\Omega = Sym(\Omega)$, all transpositions in S^Ω are fused in G^Ω, so (2) holds. Also G^Ω is generated by its transpositions, so as $t \in H_0 \trianglelefteq G$, $G^\Omega = H_0^\Omega$, completing the proof of (1).

Part (3) is a consequence of 13.1.1.1. By (1) and (3), W is an image of the permutation module for G^Ω, and $C_W(G)$ is generated by c, so (4) holds. □

NOTATION 13.2.4. Let $I = \{1, \ldots, n\}$. By 13.2.3.4, if $c \neq 1$ then we can regard W as the power set on I under symmetric difference, while if $c = 1$ then W is the image of that module modulo $\langle e_I \rangle$. Thus for $\theta \subseteq I$, we write e_θ for θ regarded as a member of W when $c \neq 1$, or its image modulo $\langle e_I \rangle$ when $c = 1$. In particular $\mathcal{Z}_S = \{e_i : i \in I\}$. Write $O(\theta)$ for $\prod_{i \in \theta} O(e_i)$.

LEMMA 13.2.5. *Let* $\emptyset \neq \theta \subseteq I$ *and* $1 \neq x = e_\theta \in \mathcal{F}^f$. *Let* $\theta_1 = \theta$, $\theta_2 = I - \theta$, $\tau_x = (\mathcal{F}_x, \Omega)$, $n_j = |\theta_j|$, $\Omega_j = \bigcup_{i \in \theta_j} O(e_i)$, $\mathcal{F}_j = \mathcal{F}_j(x) = [\Omega_j]_{\mathcal{F}_x^\circ}$, *and* $\tau_j = (\mathcal{F}_j, \Omega_j)$ *for* $j = 1, 2$.
(1) τ_x *is a quaternion fusion packet and if* 13.1.1.*3.ii holds then* τ_x *satisfies Hypothesis* 13.1.1.
(2) $\tau_x^\circ = \tau_1 * \tau_2$, *where* $\tau_2 = \emptyset$ *if* $x = c$.
(3) For $j \in \{1, 2\}$, *one of the following holds:*
(i) $\mathcal{F}_j \cong Sp_{2n_j}[m]$ *or* $SL^-_{2n_j}[m]$ *and* $\langle e_{\theta_j} \rangle = Z(\mathcal{F}_j)$.
(ii) $n_j = 1$, $m = 8$, *and* $\mathcal{F}_j = O(\theta_j)$.
(iii) $n_j = 2$, $m = 8$, *and* $\mathcal{F}_j \cong \hat{A}E_5$.
(iv) $j = 2$, $c = 1$, *and either* $\mathcal{F}_2 \cong L^-_{2n_2+1}[m]$ *or* $n_2 = 1$ *and* $\mathcal{F}_2 \cong L_2[2m]^{(1)}$.
(4) Let $\mathcal{Z}_j^m = \{t \in \mathcal{Z}^m : \theta_{3-j} \subseteq Fix(t)\}$ *and* $\mathcal{Z}'_j = \{t \in \mathcal{Z}' : \Omega_{3-j} \subseteq C_S(t)\}$. *Then* $\mathcal{Z}_j^m = \mathcal{Z}^m(\tau_j)$ *and* $\mathcal{Z}'_j = \mathcal{Z}'(\tau_j)$.

PROOF. By 2.5.2, τ_x is a quaternion fusion packet, and then (1) follows from 13.1.5.1.

If 13.1.1.3.i holds then the lemma follows from the discussion in section 5.4, so assume that 13.1.1.3.ii holds.

Let $G_x = C_G(x)$. From 13.2.3.4, $G_x^\Omega = G_1^\Omega \times G_2^\Omega$, where G_j^Ω is the pointwise stabilizer in G^Ω of θ_{3-j}, and G_j^Ω acts faithfully as $Sym(\theta_j)$ on θ_j.

If $x = c$ assume $x \notin Z(\mathcal{F})$. Then by 13.1.5.2, $\tau_x^\circ = \rho_1 * \cdots * \rho_r$ is the central product of its coconnected components $\rho_i = (\mathcal{Y}_i, \Gamma_i)$. Let $K \in \Omega_j$; then $K \in \Gamma_i$ for some i. By 13.1.5.3, $Aut_{\mathcal{Y}_i}(O(\rho_i))^{\Gamma_i} = Sym(\Gamma_i)$, so we have $\Gamma_i \subseteq \Omega_j$. On the other hand as G_j is primitive on Ω_j, it follows that $\Omega_j = \Gamma_i$. We conclude that $r = 2$ (or $x = c$ and $r = 1$) and $\Omega_j = \Gamma_i$, and then that $\rho_i = \tau_j$ and (2) holds. Further if $n_j > 1$ then by 13.1.5.3, either $\mathcal{F}_j \cong (P)Sp_{2n_j}[m]$ or $(P)SL^-_{2n_j+a}[m]$ with $a \in \{0,1\}$, or $n_j = 2$, $m = 8$, and $\mathcal{F}_j \cong \hat{A}E_5$. As $e_{\theta_1} = x \in Z(\mathcal{F}_1)$, we have $\mathcal{F}_1 \cong Sp_{2n_1}[m]$, $SL^-_{2n_1}[m]$, or $\hat{A}E_5$, and $\langle e_{\theta_1}\rangle = Z(\mathcal{F}_1)$. If $n = 1$ then $\mathcal{F}_1 = O(\theta_1)$. Similarly either $\mathcal{F}_2 \cong Sp_{2n_2}[m]$, $SL^-_{2n_2}[m]$, or $\hat{A}E_5$, and $\langle e_{\theta_2}\rangle = Z(\mathcal{F}_2)$, or $\mathcal{F}_2 \cong L^-_{2n_2+1}[m]$ (or possibly $L_2[2m]^{(1)}$ if $n_2 = 1$) and $e_{\theta_2} \notin Z(\mathcal{F}_2)$, so that $e_{\theta_2} \neq x$, and hence $c \neq 1$. Thus (3) holds.

By 13.1.10.3, $\mathcal{Z}_j^m$ centralizes $O(\theta_{3-j})$, and then by 13.1.7.3, $\mathcal{Z}_j^m \subseteq \mathcal{Z}^m(\tau_j)$. Similarly $\mathcal{Z}'_j \subseteq \mathcal{Z}'(\tau_j)$ by 13.1.7.3. As $\mathcal{Z}_j^m \subseteq \mathcal{Z}^m(\tau_j)$ and $\mathcal{Z}'_j \subseteq \mathcal{Z}'(\tau_j)$ and as (2) holds, (4) also holds.

Finally suppose $x = c \in Z(\mathcal{F})$ and set $\mathcal{F}^+ = \mathcal{F}/\langle c\rangle$. If $n = 2$ the lemma holds by 13.1.8, so take $n > 2$. Then, using 6.6.9.2, $\tau^+ = (\mathcal{F}^+, \Omega^+)$ satisfies Hypothesis 13.2.1, so by induction on the order of τ, $\mathcal{F}^{\circ +} \cong PSp_{2n}[m]$ or $PSL^-_{2n}[m]$, and then $\mathcal{F}^\circ \cong Sp_{2n}[m]$ or $SL^-_{2n}[m]$ by 3.3.16, so again the lemma holds. □

LEMMA 13.2.6.

$$\mathcal{E}(\tau) = \langle SO(\tau), N_{\mathcal{F}}(O), N_{\mathcal{F}}(O(\theta)) : \theta \in \Xi\rangle,$$

where Ξ *consists of those* $\theta \subseteq I$ *such that* $N_S(\theta)$ *is transitive on* θ, *and* $O(\theta) \in \mathcal{F}^f\rangle$.

PROOF. This follows from 3.4.12.1. □

DEFINITION 13.2.7. Define

$$\mathcal{B}(\tau) = \langle SO(\tau), N_{\mathcal{F}}(O), \mathcal{F}_{e_\theta} : \theta \subseteq I, |\theta| \text{ is a power of 2, and } e_\theta \in \mathcal{F}^f\rangle.$$

LEMMA 13.2.8. *(1)* $N_{\mathcal{F}}(W) = \langle SO(\tau), N_{\mathcal{F}}(O)\rangle \leq \mathcal{B}(\tau)$.
(2) For each $\theta \subseteq I$ *with* $|\theta|$ *a power of 2*, $\mathcal{F}_{e_\theta} \leq \mathcal{B}(\tau)$.

PROOF. By 3.3.7.2, $\mathcal{O}(\tau) \trianglelefteq N_{\mathcal{F}}(W)$. Then (1) follows from 1.3.2 in [**Asc19**].

Set $x = e_\theta$. There is $g \in G$ such that $\alpha = c_g^* \in \mathfrak{A}(x)$. By definition of $\mathcal{B} = \mathcal{B}(\tau)$, $\mathcal{F}_{x^g} \leq \mathcal{B}$. Therefore as $N_{\mathcal{F}}(O) \leq \mathcal{B}$ by (1), we can complete the proof as in 11.3.4. □

THEOREM 13.2.9. $\mathcal{F} = \mathcal{B}(\tau)$.

PROOF. Observe that if $\theta \subseteq I$ then $N_{\mathcal{F}}(O(\theta)) \leq C_{\mathcal{F}}(e_\theta)$. Therefore $\mathcal{E}(\tau) \leq \mathcal{B}(\tau)$ by 13.2.6. Observe also that Hypothesis 11.2.1 is satisfied by 13.2.3.4. Also 11.3.8.1 holds in 13.1.1.3.ii by 13.1.5.4. Hence by Theorem 11.3.10, either $\mathcal{F} = \mathcal{B}(\tau)$ or conclusion (2) of that theorem holds. However the latter case is contrary to 13.2.1.2. □

LEMMA 13.2.10. *Let $1 \neq x = e_\theta$ for some $\theta \subseteq I$ with $x \in \mathcal{F}^f$. Then one of the following holds:*
(1) $\mathcal{F}_x \leq \mathcal{E}(\tau)$.
(2) $x = c$.
(3) n is even, $c = 1$, and $|\theta| = n/2$.

PROOF. See the proof of 11.1.8 and use 13.2.5. □

LEMMA 13.2.11. *One of the following holds:*
(1) $\mathcal{F} = \mathcal{E}(\tau)$.
(2) $n = 2^k$, $c \neq 1$, and $\mathcal{F} = \langle \mathcal{E}(\tau), \mathcal{F}_c \rangle$.
(3) $n = 2^k$, $c = 1$, and $\mathcal{F} = \langle \mathcal{E}(\tau), \mathcal{F}_x \rangle$, where $x \in \mathcal{F}^f$, $x = e_\theta$, and $|\theta| = n/2$.

PROOF. By 13.2.6 and Theorem 13.2.9,

$$\mathcal{F} = \langle \mathcal{E}(\tau), \mathcal{F}_{e_\theta} : \theta \subseteq I \text{ and } |\theta| \text{ is a power of } 2 \rangle.$$

Then the lemma follows from 13.2.10. □

LEMMA 13.2.12. *Let $\theta \subseteq I$ and $x = e_\theta$ with $1 \neq x \in \mathcal{F}^f$.*
(1) $N_{\mathcal{F}}(O(\theta)) \leq \mathcal{F}_x$.
(2) $O(\theta) \in \mathcal{F}^f$ iff $x \in \mathcal{F}^f$.
(3) Either
(a) $c \in Z(\mathcal{F}_x)$, or
(b) $\mathcal{F}_2(x) \cong L^-_{2n_2+1}[m]$ or $n_2 = 1$ and $\mathcal{F}_2(x) \cong L_2[2m]^{(1)}$.

PROOF. Set $P = O(\theta)$. As $N_{\mathcal{F}}(P)$ permutes $\{O(i) : i \in \theta\}$, (1) follows. As W is the permutation module for $G^\Omega = Sym(I)$, or the image of that module modulo $\langle e_I \rangle$, it follows from 13.2.3.3 that either $C_S(x) = N_S(P)$ or n is even, $c = 1$, and $|\theta| = n/2$ with $|S : N_S(P)| = 2$. In either case (2) holds.

Next $\mathcal{F}_x^\circ \trianglelefteq \mathcal{F}_x$, and by 13.2.5, $c = e_{\theta_1} \cdot e_{\theta_2}$. Moreover by 13.2.5.3, either $\mathcal{F}_x$ permutes $\{e_{\theta_1}, e_{\theta_2}\}$, or (3b) holds so (3) follows. □

LEMMA 13.2.13. *(1) Either*
(a) $c \in Z(\mathcal{F})$, or
(b) For some $1 \neq x = e_\theta \in \mathcal{F}^f$ with $|\theta|$ a power of 2, $\mathcal{F}_2(x) \cong L^-_{2n_2+1}[m]$ or $n_2 = 1$ and $\mathcal{F}_2(x) \cong L_2[2m]^{(1)}$.
(2) If $1 \neq c \in Z(\mathcal{F})$ then either $\mathcal{F}^\circ \cong Sp_{2n}[m]$ or $SL^-_{2n}[m]$, or $n = 2$ and $\mathcal{F}^\circ \cong \hat{A}E_5$.

PROOF. First $c \in Z(SO(\tau))$. By 13.2.12.1, $N_{\mathcal{F}}(O)$ centralizes c, while by 13.2.12.3, either $\mathcal{F}_x$ centralizes c for each $1 \neq x \in W$ with $x = e_\theta \in \mathcal{F}^f$ and $|\theta|$ a power of 2, or (1b) holds. Therefore (1) follows from Theorem 13.2.9. Then (2) follows from 13.2.5.3. □

Given 13.2.13, during the remainder of the section we assume:

HYPOTHESIS 13.2.14. *Hypothesis* 13.2.1 *holds and $c = 1$ if $c \in Z(\mathcal{F})$.*

LEMMA 13.2.15. *Let $x = e_\theta \in \mathcal{F}^f$.*
(1) If $n_1 = |\theta| = 2$ then $\mathcal{F}_1(x) \cong Sp_4[m]$ or $SL^-_4[m]$.
(2) If $n_1 = n - 1$ then either $\mathcal{F}_2(x) \cong SL_2[m]$ or $L^-_3[m]$, or $m = 8$ and $\mathcal{F}_2(x) = O(n)$.

PROOF. Set $\mathcal{F}_i = \mathcal{F}_i(x)$. Assume (1) fails; then by 13.2.5.3, $\mathcal{F}_1 \cong \hat{A}E_5$, so $m = 8$. Let $z = e_1$ and set $\mathcal{G}_i = \mathcal{F}_i(z)$. By 13.2.5.3, $\mathcal{G}_2 \cong Sp_{2n-2}[8]$, $SL^-_{2n-2}[8]$, or $L^-_{2n-1}[8]$, or $n = 3$ and $\mathcal{G}_2 \cong \hat{A}E_5$, or $n = 2$ and $\mathcal{G}_2 \cong L_2[16]^{(1)}$. In the first case there is $y \in G$ of order 3 centralizing K with $O(2) = [O(2), y]$. Let $g \in G$ with $e^g_{1,2} = x$. Now y^g centralizes x so y^g acts on L the model in G for $O^2(N_{\mathcal{F}_1}(O(\theta)))$. Further y^g acts on $O_2(\mathcal{F}_1)$ and hence on $O(\theta) \cap O_2(\mathcal{F}_1) = E$. Set $\tilde{\mathcal{F}}_1 = \mathcal{F}_1/\langle x\rangle$. Then $\tilde{E} \cong E_8$, so as y^g centralizes K^g, y^g centralizes $\tilde{E}$. This contradicts $O(2)^g = [O(2)^g, y^g]$.

Therefore case two or three holds. Suppose $n = 3$; then we may take $\theta = \{2, 3\}$, and there is $h \in G$ of order 3 centralizing K with $O(\theta) = [O(\theta), h]$. Then for $a \in G$ with $(1, 2, 3)^a = (3, 1, 2)$, h^a centralizing $O(3)$ and $O(2) = [O(2), h^a]$, a contradiction as in the previous paragraph.

Finally suppose $n = 2$. There is $l \in G$ of order 3 with $O = [O, l]$. But l centralizes z, so l acts on $\mathcal{F}_2$, a contradiction as $O(2) = [O(2), l]$ and $\mathbf{Z}_4 \cong O(2) \cap O^2(\mathcal{F}_2)$. This completes the proof of (1). Moreover the argument in case three establishes (2). □

LEMMA 13.2.16. *Assume $m = 8$. Then there exists $\mathcal{O} \trianglelefteq \mathcal{F}_z$ such that K is Sylow in $\mathcal{O}$ and $\mathcal{O} \cong SL_2[8]$.*

PROOF. Set $T = C_S(K)$. We first show there exists $\mathcal{O} \leq C_{\mathcal{F}_z}(T)$ such that K is Sylow in $\mathcal{O}$ and $\mathcal{O} \cong SL_2[8]$.

Choose notation so that there is $t \in \mathcal{Z}^m$ such that $t^\Omega = (1, 2)$ and $x = e_{1,2} \in \mathcal{F}^f$. Adopt the notation of 13.2.5. By 13.2.15, $\mathcal{F}_1 \cong Sp_4[8]$ or $SL^-_4[8]$, so there is $\mathcal{O}_1 \trianglelefteq C_{\mathcal{F}_1}(e_1)$ with $K_1 = O(1)$ Sylow in $\mathcal{O}_1 \cong SL_2[8]$. Then by 2.3.5 in [**Asc19**], $S_0 O(I - \{1\})$ centralizes $\mathcal{O}_1$. As G is transitive on Ω, it follows from 2.2 in [**Asc11**] that there is a central product $\mathcal{L} = \prod_{i \in I} \mathcal{O}_i \leq N_{\mathcal{F}}(O)$ with $K_i = O(i)$ Sylow in $\mathcal{O}_i \cong SL_2[8]$ for each $i \in I$. As $N_{\mathcal{F}}(O)$ permutes Ω, $Aut_{\mathcal{L}}(O) \trianglelefteq Aut_{\mathcal{F}}(O)$, so $\mathcal{L} \trianglelefteq N_{\mathcal{F}}(O)$. Therefore there is a model L for $\mathcal{L}$ normal in G, and of course $L = \prod_{i \in I} L_i$ is a central product of groups $L_i \cong SL_2(3)$ with $K_i = O_2(L_i)$. Let $z = e_j$, so that $K = K_j$, and as S permutes $\{L_i : i \in I\}$, T acts on L_j. Then $[L_j, T] \leq C_{L_j}(K) = \langle z\rangle$, so $L_j = O^2(L_j)$ centralizes T, and therefore $\mathcal{O} = \mathcal{O}_j$ centralizes T.

Next let $\mathcal{E} = C_{\mathcal{F}}(K)$ and $\mathcal{C} = C_{\mathcal{F}_z}(\mathcal{E})$. As $K \trianglelefteq \mathcal{F}_z$, it follows from Theorem 4 in [**Asc11**] that $\mathcal{C} \trianglelefteq \mathcal{F}_z$ and $\mathcal{E} * \mathcal{C} \leq \mathcal{F}_z$. Let R be Sylow in $\mathcal{C}$ and set $Z = Z(\mathcal{E})$. Then $C_R(K) \leq R \cap T = Z$, so $KZ \in \mathcal{C}^{fc}$ is normal in $\mathcal{C}$, and hence $\mathcal{C}$ is constrained. Let G_c be a model for $\mathcal{C}$. Next $Aut_{\mathcal{O}}(KZ) \leq O^2(Aut_{C_{\mathcal{F}_z}(T)}(KZ))$, so by Theorem 4 in [**Asc11**], $Aut_{\mathcal{O}}(KZ) \leq Aut_{\mathcal{C}}(KZ)$. Therefore we may take $\mathcal{O} \leq \mathcal{C}$. Then as $\mathcal{O}$ is irreducible on $K/\langle z\rangle$, we conclude $KZ = O_2(G_c)$, so that $\mathcal{O} = O^2(\mathcal{C}) \trianglelefteq \mathcal{F}_z$, completing the proof of the lemma. □

NOTATION 13.2.17. Suppose for the moment that $t \in \mathcal{Z}_S$. If $m > 8$ set $\mathcal{O}_t = \mathcal{O}(t)$ and observe that $\mathcal{O}(t) \cong SL_2[m]$ by 13.2.15. If $m = 8$ set $\mathcal{O}_t = \mathcal{O}\alpha^{-*}$ for $\alpha \in \mathfrak{A}(t)$ with $t\alpha = z$ and $\mathcal{O}$ the system in 13.2.16. As $\mathcal{O} \trianglelefteq \mathcal{F}_z$, $\mathcal{O}_t$ is independent of α. In either case set $\mathcal{O}_\tau = \prod_{t \in \mathcal{Z}_S} \mathcal{O}_t$. Observe that $\mathcal{O}_\tau \trianglelefteq N_{\mathcal{F}}(W)$ using 13.2.16 when $m = 8$. Write $\mathcal{O}_i$ for $\mathcal{O}_{e_i}$ and for $\theta \subseteq I$ set $\mathcal{O}_\theta = \prod_{i \in \theta} \mathcal{O}_i$. Let L_0 be a model for $N_{\mathcal{O}_\tau}(OS_0)$ in G and $H = \langle \mathcal{Z}^G\rangle L_0 \leq G$. Let M be the kernel of the action of H on Ω.

DEFINITION 13.2.18. Suppose for the moment that $x = e_\theta \in \mathcal{F}^f$ with $|\theta| = 2$. Define $\mathcal{F}$ to be of *type* $PSp_{2n}[m]$ if $\mathcal{F}_1(x) \cong Sp_4[m]$, and define $\mathcal{F}$ to be of *type* $L^-_{2n}[m]$, $L^-_{2n+1}[m]$ if $\mathcal{F}_1(x) \cong SL^-_4[m]$ and $\mathcal{F}_2(z) \cong SL^-_{2n-2}[m]$, $L^-_{2n-1}[m]$, respectively.

LEMMA 13.2.19. *Let $1 \neq x = e_\theta \in \mathcal{F}^f$ with $|\theta| = n_1$ and set $\mathcal{F}_i = \mathcal{F}_i(x)$ for $i = 1, 2$.*

(1) $\mathcal{F}$ is of type $PSp_{2n}[m]$, $L^-_{2n}[m]$, or $L^-_{2n+1}[m]$.

(2) If $\mathcal{F}$ is of type $PSp_{2n}[m]$ or $L^-_{2n}[m]$ then $c = 1$, while if $\mathcal{F}$ is of type $L^-_{2n+1}[m]$ then $c \neq 1$.

(3) If $\mathcal{F}$ is of type $PSp_{2n}[m]$ then $\mathcal{Z}' = \mathcal{Z}^m$ and $\mathcal{F}_i \cong Sp_{2n_i}[m]$ for $i = 1, 2$, unless $n_i = 1$ and $m = 8$ where $\mathcal{F}_i \cong K$.

(4) If $\mathcal{F}$ is of type $L^-_{2n}[m]$ then $\mathcal{Z}' \neq \mathcal{Z}^m$, $|\mathcal{D}^o(t) \cap \mathcal{Z}_S| = 2$ for each $t \in \mathcal{Z}' - \mathcal{Z}^m$, and $\mathcal{F}_i \cong SL^-_{2n_i}[m]$ for $i = 1, 2$, unless $n_i = 1$ and $m = 8$ where $\mathcal{F}_i \cong K$.

(5) If $\mathcal{F}$ is of type $L^-_{2n+1}[m]$ then $\mathcal{Z}' \neq \mathcal{Z}^m$, there exists $t \in \mathcal{Z}'$ with $|\mathcal{D}^o(t) \cap \mathcal{Z}_S| = 1$, and $\mathcal{F}_1 \cong SL^-_{2n_1}[m]$ (unless $n_1 = 1$ and $m = 8$ where $\mathcal{F}_1 \cong K$, and $\mathcal{F}_2 \cong L^-_{2n_2+1}[m]$).

(6) If $n = 2$ then $\mathcal{F}$ is of type $L^-_5[m]$.

PROOF. By 13.2.1.2, $n \geq 2$. Thus there is $u = e_\xi \in \mathcal{F}^f$ with $|\xi| = 2$. Set $\mathcal{G}_i = \mathcal{F}_i(u)$ for $i = 1, 2$. Set $\mathcal{F}_2(z) = \mathcal{C}$. By 13.2.15.1, $\mathcal{G}_1 \cong Sp_4[m]$ or $SL^-_4[m]$.

Assume first that $n = 2$. Then $c = u \neq 1$ by 13.1.1.1. Now by 13.2.14, $c \notin Z(\mathcal{F})$, so by 13.2.13.1, $\mathcal{C}$ is isomorphic to $L^-_3[m]$ or $L_2[2m]^{(1)}$, and then the former holds by 13.2.15.2. We will show that $\mathcal{G}_1 \cong SL^-_4[m]$, so $\mathcal{F}$ is of type $L^-_5[m]$ and (1) and (6) hold. Also (2)-(4) are vacuously true and, as $\mathcal{C} \cong L^-_3[m]$ there is $t \in \mathcal{D}^o(z_2) \cap z_2^{\mathcal{C}}$ with $\mathcal{D}^o(t) \cap \mathcal{Z}_S = \{z_2\}$, so (5) holds. Therefore when $n = 2$ it remains only to show that $\mathcal{G}_1 \cong SL^-_4[m]$. Hence by 13.2.15.1 we may assume:

(a) $\mathcal{G}_1 \cong Sp_4[m]$.

Let $\Omega = \{K_1, K_2\}$ with $K = K_1$ and set $z_i = z(K_i)$. Let T_u, T_2 be Sylow in $\mathcal{G}_1$, $\mathcal{C}$, respectively, $r \in \mathcal{Z}^m \cap T_u$, $T_1 = T_2^r$, and $T = \langle T_1, T_2 \rangle$. By (a), $T_u = KK_2\langle r \rangle$, while as $\mathcal{C} \cong L^-_3[m]$, $T_i \cong SD_{2m}$. Next $T_2 \trianglelefteq N_S(K)$, so $T_1 = T_2^r \trianglelefteq N_S(K)$ and then $T_0 = T_1 \cap T_2 \trianglelefteq S$. Therefore $T_0 \leq Z(T_1) \cap Z(T_2) = \langle z \rangle \cap \langle z_2 \rangle = 1$, so

(b) $T = T_1 \times T_2$.

But $T_u \trianglelefteq S$, so $[T, r] \leq T_u \cap T = KK_2$, a contradiction as for $t \in T - K$, $[t, r] \notin KK_2$ by (b). Thus we may assume that $n > 2$.

As $n > 2$ there is $b \in u^{\mathcal{F}}$ with $b \in \mathcal{C}^f$, and a coconnected component $\mathcal{X}$ of $(C_{\mathcal{C}}(b), \Omega - \{K\})$ with $\mathcal{X}\beta^* = \mathcal{G}_1$ for $\beta \in \mathfrak{A}(b)$ with $b\beta = u$. It follows that $\mathcal{C} \cong Sp_{2n-2}[m]$ if $\mathcal{G}_1 \cong Sp_4[m]$ and $\mathcal{C} \cong SL^-_{2n-2}[m]$ or $L^-_{2n-1}[m]$ if $\mathcal{G}_1 \cong SL^-_4[m]$. Therefore (1) holds.

Similarly if $n_i \geq 2$ then there is $a \in u^{\mathcal{F}}$ with $a \in \mathcal{F}_i^f$ and a coconnected component $\mathcal{Y}$ of $(C_{\mathcal{F}_i}(a), \Omega_i)$ such that $\mathcal{Y}\alpha^* = \mathcal{G}_1$ for $\alpha \in \mathfrak{A}(a)$ with $a\alpha = u$. Therefore $\mathcal{F}_i \cong Sp_{2n_2}[m]$ iff $\mathcal{F}$ is of type $PSp_{2n}[m]$, while $\mathcal{F}_i \cong SL^-_{2n_2}[m]$ or $L^-_{2n_2+1}[m]$ otherwise.

Suppose $\mathcal{F}_2 \cong L^-_{2n_2+1}[m]$. Let $j \in \theta$ with $y = e_j \in \mathcal{F}_1^f$ and $\gamma \in \mathfrak{A}(y)$ with $y\gamma = z$. Then $\mathcal{F}_2 = C_{\mathcal{C}}(O(\theta)\gamma)^\circ$, so as $\mathcal{F}_2 \cong L^-_{2n_2+1}[m]$, it follows that $\mathcal{C} \cong L^-_{2n-1}[m]$, and therefore $\mathcal{F}$ is of type $L^-_{2n+1}[m]$. Similarly the converse is true: if $\mathcal{F}$ is of type $L^-_{2n_2+1}[m]$ then $\mathcal{F}_2 \cong L^-_{2n_1+1}[m]$. Together with the previous paragraph, these observations establish the statements in (3)-(5) about $\mathcal{F}_i$.

Suppose $\mathcal{F}$ is of type $PSp_{2n}[m]$ or $L^-_{2n}[m]$. From the previous paragraph, $\mathcal{F}_2 \cong Sp_{2n_2}[m]$ or $SL^-_{2n_2}[m]$, so $c \in Z(\mathcal{F})$ by 13.2.13.1. Therefore $c = 1$ by 13.2.14. On the other hand if $\mathcal{F}$ is of type $L^-_{2n+1}[m]$ then $x \notin Z(\mathcal{F}_2)$, so $c \neq 1$, completing the proof of (2).

If $\mathcal{G}_1$ is $SL^-_4[m]$ then there is $t \in \mathcal{Z}'(\rho_1)$ (where ρ_1 is the Lie packet of $\mathcal{G}_1$) with $e_i \in \mathcal{D}^o(t)$ for $i \in \xi$, so $\mathcal{Z}' \neq \mathcal{Z}^m$. Conversely if $t \in \mathcal{Z}'$ with $\mathcal{D}^o(t) \cap \mathcal{Z}_S = \zeta$ of order 2, then by 13.1.10.2, $\mathcal{G}_1 \cong SL^-_4[m]$.

Next if $t \in \mathcal{Z}'$ with $\mathcal{D}^o(t) \cap \mathcal{Z}_S = \{e_i\}$ of order 1, then conjugating in $\mathcal{F}$ we may take $\theta_2 = \{i\}$, and then by 13.1.7.3 and 13.2.15.2, $\mathcal{F}_2 \cong L^-_3[m]$, so $\mathcal{F}$ is of type $L^-_{2n+1}[m]$. Together with the previous paragraph, this shows $\mathcal{Z}' = \mathcal{Z}^m$ when $\mathcal{F}$ is of type $PSp_{2n}[m]$, completing the proof of (3). Similarly if $\mathcal{F}$ is of type $L^-_{2n}[m]$ then $|\mathcal{D}^o(t) \cap \mathcal{Z}_S| = 2$ for each $t \in \mathcal{Z}' - \mathcal{Z}^m$, completing the proof of (4). Finally if $\mathcal{F}$ is of type $L^-_{2n+1}[m]$ then, taking $n_2 = 1$, $\mathcal{F}_2 \cong L^-_3[m]$, so for $t \in \mathcal{Z}'(\tau_2)$, $|\mathcal{D}^o(t) \cap \mathcal{Z}_S| = 1$, completing the proof of (5). □

LEMMA 13.2.20. *Assume $x = e_{1,2} \in \mathcal{F}^f$ and let T_1 be Sylow in $\mathcal{F}_1 = \mathcal{F}_1(x)$. Then*

(1) $C_S(T_1) = C_S(\mathcal{F}_1)$.

(2) Let $t \in \mathcal{Z}^m \cap T_1$ and set $\bar{T}_1 = O(1,2)\langle t\rangle$. Then $[T_1, C_G(O(1,2))] \leq \langle e_1, e_2\rangle$, $C_S(O(1,2)) = \langle e_1\rangle \times C_S(\bar{T}_1)$, and $C_S(\bar{T}_1)$ centralizes a subsystem $\bar{\mathcal{F}}_1$ of $\mathcal{F}_1$ isomorphic to $Sp_4[m]$. In particular if $\mathcal{F}$ is of type $PSp_{2n}[m]$ then $\mathcal{F}_1 = \bar{\mathcal{F}}_1$.

(3) $N_{\mathcal{F}_1}(O(1,2)) = T_1$ if $m > 8$ and $T_1\mathcal{O}_{1,2}$ if $m = 8$.

(4) For each $r \in \mathcal{D}^m$, $r \in T_1^g$ for some $g \in G$.

(5) Let $\mathcal{D}^o = \mathcal{D}^o(\tau) = \{s \in \mathcal{Z} : \mathcal{D}^o(s) \cap \mathcal{Z}_S \neq \emptyset\}$. Then $C_{\mathcal{Z}'}(W) = \mathcal{D}^o$.

(6) For $s \in \mathcal{D}^o$, $|\mathcal{D}^o(s) \cap \mathcal{Z}_S| \leq 2$ with $s \in T_1^g$ for some $g \in G$ in case of equality. If $|\mathcal{D}^o(s) \cap \mathcal{Z}_S| = 1$ then $\mathcal{F}$ is of type $L^-_{2n+1}[m]$ and $[s, C_G(O)] \leq W$.

PROOF. By 13.2.15, $\mathcal{F}_1 \cong Sp_4[m]$ or $SL^-_4[m]$. Hence by 9.4.18, 9.4.22, and 9.4.28, together with 2.1.15, $\mathcal{F}_1 = \langle \mathcal{G}_1, O^2(Aut_{\mathcal{F}_1}(R_i)) : i \in I'\rangle$, where $\mathcal{G}_1 = T_1\mathcal{O}_{1,2}$ and $R_i = C_{T_1}(Q_i/\langle x\rangle)$ for suitable $Q_i \leq T_1$ isomorphic to $\mathbf{Z}_4 * Q_8$. Let G_2 be a model for $N_{\mathcal{F}_1}(Q_i)$ normal in a model for $N_{\mathcal{F}}(Q_iC_S(Q_i))$, and set $P = C_S(T_1)$. Then $[P, G_2] \leq C_{G_2}(R_i) \leq C_{R_i}(O^2(G_2))$, so $O^2(G_2)$ centralizes P, and therefore P centralizes $\mathcal{F}_1$, proving (1).

Part (3) follows from 5.4.8-5.4.10.

Set $Q = O(1,2)$ and $Y = C_G(Q)$. If $m > 8$ set $L_{1,2} = 1$, while if $m = 8$ let $L_{1,2}$ be a model for $\mathcal{O}_{1,2}$ in G. By (3), $[T_1, Y] \leq C_{T_1L_{1,2}}(Q) = \langle e_1, e_2\rangle$, so Y acts on $D = \langle t, e_1\rangle \cong D_8$ and on $\langle e_1, e_2\rangle$. Thus the first statement in (2) holds and $[Y, D] \leq Z(D) = \langle x\rangle$, so $|Y : C_Y(D)| = 2$. Then as $C_Y(D) = C_Y(\bar{T}_1)$, the second statement in (2) follows. The third statement follows from the second and (1) applied to the subsystem $\bar{\mathcal{F}}_1$ of $\mathcal{F}_1$ on $\bar{T}_1$ isomorphic to $Sp_4[m]$. Thus (2) holds.

Part (4) follows from 13.1.10.2 and 13.2.3, while (5) follows from 13.1.1.2.

Suppose $s \in \mathcal{D}^o$. By 13.1.10.4, $\Gamma = \mathcal{D}^o(s) \cap \mathcal{Z}_S$ is of order at most 2, and in case of equality, $s \in T_1^g$ for some $g \in G$ by 13.1.10.2. So assume $\Gamma = \{e_i\}$ is of order 1. We may assume $e_i \neq z = e_1$, so by 13.1.10.4, $s \in \mathcal{D}(z)$. Then by 13.1.7, $s \in \mathcal{Z}(\tau_z)$. Hence as Γ is of order 1 it follows from 13.2.19.1 that $\mathcal{F}$ is of type $L^-_{2n+1}[m]$. Let $H_z \trianglelefteq C_G(z)$ be a model for $N_{\mathcal{F}_2(z)}(O(1'))$. Then $s \in H_z$ so $[s, C_G(O)] \leq C_{H_z}(O(1')) \leq W$, completing the proof of (6). □

LEMMA 13.2.21. *(1) $H^\Omega = Sym(\Omega)$, $[H, S_0] \leq W$, and if $\mathcal{F}$ is of type $PSp_{2n}[m]$ then $S_0 = C_S(H)W$.*

(2) $L_0 \trianglelefteq G$, *and if* $m > 8$ *then* $L_0 = O$, *while if* $m = 8$ *then* $L_0 = \prod_{i \in I} L_i$ *where* $L_i \cong SL_2(3)$ *is the model for* $\mathcal{O}_i$.

(3) If $\mathcal{F}$ *is of type* $PSp_{2n}[m]$ *then* $M = L_0$.

(4) Suppose $\mathcal{F}$ *is not of type* $PSp_{2n}[m]$ *and set* $\mathcal{Z}^1 = C_{\mathcal{Z}'}(W)$ *and* $\mathcal{Z}^2 = \{t \in \mathcal{Z}^1 : |\mathcal{D}^o(t) \cap \mathcal{Z}_S| = 2$. *Then* $M = \langle \mathcal{Z}^1 \rangle L_0$ *and* $\tilde{M} = M/L_0$ *is the permutation module, core of the permutation module for* $H/M = Sym(I)$ *for* $\mathcal{F}$ *of type* $L^-_{2n+1}[m]$, $L^-_{2n}[m]$, *respectively, with* $\tilde{t}$ *of weight 1,2 in* $\tilde{M}$ *for* $t \in \mathcal{Z}^1 - \mathcal{Z}^2$, $t \in \mathcal{Z}^2$, *respectively.*

(5) If $\mathcal{F}$ *is of type* $PSp_{2n}[m]$ *then* $H \cong \sigma(n,m)$, *where* $\bar{\sigma}(n,m)$ *is the wreath product* L_1 *wr* S_n *with* $L_1 = K$, $SL_2(3)$ *for* $m > 8$, $m = 8$, *respectively, and* $\sigma(n,m) = \bar{\sigma}(n,m)/Z(\bar{\sigma}(n,m))$.

(6) If $\mathcal{F}$ *is of type* $L^-_{2n+1}[m]$ *then* $H \cong 2\lambda(n,m)$, *the wreath product* M_1 *wr* S_n *with* $M_1 = L_1\langle t_1 \rangle$, *where* $L_1 = O(1)$ *if* $m > 8$ *and* $t_1 \in \mathcal{Z}$ *with* $\mathcal{D}^o(t_1) \cap \mathcal{Z}_S = \{e_1\}$. *If* $\mathcal{F}$ *is of type* $L^-_{2n}[m]$ *then* $H \cong \lambda(n,m)$, *the subgroup of* $2\lambda(n,m)$ *of index 2 containing* L_0 *with* $\lambda(n,m)/L_0 \cong Weyl(C_n)$.

(7) Some complement to M *in* H *contains a member of* $\mathcal{Z}^m$.

PROOF. By 13.2.3, $m(W/\langle c \rangle) = n - 1$. Assume the setup of 13.2.20. For $1 \leq i < n$, set $\bar{T}_i = \bar{T}_1^{g_i}$ and $T_i = T_1^{g_i}$ for $g_i \in G$ with $(1,2)^{g_i^\Omega} = (i, i+1)$. Then $\bar{T}_i^\Omega = \langle (i,i+1) \rangle$, so $H^\Omega = Sym(I) = Y^\Omega$, where $Y = \langle \bar{T}_i : 1 \leq i < n \rangle$. By 13.2.20, $S_0 = C_{S_0}(\bar{T}_1) \times \langle e_1 \rangle$. Therefore $|S_0 : C_{S_0}(Y)| \leq 2^{n-1} = |W : \langle c \rangle|$. Then as $C_W(H) = \langle c \rangle$, $S_0 = WC_{S_0}(Y)$.

From 13.2.17, $\mathcal{O}_\tau \trianglelefteq N_{\mathcal{F}}(W)$, so the model L_0 for $N_{\mathcal{O}_\tau}(OS_0)$ is normal in G. Then (2) follows as $\mathcal{O}_i \cong SL_2[m]$.

Let $M_0 = \langle \mathcal{Z}^1 \rangle L_0$, N the kernel of the action of G on Ω, and $S_N = S \cap N$. Claim $M_0 \trianglelefteq G$. By a Frattini argument, $G = NN_G(S_N)$, and $N_G(S_N)$ acts on $\mathcal{Z}^1$, and hence also on M_0, so it remains to show that $M_0 \trianglelefteq N$. But $N = S_N L_0 C_G(O)$, and by 13.2.20.6, $[\mathcal{Z}^1, C_G(O)] \leq W$, so we conclude from (2) that $M_0 \trianglelefteq G$.

By 13.2.20.3, $T_1 \trianglelefteq C_G(e_{1,2})$ if $m > 8$, while $T_1 L_1 L_2 \trianglelefteq C_G(e_{1,2})$ if $m = 8$. Therefore as Y is 2-transitive on I, as $M_0 \trianglelefteq G$, and as $T_i \leq \bar{T}_i \langle \mathcal{Z}^1 \rangle$, it follows that $T_1^G \subseteq YM_0$. By parts (4)-(6) of 13.2.20, $\mathcal{Z}^G \subseteq \langle T_1^G \rangle \leq YM_0$ unless $\mathcal{F}$ is of type $L^-_{2n+1}[m]$, where $\mathcal{Z}^G \subseteq \langle (\mathcal{Z}^1)^G, T_1^G \rangle \leq YM_0$ as $M_0 \trianglelefteq G$. So in any event, $\mathcal{Z}^G \subseteq YM_0$, and hence $H = \langle \mathcal{Z}^G \rangle L_0 \leq YM_0$. As $T_1 \leq \langle \mathcal{Z} \rangle O$ and $H \trianglelefteq G$, we have $YM_0 \leq H$, so $H = YM_0$.

By 2.3.5 in [**Asc19**], L_0 centralizes S_0, and for $t \in \mathcal{Z}'$, $[S_0, t] \leq W$ by parts (2) and (6) of 13.2.20, so $[H, S_0] \leq W$. Further if $\mathcal{F}$ is of type $PSp_{2n}[m]$ then $M_0 = L_0$ by 13.2.19.3, so $H = YL_0$. Then as $S_0 = WC_{S_0}(Y)$ and L_0 centralizes S_0 by 2.3.5 in [**Asc19**], the proof of (1) is complete.

Suppose $t_1 \in \mathcal{Z}^m$ with $t_1^\Omega = (1,2)$ and $t_2 \in t_1^G$ with $t_2^\Omega = (2,3)$. Set $y = t_1 t_2$ and $u = y^3$. Then $y^\Omega = (1,3,2)$ is of order 3, so u fixes $O(i)$ for $1 \leq i \leq 3$. Suppose u is of even order and let v be the involution in $\langle u \rangle$. Then $t = t_1 v \in t_i^{\langle y \rangle}$ for $i = 1$ or 2, so $t \in \mathcal{Z}^m$. Then as $O(1)^t = O(1)^{t_1} = O(2)$, $t \in T_1$ by 13.1.10.2, so $v \in T_1$. But then y acts on $O(1,2)$ by 13.1.10.4, a contradiction. Therefore $|u|$ is odd.

Set $G^+ = G/L_0$. As u^+ is of odd order and acts on $O(i)$, u^+ centralizes t_1^+, so as t_1 inverts u we have $u^+ = 1$. Thus $u \in L_0$, so as t_i inverts u for $i = 1,2$ and centralizes L_j for $j \notin Fix_\Omega(t_i)$, we have $u \in L_1 L_2 \cap L_2 L_3 = L_2$. But now as y centralizes u, y acts on L_2, a contradiction.

We've shown that $|t_1 t_2| = 3$. Similarly if $t_i \in \mathcal{Z}^m$ with $t_i^\Omega = (i, i+1)$ and $i > 2$ then $|t_1 t_i| = 2$ by 13.1.10.7. Set $X = \langle t_j : 1 \leq j < n \rangle$. It follows that

$(X; t_j : 1 \leq j < n)$ is a Coxeter system of type A_{n-1}, so $X \cong S_n$ is a complement to M_0 in H. Therefore as $H = M_0Y$, also $H = M_0X$, so $M = M_0$, completing the proof of (3), establishing (7), and proving the first statement in (4).

Suppose $\mathcal{F}$ is of type $PSp_{2n}[m]$. Then X is a complement to L_0 in H and t_1 centralizes $\mathcal{O}_{I-\{1,2\}}$ by 13.1.10.3, so $C_X(e_1)$ centralizes $O(1)$, and hence also $\mathcal{O}_1$, so (5) follows. Thus we may assume $\mathcal{F}$ is of type $L^-_{2n}[m]$ or $L^-_{2n+1}[m]$. In the latter case for each $i \in I$ there is $t_i \in \mathcal{Z}'$ with $\mathcal{D}^o(t_i) \cap \mathcal{Z}_S = \{e_i\}$. Then $P_i = O(i)\langle t_i\rangle \cong SD_{2m}$ and by 13.1.10.4 and 13.1.10.6, $P = \langle P_i : i \in I\rangle$ is the direct product of the P_i. Further from 13.2.19.5, $P_iL_i \trianglelefteq C_G(e_i)$, so the first half of (6) follows.

Therefore we may assume $\mathcal{F}$ is of type $L^-_{2n}[m]$. There exists $t_n \in \mathcal{Z}^m \cap T_{n-1}$ with $[t_{n-1}, t_n] = 1$ and $T^+_{n-1} = \langle t^+_{n-1}, t^+_n\rangle$. As $t^\Omega_n = t^\Omega_{n-1}$, an earlier observation shows that $|t_{n-2}t_n| = 3$ and t_n centralizes t_i for $i < n-2$. Set $Y = \langle X, t_n\rangle$; it follows that $(Y, t_i : 1 \leq i \leq n)$ is a Coxeter system of type D_n, so $Y \cong \mathrm{Weyl}(D_n)$. Further $T^+_{n-1} \leq Y^+$, so as W is the image of the permutation module for X, $T_i \leq YL_0$ for each $1 \leq i < n$, and hence $H = L_0Y$, so Y is a complement to L_0 in H. Hence the second half of (6) holds, completing the proof of the lemma. □

NOTATION 13.2.22. Let $T = S \cap H$, so that T is Sylow in H. Let $\tilde{\mathcal{F}}$ be the fusion system such that $\mathcal{F}$ is of type $\tilde{\mathcal{F}}$; thus $\tilde{\mathcal{F}}$ is $PSp_{2n}[m]$, $L^-_{2n}[m]$, or $L^-_{2n+1}[m]$. Let $\tilde{\tau} = (\tilde{\mathcal{F}}, \tilde{\Omega})$ be the Lie packet for $\tilde{\mathcal{F}}$, $\tilde{S}$ the Sylow group of $\tilde{\mathcal{F}}$, etc. We see in the next lemma that $\tilde{G} \cong H$, so we may take $\tilde{S} = T$, $\tilde{\Omega} = \Omega$, and $\tilde{G} = H$.

LEMMA 13.2.23. *(1) $\tilde{G} \cong H$, so we may take $\tilde{S} = T$, $\tilde{\Omega} = \Omega$, and $\tilde{G} = H$.*
(2) $\tilde{S}\mathcal{O}_{\tilde{\tau}} = T\mathcal{O}_\tau$.
(3) Set $\mathcal{W} = \langle T\mathcal{O}_\tau, \mathcal{F}_T(H)\rangle$. Then $\tilde{\mathcal{W}} = \mathcal{W}$.
(4) $N_{\tilde{\mathcal{F}}}(\tilde{W}) = \tilde{\mathcal{W}}$.
(5) $\mathcal{W} \trianglelefteq N_{\mathcal{F}}(W)$, $\mathcal{F}_T(H) \trianglelefteq N_{\mathcal{F}}(O)$, and $T \trianglelefteq TC_{\mathcal{F}}(O)$.

PROOF. By 5.4.6-5.4.10, $\tilde{\mathcal{F}}$ satisfies Hypothesis 13.2.14, so as H is determined up to isomorphism by 13.2.21, we conclude that $\tilde{H} \cong H$. But by 5.4.8-5.4.10, $\tilde{H} = \tilde{G}$, so $\tilde{G} \cong H$. Then following the convention in 13.2.22, we may take $\tilde{S} = T$, etc, so that (1) holds.

Next by Theorem 2.7.3, $T\mathcal{O}_\tau$ is determined by T and Ω, so $T\mathcal{O}_\tau = T\mathcal{O}_{\tilde{\tau}}$, establishing (2). Then (1) and (2) imply (3).

As W is normal in $T\mathcal{O}_\tau$ and $\mathcal{F}_T(H)$, $\mathcal{W} \leq N_{\mathcal{F}}(W)$. As $\mathcal{O}_\tau \trianglelefteq N_{\mathcal{F}}(W)$ by 13.2.17, $N_{\mathcal{F}}(W) = \langle S\mathcal{O}_\tau, N_{\mathcal{F}}(O)\rangle$ by 1.3.2 in [**Asc19**]. Further $C_{\tilde{S}}(\tilde{O}) = \tilde{W}$, so $N_{\tilde{\mathcal{F}}}(\tilde{O}) = \mathcal{F}_{\tilde{S}}(\tilde{G})$ and hence (4) holds.

Set $\mathcal{G} = N_{\mathcal{F}}(W)$, $\mathcal{G}_1 = S\mathcal{O}_\tau$, $\mathcal{G}_2 = N_{\mathcal{F}}(O)$, and $\mathcal{C} = C_{\mathcal{F}}(O)$. We just saw that $\mathcal{G} = \langle \mathcal{G}_1, \mathcal{G}_2\rangle$. Set $\mathcal{E}_1 = T\mathcal{O}_\tau$ and $\mathcal{E}_2 = \mathcal{F}_T(H)$. We will show $\mathcal{E}_2 \trianglelefteq \mathcal{G}_2$. Then Hypothesis 1.4.7 in [**Asc19**] is satisfied by the tuple $\mathcal{G}_i, \mathcal{E}_i, i = 1, 2$, so $\mathcal{W} = \langle \mathcal{E}_1, \mathcal{E}_2\rangle \trianglelefteq \langle \mathcal{G}_1, \mathcal{G}_2\rangle = \mathcal{G}$ by 1.4.8 in [**Asc19**]. Similarly $\mathcal{G}_2 = \langle SC_{\mathcal{F}}(O), N_{\mathcal{G}_2}(S_0)\rangle$ by 1.3.2 in [**Asc19**], and $\mathcal{E}_2 \trianglelefteq N_{\mathcal{G}_2}(S_0)$ by construction, so if $T \trianglelefteq S\mathcal{C}$ then $\mathcal{E}_2 = \langle \mathcal{E}_2, T\rangle \trianglelefteq \mathcal{G}_2$ by 1.4.8 in [**Asc19**]. Hence to complete the proof of (5), it remains to show that $T \trianglelefteq S\mathcal{C}$.

Suppose T is strongly closed in S with respect to $S\mathcal{C}$. Then as $[T, S_0] \leq C_T(O) = W \leq Z(\mathcal{C})$, it follows from 9.5.2 in [**Asc11**] that $T \trianglelefteq T\mathcal{C}$, so the lemma holds. Therefore it suffices to show that T is strongly closed in S with respect to $S\mathcal{C}$.

Let $x = e_\theta$ be an involution in $C_W(S)$ and T_x Sylow in $\mathcal{F}_x^\circ$. As $\mathcal{F}_x^\circ \trianglelefteq \mathcal{F}_x$ and $S\mathcal{C} \leq \mathcal{F}_x$, T_x is strongly closed in S with respect to $S\mathcal{C}$. Thus it suffices to show we can choose x with $T_x = T$. If $\mathcal{F}$ is of type $L_{2n+1}^-[m]$ we choose $x = c$, so that $T = T_c$. Thus we may assume $\mathcal{F}$ is of type $PSp_{2n}[m]$ or $L_{2n}^-[m]$. If $\mathcal{F}$ is of symplectic type then (cf. section 5.4) either $T_x = T$ or n is a power of 2 with $|\theta| = n/2$, and $|T : T_x| = 2$, while if $\mathcal{F}$ is of linear type then either $|T : T_x| = 2$ and $T = T_x\langle t\rangle$ for $t \in \mathcal{Z}'$ with $\mathcal{D}^o(t) \cap Z_S = \{e_i, e_j\}$ with $i \in \theta$ and $j \notin \theta$, or n is a power of 2 with $|\theta| = n/2$ and $|T : T_x| = 4$. Suppose $\mathcal{F}$ is linear but $|\theta| \neq n/2$. Let $u = e_\xi \in \mathcal{F}^f$ with $|\xi| = 2$, $b = e_{i,j}$, and $g \in G$ with $b^g = u$ and $C_S(b)^g \leq C_S(u)$. Then a Sylow group T_u of $\mathcal{F}_u^\circ$ is strongly closed in $C_S(u)$ with respect to $C_S(u)\mathcal{C}$ and $t^g \in T_u$. Let $\beta \in \hom_{C_S(\langle b\rangle)\mathcal{C}}(t, S)$; then $\alpha = \beta^{c_g} \in \hom_{C_S(u)\mathcal{C}}(t^g, S)$ with $(t\beta)^g = t^g\alpha$, so as T_u is strongly closed we have $t^g\alpha \in T$. Therefore $t\beta \in T^{g^{-1}} \cap S \leq T$ as T is strongly closed in S with respect to G. Thus $t^{S\mathcal{C}} \subseteq T$ as S acts on T, so as T_x is strongly closed of index 2 in T, also T is strongly closed. Hence the lemma follows unless the exceptional case $|\theta| = n/2$ holds, which we now assume.

Set $\mathcal{G}_2^+ = \mathcal{G}_2/W$. As $\mathcal{C} \trianglelefteq \mathcal{G}_2$, also $\mathcal{C}^+ \trianglelefteq \mathcal{G}_2^+$ by 8.9.2 in [**Asc08a**]. Now $\mathcal{E}_2 = [T_x]_{\mathcal{E}_2} \leq C_{\mathcal{G}_2}(\mathcal{C}^+)$ with T Sylow in $\mathcal{E}_2$, so here too $T \trianglelefteq T\mathcal{C}$ with T Sylow in $\mathcal{E}_2$. □

NOTATION 13.2.24. Let $\Xi = \Xi(\tau)$ be the set of nonempty subsets ξ of I such one of the following holds:

(a) ξ is S-invariant and $\xi \neq I$, or

(b) n is a power of 2, $\mathcal{F}$ is of type $PSp_{2n}[m]$ or $L_{2m}^-[m]$, $|\xi| = n/2$, and $e_\xi \in \mathcal{F}^f$, or

(c) n is a power of 2, $\mathcal{F}$ is of type $L_{2n+1}^-[m]$, and $\xi = I$.

For $\xi \in \Xi$ and $\mathcal{F}$ of type $PSp_{2n}[m]$ or $L_{2n}^-[m]$, define $\mathcal{B}_\xi(\tau) = \langle \mathcal{F}_{e_\xi}, N_\mathcal{F}(W)\rangle$. In case (c) set $\mathcal{B}_\xi(\tau) = \langle \mathcal{F}_c, \mathcal{F}_{e_\sigma}\rangle$, where $|\sigma| = n/2$ and $e_\sigma \in \mathcal{F}^f$. Finally if $\mathcal{F}$ is of type $L_{2n+1}^-[m]$ and case (a) holds, set $\mathcal{B}_\xi(\tau) = \langle \mathcal{F}_{e_\xi}, \mathcal{F}_{ce_\xi}, N_\mathcal{F}(W)\rangle$.

LEMMA 13.2.25. *Let $\xi \in \Xi$, $f = f_\xi$, and $\mathcal{B} = \mathcal{B}_\xi(\tau)$.*

(1) $f \in Z(S)$.

(2) $N_\mathcal{F}(O) \leq \mathcal{B}$.

(3) $S\mathcal{O}_\tau \leq \mathcal{B}$.

(4) Let $\theta \subseteq I$ with $Q = O(\theta) \in \mathcal{F}^f$, and let T_Q be Sylow in $N_\mathcal{F}(Q)$. Let Σ be a nontrivial orbit of T_Q on θ and $g \in H$ such that $\alpha = c_g^* \in \mathfrak{A}(O(\Sigma))$. Then if $T_Q\alpha C_\mathcal{F}(O(\Sigma)\alpha)^\circ \leq \mathcal{B}$, we have $N_\mathcal{F}(Q) \leq \mathcal{B}$.

PROOF. In cases (a) and (c) of 13.2.24, ξ is S-invariant, so $f \in Z(S)$. In case (b), $|S : N_S(\xi)| = 2$ with $f^S = \{f, fc\}$, so as $c = 1$ by 13.2.19.2, again $f \in Z(S)$. Therefore (1) holds.

As $W \trianglelefteq N_\mathcal{F}(O)$, and as $N_\mathcal{F}(W) \leq \mathcal{F}_c$ in case (c) of 13.2.24, (2) holds. As $S\mathcal{O}(\tau) \leq \mathcal{F}_f \leq \mathcal{B}$, (3) holds.

Assume the hypothesis of (4). By 1.3.2 in [**Asc19**],

$$N_\mathcal{F}(Q) \leq \langle N_\mathcal{F}(O), T_Q C_\mathcal{F}(Q)^\circ\rangle$$

and $N_\mathcal{F}(O) \leq \mathcal{B}$ by (2), so to prove (4) it remains to show that $T_Q C_\mathcal{F}(Q)^\circ \leq \mathcal{B}$. But as α is a $\mathcal{B}$-map and $(T_Q C_\mathcal{F}(Q)^\circ)\alpha^* \leq T_Q\alpha C_\mathcal{F}(O(\Sigma)\alpha)^\circ$, $T_Q C_\mathcal{F}(Q)^\circ \leq \mathcal{B}$ if $T_Q\alpha C_\mathcal{F}(O(\Sigma)\alpha)^\circ \leq \mathcal{B}$. □

LEMMA 13.2.26. *Let $\xi \in \Xi$ and $\mathcal{B} = \mathcal{B}_\xi(\tau)$. Then $\mathcal{F} = \mathcal{B}$.*

PROOF. Set $f = f_\xi$ and assume $\mathcal{F} \neq \mathcal{B}$. Claim $N_\mathcal{F}(O(\theta)) \not\leq \mathcal{B}$ for some $\theta \subseteq \mathcal{Z}_S$ with $O(\theta) \in \mathcal{F}^f$. For if not, then by 13.2.6 and parts (2) and (3) of 13.2.25, $\mathcal{E}(\tau) \leq \mathcal{B}$, so by 13.2.11, n is a power of 2 and either $c \neq 1$ and $\mathcal{F} = \langle \mathcal{E}(\tau), \mathcal{F}_c \rangle$, or $c = 1$ and $\mathcal{F} = \langle \mathcal{E}(\tau), \mathcal{F}_x \rangle$, where $x = e_\theta \in \mathcal{F}^f$ and $|\theta| = n/2$. By 13.2.19.2, in the first case $\mathcal{F}$ is of type $L^-_{2n+1}[m]$ and in the second, $\mathcal{F}$ is of type $PSp_{2n}[m]$ or $L^-_{2n}[m]$. Thus in the first case, case (c) of 13.2.24 holds, so $\mathcal{F}_c \leq \mathcal{B}$, and hence $\mathcal{B} = \mathcal{F}$, contrary to assumption. Similarly in the second case, case (b) of 13.2.24 holds, so $\mathcal{F}_x \leq \mathcal{B}$, for the same contradiction. This establishes the claim.

By the claim and 13.2.25.4, there is $\theta \subseteq I$ such that $Q = O(\theta) \in \mathcal{F}^f$, θ is an orbit of a Sylow group T_Q of $N_\mathcal{F}(Q)$, and $\mathcal{Y} = C_\mathcal{F}(Q)^\circ \not\leq \mathcal{B}$. As $N_\mathcal{F}(O) \leq N_\mathcal{F}(W) \leq \mathcal{B}$, $\theta \neq I$. As $T_Q \leq S$, θ is contained in an orbit Σ of S. Let $\xi' = I - \xi$. In case (a) of 13.2.24, S acts on ξ, so as $T_Q \leq S$ is transitive on θ, we have $\theta \subseteq \Sigma \subseteq \zeta$, for some $\zeta \in \{\xi, \xi'\}$. If $\mathcal{F}$ is of type $PSp_{2n}[m]$ or $L^-_{2n}[m]$ then by 13.2.19.2, $c = 1$ so $f = fc$, so interchanging the roles of ξ and ξ' if necessary, we may assume $\zeta = \xi$. On the other hand if $\mathcal{F}$ is of type $L^-_{2n+1}[m]$ then as both $\mathcal{F}_f$ and $\mathcal{F}_{cf}$ are contained in $\mathcal{B}$, interchanging ξ and ξ' if necessary, we may again assume $\zeta = \xi$. Further if $\theta = \xi$ then $\mathcal{Y} \leq \mathcal{F}_f \leq \mathcal{B}$, a contradiction. So $\theta \subset \xi$ in case (a).

Similarly in cases (b), (c), conjugating in H we may take $\theta \subset \delta$, where $\delta = \xi$, σ, respectively and T_Q acts on δ.

Set $\bar{\theta} = I - \theta$, $\bar{\Omega} = \{J \in \Omega : z(J) = e_i \text{ with } i \in \bar{\theta}\}$, and $\bar{\tau} = (\mathcal{Y}, \bar{\Omega})$. Then $a = |\bar{\theta}| = n - |\theta|$ and $\bar{\tau}$ is a quaternion fusion packet.

In case (a) of 13.2.24, as $Q \in \mathcal{F}^f$, Σ is the orbit of S on I of smallest length subject to $|\Sigma| \geq |\theta|$, the T_Q-orbits and S-orbits on $I - \Sigma$ are the same, and $|\Sigma| = |S : T_Q||\theta|$. In cases (b) and (c), as $Q \in \mathcal{F}^f$, $|\delta| = |N_S(\delta) : T_Q||\theta|$.

Let $\alpha = \xi$ in (a) and (b) and $\alpha = \sigma$ in (c). In (a), S acts on ξ so T_Q acts on α. In (b) and (c), T_Q acts on $\delta = \alpha$. So in any event, T_Q acts on α.

Suppose T_Q is transitive on $\bar{\theta}$. Then as T_Q acts on α' and $\alpha - \theta$ with $\theta \neq \alpha$, we conclude $\alpha' = \emptyset$, so $\alpha = I$, contrary to 13.2.24 where α is a proper subset of I. Therefore T_Q is not transitive on $\bar{\theta}$.

Let $\bar{\alpha} = \alpha \cap \bar{\theta}$. As T_Q is Sylow in $N_\mathcal{F}(Q)$ it is also Sylow in $T_Q\mathcal{Y}$. We've seen that T_Q acts on α, so as T_Q acts on θ it also acts on $\bar{\alpha}$. Thus $\bar{\alpha} \in \Xi(\bar{\tau})$ appears in case (a), with $a = |\bar{\theta}| < n$. As T_Q is not transitive on $\bar{\theta}$, we have $a > 1$. Therefore by induction on n, $T_Q\mathcal{Y} = \mathcal{B}_{\bar{\alpha}}(\bar{\tau})$. However $\mathcal{B}_{\bar{\alpha}}(\bar{\tau}) \leq \langle \mathcal{F}^\circ_y, N_\mathcal{F}(W) \rangle \leq \mathcal{B}$, (or $\langle \mathcal{F}_y, \mathcal{F}_{cy}, N_\mathcal{F}(W) \rangle$ if $\mathcal{F}$ is of type $L^-_{2n+1}[m]$ and case (a) holds) where $y = f$ or e_σ, contradicting $T_Q\mathcal{Y} \not\leq \mathcal{B}$. This completes the proof. $\square$

13.3. Symplectic groups and systems

In this section we assume Hypothesis 5.4.1.III, except that we write L for G. Thus q is an odd prime power, $F = \mathbf{F}_q$ is the field of order q, and $m = (q^2 - 1)_2$ is the 2-share of $q^2 - 1$. Let n be a positive integer, U a $2n$-dimensional vector space over F, $\mathfrak{f}$ a symplectic form on U, and $L = O(U, \mathfrak{f}) = Sp(U)$ be the isometry group of $\mathfrak{f}$. Thus $L \cong Sp_{2n}(q)$ is the n-dimensional symplectic group over F.

We also adopt Notation 5.4.2. Thus $U = \mathfrak{U}_1 \perp \cdots \perp \mathfrak{U}_n$, where $\mathfrak{U}_i$, $i \in I = \{1, \ldots, n\}$, is a nondegenerate line, and $L_i = C_L(\mathfrak{U}_i^\perp)$ acts naturally as $Sp(\mathfrak{U}_i)$ on $\mathfrak{U}_i$. Pick $K_i \in Syl_2(L_i)$ and set $\Omega = \{K_i : i \in I\}$. Let $S \in Syl_2(L)$ with $O = K_1 \cdots K_n \leq S$ and form the fusion system $\mathcal{F} = \mathcal{F}_S(L)$. By 5.4.6.3, $\tau = (\mathcal{F}, \Omega)$

is a Lie packet of L. Set $z_i = z(K_i)$. By 5.4.6.4, $\mathcal{F}$ is transitive on Ω and for $i \in I$, $\Omega(z_i) = \{K_i\}$.

Set $c = z_1 \cdots z_n$. Then $\langle c \rangle = Z(L)$ and we set $L^+ = L/\langle c \rangle$ and $\mathcal{F}^+ = \mathcal{F}/\langle c \rangle = \mathcal{F}_{S^+}(L^+)$. Hence $L^+ \cong PSp_{2n}(q)$. By definition, $\mathcal{F} = Sp_{2n}[m]$ and $\mathcal{F}^+ = PSp_{2n}[m]$.

Let $z = z_1$ and $K = K_1$, and choose notation so that $z \in \mathcal{F}^f$. As in Notation zZ.7, $\mathcal{Z} = \mathcal{Z}(\tau)$ and $\mathcal{Z}_S = \{z_1, \ldots, z_n\}$. Set $\mathcal{Z}' = \mathcal{Z} - \mathcal{Z}_S$.

LEMMA 13.3.1. *(1)* $\{K\} = \Omega(z)$ *and if* $n > 2$ *then* $\Omega(z^+) = \{K^+\}$.

(2) $C_{\mathcal{Z}}(z) = z^{\perp}$ *and if* $n > 2$ $C_{\mathcal{Z}^+}(z^+) = (z^+)^{\perp}$.

(3) Let $\Sigma = \{\mathfrak{U}_i : i \in I\}$, $W = \langle \mathcal{Z}_S \rangle$, *and* $L_0 = \prod_i L_i$. *Then* $N_L(W) = N_L(\Sigma) = L_0 Y$ *where* Y *acts faithfully as* $Sym(\Sigma)$ *on* Σ. *Further* $N_{L^+}(W^+) = N_L(W)^+$.

PROOF. We already observed that $\{K\} = \Omega(z)$, while if $n > 2$ then $z^+ \neq z_i^+$ for $i > 1$, so also $\Omega(z^+) = \{K^+\}$ and $C_{L^+}(z^+) = C_L(z)^+$. In particular (1) holds. By 5.4.9.1, $C_{\mathcal{Z}}(z) = z^{\perp}$ and then as $C_{L^+}(z^+) = C_L(z)^+$ when $n > 2$, (2) holds.

As $c \in W$, $N_{L^+}(W^+) = N_L(W)^+$. By 3.1.5.3, $N_L(W) = N_L(\mathcal{Z}_S)$ and as z_i is the unique involution s with $[U, s] = U_i$, it follows that $N_L(W) = N_L(\Sigma)$. As $L_i = O(\mathfrak{U}_i, \mathfrak{f})$, L_0 is the kernel of the action of $N_L(\Sigma)$ on Σ, so $N_L(W)/L_0 \leq Sym(\Sigma)$. Finally let $B = \{x_i, y_i : i \in I\}$ a basis for U with $\{x_i, y_i\}$ a basis for $\mathfrak{U}_i$ and $\mathfrak{f}(x_i, y_i) = 1$ for each $i \in I$. Represent $Sym(I)$ on U via $x_i^g = x_{ig}$ and $y_i^g = y_{ig}$. Then the image Y of $Sym(I)$ in $GL(U)$ preserves $\mathfrak{f}$, so $Y \leq L$ and by construction Y acts faithfully as $Sym(\Sigma)$ on Σ, completing the proof of (3). □

NOTATION 13.3.2. Choose notation so that $O_A = K_{n-1}K_n \in \mathcal{F}^f$ and set $A = \mathfrak{U}_{n-1} + \mathfrak{U}_n$ and $L_A = C_L(A^{\perp})$. Then $L_A = Sp(A) \cong Sp_4(q)$. Further $T = S \cap L_A$ is Sylow in L_A with $T = O_A\langle t \rangle$ where $t \in \mathcal{Z}'$ with $z_{n-1}^t = z_n$. Set $\mathcal{L}_A = \mathcal{F}_T(L_A)$ and $\tau_A = (\mathcal{L}_A, \{K_{n-1}, K_n\})$, so that τ_A is a quaternion fusion packet and $\mathcal{L}_A \cong Sp_4[m]$. Set $u = z_{n-1}z_n$; then $\langle u \rangle = Z(L_A) = Z(\mathcal{L}_A)$, and we set $\tilde{L}_A = L_A/\langle u \rangle$, $\tilde{\mathcal{L}}_A = \mathcal{L}_A/\langle u \rangle$, and $\tilde{\tau}_A = (\tilde{\mathcal{L}}_A, \{\tilde{K}_{n-1}, \tilde{K}_n\})$.

Observe that $T \cong Q_m$ wr $\mathbf{Z}_2$, so T is transitive on the involutions in $T - O_A$, and this set of involutions is $\mathcal{Z}'(\tau_A)$. Further $C_{\mathcal{L}_A}(z_n)$ is transitive on the set $\mathfrak{P}(\tau_A)$ of subgroups $P = \langle t, v, z_{n-1} \rangle$ of T such that $t \in \mathcal{Z}'(\tau_A)$ and $\langle v \rangle$ is a normal subgroup of $C_{O_A}(t)$ of order 4. Set $\mathfrak{P}(\tilde{\tau}_A) = \{\tilde{P} : P \in \mathfrak{P}(\tau_A)\}$.

Observe that $P = \langle v \rangle * Q$, where $Q = \langle vt, z_n t \rangle$ is the unique Q_8-subgroup of P. If $\xi = (\mathcal{X}, \Xi)$ is a quaternion fusion packet with Sylow group X such that $\mathcal{X} = X\mathcal{X}^{\circ}$ with $\mathcal{X}^{\circ} \cong (P)Sp_4[m]$, write $\mathfrak{P}(\xi)$ for the set of $P \in \mathfrak{P}(\xi^{\circ})$ such that $\langle v, z_n \rangle$ (or $\langle \tilde{v}, \tilde{z}_n \rangle$) is normal in X. When $\mathcal{X}^{\circ}$ is $Sp_4[m]$ and $R = C_X(\tilde{P})$, define

$$Aut_{\mathcal{P}}(R) = \langle \hat{\phi} : \phi \in O^2(Aut(P)) \rangle,$$

where $\hat{\phi} \in Aut(R)$ is defined by $(cx)\hat{\phi} = c \cdot x\phi$ for $c \in C_S(Q)$ and $x \in Q$.

LEMMA 13.3.3. *Let* $\xi = (\mathcal{X}, \Xi)$ *be a quaternion fusion packet with Sylow group* X *such that* $\mathcal{X} = X\mathcal{X}^{\circ}$ *and* $\mathcal{X}^{\circ} \cong PSp_4[m]$. *Let* $P \in \mathfrak{P}(\xi)$ *and set* $R = C_X(P)$. *Then* $\mathcal{X} = \langle X\mathcal{O}_{\xi}, O^2(Aut_{\mathcal{X}}(R)) \rangle$, *where* $\mathcal{O}_{\xi} = O^2(C_{\mathcal{X}}(Z(O(\xi))))$.

PROOF. This follows from 9.4.22, 9.4.27, and 9.4.34.2. □

LEMMA 13.3.4. *Let $\xi = (\mathcal{X}, \Xi)$ be a quaternion fusion packet with Sylow group X such that $\mathcal{X} = X\mathcal{X}^\circ$ and $\mathcal{X}^\circ \cong Sp_4[m]$. Set $\langle u \rangle = Z(\mathcal{X}^\circ)$ and $\tilde{\mathcal{X}} = \mathcal{X}/\langle u \rangle$. Let X° be Sylow in $\mathcal{X}^\circ$ and $P \in \mathfrak{P}(\xi)$, and set $R = C_X(\tilde{P})$. Then*

(1) $\mathcal{X} = \langle X\mathcal{O}_\xi, O^2(Aut_{\mathcal{X}}(R)) \rangle$.

(2) $R = C_X(Q)Q$, where Q is the Q_8-subgroup of P.

(3) $C_{X^\circ}(Q) \cong D_m$.

(4) If $m > 8$ then $A_4 \cong O^2(Aut_{\mathcal{X}}(R)) = Aut_{\mathcal{P}}(R)$.

(5) Suppose $m = 8$. Then $Q^\circ = QC_{X^\circ}(Q) \cong D_8Q_8$,

$$O^2(Aut_{\mathcal{X}}(R) \cong O^2(Aut(Q^\circ))$$

via the restriction map, and is $Inn(Q^\circ)$ extended by A_5. Further $O^2(Aut_{\mathcal{X}}(R)) = \langle O^2(Aut_{N_{\mathcal{X}}(RX^\circ)}(R)), Aut_{\mathcal{P}}(R) \rangle$.

(6) $\mathcal{X} = \langle X\mathcal{O}_\xi, Aut_{\mathcal{P}}(R) \rangle$.

PROOF. As $\tilde{\mathcal{X}}$ satisfies the hypotheses of 13.3.3, we conclude from that lemma that $\tilde{\mathcal{X}} = \langle \tilde{X}\mathcal{O}_{\tilde{\xi}}, O^2(Aut_{\tilde{\mathcal{X}}}(\tilde{R})) \rangle$. Therefore by 2.1.15, $\mathcal{X}$ is generated by the preimages $X\mathcal{O}_\xi$ and $O^2(Aut_{\mathcal{X}}(R))$ of the two subcategories generating $\tilde{\mathcal{X}}$, so that (1) holds.

Recall $P = \langle v \rangle * Q$ with $Q = \langle vt, z_nt \rangle \cong Q_8$ and $R = C_X(\tilde{P})$. As $Inn(Q) = C_{Aut(Q)}(\tilde{Q})$, we have $R = C_R(Q)Q$. Observe that $C_X(Q)$ centralizes vz_n and hence also $\tilde{v}$, so $C_X(Q) \leq C_X(\tilde{P}) = R$, so $R = C_X(Q)Q$, proving (2).

Set $y = z_nt$. As $z_n \in Z(O_A)$, $C_{O_A}(y) = C_{O_A}(t) \cong K$, so $C_{O_A}(Q) = C_{O_A}(y) \cap C_{O_A}(v) = V$ is cyclic of order $m/2$. Let $l \in C_{O_A}(y) - V$. Then $i = yl$ is an involution centralizing Q and inverting V, so $Y = C_{X^\circ}(Q) = V\langle i \rangle \cong D_m$, proving (3).

Let G be a model for $N_{\mathcal{X}}(R)$ and H the normal subgroup of G which is a model for $N_{\mathcal{X}^\circ}(R)$. Observe $G = N_X(R)H$, so $O^2(G) = O^2(H)$. Also $YQ = Q^\circ \trianglelefteq G$.

Suppose $m > 8$ and set $Q_0 = \Phi(Q^\circ)$. Then $Q_0 = \Phi(Y) \cong \mathbf{Z_{m/4}}$ and $Q_1 = C_{Q^\circ}(Q_0) = VQ$. Hence Q_1 and $V = Z(Q_1)$ are normal in G, as is $\Omega_1(Q_1) = P$. Thus $O^2(G)$ acts on P and centralizes Q°/P, so $O^2(G) = O^2(H)$ and $O^2(\tilde{H}) \cong O^2(Aut(Q)) \cong A_4$, so $O^2(H) \cong SL_2(3)$. Hence $A_4 \cong O^2(Aut_{\mathcal{X}}(R))$. Let $h \in H$ be of order 3. Then $[C_X(Q), h] \leq C_{O^2(H)}(Q) = \langle u \rangle \leq C_H(h)$, so $[C_X(Q), h] = 1$. Thus completes the proof of (4).

So assume $m = 8$. Then $Q^\circ \cong Q_8D_8$, so $\Sigma = O^2(Aut(Q^\circ)$ is $Inn(Q^\circ)$ extended by A_5. Then $\Sigma = \langle N_\Sigma(\tilde{X}^\circ), \Sigma_1 \rangle$, where $\Sigma_1/Inn(Q^\circ \cong S_3$, and $O^2(\Sigma_1)$ is the image of $Aut_{\mathcal{P}}(R)$ under the restriction map. Thus (5) holds.

Finally (6) follows from (1), (4), and (5). □

DEFINITION 13.3.5. Assume $\gamma = (\mathcal{G}, \Gamma)$ is a quaternion fusion packet with Sylow group S_γ, such that $\mathcal{G} = S_\gamma\mathcal{G}^\circ$ with $\mathcal{G}^\circ = \mathcal{F}$. Let $O_* = K_1 \cdots K_{n-2} \in \mathcal{G}^f$ and set $\mathcal{P}_1 = S_\xi C_{\mathcal{G}}(O_*)^\circ$, where $S_\xi = N_{S_\gamma}(O_*)$ and $\xi = (\mathcal{P}_1, \{K_{n-1}, K_n\})$. Write $\mathfrak{P}(\gamma)$ for $\mathfrak{P}(\xi)$. Pick $P_1 \in \mathfrak{P}(\gamma)$ and set $R_1 = C_{S_\xi}(P/\langle z_{n-1}z_n \rangle))$. By 13.3.4.6, $\mathcal{P}_1 = \langle S_\xi\mathcal{O}_\xi, Aut_{\mathcal{P}}(R_1) \rangle$.

13.4. Linear and unitary groups and systems

In this section we assume case I or II of Hypothesis 5.4.1, except that we write L for G. Thus q is an odd prime power, $F = \mathbf{F}_q$ or $\mathbf{F}_{q^2}$ in case I or II, respectively, and $m = (q^2-1)_2$ is the 2-share of q^2-1. Set $\pi = \pm 1$ where $q \equiv \pi \mod 4$ and let $a \in \{0,1\}$. Let n be a positive integer, U a $(2n+a)$-dimensional vector space over F, and take $L = SL^{-\pi}(U)$. Thus U is a linear or unitary space over F and L is the subgroup of the isometry group of U contained in $SL(U)$.

We also adopt Notation 5.4.2. Thus $U = \mathfrak{U}_0 \perp \mathfrak{U}_1 \perp \cdots \perp \mathfrak{U}_n$, where $\mathfrak{U}_i$, $i \in I = \{1,\dots,n\}$, is a line with $\mathfrak{U}_i$ nondegenerate in case II, and $\mathfrak{U}_0$ is of dimension a. Further for $i \in I$, $L_i = C_L(\mathfrak{U}_i{}^\perp)$ acts naturally as $SL^{-\pi}(\mathfrak{U}_i)$ on $\mathfrak{U}_i$. Here, in case I, if A is the sum of the $\mathfrak{U}_j$ with $j \in J \subseteq I \cup \{0\}$ then $A^\perp = \langle \mathfrak{U}_k : k \notin J \rangle$. Pick $K_i \in Syl_2(L_i)$ and set $\Omega = \{K_i : i \in I\}$. Let $S \in Syl_2(L)$ with $O = K_1 \cdots K_n \leq S$ and form the fusion system $\mathcal{F} = \mathcal{F}_S(L)$. By 5.4.6.3, $\tau = (\mathcal{F}, \Omega)$ is a Lie packet of L. Set $z_i = z(K_i)$. By 5.4.6.4, $\mathcal{F}$ is transitive on Ω and for $i \in I$, $\Omega(z_i) = \{K_i\}$.

Let $z = z_1$ and $K = K_1$, and choose notation so that $z \in \mathcal{F}^f$. As in Notation zZ.7, $\mathcal{Z} = \mathcal{Z}(\tau)$ and $\mathcal{Z}_S = \{z_1,\dots,z_n\}$. Set $\mathcal{Z}' = \mathcal{Z} - \mathcal{Z}_S$.

By definition $\mathcal{F} = SL^-_{2n+a}[m]$ and we write $L^-_{2n+a}[m]$ for $\mathcal{F}/Z(\mathcal{F})$. Of course $Z(\mathcal{F})$ is cyclic of order $(q+\pi, 2n+a)_2$, so if $a=1$ then $Z(\mathcal{F}) = 1$ and $\mathcal{F} \cong L^-_{2n+1}[m]$.

Suppose on the other hand that $a = 0$. Then $Z(\mathcal{F}) = \langle c \rangle$ is of order 2, and we set $L^+ = L/\langle c \rangle$ and $\mathcal{F}^+ = \mathcal{F}/\langle c \rangle = \mathcal{F}_{S^+}(L^+)$. Thus $\mathcal{F}^+ = L^-_{2n}[m]$.

LEMMA 13.4.1. *(1) $K = \Omega(z)$ and if $n > 2$ then $\Omega(z^+) = \{K^+\}$.*

(2) $C_{\mathcal{Z}}(z) = \{z\} \cup \mathcal{D}(z) \cup \mathcal{D}^o(z)$, where $\mathcal{D}(z) = \{t \in \mathcal{Z} : [U,t] \leq \mathfrak{U}_1^\perp\}$ and $\mathcal{D}^o(z) = \{t \in \mathcal{Z} : \dim([\mathfrak{U}_1, t]) = \dim([\mathfrak{U}_1^\perp, t] = 1\}$.

(3) Let $\Sigma = \{\mathfrak{U}_j : j \in I \cup \{0\}\}$, $W = \langle \mathcal{Z}_S \rangle$, and M the kernel of the action of $N_L(\Sigma)$ on Σ. Then $N_L(W) = N_L(\Sigma) = MY$ where Y acts faithfully as $Sym(\Sigma)$ on Σ. Further if $a = 0$ then $N_{L^+}(W_0^+) = N_L(W)^+$.

(4) Set $L_0 = \prod_{i \in I} L_i$. Then $M = O(M)L_0M_0$ where M_0 is Y-invariant and generated by members of $\mathcal{Z}'$, and either

(a) $a = 0$ and M_0 is the core of the n-dimensional permutation module for Y over $\mathbf{F}_2$, or

(b) $a = 1$ and $M_0 = \langle t_i : i \in I \rangle$ is the n-dimensional permutation module with $t_i \in \mathcal{Z}'$ satisfying $[U, t_i] = [\mathfrak{U}_i, t_i] + \mathfrak{U}_0$.

PROOF. The proof of (1) is the same as that of 13.3.1.

Suppose $t \in \mathcal{Z} - \{z\}$ centralizes z. Then t acts on $\mathfrak{U}_1 = [U,z]$ and $\mathfrak{U}_1^\perp = C_U(z)$, so either $[U,t] \leq \mathfrak{U}_1^\perp$ and hence t centralizes L_1 so $t \in \mathcal{D}(z)$, or $[\mathfrak{U}_1, t] = X$ and $[\mathfrak{U}_1^\perp, t] = X'$ are 1-dimensional. In the latter case $[K,t] \neq 1$, so $t \in \mathcal{D}^o(z)$. This proves (2).

As $c \in W$, $N_{L^+}(W_0^+) = N_L(W)^+$ when $a = 0$. By 3.1.5.3, $N_L(W) = N_L(\mathcal{Z}_S)$, so as z_i is the unique involution s with $[U,s] = \mathfrak{U}_i$, and $\mathfrak{U}_i^\perp = C_U(s)$, and as $\mathfrak{U}_0 = C_U(W)$, it follows that $N_L(W) = N_L(\Sigma)$. As M is the kernel of the action of $N_L(\Sigma)$ on Σ, $N_L(W)/M \leq Sym(\Sigma)$. For $i \in I$, let $\{x_i, y_i\}$ be a basis for $\mathfrak{U}_i$, with the basis hyperbolic in case II. Represent $Sym(I)$ on U via $x_i^g = x_{ig}$ and $y_i^g = y_{ig}$ for $i \in I$, and have $Sym(I)$ centralize $\mathfrak{U}_0$. Then the image Y of $Sym(I)$ in $GL(U)$ is contained in L and by construction Y acts faithfully as the stabilizer in $Sym(\Sigma)$ of $\mathfrak{U}_0$ on Σ, completing the proof of (3).

Let N be the kernel of the action of $N_{GL^{-\pi}(U)}(\Sigma)$ on Σ. Set $\bar{N} = N/O(N)$ and for $i \in I$, choose a basis $\{x_i, y_i\}$ for U_i as in the proof of (3). Then $\bar{N} = \bar{N}_0 \times \bar{N}_1 \times \cdots \times \bar{N}_n$, where $N_j = C_N(U_j^\perp)$ for $j \in J = I \cup \{0\}$, and $\bar{N}_i = \langle \bar{s}_i \rangle \cong \mathbf{Z}_2$ for $i \in I$, and also for $i = 0$ if $a = 1$. Moreover for $i \in I$, s_i induces an outer automorphism on L_i and we may choose notation so that $[\mathfrak{U}_i, s_i] = \langle x_i \rangle$. Now if

$a = 0$ we may take $M_0 = \langle s_i s_j : i, j \in I\rangle$, so that $M_0 \cong E_{2^{n-1}}$, and, from the construction of Y in the proof of (3), M_0 is the core of the permutation module for Y. Also for $i \neq j$, $\dim([U, s_i s_j]) = 2$, so $s_i s_j \in \mathcal{Z}'$. Therefore (4a) holds in this case. Similarly if $a = 1$ we may take $M_0 = \langle t_i : i \in I\rangle$, where $t_i = s_i s_0$, so that (4b) holds. □

NOTATION 13.4.2. Choose notation so that $O_A = K_{n-1}K_n \in \mathcal{F}^f$ and set $A = \mathfrak{U}_{n-1} + \mathfrak{U}_n + \mathfrak{U}_0$ and $L_A = C_L(A^\perp)$. Then $L_A = SL^{-\pi}(A) \cong SL^{-\pi}_{4+a}(q)$. Further $T_A = S \cap L_A$ is Sylow in L_A. Set $\mathcal{L}_A = \mathcal{F}_{T_A}(L_A)$ and $\tau_A = (\mathcal{L}_A, \{K_{n-1}, K_n\})$, so that τ_A is a quaternion fusion packet and $\mathcal{L}_A \cong SL^-_{4+a}[m]$. Set $u = z_{n-1}z_n$ and $B = \mathfrak{U}_{n-1} + \mathfrak{U}_n$. Thus $B = [U, u]$, $L_B = C_L(B^\perp) \cong SL^{-\pi}_4(q)$, $T_B = S \cap L_B$ is Sylow in L_B, $\mathcal{L}_B = \mathcal{F}_{T_B}(L_B) \cong SL^-_4[m]$, and $\tau_B = (\mathcal{L}_B, \{K_{n-1}, K_n\})$ is a quaternion fusion packet. Also $\langle u\rangle = Z(L_B) = Z(\mathcal{L}_B)$, and we set $\tilde{L}_B = L_B/\langle u\rangle$, $\tilde{\mathcal{L}}_B = \mathcal{L}_B/\langle u\rangle$, and $\tilde{\tau}_B = (\tilde{\mathcal{L}}_B, \{\tilde{K}_{n-1}, \tilde{K}_n\})$.

For $i \in I$ let $\{x_i, y_i\}$ be a basis for $\mathfrak{U}_i$ chosen as in the proof of 13.4.1. We may choose our bases so that there is $t \in \mathcal{Z}^m(\tau_B)$ with $x^t_{n-1} = x_n$ and $y^t_{n-1} = y_n$, and $s \in \mathcal{Z}'(\tau_B)$ so that $[U, s] = \langle x_{n-1}, x_n\rangle$ and $C_B(s) = \langle y_{n-1}, y_n\rangle$. Then from 13.4.1.4, $T_B = O_A\langle t, s\rangle$ and $ts \in \mathcal{Z}^m(\tau_B)$.

Similarly if $a = 1$ we may choose t to centralize $\mathfrak{U}_0$, and, as in 13.4.1.4, $t_i \in \mathcal{Z}'(\tau_A)$ with $[U, t_i] = \langle y_i, \mathfrak{U}_0\rangle$ and $C_A(t_i) = \langle y_{n-1}, y_n\rangle$. Again from 13.4.1.4, $T_A = (T_{n-1} \times T_n)\langle t\rangle \cong SD_{2m}$ wr $\mathbf{Z}_2$, where $T_i = K_i\langle t_i\rangle$ and $T^t_{n-1} = T_n$.

NOTATION 13.4.3. Write $\mathfrak{P}^m(\tau_B)$ for the set of subgroups $P = \langle t, v, z_{n-1}\rangle$ of T such that $t \in \mathcal{Z}^m(\tau_B)$ and $\langle v\rangle$ is a normal subgroup of $C_{O_A}(t)$ of order 4. Set $\mathfrak{P}^m(\tilde{\tau}_B) = \{\tilde{P} : P \in \mathfrak{P}^m(\tau_B)\}$.

As in 13.3.2, $P = \langle v\rangle * Q$, where $Q = \langle vt, z_n t\rangle$ is the unique Q_8-subgroup of P.

Write $\mathfrak{P}^o(\tau_B)$ for the set of subgroups $P = \langle s, z_n, v\rangle$ such that $s \in \mathcal{D}^o(\tau_B)$. Set $\mathfrak{P}^o(\tilde{\tau}_B) = \{\tilde{P} : P \in \mathfrak{P}^o(\tau_B)\}$. From 9.4.16, $C_{\tilde{T}_B}(\tilde{P}) = \tilde{Q} \cong E_{16}$ and $Aut_{\tilde{L}_B}(\tilde{P})$ is of index 2 in $O^+_4(2)$, so it follows that $C_{T_B}(\tilde{P}) = Q \cong Q_8^2$.

Suppose $\xi = (\mathcal{X}, \Xi)$ is a quaternion fusion packet with Sylow group X such that $\mathcal{X} = X\mathcal{X}^\circ$ with $\mathcal{X}^\circ \cong (P)SL^-_4[m]$. For $! \in \{m, o\}$, write $\mathfrak{P}^!(\xi)$ for the set of $P \in \mathfrak{P}^!(\xi^\circ)$ such that when $! = m$, $\langle v, z_n\rangle$ (or $\langle\tilde{v}, \tilde{z}_n\rangle$) is normal in X, and when $! = o$, $C_X(\tilde{P}) \neq C_X(\tilde{Q})$. Write $R(P)$ for $C_X(\tilde{P})$.

LEMMA 13.4.4. *Assume $\xi = (\mathcal{X}, \Xi)$ is a quaternion fusion packet with Sylow group X such that $\mathcal{X} = X\mathcal{X}^\circ$ and $\mathcal{X}^\circ \cong L^-_4[m]$. Then either*

(1) $\mathfrak{P}^o(\xi) = \emptyset$, X has two orbits on $\mathfrak{P}^m(\xi)$ with representatives P_1 and P_2, and $\mathcal{X} = \langle X\mathcal{O}_\xi, O^2(Aut_{\mathcal{X}}(R(P_i))) : i = 1, 2\rangle$, or

(2) $\mathfrak{P}^o(\xi) \neq \emptyset$, X is transitive on $\mathfrak{P}^m(\xi)$ and $\mathfrak{P}^o(\xi)$ with representatives P_1 and P_2, and $\mathcal{X} = \langle X\mathcal{O}_\xi, O^2(Aut_{\mathcal{X}}(R(P_i))) : i = 1, 2\rangle$,

PROOF. This follows from 9.4.22, 9.4.27, and 9.4.34.2. □

LEMMA 13.4.5. *Let $\xi = (\mathcal{X}, \Xi)$ be a quaternion fusion packet with Sylow group X such that $\mathcal{X} = X\mathcal{X}^\circ$ and $\mathcal{X}^\circ \cong SL^-_4[m]$. Set $\langle u\rangle = Z(\mathcal{X}^\circ)$ and $\tilde{\mathcal{X}} = \mathcal{X}/\langle u\rangle$. Let X° be Sylow in $\mathcal{X}^\circ$. Then*

(1) X has two orbits on $\mathfrak{P}^m(\xi) \cup \mathfrak{P}^o(\xi)$ with representatives P_i, $i = 1, 2$, and with $P_1 \in \mathfrak{P}^m(\xi)$. Further, setting $R_i = R(P_i) = C_X(\tilde{P}_i)$ for $i = 1, 2$, $\mathcal{X} = \langle X\mathcal{O}_\xi, O^2(Aut_{\mathcal{X}}(R_i)) : i = 1, 2\rangle$.

(2) $R_1 = C_X(Q)Q$, where Q is the Q_8-subgroup of P_1, and $C_{X^\circ}(Q) \cong SD_{2m}$.

(3) $A_4 \cong O^2(Aut_{\mathcal{X}}(R_1)) = \{\phi \in Aut(R_1) : \phi_{|Q} \in O^2(Aut(Q))$ *and* $[\phi, C_X(Q)] = 1\}$.

(4) Assume $P_2 \in \mathfrak{P}^o(\xi)$. *Then* $Q = R_2 \cap X^\circ \cong Q_8^2$, *and* $\mathbf{Z}_3/E_{16} \cong O^2(Aut_{\mathcal{X}}(R_2))$ *consists of those* $\phi \in Aut(R_2)$ *such that* $\phi_{|Q} \in O^2(Aut(Q))$, ϕ *acts on* P_2, *and for* ϕ *of order 3,* $C_{R_2}(\phi)$ *is the preimage in* R_2 *of* $O_2(C_{Aut(Q)}(\phi_{|Q}))$ *under the conjugation map* $R_2 \to Aut(Q)$.

PROOF. As $\tilde{\mathcal{X}}$ satisfies the hypotheses of 13.4.4, we conclude from that lemma that $\tilde{X}$ has two orbits on $\mathfrak{P}^m(\tilde{\xi}) \cup \mathfrak{P}^o(\tilde{\xi})$ with representatives $\tilde{P}_i$, $i = 1, 2$, and that $\tilde{\mathcal{X}} = \langle \tilde{X}\mathcal{O}_{\tilde{\xi}}, O^2(Aut_{\tilde{\mathcal{X}}}(\tilde{R}_i)) : i = 1, 2\rangle$. Therefore by 2.1.15, $\mathcal{X}$ is generated by the preimages $X\mathcal{O}_\xi$ and $O^2(Aut_{\mathcal{X}}(R_i))$, $i = 1, 2$, of the three subcategories generating $\tilde{\mathcal{X}}$, so that (1) holds.

Adopt the notation in the proof of 13.3.4; in particular $P_1 = \langle v\rangle * Q$ with $Q = \langle vt, z_n t\rangle \cong Q_8$ and $R_1 = C_X(\tilde{P}_1)$. Then the proof of 13.3.4.2 establishes the first statement in (2), and setting $y = z_n t$, the proof of 13.3.4.3 shows $C_{O_A\langle t\rangle}(Q) = \langle V, i\rangle \cong D_m$, where $V = C_{O_A}(Q)$ is cyclic of order $m/2$, $l \in C_{O_A}(y) - V$, and $i = yl$ is an involution inverting V. Next sy centralizes Q with $(sy)^2 = u$, and $\langle s\rangle C_{O_A}(t) \cong K\langle s\rangle \cong SD_{2m}$, so $C_{X^\circ}(Q) = \langle i, sy\rangle \cong SD_{2m}$, completing the proof of (2). Then (2) and the proof of 13.3.4.4 establish (3).

Assume the hypothesis of (4), let $\mathcal{N} = N_{\mathcal{X}}(R_2)$, let G be a model for $\mathcal{N}$, and let H be the normal subgroup of G which is a model for $N_{\mathcal{X}^\circ}(R_2)$. Then $Q = O_2(H)$ and $G = N_X(R_2)H$, so $O^2(G) = O^2(H)$. From the discussion in 13.4.3, $H \cong SL_2(3) * SL_2(3)$ and $N_H(P)$ is of index 3 in H. As $Inn(Q) = C_{Aut(Q)}(\tilde{Q})$, $C_X(\tilde{Q}) = C_{R_2}(\tilde{Q}) = C_X(Q)Q$. Let $h \in H$ be of order 3. Then $[C_X(Q), h] \leq C_{O^2(H)}(Q) = \langle u\rangle = Z(H)$, so $[C_X(Q), h] = 1$. Then from the discussion in 13.4.5, $|R_2 : C_X(\tilde{Q})| = 2$ with $C_{\tilde{R}_2}(h)$ a complement to $\tilde{Q}$ in $\tilde{R}_2$, and $C_{R_2}(h)$ is the preimage in R_2 of $O_2(C_{Aut(Q)}(c_{h|Q}))$ under the conjugation map $R_2 \to Aut(Q)$. This completes the proof of (4). □

DEFINITION 13.4.6. Assume $a = 0$. Assume that $\gamma = (\mathcal{G}, \Gamma)$ is a quaternion fusion packet with Sylow group S_γ, such that $\mathcal{G} = S_\gamma \mathcal{G}^\circ$ with $\mathcal{G}^\circ = \mathcal{F}$. Let $O_* = K_1 \cdots K_{n-2} \in \mathcal{G}^f$ and for $i = 1, 2$, set $\mathcal{P} = N_{S_\gamma}(O_*)C_{\mathcal{G}}(O_*)^\circ$ and $\xi = (\mathcal{P}, \{K_{n-1}, K_n\})$. For $i = 1, 2$, write $\mathfrak{P}_i(\gamma)$ for $\mathfrak{P}_i(\xi)$, pick $P_i = P_i(\gamma) \in \mathfrak{P}(\gamma)$, and set $R_i = R_i(\gamma) = R(P_i)$. By 13.4.5.3,

$$O^2(Aut_{\mathcal{P}}(R_1)\rangle = \{\phi \in Aut(R_1) : \phi_{|Q_1} \in O^2(Aut(Q_1)) \text{ and } [\phi, C_{S_\gamma}(Q_1)] = 1\},$$

where Q_1 is the Q_8-subgroup of P_1. Further if $P_2 \in \mathfrak{P}^o(\xi)$, then by 13.4.5.4, $Q_2 = R_2 \cap X^\circ \cong Q_8^2$, where X° is Sylow in $C_{\mathcal{G}}(O_*)^\circ$, and $O^2(Aut_{\mathcal{P}}(R_2))$ consists of those $\phi \in Aut(R_2)$ such that $\phi_{|Q_2} \in O^2(Aut(Q_2))$, ϕ acts on P_2, and for ϕ of order 3, $C_{R_2}(\phi)$ is the preimage in R_2 of $O_2(C_{Aut(Q_2)}(\phi_{|Q_2}))$ under the conjugation map $R_2 \to Aut(Q_2)$.

NOTATION 13.4.7. Assume $a = 1$. Choose notation so that $K_n \in \mathcal{F}^f$ and set $C = \mathfrak{U}_n + \mathfrak{U}_0$ and $L_C = C_L(C^\perp)$. Then $L_C = SL(C) \cong SL_3^{-\pi}(q)$ with $T_C = S \cap L_C$ Sylow in L_C. Set $\mathcal{L}_C = \mathcal{F}_{T_C}(L_C)$ and $\tau_C = (\mathcal{L}_C, \{K_n\})$, so that τ_C is a quaternion fusion packet and $\mathcal{L}_C \cong L_3^-[m]$.

Let $\{x_n, y_n\}$ be a basis for $\mathfrak{U}_n$ chosen as in Rr.1, and choose the basis so that there is $t \in \mathcal{Z}'(\tau_C)$ with $[U, t] = \langle x_n\rangle + \mathfrak{U}_0$ and $C_C(s) = \langle y_n\rangle$. Then $T_C = K_n\langle t\rangle \cong SD_{2m}$.

Write $\mathfrak{P}(\tau_C)$ for the set of subgroups $P = \langle t', z_n\rangle$ of T_C such that $t' \in \mathcal{Z}'(\tau_C)$. Then K_n is transitive on $\mathfrak{P}(\tau_C)$ and for $P \in \mathfrak{P}(\tau_C)$, $P = C_{T_C}(P)$ and $\mathcal{L}_C = \langle T_C\mathcal{O}_{z_n}, O^2(Aut_{\mathcal{L}_C}(P))\rangle$.

DEFINITION 13.4.8. Assume $a = 1$. Assume that $\gamma = (\mathcal{G}, \Gamma)$ is a quaternion fusion packet with Sylow group S_γ, such that $\mathcal{G} = S_\gamma\mathcal{G}^\circ$ with $\mathcal{G}^\circ = \mathcal{F}$. Let $O_* = K_1 \cdots K_{n-1} \in \mathcal{G}^f$, set $\mathcal{P}_3 = \mathcal{P}_3(\gamma) = N_{S_\gamma}(O_*)C_{\mathcal{G}}(O_*)^\circ$. Observe that $C_{\mathcal{G}}(O_*)^\circ = \mathcal{L}_C$, pick $P_3 = P_3(\gamma) \in \mathfrak{P}(\tau_C)$, and set $R_3 = R_3(\gamma) = R(P_3) = C_{S_\gamma}(P)$.

Next $\mathfrak{U}_0 = C_U(c)$ and $[U, c] = \mathfrak{U}_0^\perp$, so $C_L(c) = C_L(\mathfrak{U}_0)\langle t\rangle$ where $P = \langle t, z_n\rangle$. Further $c \in Z(S_\gamma)$ and $\mathcal{G}_c = C_{\mathcal{G}}(c) = S_\gamma\mathcal{G}_c^\circ$, where $\mathcal{G}_c^\circ = \mathcal{F}_{C_S(\mathfrak{U}_0)}(L_c) \cong SL_{2n}^-[m]$. Of course $\gamma_c = (\mathcal{G}_c, \Omega)$ is a quaternion fusion packet. As $t \in \mathcal{S}_\gamma$, $\mathfrak{P}^\circ(\mathcal{G}_c) \neq \emptyset$. For $i = 1, 2$, define $P_i(\gamma) = P_i(\gamma_c)$, $R_i(\gamma) = R_i(\gamma_c)$, and $\mathcal{P}_i(\gamma) = \mathcal{P}(\gamma_c)$.

LEMMA 13.4.9. *Assume $a = 1$. For $i \in I$, let $\{x_i, y_i\}$ be a basis for $\mathfrak{U}_i$ chosen as in the proof of 13.4.1, and more generally adopt the notation of that proof. Let H be a model for $N_{\mathcal{F}}(O(\tau))$.*

(1) $H = \bar{L}Y_0$ where $\bar{L}_0 = \prod_{i\in I}\bar{L}_i$ and either $m > 8$ and $\bar{L}_i = K_i$ or $m = 8$, $K_i = O_2(\bar{L}_i)$, and $\bar{L}_i \cong SL_2(3)$. Further $Y_0 = M_0Y$ where $M_0 = \langle t_i : i \in I\rangle$ with $t_i \in \mathcal{Z}'$ such that $[U, t_i] = \langle x_i\rangle + \mathfrak{U}_0$ and $C_{\mathfrak{U}_i}(t_i) = \langle y_i\rangle$. Also $Y = Sym(I)$ is represented on U so that Y centralizes $\mathfrak{U}_0$ and for $g \in Y$ and $i \in I$, $x_i^g = x_{ig}$ and $y_i^g = y_{ig}$. In particular M_0 is the n-dimensional $\mathbf{F}_2$-permutation module for Y.

(2) Define $P_3 = P_3(\tau)$, $R_3 = R_3(\tau)$, and $\mathcal{P}_3 = \mathcal{P}_3(\tau)$ as in Rr.8, and choose notation so that $t = t_n$. Then $R_3 = P_3XT$ where $X = \prod_{1\leq i<n} K_i\langle t_i\rangle$ and T is the stabilizer in $S \cap Y$ of n.

(3) $O^2(Aut_{\mathcal{P}_3}(R_3))$ consists of those $\phi \in Aut(R_3)$ such that $\phi_{|P_3} \in O^2(Aut(P_3))$ and ϕ centralizes T and $K_i\langle t_iz_n\rangle$ for $1 \leq i < n$.

PROOF. Part (1) follows from parts (3) and (4) of 13.4.1.

Set $O_* = K_1 \cdots K_{n-1}$. By construction of P_3, R_3, and $\mathcal{P}_3$ in Rr.8, $\mathcal{P}_3 = S\mathcal{L}_C$, with $T_C = K_n\langle t\rangle$ Sylow in $\mathcal{L}_C$, $P_3 = \langle z_n, t\rangle$, and $R_3 = C_S(P_3)$. In particular $P_3 = C_{T_C}(P_3) = T_C \cap R_3$, so as $\mathcal{P}_3$ acts on $\mathcal{L}_C$, $N_{\mathcal{P}_3}(R_3)$ acts on $R_3 \cap T_C = P_3$. As R centralizes P_3, $R \leq Y_n\bar{L}_0M_0$, so $R \leq T\bar{L}_0M_0$. From (1), T centralizes P_3, so $R = TC_{S\cap\bar{L}_0M_0}(P_3)$. Again from (1), X centralizes P_3, so $R = TM_0C_{S\cap\bar{L}_0}(P_3)$. Then from one more application of (1), $C_{S\cap\bar{L}_0}(P_3) = O_*\langle z_n\rangle$, so (2) holds.

Next $Aut_{\mathcal{L}_C}(P_3) \lhd Aut_{\mathcal{P}_3}(R_3)$, so there is a normal model Φ for $O^2(Aut_{\mathcal{L}_C}(R_3))$ in a model Γ for $Aut_{\mathcal{P}_3}(R_3)$. Of course $\Phi \cong A_4$ with $O_2(\Phi) = P_3$, so $[\Phi, R_3] \leq \Phi \cap R_3 = P_3$. As L_C centralizes O_* and T, so does Φ. Similarly L_C centralizes t_iz_n, so t_iz_n centralizes Φ, establishing (3). □

13.5. Generating symplectic and linear systems

In this section we assume the following hypothesis:

HYPOTHESIS 13.5.1. *$\tau = (\mathcal{F}, \Omega)$ is a quaternion fusion packet such that $\mathcal{F}^\circ = O^2(\mathcal{F})$ and either*

(I) $\mathcal{F}^\circ$ is isomorphic to $(P)Sp_{2n}[m]$, $(P)SL_{2n}^-[m]$, or $L_{2n+1}^-[m]$, or

(II) $n \geq 4$ is a power of 2 and $\mathcal{F}^\circ = \mathcal{F}_1 * \mathcal{F}_2$ where $\mathcal{F}_i \cong Sp_n[m]$ or $SL_n^-[m]$ with $Z(\mathcal{F}_1) = Z(\mathcal{F}_2) = \langle f\rangle$ and the Sylow group S of $\mathcal{F}$ is transitive on $\{\mathcal{F}_1, \mathcal{F}_2\}$.

NOTATION 13.5.2. (I) Suppose for the moment that (I) holds; then we adopt the notation of section 13.3 or 13.4, except we write $\mathcal{F}^\circ = \mathcal{F}_{S_L}(L)$ where S_L is Sylow in L and $L = \Omega(U, \mathfrak{f}) \cong Sp_{2n}(q)$ if $\mathcal{F}^\circ$ is symplectic and $L \cong SL^{-\pi}_{2n+a}(q)$ if $\mathcal{F}^\circ$ is linear or unitary. (We are abusing notation in the case $\mathcal{F}^\circ = PSp_{2n}[m]$ or $PSL^-_{2n}[m]$, where L should be the projective group $PSp_{2n}(q)$ or $PSL^{-\pi}_{2n}(q)$.)

If $\mathcal{F}^\circ$ is symplectic set $I' = I'(\tau) = \{1\}$ and pick $P_1 = P_1(\tau) \in \mathfrak{P}(\tau)$ and define $R_1 = R_1(\tau)$ and $\mathcal{P}_1 = \mathcal{P}_1(\tau)$ as in 13.3.5. If $\mathcal{F}^\circ$ is $(P)SL_{2n}[m]$ define $I' = \{1, 2\}$ and for $i \in I'$ define P_i, R_i, and $\mathcal{P}_i$ as in 13.4.6. Finally if $\mathcal{F}^\circ$ is $L^-_{2n+1}[m]$ define $I' = \{1, 2, 3\}$ and for $i \in I'$, define P_i, R_i, and $\mathcal{P}_i$ as in 13.4.8.

(II) Now suppose that (II) holds. Again we adopt that notation of section 13.3 if $\mathcal{F}_i$ is symplectic or that of section 13.4 if $\mathcal{F}_i$ is linear or unitary. Here $U = U_1 \perp U_2$ with $U_1 = \mathfrak{U}_1 + \cdots + \mathfrak{U}_{n/2}$ and $U_1 = U_1^\perp$, $\mathfrak{L}_i = L_{U_i}$, and $\mathcal{F}_i = \mathcal{F}_{T_i}(\mathfrak{L}_i)$, where $T_i = S \cap \mathfrak{L}_i$ is Sylow in $\mathfrak{L}_i$. Formally $\mathcal{F}^\circ$ is the 2-fusion system of the image of $\mathfrak{L}_1\mathfrak{L}_2$ in $PSL(U)$, but it will be more convenient to abuse notation and work in $SL(U)$.

If U is symplectic then $I' = \{1\}$ and we define P_1, R_1, and $\mathcal{P}_1$ in $\tau_1 = (N_S(\mathcal{F}_1)\mathcal{F}_1, \Omega_1)$ as in 13.3.5. If U is linear or unitary, then $I' = \{1, 2\}$ and for $i \in I'$ we define P_i, R_i, and $\mathcal{P}_i$ in τ_1 as in 13.4.6.

In either (I) or (II), set $W = \langle \mathcal{Z}_S \rangle$, let T be Sylow in $\mathcal{F}^\circ$, H a model for $N_{\mathcal{F}^\circ}(O(\tau)C_T(O(\tau))$, and set $\mathcal{W} = \langle T\mathcal{O}_\tau, \mathcal{F}_T(H) \rangle$, where $\mathcal{O}_\tau = \mathcal{F}_{O(\tau)}(L_0)$.

In this section we prove:

THEOREM 13.5.3. *Assume Hypothesis* 13.5.1 *and adopt notion* 13.5.2. *Set* $\mathcal{Y}_0 = \langle S\mathcal{W}, Aut_{\mathcal{P}_j}(R_j) : j \in I'(\tau) \rangle_S$. *In case (I) set* $\mathcal{Y} = \mathcal{Y}_0$. *In case (II) let* $\mathfrak{X}$ *be the set of involutions* $e_\theta \in W$ *with* $|\theta| \neq n/2$ *or* $C_S(e_\theta) \leq N_S(\mathcal{F}_1)$, *and set* $\mathcal{Y} = \langle \mathcal{Y}_0, C_{\mathcal{F}}(x) : x \in \mathfrak{X} \rangle_S$. *Then* $\mathcal{F} = \mathcal{Y}$.

Assume the theorem is false and pick a counter example with $d = \dim(U)$ minimal.

LEMMA 13.5.4. *Assume (I) holds. Then*
(1) If $n \leq 2$ *then* $n = 2$ *and* $\mathcal{F}^\circ$ *is* $L^-_5[m]$.
(2) $\mathcal{F}^\circ \cong PSp_{2n}[m]$, $L^-_{2n}[m]$, *or* $L^-_{2n+1}[m]$.

PROOF. Suppose $n = 1$; then either $\mathcal{F}^\circ = \mathcal{O}(z)$ or $\mathcal{F}^\circ$ is $L^-_3[m]$. In the first case $\mathcal{F} = S\mathcal{O}(z) = \mathcal{Y}$, contrary to the choice of τ. In the second, $\mathcal{F} = \mathcal{Y}$ by Example 2.1.3.3, for the same contradiction.

Assume next that $n = 2$. Then $\mathcal{F}^\circ$ is $(P)Sp_4[m]$, $(P)SL^-_4[m]$, or $L^-_5[m]$. In the first case $\mathcal{F} = \mathcal{Y}$ by 13.3.3 and 13.3.4.6. In the second $\mathcal{F} = \mathcal{Y}$ by 13.4.4 and 13.4.5.1. Finally the third case is allowed in part (1). This completes the proof of (1).

Suppose $Z(\mathcal{F}^\circ) \neq 1$. Then by induction on the order of τ, $\mathcal{F}^+ = \mathcal{F}/Z(\mathcal{F}^\circ) = \langle S^+\mathcal{W}^+, Aut_{\mathcal{P}_j^+}(R_j^+) : j \in I' \rangle$, so $\mathcal{F} = \langle S\mathcal{W}, Aut_{\mathcal{P}_j}(R_j) : j \in I' \rangle$ by 2.1.15, contrary to the choice of τ. This completes the proof of (2) and the lemma. □

LEMMA 13.5.5. *Let* $\theta \subseteq \mathcal{Z}_S$ *and* $f = e_\theta = \prod_{t \in \theta} t$ *with* $1 \neq f \in \mathcal{F}^f$. *Then either*
(1) $\mathcal{F}_f \leq \mathcal{Y}$, *or*
(2) $|\theta| = n/2$ *and in case (I),* $L = PSp_{2n}(q)$ *or* $PSL^-_{2n}(q)$, *while in case (II),* $C_S(f)$ *does not act on* $\mathcal{F}_1$.

PROOF. Assume that neither (1) nor (2) holds. Assume first that (I) holds and let $|\theta| = k$ and $U_1 = [U, f]$ and $U_2 = C_U(f)$. Thus $d_1 = \dim(U_1) = 2k$ and $U_2 = U_1^\perp$ is of dimension $d_2 = 2n + a - 2k = 2(n-k) + a$. Let $L_i = L_{U_i}$ have Sylow group T_i and $\mathcal{F}_i = \mathcal{F}_{T_i}(L_i)$. Let S_f be Sylow in $\mathcal{F}_f$. Claim $C_L(f)$ acts on U_i for $i = 1, 2$. If not $d_1 = d_2$, so $2n + a = 4k \equiv 0 \mod 4$. Thus $a = 0$ and $k = n/2$. Further $f = (-I) \cdot f$, so $L = PSp_{2n}(q)$ or $PSL^-_{2n}(q)$, contrary to the assumption that (2) does not hold. This completes the proof of the claim.

Now $\mathcal{F}_f = T\mathcal{F}_1\mathcal{F}_2$, and by the claim, $\mathcal{F}_i \trianglelefteq \mathcal{F}_f$ for $i = 1, 2$. By minimality of d, $S_f\mathcal{F}_1 = \langle S_f\mathcal{W}_1, Aut_{\mathcal{P}_{1,j}}(R_{1,j}) : j \in I_1' \rangle_T$, and if $d_2 > 1$ then $S_f\mathcal{F}_2 = \langle S_f\mathcal{W}_2, Aut_{\mathcal{P}_{2,j}}(R_{2,j}) : j \in I_2' \rangle$. But $S_f\mathcal{W}_i \leq S\mathcal{W} \leq \mathcal{Y}$ and we will show that $Aut_{\mathcal{P}_{i,j}}(R_{i,j}) \leq \mathcal{Y}$, so $\mathcal{F}_f \leq \mathcal{Y}$ by 1.3.2 in [**Asc19**].

So assume $d_2 \leq 1$. By 13.5.4.2, $d_2 \neq 0$, so $d_2 = 1$. Then $\mathcal{F}_2$ is empty, so the lemma holds.

Thus we may assume that $d_2 > 1$ and it remains to show that $Aut_{\mathcal{P}_{i,j}}(R_{i,j}) \leq \mathcal{Y}$. If $d_i = 2$ then $SL_2[m] \cong \mathcal{F}_i \leq \mathcal{O}_\tau$, so we may take $d_i > 2$. Suppose first that d_i is even, so that $d_i \geq 4$. Then $K_{n_i}K_{n_i-1}$ is conjugate in $\mathcal{W}$ to K_nK_{n-1}, so $Aut_{\mathcal{P}_{i,j}}(R_{i,j}) \leq \mathcal{Y}$. So take d_i odd. Then K_{n_i} is conjugate in $\mathcal{W}$ to K_n, and hence again $Aut_{\mathcal{P}_{i,j}}(R_{i,j}) \leq \mathcal{Y}$.

So assume case (II) holds. Arguing as above, $\mathcal{F}_e \leq \mathcal{Y}_0$ if $C_S(e)$ acts on $\mathcal{F}_1$, so $C_S(e)$ does not act on $\mathcal{F}_1$. Then by definition of $\mathcal{Y}$, $|\theta| = n/2$, completing the proof of the lemma. □

LEMMA 13.5.6. *Assume (I) holds. Then*
(1) n is a power of 2.
(2) $L = PSp_{2n}(q)$ or $PSL^-_{2n}(q)$.
(3) Let $\theta \subseteq \mathcal{Z}_S$ be of order $n/2$ and $f = e_\theta \in \mathcal{F}^f$. Then $\mathcal{F}_f \not\leq \mathcal{Y}$.

PROOF. Observe $N_{\mathcal{F}}(W) = \mathcal{W}$, so $N_{\mathcal{F}}(W) \leq \mathcal{Y}$. Let $\xi \in \Xi$ and $\mathcal{B} = \mathcal{B}_\xi(\tau)$. By 13.2.26, $\mathcal{F} = \mathcal{B}$, so $\mathcal{B} \not\leq \mathcal{Y}$. But from 13.2.24, one of (a), (b), or (c) of 13.2.24 holds and $\mathcal{B} = \langle \mathcal{F}_{x_j}, N_{\mathcal{F}}(W) : j \in J \rangle$ for some set of involutions $x_j = e_{\theta_j}$, so $\mathcal{F}_{x_j} \not\leq \mathcal{Y}$ for some $j \in J$. Thus by 13.5.5, (2) holds and $|\theta_j| = n/2$, so also (3) holds. Thus it remains to show that n is a power of 2. In case (b) and (c) this holds by definition, so we may assume (a) holds. Therefore S is not transitive on I and we may choose ξ to be an orbit of S, contradicting $|\theta_j| = n/2$. □

LEMMA 13.5.7. *Case (II) holds.*

PROOF. Assume instead that (I) holds and choose θ and f as in 13.5.6.3. Set $U_1 = \langle \mathfrak{U}_j : j \in \theta \rangle$ and $\mathfrak{U}_2 = \mathfrak{U}_1^\perp$. By 13.5.6.1, n is a power of 2 so S is transitive on I and centralizes f. Set $L_i = L_{U_i}$, $T_i = S \cap L_i$ and $\mathcal{F}_i = \mathcal{F}_{T_i}(L_i)$. Then $O^2(\mathcal{F}_f) = \mathcal{F}_1 * \mathcal{F}_2$ and S is transitive on $\{\mathcal{F}_1, \mathcal{F}_2\}$. Therefore $\mathcal{F}_f$ satisfies Hypothesis 13.5.1.II. By 13.5.5, for each $x \in \mathfrak{X}$, $C_{\mathcal{F}}(x) \leq \mathcal{Y}$, so by 13.5.6.3, $\mathcal{F}_f$ is a counterexample to Theorem 13.5.3. Thus (II) holds by minimality of τ. □

NOTATION 13.5.8. Recall that we abuse notation and work in $SL(U)$ rather than in the projective group $PSL(U)$. Let $\theta_1 = \{z \in \mathcal{Z}_S : [U, z] \leq U_1\}$. Now $\Pi_1 = \{\theta_1, \mathcal{Z}_S - \theta_1\}$ is an S-invariant partition of $\mathcal{Z}_S$.

Proceeding recursively, assume Π_k is an S-invariant partition of $\mathcal{Z}_S$ into 2^k blocks of size $n/2^k$. Pick $\theta_k \in \Pi_k$ and let $\theta_{k+1} \subseteq \theta_k$ be of order $n/2^{k+1}$ such that $|N_S(\theta_k) : N_S(\theta_{k+1})| = 2$, and set $\Pi_{k+1} = \theta_{k+1}^S$, so that Π_{k+1} is an S-invariant partition of $\mathcal{Z}_S$ into 2^{k+1} blocks of size $n/2^{k+1}$. Continuing this process, we arrive at $\theta_r = \{z\}$ for some $z \in \mathcal{Z}_S$; that is $n = 2^r$.

Let $\Pi_k = \{\theta_{k,1}, \ldots, \theta_{k,2^k}\}$ and $U_{k,i} = [U, f_{\theta_{k,i}}]$. Thus $U = U_{k,1} \perp \cdots \perp U_{k,2^k}$ and we set $\mathcal{U}_k = \{U_{k,1}, \ldots, U_{k,2^k}\}$, $L_{k,i} = L_{U_{k,i}}$ and $L_k = \prod_i L_{k,i}$ the direct product of 2^k copies of $\mathfrak{L}_{2^{r-k+1}}(q)$, where $\mathfrak{L} = Sp$ or SL^-. Let $\mathcal{U}^k = \{\mathcal{U}_1, \ldots, \mathcal{U}_k\}$ and G_k the stabilizer in SL of $\mathcal{U}^k$. Set $\mathcal{D}_k = \mathcal{F}_S(G_k)$.

LEMMA 13.5.9. $G_k = SL_k O(G_k)$.

PROOF. By construction, $SL_k \leq G_k$ and L_k is contained in the kernel X of the action of G_k on $\mathcal{U}_k$. By 2.1.16, $G_k^{\mathcal{U}_k}$ is contained in a Sylow 2-group of $Sym(\mathcal{U}_k)$. For $x \in X$ of odd order, $x_{|U_{k,i}} \in L_{k,i} O(G_k)$ as $L_{k,i}$ acts faithfully as $\mathfrak{L}(U_{k,i})$ on $U_{k,i}$. Therefore $X/L_k O(G_k)$ is a 2-group, completing the proof. □

LEMMA 13.5.10. *(1)* $L_{r,i} \cong SL_2(q)$ *has Sylow 2-group* $O(s_i)$ *for that* $s_i \in \mathcal{Z}_S$ *such that* $s_i \in Z(L_{r,i})$ *and* $U_{r,i} = [U, s_i]$.

(2) $\mathcal{O}_\tau = \mathcal{F}_{O(\tau)}(L_r)$.

(3) $\mathcal{D}_r = \mathcal{F}_S(G_r) = S\mathcal{W} \leq \mathcal{Y}$.

PROOF. By construction, $U_{r,i} = [U, s_i]$ for some $s_i \in \mathcal{Z}_S$, and then (1) follows. In particular by (1), $O(\tau)$ is Sylow in L_r and (2) holds. By 13.5.9, $G_r = SL_r$, so $\mathcal{F}_S(G_r) = \mathcal{F}_S(SL_r) = S\mathcal{O}_\tau$, establishing (3). □

LEMMA 13.5.11. $\mathcal{E}(\tau) \leq \mathcal{Y}$.

PROOF. From Qq.6, $\mathcal{E} = \mathcal{E}(\tau) = \langle N_{\mathcal{F}}(O(\xi)), S\mathcal{O}_\tau, N_{\mathcal{F}}(O) : \xi \in \Xi \rangle$, where Ξ consists of those $\xi \subseteq \mathcal{Z}_S$ with $O(\xi) \in \mathcal{F}^f$ and $N_S(O(\xi))$ transitive on ξ. Now $N_{\mathcal{F}}(O) \leq \langle S\mathcal{O}_\tau, \mathcal{F}_S(SH) \rangle = S\mathcal{W} \leq \mathcal{Y}$. Thus it remains to show that $\mathcal{N} = N_{\mathcal{F}}(O(\xi)) \leq \mathcal{Y}$ for $\xi \in \Xi$. But $\mathcal{N} \leq \mathcal{F}_x$, where $x = e_\xi$, and by definition of $\mathcal{Y}$, $\mathcal{F}_x \leq \mathcal{Y}$ unless $\xi = \theta$, so we may the latter holds.

Now $\mathcal{N} = \mathcal{F}_S(X)$, where $X = N_{L_{1,1}}(O(\xi)) L_{1,2} S$. Therefore

$$\mathcal{N} = \langle \mathcal{F}_S(SL_{1,2}), N_{\mathcal{N}}(S_2) \rangle,$$

where S_2 is Sylow in $L_{1,2}$. By minimality of τ, $\mathcal{F}_S(SL_{1,2}) \leq \mathcal{Y}$. Further $N_{\mathcal{N}}(S_2) \leq N_{\mathcal{N}}(O)$, which is contained in $\mathcal{Y}$ by the previous case. □

NOTATION 13.5.12. Observe that $\mathcal{D}_1 = \mathcal{F}$, so $\mathcal{D}_1 \not\leq \mathcal{Y}$. On the other hand by 13.5.10.3, $\mathcal{D}_r \leq \mathcal{Y}$. Thus there exists a least k such that $\mathcal{D}_k \not\leq \mathcal{Y}$. Set $\mathcal{D} = \mathcal{D}_k$, $\rho = (\mathcal{D}, \Omega)$, let $S_0 = S \cap L_k$, and write $\mathcal{D}_0$ for $\mathcal{F}_{S_0}(L_k)$. Let $I = \{1, \ldots, 2^k\}$ and for $i \in I$ set $S_i = S \cap L_{k,i}$ and $\mathcal{L}_i = \mathcal{F}_{S_i}(L_{k,i})$. From 13.5.8, $L_k = \prod_{i \in I} L_{k,i}$, so $\mathcal{D}_0 = \prod_{i \in I} \mathcal{L}_i$ is a direct product. Set $\Omega_i = \Omega \cap S_i$ and $\rho_i = (\mathcal{L}_i, \Omega_i)$. Observe:

LEMMA 13.5.13. *The pair* $\mathcal{D}, \mathcal{D}_0$ *satisfies Hypotheses* 11.4.1 *and* 11.4.6.

NOTATION 13.5.14. As in 11.4.2 and 11.4.6, for $X \leq S$ and $0 \leq i \leq 2^k$, write X_i for $X \cap S_i$.

As $\mathcal{D} \not\leq \mathcal{Y}$, it follows from the Alperin-Goldschmidt Fusion Theorem that the set $\mathcal{R}$ of subgroups $R \in \mathcal{D}^e$ such that $Aut_{\mathcal{D}}(R) \not\leq Aut_{\mathcal{Y}}(R)$ is nonempty. Pick $R \in \mathcal{R}$ and let G be a model for $N_{\mathcal{D}}(R)$ and set $G^* = G/R$. By 13.5.13 and 11.4.4.1, $R_0 \in \mathcal{F}^{frc}$, so there is a model G_0 for $N_{\mathcal{D}_0}(R_0)$. By 11.4.4, we can embed G and G_0 in a certain group G_c with $G_0 \trianglelefteq G_c$ and $G = N_{G_c}(R)$. Thus we can form $H_0 = N_{G_0}(R)$ and by 11.4.4, $H_0 \trianglelefteq G$ with $R_0 = O_2(H_0)$ and $H_0^* \cong H_0/R_0$.

Set $Q = N_R(L_{k,1})$ and $\mathcal{X} = \mathcal{F}_{QS_1}(QL_{k,1})$. Let $1 \in \Sigma$ be a set of coset representatives for Q in R. Write $\mathcal{R}^*$ for the set of those $R \in \mathcal{R}$ of maximal order.

LEMMA 13.5.15. *For each $K \in \Omega$, $|K \cap R| \leq 2$.*

PROOF. Assume otherwise; then from 3.4.5, $Aut_{\mathcal{D}}(R) = Aut_{\mathcal{E}(\rho)}(R)$. But $\mathcal{E}(\rho) \leq \mathcal{E}(\tau)$ and by 13.5.11, $\mathcal{E}(\tau) \leq \mathcal{Y}$, contrary to the choice of k. □

LEMMA 13.5.16. *(1) For each $i \in I$, $R_i \neq S_i$.*
(2) R is transitive on $\{L_{k,i} : i \in I\}$.
(3) $Q \in \mathcal{X}^e$. Let G_Q be a model for $N_{\mathcal{X}}(Q)$.
(4) $G_Q/Q \cong H_0^*$.
(5) Let $W_1 = \langle(\mathcal{Z}_S(\rho_1))\rangle$ and define $\alpha : W_1 \cap Z(Q) \to W \cap R$ by $\alpha : x \to \prod_{\sigma\in\Sigma} x^\sigma$. Then $x\alpha \in Z(R)$ and if x is of weight p in W_1 then $x\alpha$ is of weight 2^kp in W.
(6) If $p \neq |\mathcal{Z}_S(\rho_1)|$ then $C_{\mathcal{D}}(x\alpha) \leq \mathcal{Y}$.

PROOF. If $R_i = S_i$ then $\Omega_i \subseteq R$, contrary to 13.5.15. Thus (1) holds.

By 13.5.13, we can appeal to results in section 11.4. In particular (2) follows from (1) and 11.4.8.1, (3) follows from 11.4.8.4, and (4) follows from 11.4.8.5.

From 13.5.14, $1 \in \Sigma$ is a set of coset representatives for Q in R. If $\sigma_i \in \Sigma$ with $R_1\sigma_i = R_i$ then $R_0 = \prod_{i\in I} R_i = \prod_{i\in I} R_1^{\sigma_i}$, so for $x \in W_1 \cap Z(Q)$ of weight p, $x\alpha$ centralizes $\langle Q, \Sigma\rangle = R$ and $x\alpha$ is of weight $|\Sigma|p = |I|p = 2^kp$ in W. Thus (5) holds.

Assume the hypothesis of (6) and set $y = x\alpha$. Observe that $|\mathcal{Z}_S(\rho_1)| = n/2^k = 2^{r-k}$. As $p \neq |\mathcal{Z}_S(\rho_1)|$, $y \neq 1$ so y is an involution. Then by definition of $\mathcal{Y}$, either $\mathcal{F}_y \leq \mathcal{Y}$ or $2^kp = n/2 = 2^{r-1}$, and we may assume the latter, so that $p = 2^{r-k-1}$. Then conjugating in $\mathcal{Y}$, we may take $x = f_{\theta_{k+1}}$, so $\Pi_{k+1} = \theta_{k+1}^R \cup (\theta_k - \theta_{k+1})^R$. But then $C_{SL_k}(y) = C_S(y)L_{k+1}$, so as $\mathcal{D}_{k+1} \leq \mathcal{Y}$, again $C_{\mathcal{D}}(y) \leq \mathcal{Y}$. □

LEMMA 13.5.17. *$\rho_1 = (\mathcal{L}_1, \Omega_1)$ satisfies Hypothesis iI.1 and if $R \in \mathcal{R}^*$ then ρ_1, Q satisfies Hypothesis iI.7.*

PROOF. Visibly ρ_1 satisfies Hypothesis 11.2.1 with $|\Omega(z)| = 1$.

So assume $R \in \mathcal{R}^*$. By 13.5.16.4, $Q \in \mathcal{L}_1^e$. From 11.2.6, $\mathcal{W}(Q)$ consists of those involutions $x \in Z(Q) \cap W_1$ such that $x = e_\alpha$ for some $\alpha \subseteq I(\rho_1)$ of weight a power of 2. Passing to $\mathcal{L}_1/Z(\mathcal{L}_1)$ if necessary we may assume $e_{I(\rho_1)} = 1$. Hence by 13.5.16.6, for each $x \in \mathcal{W}(Q)$, $C_{\mathcal{D}}(x\alpha) \leq \mathcal{Y}$.

Let $x \in \mathcal{W}(Q)$, $y = x\alpha$ and $Z = \langle y^G\rangle$. As $R \in \mathcal{R}^*$ and $C_{\mathcal{D}}(y) \leq \mathcal{Y}$, the proof of 11.3.5.1 shows that $O^{2'}(C_G(Z)) = R$. Set $Z_x = \langle x^{G_Q}\rangle$. Then Z is a full diagonal subgroup of $\prod_{\sigma\in\Sigma} Z_x^\sigma$ and the isomorphism of 13.5.16.4 maps $O^{2'}(C_{G_Q}(Z_x))/Q$ to $O^{2'}(C_G(Z))/R = 1$, so $O^{2'}(C_{G_Q}(Z_x)) = Q$. Hence ρ_1, Q satisfies 11.2.7.1.

Similarly if ρ, R satisfies the setup of 11.2.7.2 and $\mathcal{U} \neq \emptyset$ then the proof of 11.3.5.2 shows that $m_2(G^*) = 1$ and $\langle\mathcal{U}\rangle$ is not normal in G^*. Let B^* be the subgroup of G^* generated by its involutions. By 11.4.5.2, B^* is the subgroup of H^* generated by its involutions. Thus isomorphism of 13.5.16.4 maps the subgroup B_Q^* of $G_Q^* = G_Q/Q$ generated by involutions to B^*, so $m_2(G_Q^*) = 1$ and $\langle\mathcal{U}_Q\rangle \neq B_Q^*$. Therefore ρ_1, Q satisfies 11.2.7.2. This completes the proof of the lemma. □

We are now in a position to obtain a contradiction, completing the proof of Theorem 13.5.3. Namely choose $R \in \mathcal{R}^*$. By 13.5.17, the pair ρ_1, Q satisfies Hypotheses 11.2.1 and 11.2.7. Further condition (a) of Theorem 11.2.33 is satisfied, as is condition (b) by 13.5.15. Condition (c1) of 11.2.33 hold as $\mathcal{X} = \mathcal{F}_{QS_1}(SL_{k,1})$. But then by Theorem 11.2.33, $|\mathcal{Z}_S(\rho_1)| = 1$, a contradiction. This completes the proof of Theorem 13.5.3.

13.6. Finishing μ abelian

In this section we continue to assume Hypotheses 13.2.1 and 13.2.14, and adopt the various notational conventions from section 13.2. In addition we adopt the following notation:

NOTATION 13.6.1. Pick $\xi \in \Xi$ and in cases (a) and (b) of 13.2.24 set $f = e_\xi$ and $\theta_1 = \xi$. In case (c) of 13.2.24, set $f = e_\sigma$ and $\theta_1 = \sigma$. For $i = 1, 2$, set $\mathcal{F}_i = \mathcal{F}_i(f)$, $\theta_2 = I - \theta_1$, and $n_i = |\theta_i|$. Let T_i be Sylow in $\mathcal{F}_i$.

If $\mathcal{F}$ is of type $PSp_{2n}[m]$ or $L^-_{2n}[m]$ then $c = 1$ by 13.2.19.2. Thus $n > 2$ in this case by 13.1.1.1, so $n_{i_0} > 1$ for some $i_0 \in \{1, 2\}$, and, interchanging the roles of 1 and 2 if necessary, we may assume $i_0 = 2$. On the other hand if $\mathcal{F}$ is of type $L^-_{2n+1}[m]$ then $n \geq 2$ by 13.2.1.2, and (a) or (c) holds, so $\mathcal{F}_2 \cong L_{2n_2+1}[m]$ with $n_2 \geq 1$. Set $\rho_2 = (T\mathcal{F}_2, \Omega_2)$.

If $\mathcal{F}$ is of type $PSp_{2n}[m]$ or $L^-_{2n}[m]$ set $I'(\tau) = I'(\rho_2)$, so that by 13.5.2, $I'(\tau) = \{1\}$ or $\{1, 2\}$, respectively. Further set $\mathcal{Y} = \langle \mathcal{W}, Aut_{\mathcal{P}_j}(R_j) : j \in I'(\tau)\rangle$, where $R_j = R_j(\tau) = R_j(\rho_2)$ and $\mathcal{P}_j = \mathcal{P}_j(\tau) = \mathcal{P}_j(\rho_2)$. If $\mathcal{F}$ is of type $L^-_{2n+1}[m]$ then by 13.5.2, $I'(\tau_c) = \{1, 2\}$ is of order 2 and $I'(\rho_2) = \{3\}$, $\{1, 2, 3\}$ for $n_2 = 1$, $n_2 > 1$, respectively. In this case set $I'(\tau) = \{1, 2, 3\}$ and $\mathcal{Y} = \langle \mathcal{W}, Aut_{\mathcal{P}_j}(R_j) : j \in I'(\tau)\rangle$, where for $j = 1, 2$, $R_j = R_j(\tau) = R_j(\tau_c)$ and $\mathcal{P}_j = \mathcal{P}_j(\tau) = \mathcal{P}_j(\tau_c)$, while $R_3 = R_3(\tau) = R_3(\rho_2)$ and $\mathcal{P}_3 = \mathcal{P}_3(\tau) = \mathcal{P}_3(\rho_2)$.

LEMMA 13.6.2. $\mathcal{Y} = \tilde{\mathcal{F}}$.

PROOF. By 13.2.23, $\tilde{S} = T$, $\tilde{\Omega} = \Omega$, $\tilde{G} = H$, and $\tilde{\mathcal{W}} = \mathcal{W}$. By 13.5.3 applied to $\tilde{\mathcal{F}}$, we have $\tilde{\mathcal{F}} = \langle \tilde{\mathcal{W}}, Aut_{\tilde{\mathcal{P}}_j}(\tilde{R}_j) : j \in I'(\tilde{\tau})\rangle$.

Let $j \in I'(\tau)$. If $j \neq 3$, let $\theta = \{n-1, n\} \subseteq I$ with $O(\theta) \in \mathcal{F}^f$. If $j = 3$ let $\theta = \{n\}$ with $O(\theta) \in \mathcal{F}^f$. Set $\bar{\theta} = I - \theta$ and consider $\mathcal{X} = C_{\mathcal{H}}(O(\bar{\theta}))^\circ$, where $\mathcal{H} = \mathcal{F}_c$ if $\mathcal{F}$ is of type $L^-_{2n+1}[m]$ and $j = 1$ or 2, while $\mathcal{H} = \mathcal{F}_2$ otherwise.

Suppose $j \neq 3$. Then $\mathcal{X} \cong Sp_4[m]$ if $\mathcal{F}$ is of type $Sp_{2n}[m]$ and $\mathcal{X} \cong L^-_4[m]$ otherwise. A Sylow group X of $\mathcal{X}$ is $O(\theta)\langle t_i : i \in I(\tau)\rangle$, where $t_i \in \mathcal{Z}^m(\tau_{\mathcal{X}})$ with $e^{t_i}_{n-1} = e_n$ and $\tau_{\mathcal{X}} = (\mathcal{X}, \{K_{n-1}, K_n\})$. The subgroup X is determined by T, Ω, and θ by 13.1.10. Further from sections 13.3 and 13.4, $P_j \in \mathfrak{P}(\mathcal{H})$ is determined by X and $\{K_{n-1}, K_n\}$ up to conjugation in X, so we may take $\tilde{P}_j = P_j$. Then $R_j = C_T(\bar{P}_j)$, where $\bar{R} = R/\langle u\rangle$ and $u = z_{n-1}z_n$. Thus R_j is determined by T and P_j, so $\tilde{R}_j = R_j$.

Consider the case $j = 1$. By parts (4) and (5) of 13.3.5 and by 13.4.5.3,

$$Aut_{\mathcal{P}_1}(R_1) = \{\phi \in Aut(R_1) : \phi_{|Q} \in O^2(Aut(Q)) \text{ and } [\phi, C_{R_1}(Q)] = 1\},$$

where Q is the Q_8-subgroup of P_1. Then visibly $Aut_{\mathcal{P}_1}(R_1)$ is determined by P_1 and T, so $Aut_{\mathcal{P}_1}(R_1)$ is determined by T and Ω. Therefore $Aut_{\tilde{\mathcal{P}}_1}(\tilde{R}_1) = Aut_{\mathcal{P}_1}(R_1)$. In particular $\tilde{\mathcal{F}} = \mathcal{Y}$ if $\mathcal{F}$ is of type $PSp_{2n}[m]$.

So suppose $j = 2$. By 13.4.5.4, $R \cap X$ has a unique Q_8^2-subgroup Q, and $Aut_{\mathcal{P}_2}(R_2)$ consists of those $\phi \in Aut(R)$ such that $\phi_{|Q} \in O^2(Aut(Q))$, ϕ acts on

P_2, and for ϕ of order 3, $C_R(\phi)$ is the preimage in R of $O_2(C_{Aut(Q)}(\phi_{|Q}))$ under the conjugation map $R \to Aut(Q)$. Thus $Aut_{\mathcal{P}_2}(R_2)$ is determined by T and Ω, so $Aut_{\tilde{\mathcal{P}}_2}(\tilde{R}_2) = Aut_{\mathcal{P}_2}(R_2)$, and in particular $\tilde{\mathcal{F}} = \mathcal{Y}$ if $\mathcal{F}$ is of type $L^-_{2n}[m]$.

Hence we may assume $\mathcal{F}$ is of type $L^-_{2n+1}[m]$ and it remains to consider the case $j = 3$, where we must show $R_3 = \tilde{R}_3$ and $Aut_{\mathcal{P}_3}(R_3) = Aut_{\tilde{\mathcal{P}}_3}(\tilde{R}_3)$. Then $\mathcal{X} \cong L^-_3[m]$, and $X = K_n\langle t_n\rangle \cong SD_{2m}$ for $t_n \in \mathcal{D}^o(z_n)$. Further $P_3 = \langle t_n, z_n\rangle$ is determined up to conjugation in X and $R_3 = C_T(P_3)$. In particular, using 13.1.10, P_3 and R_3 are determined by H and Ω, so as $\tilde{G} = H$, P_3 and R_3 are determined by T and Ω. Thus we may take $\tilde{P}_3 = P_3$ and $\tilde{R}_3 = R_3$.

Next from the proof of 13.2.21, a Sylow group Q of the kernel M of the action of H on Ω is of the form $Q = X_1 \times \cdots \times X_n$, where $X_i = K_i\langle t_i\rangle$ and $t_i \in \mathcal{D}^o(z_i)$. Further there is $C \leq M$ acting faithfully as $Sym(I)$ on I with $X_i^h = X_{ih}$ for $h \in C$ and $QC \cong X$ wr S_n. From 13.4.9.2, $\tilde{R} = P_3X'_nB$, where $X'_n = \prod_{i<n} X_i$ and $B = S \cap C_n$ for C_n the stabilizer in C of n. Moreover $Aut_{\tilde{\mathcal{P}}_3}(\tilde{R}_3)$ consists of those $\phi \in Aut(R)$ such that $\phi_{|P_3} \in O^2(Aut(P_3))$ and ϕ centralizes B and $J_i = K_i\langle t_iz_n\rangle$ for $i < n$.

Let $Y = O(\bar{\theta})P_3$, $\mathcal{N} = N_{\mathcal{F}}(YC_S(Y))$, and G_0 a model for $\mathcal{N}$. Then $O(\bar{\theta}) = \langle \Omega \cap Y\rangle \trianglelefteq \mathcal{N}$, so $P_3 = X \cap Y \trianglelefteq \mathcal{N}$, and then a model L for $O^2(N_{\mathcal{X}}(P_3)$ is normal in G_0 with $L \cong A_4$. Further we may take $C_n \leq G_0$ and C_n centralizes P_3. It remains to show that L centralizes B and J_i. First we show C_n centralizes L. If $n = 2$ then $C_n = 1$, so we may take $n > 2$. Now $[C_n, L] \leq C_L(P_3)$, so $O^2(C_n)$ centralizes L. Then as $C_n = O^2(C_n)\langle s\rangle$ for $s \in \mathcal{Z}^m \cap C$, it remains to show that s centralizes L. Take $z_1^s = z_2$, $u = z_1z_2$, and $\alpha \in \mathfrak{A}(u)$. By 13.1.10.2, there is a coconnected component $\rho = (\mathcal{G}, \Gamma)$ of $\tau^\circ_{u\alpha}$ containing $s\alpha$ with $\Gamma = \{K_1\alpha, K_2\alpha\}$. Then $\mathcal{X}\alpha^*$ centralizes $\mathcal{G}$, so $\mathcal{X}$ centralizes s, so indeed C_n centralizes L.

Next by construction, K_i centralizes $\mathcal{X}$, so it remains to show $t = t_iz_n$ centralizes L. As t_i centralizes X, either t or t_i centralizes L, and we may assume the latter. If $n > 2$ there is $i \neq k < n$ and $\alpha \in \mathfrak{A}(z_k)$. Then working in $\mathcal{F}_2(z)$, $t\alpha$ centralizes $\mathcal{X}\alpha^*$, contradicting $[t, L] \neq 1$. Therefore $n = 2$. Let $T_2 = X$, $s \in \mathcal{Z}^m$, and $T_1 = T_2^s$. By construction, $\langle T_1, T_2\rangle = T_1 \times T_2$ with $T_i = \langle t_i\rangle K_i$. Now K_i centralizes $\mathcal{X}$ and t_1 centralizes $\langle \mathcal{F}_{P_3}(L), T_2\mathcal{O}_{z_2}\rangle = \mathcal{X}$, so T_1 centralizes $\mathcal{X}$. Set $\mathcal{C}_2 = \mathcal{X}$ and $\mathcal{C}_1 = \mathcal{C}_2c_s^*$, so that $T_2 = T_1^s$ centralizes $\mathcal{C}_1$. Therefore by 2.2 in [**Asc11**], we have $\mathcal{C}_1 \times \mathcal{C}_2 \leq \mathcal{F}$. Observe that Hypothesis 8.1.17 is satisfied. Further $\mathcal{C}_2 = \mathcal{X} \cong L^-_3[m]$, so $\mathcal{C}_2$ has one class of involutions. It follows from 8.1.21 that $\mathcal{F}^\circ = \mathcal{C}_1 \times \mathcal{C}_2$. This contradicts $K_2 \in K_1^{\mathcal{F}^\circ}$ by 13.2.3.1. □

LEMMA 13.6.3. *$N_{\mathcal{Y}}(W) = \mathcal{W} \trianglelefteq N_{\mathcal{F}}(W)$.*

PROOF. By 13.6.2, $\mathcal{Y} = \tilde{\mathcal{F}}$, so by 13.2.23, $N_{\mathcal{Y}}(W) = N_{\tilde{\mathcal{F}}}(\tilde{W}) = \mathcal{W} \trianglelefteq N_{\mathcal{F}}(W)$. □

LEMMA 13.6.4. *Let $g = f$, or if $\mathcal{F}$ is of type $L^-_{2n+1}[m]$, let $g = c$ or cf. Then*
(1) $C_T(g)N_{\mathcal{F}_g^\circ}(W) \trianglelefteq N_{\mathcal{F}_g}(W)$.
(2) $\mathcal{Y}_g \trianglelefteq \mathcal{F}_g$.

PROOF. Let H_g be the model for $N_{\mathcal{F}_g^\circ}(W)$ contained in $C_G(g)$. To prove (1), it suffices to show that $C_T(g)H_g \trianglelefteq C_G(g)$. Then as $L_0 \leq H_g$, it suffices to show that $C_T(g)L_0 \trianglelefteq C_G(g)$. We use the description of H in 13.2.21. By parts (5) and (6) of that lemma, if $\mathcal{F}$ is of type $PSp_{2n}[m]$ or $L^-_{2n+1}[m]$, then H is M_1 wr S_n, where $M_1 = L_1$ or $L_1\langle t_1\rangle$, respectively, with $L_1 = O(1)$ if $m > 8$ and $L_1 \cong SL_2(3)$ if $m = 8$. Thus M_1 is $O(1)$, $SL_2(3)$, SD_{2m}, or $SL_2(3)^{(1)}$. If $\mathcal{F}$ is of type $PSL^-_{2n}[m]$ then H is of index 2 in the group for $L^-_{2n+1}[m]$.

Next $C_G(g)$ is the stabilizer in G of a partition $\Lambda = \{\theta_1, \theta_2\}$ of I. Thus, setting $n_i = |\theta_i|$, $C_G(g)^I \cong S_{n_1} \times S_{n_2}$, unless $n_1 = n_2 = n/2$ and $c = 1$, where $C_G(g)^I \cong S_{n/2}$ wr $\mathbf{Z_2}$. Moreover $g \in Z(S)$, except in (c) where $|S : S_f| = 2$.

Next if $n_i > 1$ then the factor S_{n_i} is generated by transpositions that are induced by members of T_g, so $S^I_g = T^I_g$, except in the case $n_i = n/2$ and $c = 1$. It follows that $C_T(g) = N_{C_H(g)}(T_g) \trianglelefteq C_G(g)$, except when $\mathcal{F}$ is of symplectic type and $m = 8$, where at least $C_T(g)L_0 \trianglelefteq C_G(g)$. Therefore (1) holds.

Next by 13.6.2, $\mathcal{Y}_g = C_T(g)\mathcal{F}^\circ_g$, so by 1.5.1 in [**Asc19**], to prove (2) it suffices to show that $N_{\mathcal{Y}_g}(T_g) \trianglelefteq N_{\mathcal{F}_g}(T_g)$. But $N_{\mathcal{Y}_g}(T_g) = C_T(g)N_{\mathcal{F}^\circ_g}(T_g)$. Also $O(\tau) \le T_g$, so $N_{\mathcal{F}_g}(T_g) \le N_{\mathcal{F}_g}(W)$. Then as $\mathcal{O}_\tau \le N_{\mathcal{F}^\circ_g}(T_g)$, (2) follows from (1). □

LEMMA 13.6.5. $\mathcal{F}^\circ = \mathcal{Y} \cong \tilde{\mathcal{F}}$.

PROOF. We show $\mathcal{Y} \trianglelefteq \mathcal{F}$. Then as $\Omega \subseteq T$ and T is Sylow in $\mathcal{Y}$, $\mathcal{F}^\circ = \mathcal{Y}$, so the lemma follows from 13.6.2.

If $\mathcal{F}$ is of type $PSp_{2n}[m]$ or $PSL^-_{2m}[m]$, set $J = \{1,2\}$, $\mathcal{G}_1 = \mathcal{F}_f$, $\mathcal{E}_1 = \mathcal{Y}_f$, $\mathcal{G}_2 = N_{\mathcal{F}}(W)$, and $\mathcal{E}_2 = \mathcal{W}$. In case (c), set $J = \{1,2\}$, $\mathcal{G}_1 = \mathcal{F}_f$, $\mathcal{E}_1 = \mathcal{Y}_f$, $\mathcal{G}_2 = \mathcal{F}_c$, and $\mathcal{E}_2 = \mathcal{Y}_c$. Finally if $\mathcal{F}$ is of type $L^-_{2n+1}[m]$ and (a) holds, set $J = \{1,2,3\}$, $\mathcal{G}_1 = \mathcal{F}_f$, $\mathcal{E}_1 = \mathcal{Y}_f$, $\mathcal{G}_2 = \mathcal{F}_{fc}$, $\mathcal{E}_2 = \mathcal{Y}_{fc}$, $\mathcal{G}_3 = N_{\mathcal{F}}(W)$, and $\mathcal{E}_3 = \mathcal{W}$.

We verify that the tuple $(\mathcal{G}_j, \mathcal{E}_j : j \in J)$ satisfies Hypothesis 1.4.7 in [**Asc19**], and then appeal to 1.4.8 in [**Asc19**] to conclude that $\mathcal{Y} \trianglelefteq \mathcal{F}$, except in case (c) where we verify Hypothesis 1.4.3 in [**Asc19**] and appeal to 1.4.5 in [**Asc19**].

By 13.2.26, $\mathcal{F} = \langle \mathcal{G}_j : j \in J\rangle_S$ and $\mathcal{Y} = \langle \mathcal{E}_j : j \in J\rangle_T$. Thus in particular Hypotheses 1.4.1 and 1.4.7.3 in [**Asc19**] are satisfied. By 13.6.3 and 13.6.4.2, $\mathcal{E}_j \trianglelefteq \mathcal{G}_j$ for each $j \in J$, so Hypotheses 1.4.3.1 and 1.4.7.1 in [**Asc19**] are satisfied. By 13.6.2, $\mathcal{Y}$ is saturated, verifying 1.4.7.4 in [**Asc19**].

Suppose (c) does not hold. In this case to complete the proof by verifying Hypothesis 1.4.7 in [**Asc19**], it remains to show that 1.4.7.2 in [**Asc19**] holds. As we observed during the proof of 13.6.4, S, T is Sylow in each $\mathcal{G}_j, \mathcal{E}_j$, respectively, so it remains to show $Aut_{\mathcal{F}}(T) \le Aut(\mathcal{E}_j)$ for each $j \in J$. As $O(\tau) \le T$, $N_{\mathcal{F}}(T) \le N_{\mathcal{F}}(W)$, so as $\mathcal{W} \trianglelefteq N_{\mathcal{F}}(W)$, we have $Aut_{\mathcal{F}}(T) \le Aut(\mathcal{W})$. From the proof of 13.6.4, $Aut_{\mathcal{F}}(T) = Aut_{SL_0}(T)$, so as $g \in Z(SL_0)$ for $g \in \{f, fc\}$, $Aut_{\mathcal{F}}(T)$ centralizes g, and then as $\mathcal{E}_j \trianglelefteq \mathcal{G}_j$, we have $Aut_{\mathcal{F}}(E) \le Aut(\mathcal{E}_j)$. This completes the proof in (a) and (b), so we may assume (c) holds.

In (c) we must verify Hypothesis 1.4.3 in [**Asc19**], so it remains to check parts (2) and (3) of that hypothesis. The Sylow group of $\mathcal{E}_2$ is T, so the lifting condition 1.4.3.3 is trivial when $j = 2$. Further, as above, $Aut_{\mathcal{F}}(T) \le Aut(\mathcal{E}_2)$. Next $\mathcal{E}_1 = \mathcal{Y}_f$ and $|S : C_S(f)| = 2 = |T : C_T(f)|$ with $f^t = fc$ for $t \in T - C_T(f)$. Now, as above, $A = Aut_{\mathcal{F}}(C_T(f)) = Aut_S(C_T(f))$ so each member of A lifts to $c_s \in Aut_{\mathcal{F}}(T)$ for some $s \in S$, and hence 1.4.3.3 is satisfied when $j = 1$. Also $\mathcal{E}_1^t = \mathcal{Y}_f^t = \mathcal{Y}_{fc} \le \mathcal{Y}$, so 1.4.3.2 holds by 1.4.6 in [**Asc19**]. This completes the proof. □

THEOREM 13.6.6. *Assume Hypothesis* 13.1.1 *and* $\mathcal{F}^\circ$ *is transitive on* Ω. *Then one of the following holds:*

(1) $\mathcal{Z}_S(\tau) = \{z\}$ *is of order 1 and either* $\mathcal{F}^\circ = \mathcal{O}(z)$ *or* $\mathcal{F}^\circ \cong L_2[2m]^{(1)}$.

(2) $m = 8$, $n = 2$, *and* $\mathcal{F}^\circ \cong \hat{A}E_5$.

(3) $\mathcal{F}^\circ \cong (P)Sp_{2n}[m]$, $(P)SL^-_{2n}[m]$, *or* $L^-_{2n+1}[m]$.

PROOF. Recall $n = |\Omega|$. If $n = 1$ then (1) or (3) holds by Theorem 7.1.7. Thus we may assume $n > 1$. Hence by 13.1.11, $\mathcal{Z}' \neq \emptyset$.

If $\mathcal{F}^\circ$ is the central product of k conjugates of some subnormal subsystem $\mathcal{E} = \mathcal{E}^\circ$ then as $\mathcal{F}^\circ$ is transitive on Ω by hypothesis, it follows that $k = 1$. Therefore $\mathcal{Z}^m \neq \emptyset$ by 13.1.12. Hence Hypothesis 13.1.13 is satisfied. Then by Theorem 13.1.15, $Aut_{\mathcal{F}}(O)$ induces $Sym(\Omega)$ on Ω. Therefore Hypothesis 13.2.1 is satisfied.

If $1 \neq c \in Z(\mathcal{F})$ then (2) or (3) holds by 13.2.13.2, so we may assume otherwise. Therefore Hypothesis 13.2.14 holds. Then 13.6.5 completes the proof. □

REMARK 13.6.7. Observe that Theorem 13.6.6 and Theorem 1 imply Theorem 7. Namely Theorem 1 implies the Extended Inductive Hypothesis , while the Hypotheses of Theorem 7 imply the remaining parts of Hypothesis 13.1.1.

CHAPTER 14

More generation

Let $\tau = (\mathcal{F}, \Omega)$ be a quaternion fusion packet such that $W \in W(\tau)$ is weakly closed in S with respect to $\mathcal{F}$; for example by 3.3.14, if $\mathcal{F}$ is transitive on Ω then W is weakly closed unless possibly $m = 8$ and $\mu = \mu(\tau)$ is abelian. In particular if μ is isomorphic to Weyl(D_n) for some $n \geq 3$ or to Weyl(E_n) then W is weakly closed. Define

$$\mathcal{B}_0(\tau) = \langle N_{\mathcal{F}}(W), \mathcal{F}_x : \ x \in \mathcal{F}^f \cap W \text{ is an involution}\rangle.$$

In Chapter 14 we prove two theorems which insure that $\mathcal{F} = \mathcal{B}_0(\tau)$ under suitable hypotheses.

We begin in section 14.1 with some general results about $\mathcal{B}_0(\tau)$ and the structure and $\mathcal{F}$-automizer of essential subgroups R such that $Aut_{\mathcal{F}}(R) \neq Aut_{\mathcal{B}_0(\tau)}(R)$. Then in Hypothesis 14.1.11 we consider those τ such that $|\Omega(z)| = 1$ for $z \in \mathcal{Z}_S(\tau)$, there exists an involution $u \in C_W(S)$ with $\mathcal{B}_0(\tau) = \mathcal{F}_u$, and either $\mu \cong$ Weyl(D_n) for some $n \geq 3$ or $\mu \cong$ Weyl(E_7) and $M \in M(\tau)$ is isomorphic to $\bar{\omega}(E_7, m)$. We prove in Theorem 14.1.18 that if Hypothesis 14.1.11 is satisfied then $u \in Z(\mathcal{F})$, so that in particular $\mathcal{F} = \mathcal{B}_0(\tau)$.

Then in Theorem 14.2.4 we prove $\mathcal{F} = \mathcal{B}_0(\tau)$ when $\mu \cong$ Weyl(E_8).

14.1. A generation lemma

In this section we assume the following hypothesis:

HYPOTHESIS 14.1.1. *$\tau = (\mathcal{F}, \Omega)$ is a quaternion fusion packet such that $W \in W(\tau)$ is weakly closed in S with respect to $\mathcal{F}$.*

DEFINITION 14.1.2. Set

$$\mathcal{B}_0 = \mathcal{B}_0(\tau) = \langle N_{\mathcal{F}}(W), \mathcal{F}_x : x \in W \cap \mathcal{F}^f \text{ is an involution } \rangle.$$

LEMMA 14.1.3. *(1) For each $1 \neq X \leq W$, $N_{\mathcal{F}}(X) \leq \mathcal{B}_0$.*
(2) $\mathcal{E}(\tau) \leq \mathcal{B}_0$.

PROOF. Let $1 \neq X \leq W$. By 14.1.1, W is weakly closed in S, so $N_{\mathcal{F}}(W)$ controls fusion in W and for each $\alpha \in \mathfrak{A}(X)$, $W\alpha = W$, so $X\alpha \leq W$. As $N_{\mathcal{F}}(W) \leq \mathcal{B}_0$, α is a $\mathcal{B}_0$-map, so, replacing X by $X\alpha$, we may assume $X \in \mathcal{F}^f$. In particular if X is of order 2 then by definition of $\mathcal{B}_0$, we have $N_{\mathcal{F}}(X) \leq \mathcal{B}_0$.

Set $\mathcal{X} = N_{\mathcal{F}}(X)$. By 1.3.2 in [**Asc10**], $\mathcal{X} = \langle N_S(X)C_{\mathcal{F}}(X), N_{\mathcal{F}}(C_S(X))\rangle$. But for x an involution in $X \cap Z(N_S(X))$, we have $N_S(X)C_{\mathcal{F}}(X) \leq \mathcal{F}_x \leq \mathcal{B}_0$ from the previous paragraph, and $N_{\mathcal{F}}(C_S(X)) \leq N_{\mathcal{F}}(W) \leq \mathcal{B}_0$, completing the proof of (1).

Recall from 3.4.2 that

$$\mathcal{E}(\tau) = \langle S\mathcal{O}(\tau), N_{\mathcal{F}}(W), N_{\mathcal{F}}(O(\theta)) : \theta \subseteq \mathcal{Z}_S \text{ and } O(\theta) \in \mathcal{F}^f \rangle.$$

But $N_{\mathcal{F}}(W) \le \mathcal{B}_0$ and $S\mathcal{O}(\tau) \le N_{\mathcal{F}}(\langle \mathcal{Z}_S \rangle) \le \mathcal{B}_0$ and $N_{\mathcal{F}}(O(\theta)) \le \mathcal{B}_0$ by (1), so (2) holds. □

Until the last result of the section, in the remainder of the section we assume:

HYPOTHESIS 14.1.4. $\mathcal{F} \neq \mathcal{B}_0$.

NOTATION 14.1.5. By 14.1.4 and the Alperin-Goldschmidt Fusion Theorem, we have $\mathcal{R} \neq \emptyset$, where $\mathcal{R}$ consists of those $R \in \mathcal{F}^e$ with $Aut_{\mathcal{F}}(R) \neq Aut_{\mathcal{B}_0}(R)$. Let $\mathcal{R}^*$ consist of those members of $\mathcal{R}$ of maximal order. Pick $R \in \mathcal{R}^*$ and let G be a model for $N_{\mathcal{F}}(R)$.

Set $W_O = \langle \Omega_2(K \cap W) : K \in \Omega \rangle$ and $W_\natural = W$ or W_O. Set $G^* = G/R$, $W_R = W_\natural \cap R$, $W_G = N_{W_\natural}(R)$ and $S_G = N_S(R)$. Define

$$H = \langle N_G(X), N_G(Y) : 1 \neq X \le W_G \text{ with } R \le N_S(X), R < Y \le S_G \rangle.$$

LEMMA 14.1.6. *(1) $Aut_H(R) \le Aut_{\mathcal{B}_0}(R)$, so $H \neq G$.*
(2) H^* is strongly embedded in G^*.

PROOF. Let $R < Y \le S_G$. By maximality of $|R|$ and the Alperin-Goldschmidt Fusion Theorem, $N_{\mathcal{F}}(Y) \le \mathcal{B}_0$, so $Aut_{N_G(Y)}(R) \le Aut_{\mathcal{B}_0}(Y)$. If $1 \neq X \le W_G$ with $R \le N_S(X)$ then by 14.1.3.1, $N_{\mathcal{F}}(X) \le \mathcal{B}_0$, so $Aut_{N_G(X)}(R) \le Aut_{\mathcal{B}_0}(R)$. This proves (1).

Let $1 \neq Y_0^* \le S_G^*$, and let Y be the preimage of Y_0^* in G. Then $N_{G^*}(Y_0^*) = N_{G^*}(Y^*) = N_G(Y)^* \le H^*$, and $H^* \neq G^*$ by (1), so (2) follows. □

LEMMA 14.1.7. *(1) $W_R \neq 1$.*
(2) $W_R \neq W_G$, so $W_G^* \neq 1$.
(3) For each $X \le W_G$, $[X, R] \le W_R$, so if $W_R \le X$ then $R \le N_S(X)$.
(4) $\Phi(W_G) \le W_R$, so $\Phi(W_G^*) = 1$.

PROOF. As $W_\natural \trianglelefteq S$, $1 \neq Z(S) \cap W_\natural \le C_{W_\natural}(R) \le R \cap W_\natural = W_R$, so (1) holds.

If $W_R = W_G$ then $W_\natural = W_R \le R$, so for $K \in \Omega$, $V = \Omega_2(K \cap W) \le R$, contrary to 14.1.3.2 and 3.4.5. Thus (2) holds.

Let $X \le W_G$. Then as $W_\natural \trianglelefteq S$, we have $[X, R] \le W_\natural \cap R = W_R$, so (3) holds.

As W is abelian, W_G is quadratic on R by (3), and hence W_G is also quadratic on $\tilde{R} = R/\Phi(R)$. But $O_2(G^*) = 1$, so G^* is faithful on $\tilde{R}$ and hence as W_G^* is quadratic on $\tilde{R}$ we have $\Phi(W_G^*) = 1$, proving (4). □

LEMMA 14.1.8. *Let $g \in G$ with $W_G^g \not\le H$ and set $L = \langle W_G, W_G^g \rangle$ and $B = W_R W_R^g$.*
(1) $[L, R] \le B \le L$ so $R \le N_G(L)$.
(2) $W_R \cap W_R^g = 1$, so $B = W_R \times W_R^g$.
(3) Either
(a) $|W_G^*| = 2$ and L^* is dihedral, or
(b) $L^* \cong L_2(2^k)$ or $Sz(2^k)$ with $1 < m(W_G^*) \le k$.
(4) We may choose $g \in L$.
(5) For each $y \in W_G - W_R$, $C_B(y) = W_R = [B, y] = [R, y]$.
(6) $\Phi(B) = 1$.
(7) $m(W_G^*) \le m(W_R)$.

PROOF. By 14.1.7.3, $[W_G, R] \leq W_R$, so also $[W_G^g, R] \leq W_R^g$ and hence (1) holds.

Set $A = W_R \cap W_R^g$ and suppose $A \neq 1$. As W_G and W_G^g are abelian, $A \leq Z(L)$. As W_R and W_R^g are R-invariant, A is normal in R, so $L \leq N_G(A) \leq H$, contradicting $W_G^g \not\leq H$. Therefore (2) holds.

By 14.1.6.2, H^* is strongly embedded in G^*, so as $L^* \not\leq H^*$, $H_L^* = H^* \cap L^*$ is strongly embedded in L^*. Then (3) follows from the structure of groups with a strongly embedded subgroup. By (3) there is $l \in L$ with $W_G^{*l} \not\leq H_L^*$, so, replacing g by l if necessary, (4) holds.

Choose y as in (5) and suppose $W_R < C_B(y)$. Then $1 \neq B_0 = C_B(y^g) \cap W_R$ is centralized by $L_0 = \langle W_G, y^g \rangle$, so $B_0 \leq W_R \cap W_R^i$, where $i \in L_0$ such that $\langle W_G^{*i}, y^{g*} \rangle$ is a 2-group. But this is contrary to (2) with i in the role of g. Thus $C_B(y) = W_R$. By 14.1.7.4, $y^2 \in W_R$. By 14.1.7.3, $W_R = C_B(y) = [B, y]$, so as B is abelian by (2), it follows that $\Phi(B) = 1$, establishing (6). Then as $m(C_B(y)) = m(W_R) = m(B/W_R)$, also $W_R = [B, y]$, and then $W_R = [R, y]$ by 14.1.7.3.

For $b \in B - W_R$, W_G^* induces a group of transvections of $W_R\langle b \rangle$ with axis W_R, so (7) holds. □

LEMMA 14.1.9. *Take $W_\natural = W$ and suppose R acts on a subgroup U of W containing W_R. Then $C_{U/W_R}(R) = (U \cap W_G)/W_R$.*

PROOF. Set $V/W_R = C_{U/W_R}(R)$. By 14.1.7.3, $U \cap W_G \leq V$. Conversely $[V, R] \leq W_R \leq R$, so $V \leq N_W(R) = W_G$, so that $V \leq U \cap W_G$. □

LEMMA 14.1.10. *(1) $C_S(B) \leq R$.*

(2) $C_W(W_R^g) = C_W(B) = W_R$.

PROOF. Let $S_1 = C_S(B)$ and $S_2 = N_{S_1}(R)$. Then $[S_2, L] \leq C_S(B) \cap C_S(R/B) \leq R$, so $[S_2^*, L^*] = 1$, Then as $L^* \not\leq H^*$, it follows from 14.1.6.2 that $S_2^* = 1$, so $S_2 \leq R$, Hence also $S_1 \leq R$, completing the proof of (1).

As W is abelian and $B = W_R W_R^g$, we have $C_W(B) = C_W(W_R^g)$. By (1), $C_W(B) \leq W \cap R = W_R$, so (2) holds. □

In the remainder of the section we assume the following hypothesis:

HYPOTHESIS 14.1.11. *(1) $\tau = (\mathcal{F}, \Omega)$ is a quaternion fusion packet such that either $\mu(\tau) \cong Weyl(D_n)$ for some $n \geq 3$ or $\mu(\tau) \cong Weyl(E_7)$ and $M(\tau) \cong \bar{\omega}(E_7, m)$.*

(2) $|\Omega(z)| = 1$ for $z \in \mathcal{Z}_S$.

(3) There exists an involution $u \in C_W(S)$ such that $\mathcal{B}_0(\tau) = \mathcal{F}_u$.

NOTATION 14.1.12. Take $W_\natural = W$. Set $\Sigma = Aut_{\mathcal{F}}(B)$, $W_B = N_W(B)$, and $\Sigma_W = Aut_{W_B}(B)$.

LEMMA 14.1.13. *(1) $[\Sigma_W, B] \leq W_R$.*

(2) $\Phi(W_B) \leq W_R$, so $\Phi(\Sigma_W) = 1$.

(3) For each $1 \neq \phi \in \Sigma_W$, $C_B(\phi) = W_R$.

(4) $m(\Sigma_W) \leq m(W_R)$.

PROOF. First $[W_B, B] \leq W \cap B = W_R$, so (1) holds. As W_B is quadratic on B, $\Phi(W_B) \leq C_W(B) = W_R$ by 14.1.10.1, so (2) holds.

Choose ϕ as in (3) and assume $W_R < B_0 = C_B(\phi)$. Then for each conjugate φ of ϕ under Σ, $1 \neq W_0 = C_{W_R}(\varphi)$, so by 14.1.3.1, $\varphi \in Aut_{\mathcal{B}_0}(B) = \Sigma_0$. Therefore $\Xi = \langle \phi^\Sigma \rangle \leq \Sigma_0$.

Let $\theta = \Sigma_W \cap \Xi$ and U the preimage in W_B of θ. Claim $\theta \cap Aut_{W_G}(B) \neq 1$. For if not $U \cap W_G = W_R$. But as $\Xi \trianglelefteq \Sigma$, R acts on U and as $\phi \in \theta$, $W_R < U$, so 14.1.9 supplies a contradiction.

Let $\theta \leq \Gamma \in Syl_2(\Xi)$; by a Frattini argument, $\Sigma = \Xi N_\Sigma(\Gamma)$. By the claim and 14.1.8.5, $X = C_B(\Gamma) \leq C_B(\theta) = W_R$, so $N_\Sigma(\Gamma) \leq N_\Sigma(X)$, and by 14.1.3.1, $N_\Sigma(X) \leq Aut_{\mathcal{B}_0}(B)$. Therefore $\Sigma \leq Aut_{\mathcal{B}_0}(B)$. But by 14.1.11.3, $\mathcal{B}_0 = \mathcal{F}_u$ with $u \in C_W(S)$. Hence as $Z(S) \leq R$, we have $u \in W_R$, a contradiction as $Aut_L(B)$ does not centralize any involution in W_R. This contradiction completes the proof of (3).

The proof of (4) is the same as that of 14.1.8.7. □

LEMMA 14.1.14. *Assume $W \neq W_B$ and let $k = m(W_R)$. Then*
(1) $m(\Sigma_W) = k$.
(2) $W_R = \Omega_1(W_B)$, so $W_B \cong \mathbf{Z}_4^k$.
(3) $W_R = \Omega_1(W)$.

PROOF. Set $A = W_B B$ and let $x \in N_W(A) - W_B$. As $W_B = N_W(B)$, $B \neq B^x$. As x centralizes W_B, $W_R \leq B \cap B^x$. But by 14.1.13.3, for $a \in A - B$, $C_B(a) = W_R$, so for $y \in B^x - B$, $W_R = C_B(y)$. Therefore as $B \cap B^x \leq C_B(y)$ it follows that $W_R = B \cap B^x$. Therefore $m(A/B) \geq m(B^x/W_R) = k$, while by 14.1.13.4, $m(A/B) \leq k$, so (1) holds and $A = BB^x$. Then as $C_A(b) = B$ for $b \in B - W_R$, it follows that B and B^x are the maximal elementary abelian subgroups of A, so in particular (2) holds. Then (2) and 14.1.9, applied to $\Omega_1(W)$ in the role of U, imply (3). □

LEMMA 14.1.15. *$W_R \neq \Omega_1(W)$.*

PROOF. Assume $W_R = \Omega_1(W)$ and set $A = W_R^g$. Then A centralizes $\Omega_1(W)$, so for each $V \in \eta$, A centralizes $z(V)$ and hence A acts on V as $|\Omega(z(V))| = 1$ by 14.1.11.1. Set $V_1 = \Omega_2(V)$; then $[A, V_1] \leq \langle z(V) \rangle \leq \Omega_1(W) = W_R \leq B$, so $V_1 \leq W_B$. But $|A : C_A(V_1)| \leq 2$, so in particular $C_A(V_1) \neq 1$, contrary to 14.1.13.3. □

LEMMA 14.1.16. *$W = W_B$.*

PROOF. This follows from 14.1.14.3 and 14.1.15. □

LEMMA 14.1.17. *(1) $m = 8$.*
(2) W is not homocyclic.
(3) $\mu(\tau) \cong Weyl(D_n)$ with $n \geq 6$ even.
(4) $W_R = \Omega_1(W)$.

PROOF. By 14.1.8.6, $\Phi(W_R) = 1$ while by 14.1.16 and 14.1.13.2, $\Phi(W) \leq W_R$, so W is of exponent 4. Thus (1) holds. If W is homocyclic then $\Omega_1(W) = \Phi(W) \leq W_R \leq \Omega_1(W)$, so $W_R = \Omega_1(W)$, contrary to 14.1.15. Therefore (2) holds.

Let $\bar{M}$ be the universal group of type D_n or E_7 defined in 4.3.1. By 4.3.8 there is a surjective homomorphism $\pi : \bar{M} \to M$ with $\bar{Z} = \ker(\pi) \leq Z(\bar{M})$. By 4.3.7.2, $\bar{W} \cong \mathbf{Z}_4^n$, so $\bar{Z} \neq 1$ by (2). Hence $\bar{M}$ is of type D_n by 14.1.11.3.

As $|\Omega(z)| = 1$, $\bar{Z} = 1$ if $n \leq 4$, so $n \geq 5$. Therefore by the proof of 15.1.6, $u_0 = u(\tau)$ exists in $Z(M)$. In particular $\bar{u}_0 \notin \bar{Z}$. But if n is odd then $Z(\bar{M})$ is cyclic, so $\bar{Z} = 1$. Therefore n is even, completing the poof of (3). Also $Z(\bar{M}) \cong E_4$, so $\bar{Z} \cong \mathbf{Z}_2$.

We've shown that $\bar{Z} \cong \mathbf{Z}_2$, so $W \cong \mathbf{Z}_4^{n-1} \times \mathbf{Z}_2$. Therefore $\Phi(W) \cong E_{2^{n-1}}$ and $W/\Phi(W) \cong E_{2^n}$, so as $\Phi(W) \leq W_R$, (4) follows from 14.1.13.4, completing the proof. □

THEOREM 14.1.18. *Assume Hypothesis* 14.1.11. *Then* $u \in Z(\mathcal{F})$.

PROOF. Assume otherwise. By 14.1.11.3, $\mathcal{B}_0 = \mathcal{F}_u$, so Hypothesis 14.1.4 is satisfied. Observe however that 14.1.15 and 14.1.17.4 now supply a contradiction, completing the proof of the theorem. □

14.2. A generation lemma for E_8

In this section we assume the following hypothesis:

HYPOTHESIS 14.2.1. $\tau = (\mathcal{F}, \Omega)$ *is a quaternion fusion packet such that* $\mu(\tau) \cong$ *Weyl*(E_8).

NOTATION 14.2.2. Let $W \in \mathcal{W}(\tau)$ and $M \in \mathcal{M}(\tau)$. As in 14.1.5, set $W_O = \langle \Omega_2(K \cap W) : K \in \Omega \rangle$. Set $\Theta = \mathcal{Z}_S$ and for $t \in \Theta$ let $K_t \in \Omega(t)$ and $V_t = \langle v_t \rangle = \Omega_2(K_t \cap W)$. As $\mu(\tau) \cong \mathrm{Weyl}(E_8)$, we have $\Omega(t) = \{K_t\}$. Set $n_S = |\Theta|$.

LEMMA 14.2.3. *(1)* $n_S = 8$.
(2) $W_S \cong E_{16}$.
(3) There is a hyperplane U *of* W_S *such that* $\Theta = W_S - U$.
(4) $W_O \cong \mathbf{Z}_4^4 \times E_{16}$, $\Phi(W_O) = W_S$, *and* $\Omega_1(W) = \Omega_1(W_O) = W_\Delta$.

PROOF. See 15.8.6 for parts (1)-(3). By definition of W_O, we have $W_S = \Phi(W_O)$ and $W_O/W_S \cong E_{2^{n_S}}$, so as W_O is abelian it follows from (1) and (2) that $W_O \cong \mathbf{Z}_4^4 \times E_{16}$. In particular $m(W_O) = 8 = m(W)$, so $\Omega_1(W_O) = \Omega_1(W) = W_\Delta$, completing the proof of (4). □

Define $\mathcal{B}_0 = \mathcal{B}_0(\tau)$ as in Definition 14.1.2. The main result of this section is:

THEOREM 14.2.4. $\mathcal{F} = \mathcal{B}_0(\tau)$.

Assume that $\mathcal{F} \neq \mathcal{B}_0$. Then Hypothesis 14.1.4 is satisfied, and we adopt Notation 14.1.5 with $W_\natural = W_O$. Write $\mathfrak{Z}_0$ for the set of elementary abelian normal subgroups Z of G such that $Z \leq Z(R)$, and set $\mathfrak{Z} = \{Z \in \mathfrak{Z}_0 : W_O \cap Z \neq 1\}$.

LEMMA 14.2.5. $\Theta \cap Z(R) = \emptyset$.

PROOF. See 14.1.3.2 and 3.4.6. □

LEMMA 14.2.6. *(1) If* $Z \in \mathfrak{Z}_0$ *with* $[W_G, Z] \neq 1$ *then* $Z \in \mathfrak{Z}$.
(2) For each $Z \in \mathfrak{Z}$, $R = C_S(Z)$.

PROOF. By 14.1.7.3, $[W_G, Z] \leq W_0$, so (1) follows. Assume $Z \in \mathfrak{Z}$. From the proof of 11.3.5.1, $O^{2'}(C_G(Z)) = R$. Then $C_S(Z) = R$ by 3.4.4.1. □

LEMMA 14.2.7. *For each $Z \in \mathfrak{Z}$, we have $|Z| > 4$.*

PROOF. Assume otherwise. By 14.1.3.1, $Z \not\leq W_O$, so there is $s \in Z - W_R$. As $|Z| = 4$, $Z = \langle w, s\rangle$ with $\langle w\rangle = W_R \cap Z$, so as W_O is abelian, $C_{W_O}(s) = C_{W_O}(Z)$. By 14.2.6.2, $C_{W_O}(s) = C_{W_O}(Z) \leq R$. For $t \in \Theta$, $v = v_t v_t^s \in C_{W_O}(s) \leq W_R$, so $v^2 = 1$ by 14.1.8.6. Therefore as $\Omega(t) = \{K_t\}$ if follows that s centralizes t and inverts v_t. Therefore $W_S \leq C_{W_O}(s) \leq R$ and s inverts W_O. Hence s centralizes $\Omega_1(W_O) = W_\Delta$ by 14.2.3.4, so $W_R = W_\Delta$ by 14.1.8.6.

Let $y \in W_G - W_R$. Then s inverts y of order 4, so $[y, s] = y^2 = w$. Recall L is defined in 14.1.8 and observe $L^* = GL(Z) \cong S_3$. By 14.1.8, $O_2(L) = L \cap R = B = W_R \times W_R^g$. Let $h \in L$ be of order 3.

Set $W_t = \langle t^L\rangle$. As $C_B(h) = 1$, $E_4 \cong W_t = \langle t, t^h\rangle$. As B is abelian, t^h acts on V_a for each $a \in \Theta$. As $W_R = W_\Delta$ we can pick $y = v_1 \cdots v_4$ with $v_i = v_{z_i}$ for some $\theta = \{z_1, \ldots, z_4\} \subseteq \Theta$ independent in W_S and with $z_1 = t$ and $z_2 = a$. Now

$$t = [t^h, y] = [t^h, v_1 \cdots v_4] = \prod_i [t^h, v_i] = \prod_i z_i^{\epsilon_i},$$

so as θ is independent in W_S it follows that t^h inverts v_t and centralizes v_i for $i > 1$. In particular t^h centralizes v_a for each $a \in \Theta - \{t\}$, so as $W_O = \langle v_a : a \in \Theta - \{t\}\rangle$, it follows that t^h centralizes W_O. This is a contradiction as t^h inverts v_t. □

LEMMA 14.2.8. *Let $Z \in \mathfrak{Z}_0$ and $x \in W_G$.*
(1) x does not induce a transvection on Z.
(2) If $|[R, x]| \leq 2$ then $x \in R$.
(3) If $Y \leq W_O$ is R-invariant of order 4 then $Y \leq R$.
(4) $|U \cap R| \geq 4$.

PROOF. Assume (1) fails. Then by 2.4.1 in [**Asc19**] there is $Z_1 \in \mathfrak{Z}$ with $Z_1 \leq Z$, $m(Z_1) = 2$, and x induces a transvection on Z_1. Now 14.2.7 supply a contradiction.

Assume the setup of (2) but $x \notin R$. Let w generate $[R, x]$ and set $Z_w = \langle w^G\rangle$. As $[R, x] \trianglelefteq R$, $w \in Z(R)$, so $Z_w \in \mathfrak{Z}$. But x induces a transvection on Z_w, contrary to (1), establishing (2).

Part (3) follows from (2), and (4) follows from (3). □

LEMMA 14.2.9. *If R is transitive on Θ then $W_S \not\leq R$.*

PROOF. Assume otherwise. Let $f \in C_U(R)$ and $Z = \langle f^G\rangle$, so that $Z \in \mathfrak{Z}$. By 14.2.6.2, $R = C_S(Z)$. As $W_S \leq R = C_S(Z)$, for each $t \in \Theta$, Z acts on V_t. Then $|Z : C_Z(V_t)| = 2$ as $V_t \not\leq R$ by 14.1.8.6. As R is transitive on Θ and $Z \leq Z(R)$, we have $C_Z(V_t) = C_Z(V_a)$ for each $a \in \Theta$, so $C_Z(V_t)$ centralizes W_O, and for $s \in Z - C_Z(V_t)$, s inverts each V_a. Therefore s inverts W_O, so $W_\Delta \leq C_S(\langle s, C_Z(V_t)\rangle) = C_S(Z) = R$. Hence $W_\Delta = W_R$ by 14.1.8.6.

Pick $y \in W_G - W_R$; thus $y \in W_0$. Then y centralizes $C_Z(V_t)$ and $[y, s] = y^2 = u$, so y induces a transvection on Z, contrary to 14.2.8.1. □

LEMMA 14.2.10. *The orbits of R on Θ are of length 4 or 8.*

PROOF. Assume $\theta = t^R$ is an orbit of length less than 4. By 14.2.5, $|\Theta| \neq 1$, so $\theta = \{t_1, t_2\}$ is of order 2. By 14.2.8.3, $X = \langle\theta\rangle \leq R$. Let $v_i = v_{t_i}$, $y = v_1 v_2$, and $u = t_1 t_2$. As $[R, y] \leq X \leq R$, $y \in W_G$ and $u \in Z(R)$, so $Z = \langle u^G\rangle \in \mathfrak{Z}$ and hence $R = C_S(Z)$ by 14.2.6.2. As $|y| = 4$, $y \notin R$ by 14.1.8.6, so $[Z, y] \neq 1$. But $t_i \notin Z$ by 14.2.5, so $[Z, y] = \langle u\rangle$, contrary to 14.2.8.1. □

NOTATION 14.2.11. Let $\mathcal{T}$ be the set of subsets θ of Θ of order 4. Observe from 14.2.3 that either θ is independent and $T = \langle\theta\rangle = W_S$, or $T \cong E_8$ and $\theta = T - (T \cap U)$. In the first case we say θ is of the first kind, and in the second case we say θ is of the second kind. Write $\mathcal{T}_i$ for the members of $\mathcal{T}$ of the ith kind.

LEMMA 14.2.12. *If θ is an orbits of R on Θ of length 4 then $\theta \in \mathcal{T}_2$.*

PROOF. Assume $\theta \in \mathcal{T}_1$ and set $u = \prod_{t\in\theta} t$; then $u \in C_{W_S}(R) \leq Z(R)$ is an involution. Set $Z = \langle u^G\rangle$, and observe $Z \in \mathfrak{Z}$, so $R = C_S(Z)$ by 14.2.6.2.

Suppose first that $Z = Z_\theta$ is the pointwise stabilizer in Z of θ. Then $W_S \leq C_S(Z) = R$. For $t \in \theta$, $v_t \notin R$ by 14.1.8.6, so $|Z : C_Z(v_t)| = 2$. But as R is transitive on θ, $C_Z(v_t) = C_Z(v_a)$ for each $a \in \theta$. Thus $s \in Z - C_Z(v_t)$ inverts each v_a and hence $C_Z(v_t)$ centralizes $W_O = \langle v_a : a \in \theta\rangle$ while s inverts W_O. Therefore, as usual, $W_R = W_\Delta$. Also s inverts $y = \prod_{t\in\theta} v_t$ of order 4, and $[R, y] \leq W_S \leq R$, so $y \in W_G - W_R$. But now y induces a transvection on Z with axis $C_Z(v_t)$, contrary to 14.2.8.1. Therefore $Z \neq Z_\theta$.

Suppose next that Z^θ is of order 2 and let $s \in Z - Z_\theta$. Then $s = (t, t^s)(a, a^s) \in Sym(\Theta)$ as R is transitive on θ and $Z \leq Z(R)$. Therefore s centralizes $x = tt^s$, so x centralizes $Z = Z_\theta\langle s\rangle$ and hence $x \in R$. Set $y = ta$; then $U = \langle y, x, u\rangle$ with $U_R = \langle x, u\rangle \leq R$, so $[R, y] \leq U_R \leq R$ and hence $y \in W_G$. But $[s, y] = u \neq 1$ and y centralizes the hyperplane Z_θ of Z, contrary to 14.2.8.1.

Therefore $E_4 \cong Z^\theta \leq Z(R^\theta)$ is regular on θ, so $C_U(Z) = \langle u\rangle$. This is a contradiction as $Z \leq Z(R)$ and $|U \cap R| > 2$ by 14.2.8.4. □

LEMMA 14.2.13. *Suppose θ is an orbit of R on Θ of length 4. Then*
(1) $P = \langle\theta\rangle \leq R$.
(2) $W_S \leq R$.

PROOF. By 14.2.10, $\theta' = \Theta - \theta$ is also an orbit of R, so if (1) holds then also $\theta' \subseteq R$, and hence (2) holds. So we may assume $P \not\leq R$ and it remains to derive a contradiction.

By 14.2.12, $\theta \in \mathcal{T}_2$. Thus $\sigma = \prod_{t\in\theta} t = 1$. As R is transitive on θ and $P \not\leq R$, $t \in \theta$ is not in R. By 14.2.8.3, $U_P = U \cap P \leq C_U(Z) = U \cap R$. Let $1 \neq u \in C_U(R)$ and $Z = \langle u^G\rangle$, so that $Z \in \mathfrak{Z}$.

As R acts on P, $[R, t] \leq U_P \leq R$, so $t \in W_G$. By 14.2.8.1, $|[Z, t]| > 2$, so as $[R, t] \leq U_P$, we conclude that $U_P = [Z, t] \leq Z$. Therefore Z is transitive on $\theta = P - U_P$, so $E_4 \cong Z^\theta$ is regular on θ.

Let D be a complement to Z_θ in Z and $y = \prod_{d\in D} v_t^d$. As Z^θ is regular on θ and $Z \leq Z(R)$, we have $R = D \times R_\theta$. Also $y \in C_{W_O}(D)$ is an involution centralizing R_θ as $D \leq Z(R)$ is transitive on θ and $\sigma = 1$. Therefore $y \in C_S(R) = Z(R)$. Let $x = v_t v_t^d$ for some $d \in D^\#$. Then $[x, D] = \langle y\rangle$ and $[R_\theta, x] \leq \langle tt^d\rangle \leq U_P \leq R$, so $x \in W_G$. But if $y \notin Z$ then x induces a transvection on Z, contrary to 14.2.8.1. Therefore $[Z, x] = \langle y, tt^d\rangle$. Now L acts on $Z \cap B \geq Z_0 = \langle U_P, y\rangle \cong E_8$, so by 14.1.8.5, $2 = m([Z, x] \geq m(Z_0) = 3$, a contradiction. □

LEMMA 14.2.14. *R is transitive on Θ.*

PROOF. Assume otherwise. By 14.2.10 and 14.2.12, an orbit θ for R on Θ is in $\mathcal{T}_2$. By 14.2.13.2, $W_S \leq R$. We argue as in the proof of 14.2.9.

Let $Z \in \mathfrak{Z}$. For $t, a \in \theta$, $C_Z(V_t) = C_Z(V_a)$ is a hyperplane of Z and $s \in Z - C_Z(V_t)$ inverts $Y = \langle V_a : a \in \theta\rangle$. Thus Y acts on R. By 14.1.8.6, $\Omega_1(Y) = Y \cap R$, so there is $y \in W_G - R$ of order 4. Then y centralizes $C_Z(V_t)$, contrary to 14.2.8.1. □

LEMMA 14.2.15. $\Theta \cap R = \emptyset$.

PROOF. By 14.2.14, R is transitive on Θ, so if $t \in \Theta \cap R$ then $\Theta \subseteq R$, contrary to 14.2.9. $\square$

Pick $Z \in \mathfrak{Z}$.

LEMMA 14.2.16. $Z^{\Theta} \neq 1$.

PROOF. If so $Z = Z_{\Theta}$, so $W_S \leq C_S(Z) = R$ by 14.2.6.2, contrary to 14.2.15. $\square$

NOTATION 14.2.17. Let $1 \neq \gamma \in Z^{\Theta}$. By 14.2.14, R is transitive on Θ, so as $Z^{\Theta} \leq Z(R^{\Theta})$, we have Z^{Θ} semiregular on Θ. In particular $\gamma = (t_1, s_1) \cdots (t_4, s_4)$ is a fixed-point-free involution in $Sym(\Theta)$. Set $u_i = t_i s_i$ and $U_{\gamma} = C_U(\gamma)$. Then $U_{\gamma} \geq \langle u_1, \ldots, u_4 \rangle$. As $m(U) = 3$, U_{γ} is of rank 3 or 2. In the first case we say γ is of type 1 and in the second of type 2. Set $\sigma = \prod_i u_i$ and observe that $\sigma = 1$. If γ is of type 2 then $u_i = u_j$ for some distinct i, j, so we may choose notation such that $u_1 = u_2$. Then as $\sigma = 1$, also $u_3 = u_4$, but as $m(U_{\gamma}) = 2$, we have $u_1 \neq u_3$.

LEMMA 14.2.18. *Z^{Θ} is of order 4 or 8.*

PROOF. From 14.2.17, Z^{Θ} is semiregular on Θ, so $|Z^{\Theta}|$ divides $|\Theta| = 8$. Thus we may assume $Z^{\Theta} = \langle \gamma \rangle$ is of order 2.

Suppose γ is of type 1. Then $U = U_{\gamma} \leq C_S(Z) = R$. Hence for $t \in \Theta$, $[R, t] \leq U \leq R$, so $t \in W_G$, but $t \notin R$ by 14.2.15. However t centralizes the hyperplane Z_{Θ} of Z, contrary to 14.2.8.1.

Therefore γ is of type 2. Set $y = t_1 t_3$, so that $y \in U - U_{\gamma}$ and $[y, R] \leq U_{\gamma} \leq R$, so $y \in W_G$. But $[Z, y] = [\gamma, y] = u_1 u_3$, contrary to 14.2.8.1. $\square$

LEMMA 14.2.19. *Z^{Θ} is of order 4.*

PROOF. Assume otherwise; by 14.2.18, Z^{Θ} is regular on Θ, so $R = R_{\Theta} \times D$, where D is a complement to Z_{Θ} in Z. Now R_{Θ} centralizes W_S, and hence for each $t \in \Theta$, R_{Θ} acts on V_t. Hence $X = C_{R_{\Theta}}(V_t)$ is of index at most 2 in R_{Θ}, and as Z is transitive on Θ, we have $X = C_{R_{\Theta}}(V_a)$ for each $a \in \Theta$, so that X centralizes $W_O = \langle V_a : a \in \Theta \rangle$. Similarly if $r \in R_{\Theta} - X$ then r inverts W_O, so R_{Θ} centralizes W_{Δ}.

For $F \leq D$ set $v_F = \prod_{f \in F} v_t^f$ and $w_F = \prod_{f \in F} t^f$. Observe F centralizes v_F and w_F with $v_F^2 = w_F$. In particular D centralizes $v_D \in W_O$ and $w_D = 1$ so $v_D \in W_{\Delta}$. Therefore $v_D \in C_S(ZR_{\Theta}) = Z(R)$.

Let E be a hyperplane of D. Then E centralizes w_E and for $d \in D - E$, $[w_E, d] = w_D = 1$, so $w_E \in C_{W_{\Delta}}(D) = Z(R)$. Similarly E centralizes v_E and $[v_E, d] = v_D \in R$. Also X centralizes v_E and if $r \in R_{\theta} - X$ then $[r, v_E] = w_E \in R$. Therefore $v_E \in W_G$. But $[R, v_E] = \langle v_D, w_E \rangle$, so by 14.2.8.1 we have $w_E \neq 1$ and $[Z, v_E] = [R, v_E]$. As this holds for each hyperplane E of D, it follows that $\mathcal{V} = \{v_E^* : |D : E| = 2\}$ is of order 7, so $|\mathcal{V}| \geq 8$. This contradicts 14.1.8.7 which says $m(W_G^*) \leq m([R, w^*])$ for $1 \neq w^* \in W_G^*$. $\square$

NOTATION 14.2.20. By 14.2.18 and 14.2.19, Z has two orbits θ_1 and θ_2 on Θ, each of length 4. As usual let D be a complement to Z_{Θ} in Z, pick $t \in \Theta$, and for $F \leq D$ define v_F and w_F as in the proof of 14.2.19. Observe F centralizes v_F and w_F.

LEMMA 14.2.21. *$\theta_i \in \mathcal{T}_2$ for $i = 1, 2$.*

PROOF. Assume otherwise; then $\theta_i \in \mathcal{T}_1$ for $i = 1, 2$. But now $C_U(D) = \langle w_D \rangle$ is of order 2, so as $Z \leq Z(R)$ we conclude that $U \cap R = \langle w_D \rangle$, contrary to 14.2.8.4. □

LEMMA 14.2.22. *(1) $w_D = 1$.*
(2) $v_D \in R$.
(3) Set $\langle P_i \rangle = \langle \theta_i \rangle$. Then $U_D = U \cap P_1 = U \cap P_2 \leq R$.

PROOF. By 14.2.21, $\theta_i \in \mathcal{T}_2$. Therefore $w_D = 1$, so (1) holds. Further D centralizes v_D and as D is transitive on θ_i, $[v_D, Z_\Theta] \leq \langle w_D \rangle = 1$. Hence $v_D \in C_S(Z) = R$, so (2) holds. Calculating using 14.2.3, $U_D = U \cap P_1 = U \cap P_2$ and D centralizes U_D. Thus as Z_Θ centralizes W_S, $U_D \leq C_S(DZ_\Theta) = C_S(Z) = R$, proving (3). □

LEMMA 14.2.23. *$U_D = U \cap R$.*

PROOF. Assume otherwise; then $U \leq R$ by 14.2.22.3. Thus $[R, t] \leq U \leq R$, so $t \in W_G$. Also $v_D \in R$ by 14.2.22.2, so $X = U\langle v_D \rangle \leq W_R$ with $m(X) = 4$. This is contrary to 14.1.8.5 which says $m(W_R) = m([R, t])$. □

We are now in a position to obtain a contradiction, and hence establish Theorem 14.2.4. Let $u \in U - U_D$; then $[R, u] \leq U_D \leq R$ by 14.2.22.3, so $u \in W_G$. By 14.2.23, $u \notin R$. But $X = U_D\langle v_D \rangle \leq W_R$ by 14.2.22.2, and $m(X) = 3$. This is contrary to 14.1.8.5 which says $m([R, u]) = m(W_R)$.

THEOREM 14.2.24. *Assume Hypothesis 14.2.1. Then $\mathcal{F} = \langle \mathcal{F}_z, \mathcal{F}_u, N_{\mathcal{F}}(W) \rangle$, where $z \in \mathcal{Z}_S$ and $u \in W_S \cap \mathcal{F}^f$ is not in $\mathcal{Z}_S$.*

PROOF. By 15.8.5 and 15.8.6, W_Δ is the orthogonal space for $\tilde{\mu} = \mu/Z(\mu) \cong O_8^+(2)$. In particular M has two orbits on involutions in W: the singular and nonsingular points, with z nonsingular and $u \in W_S - \mathcal{Z}_S$ singular. Hence the theorem follows from Theorem 14.2.4. □

CHAPTER 15

$|\Omega(z)| = 1$ and μ nonabelian

Let $\tau = (\mathcal{F}, \Omega)$ be a quaternion fusion packet. In Chapter 15 we assume $|\Omega(z)| = 1$ for $z \in \mathcal{Z}_S(\tau)$ and $\mu = \mu(\tau)$ is nonabelian. In addition in Hypothesis 15.1.1 we add the assumptions that $\mathcal{F}$ is transitive on Ω and τ satisfies the Inductive Hypothesis. Later in Hypothesis 15.2.1 we strengthen this last condition and require that τ satisfy the Extended Inductive Hypothesis.

Section 15.1 is devoted to general results about τ under Hypothesis 15.1.1.

In section 15.2 we assume Hypothesis 15.2.1 with $\mathcal{F} = \mathcal{F}^{\circ}$, and prove that μ is isomorphic to Weyl(Φ), where Φ is A_n, D_n, or E_n. In Remark 15.2.30 we observe that τ is determined when Φ is A_2, the smallest case.

Then we go on to determine τ when Hypothesis 15.2.1 holds with $\mathcal{F} = \mathcal{F}^{\circ}$ for the various possibilities for Φ. Section 15.3 treats the case where Φ is D_n for $n = 3$ or $n > 4$. In section 15.4 we determine τ when Φ is D_4.

In section 15.5 we assume Φ is A_n with $n \geq 4$ and either Hypothesis 15.2.1 holds or $O^2(\mathcal{F}) = \mathcal{F}^{\circ}$ is the 2-fusion system of a central factor group of $SL_d(q)$. Various preliminary results are established, leading to the generational result in Theorem 15.6.3. Then the case $\Phi = A_n$ is completed in Theorem 15.7.7.

Finally the case $\Phi = E_n$ is handled in Theorem 15.8.16.

15.1. $|\Omega(z)| = 1$

In this section we assume the following hypothesis:

HYPOTHESIS 15.1.1. *(1) $\tau = (\mathcal{F}, \Omega)$ is a quaternion fusion packet such that for all $z \in \mathcal{Z}_S$, $|\Omega(z)| = 1$.*

(2) $\mathcal{F}$ is transitive on Ω.

(3) For $z \in \mathcal{Z}_S \cap \mathcal{F}^f$, $\mathcal{D}^*(z) \neq \mathcal{D}(z)$.

(4) The Inductive Hypothesis is satisfied by τ.

LEMMA 15.1.2. *There exists a unique $\eta \in \eta(\tau)$, so $W = \langle \eta \rangle$ is the unique member of $W(\tau)$ and W is weakly closed in S with respect to $\mathcal{F}$.*

PROOF. By 15.1.1.2, $\mathcal{F}$ is transitive on Ω, while by 15.1.1.3, $\Delta(\tau) \neq \Omega$, so the lemma follows from 3.3.14. □

NOTATION 15.1.3. Let G be a model for $N_{\mathcal{F}}(C_S(W))$ and $M = \langle \Omega^G \rangle$. Set $\mu = \mu(\tau)$, so that $\mu = M/C_M(W)$. Let $K \in \Omega$ such that $z = z(K) \in \mathcal{F}^f$. Set $D = \{d_V : V \in \eta\} \subseteq \mu$.

LEMMA 15.1.4. *(1) Let* $\{\eta_i : 1 \leq i \leq r\}$ *be the orbits of* M *on* η, $\Omega_i = \{J \in \Omega : J \cap W \in \eta_i\}$, $M_i = \langle J_i^M \rangle$ *for* $J_i \in \Omega_i$, $W_i = \langle \eta_i \rangle$, *and* $D_i = \{d_V : V \in \eta_i\}$. *Then* D *is a set of 3-transpositions of* μ, D_i *is a conjugacy class of 3-transpositions of* $\mu_i = Aut_{M_i}(W)$, *and* $M = M_1 \cdots M_r$ *is a central product.*

(2) $W = C_M(W)$ *and* $W_i = M_i \cap W = C_{M_i}(W_i)$, *so* $\mu = M/W$, $\mu_i = M_i/W_i$, *and* $\mu = \mu_1 \times \cdots \times \mu_r$.

(3) $N_G(O(\tau))$ *is transitive on* Ω *and* $\{\Omega_1, \dots, \Omega_r\}$, *and* G *is transitive on* $\{M_1, \dots, M_r\}$.

(4) For $1 \leq i \leq r$, $\tau_i = (\mathcal{F}_{S \cap M_i}(M_i), \Omega_i)$ *is a quaternion fusion packet with* $\mu_i = \mu(\tau_i)$.

(5) For each $1 \leq i \leq r$, $\mu_i \cong Weyl(\Phi)$, *where* Φ *is* A_n, D_n, *or* E_n.

PROOF. By 15.1.1.1, $\Omega(z) = \{K\}$ is of order 1 for each $z \in \mathcal{Z}_S$, so $\mathcal{A}(\tau) = \emptyset$. Hence D is a set of 3-transpositions of μ by 4.2.10.3. Next $M = M_1 * \cdots * M_r$ is a central product by 4.2.10.2, so as M is transitive on η_i, so is M_i. Then D_i is a conjugacy class of 3-transpositions of μ_i by 4.2.8.1. Thus (1) is established.

Set $O = O(\tau)$. By 15.1.1.2, $\mathcal{F}$ is transitive on Ω, so $N_G(O)$ is transitive on Ω by 6.1.6.2. Then (3) follows as $M \trianglelefteq G$, so G permutes the orbits of M on η.

We've seen that $M = M_1 * \cdots * M_r$, so $M_i \trianglelefteq M$ and hence $S_i = S \cap M_i \in Syl_2(M_i)$ and M_i has 2-fusion system $\mathcal{F}_i = \mathcal{F}_{S_i}(M_i)$. Also Ω_i is $\mathcal{F}_i$-invariant, so by 2.5.2, $\tau_i = (\mathcal{F}_i, \Omega_i)$ is a quaternion fusion packet. As M_i is transitive on η_i, $\mathcal{F}_i$ is transitive on Ω_i. Then by 4.3.8.1, $\nu_i = \mu(\tau_i) \cong \text{Weyl}(\Phi)$ for some $\Phi \in \{A_n, D_n, E_n\}$. By 4.3.8.4, $W_i = C_{M_i}(W_i)$. Therefore $\nu_i = M_i/W_i$. Then as in addition, $M = M_1 * \cdots * M_r$, also $W = C_M(W)$, so $\mu = M/W$ and $W_i = W \cap M_i$, so $\mu_i = Aut_{M_i}(W) = M_iW/W \cong M_i/W_i = \nu_i$, and $\mu = \mu_1 \times \cdots \times \mu_r$. That is (2), (4), and (5) hold. □

LEMMA 15.1.5. *Let* $\rho = (\mathcal{Y}, \Gamma)$ *be a subpacket of* τ *with* $\Gamma \subseteq \Omega$.

(1) $\rho^\circ = \rho_1 * \cdots * \rho_s$ *is a central product of quaternion fusion packets* $\rho_i = (\mathcal{Y}_i, \Gamma_i)$ *such that* $\mathcal{Y}_i = \mathcal{Y}_i^\circ$ *is transitive on* Γ_i. *Set* $n_i = |\Gamma_i|$.

(2) For each $J \in \Gamma_i$, $\Gamma_i(z(J)) = \{J\}$.

(3) If $\mu(\rho_i)$ *is abelian then one of the following holds:*

(i) $n_i = 1$ *and either* $\mathcal{Y}_i = \mathcal{O}(z_i)$ *for* $\{z_i\} = \mathcal{Z}(\rho_i)$, *or* $\mathcal{Y}_i \cong L_2[2m]^{(1)}$.

(ii) $m = 8$, $n_i = 2$, *and* $\mathcal{Y}_i \cong \hat{A}E_5$.

(iii) $\mathcal{Y}_i$ *is* $(P)Sp_{2n_i}[m]$, $(P)SL^-_{2n_i}[m]$, *or* $L^-_{2n_i+1}[m]$.

(4) Assume ρ_i *is proper in* τ *and* $\mu(\rho_i)$ *is nonabelian. Then one of the following holds:*

(i) $W(\rho_i) \trianglelefteq \mathcal{Y}_i$.

(ii) $\mu(\rho_i) \cong Weyl(D_n)$. *for some* $n \geq 3$.

(iii) $\mu(\rho_i) \cong Weyl(E_6)$ *and* $\mathcal{Y}_i \cong E_6^+[m]$.

(iv) $\mu(\rho_i) \cong Weyl(E_7)$ *and* $\mathcal{Y}_i \cong E_7[m]$ *or* $\tilde{E}_7[m]$.

(v) $\mu(\rho_i) \cong Weyl(E_8)$ *and* $\mathcal{Y}_i \cong E_8[m]$.

(vi) $\mu(\rho_i) \cong S_3$ *and* $\mathcal{Y}_i \cong L_3^+[m]$, $L_3^+[m]^{(1)}$, $G_2[m]$, *or* M_{12}, *or* $m = 8$ *and* $\mathcal{Y}_i \cong L_3(2)/E_8$.

(vii) $\mu(\rho_i) \cong S_n$ *for some* $n \geq 5$ *and* $\mathcal{Y}_i$ *is a central factor system of* $SL_n^+[m]$.

PROOF. Part (1) follows from 15.1.1.4 and 6.6.6. Part (2) follows as the condition $|\Omega(z)| = 1$ in 15.1.1.1 inherits to subpackets $(\mathcal{E}, \Theta)$ of τ with $\Theta \subseteq \Omega$. If μ is abelian then ρ_i is proper and hence appears as a conclusion in Theorem 1 with an abelian Thompson group by 15.1.1.4. Then (3) follows. (cf. Theorem 7) Thus it remains to prove (4), so we may assume $\rho = \rho_i$ is proper in τ and $\nu = \mu(\rho)$ is nonabelian. Now ρ satisfies Hypothesis 15.1.1 with $\mathcal{Y} = \mathcal{Y}^\circ$, so ρ satisfies the hypothesis of Theorem 1, and hence also satisfies one of its conclusions by 15.1.1.4. Then (4) follows (cf. 15.2.29, Remark 15.2.30, Theorem 15.8.16, and Theorem 15.7.8 for details). □

LEMMA 15.1.6. *Assume $\mu_j \cong Weyl(D_n)$.*

(1) If $n \neq 4$ there exists a unique involution $u_j = u(\tau_j) \in Z_j = Z(M_j)$ such that for each $t \in \mathcal{Z}_S(\tau_j)$, $t' = tu_j \in \mathcal{Z}_S(\tau_j)$.

(2) If $n = 4$ then $Z_j \cong E_4$ and for each $t \in \mathcal{Z}_S(\tau_j)$, $tZ_j = \mathcal{Z}_S(\tau_j)$.

(3) If n is odd then $Z_j \cong \mathbf{Z}_4$ and $M_j \cong \bar{\omega}(D_n, m)$.

PROOF. Let $\bar{G}$ be the universal group $\bar{\omega}(D_n, m)$ defined in 4.3.8.1. By 4.3.8.2 there is a surjective homomorphism $\pi : \bar{G} \to M_j$ with $\bar{W}\pi = W_j$, $\bar{K}_i\pi = K_i \in \Omega_j$, and $\ker(\pi) \leq Z(\bar{G})$ by 4.3.8.3. Set $\bar{z}_i = z(\bar{K}_i)$ and $z_i = \bar{z}_i\pi$. Set $\bar{u}_j = \bar{z}_{n-1}\bar{z}_n$ and $u_j = \bar{u}_j\pi$. By 5.8.2, $\bar{u}_j \in Z(\bar{G})$. As $\Omega(z_i) = \{K_i\}$, $\bar{u}_j \notin \ker(\pi)$ by 5.8.5, so u_j is an involution in Z_j. Also $z_n u_j = z_{n-1} \in \mathcal{Z}_S(\tau_j)$, so as $u_j \in Z_j$ and $\mathcal{F}_{S \cap M_j}(M_j)$ is transitive on Ω_j, it follows that for each $t \in \mathcal{Z}_S(\tau_j)$ we have $t' = tu_j \in \mathcal{Z}_S(\tau_j)$.

By 5.8.4, $Z(\bar{G}) \cong \mathbf{Z}_4$ or E_4 for n odd or even, respectively. In particular if n is odd then $\bar{u}_j$ is the unique involution in $Z(\bar{G})$, so as $\bar{u}_j \notin \ker(\pi)$, (3) holds. Thus we may assume n is even.

Suppose $u \in Z_j - \langle u_j \rangle$ with $tu \in \mathcal{Z}_S(\tau_j)$ for each $t \in \mathcal{Z}_S(\tau_j)$. Set $M_j^+ = M_j/Z_j$. Then $|\Omega(t^+)| > 2$, so by 10.2.13, $n = 4$. This completes the proof of (1). Part (2) follows from 5.8.7.1. □

LEMMA 15.1.7. *Assume $\rho = (\mathcal{Y}, \Gamma)$ is a proper subpacket of τ with $\Gamma \subseteq \Omega$ such that $\mathcal{Y} = \mathcal{Y}^\circ$ is transitive on Γ and $\nu = \mu(\rho)$ is nonabelian. Let $J \in \Gamma$. Then*

(1) $\nu \cong$ Weyl(Σ) for some $\Sigma \in \{A_k, D_k, E_k\}$.

(2) $J \in \Omega_j$ for some $1 \leq j \leq r$, $\Gamma \subseteq \Omega_j$, $\eta(\rho) \subseteq \eta_j$, $\mathcal{Z}(\rho) \subseteq \mathcal{Z}(\tau_j)$, $W(\rho) \leq W_j$, $M(\rho) \leq M_j$, and ν is a D_j-subgroup of μ_j.

PROOF. As $\mathcal{Y} = \mathcal{Y}^\circ$ is transitive on Γ, it follows from 15.1.5.1 that ρ is coconnected. Hence by 15.1.5.4 either (1) holds or $W(\rho) \trianglelefteq \mathcal{Y}$. In the latter case, (1) follows from 4.3.8.1 and the hypothesis that $\mathcal{Y} = \mathcal{Y}^\circ$ is transitive on Γ.

Next $\Gamma \subseteq \Omega$ by hypothesis, and then $\eta(\rho) \subseteq \eta$, so $W(\rho) \leq W$, $M(\rho) \leq M$, and ν is a D-subgroup of μ.

Let E be the set of 3-transpositions generating ν. By (1), $\mathcal{D}(E)^c$ is connected, so as $J \in \Omega_j$, (2) follows from 15.1.4.1. □

LEMMA 15.1.8. *Assume $\rho = (\mathcal{Y}, \Gamma)$ is a proper subpacket of τ with $\Gamma \subseteq \Omega$ such that $\mathcal{Y} = \mathcal{Y}^\circ$ is transitive on Γ and $\nu = \mu(\rho)$ is nonabelian. Then*

(1) For some $1 \leq j \leq r$, $\Gamma \subseteq \Omega_j$ and ν is a D_j-subgroup of μ_j.

(2) If $\mu_j \cong$ Weyl(A_n) $\cong S_{n+1}$ then $\nu \cong S_{k+1}$ for some $k \leq n$.

(3) If $\mu_j \cong$ Weyl(E_n) then $\nu \cong$ Weyl(E_k) for some $6 \leq k \leq n$; or Weyl(A_l) or Weyl(D_l) for some $l < n$; or Weyl(D_8) if $n = 8$.

(4) If $\mu_j \cong$ Weyl(D_n) then $\nu \cong$ Weyl(D_k) or S_k for some $k \leq n$.

(5) Suppose $\nu \cong$ Weyl(D_k) for some $k \geq 4$ and $\mu_j \cong$ Weyl(D_n). Set $\zeta = (N_{\mathcal{Y}}(W(\rho))^\circ, \Gamma)$. Then one of the following holds:

(i) $5 \leq k \leq n$ and $u_j = u(\zeta)$.

(ii) $k = 4$, $n \geq 5$, and $u_j \in Z(M(\rho))$.

(iii) $k = n = 4$ and $M(\rho) = M_j$.

PROOF. Part (1) follows from 15.1.7.2.

Let E be the set of 3-transpositions generating ν. By 15.1.7.1, $\nu \cong \mathrm{Weyl}(\Sigma)$, where $\Sigma \in \{A_k, D_k, E_k\}$. In particular $\mathcal{D}(E)^c$ is connected.

Suppose $\mu_j \cong S_{n+1}$. As ν is a D_j-subgroup of μ_j, ν is generated by a class E of transpositions, so as $\mathcal{D}(E)^c$ is connected, $\nu \cong S_{k+1}$ with $k \leq n$, proving (2).

Suppose $\mu_j \cong \mathrm{Weyl}(E_n)$. Then as ν is a subreflection group of μ_j, (3) follows. Similarly (4) follows.

Assume the setup of (5). By (4), $4 \leq k \leq n$. If $n = 4$ then as $\nu \leq \mu_j$, (5iii) holds, so we may assume $n \geq 5$. Set $Q = O_2(\mu_j)$, $Z = Z(\mu_j)$, $W_0 = \langle \mathcal{Z}_S(\tau_j) \rangle$, and $\mu_j^+ = \mu_j/Q$, so that $\mu_j^+ \cong S_n$ and D_j^+ is the set of transpositions in μ_j^+. Thus, as in the proof of (2), $\nu^+ \cong S_l$ for some l. In particular if $k > 4$ then $O_2(\nu) = \nu \cap Q$. But by 5.8.3.2 and 15.1.6, Q/Z is a group of transvections on W_0 with center u_j, so $\langle u(\zeta) \rangle = [O_2(\nu), W_0(\rho)] \leq [Q, W_0] = \langle u_j \rangle$, so $u(\zeta) = u_j$ and (5i) holds.

Therefore we may take $k = 4$. Here as $\mu^+ \cong S_l$ for some l, we have $l = 3$ or 4, so $P = O_2(\nu) \cap Q$ is of order 32 or 8, and $\langle u_j \rangle = [P, W_0(\rho)] \leq Z(M(\rho))$, so (5ii) holds. This completes the proof of (5). □

LEMMA 15.1.9. *Assume $\rho = (\mathcal{Y}, \Gamma)$ is a proper subpacket of τ with $\Gamma \subseteq \Omega$ such that $\mathcal{Y} = \mathcal{Y}^\circ$ is transitive on Γ and $\nu = \mu(\rho) \cong$ Weyl(D_k) for some $k \geq 3$. Then*

(1) Either

(i) there exists an involution $u \in Z(\mathcal{Y})$ such that for each $t \in \mathcal{Z}_S(\rho)$, $t' = tu \in \mathcal{Z}_S(\rho)$, or

(ii) $k = 4$ and $\mathcal{Y} \cong F_4[m]$ or $E_6^-[m]$.

(2) If $k \neq 4$ or $1 \neq Z(\mathcal{Y})$ is cyclic, then $u = u(\rho)$ is the unique such involution and one of the following holds:

(i) $W(\rho) \vartriangleleft \mathcal{Y}$.

(ii) $\mathcal{Y}$ is $Spin_{2k}^+[m]$, $HSpin_{2k}^+[m]$, $Spin_{2k+1}[m]$, or $Spin_{2k+2}^-[m]$, with $k \neq 4$ in the first case.

(iii) $m = 8$ and $\mathcal{Y} \cong H\hat{A}E_{2k}$, $H\hat{A}E_{2k}$, or $\hat{A}E_{2k+1}$, with $k \neq 4$ in the first case.

(iv) $(n, m) = (3, 8)$ and $\mathcal{Y} \cong \hat{L}_3(2)/E_{64}$ or $\hat{S}p_6(2)$.

(3) If $k = 4$ and $Z(\mathcal{Y})$ is noncyclic, then $Z(\mathcal{Y}) = Z(M(\rho)) = Z \cong E_4$ and each $u \in U^\#$ satisfies (1i). Moreover one of the following holds:

(i) $W(\rho) \vartriangleleft \mathcal{Y}$.

(ii) $\mathcal{Y} \cong Spin_8^+[m]$.

(iii) $m = 8$ and $\mathcal{Y} \cong \hat{A}E_8$, $\hat{L}_3(2)/2^{3+6}$, or $\hat{O}_8^+(2)$.

(4) $\Gamma \subseteq \Omega_j$ for some $1 \leq j \leq r$. Suppose $k \geq 4$ and $\mu_j \cong$ Weyl(D_n). Then one of the following holds:

(i) $5 \leq k \leq n$ and $u_j = u(\rho)$.

(ii) $k = 4$, $n \geq 5$, and $u_j \in Z(\mathcal{Y})$.

(iii) $k = 4$, $n \geq 5$, $u_j \notin Z(\mathcal{Y})$, and $\mathcal{Y} \cong Spin_9[m]$, $Spin_{10}^-[m]$ or $\hat{A}E_9$.

(iv) $k = n = 4$, $M(\rho) = M_j$, and $Z(M_j) \cap Z(\mathcal{Y}) \neq 1$.

(v) $k = 4$ and $\mathcal{Y} \cong F_4[m]$ or $E_6^-[m]$.

PROOF. Set $\mathcal{Y}^+ = \mathcal{Y}/Z(\mathcal{Y})$ and $\rho^+ = (\mathcal{Y}^+, \Gamma^+)$. Let $t \in \mathcal{Z}_S(\rho)$ and set $\zeta = (N_{\mathcal{Y}}(W(\rho))^\circ, \Gamma)$. As $\nu \cong \mathrm{Weyl}(D_k)$ for some $k \geq 3$ and ρ is proper in τ, it follows from 15.1.1.4 (cf. Theorems 15.3.10, 15.3.14, and 15.4.30) that one of the following holds:

(a) $Z(\mathcal{Y}) \neq 1$, $|\Gamma^+(t^+)| = 2$, and $\mathcal{Y}^+ \cong \mathcal{X}/Z(\mathcal{X})$ for some $\mathcal{X}$ listed in (2).

(b) $k = 4$, $Z(\mathcal{Y}) \cong E_4$, $|\Gamma^+(t^+)| = 4$, and $\mathcal{Y}^+ \cong \mathcal{X}/Z(\mathcal{X})$ for some $\mathcal{X}$ listed in (3).

(c) $k = 4$ and $\mathcal{Y} \cong F_4[m]$ or $E_6^-[m]$.

If (c) holds then so does (1ii) and (4v), while (2) and (3) are vacuously true. Thus we may assume (a) or (b) holds.

Suppose (a) holds. Then $\Omega(t^+) = \{K_1^+, K_2^+\}$, where $K_i = \Omega(t_i)$, $t_1 = t$, and $t_2 = tu$ for a uniquely determined involution $u \in Z(\mathcal{Y})$. Thus (1i) and (2) hold in this case. Similarly in case (b), $\Gamma^+(t^+) = \{K_i^+ : 1 \leq i \leq 4\}$, and setting $t_i = z(K_i)$, $Z(\mathcal{Y}) = Z$ with $\{t_i : 1 \leq i \leq 4\} = tZ$. Thus (1i) and (3) hold in this case. In particular (1)-(3) are established.

Finally the first remark in (4) follows 15.1.8.1. If $k \geq 5$ then 15.1.8.5.i holds, so $u_j = u(\zeta)$, while as $k \geq 5$, $u(\zeta) = u(\rho)$ by (2), so that (4i) holds. Thus we may assume $k = 4$. If $n \geq 5$ then 15.1.8.5.ii holds, so $u_j \in Z(M(\rho))$. If $u_j \in Z(\mathcal{Y})$ then (4ii) holds, so we may assume otherwise; in particular $Z(M(\rho)) \neq Z(\mathcal{Y})$, so $Z(\mathcal{Y})$ is cyclic by (3), and then (4iii) holds by (2).

It remains to treat the case $k = n = 4$, where (4iv) holds by (1). □

15.2. The case $r > 1$

In this section we assume the following hypothesis:

HYPOTHESIS 15.2.1. *Hypothesis* 15.1.1 *holds and τ satisfies the Extended Inductive Hypothesis.*

Adopt Notation 15.1.3 and the notation in Lemma 15.1.4. The main result in this section is the following theorem:

THEOREM 15.2.2. *Assume $\mathcal{F} = \mathcal{F}^\circ$. Then $r = 1$.*

Until the proof of Theorem 15.2.2 is complete, we assume that $\mathcal{F} = \mathcal{F}^\circ$ but $r > 1$.

LEMMA 15.2.3. *(1) We can choose $z \in \mathcal{Z}(\tau_1)$, W_1, and $O(\tau_1)$ in $\mathcal{F}^f$.*

(2) As $W_1 \in \mathcal{F}^f$, for each $w \in W_1$ there is a member of $w^{\mathcal{F}} \cap W_1$ in $\mathcal{F}^f$.

PROOF. See the proof of 12.3.17. □

LEMMA 15.2.4. *$\mathcal{F}_z^\circ = \mathcal{O}(z) * \mathcal{C}_z * \mathcal{C}_2 * \cdots * \mathcal{C}_r$ where*

(1) For $2 \leq i \leq r$, $\rho_i = (\mathcal{C}_i, \Omega_i)$ is a quaternion fusion packet, $\eta_i \in \eta(\rho_i)$, and $M_i \in M(\rho_i)$, so $\mu_i = \mu(\rho_i)$.

(2) If $\Phi = A_2$ then $\mathcal{C}_z = 1$, while otherwise $\rho_z = (\mathcal{C}_z, \Omega_z)$ is a quaternion fusion packet, where $\Omega_z = \Omega_1 - \Omega(z)$, $\eta_z \in \eta(\rho_z)$, and $M_z \in M(\rho_z)$, where $\eta_z = \{V \in \eta_1 : V \leq C_S(O(z))\}$ and $M_z = \langle \Omega_z^{C_{M_1}(z)} \rangle$. Further $\mu_z = \mu(\rho_z) \cong \langle C_{D_1}(d_K) - \{d_K\} \rangle$.

(3) If $\Phi = A_n$ with $n \geq 4$ or D_n with $n \geq 5$ then $\mu_z \cong$ Weyl(A_{n-2}) or Weyl(D_{n-2}), respectively. If $\Phi = D_n$ for $n = 3$ or 4 then $\mu_z \cong \mathbf{Z}_2$ or E_8, respectively. If Φ is E_n then $\mu_z \cong S_6$, Weyl(D_6), Weyl(D_7) for $n = 6, 7, 8$, respectively.

(4) If Φ is not A_2 or D_4 then ρ_z is a coconnected component of τ_z°.

PROOF. By 15.1.5.1, $\tau_z^\circ = \xi_1 * \cdots * \xi_a$ where $\xi_j = (\mathcal{F}_j, \Gamma_j)$ with $(\Gamma_j : j)$ a partition of Ω and $(\xi_j : j)$ the coconnected components of τ_z°. For $2 \leq i \leq r$, $\mathcal{F}_{S\cap M_i}(M_i)$ is coconnected and centralizes z, so by 6.6.11, $\mathcal{F}_{S\cap M_i}(M_i) \leq \mathcal{F}_{j(i)}$ for some $j(i)$. Therefore $\Omega_i \subseteq \Gamma_{j(i)}$, while $\Gamma_{j(i)} \subseteq \Omega_i$ by 15.1.8.1. We conclude $\Gamma_{j(i)} = \Omega_i$ and $\mu(\xi_{j(i)}) = \mu_i$. Thus we may choose notation such that for $2 \leq j \leq r$, $\mathcal{F}_j = \mathcal{C}_j$ and $\Gamma_j = \Omega_j$; that is $\xi_j = \rho_j$.

Now for $k \notin \{2, \ldots, r\}$, $\Gamma_k \subseteq \Omega_1$. We may choose notation so that $\xi_1 = (\mathcal{O}(z), \{K\})$, and $\mathcal{C}_z$ is the central product of the $\mathcal{F}_k$ with $r < k \leq a$. In particular (1) and (2) hold.

From 15.1.4.5, $\mu_1 \cong \text{Weyl}(\Phi)$ for some $\Phi \in \{A_n, D_n, E_n\}$. If $\Phi = A_2$ then $\mu_1 \cong S_3$, so $\mathcal{C}_z = 1$. If $\Phi = A_n$ for some $n \geq 4$ then $\mu_1 \cong S_{n+1}$ and $\langle D_{d_K}\rangle \cong S_{n-1}$, so $\mu_z \cong \text{Weyl}(A_{n-2})$. A similar argument establishes (3) when Φ is D_n with $n \geq 3$. Finally if $\Phi = E_n$ then (3) follows from 5.5.4, completing the proof of (3). Part (4) follows from (3). □

NOTATION 15.2.5. Let $I = \{1, \ldots, r\}$ and $I_1 = I - \{1\}$. For $i \in I_1$, let $z_i \in \mathcal{Z}(\rho_i)$ with $z_i \in \mathcal{C}_i^f$, and, when Φ is not A_2, set $\mathcal{D}_i = C_{\mathcal{C}_i}(O(z_i))^\circ$, $\Gamma_i = \Omega_i - \Omega(z_i)$, and $\delta_i = (\mathcal{D}_i, \Gamma_i)$.

LEMMA 15.2.6. *Assume Φ is not A_2. Then for all $i \in I_1$ and $\alpha_i \in \mathfrak{A}(z_i)$ with $z_i\alpha_i = z$, we have $\mathcal{D}_i\alpha_i^* = \mathcal{C}_z$, so $\mathcal{D}_i \cong \mathcal{C}_z$.*

PROOF. For $\beta, \gamma \in \{\delta_i : i \in I_1\} \cup \{\rho_z\}$, write $\beta \lesssim \gamma$ if β is isomorphic to a subpacket of γ.

Let $i \in I_1$. Then there is $\alpha \in \mathfrak{A}(z_i)$ with $z_i\alpha = z$. Observe $\Omega\alpha = \Omega$ and $W\alpha = W$. Then from 15.1.4.4 and 6.6.11, for $s \neq i$ we have $\tau_s\alpha^* = \tau_t$ for some $t > 1$. It follows that for $j \in J = I_1 - \{i\}$, $\rho_j\alpha^* \leq \rho_{j\sigma}$ for some injection $\sigma : J \to I_1$ by 15.2.4 and as ρ_j is coconnected. Similarly $\delta_i\alpha^* \leq \rho_z$ and $\rho_z\alpha^* \leq \delta_k$, where $I_1 = J\sigma \cup \{k\}$. Thus for each $i \in I_1$, $\delta_i \lesssim \rho_z \lesssim \delta_k$.

Let $I_0 = \{i_0 \in I_1 : \delta_{i_0} \cong \rho_z\}$ and observe that $(I_0 - \{i\})\sigma \subseteq I_0$. Suppose that $i \in I_1 - I_0$. Then $I_0\sigma \subseteq I_0$, so as σ is an injection, we have $I_0\sigma = I_0$. But this is a contradiction as also $\rho_z \lesssim \delta_k \lesssim \rho_z$, so that $k \in I_0$. This contradiction completes the proof. □

LEMMA 15.2.7. *Assume $\Phi = D_n$ for some $n \geq 3$. If $n \neq 4$ set $u_i = u(\tau_i)$ for each $i \in I$ and $u = u_1$. If $n = 4$ pick $u \in Z(M_1)^\# \cap \mathcal{F}^f$.*

(1) $\tau_u^\circ = \xi_1 * \cdots * \xi_r$, where $\xi_i = (\mathcal{U}_i, \Omega_i)$ is a quaternion fusion packet with $M_i \in M(\xi_i)$ and $\mu_i = \mu(\xi_i)$. Let P_i be Sylow in $\mathcal{U}_i$ and $P = P_1 \cdots P_r$.

(2) For $i \in I_1$, $\mathcal{U}_i = \mathcal{C}_i$, so $\rho_i = \xi_i$.

(3) If $n \neq 4$ then for $i \in I_1$, $u_i \in u^{\mathcal{F}}$, $u_i = u(\xi_i)$, and for $\alpha_i \in \mathfrak{A}(u_i)$ with $u_i\alpha_i = u$, we have $\mathcal{U}_i\alpha_i^* = \mathcal{U}_1$ and α_i permutes $\mathfrak{U} = \{\mathcal{U}_j : j \in I\}$. In particular $Aut_{\mathcal{F}}(P)$ is transitive on $\mathfrak{U}$.

(4) If $n = 4$ then for $i \in I_1$ and $\beta_i \in \mathfrak{A}(Z_i)$ with $Z_i\beta_i = Z_1$, we have $C_{\mathcal{U}_i}(Z_i)^\circ\alpha_i = C_{\mathcal{U}_1}(Z_1)^\circ$ and α_i permutes $\mathfrak{V} = \{C_{\mathcal{U}_j}(Z_j) : J \in I\}$. In particular $Aut_{\mathcal{F}}(P)$ is transitive on $\mathfrak{V}$.

(5) Suppose $n = 4$. For $i \in I_1$, $v_i \in u^{\mathcal{F}} \cap Z_i$, and $\alpha_i \in \mathfrak{A}(v_i)$ with $v_i\alpha_i = u$, we have $C_{\mathcal{C}_i}(v_i)^\circ\alpha_i^* \leq \mathcal{U}_1$, $\mathcal{U}_1\alpha_i^* \leq \mathcal{C}_{k(1)}$ for some $k(1) \in I_1$, and for each $i \neq j \in I_1$, $\mathcal{C}_j\alpha_i^* \leq \mathcal{C}_{k(j)}$ for some $k(j) \in I_1 - \{k(1)\}$.

PROOF. Part (1) follows from 15.1.5, 15.1.7, and the fact that $u \in Z(M)$; the argument essentially appears in the proof of 15.2.4.

Let $i \in I_1$. As $z \in W_1 \leq C_S(\mathcal{U}_i)$ and ξ_i is coconnected on Ω_i, it follows from 6.6.11 that $\mathcal{U}_i \leq \mathcal{C}_i$. Similarly $u \in K \cdot O(\rho_z) \leq C_S(\mathcal{C}_i)$, so $\mathcal{C}_i \leq \mathcal{U}_i$, establishing (2).

Suppose $v_i \in u^{\mathcal{F}} \cap Z_i$, so that there is $\alpha_i \in \mathfrak{A}(v_i)$ with $v_i\alpha_i = u$. Then $\Omega_i\alpha_i = \Omega_1$ and as v_i centralizes M_i, $C_{\mathcal{C}_i}(v_i)^\circ$ is coconnected, so we have $C_{\mathcal{C}_i}(v_i)^\circ\alpha_i^* \leq \mathcal{U}_1$. Similarly as v_i centralizes $\mathcal{U}_1$, we have $\mathcal{U}_1\alpha_i^* \leq \mathcal{C}_{k(1)}$ for some $k(1) \in I_1$. Also for each $i \neq j \in I_1$, v_i centralizes $\mathcal{C}_j$, so $\mathcal{C}_j\alpha_i^* \leq \mathcal{C}_{k(j)}$ for some $k(j) \in I_1 - \{k(1)\}$. In particular (5) holds.

Suppose $n \neq 4$. Then from parts (2) and (4) of 15.1.9 and the proof of 15.1.9.4 in the case $n = 3$, we have $u(\xi_i) = u_i$, and from 15.1.4.3, $N_G(W)$ is transitive on $\{u_j : j \in I\}$, so $u \in u_i^{\mathcal{F}}$. Hence from the previous paragraph, $\mathcal{C}_i\alpha_i^* \leq \mathcal{U}_1$, $\mathcal{U}_1\alpha_i^* \leq \mathcal{C}_{k(1)}$, and for $i \neq j \in I_1$, $\mathcal{C}_j\alpha^* \leq \mathcal{C}_{k(j)}$. Now, arguing as in the proof of 15.2.6, (3) holds. A similar argument establishes (4). □

LEMMA 15.2.8. *Assume $\Phi = D_4$, for $i \in I_1$ let T_i be Sylow in $\mathcal{C}_i$, and set $Q = \langle T_i : i \in I_1\rangle$, $\tau_Q = (C_{\mathcal{F}}(Q), \Omega_1)$, and $\mathcal{U} = C_{\mathcal{F}}(Q)^\circ$. Let T_1 be Sylow in $\mathcal{U}$ and $T = T_1Q$. Then*

(1) τ_Q is a quaternion fusion packet with $M_1 \in M(\tau_Q)$ and $\mu_1 = \mu(\tau_Q)$. Further $\mathcal{U}$ is transitive on Ω_1.

(2) If $Z(\mathcal{U})$ is noncyclic then $Z(\mathcal{U}) = Z_1$ and for each $i \in I_1$, $Z(\mathcal{C}_i) = Z_i$. Moreover for each $u \in Z_1^\# \cap \mathcal{F}^f$, $\{\tau_Q^\circ, \rho_i : i \in I_1\}$ is the set of coconnected components of τ_u°.

(3) If $1 \neq Z(\mathcal{U}) = \langle u\rangle$ then for each $i \in I_1$, $Z(\mathcal{C}_i) = \langle v_i\rangle$ with $v_i \in u^{\mathcal{F}}$, $\{\tau_Q^\circ, \rho_i : i \in I_1\}$ is the set of coconnected components of τ_u°, and $\alpha_i \in \mathfrak{A}(v_i)$ with $v_i\alpha_i = u$ permutes $\mathfrak{U} = \{\mathcal{U}, \mathcal{C}_j : j \in I_1\}$. In particular $Aut_{\mathcal{F}}(T)$ is transitive on $\mathfrak{U}$.

(4) If $Z(\mathcal{U}) = 1$ then for each $i \in I_1$, $\mathcal{U} \cong \mathcal{C}_i \cong F_4[m]$ or $E_6^-[m]$. Further for $\beta_i \in \mathfrak{A}(W_i)$, $\mathcal{C}_i\beta_i^* = \mathcal{U}$, so β_i permutes $\mathfrak{U}$. In particular $Aut_{\mathcal{F}}(T)$ is transitive on $\mathfrak{U}$.

(5) $C_{\mathcal{U}}(z)^\circ = \mathcal{O}(z)\mathcal{C}_z$, so $C_{\mathcal{U}}(z)^\circ \cong \mathcal{O}(z_i)\mathcal{D}_i$ for each $i \in I_1$.

PROOF. By 2.5.2, τ_Q is a quaternion fusion packet. By parts (1) and (2) of 15.2.7, Q centralizes M_1, so (1) holds as usual. Similarly by 15.2.4, Q centralizes $\mathcal{O}(z)\mathcal{C}_z$, so $\mathcal{O}(z)\mathcal{C}_z \leq C_{\mathcal{U}}(z)^\circ$, and the opposite inclusion follows from 15.2.4. Then 15.2.6 completes the proof of (5).

Let $i \in I_1$. Suppose $Z(\mathcal{U})$ is noncyclic; then by 15.1.9.3, $Z_1 = Z(\mathcal{U})$ and $\mathcal{U}$ is described in 15.1.9.3. Choose $u \in Z_1^\# \cap \mathcal{F}^f$ and define $\mathcal{U}_1$ as in 15.2.7. Then $\mathcal{U}_1 \leq \mathcal{U}$, while as $u \in Z_1 = Z(\mathcal{U})$ we also have the opposite inclusion. Thus $\mathcal{U}_1 = \mathcal{U}$ and $Z_1 = Z(\mathcal{U}_1)$. As this holds for each such choice of u, it follows from 15.2.7.5 that if $Z(\mathcal{C}_i) \neq 1$ then for v_i an involution in $Z(\mathcal{C}_i)$ and $\alpha_i \in \mathfrak{A}(v_i)$ with $v_i\alpha_i = u$, we have $\mathcal{C}_i\alpha_i^* \leq \mathcal{U}_1 = \mathcal{U}$, so $Z_i = Z(\mathcal{C}_i)$. On the other hand if $Z(\mathcal{C}_i) = 1$ then by 15.1.9.1, $\mathcal{C}_i \cong F_4[m]$ or $E_6^-[m]$. Then from 5.5.3.1, $\mathcal{D}_i \cong Sp_6[m]$ or $SL_6^-[m]$, respectively. Then by (5), $\mathcal{U} \cong \mathcal{C}_i$, a contradiction. Therefore $Z_i = Z(\mathcal{C}_i)$ for each $i \in I_1$, so that (2) holds.

Suppose next that $1 \neq Z(\mathcal{U}) = \langle u\rangle$ is of order 2. Then from 15.1.9.2, $\mathcal{U} \cong Spin_9[m]$, $Spin_{10}^-[m]$, or $\hat{A}E_9$. Hence by (5), $\mathcal{C}_z \cong Spin_5[m]$, $Spin_6^-[m]$, or $\hat{A}E_5$, as is $\mathcal{D}_i$ for each $i \in I_1$. This implies that $\mathcal{C}_i \cong \mathcal{U}$. Let v_i generate $Z(\mathcal{C}_i)$ and $\alpha_i \in \mathfrak{A}(v_i)$ with $v_i\alpha_i \in U_1$; by 15.2.7.5, $\mathcal{C}_i\alpha_i^* \leq \mathcal{U}$, so it follows that $v_i\alpha_i = u$. Now (3) holds.

This leaves the case $Z(\mathcal{U}) = 1$, where by 15.1.9.1, $\mathcal{U} \cong F_4[m]$ or $E_6^-[m]$. Then by (5), for each $i \in I_1$, $\mathcal{D}_i \cong \mathcal{C}_z$ is isomorphic to $Sp_6[m]$ or $SL_6^-[m]$, respectively, so $\mathcal{C}_i \cong \mathcal{U}$. Let $O_1' = \langle \Omega - \Omega_1\rangle$ and $\mathcal{Y} = C_{\mathcal{F}}(O_1')^\circ$. From 15.2.9.2, $\mathcal{Y} \cong \mathcal{C}_i$. But

$O_1' \leq Q$, so $\mathcal{U} \leq \mathcal{Y}$ and then as $\mathcal{U} \cong \mathcal{C}_i \cong \mathcal{Y}$, we have $\mathcal{Y} = \mathcal{U}$. Let $\beta_i \in \mathfrak{A}(W_i)$. Then β_i is also in $\mathfrak{A}(O_i)$, so 15.2.9.2 completes the proof of (4). □

LEMMA 15.2.9. *For $k \in I$ let $O_k = O(\tau_k)$ and $O_k' = \prod_{k \neq j \in I} O_j$. Set $\mathcal{Y} = C_{\mathcal{F}}(O_1')^\circ$ and $\theta = (\mathcal{Y}, \Omega_1)$. For $i \in I_1$ set $\mathcal{O}_i = O^2(C_{\mathcal{C}_i}(Z(O_i)))$ and $\mathcal{O}_1 = O^2(C_{\mathcal{Y}}(Z(O_1)))$. Then*

(1) $O_i' \in \mathcal{F}^f$ and θ is a quaternion fusion packet with $M_1 \in M(\theta)$ and $\mu_1 = \mu(\theta)$. Further $\mathcal{Y}$ is transitive on Ω_1.

(2) For $i \in I_1$ and $\alpha_i \in \mathfrak{A}(O_i)$, $\mathcal{C}_i\alpha_i^* = \mathcal{Y}$ and α_i permutes $\mathfrak{U} = \{\mathcal{Y}, \mathcal{C}_i : i \in I_1\}$.

(3) For $i \in I_1$ let T_i be Sylow in $\mathcal{C}_i$ and T_1 Sylow in $\mathcal{Y}$. Set $T = \prod_{j \in I} T_j$. Then $Aut_{\mathcal{F}}(T)$ is transitive on $\mathfrak{U}$.

(4) Let $S_k = S \cap M_k$ and $S_k' = \prod_{k \neq j \in I} S_j$. Assume S_1 centralizes $\mathcal{C}_i$ for some $i \in I_1$. Then $\mathcal{Y} = C_{\mathcal{F}}(S_1')^\circ$.

(5) Suppose $T_2 = S_2$. Then $T_j = S_j$ for each $j \in I$ and if S_1 centralizes $\mathcal{C}_i$ for some $i \in I_1$ then $\mathcal{F}$ contains a central product $\mathcal{Y} * \mathcal{C}_2 * \cdots * \mathcal{C}_r$.

(6) Suppose $T_2 = S_2$ and $\mathcal{C}_2 = \langle S_2\mathcal{O}_2, \mathcal{F}_{S_2}(M_2)\rangle$. Then $\mathcal{F}$ contains a central product $\mathcal{Y} * \mathcal{C}_2 * \cdots * \mathcal{C}_r$.

PROOF. As $W_1 \in \mathcal{F}^f$, also $O_1' \in \mathcal{F}^f$. By 2.5.2, θ is a quaternion fusion packet. As O_1' centralizes M_1, $M_1 \in M(\theta)$ and $\mu_1 = \mu(\theta)$, so (1) holds.

As in the proof of 15.2.6, for $\beta, \gamma \in \{\theta, \rho_i : i \in I_1\}$, write $\beta \lesssim \gamma$ if β is isomorphic to a subpacket of γ.

Let $i \in I_1$ and $\gamma_i \in \mathfrak{A}(O_i')$. Then $\mathcal{C}_i\gamma_i^* \leq \mathcal{Y}$, so $\rho_i \lesssim \theta$. Let $\alpha_i \in \mathfrak{A}(O_i)$. Then O_i centralizes $\mathcal{Y}$ and $\mathcal{C}_j$ for $j \in I_1 - \{i\}$, while $O_i\alpha_i = O_1$, so $\mathcal{Y}\alpha_i^* \leq \mathcal{C}_{k(1)}$ and $\mathcal{C}_j\alpha_i^* \leq \mathcal{C}_{k(j)}$ for some $k(1), k(j) \in I_1$. Therefore $\theta \lesssim \rho_{k(1)}$ and $\rho_j \lesssim \rho_{k(j)}$. Hence $\theta \cong \rho_{k(1)}$.

Let $I_0 = \{i_0 \in I_1 : \rho_{i_0} \cong \theta\}$; claim $I_0 = I_1$. Suppose that $i \in I - I_0$. Then from the previous paragraph, $k(1)$ and $k(j)$ are in I_0 for each $j \in I_0$, so $k(1) \in I_0 = k(I_0)$, a contradiction that establishes the claim.

By the claim, for each $i \in I_1$ we have $\mathcal{C}_i \cong \mathcal{Y}$, $\mathcal{Y}\alpha_i^* = \mathcal{C}_{k(1)}$, and $\mathcal{C}_j\alpha_i^* = \mathcal{C}_{k(j)}$. Also O_1' centralizes $\mathcal{C}_i\alpha_i^*$ as O_i' centralizes $\mathcal{C}_i$. Thus $\mathcal{C}_i\alpha_i^* \leq \mathcal{Y}$, and then as $\mathcal{C}_i \cong \mathcal{Y}$ we have $\mathcal{C}_i\alpha_i^* = \mathcal{Y}$. Therefore (2) holds.

By (2), α_i permutes $\mathfrak{T} = (T_j : j \in I\}$, so α_i acts on T. Further $T_i\alpha_i = T_1$, so (3) holds.

Similarly assume the setup of (4) and let $\mathcal{X} = C_{\mathcal{F}}(S_1')^\circ$. Then $\zeta = (\mathcal{X}, \Omega_1)$ is a quaternion fusion packet and as S_1 centralizes $\mathcal{C}_i$, as does S_k for each $k \in I_1 - \{i\}$, we conclude that S_i' centralizes $\mathcal{C}_i$. Therefore for $\beta_i \in \mathfrak{A}(S_i')$, $\mathcal{C}_i\beta_i^* \leq \mathcal{X}$. But $\mathcal{X} \leq \mathcal{Y} \cong \mathcal{C}_i$, so $\mathcal{X} = \mathcal{Y}$ and (4) holds.

Suppose $T_2 = S_2$. Then by (3), $T_j = S_j$ for each $j \in I$. Assume in addition that S_1 centralizes $\mathcal{C}_i$ for some $i \in I_1$. Then by (4), S_1' centralizes $\mathcal{Y}$, so by (2) and (3), S_i' centralizes $\mathcal{C}_i$ for each $i \in I_1$. Now (5) follows from 2.2 in [**Asc11**].

Assume the setup of (6). We claim that S_1 centralizes $\mathcal{C}_2$. By the transitivity in (3), it suffices to show that S_1' centralizes $\mathcal{Y}$. But S_1' centralizes both $S_1\mathcal{O}_1$ and $\mathcal{F}_{S_1}(M_1)$, so by our assumption in (6), it centralizes the subsystem $\mathcal{Y}$ they generate. This establishes the claim, and by the claim and (5), $\mathcal{F}$ contains a central product $\mathcal{C}_0 = \mathcal{Y} * \mathcal{C}_2 * \cdots * \mathcal{C}_r$. □

NOTATION 15.2.10. Recall that $\{\mathcal{C}_i : i \in I_1\}$ and $\{\rho_i : i \in I_1\}$ are defined in 15.2.4. If Φ is D_n with $n \neq 4$, define $\mathcal{C}_1 = \mathcal{U}_1$ and $\rho_1 = \xi_1$ as in aC.7.1. If Φ is D_4

define $\mathcal{C}_1 = \mathcal{U}$ and $\rho_1 = (\mathcal{C}_1, \Omega_1)$ as in 15.2.8. In the remaining cases by 15.2.4.4, Φ is A_2, A_n with $n \geq 4$, or E_n with $n \in \{6,7,8\}$. In those cases define $\mathcal{C}_1 = \mathcal{Y}$ and $\rho_1 = \theta$ as in 15.2.9. In any event for $j \in I$, let T_j be Sylow in $\mathcal{C}_i$ and set $T = T_1 \cdots T_r$.

LEMMA 15.2.11. *(1)* $\mathcal{C}_1 = \mathcal{Y}$.
(2) $Aut_{\mathcal{F}}(T)$ *is transitive on* $\{\mathcal{C}_j : j \in I\}$.
(3) $\mathcal{F}$ *contains a central product* $\mathcal{C}_0 = \mathcal{C}_1 * \cdots * \mathcal{C}_r$.
(4) If Φ *is* A_n *for some* $n \geq 4$ *or* E_n *for some* $n \in \{6,7,8\}$ *then for* $j \in I$, $S_i = T_i$ *and* $\mathcal{C}_i = \langle S_i\mathcal{O}_i, \mathcal{F}_{S_i}(M_i)\rangle$.

PROOF. Part (1) holds by definition in 15.2.10 unless Φ is D_n. In that case we will see that (3) holds, so that $\mathcal{C}_1$ centralizes $\mathcal{O}_1'$, and hence $\mathcal{C}_1 \leq \mathcal{Y}$. But from (2) and 15.2.9.3, $\mathcal{C}_1 \cong \mathcal{C}_2 \cong \mathcal{Y}$, so (1) holds in this case too.

Suppose first that Φ is D_n with $n \neq 4$. From parts (1) and (2) of 15.2.7, $u = u_1 = u(\tau_1)$ and $\tau_u^\circ = \rho_1 * \cdots * \rho_r$, so that (3) holds. Part (2) follows from 15.2.7.3.

Suppose next that Φ is D_4. In cases (2) and (3) of 15.2.8, we again have $\tau_u^\circ = \rho_1 * \cdots * \rho_r$ for some $u \in Z_1^\#$, so that (3) holds, while (2) follows from parts (2) and (3) of 15.2.8. Thus we may assume we are in case (4) of 15.2.8, so (2) holds by 15.2.8.4, and $\mathcal{C}_j \cong F_4[m]$ or $E_6^-[m]$.

By 5.5.6, $\mathcal{C}_1$ has two classes of involutions with representatives u and z, where $u \in Z_1$ and $\mathcal{E} = C_{\mathcal{C}_1}(u)^\circ \cong Spin_9[m]$ or $Spin_{10}^-[m]$, respectively. In the first case T_1 is Sylow in $\mathcal{E}$ by 5.2.7.3 and 5.2.8, so as $\tau_u^\circ = (\mathcal{E}, \Omega_1) * \rho_2 * \cdots * \rho_r$, it follows that T_1 centralizes $\mathcal{C}_i$ for each $i \in I_1$. Then (3) follows from (2) and 2.2 in [**Asc11**]. Therefore we may assume $\mathcal{C}_1$ is $E_6^-[m]$, where is remains to show that T_2 centralizes $\mathcal{C}_1$. Let P_1 be Sylow in $\mathcal{E}$; then again P_1 centralizes $\mathcal{C}_i$ for $i \in I_1$, but this time $|T_1 : P_1| = 2$ by 5.2.7 and 5.2.9. For $i \in I_1$, define v_i as in 15.2.7.5 and let P_i be Sylow in $C_{\mathcal{C}_i}(v_i)^\circ$. Then $[T_1, T_i] \leq P_1 \cap P_i = 1$. By 15.4.25.2, $\mathcal{C}_1 = \langle T_1\mathcal{E}, N_{\mathcal{C}_1}(U_1)\rangle$, where $U_j = Z(M_j)$. Let $\mathfrak{M}_j$ be a model for $N_{\mathcal{C}_j}(U_j)$ and G_0 the kernel of the action of G on $\{W_j : j \in I\}$; we may take $\mathfrak{M}_j \leq G$ and then we have $\mathfrak{M}_j \trianglelefteq G_0$. We've seen that T_2 centralizes $T_1\mathcal{E}$ and $\mathfrak{M}_2$ centralizes P_1, so $\mathfrak{M}_2$ centralizes $\langle P_1^{\mathfrak{M}_1}\rangle = \mathfrak{M}_1$ as $\mathfrak{M}_2 \trianglelefteq G_0$. Then T_2 centralizes $\mathcal{C}_1$, completing the proof of the lemma when Φ is D_4.

If Φ is A_2 then $\mathcal{C}_1 = \langle T_1\mathcal{O}_1, \mathcal{F}_{T_1}(T_1M_1)\rangle$ and $|T_1 : S_1| \leq 2$. Thus arguments above establish (3) in this case, while (2) follows from 15.2.9.3. Thus we may assume Φ is A_n for some $n \geq 4$ or E_n for some $n \in \{6,7,8\}$. Therefore by 5.2.5, 5.2.9, 5.2.10, and 5.5.8 we have $T_1 = S_1$, while by 15.6.3 and 15.8.12.3 we have $\mathcal{C}_1 = \langle T_1\mathcal{O}_1, \mathcal{F}_{T_1}(T_1M_1)\rangle$, so we can complete the proof as usual. □

LEMMA 15.2.12. *Let* $X \leq T_1$ *such that* X *is not normal in* $\mathcal{F}$. *Then either*
(1) $X^{\mathcal{F}} \cap \mathcal{F}^f \cap T_1 \neq \emptyset$, *or*
(2) $X \leq Z(\mathcal{C}_1)$ *and* $X^{\mathcal{F}} \cap \mathcal{F}^f \cap Z(\mathcal{C}_0) \neq \emptyset$.

PROOF. Let $P = O_1'$, $\alpha \in \mathfrak{A}(X)$, $Y = X\alpha$, and $\mathcal{X} = N_{\mathcal{F}}(Y)$; then by 15.2.11.3, $P\alpha \leq S_1')\alpha \leq C_S(Y)$. Let $\beta \in \mathfrak{A}(P\alpha)$ with $P\zeta = P$, where $\zeta = \alpha\beta$. Then ζ is a $N_{\mathcal{F}}(P)$-map, so as $\mathcal{C}_1 \trianglelefteq N_{\mathcal{F}}(P)$ by 15.2.11.1, we have $X\zeta \leq T_1$. Hence if $N_S(Y) \leq N_S(P\alpha)$, we have $\zeta \in \mathfrak{A}(X)$ and (1) holds. Therefore we may assume that for some $s \in N_S(Y)$, $P\alpha c_s \neq P\alpha$.

Now as X is not normal in $\mathcal{F}$, it follows from 15.1.5 that for $i \in I_1$, $\rho_i\alpha^* \leq \theta_i = (\mathcal{T}_i, \Omega_i)$ a coconnected component of $\mathcal{X}$. As P is weakly closed in a Sylow group of $\mathcal{T}_2 \cdots \mathcal{T}_r$, it follows that for some $i \in I_1$, $\theta_i c_s^* \neq \theta_k$ for any $k \in I_1$. Therefore $\theta_i c_s^* = \theta_1 = (\mathcal{T}_1, \Omega_1)$ is a component of $\mathcal{X}$. Now $\mathcal{T}_1\beta^* \leq \mathcal{Y}$ with $\mathcal{T}_1 \cong \mathcal{T}_i \geq \mathcal{C}_i\alpha^* \cong \mathcal{Y}$, so we conclude that $\mathcal{T}_1\beta^* = \mathcal{Y}$. Then as $\mathcal{T}_1\beta^*$ centralizes $Y\beta = X\zeta$, we have $X\zeta \leq Z(\mathcal{Y})$, so as ζ is a $N_{\mathcal{F}}(P)$-map and $\mathcal{Y} \trianglelefteq N_{\mathcal{F}}(P)$, we have $X \leq Z(\mathcal{Y})$. Set $U = Z(\mathcal{C}_0)$. Now S acts on U (cf. 15.2.17), so $U \in \mathcal{F}^f$. Let $\gamma \in \mathfrak{A}(X\zeta)$ with $X\zeta\gamma = Y$. As $r > 1$ and $\mathcal{C}_0\gamma^* \leq \mathcal{X}$, for each $j \in I$, $\rho_j\gamma^* = \theta_j$, so $\mathcal{C}_0\gamma^* \trianglelefteq \mathcal{X}$, and hence $N_S(Y) \leq N_S(U\gamma)$. Now for $\delta \in \mathfrak{A}(U\gamma)$ with $U\gamma\delta = U$, $\gamma\delta \in \mathfrak{A}(X)$, so that (2) holds. □

LEMMA 15.2.13. *Let $X \leq T_1$ with $X \in \mathcal{F}^f$ and assume $\mathcal{C}' = \mathcal{C}_2 \cdots \mathcal{C}_r \trianglelefteq N_{\mathcal{F}}(X)$.*

(1) $O^{2'}(N_{\mathcal{C}_1}(X)) \trianglelefteq N_{\mathcal{F}}(X)$.

(2) Each coconnected component of $N_{\mathcal{F}}(X)$ not in $\{\mathcal{C}_i : i \in I_1\}$ is a subcomponent of $(N_{\mathcal{C}_1}(X), C_{\Omega_1}(X))$.

PROOF. Let $\mathcal{X} = N_{\mathcal{F}}(X)$, $\mathcal{E} = O^{2'}(N_{\mathcal{C}_1}(X))$, and E Sylow in $\mathcal{E}$. Then $E \leq C_S(\mathcal{C}')$, so $\mathcal{E} = [E]_{\mathcal{E}} \leq \mathcal{D} = C_{\mathcal{X}}(\mathcal{C}')$. But $\mathcal{D} \leq C_{\mathcal{X}}(O_1')$ and $\mathcal{E} \trianglelefteq C_{\mathcal{X}}(O_1')$, so by 1.1.10 in [**Asc19**], we have $\mathcal{E} \trianglelefteq \mathcal{D}$. Let D be Sylow in $\mathcal{D}$; as $\mathcal{X}$ is saturated, $Aut_{\mathcal{X}}(D) = Aut_{\mathcal{G}}(D)$, where $\mathcal{G} = N_{\mathcal{X}}(DO_1'$. Then since E is strongly closed in D with respect to $\mathcal{G}$, it follows from 7.4 in [**Asc11**] that (1) holds. Then (1) implies (2). □

LEMMA 15.2.14. *Let $\mathfrak{X}$ be the set of involutions x of T_1 in $\mathcal{F}^f$ but not in $Z(\mathcal{C}_0)$. Then*

(1) for each $i \in I_1$, ρ_i is a coconnected component of $\mathcal{F}_x$.

(2) $\mathcal{C}' = \mathcal{C}_2 \cdots \mathcal{C}_r \trianglelefteq \mathcal{F}_x$.

PROOF. Assume (1) fails. Then for some $i \in I_1$, $\rho_i < \theta_i = (\mathcal{T}_i, \Omega_i)$ a component of τ_x°. Observe

(a) $x \notin z^{\mathcal{F}}$ and $r = 2$.

If $r > 2$ or $x \in z^{\mathcal{F}}$ then some $t \in z^{\mathcal{F}} \cap C_S(x)$ centralizes $\mathcal{T}_i$. Let $\alpha \in \mathfrak{A}(t)$ with $t\alpha = z$. Then $\theta_i\alpha^* \leq \rho$ for some cocomponent $\rho = (\mathcal{C}, \Gamma)$ of τ_z. This is a contradiction as $\mathcal{C}$ is isomorphic to a subsystem of $\mathcal{C}_i$.

(b) $\mathcal{C}_2$ is not $\mathcal{F}_{S_2}(M_2)$.

Assume otherwise. By 15.2.13.1, $O^{2'}(C_{\mathcal{C}_1}(z)) \trianglelefteq \mathcal{F}_z$. But $W_1 \trianglelefteq C_{\mathcal{C}_1}(z)$, so as W_1 is weakly closed in $C_S(z)$ with respect to $\mathcal{F}_z$, we conclude that $W_1 \trianglelefteq \mathcal{F}_z$. As $\mathcal{F}_{S_2}(M_2) \neq \mathcal{T}_2$, it follows from 15.1.5 that W_2 is not normal in $C_{\mathcal{T}_2}(z_2)$, a contradiction.

(c) $\mathcal{T}_2$ has more than one class of involutions.

If not, as $x^{\mathcal{F}} \cap T_2 \neq \emptyset$, we have $x \in z^{\mathcal{F}}$, contrary to (a).

(d) Φ is not A_2.

Assume otherwise. We have the proper containment $\mathcal{C}_2 < \mathcal{T}_2$. By (b), $\mathcal{C}_2$ is not $\mathcal{F}_{S_2}(M_2)$, so it follows from 15.1.5.4.vi that $\mathcal{T}_2$ is $G_2[m]$, contrary to (c).

(e) Φ is D_n for some $n \geq 3$.

If not then by (d) and 15.1.4.5, Φ is A_n for some $n \geq 4$ or E_n for some $n \in \{6, 7, 8\}$. But then by 15.1.5.4 and (b), we have no proper containment $\mathcal{C}_2 < \mathcal{T}_2$.

(f) If Φ is D_4 then $\mathcal{C}_2$ is not $F_4[m]$ or $E_6^-[m]$.

Assume otherwise. As $\mathcal{C}_2 < \mathcal{T}_2$ it follows from 15.1.9.1 that $\mathcal{C}_2$ is $F_4[m]$ and $\mathcal{T}_2$ is $E_6^-[m]$. Then by 5.5.3.1, $C_{\mathcal{T}_2}(z_2)$ has a coconnected component $\mathcal{E} \cong SL_6^-[m]$. But $\mathcal{C}_z \cong Sp_6[m]$ by 5.5.3.1, so $\mathcal{E}$ is isomorphic to no subsystem of $\mathcal{C}_z$, a contradiction.

(g) Φ is D_4.

If not then by (e), Φ is D_n for some $n \neq 4$. Then by parts (1) and (2) of 15.1.9 and 15.1.6, $u_2 = u(\rho_2) = u(\theta_2)$, whereas by 15.2.7, $\mathcal{C}_2$ is a component of $\mathcal{F}_{u_2}$, contradicting $\mathcal{C}_2 < \mathcal{T}_2$.

(h) $\mathcal{T}_2$ is $F_4[m]$ or $E_6^-[m]$.

Suppose otherwise; then by 15.1.9.1, there is an involution $u \in Z(\mathcal{T}_2) \cap Z_2$. Hence from 15.2.8, $\mathcal{T}_2$ is isomorphic to a subsystem of $\mathcal{U} \cong \mathcal{C}_2$, contradicting $\mathcal{C}_2 < \mathcal{T}_2$.

(i) $\mathcal{C}_2 = C_{\mathcal{T}_2}(u)^\circ$ for some involution u in Z_2.

By (f) and 15.1.9.1, there is an involution $u \in Z(\mathcal{C}_2)$, so $\mathcal{C}_2 \leq C_{\mathcal{T}_2}(u)^\circ$. On the other hand by aC.8, $C_{\mathcal{T}_2}(u)^\circ$ is isomorphic to a subsystem of $\mathcal{U}$, so the remark follows as $\mathcal{U} \cong \mathcal{C}_2$.

We now obtain a contradiction, establishing (1). As $x \in T_1 \leq C_1$, x is conjugate in $\mathcal{F}$ to z_2 or u, the representatives for the two classes of involutions in $\mathcal{T}$ by (h) and 5.5.6.1. By (a), x is not conjugate to z_2, so $x \in u^{\mathcal{F}}$. Therefore by (i) there is a component $\mathcal{D}$ of $\mathcal{F}_x^\circ$ isomorphic to $\mathcal{C}_2$. But now there is a conjugate of z in $\mathcal{D}$ centralizing $\mathcal{T}_2$, a contradiction.

We've established (1). Thus if (2) fails and $\mathcal{D}$ is the coconnected component of $\mathcal{F}_x$ containing $\mathcal{F}_{S_1}(M_1)$, then $\mathcal{D}$ is conjugate to some $\mathcal{C}_i$ in $\mathcal{F}_x$. Therefore $\mathcal{D} \cong \mathcal{C}_i$ and $\mathcal{D} \leq \mathcal{Y} \cong \mathcal{C}_i$, so $\mathcal{D} = \mathcal{Y}$. Therefore $x \in Z(\mathcal{C}_0)$, contrary to $x \in \mathfrak{X}$. Therefore (2) holds. □

LEMMA 15.2.15. $Z(\mathcal{C}_1) \neq 1$.

PROOF. Assume $Z(\mathcal{C}_1) = 1$; we claim that Hypothesis 8.1.8 holds. Then by 8.1.10, $\mathcal{F}^\circ = \mathcal{C}_0$, for our usual contradiction. First, Hypothesis 8.1.7 is satisfied by 15.2.11.3, so it remains to check the additional four conditions in 8.1.8. Condition (1) holds by 15.2.11.2. By 15.2.12, we may choose the involution x in 8.1.14 to be in $\mathfrak{X}$, and we must show that ρ_i is a component of τ_x for each $i \in I_1$, and all other components are subcomponents of $(N_{\mathcal{C}_1}(x)^\circ, C_{\Omega_1}(x))$. The first condition follows from 15.2.14.1. Then the second condition holds by 15.2.14.2 and 15.2.13.2. □

NOTATION 15.2.16. Set $U = Z(\mathcal{C}_0)$. Observe that by 15.2.11.3, $U = \langle Z(\mathcal{C}_j) : j \in I \rangle$, so $U \neq 1$ by 15.2.15. Set $\mathcal{U} = N_{\mathcal{F}}(U)$.

LEMMA 15.2.17. *(1)* $U \trianglelefteq S$, *so* $U \in \mathcal{F}^f$.

(2) S *permutes* $\{T_j : j \in I\}$ *and* $\{\mathcal{C}_j : j \in I\}$.

(3) $N_{\mathcal{F}}(O_1) \leq \mathcal{U}$.

PROOF. As $WO(\tau) \trianglelefteq S$, S permutes $\{O_j : j \in I\}$. In particular for $s \in S$, $O_1' c_s = O_i'$ for some $i \in I$, so $\phi = c_s \alpha_i$ is a $N_{\mathcal{F}}(O_1')$-map by 15.2.9.2. As $\mathcal{Y} \trianglelefteq N_{\mathcal{F}}(O_1')$, we have $T_1\phi = T_1$ and $\mathcal{Y}\phi^* = \mathcal{Y}$, so by 15.2.9.2, $T_1^s = T_1 c_s = T_1\alpha_i^{-1} = T_i$. As $z \in O_1$, $\{\rho_k : k \in I_1\}$ are the coconnected components of $C_{\mathcal{F}}(O_1)$, so $\mathcal{C}' = \mathcal{C}_2 \cdots \mathcal{C}_r \trianglelefteq \mathcal{N} = N_{\mathcal{F}}(O_1)$ and ϕ permutes $\{T_j : j \in I\}$ and $\{\mathcal{C}_j : j \in I\}$, as does α_i by 15.2.9.2, so $c_s = \phi\alpha_i^{-1}$ does too. This proves (2). By (2), S acts on $\langle Z(\mathcal{C}_j) : j \in I \rangle = U$, so (1) holds.

As $\mathcal{C}' \trianglelefteq \mathcal{N}$, we have $\mathcal{N} = \langle N_S(O_1)\mathcal{C}', N_{\mathcal{N}}(O_1') \rangle$ by 1.3.2 in [**Asc19**]. Thus $Z(\mathcal{Y}) \trianglelefteq \mathcal{N}$ by 15.2.11.1, so $U \trianglelefteq \mathcal{N}$, completing the proof of (3). □

LEMMA 15.2.18. *(1)* $Z(\mathcal{F}) = 1$.

(2) $O_2(\mathcal{F}) = 1$.

PROOF. Assume (1) fails and set $\mathcal{F}^+ = \mathcal{F}/Z(\mathcal{F})$. Then $\tau^+ = (\mathcal{F}^+, \Omega^+)$ is a quaternion fusion packet by 3.3.2. By 15.1.1.2, $\mathcal{F}^+$ is transitive on Ω^+, and as $\mathcal{F} = \mathcal{F}^\circ$, also $\mathcal{F}^+ = (\mathcal{F}^+)^\circ$. By 6.6.9.2, τ^+ satisfies the Extended Inductive Hypothesis. Thus if $k = |\Omega(z^+)| = 1$ then the hypotheses of Theorem 15.2.2 are satisfied, so by induction on the order of τ we have a contradiction to $r > 1$.

Therefore $k > 1$. If $k > 2$ then 10.2.13 supplies a contradiction, so $k = 2$. Then 9.3.11 and 12.3.8 supply a contradiction to $r > 1$, completing the proof of (1).

Let $Q = O_2(\mathcal{F})$. If $z \in Q$ then Theorem 2 supplies a contradiction, so we may assume otherwise. Then by 3.3.2, $\mathcal{F}^\circ$ centralizes Q, so as $\mathcal{F} = \mathcal{F}^\circ$, $Q \leq Z(\mathcal{F}) = 1$ by (1). □

LEMMA 15.2.19. *(1) For each $1 \neq X \leq U$ with $X \in \mathcal{F}^f$, $\{\rho_j : j \in I\}$ is the set of components of $N_\mathcal{F}(X)$, so $N_\mathcal{F}(X) \leq \mathcal{U}$.*

(2) $\mathcal{C}_0 \trianglelefteq \mathcal{U}$.

PROOF. By 15.1.5, for $j \in I$, $\rho_j \leq \theta_j = (\mathcal{T}_j, \Omega_j)$ a component of $N_\mathcal{F}(X)$. As $r > 1$ and $\mathcal{T}_j$ centralizes z_k for $k \neq j$, it follows that $\mathcal{T}_j = \mathcal{C}_j$, so (1) holds. Then (1) and 15.2.17.1 imply (2). □

LEMMA 15.2.20. *(1) $N_\mathcal{F}(W) \leq \mathcal{U}$.*

(2) $N_\mathcal{F}(O(\tau)) \leq \mathcal{U}$.

PROOF. Set $O = O(\tau)$, $\mathcal{W} = N_\mathcal{F}(W)$, $S_0 = S_1 \cdots S_r$, and $\mathcal{M} = \mathcal{F}_{S_0}(M)$. Then $\mathcal{M} \trianglelefteq \mathcal{W}$, so by 1.3.2 in [**Asc19**], $\mathcal{W} = \langle S\mathcal{M}, N_\mathcal{W}(S_0)\rangle$. By construction, $\mathcal{M} \leq \mathcal{U}$ so $S\mathcal{M} \leq \mathcal{U}$ by 15.2.17.1. Thus to prove (1), it suffices to show that $\mathcal{N} = N_\mathcal{F}(S_0) \leq \mathcal{U}$. Let $P = C_S(S_0)$. Next $C_\mathcal{F}(S_0) \leq N_\mathcal{F}(O_1) \leq \mathcal{U}$ by 15.2.17.3 and by 1.3.2 in [**Asc19**], $\mathcal{N} = \langle SC_\mathcal{F}(S_0), N_\mathcal{N}(P)\rangle$, so it suffices to show $N_\mathcal{N}(P) \leq \mathcal{U}$. Let H be a model for $N_\mathcal{N}(P)$; it suffices to show that $U \trianglelefteq H$. But as $O_j\alpha_j = O_1$, $N_H(U)$ is transitive on $\{O_j : j \in I\}$, so $H = N_H(O_1)N_H(U) = N_H(U)$ by 15.2.17.3, completing the proof of (1).

Set $\mathcal{D} = N_\mathcal{F}(O)$ and let E be the kernel of the action of S on Ω. By 2.8.11 in [**Asc19**] there is a normal subsystem $\mathcal{E}$ of $\mathcal{D}$ on E such that each member of Ω is normal in $\mathcal{E}$. By 1.3.2 in [**Asc19**], $\mathcal{D} = \langle S\mathcal{E}, N_\mathcal{D}(E)\rangle$. By 15.2.17, $S\mathcal{E} \leq \mathcal{U}$. As $W \leq E$, $N_\mathcal{D}(E) \leq \mathcal{W} \leq \mathcal{U}$, proving (2). □

LEMMA 15.2.21. *(1) $\mathcal{U}$ controls fusion in U.*

(2) U is tightly embedded in $\mathcal{F}$.

PROOF. To prove (2), we must verify the three conditions (T1)-(T3) of Chapter 3 of [**Asc19**]. Condition (T3) is trivial in this case. As W is weakly closed in S and abelian, $N_\mathcal{F}(W)$ controls fusion in W, so as $U \leq W$ and $N_\mathcal{F}(W) \leq \mathcal{U}$ by 15.2.20.1, it follows that $\mathcal{U}$ controls fusion in U, so that (1) and (T2) hold.

It remains to verify (T1). Let $1 \neq X \leq U$. Using 15.2.20.2 we may assume $X \in \mathcal{F}^f$. We must show that $\mathcal{E} = O^{2'}(N_\mathcal{F}(X)) \leq \mathcal{U}$. But this follows from 15.2.19.1, completing the proof. □

LEMMA 15.2.22. *(1) Φ is not A_n with n even, E_6, or E_8.*

(2) If Φ is E_7 then either $\mathcal{C}_i = \mathcal{F}_{S_i}(M_i)$ or $\mathcal{C}_i \cong \tilde{E}_7[m]$, and in either case $Z(M_i) = Z(\mathcal{C}_i) \cong \mathbf{Z}_2$.

(3) If Φ is A_n with n odd then either $\mathcal{C}_i = \mathcal{F}_{S_i}(M_i)$ or $\mathcal{C}_i$ is a central factor system of $SL^+_{n+1}[m]$ and in either case $Z(M_i) = Z(\mathcal{C}_i)$ is cyclic.

(4) If Φ is D_n with $n \neq 4$ then $u(\tau_i) = u(\rho_i) \in Z(\mathcal{C}_i)$.

(5) If Φ is D_4 then $1 \neq Z(\mathcal{C}_i) \leq Z(M_i) \cong E_4$.

(6) For $1 \neq X \leq Z(\mathcal{C}_i)$, $N_\mathcal{F}(X) \leq \mathcal{U}$.

PROOF. Part (1) follows from 15.2.15 and 15.1.5.4. Similarly if Φ is E_7 then by 15.1.5.4, either $\mathcal{C}_i = \mathcal{F}_{S_i}(M_i)$ or $\mathcal{C}_i \cong \tilde{E}_7[m]$ or $E_7[m]$, and the last case is out by 15.2.15. Therefore $Z(\mathcal{C}_i) \cong \mathbf{Z}_2$ and as S_i is Sylow in $\mathcal{C}_i$ by 5.3.10 and 5.5.8, we have $Z(\mathcal{C}_i) \leq Z(M_i)$, so we conclude that (2) holds. A similar argument using 5.7.2.6 establishes (3).

Part (4) follows from 15.2.7.3, while (5) follows from 15.2.15 and 15.2.8.

Let $1 \neq X \leq Z(\mathcal{C}_i)$. Then $X \leq U$. By 15.2.21.1 we may assume $X \in \mathcal{F}^f$. Then 15.2.19 establishes (6). □

REMARK 15.2.23. From 15.2.16, 12.3.24 is satisfied in our setup. Most of the lemmas in section 12.3 following 12.3.24 hold in our situation, and indeed usually their proofs go through verbatim, or sometimes we have already proved the lemma. In particular 12.3.25 follows from 15.2.18.1 and 12.3.26 follows from 15.2.20.1. Then the proofs of 12.3.27-12.3.29, go through verbatim. We now establish 12.3.30 in the following lemma:

LEMMA 15.2.24. *Assume $\Theta = (\mathcal{T}, \Gamma)$ is a proper subpacket of τ with $\Omega_i \subseteq \Gamma \subseteq \Omega$ and $\mathcal{C}_i \leq \mathcal{T}$. Then*

(1) ρ_i is a coconnected component of Θ°.

(2) $\mathcal{T} \leq \mathcal{U}$.

PROOF. To prove (1) we may assume $\Gamma = \Omega_i$ and $\mathcal{C}_i < \mathcal{T}$. Let $X = Z(\mathcal{C}_i)$; observe $X \neq 1$ by 15.2.15. By 15.2.22.6, $\mathcal{C}_i = C_{\mathcal{T}}(X)^\circ$. By 15.1.4.5 and 15.2.22.1, Φ is E_7, D_n for some $n \geq 3$, or A_n for some odd $n \geq 3$. In the first case $Z(M_i) = Z(\mathcal{C}_i) \cong \mathbf{Z}_2$ by 15.2.22.2, so as $M_i \in M(\Theta)$ it follows that $\mathcal{T}$ is not $E_7[m]$ and hence $X = Z(\mathcal{T})$ by 15.1.5.4, contradicting $\mathcal{T} \neq \mathcal{C}_i = C_{\mathcal{T}}(X)^\circ$. Similarly if Φ is A_n then $X = Z(\mathcal{T})$ and we have the same contradiction. Therefore Φ is D_n. By 15.1.9.1, either there is an involution $u = u(\Theta) \in Z(\mathcal{T})$ or Φ is D_4 and $\mathcal{T}$ is $F_4[m]$ or $E_6^-[m]$. However in the latter case, $N_{\mathcal{T}}(W_i)$ is transitive on $Z_i^\#$, so as $Z(\mathcal{C}_i) \neq 1$ is cyclic by 5.5.6, and as $N_{\mathcal{F}}(W_i) \leq \mathcal{U}$ by 12.3.28.2, we have a contradiction.

Therefore $u = u(\Theta) \in Z(\mathcal{T})$. If $n \neq 4$ then by 15.2.7.3, $u = u_i \in Z(\mathcal{C}_i) \leq U$, contrary to $\mathcal{C}_i \neq \mathcal{T}$ and 15.2.22.6. Therefore $n = 4$ and at least $u \in Z_i$, so $u \in Z(\mathcal{C}_i) \leq U$, yielding our usual contradiction. This completes the proof of (1). Then (2) follows from (1) as in the proof of 12.3.30.2. □

REMARK 15.2.25. By 15.2.18.2, $O_2(\mathcal{F}) = 1$, so as $U \leq O_2(\mathcal{U})$ we have $\mathcal{U} \neq \mathcal{F}$. This is 12.3.31.1, while the remaining two parts of 12.3.31 follow from 15.2.19.2. Observe 12.3.32 is 15.2.21.2. Given that 12.3.25-12.3.32 hold, the proofs of 12.3.33 and 12.3.34 go through verbatim. Lemma 12.3.35 is unnecessary, while the proofs of 12.3.36-12.3.42 go through verbatim. Indeed 12.3.38 is 15.2.18.2. The proofs of 12.3.43 and 12.3.44 go through verbatim. Lemma 12.3.45 is also unnecessary. The proof of 12.3.46 goes through verbatim. We only need parts of 12.3.47, which we establish in the next lemma.

LEMMA 15.2.26. *Assume $A \in \Xi_P$ and let $J \in \Omega$. Then*

(1) $|N_J(P)| \leq 2$.

(2) AP acts on J.

PROOF. We follow part of the proof of 12.3.47. By 12.3.46.2, $C_U(P)\alpha \leq Q \in \hat{\mathcal{P}}$. Assume (1) fails. Then there is x of order 4 in $N_J(P)$. By 12.3.46.5, $x\alpha \in \Sigma_Q$. But $x\alpha \in D \in \Delta$, so $x\alpha \in S_j$ for some $j \in I$, contrary to 12.3.42.3. This proves (1).

To prove (2) we proceed as in the proof of 12.3.47.1. Let $t = z(J)$. If A centralizes t then so does AP by 12.3.44.3, so (2) holds. Thus we may assume for some $a \in A$, $s = t^a \neq t$. Let $L \in \Omega(s)$, so that $L = J^a$. Then $Y = JL = J \times L$, and arguing as in the proof of 12.3.47.1, $B = N_A(J)$ is of index 2 in A and centralizes Y. But then $J = N_J(P)$, contrary to (1). This completes the proof of (2). □

THEOREM 15.2.27. $\Xi_P = \emptyset$.

PROOF. Assume $A \in \Xi_P$. By 15.2.26.2, AP acts on K and hence on $V = K \cap W$. By 12.3.44.3, P centralizes the subgroup V_0 of V of order 4, so $V_0 \leq N_K(P)$, contrary to 15.2.26.1. □

LEMMA 15.2.28. *(1)* $P = \langle a \rangle \cong \mathbf{Z}_2$.
(2) $r = 2$.
(3) $W_1^a = W_2$.

PROOF. By 12.3.46.4, each nontrivial subgroup of P is in Σ_P, so P is cyclic by 15.2.27. Let a be the involution in P.

By 12.3.46.2, $C_U(P)\alpha \leq Q \in \hat{\mathcal{P}}$. Suppose $\emptyset \neq J \subset I$ such that $Y = \prod_{j \in J} W_j$ is a-invariant. Then $Y \leq C_S(\mathcal{C}_i)$ for $i \in I - J$, so by 12.3.46.5, each nontrivial subgroup of $X = C_Y(a)\alpha$ is in Σ_Q. Hence X is cyclic by 15.2.27, so $m(Y) \leq 2$. But if a acts on W_1 we may take $J = \{i\}$ and $Y = W_1$, whereas $m(W_1) \geq 3$. Therefore we may assume $W_1^a = W_2$, so (3) holds. Similarly (2) holds or else we could choose $Y = W_1W_2$. Finally as $r = 2$, $\Phi(S)$ acts on W_1, so $a \notin \Phi(P)$ and therefore $P = \langle a \rangle$, establishing (1). □

We are now in a position to obtain a contradiction, and hence finally establish Theorem 15.2.2. Namely Lemma 15.2.28 is just 12.3.49, and the last paragraph of the proof of Theorem 12.3.8 (which appears just after the proof of 12.3.49) supplies our contradiction. Thus the proof of Theorem 15.2.2 is finally complete.

THEOREM 15.2.29. *Assume Hypothesis* 15.2.1 *with* $\mathcal{F} = \mathcal{F}^\circ$. *Then* $\mu(\tau) \cong$ *Weyl*(Φ), *where* Φ *is* A_n, D_n, *or* E_n.

PROOF. By Theorem 15.2.2, $r = 1$, so $W = W_1$ and $\mu(\tau) = \mu_1$. Thus the theorem follows from 15.1.4.5. □

REMARK 15.2.30. Assume Hypothesis 15.2.1 with $\mathcal{F} = \mathcal{F}^\circ$, and suppose $\Phi = A_2$. We claim that either

(a) $z \in O_2(\mathcal{F})$ and $\mathcal{F} = \mathcal{F}_S(M)$ or $\mathcal{F} \cong L_3(2)/E_8$, or

(b) $\mathcal{F} \cong L_3^+[m]$, $L_3^+[m]^{(1)}$, $G_2[m]$, or M_{12}.

For $\Omega = \{K\}$ is of order 1 and $\mu = \mu(\tau) \cong S_3$. Hence Hypothesis 7.1.1 holds, so we can appeal to the results in section 7.1. In particular if $z \in O_2(\mathcal{F})$ then (a) holds by Theorem 7.1.6, while if $a \notin O_2(\mathcal{F})$ then (b) holds by Theorem 7.1.29.

15.3. $\Phi = D_n$

In this section we assume the following hypothesis:

HYPOTHESIS 15.3.1. *Hypothesis 15.2.1 holds with $\mathcal{F} = \mathcal{F}^\circ$ and $\mu = \mu(\tau) \cong Weyl(\Phi)$, where Φ is of type D_n for some $n \geq 3$.*

In addition we adopt Notation 15.1.3, set $S_M = S \cap M$, $\mathcal{M} = \mathcal{F}_{S_M}(M)$, and we adopt the following notation:

NOTATION 15.3.2. We adopt a modified version of the notation in section 5.3. Let q be a prime power with $(q-1)_2 = m/2$ and let $\bar{L} = Spin_{2n}^+(q)$. There is a representation $\bar{\pi} : \bar{L} \to L = \Omega(U, \mathfrak{q})$ of $\bar{L}$ on an orthogonal space $(U, \mathfrak{q})$ over $F = \mathbf{F}_q$ of sign $+1$ and dimension $2n$. Let $\bar{S} \in Syl_2(\bar{L})$, $\bar{\mathcal{F}} = \mathcal{F}_{\bar{S}}(\bar{L})$, and $\bar{\tau} = (\bar{\mathcal{F}}, \bar{\Omega})$ a Lie packet for $\bar{L}$. Then $\bar{\pi}$ induces a morphisms $\bar{\pi}_L : \bar{\tau} \to \tau_L = (\mathcal{F}_L, \Omega_L)$, where τ_L is a Lie packet for L with Sylow group $S_L = \bar{S}\bar{\pi}_L$. Let $\bar{W} \in W(\bar{\tau})$ and $\bar{M} \in M(\bar{\tau})$. By 4.3.8 there is a surjective homomorphism $\pi : \bar{M} \to M$ with $\ker(\pi) \leq Z(\bar{M})$. Similarly $\bar{\pi}_L$ induces a surjective homomorphism $\pi_L : \bar{M} \to M_L$, where $W_L = \bar{W}\bar{\pi}_L$ and $M_L \in M(\tau_L)$. We use these maps to study M and τ, via the representation of M_L on U.

Let $I = \{1, \ldots, n\}$. In particular, as in 5.3.4 and 5.3.5, there is an M_L-invariant decomposition $U = U_1 \perp \cdots \perp U_n$ of U such that for $i \in I$, U_i is a line of sign $+1$ and W_L is the kernel of the action of M_L on $\Gamma = \{U_i : i \in I\}$.

NOTATION 15.3.3. If $n \neq 4$ then by 15.1.6 there exist unique involutions $\bar{u} = u(\bar{\tau}) \in Z(\bar{M})$ and $u = u(\tau) \in Z(M)$ such that for each t in $\mathcal{Z}_S(\bar{\tau})$ or $\mathcal{Z}_S(\tau)$, $\bar{t}' = t\bar{u}$ or $t' = tu$ is also in $\mathcal{Z}_S(\bar{\tau})$ or $\mathcal{Z}_S(\tau)$, respectively. If $n = 4$ then $\ker(\pi_L)$ is of order 2 and we write $\bar{u}$ for a generator of the kernel.

For $1 \leq k \leq n$, write $i(2k, +)$ for the set of involutions $j \in L$ such that $[U, j]$ is nondegenerate of dimension $2k$ and sign $+1$, and write $\bar{i}(2k, +)$ for the set of preimages in $\bar{L}$ of such involutions. A *projective involution* in L is an element x of order 4 such that the eigenspaces V_λ and $V_{-\lambda}$ for x on V are totally singular of dimension n, where $\lambda \in F$ is of order 4.

For w in W or W_L, write $\bar{w}$ for a preimage in $\bar{W}$ of w under π or π_L.

LEMMA 15.3.4. *(1) If $n \neq 4$ then $\bar{u}\pi = u$, so $\bar{u} \notin \ker(\pi)$.*
(2) $\langle \bar{u} \rangle = \ker(\pi_L)$, so for $\bar{t} \in \mathcal{Z}_S(\bar{\tau})$, $\bar{t}'\pi_L = \bar{t}$.
(3) For $t_L \in \mathcal{Z}_\Delta(\tau_L)$, $[U, t_l] = U_i + U_j$ for some distinct $i, j \in I$.
(4) If n is odd or $n = 4$ then π is an isomorphism.
(5) If $n > 4$ is even then either π is an isomorphism or $\ker(\pi) \cong \mathbf{Z}_2$.

PROOF. The uniqueness of u and $\bar{u}$ imply (1).

If n is odd then $Z(\bar{M})$ is cyclic by 5.8.4, so as $\bar{u} \notin \ker(\pi)$ by (1), it follows that $\ker(\pi) = 1$, so that π is an isomorphism. If $n = 4$ then π is an isomorphism as $|\Omega(z)| = 1$. Thus (4) holds. Similarly if $n > 4$ is even then $Z(\bar{M}) \cong E_4$ by 5.8.4, so (1) implies (5).

If $n = 4$ then (2) holds by construction. For $t_L \in \mathcal{Z}_S(\tau_L)$, we have $\Omega(t_L) = \{K_L, K_L'\}$ of order 2 with $[U, K_L] = [U, K_L'] = [U, t_L] = U_i + U_j$ for some distinct $i, j \in I$. Then (2) follows from the uniqueness of $\bar{u}$. Also as M_L is transitive on $\mathcal{Z}_S(\tau_L)$, (3) follows. $\square$

LEMMA 15.3.5. *(1) Members $\bar{j}$ of $\bar{i}(2k,+)$ are of order 2,4, for k even, odd, respectively, with $\bar{j}^2 = \bar{u}$ if k is odd.*

(2) If j is an involution in W_L then $[U,j]$ is the sum of k members of Γ, so $j \in i(2k,+)$.

(3) If t is an involution in W then one of the following holds:

(a) $\bar{t}\pi_L \in i(2k,+)$ for some even k.

(b) $\bar{t}\pi_L$ is a projective involution.

(c) $t = u$.

(4) If x is a projective involution in L then $C_L(x) \cong GL_n(q)$, and if $x \in \mathcal{F}_L^f$ then $\tau(C_{\mathcal{F}_L}(x), C_{\Omega_L}(x))$ is a Lie packet of $C_L(x)$ with Thompson group Weyl(A_{n-1}).

PROOF. For $i \in I$ let j_i be the involution in L with $[U, j_i] = U_i$. Then j_1 is a representative for $i(2,+)$. Let $B = U_1 + U_2$ and regard j_1 as a member of $\Omega(B,\mathfrak{q}) \cong SL_2(q) * SL_2(q)$. As $Spin(B,\mathfrak{q}) \cong SL_2(q) \times SL_2(q)$, $\bar{j}_1$ is of order 4 with $\bar{j}_1^2 = \bar{u}$. Now a representative x for $i(2k,+)$ is $x = j_1 \cdots j_k$, so as $[\bar{j}_r, \bar{j}_s] = 1$ for $r,s \in I$, (1) follows.

Next from 15.3.2, W_L is the kernel of the action of M_L on Γ, so each involution y in W_L is the product $y = \prod_{j_i \in I(y)} j_i$ for some $I(y) \subseteq I$. Hence (2) holds.

Let t be an involution in W and $t_L = \bar{t}\pi_L$. Then $t_L \in W_L$, so if t_L is an involution then $t_L \in i(2k,+)$ for some k by (2). But if k is odd then $\bar{t}^2 = u$ by (1), so as $\bar{u} \notin \ker(\pi)$ by 15.3.4.1, we have a contradiction to t an involution. Thus (3a) holds if t_L is an involution, so we may assume otherwise. If $t_L = 1$ then $\bar{t}_L = \bar{u}$ by 15.3.4.2, so (c) holds by 15.3.4.1; thus we may assume $t_L \neq 1$, so $|t_L| \geq 4$ and hence $|\bar{t}| = |\bar{t}_L| \geq 4$. Then as t is an involution, π is not an isomorphism, and therefore $n > 4$ is even and $\ker(\pi) \cong \mathbf{Z}_2$ by 15.3.4. Then $\bar{t}^2 = \bar{y}$ generates $\ker(\pi)$, but $\bar{y}\pi_L = y \neq 1$ is in $Z(M_L)$, so $y = -1_U$. Therefore t_L is a projective involution, completing the proof of (3).

Part (4) is an exercise; see for example 15.12 in [**Asc77**]. □

LEMMA 15.3.6. *Let t be an involution in $W \cap \mathcal{F}^f$ with $t_L = \bar{t}\pi_L \in i(2k,+)$. Set $U(1) = [U, t_L]$, $U(2) = C_U(t_L)$, and $2k_i = \dim(U_i)$.*

(1) $\theta_t = (C_{\mathcal{M}}(t)^\circ, C_\Omega(t))$ is a quaternion fusion packet.

(2) $\theta_t = \theta_1 * \theta_2$, where $\theta_i = (\mathcal{M}_i, \Omega_i)$, $\Omega_i = \{\bar{K}\pi : K \in \Omega_L$ and $[U,K] \leq U(i)\}$, and $\mathcal{M}_i = [\Omega_i]_{C_{\mathcal{M}}(t)^\circ}$.

(3) $\mu(\theta_i) \cong$ Weyl(Φ_i) where $\Phi = D_{k_i}$, unless $k_2 \leq 1$ where $\tau_2 = \emptyset$.

(4) If $n \geq 5$ and $k_i \geq 2$, then $u \in Z(\mathcal{M}_i)$. If in addition $k_i \neq 4$ then $u = u(\theta_i)$.

PROOF. Part (1) follows from 2.5.2.

As $t \in \mathcal{F}^f$ also $t_L \in \mathcal{F}_L^f$. As $t_L \in i(2k,+)$, k is even by 15.3.5.3, so in particular $k_1 \geq 2$. Now $C_L(t)^\circ = L_1 * L_2$, where $L_i = \Omega(U(i),\mathfrak{q}) \cong \Omega_{2k_i}^+(q)$, except if $k_2 \leq 1$ then $L_2 = 1$. Let $S_i = S_L \cap L_i$ and $\xi_i = (\mathcal{F}_{S_i}, \Lambda_i)$, where $\Lambda_i = \{K \in \Omega_L : [U,K] \leq U(i)\}$. Then as $t_L \in W_L \cap \mathcal{F}_L^f$, ξ_i is a Lie packet of L_i with $\mu(\xi_i) = D_{k_i}$. Let $\mathcal{N} = \mathcal{F}_{S_L}(M_L)$, $N_i \in M(\xi_i)$, and $\mathcal{N}_i = \mathcal{F}_{S_i}(N_i)$. Then $C_{\mathcal{N}}(t_L)^\circ = \mathcal{N}_1 * \mathcal{N}_2$, so pulling back to $\bar{M}$ using π_L and projecting onto $\bar{M}$ using π, we conclude that (2) and (3) hold.

Assume $n \geq 5$ and $k_i \geq 2$, so that u is defined. Then there exist $K, K' \in \Lambda_i$ with $z(K) = z(K')$, so $z(\bar{K}') = z(\bar{K})\bar{u}$ is in $Z(\bar{\mathcal{N}}_i)$, and hence $u = \bar{u}\pi \in Z(\mathcal{M}_i)$. Then if $k_i \neq 4$, $u(\theta_i)$ is defined and $u = u(\theta_i)$ by uniqueness of $u(\theta_i)$. This completes the proof of (4) and the lemma. □

LEMMA 15.3.7. *Assume $n \geq 5$ and t is an involution in $W \cap \mathcal{F}^f$ with $t_L = \bar{t}\pi_L \in i(2k,+)$. Then $u \in Z(\mathcal{F}_t)$.*

PROOF. Adopt the notation from 15.3.6. If $u \in Z(\mathcal{F})$ the lemma is trivial, so we may assume otherwise.

Suppose $t \in Z(\mathcal{F})$ and set $\mathcal{F}^+ = \mathcal{F}/\langle t\rangle$. Then, using 6.6.9, $\tau^+ = (\mathcal{F}^+, \Omega^+)$ satisfies Hypothesis 15.3.1, so $u^+ \in Z(\mathcal{F}^+)$ by induction on the order of τ (cf. 15.3.10). But then pulling back to $\mathcal{F}$, the lemma holds in τ. Therefore $t \notin Z(\mathcal{F})$, so $\tau_t = (\mathcal{F}_t^\circ, \Omega)$ is proper in τ. Then by 15.1.5 and an argument in the proof of 15.2.4, $\tau_t = \tau_1 * \tau_2$ where $\tau_i = (\mathcal{F}_i, \Omega_i)$ and for $M_i \in M(\tau_i)$, $\mathcal{M}_i = \mathcal{F}_{S_i}(M_i)$, so that $\mu(\tau_i) = \mathrm{Weyl}(\Phi_i)$ with $\Phi_i = D_{k_i}$ by 15.3.6.

(a) If $4 \neq k_i \geq 3$ then $u = u(\tau_i)$.

Namely by 15.3.6.4, $u = u(\theta_i)$. But by 15.1.9.1, $u(\theta_i) \in Z(\mathcal{F}_i)$, so (a) holds.

(b) For $i = 1, 2$, either $k_i \leq 2$ or $k_i = 4$.

If not then we can choose i with $4 \neq k_i \geq 3$. Then by (a), $u = u(\tau_i) \in Z(\mathcal{F}_i)$. But observe that either $\mathcal{F}_i \trianglelefteq \mathcal{F}_t$, or $k_1 = k_2 = n/2$ and $\mathcal{F}_i$ is normal in a subsystem of $\mathcal{F}_t$ of index 2. Then as $u \in Z(S)$, in any event $u \in Z(\mathcal{F}_t)$, contrary to assumption. This proves (b). Then as $k_1 + k_2 = n \geq 5$ and k_1 is even, we conclude from (b) that:

(c) (k_1, k_2) is $(2,4)$, $(4,1)$, $(4,2)$, or $(4,4)$.

Observe that by 15.3.6.4:

(d) $u \in Z(\mathcal{M}_i)$ for each i with $k_i \neq 1$.

(e) If $(k_1, k_2) \neq (4,1)$ then $u \in Z(\mathcal{F}_i)$ for $i = 1, 2$.

By (d), $u \in S_j$ for $j = 1, 2$, so as S_{3-i} centralizes $\mathcal{F}_i$, part (e) follows.

(f) $u \notin Z(\mathcal{F}_i)$ for $i \in \{1, 2\}$.

Assume otherwise and set $Z_i = Z(\mathcal{F}_i)$. If $Z_i = \langle u\rangle$ then the proof of (b) shows that $u \in Z(\mathcal{F}_t)$, contrary to assumption. Then from (c), $Z_i \cong E_4$, and the proof of (b) shows that there is a 3-element $\phi \in Aut_{\mathcal{F}_t}(S_1S_2)$ irreducible on Z_i. Then ϕ acts on W_1W_2 and is induced in $N_{\mathcal{F}}(C_S(W_1W_2)$, so as W is weakly closed in $C_S(W_1W_2)$, ϕ acts on W. This is a contradiction as $u \in Z(N_{\mathcal{F}}(W))$.

Observe that by (e) and (f):

(g) $n = 5$ and $(k_1, k_2) = (4, 1)$.

(h) $\mathcal{F}_1$ is isomorphic to $Spin_9[m]$, $Spin_{10}^-[m]$, or $\hat{A}E_9$.

This follows from (g) and 15.1.9.2.

We now obtain a contradiction, establishing the lemma. By (g), $n = 5$, so $Z(M) = \langle y\rangle \cong \mathbf{Z}_4$. Now by (f), y acts faithfully on $\mathcal{F}_1$ centralizing $\mathcal{M}_1$. But then (h) and 11.5.17.3 supply a contradiction. □

LEMMA 15.3.8. *Assume $n \geq 5$ and t is an involution in $W \cap \mathcal{F}^f$ with $t_L = \bar{t}\pi_L$ a projective involution. Then*

(1) $\tau_t = (\mathcal{F}_t^\circ, C_\Omega(t))$ is a quaternion fusion packet with $\mu(\tau_t) \cong Weyl(A_{n-1})$.

(2) Either

(a) $W(\tau_t) \trianglelefteq \mathcal{F}_t$, or

(b) $\mathcal{F}_t^\circ$ is a central factor system of $SL_n^+[m]$.

(3) $u \in Z(\mathcal{F}_t)$.

PROOF. By 2.5.2, τ_t is a quaternion fusion packet, and by 15.3.5.4, $\mu(\tau_t) \cong Weyl(A_{n-1})$, so (1) holds. Then (2) follows from (1) and 15.1.5.

Assume $u \notin Z(\mathcal{F}_t)$. Claim u centralizes $\mathcal{Y} = \mathcal{F}_t^\circ$. In case (a) of (2) this is clear, so assume case (b) holds. Then by (b) and 15.7.1 applied in an inductive setting,

$\mathcal{Y} = \langle \mathcal{F}_T(N), T\mathcal{O}_{\tau_t}\rangle$, where $N \in M(\tau_t)$, so as u centralizes N and $\mathcal{O}_{\tau_t}$, the claim also holds in case (b).

Let $\mathcal{X} = C_{\mathcal{F}_t}(\mathcal{Y})$ and X Sylow in $\mathcal{X}$. We may take $K \in C_\Omega(t)$, so $\mathcal{X} \leq C_{\mathcal{F}_t}(z) \leq C_{\mathcal{F}_t}(u)$ by 15.3.7. Let R be Sylow in $\mathcal{XY}$ and $S_t = C_S(t)$. By 1.3.2 in [**Asc19**], $\mathcal{F}_t = \langle S_t\mathcal{XY}, N_{\mathcal{F}_t}(R)\rangle$, so it suffices to show that $\mathcal{N} = N_{\mathcal{F}_t}(R)$ centralizes u. Let $P \leq S_t$ and $\phi \in \hom_{\mathcal{N}}(P, S_t)$; we may take $R \leq P$ and it suffices to show $u\phi = u$. Now ϕ acts on $O = O(\tau_t)$ and X, so it suffices to show that $\alpha = \phi_{|XO} \in Aut_{\mathcal{N}}(XO)$ fixes u. But $N_{\mathcal{Y}}(O)$ is transitive on $C_\Omega(t)$, so there is $\psi \in Aut_{\mathcal{Y}}(XO)$ with $z\alpha\psi = z$. As $C_{\mathcal{F}_t}(z)$ centralizes u, we have $u\alpha\psi = u$, and then as $\mathcal{Y}$ centralizes u, $u\alpha = u\psi^{-1} = u$, completing the proof of (3) and the lemma. □

LEMMA 15.3.9. *If $n \geq 5$ then $u \in Z(\mathcal{F})$.*

PROOF. Let $\mathfrak{T}$ be the set of fully centralized involutions in W. Let $t \in \mathfrak{T}$ and set $t_L = \bar{t}\pi_L$. By 15.3.5.3, either $t_L \in i(2k, +)$ for some even k or t_L is a projective involution, or $t = u$. Hence by 15.3.7 and 15.3.8.3, $u \in Z(\mathcal{F}_t)$. Also $u \in Z(N_{\mathcal{F}}(W))$. Therefore $\mathcal{F}_u = \langle N_{\mathcal{F}}(W), \mathcal{F}_t : t \in \mathfrak{T}\rangle$, so the lemma follows from Theorem 14.1.18. □

THEOREM 15.3.10. *Assume Hypothesis* 15.3.1 *with $n \geq 5$. Then $u \in Z(\mathcal{F})$ and one of the following holds:*

(1) $W \trianglelefteq \mathcal{F}$.

(2) $\mathcal{F} \cong Spin^+_{2n}[m]$, $Spin_{2n+1}[m]$, or $Spin^-_{2n+2}[m]$.

(3) n is even and $\mathcal{F} \cong HSpin^+_{2n}[m]$.

(4) $\mathcal{F} \cong \hat{A}E_{2n}$ or $\hat{A}E_{2n+1}$.

(5) n is even and $\mathcal{F} \cong H\hat{A}E_{2n}$.

PROOF. By 15.3.9, $u \in Z(\mathcal{F})$. Set $\mathcal{F}^+ = \mathcal{F}/\langle u\rangle$; then $\tau^+ = (\mathcal{F}^+, \Omega^+)$ is a quaternion fusion packet with $\mathcal{F}^+ = (\mathcal{F}^+)^\circ$ transitive on Ω^+. By 6.6.9.2, τ^+ satisfies the Extended Inductive Hypothesis. Further $z^+ = z_2^+$ where $z_2 = zu$, so $\Omega(z^+) = \{K^+, K_2^+\}$ is of order 2. Therefore $\mathcal{F}^+$ appears in one of cases (1)-(3) of Theorem 12.3.50, using Theorem 9.3.24 to verify that $\mathcal{Z} \cap O(z) = \{z\}$. Then lifting back to $\mathcal{F}$ we conclude the theorem holds using 5.1.21 and 3.3.16. □

NOTATION 15.3.11 (aE.11). In the remainder of the section we assume that $n = 3$. Hence by 15.3.4.4, $\pi : \bar{M} \to M$ is an isomorphism, so we identify M with the universal group $\bar{M}$ of 4.3.1 via this isomorphism and adopt the notation found there, except we suppress the "bars". Thus $M = \langle K_1, K_2, K_3\rangle$ with $K_i = \langle k_i, v_i\rangle$ and $V_i = \langle v_i\rangle = W \cap K_i$. Set $z_i = z(K_i)$; we choose notation so that $z = z_3$, so that $K = K_3$ and $u = z_2z_3$.

LEMMA 15.3.12. *Assume $n = 3$.*

(1) $\tau_z = (\mathcal{F}_z, \Omega)$ is a quaternion fusion packet with $\tau_z = \tau_1 * \tau_2$, where $\tau_1 = (\mathcal{O}(z), \{K\})$ and $\tau_2 = (\mathcal{F}_2, \{K_2\})$.

(2) Either

(a) $\mathcal{F}_2 = \mathcal{O}(z_2)$ and $u \in Z(\mathcal{F})$, or

(b) $\mathcal{F}_2 \cong L_3^-[m]$ or $L_2[2m]^{(1)}$.

PROOF. By 3.3.1, τ_z is a quaternion fusion packet. As $n = 3$, $\Omega = \{K, K_2\}$, so (1) follows.

From 14.1.2, $\mathcal{B}_0 = \langle N_{\mathcal{F}}(W), \mathcal{F}_t : t \in \mathfrak{T}\rangle$, where $\mathfrak{T}$ is the set of fully centralized involutions in W. As $n = 3$ and $M = \bar{M}$, up to conjugation in M, $\mathfrak{T} = \{u, z\}$. Also $N_{\mathcal{F}}(W) \leq \mathcal{F}_u$, so $\mathcal{B}_0 = \langle \mathcal{F}_u, \mathcal{F}_z\rangle$.

If $\mathcal{F}_2 = \mathcal{O}(z_2)$ then z_2 is in the center of $\mathcal{F}_z$, so $u = zz_2$ is too, and hence $\mathcal{B}_0 = \mathcal{F}_u$. But then (2a) holds by Theorem 14.1.18. Thus we may assume z_2 is not in the center of $\mathcal{F}_2$. But as $\mu \cong S_4$, $\mu(\tau_z) \cong E_4$ and hence $\mu(\tau_2) \cong \mathbf{Z}_2$, so (3b) holds by Theorems 7.1.6 and 7.1.7. □

LEMMA 15.3.13. *Assume $n = 3$. Then $u \in Z(\mathcal{F})$.*

PROOF. Assume otherwise; by 15.3.12.2, $\mathcal{F}_2 \cong L_2[2m]^{(1)}$ or $L_3^-[m]$. In either case a Sylow subgroup S_2 of $\mathcal{F}_2$ is semidihedral of order $2m$, with $S_2 = \langle x, y\rangle$ where $|y| = m$ with $y^2 = v_2$, x is an involution, and $y^x = y^{-1}z_2$.

Now $E = \Omega_1(W) = \langle z_1, z_2, z\rangle$ with V_3 centralizing S_2 and z_1 centralizing V_3 and acting on S_2. Thus $[x, z_1] \in W \cap S_2 = V_2$, so $[x, z_1] = z_2^a$ for some $a \in \{0, 1\}$. If $a = 0$ set $t = z_1$, and observe t centralizes z_1. If $a = 1$ set $t = jx$ where $j \in K_2 - V_2$ with $j = yx$; then as $[k_2, z_1] = z_2$, again t centralizes z_1 and this time $t = y$.

Define G as in 15.1.3 and form the subgroups $U = W\langle t\rangle$ and $X = M\langle t\rangle$ of G. Then $U = C_X(E)$, so $U \trianglelefteq X$. Then as t centralizes V_3 and M is transitive on η, it follows that U, and hence also t, centralizes W. Hence as x does not centralize V_2, we conclude that $t = y$. But then $U^{m/2} = \langle y^{m/2}\rangle = \langle z_2\rangle$ is normal in M, a contradiction. □

THEOREM 15.3.14. *Assume Hypothesis 15.3.1 with $n = 3$. Then $u \in Z(\mathcal{F})$ and one of the following holds:*

(1) $W \trianglelefteq \mathcal{F}$.

(2) $\mathcal{F} \cong Spin_6^+[m]$, $Spin_7[m]$, or $Spin_8^-[m]$.

(3) $\mathcal{F} \cong \hat{A}E_6$ or $\hat{A}E_7$.

(4) $m = 8$ and $\mathcal{F} \cong \hat{L}_3(2)/E_{64}$ or $\hat{S}p_6(2)$.

PROOF. The proof is much like that of Theorem 15.3.10. By 15.3.13, $u \in Z(\mathcal{F})$. Set $\mathcal{F}^+ = \mathcal{F}/\langle u\rangle$; then $\tau^+ = (\mathcal{F}^+, \Omega^+)$ is a quaternion fusion packet with $\mathcal{F}^+ = (\mathcal{F}^+)^\circ$ transitive on Ω^+. By 6.6.9.2, τ^+ satisfies the Extended Inductive Hypothesis. Further $z^+ = z_2^+$, so $\Omega^+ = \Omega(z^+) = \{K^+, K_2^+\}$ is of order 2. Therefore $\mathcal{F}^+$ appears in one of cases (1)-(3) or (5) of Theorem 12.3.50, using Theorem 9.3.24 to verify that $\mathcal{Z} \cap O(z) = \{z\}$. Then lifting back to $\mathcal{F}$ we conclude the theorem holds using 3.3.16, 5.10.7.2, and 5.1.21. □

15.4. $\Phi = D_4$

In this section we assume the following hypothesis:

HYPOTHESIS 15.4.1. *Hypothesis 15.2.1 holds with $\mathcal{F} = \mathcal{F}^\circ$ and $\mu = \mu(\tau) \cong$ Weyl(Φ), where Φ is of type D_4.*

NOTATION 15.4.2. As usual let S_M be Sylow in M and adopt the notation from section 15.3, particularly Notation 15.1.3, 15.3.2, and 15.3.3. By 15.3.4.4, $\pi : \bar{M} \to M$ is an isomorphism, so, just as in 15.3.11, identify M with $\bar{M}$, except we suppress the "bars". Thus $M = \langle K_i : i \in I\rangle$ with $K_i = \langle k_i, v_i\rangle$, and $V_i = \langle v_i\rangle = K_i \cap W$.

Set $z_i = z(K_i)$. We choose notation so that $z = z_4$ and hence $K = K_4$. Let K_0 be the fourth member of Ω other than K_1, K_3, K_4. Set $U = Z(M)$.

LEMMA 15.4.3. *(1) $U = \langle zz_3, zz_1\rangle \cong E_4$.*

(2) Let $\mathfrak{T}$ be the set of fully centralized involutions in W. Then up to conjugation in M, $\mathfrak{T} = U^\# \cup \{z\}$.

(3) $\mathcal{F} = \langle N_{\mathcal{F}}(U), \mathcal{F}_t : t \in \mathfrak{T}\rangle$.

(4) Assume $\tilde{\tau} = (\tilde{\mathcal{F}}, \tilde{\Omega})$ is the Lie packet with $\tilde{\mathcal{F}}$ isomorphic to $F_4[m]$ or $E_6^-[m]$. Then $\tilde{\mathcal{F}} = \langle N_{\tilde{\mathcal{F}}}(Z(\tilde{M})), \tilde{\mathcal{F}}_{\tilde{z}}, \tilde{\mathcal{F}}_{\tilde{u}}\rangle$, where $\tilde{M} = M(\tilde{\tau})$, $\tilde{z} \in \mathcal{Z}_{\tilde{S}}$, and $\tilde{u}$ is an involution in $Z(\tilde{M})$.

PROOF. Part (1) follows from 5.8.7.2, while (2) follows from 5.8.3. Observe Hypothesis 11.3.1 is satisfied with $n_S = 4$ and $W_S = U\langle z\rangle$ of rank 3. Then from 11.3.2,

$$\mathcal{B} = \mathcal{B}(\tau) = \langle \mathcal{E}(\tau), \mathcal{F}_x : x \in \mathfrak{T}\rangle.$$

From 14.1.3.1, $\mathcal{E}(\tau) \le \mathcal{B}_0 = \langle N_{\mathcal{F}}(W), \mathcal{F}_x : x \in \mathfrak{T}\rangle$, so as $N_{\mathcal{F}}(W) \le N_{\mathcal{F}}(U)$, (3) follows from 11.3.10.

Assume the hypothesis of (4). From 5.5.2.1 and 5.5.4.1, Hypothesis 15.4.1 is satisfied by $\tilde{\tau}$, with the exception of 15.1.1.4. Similarly by 5.5.4.1, $\tilde{\tau}$ satisfies Hypothesis 11.3.8.3. By 5.5.6.1, $\tilde{\mathcal{F}}$ has two classes of involutions with representatives $\tilde{z}$ and $\tilde{u}$. Now the proof of (3) establishes (4). □

LEMMA 15.4.4. *Let $u \in U^\# \cap \mathcal{F}^f$. Then $\tau_u^\circ = (\mathcal{F}_u^\circ, \Omega)$ is a quaternion fusion packet and one of the following holds:*

(1) $W \trianglelefteq \mathcal{F}_u$.

(2) $\mathcal{F}_u^\circ \cong Spin_8^+[m]$.

(3) $m = 8$ and $\mathcal{F}_u^\circ \cong \hat{\Omega}_8^+(2)$, $\hat{L}_3(2)/2^{3+6}$, or $\hat{A}E_8$.

(4) $\mathcal{F}_u^\circ \cong Spin_9[m]$ or $Spin_{10}^-[m]$.

(5) $m = 8$ and $\mathcal{F}_u^\circ \cong \hat{A}E_9$.

Moreover in cases (1)-(3), $U = Z(\mathcal{F}_u^\circ)$, while in cases (4) and (5), $\langle u\rangle = Z(\mathcal{F}_u^\circ)$.

PROOF. First, τ_u° is a quaternion fusion packet by 2.5.2. Then if τ_u is proper in τ, the lemma follows from parts (2) and (3) of 15.1.9, so assume otherwise; then $u \in Z(\mathcal{F})$. Set $\mathcal{F}^+ = \mathcal{F}/\langle u\rangle$; then $\tau^+ = (\mathcal{F}^+, \Omega^+)$ is a quaternion fusion packet by 3.3.2, and as $\mathcal{F}$ is transitive on Ω, we have $\mathcal{F}^+$ transitive on Ω^+. By 6.6.9.2, τ^+ satisfies the Extended Inductive Hypothesis. Further $z = z_i^+$, where $i \in \{0, 1, 3\}$ with $u = zz_i$. Hence $|\Omega(z^+)| = 2$, so $\mathcal{F}^+$ appears as one of conclusions (1)-(4) of Theorem 12.3.50, using Theorem 12.3.24 to verify that $\mathcal{Z}^+ \cap O(z^+) = \{z^+\}$. Now, lifting back to $\mathcal{F}$ and appealing to 3.3.16, 5.10.12, and 5.1.16, the lemma follows. □

LEMMA 15.4.5. *(1) $\tau_z^\circ = (\mathcal{F}_z^\circ, \Omega)$ is a quaternion fusion packet and $\tau_z = \tau_1 * \tau_2$, where $\tau_i = (\mathcal{T}_i, \Omega_i)$, $\mathcal{T}_1 = \mathcal{O}(z)$, $\Omega_1 = \{K\}$, and $\Omega_2 = \{K_0, K_1, K_3\}$. Further $\mu(\tau_2) \cong E_8$ and $z \in Z(\mathcal{T}_2)$.*

(2) Up to choice of notation, one of the following holds:

(i) $z_i \in Z(\mathcal{T}_2)$ for each $i \in \{0, 1, 3, 4\}$ and $U \trianglelefteq \mathcal{F}_z$.

(ii) $\tau_2 = \xi_0 * \xi_1$, where $\xi_0 = (\mathcal{O}(z_0), \{K_0\})$ and $\xi_1 = (\mathcal{G}, \{K_1, K_3\}$ with $\mu(\xi_1) \cong E_4$ and either $\mathcal{G} \cong Sp_4[m]$ or $SL_4^-[m]$, or $m = 8$ and $\mathcal{G} \cong \hat{A}E_5$. Moreover $zz_0 \in Z(\mathcal{F}_z)$.

(iii) $\mathcal{T}_2 \cong Sp_6[m]$ or $SL_6^-[m]$ and $\langle z\rangle = Z(\mathcal{T}_2)$.

PROOF. By 3.3.1, τ_z° is a quaternion fusion packet; then (1) follows. Observe that $1 = z_0z_1z_3z$, so $z = z_0z_1z_3 \in \mathcal{T}_2$ and hence $z \in Z(\mathcal{T}_2)$.

If $z_i \in Z(\mathcal{T}_i)$ for each $i \in \{0,1,3\}$ then (2i) holds, so we may assume $z_1 \notin Z(\mathcal{T}_2)$. Then as $z_1 = zz_0z_3$ and $z \in Z(\mathcal{F}_z)$, we may also assume $z_3 \notin Z(\mathcal{T}_2)$.

Suppose that $z_0 \in Z(\mathcal{T}_2)$. Then $\tau_2 = \xi_0 * \xi_1$, where $\xi_0 = (\mathcal{O}(z_0), \{K_0\})$ and $\xi_1 = (\mathcal{G}, \{K_1, K_3\})$ with $\mu(\xi_1) \cong E_4$. Then (2ii) holds by 15.1.5.3, observing that $zz_0 = z_1z_3 \in Z(\mathcal{G})$.

This leaves the case where for each $i \in \{0,1,3\}$, $z_i \notin Z(\mathcal{T}_2)$. Here (2iii) holds by 15.1.5.3, again keeping in mind that $z \in Z(\mathcal{T}_2)$. □

LEMMA 15.4.6. *Assume $U \trianglelefteq \mathcal{F}_u$ for each $u \in U^\#$. Then $U \leq Z(\mathcal{F})$.*

PROOF. If $U \trianglelefteq \mathcal{F}$ then as U centralizes Ω and $\mathcal{F} = \mathcal{F}^\circ$, also $U \leq Z(\mathcal{F})$. So it suffices to show that $U \trianglelefteq \mathcal{F}$.

By 15.4.3.3, $\mathcal{F} = \langle N_{\mathcal{F}}(W), \mathcal{F}_t : t \in \mathfrak{T}\rangle$, so as $U \trianglelefteq N_{\mathcal{F}}(W)$ it remains to show $U \trianglelefteq \mathcal{F}_t$ for each $t \in \mathfrak{T}$. This holds by hypothesis if $t \in U$, so by 15.4.3.2 it suffices to show $U \trianglelefteq \mathcal{F}_z$. Thus we may assume case (i) of 15.4.5.2 does not hold. If case (ii) of that lemma holds, then $u = zz_0 \in Z(\mathcal{F}_z)$, so as $U \trianglelefteq \mathcal{F}_u$, our lemma holds in that case. Thus we may assume 15.4.5.2.iii is satisfied. Let $u = zz_0$; we may assume $u \in \mathcal{F}^f$. By hypothesis, $U \trianglelefteq C_{\mathcal{F}_u}(z)$. But $C_{\mathcal{F}_u}(z) = C_{\mathcal{F}_z}(u)$ and U is not normal in $C_{\mathcal{F}_z}(u) = C_{\mathcal{F}_z}(z_0)$ in 15.4.5.2.iii, completing the proof of the lemma. □

In light of 15.4.6, in the remainder of the section we assume:

HYPOTHESIS 15.4.7. *Hypothesis* 15.4.1 *holds and U is not normal in $\mathcal{F}_{u_0}$ for some $1 \neq u_0 \in U \cap Z(S)$.*

LEMMA 15.4.8. *We have $\langle u_0\rangle = Z(\mathcal{F}_{u_0}\rangle$ and either*
(1) $\mathcal{F}_{u_0}^\circ \cong Spin_9[m]$ or $Spin_{10}^-[m]$, or
(2) $m = 8$ and $\mathcal{F}_{u_0}^\circ \cong \hat{A}E_9$.

PROOF. This is a consequence of 15.4.7 and 15.4.4. □

LEMMA 15.4.9. *(1) For T_0 Sylow in $\mathcal{F}_{u_0}^\circ$, $\langle u_0\rangle = C_U(T_0)$.*
(2) Either
(a) $Aut_{N_{\mathcal{F}}(W)}(U) = GL(U)$, or
(b) u_0 is weakly closed in U with respect to $\mathcal{F}$ and for each $u \in U - \langle u_0\rangle$, $u \in \mathcal{F}^f$ and $u_0 \in Z(\mathcal{F}_u)$.

PROOF. Part (1) follows from 15.4.8 and 11.5.17.2.

By (1), either $Aut_{\mathcal{F}}(U) = Aut_{T_0}(U) \cong \mathbf{Z}_2$, or $Aut_{\mathcal{F}}(U) = GL(U)$. As $\mathcal{N} = N_{\mathcal{F}}(W)$ controls fusion in W, (2a) holds in the latter case, so we may assume the former, where u_0 is weakly closed in U. Let $u \in U - \langle u_0\rangle$; then $u \notin u_0^{\mathcal{F}}$ and $u \in \mathcal{F}^f$. Let T_u be Sylow in $\mathcal{F}_u^\circ$. As u_0 is weakly closed in U, T_u centralizes u_0, so by 11.5.17.2, $\mathcal{F}_u^\circ$ does not satisfy one of cases (4) or (5) of 15.4.4. Hence $U \trianglelefteq \mathcal{F}_u$ by 15.4.4, so $u_0 \in Z(\mathcal{F}_u)$ as u_0 is weakly closed in U, and hence (2b) holds. □

LEMMA 15.4.10. *Assume u_0 is weakly closed in U with respect to $\mathcal{F}$. Then $u_0 \in Z(\mathcal{F})$ and either*
(1) $\mathcal{F} \cong Spin_9[m]$ or $Spin_{10}^-[m]$, or
(2) $m = 8$ and $\mathcal{F} \cong \hat{A}E_9$.

PROOF. If $u_0 \in Z(\mathcal{F})$ then (1) or (2) holds by 15.4.4 and 15.4.7. Thus it remains to show $u_0 \in Z(\mathcal{F})$. As u_0 is weakly closed in U, $N_{\mathcal{F}}(W) \leq \mathcal{F}_{u_0}$, so by 15.4.3 and 15.4.9.2, it suffices to show $u_0 \in Z(\mathcal{F}_z)$. Assume otherwise; then by 15.4.9.2, $U \cap Z(\mathcal{F}_z) = 1$, and as u_0 is weakly closed in U, U is not normal in $\mathcal{F}_z$. Therefore by 15.4.5.2, conclusion (iii) of that lemma holds. But now u_0 is not weakly closed in U with respect to $\mathcal{T}_2$, a contradiction. □

Given lemmas 15.4.9.2 and 15.4.10, in the remainder of the section we assume:

HYPOTHESIS 15.4.11. *Hypothesis* 15.4.1 *holds,* $Aut_{N_{\mathcal{F}}(W)}(U) = GL(U)$, *and for* $1 \neq u_0 \in Z(S) \cap U$, U *is not normal in* $\mathcal{F}_{u_0}$.

LEMMA 15.4.12. $\mathcal{F}_{u_0}$ *is* $Spin_9[m]$ *or* $Spin_{10}^-[m]$.

PROOF. Assume otherwise; then by 15.4.8, $m = 8$ and $\mathcal{Y} = \mathcal{F}_{u_0}^\circ$ is $\hat{A}E_9$. Recall Notation 15.4.2, and choose $u_0 = zz_0$. Let $u = zz_3$. As $Aut_{\mathcal{F}}(U) = GL(U)$ there is $\alpha \in \mathfrak{A}(u)$ with $u\alpha = u_0$. Then α acts on W and hence also on U and $C_S(U)$. Then as $O = O(\tau) \leq C_S(U)$, we have $O\alpha = O$. Indeed as $Aut_{N_{\mathcal{F}}(W)}(U) = GL(U)$, we can choose α to centralize $\langle z, z_1 \rangle$ and interchange u and u_0, so $\alpha = (z_0, z_3)$ as a permutation of $\mathcal{Z}_S$. There is $\phi \in Aut_{\mathcal{Y}}(O)$ of order 3 centralizing K_1K_3 with $C_{KK_0}(\phi) = \langle z, z_0 \rangle$. Let $\varphi = \phi\alpha^*$; then as $\alpha = (z_0, z_3)$, φ centralizes K_1K_0 with $C_{KK_3}(\varphi) = \langle z, z_3 \rangle$. Now $\mathcal{Y} \trianglelefteq \mathcal{F}_{z_0}$, so and argument in the proof of 5.10.13 supplies a contradiction. □

DEFINITION 15.4.13. Define $\mathcal{F}$ to be of *type* $F_4[m]$ if $\mathcal{F}_{u_0}^\circ$ is $Spin_9[m]$ and $\mathcal{F}_z^\circ$ is $\mathcal{O}(z) * Sp_6[m]$. Define $\mathcal{F}$ to be of *type* $E_6^-[m]$ if $\mathcal{F}_{u_0}^\circ$ is $Spin_{10}^-[m]$ and $\mathcal{F}_z^\circ$ is $\mathcal{O}(z) * SL_6^-[m]$.

LEMMA 15.4.14. $\mathcal{F}$ *is of type* $\Psi[m]$, *where* Ψ *is* F_4 *or* E_6^-.

PROOF. Set $\mathcal{Y} = \mathcal{F}_{u_0}^\circ$; by 15.4.12, $\mathcal{Y}$ is $Spin_9[m]$ or $Spin_{10}^-[m]$. As usual take $u_0 = zz_0$. Let $\mathcal{X} = C_{\mathcal{F}_{u_0}}(z)^\circ$; then $\mathcal{X} = (\mathcal{O}(z) \times \mathcal{O}(z_0)) * \mathcal{C}$, where $\mathcal{C}$ is $Sp_4[m]$ or $SL_4^-[m]$, respectively. Now $\mathcal{X} = C_{\mathcal{F}_z}(u_0)^\circ$, so we conclude from 15.4.5.2 that case (ii) or (iii) of that lemma holds. By 15.4.11, and arguing as in the proof of 15.4.12, there is $\gamma \in Aut_{\mathcal{F}}(C_S(U))$ centralizing z with $|\gamma_{|U}| = 3$. Then γ is transitive on $\{K_0, K_1, K_3\}$, so 15.4.5.2.iii holds. But this implies the lemma. □

LEMMA 15.4.15. *Let* $\mathcal{Y} = C_{\mathcal{F}}(U)^\circ$ *and* $\bar{\mathcal{Y}} = \mathcal{Y}/U$.

(1) $\mathcal{Y} \cong Spin_8^+[m]$ *and* $\bar{\mathcal{Y}} \cong P\Omega_8^+[m]$ *is tamely realized by* $\bar{L}_U \cong P\Omega_8^+(q)$ *for suitable* q.

(2) $\mathcal{Y} = \mathcal{F}_{S_M}(L_U)$ *where* $L_U \cong Spin_8^+(q)$ *is the universal covering group of* $\bar{L}_U$.

(3) $C_S(S_M) = C_S(\mathcal{Y})$.

(4) $O^2(C_{Aut_{\mathcal{F}}(S_M)}(U)) = 1$.

PROOF. As $\mathcal{Y} = C_{\mathcal{F}}(U)^\circ$, $\mathcal{Y} \cong Spin_8^+[m]$ by 15.4.14. Therefore $\bar{\mathcal{Y}} \cong P\Omega_8^+[m]$, and then (1) holds by 3.5 in [**AO16**]. Now (1) and 3.3.16 imply (2).

Next $S\mathcal{Y}$ reduces to $\bar{\mathcal{Y}}$, so by (1) and 2.22.3 in [**AO16**], $\bar{S\mathcal{Y}} = S\mathcal{Y}/C_S(\mathcal{Y}) = \mathcal{F}_{\bar{S}}(\bar{X})$ for some $\bar{L}_U \leq \bar{X} \leq Aut(\bar{L}_U)$. Hence if $s \in S - C_S(\mathcal{Y})$ centralizes S_M, then $\bar{s} = \bar{z}$, so $s = zc$ for some $c \in C_S(S_M)$. This is a contradiction as $z \notin Z(S_M)$, so (3) is established.

Let $\Sigma = O^2(C_{Aut_{\mathcal{F}}(S_M)}(U))$ and σ of odd order in Σ. Set $E = \langle \mathcal{Z}_S \rangle$. Then $E \cong E_8$ and $\mathcal{Z}_S = zU$, so as σ centralizes U and is of odd order, we conclude that σ centralizes E. Therefore σ acts on each $K_i \in \Omega$ and then on the cyclic subgroup $K_i \cap W$ of index 2 in K_i, so σ centralizes K_i. Thus σ centralizes $O = O(\tau)$. Now $W/(W \cap O(\tau))$ is cyclic, so σ centralizes W. Then $[S_M, \sigma] \leq C_{S_M}(W) = W \leq C_{S_M}(\sigma)$, so $\sigma = 1$, proving (4). □

NOTATION 15.4.16. For $x \in \{z, u_0\}$ let T_x be Sylow in $\mathcal{F}_x^\circ$. Set $T_0 = T_{u_0}$ and $T = T_0 T_z$.

LEMMA 15.4.17. *Assume* $\Psi = F_4$. *Then*
(1) $|T_0| = (m/2)^4 \cdot 2^7$.
(2) $|T_z| = (m/2)^4 \cdot 2^4$.
(3) $T_z \leq T_0$ *so* $T = T_0$.

PROOF. Part (1) follows from 5.2.7.3. Next $\mathcal{F}_z^\circ = \mathcal{O}(z) * \mathcal{T}_2$ with $\mathcal{T}_2 \cong Sp_6[m]$, so $T_z = K * T_2$ with T_2 Sylow in $\mathcal{T}_2$. By 5.2.6, $|T_2| = (m/2)^3 \cdot 2^4$, so (2) follows. Indeed $C_{\mathcal{F}_z}(u_0)^\circ = (\mathcal{O}(z) \times \mathcal{O}(z_0)) * \mathcal{C}$ where $\mathcal{C} \cong Sp_4[m]$, so a Sylow group of $C_{\mathcal{F}_z}(u_0)^\circ$ is of order $(m/2)^4 \cdot 2^4$, and hence is T_z by (2). Therefore (3) holds. □

LEMMA 15.4.18. *Assume* $\Psi = E_6^-$. *Then*
(1) $|T_0| = (m/2)^4 \cdot 2^8$.
(2) $|T_z| = (m/2)^4 \cdot 2^6$.
(3) $|T : T_0| = 2$ *and* $C_T(S_M) = U$.

PROOF. Part (1) follows from 5.2.7.5. By 5.2.5.2 a Sylow group of $\mathcal{T}_2$ is of order $(m/2)^3 \cdot 2^6$, so (2) follows.

This time $C_{\mathcal{F}_z}(u_0)^\circ = (\mathcal{O}(z) \times \mathcal{O}(z_0)) * \mathcal{C}$ with $\mathcal{C} \cong SL_4^-[m]$, so its Sylow group P is $(K \times K_0) * P_1$ where P_1 is Sylow in $\mathcal{C}$ of order $(m/2)^2 \cdot 2^4$. Hence $|P| = (m/2)^4 \cdot 2^5$, so $|T_z : P| = 2$. Indeed $T_z = P\langle s \rangle$ where $s \in \mathcal{Z}(\tau_2)$ induces an outer automorphism on K_0 and K_1 and centralizes KK_3. Then $s \notin T_0$, as involutions in T_0 centralizing K also centralize K_0, Therefore s induces an outer automorphism on $\mathcal{F}_{u_0}^\circ$, so $|T_z : T_z \cap T_0| = 2$, proving (3). □

LEMMA 15.4.19. *(1)* $\mathcal{F}_z = \langle C_S(z)\mathcal{F}_z^\circ, C_{\mathcal{F}_z}(u_0) \rangle$.
(2) $C_T(z)\mathcal{F}_z^\circ \trianglelefteq \mathcal{F}_z$.

PROOF. Set $S_z = C_S(z)$ and $\mathcal{Y} = \mathcal{F}_z^\circ$. By 1.3.2 in [**Asc19**],

$$\mathcal{F}_z = \langle S_z \mathcal{Y}, N_{\mathcal{F}_z}(T_z) \rangle,$$

so to prove (1) it suffices to show $N_{\mathcal{F}_z}(T_z) \leq \mathcal{F}_{u_0}$. But this holds as K_0 is the unique member of $\Omega - \{K\}$ normal in T_z, so $N_{\mathcal{F}_z}(T_z) \leq C_{\mathcal{F}_z}(z_0) = C_{\mathcal{F}_z}(u_0)$. Hence (1) holds.

Let $\mathcal{X} = T\mathcal{F}_{u_0}^\circ$, $\mathcal{X}_z = C_{\mathcal{X}}(z)$, and $T_1 = C_T(z)$. Then $N_{\mathcal{X}_z}(T_z) = T_1$ if $m > 8$, while if $m = 8$ at least $N_{\mathcal{X}_z}(T_z) = T_1 N_{\mathcal{Y}}(T_z)$. Hence as $N_{\mathcal{F}_z}(T_z) \leq C_{\mathcal{F}_z}(u_0)$, we conclude that $T_1 N_{\mathcal{Y}}(T_z) \trianglelefteq N_{\mathcal{F}_z}(T_z)$. Then (2) follows from 1.5.2 in [**Asc19**]. □

LEMMA 15.4.20. $T \trianglelefteq S$.

PROOF. As $u_0 \in Z(S)$, we have $T_0 \trianglelefteq S$. Similarly $T_z \trianglelefteq C_S(z)$, so $C_S(z)$ acts on $T_0 T_z = T$. As $S_M \leq T_0 \leq T$, S_M acts on T. Finally as S_M is transitive on $\mathcal{Z}_S$, $S = S_M C_S(z)$, completing the proof. □

LEMMA 15.4.21. $T\mathcal{F}_{u_0}^\circ \trianglelefteq \mathcal{F}_{u_0}$.

PROOF. Set $\mathcal{Y} = \mathcal{F}_{u_0}^{\circ}$ and $\mathcal{X} = \mathcal{F}_{u_0}$. If $\Psi = F_4$ then $T = T_0$ so $T\mathcal{Y} = \mathcal{Y} \trianglelefteq \mathcal{X}$, so we may assume that $\Psi = E_6^-$. By 1.5.2 in [**Asc19**], it suffices to show that $TN_{\mathcal{Y}}(T_0) \trianglelefteq N_{\mathcal{X}}(T_0) = \mathcal{N}$. Observe that $N_{\mathcal{Y}}(T_0) = T_0$, so we must show that $T \trianglelefteq \mathcal{N}$. Now $C_T(z) \trianglelefteq C_{\mathcal{N}}(z)$ by 15.4.19.2. Also S_M is transitive on $\mathcal{Z}_S$, so $\mathcal{N} = T_0C_{\mathcal{N}}(z)$. Therefore $T = T_0C_T(z) \trianglelefteq T_0C_{\mathcal{N}}(z) = \mathcal{N}$. □

LEMMA 15.4.22. *T is strongly closed in S with respect to $N_{\mathcal{F}}(U)$.*

PROOF. Set $\mathcal{N} = N_{\mathcal{F}}(U)$ and $\mathcal{U} = C_{\mathcal{F}}(U)^{\circ}$. By 1.3.2 in [**Asc19**], $\mathcal{N} = \langle S\mathcal{U}, N_{\mathcal{N}}(S_M)\rangle$. As $S\mathcal{U} \leq \mathcal{F}_{u_0}$, T is strongly closed in S with respect to $S\mathcal{U}$ by 15.4.21. Therefore it suffices to show that T is strongly closed in S with respect to $\mathcal{M} = N_{\mathcal{N}}(S_M)$.

Suppose $t \in T - S_M$ and $\phi \in \hom_{\mathcal{M}}(t, S)$; we must show $t\phi \in T$. Set $P = S_M\langle t\rangle$; we may assume $\phi \in \hom_{\mathcal{M}}(P, S)$. As S_M is transitive on $\mathcal{Z}_S$, we may take $t \in C_T(z)$. Now $z\phi \in \mathcal{Z}_S$ so there is $s \in S_M$ with $z\phi c_s = z$. By 15.4.19.2, $C_T(z)$ is strongly closed in $C_S(z)$ with respect to $\mathcal{F}_z$, so $t\phi c_s \in T$. Then by 15.4.20, $t\phi \in Tc_s^{-1} = T$, completing the proof. □

NOTATION 15.4.23. Set $\mathcal{F}_1 = \mathcal{F}_{u_0}$, $\mathcal{E}_1 = T\mathcal{F}_{u_0}^{\circ}$, $\mathcal{F}_2 = N_{\mathcal{F}}(U)$, and $\mathcal{E}_2 = [T]_{\mathcal{F}_2}$. Set $\mathcal{E} = \langle \mathcal{E}_1, \mathcal{E}_2\rangle$.

LEMMA 15.4.24. *(1) $\bar{\mathcal{E}}_2 = \mathcal{E}_2/U = \mathcal{F}_{\bar{T}}(\bar{L}_2)$, where $P\Omega_8^+(q) \cong \bar{L}_U \leq \bar{L}_2 \leq Aut(\bar{L}_U)$ and $\bar{L}_2$ is generated by conjugates of reflections under $Aut(\bar{L}_U)$.*

(2) If $\Psi = F_4$ then $\bar{L}_2/\bar{L}_U \cong S_3$.

(3) If $\Psi = E_6^-$ then $\bar{L}_2/\bar{L}_U$ is the subgroup of $Out(\bar{L}_U)$ generated by all $Out(\bar{L}_U)$-conjugates of images of reflections.

(4) T is Sylow in $\mathcal{E}_2$.

(5) $\mathcal{E}_2 = \mathcal{F}_T(L_2)$ where L_2 is the pull back of $\bar{L}_2$ via the universal property of the universal cover L_U of $\bar{L}_U$.

PROOF. We appeal to 2.1.14 applied to $\mathcal{N} = N_{\mathcal{F}}(U)$, $\mathcal{Y} = C_{\mathcal{F}}(U)^{\circ}$, S_M, T in the roles of $\mathcal{F}$, $\mathcal{F}_0$, S_0, T. Hypotheses (a), (c), and (d) of 2.1.14 are satisfied by 15.4.15. Hypothesis (b) holds by 15.4.11. Finally we check that (e) holds: First, T is strongly closed by 15.4.22. By 15.4.9.1, $[T, U] \neq 1$. Recall S_M is Sylow in $\mathcal{Y}$, so to verify (e) it remains to show that $C_T(S_M) = U$. If $\Psi = E_6^-$ this follows from 15.4.18.3, while if $\Psi = F_4$ it follows from 15.4.17.3 and the structure of $Spin_9[m]$. This completes the verification of the hypotheses of 2.1.14.

By 2.1.14.1, (4) holds. By 2.1.14.2, $F^*(\mathcal{E}_2) = \mathcal{Y}$. By 2.1.14.5, $\mathfrak{red}(\mathcal{E}_2) = \bar{\mathcal{Y}} = \mathcal{Y}/U$, so by 2.22.3 in [**AO16**], $\mathcal{E}_2 = \mathcal{F}_T(L_2)$ for some group L_2 with $F^*(L_2) = L_U$. Then, using 15.4.15, $P\Omega_8^+(q) \cong \bar{L}_U \leq \bar{L}_2 \leq Aut(\bar{L}_U)$. Now (5) follows from the universal property of the universal covering group L_U of $\bar{L}_U$.

From 15.4.14 and 5.9.1, there is $r \in T - S_M$ with $r^2 = u_0$ and $\bar{r}$ induces a reflection on $\bar{L}_U$. Set $H = N_{L_2}(S_2)$. From 2.1.14.4, $C_H(U) = C_T(U)$ and $H/C_T(U) \cong S_3$. If $\Psi = F_4$ then $S_M = C_T(U)$, so as $r \in T - S_M$ induces a reflection, (1) and (2) hold, completing the proof in this case. So we may take $\Psi = E_6^-$. Then $Out(\bar{L}_U) = \Sigma \times \Gamma$ where $\Sigma \cong S_4$ is the subgroup described in (3) and Γ is the group of field automorphisms. Now $T/S_M = \Omega_1(T/S_M)$ is of order 8, so we conclude $T/S_M \cong D_8$ and $\bar{L}_2/\bar{L}_U \cong S_4$. As $\bar{r}$ is a reflection, we have $\bar{r}\bar{L}_U \in \Sigma$, so $\bar{L}_2/\bar{L}_U = \Sigma$. So in this case (1) and (3) hold, completing the proof. □

LEMMA 15.4.25. *(1) $\mathcal{F} = \langle \mathcal{F}_1, \mathcal{F}_2\rangle$.*

(2) Assume the setup of 15.4.3.4. Then $\tilde{\mathcal{F}} = \langle \tilde{\mathcal{F}}_1, \tilde{\mathcal{F}}_2 \rangle$, *where* $\tilde{\mathcal{F}}_1 = \tilde{\mathcal{F}}_{\tilde{u}}$ *and* $\tilde{\mathcal{F}}_2 = N_{\tilde{\mathcal{F}}}(Z(\tilde{M}))$.

PROOF. We first prove (1). By 15.4.3 and 15.4.11, $\mathcal{F} = \langle \mathcal{F}_1, \mathcal{F}_2, \mathcal{F}_z \rangle$, so it remains to show that $\mathcal{F}_z \leq \langle \mathcal{F}_1, \mathcal{F}_2 \rangle$.

Set $S_z = C_S(z)$. From Theorem 13.6.3, $S_z\mathcal{T}_2 = \langle S_z C_{\mathcal{T}_2}(z_0), N_{S_z\mathcal{T}_2}(\langle z_0, z_1, z_3 \rangle) \rangle$. By 1.3.2 in [**Asc19**], $\mathcal{F}_z = \langle S_z\mathcal{T}_2, N_{\mathcal{F}_z}(T_2) \rangle$, where T_2 is Sylow in $\mathcal{T}_2$. Now $\mathcal{F}_1$ contains $S_z C_{\mathcal{T}_2}(z_0)$ and $N_{\mathcal{T}_2}(\langle z_0, z_1, z_3 \rangle)$, so $S_z\mathcal{T}_2 \leq \mathcal{F}_1$. Further $N_{\mathcal{F}_z}(T_2) \leq \mathcal{F}_2$, so (1) holds.

The proof of (2) is essentially the same, given 15.4.3.4. □

NOTATION 15.4.26. Let $\tilde{\mathcal{F}} = \Psi[m]$. Then $\tilde{\mathcal{F}}$ satisfies our hypotheses (with the exception of aB.1.4), with $\tilde{\mathcal{F}}$ of type $\Psi[m]$. We will show that $\mathcal{F} \cong \tilde{\mathcal{F}}$. By 15.4.14, $\mathcal{F}^\circ_{u_0} = \mathcal{F}_{T_0}(L_0)$ for $L_0 \cong Spin_9(q)$ or $Spin^-_{10}(q)$ for suitable q. The same is true for $\tilde{\mathcal{F}}$ by 5.5.6.2, so we may take $\mathcal{F}^\circ_{u_0} = \tilde{\mathcal{F}}^\circ_{\tilde{u}_0}$, $S_M = \tilde{S}_M$, etc.

LEMMA 15.4.27. *If* $\Psi = F_4$ *then* $\mathcal{E} = \tilde{\mathcal{F}}$.

PROOF. From 15.4.17.3, $T = T_0 = \tilde{S}$, so $\mathcal{E}_1 = \mathcal{F}_T(L_0) = \tilde{\mathcal{F}}_{\tilde{S}}(L_0) = \tilde{\mathcal{F}}_1$.

Also $L_U = C_{L_0}(U)^\infty$ and $T = \langle r \rangle S_M$, where $r^2 = u_0$ and $\bar{r}$ acts as a reflection on $\bar{L}_U$. From 15.4.24.2, $L_2/L_U \cong S_3$ and $\bar{L}_2$ is generated by reflections. Therefore $\bar{L}_2$ is determined up to conjugacy under $C_{Aut(\bar{L}_U)}(\bar{r})$. Hence L_2 is determined up to conjugation via $\alpha \in N_{Aut(\bar{L}_U)}(L_U\langle r \rangle)$, and α is induced by $\beta \in Aut(L_0)$, so, adjusting our identification of $\mathcal{E}_1$ with $\tilde{\mathcal{F}}_1$ via β if necessary, we may assume $\mathcal{E}_2 = \mathcal{F}_T(L_2) = \tilde{\mathcal{F}}_2$. By 15.4.25.2 we have $\tilde{\mathcal{F}} = \langle \tilde{\mathcal{F}}_1, \tilde{\mathcal{F}}_2 \rangle$, so the lemma holds. □

LEMMA 15.4.28. $\mathcal{E} = \tilde{\mathcal{F}}$.

PROOF. By 15.4.27, we may assume $\Psi = E^-_6$. By 15.4.24, $\mathcal{E}_2 = \mathcal{F}_T(L_2)$ with $\bar{L}_2/\bar{L}_U = \Sigma \cong S_4$, where Σ is the normal closure in $Aut(\bar{L}_U)$ of the image of a reflection. As the same holds for $\tilde{\mathcal{F}}$, it follows that $T = \tilde{S}$ and $\mathcal{E}_2 = \tilde{\mathcal{F}}_2$.

Next $T_0/S_M \cong E_4$ contains rS_M where $\bar{r}$ is a reflection on $\bar{L}_U$, so T_0/S_M is the 4-subgroup of T/S_M distinct from $O_2(N_{L_2}(S_M)/S_M)$. Let hS_M generate $\Phi(T/S_M)$. Then $\bar{h}$ can be chosen to be an involution of type $i(2,-)$ on $\bar{L}_U$. Hence $C_{\mathcal{E}_2}(h)^\circ \cong Spin^-_6[m]$, and of course a conjugate g of h under $N_{L_2}(S_M)$ in $T - T_0$ also satisfies $C_{\mathcal{E}_2}(g)^\circ \cong Spin^-_6[m]$. Hence g induces an outer diagonal automorphism on L_0, so $TL_0/\langle u_0 \rangle$ is determined as a subgroup of $Aut(L_0/\langle u_0 \rangle)$. Then as L_0 is the universal covering group of $L_0/\langle u_0 \rangle$, TL_0 is also determined. Thus $\mathcal{E}_1 = \mathcal{F}_T(TL_0) = \mathcal{F}_{\tilde{S}}(\tilde{S}L_0) = \tilde{\mathcal{F}}_1$. Then, as in the proof of the previous lemma, $\mathcal{E} = \langle \mathcal{E}_1, \mathcal{E}_2 \rangle = \langle \tilde{\mathcal{F}}_1, \tilde{\mathcal{F}}_2 \rangle = \tilde{\mathcal{F}}$. □

THEOREM 15.4.29. *Assume Hypothesis* 15.4.11. *Then* $\mathcal{F} \cong F_4[m]$ *or* $E^-_6[m]$.

PROOF. We show that the tuple $\mathcal{F}_1, \mathcal{F}_2, \mathcal{E}_1, \mathcal{E}_2$ satisfies Hypothesis 1.4.7 in [**Asc19**]. Then we appeal to 1.4.8 in [**Asc19**] to conclude that $\mathcal{E} \trianglelefteq \mathcal{F}$. Hence $\mathcal{F} = \mathcal{F}^\circ = \mathcal{E}$, so the theorem follows from 15.4.28.

Hypothesis 1.4.1 in [**Asc19**] is satisfied by 15.4.25. By 15.4.21, $\mathcal{E}_1 \trianglelefteq \mathcal{F}_1$, while by construction in 15.4.23, $\mathcal{E}_2 \trianglelefteq \mathcal{F}_2$. Hence condition (1) of 1.4.7 in [**Asc19**] holds.

Next $N_{\mathcal{F}}(T) \leq N_{\mathcal{F}}(W) \leq N_{\mathcal{F}}(U) = \mathcal{F}_2$, so as $\mathcal{E}_2 \trianglelefteq \mathcal{F}_2$, we have $Aut_{\mathcal{F}}(T) \leq Aut(\mathcal{E}_2)$. Similarly $\langle u_0 \rangle = C_U(T)$, so as $\mathcal{E}_1 \trianglelefteq \mathcal{F}_1$, also $Aut_{\mathcal{F}}(T) \leq Aut(\mathcal{E}_1)$. That is condition (2) of 1.4.7 in [**Asc19**] holds.

Condition (3) follows from the definition of $\mathcal{E}$ as $\langle \mathcal{E}_1, \mathcal{E}_2 \rangle$, and finally $\mathcal{E}$ is saturated by 15.4.28, so condition (4) holds. This completes the verification of Hypothesis 1.4.7 in [**Asc19**] and hence also the proof of the theorem. □

THEOREM 15.4.30. *Assume $\tau = (\mathcal{F}, \Omega)$ is a quaternion fusion packet such that $\mathcal{F} = \mathcal{F}^\circ$ and $\mu(\tau) \cong Weyl(D_4)$. Assume for $z \in \mathcal{Z}_S$ that $|\Omega(z)| = 1$, and let $M \in M(\tau)$ and $U = Z(M)$. Assume the Extended Inductive Hypothesis. Then $Z(\mathcal{F}) \leq U$ and one of the following holds:*

(1) $U = Z(\mathcal{F})$ and one of the following holds:

(i) $W \trianglelefteq \mathcal{F}$.

(ii) $\mathcal{F} \cong Spin_8^+[m]$.

(iii) $m = 8$ and $\mathcal{F} \cong \hat{O}_8^+(2)$, $\hat{L}_3(2)/2^{3+6}$, or $\hat{A}E_8$.

(2) $Z(\mathcal{F}) \cong \mathbf{Z_2}$ and either

(a) $\mathcal{F} \cong Spin_9[m]$ or $Spin_{10}^-[m]$, or

(b) $m = 8$ and $\mathcal{F} \cong \hat{A}E_9$.

(3) $Z(\mathcal{F}) = 1$ and $\mathcal{F} \cong F_4[m]$ or $E_6^-[m]$.

PROOF. By assumption, conditions (1) and (4) of Hypothesis 15.1.1 hold, and as $\mu(\tau) \cong Weyl(D_4)$, conditions (2) and (3) of that hypothesis are also satisfied. That is Hypothesis 15.1.1 holds, and then also Hypothesis 15.4.1 is satisfied.

If $U \trianglelefteq \mathcal{F}_u$ for each $u \in U^\#$ then by 15.4.6, $U \leq Z(\mathcal{F})$. Then (1) holds by 15.4.4 applied to any $u \in U^\#$. Therefore we may assume Hypothesis 15.4.7 is satisfied.

If u_0 is weakly closed in U with respect to $\mathcal{F}$ then by 15.4.10, $u_0 \in Z(\mathcal{F})$ and then (2) holds by that lemma. Therefore we may assume u_0 is not weakly closed, so by 15.4.9, Hypothesis 15.4.11 holds. Now Theorem 15.4.29 says that (3) holds, completing the proof. □

15.5. $\Phi = A_n$

In this section we assume the following hypothesis:

HYPOTHESIS 15.5.1. *Either*

(1) $\tau = (\mathcal{F}, \Omega)$ is a quaternion fusion packet with $\mathcal{F}^\circ = O^2(\mathcal{F})$ and τ° is the Lie packet of $\bar{L}/\bar{Z}$, where $\bar{L} = SL_d(q)$ with $(q-1)_2 = m/2$, $d = n+1 \geq 5$, and $\bar{Z} \leq Z(\bar{L})$, or

(2) Hypothesis 15.2.1 *holds with $\mathcal{F} = \mathcal{F}^\circ$ and $\mu = \mu(\tau) \cong Weyl(\Phi)$, where Φ is of type A_n for some $n \geq 4$.*

In addition we adopt Notation 15.1.3, set $S_M = S \cap M$, $\mathcal{M} = \mathcal{F}_{S_M}(M)$, and we adopt the following notation:

NOTATION 15.5.2. We adopt a modified version of the Notation 5.4.12 and 5.7.1. Let q be a prime power with $(q-1)_2 = m/2$, $d = n+1$, and U a d-dimensional vector space over $F = \mathbf{F}_q$. Let $\bar{L} = SL(U) = SL_d(q)$, $\bar{S} \in Syl_2(\bar{L})$, $\bar{\mathcal{F}} = \mathcal{F}_{\bar{S}}(\bar{L})$, and $\bar{\tau} = (\bar{\mathcal{F}}, \bar{\Omega})$ a Lie packet for $\bar{L}$. Let $\bar{W} \in W(\bar{\tau})$ and $\bar{M} \in M(\bar{\tau})$. By 4.3.8 there is a surjective homomorphism $\pi : \bar{M} \to M$ with $\ker(\pi) \leq Z(\bar{M})$. We use π to study M and τ, via the representation of $\bar{M}$ on U.

Let $I = \{1,\dots,d\}$, $\mathfrak{B} = \{u_i : i \in I\}$ a basis for U, and $\Gamma = \{Fu_i : i \in I\}$. Let $\mathfrak{M} = N_{\bar{L}}(\Gamma)$ and $\mathfrak{W}$ the kernel of the action of $\mathfrak{M}$ on Γ. From 5.4.13 and 5.7.2, $\mathfrak{M} = \bar{M}O(\mathfrak{M})$, $\bar{W} = \bar{M} \cap \mathfrak{W}$, and $\bar{M}$ and $\bar{W}$ are described in 5.7.2. In particular $\bar{M}$ is represented as $Sym(\Gamma) \cong Sym(I)$ on Γ with kernel $\bar{W}$.

LEMMA 15.5.3. *Set $\bar{W}_0 = \Omega_1(\bar{W})$.*
(1) $\bar{W}_0$ is the core of the permutation module for $\bar{M}/\bar{W} \cong Sym(I)$.
(2) For $\bar{t} \in \bar{W}_0$ of weight k, $\dim([U,\bar{t}]) = k$.
(3) $\mathcal{Z}_\Delta$ is the set of vectors e_α, $\alpha \in \binom{I}{2}$, of weight 2.

PROOF. This is essentially 5.7.3. □

NOTATION 15.5.4. Recall from 3.4.11 that

$$\mathcal{E}(\tau) = \langle S\mathcal{O}(\tau), N_{\mathcal{F}}(O(\tau)), N_{\mathcal{F}}(O(\xi)), N_{\mathcal{F}}(W) : \xi \in \Xi \rangle,$$

where Ξ consists of those $\emptyset \neq \xi \subseteq \mathcal{Z}_S$ such that $N_S(\xi)$ is transitive on ξ and $O(\xi) = \prod_{s\in\xi} O(s) \in \mathcal{F}^f$. Further from Definition 11.3.2, $\mathcal{W}(\tau)$ consists of those $1 \neq f_\theta = \prod_{t\in\theta} t$ for $\theta \subseteq \mathcal{Z}_S$ such that $|\theta|$ is a power of 2, unless $n_S = 4$ and $m(W_S) = 3$, where $\mathcal{W}(\tau)$ also includes the involutions in $\mathcal{W}_*(\tau)$. In the latter case by iJ.3.2 we have $n = 7$ and the members of $\mathcal{W}_*(\tau)$ are the M-conjugates of c_0x_0 where $c_0 \in Z(M)$ with $\bar{c}_0 = c_0\pi^{-1}$ of order 4 in $Z(\bar{M})$ and $x_0 = f_\theta$ for some 2-subset θ of $\mathcal{Z}_S$. In particular either $c_0 = 1$ and $x = f_\theta \in W_S$ or $c_0 \neq 1$ and $x \notin W_S$. Moreover

$$\mathcal{B}(\tau) = \langle \mathcal{E}(\tau), \mathcal{F}_x : x \in \mathcal{W}(\tau) \cap \mathcal{F}^f \rangle.$$

LEMMA 15.5.5. *(1) $n_S = |\mathcal{Z}_S| = d/2$ if d is even and $(d-1)/2$ of d is odd.*
(2) Let $\Theta = \{\{2i-1, 2i\} : 1 \leq i \leq n_S\}$. We can choose notation so that $\mathcal{Z}_S = \{e_\theta : \theta \in \Theta\}$.
(3) Let $W_S = \langle \mathcal{Z}_S \rangle$. Then either $m(W_S) = n_S$, or d is even, $e_I \in \ker(\pi)$, so that $f_{\mathcal{Z}_S} = e_I\pi = 1$, and $m(W_S) = n_S - 1$.

PROOF. This follows from 15.5.3. □

LEMMA 15.5.6. *$\mathcal{F} = \mathcal{B}(\tau)$.*

PROOF. The lemma follows from 11.3.10, once we verify that Hypothesis 11.3.8 holds; note that conclusion (2) of 11.3.10 does not hold as $\mu \cong \mathrm{Weyl}(A_n)$.

We first observe that Hypothesis 11.2.1 holds; this is immediate, except for 11.2.1.2, which follows from 15.5.5.3. Similarly 11.3.8.2 holds in 15.5.1.1, while 11.3.8.1 holds in 15.5.1.2 by 15.1.1.4. □

LEMMA 15.5.7. *$\mathcal{F} = \langle N_{\mathcal{F}}(W), \mathcal{F}_x : x \in \mathcal{W}(\tau) \cap \mathcal{F}^f \rangle$.*

PROOF. Set $\mathcal{B}_1 = \langle N_{\mathcal{F}}(W) : \mathcal{F}_x : x \in \mathcal{W}(\tau) \cap \mathcal{F}^f \rangle$. By 15.5.6, it suffices to show that $\mathcal{E}(\tau) \leq \mathcal{B}_1$.

For $x \in \mathcal{W}(\tau)$, x is conjugate under $\mathcal{M} \leq N_{\mathcal{F}}(W) \leq \mathcal{B}_1$ to a member of $\mathcal{W}(\tau) \cap \mathcal{F}^f$, so $\mathcal{F}_x \leq \mathcal{B}_1$.

Let $\xi \in \Xi$. If $f_\xi \neq 1$ then $f_\xi \in \mathcal{W}(\tau)$ and $N_{\mathcal{F}}(O(\xi)) \leq C_{\mathcal{F}}(f_\xi) \leq \mathcal{B}_1$. On the other hand if $f_\xi = 1$ then d is even, $\xi = \mathcal{Z}_S$, and it preimage in $\bar{W}_0$ is $\bar{f}_\xi = -1_U \in \ker(\pi)$. As $\xi = \mathcal{Z}_S$, $O(\xi) = O(\tau) = O$, so it remains to show that $S\mathcal{O}(\tau)$ and $N_{\mathcal{F}}(O)$ are contained in $\mathcal{B}_1$.

Observe that $W_S \leq Z(\mathcal{O}(\tau))$ and there exists $x \in \mathcal{W}(\tau) \cap Z(S) \subseteq C_{W_S}(S)$, so $S\mathcal{O}(\tau) \leq \mathcal{F}_x \leq \mathcal{B}_1$.

The same argument shows that $SC_{\mathcal{F}}(W_S) \leq \mathcal{F}_x \leq \mathcal{B}_1$. By 1.3.2 in [**Asc19**], $\mathcal{N} = N_{\mathcal{F}}(O) = \langle SC_{\mathcal{F}}(W_S), N_{\mathcal{N}}(T)\rangle$, where $T = C_S(W_S)$. But $W \leq T$ is weakly closed, so $N_{\mathcal{N}}(T) \leq N_{\mathcal{F}}(W) \leq \mathcal{B}_1$, completing the proof. □

LEMMA 15.5.8. *For each involution $t \in W_0$ and each $t_0 \in t^{\mathcal{F}} \cap \mathcal{F}^f$, we have $t_0 \in t^M \cap W_S$.*

PROOF. By 15.5.3, $\bar{W}_0$ is the core of the permutation module for $\bar{M}/\bar{W} = Sym(I)$. Therefore $t = e_\theta$ for some $\theta \subseteq I$ of even order k. Set $\beta_i = \{2i-1, 2i\} \in \Theta$; from 15.5.5.3, $t_i = e_{\beta_i} \in \mathcal{Z}_S$. Let $\beta = \beta_1 \cup \cdots \cup \beta_{k/2}$; then $s = e_\beta = t_1 \cdots t_{k/2} \in W_S$, and as $\bar{W}_0$ is the core of the permutation module, $s \in t^M$, so we may take $t = s \in W_S$. In particular, $O = O(\tau) \leq C_S(t)$. Let $\alpha \in \mathfrak{A}(t)$; then $W = W\alpha$ and $O = O\alpha$, so $t\alpha \in t^M$ and $t\alpha \in W_S$, completing the proof. □

REMARK 15.5.9. From 5.7.2.6, $Z(\bar{M})$ is cyclic of order $(d, m/2)$. In particular if d is odd then $Z(\bar{M}) = 1$ so $\pi : \bar{M} \to M$ is an isomorphism, and we identify the two groups via this isomorphism. If d is even then $\bar{W}_0 \cap Z(\bar{M}) = e_I = \bar{f}_{\mathcal{Z}_S}$, which we denote by $\bar{c}$. Set $c = \bar{c}\pi$; if $c \neq 1$ then as $Z(\bar{M})$ is cyclic, again π is an isomorphism. Finally if $c = 1$ then each $x \in W_0$ has two preimages; we pick one and call it $\bar{x}$; the second preimage is $\bar{x}\bar{c}$.

Recall the exceptional case from 15.5.4, where $n = 7$ and $m(W_S) = 3$. In this case $c = 1$, and, as in 15.5.4 let $\bar{c}_0 \in Z(\bar{M})$ be of order 4 and set $c_0 = \bar{c}_0\pi$. From 15.5.4, $\mathcal{W}_*(\tau) \subseteq \mathcal{W}(\tau)$ consists of the W-conjugates of $c_0 f_\theta$, where $\theta \subseteq \mathcal{Z}_S$ is of order 2. Recall that if $c_0 = 1$ then $c_0 f_\theta = f_\theta \in W_S$.

LEMMA 15.5.10. *Let t be an involution in $W_0 \cap \mathcal{F}^f$, set $U(1) = [U, \bar{t}]$, $U(2) = C_U(\bar{t})$, and $k_i = \dim(U_i)$.*

(1) k_1 is the weight of $\bar{t}$ in $\bar{W}_0$, so k_1 is even. Further $k_2 = d - k_1$.

(2) $\theta_t = (C_{\mathcal{M}}(t)^\circ, \Omega)$ is a quaternion fusion packet.

(3) $\theta_t = \theta_1 * \theta_2$, where $\theta_i = (\mathcal{M}_i, \Omega_i)$, $\Omega_i = \{K \in \Omega : [U, \bar{K})] \leq U(i)\}$, and $\mathcal{M}_i = [\Omega_i]_{C_{\mathcal{M}}(t)^\circ}$.

(4) $\mu(\theta_i) \cong S_{k_i} \cong Weyl(A_{k_i - 1})$, unless $k_2 \leq 1$ where $\tau_2 = \emptyset$.

PROOF. By 15.5.8, $t \in W_S$, so t centralizes Ω and $\bar{t}$ is of even weight k in $\bar{W}_0$. By 15.5.3.2, $k = k_1$, so (1) holds. As t centralizes Ω, (2) follows from 2.5.2.

As k is even, $k \geq 2$. Now $C_{\bar{L}}(\bar{t})^\circ = L_1 * L_2$, where $L_i = C_{\bar{L}}(U(3-i))^\circ \cong SL_{k_i}(q)$, except if $k_2 \leq 1$ where $L_2 = 1$. Let $S_i = S_L \cap L_i$ and $\xi_i = (\mathcal{F}_{S_i}, \Lambda_i)$. As $t \in W_0 \cap \mathcal{F}^f$, ξ_i is a Lie packet of L_i with $\mu(\xi_i) \cong S_{k_i}$. Let $M_i \in M(\xi_i)$ and $\mathcal{M}_i = \mathcal{F}_{S_i}(M_i)$. Then $C_{\mathcal{M}}(t)^\circ = \mathcal{M}_1 * \mathcal{M}_2$, so we conclude that (3) and (4) hold. □

LEMMA 15.5.11. *Let t be an involution in $W_0 \cap \mathcal{F}^f$ of weight k_1 and set $U(1) = [U, \bar{t}]$, $U(2) = C_U(\bar{t})$, and $k_2 = d - k_1$.*

(1) $\tau_t^\circ = \tau_1 * \tau_2$ is a central product of quaternion fusion packets $\tau_i = (\mathcal{Y}_i, \Omega_i)$ such that $\mathcal{Y}_i = \mathcal{Y}_i^\circ$ is transitive on $\Omega_i = \{K \in \Omega : [U, \bar{K})] \leq U(i)\}$.

(2) $M_i \in M(\theta_i)$ is in $M(\tau_i)$ and $\mu_i = \mu(\tau_i) \cong S_{k_i}$, unless $k_2 \leq 1$ where $\tau_2 = \emptyset$.

(3) One of the following holds:

(i) $W(\tau_i) \trianglelefteq \mathcal{Y}_i$.

(ii) $\mathcal{Y}_i \cong SL_{k_i}[m]$, or $k_i = 2$ and $\mathcal{Y}_i \cong Q_m$.

(iii) $k_i = 3$ and $\mathcal{Y}_i \cong L_3^+[m]^{(1)}$ or $G_2[m]$, or $m = 8$ and $\mathcal{Y}_i \cong M_{12}$ or $L_3(2)/E_8$.

(iv) $k_i = 4$ and $\mathcal{Y}_i \cong Spin_7[m]$ or $Spin_8^-[m]$, or $m = 8$ and $\mathcal{Y}_i \cong \hat{A}E_6$, $\hat{A}E_7$, $\hat{L}_3(2)/E_{64}$, or $\hat{S}p_6(2)$.

PROOF. Adopt the notation from 15.5.10. By 15.5.10.3, $\theta_t = \theta_1 * \theta_2$ with $\mu_i = \mu(\theta_i) \cong S_{k_i}$, unless $k_2 \leq 1$, where $\theta_2 = \emptyset$.

If 15.5.1.1 holds then the lemma follows from the discussion in section 5.4, so we may assume 15.5.1.2 holds.

Set $\rho = \tau_t^\circ$; by 15.1.5.1, $\rho = \rho_1 * \cdots * \rho_s$ is a central product of quaternion fusion packets $\rho_i = (\mathcal{Y}_i, \Gamma_i)$ such that $\mathcal{Y}_i = \mathcal{Y}_i^\circ$ is transitive on Γ_i. As $\mathcal{M}_i$ is transitive on Ω_i, we can choose notation so that $\mathcal{M}_i \leq \mathcal{Y}_i$ and $\Omega_i \subseteq \Gamma_i$, for $i = 1, 2$.

If $t \in Z(M)$ then $k_2 \leq 1$, so $\theta_2 = \emptyset$, $\mathcal{M}_1 = \mathcal{M}$, and $\Omega_1 = \Omega$. Thus $s = 1$ and $\rho = \rho_1 = \tau_1$, so that (1) and (2) hold in this case. On the other hand if $t \notin Z(M)$ then $t \notin Z(\mathcal{F})$, so by induction on the order of τ, $\mu(\mathcal{Y}_i)$ is coconnected, and hence $\mu(\rho_i) = \mu(\theta_i)$ and $M_i \in M(\theta_i)$ is in $M(\rho_i)$. As $\Omega = \Omega_1 \cup \Omega_2$, we conclude that $s = 2$ (unless $\Omega = \Omega_1$ where $s = 1$). This completes the proof of (1) and (2).

It remains to prove (3). Suppose first that $t \in Z(\mathcal{F})$ and set $\mathcal{F}^+ = \mathcal{F}/\langle t \rangle$. By induction on the order of τ, either $W^+ \trianglelefteq \mathcal{F}^+$ or $\mathcal{F}^+$ is the image of $SL_d[m]$ modulo its central subgroup of order 2. But then, pulling back to $\mathcal{F}$ and using 3.3.16, either $W \trianglelefteq \mathcal{F}$ or $\mathcal{F} \cong SL_d[m]$. That is (3i) or (3ii) holds in this case.

Therefore we may assume that ρ is proper in τ. Then (3) follows from 15.1.5.4 and 15.1.9.2. □

THEOREM 15.5.12. *Assume d is even. Then*
(1) $c \in Z(\mathcal{F})$, and
(2) if $c \neq 1$ then either $\mathcal{F} \cong SL_{n+1}[m]$ or $W \trianglelefteq \mathcal{F}$.

PROOF. If $c = 1$ then the lemma is trivial, so we may assume $c \neq 1$. In particular from 15.5.9 the exceptional case $n = 7$ and $m(W_S) = 3$ does not hold.

From 15.5.9, $c = e_I = f_\zeta$, where $\zeta = \mathcal{Z}_S$, and $\bar{c}$ is the involution in $Z(\bar{M})$, so $c \in Z(N_{\mathcal{F}}(W))$. Let $t \in \mathcal{W}(\tau) \cap \mathcal{F}^f$ with $t \neq c$. As the exceptional case does not hold, it follows from 15.5.8 and 15.5.9 that $t \in W_S$, and we adopt the notation of 15.5.11. Then $c = t \cdot s$, where s is the involution in $Z(\mathcal{Y}_2)$. Now either $\mathcal{Y}_2 \trianglelefteq \mathcal{F}_t$, so $s \in Z(\mathcal{F}_t)$, or t is of weight $d/2$ and $t = s$. But in the latter case $c = 1$, so that (1) holds; thus we may assume the former holds, so that $c = ts \in Z(\mathcal{F}_t)$.

We've shown that $c \in Z(\mathcal{F}_t)$ for each $t \in \mathcal{W}(\tau) \cap \mathcal{F}^f$. Therefore by 15.5.7, $c \in Z(\mathcal{F})$, completing the proof of (1). Then (2) follows from 15.5.11 (and its proof) applied to c in the role of t. □

NOTATION 15.5.13. By 15.5.11, $\tau_z^\circ = \tau_K * \tau'$ where $\tau_K = (\mathcal{O}(z), \{K\})$ and $\tau' = (\mathcal{T}, \Omega - \{K\})$ with $\mu(\tau') \cong S_{d-2}$.

LEMMA 15.5.14. *Assume $n = 7$, $m(W_S) = 3$, and $c_0 \neq 1$.*
(1) $\tau_{c_0}^\circ = (\mathcal{Y}, \Omega)$ is a quaternion fusion packet with $\mu(\tau_{c_0}^\circ) \cong S_8$.
(2) $c_0 \in Z(S)$.
(3) Either $W \trianglelefteq \mathcal{Y}$ or $\mathcal{Y}$ is the 2-fusion system of $\bar{L}/\langle -1_U \rangle$.
(4) Either $W(\tau') \trianglelefteq \mathcal{T}$ or $\mathcal{T} \cong SL_6^+[m]$. Further $W \trianglelefteq \mathcal{Y}$ iff $W(\tau') \trianglelefteq \mathcal{T}$.
(5) There is a 2-subset θ of $\mathcal{Z}_S$ with $f = f_\theta \in Z(S)$. Set $x = c_0 f$.
(6) c_0 is in the center of $\mathcal{F}_f$ and $\mathcal{F}_x$.
(7) If $c_0 \in Z(\mathcal{F}_z)$ then $c_0 \in Z(\mathcal{F})$.

PROOF. From 15.5.9, c_0 is the involution in $Z(M)$, so (1) and (2) hold. In 15.5.1.1, (3), (6), and (7) follow as $\bar{c}_0 \in Z(\bar{L})$, while in 15.5.1.2, (3) follows from (1) and 15.1.5, unless $c_0 \in Z(\mathcal{F})$. But in the latter case, passing to $\mathcal{F}/\langle c_0 \rangle$, inducting on the order of τ, and appealing to 3.3.16, again (3) follows.

From 15.5.9, $c = 1$, so (5) follows.

The first sentence in (4) follows from 15.5.11.3. Suppose $\mathcal{Y}$ is the 2-fusion system of $\bar{L}/\langle -1_U \rangle$, and let $\mathcal{X} = C_{\mathcal{Y}}(K)^\circ$. Then $\mathcal{X} \cong SL_6[m]$ and $\mathcal{X} \leq \mathcal{T}$, so $\mathcal{X} = \mathcal{T}$. Conversely assume $\mathcal{T} \cong SL_6 + [m]$ and write $f = zt$ for some $t \in \mathcal{Z}_S$. Then $C_{\mathcal{T}}(f)^\circ = \mathcal{O}(z) * \mathcal{O}(t) * \mathcal{E}$ with $\mathcal{E} \cong SL_4^+[m]$. In a moment we will use this fact to show W is not normal in $\mathcal{Y}$, completing the proof of (4).

Next, using 15.5.11.2, we have $M(\tau_f) = M(\tau_x) = M_1 \times M_2$, with $\mu(M_i) \cong S_4$ and $Z(M_i) = \langle c_i \rangle \cong \mathbf{Z}_4$, where $f = c_1^2$ and $c_0 = c_1c_2$. From 15.5.11.1, $\tau_f^\circ = \tau_1 * \tau_2$ with $\tau_i = (\mathcal{Y}_i, \Omega_i)$ and $M_i \in M(\tau_i)$. Some member of S interchanges τ_1 and τ_2, so $\mathcal{Y}_1 \cong \mathcal{Y}_2$. By 15.5.11.3, one of the following holds:

(a) $W(\tau_i) \trianglelefteq \mathcal{Y}_i$.

(b) $\mathcal{Y}_i \cong SL_4^+[m]$.

(c) $m = 8$ and $\mathcal{Y}_i$ is $\hat{A}E_6$, $\hat{L}_3(2)/E_{64}$, or $\hat{S}p_6(2)$.

Here we eliminate the possibilities $\mathcal{Y}_i \cong Spin_7[m]$ or $Spin_8^-[m]$ using $f \in Z(\mathcal{Y}_i)$. In case (c), $z, t \in \mathcal{Z}_S(\tau_1)$, so $\mathcal{Y}_2 \leq C_{\mathcal{F}}(O(z)O(t))^\circ = \mathcal{E}$, so $\mathcal{Y}_2 \cong \hat{A}E_6$ as $\mathcal{E} \cong SL_4^+[m]$. But similarly $\mathcal{E} \leq C_{\mathcal{F}}(f)^\circ$, so $\mathcal{E} \leq \mathcal{Y}_2$, a contradiction. Therefore (a) or (b) holds. Further if $\mathcal{T} \cong SL_6^+[m]$ then $\mathcal{E} \leq C_{\mathcal{F}}(O(z)O(t)) = \mathcal{Y}_2$, so (b) holds. Then as $c_0 = c_1c_2$, we have $c_0 \in Z(\mathcal{F}_f)$, so c_0 centralizes $\mathcal{E}$, and hence W is not normal in $\mathcal{Y}$, completing the proof of (4). Notice we've also shown that $c_0 \in Z(\mathcal{F}_f)$.

Recall (6) and (7) hold in 15.5.1.1, so we may assume 15.5.1.2 holds. As $c_0 \in Z(\mathcal{F}_f)$ we have $\mathcal{F}_f \leq \mathcal{F}_x$. By 15.1.5 and 15.1.9.2, $\tau_x^\circ = \rho_1 * \rho_2$ with $\rho_i = (\mathcal{X}_i, \Omega_i)$, $M_i \in M(\rho_i)$, $\mathcal{Y}_i \leq \mathcal{X}_i$, and either $\mathcal{Y}_i = \mathcal{X}_i$ or $\mathcal{X}_i$ appears in 15.5.11.3.iv. As $x = c_0 f \in Z(M_1M_2)$, we have $x \in Z(\mathcal{X}_1\mathcal{X}_2)$, and then arguing as above, $\mathcal{X}_i = \mathcal{Y}_i$, completing the proof of (6).

Finally if $c_0 \in Z(\mathcal{F}_z)$ then $c_0 \in Z(\mathcal{F})$ by 15.5.7 and (6), establishing (7). □

LEMMA 15.5.15. *Let t be an involution in $W_0 \cap \mathcal{F}^f$ and for $i = 1, 2$, define $\tau_i = (\mathcal{Y}_i, \Omega_i)$ and k_i as in aG.11. Assume $W(\tau') \trianglelefteq \mathcal{T}$. Then for $i = 1, 2$, $W(\tau_i) \trianglelefteq \mathcal{Y}_i$.*

PROOF. Assume $W' = W(\tau') \trianglelefteq \mathcal{T}$ but $W_i = W(\tau_i)$ is not normal in $\mathcal{Y}_i$ for some i.

Suppose first that $J \in \Omega_{3-i}$ and let $j = z(J)$ and $\alpha \in \mathfrak{A}(j)$ with $j\alpha = z$. Then $\mathcal{X} = \mathcal{Y}_i\alpha^* \leq \mathcal{F}_z$, and as $\mathcal{Y}_i$ is coconnected, it follows that either $\mathcal{X} \leq \mathcal{O}(z)$ or $\mathcal{X} \leq \mathcal{T}$, while as $j\alpha = z$ and $J \notin \Omega_i$, it must be the latter. Therefore $\mathcal{X} \leq \mathcal{T}$, so $W(\tau_i)\alpha \leq W' \trianglelefteq \mathcal{T}$ and hence $W(\tau_i) \trianglelefteq \mathcal{Y}_i$, contrary to assumption.

Therefore we may assume $\Omega_{3-i} = \emptyset$, so $\Omega = \Omega_i$ and hence $i = 1$ and either $k_1 = d$ is even or d is odd and $k_1 = d - 1$. In the first case $t = c$, and then 15.5.12.2 says that either $\mathcal{F} \cong SL_d[m]$, contradicting $W' \trianglelefteq \mathcal{T}$, or $W \trianglelefteq \mathcal{F}$, where the lemma holds. Therefore we may assume that the second case holds.

Now $t \in Z(S)$, so $z \in \mathcal{F}_t^f$ and $C_{\mathcal{Y}_1}(z)^\circ = \mathcal{O}(z) * \mathcal{C}$, where $\rho = (\mathcal{C}, \Omega - \{K\})$ is a quaternion fusion packet with $\mu(\rho) \cong S_{d-3}$. Observe $\mathcal{C} \leq \mathcal{T}$, so $W(\rho) \leq W'$, and then as $W' \trianglelefteq \mathcal{T}$, also $W(\rho) \trianglelefteq \mathcal{C}$. Then as $W(\tau_1)$ is not normal in $\mathcal{Y}_1$ and $d \geq 5$, we conclude from 15.5.11.3 that $d = 5$, $m = 8$, and $\mathcal{Y}_1$ appears in 15.5.11.3.iv.

Set $O = O(\tau)$, $R = OC_S(O)$, and let G_R be a model for $N_{\mathcal{F}}(R)$. As $d = 5$, $\Omega = \{K, K_0\}$ is of order 2, and from 15.5.11.3.iv, there exists X of order 3 in the normal submodel of G_R for $N_{\mathcal{Y}_1}(R)$ with $C_O(X) = Z(O)$. But as $W' \trianglelefteq \mathcal{T}$, X acts on $W' \cap K_0$ of index 2 in K_0, contradicting $C_{K_0}(X) = Z(K_0)$. The proof is complete. □

THEOREM 15.5.16. *If $W(\tau') \trianglelefteq \mathcal{T}$ then $W \trianglelefteq \mathcal{F}$.*

PROOF. Assume $W' = W(\tau') \lhd \mathcal{T}$. Let $r \in z^{\mathcal{F}} \cap W'$ with $r \in \mathcal{F}_z^f$, and let $\alpha \in \mathfrak{A}(r)$ with $r\alpha = z$. By 15.5.15, $W_z = W(\tau_z) \lhd \mathcal{F}_z^{\circ}$, so $W_z \lhd \mathcal{F}_z$. As $W \leq C_S(r)$, we have $W\alpha = W$. Then as $W_z \lhd \mathcal{F}_z$, we have $W_r = W_z\alpha^{-1} \lhd C_{\mathcal{F}_z}(r)$. Therefore $W = W_zW_r \lhd C_{\mathcal{F}_z}(r)$. However $r \in W'$, so $C_{\mathcal{F}_z}(W_z) \leq C_{\mathcal{F}_z}(r)$, and hence $W \lhd C_{\mathcal{F}_z}(W_z)$. Thus $W \leq O_2(\mathcal{F}_z)$, so as W is weakly closed, we have $W \lhd \mathcal{F}_z$.

Now choose t as in 15.5.11 and adopt the notation of that lemma. By 15.5.15, $W_t = W(\mathcal{F}_t^{\circ}) \lhd \mathcal{F}_t$. Let $s \in z^{\mathcal{F}} \cap \mathcal{F}_t^f$ with $s \in \mathcal{Z}_S(\tau_t)$, and let $\beta \in \mathfrak{A}(s)$ with $s\beta = z$. As above, $W_s = W_z\beta^{-1} \lhd C_{\mathcal{F}_t}(s)$. Then $W = W_tW_s \lhd C_{\mathcal{F}_t}(W_t)$, so as above, $W \lhd \mathcal{F}_t$.

Now, if the exceptional case in 15.5.14 does not hold, then $\mathcal{W}(\tau) \subseteq W_S$, so by 15.5.7, $\mathcal{F} \leq N_{\mathcal{F}}(W)$, completing the proof. So assume the exceptional case holds and adopt the notation of 15.5.14; in this case we must also show that $\mathcal{F}_x \leq N_{\mathcal{F}}(W)$. But $W(\tau') \lhd \mathcal{T}$, so by 15.5.14.4, $W \lhd \mathcal{F}_{c_0}$, and by 15.5.14.6. $\mathcal{F}_x \leq \mathcal{F}_{c_0}$, completing the proof. □

THEOREM 15.5.17. *Assume $n = 7$, $m(W_S) = 3$, and $c_0 \neq 1$. Then $c_0 \in Z(\mathcal{F})$.*

PROOF. By 15.5.14.3 either $W \lhd \mathcal{Y}$ or $\mathcal{Y}$ is the 2-fusion system of $\bar{L}/\langle -1_U\rangle$. In the first case by 15.5.14.4 and Theorem 15.5.16, we have $W \lhd \mathcal{F}$, so the lemma holds as c_0 is the involution in M. Thus we may assume $\mathcal{Y}$ is the 2-fusion system of $\bar{L}/\langle -1_U\rangle$. Hence by 15.5.14.4, $\mathcal{T} \cong SL_6^+[m]$. By 15.5.14.7, it suffices to show that $c_0 \in Z(\mathcal{F}_z)$.

Set $W' = W(\tau')$. Applying 15.5.7 to $\mathcal{T}$ is the role of $\mathcal{F}$, we have $\mathcal{T} = \langle C_{\mathcal{T}}(f), N_{\mathcal{T}}(W')\rangle$. By 15.5.14.6, $\mathcal{F}_f \leq \mathcal{F}_{c_0}$. As W' is selfcentralizing in $\mathcal{T}$, we have $W = C_{W\mathcal{T}}(W') \lhd N_{W\mathcal{T}}(W')$, so $N_{\mathcal{T}}(W') \leq \mathcal{F}_{c_0}$. Therefore $\mathcal{T}$ centralizes c_0.

By 1.3.2 in [**Asc19**], $\mathcal{F}_z = \langle C_S(z)\mathcal{T}, N_{\mathcal{F}_z}(T)\rangle$ for T Sylow in $\mathcal{T}$. Thus it remains to show $N_{\mathcal{F}_z}(T) \leq \mathcal{F}_{c_0}$. But $N_{\mathcal{F}_z}(T) \leq C_{\mathcal{F}_z}(f) \leq \mathcal{F}_{c_0}$ by 15.5.14.6, completing the proof. □

In light of theorems 15.5.12, 15.5.16, and 15.5.17, during the remainder of the section we assume:

HYPOTHESIS 15.5.18. *Hypothesis* 15.5.1 *holds. Further*
(1) $W(\tau')$ is not normal in $\mathcal{T}$.
(2) If n is odd then $e_I = 1$.
(3) If $n = 7$ then $c_0 = 1$.

LEMMA 15.5.19. *Suppose $m > 8$, t is in the setup of* 15.5.11, *and for some i with $k_i \geq 3$, we have $\mathcal{Y}_i \cong SL_{k_i}[m]$. Then $\mathcal{O}(z) \cong SL_2[m]$.*

PROOF. Let $J \in \Omega_i$ and $j = z(J)$. As $\mathcal{Y}_i \cong SL_{k_i}[m]$ with $k_i \geq 3$, it follows that $\mathcal{O}(j) \cong SL_2[m]$, so the lemma follows. □

LEMMA 15.5.20. *Assume $d \geq 7$. Then*
(1) $\mathcal{T} \cong SL_{d-2}[m]$.
(2) Assume the setup of 15.5.11 *with $k_i \geq 3$. Then $\mathcal{Y}_i \cong SL_{k_i}[m]$.*

PROOF. Part (1) follows from 15.5.18.1 and 15.5.11.3.

Assume the hypothesis of (2). Suppose first that $J \in \Omega_{3-i}$, and choose J so that $j = z(J) \in \mathcal{F}_t^f$. Let $\alpha \in \mathfrak{A}(j)$ with $j\alpha = z$. Then $\mathcal{Y}_i\alpha^*$ is a coconnected component of $C_{\mathcal{T}}(t\alpha)^\circ$, so by (1), $\mathcal{Y}_i \cong \mathcal{Y}_i\alpha^* \cong SL_{k_i}[m]$. Thus we may assume $\Omega_{3-i} = \emptyset$, so $\Omega = \Omega_i$ and $k_i = d$ or $d-1$. Indeed $t \in Z(\mathcal{Y}_i)$, so k_i is even, and hence by aG.16, d is odd and $k_i = d-1$. Hence as $d \geq 7$ we have $k_i \geq 6$, so (2) follows from 15.5.11. □

LEMMA 15.5.21. *Assume $d = 6$. Then*
(1) $z^M = W_0^\#$, and
(2) $\mathcal{T} \cong SL_4^+[m]$.

PROOF. By 15.5.18.2, $e_I = 1$ in W_0, so W_0 is a natural module for $M/W \cong S_6$. This implies (1).

Assume (2) fails. Then by 15.5.18.1 and 15.5.11.3, $\mathcal{T}$ appears in case (iv) of the latter lemma. As $d = 6$, $\Omega = \{K_1, K_2, K_3\}$ is of order 3. Set $z_i = z(K_i)$ and choose $z = z_1$. Let $\alpha \in \mathfrak{A}(z_2)$ with $z_2\alpha = z$. Set $O = O(\tau)$, $\mathcal{D} = O^2(C_{\mathcal{T}}(O))$, and $\mathcal{B} = \mathcal{D}\alpha^*$. Then $O = O\alpha$, so $\mathcal{B} \leq C_{\mathcal{F}}(O)$.

Suppose first that $\mathcal{T} \cong Spin_7[m]$ or $Spin_8^-[m]$. Then $\mathcal{D} \cong SL_2[m]$ or $SL_2[2m]$, respectively, with $z = z(\mathcal{D})$. Therefore $z(\mathcal{B}) = z\alpha \neq z$, so $\mathcal{B} \neq \mathcal{D}$. But $\mathcal{B}$ is a component or solvable component of $C_{\mathcal{F}_z}(z\alpha)$, so by 2.2.6.4, either $\mathcal{B}$ centralizes $\mathcal{T}$ or the Sylow group B of $\mathcal{B}$ is contained in the Sylow group T_z of $\mathcal{T}$. As $z\alpha = z(\mathcal{B})$ does not centralize $\mathcal{T}$, we conclude that $B \leq T_z$, so $B \leq C_{T_z}(O) = D\langle z_2\rangle$, where D is Sylow in $\mathcal{D}$. But then $z(B) = z(D)$, contradicting $z \neq z\alpha$.

In the remaining cases, $m = 8$. Let $R = OC_S(O)$ and G_R a model for $N_{\mathcal{F}}(R)$. Then G_R is transitive on Ω. Let $H \lhd C_{G_R}(z)$ be the model for $N_{\mathcal{T}}(R)$. If $\mathcal{T} \cong \hat{A}E_7$ then $\mathcal{D} \cong SL_2(3)$ with $z(\mathcal{D}) = z$, and then the argument in the previous paragraph supplies a contradiction. Suppose $\mathcal{T} \cong \hat{A}E_6$ or $\hat{L}_3(2)/E_{64}$. Then $H = T_zX$ where $X \cong \mathbf{Z}_3$ with $C_O(X) = K\langle z_2\rangle$. As G_R is transitive on Ω, there is a G_R-conjugate X_2 of X with $C_O(X_2) = K_2\langle z\rangle$. But by 5.10.13.3, R admits no such group of automorphisms when $\mathcal{T}$ is $\hat{A}E_6$ or $\hat{L}_3(2)/E_{64}$.

This leaves the case where $\mathcal{T} \cong \hat{S}p_6(2)$. Set $Q_1 = [C_{T_z}(O), O^2(H)]$; then $Q_1 \cong Q_8$ with $z(Q_1) = z_1 = z$. Further $Q_1 \lhd C_{G_R}(z)$ with $C_H(z)$ irreducible on $Q_1/\langle z\rangle$, so $[R, Q_1] \leq \langle z\rangle$. Let $Q_i = Q_1^{g_i}$ for $g_i \in G_R$ with $z^g = z_i$ for $i = 2, 3$. Then $[Q_1, Q_i] \leq \langle z\rangle \cap \langle z_i\rangle = 1$, so $Q = \langle Q_1^{G_R}\rangle = Q_1 * Q_2 * Q_3$. Let $t \in \mathcal{Z}(\tau')$ with $Q_2^t = Q_3$; then $[Q_2, t] \leq Q_2Q_3 \cap T_z \leq C_{T_z}(O) = Q_1\langle z_2\rangle$, so $[Q_2, t] = Q_1\langle z_2\rangle$, whereas $Q_2Q_3 \cap Q_1 = \langle z\rangle$. □

LEMMA 15.5.22. *Assume $d = 5$. Then $\Omega = \{K, K_0\}$ is of order 2, and:*
(1) M has two orbits on $W_0^\#$ with representatives z and $t = zz(K_0)$.
(2) $\mathcal{T} \cong L_3^+[m]$.
(3) $\mathcal{F}_t^\circ \cong SL_4^+[m]$.

PROOF. As $d = 5$, $\Omega = \{K, K_0\}$ is of order 2; set $z_0 = z(K_0)$ and $t = zz_0$. Then $t \in Z(S)$, so $t \in \mathcal{F}^f$. By 15.5.3.1 and 15.5.9, W_0 is the core of the permutation module for $M/W \cong Sym(I) \cong S_5$, so (1) holds.

By 15.5.11, $\tau_t^\circ = (\mathcal{Y}, \Omega)$ and either (3) holds or $\mathcal{Y}$ appears in 15.5.11.3.iv. Similarly either (2) holds or $\mathcal{T}$ appears in 15.5.11.3.iii. Set $O = O(\tau) = KK_0$ and $\mathcal{D} = O^2(C_{\mathcal{Y}}(O))$.

Suppose $\mathcal{T} \cong L_3^+[m]^{(1)}$ or $G_2[m]$. Then $\mathcal{C} = O^2(C_{\mathcal{T}}(O)) \cong SL_2[m]$ is a component or solvable component of $C_{\mathcal{F}}(\langle z, t\rangle) = C_{\mathcal{F}_t}(z_0)$ with $z_0 = z(\mathcal{D})$. By 2.2.6.4, either $\mathcal{C}$ centralizes $\mathcal{Y}$ or the Sylow group T of $\mathcal{C}$ is contained in the Sylow group T_t of $\mathcal{Y}$. The former is impossible as z_0 does not centralize $\mathcal{Y}$. Therefore $T \leq T_t$, so

$T \leq C_{T_t}(O)$, so by inspection of the list in 15.5.11.3.iv, $\mathcal{Y}$ is $Spin_7[m]$, $Spin_8^-[m]$, $\hat{A}E_7$, $\hat{L}_3(2)/E_{64}$, or $\hat{S}p_6(2)$, and $z_0 = z(T) = t$, a contradiction. Hence by 15.5.11.3, either $\mathcal{T}$ is $L_3^+[m]$ or $m = 8$ and $\mathcal{T}$ is M_{12} or $L_3(2)/E_8$. In particular if $m > 8$ then (2) holds.

Suppose $\mathcal{Y}$ is $Spin_7[m]$, $Spin_8^-[m]$, or $\hat{A}E_7$. Then $\mathcal{D} \cong SL_2[m]$ or $SL[2m]$ with $z(\mathcal{D}) = t$ and $\mathcal{D}$ is a component or solvable component of $C_{\mathcal{F}}(\langle z, t\rangle) = C_{\mathcal{F}_z}(t)$. Therefore by 2.2.6.4, either $\mathcal{D}$ centralizes $\mathcal{T}$ or the Sylow group D of $\mathcal{D}$ is contained in the Sylow group T_z of $\mathcal{T}$. As t does not centralize $\mathcal{T}$, we conclude that $D \leq T_z$, contradicting $t \notin T_z$. In particular if $m > 8$ then $\mathcal{Y} \cong SL_4^+[m]$, and hence the lemma holds. Thus we may assume $m = 8$. Also by 15.5.11.3, $\mathcal{Y}$ is $SL_4^+[8]$, $\hat{A}E_6$, $\hat{L}_3(2)/E_{64}$, or $\hat{S}p_6(2)$.

Set $R = OC_S(O)$ and let G_R be a model for $N_{\mathcal{F}}(R)$ and $H \trianglelefteq G_R$ a model for $N_{\mathcal{Y}}(R)$. Observe there is X of order 3 in G_R induced in $\mathcal{T}$ such that X centralizes K and $XK_0 \cong SL_2(3)$. By 5.10.13.3, R admits no such group of automorphisms when $\mathcal{Y}$ is $\hat{A}E_6$ or $\hat{L}_3(2)/E_{64}$, so $\mathcal{Y}$ is $SL_4^+[8]$ or $\hat{S}p_6(2)$. Assume the latter and set $Q = [C_{T_t}(O), O^2(H)]$. As in the proof of 15.5.21, $Q \cong Q_8$ with $t = z(Q)$. As t does not centralize $\mathcal{T}$, Q is faithful on $\mathcal{T}$. It follows that $\mathcal{T}$ is M_{12} or $L_3(2)/E_8$ and $Q \leq K_1 \times C_S(\mathcal{T})$ with $K_1 = C_{T_z}(O) \cong Q_8$ and $z(K_1) = z_0$. But we can choose Y of order 3 in H centralizing K_0 with $Q = [Q, Y]$, whereas by 5.10.13.3, R admits no such group of automorphisms when $\mathcal{T}$ is M_{12} or $L_3(2)/E_8$. Therefore $\mathcal{Y} \cong SL_4^+[8]$, completing the proof of (3). Moreover we may assume that (2) fails, so $\mathcal{T} \cong M_{12}$ or $L_3(2)/E_8$. Now H contains A of order 3 with $[R, A] = K_0$. Again by 5.10.13.3, R admits no such group of automorphisms when $\mathcal{Y}$ is M_{12} or $L_3(2)/E_8$. This finally completes the proof of the lemma. □

LEMMA 15.5.23. *Assume the setup of* 15.5.11 *with* $k_i \geq 3$. *Then* $\mathcal{Y}_i \cong SL_{k_i}[m]$.

PROOF. If $d \geq 7$ this is a consequence of 15.5.20.2, while if $d = 6$ it follows from 15.5.21 and if $d = 5$ it follows from 15.5.22. □

LEMMA 15.5.24. *Let* $\mathcal{Z}_S = \{z_i : 1 \leq i \leq n_S\}$. *Then for* $1 \leq i \leq n_S$ *there exists* $SL_2[m] \cong \mathcal{O}_i = \mathcal{O}_{z_i} \leq \mathcal{F}$ *with Sylow group* $O(z_i)$ *such that*

(1) $\mathcal{O}_i = \mathcal{O}(z_i)$ *if* $m > 8$.

(2) $\mathcal{O}_\tau = \mathcal{O}_1 * \cdots * \mathcal{O}_{n_S} \trianglelefteq N_{\mathcal{F}}(W_S)$ *and* $N_{\mathcal{F}}(W_S)$ *permutes* $\Lambda = \{\mathcal{O}_i : 1 \leq i \leq n_S\}$ *as* $Sym(\Lambda)$.

(3) $\mathcal{O}_z \trianglelefteq \mathcal{F}_z$.

PROOF. If $m > 8$ then $\mathcal{O}(z_i) \cong SL_2[m]$ by 15.5.19, and the remainder of the lemma follows from 2.6.9 and 3.3.7. Therefore (1) holds and we may assume that $m = 8$.

Let $O = O(\tau)$ and G_0 be a model for $\mathcal{N} = N_{\mathcal{F}}(OC_S(O))$. Choose notation so that $z = z_{n_S}$. By 15.5.23, $\mathcal{T} \cong SL_{d-2}[8]$, so there is a model H for $N_{\mathcal{F}_z^\circ}(E)$ normal in $C_{G_0}(z)$ such that $H = K * H_0$ and $H_0 = H_1 \times \cdots \times H_k$ where $k = n_S - 1$, $O(z_i) = O_2(H_i)$, and $H_i \cong SL_2(3)$. As G_0 acts transitively on $\mathcal{Z}_S$ as $Sym(\mathcal{Z}_S)$ and $K \trianglelefteq C_{G_0}(z)$, it follows that there is $SL_2(3) \cong H_{n_S} \leq G_0$ with $K = O_2(H_{n_S})$ centralizing H_0, and G_0 permuting $\Gamma = \{H_i : 1 \leq i \leq n_S\}$ as $Sym(\Gamma)$. This establishes (1) and (2) with $\mathcal{O}_i = \mathcal{F}_{O(z_i)}(H_i)$.

Let $X \in Syl_3(H_{n_S})$, T Sylow in $\mathcal{T}$, $W_1 = W(\tau')$, and $M_1 = M(\tau')$. Observe that W is in the kernel of the action of G_0 on Γ and $[H_{n_S}, W_1] \leq C_{H_{n_S}}(K) = \langle z\rangle$, so W_1 centralizes $H_{n_S} = O^2(H_{n_S})$. Then X acts on M_1 with $[M_1, X] \leq C_{M_1}(W_1) =$

$W_1 \leq C_S(X)$, so X centralizes the Sylow group T of M_1 and $\mathcal{T}$. Therefore by 9.5 in [**Asc11**], X centralizes $\mathcal{T}$.

We've shown that $\mathcal{O}_z \leq \mathcal{C} = C_{\mathcal{F}_z}(\mathcal{T})$. Then $\mathcal{C}$ centralizes W_S, so by (2), we have $\mathcal{O}_z \trianglelefteq \mathcal{C}$. Now as $K \trianglelefteq \mathcal{F}_z$, we conclude from 7.4 in [**Asc11**] that (3) holds, completing the proof. □

LEMMA 15.5.25. *Assume the setup of* 15.5.11 *with* $k_1 \neq d/2$. *Then*

(1) $W \trianglelefteq N_{\mathcal{F}_t}(W(\tau_t))$.

(2) $\mathcal{F}_t = \langle \mathcal{X}_i, N_{\mathcal{F}_t}(W) : i = 1, 2\rangle$, *where* $\mathcal{X}_i = C_S(t)\mathcal{Y}_i$, *unless* $t = z$ *and* $m = 8$ *where* $\mathcal{X}_1 = C_S(z)\mathcal{O}_z$.

(3) Suppose for $i = 1, 2$ *that* $\mathcal{X}_i = \langle C_S(t)\mathcal{O}_{\tau_i}, N_{\mathcal{X}_i}(W(\tau_i))\rangle$. *Then* $\mathcal{F}_t \leq \langle C_S(t)\mathcal{O}_\tau, N_{\mathcal{F}_t}(W)\rangle$.

PROOF. As $k_1 \neq d/2$, we have $\mathcal{Y}_i \trianglelefteq \mathcal{F}_t$ for $i = 1, 2$. Let $S_t = C_S(t)$, T_i Sylow in $\mathcal{Y}_i$, $W_i = W(\tau_i)$, and $W_t = W(\tau_t)$. Then $T = T_1T_2$ is Sylow in $\mathcal{F}_t^\circ$ and $W_t = W_1W_2$. By 1.3.2 in [**Asc19**], $\mathcal{F}_t = \langle \mathcal{X}_1, N_{\mathcal{F}_t}(T_1)\rangle$ and $N_{\mathcal{F}_t}(T_1) = \langle \mathcal{X}_2, N_{\mathcal{F}_t}(T)\rangle$. But unless $t = z$ and $m = 8$, W_t is weakly closed in T, and in the exceptional case $N_{\mathcal{F}_t}(T) = \langle \mathcal{X}_1, N_{\mathcal{F}_t}(W_t)\rangle$, so:

(a) $\mathcal{F}_t = \langle \mathcal{X}_i, N_{\mathcal{F}_t}(W_t) : i = 1, 2\rangle$.

Hence (1) implies (2). Further for $i = 1, 2$, $S_t\mathcal{O}_{\tau_i} \leq S_t\mathcal{O}_\tau$ and as $W_{3-i} \trianglelefteq N_{\mathcal{X}_i}(W_i)$, we have $N_{\mathcal{X}_i}(W_i) \leq N_{\mathcal{F}_t}(W_t)$, so (1) and (2) imply (3). Therefore it remains to prove (1).

Pick $r \in \mathcal{Z}(\tau_t) \cap \mathcal{F}_t^f$ and let $\alpha \in \mathfrak{A}(r)$ with $r\alpha = z$. If $t = z$ pick $r \in \mathcal{Z}(\tau')$. Set $W_z = W(\tau_z)$. As in the proof of 15.5.16, $W_r = W_z\alpha^{-1} \leq W$ with $W = W_rW_t$.

Let $\mathcal{C} = C_{\mathcal{F}_t}(\mathcal{F}_t^\circ)$ and $P = C_S(\mathcal{F}_t^\circ)$. Then $\mathcal{C}\alpha^*$ centralizes $\mathcal{O}_\tau$, so by 15.5.23 and x.y, $\mathcal{C}\alpha^* \leq C_{\mathcal{F}_z}(\mathcal{F}_z^\circ)P\alpha$, so $\mathcal{C} \leq C_{\mathcal{F}_r}(W_r)P$. Therefore $W = W_rW_t \leq O_2(W\mathcal{C})$ and hence:

(b) $W \trianglelefteq S_t\mathcal{C}$.

Let $\mathcal{M}_t$ be the 2-fusion system of $M(\tau_t)$. Then $\mathcal{M}_t \trianglelefteq N_{\mathcal{F}_t}(W_t)$, so by 1.3.2 in [**Asc19**], $N_{\mathcal{F}_t}(W_t) = \langle S_t\mathcal{M}_t, N_{\mathcal{F}_t}(T)\rangle$ as T is Sylow in $\mathcal{M}_t$ by 5.3.5. Now $W \leq C_{S_t\mathcal{M}_t}(W_t) \trianglelefteq S_t\mathcal{M}_t$, so as W is weakly closed in S_t we have $W \trianglelefteq S_t\mathcal{M}_t$. Finally $N_{\mathcal{F}_t}(T) = S_t\mathcal{C}$ as $N_{Aut(L_1L_2)}(T) = RO(N_{Aut(L_1L_2)}(T))$, where $R \in Syl_2(Aut(L_1L_2))$ and $L_i \cong SL_{k_i}(q)$ tamely realizes $\mathcal{Y}_i$. Therefore $W \trianglelefteq N_{\mathcal{F}_t}(W_t)$ by (b), so (1) follows, completing the proof. □

LEMMA 15.5.26. *Let* $\mathfrak{X}$ *consist of those* $x \in \mathcal{W}(\tau)$ *which are not of weight* $d/2$. *Assume*

(a) For each $t \in \mathfrak{X} \cap \mathcal{F}^t$, *and for each* $i = 1, 2$,

$$\mathcal{X}_i = C_S(t)\mathcal{Y}_i = \langle C_S(t)\mathcal{O}_{\tau_i}, N_{\mathcal{X}_i}(W(\tau_i))\rangle.$$

(b) If d *is a power of 2 and* $t \in \mathcal{W}(\tau) \cap Z(S)$ *is of weight* $d/2$ *then* $\mathcal{X} = S\mathcal{F}_t^\circ \leq \langle C_{\mathcal{X}}(x), N_{\mathcal{X}}(W(\tau_t)), S\mathcal{O}_\tau : x \in \mathfrak{X}\rangle$.

Then $\mathcal{F} = \langle S\mathcal{O}_\tau, N_{\mathcal{F}}(W)\rangle$.

PROOF. Set $\mathcal{A} = \langle S\mathcal{O}_\tau, N_{\mathcal{F}}(W)\rangle$. By (a) and 15.5.25.3, for each $x \in \mathfrak{X} \cap \mathcal{F}^f$, we have $\mathcal{F}_x \leq \mathcal{A}$. Then as $N_{\mathcal{F}}(W) \leq \mathcal{A}$ and M controls fusion in W, for each $x \in \mathfrak{X}$ we have $\mathcal{F}_x \leq \mathcal{A}$.

Next suppose d is a power of 2 and let $t \in \mathcal{W}(x) \cap \mathcal{F}^f$ be of weight $d/2$. Then $t \in Z(S)$. Let $W_t \in W(\tau_t)$, T Sylow in $\mathcal{F}_t^\circ$, and $\mathcal{M}_t$ the 2-fusion system of $M_t \in M(\tau_t)$. By 1.3.2 in [**Asc19**], $\mathcal{F}_t = \langle \mathcal{X}, N_{\mathcal{F}_t}(W_t)\rangle$. Then the proof of 15.5.25.3 shows that $W \trianglelefteq N_{\mathcal{F}_t}(W_t)$, so $N_{\mathcal{F}_t}(T) \leq N_{\mathcal{F}}(W) \leq \mathcal{A}$. Next $N_{\mathcal{X}}(W_t) = S\mathcal{M}_t$

and $W \leq C_S(W_t) \leq O_2(S\mathcal{M}_t)$, so $W \trianglelefteq N_{\mathcal{X}}(W_t)$. Finally we've shown that $C_{\mathcal{X}}(x) \leq \mathcal{A}$ for each $x \in \mathfrak{X}$, so it follows from (b) that $\mathcal{X} \leq \mathcal{A}$. But now by 15.5.7, $\mathcal{F} = \langle N_{\mathcal{F}}(W), \mathcal{F}_x : x \in \mathcal{W}(\tau) \cap \mathcal{F}^f \rangle \leq \mathcal{A}$, completing the proof. □

15.6. Generating linear systems

In this section we assume the following hypothesis:

HYPOTHESIS 15.6.1. *$\tau = (\mathcal{F}, \Omega)$ is a quaternion fusion packet such that $\mathcal{F}^\circ = O^2(\mathcal{F})$ and either*

(I) $\mathcal{F}^\circ$ is isomorphic to the 2-fusion system of $\bar{L}/\bar{Z}$, where $\bar{L} \cong SL_d(q)$ with $(q-1)_2 = m/2$, $d \geq 5$, and $\bar{Z} \leq Z(\bar{L})$, or

(II) $d \geq 8$ is a power of 2 and $\mathcal{F}^\circ = \mathcal{F}_1 * \mathcal{F}_2$ where $\mathcal{F}_i = \mathcal{F}_{\bar{T}_i}(\bar{L}_i)$ for $\bar{L}_i \cong SL_{d/2}(q)$ and $\bar{T}_i \in Syl_2(\bar{L}_i)$, and $Z(\mathcal{F}_1) \cap Z(\mathcal{F}_2) \neq 1$. Further the Sylow group S of $\mathcal{F}$ is transitive on $\{\mathcal{F}_1, \mathcal{F}_2\}$.

NOTATION 15.6.2. Suppose for the moment that (I) holds; then we adopt Notation 15.5.2 in discussing $\bar{L}$. In particular $\bar{L} = SL(U)$ for U a d-dimensional vector space over $\mathbf{F}_q$.

Suppose on the other hand that (II) holds. We continue to take $\bar{L} = SL(U)$ and choose a direct sum decomposition $U = U_1 \oplus U_2$ with $\dim(U_i) = d/2$. We take $\bar{L}_i = C_{\bar{L}}(U_{3-i})$. Let $\bar{\mathcal{F}}^\circ = \mathcal{F}_{\bar{T}}(\bar{L}_1 \times \bar{L}_2) = \bar{\mathcal{F}}_1 \times \bar{\mathcal{F}}_2$, with $\bar{\mathcal{F}}_i = \mathcal{F}_{\bar{T}_i}(\bar{L}_i)$, $\bar{T}_i \in Syl_2(\bar{L}_i)$, and $\bar{T} = \bar{T}_1\bar{T}_2$. Then $\mathcal{F}^\circ$ is a central factor system of $\bar{\mathcal{F}}^\circ$ determined by a central factor group of $\bar{L}_1 \times \bar{L}_2$.

As in 15.5.11, $\tau = \tau_1 * \tau_2$, where $\tau_i = (\mathcal{F}_i, \Omega_i)$ and $\Omega_i = \{K : [U, K] \leq U_i\}$.

In either case let $W \in W(\tau)$ and $M \in M(\tau)$. In case (II), $W = W_1 * W_2$ and $M = M_1 * M_2$, where $W_i \in W(\tau_i)$ and $M_i \in M(\tau_i)$. Define $\mathcal{O}_\tau$ as in 15.5.24; more precisely in (II), $\mathcal{O}_\tau = \mathcal{O}_{\tau_1} * \mathcal{O}_{\tau_2}$.

As in 15.5.3, $\bar{W}_0 = \Omega_1(\bar{W})$, and W_0 is the image of $\bar{W}_0$ in the Sylow group T of $\mathcal{F}^\circ$. Thus in (II), $W_0 = W_{0,1} * W_{0,2}$, and we can view $\bar{W}_0$ has a subgroup of $\bar{W} \in W(\bar{\tau})$ from (I), and calculate weights in the latter module.

In (II) define $\mathfrak{X}_0$ to consist of those involutions $x \in \mathcal{W}(\tau)$ such that x is of weight $wt(x)$ a power of 2 and either $wt(x) \neq d/2$ or $C_S(x) \leq N_S(\mathcal{F}_1)$.

In this section we prove:

THEOREM 15.6.3. *Assume Hypothesis* 15.6.1 *and adopt notion* 15.6.2. *Set $\mathcal{A}_0 = \langle S\mathcal{O}_\tau, N_{\mathcal{F}}(W)\rangle_S$. In case (I) set $\mathcal{A} = \mathcal{A}_0$, and in case (II) set $\mathcal{A} = \langle \mathcal{A}_0, C_{\mathcal{F}}(x) : x \in \mathfrak{X}_0\rangle_S$. Then $\mathcal{F} = \mathcal{A}$.*

Assume the theorem is false and pick a counter example with τ of minimal order.

LEMMA 15.6.4. *Assume (I) holds. Then $\bar{Z} = Z(\bar{L})$, so $\mathcal{F}^\circ \cong L_d^+[m]$.*

PROOF. Assume otherwise and set $\mathcal{F}^+ = \mathcal{F}/\langle c\rangle$ where c is the involution in $Z(\mathcal{F}^\circ)$. Then $\tau^+ = (\mathcal{F}^+, \Omega^+)$ is a quaternion fusion packet satisfying Hypothesis 15.6.1, so by minimality of τ, $\mathcal{F}^+ = \langle S^+\mathcal{O}_{\tau^+}, N_{\mathcal{F}^+}(W^+)\rangle$, where $W^+ \in W(\tau^+)$. But now $\mathcal{F} = \mathcal{A}$ using 2.1.15, contrary to the choice of τ as a counterexample. □

LEMMA 15.6.5. *Let $t \in \mathcal{W}(\tau)$. Then either*
(1) $\mathcal{F}_t \leq \mathcal{A}$, or
(2) d is a power of 2, $wt(t) = d/2$, and in case (II), $C_S(t)$ does not act on $\mathcal{F}_1$.

PROOF. Assume that neither (1) nor (2) holds. Assume first that (I) holds. By 15.5.8 and as $N_{\mathcal{F}}(W) \leq \mathcal{A}$, we may assume $t \in \mathcal{F}^f$. Adopt the notation of 15.5.11. As (2) fails, $\mathcal{Y}_i \trianglelefteq \mathcal{F}_t$ for $i = 1, 2$. If $k_i = 3$ then the hypothesis of 15.5.25.3 is satisfied by 15.5.7, while if $k_i = 4$ it holds by 9.3.20. If $k_i \geq 5$ it holds by minimality of τ. Hence $\mathcal{F}_t \leq \mathcal{A}$ by 15.5.25.3, contrary to assumption.

Therefore (II) holds. By definition of $\mathcal{A}$, $wt(t) = d/2$ and $C_S(t)$ does not act on $\mathcal{F}_1$, so t satisfies (2), again contrary to assumption. □

LEMMA 15.6.6. *Assume (I) holds. Then*
(1) d is a power of 2.
(2) Let $f \in Z(S) \cap W_0$ be of weight $d/2$. Then $\mathcal{F}_t \not\leq \mathcal{A}$.

PROOF. By 15.5.7, $\mathcal{F} = \langle N_{\mathcal{F}}(W), \mathcal{F}_x : x \in \mathcal{W}(\tau) \cap \mathcal{F}^f\rangle$. By definition of $\mathcal{A}$, $N_{\mathcal{F}}(W) \leq \mathcal{A}$, and by 15.6.5, and 15.5.14.6 when $d = 8$, for each $x \in \mathcal{W}(\tau)$, either $\mathcal{F}_x \leq \mathcal{A}$ or d is a power of 2 and we may choose $x \in W_0$ with $wt(x) = d/2$. In particular if d is not a power of 2 then $\mathcal{F} = \mathcal{A}$, contrary to the choice of τ as a counter example. Therefore (1) holds. Similarly (using 15.5.14.6 when $d = 8$) there is $x \in \mathcal{W}(\tau) \cap \mathcal{F}^f$ of weight $d/2$ such that $\mathcal{F}_x \not\leq \mathcal{A}$. But as d is a power of 2 and $x \in \mathcal{F}_x$, we have $x \in Z(S)$, so (2) holds. □

LEMMA 15.6.7. *Case (II) holds.*

PROOF. Assume instead that (I) holds and choose f as in 15.6.6.2. Then by 15.5.11 and 15.6.4, $\mathcal{F}_f = S\mathcal{F}_f^\circ$ satisfies Hypothesis 15.6.1.II. Therefore (II) holds by minimality of τ. □

NOTATION 15.6.8. We will abuse notation and work in $SL(U)$ rather than in the projective group $PSL(U)$. Let $\theta_1 = \{z \in \mathcal{Z}_S : [U, z] \leq U_1\}$. Now $\Pi_1 = \{\theta_1, \mathcal{Z}_S - \theta_1\}$ is an S-invariant partition of $\mathcal{Z}_S$.

Proceeding recursively, assume Π_k is an S-invariant partition of $\mathcal{Z}_S$ into 2^k blocks of size $n/2^k$. Pick $\theta_k \in \Pi_k$ and let $\theta_{k+1} \subseteq \theta_k$ be of order $n/2^{k+1}$ such that $|N_S(\theta_k) : N_S(\theta_{k+1})| = 2$, and set $\Pi_{k+1} = \theta_{k+1}^S$, so that Π_{k+1} is an S-invariant partition of $\mathcal{Z}_S$ into 2^{k+1} blocks of size $n/2^{k+1}$. Continuing this process, we arrive at $\theta_r = \{z\}$ for some $z \in \mathcal{Z}_S$; that is $n = 2^r$.

Let $\Pi_k = \{\theta_{k,1}, \ldots, \theta_{k,2^k}\}$ and $U_{k,i} = [U, f_{\theta_{k,i}}]$. Thus $U = U_{k,1} \oplus \cdots \oplus U_{k,2^k}$ and we set $\mathcal{U}_k = \{U_{k,1}, \ldots, U_{k,2^k}\}$, $L_{k,i} = C_{U_{k,i}}$ and $L_k = \prod_i L_{k,i}$ the direct product of 2^k copies of $SL_{2^{r-k}}(q)$. Let $\mathcal{U}^k = \{\mathcal{U}_1, \ldots, \mathcal{U}_k\}$ and G_k the stabilizer in $S\bar{L}$ of $\mathcal{U}^k$. Set $\mathcal{D}_k = \mathcal{F}_S(G_k)$.

LEMMA 15.6.9. $G_k = SL_kO(G_k)$.

PROOF. By construction, $SL_k \leq G_k$ and L_k is contained in the kernel X of the action of G_k on $\mathcal{U}_k$. By 2.1.16, $G_k^{\mathcal{U}_k}$ is contained in a Sylow 2-group of $Sym(\mathcal{U}_k)$.

For $x \in X$ of odd order, $x_{|U_{k,i}} \in L_{k,i}O(G_k)$ as $L_{k,i}$ acts faithfully as $SL(U_{k,i})$ on $U_{k,i}$. Therefore $X/L_kO(G_k)$ is a 2-group, completing the proof. □

LEMMA 15.6.10. *(1)* $L_{r,i} \cong SL_2(q)$ *has Sylow 2-group* $O(s_i)$ *for that* $s_i \in \mathcal{Z}_S$ *such that* $s_i \in Z(L_{r,i})$ *and* $U_{t,i} = [U, s_i]$.
(2) $\mathcal{O}_\tau = \mathcal{F}_{O(\tau)}(L_r)$.
(3) $\mathcal{D}_r = \mathcal{F}_S(G_r) = S\mathcal{O}_\tau \leq \mathcal{A}$.

PROOF. By construction, $U_{r,i} = [U, s_i]$ for some $s_i \in \mathcal{Z}_S$, and then (1) follows. In particular by (1), $O(\tau)$ is Sylow in L_r and (2) holds. By 15.6.9, $G_r = SL_rO(G_r)$, so $\mathcal{F}_S(G_r) = \mathcal{F}_S(SL_r) = S\mathcal{O}_\tau$, establishing (3). □

LEMMA 15.6.11. $\mathcal{E}(\tau) \leq \mathcal{Y}$.

PROOF. From 15.5.4, $\mathcal{E} = \mathcal{E}(\tau) = \langle N_{\mathcal{F}}(O(\xi)), S\mathcal{O}_\tau, N_{\mathcal{F}}(O), N_{\mathcal{F}}(W) : \xi \in \Xi \rangle$, where Ξ consists of those $\xi \subseteq \mathcal{Z}_S$ with $O(\xi) \in \mathcal{F}^f$ and $N_S(O(\xi))$ transitive on ξ. Now $N_{\mathcal{F}}(O) \leq \langle S\mathcal{O}_\tau, N_{\mathcal{F}}(W) \rangle \leq \mathcal{A}$. Thus it remains to show that $\mathcal{N} = N_{\mathcal{F}}(O(\xi)) \leq \mathcal{A}$ for $\xi \in \Xi$. But $\mathcal{N} \leq \mathcal{F}_x$, where $x = f_\xi$, and by definition of $\mathcal{A}$, $\mathcal{F}_x \leq \mathcal{Y}$ unless $\xi = \theta_1$ and $U_1 = [U, x]$, so we may the latter holds.
Now $\mathcal{N} = \mathcal{F}_S(X)$, where $X = N_{L_{1,1}}(O(\xi))L_{1,2}S$. Therefore

$$\mathcal{N} = \langle \mathcal{F}_S(SL_{1,2}), N_{\mathcal{N}}(S_2) \rangle,$$

where S_2 is Sylow in $L_{1,2}$. By minimality of τ, $\mathcal{F}_S(SL_{1,2}) \leq \mathcal{A}$. Further $N_{\mathcal{N}}(S_2) \leq N_{\mathcal{N}}(O)$, which is contained in $\mathcal{A}$ by the previous case. □

NOTATION 15.6.12. Observe that $\mathcal{D}_1 = \mathcal{F}$, so $\mathcal{D}_1 \not\leq \mathcal{A}$. On the other hand by 15.6.10.3, $\mathcal{D}_r \leq \mathcal{Y}$. Thus there exists a least k such that $\mathcal{D}_k \not\leq \mathcal{A}$. Set $\mathcal{D} = \mathcal{D}_k$, $\rho = (\mathcal{D}, \Omega)$, let $S_0 = S \cap L_k$, and write $\mathcal{D}_0$ for $\mathcal{F}_{S_0}(L_k)$. Let $I = \{1, \ldots, 2^k\}$ and for $i \in I$ set $S_i = S \cap L_{k,i}$ and $\mathcal{L}_i = \mathcal{F}_{S_i}(L_{k,i})$. From 15.6.8, $L_k = \prod_{i \in I} L_{k,i}$, so $\mathcal{D}_0 = \prod_{i \in I} \mathcal{L}_i$ is a direct product. Set $\Omega_i = \Omega \cap S_i$ and $\rho_i = (\mathcal{L}_i, \Omega_i)$. Observe:

LEMMA 15.6.13. *The pair* $\mathcal{D}, \mathcal{D}_0$ *satisfies Hypotheses* 11.4.1 *and* 11.4.6.

NOTATION 15.6.14. As in 11.4.2 and 11.4.6, for $X \leq S$ and $0 \leq i \leq 2^k$, write X_i for $X \cap S_i$.
As $\mathcal{D} \not\leq \mathcal{Y}$, it follows from the Alperin-Goldschmidt Fusion Theorem that the set $\mathcal{R}$ of subgroups $R \in \mathcal{D}^e$ such that $Aut_{\mathcal{D}}(R) \not\leq Aut_{\mathcal{Y}}(R)$ is nonempty. Pick $R \in \mathcal{R}$ and let G be a model for $N_{\mathcal{D}}(R)$ and set $G^* = G/R$. By 15.6.13 and 11.4.4.1, $R_0 \in \mathcal{F}^{frc}$, so there is a model G_0 for $N_{\mathcal{D}_0}(R_0)$. By 11.4.4, we can embed G and G_0 in a certain group G_c with $G_0 \trianglelefteq G_c$ and $G = N_{G_c}(R)$. Thus we can form $H_0 = N_{G_0}(R)$ and by 11.4.4, $H_0 \trianglelefteq G$ with $R_0 = O_2(H_0)$ and $H_0^* \cong H_0/R_0$.
Set $Q = N_R(L_{k,1})$ and $\mathcal{X} = \mathcal{F}_{QS_1}(QL_{k,1})$. Let $1 \in \Sigma$ be a set of coset representatives for Q in R. Write $\mathcal{R}^*$ for the set of those $R \in \mathcal{R}$ of maximal order.

LEMMA 15.6.15. *For each* $K \in \Omega$, $|K \cap R| \leq 2$.

PROOF. Assume otherwise; then from 3.4.5, $Aut_{\mathcal{D}}(R) = Aut_{\mathcal{E}(\rho)}(R)$. But $\mathcal{E}(\rho) \leq \mathcal{E}(\tau)$ and by 15.6.11, $\mathcal{E}(\tau) \leq \mathcal{A}$, contrary to the choice of k. □

LEMMA 15.6.16. *(1) For each* $i \in I$, $R_i \neq S_i$.
(2) R *is transitive on* $\{L_{k,i} : i \in I\}$.
(3) $Q \in \mathcal{X}^e$. *Let* G_Q *be a model for* $N_{\mathcal{X}}(Q)$.
(4) $G_Q/Q \cong H_0^*$.

(5) Let $E_1 = \langle(\mathcal{Z}_S(\rho_1))\rangle$ *and define* $\alpha : W_1 \cap Z(Q) \to W_S \cap R$ *by* $\alpha : x \to \prod_{\sigma\in\Sigma} x^\sigma$. *Then* $x\alpha \in Z(R)$ *and if* x *is of weight* p *in* E_1 *then* $x\alpha$ *is of weight* $2^k p$ *in* W_S.

(6) If $p \neq |\mathcal{Z}_S(\rho_1)|$ *then* $C_{\mathcal{D}}(x\alpha) \leq \mathcal{Y}$.

PROOF. If $R_i = S_i$ then $\Omega_i \subseteq R$, contrary to 15.6.15. Thus (1) holds.

By 15.6.13, we can appeal to results in section 11.4. In particular (2) follows from (1) and 11.4.8.1, (3) follows from 11.4.8.4, and (4) follows from 11.4.8.5.

From 15.6.14, $1 \in \Sigma$ is a set of coset representatives for Q in R. If $\sigma_i \in \Sigma$ with $R_1\sigma_i = R_i$ then $R_0 = \prod_{i\in I} R_i = \prod_{i\in I} R_1^{\sigma_i}$, so for $x \in E_1 \cap Z(Q)$ of weight p, $x\alpha$ centralizes $\langle Q, \Sigma\rangle = R$ and $x\alpha$ is of weight $|\Sigma|p = |I|p = 2^k p$ in W_S. Thus (5) holds.

Assume the hypothesis of (6) and set $y = x\alpha$. Observe that $|\mathcal{Z}_S(\rho_1)| = n/2^k = 2^{r-k}$. As $p \neq |\mathcal{Z}_S(\rho_1)|$, $y \neq 1$ so y is an involution. Then by definition of $\mathcal{A}$, either $\mathcal{F}_y \leq \mathcal{A}$ or $2^k p = d/2 = 2^{r-1}$, and we may assume the latter, so that $p = 2^{r-k-1}$. Then conjugating in $\mathcal{A}$, we may take $x = f_{\theta_{k+1}}$, so $\Pi_{k+1} = \theta_{k+1}^R \cup (\theta_k - \theta_{k+1})^R$. But then $C_{SL_k}(y) = C_S(y)L_{k+1}$, so as $\mathcal{D}_{k+1} \leq \mathcal{Y}$, again $C_{\mathcal{D}}(y) \leq \mathcal{Y}$. □

LEMMA 15.6.17. *$\rho_1 = (\mathcal{L}_1, \Omega_1)$ satisfies Hypothesis* 11.2.1 *and if* $R \in \mathcal{R}^*$ *then* ρ_1, Q *satisfies Hypothesis* 11.2.7.

PROOF. Visibly ρ_1 satisfies Hypothesis 11.2.1 with $|\Omega(z)| = 1$.

So assume $R \in \mathcal{R}^*$. By 15.6.16.4, $Q \in \mathcal{L}_1^e$. From 11.2.6, $\mathcal{W}(Q)$ consists of those involutions $x \in Z(Q) \cap E_1$ such that $x = f_\alpha$ for some $\alpha \subseteq \mathcal{Z}_S(\rho_1)$ of weight a power of 2. Passing to $\mathcal{L}_1/Z(\mathcal{L}_1)$ if necessary we may assume $f_{\mathcal{Z}_S(\rho_1)} = 1$. Hence by 15.6.16.6, for each $x \in \mathcal{W}(Q)$, $C_{\mathcal{D}}(x\alpha) \leq \mathcal{A}$.

Let $x \in \mathcal{W}(Q)$, $y = x\alpha$ and $Z = \langle y^G\rangle$. As $R \in \mathcal{R}^*$ and $C_{\mathcal{D}}(y) \leq \mathcal{Y}$, the proof of 11.3.5.1 shows that $O^{2'}(C_G(Z)) = R$. Set $Z_x = \langle x^{G_Q}\rangle$. Then Z is a full diagonal subgroup of $\prod_{\sigma\in\Sigma} Z_x^\sigma$ and the isomorphism of 15.6.16.4 maps $O^{2'}(C_{G_Q}(Z_x))/Q$ to $O^{2'}(C_G(Z))/R = 1$, so $O^{2'}(C_{G_Q}(Z_x)) = Q$. Hence ρ_1, Q satisfies 11.2.7.1.

Similarly if ρ, R satisfies the setup of 11.2.7.2 and $\mathcal{U} \neq \emptyset$ then the proof of 11.3.5.2 shows that $m_2(G^*) = 1$ and $\langle\mathcal{U}\rangle$ is not normal in G^*. Let B^* be the subgroup of G^* generated by its involutions. By 11.4.5.2, B^* is the subgroup of H^* generated by its involutions. Thus the isomorphism of 15.6.16.4 maps the subgroup B_Q^* of $G_Q^* = G_Q/Q$ generated by involutions to B^*, so $m_2(G_Q^*) = 1$ and $\langle\mathcal{U}_Q\rangle \neq B_Q^*$. Therefore ρ_1, Q satisfies 11.2.7.2. This completes the proof of the lemma. □

We are now in a position to obtain a contradiction, completing the proof of Theorem 15.6.3. Namely choose $R \in \mathcal{R}^*$. By 15.6.17, the pair ρ_1, Q satisfies Hypotheses 11.2.1 and 11.2.7. Further condition (a) of Theorem 11.2.33 is satisfied, as is condition (b) by 15.6.15. Condition (c) of 11.2.33 holds as $\mathcal{X} = \mathcal{F}_{QS_1}(SL_{k,1})$. But then by Theorem 11.2.33, $|\mathcal{Z}_S(\rho_1)| = 1$, a contradiction. This completes the proof of Theorem 15.6.3.

15.7. Wrapping up $\Phi = A_n$

In this section (until the final theorem) we assume Hypothesis 15.5.18 and adopt the notation from section 15.5. We complete our treatment of the case $\Phi = A_n$.

Define $\mathcal{O}_\tau$ as in 15.5.24.

LEMMA 15.7.1. $\mathcal{F} = \langle S\mathcal{O}_\tau, N_\mathcal{F}(W)\rangle$.

PROOF. This follows from 15.5.26 and 15.6.3. We verify condition (a) of 15.5.26 using aH.3 in case (I) applied to $\mathcal{X}_i$ for $i = 1, 2$. By 15.5.23, $\mathcal{Y}_i \cong SL_{k_i}[m]$. Then 15.6.3 implies (a) as long as $k_i \geq 5$. When $k_i = 2$, condition (a) is trivial, while if $k_i = 3$ or 4 then (a) follows from 15.5.7 or 9.3.20, respectively.

Similarly condition (b) of 15.5.26 follows from an application of 15.6.3 in case (II). □

LEMMA 15.7.2. *If $Z(M) \neq 1$ then $\mathcal{F}$ is an image of $SL_d^+[m]$ modulo some central subgroup.*

PROOF. Assume $Z(M) \neq 1$ and let x be the involution in $Z(M)$. Then x is in the center of $N_\mathcal{F}(W)$ and $S\mathcal{O}_\tau$, so $x \in Z(\mathcal{F})$ by 15.7.1. Set $\mathcal{F}^+ = \mathcal{F}/\langle x\rangle$, and observe that, using 6.6.9.2, $\tau^+ = (\mathcal{F}^+, \Omega^+)$ satisfies Hypothesis 15.5.18. Therefore by induction on the order of τ, $\mathcal{F}^+$ is a central factor system of $SL_d^+[m]$, so by 3.3.16, $\mathcal{F}$ is also such a factor system. □

Because of 15.7.2, in the remainder of the section (until the final theorem) we assume:

HYPOTHESIS 15.7.3. *Hypothesis* 15.5.18 *holds with* $Z(M) = 1$.

NOTATION 15.7.4. Set $\tilde{\mathcal{F}} = L_d^+[m]$, $T = S_M$, $\mathcal{F}_1 = S\mathcal{O}_\tau$, $\mathcal{E}_1 = T\mathcal{O}_\tau$, $\mathcal{F}_2 = N_\mathcal{F}(W)$, $\mathcal{E}_1 = \mathcal{F}_T(M)$, and $\mathcal{E} = \langle \mathcal{E}_1, \mathcal{E}_2\rangle_T$. By 5.4.13, the Lie packet $\tilde{\tau} = (\tilde{\mathcal{F}}, \tilde{\Omega})$ of $L_d(q)$ satisfies Hypothesis 15.7.3 for suitable q with $(q-1)_2 = m/2$, and by 5.2.5, $\tilde{M} \in M(\tilde{\tau})$ contains the Sylow group $\tilde{S}$ of $\tilde{\mathcal{F}}$. As $Z(M) = 1$, we conclude from 4.3.8 that $M = \tilde{M}$. Therefore, identifying M with $\tilde{M}$, we may take $M = \tilde{M}$, $T = \tilde{S}$, etc. Subject to this convention, $\mathcal{E}_2 = \mathcal{F}_{\tilde{S}}(\tilde{M})$.

LEMMA 15.7.5. *(1) $\mathcal{E}_1 = \tilde{S}\mathcal{O}_{\tilde{\tau}}$.*
(2) $\mathcal{E} = \tilde{\mathcal{F}}$.

PROOF. By Theorem 2.7.3, $T\mathcal{O}_\tau$ is uniquely determined by T and Ω, so as $T = \tilde{S}$ and $\Omega = \tilde{\Omega}$, we conclude that (1) holds. From 15.7.4, $\mathcal{E}_2 = \mathcal{F}_{\tilde{S}}(\tilde{M})$, so by (1) we have $\mathcal{E} = \langle \tilde{S}\mathcal{O}_{\tilde{\tau}}, \mathcal{F}_{\tilde{S}}(\tilde{M})\rangle$. Finally $\tilde{\mathcal{F}} = \langle \tilde{S}\mathcal{O}_{\tilde{\tau}}, \mathcal{F}_{\tilde{S}}(\tilde{M})\rangle$ by 15.7.1 applied to $\tilde{\mathcal{F}}$, since $N_{\tilde{\mathcal{F}}}(\tilde{W}) = \mathcal{F}_{\tilde{S}}(\tilde{M})$. □

LEMMA 15.7.6. *For $i = 1, 2$, $\mathcal{E}_i \trianglelefteq \mathcal{F}_i$.*

PROOF. By construction, $\mathcal{E}_2 \trianglelefteq \mathcal{F}_2$. As T is Sylow in $\mathcal{E}_2 \trianglelefteq \mathcal{F}_2$, we have $T \trianglelefteq S$, so $\mathcal{E}_1 = T\mathcal{O}_\tau \trianglelefteq S\mathcal{O}_\tau = \mathcal{F}_2$ by Theorem 5 in [**Asc11**]. □

THEOREM 15.7.7. *Assume Hypothesis* 15.7.3. *Then $\mathcal{F} \cong L_d^+[m]$.*

PROOF. We show that the tuple $\mathcal{F}_1, \mathcal{F}_2, \mathcal{E}_1, \mathcal{E}_2$ satisfies Hypothesis 1.4.7 in [**Asc19**]. Then we appeal to 1.4.8 in [**Asc19**] to conclude that $\mathcal{E} \trianglelefteq \mathcal{F}$. Hence $\mathcal{F} = \mathcal{F}^\circ = \mathcal{E}$, so the theorem follows from aI.5.2.

Hypothesis 1.4.1 in [**Asc19**] is satisfied by 15.7.1. By 15.7.6, $\mathcal{E}_i \trianglelefteq \mathcal{F}_i$ for $i = 1, 2$. Hence condition (1) of 1.4.7 in [**Asc19**] holds.

Next $N_\mathcal{F}(T) \leq N_\mathcal{F}(W) = \mathcal{F}_2$, so as $\mathcal{E}_2 \trianglelefteq \mathcal{F}_2$, we have $Aut_\mathcal{F}(T) \leq Aut(\mathcal{E}_2)$. Similarly, setting $E = \langle \mathcal{Z}_S\rangle$, $\mathcal{O}_\tau \trianglelefteq N_\mathcal{F}(E)$, so $Aut_\mathcal{F}(T) \leq Aut(\mathcal{E}_1)$. That is condition (2) of 1.4.7 in [**Asc19**] holds.

Condition (3) follows from the definition of $\mathcal{E}$ as $\langle \mathcal{E}_1, \mathcal{E}_2 \rangle$, and finally $\mathcal{E}$ is saturated by 15.7.5.2, so condition (4) holds. This completes the verification of Hypothesis 1.4.7 in [**Asc19**] and hence also the proof of the theorem. $\square$

THEOREM 15.7.8. *Assume $\tau = (\mathcal{F}, \Omega)$ is a quaternion fusion packet such that $\mathcal{F} = \mathcal{F}^\circ$ and $\mu(\tau) \cong$ Weyl(A_n) for some $n \geq 4$. Assume the Extended Inductive Hypothesis is satisfied. Then for $z \in \mathcal{Z}_S$ we have $|\Omega(z)| = 1$, and either*

(1) $W \in W(\tau)$ is normal in $\mathcal{F}$, or

(2) $\mathcal{F}$ is a central factor system of $SL^+_{n+1}[m]$.

PROOF. Let $d = n + 1$, $W \in W(\tau)$, and $M \in M(\tau)$. Then $\mu = M/W \cong S_d$. Let $K \in \Omega$ and $z = z(K)$. Then $a = d_K$ is a transposition in μ and $D = a^\mu$ is the set of all transpositions. As $\mu \cong S_d$ with $d \geq 5$, $V_a = \{a\}$, so $\Omega(z) = \{K\}$ as for each $J \in \Omega(z)$, $d_J \in V_a$. Therefore 15.1.1.1 is satisfied. By hypothesis the Extended Inductive Hypothesis is satisfied, so 15.1.1.4 holds, and as $\mu(\tau) \cong S_{n+1}$, conditions (2) and (3) of 15.1.1 hold. Hence Hypothesis 15.1.1 is satisfied, so as $\mu \cong \text{Weyl}(A_n)$, also Hypothesis 15.5.1 holds.

Define $c \in W_0$ as in 15.5.9. If $c \neq 1$ then conclusion (2) of our theorem holds by 15.5.12.2, so we may assume $c = 1$. Similarly if $W(\tau') \lhd \mathcal{T}$ then conclusion (1) of our theorem holds by 15.5.16, so we may assume otherwise. Hence Hypothesis 15.5.18 is satisfied. By 15.7.2 we may assume that $Z(M) = 1$. Therefore Hypothesis 15.7.3 holds, so conclusion (2) of our theorem holds by Theorem 15.7.7. This completes the proof. $\square$

15.8. $\Phi = E_n$

In this section we assume the following hypothesis:

HYPOTHESIS 15.8.1. *Either*

(1) $\tau = (\mathcal{F}, \Omega)$ is the Lie packet with $\mathcal{F}$ isomorphic to $E_6^+[m]$, $E_7[m]$, $\tilde{E}_7[m]$, or $E_8[m]$, or

(2) Hypothesis 15.2.1 *holds with $\mathcal{F} = \mathcal{F}^\circ$ and $\mu = \mu(\tau) \cong$ Weyl(Φ), where Φ is of type E_n for some $n \in \{6, 7, 8\}$.*

NOTATION 15.8.2. We adopt Notation 15.1.3, and as usual let S_M be Sylow in M.

LEMMA 15.8.3. *Assume $n = 7$. Then*

(1) $\mu \cong \mathbf{Z}_2 \times Sp_6(2)$.

(2) Let $d = d_{K \cap W}$. Then $C_\mu(d) = \langle d^\perp \rangle = \langle d \rangle \times \langle D_d \rangle$, with $\langle D_d \rangle \cong Weyl(D_6)$.

(3) $\mu(\tau_z) \cong$ Weyl$(A_1 D_6)$.

(4) $W_0/Z(M)$ is centralized by $Z(\mu)$ and is the natural module for $\mu/Z(\mu) \cong Sp_6(2)$.

(5) If $Z(M) \neq 1$ then $Z(M) = Z(\mathcal{F})$ and either $W \lhd \mathcal{F}$ or $\mathcal{F} \cong \tilde{E}_7[m]$.

PROOF. By hypothesis, $\mu \cong Weyl(E_7)$, so (1) holds. Let σ generate $Z(\mu)$ and $\nu = E(\mu)$. Then $d = \sigma \cdot \delta$, where $\delta \in \nu$ is a transvection, so $C_\nu(d) = C_\nu(\delta) \cong Weyl(D_6)$. Further for $a \in D_d$, $a = \sigma\alpha$ for some transvection $\alpha \in C_\nu(d)$, so $\langle D_d \rangle \cong C_\nu(d)$ with $\langle D_d \rangle \cap C_\nu(d) = C_\mu(d)^\infty$ of index two in each of the two subgroups. As

$\delta \in C_\mu(d)^\infty$, it follows that $\sigma \in \langle d^\perp \rangle$, and then that (2) holds. Then (2) implies (3).

Let $\bar{M}$ be the universal group $\bar{\omega}(E_7, m)$ defined in 4.3.1. By 4.3.8 there is a surjective homomorphism $\pi : \bar{M} \to M$ with $\ker(\pi) \leq Z(\bar{M}) = \bar{Z}$. By 4.3.7.2, $\bar{W}_0$ is of rank 7, and by 5.5.5 there is $\bar{Z}_0 \leq Z(\bar{M})$ of order 2. Then as the natural module for $Sp_6(2)$ is its only nontrivial 6-dimensional module, we conclude that $\bar{Z} = \bar{Z}_0$ and (4) holds.

Finally suppose $Z(M) \neq 1$. Then π is an isomorphism and we identify M with $\bar{M}$ via that map. Let c generate $Z = Z(M)$. From (4), M has three orbits on $W_0^\#$ with representatives c, z, cz. As $\langle c \rangle = Z(M)$, $c \in Z(N_{\mathcal{F}}(W))$. We show c is in the center of $\mathcal{F}_t$ for $t \in \{z, cz\}$, so that Hypothesis 14.1.11 is satisfied and then $c \in Z(\mathcal{F})$ by 14.1.18. Set $\mathcal{F}^+ = \mathcal{F}/\langle c \rangle$. By 6.6.9.2 and induction on the order of τ, either $W^+ \trianglelefteq \mathcal{F}^+$ or $\mathcal{F}^+ \cong E_7[m]$; then (5) follows from 3.3.16.

It remains to show $\mathcal{F}_t \leq \mathcal{F}_c$. In 15.8.1.1 this follows as $c \in Z(\mathcal{F})$, so assume 15.8.1.2 holds. By (3), $\mu(\tau_t) \cong \text{Weyl}(A_1 D_6)$, so by 15.1.5 and 15.1.9, $\tau_t^\circ = \tau_1 * \tau_2$, where $\tau_1 = (\mathcal{T}_1, K)$ and $\tau_2 = (\mathcal{T}_2, \Omega - \{K\})$ with $\mu(\tau_1) \cong \mathbf{Z}_2$ and $\mu(\tau_2) \cong \text{Weyl}(D_6)$. Let $M_2 \in M(\tau_2)$. Then $M_2 = \bar{\omega}(D_6, m)$ is the universal group with $Z(M_2) = \langle z, c \rangle$.

If $Z(M_2) = Z(\mathcal{T}_2)$ then as c is weakly closed in W, we have $c \in Z(\mathcal{F}_t)$. Thus we may assume $Z(\mathcal{T}_2) = \langle t \rangle$, so $\mathcal{T}_2$ is not by 15.1.9, $\mathcal{T}_2 \cong Spin_{13}[m]$, $Spin_{14}^-[m]$, or $\hat{A}E_{13}$, But then, as in the proof of 11.5.17.2, c is fused to ct in $\mathcal{T}_2$, contradicting c weakly closed in W. □

In light of 15.8.3.5, during the remainder of the section we assume:

HYPOTHESIS 15.8.4. *Hypothesis* 15.8.1 *holds with* $Z(M) = 1$ *if* $n = 7$.

LEMMA 15.8.5. *(1)* $\mu \cong O_6^-(2)$, $\mathbf{Z}_2 \times Sp_6(2)$, *or* $\hat{O}_8^+(2)$, *for* $n = 6$, *7, or 8, respectively.*

(2) $Z(\mu)$ *centralizes* W_0 *and* W_0 *is the natural module for* $\mu/Z(\mu)$.

(3) If $n = 7$ *then* M *is transitive on* $W_0^\#$.

(4) If $n = 6$ *or 8 then* M *has two orbits on* $W_0^\#$ *with representatives* z *and* t, *with* z *nonsingular and* t *singular in the orthogonal space* W_0.

(5) If $n = 6$ *then* $\mu(\tau_z) \cong$ *Weyl*$(A_1 A_5)$ *and* $\mu(\tau_t) \cong$ *Weyl*(D_5).

(6) If $n = 8$ *then* $\mu(\tau_z) \cong$ *Weyl*$(A_1 E_7)$ *and* $\mu(\tau_t) \cong$ *Weyl*(D_8).

(7) If $n = 7$ *then* $\mathcal{F} = \langle N_{\mathcal{F}}(W), \mathcal{F}_z \rangle$.

(8) If $n = 6$ *or 8 then* $\mathcal{F} = \langle N_{\mathcal{F}}(W), \mathcal{F}_z, \mathcal{F}_t \rangle$.

PROOF. As $\mu \cong \text{Weyl}(E_n)$, (1) holds.

Suppose $n = 7$. Then $Z(M) = 1$ by 15.8.4, so (2) and (3) follow from parts (3) and (4) of 15.8.3. Now $n_S = 7$ is odd (cf. 15.8.6.2), so $\mathcal{F} = \mathcal{E}(\tau)$ by 3.4.6. By 3.4.11,

$$\mathcal{E}(\tau) = \langle S\mathcal{O}(\tau), N_{\mathcal{F}}(O(\xi), N_{\mathcal{F}}(W) : \xi \in \Xi \rangle,$$

where Ξ consists of the subsets ξ of $\mathcal{Z}_S$ with $N_S(\xi)$ transitive on ξ and $O(\xi) \in \mathcal{F}^f$. Now $S\mathcal{O}(\tau) \leq \mathcal{F}_z$, and from 15.8.6.2, $N_{\mathcal{F}}(O(\xi)) \leq \mathcal{F}_z$ if $\xi \in \Xi$ is of order 1 or 2. Finally if $|\xi| = 4$ then $N_{\mathcal{F}}(O(\xi)) \leq N_{\mathcal{F}}(W_S) = \mathcal{N}$, and by 1.3.2 in [**Asc19**], $\mathcal{N} = \langle SC_{\mathcal{F}}(W_S), N_{\mathcal{N}}(T) \rangle$ where T is Sylow in $C_{\mathcal{F}}(W_S)$. But $SC_{\mathcal{F}}(W_S) \leq \mathcal{F}_z$ and $N_{\mathcal{F}}(T) \leq N_{\mathcal{F}}(W)$, so (7) holds. Therefore we may assume n is 6 or 8.

Let $d = d_{K \cap W} \in D \subseteq \mu$. By (1), $\bar{\mu} = \mu/Z(\mu) \cong O_n^\epsilon(2)$ where $(n, \epsilon) = (6, -1)$ or $(8, +1)$. By 3.1.20.3, $\bar{d}$ induces a transvection on W_0 with center z. By 4.3.7.2,

$m(\bar{W}_0) = n$, so as the natural module is the only nontrivial n-dimensional module for $\bar{\mu}$, it follows that (2) holds. Further as z is the center of d, z is nonsingular in the orthogonal space W_0; then (4) follows from (2).

As $\bar{d}$ is a transvection in $\bar{\mu}$, we conclude that $C_\mu(d) \cong \mathbf{Z}_2 \times S_6$, $E_4 \times Sp_6(2)$ for $n = 6, 8$, respectively, so $\mu(\tau_z) \cong \mathrm{Weyl}(A_1\Sigma)$ for $\Sigma = A_5, E_7$, respectively. Similarly by (2) and (4), $C_\mu(t)$ is the parabolic S_5/E_{16} or S_8/E_{2^7}, so $\mu(\tau_t)$ is $\mathrm{Weyl}(D_5)$ or $\mathrm{Weyl}(D_8)$, completing the proof of (5) and (6).

If remains to prove (8). If $n = 8$ then (8) follows from (4) and 14.2.4, so we may assume $n = 6$. Then $m(W_S) = 3$ by 15.8.5.3, so Hypothesis 11.3.8 is satisfied, and then (8) follows from Theorem 11.3.10. □

LEMMA 15.8.6. *(1) W_S is a maximal totally isotropic subspace of the symplectic or orthogonal space W_0, so $m(W_S) = m(W_0)/2$.*

(2) If $n = 7$ then $\mathcal{Z}_S = W_S^\#$, so $n_S = |\mathcal{Z}_S| = 7$.

(3) If $n = 6$ or 8 then $\mathcal{Z}_S = W_S - E_0$, where E_0 is the totally singular hyperplane of W_S, so $n_S = 4$ or 8.

(4) $\prod_{s\in\mathcal{Z}_S} s = 1$.

PROOF. For $a, b \in \mathcal{Z}_\Delta$, $b \in \mathcal{D}(a)$ iff $[d_{O(a)\cap W}, d_{O(b)\cap W}] = 1$ iff a is orthogonal to b in W_0. Therefore W_S is totally isotropic, so $m(W_S) \leq m(W_0)/2$. As n_S is the width of the 3-transposition group μ, we have $n_S = 4, 7, 8$, respectively, so $m(W_S) \geq m(W_0)/2$. The lemma follows. □

LEMMA 15.8.7. *(1) $\tau_z^\circ = \tau_K * \tau'$ is a central product of packets, where $\tau_K = (\mathcal{O}(z), K)$ and $\tau' = (\mathcal{T}, \Omega - \{K\})$ with $\mu(\tau') \cong$ Weyl(Φ') and Φ' is A_5, D_6, E_7 for $n = 6, 7, 8$, respectively.*

(2) $Z(\mathcal{T}) = \langle z \rangle$.

(3) One of the following holds:

(i) $W' \in W(\tau')$ is normal in $\mathcal{T}$.

(ii) $n = 6$ and $\mathcal{T} \cong SL_6^+[m]$.

(iii) $n = 7$ and $\mathcal{T} \cong HSpin_{12}^+[m]$.

(iv) $n = 8$ and $\mathcal{T} \cong \tilde{E}_7[m]$.

PROOF. By 15.8.3 and 15.8.5, $\mu(\tau_z) \cong \mathrm{Weyl}(A_1\Phi')$, where Φ' is A_5, D_6, or E_7. Then in 15.8.1.2, (1) follows from 15.1.5 and 15.1.9, while in 15.8.1.1 it follows from 5.5.3.1.

By 15.8.6.4, $z = p$ where $p = \prod_{z\neq z\in\mathcal{Z}_S} s$. Therefore $z \in \mathcal{T}$, so $z \in Z(\mathcal{T})$. Let $M_2 \in M(\tau')$. If $n = 7$ then from the proof of 15.8.3.5 and 15.8.4, we have $Z(M_2) = \langle z \rangle$, so (2) holds in this case. If $n = 6$ or 8 then by 5.7.2.6 and 15.8.3.4, $Z(\bar{M}_2) \cong \mathbf{Z}_2$, so again (2) holds.

In 15.8.1.1, (3) follows from 5.5.3.1, so assume 15.8.1.2 holds. Assume (3i) does not hold. If $n = 6$ or 8 then (3) holds by (1), (2), and 15.1.5.4, so take $n = 7$. Then by 15.1.9.2, $\mathcal{T}$ is $Spin_{12}^+[m]$, $HSpin_{12}^+[m]$, $Spin_{13}[m]$, or $Spin_{14}^-[m]$, or $m = 8$ and $\mathcal{T}$ is $\hat{A}E_{12}$, $H\hat{A}E_{12}$, or $\hat{A}E_{13}$. As $Z(M_2) \cong \mathbf{Z}_2$, we conclude that $\mathcal{T}$ is $HSpin_{12}^+[m]$ or $H\hat{A}E_{12}$, and we may assume the latter and it remains to derive a contradiction.

Let $z \neq z_1 \in \mathcal{Z}_S$ and $E_1 = \langle z, z_1 \rangle$; by 15.8.6.2, $z_2 = zz_1 \in \mathcal{Z}_S$. Let $K_i \in \Omega(z_i)$. Now $\mathcal{F}_z^\circ$ is constrained, so it has a model G_z and $G_z = K * H$ where $H \cong H\hat{A}E_{12}$. By (2), $\langle z \rangle = Z(H)$, so there is X of order 3 in H with $C_{KK_1K_2}(X) = KE_1$. Let $R = O_2(G_z)C_S(O_2(G_z))$ and G_0 a model for $N_\mathcal{F}(R)$. By 15.8.5, $Aut_\mathcal{F}(E_1) = GL(E_1)$, so there is X_1 of order 3 in G_0 with $C_{KK_1K_2}(X_1) = K_1E_1$. Then X_1 acts on G_z, a contradiction as H has no group X_1 of automorphisms with $C_{K_1K_2}(X_1) = K_1E_1$. This completes the proof of (3) and the lemma. □

LEMMA 15.8.8. *Assume $n = 6$ or 8. Then*
(1) $\tau_t^\circ = (\mathcal{Y}, \Omega)$ with $\mu(\tau_t^\circ) \cong Weyl(D_k)$ and $(n,k) = (6,5)$ or $(8,8)$.
(2) $t = u(\tau_t^\circ) \in Z(\mathcal{Y})$.
(3) One of the following holds:
(i) $W_x \in W(\tau_x)$ is normal in $\mathcal{F}_x$ for $x \in \{z,t\}$.
(ii) $n = 6$, $\mathcal{T} \cong SL_6^+[m]$, and $\mathcal{Y} \cong Spin_{10}^+[m]$.
(iii) $n = 8$, $\mathcal{T} \cong \tilde{E}_7[m]$, and $\mathcal{Y} \cong HSpin_{16}^+[m]$.

PROOF. Part (1) follows from 15.8.5.

In 15.8.1.1, (2) and (3) follow from 5.5.6, augmented by arguments in the proof of 15.1.6. Therefore we may assume 15.8.1.2 holds.

By 15.8.6, for each $s \in \mathcal{Z}_S$, $st \in \mathcal{Z}_S$, so $t = u(\tau_t^\circ)$ by 15.1.6.1. Thus (2) holds.

By 15.1.9.2, one of the following holds:

(a) $W_t \in W(\tau_t)$ is normal in $\mathcal{Y}$.

(b) $\mathcal{Y} \cong Spin_{2k}^+[m]$, $HSpin_{2k}^+[m]$, $Spin_{2k+1}[m]$, or $Spin_{2k+2}^-[m]$.

(c) $m = 8$ and $\mathcal{Y}$ is $\hat{A}E_{2k}$, $H\hat{A}E_{2k}$, or $\hat{A}E_{2k+1}$.

Let $\mathcal{X} = C_{\mathcal{F}}(\langle z,t\rangle)^\circ$ and $\xi = (\mathcal{X}, \Omega)$. Then ξ is a quaternion fusion packet such that $\xi = C_{\mathcal{F}_z^\circ}(t)^\circ$, so from 15.8.7.3, $\xi = \tau_K * \tau_0 * \zeta$, where $t = zz_0$ with $z_0 \in \mathcal{Z}_S$, $K_0 \in \Omega(z_0)$, $\tau_0 = (\mathcal{O}(z_0), K_0)$, and $\zeta = (\mathcal{D}, \Omega - \{K, K_0\})$ with $\mu(\zeta) = \text{Weyl}(\Sigma)$ and $\Sigma = A_3$ or D_6 for $n = 6$ or 8, respectively. In particular if 15.8.7.3.i does not hold then $\mathcal{D} \cong SL_4^+[m]$ or $Spin_{12}^+[m]$, respectively, so as ξ is the packet for $C_{\mathcal{Y}}(z)^\circ$, we conclude that (b) holds with $\mathcal{Y} \cong Spin_{2k}^+[m]$ or $HSpin_{2k}^+[m]$. If $n = 6$ then $k = 5$ is odd, so the half spin system does not exist, and hence (3ii) holds in this case. If $n = 8$ then from 15.8.6, $W_S \cong E_{16}$, so $\mathcal{Y} \cong HSpin_{16}^+[m]$ and therefore (3iii) holds in this case.

Therefore we may assume that 15.8.7.3.i holds, so $W' \trianglelefteq \mathcal{T}$. Let $J \in \Omega - \{K\}$ and $j = z(J)$; then $J \cap W = J \cap W' \trianglelefteq C_{\mathcal{G}}(W_S)$. But if $\alpha \in \mathfrak{A}(z_0)$ with $z_0\alpha = z$ then $W\alpha = W$ and $W_S\alpha = W_S$ with $K\alpha \in \Omega - \{K\}$, so also $K \cap W \trianglelefteq C_{\mathcal{F}}(W_S)$. Hence (a) holds, completing the proof. □

THEOREM 15.8.9. *If $W_z \in W(\tau_z)$ is normal in $\mathcal{F}_z^\circ$ then $W \trianglelefteq \mathcal{F}$.*

PROOF. The proof is much the same as that of 15.5.16. If $n = 6$ or 8 let $\{x,y\} = \{z,t\}$ and define W_x as in 15.8.8.3.i, so that $W_x \trianglelefteq \mathcal{F}_x$. If $n = 7$ let $(x,y) = (z,z_0)$ with $z \neq z_0 \in \mathcal{Z}_S$, let $\alpha \in \mathfrak{A}(y)$ with $y\alpha = x$, and let $W_y = W_x\alpha^{-1}$. Then $W_x \trianglelefteq \mathcal{F}_x$ and $W_y \trianglelefteq C_{\mathcal{F}_x}(y)$.

Observe next that, in any event, $W = W_xW_y$ and $W_y \trianglelefteq C_{\mathcal{F}_x}(y)$. Hence as $y \in W_x$, we have $W = W_xW_y \trianglelefteq C_{\mathcal{F}_x}(W_x)$, so $W \leq O_2(C_{\mathcal{F}_x}(W_x))$ and therefore $W \trianglelefteq \mathcal{F}_x$. Then by parts (7) and (8) of 15.8.5, $W \trianglelefteq \langle N_{\mathcal{F}}(W), \mathcal{F}_r : r \in \{x,y\}\rangle = \mathcal{F}$, completing the proof. □

Given Theorem 15.8.9, in the remainder of the section we assume:

HYPOTHESIS 15.8.10. *Hypothesis 15.8.1 holds, $W_z \in W(\tau_z)$ is not normal in $\mathcal{F}_z$, and if $n = 7$ then $Z(M) = 1$.*

LEMMA 15.8.11. *Let $\mathcal{Z}_S = \{z_i : 1 \le i \le n_S\}$. Then for $1 \le i \le n_S$ there exists $SL_2[m] \cong \mathcal{O}_i = \mathcal{O}_{z_i} \le \mathcal{F}$ with Sylow group $O(z_i)$ such that*

(1) $\mathcal{O}_i = \mathcal{O}(z_i)$ if $m > 8$.

(2) $\mathcal{O}_\tau = \mathcal{O}_1 * \cdots * \mathcal{O}_{n_S} \trianglelefteq N_{\mathcal{F}}(W_S)$ and $N_{\mathcal{F}}(W_S)$ is transitive on $\Lambda = \{\mathcal{O}_i : 1 \le i \le n_S\}$.

(3) $\mathcal{O}_z \trianglelefteq \mathcal{F}_z$.

PROOF. The proof is much like that of 15.5.24, and as such is omitted. □

LEMMA 15.8.12. *Let $x \in \{z, t\}$. Then*

(1) $W \trianglelefteq N_{\mathcal{F}_x}(W_x)$.

(2) $\mathcal{F}_x = \langle C_S(x)\mathcal{O}_\tau, N_{\mathcal{F}_x}(W)\rangle$.

(3) In aJ.1.1, $\mathcal{F} = \langle S\mathcal{O}_\tau, \mathcal{F}_S(M)\rangle$.

PROOF. The proof of (1) and (2) is much like that of 15.5.25. Define x and y as in the proof of 15.8.9, and let $S_x = C_S(x)$, $\mathcal{X} = S_x\mathcal{F}_x^\circ$ if $x \ne z$, $\mathcal{X} = S_z\mathcal{O}_z\mathcal{T}$ if $x = z$, and T Sylow in $\mathcal{F}_x^\circ$. By 1.3.2 in [**Asc19**], $\mathcal{F}_x = \langle \mathcal{X}, N_{\mathcal{F}_x}(T)\rangle$, so as W_x is weakly closed in T unless $x = z$ and $m = 8$, we conclude:

(a) $\mathcal{F}_x = \langle \mathcal{X}, N_{\mathcal{F}_x}(W_x)\rangle$.

We next observe that:

(b) $\mathcal{X} = \langle S_x\mathcal{O}_\tau, N_{\mathcal{X}}(W_x)\rangle$.

For $\mathcal{T} \cong SL_6^+[m]$, $HSpin_{12}^+[m]$, or $\tilde{E}_7[m]$, while $\mathcal{Y} \cong Spin_{10}^+[m]$ or $HSpin_{16}^+[m]$. Therefore (b) follows from 11.5.3 if $\mathcal{T}$ or $\mathcal{Y}$ is a spin system, from 15.6.3 if $\mathcal{T} \cong SL_6^+[m]$, and by induction on the order of τ if $\mathcal{T} \cong \tilde{E}_7[m]$.

Observe that (1), (a), and (b) imply (2), so it remains to prove (1).

As in the proof of 15.8.9, $W = W_xW_y$. Let $\mathcal{C} = C_{\mathcal{F}_x}(\mathcal{F}_x^\circ)$; as in the proof of 15.5.25:

(c) $W \trianglelefteq S_x\mathcal{C}$.

Let $\mathcal{M}_t$ be the 2-fusion system of $M(\tau_x)$; as in the proof of 15.5.25, $N_{\mathcal{F}_x}(W_x) = \langle S_x\mathcal{M}_x, N_{\mathcal{F}_x}(T)\rangle$, with $W \trianglelefteq S_x\mathcal{M}_x$ and $N_{\mathcal{F}_x}(T) = S_x\mathcal{C}$, so (1) follows from (c).

Finally we prove (3), so assume that 15.8.1.1 holds. We show $\mathcal{F} = \mathcal{G}$, where $\mathcal{G} = \langle N_{\mathcal{F}}(W), \mathcal{F}_z\rangle$. Then as $N_{\mathcal{F}}(W) = \mathcal{F}_S(M)$ and as $\mathcal{F}_z = \langle S\mathcal{O}_\tau, N_{\mathcal{F}_z}(W)\rangle$ by (2), the lemma holds. If $n = 7$ then $\mathcal{F} = \mathcal{G}$ by 15.8.5.7, so take $n = 6$ or 8. Then $\mathcal{F} = \langle \mathcal{G}, \mathcal{F}_t\rangle$ by 15.8.5.8, so it suffices to show that $\mathcal{F}_t \le \mathcal{G}$. But this follows from (2). □

NOTATION 15.8.13. Set $\tilde{\mathcal{F}} = E_n[m]$, or more precisely $E_6^+[m]$ if $n = 6$. Set $T = S_M$, $\mathcal{F}_1 = S\mathcal{O}_\tau$, $\mathcal{E}_1 = T\mathcal{O}_\tau$, $\mathcal{F}_2 = N_{\mathcal{F}}(W)$, $\mathcal{E}_2 = \mathcal{F}_T(M)$, and $\mathcal{E} = \langle \mathcal{E}_1, \mathcal{E}_2\rangle_T$. By 5.5.2 and 5.5.3, the Lie packet $\tilde{\tau} = (\tilde{\mathcal{F}}, \tilde{\Omega})$ of $E_n(q)$ satisfies Hypothesis 15.8.10 for suitable q, and by 5.2.9 and 5.2.10, $\tilde{M} \in M(\tilde{\tau})$ contains the Sylow group $\tilde{S}$ of $\tilde{\mathcal{F}}$. We conclude from 4.3.8 that $M = \tilde{M}$. Therefore, identifying M with $\tilde{M}$, we may take $M = \tilde{M}$, $T = \tilde{S}$, etc. Subject to this convention, $\mathcal{E}_2 = \mathcal{F}_{\tilde{S}}(\tilde{M})$.

LEMMA 15.8.14. *(1) $\mathcal{E}_1 = \tilde{S}\mathcal{O}_{\tilde{\tau}}$.*

(2) $\mathcal{E} = \tilde{\mathcal{F}}$.

(3) For $i = 1, 2$, $\mathcal{E}_i \trianglelefteq \mathcal{F}_i$.

PROOF. See the proofs of 15.7.5 and 15.7.6. □

THEOREM 15.8.15. *Assume Hypothesis* 15.8.10. *Then $\mathcal{F} \cong E_n[m]$ if $n = 7$ or 8 and $\mathcal{F} \cong E_6^+[m]$ if $n = 6$.*

PROOF. See the proof of 15.7.7. □

THEOREM 15.8.16. *Assume* $\tau = (\mathcal{F}, \Omega)$ *is a quaternion fusion packet such that* $\mathcal{F} = \mathcal{F}^{\circ}$ *and* $\mu(\tau) \cong Weyl(E_n)$ *for some* $n \in \{6, 7, 8\}$. *Assume the Extended Inductive Hypothesis holds. Then for* $z \in \mathcal{Z}_S$ *we have* $|\Omega(z)| = 1$, *and either*

(1) $W \in W(\tau)$ *is normal in* $\mathcal{F}$, *or*

(2) $\mathcal{F} \cong E_6^+[m]$, $E_7[m]$, $\tilde{E}_7[m]$, *or* $E_8[m]$.

PROOF. Let $K \in \Omega$ and $z = z(K)$. Arguing as in the proof of 15.7.8, Hypothesis 15.8.1 is satisfied.

If $n = 7$ and $Z(M) \neq 1$ then the theorem follows from 15.8.3.5, so we may assume otherwise. Therefore Hypothesis 15.8.4 holds. Similarly if $W_z \trianglelefteq \mathcal{F}_z^{\circ}$ then (1) holds by 15.8.9, so we may assume Hypothesis 15.8.10 is satisfied. Now Theorem 15.8.15 completes the proof. □

Theorem 1 and the Main Theorem

CHAPTER 16

Proofs of four theorems

16.1. The proof of Theorem 1

In this section we prove Theorem 1. Thus we assume the hypotheses of that theorem: $\tau = (\mathcal{F}, \Omega)$ is a quaternion fusion packet such that $\mathcal{F} = \mathcal{F}^\circ$ is transitive on Ω. Let $K \in \Omega$ and $z = z(K) \in \mathcal{F}^f$. Set $\mu = \mu(\tau)$. We may assume $z \notin O_2(\mathcal{F})$. Further, proceeding by induction on the order of τ:

LEMMA 16.1.1. *τ satisfies the Extended Inductive Hypothesis.*

LEMMA 16.1.2. *Assume $\mathcal{F}$ is isomorphic to $\mathcal{F}_S(G)$ for some group G of Lie type and odd characteristic. Set $\tilde{\mathcal{F}} = \mathcal{F}_S(G)$ and let $\tilde{\tau} = (\tilde{\mathcal{F}}, \tilde{\Omega})$ be the Lie packet of G. Then there exists an isomorphism $\delta : \mathcal{F} \to \tilde{\mathcal{F}}$ with $\Omega\delta = \tilde{\Omega}$, so τ is isomorphic to a Lie packet.*

PROOF. Assume first that $\mu = \mathrm{Weyl}(\Phi)$, where $\Phi = A_n$ for some $n \geq 2$, or $\Phi = D_n$ for some $n \geq 4$, or Φ is E_6, E_7, or E_8. Let $M \in M(\tau)$, $\tilde{M} \in M(\tilde{\tau})$, and $\bar{M} = \bar{\omega}(\Phi, m)$. Sift through the proof below for the lemma (Lemma X) where $\mathcal{F} \cong \tilde{\mathcal{F}}$ arises; the isomorphism is established in a theorem (Theorem Y) proven earlier in the paper. Inspecting the proof of Theorem Y, we find that $M \cong \bar{M}/\bar{U} \cong \tilde{M}$ for a certain $\bar{U} \leq Z(\bar{M})$. Now $\bar{M}/\bar{U}$ is a model for a system $\mathcal{M}$ where $\rho = (\mathcal{M}, \Gamma)$ is the quaternion fusion packet described in 4.3.7.5. We have isomorphisms $\alpha : M \to \bar{M}/\bar{U}$ and $\beta : \tilde{M} \to \bar{M}/\bar{U}$ such that $\Omega\alpha = \Gamma = \tilde{\Omega}\beta$. Hence $\gamma = \alpha\beta^{-1} : M \to \tilde{M}$ is an isomorphism with $\Omega\gamma = \tilde{\Omega}$. Then the proof of Theorem Y shows that γ extends to an isomorphism $\delta : \mathcal{F} \to \tilde{\mathcal{F}}$, and of course $\Omega\delta = \tilde{\Omega}$.

So assume μ is not $\mathrm{Weyl}(\Phi)$. Then from the lemmas below, μ is elementary abelian. Hence from 16.1.3, and as μ is abelian, we get $|\Omega(z)| = 1$ or 2. Suppose $|\Omega(z)| = 2$. Then as μ is abelian, we conclude from 16.1.8 that $\mathcal{F}$ is $PSp_4[m]$ or $L_4^-[m]$. But then $Z(S) = \langle z \rangle$ and Ω is the pair of Sylow groups of the factors $O^2(\mathcal{F}_z) \cong SL_2[m] * SL_2[m]$.

So we may take $|\Omega(z)| = 1$. Then $\mathcal{F}$ is one of the conclusions in 16.1.11. Hence $\mathcal{F}$ makes its appearence in Theorem 13.6.6. Tracing through the proof of that theorem, we find that in lemmas 13.2.21 and 13.2.23, an isomorphism $\delta : S \to \tilde{S}$ is constructed mapping Ω to $\tilde{\Omega}$, and δ is shown to define an iomorphism $\delta; \mathcal{F} \to \tilde{\mathcal{F}}$ in 13.6.2. □

LEMMA 16.1.3. *If $|\Omega(z)| > 2$ then $\Omega = \Omega(z)$ is of order 4 and $\mathcal{F} \cong \Omega_8^+(2)$ or $P\Omega_8^+[m]$.*

PROOF. This is a consequence of 16.1.1, Theorem 10.2.13, and the assumption that $z \notin O_2(\mathcal{F})$. □

By 16.1.3 we may assume:

LEMMA 16.1.4. *$|\Omega(z)| = 1$ or 2.*

LEMMA 16.1.5. *$\mathcal{A}(\tau) = \emptyset$.*

PROOF. Assume otherwise. By Theorem 7.2.3, as $\mathcal{F} = \mathcal{F}^\circ$, $\mathcal{F}$ is not transitive on Ω, contrary to the hypothesis of Theorem 1. □

LEMMA 16.1.6. *D is a set of 3-transpositions of μ.*

PROOF. This follows from 16.1.5 and 4.2.10.3. □

LEMMA 16.1.7. *Assume*
(1) $|\Omega(z)| = 2$, and
(2) $\mathcal{Z} \cap O(z) \neq \{z\}$.
Then $\mu \cong S_4$ and $\mathcal{F} \cong L_4^+[m]$.

PROOF. We first observe that Hypothesis 9.3.1 is satisfied. For example $\mathcal{A}(\tau) = \emptyset$ by 16.1.5 and 9.3.1.4 holds by 16.1.1. Now appeal to Theorem 9.3.24 to complete the proof. □

LEMMA 16.1.8. *Assume $|\Omega(z)| = 2$ and $\mathcal{D}^*(z) = \mathcal{D}(z)$. Then $\mathcal{F} \cong PSp_4[m]$ or $L_4^-[m]$.*

PROOF. Observe that Hypothesis 9.4.1 is satisfied and appeal to Theorem 9.4.33. □

LEMMA 16.1.9. *Assume $|\Omega(z)| = 2$ and $\mathcal{D}^*(z) \neq \mathcal{D}(z)$. Then $\mu \cong Weyl(D_n)$ for some $n \geq 3$ and one of the following holds:*
(1) $\mathcal{F} \cong (P)\Omega_{2n}^+[m]$, $\Omega_{2n+1}[m]$, or $\Omega_{2n+2}^-[m]$.
(2) $(n, m) = (4, 8)$ and $\mathcal{F} \cong \check{\Omega}_8^+(2)$.
(3) $(n, m) = (3, 8)$ and $\mathcal{F} \cong Sp_6(2)$.

PROOF. If $\mathcal{Z} \cap O(z) \neq \{z\}$ then the lemma follows from 16.1.7 since $A_3 = D_3$ and $L_4^+[m] = P\Omega_6^+[m]$. Therefore we may assume that $\mathcal{Z} \cap O(z) = \{z\}$, so Hypothesis 12.3.1 is satisfied. Now the lemma follows from Theorem 12.3.50. □

LEMMA 16.1.10. *We may assume that $\Omega(z) = \{K\}$ is of order one.*

PROOF. Assume otherwise; then by 16.1.4, $|\Omega(z)| = 2$. We will show that one of the conclusions of Theorem 1 holds in this case: namely we will show that either τ is a Lie packet or $\mathcal{F}$ is $\check{\Omega}_8^+(2)$ or $Sp_6(2)$. If $\mathcal{D}^*(z) = \mathcal{D}(z)$ this follows from 16.1.8 and 16.1.2, while if $\mathcal{D}^*(z) \neq \mathcal{D}(z)$, it follows from 16.1.9 and 16.1.2. □

LEMMA 16.1.11. *Assume μ is abelian. Then $\mathcal{F} \cong (P)Sp_{2n}[m]$, $(P)SL_{2n}^-[m]$, $L_{2n+1}^-[m]$, or $L_2[2m]^{(1)}$.*

PROOF. Using 16.1.1 and 16.1.10, Hypothesis 13.1.1 is satisfied. Then Theorem 13.6.6 completes the proof. □

LEMMA 16.1.12. *We may assume that $\mathcal{D}^*(z) \neq \mathcal{D}(z)$.*

PROOF. If not, μ is abelian, so Theorem 1 holds by 16.1.11 and 16.1.2. □

LEMMA 16.1.13. *We may assume that Hypothesis* 15.2.1 *holds with μ isomorphic to Weyl(Φ) where Φ is A_n for some $n \geq 4$, D_n for some $n \geq 3$, or E_n for some $n \in \{6, 7, 8\}$.*

PROOF. By 16.1.1, 16.1.10, and 16.1.12, Hypothesis 15.2.1 holds. Then by Theorem 15.2.29, $\mu \cong$ Weyl(Φ) where Φ is A_n, D_n, or E_n. As μ is nonabelian, Φ is not A_1. If Φ is A_2 then Theorem 1 holds by Remark 15.2.30. Similarly as μ is nonabelian we have $n \geq 3$ if Φ is D_n. As $D_3 = A_3$, we may take $n \geq 4$ if Φ is A_n. Therefore the lemma holds. □

LEMMA 16.1.14. *If $\Phi = D_n$ for some $n \geq 3$ then Theorem 1 holds.*

PROOF. If $n \geq 5$ this follows from Theorem 15.3.10. If $n = 4$ it follows from Theorem 15.4.30. Finally if $n = 3$ it follows from Theorem 15.3.14. □

LEMMA 16.1.15. *If $\Phi = A_n$ for some $n \geq 4$ then Theorem 1 holds.*

PROOF. See Theorem 15.7.8. □

LEMMA 16.1.16. *If $\Phi = E_n$ for some $n \in \{6, 7, 8\}$ then Theorem 1 holds.*

PROOF. This is a consequence of Theorem 15.8.16. □

Observe that we have completed the proof of Theorem 1. Namely by 16.1.13, $\mu =$ Weyl(Φ) where Φ is A_n for some $n \geq 4$, D_n for some $n \geq 3$, or E_n for some $n \in \{6, 7, 8\}$. In each case, Theorem 1 holds by one of 16.1.14-16.1.16.

16.2. Proofs of the Main Theorem and Theorems 6, 7, and 8

In this short section we complete the proofs of the Main Theorem and of Theorems 6, 7, and 8.

First, Theorem 6 follows from Theorem 1 and the discussion in section 12.4. Namely from section 12.4, Theorem 6 holds under the extra assumption that the Extended Inductive Hypothesis holds. But now that we have established Theorem 1 in section 16.1, this extra assumption is indeed satisfied, completing the proof of Theorem 6.

Similarly Theorem 7 follows from Theorem 1 and Remark 13.6.7.

We next prove Theorem 8, so assume the hypothesis of that theorem. We claim that Hypothesis 15.2.1 holds. By the hypothesis of Theorem 8, parts (1) and (2) of 15.1.1 hold, and as μ is nonabelian, 15.1.1.3 holds. By Theorem 1, 15.1.1.4 holds, completing the proof of the claim.

By the claim and the proof of 16.1.13, the conclusion of 16.1.13 holds. Then, as in the proof of 16.1.14-16.1.16, Theorem 8 holds.

Finally here is the proof of the Main Theorem. Assume the hypothesis of that theorem. Then there is an involution z in S and a subnormal subsystem $\mathcal{E}$ of $\mathcal{F}_z$ isomorphic to $SL_2[m]$ for some m such that z is contained in the Sylow group K of $\mathcal{E}$. Then $\mathcal{E}$ is a component, or a solvable component if $m = 8$, of $\mathcal{F}_z$, so, setting $\Omega = K^{\mathcal{F}}$, $\tau = (\mathcal{F}, \Omega)$ is a quaternion fusion packet by 2.4.1. As $\Omega = K^{\mathcal{F}}$, $\mathcal{F}$ is transitive on Ω, and by the hypothesis of the Main Theorem, $\mathcal{F} = [K]_{\mathcal{F}}$, so $\mathcal{F} = \mathcal{F}^\circ$.

Therefore τ satisfies the hypothesis of Theorem 1, so by that theorem, τ satisfies one of its conclusions.

Suppose that $z \in Z(\mathcal{F})$. Then $\mathcal{E}$ is subnormal in $\mathcal{F}_z$, so as $\mathcal{F} = \mathcal{F}^\circ$ is transitive on Ω, it follows that $\mathcal{F} = \mathcal{E}$. Hence τ is the Lie packet of $SL_2(q)$ for suitable q, so the Main Theorem holds in this case. Therefore we may assume that $z \notin Z(\mathcal{F})$. Thus if $z \in O_2(\mathcal{F})$ then the hypotheses of Theorem 2 are satisfied, so τ satisfies one of its conclusions. This contradicts $SL_2[m] \cong \mathcal{E}$ subnormal in $\mathcal{F}_z$.

Therefore $z \notin O_2(\mathcal{F})$, so τ satisfies one of conclusions (2), (3), or (4) of Theorem 1. Now if conclusion (2) holds then so does the Main Theorem, so we may assume conclusion (3) or (4) holds. But this contradicts $SL_2[m] \cong \mathcal{E}$ subnormal in $\mathcal{F}_z$, and this contradiction completes the proof of the Main Theorem.

16.3. Lie fusion packets

In this last section we tie up a loose end. From Theorem 1, if $\mathcal{F}$ is the 2-fusion system of a group G of Lie type and odd characteristic, and $\tau = (\mathcal{F}, \Omega)$ is a quaternion fusion packet with $\mathcal{F} = \mathcal{F}^\circ$ transitive on Ω, then τ is isomorphic to a Lie fusion packet. But is τ *the* Lie packet for G? The main result of this section shows that this is almost always the case.

THEOREM 16.3.1. *Let $\tau = (\mathcal{F}, \Omega)$ be a Lie fusion packet. Then*

(1) $\mathcal{F}$ is tamely realized by some group G of Lie type and odd characteristic.

(2) Let $\rho = (\mathcal{F}, \Gamma)$ be the Lie packet of G. Then there exist $\alpha \in Aut(\mathcal{F})$ with $\Gamma\alpha = \Omega$.

(3) There exists $\beta \in Aut(G)$ with $\alpha = \beta_{|S}$.

(4) Either $\Omega = \Gamma$ or $G = G_2(q)$ with $q = 3^e$ and β interchanges long and short root subgroups of G. Thus $\Gamma = \{K\}$ where $K = S \cap L_l$ for some long fundamental subgroup L_l of G and $K\alpha = C_S(K) = S \cap L_s$, where L_s is a short fundamental subgroup of G with $O^2(C_G(z)) = L_l * L_s$.

The proof of Theorem 16.3.1 involves a short series of reductions. Assume $\tau = (\mathcal{F}, \Omega)$ is a Lie fusion packet.

LEMMA 16.3.2. *(1) $\mathcal{F} = \mathcal{F}_S(G)$ for some group G of Lie type and odd characteristic tamely realizing $\mathcal{F}$.*

(2) Let $\rho = (\mathcal{F}, \Gamma)$ be the Lie packet of G. Then $\Gamma = \{S \cap L : L \in \mathfrak{L}$ and $S \cap L \in Syl_2(L)\}$, where $\mathfrak{L}$ is the set of fundamental subgroups of G.

(3) There exists $\alpha \in Aut(\mathcal{F})$ with $\Gamma\alpha = \Omega$.

(4) $\alpha = \beta_{|S}$ for some $\beta \in Aut(G)$.

PROOF. As τ is a Lie packet, $\mathcal{F} = \mathcal{F}_S(G)$ for some group G of Lie type and odd characteristic. By 3.5 in [**AO16**] we may choose G to tamely realize $\mathcal{F}$; that is (1) holds. Then (2) follows directly from the definition of "Lie packet" in the introduction. From Theorem 1, all quaternion fusion packets $\zeta = (\mathcal{F}, \Sigma)$ with $\mathcal{F} = [\Sigma]_{\mathcal{F}}$ and $\mathcal{F}$ transitive on Σ are isomoprhic so (3) holds. Then (1) and 2.22 in [**AO16**] implies (4). □

LEMMA 16.3.3. *If β permutes the long root subgroups of G then $\Gamma\alpha = \Gamma$.*

PROOF. Let $\mathfrak{U}$ denote the set of long root subgroups of G, and suppose β acts on $\mathfrak{U}$. Then β also acts on the set $\mathfrak{P}$ of pairs (U, V) such that $U, V \in \mathfrak{U}$ and U and V are opposites. Hence β permutes $\mathfrak{L}$. Therefore for $K \in \Gamma$, $K = S \cap L \in Syl_2(L)$ for some $L \in \mathfrak{L}$, so $K\alpha = K\beta = (S \cap L)\beta = S \cap L\beta \in \Gamma$, completing the proof. □

LEMMA 16.3.4. *If $\Gamma\alpha \neq \Gamma$ then $G = G_2(q)$, where $q = 3^e$, and β interchanges long and short root subgroups of G.*

PROOF. By 16.3.3, the root system of G has two root lengths and β interchanges long and short root subgroups. Then the lemma follows from 2.5.1 in [**GLS98**]. □

Observe that Theorem 16.3.1 follows from lemmas 16.3.2-16.3.4, together with the fact that $G_2(q)$ has one class z^G of involutions and $O^2(C_G(z)) = L_l * L_s$ where L_l is a long fundamental subgroup and L_s is a short fundamental subgroup.

References and Index

Bibliography

[AKO11] Michael Aschbacher, Radha Kessar, and Bob Oliver, *Fusion systems in algebra and topology*, London Mathematical Society Lecture Note Series, vol. 391, Cambridge University Press, Cambridge, 2011. MR2848834

[Alp67] J. L. Alperin, *Sylow intersections and fusion*, J. Algebra **6** (1967), 222–241, DOI 10.1016/0021-8693(67)90005-1. MR215913

[AO16] Michael Aschbacher and Bob Oliver, *Fusion systems*, Bull. Amer. Math. Soc. (N.S.) **53** (2016), no. 4, 555–615, DOI 10.1090/bull/1538. MR3544261

[AS04] Michael Aschbacher and Stephen D. Smith, *The classification of quasithin groups. I*, Mathematical Surveys and Monographs, vol. 111, American Mathematical Society, Providence, RI, 2004. Structure of strongly quasithin K-groups. MR2097623

[Asc73] Michael Aschbacher, *A condition for the existence of a strongly embedded subgroup*, Proc. Amer. Math. Soc. **38** (1973), 509–511, DOI 10.2307/2038941. MR318308

[Asc77] Michael Aschbacher, *A characterization of Chevalley groups over fields of odd order. II*, Ann. of Math. (2) **106** (1977), no. 3, 399–468, DOI 10.2307/1971063. MR498829

[Asc80] Michael Aschbacher, *On finite groups of Lie type and odd characteristic*, J. Algebra **66** (1980), no. 2, 400–424, DOI 10.1016/0021-8693(80)90095-2. MR593602

[Asc86] Michael Aschbacher, *Finite group theory*, Cambridge Studies in Advanced Mathematics, vol. 10, Cambridge University Press, Cambridge, 1986. MR895134

[Asc87] Michael Aschbacher, *Chevalley groups of type G_2 as the group of a trilinear form*, J. Algebra **109** (1987), no. 1, 193–259, DOI 10.1016/0021-8693(87)90173-6. MR898346

[Asc94] Michael Aschbacher, *Sporadic groups*, Cambridge Tracts in Mathematics, vol. 104, Cambridge University Press, Cambridge, 1994. MR1269103

[Asc97] Michael Aschbacher, *3-transposition groups*, Cambridge Tracts in Mathematics, vol. 124, Cambridge University Press, Cambridge, 1997. MR1423599

[Asc08a] Michael Aschbacher, *Normal subsystems of fusion systems*, Proc. Lond. Math. Soc. (3) **97** (2008), no. 1, 239–271, DOI 10.1112/plms/pdm057. MR2434097

[Asc08b] Michael Aschbacher, *Standard components of alternating type centralized by a 4-group*, J. Algebra **319** (2008), no. 2, 595–615, DOI 10.1016/j.jalgebra.2006.09.033. MR2381797

[Asc10] Michael Aschbacher, *Generation of fusion systems of characteristic 2-type*, Invent. Math. **180** (2010), no. 2, 225–299, DOI 10.1007/s00222-009-0229-z. MR2609243

[Asc11] Michael Aschbacher, *The generalized Fitting subsystem of a fusion system*, Mem. Amer. Math. Soc. **209** (2011), no. 986, vi+110, DOI 10.1090/S0065-9266-2010-00621-5. MR2752788

[Asc15] Michael Aschbacher, *Classifying finite simple groups and 2-fusion systems*, ICCM Not. **3** (2015), no. 1, 35–42, DOI 10.4310/ICCM.2015.v3.n1.a5. MR3385504

[Asc19] Michael Aschbacher, *On fusion systems of component type*, Mem. Amer. Math. Soc. **257** (2019), no. 1236, v+182, DOI 10.1090/memo/1236. MR3898993

[BCG+05] Carles Broto, Natàlia Castellana, Jesper Grodal, Ran Levi, and Bob Oliver, *Subgroup families controlling p-local finite groups*, Proc. London Math. Soc. (3) **91** (2005), no. 2, 325–354, DOI 10.1112/S0024611505015327. MR2167090

[BLO03] Carles Broto, Ran Levi, and Bob Oliver, *The homotopy theory of fusion systems*, J. Amer. Math. Soc. **16** (2003), no. 4, 779–856, DOI 10.1090/S0894-0347-03-00434-X. MR1992826

[Cra11] David A. Craven, *Normal subsystems of fusion systems*, J. Lond. Math. Soc. (2) **84** (2011), no. 1, 137–158, DOI 10.1112/jlms/jdr004. MR2819694

[Dic58] Leonard Eugene Dickson, *Linear groups: With an exposition of the Galois field theory*, with an introduction by W. Magnus, Dover Publications, Inc., New York, 1958. MR0104735

[Foo97] Richard Foote, *A characterization of finite groups containing a strongly closed* 2-*subgroup*, Comm. Algebra **25** (1997), no. 2, 593–606, DOI 10.1080/00927879708825876. MR1428800

[Gla66] George Glauberman, *Central elements in core-free groups*, J. Algebra **4** (1966), 403–420, DOI 10.1016/0021-8693(66)90030-5. MR202822

[GLS98] Daniel Gorenstein, Richard Lyons, and Ronald Solomon, *The classification of the finite simple groups*, Mathematical Surveys and Monographs, vol. 40, American Mathematical Society, Providence, RI, 1994. MR1303592

[GLS99] Daniel Gorenstein, Richard Lyons, and Ronald Solomon, *The classification of the finite simple groups*, Mathematical Surveys and Monographs, vol. 40, American Mathematical Society, Providence, RI, 1994. MR1303592

[Gol74] David M. Goldschmidt, 2-*fusion in finite groups*, Ann. of Math. (2) **99** (1974), 70–117, DOI 10.2307/1971014. MR335627

[Gol75] David M. Goldschmidt, *Strongly closed* 2-*subgroups of finite groups*, Ann. of Math. (2) **102** (1975), no. 3, 475–489, DOI 10.2307/1971040. MR393223

[Lyn14] Justin Lynd, *The Thompson-Lyons transfer lemma for fusion systems*, Bull. Lond. Math. Soc. **46** (2014), no. 6, 1276–1282, DOI 10.1112/blms/bdu083. MR3291264

[McL69] Jack McLaughlin, *Some subgroups of* $\mathrm{SL}_n(\mathbf{F}_2)$, Illinois J. Math. **13** (1969), 108–115. MR237660

[Pui09] Lluís Puig, *Frobenius categories versus Brauer blocks*, Progress in Mathematics, vol. 274, Birkhäuser Verlag, Basel, 2009. The Grothendieck group of the Frobenius category of a Brauer block. MR2502803

Index

Selected Published Titles in This Series

765 **Michael Aschbacher,** Quaternion Fusion Packets, 2021
763 **Tyler J. Jarvis and Nathan Priddis, Editors,** Singularities, Mirror Symmetry, and the Gauged Linear Sigma Model, 2021
762 **Atsushi Ichino and Kartik Prasanna,** Periods of Quaternionic Shimura Varieties. I., 2021
761 **Ibrahim Assem, Christof Geiß, and Sonia Trepode, Editors,** Advances in Representation Theory of Algebras, 2021
760 **Olivier Collin, Stefan Friedl, Cameron Gordon, Stephan Tillmann, and Liam Watson, Editors,** Characters in Low-Dimensional Topology, 2020
759 **Omayra Ortega, Emille Davie Lawrence, and Edray Herber Goins, Editors,** The Golden Anniversary Celebration of the National Association of Mathematicians, 2020
758 **Jan Šťovíček and Jan Trlifaj, Editors,** Representation Theory and Beyond, 2020
757 **Kaïs Ammari and Stéphane Gerbi, Editors,** Identification and Control: Some New Challenges, 2020
756 **Joeri Van der Veken, Alfonso Carriazo, Ivko Dimitrić, Yun Myung Oh, Bogdan D. Suceavă, and Luc Vrancken, Editors,** Geometry of Submanifolds, 2020
755 **Marion Scheepers and Ondřej Zindulka, Editors,** Centenary of the Borel Conjecture, 2020
754 **Susanne C. Brenner, Igor Shparlinski, Chi-Wang Shu, and Daniel B. Szyld, Editors,** 75 Years of Mathematics of Computation, 2020
753 **Matthew Krauel, Michael Tuite, and Gaywalee Yamskulna, Editors,** Vertex Operator Algebras, Number Theory and Related Topics, 2020
752 **Samuel Coskey and Grigor Sargsyan, Editors,** Trends in Set Theory, 2020
751 **Ashish K. Srivastava, André Leroy, Ivo Herzog, and Pedro A. Guil Asensio, Editors,** Categorical, Homological and Combinatorial Methods in Algebra, 2020
750 **A. Bourhim, J. Mashreghi, L. Oubbi, and Z. Abdelali, Editors,** Linear and Multilinear Algebra and Function Spaces, 2020
749 **Guillermo Cortiñas and Charles A. Weibel, Editors,** K-theory in Algebra, Analysis and Topology, 2020
748 **Donatella Danielli and Irina Mitrea, Editors,** Advances in Harmonic Analysis and Partial Differential Equations, 2020
747 **Paul Bruillard, Carlos Ortiz Marrero, and Julia Plavnik, Editors,** Topological Phases of Matter and Quantum Computation, 2020
746 **Erica Flapan and Helen Wong, Editors,** Topology and Geometry of Biopolymers, 2020
745 **Federico Binda, Marc Levine, Manh Toan Nguyen, and Oliver Röndigs, Editors,** Motivic Homotopy Theory and Refined Enumerative Geometry, 2020
744 **Pieter Moree, Anke Pohl, L'ubomír Snoha, and Tom Ward, Editors,** Dynamics: Topology and Numbers, 2020
743 **H. Garth Dales, Dmitry Khavinson, and Javad Mashreghi, Editors,** Complex Analysis and Spectral Theory, 2020
742 **Francisco-Jesús Castro-Jiménez, David Bradley Massey, Bernard Teissier, and Meral Tosun, Editors,** A Panorama of Singularities, 2020
741 **Houssam Abdul-Rahman, Robert Sims, and Amanda Young, Editors,** Analytic Trends in Mathematical Physics, 2020
740 **Alina Bucur and David Zureick-Brown, Editors,** Analytic Methods in Arithmetic Geometry, 2019

For a complete list of titles in this series, visit the
AMS Bookstore at **www.ams.org/bookstore/conmseries/**.